STATISTICS
FOR BUSINESS AND ECONOMICS
Second Edition

Heinz Kohler
Amherst College

Scott, Foresman and Company
Glenview, Illinois   London, England

To my daughter, Vicky

Acknowledgments
Acknowledgments for literary selections and tables appear at the back of the book on page C-1, an extension of the copyright page.

Library of Congress Cataloging-in-Publication Data

Kohler, Heinz, 1934-
  Statistics for business and economics.

  Includes indexes.
  1. Commercial statistics.   2. Statistics.   I. Title.
HF1017.K64   1988   519.5′02433   87-11781
ISBN 0-673-18444-7

PREFACE

TO THE INSTRUCTOR

This text is designed for an introductory course in statistics that is, above all else, *applications-oriented*. As the detailed listing in the front endpapers shows, the applications found in this book are drawn from such diverse fields as accounting, advertising, finance, general management, marketing, personnel, production and operations management, quality control, public administration, economics, and a variety of other fields. Many of these applications have been integrated into the basic structure of chapters as "Practice Problems" or "Applications." Others have been placed, near the end of each chapter, as self-contained "Analytical Examples" or somewhat less rigorous "Close-Ups." Altogether, well over 200 examples illustrate the usefulness of statistics in numerous aspects of our lives. The end-of-chapter problems (which have grown to sixty per chapter in this second edition), the *Student Workbook,* and the *Instructor's Manual* abound with almost 1500 additional applications still. Also, most chapters contain one or more biographical sketches of the scholars who developed the statistical theory being discussed. The 20 "Biographies," also found near the ends of chapters, provide a sense of history; they also lend human interest to the theoretical material and highlight some of the controversies that have divided statisticians in the past and, in some cases, still do so in the present.

Following the brief introductory chapter, this book is divided into five parts. Part One, "Descriptive Statistics," deals with the *collection* of data by means of surveys and experiments (Chapter 2) and with the *presentation* of data by means of tables, graphs, and numerical summary measures (Chapters 3 and 4). Part Two, "Analytical Statistics: The Foundations of Inference," introduces the theory of probability (Chapter 5) and then discusses a number of discrete and continuous probability distributions (Chapters 6 and 7). Part Three, "Analytical Statistics: Basic Inference," introduces sampling distributions and estimation (Chapter 8) and hypothesis testing using classical, chi-square, and other nonparametric techniques (Chapters 9–11). Part Four, "Analytical Statistics: Advanced Inference," introduces the analysis of variance (Chapter 12) and develops the techniques of simple and multiple regression and correlation (Chapters 13 and 14). Part Five, "Special Topics for Business and Economics," focuses on time series and forecasting (Chapter 15), index numbers (Chapter 16), and decision theory (Chapter 17).

This text assumes a knowledge of only college-entrance algebra, not of calculus. Certain additional skills (such as the use of the summation sign, the use of factorials, and the use of computer printouts) are introduced when needed, but the derivation of formulas is left to more advanced texts in mathematical statistics and, in part, to supplementary sections in the *Student Workbook* that accompanies this text. This approach has been dictated in part by space considerations: the sheer volume of important statistical techniques that are in common use and, thus, make a legitimate claim on the inevitably limited number of pages available for a standard-sized text leaves little room for extended mathematical proofs. A pedagogical consideration, however, has played a role, too: the text aims to prepare students for their future careers, and most students in business-administration and economics programs do not go on to careers in theoretical mathematics, but they do take up jobs that often require a fine appreciation of the power and applicability of statistical methods. Above all else, they must be taught sound statistical reasoning so that they know which techniques can be used under which assumptions and how the results, properly interpreted, can aid their decision making. The applications approach, which is also very space-consuming, is considered crucial for achieving this goal.

Course Outlines

This text has been written to provide maximum flexibility to instructors who wish to fashion their own courses suitable to their own tastes and to the needs of their particular students. Most of the chapters of this book (and even many of the major sections within chapters) are self-contained units. It is not necessary to cover them all or even to assign them in the order in which they appear. In general, instructors can feel free to omit sections designated by an asterisk as an "optional section," or to move sections or chapters to other positions in the Table of Contents. When one section builds upon optional material in other chapters, this fact is footnoted on the first page of the section. Given this flexibility, this text is suitable for a statistics course as short as a quarter or as long as an entire academic year. The following are recommended sequences (with possible omissions of optional sections within chapters):

ONE-YEAR COURSES

First quarter: Chapters 1–7 (descriptive statistics and probability)
Second quarter: Chapters 8–11 (estimation and hypothesis testing)
Third quarter: Chapters 12–14 (ANOVA, regression, and correlation)
Fourth quarter: Chapters 15–17 (special topics for business and economics)

First semester: Chapters 1–11
Second semester: Chapters 12–17

ONE-SEMESTER COURSES (exclude all optional sections)

Week 1: Chapters 1–3 (focus on frequency distributions)
Week 2: Chapter 4
Week 3: Chapter 5
Week 4: Chapter 6 (binomial distribution only)
Week 5: Chapter 7 (normal distribution only)
Week 6: Chapter 8
Week 7: Chapter 9
Week 8: Chapter 10 (only applications to binomial and normal distributions)
Week 9: Chapter 12 (one-way ANOVA only)
Week 10: Chapter 13
Week 11: Chapter 14
Weeks 12–14: Review above material *or* expand above material with optional sections *or* introduce special topics (from Chapters 15–17), depending on previous experience with students in that class.

Supporting Materials

Three types of supporting materials are available: an *Instructor's Manual,* a *Student Workbook,* and a set of two personal-computer diskettes, called DAYSAL-2.

The *Instructor's Manual* contains responses to or solutions for the even-numbered end-of-chapter Questions and Problems in this text. It also contains, for Chapters 2–17 of the text, numerous true/false questions, multiple-choice questions, and problems (with solutions). These questions and problems do not duplicate those of the *Student Workbook* that is described in the Preface to the Student.

The DAYSAL-2 disks, designed to be used with IBM microcomputers and compatible machines, contain numerous mathematical programs, graphics programs, and educational games and simulations, as well as 26 files of data on the U.S. economy. Jointly, these programs cover almost every aspect of this text. Note the descriptions of relevant programs that follow each chapter's Questions and Problems Section.

One final note: The most important skill, perhaps, that students in statistics learn is *the ability to discern which particular technique is applicable to the solution of a given problem.* Yet, inevitably, a text that introduces each technique in the context of a problem (or that lists problems at the end of a given chapter) gives away the answer to this most difficult task of fitting the technique to the problem at hand. Instructors, therefore, are advised to consider the text and *Workbook* problems as exercises in technique and to use the *Instructor's Manual* problems (as well as the *Workbook* "Achievement Tests") as a means of testing the additional student skill just noted.

TO THE STUDENT

Statistics is an important subject. Consider the business executive who claims that a 20 percent reduction in product price will so stimulate sales as to bring about a significant gain in the firm's market share. Consider the economist who argues that a $20 billion tax cut will induce a rise in national income by a multiple, will bring about a significant drop in the unemployment rate, and will actually increase government revenue. Decision makers about to act on the basis of such beliefs may, however, have second thoughts: Might the quantity sold rise so little that the firm's dollar market share actually *drops?* Might the tax cut fail to end the recession and produce soaring deficits instead of revenue gains? In quandaries such as these, decision makers can use statistical methods to help them choose a proper course of action because these methods can help them *test* the accuracy of mere opinions and introduce precision where ambiguity reigns supreme. A proper statistical test, for example, might conclude that the 20 percent price reduction will raise quantity sold by only 10 percent and will actually reduce the firm's market share by 5 percent. Similarly, after gathering and analyzing data, the economist might discover that the $20 billion tax cut will raise national income by $60 billion and will lower the unemployment rate by half a percentage point but will raise the government deficit by $15 billion. Those who master the theoretical knowledge of statistics are, thus, able to wield a powerful set of tools capable of unearthing the likely truth about matters merely hypothesized and of, thus, facilitating wise decision making in the face of uncertainty.

Statistics is also an exciting subject. This fact is evidenced throughout this book by a continuous link between theory and applications. Although the discipline of statistics is crucial for all sciences, this introductory book, being designed for students of business administration and economics, draws examples primarily from these two fields. Again and again, this book shows how the careful collection, effective presentation, and proper analysis of numerical information enables people to draw important inferences from inevitably limited information and, thus, to build windows that afford an occasional glance into the world of the unknown. The front endpapers of this book contain a partial list of the examples found throughout this text and its supplements.

Students will find a number of aids to the study of the material presented in this book. Some of these aids have been built into the text: the figures made self-contained by carefully worded captions, the formulas highlighted in boxed spaces and followed by numbered "Practice Problems" or "Applications," the free-standing "Analytical Examples" and "Close-Ups" that give additional real-world applications, the chapter "Summaries," the end-of-chapter listings of "Key Terms" (boldfaced in the text), the end-of-chapter "Questions and Problems," and the "Selected Readings." In addition, a "Glossary" of all key terms and answers to odd-numbered Questions and Problems appear at the end of the book. Finally, a "Glossary of Symbols" appears in the back endpapers. A separate *Student Workbook* is also available. For Chapters 2–17, the *Student Workbook* contains true/false questions, multiple-choice questions, and numerous problems, along with answers to all of these. For selected chapters, the *Workbook* also contains "Supplementary Topics" that expand upon the materials presented in the text. For each of the five parts of the text, the *Workbook* contains a major "Achievement Test" along with all the answers.

Students who have access to IBM microcomputers or compatible machines may wish to acquire DAYSAL-2, a set of diskettes that contain numerous mathematical programs, graphics programs, and educational games and simulations, as well as 26 files of data on the U.S. economy. Jointly, these programs cover almost every aspect of this text and can, thus, aid mightily in the number crunching that the application of formulas inevitably requires. Note the description of relevant programs that follow each chapter's Questions and Problems section.

Another kind of aid that is built into the text may appear like a hindrance at first. Many important concepts appear under more than one name, and many symbols take on more than one form. For example, a particular kind of correlation may be described as both *nonsense correlation* and *spurious correlation,* and a particular concept may be symbolized as both $s_{Y \cdot X_1 X_2}$ and $s_{Y \cdot 12}$. Both forms are given, not to make life hard for you (or because the author couldn't make up his mind), but because both versions are in common use. A statistics textbook should make the student aware of the fact that different books or journals may describe a concept in different ways.

Acknowledgments

I would like to express my sincere gratitude to many who have helped me in the creation of this text. Many reviewers took the time to examine at least a part of the project and gave me good advice:

Richard E. Beckwith	*California State University at Sacramento*
Harry Benham	*University of Oklahoma*
Stephen G. Cecchetti	*New York University*
Georgio Canarella	*California State University at Los Angeles*
John S. Gafar	*Long Island University*

Tarsaim L. Goyal	*University of the District of Columbia*
Paul W. Guy	*California State University at Chico*
Stephen Grubaugh	*Bentley College*
Hannah Hiles	*Wesleyan University*
Peter Hoefer	*Pace University*
Patrick R. Huntley	*University of Arkansas*
Mel Jameson	*University of Rochester*
Denis F. Johnston	*Georgetown University*
Allen Lee	*Northeastern University*
Robert C. Marshall	*Duke University*
Peter H. Michael	*California State University at Sacramento*
Keith T. Poole	*Carnegie-Mellon University*
Harry O. Posten	*University of Connecticut*
Lawrence B. Pulley	*Brandeis University*
David P. Rogers	*Purdue University*
John D. Scholl	*Rio Grande College*
Saul Schwartz	*Tufts University*
C. B. Sowell	*Berea College*
Leonard W. Swanson	*Northwestern University*
Stanley A. Taylor	*California State University at Sacramento*
Willbann D. Terpening	*University of Notre Dame*
Bruce Vermeulen	*Colby College*

I am equally grateful to acquiring editor George Lobell, developmental editor Bruce Kaplan, project editor Carol Karton, designer Jacqueline Kolb, and layout artist Cathy Wacaser. They have guided this project through the long process of production and have created, as most will agree, a beautiful book. Permission to use selected materials from other books is gratefully acknowledged on the copyright page.

Heinz Kohler
Amherst College

CONTENTS

*Asterisks indicate optional sections.

||| Part Two

ANALYTICAL STATISTICS: THE FOUNDATIONS OF INFERENCE 135

Part Five

SPECIAL TOPICS FOR BUSINESS AND ECONOMICS   645

STATISTICS
FOR BUSINESS AND ECONOMICS

Second Edition

CHAPTER ONE

The Nature of Statistics

Most people associate the term *statistics* with masses of numbers or, perhaps, with the tables and graphs that display them and with the averages and similar measures that summarize them. This mental image is reinforced daily as people encounter an abundance of numerical information in newspapers, magazines, and on television screens: on the prices of bonds and stocks, on the performance of businesses and sports teams, on the movements of exchange rates and commodity futures, on the rates of unemployment and inflation, on the incidence of poverty and disease, on accidents, crime, and the weather—the list goes on. It is not surprising, therefore, that people who think about statistics not in the plural sense of the word, but in the singular (that is, as a field of study) imagine it to deal with the collection and presentation of numbers. As a matter of fact, this widely held view quite accurately depicts the original concern of the discipline as well as the continuing concern of one of its modern branches.

DESCRIPTIVE STATISTICS

Early statisticians were typically interested in collecting and displaying information useful to the *state*—and that is how *statistics* got its name. They collected, for example, data on births and deaths to aid officials in charge of the military draft, they collected data on diseases to aid others concerned with the public health, and they collected data on exports, imports, incomes, and expenditures in order to facilitate the collection of taxes. A look at present-day government publications, such as the *Statistical Abstract of the United States* or the *Economic Report of the President,* suggests that the tradition of statisticians-aiding-government has certainly not died out.

Accordingly, Part One of this book deals with **descriptive statistics,** a branch of the discipline that is concerned with developing and utilizing techniques for the careful collection and effective presentation of numerical information. Some statisticians prefer a more narrow definition of descriptive statistics, confining the term to the *presentation* of data that have already been collected by someone else. The broader definition used in this text, however, is a more accurate reflection of the history of the discipline (early statisticians were equally interested in collection and presentation) and, perhaps more important, it is more logical (the careful collection of information, inevitably, is the first step toward its effective presentation). As will be shown in the next chapter, statistical data can be collected with the help of complete or sample surveys, or they can be acquired through experiments. Subsequently, these data can be presented effectively by organizing and condensing them in tables, graphs, or numerical summary measures. Unlike a mere listing of masses of raw data, this type of presentation alone makes the data understandable and often reveals patterns otherwise hidden in them.

The importance of descriptive statistics is illustrated vividly by the examples found in Part One. Consider the matter of *collecting* data. Accountants who wish to acquire accurate information about a firm's accounts receivable, accounts payable, or goods in process could organize a complete count and spend vast amounts of money in the process. As will be shown, however, they can get the same information at a tiny fraction of the cost by carefully collecting small amounts of sample data only. Similarly, researchers who wish to determine the effectiveness of drugs or vaccines or the safety of anesthetics have the option of deriving their conclusions from a few data drawn from carefully designed experiments; if they follow proper procedures of data collection, they can avoid the cost of a massive and expensive study while learning precisely the same things.

Consider, in turn, the matter of *presenting* data. As will be shown, the effective presentation of data can lead a production engineer to discover the secret behind the recent breakdowns of motors produced by a firm, can help a historian unravel the mystery of a disputed authorship, and can aid spies in the cracking of a secret code.

Useful as it may be, descriptive statistics, however, comprises only a small part of the modern discipline of statistics (and of this book). Contrary to the common view that identifies descriptive statistics with the entire field of study, another branch of the discipline is nowadays considerably more important.

ANALYTICAL STATISTICS

Modern statisticians direct most of their effort not toward the collection and presentation but toward the analysis of numerical information. They focus on applying reason to data in order to draw sensible conclusions from them; their chief concern is with the making of reasonable inferences, from the limited information that is available, about matters that are not known. Accordingly, Parts Two through Four of this book deal exclusively with **analytical statistics,** a branch of the discipline that is concerned with developing and utilizing techniques for properly analyzing numerical information. Because analyzing data means drawing inferences from them, this branch of statistics is also referred to as **inferential statistics.**

Sometimes a general truth is inferred from particular instances; at other times, statisticians reverse the process and draw conclusions about the particular from a knowledge of the general. The drawing of inferences about an unknown whole from a known part is **inductive reasoning.** Inductive reasoning is at work, for example, when a statistician concludes, with a 95-percent chance of being right and a 5-percent chance of being wrong, that between 1 and 3 percent of a firm's total output is defective because 2 percent of an output sample did in fact not meet quality standards. On the other hand, the drawing of inferences about an unknown part from a known whole is **deductive reasoning.** Deductive reasoning is at work, for example, when a statistician concludes that a particular unit of a firm's output (namely, a unit produced at plant 7, during the night shift, and with the help of components supplied by firm X) has 5 chances in 100 of being included in a quality test because (a) 5 percent of the firm's total output meets the above criteria and (b) the portion of output to be tested is being selected from total output by a random process and is, therefore, likely to reflect the characteristics of total output. As a matter of fact, deductive reasoning and inductive reasoning are complementary. Before statisticians can safely generalize from the part to the whole, they must study how the part has been generated from the whole.

The importance of analytical statistics is illustrated by the rich array of examples found in Parts Two through Four. As we shall see, statistical techniques help firms screen job applicants, budget research and development expenditures, determine the quality of raw materials received or of output produced, and decide whether sales personnel are better motivated by salary or commission, whether one shipping company makes faster deliveries than another, whether one production method is less time-consuming than another, and even whether people prefer brand A to brand B. Statistical techniques can, similarly, help firms choose the best one among several product designs, leasing arrangements, oil-drilling sites, fertilizer types, and advertising media. And statistics can tell firms precisely how the quantity of their product that is demanded relates to the product's price, to the prices of substitutes and complements, to consumer income, and, perhaps, even to the consumer's sex.

Government officials are equally avid users of what this book has to teach. As will be shown, analytical statistics plays an important role in assuring the reliability of space missions and of more mundane airport lighting systems. And statistical techniques help answer questions such as: Do motorcycle helmets really reduce accident fatalities? Do nursing homes discriminate against Medicaid recipients? Do the boxes of raisins marketed by this firm truly contain 15 ounces as claimed? Do the firms in this state meet antipollution standards? Have the rents in this city risen over the past year? Are wages in New York higher than in Chicago? Do different IRS offices perform work of equal quality? Should this hurricane be seeded to reduce its destructiveness? Is that defendant guilty? Where should the political candidate concentrate further campaign efforts? Do entry barriers explain the degree of dominance achieved by a few firms in this industry? What will happen to tax revenues, the level of employment, and the rate of inflation next year? How are the incomes of people related to their education, job experience, age, sex, or even race? This list, too, can be expanded at will.

All of the previous examples have one thing in common. They illustrate dramatically how analytical statistics can facilitate decision making in the face of uncertainty. Indeed, one can argue that facilitating decision making under uncertainty is the main purpose of the entire discipline.

STATISTICS—A UNIVERSAL GUIDE TO THE UNKNOWN

According to one widely held view, **statistics** is best defined as a branch of mathematics that is concerned with facilitating wise decision making in the face of uncertainty and that, therefore, develops and utilizes techniques for the careful collection, effective presentation, and proper analysis of numerical information. This definition clearly incorporates both descriptive and analytical statistics. Jointly, the two branches help decision makers extract the maximum usefulness from limited information. On the one hand, tables, graphs, and summary calculations highlight patterns otherwise buried in unorganized data. On the other hand, proper inferences provide reasonable estimates of things unknown, along with clearly stated probabilities of such estimates being right or wrong.

Note: By failing to specify who the decision makers are, the above definition quite correctly suggests the universal applicability of what statistics has to offer. As the earlier examples have shown, modern statistical techniques routinely guide business executives, as well as government economic-policy makers, in making reasonable decisions in the face of uncertainty. It may even be true, as some have argued, that modern statistics is particularly useful in the fields of business and economics because controlled experiments can rarely be performed in these fields. Therefore, relationships between variables (as between advertising expenditures and sales or between interest rates and investment) that are of interest to economic decision makers can only be established by applying statistical techniques to raw data that reflect the complicated and simultaneous interaction of large numbers of variables. Nevertheless, the same techniques are just as useful, and are just as frequently applied, outside the fields of business and economics. Like mathematics in general, statistics is a universal type of language that is regularly used by all sciences. The drawing of valid inferences from limited information is just as important to historians and psychologists, to geneticists and medical researchers, to astronomers and engineers as it is to business executives and economists. Of this universal use of statistics we shall see examples as well. Among others, questions such as these will be posed and answered: Do members of Congress influence the actions of the regulatory agencies that they oversee in order to benefit their home districts? Is ESP real? Does age influence IQ? Is toothpaste A better than toothpaste B? Does smoking cause heart disease? Can rats (and maybe people) expect to live lives of normal length with nothing but artificial blood?

Summary

1. The academic discipline of statistics is a branch of mathematics and is itself divided into two branches. *Descriptive statistics* is concerned with developing and utilizing techniques for the careful collection and effective presentation of numerical information.

2. *Analytical statistics,* in contrast, is concerned with developing and utilizing techniques for properly analyzing numerical information. Because analyzing data means drawing inferences from them, this branch of statistics is also referred to as *inferential*

statistics. Inferences can be made by inductive or deductive reasoning. The degree of their validity can be indicated by attaching measures of reliability to them (for example, a 95-percent chance of being right, a 5-percent chance of being wrong).

3. By developing and utilizing techniques for the careful collection, effective presentation, and proper analysis of numerical information, statistics facilitates wise decision making in the face of uncertainty. Statistics can be viewed as a universal guide to the unknown.

Key Terms

analytical statistics
deductive reasoning
descriptive statistics

inductive reasoning
inferential statistics
statistics

Selected Readings

American Statistical Association and Institute of Mathematical Statistics. *Careers in Statistics.* Washington, D.C.: 1974.

Mosteller, Frederick, et al., eds. *Statistics By Example.* Reading, Mass.: Addison-Wesley, 1973. A good introduction to the range of issues dealt with by modern statistics: vol. 1, *Exploring Data;* vol. 2, *Weighing Chances;* vol. 3, *Detecting Patterns;* vol. 4, *Finding Models.*

Tanur, Judith M., et al., eds. *Statistics: A Guide to the Unknown.* 2nd ed. San Francisco: Holden-Day, 1978. A superb volume of 46 essays that describe important applications of statistics in many fields of endeavor. Prepared by a joint committee of the American Statistical Association and the National Council of Teachers of Mathematics.

U.S. Department of Commerce, Office of Federal Statistical Policy and Standards. *Revolution in United States Government Statistics, 1926–1976.* Washington, D.C.: U.S. Government Printing Office, 1978. A review of 200 years of federal government statistics gathering, with particular emphasis on such recent developments as probability sampling and the use of computers.

PART ONE

Descriptive Statistics

CHAPTER TWO

The Collection of Data: Surveys and Experiments

S tatistical work cannot be performed in a vacuum. Before all else, such work requires the acquisition of a crucial type of raw material: information relevant to the subject matter under study. This chapter introduces a number of basic concepts employed in data gathering and then discusses the major methods of accomplishing that task.

BASIC CONCEPTS

A statistical investigation invariably focuses on people or things with characteristics in which someone is interested. The persons or objects possessing the characteristics that interest the statistician are referred to as **elementary units.** Thus, someone who wanted to learn about the racial composition of a firm's labor force would quickly identify the individual employees of that firm as the elementary units, but someone concerned about the amount of credit extended by that firm might view the individual credit accounts issued by it as the elementary units to be investigated. Even the flash cubes produced by the firm, the light bulbs installed in its plants, or the boxes of cereal shipped by one of its divisions could be regarded as elementary units—provided someone was interested in discovering, respectively, the percentage of defective flash cubes produced, the lifetimes of light bulbs used, or the content weights of boxes of cereal sold. A complete listing of all elementary units relevant to a statistical investigation is called a **frame.**

Any one elementary unit may possess one or more characteristics that interest the statistician. An investigator may, indeed, be interested only in the race of each employee, but it would be just as possible to observe, in addition, each employee's sex or salary. The characteristics of elementary units are themselves called **variables,** presumably because observations about these characteristics are likely to vary from one elementary unit to the next.

Any single observation about a specified characteristic of interest is called a **datum;** it is the basic unit of the statistician's raw material. Any collection of observations about one or more characteristics of interest, for one or more elementary units, is called a **data set.** A data set is said to be **univariate, bivariate,** or **multivariate** depending on whether it contains information on one variable only, on two variables, or on more than two.

An example can help us assimilate these new concepts. Consider a statistician who was hired to evaluate charges of racial and sex discrimination in a firm. Using the firm's personnel records, the statistician might record observations about various characteristics (such as race, sex, job title, and the like) associated with each one of the firm's employees. The resultant frame and data set is shown as Table 2.1. Any one entry in column 1 is an elementary unit; all the entries in the shaded portion of that column jointly represent the frame. The headings of columns 2–6 designate the variables; any one entry in these columns is a datum; all entries in these columns together constitute the entire (multivariate) data set. Indeed, we can learn more than this from Table 2.1.

Table 2.1   Selected Characteristics of All the Full-Time Employees of Mountain Aviation, Inc.; July 1, 1987

This table contains a statistical frame and the multivariate data set derived from it. The table illustrates the meaning of a number of basic statistical concepts. Thus, column 1 lists 9 *elementary units* that jointly constitute the *frame* (shaded). The headings of columns 2–6 show various characteristics of the elementary units that are called *variables* and that can be qualitative (race, sex, job title) or quantitative (years of service, annual salary). All possible observations about a given variable constitute a statistical *population*—the shaded entries in column 3 and in column 6 are two examples of populations; any single observation is a *datum;* any subset of a population or of the frame is a *sample.*

Qualitative variable	Employees (1)	Race (2)	Sex (3)	Job Title (4)	Years of Service (5)	Annual Salary (6)	Quantitative variable
	Abel	White	Male	Pilot	2	$17,000	
	Cruz	White	Male	Chief mechanic	10	30,000	
	Dunn	Black	Male	Chief pilot	23	35,000	←Datum
Elementary unit	Hill	Black	Female	Secretary	5	7,000	
	King	White	Male	Janitor	8	8,500	
	Otis	White	Male	Grounds keeper	10	10,000	
	West	Black	Male	Mechanic	2	18,000	Sample of
	Wolf	White	Female	Pilot	7	18,000	←employee
	Zorn	White	Female	Mechanic	7	20,000	salaries
	↑ Frame		↑ Population of employee sexes			↑ Population of employee salaries	

Population Versus Sample

Statisticians call the set of all possible observations about a specified characteristic of interest a **population** or **universe.** As Table 2.1 illustrates, it is possible to draw several such statistical populations from a given frame. Because there are five variables—the headings of columns 2–6—our table contains five populations. The shaded entries in column 3, for example, make up the population of employee sexes; those in column 6 make up the population of employee salaries; and so on for columns 2, 4, and 5.

It should be noted that a statistical population consists of *all possible* observations about a variable. It is irrelevant how many observations, if any, are actually being made. Because they correspond to *all* the employees of our firm, the entries in column 6 make up the population of employee salaries in our hypothetical study, regardless of whether anyone actually gathers these data.

In a firm of only nine employees, it is, of course, easy to observe the entire salary population (as we assumed), but imagine the difficulty of such an undertaking if our statistician attempted to carry out a similar study for the entire aviation industry or even for the entire labor force of the United States! Under such circumstances, statisticians often make observations concerning only selected elementary units. They observe only n such units out of the larger number N that exist. Naturally, they end up with a subset of all the possible observations about the characteristic of interest. Such a subset of a statistical population, or of the frame from which it is derived, is called a **sample.** The boxed entries in column 6 of Table 2.1, for instance, make up one of many possible samples of employee salaries—namely, the sample based on observing the salary characteristics of only West, Wolf, and Zorn. These three names themselves can, in turn, be viewed as a sample of the frame.

But note: What constitutes a population or a sample of it depends entirely on the context in which the question arises. If the goal is to study salaries only at Mountain Aviation, Inc., the data in shaded column 6 of Table 2.1 do, indeed, as a group, make up the relevant population. If the goal were to study salaries in the entire aviation industry, however, the identical column 6 data, even as a group, would constitute only a sample of the now much larger population of salaries throughout the industry.

Qualitative Versus Quantitative Variables

Table 2.1 teaches us something else: Any given characteristic of interest to the statistician can differ in kind or in degree among various elementary units. A variable that is normally not expressed numerically (because it differs in kind rather than degree among elementary units) is called a **qualitative variable.** Table 2.1 contains three qualitative variables: race, sex, and job title. Qualitative variables can, in turn, be *dichotomous* or *multinomial*. Observations about a **dichotomous qualitative variable** can be made only in two categories: for example, male or female, employed or unemployed, correct or incorrect, defective or satisfactory, elected or defeated, absent or present. Observations about a **multinomial qualitative variable** can be made in more than two categories; consider job titles, colors, languages, religions, or types of businesses.

On the other hand, a variable that is normally expressed numerically (because it differs in degree rather than kind among elementary units) is called a **quantitative variable.** Table 2.1 contains two of them: years of service and annual salary. Quantitative variables can, in turn, be *discrete* or *continuous.* Observations about a **discrete quantitative variable** can assume values only at specific points on a scale of values, with gaps between them. Such data differ from each other by clearly defined steps. Consider observing the number of children in families, of employees in firms, of students in classes, of rooms in houses, of cars in stock, of cows in pastures. Invariably, the individual data will be disconnected from each other by gaps on the scale of values; in the above instances, they will look like 1, 2, 3, . . . and 49; never like 3.28 or 20.13. But note: the gaps representing impossible values need not span the entire space between *whole* numbers. Stock prices, for example, are customarily reported in eighths of dollars (that is, in units of $0.125). These discrete figures can take on values of 67⅛, 67⅜, 67⅜ (in dollars per share) but cannot take on values between these. (The quoting of prices by eighths is a throwback to the old pirate days and the Spanish gold "pieces of eight.")

Observations about a **continuous quantitative variable** can, in contrast, assume values at all points on a scale of values, with no breaks between possible values. Consider height, temperature, time, volume, or weight. Weight, for instance, might be reported as 7 pounds or 8 pounds but also as 7.3 pounds or even 7.3425 pounds, depending entirely on the sensitivity of the measuring instrument involved. No matter how close two values are to each other, it is always possible for a more precise device to find another value between them.

Note: The distinction between qualitative and quantitative variables is visually obvious in Table 2.1. The observations about one type of variable are recorded in words; those about the other type in numbers. Yet that distinction can easily be blurred. Quantitative variables can be converted into seemingly qualitative ones, and the opposite is also true. Thus, a statistician could replace the column 5 and 6 data by words, such as *low, intermediate,* or *high,* although, presumably, nobody would in this way wish to give up the more precise information presently recorded in Table 2.1. On the other hand, it is very common to *code* observations about qualitative variables with the help of numbers. Thus, a statistician might turn the verbal entries of Table 2.1 into *numbers* by recording "white" as 1 and "black" as 2 in column 2, by recording "male" as 0 and "female" as 1 in column 3, and by assigning numbers between 0 and 6 to the seven job titles in column 4. Nevertheless, the distinction between qualitative and quantitative variables, although then hidden, would remain. Even when they are encoded numerically, one cannot perform arithmetic operations with qualitative data, while one can perform them with quantitative ones. Thus, it would make no sense to report the "sum of races" in our firm as 12 (using the code just noted), but one can report the sum of annual salaries as $163,500.

Types of Data

The preceding discussion makes clear why statisticians, who inevitably work with numbers, must be keenly aware of what, if anything, is measured by these numbers. The assignment of numbers to characteristics that are being observed—which is **measurement**—can yield four types of data of increasing sophistication. It can produce nominal, ordinal, interval, or ratio data, and different statistical concepts and techniques are appropriately applied to each type.

The weakest level of measurement produces **nominal data,** which are numbers that merely *name* or label differences in kind and, thus, can serve the purpose of classifying observations about qualitative variables into mutually exclusive groups. As noted, "male" might be numbered as 0, "female" as 1, but alternative labels of "male" = 100 and "female" = 17 would serve as well. Other examples of creating nominal data are classifying defective units of a product as 1, satisfactory units as 2; labeling rooms on first, second, or third floors by numbers in the 100s, 200s, or 300s, respectively; or designating rooms on the north or south side of buildings by even or odd last digits in room numbers: 102, 104, 106 for first-floor rooms to the north; 301, 303, 305 for third-floor rooms facing south. As these examples confirm, it never makes sense to add, subtract, multiply, or divide nominal data, but one can *count* them: The presence of five 17s, by one of the above codes, denotes the presence of five females; 25 odd room numbers mean there are 25 rooms facing south.

The next level of measurement produces **ordinal data,** which are numbers that by their size *order* or rank observations on the basis of importance, while intervals between the numbers, or ratios of such numbers, are meaningless. Assessments of a product as *superb, average,* or *poor* might be recorded as 2, 1, 0, as 250, 10, 2, or even as 10, 9, 4.5—the important thing is that larger numbers denote a more favorable assessment, or a higher ranking, while smaller ones do the opposite. Yet ordinal data make no statement about *how much* more or less favorable one assessment is compared to another. A 2 is deemed better than a 1 but not necessarily twice as good; a 250 is deemed better than a 10 but not necessarily 25 times as good; a 4.5 is deemed worse than a 9 but not necessarily half as good; and that is all. Thus, an *ordinal* coding of "professor," "associate professor," and "assistant professor" as 3, 2, and 1 does imply that, in some sense, professors are viewed as more important than assistant professors, while a *nominal* coding of male = 100, female = 17 does not imply the superiority of males over females any more than a coding of male = 0, female = 1 denotes the opposite. In either case, arithmetic operations with the data are out of the question.

Somewhat more sophisticated are **interval data** that permit at least addition and subtraction. Interval data are numbers that by their size rank observations in order of importance and between which *intervals* or distances are comparable, while their ratios are meaningless. Because these data do not possess an intrinsically meaningful origin, their measurement starts from an arbitrarily located zero point and utilizes an equally arbitrary unit distance for expressing intervals between numbers. Commonly used temperature scales are examples of interval data. The Celsius scale places zero at the water-freezing point; the Fahrenheit scale places it far below that. Within the context of either scale, the unit distance (degree of temperature) has a consistent meaning: each degree Celsius equals 1/100 of the distance between water's freezing and boiling points; each degree Fahrenheit equals 1/180 of that distance. Note that the zero point, being arbitrarily located, does not denote the absence of the characteristic being measured. Unlike 0° on the absolute (Kelvin) temperature scale that is familiar to astronauts, neither 0°C nor 0°F expresses a complete absence of temperature. Note also that any ratio of interval data is meaningless: 90°F is *not* twice as hot as 45°F. Indeed, the ratio of the corresponding Celsius figures (32.2° and 7.2°) does not equal 2:1 but well over 4:1. This discussion brings us to the most sophisticated types of numbers of all.

The most useful types of data, such as those in columns 5 and 6 of Table 2.1, are **ratio data,** which are numbers that by their size rank observations in order of importance and between which intervals as well as *ratios* are meaningful. All types of arithmetic operations can be performed with such data because these types of numbers have a natural or true zero point that denotes the complete absence of the characteristic they measure and makes the ratio of any two such numbers independent of the unit of measurement (unlike in the Fahrenheit-Celsius example). Thus, the measurement of salaries, age, distance, height, volume, or weight produces ratio data. For example, because it is meaningful to rank salary data and to say that a salary of $15,000 is larger than one of $10,000, which is larger than one of $5,000, salary data give the kind of information provided by ordinal data. Yet in addition, because it is meaningful to compare intervals between salary data and to say that the distance between $15,000 and $10,000 equals the distance between $10,000 and $5,000, these data also give the type of information provided by interval data. Further, these data are ratio data because we can safely describe $15,000 as three times as much money as $5,000; even a change in the unit of measurement, as from dollars to cents or from dollars to francs (at an exchange rate of, say, 6 francs to the dollar) does not change this conclusion: 1,500,000 cents still is three times as much money as 500,000 cents; 90,000 francs still is three times as much as 30,000 francs. The fact that ratios of numbers convey meaningful information is the advantage of ratio data over interval data. No wonder that statisticians, when they have a choice, prefer ratio data to interval data, interval data to ordinal data, and ordinal data to nominal data.

Having discussed the major types of data, we introduce, finally, the two basic methods of generating them.

Surveys Versus Experiments

New data can be generated either by conducting a survey or by performing an experiment. A **survey,** or **observational study,** is the collection of data from elementary units without exercising any particular control over factors that may make these units different from one another and that may, therefore, affect the characteristic of interest being observed. A characteristic such as the annual salary of workers, for example, is simply being observed and recorded for different workers without regard to factors, such as education, work experience, or length of service, that make workers different from one another and that may be responsible for differences in their salaries. In contrast, an **experiment** involves the collection of data from elementary units, while exercising control over some or all factors that may make these units different from one another and that may, therefore, affect the characteristic of interest being observed. Thus, a firm may divide its 40 new employees into two similar groups of equal size (with the help of some random device) and then administer an obligatory special training program to one of the groups only. If the 20 employees who went through the program exhibited superior productivity later on, the training program might justifiably be credited because other factors that could account for this result (such as group differences in age, motivation, or prior work experience) were eliminated by the random division of the original group of 40 newcomers. Note: If the same data on superior productivity had been generated from a group of volunteers, the data would have to be viewed as mere observational ones because of the lack of special controls. Without the random assignment, some systematic difference might have emerged between the two groups of workers even prior to the training program. If only more mature, more motivated, or more experienced workers volunteered for the training, they would probably have shown greater productivity even without the program. Thus, the higher productivity later on could not be credited to that program with absolute confidence. Experimental data tend to be stronger than survey data.

Unfortunately, most data in business and economics, and in many other fields, are generated not by experiments but by surveys because it is often impossible, or extremely costly, to exercise proper experimental controls. One cannot easily divide the country's labor force (or even a segment of it) at random into three groups and then subject each group to different tax rates in order to study the effect of taxes on labor supply. One cannot simply divide the country's newly born (or even a segment of them) at random into two groups and then subject only one of the groups to a lifetime of smoking in order to study the effect of smoking on health. Nevertheless, with proper statistical techniques, one can learn a great deal about such matters even from survey data. Accordingly, the remainder of this chapter will first deal with surveys (this being the most common method of generating data); it will then point out the errors likely to be associated with survey procedures and will, finally, return to the subject of data generation by means of experiments.

SURVEYS: CENSUS TAKING OR SAMPLING*

Two types of surveys exist: complete and partial ones. In a complete survey, or **census,** observations about one or more characteristics of interest are being made for every elementary unit that exists. Accordingly, the resultant data set consists of at least one statistical population but possibly several of them, as in columns 2–6 of Table 2.1. But note: When the number of elementary units is very large, complete success in observing all of them is likely to elude the census takers. Nevertheless, any reasonably successful attempt to collect data on an entire statistical population is referred to as a census.

The precise procedure of data collection may vary. It may involve *direct observation* of some ongoing activity by an investigator who might record, say, the net weights of boxes of cereal as they are being filled by a machine or the numbers and brands of such boxes as they are being picked up by customers from a supermarket shelf. A census may instead involve a *personal or telephone interview* in which the investigator asks questions printed on a carefully prepared list and records the verbal answers given by people. Finally, a census may proceed by *self-enumeration,* as when people, having read a set of instructions, make written replies on questionnaires that they received in the mail, on the street corner, or, perhaps, when purchasing an appliance. More often than not, however, these same procedures are applied to a different type of survey. In a partial or **sample survey,** observations about one or more characteristics of interest are being made for only a subset of all existing elementary units. The resultant data set consists of one or more samples, as noted in column 6 of Table 2.1, and it is hoped that the sample data mirror the population data in the sense that valid inferences can be drawn from any given sample about the corresponding, but unknown, statistical population. There are good reasons why sample surveys are often undertaken in place of censuses.

———————
*Optional section.

The Reasons for Sampling

In one way or another, all the reasons typically cited for preferring sampling to census taking have something to do with reducing the cost of getting a given type of information or with increasing the quantity or quality of information received with a given cost.

First, the cost of collecting and processing data is obviously lower the fewer are the elementary units that have to be contacted. This becomes a crucial consideration whenever the number of relevant elementary units is large. The U.S. Bureau of Labor Statistics, for example, gathers its monthly unemployment data from a mere 60,000 households, not by questioning the more than 100 million members of the labor force. Yet—and this is the crucial point—the information so received is, nevertheless, a reliable indicator of what would be known if a complete census were taken. When first encountered, this statement is hard to believe, but consider an analogy. Suppose we wanted to determine the proportion of green and yellow peas in a sackful of them. If the peas were thoroughly mixed, we could surely take a mere cupful from the sack and get our answer by counting the peas in that cup. The result would be almost exactly the same as when all the peas in the sack were counted. And the same would continue to be true if we had not a sackful, but a truckload of peas, or even an entire trainload of them. Even then a sample of a single cupful would answer our question, *provided* the peas were still thoroughly mixed. In the same way, one can count the employed (green peas) and the unemployed (yellow peas) in the labor force, provided the sample is properly selected (contains a thorough mixture of the employed and unemployed). Thus, desired information can, indeed, be gained at a fraction of what a census would cost. For the same reason, business firms that are interested in consumer preferences concerning old or new products never survey all existing consumers but only a tiny percentage of them. Firms even answer questions about their *internal* operations by sampling whenever a census would be too unwieldy and costly. Consider a bank that wishes to ascertain the percentage of errors made when crediting monthly interest to some 3 *million* savings accounts or when billing some 5 *million* credit accounts. The clerical cost of conducting a census of all accounts would be enormous.

Second, a census is sometimes physically impossible (that is, infinitely costly), as when the number of elementary units is infinitely large or when some of them are totally inaccessible. Any *process* that is expected to operate indefinitely under identical conditions, for example, generates an infinite number of outcomes. Thus, a census can never observe all the outcomes of this process. It can never record, for example, all the defective memory chips likely to be produced by a new and ongoing production process or all the effects ever to be produced by a new drug. In such situations, sampling is inevitable. The same is true when some elementary units are practically impossible to contact, such as airplanes presumed to have crashed in the ocean or in remote mountain regions or people with addresses unknown or temporarily hospitalized in intensive care. Third, a census is senseless (that is, again infinitely costly) whenever the acquisition of the desired information destroys the elementary units of interest. Consider questions about the lifetime of batteries or about the quality of flash cubes produced by a firm. If every battery or flash cube were tested, all of the output would be used up and the answers to the original questions would be useless. Fourth, a census is senseless whenever it produces information that comes too late. Consider political opinion polling undertaken prior to an election. A census of many millions of registered voters (or even repeated censuses in the face of rapidly fluctuating preferences) would take too long to yield results; only sampling can provide the desired information in time. Fifth, for a given cost, sampling can provide *more detailed* information than a census. Consider how many more questions the Bureau of Labor Statistics could ask those 60,000 households if it spent the same amount of money, or even a quarter of the money, that it would have to spend on a more-than-100-million-person census of the entire labor force. Sixth, sampling can provide *more accurate* data than a census. This seems paradoxical but is true because fewer statistical workers are needed, and they can be better trained and more effectively supervised. Thus, for a given cost, higher quality information is received.

Types of Samples

Different types of samples are obtained depending on the method by which elementary units are selected for observation. These include convenience samples, judgment samples, and various forms of random samples. When expediency is the primary consideration and only the most easily accessible elementary units are chosen for observation, the resultant subset of the frame, or of an associated population, constitutes a **convenience sample.** It is unlikely that a convenience sample of a statistical population is representative of it in the sense that valid inferences can be drawn from the sample about the population. The selection procedure based on convenience almost assures the opposite. The boxed salary figures in column 6 of Table 2.1 provide an example. For purposes of illustrating the concept of a sample, it was simply *convenient* to combine the last three numbers in that column, but no serious thought was given about their representativeness of the entire column 6 salary population. In the same way, a member of Congress who judges constituent attitudes on the basis of mail received is taking a convenience sample; so is the aide who telephones constituents about their opinions—without regard to the opinions of those who happen not to answer the phone, who do not have a phone, or who have unlisted telephone numbers.

A somewhat more sophisticated type of sample emerges when personal judgment, presumably based on prior experience, plays a major role in selecting elementary units for observation. Because the "expert" judgment is believed to make the sample representative of the whole, such a subset of the frame, or of an associated population, is called a **judgment sample.**

Making such a judgment can be next to impossible, however, especially when the elementary units are heterogeneous and when the desired sample is small. Consider, for example, selecting a "representative" group of 10 persons from among 1,000 people about whom we already know a great deal. If there were 400 males and 600 females, we could select 4 males and 6 females for our sample, and in this sense our sample would be a miniature of the population. But what about all the other characteristics people have, such as age, education, income, and race? If half the males and two thirds of the females were over 50 years of age, we could, of course, make sure that two of our 4 males were under 50 and two of them were over 50 and that two of our 6 females were under and four of them over that age. If the population also contained 20 percent blacks among males and 30 percent blacks among females, however, we couldn't also arrange for 0.8 black males and 1.8 black females to enter our sample. We might choose 1 black male and 2 black females, of course, but already our sample would have ceased being a miniature of the whole. The more characteristics we considered, the worse our problem would get. The more we tried, the more our judgment sample would look like a convenience sample, as if it were determined solely by our personal whim. Nevertheless, "expert" judgment is oftentimes used to select samples.

A good example of such judgment at work is the construction of the monthly Consumer Price Index (which will be discussed more fully in Chapter 16). Someone decides, simply on the basis of personal judgment, which prices, among literally billions of prices, are to be sampled and what weights are to be assigned to them. Determining the CPI involves complex decisions on which stores in which geographic areas are to be surveyed on which days concerning the prices of which products. Clearly, a price charged many customers in a popular store is more important than one charged few customers in a nearly abandoned one; prices in tiny Hadley, Massachusetts, affect fewer people than prices in Chicago; more people shop on Saturdays than on Mondays (and then take advantage of weekend specials); and automobile tires, fruit, and meat somehow *seem* more important than horseshoes, tea, and matches. On all these matters and more, "expert" judgment is brought to bear. When it is wrong, as is easily possible, the sample ends up unrepresentative of the associated whole.

Most important, because it avoids the problem of unrepresentativeness, is the **random sample,** or **probability sample,** which is a subset of a frame, or of an associated population, chosen by a random process that gives each unit of the frame or associated population a known positive (but not necessarily equal) chance to be selected. If properly executed, the random selection process allows no discretion to the investigator as to which particular units in the frame or population enter the sample. As a result, such a sample tends to maximize our chances of making valid inferences about the totality from which it is drawn. Because random samples are so important, we must look at them in some detail. Four major types of random samples exist. We will consider each of them in turn, always assuming that the sample is to be taken (a) from a *finite* frame or population (containing a countable number of units) and (b) without replacement (so that any given unit can enter the sample only once).

The Simple Random Sample

A **simple random sample** is a subset of a frame, or of an associated population, chosen in such a fashion that every possible subset of like size has an equal chance of being selected. This procedure implies that each individual unit of the frame or population has an equal chance of being selected, but the converse is not true (giving each individual unit an equal chance of being selected does not assure that every possible subset of like size has an equal chance). Consider a frame consisting of six elementary units, A, B, C, D, E, and F. Suppose we divided these units into two categories, a subset of larger units (A, D, E) and a subset of small units (B, C, F). One *could* decide to select one of these two subsets as the sample, if a sample of three were desired, and the actual sample could then be selected by simply tossing a coin. Each of the two listed samples would have an equal chance of being selected—namely, 1 out of 2, depending on whether heads or tails appears. The fact that each of these samples would have an equal chance of being selected would imply that each individual unit contained in either one of these two samples, and, therefore, each unit in the frame, would also have an equal chance (1 out of 2) of being selected. Yet this procedure would not satisfy the above prescription for selecting a simple random sample. There would be some possible samples of size three—namely, combinations of large *and* small elementary units (such as A, B, C or C, E, F)—that would have a *zero* chance of being selected. In order to select a simple random sample of three from the above list of six elementary units, each conceivable subset of three must be given an equal chance of being selected. There are 20 such conceivable subsets[1]:

ABC	ACD	(ADE)	AEF	BCD	BDE	BEF	CDE	CEF	DEF
ABD	ACE	ADF		BCE	BDF		CDF		
ABE	ACF			(BCF)					
ABF									

In order for the sample to qualify as a simple random one, all 20 of these subsets, and not just the encircled ones, must have an equal chance (here 1 in 20) of being selected.

A simple random sample of a city's residents, obviously, cannot be obtained by interviewing people on Main Street at 11 A.M. on a Monday. Samples including people at work or in hospitals, for example, would have no chance of being selected. Indeed, the most common procedure of obtaining a simple random sample involves a lottery process applied to the relevant frame or population. Consider the alphabetical listing, in Table 2.2, of the 100 largest multinational companies headquartered in the United States. For someone who wanted to investigate characteristics of these companies, such as their foreign and domestic sales, their profits, or their assets, this listing represents the frame. If a census were to be undertaken, relevant data could be collected for every one of the companies, and these would represent the statistical populations of sales, profits, or assets. Yet how could one acquire simple random samples of such populations corresponding to, say, a mere 10 of these companies?

One procedure is akin to a raffle. One records each of the 100 company names on a separate slip of paper, places the slips of paper in a container, mixes thoroughly, and draws 10 slips. The firms named on the slips drawn are then investigated with respect to the variables of interest. This procedure is, however, troublesome when there is a large number of elements because it is difficult to mix them properly. (Close-Up 2.1, "The 1970 Draft Lottery Fiasco," provides a vivid illustration of this fact.)

A more common procedure nowadays makes use of a **random-numbers table;** that is, a listing of numbers generated by a random process in such a way that each possible digit is equally likely to precede or follow any other one. In such a table, there is no discernible pattern in the way the digits appear. Suppose we wanted to construct our own table of five-digit random numbers. All we would have to do is write the numbers from 0 to 9 separately on 10 slips of paper, mix the slips in a bowl, and pull out one. We would have the first digit for our first number. After returning the slip we pulled to the bowl, we would once more mix the 10 slips, pull out one, and

---

[1]The number of different samples of size n that can be drawn without replacement from a list of N elements equals $\dfrac{N!}{(N-n)!\,n!}$ which, in this example, equals

$$\frac{6!}{(6-3)!3!} = \frac{6!}{3!3!} = \frac{6\cdot5\cdot4\cdot3\cdot2\cdot1}{(3\cdot2\cdot1)(3\cdot2\cdot1)} = \frac{720}{36} = 20.$$

The expression $N!$ is pronounced "N factorial"; its meaning is obvious from the example. But note: By definition, $1! = 1$ and $0! = 1$. Factorials are discussed further in Chapter 5.

Table 2.2   The 100 Largest U.S. Multinationals

This table shows the names of those 100 U.S.-based multinational companies that had the highest foreign revenues in 1981. Such revenues ranged from $75.8 billion (Exxon) to $1.1 billion (Ingersoll-Rand). Company names have been coded with two-digit numbers from 00 to 99.

00 Allied Corporation	50 Honeywell
01 American Brands	51 IBM
02 American Cyanamid	52 Ingersoll-Rand
03 American Express	53 International Harvester
04 American Home Products	54 International Telephone & Telegraph
05 American International Group	55 Irving Bank
06 Atlantic Richfield	56 Johnson and Johnson
07 Avon Products	57 K mart
08 Bank America	58 Litton Industries
09 Bankers Trust New York	59 Manufacturers Hanover
10 Beatrice Foods	60 Marine Midland Banks
11 Bendix	61 Merck
12 Bristol-Myers	62 Minnesota Mining & Manufacturing
13 Burroughs	63 Mobil
14 Caterpillar Tractors	64 Monsanto
15 Chase Manhattan	65 Morgan, J. P.
16 Chemical New York	66 Motorola
17 Chrysler	67 Nabisco Brands
18 Citicorp	68 NCR
19 Coca-Cola	69 Occidental Petroleum
20 Colgate-Palmolive	70 Pan American
21 Consolidated Foods	71 Pepsi Co.
22 Continental Group	72 Pfizer
23 Continental Illinois	73 Phibro-Salomon
24 CPC International	74 Philip Morris
25 Dart and Kraft	75 Phillips Petroleum
26 Deere	76 Procter and Gamble
27 Digital Equipment	77 Ralston Purina
28 Dow Chemical	78 RCA
29 Du Pont de Nemours	79 R. J. Reynolds Industries
30 Eastman Kodak	80 Safeway Stores
31 Exxon	81 Scott Paper
32 Firestone	82 Sears, Roebuck
33 First Chicago	83 Singer
34 First National Boston	84 Sperry
35 Fluor	85 Standard Oil California
36 Ford Motor	86 Standard Oil Indiana
37 General Electric	87 Sun Co.
38 General Foods	88 Tenneco
39 General Motors	89 Texaco
40 General Telephone and Electric	90 Texas Instruments
41 Getty Oil	91 Trans World
42 Gillette	92 TRW
43 Goodyear	93 Union Carbide
44 Grace, W. R.	94 Union Oil California
45 Gulf and Western Industries	95 United Technologies
46 Gulf Oil	96 Warner-Lambert
47 Halliburton	97 Westinghouse Electric
48 Heinz, H. J.	98 Woolworth, F. W.
49 Hewlett-Packard	99 Xerox

Source: Forbes, July 5, 1982, pp. 126–28.

thus get the second digit; then, repeating the process, we would determine the third, fourth, and fifth digits. We would, thus, have created our first five-digit random number, and if we cared to we could create 500 of them, such as those listed in Appendix Table A. Actually, the numbers listed there come from a list of 1 million random numbers generated by a computer at the Rand Corporation. Because the computer, like an electronic roulette wheel, followed the selection principle outlined above, each value between 00000 and 99999 had an equal chance of appearing at each of the 500 locations given in the table. (See Close-Up 2.2, "The Making of Random Numbers".)

It is easy to select a simple random sample of the multinational companies with the help of our random-numbers table. For this purpose, the 100 company names have already been coded by two-digit numbers ranging from 00 to 99. (These numbers are a good example of nominal data.) We decide which 10 of the companies are to be in the sample simply by entering the random-numbers table at random and reading off two-digit numbers in any predetermined, systematic way. The first 10 two-digit numbers encountered that correspond to previously unused company codes tell us what the sample is. Let us elaborate.

Many random-numbers tables, unlike Appendix Table A, consist of many pages. It is a good idea to begin the process of selecting a simple random sample by first selecting a page of the table and then selecting a number on that page, at random. Any nonpatterned procedure will do. For example, we might simply open the 400-page Rand Corporation table at random and select page 377. (This page has, in fact, been reproduced as Appendix Table A.) Closing our eyes, we might then place a finger on the chosen page and thus determine number 00717 (row 24, column 5) as our starting number. Actually though, we should not even go about finding this number until we have decided how we will use it and how we will then proceed to find other numbers. If, as in our case, the data consist of two digits, while the random numbers contain five, we must decide how we will extract two-digit numbers from our chosen page. We could read only the first two digits of each five-digit number, we could read the last two digits only, or even the second and last digits in each number. A multitude of possibilities exist. And having found our first *two-digit* number, we can go about finding additional ones (up to the desired sample size) in any systematic fashion we like: by proceeding along our initial row to the right or to the left, by moving along our initial column down or up, even by reading numbers diagonally. We can follow any self-imposed plan we like, as long as we never read a given location twice.

Suppose, for example, that we had decided to select our sample of 10 companies by reading the last two digits of the five-digit random numbers and by proceeding down along (possibly successive) columns. Since our initial number was the 24th one in the fifth column of Appendix Table A, our simple random sample would quickly be determined as follows:

007⃝17 Chrysler
107⃝97 Westinghouse Electric
739⃝35 Fluor
53497 omitted (97 used above)
692⃝14 Caterpillar Tractors
747⃝74 Philip Morris

945(82) Sears, Roebuck
151(07) Avon Products
807(56) Johnson and Johnson
894(27) Digital Equipment
43335 omitted (35 used on p. 20)
830(01) American Brands

Two things should be noted: First, if one encounters a given value a second time during the above procedure (as happened with numbers 97 and 35 in our case), one skips to the next number. The same is recommended when the value encountered has no counterpart in the frame or population being sampled (as when one is choosing from a list of 60 companies only and one encounters numbers between 61 and 99 in the random-numbers table).

Second, there are those who argue that the above procedure for finding a starting position (our number 00717) may not be sufficiently random for *frequent* users of a random-numbers table. This may be so because a given book has the tendency to open to the same page repeatedly (and, thus, a frequent user might again and again select page 377), while a person placing a finger on a page has a tendency to choose a number toward the center of the page (as we did). In order to guard against these dangers, one can, of course, select a starting position in any number of truly random ways. For example, if we were working with an eight-page table, we might select a page by tossing a coin three times. There are two possible outcomes for the first toss, heads (H) or tails (T). There are two possible follow-ups for each of these two contingencies on the second toss, making for sequences of HH, HT, TH, or TT. And there are two possible follow-ups for each of these four contingencies on the third toss, making for eight conceivable sequences for the three tosses. These sequences might be linked to the choice of a page in the random-numbers table, as shown in the accompanying table.

Possible Sequence	Page Number Picked
H H H	1
H H T	2
H T H	3
H T T	4
T H H	5
T H T	6
T T H	7
T T T	8

Having thus found the page by tossing a coin three times, we might determine the starting position by reading two digits at a time along successive rows and letting the first usable set of two digits point us to a row and the second such set to a column.

(Can you see why, in Appendix Table A, this procedure would have us start with row 27 and column 01; that is, with number 20545?) The possibilities of random selection are truly limitless.

The Systematic Random Sample

Another commonly used type of sample, the **systematic random sample,** is a subset of a frame, or of an associated population, chosen by randomly selecting one of the first k elements and then including every kth element thereafter. If this procedure is employed, k is determined by dividing population size, N, by desired sample size, n. Consider again Table 2.2. To select a systematic random sample of, say, five companies, one would wish to include every 20th company on the list (there being a total of 100 firms); hence $k = 20$. The procedure, however, requires that the starting point (the first company included) be chosen at random. By drawing a number between 00 and 19 out of a hat, or by using a table of random numbers, we might decide that company number 09 is to be selected as the starting point. Then our entire sample would consist of the companies coded as 09, 29, 49, 69, 89. Had our initial choice been 01 or 17, the sample would have been different, of course—namely, 01, 21, 41, 61, 81 or 17, 37, 57, 77, 97, respectively.

The Stratified Random Sample

Sometimes the frame or population to be sampled is known to contain two or more mutually exclusive and clearly distinguishable subgroups or *strata* that differ greatly from one another with respect to some characteristic of interest, while the elements within each stratum are fairly homogeneous. In such circumstances, one can select a **stratified random sample,** which is a subset of a frame, or of an associated population, chosen by taking separate (simple or systematic) random samples from every stratum in the frame or population, often in such a way that the sizes of the separate samples vary with the importance of the different strata. If we knew that 10 of the 100 companies listed in Table 2.2 accounted for 70 percent of the 100 companies' sales, while the other 90 companies accounted for the remaining 30 percent, and if sales were the characteristic of interest to us, we might wish to ensure that the 10 giants were not missed by our sampling procedure (as might well happen if a simple or systematic random sample were taken). We could divide our list into two strata (10 giants and 90 dwarfs) and then create our overall sample by selecting some firms from each of these two groups. For a 10-percent overall sample, we might select 8 firms from the giant stratum and 2 firms from the dwarf stratum, and these 10 firms could be expected to account for more than half of all sales.

This type of procedure is, in fact, quite common. Some examples from the business world are summarized in Analytical Example 2.1, "How Accountants Save Money by Sampling." Other examples of stratified sampling include a study of banking (wherein strata were formed to ensure representation of very large banks that may be few in number but that hold a high percentage of all deposits), a study of affirmative-action programs (concerned with the representation of all ethnic or racial groups), and a study of teacher qualifications (designed to ensure the proportional representation of elementary-school, high-school, and college teachers). Stratified random sampling has also become crucially important for pollsters who wish to predict the outcome of U.S. presidential elections. It is not good enough to estimate correctly which candidate will win the popular vote in the nation as a whole. Such a candidate can still lose the election because the president is elected by the Electoral College in which representatives traditionally vote as a block *by state,* and electoral votes are rationed unequally to each state in accordance with its representation in Congress. To be successful, a candidate must, therefore, get popular vote margins in the large states. Hence, a pollster must treat each state as a separate stratum, must determine the winner in each state, and must combine the results in the same proportion as the electoral votes.

The Clustered Random Sample

Finally, there are occasions when the frame or population to be sampled is naturally subdivided into *clusters* on the basis of physical accessibility. In such circumstances, one can select a **clustered random sample,** which is a subset of a frame, or of an associated population, chosen by taking separate censuses in a randomly chosen subset of geographically distinct clusters. Someone who wanted to sample the residents or shops of a city, for example, might divide the city into blocks, randomly select a few of these (by any of the methods previously discussed), and then interview every resident or shop owner within the chosen blocks. Because of the geographic proximity of those interviewed, this procedure would save considerable transportation expenses and time in comparison to a citywide simple random sample of individual residents or shop owners (who would almost certainly be located in a multitude of different places). Acceptance sampling by business firms is likely to follow a similar route. Imagine a firm that had just received a shipment of 1 million coffee cups. Its warehouse would be filled, perhaps, with 10,000 sealed cartons, each containing 100 cups. An inspector who wanted to determine the quality of the cups by taking a 10,000-cup sample could take a simple or systematic random sample of them, but might then have to open and unpack nearly all of the 10,000 cartons in order to find the individual cups selected. It would be much easier to regard each carton as a geographic cluster, to select 100 cartons randomly, and to inspect every single cup in these cartons only. Some 9,900 of the cartons would not have to be opened at all.

Note: The procedure described here is also called *single-stage* cluster sampling. On occasion, this type of sampling is replaced by the more complex *multistage* cluster sampling. An example of the latter might be a nationwide household survey conducted as follows: first, a subset of states (primary clusters) is chosen at random; second, a subset of cities (secondary clusters) is randomly chosen within the previously selected states; third, a subset of city blocks (tertiary clusters) is randomly selected within the cities chosen before. As in the earlier example, censuses occur in the city blocks finally selected.

Note also that the random selection of clusters need not necessarily involve simple or systematic random sampling. By an appropriately designed type of stratified sampling, different clusters might, for example, be given unequal chances of selection, depending, perhaps, on the proportions of relevant units (such as people, shops, or coffee cups) that are located in different clusters (such as cities, city blocks, or cartons).

ERRORS IN SURVEY DATA*

We have noted how data can be acquired by taking a census or by engaging in various forms of sampling. Inevitably, all survey data are subject to error; errors can arise from innumerable, often unexpected, sources. At best, errors obscure the truth only slightly; at worst, they can reduce the value of a survey below zero, for nothing is more unfortunate than "knowing" something that isn't so. Errors can be generated during the planning stage of a survey; more errors can be created during the later stages of data recording and processing. Indeed, it is a hopeless task to list all the ways in which surveys can go wrong, and we won't attempt it. It is possible, however, to raise our awareness of the problem by focusing on two broad categories of error.

Random Error Versus Systematic Error

Suppose we were interested in the average annual salary received by the employees of a particular firm, such as those of Mountain Aviation, Inc. discussed earlier. If we had none of the information given in Table 2.1, except the listing of employee names, we could take a random sample of this frame, interview the employees so selected about their salaries, and let the average of these salaries serve as an estimate of the average salary of all employees.(The choice of such a sampling procedure would be all the more likely, of course, the greater was the number of employees of the firm.) A random sample of three might consist of Cruz, Hill, and West, and we could estimate the average annual salary of all employees by the average of the Cruz, Hill, and West salaries listed in column 6; that is, as $18,333 per year. Yet it is clear that this result depends entirely on the particular elementary units that happened to be selected for our sample. One can, in fact, select 84 different samples of three from among our nine hypothetical employees (see footnote 1), and almost every one of these 84 samples would provide us with a different average-salary estimate, if the above procedure were followed. The sample consisting of the salaries of West, Wolf, and Zorn—see the box in column 6 of Table 2.1—for example, would give us an average-salary estimate of $18,667 per year, but a sample consisting of the salaries of Hill, Otis, and Zorn would yield an estimate of $12,333 per year. All these estimates differ from the average-annual-salary figure a census would reveal, for the average of *all* the column 6 data is $18,167 per year. But note: The individual-sample estimates are just as likely to depart from the census figure in one direction as the other. In the long run, if one took repeated samples and averaged the resulting estimates of average salary made by all possible samples of three, the same census-derived figure would be found.

Thus, the first major type of error of which statisticians must be aware is called **random error,** or **chance error,** or **sampling error,** and equals the difference between the value of a variable obtained by taking a single random sample and the value obtained by taking a census (or by averaging the results of all possible random samples of like size). This definition is summarized in the left half of Figure 2.1.

This type of error, as one of its names suggests, is associated only with sample surveys. It arises from the operation of chance that determines which particular units of the population happen to be included in the sample. This error can, of course, be positive or negative, tiny or huge, but it can always be reduced by increasing the size

---

*Optional section.

Figure 2.1 Random Error and Systematic Error

This graph illustrates how values obtained by data gatherers can differ from true values as a result of random error (in the case of random sampling) or of systematic error (in the case of all types of surveys). Either type of error can be positive or negative.

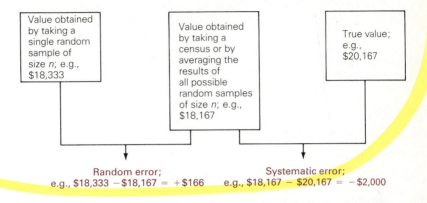

Value obtained by taking a single random sample of size *n*; e.g., $18,333

Value obtained by taking a census or by averaging the results of all possible random samples of size *n*; e.g., $18,167

True value; e.g., $20,167

Random error; e.g., $18,333 − $18,167 = +$166

Systematic error; e.g., $18,167 − $20,167 = −$2,000

or number of random samples taken (and it is zero in a census). Most important, as later chapters will show, the size of this error can itself be estimated (and reported alongside the observed data). As a result, statisticians can state, for example, that the average annual salary in a firm equals $18,333 and that there are 95 chances in 100 that a census would reveal a number within $205 (plus or minus) of the stated one.

Unfortunately, random errors are the least of our problems. Even if we had the complete census information given in column 6 of Table 2.1, we could not be *certain* that the average annual salary in our firm was $18,167. The true value might well equal $20,167, as noted on the right-hand side of Figure 2.1. This surprising statement is true because every type of survey (sampling or census taking) could produce incorrect data if it was subject to some fundamental flaw in the manner in which observations were made or recorded. Such a flaw might affect every single measurement in the same way, systematically pushing each one either below or above its true value.

Consider the story of the sales clerk who measured yards of cloth with an old, stretched-out cloth tape, thus causing all the recorded lengths to be lower than the true lengths. A similar predicament could befall the survey taker. Unbeknownst to anyone, an incorrectly programmed computer might, for example, deduct $2,000 from every true salary figure entered in it, and it might thus print out an incorrect set of column 6 data (in the case of a census) or an incorrect subset of them (in the case of sampling). The data, mind you, would look precisely like those in Table 2.1, but they would be wrong, nevertheless.

The people who are being interviewed can be at fault, too. Consider the story of the butcher who weighed each piece of meat with a thumb on the scales, thus causing all the recorded weights to be higher than the true weights. An analogous misfortune could befall the survey taker when "shopping" for information. Like the scheming butcher, the employees being interviewed about their salaries might not tell the truth. They might understate their salaries if they suspected the information was being channeled to the tax collector; they might overstate their salaries if they thought the information was about to be used to compute a 20-percent raise.

This sort of problem is the basis for defining the second major type of error that statisticians must consider. It is called **systematic error,** or **bias,** or **nonsampling error,** and equals the difference between the value of a variable obtained by taking a census (or by averaging the results of all possible random samples of a given size) and the true value. This definition is summarized in the right half of Figure 2.1.

Unfortunately, bias can be hard to detect, and its size—unlike that of random error—cannot be estimated. For this reason, statisticians who seek to discover the truth must become aware of the major sources of bias and try their best to neutralize them. To that issue we now turn.

How Bias Can Enter Surveys

Bias can enter a census or sampling procedure at the planning stage, during the data-collection stage, or even while data are being processed.

In the planning stage. If they are not careful, statisticians can literally build systematic error into the very design of their surveys. This can happen, for example, as a result of **selection bias,** a systematic tendency to favor the inclusion in a survey of selected elementary units with particular characteristics, while excluding other units with other characteristics. As a result of selection bias, any data that are eventually collected are bound to overrepresent the former characteristics. The use of a faulty frame or the selection of a *nonrandom* sample can easily result in a selection bias.

Consider what would happen in our quest to determine the average salary earned in a firm if we planned our survey on the basis of a list of those employees who were seen walking through the firm's gates last Friday around 5 P.M. Regardless of whether we were planning a census or a random sample, the accuracy of our survey results would be compromised because we would not be working the proper frame (such as an up-to-date listing of all employees culled from the firm's personnel records). Many employees—such as those on vacation, those working part-time, those who are ill, and, perhaps, executives working late—would have no chance of being included in our survey. It would be hazardous to assume that their salaries would, on average, equal those of the people on our list. We would be studying an *actual* population (salaries of employees who were seen leaving the firm last Friday around 5 P.M.) that was almost certainly quite different from our *target* population (salaries of all employees). Unfortunately, the use of faulty frames is quite common; sometimes, for lack of better alternatives, it is even unavoidable. Examples include the firm that surveys its actual customers (instead of its potential customers) to assess the prospects of a new product and the political pollster who surveys a group of people whose names were taken from the telephone directory or from car-registration rolls or even from a list of *eligible* voters in order to assess the chances of a candidate (when the people who really matter are those *registered* voters who are sure to vote on election day).

As noted, selection bias can also result from the nonrandomness of sampling, even though a perfect frame is used. The use of any convenience or judgment sample is automatically suspect, but even an attempted random sample can turn out to be non-random in fact. Consider the attempted selection of a systematic random sample from a perfect frame that lists elementary units with some type of rhythm. If the frame were a 50-year listing of monthly stock prices, and we selected every 12th number (or some multiple thereof), we would end up with a set of data all of which would pertain to the same month—say December; these selected data could easily differ in some systematic way from all the excluded ones. Or consider an alphabetical listing of names. If most of the members of some ethnic group had names that began with certain letters (such as Irish names that begin with *Mc* or *O'*), a systematic random sample could easily miss every one of these people, and this would be disastrous if the survey were designed to study ethnic questions. The same sort of problem can occur in clustered random sampling. Because people with similar characteristics (such as education level, ethnic grouping, or income) tend to live close to each other, it is not impossible for cluster sampling to miss a whole range of backgrounds altogether, and this omission could seriously bias the survey results.

Another common type of systematic error is **response bias,** a tendency for answers to survey questions to be wrong in some systematic way. Nothing, probably, contributes more to this problem than the faulty design of questionnaires. As students who take exams know too well, it is not easy to ask good questions that elicit the answers really sought. Consider the frequently told story of the student who answered (c) when confronted with the following question:

> 4. Wind-eroded rocks are most commonly found
>   a. on mountain tops.   c. in fertile valleys.
>   b. on river bottoms.   d. in outer space.

The teacher had hoped for (a), but the student argued that hardly any people ever *were* on mountain tops, river bottoms, or in outer space; hence, if such rocks were to be *found,* it would most likely happen in (heavily populated) fertile valleys.

Nothing seems easier than to formulate similarly inept questions on survey questionnaires. When this happens, people cannot be expected to give reliable answers; data gatherers who imagine themselves measuring something may, in fact, end up measuring something else or nothing at all. When asked about their income, for instance, people give different answers depending on whether they suspect that their answer will affect their taxes, their pension, or their bank credit. (A good questionnaire, therefore, contains instructions that motivate respondents to tell the truth by providing them with reasons for the survey). The use of clearly defined concepts is crucial, too. It is not obvious what it means to live in poverty or in a substandard home or in a crime-ridden neighborhood. Such terms, therefore, should not be placed on a questionnaire unless they are carefully defined. Nor can people be expected to know what to do when asked to list their home town (what about soldiers or students?), their employment status (is a retired person *unemployed?*), their trips abroad (does Canada qualify?), their weight (with or without clothes? at morning or at night?), their age (last birthday or nearest birthday?), or their spending plans for next year (next *calendar* year or the next 12 months?).

Most worrisome of all, however, is the leading question that is likely to lead the respondent, inexorably, to a particular, predictable answer. Leading questions can be obvious or extremely subtle. Here are some obvious ones:

1. As usual, today's food in the dining hall was rotten.	Yes	No
2. Should we waste further billions to send people into outer space, while there is poverty on earth?	Yes	No
3. In light of their $3 billion profit, should we reject management's skimpy wage offer?	Yes	No
4. Does the name *Apple* come to mind when thinking about high-quality computers?	Yes	No

One can safely bet that most answers would be Yes, No, Yes, Yes. Such questions are sure to invite response bias. There are, however, more subtle ways to ask leading questions.

Response bias can result from the *sequencing* of different questions in such a way that people answer one question within a frame of reference provided by a (possibly unrelated) earlier question. Consider this:

10. Which of the following is responsible for making "Fruit and Fibre" such a superior cereal?
    a. It is fruitier.　　　c. It is sweeter.
    b. It is nuttier.　　　d. It is crunchier.
11. Which of the following do you consider superior?
    a. Fruit and Fibre.　　c. Granola.
    b. Bran Chex.　　　　d. Rice Krispies.

In light of question 10, most people will answer question 11 by choosing (a).

Faulty questionnaire design can, finally, be responsible for **nonresponse bias,** a systematic tendency for selected elementary units with particular characteristics not to contribute data in a survey while other such units, with other characteristics, do. In the presence of this problem, even a census based on a perfect frame, or a perfectly selected random sample, will fail. They will yield faulty conclusions because the data actually collected will in fact constitute a convenience sample (for example, of the most strongly opinionated people among all the people that were supposed to be in the survey). Questionnaire features that contribute to nonresponse bias include a physically unattractive design; hard-to-read print; questions that are boring, unclear, or long and involved; an excessive number of questions; bad sequencing of questions (so that respondents are forced to jump back and forth from topic to topic), and, in the case of multiple-choice questions, the specification of answers that are not mutually exclusive or are excessively restricted to particular points of view, while omitting other possible views. Experience shows that high-income people and low-income people (unlike middle-income people) tend not to respond to surveys; it is easy to see how the exclusion of either group is apt to bias survey results.

Close-Up 2.3, "Dubious Sampling: The Literary Digest Case," gives an all too vivid picture of a survey gone wrong even at the planning stage.

In the collection stage. Even when the designers of surveys successfully avoid the emergence of bias by the use of proper frames, the careful selection of random samples, the meticulous creation (and pretesting) of questionnaires, and the like, those who execute surveys can still allow bias to enter the process.

Selection bias is apt to enter a survey when interviewers are instructed to select, within broad guidelines, the particular individuals they will question. Interviewers are routinely allowed to select their interviewees in the case of judgment sampling. If it is known that the population of a city contains 10 percent black men, 20 percent black women, 30 percent white men, and 40 percent white women, for example, an interviewer who is to question 100 persons may be given *quotas* of 10 black men, 20 black women, 30 white men, and 40 white women in order to assure that the questions are posed to a proper cross section of the population. Given the quotas, the interviewer, however, may be free to select particular individuals. Naturally, interviewers will select such people in some convenient way; perhaps, they will choose interviewees by standing at the nearest street corner (provided they can find people whose appearance is pleasant enough and whose behavior is conventional) or by knocking on doors (provided the neighborhood is attractive enough, there are no dogs in the front yard, and the time of day is right). This procedure will, of course, miss all sorts of people, such as those on vacation, those in hospitals or prisons, those at work, or those residing in unpleasant parts of town, in trailer camps, or in dormitories. Indeed, experience shows that quota sampling invariably leads to the selection of a disproportionate number of highly educated, high-income individuals. (Other types of quotas lead to similarly biased selections. When told to select 60 people over 50 years of age and 40 people under 50 years, the typical interviewer, understandably, selects lots of people who look like 70, lots of people who look like 20, and hardly anyone who is close to 50 years old). Indeed, selection bias can occur even in the execution stage of a properly selected *random* sample. Suppose a random-numbers table were used to select, from a complete list of a city's residents, a simple random sample of 100 persons (in the same way as a sample of 10 multinational firms was selected earlier in this chapter). Such a procedure would charge an interviewer with finding 100 specifically named individuals, no matter how hard the task. Yet actual interviewers have been known to substitute other persons for the randomly selected ones if the latter were hard to catch during (the interviewer's) working hours or lived in dangerous or out-of-the-way places. Naturally, such a failure to follow directions on the part of the interviewer dooms the entire survey by turning a probability sample into a convenience sample. Imagine the biased results in a survey about cosmetics use if the interviewer substituted grandmothers for all the teenagers because the latter were never around. Imagine the misleading results if elderly or working people or vacationers were excluded from surveys that sought to ascertain the demand, respectively, for household help or frozen food or leisure-time products.

Response bias can arise for a number of reasons during data collection. Both interviewers and respondents can be at fault. Interviewers, often inadvertently, may solicit a particular "acceptable" answer to a question by their dress and choice of words (which may betray their social class) or by their tone of voice and demeanor (a gesture of disapproval or surprise at some answers will almost surely affect other answers). Interviewers can, of course, also make systematic mistakes when recording answers—for example, by consistently categorizing part-time income or work as full-time. And interviewers have been known to fake answers completely, as happened during the 1980 census in New York City when drug addicts on methadone maintenance worked for the Census Bureau in ghetto areas. (Some individuals just wanted to get the work done fast; the lives of others had been threatened by unwilling respondents.) Even the timing of a survey can create response bias. What do you think would happen to all the answers if a survey on plant safety were conducted precisely one day after a rare but major accident or if a survey on hours recently worked were conducted shortly after a major holiday? Interviewers must not, however, receive all the blame.

Even when questionnaires are designed well, respondents often give false answers for a variety of reasons. Respondents may simply not know the answer but may give one anyway to conceal their ignorance. They may give whatever answer they think will please the interviewer. (It is well known that white and black interviewers get different answers from identical respondents when asking identical questions on racial issues.) Respondents may give distorted answers in line with some current fad toward optimism or pessimism. And they may tell deliberate falsehoods to mislead competitors or impress the interviewer. (People are particularly likely to boast of successes and hide failures on so-called "prestige questions" concerning, say, their knowledge of current events or famous people, their reading of books or level of education, their grade-point average or wealth, and even their brushing of teeth and taking of baths.)

The data-collection stage also generates its share of *nonresponse bias.* Respondents often refuse to answer questions they consider too personal, such as those about their age, health, and sexual habits, about their income, or about illegal and immoral activities. Indeed, many respondents refuse to answer *any* questions, particularly in surveys conducted by phone or mail. (The 1980 Census, claimed the city of New York, failed to count some 800,000 residents who refused to be counted; Chicago, Detroit, Philadelphia, and other cities made similar allegations.) Often people simply don't want to be bothered. What may be terribly important to the questioner is of no concern to them. See Close-Up 2.4, "Firm *X* versus the California Tax Authority."

In the processing stage.   The emergence of bias during data collection can conceivably be minimized by the careful training and supervision of interviewers and by a variety of techniques designed to forestall incorrect or refused responses. (These techniques include prominent and favorable coverage by the news media about surveys to be undertaken, personal special-delivery letters to respondents on the stationery of well-known organizations or over the signature of respected individuals, credible promises of anonymity or of rewards, and more.) Nevertheless, bias can enter even at the data-processing stage. Multitudes of people who code, edit, keypunch, tabulate, print, and otherwise manipulate data have multitudes of opportunities for making non-canceling errors. Those who code answers to *open-ended* questions (which elicit answers in people's own words) may consistently miscategorize them; editors may consistently introduce high values (or low ones) when encountering incomplete or illegible responses; they may fail to eliminate **outliers,** or "wild values" (maverick responses that are not believable because they differ greatly from the majority of observed values); keypunchers may consistently misread 7s as 9s; they may ignore all the decimal points, thus turning every 8.1 into an 81—the list goes on. Close-Up 2.5, "Of Teenage Widowers, American Indians, and a Billion Chinese," provides food for thought. Close-Up 2.6, "Sampling as Legal Evidence," is an illuminating contrast.

EXPERIMENTS*

The major alternative to data generation by means of surveys is data generation through controlled experiments. As noted earlier, compared to surveys, properly conducted experiments are more likely to elucidate cause-and-effect relationships. They provide stronger data. Here we will focus on *comparative* experiments in which, all else being held equal, different stimuli, or **treatments,** are applied to elementary units, now called **experimental units,** whose responses are observed and compared. (By custom, not only the deliberate exposure of an experimental unit to something new, but also the deliberate exposure of it to *nothing* new is termed a *treatment.*) All experiments have in common the need to define the treatments, to select experimental units, and to divide these units into at least one **experimental group** (that is exposed to something new) and one **control group** (that is not subjected to anything new), while

_____
*Optional section.

somehow controlling extraneous factors (other than the treatments) that might influence the outcome. As Analytical Example 2.2, "On Curing the Common Cold and Other Diseases," indicates, one could come to extremely wrong conclusions in the absence of a control group.

Randomization and Blocking

Two major devices are used to control for extraneous factors. One of these devices is **randomization,** a procedure that lets extraneous factors operate but assures, by virtue of the random assignment of experimental units to experimental and control groups, that each treatment has an equal chance to be enhanced or handicapped by these factors. The other device is **blocking,** a procedure that eliminates the effects of extraneous factors by forming blocks of experimental units within each of which all units are as alike as possible with respect to those factors. These two devices have given rise to a number of different **experimental designs,** or plans for assigning treatments to experimental units under controlled conditions and, thus, for generating valid data.

The Randomized Group Design

Perhaps the simplest experimental plan is the **randomized group design,** or **completely randomized design.** This plan creates one treatment group for each treatment and assigns each experimental unit to one of these groups by a random process. If there were 60 experimental units and two treatments (A and B), the experimenter might, for example, simply pick the first 30 numbers in the 01 to 60 range that are encountered in a table of random numbers and assign similarly numbered experimental units to treatment group A. The remaining 30 units would be given treatment B.

To be specific, consider an experiment about the effectiveness of a new foreign-language teaching device. Treatment A would be the use of the device, treatment B would be traditional teaching methods. The random assignment of 30 students to each of these treatments would control for extraneous factors in the sense that each group would probably contain similar proportions of very able and extremely dull students, of highly motivated and totally indifferent students, of well-prepared and ill-prepared students. As a result, these very relevant factors of ability, motivation, and preparation could be expected to be at work equally in both groups so that later differences in performance might, indeed, be attributed to the difference in teaching methods.

The same type of procedure could, of course, be used for a larger number of treatments as well. If there were three treatments (A, B, and C), the first 20 numbers drawn from the given range of 60 numbers would determine experimental units for treatment A; the next 20 numbers drawn (and not used before) would designate experimental units for treatment B, and the remaining 20 units would receive treatment C.

The Randomized Block Design

Another favorite type of experiment is the **randomized block design.** This plan divides the available experimental units into blocks of fairly homogeneous units (each block containing as many units as there are treatments or some multiple of that number) and then matches each treatment with one or more units within each block by a random process. Recall the above experimentation, by means of the randomized group design, with a new foreign-language teaching device. Under the randomized *block* design, the procedure would be quite different: The language test scores of the 60 students might be examined, and the top two students might be placed in block 1. Their high scores would suggest that they were very much alike with respect to such relevant factors as ability, motivation, and level of preparation. They would be like a pair of near-twins. If, by the toss of a coin, treatment A (use of the new device) were to be assigned to one of them, and treatment B (continued use of old methods) to the other, one could hardly argue that later differences in test scores were attributable to anything else but the difference in teaching approach. Now imagine that the third- and fourth-ranking students were, similarly, combined to form block 2, the fifth- and sixth-ranking ones to form block 3, and so on down the line until the worst two students were placed into block 30. In each case, this new experimental design would control for extraneous influences.

Because it involves blocks of two, this particular randomized block design is also referred to as a **matched-pairs design.** As the above definition indicates, there is, however, no necessity to create blocks of two. Even in this example, one could probably have just as well created 15 "homogeneous" blocks of four students (rather than 30 blocks of two students), and assigned each treatment at random to two students (rather than one student) within each block.

This type of block design need not, however, be confined to human experimental units. Indeed, the term *blocking* originated in agricultural experiments and first referred to the creation of pieces of land that were as homogeneous as possible with respect to soil, sun, rain, and more. See Analytical Example 2.3, "Confounding and Blocking: The Fluorescein Experiment."

Note: Many other experimental designs exist. We will not discuss them here because the generation of data by means of experiments rather than by means of surveys is relatively rare in the fields of business and economics.[2]

ERRORS IN EXPERIMENTAL DATA*

Experimental data are also subject to error, including the same types of error to which survey data are subject. As Figure 2.2 illustrates, an observed difference between the effects of two treatments will be the true difference—attributable to nothing else but the difference in treatments—only in the absence of random or systematic error. Consider the experiment, noted above, concerning two approaches to foreign-language teaching that were compared on the basis of the randomized group design. If there were no bias and the true difference in test scores were $+10$, a single experiment might, nevertheless, yield a difference of $+16$. Another experiment might yield a difference of $+5$ only, and so on for additional ones. These chance variations around the true value of $+10$ could be expected to occur, with equal frequency and magnitude, in both directions around the true value; therefore, the true difference could be dis-

---

[2]Those who are, nevertheless, intent on pursuing the subject will find an extended discussion of experimental design in Chapter 2 of the *Student Workbook* that accompanies this text.

*Optional section.

Figure 2.2 Errors in Experiments

Experiments are subject to the same type of errors as surveys are. Either type of error can be positive or negative.

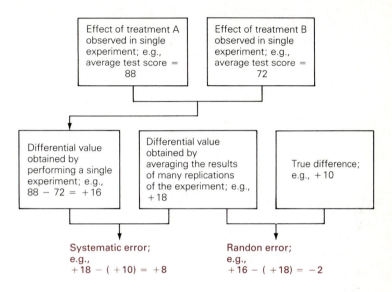

covered by averaging the results of a sufficiently large number of repetitions of the experiment. Like sampling procedures, experimental ones are thus subject to *random error,* now also called **experimental error,** which equals the difference between the value of a variable obtained by performing a single experiment and the value obtained by averaging the results of a large number of identical experiments.

Unfortunately, bias can ruin comparative experiments as well as surveys; such *systematic error* equals the divergence between the differential value obtained by averaging the results of many replications of an experiment and the true difference. If the experimenter is not careful, the differential effects of treatments that are to be measured (say, the difference between two teaching devices) may remain inextricably mixed up with (or *confounded* with) the effects of other factors (say, teacher experience or student preparation) throughout the experiment. As a result, more than one cause might be responsible for the observed experimental result. As indicated in Figure 2.2, fully half of a 16-point difference observed by comparing test scores in a single experiment might be due to factors other than the treatment and would thus represent a systematic error.

As in the case of surveys, bias can enter experiments at various stages and in many guises; nothing but eternal vigilance can prevent it. As noted before, if volunteers are used to form experimental groups and others are used to form control groups, bias is likely to occur because volunteers are rarely like nonvolunteers with respect to relevant factors other than treatments. Analytical Example 2.4, "Bias in the Salk Polio Vaccine Trial," provides another, quite spectacular example of selection bias.

Even when randomization and blocking have been used to best advantage to avoid selection bias, response bias can be as deadly. To cite one example, experimental units or experimenters who are aware of who gets what type of treatment can easily let their prejudices determine the responses they make or record. Thus, people who know they are receiving a new drug that supposedly reduces certain symptoms may report this effect even though it is wholly imagined, or outside observers who are "convinced," even before the experiment, that the drug will work may unconsciously observe what they want to observe and note a reduction in symptoms even in the absence of such an effect. This is why **double-blind experiments** are preferred in cases where human frailty is likely to play a role. In such experiments, response bias is controlled by letting neither the subjects nor the judges know who is receiving which type of treatment. (In the case of drug experiments, all experimental units might receive look-alike pills, half of which are active but half of which are *placebos*—that is, inactive and harmless substances.)

ANALYTICAL EXAMPLES

2.1 How Accountants Save Money by Sampling

Business firms and other organizations used to spend a great deal of time and money on censuses to determine their accounts receivable (or payable) and to take physical inventories of equipment, raw materials, goods in process, and the like. Recent examples have shown, however, that census taking is often unnecessary and that relatively small samples of the relevant populations, if carefully drawn, can produce high-quality results, along with dramatic savings in time and money.

Case 1. The Chesapeake and Ohio Railroad faced the problem of allocating total freight and passenger revenues among several carriers when freight or passengers had traveled over lines owned by different companies, a common occurrence. In one experiment, it examined the waybills for interline freight shipments that had occurred over a six-month period. (A *waybill* is a document describing the nature of goods shipped, their routing and the total freight charges.) As the accompanying table shows, nearly 23,000 waybills were examined, and the amount that the shippers of freight owed the Chesapeake and Ohio (rather than other railroads) was determined to be $64,651.

At a much lower cost, a sample was also taken. It was a stratified sample, containing only 1 percent of waybills totaling $5 or less, 10 percent of waybills totaling between $5.01 and $10; 20 percent of waybills totaling between $10.01 and $20; 50 percent of waybills totaling between $20.01 and $40; and 100 percent of waybills in excess of $40. The particular waybills

	Census	Sample
Number of waybills examined	22,984	2,072
Cost of examination	$5,000	$1,000
Amount found due Chesapeake and Ohio Railroad	$64,651	$64,568

included in the sample were selected with the help of a random-numbers table on the basis of the last two digits of their serial numbers.

The result was amazing: The amount due was determined within $83 or $1/10$ of 1 percent of the much costlier census figure. Similar accuracy was achieved when the railroad studied five months' worth of interline passenger tickets. A census of the 14,109 tickets revealed the amount due the railroad to be $212,164. A 5 percent sample of these tickets yielded a figure of $212,063. The $101 difference was even less than $1/10$ of 1 percent of the census figure.

Case 2. Airlines face the same problem of settling accounts with each other concerning passengers traveling on several different airlines during a given trip. Each major airline picks up hundreds of thousands of tickets per month that were issued by and initially paid to other airlines. Clerks used to figure out the revenue allocation for each ticket, but this process could be extremely cumbersome and costly. (At one time, there existed 60 possible fares between Chicago and New York!) Three airlines (Northwest, United, and TWA), therefore, conducted a four-month test similar to the one by the Chesapeake and Ohio described above. A census was taken, then a stratified sample. (The strata used were first-class, coach, military, and other types of tickets. Individual tickets were pulled on the basis of the last digit in their serial numbers with the help of a random-numbers table.) The difference in the census and sample results came to less than $700 per $1 million of tickets. After introducing sampling procedures, one major carrier cut clerical expenses by $75,000 per year, the industry as a whole saved $500,000 per year.

Case 3. One firm, Minneapolis-Honeywell, used to take 100-percent inventories of its goods in process. Then it conducted two experiments. First, it took a simple random sample of 500 items out of 5,000 items in one department. The estimated size of the inventory came within 1/10 of 1 percent of the full count. Second, it took a stratified random sample, dividing work-in-

process lots from all departments into those of high value ($500 or more) and those of low value. All of the former were counted; only 10 percent of the latter were counted. The sample of 4,200 out of 40,000 items yielded an estimate within 8/10 of 1 percent of the full count.

Conclusion. Sampling instead of census taking for accounting purposes is now widespread. Firms use it to estimate their bad debts (by estimating the total of accounts owed for more than 60 days) and to gauge the accuracy of their clerical workers (by contacting a sample of customers concerning their billings); the U.S. Air Force uses it when taking a worldwide audit of its motor-vehicle pool. The list could easily be lengthened.

Sources: John Neter, "How Accountants Save Money by Sampling," in Judith M. Tanur, et al., eds., *Statistics: A Guide to the Unknown* (San Francisco: Holden Day, 1972), pp. 203–11; Theodore J. Sielaff, *Statistics in Action: Readings in Business and Economic Statistics* (San Jose, Calif.: Lansford Press, 1963), pp. 20–21 and 26–27.

2.2 On Curing the Common Cold and Other Diseases

The importance of having a control group in an experiment was highlighted by a three-year study in the 1930s that evaluated a vaccine against the common cold. A homogeneous group of especially susceptible individuals was randomly separated into an experimental group (which received the vaccine) and a control group (which received a saline solution without knowing this fact). The accompanying table shows the results.

While the first row seems to indicate that the vaccine had an impressive effect, a comparison with the second row suggests that all the experimental units also received a heavy dose of imagination—regardless

of the treatment type, everyone's colds were reduced substantially—a factor that would not have been obvious in the absence of the control group.

Yet, unfortunate as it is, even modern-day medical decisions are often made on the basis of evidence like that in row 1 alone. Such was the case with the testing of diethylstilbesterol, or DES, which was claimed to be effective in preventing miscarriages, all on the basis of five uncontrolled trials between 1948 and 1955. Later controlled studies, however, showed no difference in the rate of live births between treated and control groups, but the drug was only abandoned in 1973 when children born to DES-treated mothers devel-

| Group | Treatment | Number of Persons | Average Number of Colds per Year | | Reduction in Colds |
			Before Vaccination	After Vaccination	
1. Experimental	Cold vaccine	272	5.9	1.6	73%
2. Control	Saline solution	276	5.6	2.1	63%

oped cancer. Another case is the testing of isoprinosine, introduced in 1972 to treat encephalitis. There were some apparent successes but also 30 deaths. Even 10 years later, in 1982, no one knew whether to praise or blame the drug because no controlled study had been made.

Sources: H. S. Diehl, A. B. Baker, and D. W. Cowan, "Cold Vaccines: An Evaluation Based on a Controlled Study," *Journal of the American Medical Association* 2 (1938): 1168–73, as reported by Austin Bradford Hill, *Principles of Medical Statistics,* 8th ed. (New York: Oxford University Press, 1967), p. 276; Jerry E. Bishop, "Doctors Debate Use of Controlled Studies To Test Effectiveness of New Treatments," *The Wall Street Journal,* August 8, 1982, p. 21.

2.3 Confounding and Blocking: The Fluorescein Experiment

In the 1930s, some alleged that plants grew better when watered with a dilute solution of fluorescein rather than plain water. Experimenters predicted the connection because a solution of this yellowish-red crystalline compound ($C_{20}H_{12}O_5$), like plants themselves, was responsive to light. In one experiment, 54 plants were watered with plain water; another such group was given fluorescein water (see Figure A). A large effect was observed, as predicted.

Yet no one could replicate the experiment! It turned out that the original experimenter had widely separated the plots of plants to avoid contamination,

yet the fluorescein plot differed not only in location but also in fertility. The effect of the treatment with fluorescein water had been confounded with that of soil fertility.

In a subsequent experiment, six homogeneous blocks were created, and each was planted with 18 plants. Within each block, half the plants were given plain water, the other half fluorescein water—and even that allocation was determined at random, by the toss of a coin. The experimental design is illustrated in Figure B. When the six differences within blocks were compared, no significant effect was noted.

Figure A

Plants watered with plain water

Plants watered with fluorescein water

Figure B

Source: William J. Youden, "Chance, Uncertainty, and Truth in Science," *Journal of Quality Technology,* January 1972, p. 8.

2.4 Bias in the Salk Polio Vaccine Trial

In 1954, the biggest public-health experiment to date was carried out in the United States to determine the effectiveness of the Salk poliomyelitis vaccine. Seemingly, the study was carefully designed. It included nearly 1 million children in grades 1–3. Such a large-scale approach was necessary because the incidence of the disease in the United States then equaled only 50 per 100,000 making an experiment with, say, 10,000 children necessarily inconclusive (because only 5 of these could be expected to contract polio and *any* observed allocation of this number between those vaccinated and those not vaccinated could then be attributed to mere chance.) The experiment involved the administration of the Salk vaccine to about 222,000 second-graders who had their parents' permission and the comparison of this group's experience with that of some 725,000 first- and third-graders who were not vaccinated. The results are shown in Table A.

A comparison of the encircled numbers seemed to indicate an amazing effectiveness of the vaccine. Yet the study was disastrously biased; this comparison of the experimental and control groups was highly unfair! *Only* those second-graders with parental permission had received the vaccine, but *all* first- and third-graders in their schools had been used as controls. Unfortunately, this was a mistake, and here is why:

Poliomyelitis always hit hardest those children who were better off hygienically and who had better nutrition, better housing, and higher-income, better-educated parents. Polio was almost unknown in places with the poorest of hygiene. (Polio is caused by one of several viruses that are ubiquitous.) Children living in poor hygienic conditions are exposed to polio early on while they are still protected by immunity passed on by their mothers. At that time, they develop their own immunity, and, henceforth, they contract polio less often than their hygienically more fortunate peers. As it turned out, higher-income, better-educated parents of second-graders freely gave permission for their children to receive the vaccine; the lesser-educated parents did not. As a result, the second-graders receiving the vaccine were unusually *likely* to contract polio, much more so than their first- and third-grade peers who belonged to all types of parents.

A second experiment was designed, involving *only* children with parental permission (and, thus, presumably, children equally at risk). In this double-blind experiment, each of the volunteers was assigned at random to receiving the Salk vaccine or a placebo (a saline solution), but neither the children nor the evaluating physicians knew who got which. (To forestall favoritism in the assignment of treatments or prejudice in the appraisal of symptoms, vaccine and placebo vials were identical in appearance and distinguished only by secret code numbers.) The results are shown in the first two rows of Table B.

Table A

School Grade	Treatment	Number of Children	Number Afflicted with Paralytic Polio	Number Afflicted per 100,000
2	Salk vaccine	221,998	38	17
1 and 3	None	725,173	330	46

Table B

Permission Given	Treatment	Number of Children	Number Afflicted with Paralytic Polio	Number Afflicted per 100,000
Yes	Salk vaccine	200,745	33	16
Yes	Placebo	201,229	115	57
No	None	338,778	121	36

A comparison of the encircled numbers in Table B with those in Table A clearly shows that the Salk vaccine was much *more* effective than the first experiment had demonstrated.

Note: As the last row of Table B indicates, the second experiment also provided evidence for the home-hygiene effect noted above. Unvaccinated children from homes with probably better hygiene had a greater incidence of polio (57) than those from homes with probably poor hygiene (36).

Sources: William J. Youden, "Chance, Uncertainty, and Truth in Science," *Journal of Quality Technology,* January 1972, pp. 7–10; Paul Meier, "The Biggest Public Health Experiment Ever," in Judith M. Tanur, et al., eds., *Statistics: A Guide to the Unknown* (San Francisco: Holden-Day, 1972), pp. 2–13.

CLOSE-UPS

2.1 The 1970 Draft Lottery Fiasco

During both World Wars I and II, it became necessary to establish an order in which men were to be drafted into the U.S. military. In 1917, accordingly, 10,500 black capsules, containing numbers previously assigned to eligible men, were drawn from a glass fishbowl. In 1940, a similar procedure was adopted to draw 9,000 numbers, but there were criticisms: The small, wooden paddle (which was made from a piece of rafter traceable to Independence Hall in Philadelphia) used to stir the capsules in the bowl would not reach deep enough into the bowl. It also broke open some of the capsules, impeding the mixing process further. In the end, the numbers drawn looked like anything but random ones; they were concentrated in certain clusters of hundreds, apparently reflecting the fact that the numbers had been poured into the bowl in lots of 100 each. The lesson was clear: thorough physical mixing of capsules in a bowl is difficult.

Apparently the lesson had been forgotten when the 1970 draft lottery was instituted. Some 366 capsules, containing all the possible birthdates in a year, were poured into a bowl but not stirred. (There had been some mixing during the process of inserting dated slips of paper into the capsules). Then capsules were drawn out, the order of their withdrawal determining the priorities for the draft. The observed sequence strongly reflected the order in which the capsules were created (one month at a time), with late-in-the-year birthdays (that were encapsulated last) being drawn first and early-in-the-year birthdays (that were encapsulated first) being drawn last and, therefore, less subject to the draft. Several young men filed suit in federal court seeking to have the 1970 lottery voided on the basis of the apparent lack of randomization. Note: A 1971 draft lottery, in response to widespread criticism of the 1970 lottery, made use of random-numbers tables, discussed in the text. A discussion of the 1971 lottery appears as Analytical Example 11.1.

Source: Stephen E. Fienberg, "Randomization and Social Affairs: The 1970 Draft Lottery," *Science,* January 22, 1971, p. 255–61.

2.2 The Making of Random Numbers

Random numbers, such as those in Appendix Table A, provide important protection against selection bias in sampling and experimentation. Such numbers have been generated by physical processes, by mathematical ones, and by a combination of the two. Physical processes have included

a. drawing a capsule from a bowl, replacing the drawn capsule, remixing, and drawing again;
b. flashing a beam of light at irregular intervals onto a rotating disk and reading off numbers thereon;
c. recording digitized electronic pulses that occur at random times; and even
d. counting the number of output pulses generated by the radioactive decay of cobalt 60 during given time intervals.

The trouble with these physical processes is that a specific sequence of numbers cannot be regenerated; therefore, one cannot repeat any procedure that made use of such numbers, unless the numbers are somehow stored.

Mathematical generating processes have included:

e. programming a digital computer by means of a recurrence relation such that the next random number is always derived from one or more prior numbers and

f. a variety of more complicated procedures, such as the "middle-square" and "multiplicative-congruence" methods.

However, mathematical processes cannot generate sequences of truly random numbers that are independent of prior numbers and that cannot be predicted with certainty. Their output is often referred to as *pseudo* random numbers. Yet this type of generation has been found useful because it can avoid certain undesirable, although unlikely, outcomes of truly random processes, such as repetitive sequences of numbers or unequal proportions of odd and even numbers that can give rise to questionable samples or experimental plans.

Amazingly, random numbers had never been stored in tabular form until the appearance, in 1927, of Leonard H. C. Tippett's *Random Sampling Numbers*

(London: Cambridge University Press). By randomly taking digits out of a 1925 British census report, Tippett produced a table of 41,600 such digits. Subsequently, in 1938, Ronald A. Fisher and Frank Yates pulled 15,000 random digits from a table of logarithms and published *Statistical Tables for Biological, Agricultural, and Medical Research* (Edinburgh: Oliver & Boyd). One year later, in 1939, M. G. Kendall and B. B. Smith employed method (b) above to create 100,000 random digits, published as *Tables of Random Sampling Numbers* (London: Cambridge University Press). They also introduced a variety of tests to check the randomness of the numbers generated. The most ambitious project of all, however, employed method (c), along with so-called compound-randomization techniques "to improve the numbers." This 1955 project yielded the Rand Corporation's *A Million Random Digits* from which Appendix Table A has been excerpted.

Source: Mervin E. Muller, "Random Numbers," in William H. Kruskal and Judith M. Tanur, eds., *International Encyclopedia of Statistics,* vol. 2 (New York: The Free Press, 1978), pp. 839–47.

2.3 Dubious Sampling: The Literary Digest Case

A classic case of both selection bias and nonresponse bias occurred in 1936. The *Literary Digest* magazine, which had correctly predicted the winner in every U.S. presidential election since 1916, predicted confidently a comfortable victory for Alfred M. Landon, the Republican candidate, over Franklin D. Roosevelt, the Democratic candidate, by a margin of 57 to 43. Yet Roosevelt won by a landslide never before seen in U.S. history, receiving 62 percent of the votes cast. What had gone wrong?

First, there was selection bias. The *Digest* had mailed questionnaires to 10 million people whose names had been taken from various lists such as its own subscribers, telephone directories, and automobile-registration rolls. During the Great Depression, higher-income people typically voted Republican, and these people were well represented in the *Digest's* sample. On the other hand, lower-income people, who heavily favored the Democrats, were underrepresented because a lower percentage could afford magazine subscriptions, telephones, and automobiles.

Second, there was nonresponse bias. Only 2.4 million of the 10 million questionnaires were mailed back. Although this made the survey the largest sample ever taken, more educated people are more likely to

respond to mail questionnaires than less educated ones. The former, again, tended to favor Republicans; the latter, the Democrats. Thus, a much larger percentage of the nonrespondents than of the respondents were for Roosevelt. This bias reinforced the selection bias.

Note: The *Digest* never survived the debacle and folded shortly thereafter. At the same time, George Gallup was setting up his survey organization, and he correctly forecast the Roosevelt victory from a mere sample of 50,000 people. Yet in 1948, using another dubious procedure (a form of judgment sampling, called *quota sampling*), Gallup's organization (along with Crossley's and Roper's) incorrectly predicted the victory of Thomas Dewey over Harry S. Truman.

Sources: Frederick Mosteller, et al., *The Pre-Election Polls of 1948* (New York: Social Science Research Council, 1949); Mildred B. Parten, *Surveys, Polls, and Samples* (New York: Harper, 1950); Frederick F. Stephan and Philip J. McCarthy, *Sampling Opinions* (New York: Wiley, 1958); and George H. Gallup, *The Sophisticated Poll-Watcher's Guide* (Princeton: Opinion Research, 1976). See also Maurice C. Bryson, "The *Literary Digest* Poll: Making a Statistical Myth," *The American Statistician,* November 1976, pp. 184–85, which lays the entire blame for the debacle on nonresponse bias.

2.4 Firm X versus the California Tax Authority

Two Colorado corporations (a wholesale seller of Christmas cards and a mail-order house for stationery) had merged in 1967. They continued to operate independently in California (among other places) where sales at wholesale as well as mail-order sales by *out-of-state* corporations are not ordinarily subject to sales tax. A ruling by the California tax authority, however, held that the wholesale operations gave the merged firm a "substantial business presence" in California; therefore, the firm's mail-order sales were subject to sales tax.

The firm paid the tax but sued to get it back, arguing that the mail-order business was wholesale, in part, and, thus, tax-exempt. To prove the point (and with the tax authority's agreement), the firm made a sample survey of its mail orders, picking every thirty-fifth order placed in 1975 by California residents, and asking each buyer whether the purchase had been for resale or personal use. The sample contained 5,121 orders, but only 2,966 buyers responded, of whom 282 had resold (9.5 percent). Yet the *value* of all the respon-

dents' purchases came to $38,160 and 21 percent of the *value* had been resold. Thus, the firm argued, 21 percent of California mail-order sales should be tax exempt. The tax authority, however, citing the low response rate, argued that all nonrespondents, by inference, had bought for personal use, and only agreed to a 12-percent exemption.

The firm, in turn, presented evidence from its records that there was little difference between respondents and nonrespondents: the distribution of orders by size (under $10, $10 to $50, over $50) as well as the distribution of orders by residence (San Francisco, Los Angeles, other) was identical for the two groups. Thus, there was reason to believe that some of the nonrespondents had bought for resale. The firm won its case.

Source: David A. Freedman, "A Case Study in Nonresponse: Plaintiff vs. California State Board of Equalization," *Journal of Business and Economic Statistics,* January 1986, pp. 123–24.

2.5 Of Teenage Widowers, American Indians, and a Billion Chinese

Contrary to common belief, a census does not necessarily ensure more reliable results than do sampling techniques, especially when large numbers of investigators are involved. The decennial U.S. Census of Population, which uses hundreds of thousands of workers, is a case in point. It is simply impossible to train and supervise such numbers properly and, thus, to ensure quality control in the data-collection and data-processing stages. No wonder that errors abound. To cite just one, between 1940 and 1950, the U.S. Census yielded an enormous increase in the number of 14-year-old widowers and divorcés and an equally impressive rise in that of young Indians. These data could not possibly have been correct and, as was later discovered, resulted from a column shift during key-punching.

Now consider the likelihood of error in the most ambitious census ever undertaken in the world's his-

tory, that of mainland China in 1982. In a mere 10 days, some 1,000 computer technicians, 4,000 data-entry workers, 100,000 coders, 1 million supervisors, and 4 million enumerators supposedly counted and categorized an estimated 1 *billion* people! Many of the respondents had good reasons not to cooperate: by underreporting births, they avoided criticism of their attitude toward family planning; by not reporting deaths or migrations of family members, they assured themselves larger rations of grain, cloth, and housing space. Nevertheless, the Chinese government reported the population at precisely 1,008,175,288.

Sources: Ansley J. Coale and Frederick F. Stephan, "The Case of the Indians and the Teen-Age Widows," *Journal of the American Statistical Association,* June 1962, pp. 338–47; Frank Ching, "Abacuses Are Out, Computers Are In In Chinese Census," *The Wall Street Journal,* July 7, 1982, pp. 1 and 15.

2.6 Sampling as Legal Evidence

Judicial procedures in the United States are designed to establish guilt or innocence "beyond a reasonable doubt" (in criminal cases) or by a "preponderance of evidence" (in civil cases). One would think

that statistical data would play a large role in this process. Indeed, official publications of census-like counts, such as those published by the Bureau of the Census, have usually been considered admissible evidence, but counts based on sampling have often encountered hurdles. This negative court attitude toward sampling has changed recently, partly because modern election polls have established the power of sampling and partly because it has been recognized that crucial knowledge can oftentimes not be secured in any other way. Thus, the courts have accepted estimates of market shares based on store samples, of mineral deposits based on test drillings, and of depreciation based on samples of plant facilities. (In one case, involving a tax refund, a court refused to accept sample evidence based on several hundred thousand sales slips; it insisted on a complete count. Yet the result of the census was within 1 percent of the sample estimate.)

Courts are still more reluctant to accept evidence secured through sampling where interviewing is involved because these interviews are considered *hearsay evidence* (evidence based not on a witness's personal knowledge but on matters told him or her by another). Courts have, however, allowed exceptions even here, as long as the interview response did not concern the truth of the matter but only the respondents' state of mind. If respondents are asked whether two trademarks represent the same manufacturer, the interviewers know the true facts already; their goal is to find out whether the respondents know them. The interviewers in such cases are deemed competent witnesses. Sample surveys conducted under this rubric have been concerned with consumer awareness to establish (1) whether trademarks or advertising slogans were sufficiently established in consumers' minds to grant them continued legal protection, (2) whether a new trademark was sufficiently different from an existing one to allow its registration, (3) whether certain words (such as "English lavender") were taken literally, thus possibly violating truth-in-labeling acts, and (4) whether certain brand names (such as "thermos") were still mentally connected with a specific manufacturer or had become generic names that could not be protected.

Courts have also allowed sample evidence secured by interviewing to establish the existence or absence of community prejudice against a defendant (and the need for postponing the trial or securing a fair trial in another community).

Source: Hans Zeisel, "Statistics as Legal Evidence," *International Encyclopedia of Statistics,* vol. 2 (New York: Free Press, 1978), pp. 1118–19.

BIOGRAPHY

2.1 Adolphe Quetelet

Lambert Adolphe Jacques Quetelet (1796–1874) was born in Ghent, Belgium. He had an early interest in the fine arts (he painted, wrote poems, and even produced an opera), but this interest was soon overshadowed by his attraction to mathematics. His was the first doctoral dissertation at the newly established University of Ghent, and it was widely acclaimed as an original contribution to analytical geometry. The dissertation resulted in his election, at age 24, to the Brussels Académie Royale des Sciences et des Belles-Lettres (in which he soon became the dominant spirit) and to a position of teaching mathematics, physics, and astronomy at the Brussels Athenaeum. He was a great teacher (students and visitors from all of Europe crowded his lectures), and he was a prodigious writer (producing a vast array of essays and books and editing a leading journal). Yet the seemingly undefatigable energy that he poured into his career also changed its orientation. His enthusiasm for astronomy that led, eventually, to the building of an observatory in Brussels, and his directorship of it, also brought him in contact with illustrious French mathematicians, such as Fourier, Laplace (whose Biography appears in Chapter 5 of the *Student Workbook*), and Poisson (Biography 6.2). Their interest in probability theory and its applications to social phenomena excited Quetelet. His subsequent active encouragement of the collection of empirical social data led to the first national census in Belgium and Holland (in 1829), the formation (in 1834) of the Statistical Society of London (now named the Royal Statistical Society), and the organization (in 1841) of the Belgian Central Statistical Commission, a central agency responsible for collecting statistics. As president of the latter, Quetelet did much to inspire the creation of statistical bureaus all over Europe and labored unstintingly to promote internationally uniform methods and terminology in data collection and presentation. Under his leadership, the first of a long series of International Statistical Congresses was held in Brussels in 1853.

Although there had been forerunners in England, France, and Germany, Quetelet earned the honor of being called the father of modern statistics by the publication of his *Sur l'Homme et le Développement de Ses Facultés* in 1835. In this book, Quetelet noted how social phenomena (such as crime or suicides) reproduced themselves with amazing regularity; he argued that such regularities were discoverable only by statistical techniques and, even more important, could also be linked to causes with the help of such techniques. Unlike earlier writers who had given a theological interpretation to social regularities (seeing in them evidence of a divine presence), Quetelet pointed to social conditions as causes and suggested that legislation could also ameliorate their effects (such as crime or suicides). As he put it[1]:

The constancy with which the same crimes repeat themselves every year with the same frequency and provoke the same punishment in the same ratios, is one of the most curious facts we learn

from the statistics of the courts. . . . And every year the numbers have confirmed my prevision in a way that I can even say: there is a tribute man pays more regularly than those owed to nature or to the Treasury; the tribute paid to crime! Sad condition of human race! We can tell beforehand how many will stain their hands with the blood of their fellow-creatures, how many will be forgers, how many poisoners, almost as one can foretell the number of births and deaths.

Quetelet believed, however, that masses of numbers had to be studied before one could reach any reliable conclusions about causes[2]:

It seems to me that *that which relates to the human species, considered en masse, is of the order of physical facts;* the greater the number of individuals, the more the influence of the individual will is effaced, being replaced by the series of general facts that depend on the general causes according to which society exists and maintains itself. These are the causes we seek to grasp, and when we do know them, we shall be able to ascertain their effects in social matters, just as we ascertain effects from causes in the physical sciences.

[1] From *Sur l'Homme . . .*, as cited in *Dictionary of Scientific Biography* vol. XI (New York: Charles Scribner's, 1975), p. 237.

[2] *Recherches sur le Penchant au Crime aux Différens Âges,* 2nd ed. (Brussels: Hayez, 1833), pp. 80–81.

Summary

1. The process of data collection employs a number of basic concepts. Thus, a statistical investigation focuses on persons or objects, which are called *elementary units;* these elementary units possess characteristics of interest, called *variables;* observations about them are called *data.* The set of all possible observations about a specified characteristic of interest is called a *population;* a subset of it (or of the frame from which the population is derived) is referred to as a *sample.* Variables themselves can be qualitative or quantitative; the assignment of numbers to observations about them, which is called *measurement,* can yield data of varying quality, such as nominal, ordinal, interval, and ratio data. Whatever their type, new data can be generated either by conducting a *survey* or by performing an *experiment.* A survey is the collection of data from elementary units without exercising any particular control over factors that may make these units different from one another and that may, therefore, affect the characteristic of interest being observed. In an experiment, in contrast, such control is exercised.

2. Surveys are the most common method of generating data in business and economics. In a complete survey, or *census,* observations are made about one or more characteristics of interest for every elementary unit that exists—by direct observation, personal or telephone interview, or self-enumeration. In a partial or *sample survey,* such observations are made only for a subset of existing elementary units. A number of reasons for preferring sampling to census taking can be cited; in one way or another, these reasons have something to do with reducing the cost of getting a given type of information (or with raising the quantity or quality of information received at a given cost). Samples of different quality are obtained depending on the method by which elementary units are selected for observation; convenience samples, judgment samples, and probability samples are the major types. Among the latter, which are the most important, are simple random samples, systematic random samples, stratified random samples, and clustered random samples.

3. Whether collected by census or sample, all survey data are subject to error. Values obtained by data gatherers can differ from true values as a result of *random error* (in the case of random sampling) or of *systematic error* (in the case of all types of surveys). The size of the former can be estimated, but not so the size of the latter. Systematic error or bias can enter surveys at the planning stage, during the data-collection stage, and even while data are being processed, and such error can take the form of selection bias, response bias, or nonresponse bias.

4. Controlled (comparative) experiments provide the major alternative to surveys as a means of data generation. All such experiments have in common the need to define treatments, to select experimental units, and to divide these units into at least one experimental group (that is exposed to something new) and one control group (that is not subjected to anything new), while somehow controlling extraneous factors (other than treatments) that might influence the outcome. Extraneous factors are controlled either by randomization or by blocking, and these two devices have given rise to a number of experimental designs, including the randomized group design and the randomized block design.

5. Just like survey data, experimental data are subject to different types of error, including random error (now also called *experimental error*) and systematic error.

Key Terms

bias
bivariate data set
blocking
census
chance error
clustered random sample
completely randomized design
continuous quantitative variable
control group
convenience sample
data set
datum
dichotomous qualitative variable
discrete quantitative variable
double-blind experiments
elementary units
experiment
experimental designs
experimental error
experimental group

experimental units
frame
interval data
judgment sample
matched-pairs design
measurement
multinomial qualitative variable
multivariate data set
nominal data
nonresponse bias
nonsampling error
observational study
ordinal data
outliers
population
probability sample
qualitative variable
quantitative variable
random error
randomization

randomized block design
randomized group design
random-numbers table
random sample
ratio data
response bias
sample
sample survey
sampling error
selection bias
simple random sample
stratified random sample
survey
systematic error
systematic random sample
treatments
univariate data set
universe
variables

Questions and Problems

The computer programs noted at the end of the chapter can be used to work many of the subsequent problems.

1. The tables at the top of the following page represents the 1985 mutual funds "honor roll" published by *Forbes* magazine (September 16, 1985, pp. 80–81). The twenty-three companies finished in the top half of the pack in both bull (up) and bear (down) markets; they also delivered good performance in the long run (over almost a nine-year period).

Table 2.A

Fund (1)	Performance UP (2)	Performance DOWN (3)	Manager (4)	Net Assets 6/30/85 (millions) (5)	Annual Expenses per $100 Assets (6)	Portfolio Strategy (7)	Average Annual Total Return (8)
01 Acorn Fund	B	A	Ralph Wanger	$ 269	$0.85	Small CO Growth	20.6%
02 Amcap Fund	B	A	Marjorie Fisher	1,209	0.69	Growth	21.5
03 American Capital Pace Fund	B	A+	John Doney	1,521	0.62	Aggressive Growth	27.5
04 American Capital Venture Fund	B	A	Steve Hayward	438	0.72	Aggressive Growth	24.9
05 Claremont Capital Corp.	B	A+	Erik Bergstrom	44	1.03	Growth	23.1
06 Evergreen Fund	A+	B	Stephen Lieber	323	1.10	Growth	26.4
07 Explorer Fund	A	B	Frank Wisneski	355	1.00	Small Co Growth	20.7
08 Fidelity Destiny Fund	B	A	George Vanderheiden	764	0.70	Growth	22.1
09 Fidelity Magellan Fund	A+	A	Peter Lynch	2,879	1.12	Growth	33.3
10 Growth Fund of America	B	A	James Rothenberg	606	0.68	Growth	22.3
11 Janus Fund	B	A	Thomas Bailey	412	1.06	Aggressive Growth	20.9
12 Loomis-Sayles Capital Development Fund	A	B	Kenneth Heebner	169	0.70	Growth	22.7
13 Mass Cap Development Fund	A+	B	William S. Harris	785	0.75	Growth	23.8
14 NEL Growth Fund	A	B	Kenneth Heebner	216	0.75	Growth	22.2
15 Nicholas Fund	B	A+	Albert Nicholas	462	0.82	Growth	25.4
16 Over-the-Counter Securities Fund	B	A	Binkley Shorts	114	1.28	Small Co Growth	23.4
17 Pennsylvania Mutual Fund	B	A	Charles Royce	292	1.18	Small Co Growth	22.6
18 Pioneer II	B	A	David Tripple	1,834	0.71	Growth and Income	20.7
19 Scudder Development Fund	B	A	Edmund Swanberg	254	1.34	Small Co Growth	20.8
20 Sigma Venture Shares	A	B	Richard King	68	0.99	Small Co Growth	22.7
21 Tudor Fund	A	B	Melville Straus	130	1.59	Aggressive Growth	21.5
22 Twentieth Century Select	A+	B	Fund Has Four Managers	1,076	1.01	Growth	28.0
23 United Vanguard Fund	B	B	Hank Herrmann	409	1.04	Growth	21.0

a. How many *variables* are listed for these 23 mutual funds?

b. List the qualitative variables, then the quantitative ones.

c. Separate the variables into dichotomous/multinomial and discrete/continuous, respectively.

2. In Table 2.A, identify the elements on page 45.

a. the elementary units.

b. the population of managers.

c. the population of net assets.

3. Classify the following variables, first, as qualitative or quantitative and, second, as dichotomous/multinomial or discrete/continuous:

   a. the number of telephone calls made by someone during a day.

   b. the dollar figures listed on a sheet of paper.

   c. the sexes of corporate executives.

   d. the running times of participants in a race.

   e. the employment/unemployment status of workers.

   f. the types of hair coloring sold in a drugstore.

4. Classify the following variables, first, as qualitative or quantitative and, second, as dichotomous/multinomial or discrete/continuous:

   a. the weight lost by a dieter.

   b. the types of skills found among a firm's employees.

   c. the attendance record of students in a class.

   d. the ages of applicants for a marriage license.

   e. the types of cars seen in a parking lot.

   f. the pressure required to fracture a casting.

5. Classify the following variables as qualitative or quantitative:

   a. a firm's average cost of production.

   b. a town's tax rate.

   c. the religious affiliations of a firm's employees.

   d. the national unemployment rate.

   e. the brands of gasoline for sale in a city.

   f. a listing of the states in which 50 firms achieved their highest sales.

6. Classify the following variables as qualitative or quantitative:

   a. a list of foreign exchange rates.

   b. the number of black executives in an industry.

   c. the depth of tread remaining on aircraft tires after 1,000 landings.

   d. the Dow-Jones Industrial Average.

   e. the political-party affiliations of a firm's employees.

   f. the types of sports practiced by a group of people.

7. Classify the variables in problem 5 also as dichotomous/multinomial or discrete/continuous.

8. Classify the variables in problem 6 also as dichotomous/multinomial or discrete/continuous.

9. Classify numbers describing the following as nominal, ordinal, interval, or ratio data:

   a. the location of voters by district.

   b. the ages of employees.

   c. the order in which cars finish a race.

   d. models of computers.

   e. white blood cells found in a cubic centimeter.

   f. the colors of new cars.

10. Classify numbers describing the following as nominal, ordinal, interval or ratio data:

    a. ratings of colleges.

    b. temperature readings at the airport.

    c. the daily receipts of a supermarket.

    d. consumer brand preferences concerning types of coffee.

    e. army ranks.

    f. a corporate hierarchy from president to janitor.

    g. calendar years.

11. Among numbers describing the following, which are nominal data?

    a. distances traveled

    b. student I.D. numbers

    c. net assets

    d. room numbers

    e. sound levels inside different airplanes

    f. drivers' ratings of handling characteristics of cars

12. Are there interval data in the listing of problem 11? If so, explain.

13. Consider the following situations and determine whether a census or a sample would be more appropriate:

    a. A personnel director, who wants to improve labor relations, seeks to ascertain the attitudes of the firm's employees.

    b. A pharmaceutical firm wants to ascertain the side effects, if any, of a drug that has proven effective against some types of cancer.

    c. NASA wants to check the quality of all space-shuttle components.

14. Consider the following situations and determine whether a census or a sample would be more appropriate:

    a. A manufacturer of matches has discovered that a certain percentage of production always seems to be defective and wants to ascertain what the percentage is.

    b. A retail hardware store wants to determine the value of its inventory.

    c. Someone wants to predict the outcome of a planned vote on the establishment of no-smoking zones in all public places.

15. If you had to do it, how would you go about measuring the popularity of next fall's new TV shows?

16. Which type of sample was actually taken in each of the following situations?

    a. A newscaster reports having taken a random sample of city residents by interviewing people coming out of a supermarket.

b. Between 10 A.M. and 4 P.M., Mondays to Fridays, a pollster calls every 100th number in a city's telephone directory and asks whomever answers to rate various types of detergents. (No effort is made to contact people who do not answer the phone.)

c. The Dow-Jones Industrials stock index is constructed on the basis of observing the performance of a mere 30 stocks selected by experts as representative of all industrial stocks.

17. Which type of sample was actually taken in each of the following situations?

a. The state tourism bureau creates a profile of the typical out-of-state tourist by interviewing one evening all out-of-state guests at a single hotel chosen at random.

b. In order to ascertain how well a new product is doing, a pollster checks at known trendsetters among retail stores.

c. Using a random-numbers table, a quality inspector selects 20 from among 1000 numbered boxes containing paper napkins produced during a day, then inspects all the napkins in these boxes.

18. Which of the following are probability samples? How would you characterize the others?

a. Using a table of random numbers, an interviewer selects 10 shopping malls from a list of 200, then interviews the owners of every store in these 10 malls.

b. A farmer throws a pair of dice, comes up with a seven, and then, starting with tree number 7 inspects every peach on every twelfth tree in an orchard to ascertain frost damage.

c. A manufacturer sends a questionnaire to all new-car purchasers. The results, based on a 20 percent response, are then tabulated.

d. Representatives of the mayor interview three blacks and seven whites prior to a town meeting concerning a proposed ordinance to close bars at 10 P.M. The proportions chosen reflect the racial makeup of the town.

19. Consider all the probability samples you found when answering problems 16–18. What type were they?

20. How many different samples of three might one select from four units (A, B, C, D), and what probability of selection must each of these have to make the sample a simple random sample?

21. Consider Table 2.2. How many different simple random samples of size $n = 5$ can be drawn without replacement from the list of multinationals? Of size $n = 50$?

22. Consider Table 2.A given in problem 1. How many different simple random samples of size $n = 4$ can be drawn without replacement from the list of mutual funds? Of size $n = 10$?

23. Consider Table 2.1. How many different simple random samples of size $n = 3$ can be drawn without replacement from the list of employees? Of size $n = 1$?

24. Start at the bottom of column 7 of Appendix Table A. By using the first two digits of numbers and reading upwards, select a simple random sample of 10 multinationals from Table 2.2. What are their names?

25. Repeat the exercise given in problem 24, but assume that there are only 50 such multinationals; namely, the ones numbered 00 through 49. What are the names of sampled firms now?

26. Appendix Table A has 50 rows and 10 columns of five-digit numbers. Starting at the upper left-hand corner of the table, read two digits at a time along the first row (and possibly along subsequent rows), letting the first usable set of two digits point you to a row and the second usable set to a column where random-number selection is to start. What is your starting number?

27. Given the starting number found in problem 26, proceed to the right along successive rows, reading the first and last digits of the 5-digit numbers in order to select a sample of six mutual funds from Table 2.A. What is your sample?

28. Given the starting number found in problem 26, proceed downward along successive columns, reading the second and third digits of the 5-digit numbers in order to select a sample of four mutual funds from Table 2.A. What is your sample?

29. Repeat problem 27, but move upward along successive columns.

30. Repeat problem 28, but proceed diagonally to the right and down. If you run out of numbers, read successive diagonals below the initial one.

31. What would your sample be if, in problem 27, you selected six multinationals from Table 2.2?

32. What would your sample be if, in problem 28, you selected four multinationals from Table 2.2?

33. Use the table of random numbers to select a simple random sample of three doctors from among those listed in the yellow pages of your local telephone directory.

34. Select a systematic random sample of five firms from Table 2.2. Assume the first number selected is 07. List the names.

35. Select a systematic random sample of four firms from Table 2.2. What is the value of k? Find the starting point for your sample by reading the last row of Appendix Table A and picking the first usable number between 00 and k. List your sample.

36. Select a systematic random sample of four firms from Table 2.A, using the procedure of problem 35. If necessary, read successive rows above the last one.

37. How might one select a systematic random sample of 200 from among 10,000 bills?

38. How might one take a 10-percent systematic random sample of a firm's 200 employees?

39. How big a sample would one have to take from each stratum if a population was stratified in such a way that each stratum contained completely homogeneous elements and if one desired information equivalent to a census?

40. Imagine that the 100 firms listed in Table 2.2 could meaningfully be subdivided into 10 strata, the members of which were, respectively, the firms numbered 00–09, 10–19, 20–29, and so on. Select a stratified random sample of one firm from each group, assuming that you picked the following random numbers to represent the last digits of firms to be picked from the strata: 0   8   2   3   3   5   9   5   0   6.   List the names.

41. Repeat problem 40, but find the ten random numbers in the top two numbers given in the last column of Appendix Table A.

42. Both stratified and cluster sampling separate a population into groups, yet there are major differences between the two approaches. Explain.

43. Once again consider Table 2.2, but now imagine that each successive five firms were a meaningful cluster. Select a sample of two from the 20 clusters in which to take censuses. Assume you assigned numbers 00 to 19 to the clusters, then picked the random numbers 17 and 05 from a table. List the firms to be sampled.

44. Once again consider Table 2.A, but now imagine that each successive two firms and the last three firms were a meaningful cluster. Select a sample of two from the 11 clusters in which to take censuses. Assume you picked clusters 5 and 9 at random. List your sample.

45. What kinds of errors are likely to occur as a result of each of the following:
   a. consistently keypunching a wrong column on census cards used as computer input.
   b. the inability to interview a family selected as part of a simple random sample.
   c. having only 21 percent of questionnaires returned from a mail survey.

d. judging public opinion from a radio talk show.
   e. asking people about their age rather than their birthday.

46. What kinds of errors are likely to occur as a result of each of the following:
   a. judging consumer preferences for a product by interviewing people stopped in cars at a red light.
   b. calling 100 people listed in the phone book to rate a new TV show.
   c. taking a random sample of 10 pages of this book to estimate the fractions of the book devoted to text, boxed examples, biographies, summaries, questions, and the like.
   d. taking a random sample of 10 sentences from the population of sentences found on the second page of Chapter 1 in order to estimate the length of sentences in this book.
   e. asking people how many phone calls they have made (or bars of soap they have used) in the past 24 months.

47. An economist randomly samples some small trucking firms, some medium-sized firms, and some large ones and finds their average annual insurance expenses to be $5,503 per truck. A census shows the figure to be $4,973. What kind of sample was taken, what kind of error was made?

48. The manager of a fast-food chain determined the average age of the chain's employees by making censuses in randomly chosen establishments 5, 29, and 172. The average age found was 25.1; yet data from all establishments showed it to be 21.7. What kind of sample was taken, what kind of error was made?

49. Develop a short questionnaire that is to be mailed to all students at your university to solicit their opinions about a planned fee for computer use. Then analyze your questionnaire: Does it avoid response bias?

50. Do you think the project noted in problem 49 will encounter nonresponse bias? How might you reduce it?

51. Modify the questionnaire in problem 49 so it can be used for a sample telephone survey. How would you select the telephone numbers to be called?

52. What problems might you encounter if you decided to carry out the project in problem 49 by means of personal interviews?

53. You are a manager about to expand into a new market (your home town). You want to study people's preferences for the goods you want to offer. Prepare a sampling plan.

54. Repeat problem 53, but assume that your new market is the United States east of the Mississippi.

55. Students are forever being polled: about the quality of their instructors, about library hours, about athletics-program fees, about the new dorms. Consider the last poll taken at your university; analyze its validity.

56. Devise an experimental plan to test the relative effectiveness of three types of toothpaste, using a total of 15 people over a 10-year period.

57. Develop a plan like problem 56 that also controls for ethnic backgrounds of participants.

58. A firm receives a certain component from four different contractors. How might one test whether the average component-lifetimes differ?

59. An economist wants to test whether the average annual income of workers in four occupations is the same regardless of their education levels (low, medium, or high). Can you help?

60. Reread Analytical Example 2.4. Comment on an experiment that compared later driving records of two groups: those who had volunteered for and taken a high-school driver-training course (mostly females) and all others who had not taken such a course. (The record of the former group was superior, allegedly because of the training course.)

Selected Readings

Cochran, William G. *Sampling Techniques,* 3rd ed. New York: Wiley, 1977. A lucid and clear account of sample survey techniques.

Fisher, Ronald A. *The Design of Experiments,* 9th ed. New York: Hafner, 1971. A classic, first published in 1935.

Hankins, Frank H. "Adolphe Quetelet as Statistician." In *Studies in History, Economics, and Public Law,* vol. 31. New York: Columbia University, 1908, pp. 443–576.

Moore, D. S. *Statistics: Concepts and Controversy.* San Francisco: Freeman, 1979. On how data are collected; the focus is on ideas and their impact on everyday life.

Morgenstern, Oskar. *On the Accuracy of Economic Observations,* 2nd ed. Princeton: Princeton University Press, 1963. A classic book about error in statistics.

Riecken, Henry W., and Boruch, Robert F. *Social Experimentation: A Method for Planning and Evaluating Social Intervention.* New York: Academic Press, 1974. Reports on the use of randomized experiments in social programs dealing with fertility control, mental-health rehabilitation, negative income taxation, nursing-home patient self-care, police training, and more. See also a companion volume, *Experimental Testing of Public Policy: Proceedings of the Social Science Research Council Conference on Social Experiments* (Boulder, Colo.: Westview, 1976), edited by the above authors.

Sielaff, Theodore J. *Statistics in Action: Readings in Business and Economic Statistics.* San Jose, Calif.: Lansford Press, 1963. Parts II–IV contain a dozen case studies on the use of sampling techniques.

Tanur, Judith M., et al., eds. *Statistics: A Guide to the Unknown,* 2nd ed. San Francisco: Holden-Day, 1978. Pages xx to xxii list numerous articles, contained in this volume, that deal with census taking, sampling, and experiments.

U.S. Department of Commerce. *Revolution in United States Government Statistics, 1926–1976.* Washington, D.C.: U.S. Government Printing Office, 1978. A superb review of the history of the U.S. federal statistical system, including a discussion of the role of sampling in the collection of government statistics.

Computer Programs

The DAYSAL-2 personal-computer diskettes that accompany this book contain four programs of interest to this chapter:

1. *Random Numbers* lets you generate your own random-numbers tables.

2. *Secret Code* illustrates how random numbers can help create secret codes. You can write and decipher your own messages.

17. *Random Walk* uses random numbers to illustrate the random-walk hypothesis about stock prices or futures prices.

18. *Cointoss* uses random numbers to simulate the tossing of a coin and determines the ever-changing empirical probability of getting heads.

Note: Programs 17 and 18 require graphics capability.

CHAPTER THREE

The Presentation of Data: Tables and Graphs

After collecting their data, descriptive statisticians turn to an equally important task: the effective presentation of data. This task is particularly crucial when the data collection is large. No human mind is capable of grasping the meaning of any considerable quantity of numerical data, unless their mass is somehow reduced to relatively few convenient categories or is condensed with the help of some kind of visual aid. Yet it is not easy to cast masses of raw data into a form that is not only succinct or visually attractive, but also successful in weeding out the irrelevant and in then conveying all the relevant information in an accurate and unambiguous manner. This chapter discusses the two common procedures for making data collections intelligible: the construction of tables and the drawing of graphs.

TABLES

The construction of a good statistical table is an art. It involves a great deal more than simply presenting data in rows and columns. Imagine for example, that a census had been taken of the 100 multinational companies listed in Table 2.2 on pp. 19 and that for each of them data had been collected on the combined profit from domestic and foreign operations, as shown in Table 3.1. Even though Table 3.1 contains a mere 100 numbers, it is pretty confusing. A listing of these data in order of ascending or descending magnitude would certainly be a major improvement. Such an **ordered array,** in this case going from the smallest to the largest number, is provided by Table 3.2. More likely than not, however, most users would not be interested in the detail contained in either table and would prefer a considerable condensation of the data. The table maker obliges by carefully dividing the range of available data into nonoverlapping classes, noting how many observations fall into each, and constructing a table on that basis.

Table 3.1　A Population of Profits

This table shows 1981 combined profits from domestic and foreign operations, in millions of dollars, of the 100 largest U.S.-based multinationals listed in Table 2.2. Successive columns of numbers correspond to the alphabetical listing of companies in the earlier table.

1,071	724	457	2,060	722	499	81	−353	115	108
784	447	283	5,423	2,276	3,308	579	708	133	165
197	600	405	258	312	430	1,169	524	592	489
835	448	254	119	772	60	2,433	535	145	1,177
960	1,159	473	119	803	1,576	754	1,369	749	791
441	412	863	370	512	97	348	2,762	2,380	458
1,671	215	522	−1,060	1,231	838	339	668	1,992	580
403	−476	536	1,652	1,181	220	627	409	1,281	729
445	531	916	532	338	519	437	98	1,851	257
188	918	1,467	333	580	252	2,227	1,567	2,310	598

Source: Forbes, July 5, 1982, pp. 126-128.

The Absolute Frequency Distribution

It is a matter of personal judgment how many such collectively exhaustive and mutually exclusive groupings of data, called **classes,** best serve the goal of clarity. Obviously, this judgment will vary with the nature of the raw data in question and, thus, with the possible number of different qualitative observations or the possible range that quantitative observations can assume. If data pertain to a *qualitative* variable (such as types of businesses), the number of qualitative categories (such as proprietorships, partnerships, or corporations) may well point to the most appropriate number of classes, but even then alternatives exist. (One could combine observations about proprietorships and partnerships, for example, or one may wish to list data separately for small, medium-sized, and large corporations.) Data pertaining to *discrete quantitative* variables (such as the number of television sets owned by households) can, similarly, be classified on the basis of the number of discrete possibilities (households owning zero television sets can be counted separately from those owning one set, two sets, and so on), but once more different classifications are possible. (One could combine the counts of households owning two or more sets, for example.) Data concerning *continuous quantitative* variables (such as the tonnage produced by firms in the steel industry), finally, can be classified in any convenient fashion. Whenever the choice on the matter is wide open, most practitioners recommend between 5 and 20 classes; a smaller number sacrifices too much detail, a larger one retains too much of it.[1]

In the case of quantitative data, it is also considered desirable to make the **class width**—the difference between the numerical lower and upper limit of a class—the

---

[1]Some statisticians like to use *Sturgess's rule* to determine the desirable number of classes, k. According to this rule, $k = 1 + 3.3 \log n$, where n equals the size of the data set. In our example, with $n = 100$ and $\log 100 = 2$, this comes to $k = 1 + 3.3(2) = 7.6$, which might be rounded to 7 or 8. As will be shown, it is rarely desirable, however, to follow such a fixed rule blindly.

Table 3.2   The Profit Population Rearranged into an Ordered Array

When the data of Table 3.1 are turned into an *ordered array* from the lowest to the highest number, better comprehension is the result. It is instantly clear, for example, that profits ranged from − 1,060 million to + 5,423 million. (These extremes referred to Ford and Exxon, respectively. The other two losers were Chrysler and Pan Am.)

− 1,060	119	257	405	458	535	708	835	1,181	1,992
− 476	133	258	409	473	536	722	838	1,231	2,060
− 353	145	283	412	489	579	724	863	1,281	2,227
60	165	312	430	499	580	729	916	1,369	2,276
81	188	333	437	512	580	749	918	1,467	2,310
97	197	338	441	519	592	754	960	1,567	2,380
98	215	339	445	522	598	772	1,071	1,576	2,433
108	220	348	447	524	600	784	1,159	1,652	2,762
115	252	370	448	531	627	791	1,169	1,671	3,308
119	254	403	457	532	668	803	1,177	1,851	5,423

Source: Table 3.1.

same for all classes because this fact facilitates comparisons between classes, as well as the calculation of summary values (to be discussed in Chapter 4). Having decided on the desired *number* of classes, one can quickly approximate an appropriate class *width* by dividing the difference between the largest and smallest observation in one's data set by the desired number of classes.

If we tentatively decided to condense the Table 3.2 data into seven classes of equal width (a number, perhaps, suggested by Sturgess's rule noted in footnote 1), we would calculate

$$\text{approximate class width} = \frac{\text{largest value} - \text{smallest value in data set}}{\text{desired number of classes}}$$

$$= \frac{5,423 - (-1,060)}{7} = 926.1.$$

In order to set up nonoverlapping classes for the entire range of data, we might round this number to 900 and arrive at the following:

1st class:	− 1,100 to under − 200
2nd class:	− 200 to under 700
3rd class:	700 to under 1,600
4th class:	1,600 to under 2,500
5th class:	2,500 to under 3,400
6th class:	3,400 to under 4,300
7th class:	4,300 to under 5,200
8th class:	5,200 to under 6,100

We would end up with eight classes of equal width and would have satisfied both of the recommendations (concerning the desirable number of classes and the equality of widths). We would then be able to place each of the Table 3.2 data into one, and only one, of these eight categories. (The use of the term *under* in the previous classification avoids ambiguity by clearly distinguishing between the upper limit of one class and the lower limit of the class above it. Thus, an observation of 699 would be placed in the second class, as would a 699.99. Yet an observation of 700 would clearly belong to the third class, a fact that would not be obvious if the term *under* were missing.)

Yet our creation of classes would hardly be an appropriate one in light of the distribution of values found in Table 3.2. We would end up with 3 observations in the 1st class, then lump fully 57 observations into the 2nd class, 27 more into the 3rd class, and place a mere 13 observations into the last five classes (none of them into classes 6 and 7). This is where good judgment must come to play.

Noting that most observations fall into the range between 0 and 2,500, we might divide that range into five equal classes and then place two larger classes at either end of the distribution, as column 1 of Table 3.3 illustrates. Then we would be ready to note the total number of observations that fall into each of our classes, a procedure that would be easy with the help of Table 3.2 and would yield the numbers in column 3 of Table 3.3. (Note: If we had worked with unordered raw data, such as those in Table 3.1, we would have first placed a tally mark in the appropriate row of column 2 for every value found in Table 3.1, while checking off the Table 3.1 data so that none of them were counted twice. At the end of this process, we would also have found the values in column 3 of Table 3.3 by counting the tally marks.)

As Table 3.3 indicates, the absolute number of observations that fall into a given class is referred to as the **absolute class frequency.** In turn, the tabular summary of a data set showing the absolute class frequencies in each of several collectively exhaustive and mutually exclusive classes is called the data set's **absolute frequency distribution.** The column 1 and 3 data as a group are an example. Consider how this presentation tells us at a glance that three among the 100 largest U.S. multinationals made losses in 1981, that 24 of them made profits of $1 billion or more; while most of them (41 and 32, respectively) earned profits between 0 and under $500 million or between $500 million and under $1 billion. Such succinct information could hardly have been derived, equally rapidly, from the raw data of Table 3.1 and would have been even less possible to derive if the raw data had been more massive in number, as often is the case (except in textbooks).

Note: Although it is rarely done, the usefulness for later analysis of a summary such as Table 3.3 could be vastly improved by the addition of another column, showing the *sum* of data pertaining to each class. Thus, the sum of losses made by the 3 companies in class 1 might be given as $-\$1,889$ million (as Table 3.2 attests), the sum of profits of the 41 companies in class 2 might be given as $12,057 million, and similar totals (of $21,901 million, $11,105 million, $10,309 million, $13,686 million, and $11,493 million) might be given for the other five classes. This sort of information could, for example, be used to calculate precise class averages (such as $-\$1,889$ million divided by 3, or $-\$629.67$ million for class 1) and would obviate the need for estimating them. Such estimates are often made by assuming (perhaps incorrectly) that the average pertaining to each class equals the midpoint of the class interval (such as $-\$750$ million for class 1). The listing of class sums is particularly crucial for **open-ended classes** that have

Table 3.3 An Absolute Frequency Distribution of Profits of the 100 Largest U.S.-Based Multinationals in 1981

An absolute frequency distribution, such as columns 1 and 3, can highlight meaningful patterns hidden in raw data (such as Tables 3.1 or 3.2). Note: The sum of absolute class frequencies necessarily equals the total number of observations.

Class (size of profit in millions of dollars) (1)	Absolute Class Frequency (number of companies in class[a]) (2)	(3)
− 1,500 to under 0		3
0 to under 500		41
500 to under 1,000		32
1,000 to under 1,500		9
1,500 to under 2,000		6
2,000 to under 2,500		6
2,500 to under 5,500		3
		Total 100

[a]Strictly speaking, each class frequency refers to the number of *profit figures* from the profit population of Table 3.1 that fit into the designated class, but one can just as well talk about the number of companies because each profit figure is linked to a particular (different) company.

Source: Table 3.1.

only one stated end point, the upper or the lower limit (examples are "under 0" or "above $2,500 million"). In that case, users do not even have the option of approximating the class average by the class midpoint. Because this desirable practice (of listing class sums) is rarely followed, anyone who prepares an absolute frequency distribution, such as Table 3.3, can render subsequent table users a great favor by selecting classes in such a fashion that the average of values in a class comes close to the class midpoint.[2]

The Relative Frequency Distribution

The frequency distribution discussed so far in this chapter is termed an absolute one because it is based on counting absolute *numbers* of observations for each class. One can, however, also determine the *fractions* or *proportions* of all observations that fall into various classes; any such proportion is a **relative class frequency** and is the ratio of the number of observations in a particular class to the total number of observations made. Accordingly, the tabular summary of a data set showing the relative class frequencies in each of several collectively exhaustive and mutually exclusive classes is called the data set's **relative frequency distribution.** Table 3.4 illustrates how a

[2]The sets of class midpoints and precise class averages for the seven classes in Table 3.3 are − 750 and − 629.67; 250 and 294; 750 and 684; 1,250 and 1,234; 1,750 and 1,718; 2,250 and 2,281; and 4,000 and 3,831.

Table 3.4   Deriving the Relative Frequency Distribution of Profits of the 100 Largest U.S.-Based Multinationals in 1981

An absolute frequency distribution such as columns 1 and 2, is easily converted into a relative one—columns 1 and 3—by successively dividing each absolute frequency—such as the number 3 in row 1 of column 2—by the total number of observations (here $N = 100$) to yield the corresponding relative frequency (such as .03). *Caution:* It is an accident that the total number of observations equals 100. Only because of that fact do the relative frequencies look so similar to the absolute ones. They would not if the column 2 total added to a number other than 100 (as a look at Table 3.5 can confirm). It is no accident, however, that the sum of relative frequencies equals 1.00; such is always necessarily the case.

Class (size of profit in millions of dollars) (1)	Absolute Class Frequency (number of companies in class) (2)	Relative Class Frequency (proportion of all companies in class) (3)
−1,500 to under 0	3	.03
0 to under 500	41	.41
500 to under 1,000	32	.32
1,000 to under 1,500	9	.09
1,500 to under 2,000	6	.06
2,000 to under 2,500	6	.06
2,500 to under 5,500	3	.03
Totals	100	1.00

Source: Table 3.1.

relative frequency distribution is derived from an absolute one, but as the table caption cautions, it is an accident that the total number of observations equals 100. As a result of this accident, the column 2 and column 3 entries look very similar in this case. In all cases, however, proportions can be read as *percentages* by mentally multiplying them by 100. Thus, .03(100) = 3 percent of the multinationals in question made losses, .41(100) = 41 percent earned profits between 0 and under 500 million dollars, and so on.

Relative frequency distributions are particularly useful for comparisons of populations that are similar in type but vastly different in size. A most dramatic illustration of this fact is provided by Analytical Example 3.1, "Deciphering Secret Codes," wherein a small population of letters from the words of a secret message is compared with the vast population of letters in all the words of the English language. For a more mundane example, consider Table 3.5, which shows hypothetical frequency distributions of the racial and sexual composition of the labor force in a single firm and in the entire industry. It would be absurd to argue, on the basis of absolute numbers, that Mountain Aviation, Inc. was discriminating against blacks or females in hiring because it had hired so few of them compared to the industry as a whole. Relative frequencies alone tell a meaningful story here: 33 percent of the employees of Mountain Aviation, Inc. are black, while only 20 percent of the entire industry's employees are—note the first two entries in columns 4 and 5. On the other hand, 33 percent of that firm's employees are also female, while the industrywide percentage is 49. [Add the first and third entries in columns 4 and 5 for this comparison.]

Table 3.5 Absolute and Relative Frequency Distributions of Selected Characteristics of Aviation Industry Employees

As these hypothetical data indicate, comparisons of absolute numbers—columns 2 vs. 3—can sometimes be less insightful than comparisons of relative numbers—columns 4 vs. 5.

Class (employee race and sex) (1)	Absolute Class Frequencies (number of employees in class)		Relative Class Frequencies (proportion of all employees in class)	
	Mountain Aviation, Inc. (2)	Entire Aviation Industry (3)	Mountain Aviation, Inc. (4)	Entire Aviation Industry (5)
Black female	1	69,000	.111	.100
Black male	2	69,000	.222	.100
White female	2	269,000	.222	.390
White male	4	283,000	.444	.410
Totals	9	690,000	1.000	1.000

Note: As a comparison of the column 1 entries in Table 3.5 with those in Table 3.4 confirms, absolute and relative frequency distributions can be produced for qualitative variables as well as quantitative ones. Such is not the case, however, for another type of distribution.

The Cumulative Frequency Distribution

Only when variables are quantitative can one determine a **cumulative class frequency** as the sum of the (absolute or relative) class frequencies for all classes up to and including the class in question, beginning at either end of the frequency distribution. If the cumulation process proceeds from lesser to greater classes, the result is a "less-than-or-equal," or LE, type of distribution because the frequencies counted pertain to the class in question or lower ones. If cumulation proceeds from greater to lesser classes, a "more-than-or-equal," or ME, type of distribution emerges because the frequencies counted pertain to the class in question or higher ones. Accordingly, a **cumulative frequency distribution** is a tabular summary of a data set showing for each of several collectively exhaustive and mutually exclusive classes the absolute number or proportion of observations that are less than or equal to the upper limits of the classes in question (LE type) or that are more than or equal to their lower limits (ME type). All this is illustrated in Tables 3.6 and 3.7, with the help of the Table 3.4 data.

Consider Table 3.6. Note how the third (encircled) entry in column 3 tells us that 76 companies had profits that were less than $1,000 million (and at most equal to the upper class limit of just under $1,000 million) while the fourth (encircled) entry in column 5 tells us that 85 *percent* of the companies had profits that were less than $1.5 billion.

Now consider Table 3.7. Note how the fourth (encircled) entry in column 3 tells us that 24 companies had profits that were more than, or at least equal to, $1 billion, while the third (encircled) entry in column 5 tells us that 56 *percent* of the companies had profits of at least $500 million.

Table 3.6 Deriving the LE Type of Cumulative Frequency Distribution of Profits of the 100 Largest U.S.-Based Multinationals in 1981

An absolute frequency distribution—columns 1 and 2—is easily converted into a cumulative absolute one—columns 1 and 3—just as a relative frequency distribution—columns 1 and 4—can easily become a cumulative relative one—columns 1 and 5. Note how column 2 has been turned into column 3 and how 4 has been turned into 5. When cumulation proceeds from lesser to greater classes, a "less-than-or-equal," or LE, type of cumulative frequency distribution emerges. Each member of such a distribution tells us the number or proportion of observations that are less than, or at most equal to, the *upper* limit of the class in question. Note: While it is an accident that the largest cumulative absolute frequency equals 100 in this example, the largest cumulative relative frequency will always equal 1.00.

Class (size of profit in millions of dollars) (1)	Absolute Class Frequency (number of companies in class) (2)	Cumulative Absolute Class Frequency (number of companies in class or lower ones) (3)	Relative Class Frequency (proportion of all companies in class) (4)	Cumulative Relative Class Frequency (proportion of all companies in class or lower ones) (5)
		Starting Point		Starting Point
−1,500 to under 0	3	3	.03	.03
0 to under 500	41	3 + 41 = 44	.41	.03 + .41 = .44
500 to under 1,000	32	44 + 32 = ⑦⑥	.32	.44 + .32 = .76
1,000 to under 1,500	9	76 + 9 = 85	.09	.76 + .09 = ⑧⑤
1,500 to under 2,000	6	85 + 6 = 91	.06	.85 + .06 = .91
2,000 to under 2,500	6	91 + 6 = 97	.06	.91 + .06 = .97
2,500 to under 5,500	3	97 + 3 = 100	.03	.97 + .03 = 1.00

Source: Table 3.4.

Table 3.7 Deriving the ME Type of Cumulative Frequency Distribution of Profits of the 100 Largest U.S.-Based Multinationals, in 1981

An absolute frequency distribution—columns 1 and 2—is easily converted into a cumulative one—columns 1 and 3. The same is true of a relative frequency distribution; note how column 4 has been turned into column 5. When cumulation proceeds from greater to lesser classes, a "more-than-or equal," or ME, type of cumulative frequency distribution emerges. Each member of such a distribution tells us the number or proportion of observations that are more than, or at least equal to, the *lower* limit of the class in question.

Class (size of profit in millions of dollars) (1)	Absolute Class Frequency (number of companies in class) (2)	Cumulative Absolute Class Frequency (number of companies in class or higher ones) (3)	Relative Class Frequency (proportion of all companies in class) (4)	Cumulative Relative Class Frequency (proportion of all companies in class or higher ones) (5)
−1,500 to under 0	3	97 + 3 = 100	.03	.97 + .03 = 1.00
0 to under 500	41	56 + 41 = 97	.41	.56 + .41 = .97
500 to under 1,000	32	24 + 32 = 56	.32	.24 + .32 = ⑤⑥
1,000 to under 1,500	9	15 + 9 = ㉔	.09	.15 + .09 = .24
1,500 to under 2,000	6	9 + 6 = 15	.06	.09 + .06 = .15
2,000 to under 2,500	6	3 + 6 = 9	.06	.03 + .06 = .09
2,500 to under 5,500	3	3	.03	.03
		Starting Point		Starting Point

Source: Table 3.4.

A Final Note

The preceding sections have focused on how statisticians condense masses of raw data in order to highlight the most important information contained in them. We have paid no particular attention, however, to the proper construction of the tables themselves that are to convey the condensed data to their ultimate users. Good table making requires that attention be paid to a number of important factors. Table 3.8 can serve as an example of what it takes to make a table publishable. The overriding goal must be to help the user.

Number and title. When published along with other tables, the table should have a number so it can be referred to easily. Often, as in this case, it is convenient to use a chapter number for the purpose; Table 3.8 then refers to the eighth table in Chapter 3. Sometimes it is useful to use page numbers instead; Table 361a then refers to the first table on p. 361. The table number must be followed by a title that is as short as possible but also complete. It must focus on what, where, and when: in this case, sales of household washing machines (what), in Boston and Massachusetts (where), during 1976–85 (when). Consider how useless the entire table would be if any one of these ingredients were left out of the title.

Footnotes. Additional information can be extremely valuable to a user and can often easily be provided in a footnote (designated by letter or symbol to avoid confusion with numbers in the body of the table). When the information contained in the table is based on a sample survey, details about the size and type of sample can be given, along with the size of the sampling error. (A later chapter will note how the sampling error is calculated.) Information useful in determining the likely extent of bias can also be added, such as the exact wording of the question that was asked or the rate of response to a mail questionnaire. (Anyone noting the low response rate in footnote *a* is likely to suspect a significant nonresponse bias.) In other instances, other types of information may be equally useful. Anyone who wanted to publish a table based on columns 1, 4, and 5 of Table 3.5, for example, would serve the user immensely by footnoting the absolute totals (9 and 690,000) to which the relative frequencies refer. Any user could then easily reconstruct columns 2 and 3.

Table 3.8 Sales of Household Washing Machines in Boston and Massachusetts, 1976–1985[a]

| Year | Thousands of Units Sold | | Year | Thousands of Units Sold | |
	Boston	Massachusetts		Boston	Massachusetts
1976	60	300	1981	61	295
1977	51	255	1982	67	345
1978	42	210	1983	72	360
1979	37	200	1984	75	375
1980	59	245	1985	79	395

[a]Based on questionnaire mailed to a 5 percent simple random sample of retail outlets. ("How many household washing machines did you sell in each of the stated years?") Sampling error: xxx. Sample size: Boston 50, Massachusetts 500. Response: Boston 21, Massachusetts 101. Massachusetts data include Boston.

Source: Hypothetical data.

Decimals and rounding. A decision must inevitably be made on how many decimals to retain in the numbers appearing in the table. Should an absolute value of 600.493 be reported as such, as 600.49, or even as 600, with all the decimals dropped? Should a proportion of .389855 be reported as such or would .39 do just as well? An advantage of many decimals in proportions is that the user can more accurately recalculate absolute numbers, provided the total on which the proportions are based has been provided. Thus, a reader who is given the column 3 total of 690,000 in Table 3.5 in a footnote, but is not given the other column 3 values, can calculate the third entry of 269,000 in that column precisely when multiplying the given total by a column 5 proportion of .389855. Yet the calculation would yield 269,100 if a proportion of .39 were used. Still, this less precise calculation would lie within a few hundredths of 1 percent of the correct one. More often than not, the retention of many decimals serves no good purpose (in that it adds little information) and may, in fact, do harm (by giving the published data a totally unwarranted air of precision). Having decided on a reasonable number of decimals to retain, some systematic procedure for rounding must be employed. Typically, decimals from 1 to 4 are eliminated by rounding down (a 5.42 becomes a 5.4); those between 6 and 9 are eliminated by rounding up (a 5.47 becomes a 5.5); and an arbitrary rule is followed with the decimal 5, such as always rounding it to the *even* possibility (a 5.45 becomes a 5.4 and not a 5.5, but a 5.35 also becomes a 5.4 and not a 5.3). This arbitrary rule serves the purpose of avoiding bias from rounding up more often than down. Finally, note that numbers rounded to zero, as from .03, are listed as 0.0, while an original observation of zero is recorded as 0.0e (e = exact).

GRAPHS

Graphs, undoubtedly, are the most popular way of presenting statistical data. In one way or another, the information condensed into any type of table can also be presented with the help of a graph.

The Frequency Histogram

Consider graphing the information contained in Table 3.3. A graphical portrayal of an absolute or relative frequency distribution of a continuous quantitative variable is called a **frequency histogram.** As Figure 3.1 indicates, a frequency histogram is drawn on a set of rectilinear coordinates and consists of a series of contiguous rectangles (also referred to as bars or columns) placed on the horizontal axis so their bases correspond to classes of possibly different widths and their *areas* to class frequencies.

It is easy to see how the various classes of profit from column 1 of Table 3.3 can be represented on the horizontal axis by starting at the lower limit of the first class and going to the upper limit of the last one. (Hence, the first rectangle of a histogram need not start at the zero point.) The widths of the various classes can be measured in terms of any convenient standard. If, as in Table 3.3, most classes span intervals of $500 million, that range might just as well be used as the unit of measurement. Note the tick marks on the horizontal axis of Figure 3.1 that clearly show the first class to be three units wide, classes 2 to 6 to be one unit wide, and the last class to be six units wide.

It is much more difficult to determine what to measure on the vertical axis. Vertical lines are to be erected at the class limits to form the sides of rectangles the *areas* of which are to represent class frequencies, but how long should these lines be? A reminder from basic geometry quickly solves the puzzle. The area of any rectangle equals

Figure 3.1 The Frequency Histogram

A frequency histogram is a graphical portrayal of a frequency distribution for continuous quantitative data. While classes are shown horizontally, frequency densities (here the absolute numbers of observations per $500 million class range) are plotted vertically. As a result, the *area* of each rectangle, rather than its height, equals the frequency (here the absolute number of observations) for the corresponding class. Notice how the area of the first rectangle equals *three* $500 million units (base) times *one* observation per $500 million range (height), or *three* observations altogether. Similarly, the areas of the second or last rectangles (and, therefore, the absolute frequencies of the second and last classes) equal, respectively, $1 \times 41 = 41$ or $6 \times \frac{1}{2} = 3$ (just as noted in Table 3.9).

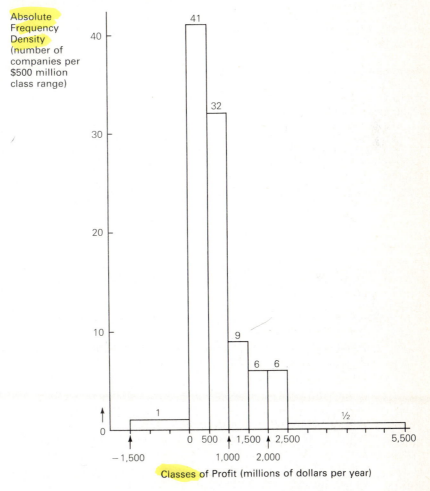

base times height. Height, therefore, equals area divided by base—or, in our case, class frequency (area) divided by class width (base). This ratio of class frequency to class width is also called **frequency density** and is, inevitably, what must be measured on the vertical axis if the *areas* of histogram rectangles are to correspond to class frequencies. These relationships are summarized in Table 3.9.

Table 3.9   Preparing to Draw a Frequency Histogram

The data of columns 2 through 4 shown here become, respectively, the area, base, and height of the corresponding histogram (Figure 3.1). Note how the concept of frequency density is analogous to the familiar one of human population density. Just as one commonly speaks of so many people living per square mile in Texas, Michigan, or New York, the column 4 data tell us of so many companies "living" per $500 million range in the "countries" of column 1. The total "populations" of these "countries" are shown in column 2; the sizes of these "countries" are given in column 3.

Basic Data			Absolute Frequency Density
Class Interval (size of profit in millions of dollars) (1)	Absolute Class Frequency (number of companies in class) (2)	Class Width (number of $500 million units in class) (3)	(number of companies per $500 million class range) (4) = (2)/(3)
−1,500 to under 0	3	3	3:3 = 1
0 to under 500	41	1	41:1 = 41
500 to under 1,000	32	1	32:1 = 32
1,000 to under 1,500	9	1	9:1 = 9
1,500 to under 2,000	6	1	6:1 = 6
2,000 to under 2,500	6	1	6:1 = 6
2,500 to under 5,500	3	6	3:6 = ½
Area	=	Base	×	Height

Source: Table 3.3.

Note how the column 4 data—and not the column 2 data—have been plotted on the vertical axis of Figure 3.1. The first class, for example, spans an interval of $1,500 million; it is *three* times as wide as the standard unit of measurement employed. Hence, the absolute class frequency of 3 is properly represented by the first rectangle's *area* if that rectangle's height is plotted as the frequency density of 1. In this large class, one is likely to find only one company per $500 million range. Similarly, the last class spans $3,000 million (or $3 billion); it is *six* units wide. Therefore, the absolute frequency of 3 is properly represented by the last rectangle's area only if the rectangle's height equals 1/2.

There is, of course, good reason for all this adjustment. Failure to draw the heights of all histogram columns with respect to the same standard (such as a $500 million range) would make these columns not comparable and would present a false visual impression. In our case, drawing the first and last rectangles at a height of 3 [as the first and last entries of column 2, Table 3.9, might suggest] would make these two classes of profit seem much more important than they really are. Per $500 million range, these two "countries" are sparsely populated, indeed.

Histograms for relative frequency distributions. Histograms for relative frequency distributions are constructed on the same principles as those for absolute distributions. A histogram constructed on the basis of columns 1 and 3 of Table 3.4, for example, would have the same shape as that shown in Figure 3.1; however, the labeling on the vertical axis (and, therefore, the tops of the columns) would change to reflect the fact that *proportions* of companies, rather than absolute numbers of the companies, were being measured per 500-million dollar range.

Common types of histograms. All sets of quantitative data can be fitted into some form of frequency distribution that is likely to yield one or another of a relatively few common types of histogram. Figure 3.2 shows some of the more common shapes of histograms. Histogram (a) pictures the relative frequency distribution of the actual diameters of supposedly 1-foot-wide water pipes produced during a day. This histogram is symmetrical about the interval from .998 to 1.002 feet, with the histogram columns to the left and right becoming progressively shorter. The larger the divergence of the actual diameter, in either direction, from the desired diameter of 1 foot, the smaller is the proportion of such pipes produced. Thus, 30 percent of all pipes produced fit in the class on which the central column stands, 20-percent each fit in the two adjacent classes, 10-percent each into the next two, and 5-percent each into the two extreme classes (making for a total of 100 percent). This distribution follows from the fact that the area of the central column takes up .3 times the entire histogram area; the areas of the two adjacent columns each take up .2 times the total area, and so on, as the numbers inside the columns indicate. Naturally, these numbers must add to 1.00. This type of histogram describes a *normal distribution* that is typical of populations of physical measurements.

Histogram (b) once more shows the profits of multinationals from Figure 3.1. This time the proportion of the total area of the histogram taken by each column is indicated (which is not to be confused with the column heights given in Figure 3.1). Again, these proportions add to 1.00, and they tell us this: while 41 percent of the companies fall in the 0 to $500 million category and only 3 percent of them are found below that range, 56 percent of companies are above that range, which gives the distribution a long "tail" to the right. This shape corresponds to a general class of *skewed* distributions in which there are relatively few small observations concentrated in a narrow range below a dominant class, while relatively more large observations are spread over a wide range of values above that class.

Histogram (c) pictures gasoline mileages of a year's new-car models and represents an opposite type of skewed distribution. In this case, the smaller values are widely spread out below a relatively important range, but there are hardly any observations above it. This distribution is skewed to the left.

Histogram (d) depicts the *exponential distribution.* Such a picture is typical of populations that exhibit changes over time, such as the lifetimes of machines. A relatively large proportion of machines last one year, for example, but the proportion of machines that exceeds that number by one, two, three, and more years gets rapidly smaller and smaller.

Histogram (e), finally, depicts the *rectangular* or *uniform distribution.* For a construction job, lumber may have to be purchased at lengths of 8 feet, while pieces of all sizes from 14 inches to 8 feet are needed. As a result, lots of scrap is produced, and every one of the scrap pieces measures less than 14 inches. About equal proportions of them are found in each of the classes between 0 and 14 inches.

Vivid illustrations of the usefulness of histograms are provided by Analytical Examples 3.2, "Deciding Authorship," and 3.3, "Quality Control in Manufacturing."

Figure 3.2   Common Types of Histograms

This graph pictures some of the relative frequency distributions that will be discussed in detail in Chapters 6 and 7. Histograms for relative frequency distributions such as these are always constructed in such a way that the total area covered by a given set of columns equals unity. Therefore, the proportion of the total area taken by any one column (which is indicated throughout) can be readily converted into the percentage of observations falling into the class above which this column has been erected. Note: As was true of the absolute frequency densities in Figure 3.1, the relative frequency densities measured here are expressed with respect to some unit that is defined on the horizontal axis. This unit equals .004 feet in panel (a), $500 million per year in panel (b), 5 miles per gallon in panel (c), 1 year in panel (d), and 2 inches in panel (e). In all cases, multiplying a column's vertical distance, or frequency density, by the number of such units on which the column stands yields its share of the total area of the histogram. The numbers shown inside the columns refer to these areas.

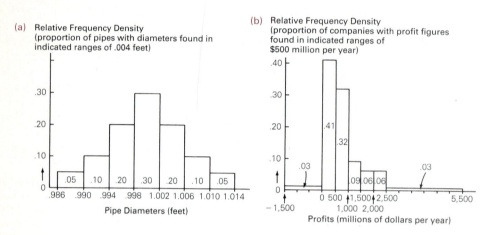

The Frequency Polygon and the Frequency Curve

An alternative graphical way of portraying an absolute or relative frequency distribution is provided by the **frequency polygon.** To draw this many-sided figure, the same set of coordinates is used as for the histogram, but this time the **class mark,** or midpoint of each class width, is identified (as the average of the two class limits) and a dot is positioned above it at a height equal to the (absolute or relative) frequency density. The dots are then connected by straight lines. To complete the polygon, straight lines are also drawn from the dots above the first and last class marks, respectively, to points on the horizontal axis that lie one-half the length of the standard unit of measurement below the lowest or above the highest class limit. Figure 3.3 indicates how histogram (a) of Figure 3.2 can be converted into a polygon.

The shape of a histogram can also be approximated by a smoothed **frequency curve.** This approximation can be accomplished by mathematical or graphical techniques and typically serves the purpose of removing irregularities as a result of sampling error from a histogram depicting information gathered in a sample survey. The frequency curve can then be viewed as an estimate of the unknown histogram that would emerge if census information were graphed with many classes of tiny widths. Figure 3.4 indicates how histogram (b) of Figure 3.2 might be converted into a smoothed frequency curve.

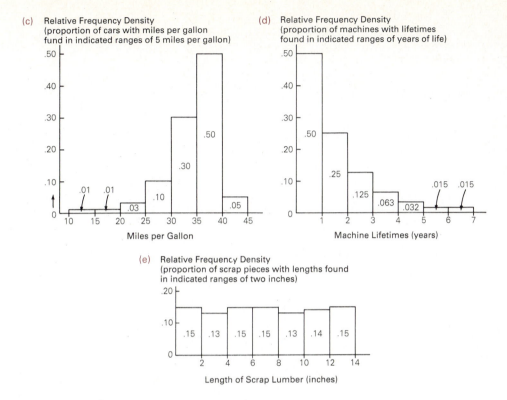

(c) Relative Frequency Density
(proportion of cars with miles per gallon
fund in indicated ranges of 5 miles per gallon)

Miles per Gallon

(d) Relative Frequency Density
(proportion of machines with lifetimes
found in indicated ranges of years of life)

Machine Lifetimes (years)

(e) Relative Frequency Density
(proportion of scrap pieces with lengths found
in indicated ranges of two inches)

Length of Scrap Lumber (inches)

Figure 3.3 Turning a Histogram into a Polygon

Panel (a) depicts the histogram of pipe diameters (now dashed) that was first depicted in Figure 3.2. Fat dots *b* through *h* have been placed over the midpoints of every class width at heights equal to that of the histogram columns (and thus measuring the relative frequency densities). Two additional dots, *a* and *i,* have been placed on the horizontal axis, one-half the length of the .004 foot standard unit of measurement, below the lowest and above the highest class limit, respectively. The polygon is created by connecting all these dots *(a* through *i)* by straight lines. Panel (b) shows the polygon standing alone, but class marks have replaced class limits on the horizontal axis.

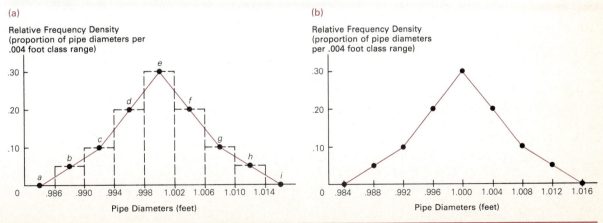

(a)

Relative Frequency Density
(proportion of pipe diameters per
.004 foot class range)

Pipe Diameters (feet)

(b)

Relative Frequency Density
(proportion of pipe diameters
per .004 foot class range)

Pipe Diameters (feet)

Figure 3.4   Turning a Histogram into a Frequency Curve

Panel (a) depicts the histogram of miles per gallon (now dashed) that was first noted in Figure 3.2. A smooth curve is drawn to approximate the shape of this histogram. This frequency curve is shown separately in panel (b).

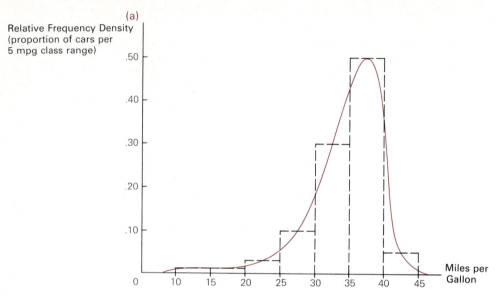

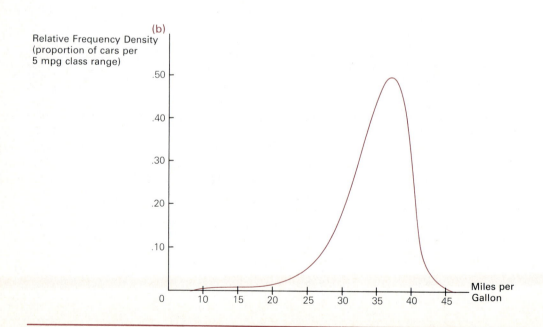

Ogives

Any cumulative frequency distribution, such as those noted in Table 3.6 or 3.7, can be shown graphically as well. The resulting curve is called an **ogive** (pronounced ōjive).

In the case of a less-than-or-equal type of cumulative frequency distribution, each cumulative class frequency is plotted vertically above the *upper* limit of the corresponding class (which, in turn, is measured horizontally). Panel (a) of Figure 3.5, for example, shows an ogive based on columns 1 and 3 of Table 3.6. The encircled entry in column 3 now appears as point *b:* 76 companies had profits of less than $1,000 million ($1 billion).

In the case of a more-than-or-equal type of cumulative frequency distribution, each cumulative class frequency is plotted vertically above the *lower* limit of the corresponding class. Panel (b) of Figure 3.5, for example, shows an ogive based on columns 1 and 3 of Table 3.7. The encircled entry in column 3 now appears as point *e:* 24 companies had profits of $1,000 million ($1 billion) or more.

Note: In the case of discrete quantitative data, an ogive is typically drawn in stair-step fashion, the horizontal portions of the line being placed in the gaps between possible values. Panels (c) and (d) of Figure 6.1 on p. 193 provide examples.

OTHER TYPES OF GRAPHS*

Bar and Column Charts

Bar charts and **column charts** are particularly effective in presenting data about qualitative or discrete quantitative variables. Such data are portrayed by a series of noncontiguous horizontal bars or vertical columns, the lengths of which are proportional to the values that are to be depicted. Many variations of these types of charts exist; Figure 3.6 on page 67, which is based on Table 3.5, shows just some of them.

Time-Series Line Graphs

Some data, such as those of Table 3.8, that are linked with specific points in time or intervals of time can be presented in bar or column charts but are also suited to being plotted on a rectilinear coordinate surface. Successive points are then joined by a continuous line, yielding a **time-series line graph.** On page 68 panel (a) of Figure 3.7 provides the simplest example; panel (b) provides a slightly more complex one. Panel (a) is called an *arithmetic* chart because it uses the regular arithmetic scale on both axes. It has a disadvantage, however, when, as in this case, two series of numbers are being compared that move within rather different ranges of values. The series within the higher range of values (in this case, statewide sales) will appear to be considerably more volatile, but this volatility can be an illusion. The illusion can be removed by

---

*Optional section.

Figure 3.5 Two Types of Ogives

Just as there are two types of cumulative frequency distributions, there are two types of graphs to portray them. Panel (a) is based on columns 1 and 3 of Table 3.6 and represents a less-than type ogive. As did Table 3.6, it tells us the number of companies with 1981 profit less than a certain figure. Thus point *a* indicates that 0 companies earned less than − $1,500 million, point *b* shows that 76 companies earned less than $1,000 million, and point *c* tells us that 100 companies earned less than $5,500 million. Panel (b) is based on columns 1 and 3 of Table 3.7 and represents a more-than-or-equal type ogive. As did Table 3.7, it gives us the number of companies with 1981 profit of a certain figure or more. Point *d*, for example, indicates that all 100 companies earned − $1,500 million or more, point *e* shows that 24 companies earned $1,000 million or more, and point *f* tells us that 0 companies earned $5,500 million or more. All this can, of course, be shown for relative frequency distributions as well.

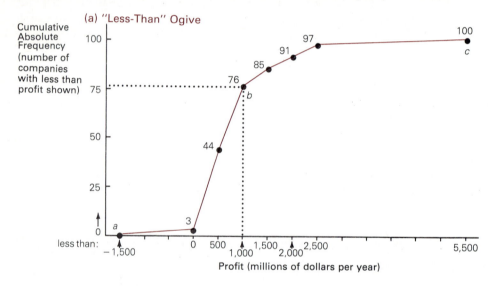

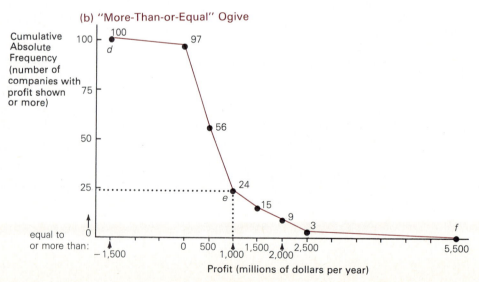

Figure 3.6 Bar and Column Charts

Panel (a) is a bar chart; panels (b) through (d) are various forms of column charts (an ordinary column chart, a component-column chart, and a grouped-column chart). Note that the identical information is conveyed by panels (a) through (c); panel (d), unlike the other panels, portrays relative rather than absolute frequencies.

(a) Mountain Aviation, Inc.
    Composition of Employees by
    Race and Sex, 7/1/85

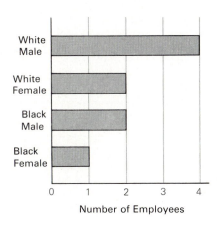

(b) Mountain Aviation, Inc.
    Composition of Employees by
    Race and Sex, 7/1/85

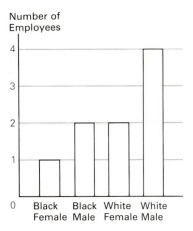

(c) Mountain Aviation, Inc.
    Composition of Employees by
    Race and Sex, 7/1/85

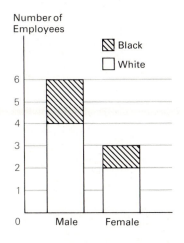

(d) Mountain Aviation, Inc. vs. Aviation Industry:
    Composition of Employees by
    Race and Sex, 7/1/85

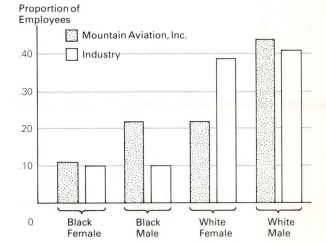

Source: Table 3.5.

Figure 3.7 Time-Series Line Graphs

Line graphs can be drawn on a regular arithmetic scale on both axes, as in panel (a). They can also be drawn on a semilogarithmic scale, as in panel (b). In the latter case, equal percentage changes in different series of data produce lines of equal slope. Note: A logarithmic scale does not have a zero point because the logarithm of zero equals minus infinity. Such a scale is calibrated in log values in various "cycles" from 1–10, 10–100, 100–1,000 and so on (or, in the other direction, from 1–10, .1 to 1, .01 to .1, and so forth).

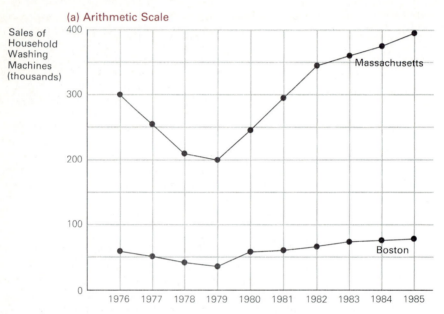

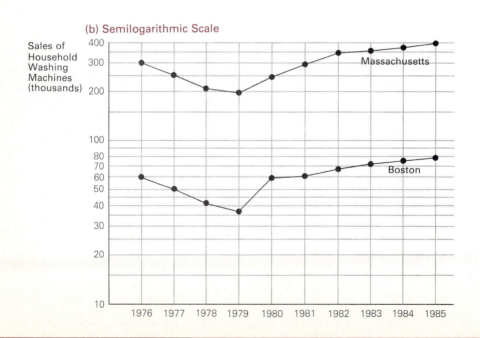

using a *semilogarithmic* chart, as in panel (b). The horizontal scale remains arithmetic, the vertical one becomes logarithmic. As a result, equal percentage changes on two series of data (as between Boston sales and statewide sales until 1978 and again from 1983 onwards) produce lines of equal slope; the illusion disappears.

Maps, Pie Charts, and Pictograms

A great number of other devices can be used to display data visually. Some of them are shown in Figure 3.8. Panel (a) pictures a **statistical map,** which portrays data for areal units, such as world regions, nations, states, counties, or even census tracts, by crosshatching or shading these units in different ways on a geographic map. Such a map can be very effective, as in panel (a).

Note how anyone with the proper geographical knowledge can tell instantly, for example, that 1972 per capita personal income reached $5,000 or more in only 6 among the lower 48 states of the United States (namely, in California, Nevada, Illinois, New York, New Jersey, and Connecticut), while such per capita income in Maine, Louisiana, and many other states was below $4,000. Other statistical maps are, however, less easy to read—particularly three-dimensional ones that attempt, for example, to depict the extent of urban sprawl or air pollution by drawing pictures of buildings or billowing chimneys on top of a geographic map.

Panel (b) is a typical **pie chart,** a segmented circle that is amenable to portraying divisions of some aggregate, provided there are not too many divisions. The circular pie is cut into slices by radii in such a way that the central angles (and, therefore, the circumference arcs and sector areas) are proportional to the sizes of the divisions being displayed.[3] The pie chart shown in panel (b) of Figure 3.8 demonstrates how Tylenol dominated the 1981 U.S. market for over-the-counter pain relievers, which at the time gave aspirin makers a headache that even extra-strength Tylenol couldn't cure.

A pictorial chart or **pictogram** depicts data with the help of symbols, such as those in panel (c). All kinds of symbols can be used, from stars to stacks of silver coins, from tanks to trees, from crosses to castles, from smiling faces to billowing smokestacks. Typically, each symbol represents a definite and uniform value, as is the case here, but sometimes the size of the pictorial symbol is made proportional to the values that are to be portrayed. Making the size of the symbol proportional to values is tricky business, however, and can easily lead to confusion. How, for instances, does one double the

---

[3]Suppose the aggregate in question were industry sales of $132 million, of which firm *X* sold $48.84 million, or 37 percent. This market share would be represented in a pie chart by a central angle of .37(360°) = 133.2°, since a circle has a total of 360 degrees and 133.2° equal 37 percent of this total, as the accompanying graph illustrates.

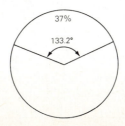

Figure 3.8   Maps, Pie Charts, and Pictograms

These are just three examples of a near-infinite variety of graphical devices commonly used to portray statisticsl data.

(a) Per Capita Personal Incomes, 1972

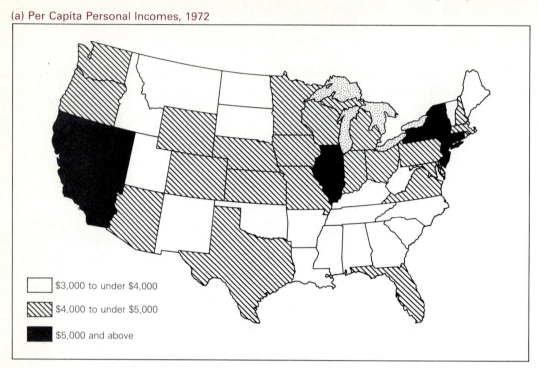

☐ $3,000 to under $4,000

▨ $4,000 to under $5,000

■ $5,000 and above

(b) The U.S. Analgesic Market, 1981   (c) Employment at City Motors, 1950 vs. 1985

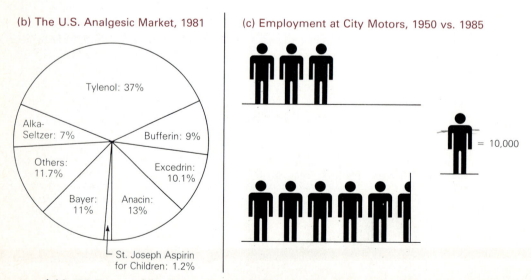

Sources: panel (a): U.S. Bureau of the Census, *Statistical Abstract of the United States: 1973* (Washington, D.C.: U.S. Government Printing Office, 1973), p. 326; panel (b): Paine Webber Mitchell Hutchins, Inc.; panel (c): hypothetical data.

size of a three-dimensional symbol, such as a box? By doubling each length, one inevitably quadruples the areas portrayed, and one raises the volume eightfold. (For an illustration of this problem see Close-Up 3.1, "How to Lie with Statistics.")

The problem just described haunts many of the three-dimensional perspective charts that nowadays are so speedily produced by computers. Their pictures inevitably distort the underlying data. In spite of the computer's popular appeal, it pays to remember that complex technology cannot replace personal judgment and expertise in designing graphs. At least at the time of this writing, computer graphics were still crude compared to the artistry often exhibited by handmade graphs. Serious computer graphics have been around only since about 1970; in the early 1980s, there was hope that the advent of "micrographics" would improve their quality and even that television and motion-picture technology could be put in the service of a new kind of descriptive "kinostatistics."[4]

Stem-and-Leaf Diagrams and Residual Plots

A somewhat unusual type of diagram, the **stem-and-leaf diagram,** combines the features of an ordered array of numbers (such as Table 3.2) and a frequency histogram (such as Figure 3.1). Consider selecting a simple random sample of 80 persons from a complete list of subscribers to an economic newsletter. If each of them were interviewed, a variety of information might be acquired, including the data found in Table 3.10.

One way of displaying these data is the stem-and-leaf diagram of Figure 3.9. The possible decade numbers are arranged in a column to the left of a vertical line. These first digits of the two-digit numbers are called the *stems.* (When large numbers are involved, the stems can consist of several digits, however.) Subsequently, each of the numbers in Table 3.10 is inspected and its *second* digit only is recorded as a *leaf* to the right of the vertical line next to the appropriate stem. With some skill, the leaves belonging to a given stem can even be put in order at the same time (0s being placed before 1s, 1s before 2s, and so on). The result is Figure 3.9, and its interpretation is easy: As row 1 shows, only one subscriber in the sample is a teenager and is 19 years

Table 3.10   A Population of Ages

This table shows the ages of a hypothetical random sample of 80 subscribers to an economic newsletter.

40	50	42	40	41	54	47	55	30	45
21	70	60	31	45	50	54	50	30	35
30	19	50	52	29	25	60	60	34	47
50	45	60	55	30	35	40	48	43	56
70	58	50	65	32	41	48	40	55	53
51	68	65	85	49	75	45	52	40	42
47	66	58	20	48	37	69	55	65	53
49	46	40	51	55	73	20	50	75	52

---

[4]A "Close-Up" example called "The Faces of American Cities," which appears in Chapter 3 of the *Student Workbook* that accompanies this text, shows computer-produced pictograms of multivariate data.

Figure 3.9   A Stem-and-Leaf Diagram

This unusual diagram involves no dots or lines and can be read simply as an ordered array of the raw data given in Table 3.10. Yet when looked at from the side, the array serves the same function as a histogram. (Suggestion: Rotate this page 90 degrees so that the askerisk appears at the bottom. Then note its similarity to a graph such as Figure 3.1.)

Stem	Leaf	Frequency
1	9	1
2	0 0 1 5 9	5
3	0 0 0 0 1 2 4 5 5 7	10
4	0 0 0 0 0 0 1 1 2 2 3 5 5 5 5 6 7 7 7 8 8 8 9 9	24
5	0 0 0 0 0 0 0 1 1 2 2 2 3 3 4 4 5 5 5 5 5 6 8 8	24
6	0 0 0 0 5 5 5 6 8 9	10
7	0 0 3 5 5	5
8	5	1

old. Row 2 shows that there are five subscribers in their twenties; they are aged 20, 20, 21, 25, and 29, respectively. And so it goes. As the caption indicates, the diagram can be viewed as an array of raw data but also as a histogram.

The arrangement of data in Figure 3.9 highlights a curious fact, totally hidden among the raw data of Table 3.10: fully 43 of the 80 persons interviewed claimed to have an age ending precisely with a zero or a five. Considering that the sample was a random one, this distribution seems difficult to accept as true but strongly suggests that

Table 3.11   Residuals Implied by the Stem-and-Leaf Diagram

The residuals in column 4 are calculated as the difference between an observed value—column 2— and an anticipated value—column 3. It is no accident that the sums of actual and expected frequencies are alike (80 people were in fact interviewed) nor that the sum of residuals equals zero (to the extent that some digits appear too often, others must appear not often enough).

Second Digit of Age (1)	Actual Frequency (2)	Expected Frequency (3)	Residual (4) = (2) − (3)
0	25	8	+17
1	6	8	−2
2	6	8	−2
3	4	8	−4
4	3	8	−5
5	18	8	+10
6	3	8	−5
7	4	8	−4
8	6	8	−2
9	5	8	−3
Totals	80	80	0

many of the respondents were simply rounding their true ages to the nearest zero or five. This is where another unusual type of graph comes in. Statisticians define the difference between a value obtained by observation and the corresponding value anticipated on the basis of a theoretical model or previous experience as a *residual*; they refer to a graphical display of such residuals as a set of **residual plots.** In this example, some 80 randomly chosen people were interviewed about their ages and one would expect that each of the 10 possible second digits (0 through 9) would appear just about equally often—that is, 8 times. Yet a 0 appears 25 times, not 8 times, so the residual here is $25 - 8 = +17$. Similarly, a 1 appears 6 times, not 8 times, making the residual $6 - 8 = -2$. Similar residuals can be calculated for the other digits by closely inspecting Figure 3.9. The results are shown in Table 3.11. The unusual divergence of the actual from the expected that is noted in columns 3 and 4 of Table 3.11 can also be graphed, as in Figure 3.10.

Note: Chapter 3 of the *Student Workbook* that accompanies this text introduces another unusual type of data display called the *box-and-whisker diagram.*

Figure 3.10 Residual Plots

When actual observations coincide with expected ones, residuals are zero, and all residual plots would lie on the horizontal line in the center of the graph. The fact that actual plots are not found on this line may suggest that something is amiss. In this case, the frequencies with which digits 0 and 5 appear are excessive; not the two plots above the horizontal zero line. Correspondingly, the frequencies with which the other digits appear are too low, as the plots below the zero line show.

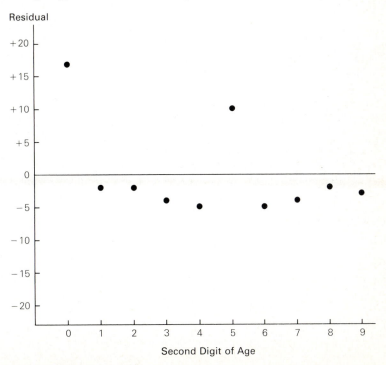

Source: Table 3.11, columns (1) and (4).

ANALYTICAL EXAMPLES

3.1 Deciphering Secret Codes

Cryptanalysis, the scientific study of converting ciphers or codes to which the key is not known into plain text, provides a dramatic illustration of how frequency distributions shed light on masses of raw data. Consider the secret message (presumed to be written in English) contained in Table A.

To decipher the message, cryptanalysis first establishes the absolute frequency distribution of the letters in the secret message. Table B provides a count of the number of time each letter occurs and also shows the implied relative frequency distribution. Next, the relative frequency distribution is compared with that of the letters in a normal English language text, shown in Table C (small letters are used in the list of normal English letters to avoid confusion with the capital letters in the coded message). To make the comparison,

Table A

IMEUX	ANEMO	MNPKN	EONUC	MOMXB
MJDAM	VNNES	NSXXW	MVSPM	RPMSN
MAMQK	SXNES	NNEMG	SPMMV	AUIMA
CGNEM	DPRPM	SNUPI	DNERM	PNSDV
KVSXD	MVSCX	MPDLE	NONES	NSWUV
LNEMO	MSPMX	DBMXD	CMPNG	SVANE
MTKPO	KDNUB	ESTTD	VMOON	ESNNU
OMRKP	MNEMO	MPDLE	NOLUJ	MPVWM
VNOSP	MDVON	DNKNM	ASWUV	LWMVA
MPDJD	VLNEM	DPZKO	NTUIM	POBPU
WNEMR	UVOMV	NUBNE	MLUJM	PVMA

Table B Frequency Distribution of 274 Letters in Coded Message

Letter	Absolute Frequency (number)	Relative Frequency (percent)	Letter	Absolute Frequency (number)	Relative Frequency (percent)
A	9	3.3	N	36	13.1
B	5	1.8	O	16	5.8
C	4	1.5	P	21	7.7
D	16	5.8	Q	1	.4
E	19	6.9	R	5	1.8
F	0	0	S	19	6.9
G	3	1.1	T	4	1.5
H	0	0	U	14	5.1
I	4	1.5	V	18	6.6
J	4	1.5	W	6	2.2
K	8	2.9	X	9	3.3
L	7	2.5	Y	0	0
M	45	16.4	Z	1	.4
			Totals	274	100.0

the letters of normal English and of the secret message are arranged in Table D in order of decreasing relative frequency.

Although one can hardly expect an immediate letter-for-letter matching in Table D, the juxtaposition can be very helpful in breaking the code. After some trying

Table C Frequency Distribution of 200 Letters of a Normal English-Language Text

Letter	Absolute Frequency (number)	Relative Frequency (percent)	Letter	Absolute Frequency (number)	Relative Frequency (percent)
a	16	8	n	14	7
b	3	1.5	o	16	8
c	6	3	p	4	2
d	8	4	q	.5	.25
e	26	13	r	13	6.5
f	4	2	s	12	6
g	3	1.5	t	18	9
h	12	6	u	6	3
i	13	6.5	v	2	1
j	1	.5	w	3	1.5
k	1	.5	x	1	.5
l	7	3.5	y	4	2
m	6	3	z	.5	.25
			Totals	200	100.0

Source: David Kahn, *The Codebreakers: The Story of Secret Writing* (New York: Macmillan, 1967), p. 100.

Table D Comparative Relative Frequency Distributions of Plain and Coded English

Plain English		Coded English		Plain English		Coded English	
Letter	Relative Frequency	Letter	Relative Frequency	Letter	Relative Frequency	Letter	Relative Frequency
e	13	M	16.4	u	3	W	2.2
t	9	N	13.1	f	2	B	1.8
a	8	P	7.7	p	2	R	1.8
o	8	E	6.9	y	2	C	1.5
n	7	S	6.9	b	1.5	I	1.5
i	6.5	V	6.6	g	1.5	J	1.5
r	6.5	D	5.8	w	1.5	T	1.5
h	6	O	5.8	v	1	G	1.1
s	6	U	5.1	j	.5	Q	.4
d	4	A	3.3	k	.5	Z	.4
l	3.5	X	3.3	x	.5	F	0
c	3	K	2.9	q	.25	H	0
m	3	L	2.5	z	.25	Y	0
				Total	100.0	Total	100.0

about, some letters can be identified (for example, *M* as *e* and *N* as *t*), then portions of words, and finally, the whole message. (Note: A good codebreaker looks not only at general letter frequencies, but also considers the preferred associations of one letter with other letters, the order of frequency of the most common doubles, of initial letters, of final letters, of the most frequent one-letter words, and much more.) The key for this particular message is a simple one: the word *SCRAMBLED* is used for the first nine letters of the alphabet, and the remaining code letters (now excluding the letters in the word *SCRAMBLED*) are listed in reverse alphabetical sequence. The first line below shows the plain text and the second line below shows the coded text.

a b c d e f g h i j k l m n o p q r s t u v w x y z
S C R A M B L E D Z Y X W V U T Q P O N K J I H G F

As an analysis of Table D can show, even this simple table correctly identifies the meanings of five code letters *(M, N, A, X,* and *B),* and it provides strong hints in the case of a dozen others, the meanings of which are found within a line or two of the code letter. (Thus a *V* is not an *i,* but an *n;* a *D* is not an *r,* but an *i,* and so on). In fact, all but two of the alphabet's 26 letters are decipherable by looking within four lines of the code letter in Table D. The decoded message is an excerpt from the Declaration of Independence.

We hold these truths to be self-evident, that all men are created equal, that they are endowed by their Creator with certain unalienable rights, that among these are life, liberty and the pursuit of happiness, that to secure these rights, governments are instituted among men, deriving their just powers from the consent of the governed.

3.2 Deciding Authorship

On occasion, an important question arises concerning the authorship of a piece of writing. A well-known case concerned the authorship of the Paulines, a set of Christian religious writings possibly written by St. Paul. A second famous case was the Shakespeare-Bacon-Marlowe controversy over who wrote certain plays traditionally attributed to Shakespeare. And then there was the matter of the *Federalist* papers, some of which were claimed simultaneously by Alexander Hamilton and James Madison. Let us consider the latter case to show how descriptive statistics can play an important role in solving the puzzle simply by counting something in a systematic way.

After most of them had appeared anonymously in various newspapers, some 85 essays written by Alexander Hamilton, John Jay, and James Madison appeared in book form in 1788. *The Federalist,* as the book was called, has ever since been a leading source of information concerning the intent of the framers of the U.S. Constitution, but between 1807 and 1818 a variety of lists appeared with conflicting information about which authors wrote which papers. In particular, 12 papers were claimed by both Hamilton and Madison after they had become bitter political enemies. Historians have tried in vain to decide authorship on the basis of political content. This line of inquiry led nowhere partly because both authors were at the time writing these papers as lawyers' briefs in favor of ratifying the Constitution (and not every argument put forward was necessarily their own), partly because the authors' political opinions changed over time (and one couldn't infer their earlier positions from later ones). Nor could average sentence length be used to differentiate the disputed papers. It was 34.5 words in 51 undisputed Hamilton papers and 34.6 words in 14 undisputed Madison papers—the authors wrote in similar styles.

Statisticians were able, however, to employ as discriminators *rates of word use.* They ignored contextual words, such as *law* and *liberty,* that were likely to appear in all the papers to a great extent, simply because of the topic discussed. They focused instead on filler words, such as *by, to,* and *upon,* that were unrelated to the topic. Three sets of histograms depicting rates of word use are shown in panels (a)–(c) of the accompanying figure. Focus on the tallest columns in panel (a): In 18 of 48 undisputed Hamilton papers, the word *by* occurred between 7 and 9 times per 1,000 words; yet in 16 of 50 undisputed Madison papers, it occurred between 11 and 13 times per 1,000 words, just as it did in 4 of 12 disputed papers. Compared to Hamilton's, the entire Madison histogram is displaced to the right (as is that of the disputed papers). The more frequent use of *by* suggested Madison authorship.

(a) Occurrence of *by*   (b) Occurrence of *to*   (c) Occurrence of *upon*

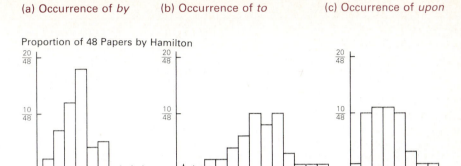

Proportion of 48 Papers by Hamilton

Rate per 1,000 Words

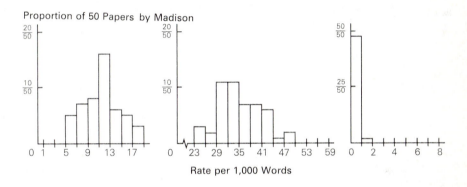

Proportion of 50 Papers by Madison

Rate per 1,000 Words

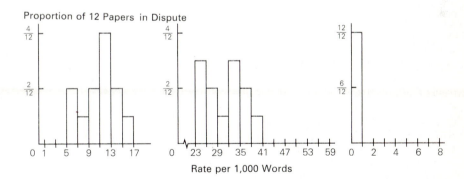

Proportion of 12 Papers in Dispute

Rate per 1,000 Words

Similarly, the *less* frequent use of the words *to* and *upon* in Madison's undisputed papers and in the disputed papers lead to the same conclusion. Indeed, the three histogram sets shown here, along with 27 other sets like these, provided overwhelming evidence in favor of *Madison* being the author of each one of the disputed papers.

Source: Frederick Mosteller and David L. Wallace, "Deciding Authorship," in Judith M. Tanur et al., eds., *Statistics: A Guide to the Unknown* (San Francisco: Holden-Day, 1972), pp. 164–75.

77

3.3 Quality Control in Manufacturing

A manufacturer of small motors was in trouble. Recently produced motors were breaking down at an unexpected rate. A close look at the manufacturing process was in order. The investigation centered on steel rods that might be too loose in their bearings.

The investigator noted that quality inspectors had been instructed to inspect the inside diameters of the rods and to reject all those with measurements of .9995 centimeters or less. The inspection records of 500 such rods were requisitioned and the physical measurements were plotted as a histogram. The investigator expected to find a symmetrical histogram, such as histogram (a) in Figure 3.2 because in the past, errors in either direction from the standard had been equally frequent and smaller errors had been more frequent than larger errors. Yet the investigator found the histogram reproduced in the accompanying figure, showing absolute frequencies of 10, 30, 0, 80, 60, 100, 90, 60, 40, 20, and 10 for the eleven classes of diameters. (Note: Because all the classes are of equal size, absolute frequencies and absolute frequency densities

coincide in this case.) The unusual gap just below the lower tolerance limit and the unusually tall column just above it were too obvious to be ignored.

As it turned out, inspectors had felt pressure to keep scrappage low (because any rod thrown out meant labor, materials, and other costs were wasted in its manufacture). As a result, they had misclassified borderline defective rods (with diameters, say, of .9993 centimeters) as borderline acceptable (as if they had diameters of .9995 centimeters). No wonder that too few rods appeared in class 3 and too many in class 4. When their inspection procedures were corrected, not 40, but 105 rods had to be scrapped from the next batch of 500.

The high number of defective rods itself was subsequently traced to a faulty machine setting and corrected. The breakdowns of the motors disappeared.

Source: W. Edwards Deming, "Making Things Right," in Judith M. Tanur et al., eds., *Statistics: A Guide to the Unknown* (San Francisco: Holden-Day, 1972), pp. 229–31.

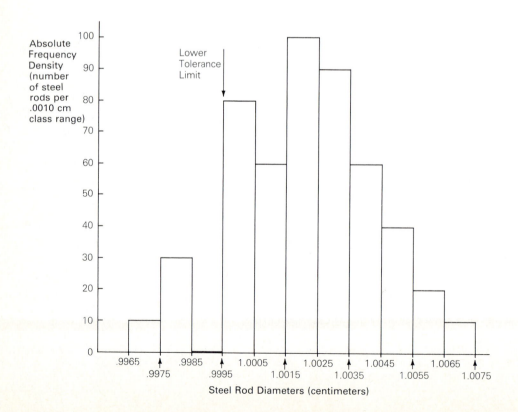

3.1 How to Lie with Statistics

Benjamin Disraeli, the British Prime Minister from 1874–80, purportedly said: "There are three kinds of lies: ordinary lies, damned lies, and statistics." Regardless of whether this tale is true, it is a fact that one can lie with statistics. Such distortion may happen accidentally, out of ignorance; it may also happen deliberately, out of malice, in order to defraud the gullible. The ways of doing so are legion. Consider just some of the vices illustrated in the accompanying graphs.

Figure A graphs the identical data about Boston washing-machine sales (Table 3.8) in three ways. Every one of the graphs is technically correct, but consider the different impressions one can make just by stretching the horizontal or vertical axis.

Figure B is a worse case, although commonly encountered. By not labeling the vertical axis at all, the desired impression is achieved, yet the graphs are totally meaningless. Does the panel (b) change from 1980–81, for example, depict an annual increase of 50¢ or of $5 million?

Finally, consider Figure C. Panel (a) pictures a 3.5-fold increase in the number of a firm's employees with the help of two human figures, the second of which is 3.5 times as tall as the first. Note the false impression this type of pictogram is likely to give. Compare it with the alternative version given in panel (b).[1]

[1]For further discussion of graphical fallacies, see Darrell Huff and Irving Geis, *How to Lie with Statistics* (New York: W. W. Norton, 1954), which inspired this example.

Figure A

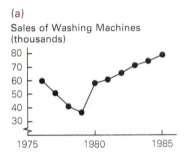

(a)
Sales of Washing Machines (thousands)

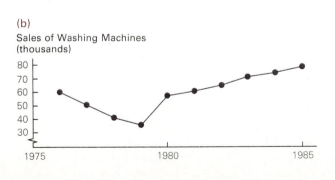

(b)
Sales of Washing Machines (thousands)

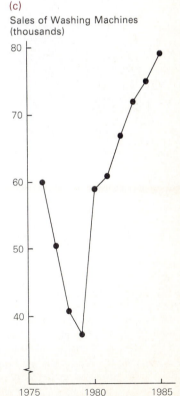

(c)
Sales of Washing Machines (thousands)

Figure B

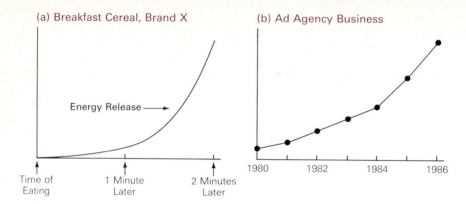

(a) Breakfast Cereal, Brand X

Energy Release ⟶

Time of Eating | 1 Minute Later | 2 Minutes Later

(b) Ad Agency Business

1980   1982   1984   1986

Figure C

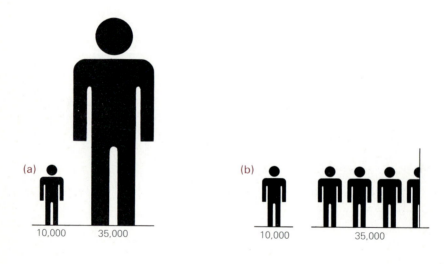

(a)   10,000   35,000

(b)   10,000   35,000

Summary

1. This chapter discusses the two most common procedures for making masses of data intelligible: the construction of tables and the drawing of graphs. Any considerable quantity of unorganized numerical data is likely to be confusing; even an ordered array is an improvement. Greater clarity still is provided by dividing the range of available data into a relatively small number of collectively exhaustive and mutually exclusive classes, noting how many observations fall into each, and constructing a table on that basis. If desired, such an absolute frequency distribution can easily be converted into a relative one that shows the proportions of all observations in the chosen classes (rather than the absolute number of observations). In the case of quantitative data, further insight can be gained from the construction of cumulative frequency distributions (of the LE or ME type). Prior to the publication of any such tabular summary, however, a number of factors concerning table title, footnotes, and the like should be designed to provide maximum help to the user.

2. Some of the most important devices for presenting absolute or relative frequency distributions of continuous quantitative data graphically include frequency histograms, frequency polygons, and frequency curves. Cumulative frequency distributions are, in turn, graphed as ogives.

3. Data about qualitative or discrete quantitative variables are most effectively presented in bar or column charts. Other key types of graphs include time-series line graphs, statistical maps, pie charts, and pictograms. Among the more unusual data displays are stem-and-leaf diagrams and residual plots.

Key Terms

absolute class frequency
absolute frequency
distribution
bar charts
classes
class mark
class width
column charts
cumulative class frequency

cumulative frequency
distribution
frequency curve
frequency density
frequency histogram
frequency polygon
ogive
open-ended classes
ordered array

pictogram
pie chart
relative class frequency
relative frequency
distribution
residual plots
statistical map
stem-and-leaf diagram
time-series line graph

Questions and Problems

The computer programs noted at the end of this chapter can be used to work many of the subsequent problems.

1. An automated filling machine filled 80 cans with supposedly 16 ounces of peas each. Upon inspection, the actual net weights, however, were found to be those listed in Table 3.A. Present these data as an ordered array. Use a stem-and-leaf diagram to produce the array.

Table 3.A

15.83	15.39	15.93	15.98	15.98	15.85	16.00	15.98	16.04	15.79
16.01	15.91	16.21	15.89	15.77	16.05	15.77	15.96	15.75	16.37
16.24	15.71	15.84	16.09	15.75	16.38	16.43	16.49	16.60	16.63
16.42	16.01	16.17	15.91	16.13	15.82	15.91	16.27	15.92	15.92
15.33	15.90	16.01	16.08	15.57	16.12	16.11	16.81	16.65	16.68
15.44	15.51	15.92	15.72	16.25	15.90	15.53	16.00	16.29	15.69
16.88	16.10	16.20	15.94	15.93	15.73	16.47	16.07	16.00	16.53
16.31	15.62	15.81	15.99	16.40	15.84	15.95	16.19	15.74	15.90

Table 3.B.

42	58	41	9	16	25	48	53
32	11	16	21	8	29	10	12
18	27	31	18	36	50	38	38
60	33	40	18	40	51	42	47
17	19	45	11	58	13	29	34

2. Table 3.B lists the miles (in thousands) driven by the 40 sales representatives of a publishing firm during the past 12 months. Present these data as an ordered array, using a stem-and-leaf diagram to produce it.

3. Table 3.C lists the average salaries of public school teachers during the 1984–85 school year (*The New York Times,* August 31, 1985, p. 15). Present these data as an ordered array.

Table 3.C

Alabama	$20,209	Kentucky	$21,486	North Dakota	$19,900
Alaska	39,751	Louisiana	19,690	Ohio	22,737
Arizona	23,380	Maine	18,329	Oklahoma	18,954
Arkansas	19,180	Maryland	25,861	Oregon	24,378
California	26,300	Massachusetts	24,110	Pennsylvania	24,435
Colorado	24,456	Michigan	28,401	Rhode Island	27,384
Connecticut	24,520	Minnesota	25,920	South Carolina	19,800
Delaware	22,924	Mississippi	15,971	South Dakota	17,356
D.C.	28,621	Missouri	20,452	Tennessee	22,141
Florida	21,057	Montana	21,705	Texas	22,600
Georgia	20,607	Nebraska	20,153	Utah	21,307
Hawaii	24,628	Nevada	22,520	Vermont	19,014
Idaho	19,700	New Hampshire	18,577	Virginia	21,536
Illinois	25,829	New Jersey	25,125	Washington	25,610
Indiana	23,089	New Mexico	22,064	West Virginia	19,563
Iowa	20,934	New York	29,000	Wisconsin	24,577
Kansas	21,208	North Carolina	20,691	Wyoming	26,709
				U.S.	23,546

4. Table 3.D lists the price-earnings ratios of selected stocks. Present these data as an ordered array.

Table 3.D

28	9	11	8	14	14	10	4	13
13	7	8	13	12	57	15	8	19
13	9	35	18	14	9	15	8	48
10	21	7	19	9	26	29	6	17
14	23	22	10	9	11	13	24	37

5. Following Sturgess's rule, what is the recommended number of classes, k, for a frequency distribution with $n = 80$ observations? For $n = 400$?

6. How many classes would Sturgess's rule recommend for the data sets given in problems 1 through 4? Explain.

7. Comment on the following two arrangements for dividing the net weights of vegetable cans into classes.

Arrangement 1:	Arrangement 2:
15.3 to under 15.5	15.3 to 15.5
15.6 to under 15.8	15.5 to 15.7
15.8 to under 16.0	15.7 to 15.9
and so forth	and so forth

8. Contrary to Sturgess's rule, create an absolute frequency distribution from the Table 3.B data, starting at 8,000 miles and using
   a. classes of 4,000 miles.
   b. classes of 25,000 miles.

9. Contrary to Sturgess's rule, create an absolute frequency distribution from the Table 3.A data, starting at 15.3 ounces and using
   a. classes of 0.1 ounce.
   b. classes of 0.8 ounce.

10. Follow Sturgess's rule and create an absolute and relative frequency distribution for the Table 3.B data. Start the first class with the minimum value in the data set.

11. Divide the Table 3.A data set into 0.2-ounce classes, beginning at 15.3 ounces. Construct an absolute and relative frequency distribution.

12. Follow Sturgess's rule and create an absolute and relative frequency distribution for the Table 3.D data. Start the first class with a value of 4.

13. Follow Sturgess's rule and create an absolute and relative frequency distribution for the Table 3.C data and a minimum value of $15,500. Round the class width suggested by Sturgess's rule upwards to the nearest $500.

14. Consider your answer to problem 10. Use it to create the cumulative absolute frequency distribution and the cumulative relative frequency distribution, each of the "less-than-or-equal" type.

15. Consider your answer to problem 11. Use it to create the cumulative absolute frequency distribution and the cumulative relative frequency distribution, each of the "less-than-or-equal" type.

16. Consider your answer to problem 12. Use it to create the cumulative absolute frequency distribution and the cumulative relative frequency distribution, each of the "more-than-or-equal" type.

17. Consider your answer to problem 13. Use it to create the cumulative absolute frequency distribution and the cumulative relative frequency distribution, each of the "more-than-or-equal" type.

18. Consider Table 3.E concerning the households in a suburban development.
   a. Fill in the missing values (denoted by dots).
   b. What percentage of these households has at least one TV set? Has at most three sets? Has three or four sets?

19. In December of 1984, U.S. commercial banks held loans and investments valued at $1,713.6 billion (*Economic Report of the President,* February 1985, p. 307). The relative class frequencies of four categories of these assets are shown in column 2 of Table 3.F. Calculate the absolute class frequencies that are missing in column 3.

20. On the basis of your answer to problem 8a, draw the two frequency histograms (absolute and relative).

21. Depict your answers to problem 9 as absolute frequency histograms.

22. Depict your problem 10 answers as histograms.

23. Consider whether each of the following variables (measured on the horizontal axis) is likely to be described by a histogram approximating a frequency distribution that is normal (symmetrically bell-shaped), skewed to the right, skewed to the left, exponential, or rectangular:
   a. family money incomes in the United States.
   b. random numbers.
   c. the dollar values of mail-order-house sales.
   d. the height of boats on the Mississippi.
   e. the ages of pinball machine players.
   f. the ages of people with false teeth or bald heads.
   g. the last digits of all New York City telephone numbers.
   h. the grades of all students at midterm.
   i. the ages of children in elementary school.

Table 3.E

Class (number of TV sets per household) (1)	Absolute Class Frequency (number of households) (2)	Relative Class Frequency (proportion of all households in class) (3)	Cumulative Relative Class Frequency	
			"less than or equal" (proportion of all households in class or lower ones) (4)	"more than or equal" (proportion of all households in class or higher ones) (5)
0	60	•	•	•
1	•	.28	•	•
2	•	•	.72	•
3	•	•	•	•
4	24	•	•	•
5	•	•	•	.04
Totals	300	•		

Table 3.F

Loans and Investments (1)	Relative Frequency (2)	Absolute Frequency (3)
Commercial loans	.274	
Other loans	.492	
U.S. gov't securities	.152	
Other securities	.082	
Total	1.000	

24. Consider whether each of the following variables (measured on the horizontal axis) is likely to be described by a histogram approximating a frequency distribution that is normal (symmetrically bell-shaped), skewed to the right, skewed to the left, exponential or rectangular:
  a. the number of printing errors found on a book's pages.
  b. the number of minutes people have to wait at the bank, doctor's office, post office, supermarket checkout counter.
  c. the number of days people spend in a hospital.
  d. the size of errors made by a supermarket checkout clerk.
  e. the number of days it takes people to pay their credit-card bills.
  f. the response times of a fire or police department.
  g. the yields per acre of U.S. wheat farmers.
  h. the ages at which famous authors wrote their best books.
  i. the mileages driven in a year by all cars registered in the United States.

25. A firm was using two machines to produce metal parts, but customers complained about erratic changes in the thickness of parts in different shipments. The firm's statistician developed the histogram at the bottom of the page from a random sample of a week's output. What does it tell you?

26. Consider the cumulative absolute frequency distribution developed in problem 14. Plot these data as a "less-than" ogive.

27. Consider the absolute frequency distribution developed in problem 11. Plot it as
  a. an absolute frequency histogram.
  b. a frequency polygon.
  c. a frequency curve.

28. Consider the cumulative relative frequency distribution developed in problem 16. Plot these data as a "more-than-or-equal" ogive.

29. Consider the cumulative absolute frequency distribution developed in problem 15. With appropriate changes, plot these data as a "more-than-or-equal" ogive.

30. Consider the first of the two cumulative relative frequency distributions developed in problem 18. Plot the corresponding ogive.

31. Graph Table 3.G as an *ordinary horizontal bar chart*.

Table 3.G   Net Import Reliance of Selected Minerals in the United States, 1981

Mineral	Import as Percent of Consumption
Chromium	90
Cobalt	91
Columbium	100
Fluorspar	85
Iron ore	28
Manganese	98
Nickel	72
Tin	80
Tungsten	52
Vanadium	42
Zinc	67

Source: American Iron and Steel Institute, *Steel and America: An Annual Report* (Washington, D.C.: May 1982), p. 31.

32. Graph Table 3.H as a *duo-directional* horizontal bar chart, with negative values graphed toward the left and positive values toward the right of a vertical zero line.

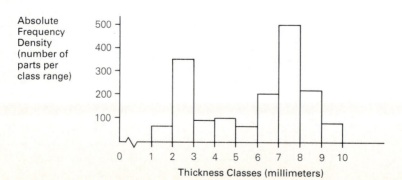

Absolute Frequency Density (number of parts per class range)

Thickness Classes (millimeters)

Table 3.H Percentage Change in U.S. GNP
(in 1972 dollars)

1978	+4.8%
1979	+3.2
1980	−0.2
1981	+2.0
1982	−1.8

Source: Federal Reserve Bulletin, various issues.

33. Graph Table 3.I as a *component bar chart.*

Table 3.I State and Local Government Taxes in the United States—Percent Distribution by Type, 1965–1982, Selected Years

Year	Property	Individual Income	Sales and Gross Receipts	Other
1965	44.1	8.0	33.4	14.5
1970	39.2	12.5	34.9	13.4
1975	36.4	15.2	35.2	13.2
1979	31.6	18.0	36.1	14.3
1980	30.7	18.8	35.8	14.7
1981	30.7	19.0	35.2	15.1
1982	30.8	19.1	35.2	14.9

Source: Statistical Abstract of the United States, 1985, p. 265.

34. From Table 3.J create a *grouped bar chart,* titled "Foreign Direct Investment in the United States—Percent Distribution by Area: 1970 and 1983."

Table 3.J Foreign Direct Investment in the United States (in millions of dollars)

	1970	1983
United Kingdom	$ 4,127	$ 32,512
Netherlands	2,151	28,817
Canada	3,117	11,115
Japan	229	11,145
Germany	680	10,482
Switzerland	1,545	7,132
Other Europe	1,051	13,747
Other Areas	370	20,573
Total	13,270	135,523

Source: Statistical Abstract of the United States, 1985, p. 804.

35. From Table 3.K create a *grouped bar chart.*
36. From Table 3.L create an *ordinary column chart.*
37. From Table 3.M create an *ordinary column chart.*

Table 3.K Retail Store Sales in the United States by Kind of Business: 1982 and 1983 (billions of dollars)

	1982	1983
Food Stores	$249.3	$259.4
Automotive Dealers	182.6	221.7
General Merchandise Group	132.6	143.0
Gasoline Service Stations	103.5	103.1
Eating and Drinking Places	104.7	115.7
Building Material and Hardware	51.3	59.9
Apparel and Accessory Stores	51.4	54.0
Furniture, Furnishings, and Equipment Stores	46.1	51.8
Drug and Proprietary Stores	36.0	38.8
Liquor Stores	19.4	19.7
Mail-order Houses	4.2	4.4

Source: Statistical Abstract of the United States, 1985, p. 782.

Table 3.L Health Expenditures in the United States as Percent of GNP, 1970–83

1970	7.6
1975	8.6
1980	9.4
1981	9.7
1982	10.5
1983	10.8

Source: Statistical Abstract of the United States, 1985, p. 96.

Table 3.M Annual Percentage Change in U.S. Consumer Price Index, 1970–1984

1970	5.9	1977	6.5
1971	4.3	1978	7.7
1972	3.3	1979	11.3
1973	6.2	1980	13.5
1974	11.0	1981	10.4
1975	9.1	1982	6.1
1976	5.8	1983	3.2
		1984	4.2

Source: Statistical Abstract of the United States, 1985, p. 467.

38. From Table 3.N create an *ordinary column chart.*

Table 3.N Business Failure Rate in the United States, 1970–1983 (per 10,000 firms)

1970	44	1977	28
1971	42	1978	24
1972	38	1979	28
1973	36	1980	42
1974	38	1981	61
1975	43	1982	88
1976	35	1983	110

Source: Statistical Abstract of the United States, 1985, p. 519.

39. Graph Table 3.O as a *component-column chart.*

Table 3.O U.S. Business Inventories (in billions of 1972 dollars)

Year	Total	Farm	Nonfarm
1961	$171.8	$33.2	$138.6
1971	267.0	39.2	227.8
1981	348.7	44.2	304.5

Source: Economic Report of the President, February 1982, p. 253.

40. U.S. federal receipts in 1970 were distributed as follows: individual income taxes 47 percent, social insurance taxes and contributions 23 percent, corporate income taxes 17 percent, other sources 13 percent. The corresponding 1983 figures were 48.1 percent, 34.8 percent, 6.2 percent, 10.9 percent. Illustrate these data by means of a *component-column chart.*

41. Graph Table 3.P as a *grouped-column chart.*

Table 3.P Personnel Problems at City Motors, 1986 (hypothetical data)

	Atlanta Plant	Chicago Plant
Average daily percent absent	15	4
Average daily percent tardy	10	1

42. Graph Table 3.Q as a *duo-directional column chart.*

Table 3.Q U.S. Air Carriers Net Operating Income (in millions of dollars), 1975–83

1975	$129	1980	$ −6
1976	576	1981	−264
1977	654	1982	−750
1978	995	1983	−176
1979	123		

Source: Statistical Abstract of the United States, 1985, p. 614.

43. Graph Table 3.R as a *pie chart.*

Table 3.R Car-Rental Market Shares, March 1982

Hertz	39.7%
Avis	23.4
National	17.0
Budget	14.5
Others	5.4
Total	100.0%

Source: Hertz Corporation.

44. Draw a *pie chart* to illustrate the 1984 U.S. beer-industry market shares: Anheuser-Busch 35 percent, Miller 20 percent, Stroh 13 percent, Heileman 9 percent, Coors 7 percent, Pabst 6 percent, others 10 percent. (*The New York Times,* September 13, 1985, p. D1.) What are the required central angles?

45. Draw three *pie charts* to illustrate the rapid comeback of Tylenol after the 1982 scare following the discovery of cyanide-laced Tylenol capsules. Use these market-share data for Tylenol, aspirin, and other pain relievers: September 1982: 37 percent, 47 percent, 16 percent; October 1982: 12 percent, 75 percent, 13 percent; April 1983: 30 percent, 50 percent, 20 percent. (*The New York Times,* September 17, 1983, p. 33.)

46. Draw a *pie chart* to illustrate the 1981 U.S. soft-drink industry market shares: Coca-Cola 35.9 percent, Pepsi 24.6 percent, Dr. Pepper 6.5 percent, Seven-Up 5.9 percent, Royal-Crown 4.3 percent, others 22.8 percent. (*The New York Times,* May 24, 1982, p. 23.) What are the required central angles?

47. Draw a *pie chart* to illustrate the 1984 division of Fairchild's $898.9 million sales as follows: government aerospace 27 percent; general industry 21 percent; commercial aerospace 19 percent; tooling for plastics 11 percent; communications, electronics, and space 11 percent, aerospace fasteners 11 percent. (*The New York Times,* September 5, 1985, p. D1.) What are the required central angles?

48. Draw two *pie charts* to compare the divisional sales in 1968 and 1984 of Boeing. 1968: commercial 69 percent; missiles, space 17 percent; military 13 percent; other 1 percent. 1984: commercial 52 percent; military 29 percent; missiles, space 13 percent; other 6 percent. (*The New York Times,* September 8, 1985, p. F1.)

49. Draw a *pie chart* to illustrate the 1984 division of Richardson-Vicks $1,280.5 million sales as follows: personal care products 47 percent, health care and nutritional products 40 percent, home care products, chemicals, and instruments 13 percent. (*The New York Times,* September 11, 1985, p. D1.) What are the required central angles?

50. Draw two *pie charts* that compare 1970 and 1983 American electric energy production by source of energy, as reported by the U.S. Bureau of the Census. 1970: coal 46 percent, natural gas 24.3 percent, nuclear 1.4 percent, hydro 16.2 percent, oil 12.1 percent. 1983: coal 54.8 percent, natural gas 11.9 percent, nuclear 12.6 percent, hydro 14.4 percent, oil 6.3 percent.

51. Draw two *pie charts* that indicate 1981 American steel shipments by market and product group. By market: steel service centers 20.3 percent, automotive 15.1 percent, capital goods 8.3 percent, converting and processing 5.8 percent, construction and contractors products 13.4 percent, containers 6.1 percent, oil and gas 7.2 percent, all others 23.8 percent. By product group: sheets and strip 42.4 percent, pipe and tubing 11.8 percent, bars and tool steel 15.9 percent, structural shapes and plates 14.1 percent, all other products 15.8 percent. (American Iron and Steel Institute, *Steel and America, An Annual Report,* 1982, p. 10.)

52. Draw a *time-series line graph* depicting U.S. pig iron production from 1975 to 1985 (in millions of net tons): 79.9, 86.9, 81.3, 87.7, 87.0, 68.7, 73.6, 43.3, 48.7, 51.9, 50.4 (American Iron and Steel Institute, *An Annual Report,* May 1986).

53. Draw a *time-series line graph* depicting the average number of employees (in thousands) of the U.S. iron and steel industry from 1975 to 1985: 457, 454, 452, 449, 453, 399, 391, 289, 243, 236, 208 (American Iron and Steel Institute, *An Annual Report,* May 1986).

54. Draw a *time-series line graph* depicting the percent return on stockholders' equity in the U.S. iron and steel industry from 1975 to 1984: 9.8, 7.8, 0.1, 7.3, 6.5, 9.6, 13.4, -15.0, -19.7, -1.3 (American Iron and Steel Institute, *An Annual Report,* May 1985).

55. Draw a *time-series line graph* depicting the U.S. iron and steel industry's capital expenditures for air and water quality control from 1975 to 1984 (in millions of dollars). Total: 453.1, 489.2, 534.8, 458.0, 650.8, 510.5, 489.2, 261.9, 140.2, 132.9. Air: 321.3, 330.5, 329.1, 277.2, 449.6, 342.3, 369.8, 157.3, 84.6, 79.6. Water: 131.8, 158.7, 205.7, 180.8, 201.2, 168.2, 119.4, 104.6, 55.6, 53.3 (American Iron and Steel Institute, *An Annual Report,* May 1985).

56. During the first quarter of a year, five packers in a mail-order house filled eight thousand orders, but 1 percent (or 80) of the orders were returned to the mail-order house because they were filled incorrectly and the customers could not use what they ordered. The packers involved could be identified by their initials, A through E. The 80 orders returned were packed by the packers listed in Table 3.S below. Construct a *residual plot diagram* to assess the situation.

Table 3.S

A	C	C	A	D	A	C	A	D	C
A	D	D	D	D	C	C	B	D	D
B	A	E	E	E	B	C	C	B	B
D	B	A	B	A	D	D	D	A	B
D	D	B	B	C	E	A	D	D	C
E	E	C	D	E	C	D	E	C	D
E	D	D	D	E	E	B	E	E	D
C	A	D	D	B	C	D	A	C	A

57. Graph Table 3.T as a pictogram.

Table 3.T   **Fewer Farmers Feeding More People**

Year	Population (of the United States in millions)	
	Farm	Total
1940	31	132
1970	10	203

Source: U.S. Bureau of the Census, *Statistical Abstract of the United States, 1973,* pp. 5 and 584.

58. Using Appendix Table A, "Random Numbers," take random samples of 20 pages of this book and one other book. Then establish and compare relative frequency distributions for
a. the occurrence of printing errors.
b. the occurrence of the word *of.*

59. The age distribution of subscribers to two magazines are shown in Table 3.U.
a. Set up the two relative frequency distributions.
b. Use them to make a recommendation to a client about placing an advertisement for baby food. For denture cleaners.

60. The salaries (in thousands of dollars per year) of family heads in two towns are shown in Table 3V. Set up a relevant frequency distribution that would help a client decide on the location (in Town A or Town B) of a new fancy restaurant that the client would like to operate.

Table 3.U

	10–19	20–29	30–39	40–49	50–59	60–69	70–79
Magazine A:	9	533	827	208	59	13	2
Magazine B:	29	533	827	1,208	2,000	2,900	73

Table 3.V

Town A:	12, 37, 15, 23, 18, 22, 31, 19, 22, 22, 51, 40, 27, 24, 21, 19, 22, 20, 14, 9, 8, 19, 26, 31, 24.
Town B:	12, 37, 61, 54, 49, 47, 45, 26, 31, 33, 9, 15, 62, 58, 29, 23, 14, 51, 47, 47, 46, 23, 56, 58, 17, 29.

Selected Readings

Brainerd, B. *Weighing Evidence in Language and Literature: A Statistical Approach.* Toronto: University of Toronto Press, 1974. Very readable.

Broome, R. "Micrographics: A New Approach to Cartography at the Census Bureau." In U.S. Bureau of the Census, *Geographic Base (DIME) System: A Local Program.* Washington, D.C.: U.S. Government Printing Office, 1975. Series GE60, No. 6, pp. 28–39. Discusses attempts to overcome the traditional crudity of computer-produced charts.

Edwards, J. A., et al. *Graphic Communication Through Isotype.* Reading, England: University of Reading, 1975. About the development, by Otto Neurath and others, of an international picture language.

Feinberg, Barry M. *Kinostatistics: Communicating a Social Report to the Nation.* Washington, D.C.: Bureau of Social Science Research, 1976. A proposal for improved graphic reporting by means of techniques, such as television, that transcend the limits of the printed page.

Hall, Ray O. *Handbook of Tabular Presentation: How to Design and Edit Statistical Tables.* New York: Ronald Press, 1946. A style manual and case book.

Huff, Darrell, and Irving Geis. *How to Lie with Statistics.* New York: W. W. Norton, 1954. A discussion of graphical fallacies.

Kjetsaa, Geir. "The Battle of the Quiet Don." *Computers and the Humanities.* Pergamon Press, 1977, vol. 11, pp. 341–46. A statistical study of disputed authorship of *The Quiet Don*. Was it written by Kryukov or Sholokhov (who received the 1965 Nobel Prize for Literature for it)?

Mosteller, Frederick, and David L. Wallace. *Inference and Disputed Authorship: The Federalist.* Reading, Mass.: Addison-Wesley, 1964. A more extensive discussion, including historical details and a variety of alternative analyses, of Analytical Example 3.2., "Deciding Authorship."

Pratt, Fletcher. *Secret and Urgent: The Story of Codes and Ciphers.* Indianapolis: Bobbs-Merrill, 1939. Contains a discussion of the Shakespeare-Bacon controversy noted in Analytical Example 3.2, "Deciding Authorship."

Smith, Laurence Dwight. *Cryptography: The Science of Secret Writing* (New York: W. W. Norton, 1943). More about codes and ciphers, including word-frequency listings for English, French, German, Italian, and Spanish.

Computer Programs

The DAYSAL-2 personal-computer diskettes that accompany this book contain four programs of interest to this chapter:

2. Secret Code illustrates how random numbers can help create secret codes. Lets you write and decipher your own secret messages.

3. Frequency Distributions From Raw Data lets you enter a set of unorganized, raw data; then produces an ordered array plus six types of frequency distributions: absolute, relative, cumulative absolute LE-type and ME-type, cumulative relative LE-type and ME-type. You can specify (a) only the desired number of classes, (b) only the desired class width, or (c) the precise limits of all classes. The program also produces six horizontal histograms, like Figure 3.9, corresponding to the six frequency distributions just noted, as well as statistical summary measures, to be discussed in Chapter 4.

4. Summary Measures From Grouped Data takes an absolute frequency distribution that you supply and produces, among other things, five other frequency distributions: cumulative absolute LE-type and ME-type, relative, cumulative relative LE-type and ME-type.

19. General Graphics lets you draw line graphs, bar charts, column charts, histograms like Figures 3.1 and 3.2, and pie charts. Numerous designs are available; for example, up to 18 bars with as many as 4 subdivisions and 3 different widths, or pies with up to 10 slices, 3 colors, and 7 patterns.

CHAPTER FOUR

The Presentation of Data: Summary Measures

As Chapter 3 has shown, properly constructed tables and graphs can be of considerable help in making otherwise confusing masses of data intelligible and in revealing the secrets hidden in such unorganized data. Yet an even more radical approach to condensing collections of data is the calculation of *arithmetic summary measures* that are designed to express the most salient features of data sets in the most compact fashion imaginable. Data about quantitative variables can be neatly summarized in three ways: by measures of central tendency, measures of dispersion, and measures of shape.

Summary **measures of central tendency** (or **location**) are values around which observations tend to cluster and that describe the location of what in some sense might be called the "center" of a data set. Consider the two frequency curves in panel (a) of Figure 4.1. They describe two hypothetical data sets on the sizes of life-insurance policies taken out by women and men. It is immediately obvious that men in general take out larger policies than women do, in spite of the fact that some women take out policies for as much as $100,000, while some men take out policies for as little as $50,000. If there were to be a single and "typical" measure of the size of women's policies, it would be $50,000; such a measure for men would be $100,000. Each of these would be a measure of central tendency because it would locate, in this case on the horizontal axis, the general center of the respective data sets. What, however, are the centers of the asymmetrical data sets in panel (c)? The answer is less obvious and, as will be shown, depends very much on how the term *center* is defined.

Summary **measures of dispersion** (or **variability**) are numbers that indicate the spread or scatter of observations; they show the extent to which individual values in a data set differ from one another and, hence, differ from their central location. The two data sets on life-insurance policies that are visually summarized in panel (a) of Figure 4.1 do not differ from each other with respect to variability. Note, for example, how both women's and men's policies vary ± $50,000 from their respective centers. Women's policies cover a $100,000 range (from 0 to $100,000); men's policies also cover a

Figure 4.1 Summary Measures and Frequency Curves

Different types of summary measures describe different aspects of frequency curves, such as the curves shown here. Measures of *central tendency* locate the center of a data set, as at 50 (women) or 100 (men) in panel (a). Measures of *dispersion* focus on the spread of data around their center, as on the ranges from 0 to 100 (women) or 50 to 150 (men) in panel (a) or from 9.7 to 10.3 (pounds of cookies) or 9.9 to 10.1 (pounds of sugar) in panel (b). Measures of *shape* describe how symmetrical frequency curves are—as are the curves in panels (a) and (b)—or how asymmetrical they are—as are the curves in panel (c). Measures of shape also describe how peaked (sugar) or how flat (cookies) frequency curves are.

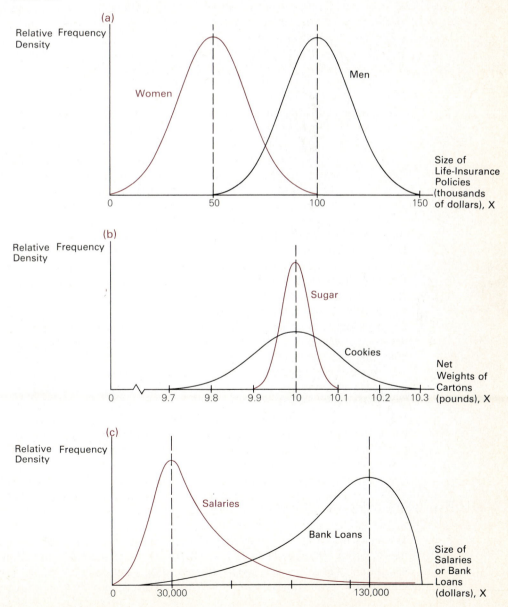

$100,000 range (from $50,000 to $150,000). But now consider the two frequency curves in panel (b), which describe two data sets on the actual net weights of supposedly 10-pound cartons of cookies and of sugar. Even though both data sets have the identical central tendency (of 10 pounds per carton), they differ significantly as far as data dispersion is concerned. The sugar-carton net weights are tightly packed around the center of the distribution. All the weights are found within the 0.2-pound range from 9.9 to 10.1 pounds; that is, within 1 percent of the stated weight. The net weights of cookie cartons, on the other hand, are more widely spread; they range over 0.6 pounds; that is, within ± 3 percent of the desired weight. (This greater dispersion may occur in the case of the cookie-carton weights because their final weight must be determined by adding or subtracting an entire cookie rather than a few tiny grains of sugar.) Clearly, the two net-weight data sets could quite nicely be described by some measure of central tendency *along with* appropriate measures of dispersion.

Summary **measures of shape** are numbers that indicate either the degree of asymmetry or the degree of peakedness in a frequency distribution. Note how the four distributions pictured in panels (a) and (b) of Figure 4.1 are symmetrical about the dashed vertical lines; the left and right halves of the distributions in each case are mirror images of each other. This symmetry need not be present, however, as is indicated in panel (c). The salaries received by the employees in an industry might cluster near a dominant value of, say, $30,000. None being below zero, the spread below $30,000 would be limited, while the spread above that number may be extreme. As a result, the frequency curve comes to look like the side view of a child's slide. Asymmetry may be found in the other direction as well: the sizes of loans made by a bank may be concentrated near $130,000, there being none much larger because of the bank's lending limit, perhaps. Yet there may be many scattered loans of much lower amount. Measures of shape try to capture this type of asymmetry, as well as the flatness or peakedness of frequency curves such as those depicted graphically in panel (b) of Figure 4.1.

Finally, in the case of qualitative data, the only summary measure available is the **proportion,** a number that describes the frequency of observations in a particular category as a fraction of all observations made. (As the Chapter 3 discussion of pie charts indicates, proportions can, however, be calculated for quantitative data as well.)

The various types of summary measures will be discussed in turn, but note: any summary measure can be calculated either for a statistical population or for a sample. If it is based on population data, the summary measure is called a **parameter;** it is referred to as a **statistic** if it is based on sample data only. In addition, the calculations can proceed from sets of original, raw, or ungrouped data, or they can be based, with some loss of accuracy as a result, on previously processed or grouped data, such as those data found in a frequency distribution.

SUMMARY MEASURES OF CENTRAL TENDENCY

The most important summary measures of central tendency are the following: (1) the *mean,* (2) the *median,* and (3) the *mode.* Because many people refer to all of them using the same word—*averages*—it is important to keep their meanings distinct so that you can use each summary measure of central tendency in an appropriate way.

The Arithmetic Mean

Most often encountered and most widely known among measures of central tendency is, no doubt, the *mean*—more precisely called the *arithmetic mean* (and in everyday usage simply referred to as the *average*).

Calculation from ungrouped data. When ungrouped data are available, the **arithmetic mean** can be calculated with precision by adding all the values observed and dividing their sum by the number of observations. If the annual profits of five firms (in millions of dollars) were 2, 2, 4, 7, and 15, the arithmetic mean would equal

$$\frac{2 + 2 + 4 + 7 + 15}{5} = \frac{30}{5} = 6.$$

This number (6) would be the *population mean* if the frame of interest contained only five firms (say, all the makers of airplanes in the United States or all the makers of beer in Detroit); the number would be a *sample mean* if it referred to five firms among a much larger group of interest (say, five among dozens of airplane manufacturers in the world or five among hundreds of breweries in the United States). The procedure just followed can also be expressed symbolically, as in Box 4.A.

Symbolic expression. By tradition, the small or the capital letter X is used to represent an observed value. Different values are distinguished by subscripts 1, 2, 3, . . . and so on, yielding the series $X_1, X_2, X_3, . . .$ and so on, for a set of data. The symbol X_1 is pronounced "X sub one," and, in the example discussed before, $X_1 = 2$. Similarly, $X_2 = 2, X_3 = 4, X_4 = 7$, and $X_5 = 15$. Also by tradition, the total number of observations (here 5) is referred to as N in the case of a population and as n in the case of a sample. Finally, *parameters* are always symbolized by Greek letters, *statistics* by Roman ones (Appendix Table B lists the Greek alphabet). The population mean, in particular, is represented by μ (which is the lowercase Greek m, is pronounced "mu," and stands for *mean*); the sample mean, in turn, is represented by capital $\overline{X}$ (pronounced "X bar"). Thus, the calculation of the arithmetic mean from ungrouped data can be symbolized by the formulas in Box 4.A.

4.A Arithmetic Mean from Ungrouped Data

For a population:

$$\mu = \frac{X_1 + X_2 + X_3 + . . . X_N}{N}$$

For a sample:

$$\overline{X} = \frac{X_1 + X_2 + X_3 + . . . + X_n}{n}$$

where the X's are observed population (or sample) values, N is the number of observations in the population, and n is the number of observations in the sample.

A considerably more succinct version of these formulas can, however, be developed with the help of the *summation sign,* Σ (which is the uppercase Greek *S,* is pronounced "sigma," and stands for *sum*). Thus, we can write

$$X_1 + X_2 + X_3 + \ldots + X_N \text{ as } \sum_{i=1}^{N} X_i$$

which means "the sum of all the individual values of X with subscripts ranging from 1 to N." Similarly, if it were desirable to sum not all of the values, but only some of them, one could indicate this fact by appropriate subscripts and superscripts connected to the summation sign. Thus,

$$\sum_{i=1}^{3} X_i$$

would indicate that only the values of X with subscripts from 1 to 3 should be added, such as $X_1 + X_2 + X_3$. Analogously,

$$\sum_{i=5}^{8} X_i$$

would equal $X_5 + X_6 + X_7 + X_8$.

As it turns out, however, all the summations in this book, with a single exception that will be duly noted, involve the summing of *all* the values in question (of *all* the individual values of X_i between 1 and N, for example). It is, therefore, possible to simplify the summation notation by dropping the subscripts and superscripts connected to the summation sign. Thus, we will write

$$\frac{X_1 + X_2 + X_3 + \ldots + X_N}{N} \text{ not as } \frac{\sum_{i=1}^{N} X_i}{N} \text{ but simply as } \frac{\Sigma X}{N}.$$

And we will, similarly, write

$$\frac{X_1 + X_2 + X_3 + \ldots + X_n}{n} \text{ not as } \frac{\sum_{i=1}^{n} X_i}{n} \text{ but as } \frac{\Sigma X}{n}.$$

In each case, the invisible message attached to the summation sign will be the same: *all* the available values are to be summed. The context in which the simplified notation appears will indicate what these values are. (In the case of ungrouped population data, the values will range from 1 to N, for example; in the case of sample data, from 1 to n.)[1]

---

[1]An extended discussion of the rules of summation appears in Chapter 4 of the *Student Workbook* that accompanies this text.

Thus, the formulas in Box 4.A can be simplified to those in Box 4.B.

4.B Arithmetic Mean from Ungrouped Data—Alternative Formulation

For a population:

$$\mu = \frac{\Sigma X}{N}$$

For a sample:

$$\overline{X} = \frac{\Sigma X}{n}$$

where ΣX is the sum of all observed population (or sample) values, N is the number of observations in the population, and n is the number of observations in the sample.

Calculation from grouped data. Sometimes raw data are unavailable, and the mean, if it is to be calculated at all, must be calculated from a frequency distribution. At other times, raw data are available but are so massive in number that it is preferable to use grouped data to speed up the calculations. Calculating the mean from grouped data can be accomplished only by assuming (perhaps incorrectly) that the observations falling into a given class are equally spaced within it and are, therefore, on the average, equal to the midpoint of the class interval. (In the case of open-ended classes, the procedure must be abandoned or a wild estimate of the midpoint must be made.) Each midpoint is then multiplied by the absolute class frequency, and the sum of these products is divided by the population size, N (or the sample size, n). If each midpoint is represented by X, and the corresponding absolute class frequency by f, the procedure can be described by the formula in Box 4.C.

4.C Arithmetic Mean from Grouped Data

For a population:

$$\mu = \frac{\Sigma f X}{N}$$

For a sample:

$$\overline{X} = \frac{\Sigma f X}{n}$$

where $\Sigma f X$ is the sum of all class-frequency (f) times class-midpoint (X) products, N is the number of observations in the population, and n is the number of observations in the sample.

Consider, for example, calculating the arithmetic mean of the 1981 profits of the 100 largest U.S.-based multinationals. If the original data were available, such as those in Table 3.2 on page 51, one could sum them, divide by $N = 100$, and find the answer to be \$786.62 million. Suppose, however, that only the grouped data of Table 3.3 (page 53) were available. The calculation would then have to proceed as shown in Table 4.1. The result—\$792.5 million—is only an approximation but a close one; it is off the true value by less than 1 percent. (When discussing the construction of absolute frequency distributions in Chapter 3, it was noted how helpful it would be for subsequent users if the maker of the distribution chose class intervals with midpoints close to the average of the observations falling into the class or even reported the sums of observations in each class. These assertions are now vindicated. If each class midpoint equaled the average of all the observations in the class, each fX would equal the precise sum of class observations, and the mean calculated on that basis would be precise also. The same would, of course, be true if the precise sums of class values were provided in the first place.)

The nature of the mean. Knowing how to calculate the mean is one thing; knowing what it tells us is another. The mean is best viewed as a point of balance in a data set, very much like the fulcrum of a seesaw. In the case of a seesaw, or teeterboard, the number of (equal-sized) weights on one side of the fulcrum multiplied by their respective distances from it must equal the number of such weights times their distances on the other side in order for the board to be horizontal. Like the fulcrum, the arithmetic mean similarly balances the number of observations on one side of the mean

Table 4.1 Approximation of the Arithmetic Mean from a Frequency Distribution

When the arithmetic mean is to be calculated from a frequency distribution, and (as is usually the case) the actual sums of values in each class are not known, an approximation can still be derived by weighting each class midpoint, X, by the corresponding class frequency, f, and dividing the sum of the weighted products, ΣfX, by the population size, N (or the sample size, n).

Class (size of profit in millions of dollars)	Absolute Class Frequency (number of companies in class) f	Class Midpoint X	fX	Calculation
−1,500 to under 0	3	−750	−2,250	
0 to under 500	41	250	10,250	$\mu = \dfrac{\Sigma fX}{N} =$
500 to under 1,000	32	750	24,000	
1,000 to under 1,500	9	1,250	11,250	$\dfrac{79,250}{100} =$
1,500 to under 2,000	6	1,750	10,500	
2,000 to under 2,500	6	2,250	13,500	\$792.5 million
2,500 to under 5,500	3	4,000	12,000	
Totals: $\Sigma f = N = 100$			$\Sigma fX = 79,250$	

Source: Table 3.3.

Figure 4.2   The Nature of the Arithmetic Mean

The mean of a data set can be compared to the fulcrum of a seesaw. The mean occupies a central position in the sense that the sum of negative and positive deviations of individual observations from the mean equals zero. Thus, for the data set represented by the blocks in panel (a), the mean equals 6, and the sum of deviation equals $-4 - 4 - 2 + 1 + 9 = 0$. For the data set represented by the blocks in panel (b), the mean equals 8, and the sum of deviations equals $-6 - 6 - 4 - 1 + 17 = 0$. Note how the arithmetic mean responds to the change in any single observation (as when 15 becomes 25) and how it is quite possible for the vast majority of all observations to be on one side of the mean—as in panel (b).

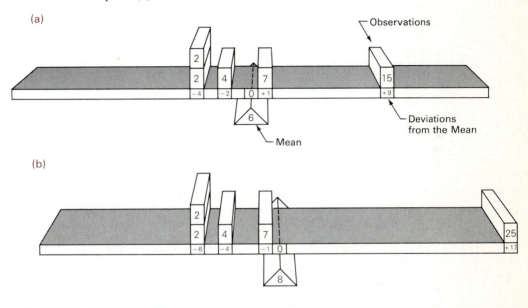

times their respective deviations from this mean with the number of observations times their deviations on the other side. This role of the mean is illustrated in Figure 4.2.

Panel (a) illustrates the relationship between the five hypothetical profit figures noted above and the mean calculated from them. The five original observations are represented by the numbered blocks on top of the teeterboard; the $6 million mean is represented by the numbered fulcrum. The deviations from the mean are shown on the side of the board. Note how one observation of $4 million lies $2 million below the mean, while two observations of $2 million lie $4 million below the mean; hence, the negative deviations add to $- \$2 \text{ million} - \$4 \text{ million} - \$4 \text{ million} = -\$10 \text{ million}$. They are perfectly matched by the positive deviations from the mean of the $7 million and $15 million profit figures, the deviations of which come to $+ \$1$ million $+$ $9 million $= +\$10$ million.

Panel (b) shows how the arithmetic mean instantly responds to a change in even a single observation (such as a change from $15 million to $25 million). Indeed, the mean is so responsive to extreme values that it can easily lie above (or below) the vast majority of individual observations! (Contrary to what most people think, it is, there-fore, quite possible for *most* people to weigh more than the average, be less intelligent than the average, or even be shorter than the average.) The mean always remains a *central* value, however, in the sense that the sum of deviations of all observations from the mean equals zero.[2] Nevertheless, and quite understandably, many people are con-fused when confronted with situations, such as that in panel (b), in which the vast majority of observations are below (or above) the average. This situation is not what *average* denotes to them; they prefer to use an entirely different concept of central tendency, a concept to which we now turn.

The Median

Another important measure of central tendency, quite different in nature from the mean, is the **median**—the value in an ordered array of data above and below which an equal number of observations can be found. This *middle value,* too, can be calcu-lated precisely from ungrouped data, or it can be approximated from grouped ones.

Calculation from ungrouped data. For the profit figures (in millions of dollars) of 2, 2, 4, 7, and 15, the median equals 4 because it is located at the center of the array. Two values lie above it; two values lie below it—their sizes do not matter. Thus, if the array had been 2, 2, 4, 7, and 25, as in panel (b) of Figure 4.2, the median would still have been 4; and so it would have been for the array of 4, 4, 4, 4, and 25. But note: when there is an even number of observations, as in the series 2, 2, 4, 7, 15, and 25, there would seem to be two central values (of 4 and 7, in this example). In such a case, the two central values are averaged, and that average (here 5.5) is designated as the me-dian. In our example, as the definition requires, three firms would earn less and three firms would earn more than the median (the values of 2, 2, and 4 would lie below 5.5, and those of 7, 15, and 25 would lie above it). Can you see with the help of Table 3.2 on page 51 that the median 1981 profit of the 100 multinationals equals $533.5 million, the average of the $532 million and $535 million values appearing in the center of the ordered array?

Symbolic expression. Denoting the median by M (the Greek *capital* mu) for a pop-ulation and by m for a sample, we can express the definition of median symbolically by the formulas in Box 4.D.

The median, these formulas tell us, always equals the value of the middle obser-vation in an ordered array. In our sample of five profit figures, it equals the value of

$$X_{\frac{5+1}{2}} = X_{\frac{6}{2}} = X_3.$$

The value of the third observation is 4. Similarly, in our population of 100 multina-tional-company profits, the median equals the value of

---

[2]A mathematical proof of this assertion appears in Chapter 4 of the *Student Workbook* that accompanies this text.

$$X_{\frac{100 + 1}{2}} = X_{50.5};$$

that is, the arithmetic mean of $X_{50} = 532$ and $X_{51} = 535$, or 533.5.

4.D Median from Ungrouped Data

For a population:

$$M = X_{\frac{N + 1}{2}} \text{ in an ordered array}$$

For a sample:

$$m = X_{\frac{n + 1}{2}} \text{ in an ordered array}$$

where X is an observed population (or sample) value, N is the number of observations in the population, and n is the number of observations in the sample.

Calculating the median from grouped data. Just as the arithmetic mean can be approximated from grouped data, so can the median. Consider again Table 4.1. It is fairly easy to find the class that contains the median and is, therefore, called the **median class.** Because the size of the population equals 100, the median is the average of the 50th and 51st observation in an ordered array of the population. There being $3 + 41 = 44$ observations in the first two classes, the median cannot lie in either of them, but it is bound to be found among the next 32 observations in the third class. Indeed, the median must be the average of the sixth and seventh observations in the median class (which would correspond to the 50th and 51st observations among all the data). If one imagined the 32 observations in the third or median class to be equally spaced within the $500 million width of that class, each observation would be 15.625 units apart from the next one. If one further assumed that the first observation within the class was one half this distance above the lower class limit (and that the last observation within the class was one half this distance below the upper limit), the sixth and seventh observations would, respectively, be found at 585.9375 and 601.5625, making their arithmetic mean equal to $M = \$593.75$ million. Other interpolation procedures are possible also, but note again how the value determined in this way is bound to be an approximation only (this time within roughly 11 percent) of the correct figure of $533.5 million determined above.

This procedure, too, can be expressed by formulas, as shown in Box 4.E. In Box 4.E, L designates the lower limit of the median class; N or n is the size of the population or sample; F is the sum of frequencies up to, but not including, the median class; f is the frequency of the median class; and w is the width of the median class. In our case, this formula produced a median estimate of

$$M = 500 + \frac{(100/2) - 44}{32}(500) = 500 + \frac{6}{32}(500) = 593.75.$$

4.E Median from Grouped Data

For a population:

$$M = L + \frac{(N/2) - F}{f} w$$

For a sample:

$$m = L + \frac{(n/2) - F}{f} w$$

where L is the lower limit of the median class, f is its absolute frequency, and w is its width, while F is the sum of frequencies up to (but not including) the median class, N is the number of observations in the population, and n is the number of observations in the sample.

Median vs. mean. Although both median and mean are often referred to as *averages,* they are such in very different senses of the term. The arithmetic mean is an average of the observed values; the median is whatever value happens to be found at the average of all the positions in an ordered array. [It is, for example, the value found at position $(1 + 100) \div 2 = 50.5$ among 100 ordered numbers]. As a result, the mean can be algebraically manipulated in many ways; the median cannot. For example, if we know the means and the sizes of two populations, we can calculate their combined mean; yet we cannot calculate the combined median from a knowledge of the medians and sizes of two populations. Consider the two populations, A and B:

A: 10   17   39
B: 6   20   31   57   82   97   99

The arithmetic means are 22 and 56, respectively; the medians are 17 and 57. One can easily calculate the combined mean as $[(3 \times 22) + (7 \times 56)] \div 10 = 45.8$, knowing only the sizes of the two populations and their means. One *cannot* calculate the combined median by a similar procedure, such as $[(3 \times 17) + (7 \times 57)] \div 10 = 45$. The combined median is, in fact, not 45, but the average of the fifth and sixth value in an ordered array of the combined populations, or $(31 + 39) \div 2 = 35$.

In spite of this disadvantage, the median is, on the other hand, a better measure of central tendency than the mean when the data set contains a few extreme values, high or low. Consider these annual income figures of the families living on a certain street:

$15,000 $19,000
$17,000 $19,000
$17,000 $301,000
$18,000

The arithmetic mean of these incomes equals $58,000, which certainly does not look like a very good summary of neighborhood incomes. The median of $18,000, on the other hand, is much more representative. Indeed, the median is often called "democratic" because in a sense it gives each value, regardless of its size, an equal "vote" in determining the central location. In the above example, the three values below $18,000 have one vote each, the three values above $18,000 have one vote each, so $18,000 is the "winner." Unlike in the case of the arithmetic mean, the sizes of the individual values below or above the median are irrelevant for determining the median. [The same issue is visually illustrated in Figure 4.2. Under the influence of one extreme observation, the mean changes from panel (a) to panel (b), and it hardly seems representative of the panel (b) data. Yet the median in both panels equals 4.]

The Mode

A third measure of central tendency—the **mode**—is defined as the most frequently occurring value in a set of data. In both panels (a) and (b) of Figure 4.2, for example, the mode equals 2. Just as the expression to be "à la mode" means to be in fashion, the mode is the most "fashionable" value in a data set. Yet, as the examples of Figure 4.2 suggest, the mode can easily be found at either extreme of a data set and can be quite atypical of the majority of observations. Sometimes a population contains two modes or even more; consider the profit figures of $119 million and $580 million in Table 3.2 on page 51. In practice, the mode is seldom used for describing ungrouped data; much more frequent is its designation for grouped data.

Calculation from grouped data. In the case of grouped data, the class containing the mode is the **modal class** and is the class with the highest frequency density. Consider Table 4.2. The modal class clearly is the second one, but there exist several alternative ways of approximating the mode. One possibility is simply to designate the midpoint

Table 4.2 Approximation of the Mode from a Frequency Distribution

The mode can be estimated from a frequency distribution in several ways, one of which is illustrated here. This method assumes that the mode is found in the modal class but is also "attracted" to the two adjacent classes by forces that are proportional to the frequency densities of these classes. But note: the mode is highly unstable. It is likely to change with every change in the grouping of data.

Class (size of profit in millions of dollars)	Absolute Frequency Density (number of companies per $500 million class range)	Calculation
−1,500 to under 0	1	$Mo = L + \dfrac{d_1}{d_1 + d_2} \cdot w =$
0 to under 500	41	
500 to under 1,000	32	
1,000 to under 1,500	9	$0 + \dfrac{41 - 1}{(41 - 1) + (41 - 32)} (500) =$
1,500 to under 2,000	6	
2,000 to under 2,500	6	$\dfrac{40}{49} (500) = \$408.16$ million
2,500 to under 5,500	½	

Source: Table 3.9.

of the modal class, here \$250 million, as the mode. Another possibility proceeds from the assumption that the mode will be closer to the adjacent class that has the greater frequency density (here closer to the third class than the first one). A favorite formula for finding the mode (*Mo* for population, *mo* for sample) based on the second alternative just described is given in Box 4.F and is applied in Table 4.2.

4.F Mode from Grouped Data

For a population or a sample:

$$Mo \text{ or } mo = L + \frac{d_1}{d_1 + d_2} w$$

where L is the lower limit of the modal class, w is its width, and d_1 and d_2, respectively, are the differences between the modal class frequency *density* and that of the preceding or following class.

In Box 4.F, L designates the lower limit of the modal class; d_1 and d_2 are the respective differences between frequency densities (*not* absolute frequencies) of the modal class and that of the preceding (or following) class; and w is the modal class width.

The same procedure can also be shown graphically, as Figure 4.3 illustrates. Note how the formula gives the same result as the graph:

$$\text{For panel (a), } Mo = 3 + \frac{2}{2 + 6}(2) = 3.5.$$

$$\text{For panel (b), } Mo = 3 + \frac{2}{2 + 2}(2) = 4.$$

$$\text{For panel (c), } Mo = 3 + \frac{6}{6 + 2}(2) = 4.5.$$

The mode and the frequency curve. The mode can also be found on a smooth frequency curve as the value lying underneath its highest point. However, if two or more different values in a data set occur with the highest frequency, or almost that, two or more modes exist, and the data set is said to have a **multimodal frequency distribution.** This distribution shows up as two or more peaks on a frequency curve. Such a situation usually arises because the population or sample in question contains two or more groups that by some criterion are fairly homogeneous internally but differ significantly from each other. For purposes of statistical analysis, it is often wiser to study such groups separately. Consider Figure 4.4 on page 104. Panels (a) and (b) give two examples of the **bimodal frequency distribution,** so called because it contains

Figure 4.3   Approximating the Mode from a Histogram

The graphical procedure for finding the mode that is shown here corresponds to the formula given in Box 4.F. In each case, the second class represents the modal class, but the mode equals the midpoint of that class only in panel (b) where both adjacent classes have equal frequency densities. Thus, the two adjacent classes exert equal pull on the midpoint of the modal class. In contrast, the mode in panel (a) is more "attracted" to the preceding class (which has a higher frequency density than the following one), while in panel (c) the mode is more "attracted" to the following class (because it has a higher frequency density than the preceding one). Thus, the mode equals 3.5 in panel (a), 4.0 in panel (b), and 4.5 in panel (c).

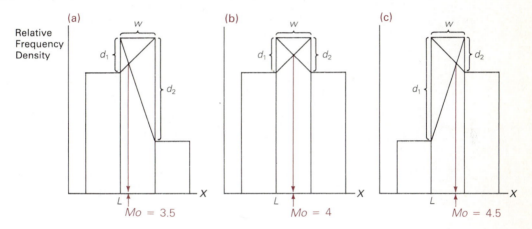

two modes. Panel (a) portrays the frequency distribution of the heights of all the production workers in a large firm. Two modes are visible, at 5'2" and 5'11", but they reflect nothing more mysterious than the fact that the firm has an equal number of female and male workers (and women tend to be shorter). To illustrate why it is often wiser to study such groupings separately, imagine that someone compared the mean heights of production workers now and 50 years ago (when almost all the workers were male). The apparent conclusion that workers had become shorter would be false, a fact that would become obvious if female workers of 50 years ago were compared with female workers now, while male workers then and now were also compared separately. The presence of more than one mode, therefore, is a warning signal to the investigator that suggests the utmost of caution.

Panel (b) presents another example. It shows the frequency distribution of the high-school grade-point averages of newly hired production workers. Humps in the distribution occur at 2.1 and 3.3; although the frequency of the former is below that of the latter (which, strictly speaking, makes 3.3 the mode), one still refers to this distribution as bimodal. The reason might be that two groups of workers were hired, a large one of apparently highly qualified ones, a small one of seemingly unqualified ones. Yet the latter group—hired, perhaps, under affirmative-action rules—may be given special training and be expected, eventually, to perform as well as the other on the production line. Once again, the presence of these heterogeneous groups suggests to an investigator (who might wish to compare early and later work performance) that these two groups should be studied separately. The mean performance at work of the entire

Figure 4.4   Multimodal Frequency Distributions

Multipeaked frequency curves illustrate the presence of two or more modes and strongly suggest the existence of some nonhomogeneous factor in the underlying data set. Under such circumstances, any single measure of central location is likely to be misleading; it is advisable to describe the two or more underlying data sets separately.

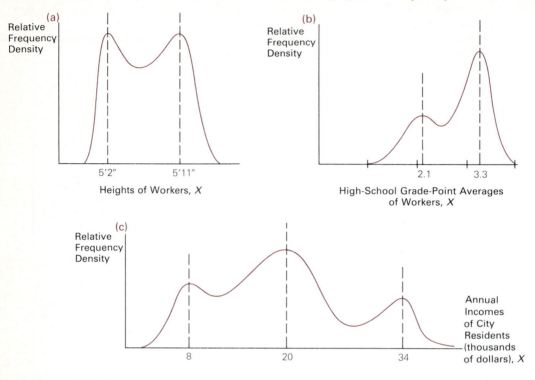

group may hardly change over time; the mean performance at work of the small special-training group may improve dramatically over time.

Panel (c), finally, gives an example of an even more heterogeneous statistical population. The city's residents may, for example, fall into three distinct groups: unskilled laborers, skilled workers, and professionals. Their respective modal incomes of $8,000, $20,000, and $34,000 per year may produce the three-peaked frequency curve when all of them are included in a single data set.

Other Measures

Additional but less frequently used measures of central tendency include (1) the *weighted mean,* (2) the *geometric mean,* and (3) the *harmonic mean.* We will here illustrate the meaning of the first of these; the second and third are discussed in Chapter 4 of the *Student Workbook* that accompanies this text.

The **weighted mean, μ_w** or **$\overline{X}_w$,** is calculated when different observations must be given unequal weights in accordance with their unequal relative importance. The weighted mean equals the sum of the products of observed values and their respective weights, divided by the sum of weights. If a firm pays hourly wages of $5, $10, and $15 to different groups of workers, it would be unwise to conclude that its workers were earning, on the average, $10 per hour, unless equal numbers of workers were found in the three categories. If, however, there were 100 unskilled workers earning $5, 50 semiskilled workers earning $10, and 10 skilled workers earning $15, these numbers of workers should be used as weights to count $5 one hundred times, $10 fifty times, and $15 ten times. Thus, the formulas for the unweighted arithmetic mean noted in Box 4.B are adjusted to read like those in Box 4.G.

4.G Weighted Mean from Ungrouped Data

For a population:

$$\mu_w = \frac{\Sigma wX}{\Sigma w}$$

For a sample:

$$\overline{X}_w = \frac{\Sigma wX}{\Sigma w}$$

where ΣwX is the sum of all weight (w) times observed-value (X) products, while Σw equals N (the number of observations in the population) or n (the number of observations in the sample).

In our case, this procedure yields a weighted mean,

$$\mu_w = \frac{(100 \times \$5) + (50 \times \$10) + (10 \times \$15)}{100 + 50 + 10} = \frac{\$1,150}{160} = \$7.19,$$

a far cry from the unweighted average hourly wage of $10.

Note: The calculation of the arithmetic mean from grouped data, in fact, produces a weighted mean; the class frequencies, f, then serve the function of weights, w. The sum of these weights, Σf, is simply designated as the total number of observations, N or n. Thus, the formulas in Box 4.C are a special case of those in Box 4.G.

SUMMARY MEASURES OF DISPERSION

A knowledge of the dispersion or variability of data around their center of location is crucial in many situations. Consider someone about to decide which of two types of business to enter. Both types may promise the same median income, yet the incomes observed in one type of activity may be scattered widely about the median while the incomes in the other type may fall within a relatively narrow range of the median. Depending on their attitudes toward risk (a matter that is discussed at length in Chapter 17), different people will evaluate these two business prospects differently. Risk seekers will welcome a high variability of income; at the risk of getting a very low income, this high variability of income offers them a chance to get an exceptionally high income. Risk-averse people, on the other hand, will prefer the activity with a low variability of income; it offers them a middle-level income with near-certainty and offers them no chance for very low or very high income. The importance of variability is reviewed in Figure 4.5.

Dispersion is measured in two ways: as distances between selected observations or as average deviations of individual observations from a central value. We consider these possibilities in turn.

Distance Measures of Dispersion

The most popular distance measures of dispersion include (1) the (overall) range and (2) various interfractile ranges.

The range.   The overall range is usually called just the **range** and is the difference between the largest and smallest observation in a set of ungrouped data. It is the difference between the upper limit of the largest class and the lower limit of the smallest class for grouped data (and, therefore, cannot be determined at all if grouped data have open-ended classes). Thus, the range in our set of ungrouped profit figures listed in Table 3.2 on page 51 is $5,423 million − (−$1,060 million) = $6,483 million. The 1981 profits of U.S. multinationals, this figure tells us, were spread out over a vast range of almost $6.5 billion. Yet the major disadvantage of this measure is obvious: it ignores all the values except the two extremes. These extremes may be atypical not only among all the values in the data set, but even among the higher and lower ones—which surely is the case here. As another look at Table 3.2 confirms, most profits are found in the much narrower $2.5 billion-wide range between $0 and $2.5 billion. This problem is even worse when the range is calculated from the corresponding grouped data, found in Table 4.1. Here the range appears to equal $5,500 million − (−$1,500 million) = $7,000 million.

Interfractile ranges.   A definite improvement over using the overall range is to use various **interfractile ranges,** which measure differences between two values called *fractiles*. A **fractile** is a value below which a specified proportion of all values is found. The same information can also be given as a **percentile,** which is a value below which a specified *percentage* of all values is found. Thus, the median is the .50 fractile (or the 50th percentile) because half (or 50 percent) of all observations lie below it. After the median, the most common fractiles or percentiles used are the **quartiles,** which divide the total of observations into four quarters, each of which contains .25 (or 25 percent) of the observed values. The .25 fractile (or 25th percentile) is called the **first quartile,** and a quarter (or 25 percent) of all observations lie below it. The median itself is the

Figure 4.5   The Importance of Variability

Even though two populations may have identical medians, the spread of observations around them may differ significantly. This spread of observations is likely to be a crucial factor in decision making. Thus, a risk-averse individual may prefer the Type A Business (that promises a less varied income), while a risk-seeking person may choose to enter the Type B Business (just because it offers a chance of fairly large incomes, along with the danger of extremely low incomes, of course).

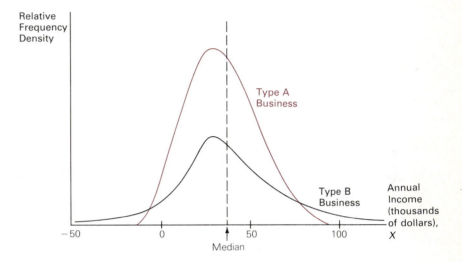

second quartile, and the .75 fractile (or 75th percentile) is referred to as the **third quartile.** Can you see in Table 3.2. on page 51 that the first, second, and third quartiles equal $335.5 million, $533.5 million, and $939 million, respectively? (In each case, the average of two adjacent values has been taken to ensure that 25, 50, and 75 percent of all observations lie below the quartile in question.)

Most popular among interfractile ranges is the **interquartile range,** which is the difference between the third and first quartiles and, hence, is the range containing the middle 50 percent of all values observed.

Although they are seldom used, other interfractile ranges can, of course, also be constructed, such as those between various **deciles**—the values in a data set that divide the total of observations into 10 parts rather than into 4 parts, as quartiles do, or into 100 parts, as percentiles do). Thus, the range between the 1st and 9th deciles (or the .10 and .90 fractiles or the 10th and 90th percentiles) contains the middle 80 percent of all values observed.

Can you see in Table 3.2 on p. 51 that the interquartile range equals $939 million − $335.5 million = $603.5 million? Can you see that the .10 to .90 interfractile range equals $1,802.5 million?

Figure 4.6 explains how fractile values can be read from a graph of a less than type ogive.

Figure 4.6   Reading Fractiles from an Ogive

Fractile values can be approximated from an ogive. The first quartile is about \$336 million (below *b*), and the third quartile is \$939 million (below *c*); thus, the interquartile range is about \$603 million wide. By the same token, the .10 to .90 interfractile range is seen to equal about \$1,803 million (going from below *a* to below *d* on the horizontal axis).

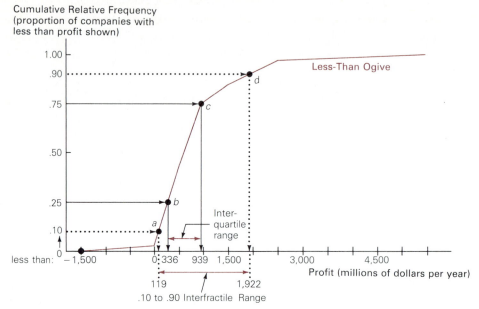

Source: Panel (a) of Figure 3.5.

Average Deviation Measures of Dispersion

All the distance measures of dispersion share a common disadvantage: they do not take into account all observations. Average deviation measures of dispersion—most important, (1) the mean absolute deviation, (2) the variance, and (3) the standard deviation— do take all observations into account.

The mean absolute deviation. To accomplish the feat of not ignoring any observed value, one might be tempted to calculate how much each individual observation, X, deviates from some central value, such as μ or $\overline{X}$, and then to combine these deviations by averaging them. However, averaging deviations from the arithmetic mean necessarily produces a result of zero, as a review of Figure 4.2 (the seesaws) can quickly remind us. Consider the five numbers 2, 2, 4, 7, and 15. Their arithmetic mean equals 6. Subtracting this mean from each value yields deviations of -4, -4, -2, $+1$, and $+9$, as panel (a) of Figure 4.2 shows. These deviations necessarily average to 0. The difficulty is overcome by realizing that the objective is merely to find the average distance of all the data from their center; for this purpose, it is irrelevant whether the individual data lie above or below the center. Hence, positive and negative deviations can be treated alike; one can ignore all the above plus and minus signs, and one can average all the *absolute* values of the deviations of individual observations from the mean (a fact that is indicated symbolically by vertical lines before and after each value). This procedure yields the **mean absolute deviation, MAD,** here equal to $[(|4| + |4| + |2| + |1| + |9|)/5] = (20/5) = 4$. Can you confirm that this measure of dispersion comes to $566.70 million for the profit data of Table 3.2 (the arithmetic mean of which came to $786.62 million)? Symbolically, the mean absolute deviation can be expressed as shown in Box 4.H (with vertical lines denoting absolute value).

4.H Mean Absolute Deviation

From Ungrouped Data
For a population:

$$MAD = \frac{\Sigma|X - \mu|}{N}$$

For a sample:

$$MAD = \frac{\Sigma|X - \overline{X}|}{n}$$

where $\Sigma|X - \mu|$ is the sum of the absolute differences between each observed population value, X, and the population mean, μ, while N is the number of observations in the population, and where $\Sigma|X - \overline{X}|$ is the sum of the absolute differences between each observed sample value, X, and the sample mean, $\overline{X}$, while n is the number of observations in the sample.

From Grouped Data
Denoting absolute class frequencies by f and class midpoints by X, substitute $\Sigma f|X - \mu|$ or $\Sigma f|X - \overline{X}|$ for the numerators given here.

Note: Occasionally, absolute deviations from the median rather than from the mean are calculated; in which case μ is replaced by **M**, and $\overline{X}$ is replaced by m.

The variance. A much more important measure of dispersion, however, which is calculated by averaging the *squares* of the individual deviations from the mean, is the **mean of squared deviations,** or the **variance.**[3] Using the five values from panel (a) of Figure 4.2, the variance is calculated in Table 4.3.

---

[3]The arithmetic mean has the property of minimizing the sum of the squared deviations of all the individual observations from itself, as shown in Chapter 4 of the *Student Workbook* that accompanies this text.

Table 4.3   Calculating the Variance for a Population of $N = 5$ Values

The population variance is calculated as the arithmetic mean of the individually squared deviations of each observed value from the population mean. Note how the process of squaring deviations (a) eliminates negative values (and thus produces a measure of dispersion that focuses on the existence of deviations rather than their direction) and (b) emphasizes large deviations more than small ones (a 9 squared counts not 9 times, but 81 times as much as a 1 squared).

Observed Values X	Population Mean $\mu = \dfrac{\Sigma X}{N}$	Deviations $X - \mu$	Squared Deviations $(X - \mu)^2$
2	6	−4	16
2	6	−4	16
4	6	−2	4
7	6	+1	1
15	6	+9	81

$\Sigma X = 30$ Sum of squared deviations $= \Sigma (X - \mu)^2 = 118$

$$\text{Mean of squared deviations} = \text{variance} = \frac{\Sigma (X - \mu)^2}{N} = \frac{118}{5} = 23.6$$

When the variance is calculated from a statistical population, it is symbolized by σ^2 (the letter σ is the lowercase Greek *sigma* and σ^2 is pronounced "sigma squared"). When the variance refers to sample data, it is denoted by s^2 instead (in line with the common procedure of labeling parameters by Greek letters and statistics by Roman ones). The variance is calculated as shown in Box 4.I.

4.I Variance from Ungrouped Data

For a population:

$$\sigma^2 = \frac{\Sigma(X - \mu)^2}{N}$$

For a sample:

$$s^2 = \frac{\Sigma(X - \overline{X})^2}{n - 1}$$

Where $\Sigma(X - \mu)^2$ is the sum of squared deviations between each population value, X, and the population mean, μ, with N being the number of observations in the population, while $\Sigma(X - \overline{X})^2$ is the sum of squared deviations between each sample value, X, and the sample mean, $\overline{X}$, with n being the number of observations in the sample.

Note that this formulation contains only one surprise: the sample variance is obtained by dividing the sum of the squared deviations of individual sample values, X, from the sample mean, $\overline{X}$, by $n - 1$ rather than by the sample size, n. The reason, to be explained in Chapter 8, is that this procedure makes the sample variance, s^2, a more accurate estimator of the usually unknown population variance, σ^2.

The variance, too, can be estimated from grouped data by the now familiar procedure of substituting class midpoints for actually observed values, X, and replacing the numerators in Box 4.I by $\Sigma f(X - \mu)^2$ and $\Sigma f(X - \overline{X})^2$, respectively, as shown in Box 4.J.

4.J Variance from Grouped Data

For a population:

$$\sigma^2 = \frac{\Sigma f(X - \mu)^2}{N}$$

For a sample:

$$s^2 = \frac{\Sigma f(X - \overline{X})^2}{n - 1}$$

where absolute class frequencies are denoted by f, class midpoints of grouped population (or sample) values by X, the population (or sample) mean by μ (or $\overline{X}$), and the number of observations in the population (or sample) by N (or n).

Practical problems. Unfortunately, there are two practical problems associated with the use of the variance concept. First, the variance tends to be a large number compared to the observations the spread of which it is to describe. When the original observations, as in Table 3.2 on p. 51 are at most equal to a few billions, their variance can equal many hundreds of billions. Second and worse yet, the variance, being a squared number, is not expressed in the same units as the observed values themselves. Thus, the variance of the Table 3.2 profit population is not expressed in hundreds of billions of *dollars,* but in hundreds of billions of *squared* dollars; in fact, it equals 704.4 billion squared dollars (and who knows the meaning of that?). But there is good news as well: both of the conceptual difficulties just cited can be overcome in one fell swoop by working with the square root of the variance, a concept to which we turn.

The standard deviation. The positive square root of the variance is called the **standard deviation** and is, without doubt, the most important measure of dispersion. It falls in the same range of magnitude as, and is expressed in the same units as, the observations themselves. Thus, the standard deviation to the data in Table 4.3 equals $\sqrt{23.6} = 4.86$; the standard deviation for the Table 3.2 profit data equals $\sqrt{704,408.34}$ million squared dollars $= \$839.29$ million. The definition of the standard deviation is expressed symbolically in Box 4.K.

4.K Standard Deviation from Ungrouped Data

For a population:

$$\sigma = \sqrt{\frac{\Sigma(X - \mu)^2}{N}}$$

For a sample:

$$s = \sqrt{\frac{\Sigma(X - \overline{X})^2}{n - 1}}$$

where $\Sigma(X - \mu)^2$ is the sum of squared deviations between each population value, X, and the population mean, μ, with N being the number of observations in the population, while $\Sigma(X - \overline{X})^2$ is the sum of squared deviations between each sample value, X, and the sample mean, $\overline{X}$, with n being the number of observations in the sample.

Once again, the standard deviation is estimated from grouped data by substituting class midpoints for actually observed values, X, and replacing the numerators under the square roots in Box 4.K by $\Sigma f(X - \mu)^2$ and $\Sigma f(X - \overline{X})^2$, respectively, as shown in Box 4.L.

4.L Standard Deviation from Grouped Data

For a population:

$$\sigma = \sqrt{\frac{\Sigma f(X - \mu)^2}{N}}$$

For a sample:

$$s = \sqrt{\frac{\Sigma f(X - \overline{X})^2}{n - 1}}$$

where absolute class frequencies are denoted by f, class midpoints of grouped population (or sample) values by X, the population (or sample) mean by μ (or $\overline{X}$), and the number of observations in the population (or sample) by N (or n).

Shortcut calculations of variance and standard deviation. Unless the variance or standard deviation is to be calculated by a computer, it is helpful to use a shortcut procedure that is mathematically equivalent to the boxed formulas encountered in the book so far.[4] The shortcut formulas for the variance appear in Box 4.M; the standard deviation is, of course, always the square root of the variance. A numerical example, based on the data in Table 4.3, is given in Table 4.4.

4.M Variance from Ungrouped Data—Shortcut Method

For a population:

$$\sigma^2 = \frac{\Sigma X^2 - N\mu^2}{N}$$

For a sample:

$$s^2 = \frac{\Sigma X^2 - n\overline{X}^2}{n - 1}$$

where ΣX^2 is the sum of squared population (or sample) values, μ^2 is the squared population mean and $\overline{X}^2$ the squared sample mean, N is the number of observations in the population, and n is the number of observations in the sample.

Table 4.4 Calculating the Variance and Standard Deviation for a Population of Ungrouped Data of $N = 5$ Values—Shortcut Method

This shortcut method of calculating the variance and standard deviation is particularly recommended whenever the calculation proceeds with pencil and paper or by simple calculator.

Observed Values X	X^2	
2	4	$\mu = \dfrac{\Sigma X}{N} = \dfrac{30}{5} = 6$
2	4	$\sigma^2 = \dfrac{\Sigma X^2 - N\mu^2}{N} = \dfrac{298 - 5(6)^2}{5} =$
4	16	
7	49	$\dfrac{298 - 180}{5} = 23.6$
15	225	
$\Sigma X = 30$	$\Sigma X^2 = 298$	$\sigma = \sqrt{23.6} = 4.86$

---

[4]Consider this proof for the population variance from ungrouped data: $\sigma^2 = \dfrac{\Sigma(X - \mu)^2}{N} = \dfrac{\Sigma(X^2 - 2X\mu + \mu^2)}{N} = \dfrac{\Sigma X^2 - 2\mu\Sigma X + N\mu^2}{N} = \dfrac{\Sigma X^2}{N} - 2\mu^2 + \mu^2 = \dfrac{\Sigma X^2 - N\mu^2}{N}$

A shortcut method is equally helpful when making approximations from grouped data, such as those in Table 4.1. The formulas appear in Box 4.N, and Table 4.5 gives an example of how to use the formulas.

4.N Variance from Grouped Data—Shortcut Method

For a population:

$$\sigma^2 = \frac{\Sigma f X^2 - N\mu^2}{N}$$

For a sample:

$$s^2 = \frac{\Sigma f X^2 - n\bar{X}^2}{n - 1}$$

where $\Sigma f X^2$ is the sum of absolute-class-frequency (f) times squared-class-midpoint (X) products, μ^2 is the squared population mean and $\bar{X}^2$ the squared sample mean, N is the number of observations in the population and n is the number of observations in the sample.

Table 4.5   Approximating the Variance and Standard Deviation for a Population from Grouped Data—Shortcut Method

This table illustrates how closely one can approximate the true variance and standard deviation of ungrouped data (such as those in Table 3.2) with the help of the corresponding grouped data (shown here). Compare the approximations calculated at the bottom of this table with the precise calculations from the underlying ungrouped data: The *variance from ungrouped data* is

$$\sigma^2 = \frac{132,317,936 - 61,877,102}{100} = \$704,408.34 \text{ million squared.}$$

The *standard deviation from ungrouped data* is $\sigma = \$839.29$ million.

Class (size of profit in millions of dollars)	Absolute Class Frequency (number of companies in class), f	Class Midpoint, X	fX	X^2	fX^2
−1,500 to under 0	3	−750	−2,250	562,500	1,687,500
0 to under 500	41	250	10,250	62,500	2,562,500
500 to under 1,000	32	750	24,000	562,500	18,000,000
1,000 to under 1,500	9	1,250	11,250	1,562,500	14,062,500
1,500 to under 2,000	6	1,750	10,500	3,062,500	18,375,000
2,000 to under 2,500	6	2,250	13,500	5,062,500	30,375,000
2,500 to under 5,500	3	4,000	12,000	16,000,000	48,000,000
Totals:	$N = 100$		$\Sigma fX = 79,250$		$\Sigma fX^2 = 133,062,500$

$$\mu = \frac{\Sigma fX}{N} = \frac{79,250}{100} = \$792.5 \text{ million}$$

$$\sigma^2 = \frac{\Sigma fX^2 - N\mu^2}{N} = \frac{133,062,500 - 100(792.5)^2}{100}$$

$$= \frac{133,062,500 - 62,805,625}{100} = \frac{70,256,875}{100}$$

$$= \$702,568.75 \text{ million squared}$$

$$\sigma = \sqrt{\$702,568.75 \text{ million squared}} = \$838.19 \text{ million}$$

Source: Tables 3.2 and 4.1.

The Special Significance of the Standard Deviation

Together with the arithmetic mean, the standard deviation is, perhaps, the most significant parameter of a statistical population. Consider just two of its many uses: describing the normal frequency distribution and comparing the degree of dispersion of different data sets.

Describing the normal frequency distribution. When a population is correctly described by the perfectly symmetrical *normal frequency distribution* shown in the graph on p. 121 (and to be discussed at length in Chapter 7), one can predict the precise percentage of observations falling within ranges from the mean defined by any desired number of standard deviations. Some 68.3 percent of all observations will then lie within the range of $\mu \pm 1\sigma$; that is, within one standard deviation of the mean. Similarly, some 95.4 percent of the population will lie within $\mu \pm 2\sigma$, and 99.7 percent will lie within $\mu \pm 3\sigma$. Thus, if the mean height of all male production workers in a large firm were $\mu = 5'8''$, if the standard deviation was $\sigma = 2$ inches, and if the heights were distributed like a normal curve, we could confidently assert that 68.3 percent of these men were between $5'6''$ and $5' 10''$ tall or that 99.7 percent of them had heights between $5'2''$ and $6'2''$.

It is difficult to overemphasize the significance of these facts. Think of it, a normally distributed population can be fully described by a mere two numbers—the parameters μ and σ. From these two numbers, the entire frequency distribution can be reconstructed. Analytical Examples 4.1, "Standard Scores," and 4.2, "Control Charts," illustrate just two of a multitude of applications. Indeed, as the discussion of *Chebyshev's theorem* in Biography 4.1 at the end of the chapter attests, the standard deviation can tell us important things about all types of populations, even those that are not normally distributed.

Comparing the degree of dispersion of different data sets. Sometimes the ratio of the standard deviation to the arithmetic mean is used as an indicator of *relative* dispersion. This ratio is called the **coefficient of variation** and is symbolized by V (or v), as shown in Box 4.O.

4.O Coefficient of Variation

For a population:

$$V = \frac{\sigma}{\mu}$$

For a sample:

$$v = \frac{s}{\overline{X}}$$

where σ is the population standard deviation, μ the population mean, s the sample standard deviation, and $\overline{X}$ the sample mean.

For the profit population of Table 3.2, the coefficient of variation comes to

$$V = \frac{\$839.29 \text{ million}}{\$786.62 \text{ million}} = 1.067.$$

Put differently, the standard deviation equals 106.7 percent of the mean.

Note that the coefficient of variation, unlike all the measures of *absolute* dispersion, is expressed as a pure number without any units. As a result, the coefficient of variation can be used to compare the relative dispersion of two or more distributions that are expressed in different units. For example, if the employees of a firm have a mean height of 66 inches (with a standard deviation of 4 inches), but a mean weight of 150 pounds (with a standard deviation of 30 pounds), one can say that heights are less varied than weights because

$$\text{for heights, } V = \frac{4''}{66''} = .06,$$

$$\text{but for weights, } V = \frac{30 \text{ lb.}}{150 \text{ lb.}} = .20.$$

Analytical Example 4.3, "On the Accuracy of National Income Statistics," shows another application of the concept.

SUMMARY MEASURES OF SHAPE*

The *shape* of a frequency distribution can be described by (1) its symmetry or lack of it *(skewness)* and (2) its peakedness *(kurtosis)*.

Skewness

The matter of **skewness,** or a frequency distribution's degree of distortion from horizontal symmetry, is illustrated in Figure 4.7. Panel (a) depicts the bell-shaped normal curve that possesses *zero* skewness because the frequency density is tapering off equally in both directions from the mode. Panel (b) shows the case of *positive* skewness, so called because the frequency density is tapering off more slowly toward the right of the mode than toward the left, and the long right tail of the frequency curve points in the *positive* direction along the horizontal axis—toward larger values of X. Panel (c), finally, illustrates *negative* skewness, so called because the frequency density is tapering off more slowly toward the left of the mode than toward the right, and the long left tail of the frequency curve points in the *negative* direction along the horizontal axis—toward smaller values of X.

Interestingly, the type of skewness has certain implications for the positions of the mean, median, and mode. In the case of zero skewness, as in panel (a), mean, median,

---

*Optional section.

Figure 4.7   The Shape of a Frequency Curve: Skewness

These graphs illustrate the three basic types of skewness. Note their implications for the positions of the mean, the median, and the mode. While these measures of central tendency coincide in the case of zero skewness, the mean is pulled toward the extreme values in the case of positive or negative skewness. Moving away from the tail, and just as in the dictionary, the mean is followed by median and mode. Also as in the dictionary, the median is typically closer to the mean than to the mode; for moderately skewed distributions, the median lies about one third of the way between mean and mode.

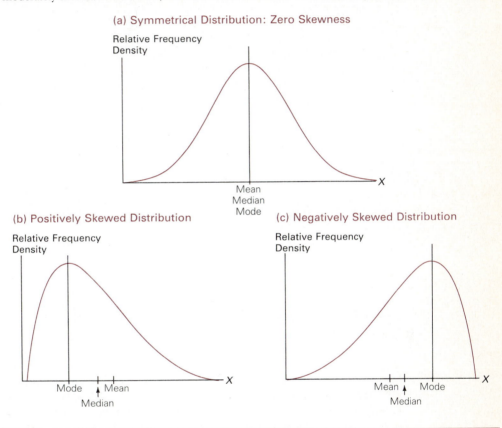

and mode coincide. When the frequency distribution exhibits positive skewness, as in panel (b), a relatively few extremely large values raise the mean considerably above the mode; the median (which is less sensitive to extreme values) often ends up somewhere between the mean and the mode. When the frequency distribution is one of negative skewness, as in panel (c), a relatively few unusually small values lower the mean considerably below the mode; the median, again, often ends up between the mean and mode. By tradition, statisticians prefer to use the median as the measure of central tendency whenever the underlying distribution is skewed.

The coefficient of skewness. The positional differences among mean, median, and mode can be used to create arithmetic measures of skewness. Of the several such measures that exist, the most useful, perhaps, is **Pearson's coefficient of skewness,**

Sk or *sk,* which is a measure of skewness that focuses on the difference between the mode and the mean (and then relates it to the standard deviation), as in Box 4.P. (The coefficient is named after Karl Pearson, who is featured in Biography 11.1.)

4.P Pearson's Coefficient of Skewness

For a population:

$$Sk = \frac{\mu - Mo}{\sigma}$$

For a sample:

$$sk = \frac{\overline{X} - mo}{s}$$

where μ or $\overline{X}$ are population or sample mean, while Mo or mo are population or sample mode and σ or s are population or sample standard deviation.

As panel (a) of Figure 4.7 indicates, the mean equals the mode for a symmetrical distribution; thus, Pearson's coefficient of skewness equals zero because $\mu - Mo = 0$. As panel (b) shows, the mean exceeds the mode for a positively skewed distribution; thus, the coefficient of skewness ends up positive because $\mu > Mo$. Similarly, as shown in panel (c), the mean is below the mode of negatively skewed distributions; the coefficient of skewness is, accordingly, negative because $\mu < Mo$.

Note: Sometimes the difference, multiplied by 3, between the mean and the *median* is substituted in the formulas of Box 4.P for the difference between the mean and the mode. For moderately skewed distributions, this substitution gives roughly the same result as the formula in Box 4.P because, as Figure 4.7 shows, the distance between mean and median then equals about one third the distance between mean and mode. Thus, for the population of profit figures in Table 3.2, the coefficient of skewness might be calculated as

$$Sk = \frac{3(\$786.62 \text{ million} - \$533.5 \text{ million})}{\$839.29 \text{ million}} = \frac{\$759.36 \text{ million}}{\$839.29 \text{ million}} = .90.$$

The result indicates positive skewness. (For a visual impression, have another look at panel (b) of Figure 3.2 on page 62, which is based on the same data.)

Kurtosis

A frequency curve's degree of peakedness, or **kurtosis,** is illustrated in Figure 4.8. Kurtosis is measured by the **coefficient of kurtosis,** described in Box 4.Q on the next page.

Figure 4.8   The Shape of a Frequency Curve: Kurtosis

This graph illustrates the three basic types of kurtosis.

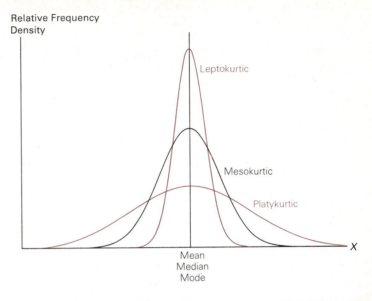

4.Q Coefficient of Kurtosis

From Ungrouped Data

For a population:

$$K = \frac{\dfrac{\Sigma(X - \mu)^4}{N}}{\sigma^4}$$

For a sample:

$$k = \frac{\dfrac{\Sigma(X - \overline{X})^4}{n}}{s^4}$$

where Xs are observed population or sample values (with μ being the population mean and $\overline{X}$ the sample mean), while N is population size, n is sample size, σ^4 is the squared population variance, and s^4 is the squared sample variance.

From Grouped Data

Replace numerators by $\dfrac{\Sigma f(X - \mu)^4}{\Sigma f}$ or $\dfrac{\Sigma f(X - \overline{X})^4}{\Sigma f}$, respectively, where X and f represent class midpoints and class frequencies.

Figure 4.8 illustrates types of kurtosis. Frequency curves with a kurtosis of 3 are called *mesokurtic;* those with larger values are more peaked and are called *leptokurtic;* those with smaller values are flatter and are called *platykurtic.* [If one approximates the value of K from the grouped profit data of Table 4.5, the result is 7.51, suggesting a leptokurtic distribution, as panel (b) of Figure 3.2 confirms.] Like the coefficient of variation, both the coefficient of skewness and that of kurtosis are expressed as pure numbers; hence, different distributions can easily be compared with respect to their degree of skewness or kurtosis.

THE PROPORTION

For qualitative data, the *proportion* of observations falling into a given category is the only summary measure available. It is calculated as the number of observations in the category divided by the total of all observations. Thus, if 138 of 200 firms sampled are single proprietorships, if 18 are partnerships, and if the rest are corporations, one could designate the first group as the modal type of firm, but one could hardly calculate the arithmetic mean, the median, or the standard deviation of firm types. It would be meaningful, however, to state that the *proportion* of single proprietorships and partnerships among all types of firms equaled (138 + 18)/200, or .78. By the same token, corporations could be reported as making up a proportion of .22. Naturally, these numbers, like all proportions, can instantly be converted to percentages, here of 78 and 22 percent.

When referring to a population, the proportion is denoted by π (which is the lowercase Greek *p,* is pronounced "pī," and stands for *proportion* and *not*—here—for the geometric constant 3.1416, which equals the ratio of the circumference to the diameter of a circle). When referring to a sample, the proportion is represented by the capital Roman letter *P.* Analytical Examples 4.4, "The Safety of Anesthetics," and 4.5, "Does Inheritance Matter in Disease?" illustrate the important uses to which the simple concept of proportion can be put.

||| ANALYTICAL EXAMPLES

4.1 Standard Scores

As this chapter indicates, one of the significant features of the standard deviation is the fact that it can be used to describe the precise percentage of observations falling within various ranges from the mean, provided the frequency distribution fits the so-called *normal distribution,* such as the one first noted in panel (a) of Figure 3.2 on pp. 62. Consider the accompanying figure. This graph shows a hypothetical normal distribution of aptitude test scores (such as the familiar SAT scores) with a mean score of $\mu = 500$ and a standard deviation of $\sigma = 100$. As a result of the normal distribution of the population of scores, 68.3 percent of all scores lie between 400 and 600 (that is, within 1 standard deviation of the mean), 95.4 percent of all scores lie between 300 and 700 (within 2 standard deviations of the mean), and 99.7 percent of all scores lie between 200 and 800 (within 3 standard deviations of the mean). As will be discussed at length in Chapter 7, it is also true that 68.3 percent of the *area* under a normal curve lies within 1 standard deviation of the mean, while 95.4 and 99.7 percent of the area, respectively, lie within 2 and 3 standard deviations.

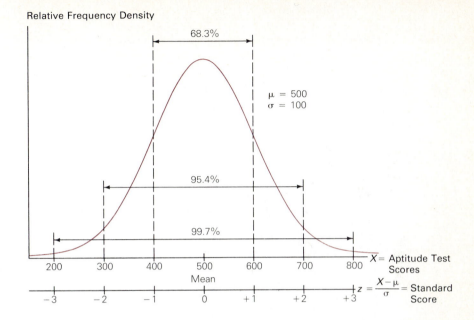

In order to make possible easy comparisons among normal distributions of different types of populations (be they test scores, heights, weights, or wages), one can convert raw data into **standard scores** that indicate the numbers of standard deviations between particular observations and the mean of all observations in a data set. As is indicated underneath the graph, the standard score, z, equals the difference between an observed value, X, and the mean, μ, divided by the standard deviation, σ; thus, a z score always shows how many standard deviations *away from the mean* a corresponding X score is located.

In this example, with $\mu = 500$ and $\sigma = 100$, someone scoring $X = 300$ has a standard score of $z = -2$

because $(300 - 500) \div 100 = -2$. Such a score lies below 50 percent of all scores (the area under the right half of the bell-shaped curve) and also below an additional $(95.4/2) = 47.7$ percent of all scores (the area under the left side of the bell between the mean and $X = 300$). Thus, a score of $X = 300$ lies below 97.7 percent of all scores.

Can you see, similarly, why a score of 800 would correspond to a standard score of $+3$ and would lie above 99.85 percent of all scores? Why someone with standard scores of $+.5$ on the weight distribution and -1 on the wage distribution would be heavier and less well paid than the average person whose weight and wages were plotted?

4.2 Control Charts

Managers of businesses know that there exists an inherent amount of variability in any data series, whether the series describes the sugar content of bottled drinks, the extent of clerical errors, the size of inventories, of labor costs, or advertising expenses, or even the loss rate of old customers. **Control charts** are graphical devices, such as those shown in the accompanying figure, that highlight the average performance of data series and the dispersion around this average. (They were introduced in 1924 by Walter A. Shewhart of the Bell Telephone Laboratories.) The av-

erage performance of the past is viewed as a standard with which to compare current performance, and the typical dispersion of the past is used to set allowable limits for performance now. When plotting new data, it becomes immediately obvious whether they fall within the range of the expected or whether there are undesirable trends that need the immediate attention of the manager.

Note how all the data series in the figure, except panels (c) and (e), are proceeding within normal limits of, say, ± 2 standard deviations from the average

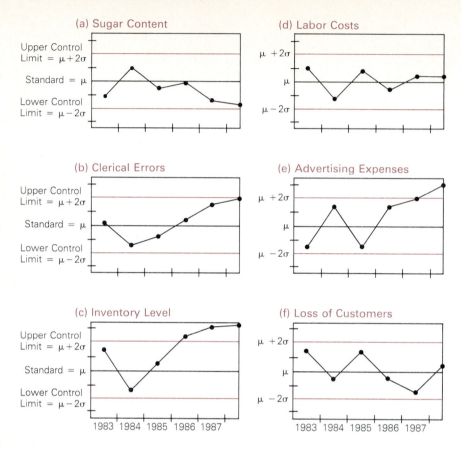

observed over some past period. Managers would waste their time if they worried about the recent decrease observed in series (a) or the recent increase in series (f). Both these changes are still within the upper and lower control limits of $\mu \pm 2\sigma$. Managers would be well advised, however, to pay special attention to series (c) and (e) that have broken through the limits and are, thus, "out of control." Managers might even consider checking up on series (b) that seems to be exhibiting a lengthy undesirable trend.

Note: For our purposes it is unimportant how the original data of the various series are measured—whether, for example, sugar content is measured in milligrams per bottle or milligrams per cubic inch and whether advertising expenses are expressed in dollars per million bottles sold or dollars per year. Important only is the fact that the data of each series, however originally measured, can be converted into "standard scores" defined by their own means and standard deviations.

4.3 On the Accuracy of National Income Statistics

People pay almost religious attention to statistics of national income. Every quiver of a decimal point is reported and widely analyzed, but rarely are people aware of the large margins of error associated with the data. Simon Kuznets, the great pioneer in national income estimation, was quite aware of these error margins and once examined the errors pertaining to U.S.

national income between 1919 and 1935. The accompanying table gives some of his results.

Kuznets found the weighted margin of error of national income to be 20 percent but thought this to be exaggerated for a variety of reasons. Yet he did not hesitate to view a 10 percent error in national income data as rather likely.

Sector Contributing to National Income	Mean Error μ (in percent of stated figure)	Standard Deviation of Error σ (percent)	Coefficient of Variation σ/μ
Manufacturing	9.45	3.40	.36
Agriculture	12.40	4.96	.40
Mining	13.10	5.90	.45
Trade	20.50	12.71	.62
Government	17.66	3.18	.18
Construction	26.91	3.23	.12
Services	27.27	3.82	.14

Consider the implications. If the 1986 U.S. GNP data contained a similar margin of error, this margin would amount to ± $420 billion, an amount roughly equal to all the durable-goods purchases of consumers, or all nonresidential private domestic investment, or all of exports and just about equal to half of the spending of the entire government sector (federal, state, and local). Yet ordinary citizens and government officials alike seriously debate the meaning of changes of even 1/10 of 1 percent of the GNP and continually draw what can only be called totally unwarranted conclusions therefrom.

Source: Simon Kuznets, *National Income and Its Composition, 1919–1938,* vol. 2 (New York: National Bureau of Economic Research, 1941), pp. 512–13.

4.4 The Safety of Anesthetics

In 1958, U.S. hospitals began using a new type of anesthetic called *halothane*. Unlike other anesthetics traditionally used, it was not flammable (which reduced explosion hazards during surgery), and it was agreeable to patients (who recovered faster and suffered fewer aftereffects). Before long, halothane was used in half the operations, but then a suspicion arose: many hospitals reported strikingly unusual, but similar deaths of patients after the use of halothane. Was halothane responsible for these unusual deaths?

A search of 850,000 records in 34 hospitals revealed 17,000 postoperative deaths within six weeks after surgery, a proportion of .02. Naturally, these deaths were attributable to all kinds of causes and not necessarily to the use of halothane. In fact, the proportion of postoperative deaths associated with various anesthetics (but not necessarily caused by them) were found to be as shown in row 1 of the accompanying table.

Contrary to the aforementioned suspicion, was halothane *twice as safe* as cyclopropane and certainly no worse than any other anesthetic? The statistical investigators were hesitant to argue on the basis of row 1 alone. They realized that (a) some anesthetics were used more often in difficult operations (cyclopropane) than in easy ones (pentothal); (b) some types of operations had vastly different death rates than other types (the range went from .0025 to .14); (c) the physical status of patients affected death rates (the range went from .005 for young patients to .26 for old ones); (d) women patients had lower death rates than men; and (e) different hospitals had different death rates as well. Conceivably, all the favorable factors (simple operations, young or female patients, and so forth) could have been associated with the use of halothane, and all the unfavorable factors could have been associated with the use of other anesthetics, thus masking in row

	Halothane	Pentothal	Cyclopropane	Ether	Others	Total
(1) Raw data	.017	.017	.034	.019	.030	.020
(2) Adjusted data	.021	.020	.026	.020	.025	.020

1 of the table the bad effect of the suspicious anesthetic. Accordingly, all the proportions were adjusted to account for these other factors. The result is given in row 2. The masking effect just referred to was, indeed, present, but even after adjusting the data, halothane proved to be at least as safe as any of the other anesthetics. (The suspicious deaths were also present with other anesthetics but had not been noticed because the traditional anesthetics had not been watched as closely as the new one.)

Source: Lincoln E. Moses and Frederick Mosteller, "Safety of Anesthetics," in Judith M. Tanur et al., eds., *Statistics: A Guide to the Unknown* (San Francisco: Holden-Day, 1972), pp. 14–22.

4.5 Does Inheritance Matter in Disease?

Do people develop certain diseases because they inherit a special susceptibility to them from their parents, or are these diseases the result of their environment? In most cases, it is difficult to answer this age-old "nature vs. nurture" question because most people differ in both their genetic endowment and in their upbringing. Yet an unusual opportunity for answering this question is provided by the existence of twins who can be identical (come from the same fertilized egg and, therefore, have the same set of genes) or nonidentical (come from separate eggs and have gene patterns that are no more or less alike than those of ordinary siblings). Both types of twins typically share the same environment during their upbringing, however. If one of the pair then develops a disease that is transmitted through the genes, it is more likely to appear in an identical twin than in a nonidentical twin. By the same token, diseases that are not genetic should affect identical and nonidentical twin siblings to the same degree.

Consider Table A, which shows the results of a Danish study of 4,368 same-sex pairs of twins. The data show *affected concordance rates*—that is, the ratio of the total number of people affected as pairs (when

Table A

Disease	Affected Concordance Rate for	
	Identical Twins	Nonidentical Twins
Bronchial asthma	.63	.38
Epilepsy	.54	.24
Tuberculosis	.54	.27
Rheumatoid arthritis	.50	.05
Cerebral apoplexy	.36	.19
Coronary occlusion	.33	.27
Rheumatic fever	.33	.10
Death from acute infection	.14	.11

Table B

Disease	Affected Concordance Rate for			
	Identical Twins		Nonidentical Twins	
	Men	Women	Men	Women
Chronic coughs				
Smokers	.146	.136	.123	.145
Nonsmokers	.077	.076	.055	.057

Source: D. D. Reid, "Does Inheritance Matter in Disease? The Use of Twin Studies in Medical Research," in Judith M. Tanur et al., eds., *Statistics: A Guide to the Unknown* (San Francisco: Holden-Day, 1972), pp. 77–83.

both members of a pair got the disease) to the total number of affected people (regardless of whether one or both members of a pair got the disease). Surely, genetics seems to have a lot to do with at least the first four of the diseases and very little with the last one.

A similar Swedish study tested the likely association between lung diseases and smoking, as shown in Table B. Clearly, these proportions suggest strongly that smoking rather than genetics is the *likely* cause of chronic coughs. (While the mere observation of differences in proportions can, thus, be suggestive of a cause-and-effect relationship, techniques discussed in later chapters of this book provide more reliable ways of establishing such a relationship.)

| BIOGRAPHY

4.1 Pafnuty Chebyshev

Pafnuty Lvovich Chebyshev (1821–1894) was born in Okatovo, Russia. His parents, who belonged to the gentry, had him privately tutored; he quickly became fascinated by mathematics and eventually studied mathematics and physics at Moscow University. Even as a student, he won a silver medal for a now famous paper on calculating the roots of equations. It was only the first of many brilliant papers that he wrote while teaching mathematics at St. Petersburg University and pursuing a keen interest in mechanical engineering. (Among other things, he contributed significantly to ballistics, which gave rise to various innovations in artillery, and he invented a calculating machine.) Always, he stressed the unity of theory and practice, saying[1]:

> Mathematical sciences have attracted especial attention since the greatest antiquity; they are attracting still more interest at present because of their influence on industry and arts. The agreement of theory and practice brings most beneficial results; and it is not exclusively the practical side that gains; the sciences are advancing under its influence as it discovers new objects of study for them, new aspects to exploit in subjects long familiar.

[1]*Dictionary of Scientific Biography,* vol. 3 (New York: Charles Scribner's, 1971), p. 226.

Chebyshev typically worked toward the effective solution of problems by establishing *algorithms* (methods of computation) that gave either an exact numerical answer or an approximation correct within precisely defined limits. A most important example of this approach in the field of statistics is his formulation of what is now called **Chebyshev's theorem.** It states that "regardless of the shape of a population's frequency distribution, the proportion of observations falling within k standard deviations of the mean is at least

$$1 - \frac{1}{k^2}$$

given that k equals 1 or more." Thus, we can make the predictions listed in the accompanying table.

Notice how this theorem is vindicated by the profit data of Table 3.2 on page 51. The mean profit equals $786.62 million; the associated standard deviation equals $839.29 million. Can at least 75 percent of the 100 profit figures be found within $\mu \pm 2\sigma$—that is, within the range of −$891.96 million and +$2,465.2 million? Can at least 89 percent be found within the range of −$1,731.25 million and +$3,304.49 million?

If the number of standard deviations, k, equals	then the proportion of all observations lying within the range of $\mu \pm k\,\sigma$ equals at least
1	$1 - (1/1^2) = 0$
2	$1 - (1/2^2) = 0.75$
3	$1 - (1/3^2) = 0.89$
4	$1 - (1/4^2) = 0.94$

Can at least 94 percent be found within the range of − $2,570.54 million and + $4,143.78 million? As a look at Table 2.2 attests, the answers are yes in all cases. In fact, the answers provided by Chebyshev's theorem are most conservative; the actual percentages of observations found within these ranges are 96, 98, and 99, respectively, in this example.

For practical purposes, many a frequency distribution that is only slightly skewed (with a coefficient of skewness between, say, − .5 and + .5) can be treated like a perfectly symmetrical one, and the higher percentages applicable to a normal curve (and shown in Analytical Example 4.1, "Standard Scores") can be ap-

plied to estimate the proportions of observations falling within specified distances from the mean.

Chebyshev's theorem, however, demonstrates well a radical change brought about by him: He was the first mathematician to insist on absolute accuracy in limit theorems. In the words of A. N. Kolmogorov, another eminent Russian mathematician, "he always aspired to estimate exactly in the form of inequalities absolutely valid under any number of tests the possible deviations from limit regularities."[2]

_____

[2]Ibid., p. 231.

Summary

1. This chapter discusses an approach to condensing collections of data that is even more radical than the construction of tables and graphs: the calculation of arithmetic summary measures. The most important summary measures are the measures of central tendency, of dispersion, and of shape for quantitative variables and the proportion for qualitative ones.

2. Measures of central tendency or location are values around which observations tend to cluster and that describe the location of the "center" of a data set. They can be calculated with precision from ungrouped data and can be approximated from grouped data. The most commonly used measures of central tendency are the mean, the median, and the mode; the mean comes in several versions, including the unweighted mean and weighted mean (among others).

3. Measures of dispersion or variability are numbers that indicate the spread or scatter of observations; they show the extent to which individual values in a data set differ from one another and, hence, differ from their central location. Dispersion can be measured as distances between selected observations or as average deviations of individual observations from a central value. The most important

distance measures of dispersion include the (overall) range and a variety of interfractile ranges, usually defined by quartiles, deciles, or percentiles. The most important average deviation measures of dispersion include the mean absolute deviation, the variance, and the standard deviation. There are a number of important uses for the standard deviation (and its derivative coefficient of variation).

4. Measures of shape are numbers that indicate either the degree of asymmetry (skewness) or that of peakedness (kurtosis) in a frequency curve. The bell-shaped normal curve possesses zero skewness because the frequency density is tapering off equally in both directions from the mode. This contrasts with positive (or negative) skewness when the frequency density is tapering off more slowly toward the right (or left) of the mode. Depending on their peakedness, frequency curves are called platykurtic, mesokurtic, or leptokurtic.

5. The only summary measure available for qualitative data is the proportion, a number that describes the frequency of observations in a particular category as a fraction of all observations made. Even this humble measure can lead to important insights that are hidden in raw data.

Key Terms

arithmetic mean (μ or $\overline{X}$)
bimodal frequency distribution
Chebyshev's theorem
coefficient of kurtosis (*K* or *k*)

coefficient of variation (*V* or *v*)
control charts
deciles
first quartile

fractile
interfractile ranges
interquartile range
kurtosis
mean absolute deviation *(MAD)*
mean of squared deviations
measures of central tendency (or location)
measures of dispersion (or variability)
measures of shape
median (M or *m*)
median class
modal class
mode *(Mo or mo)*
multimodal frequency distribution

parameter
Pearson's coefficient of skewness *(Sk or sk)*
percentile
proportion (π or P)
quartiles
range
second quartile
skewness
standard deviation (σ or s)
standard scores *(z)*
statistic
third quartile
variance (σ^2 or s^2)
weighted mean (μ_w or $\overline{X}_w$)

Questions and Problems

The computer programs noted at the end of this chapter can be used to work many of the subsequent problems.

1. Consider the accompanying sample data on the 1981 research-and-development spending of fifteen U.S. firms.
  a. Calculate the mean, median, and mode for each of the listings in Table 4.A.

b. Would your answers change, if these were population data? If so, how?

2. Consider columns 5 and 8 of Table 2.A that accompanies problem 1 of Chapter 2. For each data set, calculate the sample mean, median, and mode.

Table 4.A   Fifteen U.S. Companies: Research-and-Development Spending in 1981

In Total Dollars (millions) (1)		In Percent of Sales (2)		In Dollars per Employee (3)	
1. General Motors ..	$2,250	1. Telesciences	22.1	1. Cray Research	$15,060
2. Ford Motor	1,718	2. Kulicke & Soffa.......	18.9	2. Amdahl	14,851
3. AT&T	1,686	3. Computer Consoles ..	17.8	3. Auto-Trol	
4. IBM.............	1,612	4. Auto-Trol		Technology..........	14,760
5. Boeing..........	844	Technology	17.2	4. Telesciences	11,130
6. General Electric..	814	5. Amdahl	17.0	5. Computer Consoles...	10,677
7. United		6. Cray Research	16.0	6. Applied Materials	9,722
Technologies.....	736	7. Floating Point		7. Intergraph............	9,393
8. Du Pont.........	631	Systems..............	15.3	8. Onyx & IMI	9,039
9. Exxon	630	8. Dysan...............	15.2	9. Apple Computer......	8,532
10. Eastman Kodak...	615	9. Intel	14.8	10. Merck...............	8,462
11. Xerox...........	526	10. Applied Materials.....	14.4	11. Floating Point	
12. ITT	503	11. Cordis	14.0	Systems	8,418
13. Dow Chemical ...	404	12. Intergraph	13.1	12. Boeing..............	8,357
14. Honeywell	369	13. Teradyne............	12.7	13. International Flavors &	
15. Hewlett-Packard ..	347	14. Genrad	12.6	Fragrances	8,297
		15. Anderson Jacobson ...	11.8	14. Cado Systems........	8,226
				15. Eli Lilly..............	8,210

Source: Business Week, July 5, 1982, p. 54.

3. Consider Table 3.A that accompanies problem 1 of Chapter 3. Calculate the sample mean, median, and mode.

4. Consider Table 3.B that accompanies problem 2 of Chapter 3. Calculate the sample mean, median, and mode.

5. Consider Table 3.C that accompanies problem 3 of Chapter 3.
   a. Calculate the population mean, median, and mode.
   b. Why does your answer for the mean differ from that given in the table (that lists an average of $23,546)?

6. Consider the absolute frequency distribution found in the answer to problem 11 of Chapter 3 at the back of the book (Table A.6). Calculate the sample mean, median, and mode. Show your calculations.

7. Consider the absolute frequency distribution found in the answer to problem 13 of Chapter 3 at the back of the book (Table A.7). Calculate the population mean, median, and mode.

8. Consider the absolute frequency distribution found in Table 4.B. Calculate the sample mean, median, and mode.

Table 4.B   Sample of Orders Received By Aircraft-Parts Supplier

Price Range	Absolute Class Frequency
$ 0 to under $50	200
$ 50 to under $100	700
$ 100 to under $200	1,953
$ 200 to under $500	2,311
$ 500 to under $1,000	582
$1,000 to under $10,000	320

9. Consider the absolute frequency distribution found in Table 4.C. Calculate the sample mean, median, and mode.

Table 4.C   Sample of Household Monthly Telephone Bills

Size of Bill	Absolute Class Frequency
$ 0 to under $10	11
$ 10 to under $30	67
$ 30 to under $50	129
$ 50 to under $75	732
$ 75 to under $150	82
$150 to under $500	11
$500 to under $1,000	3

10. Consider the absolute frequency distribution found in Table 4.D. Calculate the population mean, median, and mode.

Table 4.D   Age Distribution of an Aircraft Fleet

Age Class (months)	Absolute Class Frequency
11 to under 24	513
24 to under 48	1,607
48 to under 80	192
80 to under 150	33
150 to under 240	12
240 to under 480	2

11. Reconsider problem 7. Determine the mode graphically by drawing an appropriate histogram.

12. Reconsider problem 9. If you wanted to determine the mode graphically, what kind of calculations would you have to perform prior to drawing the appropriate histogram?

13. Invent a set of 10 figures on family incomes that are *simultaneously* consistent with each of the following statements: (i) On the average, the family income in my neighborhood equals $35,000 per year. (ii) On the average, the family income in my neighborhood equals $10,000 per year. (iii) On the average, the family income in my neighborhood equals $2,000 per year.

14. Invent a set of 10 figures on the tar content of cigarettes that are *simultaneously* consistent with each of the following statements: (i) Brand A cigarettes contain 40 percent less tar than the arithmetic average of all the other brands. (ii) Eighty percent of all brands of cigarettes contain less tar than Brand A.

15. In each case determine the measure of central tendency that is most appropriate: (i) The manager of a clothing store wants to order those sizes of different types of clothing that are most likely to promote low markup, high volume. (ii) An economist wants to compare the average family income in the 50 states. (iii) A manager wants to distribute a profit-sharing fund equally among all the workers. (iv) The personnel director of a firm argues that there are as many people with above-average intelligence as with below-average intelligence. (v) The personnel director's assistant argues that it is possible, in the presence of a single genius (or moron) for most people to have below-average (or above-average) intelligence. (vi) A reporter wants to know the price most commonly paid for houses.

Table 4.E

	Labor Force (millions)	Unemployment Rate (percent)
Both sexes, 16–19 years	8.62	21.5
Males, 20 years and over	57.51	7.9
Females, 20 years and over	42.91	7.4

Source: Economic Report of the President (Washington, D.C.: U.S. Government Printing Office, 1982), pp. 268–69.

Table 4.F

	Stockholders' Equity (billions of dollars)	Profit as Percent of Equity (percent)
Royal Dutch Shell	25.2	14.4
British Petroleum	14.8	13.9
ENI	2.2	17.5
Unilever	4.8	16.5
Française des Pétroles	3.7	4.7

Source: Fortune, August 23, 1982, p. 183.

(vii) At a New York conference of 20 scientists (of whom 18 are from New York and one each is from Sweden and Japan), someone wants to determine the average distance of the conference site from the participants' homes. (viii) A manager who wants to calculate the total of accounts receivable asks for the number of accounts and the average balance.

16. Consider Table 2.A that accompanies problem 1 of Chapter 2. Calculate a weighted mean of the col. 8 data, using the col. 5 data as weights. Why does your answer differ from the answer in problem 2 of these questions and problems?

17. The U.S. labor-force and unemployment data listed in Table 4.E were reported for December, 1981. Calculate the weighted average unemployment rate for the entire U.S. labor force; compare it with an unweighted arithmetic average. Show your calculations.

18. In 1981, the five largest industrial corporations outside the United States reported the data listed in Table 4.F. Calculate the weighted average profit rate; compare it with the unweighted average. Show your calculations.

19. From the data in Table 4.G, calculate the weighted mean rate of return from public electric utilities in Canada between 1967 and 1981. Compare your answer with an unweighted mean.

Table 4.G

Year	Rate of Return (percent)	Total Invested Capital[1] ($ millions)
1967	3.66	12,081.67
1968	3.71	13,025.41
1969	3.84	14,390.25
1970	3.74	16,406.42
1971	3.62	18,266.16
1972	3.51	20,057.25
1973	3.68	23,027.77
1974	3.19	30,277.61
1975	2.32	39,361.00
1976	3.00	46,523.68
1977	3.01	52,066.63
1978	3.94	62,785.06
1979	3.95	74,715.24
1980	3.69	85,349.86
1981	3.79	99,643.62

Source: Glenn P. Jenkins, "Public Utility Finance and Economic Waste," *Canadian Journal of Economics,* August 1985, p. 493.

Table 4.H

Country	Percent Unemployed	Civilian Labor Force (millions)
Australia	7.1	6.9
Canada	11.0	11.9
France	8.5	22.9
Great Britain	12.3	25.7
Italy	4.8	21.4
Japan	2.4	57.0
Sweden	3.1	4.4
United States	9.7	110.2
West Germany	5.9	26.6

Source: Statistical Abstract of the United States 1984, p. 872.

20. From the data in Table 4.H calculate the weighted mean unemployment rate of the nine listed countries in 1982. Compare your answer with an unweighted mean.

21. Consider the answer at the back of this book for Chapter 3's problem 1. Then determine
   a. the range of the data.
   b. the first and third quartiles.
   c. the second and seventh deciles.
   Finally, determine *graphically*
   d. the median and the 87th percentile. [For this purpose, convert the column 3 relative frequency distribution of Chapter 3's answer to problem 11 into a *cumulative* relative frequency distribution and plot an appropriate ogive.]

22. Consider columns 5 and 8 of Table 2.A that accompanies problem 1 of Chapter 2. For each data set, calculate
   a. the range.
   b. the interquartile range.

23. Consider Table 4.A. For each of the three data sets, calculate
   a. the range.
   b. the interquartile range.

24. Calculate (a) the range and (b) the interquartile range for the data sets in Chapter 3's Tables 3.B and 3.C.

25. Calculate (a) the range and (b) the interquartile range for the percentage data sets in Tables 4.F, 4.G, and 4.H.

26. Consider the first and third of the three sets of sample data in Table 4.A and determine in each case
   a. the mean absolute deviation.
   b. the variance.
   c. the standard deviation.

27. Consider the second of the three sets of figures in Table 4.A. Treat it as a sample and determine
   a. the mean absolute deviation.
   b. the variance (long and short method).
   c. the standard deviation.
   In each case, show your calculations.

28. Consider columns 5 and 8 of Table 2.A that accompanies problem 1 of Chapter 2. For each of these sets of sample data, determine
   a. the mean absolute deviation.
   b. the variance.
   c. the standard deviation.

29. Consider the data sets in Chapter 3's Tables 3.A, 3.B, and 3.C. In each case, determine
   a. the mean absolute deviation.
   b. the variance.
   c. the standard deviation.

30. Consider the percentage data sets in Tables 4.F, 4.G, and 4.H. In each case, determine
   a. the mean absolute deviation.
   b. the variance.
   c. the standard deviation.

31. Consider the answer at the back of the book to Chapter 3's problem 11 (Table A.6). From the absolute frequency distribution shown there, esti-

mate the variance and standard deviation, treating the data as sample data and using the shortcut method. Show your calculations.

32. Consider the answer at the back of the book to Chapter 3's problem 13 (Table A.7). From the absolute frequency distribution shown there, estimate
a. the mean absolute deviation.
b. the population variance.
c. the population standard deviation.
Would any of your answers change if these were sample data? If so, how?

33. Consider the sample data in Table 4.B. Use them to estimate
a. the mean absolute deviation.
b. the variance.
c. the standard deviation.
Would any of your answers change if these were population data? If so, how?

34. Consider the sample data in Table 4.C. Use them to estimate
a. the mean absolute deviation.
b. the variance.
c. the standard deviation.
Would any of your answers change if these were population data? If so, how?

35. Consider the population data in Table 4.D. Use them to estimate
a. the mean absolute deviation.
b. the variance.
c. the standard deviation.
Would any of your answers change if these were sample data? If so, how?

36. A computer firm offers a paid leave of absence to its engineers who wish to get an advanced degree. However, applicants are tested and their aptitude test scores must indicate a superb chance of success (as defined by z-scores of $+2$ or better). In one test, applicants A through F earned raw scores of 500, 631, 760, 438, 598, and 720. The mean score of all 200 applicants was 520, the standard deviation was 60. Who among the six will go back to school?

37. The quality controller of a vegetable-canning firm is constructing a control chart. The production process fills cans of peas with an average of $\mu = 16.02763$ ounces, the standard deviation being $\sigma = .3168132$ ounces. Any can filled with $\mu \pm 2\sigma$ is considered acceptable. A sample reveals net weights of 15.33, 15.53, 15.77, 15.94, 16.05, 16.24, 16.43, 16.63, 16.88. Is the production process "out of control"?

38. Make a list of 10 numbers such that each number is a 1 or a 9, while the standard deviation of the set is as small (or as large) as possible.

39. The manager of a firm announced: "The light bulbs produced by our firm have a mean lifetime of 1,500 hours, with a standard deviation of 150 hours; thus, it follows that almost all of our light bulbs last between 1,200 and 1,800 hours." What was the manager assuming? What if the assumption did not hold?

40. Consider Table 3.K in Chapter 3's problem 35. In which of the two years did the sample of retail businesses show the greater variability of sales?

41. Consider again the three alternative listings of the top R&D spenders given in Table 4.A. By which of the three criteria does R&D spending show the greatest (or smallest) variability? Show your calculations.

42. Consider Table 3.V in Chapter 3's problem 60. In which of the two towns does the census reveal the greater variability of salaries?

43. In 1982, the thirty-three public-school districts of New York City reported the mathematics and reading scores given in Table 4.I. For which of the two subjects do these census data reveal the greater variability?

Table 4.I

Mathematics Scores		Reading Scores	
44.0	57.9	37.6	60.5
69.9	50.8	57.0	39.7
48.0	63.5	42.7	59.0
41.8	62.9	48.5	57.3
46.3	67.6	42.4	63.8
43.7	38.9	32.8	38.5
43.3	61.3	34.2	52.7
46.3	78.7	43.1	71.0
41.2	81.4	34.4	79.7
46.1	60.2	37.9	54.5
54.2	63.7	56.3	58.5
49.9	63.6	37.4	59.9
50.3	67.0	47.5	57.9
55.3	75.9	41.1	74.1
54.4	44.9	46.1	39.1
43.9	65.9	38.1	74.3
44.2		42.5	

Source: The New York Times, June 20, 1982, p. 38.

44. Over a ten-year period, the earnings per share for two stocks, A and B, were as shown in Table 4.J. For which of the two stocks do these sample data reveal the greater variability?

45. Find the coefficient of skewness and of kurtosis for the list 2 sample data in Table 4.A. Show your calculations.

46. Consider the sample data in lists 1 and 3 of Table 4.A. For each determine
  a. the coefficient of skewness.
  b. the coefficient of kurtosis.

47. Consider the sample data in columns 5 and 8 of Table 2.A that accompanies problem 1 of Chapter 2. For each data set, determine
  a. the coefficient of skewness.
  b. the coefficient of kurtosis.

48. Consider the sample data in Chapter 3's Tables 3.A and 3.K (1982 and 1983). For each of the three data sets, determine
  a. the coefficient of skewness.
  b. the coefficient of kurtosis.

49. Consider the population data in Chapter 3's Tables 3.C and 3.V (town A and town B), as well as in this chapter's Table 4.I (math and reading). For each of the five data sets, determine
  a. the coefficient of skewness.
  b. the coefficient of kurtosis.

50. Consider the absolute frequency distribution in Table A.7 (which helps answer Chapter 3's problem 13). Use it to determine the coefficient of skewness.

51. Consider the absolute frequency distribution in Table A.6 (which helps answer Chapter 3's problem 11). Use it to determine

  a. the coefficient of skewness.
  b. the coefficient of kurtosis.
  Show your calculations.

52. Consider the absolute frequency distribution in Table 4.B. Use it to determine the coefficient of skewness.

53. Consider the absolute frequency distribution in Table 4.C. Use it to determine the coefficient of skewness.

54. Consider the absolute frequency distribution in Table 4.D. Use it to determine the coefficient of skewness.

55. An inspector of incoming materials is supposed to reject all shipments containing a proportion of defective items in excess of .04. Which of the shipments in Table 4.K would she reject?

56. Consider Analytical Example 3.1, "Deciphering Secret Codes." Determine the proportions of vowels and consonants in 200 letters of a normal English language text.

57. Consider Table 2.A that accompanies Chapter 2's problem 1. What proportions of the sampled mutual funds pursued a portfolio strategy of "aggressive growth" or "small company growth"?

58. Consider the frequency curve in Figure A. What is the *minimum* proportion of observations falling in the crosshatched area? In the dotted area?

59. Reconsider the ordered array of Table A.2 that constitutes the answer to Chapter 3's Problem (1), along with your answers to problems 3 and 29 of the current chapter. What percentage of the data fall within ± 2.5 standard deviations of the mean? Is this number consistent with Chebyshev's theorem?

Table 4.J

Stock A:	$0.75,	0.80,	1.00,	1.20,	1.40,	1.00,	1.50,	1.70,	0.80,	1.80
Stock B:	$6.30,	5.90,	7.60,	8.50,	6.90,	7.30,	5.10,	9.30,	7.85,	6.30

Table 4.K

Shipment	Number of Items Sampled	Number of Items Found Defective
A	500	5
B	1,000	37
C	200	3
D	750	31
E	900	47

Figure A

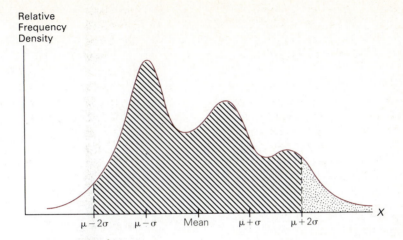

60. An airline announces that the mean time loss to passengers from transfers between planes or flight delays equals 16.5 minutes per trip, with a standard deviation of 3.5 minutes. If the frequency distribution of time losses fit a normal curve, what range of time losses would fit 95.4 percent of all delays? What if the frequency distribution did not fit a normal curve? Finally, in light of the foregoing, how confident would you be in making a connecting flight that left you only 33 minutes for transferring between planes?

Selected Readings

Campbell, Stephen K. *Flaws and Fallacies in Statistical Thinking.* Englewood Cliffs, N.J.: Prentice-Hall. 1974. An excellent and amusing discussion aimed at helping nonstatisticians recognize (intentional and unintentional) abuses of statistical tools and gain the ability to judge the quality of statistical evidence.

Computer Programs

The DAYSAL-2 personal-computer diskettes that accompany this book contain three programs of interest to this chapter.

3. Frequency Distributions from Raw Data calculates all major summary measures as a by-product of setting up frequency distributions.

4. Summary Measures from Grouped Data takes an absolute frequency distribution that you supply and computes all major summary measures, along with other types of frequency distributions.

5. Summary Measures from Raw Data calculates all major summary measures of central tendency, dispersion, and shape.

PART TWO

Analytical Statistics: The Foundations of Inference

CHAPTER FIVE

The Theory of Probability

Every day, each of us makes decisions the outcomes of which lie in the realm of the unknown, but we do not make them blindly. Figuring the *chances* of various possible outcomes helps us make the right choices. Without even knowing it, we rely on a special kind of calculus—"the calculus of the likelihood of specific occurrences"—which is what the **theory of probability** has been called. Consider how we cancel the planned picnic or move the graduation indoors when the chance of rain seems too high; how we proceed with the wedding or introduce the new product when success seems more likely than failure. Wherever we turn, we are weighing the chances. How likely is it, we ask, that the price hike will reduce sales, that the new process will raise productivity, that the merger will raise profit, that the inspection will turn up defective parts, that the tax cut will put an end to the recession? What are the chances, we ask, for life on other planets, for nuclear war in this century, for our candidate's election, for this horse to win, for this item to be on the quiz? The list goes on.

As we noted in Chapter 1, statistics as a discipline can be viewed as a guide to the unknown. Is it any wonder that all of analytical statistics is built upon the theory of probability? As the Latin origin of the word suggests (*probabilis* means "likely" or "like truth"), the theory helps us find the likely truth when we cannot know it with certainty. But we must be patient. The content of the current chapter may seem to be remote from business and economics applications; yet it will turn out to be the very foundation for the development, in later chapters, of a host of statistical techniques that are routinely and successfully used to solve problems arising in these very fields.

BASIC CONCEPTS

Probability theorists employ a number of important concepts; crucial among them are (1) the *random experiment,* (2) the *sample space,* and (3) *random events.* We discuss each of these concepts in turn.

The Random Experiment

Any activity that will result in one and only one of several well-defined outcomes but that does not allow us to tell in advance which of these will prevail in any particular instance is called a **random experiment.** The tossing of a coin, for example, is such an experiment. Assuming that tosses with such unlikely outcomes as the coin landing on its edge or disappearing in a hole do not count, the *type* of outcome is known with certainty: it must be heads or tails. Yet the *actual* outcome associated with a given toss is bound to vary from one trial to the next and, therefore, cannot be predicted with certainty. The rolling of a die is another such random experiment; so is the drawing of a card from a deck of cards or the spinning of the wheel in the game of roulette. These examples come most easily to mind because probability was first studied some 300 years ago in connection with gambling. However, the concept of probability is just as applicable to matters outside the gambling situation. Pulling selected items from the production line for inspection or selecting invoices for an audit represent other examples of random experiments (with such possible alternative outcomes as defective or satisfactory in the first case and correct or erroneous in the other). Even investing a billion dollars in a new type of car can be viewed as a random experiment (leading to profit, loss, or even bankruptcy), and so can the placing of a coin into a vending machine (leading to the desired coffee, to money returned, to an infuriating neither, or even to a quite unwelcome chicken soup).

The Sample Space

Any one of a random experiment's possible outcomes, the occurrence of which rules out the occurrence of all the alternative outcomes, is called a **basic outcome.** In turn, a listing of all the basic outcomes of a random experiment is generally referred to as the **sample space,** although other terms, such as **outcome space** and **probability space,** are occasionally used as well. The basic outcomes that constitute the sample space can be represented symbolically, pictorially, and in other ways still; let us consider some of them.

Tossing a single coin once. The single toss of a single coin involves a sample space *(S)* of two basic outcomes only, heads (H) and tails (T). This sample space might be represented symbolically as

$$S = \{H, T\}.$$

This sample space can also be shown pictorially, as in panel (a) of Figure 5.1. In any single toss, one (and only one) of the two outcomes shown must occur.

Rolling a single die once. The single roll of a single die involves a slightly larger sample space because six basic outcomes exist. Symbolically,

$$S = \{1, 2, 3, 4, 5, 6\}.$$

This sample space is shown pictorially in panel (b) of Figure 5.1.

Drawing a single card once. Larger still is the sample space describing the possible consequences of randomly selecting a single card from a standard deck of 52 playing cards. Such a deck has four colored *suits*—namely, black clubs, red diamonds, red

Figure 5.1   Sample Spaces for Simple Experiments

The sample spaces of simple random experiments can be represented pictorially.

(a) Tossing a Single Coin Once

(b) Rolling a Single Die Once

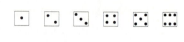

(c) Drawing a Single Card Once

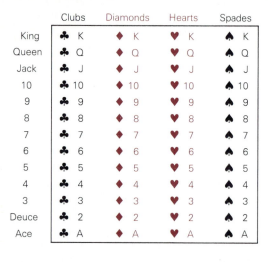

	Clubs	Diamonds	Hearts	Spades
King	♣ K	♦ K	♥ K	♠ K
Queen	♣ Q	♦ Q	♥ Q	♠ Q
Jack	♣ J	♦ J	♥ J	♠ J
10	♣ 10	♦ 10	♥ 10	♠ 10
9	♣ 9	♦ 9	♥ 9	♠ 9
8	♣ 8	♦ 8	♥ 8	♠ 8
7	♣ 7	♦ 7	♥ 7	♠ 7
6	♣ 6	♦ 6	♥ 6	♠ 6
5	♣ 5	♦ 5	♥ 5	♠ 5
4	♣ 4	♦ 4	♥ 4	♠ 4
3	♣ 3	♦ 3	♥ 3	♠ 3
Deuce	♣ 2	♦ 2	♥ 2	♠ 2
Ace	♣ A	♦ A	♥ A	♠ A

(d) Spinning a Roulette Wheel Once

---

hearts, and black spades; each suit has 13 *denominations,* from ace to 10 plus three face cards (jack, queen, and king). Symbolically,

$S = \{$ace of clubs, deuce of clubs, 3 of clubs, . . ., king of spades$\}$.

Pictorially, we can represent the same set of alternative outcomes as in panel (c) of Figure 5.1.

Spinning a roulette wheel once. Also large is the set of basic outcomes from spinning a roulette wheel. The Monte Carlo wheel that is shown in panel (d) of Figure 5.1 has 37 numbers, ranging from 0 to 36. Apart from the zero, half of them are black, half red. Symbolically,

$S = \{$red 0, red 1, black 2, . . ., red 36$\}$.

Note: The Monte Carlo wheel shown here should not be confused with the American wheel that adds a double zero or even a triple zero to the numbers shown (a version of which appears in Question 1 at the end of this chapter).

Tossing a single coin twice. More complicated experiments, such as tossing a coin twice or tossing two coins at a time, can be depicted similarly, although various other ways are often preferred. Among the favorite devices for defining the sample space of random experiments involving multiple steps is the **tree diagram,** so called because it reminds one of the branches of a tree. Consider panel (a) of Figure 5.2. The tree diagram shown there pictures the two possible outcomes of the first toss of a coin: heads (H_1) or tails (T_1), where the subscripts denote the number of the toss. For either contingency, two further possibilities exist on the second toss: H_2 or T_2. Thus, one can

Figure 5.2   Sample Spaces for Multiple-Step Experiments

Sample spaces for multiple-step experiments are typically represented by the devices shown here.

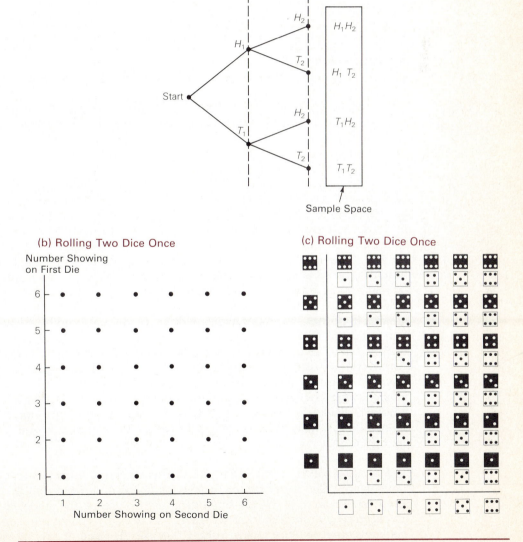

(a) Tossing a Coin Twice

(b) Rolling Two Dice Once

(c) Rolling Two Dice Once

follow the various paths along the branches from the starting position prior to the first toss and trace out the four basic outcomes of the entire two-stage experiment, and these outcomes are conveniently summarized in the box to the right of the tree diagram. Note how the identical diagram could be used to identify the possible consequences of tossing two coins at once. One would only have to replace "First Toss" by "First Coin" and "Second Toss" by "Second Coin" to see the perfect analogy. Naturally, more complex trees can be drawn, too, in order to find, say, the sample space associated with tossing a coin three times or even with tossing it more often than that.

Rolling two dice once. Finally, consider panels (b) and (c) of Figure 5.2. They depict other ways still of determining the sample space in multiple-step experiments. In this case, both panels tell us immediately that there are 36 basic outcomes of an experiment that rolls two dice at once (or, for that matter, that rolls a single die twice).

Random Events and Their Relationships

We have now seen that any random experiment will yield one of two or more possible outcomes and that these basic outcomes, as a group, make up the sample space. Any subset of this sample space is called a **random event,** of which there are two types.

Simple vs. composite events. Any single one of a random experiment's basic outcomes is called a **simple event.** Thus, the sample space for experiment (a) in Figure 5.1 consists of two simple events: the appearance of a head and the appearance of a tail. The sample spaces for experiments (b) through (d) are, similarly, made up of 6, 52, and 37 simple events, respectively.

Finally, in Figure 5.2, we find four simple events in panel (a) and 36 simple events each in panels (b) and (c), although each of these events (unlike those in Figure 5.1) inextricably links together two elements, such as "head on the first toss" with "head on the second toss" or "6 on the black die" with "1 on the white die." Some people distinguish basic outcomes or simple events that are *univariate* (such as those in Figure 5.1 where each outcome is defined by the appearance of a single element) from those that are *bivariate* (such as those in Figure 5.2 where each outcome is defined by the joint appearance of two elements), or, perhaps, even *multivariate* (not shown). On occasion, basic outcomes or simple events are also referred to as **elementary events,** or **sample points**—the latter name, perhaps, being suggested by the type of graphical representation of the sample space shown in panel (b) of Figure 5.2.

It is often useful, however, to consider subsets of the sample space that contain more than one basic outcome or simple event. Any such combination of two or more basic outcomes is called a **composite event.** Consider the event "getting an even number" when rolling a single die once. Consider the event "getting a face card" when drawing a single card from a deck. Consider, finally, "getting a sum of 10 or more" when rolling two dice. Such composite events are pictured by the shaded areas in Figure 5.3.

Mutually exclusive events. Different random events that have no basic outcomes in common are called **mutually exclusive,** or **disjoint,** or **incompatible events.** Such events cannot occur at the same time; the occurrence of one automatically precludes the occurrence of the other. As an example, consider "getting an even number" and "getting an odd number" when rolling a single die once. If one of these composite

Figure 5.3 Composite Events

The shaded subsets of the three boxed sample spaces shown here illustrate the nature of composite events. As the shaded portion of panel (a) shows, "getting an even number" when rolling a single die once is a composite event that consists of three basic outcomes or simple events (getting a 2, a 4, or a 6). The shaded portion of panel (b) shows, similarly, that "getting a face card" when drawing a single card from a deck is a composite event that consists of 12 basic outcomes or simple events (getting a jack, a queen, or a king from any one of the four suits). Finally, panel (c) tells us that "getting a sum of 10 or more" on a single roll of two dice is a composite event that consists of six simple (but bivariate) events; namely, those depicted by the six combinations of two faces of dice in the shaded area.

(a) Getting an Even Number

(b) Getting a Face Card

	Clubs	Diamonds	Hearts	Spades
King	♣ K	♦ K	♥ K	♠ K
Queen	♣ Q	♦ Q	♥ Q	♠ Q
Jack	♣ J	♦ J	♥ J	♠ J
10	♣ 10	♦ 10	♥ 10	♠ 10
9	♣ 9	♦ 9	♥ 9	♠ 9
8	♣ 8	♦ 8	♥ 8	♠ 8
7	♣ 7	♦ 7	♥ 7	♠ 7
6	♣ 6	♦ 6	♥ 6	♠ 6
5	♣ 5	♦ 5	♥ 5	♠ 5
4	♣ 4	♦ 4	♥ 4	♠ 4
3	♣ 3	♦ 3	♥ 3	♠ 3
Deuce	♣ 2	♦ 2	♥ 2	♠ 2
Ace	♣ A	♦ A	♥ A	♠ A

(c) Getting a Sum of 10 or More

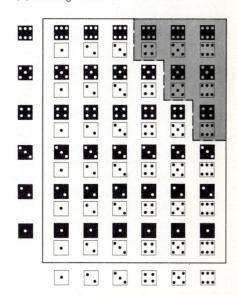

---

events occurs, the other one cannot possibly occur. The event in question need not, however, be a composite one. The same random experiment can yield *simple* events that are mutually exclusive as well, such as "getting a 6" (when rolling a single die) and "getting a 4." Equally incompatible is "getting the king of hearts" and "getting the ace of hearts" when drawing a single card from a deck or "getting a defective part" and "getting a perfect part" when pulling an item off the production line. In all these cases, it is one *or* the other.

As another look at Figure 5.3 can confirm, mutually exclusive events always occupy entirely different parts of the sample space. Note how, in panel (a), the even numbers are found in some areas (the shaded ones) and the odd ones are found in other areas (the unshaded ones). Equally distinct parts of the sample space are occupied by the six face and the four face in panel (a), or by the king of hearts and the ace of hearts in panel (b).

Collectively exhaustive events. Different random events that jointly contain all the basic outcomes in the sample space are called **collectively exhaustive events.** When the appropriate random experiment is conducted, one of these events is bound to occur. Again consider "getting an even number" and "getting an odd number" when rolling a single die once. These two events, as we saw, are mutually exclusive, but they are collectively exhaustive as well. Although they cannot occur at the same time, one of them is bound to occur. This characteristic is shown visually in panel (a) of Figure 5.3 by the fact that the shaded areas (even numbers) and the unshaded areas (odd numbers) jointly cover the entire sample space represented by the box.

But note: Not every set of mutually exclusive events is collectively exhaustive. In the same experiment, the events "getting a six" and "getting a four" are mutually exclusive, but they hardly exhaust the sample space. On the other hand, not every set of collectively exhaustive events need consist of mutually exclusive ones, either. Consider the three events of "getting a six," "getting an even number," and "getting an odd number." In our experiment, these events are surely collectively exhaustive; they cover the entire sample space in panel (a). Yet these events are not mutually exclusive because "getting a six" is a simple event that is already included in the composite event of "getting an even number."

Complementary events. There exists one relationship, however, that always describes events that are both mutually exclusive and collectively exhaustive at the same time. Two random events such that precisely all those basic outcomes that are not contained in one are contained in the other are called **complementary events.** Once more, Figure 5.3 provides perfect examples if we contrast, in each panel, the event pictured by the shaded area with that described by the unshaded one. Thus, in panel (a), one event is "getting an even number"; its complement of "not getting an even number" is, of course, equivalent to "getting an odd number." Together, these two mutually exclusive events exhaust the sample space. Similarly, in panel (b), one event is "getting a face card" (shaded area); its complement is "not getting a face card" (unshaded area). Finally, in panel (c), one event is "getting a sum of 10 or more"; its complement is "not getting a sum of 10 or more" or "getting a sum of 9 or less." Again, the two mutually exclusive events precisely exhaust the sample space.

Compatible events. Just as there are incompatible events that have no basic outcomes in common, there are **compatible events** that have at least some basic outcomes in common. Such events can occur at the same time; the occurrence of one does not rule out the occurrence of the other. The two events, "getting a face card" and "getting a heart," provide an example. This situation is depicted in Figure 5.4. The shaded area in panel (a) depicts the 12 simple events associated with the composite event of "getting a face card"; the shaded area in panel (b) pictures the 13 simple events described by the composite event "getting a heart."

Sometimes one is interested in the occurrence of *either* of two such composite events; for this purpose one defines the **union of two events** as all the basic outcomes contained in either one random event *or* another or, conceivably, in both. Such a union of two events is illustrated by the shaded area in panel (c).

On other occasions, one might be interested only in those occurrences that two composite events have in common—in our example, all the possibilities of getting a

Figure 5.4 Union and Intersection

Two composite events, such as those pictured by the shaded areas in panels (a) and (b), can be considered together either by focusing on their union—panel (c)—or on their intersection—panel (d). The union contains all the simple events found in *either one* of the composite events—found, that is, in one of the composite events only *or* in the other only *or* in both at the same time. The key word describing such a situation is *or*. In contrast, the intersection contains only the simple events found in *both* of the composite events—found, that is, in one of the composite events *and* also in the other. The key word for this situation is *and*.

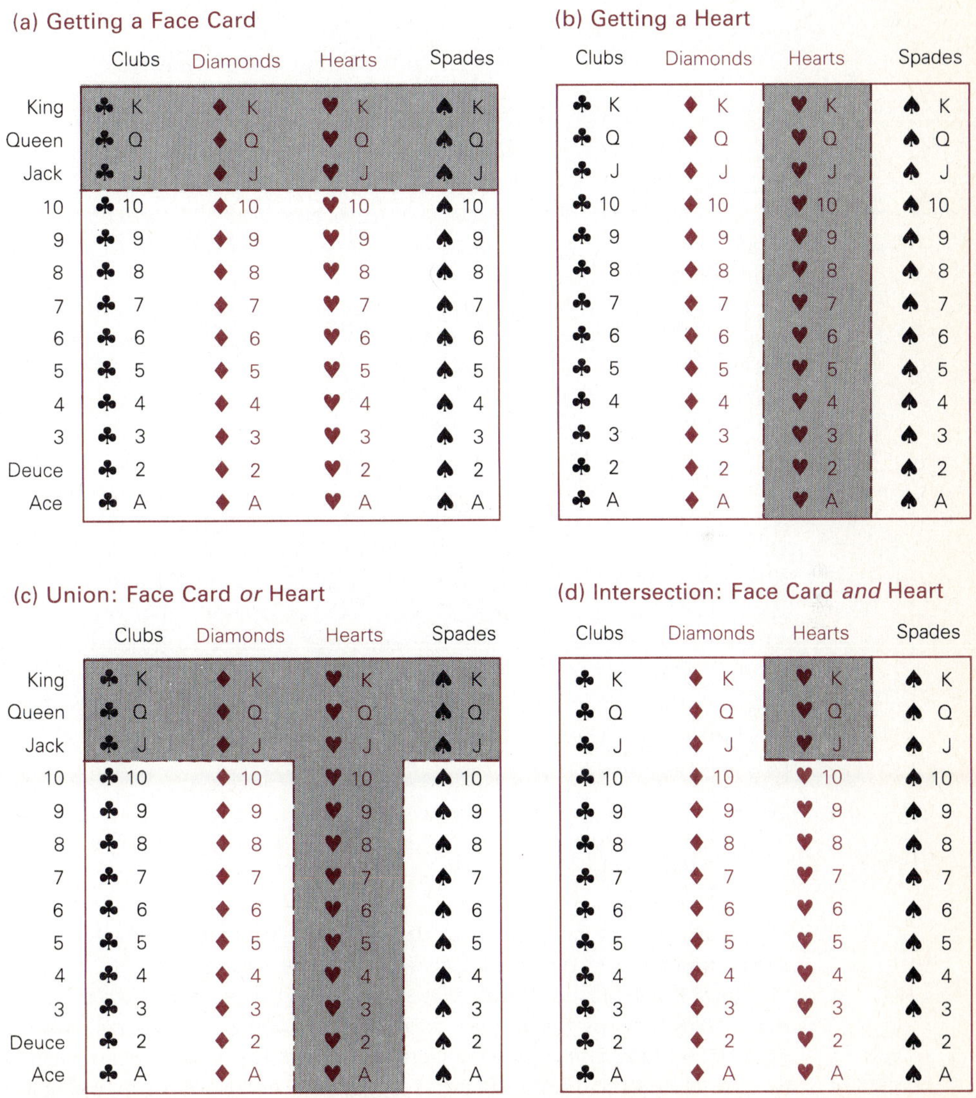

(a) Getting a Face Card

(b) Getting a Heart

(c) Union: Face Card *or* Heart

(d) Intersection: Face Card *and* Heart

face card and heart simultaneously. For this purpose, one defines the **intersection of two events** as all the basic outcomes simultaneously contained in both one random event *and* another. Such an intersection is illustrated by the shaded area in panel (d).

Summary. All the foregoing event relationships can be nicely summarized with the help of **Venn diagrams,** which are graphical devices, such as those in Figure 5.5, that depict sample spaces and random events symbolically (and that are named after their originator, John Venn). In each panel, the box represents the sample space containing all the basic outcomes of a random experiment. Particular events are depicted by circles or sections within the sample space. Thus, in panel (a), events A and B are mutually exclusive because, when performing the random experiment of "running the economy for a year," it is impossible to find inflation simultaneously at 0–5 percent and also at 10–15 percent. On the other hand, these two events are not collectively exhaustive; other outcomes, such as inflation above 5 but below 10 percent or inflation above 15 percent, are possible also.

Panel (b), on the other hand, depicts a complete *partition* of the sample space. When performing the random experiment of "running a firm for a year," one of the four noted events is bound to occur.

Panel (c) depicts the sample space for the random experiment of "checking the quality of items produced." Only two events are possible; they are complementary. Such events will be described symbolically as A vs. *not* A, or A vs. $\overline{A}$, in this book; another symbol traditionally used for the complement of A is A'.

Panels (d) and (e) once more relate to the experiment of "running the economy for a year." Note that the two events cited do not exhaust all the possibilities; therefore, they do not cover all of the sample space. On the other hand, the two events may or may not happen simultaneously. To the extent that they do, the two circles overlap. Thus, in panel (d), any outcome that fits *either* event A *or* B (and, perhaps, even both) is found in the crosshatched area, and such a *union* of the two events will be described symbolically as A *or* B in this book. (Other writers also use the designation A ∪ B, where ∪ stands for "union.") Finally, the smaller number of outcomes that fit *both* event A *and* B is found in the smaller crosshatched area of overlap in panel (e), and such an intersection of the two events will be described symbolically as A *and* B throughout this book. (Other writers also use the designation A ∩ B, where ∩ stands for "intersection.")

THE MEANING OF PROBABILITY

There are two types of random experiments, those that can be repeated over and over again under (essentially) identical conditions and those that are unique and cannot be repeated because the conditions surrounding them are forever changing. Correspondingly, there are two types of probability numbers, although both seek to measure the degree of likelihood that a particular event will occur. A numerical measure of chance that estimates the likelihood of a specific occurrence in a repeatable random experiment is called an **objective probability;** a corresponding value for a unique random experiment is called a **subjective probability.** We will discuss each of these in turn.

Figure 5.5   Venn Diagrams

Venn diagrams depict symbolically various relationships among random events.

(a) **Mutually Exclusive, but Not Collectively Exhaustive Events**

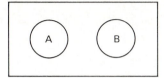

Examples:

Event A = Inflation between 0 and 5% per year
Event B = Inflation between 10 and 15% per year

(b) **Mutually Exclusive and Collectively Exhaustive Events**

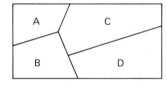

Event A: Loss
Event B: Profit between $0 and under $10 million
Event C: Profit between $10 million and under $100 million
Event D: Profit of $100 million or more

(c) **Complementary Events: A vs. *Not* A**

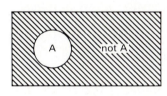

Event A: Defective items
Event not-A: Satisfactory items

(d) **Union of Two Events: A *or* B**

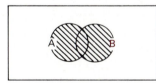

Event A: Inflation between 0 and 5% per year
Event B: Unemployment between 5 and 10% of the labor force

(e) **Intersection of Two Events: A *and* B**

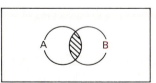

Event A: Inflation between 0 and 5% per year
Event B: Unemployment between 5 and 10% of the labor force

Objective Probability

An objective probability value can be determined in one of two ways: theoretically, by conducting repeated *thought* experiments, or empirically, by conducting repeated *actual* experiments.

The theoretical approach. The theoretical approach to determining probability is also called the *classical* approach because it was used by the originators of probability theory, such as Pierre Simon de Laplace.[1] This approach relies entirely on abstract reasoning; it does not resort to performing actual experiments because logic is deemed sufficient for providing all the answers. As Laplace saw it, if a hypothetical experiment could have N equally likely basic outcomes, and if n of these outcomes were favorable to event A, then the probability of this event, $p(A)$, equaled the ratio of n to N.

The application of this definition inevitably produces a number between 0 and 1, as Figure 5.6 illustrates. But note: while probability will usually be measured in this fashion throughout this book, there are alternative ways of expressing it—for example, in terms of percentages, chances, or odds. Thus, a probability of 7/10 or .7 might be expressed as a probability of 70 percent or as 70 chances in 100 or as odds of 70:30 for or even as odds of 30:70 against.

The classical formula for measuring probability is easily applied to games of chance, such as those pictured in Figures 5.1 to 5.3. Regardless of whether the event is a simple or composite one, the juxtaposition of the number of favorable basic out-comes to that of all possible basic outcomes immediately produces a probability value. Consider Figure 5.1. The probability of getting a head when tossing a fair coin is easily seen as $p(H) = 1/2$; that of getting a 6 when rolling a balanced die is $p(6) = 1/6$; and that of getting the king of hearts when randomly drawing a card from a well-shuffled deck is $p(\text{king of hearts}) = 1/52$, always by reference to the Box 5.A formula. Similarly, the probability of getting a 15 when drawing a card is

$$p(15) = \frac{0}{52} = 0$$

(because there is no such card), while that of getting a black card is

$$p(\text{black}) = \frac{26}{52} = \frac{1}{2}$$

In contrast, the probability of getting black when spinning the Monte Carlo wheel is

$$p(\text{black}) = \frac{18}{37}, \text{ or somewhat less than } \frac{1}{2}.$$

5.A The Probability of Event A: Classical Approach

$$p(A) = \frac{\text{number of equally likely basic outcomes favorable to the occurrence of event A}}{\text{number of equally likely basic outcomes possible}} = \frac{n}{N}$$

---

[1]A biographical sketch of Laplace (1749–1847) is contained in Chapter 5 of the *Student Workbook* that accompanies this text.

Figure 5.6 Measuring Probability

Probability as defined in Box 5.A is inevitably measured on a scale from 0 to 1. Because there can be no negative basic outcomes of a random experiment, the lowest possible number of basic outcomes favorable to an event is zero, and such an occurrence would make the probability of that event zero as well. [As a look at Box 5.A will confirm, when $n = 0$ and N is positive, $p(A) = (n/N)$ must be zero.] On the other hand, the highest possible number of basic outcomes favorable to an event equals the total number of basic outcomes that are possible, and such an occurrence would make the probability of that event equal to one. [As Box 5.A confirms, when $n = N$, their ratio $p(A) = (n/N)$ must equal one.]

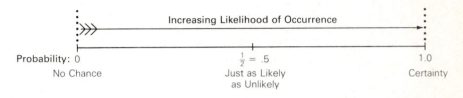

The same procedure of contrasting favorable basic outcomes with all possible ones is applied just as easily to the slightly more complex games of Figures 5.2 to 5.4.

You should be able to see from Figure 5.2 that getting two heads when tossing a coin twice carries a probability of

$$p(2H) = \frac{1}{4}$$

that getting two heads *or* two tails in the same game has a probability of

$$p(2H \text{ } or \text{ } 2T) = \frac{2}{4} = \frac{1}{2}$$

and that getting two fives when rolling two dice has a probability of

$$p(2 \text{ fives}) = \frac{1}{36}.$$

You should be able to confirm from Figure 5.3 that getting an even number when rolling a die once carries a probability of

$$p(\text{even number}) = \frac{3}{6} = \frac{1}{2}$$

that getting a face card when drawing one card has a probability of

$$p(\text{face card}) = \frac{12}{52} = \frac{3}{13}$$

and that getting a sum of 10 or more when rolling two dice once has a probability of

$$p(\text{sum of 10 or more}) = \frac{6}{36} = \frac{1}{6}.$$

Finally, it should be evident from Figure 5.4 that getting a face card *or* a heart carries a probability of

$$p(\text{face card } or \text{ heart}) = \frac{22}{52} = \frac{11}{26}$$

while getting a face card *and* a heart has a probability of only

$$p(\text{face card } and \text{ heart}) = \frac{3}{52}.$$

How easy! The classical approach allows the determination of probability values without ever *actually* tossing a coin, rolling a die, and the like. Yet there are those who have trouble with this definition of probability: the definition seems circular, for by specifying *equally likely* basic outcomes it employs the notion of probability in order to define probability! In addition, there exist repeatable random experiments (such as tossing a biased coin, rolling a loaded die, or sampling the quality of items coming off the production line) for the various possible outcomes of which no amount of arm-chair reasoning can ever discover probability values. In such cases, numerical assignments require *experience*.

The empirical approach. In order to counter the criticisms of the classical approach, many prefer to think of probability values as derived from experience, the more of it the better. They suggest that the probability of an event be considered equal to the relative frequency with which it has actually been observed in the past over the course of a large number of experiments.

The probability value is still a ratio between 0 and 1, but it is now based on empirical data rather than on *a priori* reasoning. Thus, the probability of getting a head with a *biased* coin might be established by experiment as $p(\text{H}) = (9/10)$ and that of finding a defective item on the production line as $p(\text{D}) = (3/100)$, both of these being values that could not be derived by abstract reasoning alone.

Yet here, too, controversy exists. What is a "large number" of experiments? Do there have to be a thousand of them, or even a million, or must M, as some purists suggest, approach infinity? The latter would certainly not be practical, but it is true that the probability of an event, calculated by the empirical formula in Box 5.B, will vary with the number of experiments performed and will hardly be very meaningful if that number is small. To appreciate this fact, consider tossing a (presumably unbiased) coin to determine *empirically* the probability of getting a head.

5.B The Probability of Event A: Empirical Approach

$$p(A) = \frac{\text{number of times event A did occur in the past during a large number of experiments}}{\text{maximum number of times event A could have occurred during these experiments}} = \frac{k}{M}$$

At the time of this writing, such an experiment was performed by the author. Fifty tosses were made; some of the results are shown in Table 5.1, and all of them are graphed in Figure 5.7. Note how the probability measure so determined varies with the number of experiments performed. What number of trials, therefore, reveals the

Table 5.1 Determining Probability Empirically: Tossing a Coin

When determining by actual experiment the probability of getting a head when tossing a coin, the computed value differs depending on the number of trials.

Toss Number	Result	Cumulative Number of Heads, k	Cumulative Number of Tosses, M	Probability of Head Implied by Past Experience $p(H) = \dfrac{k}{M}$
1	T	0	1	0
2	T	0	2	0
3	H	1	3	.33
4	H	2	4	.50
5	H	3	5	.60
•	•	•	•	•
•	•	•	•	•
•	•	•	•	•
10	T	6	10	.60
•	•	•	•	•
•	•	•	•	•
•	•	•	•	•
20	T	10	20	.50
•	•	•	•	•
•	•	•	•	•
•	•	•	•	•
30	H	14	30	.47
•	•	•	•	•
•	•	•	•	•
•	•	•	•	•
40	T	20	40	.50
•	•	•	•	•
•	•	•	•	•
•	•	•	•	•
50	H	26	50	.52

Figure 5.7   Determining Probability: The Classical versus the Empirical Approach

The classical approach to probability determines the probability of getting a head when tossing a fair coin by reasoning alone, and the value equals .5. The empirical approach requires actual experiments; its measurement varies with the number of trials. Using the empirical approach raises the question of what number of trials can be relied upon to reveal the "true" probability—a question that is particularly acute when the alternative, classical approach is unavailable.

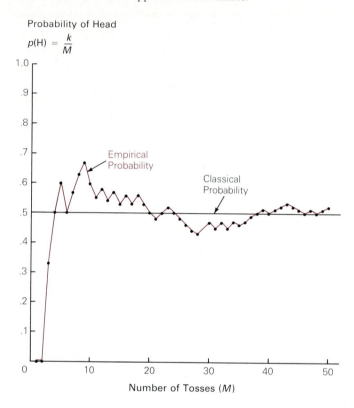

"true" probability? This question is particularly acute in situations in which the empirical approach is the only one available.

Note: Contrary to popular belief (among nonstatisticians), and contrary to what Figure 5.7 might suggest, it is not *certain* that the empirical probability value will cease to fluctuate after a "large enough" number of trials (and will thus reveal which number can be relied upon). Even if the experiment described here were continued for 20,000 tosses, the probability value may well rise again to the high of .67 observed after the ninth toss or may even fall by an equivalent amount below the classical .5. However, as the **law of large numbers** suggests, the probability of a significant deviation of an empirically determined probability value from a theoretically determined one is smaller the larger is the number of repetitions of the random experiment in question.

Subjective Probability

Considerably more controversial than objectively determined probability values are subjective ones. Such numbers reflect a purely personal degree of belief in the likelihood of the occurrence of some event. They reflect the hunches people have, their "feelings in the bone," and these vary over time and from one person to the next.

Consider how people speculate about the likehihood of a recession this year, of General Motors being in the red next fall, of research finding a source of abundant energy, or of oil being found at a given site. "There are 3 chances out of 10," they might say, yet the named experiments have never been performed before and will never be repeated. Such a probability value of .3 is based neither on logical deduction nor on repeated observations of actual events—that is, historical experience. In the view of some probability theorists, therefore, such values are totally worthless. Yet it is a fact that subjective estimates of the chances for various outcomes of unique experiments play an important role in millions of decisions every day. This fact will be noted in the last section of this chapter and, at greater length, in the last chapter of this book.

COUNTING TECHNIQUES

All the random experiments discussed so far in this chapter have been easy ones in the sense that the numbers of favorable outcomes and the numbers of possible outcomes were invariably small, and counting them presented no obstacle to the calculation of probability. Such simplicity is not always present, however. Some random experiments involve truly huge sample spaces; in such cases, the determination of probabilities would be severely hampered without special counting techniques. In order to be prepared for such occasions, we must learn a special way of counting and become familiar with the concepts of (1) factorials, (2) permutations, and (3) combinations.

Factorials

Suppose we had to know in how many different ways n distinct items can be sequenced. If the number of such items were small enough and involved, say, only the three letters A, B, and C, we could, of course, do the counting rather quickly, perhaps with the help of the tree diagram in Figure 5.8. If we started out with the three letters in our hands, we would have three alternative choices for placing a letter in the first position: A, B, or C. Having made that decision, only two choices would remain for the second position. (If the first choice were A, for example, only B or C would be available for the second position.) And having made the second choice, our hands would be tied; only one possibility would remain for the third position. (If the first choice were A and the second B, for example, we would *have* to place C in the third place.) By following the various branches of the tree diagram, we would discover the six possible sequences shown in the box and, thus, the answer to our question: 3 distinct items can be sequenced in 6 possible ways.

Now consider this: our answer can also be found by multiplying together the number of choices for the first, second, and third positions; that is, our 6 possible ways equal $3 \cdot 2 \cdot 1$. Such a product of a series of positive whole numbers that descends from a given number n down to 1 is called a **factorial product** or simply a **factorial.**

Figure 5.8   Alternative Sequences of Three Distinct Items

As this tree diagram indicates, three distinct items can be sequenced in six possible ways.

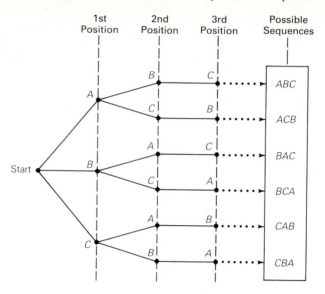

A factorial always tells us in how many different ways n distinct items can be sequenced. In our example, where $n = 3$, the answer is written as 3! and pronounced "three factorial." The more general case is given in Box 5.C.

5.C The Factorial

$$n! = n \cdot (n - 1) \cdot (n - 2) \cdot \ldots \cdot 3 \cdot 2 \cdot 1$$

where 0! and 1! are defined to equal 1.

Consider an advertising agency that wants to place 6 different items in a sales brochure. In how many different ways can they be sequenced? The answer is $6! = 6 \cdot 5 \cdot 4 \cdot 3 \cdot 2 \cdot 1 = 720$. Yet, even if there were only 10 items, the answer would already be $10! = 3,628,800$; for 20 items, it would exceed 2.4 quintillion. (A *quintillion* is a 1 followed by 18 zeros.) Can you imagine in how many different ways one could sequence the thousands of items in the Sears, Roebuck catalogue?

As these examples show, factorials of rather small numbers can take the place of extremely large ordinary numbers.

Permutations

The previous section has in fact discussed special cases of **permutations,** which are distinguishable ordered arrangements of items all of which have been drawn from a given group of items. Thus, the box in Figure 5.8 contains 6 permutations of the letters *A, B, C* because (a) it shows 6 distinct alternative sequences of these letters and (b) all of these sequences have been created out of the same group of letters: *A, B,* and *C.* Nevertheless, the letter-sequencing example of Figure 5.8 constitutes a special case of permutations for two reasons: First, the example involves a selection of *n* items out of *n,* but one could select fewer than *n.* Second, in this example, selections are made out of a group of *clearly distinguishable* items, but all of the elements of some groups may not be this distinct from one another. Both of these matters are discussed in the remainder of this section.

Selecting x *out of* n *distinct items.* Our letter-sequencing example involved finding the permutations for all the items in existence—that is, they involved finding all the different ways of arranging *n* distinct items into groups of *n.* Oftentimes, however, one is interested in the possible number of ordered sequences of only *x* items out of *n,* where *x* is smaller than *n.* In how many different ways, we may ask, can one sequence *x* = 3 items taken out of *n* = 10 distinct items? The answer cannot be *n*! because that number also counts the ways of sequencing the last (*n* − *x*) = 7 items in the sequences of which we are not now interested. Thus, the (*n* − *x*)! portion of the expression *n*! must be eliminated; this can most easily be done by dividing *n*! by (*n* − *x*)! The resultant formula for the possible number of permutations, *P,* when *x* items at a time are taken out of *n* distinct items is given in Box 5.D.

5.D Permutations for x at a Time out of n Distinct Items (with no repetitions allowed among the x items)

$$P^n_x = \frac{n!}{(n-x)!} \text{ where } x \le n$$

Applying the new formula in Box 5.D to the selection of 3 items out of 10, we find the answer as

$$P^{10}_3 = \frac{10!}{(10-3)!} = \frac{10!}{7!} = \frac{10 \cdot 9 \cdot 8 \cdot 7!}{7!} = 10 \cdot 9 \cdot 8 = 720.$$

Out of a total of 10 items, one can create 720 different ordered sequences of 3 items. For example, a firm that wants to place three new managers in three of its ten plants can do so in 720 different ways!

Similarly, applying the Box 5.D formula to the selection of *x* = 2 letters from *n* = 3 distinct letters (*A, B,* and *C*),

$$P^3_2 = \frac{3!}{(3-2)!} = \frac{3 \cdot 2 \cdot 1}{1} = 6.$$

These six sequences are *AB, AC, BA, BC, CA,* and *CB.*

Note: If one wishes, one can apply the Box 5.D formula to the special case of *x* = *n,* but one must realize that *n* − *x* then equals 0 and that 0! = 1 *by definition.* Thus, if *x* = *n,* the entire formula for P^n_n collapses to *n*!, which returns us to the formula given in Box 5.C.

Selecting n out of n nondistinct items. In our letter-sequencing example of Figure 5.8, the elements contained in the given group of items (namely, the letters *A, B, C*) were *distinct* in the sense that every one of the $n = 3$ items in the group differed from every other item in the group. Yet the elements contained in a given group of items can also be *nondistinct* in the sense that x_1 out of n items are of one kind, x_2 are of another kind, and so on for k different kinds, such that $x_1 + x_2 + \ldots + x_k = n$. (Consider the group *A, A, A, B, B, C, C, C, C, C,* for example.) In such a case, one cannot distinguish a sequence such as the one printed here from another one in which only the first and second *A*'s have exchanged places. In how many distinct ways, we may ask, can one sequence the above 10 nondistinct items? The answer cannot be $n!$ because of all the unrecognizable ways in which one can internally rearrange the $x_1 = 3$ *A*s, the $x_2 = 2$ *B*'s, and the $x_k = 5$ *C*'s in our example. These rearrangement possibilities, however, number $x_1!$ and $x_2!$ and $x_k!$, respectively, and we can get our answer by dividing $n!$ by the product of all these factorials, as shown in Box 5.E.

5.E Permutations for *n* Out of *n* Items, Given *k* Distinct Types of Items and *k* < *n*

$$P^n_{x_1, x_2, \ldots x_k} = \frac{n!}{x_1! \, x_2! \ldots \ldots x_k!}$$

where x_1 items are of one kind, x_2 items are of a second kind, . . . and x_k items are of a *k*th kind and where $x_1 + x_2 + \ldots + x_k = n$.

When applying this formula, we discover that the above 10 nondistinct items can be put into 2,520 distinct sequences because

$$P^{10}_{3,2,5} = \frac{10!}{3!2!5!} = 2,520.$$

What if our initial group contained only two kinds of items, such as $x = 2$ *A*'s and $(n - x) = 1$ *B*? From this group of $n = 3$ (consisting of *A, A, B*), we could form only 3 distinct permutations:

$$P^n_{x,n-x} = \frac{n!}{x!(n - x)!} = \frac{3!}{2!(2 - 1)!} = 3.$$

These permutations are *AAB, ABA,* and *BAA.*

Combinations

Sometimes one is interested not in permutations, but in **combinations,** which are different selections of items such that possible alternative sequences among the components of any one such selection are considered immaterial. Thus, the boxed list in Figure 5.8 shows six *permutations* but represents a single *combination* of the letters *A, B, C.* If one doesn't care about the order of the items, there is only one way to select three items out of three. While the sequence of the elements in the selected group matters in the case of permutations (*ABC* is not considered the same as *ACB,* for example), sequence does not matter in the least for combinations (combination *ABC* is considered the same as combination *ACB* because each grouping contains the same elements). Sequence does not matter in the case of card hands or of committee members or of samples of all sorts. The fact that a hand of cards contains the king of hearts, the ace of diamonds, and the 2, 8, and 10 of clubs is all that matters. The identical elements listed in a different order —ace of diamonds, 2 of clubs, king of hearts, 8 and 10 of clubs—are viewed as the same hand. They represent a different permutation but the same combination.

When calculating probabilities, one is often interested in this question: in how many different ways can one select a given number of items (say $x = 5$) from a group of distinct items (say, $n = 10$) without any concern whatsoever for the sequence in which the selected items appear? The answer cannot be $n!/(n - x)!$ precisely because the permutation formula counts separately all the different orderings of the group of x items—orderings that are now irrelevant to us. In order to count these once only, one can divide $n!$ by $x!$ as well (for $x!$ shows all the different and now irrelevant ways in which x items can be sequenced), which yields the formula in Box 5.F.

5.F Combinations for *x* at a Time out of *n* Distinct Items (with no repetitions allowed among the *x* items)

$$C_x^n = \frac{n!}{x! \, (n - x)!}, \text{ where } x \leq n$$

Note how this formula collapses to 1 for the limiting case of $x = n$:

$$C_n^n = \frac{n!}{n! \, (n - n)!} = \frac{n!}{n! \, 0!} = \frac{n!}{n! \, (1)} = 1$$

When the order of selection is irrelevant, there exists only one way to select n items out of n items.

In how many ways, however, can one select 5 cards from a deck of 52 when the order of selection does not matter? Our formula gives us the answer almost instantly:

$$C_5^{52} = \frac{52!}{5!(52 - 5)!} = \frac{52!}{5! \, 47!} = \frac{52 \cdot 51 \cdot 50 \cdot 49 \cdot 48 \cdot 47!}{5 \cdot 4 \cdot 3 \cdot 2 \cdot 1 \cdot 47!}$$

$$= \frac{52 \cdot 51 \cdot 50 \cdot 49 \cdot 48}{5 \cdot 4 \cdot 3 \cdot 2 \cdot 1} = \frac{311,875,200}{120} = 2,598,960.$$

One can select 2,598,960 different hands of 5 cards from a deck of a mere 52 cards. Imagine how long it would take you to find this answer with the help of a tree diagram such as the one in Figure 5.8. Now ask yourself this: what is the probability of picking the top 5 cards from a deck of 52 well-shuffled cards and ending up with a *straight* (an ordered denominational sequence of 5 cards in which the deuce counts as the lowest and the ace as the highest card acceptable)? Although the lowest possible card in a 5-card straight can be a 2 (or deuce), it can belong to any one of the four different suits.

The 2 can be clubs, diamonds, hearts, or spades, and the same possibilities exist for the next four cards (which would be 3, 4, 5, and 6). Thus there can be $4 \cdot 4 \cdot 4 \cdot 4 \cdot 4$ = 1,024 different straights starting with 2. Yet there are nine possible low cards—namely, all the numbers between 2 and 10. (A 5-card straight starting with 10 would reach to the end of the series, going 10, jack, queen, king, and ace.) Hence there can be $9 \cdot 1,024$ = 9,216 such straights altogether. Contrast this number with the entire sample space of 5-card sets, and you find the (low) probability of a 5-card straight:

$$p(\text{5-card straight}) = \frac{9,216}{2,598,960} = .003546.$$

Analytical Examples 5.1, "The ESP Mystery," and 5.2, "Cosimo's Counting Problem," provide two interesting cases in which special counting techniques were applied.

THE LAWS OF PROBABILITY

Deriving probability values for single events is one thing; manipulating these values to determine the combined probability for the union or intersection of several events is another. Such manipulation is often useful in that it obviates the necessity for repeated counting of favorable and possible basic outcomes once some probability values have been derived in this way. Just as a person who has separately counted the fingers on the left hand and on the right hand can determine the number of fingers on both hands by simply adding 5 + 5, rather than by looking at both hands simultaneously and counting again, so a person who has access to some probability values can often combine them to gain further knowledge. We will consider two important laws for doing just that: (1) the addition law and (2) the multiplication law. Both were first stated by Abraham de Moivre (see Biography 5.1 at the end of the chapter).

The Addition Law

The **addition law** is a law of probability theory that is used to compute the probability for the occurrence of a *union* of two or more events. Such an occurrence was first illustrated (for the case of two events) in panel (c) of Figure 5.4 and then was shown symbolically in panel (d) of Figure 5.5. For the time being, we will continue to focus on two events, but, as will be shown later, the law can easily be extended to a larger number of events. Regardless of the *number* of events, a general addition law can be formulated that applies to all *types* of events; a special and simpler law can be stated for mutually exclusive events.

The general law. The general addition law for two events is given in Box 5.G, and it is easy to interpret. The law states that the probability of event A *or* B happening equals the probability of A alone plus the probability of B alone minus the probability of A *and* B happening at the same time. All this is most easily understood by reviewing panel (d) of Figure 5.5. The probability of A alone, $p(A)$, is the probability of getting the basic outcomes contained in the black circle, labeled A. The probability of B alone, $p(B)$, is the probability of getting the basic outcomes contained in the colored circle, labeled B. If one now determined the probability of either A *or* B happening by adding the separate probabilities of the two events, the probabilities of all the basic outcomes contained in the intersection of the two events and shown separately by the cross-hatched area in panel (e) would be counted twice: once as a part of circle A and again as a part of circle B. To eliminate the double counting, the probabilities of the basic outcomes found in this overlapping section are deducted once; hence, the *negative* term $p(A$ *and* $B)$ in Box 5.G. As a whole, the formula gives us precisely the probability of all the basic outcomes inside the crosshatched area of panel (d).

5.G General Addition Law

$$p(A \text{ } or \text{ } B) = p(A) + p(B) - p(A \text{ } and \text{ } B)$$

Consider the concrete example of Figure 5.4. The probability of event A (getting a face card) equals

$$p(A) = \frac{12}{52}.$$

This calculation requires nothing more difficult than applying the classical formula: counting the basic outcomes favorable to this event, of which there are 12 in the shaded area of panel (a), and relating the count to all the possible basic outcomes, of which there are 52 in the entire boxed sample space. Similarly, in panel (b), we can find the probability of event B (getting a heart) as

$$p(B) = \frac{13}{52}.$$

What is the probability of the union of these events, of getting *either* a face card *or* a heart? Does it equal

$$\frac{12}{52} + \frac{13}{52} = \frac{25}{52}?$$

The answer is no, as a simple count of favorable to possible basic outcomes in panel (c) reveals. The probability of getting a face card *or* a heart, $p(A$ *or* $B)$, equals only 22/52, the number of panel (c) favorable basic outcomes (shaded) divided by the number of all possible basic outcomes (shaded plus unshaded). The difference of 3/52 between the incorrect result of 25/52 and the correct one of 22/52 is easily accounted for. It arises from the double counting of the jack, queen, and king of hearts when the probabilities implied by the shaded areas of panels (a) and (b) are thoughtlessly added together. The error can be eliminated by deducting the probability of getting a face card *and* a heart, $p(A$ *and* $B)$, as the formula demands. That value, of course, equals 3/52, as panel (d) so vividly illustrates. Thus,

$$
\begin{array}{ccccccc}
p(A) & + & p(B) & - & p(A \text{ } and \text{ } B) & = & p(A \text{ } or \text{ } B) \\
p(\text{face card}) & + & p(\text{heart}) & - & p(\text{face card } and \text{ heart}) & = & p(\text{face card } or \text{ heart}) \\
\dfrac{12}{52} & + & \dfrac{13}{52} & - & \dfrac{3}{52} & = & \dfrac{22}{52}
\end{array}
$$

If we know the probabilities implied by panels (a), (b), and (d) in Figure 5.4, we can calculate the probability implied by panel (c).

Note: If there were three events, we would simply expand the formula to read

$$p(A) + p(B) + p(C) - p(A \text{ and } B) - p(A \text{ and } C) - p(B \text{ and } C)$$
$$+ p(A \text{ and } B \text{ and } C) = p(A \text{ or } B \text{ or } C).$$

Further expansion is carried out similarly.

The special law. The general addition law can be used to calculate the probability for a union of all types of events, but it reduces to a simpler form for mutually exclusive events. The reason is simple: In the case of mutually exclusive events, such as those symbolized in panel (a) of Figure 5.5, there is no intersection, hence no double counting, hence no need for the negative term in our formula. The formula in Box 5.G can still be used, but the expression $p(A \text{ and } B)$ will equal zero; hence, we can just as well use the simpler version in Box 5.H.

5.H Special Addition Law for Mutually Exclusive Events

$$p(A \text{ or } B) = p(A) + p(B)$$

Consider calculating the probability of getting a face card (event A) *or* getting an ace (event B) for the experiment of drawing one card at random from a well-shuffled deck. As a quick look at panel (a) of Figure 5.4 confirms, the separate probabilities of these mutually exclusive events equal

$$p(A) = \frac{12}{52} \text{ and } p(B) = \frac{4}{52}$$

and the probability of their union is

$$
\begin{array}{ccccc}
p(A) & + & p(B) & = & p(A \text{ or } B) \\
p(\text{face card}) & + & p(\text{ace}) & = & p(\text{face card } \text{or} \text{ ace}) \\
\dfrac{12}{52} & + & \dfrac{4}{52} & = & \dfrac{16}{52}
\end{array}
$$

Naturally, the probability for any union of mutually exclusive events that are also collectively exhaustive must equal 1, and this fact, too, can easily be confirmed. Consider panel (c) of Figure 5.3. In the experiment shown there, the probability of getting a sum of 10 or more (event A) *or* getting a sum of 9 or less (event B) exhausts the sample space. Accordingly, our special formula tells us that

$$
\begin{array}{ccccc}
p(A) & + & p(B) & = & p(A \text{ or } B) \\
p(\text{sum of 10 or more}) & + & p(\text{sum of 9 or less}) & = & p(\text{sum of} \geq 10 \text{ or sum of} \leq 9) \\
\dfrac{6}{36} & + & \dfrac{30}{36} & = & 1
\end{array}
$$

Thus, the probability for a union of complementary events always equals 1. When event B is complementary to event A, one can write the above also as

$$p(A) + p(\overline{A}) = p(A \ or \ \overline{A}) = 1.$$

An awareness of this truism is often helpful when it is desired to calculate the probability of one event, say A, but when it is easier to calculate that of its complement, *not* $A = \overline{A}$. In that case $p(A)$ can simply be found as $1 - p(\overline{A})$. Similarly, of course, $p(\overline{A})$ can always be found by $1 - p(A)$.

Again note that the special formula in Box 5.H can easily be expanded to any number of mutually exclusive events. For example,

$$p(A \ or \ B \ or \ C \ or \ D) = p(A) + p(B) + p(C) + p(D)$$

in a case such as the one illustrated in panel (b) of Figure 5.5.

The Multiplication Law

The **multiplication law** is a law of probability theory that is used to compute the probability for an *intersection* of two or more events. Such an occurrence was first illustrated (for the case of two events) in panel (d) of Figure 5.4 and then shown symbolically in panel (e) of Figure 5.5. For the time being, we will once again continue to focus on two events, but, as will be shown later, this law, too, can easily be extended to a larger number of events. As was true for the addition law, a general multiplication law can be formulated that applies to all types of events; a special and simpler one can be stated for a more restricted group of events. Before we can understand these laws, however, we must become familiar with a number of new concepts.

Unconditional probability. Consider an appliance dealer who has been promoting a certain product by means of a major television advertising campaign. In order to evaluate the effectiveness of the campaign, let us suppose, all of the 600 customers who visit the dealership during a certain period are asked whether they remember the TV ad; records are kept of their answers as well as their possible purchase of the product in question. The results of this survey are given in Table 5.2.

Now consider the following four events: the purchase of the product, P; its nonpurchase, $\overline{P}$; remembering the TV ad, R; not remembering it, $\overline{R}$. Clearly, a person can

Table 5.2   Summary of Customer Survey

These hypothetical figures represent numbers of customers classified in two ways: whether they purchased a certain product and whether they remembered a TV ad promoting it.

Product \ TV Ad	Remembered, R	Not Remembered, $\overline{R}$	Total
Purchased, P	120	60	180
Not Purchased, $\overline{P}$	80	340	420
Total	200	400	600

calculate regular empirical probability values for these four events by using the numbers in the shaded last column and shaded last row of Table 5.2:

$$p(P) = \frac{180}{600} = .30 \qquad p(R) = \frac{200}{600} = .33$$

$$p(\overline{P}) = \frac{420}{600} = .70 \qquad p(\overline{R}) = \frac{400}{600} = .67$$

Experience thus reveals a probability of .30 that a customer buys the product, a probability of .70 that a customer doesn't, a probability of .33 that a customer remembers the ad, and a probability of .67 that a customer doesn't remember the ad. Because each of these values measures the likelihood that a particular event occurs, regardless of whether another event occurs, each of these values is called an **unconditional probability.** Thus .30 equals the probability of purchase by a customer without regard to the question of whether that customer does or does not remember the ad. Similarly, the value of .67 equals the probability of not remembering by a customer without regard to the question of whether that customer does or does not purchase the product. Because these unconditional probabilities are calculated from the margins of our frequency table, they are also referred to as **marginal probabilities.**

Conditional probability. It will come as no surprise that the concept of unconditional probability has a twin—namely, that of **conditional probability,** which measures the likelihood that a particular event occurs, given the fact that another event has already occurred or is certain to occur. For two events, A and B, such a probability is always denoted by $p(A|B)$ or $p(B|A)$, which is read as "the probability of A, given B" or "the probability of B, given A"—the vertical line standing for "given."

By focusing, successively, on the values in the first row only, then on those in the second row only, next on those in the first column only, and finally on those in the second column only, we can calculate the following conditional probabilities.

1. From the first row, we can calculate the probability of remembering, given the fact that a purchase has occurred, and also the probability of not remembering, given that a purchase has occurred:

$$p(R|P) = \frac{120}{180} = .67$$

$$p(\overline{R}|P) = \frac{60}{180} = .33$$

2. From the second row, we can calculate the probability of remembering, given the fact that no purchase has occurred, and also the probability of not remembering, given that no purchase has occurred:

$$p(R|\overline{P}) = \frac{80}{420} = .19$$

$$p(\overline{R}|\overline{P}) = \frac{340}{420} = .81$$

3. From the first column, we can calculate the probability of purchase, given the fact that the ad is remembered, and also the probability of no purchase, given that the ad is remembered:

$$p(P|R) = \frac{120}{200} = .60$$

$$p(\overline{P}|R) = \frac{80}{200} = .40$$

4. From the second column, we can calculate the probability of purchase, given the fact that the ad is not remembered, and also the probability of no purchase, given that the ad is not remembered:

$$p(P|\overline{R}) = \frac{60}{400} = .15$$

$$p(\overline{P}|\overline{R}) = \frac{340}{400} = .85$$

Note: It is no accident that each pair of these conditional probabilities sums to 1.00, just as the two pairs of unconditional probabilities given in the previous section do. The reason is that the events in question are in each case mutually exclusive and collectively exhaustive (either there was a purchase or there was no purchase, either the ad was remembered or it was not remembered), and the probabilities of such events always add to 1.

Joint probability. We must, finally, turn to the concept of **joint probability,** a measure of the likelihood of the simultaneous occurrence of two or more events. Four such values can be calculated from Table 5.2 because the events P, $\overline{P}$, R, and $\overline{R}$ intersect in the four ways shown in the unshaded portion of our table. It is customary to present this information in a *joint probability table,* such as Table 5.3. Note that each probability

Table 5.3   A Joint Probability Table

The joint probabilities for the designated events appear inside the colored box; the remaining values are the marginal probabilities, which equal the respective sums of the joint probabilities in the various rows and columns.

Product / TV Ad	Remembered, R	Not Remembered, $\overline{R}$	Total
Purchased, P	$\frac{120}{600} = .20$	$\frac{60}{600} = .10$	$\frac{180}{600} = .30$
Not Purchased, $\overline{P}$	$\frac{80}{600} = .13$	$\frac{340}{600} = .57$	$\frac{420}{600} = .70$
Total	$\frac{200}{600} = .33$	$\frac{400}{600} = .67$	$\frac{600}{600} = 1.00$

Source: Table 5.2

value shown therein simply equals the number of relevant observations given in Table 5.2, divided by the total number of observations (which was 600). The relative frequencies so computed are, of course, nothing else but probability numbers determined by the empirical approach (Box 5.B). The numbers within the colored box in Table 5.3 are the joint probabilities, and they are symbolized as follows:

$$p(\text{P } and \text{ R}) = .20 \qquad p(\text{P } and \text{ } \overline{\text{R}}) = .10$$
$$p(\overline{\text{P}} \text{ } and \text{ R}) = .13 \qquad p(\overline{\text{P}} \text{ } and \text{ } \overline{\text{R}}) = .57$$

Thus, we are told, the probability of finding a customer who purchases the product and also remembers the ad is .20, and the probability of finding a customer who does neither is .57. Again, it is no accident that the sum of the four joint probabilities of the four mutually exclusive and collectively exhaustive joint events equals one. (The special addition law applies.)

Two things should be noted: First, the margins of the joint probability table once again show the marginal or unconditional probabilities computed earlier. Second, by dividing any given joint probability by the marginal probability of its row (or column), the conditional probability for the given row (or column) event can be calculated. Thus,

$$\frac{.20}{.30} = .67, \text{ which means } \frac{p(\text{P } and \text{ R})}{p(\text{P})} = p(\text{R}|\text{P}).$$

The joint probability of purchase *and* remembering, this expression tells us, when divided by the unconditional probability of purchase, equals the conditional probability of remembering, given the fact of purchase. Similarly, it is true that

$$\frac{.20}{.33} = .60, \text{ which means } \frac{p(\text{P } and \text{ R})}{p(\text{R})} = p(\text{P}|\text{R}).$$

By making similar calculations based on the other three joint probabilities in Table 5.3, you can confirm the remaining six conditional probability values given in the earlier section on conditional probability.

The general law. The relationships just discovered between joint, unconditional, and conditional probabilities can be rewritten for any two events, A and B, as shown in Box 5.I, the general multiplication law.

5.I General Multiplication Law

$$1. \ p(\text{A } and \text{ B}) = p(\text{A}) \cdot p(\text{B}|\text{A})$$

and also

$$2. \ p(\text{A } and \text{ B}) = p(\text{B}) \cdot p(\text{A}|\text{B})$$

The law states that the joint probability of two events happening at the same time equals the unconditional probability of one event times the conditional probability of the other event, given that the first event has aleady occurred (or is certain to occur).

Consider again the example of Figure 5.4. The unconditional probability of event A (getting a face card) is

$$p(A) = \frac{12}{52}.$$

The conditional probability of event B (getting a heart, assuming a face card is being picked) is

$$p(B|A) = \frac{3}{12}$$

because there are 3 face cards with hearts among a total of 12 face cards. Version 1 of our formula tells us that the joint probability should be

$$p(A \ and \ B) = p(A) \cdot p(B|A) = \frac{12}{52} \cdot \left(\frac{3}{12}\right) = \frac{3}{52}$$

which is precisely what panel (d) of Figure 5.4 suggests (there are 3 basic outcomes in the shaded intersection and 52 basic outcomes in the entire sample-space box).

Note that the same result could have been reached by the use of version 2 of the formula: the unconditional probability of event B (getting a heart) is

$$p(B) = \frac{13}{52}.$$

The conditional probability of event A (getting a face card, assuming a heart is being picked) is

$$p(A|B) = \frac{3}{13}$$

because there are 3 face cards with hearts among a total of 13 cards with hearts. Thus, version 2 of our formula tells us that

$$p(A \ and \ B) = p(B) \cdot p(A|B) = \frac{13}{52} \cdot \frac{3}{13} = \frac{3}{52}.$$

Note also that the general multiplication law can be extended to more than two events such that (for three events)

$$p(A \ and \ B \ and \ C) = p(A) \cdot p(B|A) \cdot p(C|A \ and \ B).$$

If the probability of a person seeing a TV ad is .3 (event A) and if the probability of such a person walking into the advertising dealership's showroom is .1 (event B, given A) and if the probability of such a visitor making a purchase is .5 (event C, given A *and* B), the joint probability of someone seeing the ad, visiting the showroom, and making a purchase (event A *and* B *and* C) is .3(.1)(.5) = .015. Obviously, this type of calculation gets more complicated as the number of events increases. For a fascinating proof of this statement, see Analytical Example 5.3, "The Miracle of the Matching Birthdays."

Fortunately, a special and simpler version of the multiplication law can be applied whenever the events in question are independent rather than dependent.

Dependent versus independent events. Two random events are said to be **dependent events** when the probability of one event is affected by the occurrence of the other event. In contrast, two events are called **independent events** when the probability of one event is *not* affected by the occurrence of the other event.

A good example of dependent events is the drawing of more than one card from a deck *without replacing* the cards drawn. Prior to the first draw, the probability of getting the king of hearts is 1/52, and so is the probability of getting the ace of diamonds or the queen of spades. Now imagine drawing the king of hearts on the first try *and setting it aside*. Notice how all the probabilities will now differ because the deck is reduced to 51 cards, and the king of hearts is gone. The subsequent probabilities are zero for the king of hearts and 1/51 for all the other cards, and so it goes until, when only one card is left, the probability of drawing it equals 1, at which point the probability is zero for drawing any of the other cards already drawn.

Consider, in contrast, the same experiment, with the immediate replacement of any card drawn. The probability of drawing any one card will then remain unaffected by any of the preceding events; it will always equal 1/52. Even if the king of hearts were drawn on the first try, after replacing it and reshuffling the deck, the probability of drawing it on the second try would again be 1/52, just as for any other card. In this case, different drawings are independent events.

Note: While one can, if one wishes, set aside the king of hearts and draw again, it is physically impossible to set aside a head (or tail) and then toss a coin again; hence, different tosses of a coin are always independent events. Nevertheless, many gamblers refuse to believe it. After observing a series of heads, they bet on a tail coming up next because, as they put it, "it is the tail's turn" or "the tail's chances are now mature." In fact, this is nonsense. The probability of getting a head (or a tail) remains 1/2 for any toss of a fair coin, even when 37 heads have just been tossed in a row. The coin has no conscience or memory and doesn't care or know whose turn is next.

As is true throughout this chapter, the many examples from the world of gambling have innumerable analogies in the world of business and economics. Thus, an inspector who randomly picks selected items out of a warehouse in order to check quality (and who wants to determine the probability of the existence of defective items within the population of all those produced) must be aware of whether successive pickings are dependent events or independent ones. If each item inspected is set aside, as is usually the case, dependent events are occurring; if each item is replaced before the next one is picked, the events are independent.

It is easy to test for the dependence or independence of any two events, such as the remembering of an ad and the purchase of a product or the drawing of a king and a red card from a deck. Three simple steps are involved:

1. *Determine the unconditional probability for one event in the pair.* For example, note in Table 5.3 that the unconditional probability of remembering is $p(R) = .33$, or note in panel (c) of Figure 5.1 that the unconditional probability of drawing a king from a deck is $p(king) = 4/52$.

2. *Determine the conditional probability for the same event, given the occurrence of the second event in the pair.* For example, note that the conditional probability of remembering, given purchase, is $p(R \mid P) = .67$, or note that the conditional probability of drawing a king, given that a red card is drawn, equals $p(\text{king}|\text{red}) = 2/26$ (there are 2 kings among 26 red cards).
3. *Compare the unconditional with the conditional probability.* If the values differ, the events are *statistically dependent;* if they do not differ, the events are *statistically independent.* In our case, $.33 \neq .67$; remembering and purchase are dependent events. On the other hand, $4/52 = 2/26$; getting a king and getting a red card are independent events.

Symbolically, dependence implies $p(A) \neq p(A \mid B)$, while independence implies $p(A) = p(A \mid B)$. This discovery brings us directly to the simplification of the general multiplication law.

The special law. The general multiplication law can be used to calculate the probability for an intersection of all types of events, but it reduces to an easier form for independent events. The reason is simple: in the case of independent events, the unconditional and the conditional probabilities are identical; hence, the conditional probabilities, included in Box 5.I, can be replaced by the unconditional probabilities, as shown in Box 5.J.

5.J Special Multiplication Law for Independent Events

$$p(A \text{ and } B) = p(A) \cdot p(B)$$

In the case of independent events, the joint probability of A *and* B (or of any number of such events) simply equals the product of the unconditional probabilities of these events. For example, as we noted above, drawing a king and drawing a red card from a deck are independent events. Hence, their joint probability

$$p \text{ (king } and \text{ red card)} = \frac{4}{52}\left(\frac{26}{52}\right) = \frac{1}{26}.$$

This probability is, indeed, obvious by looking at panel(c) of Figure 5.1: there are 2 red kings in the deck of 52 cards.

Or consider the quality inspector who selects 3 items (like kings) from a group of 100 (like a deck of cards). If the inspected item is replaced after each selection and the whole lot is remixed, what is the probability that the inspector will find 3 defective items if (unbeknownst to the inspector) there are only 10 defective items in the lot? The answer is simple: the unconditional probability of finding a defective item, $p(D)$, equals $(10/100) = .1$. Because replacement makes the events of finding a defective item on the first, second, and third try (D_1, D_2, D_3) independent ones, the special multiplication law applies: $p(D_1 \text{ and } D_2 \text{ and } D_3) = p(D_1) \cdot p(D_2) \cdot p(D_3) = .1(.1)(.1) = .001$. There is just one such chance in 1,000 tries.

FAMOUS APPLICATIONS OF THE PROBABILITY LAWS

When properly applied, the laws of probability can tremendously enhance people's ability to make proper decisions in the face of uncertainty. This fact is vividly illustrated by Application 1, "Striving for Reliability: The Case of the Apollo Mission." On the other hand, disastrous consequences await those who have failed to do their homework. Consider the mix-up of the general multiplication law (that should have been applied) with the special multiplication law (that was applied), as reported in Application 2, "Probability in Court." Or study the equally famous mix-up of the special multiplication law (that should have been applied) with the special addition law (that was applied), a case reported in Application 3, "The Paradox of the Chevalier de Méré." That case was solved jointly by two great mathematicians, Pierre de Fermat and Blaise Pascal (see Biographies 5.2 and 5.3 at the end of the chapter).

Application 1: Striving for Reliability—The Case of the Apollo Mission*

The Apollo mission that was to land Americans on the moon required the creation of a complex artificial world to sustain three astronauts in a hostile environment for two weeks. The failure of any component of the Apollo system could lead to failure of the entire mission. The reliability problem was the greatest challenge facing the program. Because *reliability* of components is closely linked with the *probability* of their success or failure, the entire issue had to be dealt with by probability theory. First, the mission planners developed a mathematical model of the Apollo system showing the location in various subsystems of 2 million functional parts, miles of wiring, and thousands of joints. Such a simplified model is given in panel (a) of Figure A; an even more simplified model appears in panel (b).

Second, mission planners estimated the reliability of each subsystem from experimental data and then used the laws of probability to predict the reliability of the Apollo system as a whole.

Figure A is a typical example of a *series system:* if any component fails, the entire system fails (just as, in the case of old-fashioned strings of Christmas tree lights, the entire tree was darkened whenever a single bulb failed). Assuming that the components operate independently and that the individual probability-of-success values are $p(A)$, $p(B)$, $p(C)$, $p(D)$, and $p(E)$, respectively, the reliability of the entire system, $p(S)$, can be calculated by the special multiplication law for independent events (see Box 5.J) as $p(S) = p(A) \cdot p(B) \cdot p(C) \cdot p(D) \cdot p(E)$. Thus, if each component has a reliability of .95, the probability of success for the entire system shown in Figure A is given by $p(S) = (.95)^5 = .774$, which is not too encouraging. If the performance of each subsystem is not independent of that of the others (and, say, the successful performance of A portends the successful performance of E or if the failure of B puts a high load on other subsystems and causes the failure of D), the *general* multiplication law must, of course, be used to calculate $p(S)$.

The Apollo planners were also aware, however, of the implications of *parallel configurations*—of replacing, say, the large main engine A by two smaller engines, A_1 and A_2. If, then, failure occurs only if *both* A_1 and A_2 fail (or only if an intersection of two

---

*Adapted from Gerald J. Lieberman, "Striving for Reliability," in Judith M. Tanur et al., eds., *Statistics: A Guide to the Unknown* (San Francisco: Holden-Day, 1972), pp. 400–406.

Figure A

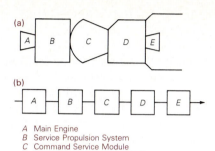

(a)

(b)

A Main Engine
B Service Propulsion System
C Command Service Module
D Lunar Excursion Module (LEM)
E LEM Engine

Figure B

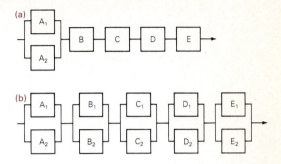

(a)

(b)

events occurs), reliability might be increased. Such a possible new engine configuration is shown in panel (a) of Figure B.

Under this scheme, the probability of success for the engine function now equals the probability of success for engine A_1 *or* A_2; that is,

$$p(A) = p(A_1 \text{ or } A_2).$$

According to the general addition law, the above equation can also be written as:

$$p(A) = p(A_1) + p(A_2) - p(A_1 \text{ and } A_2).$$

Assuming again independence of engines A_1 and A_2, the latter term can be replaced by $p(A_1) \cdot p(A_2)$ in accordance with the special multiplication law, yielding

$$p(A) = p(A_1) + p(A_2) - [p(A_1) \cdot p(A_2)].$$

Assuming the same probability of success for the smaller engine as for the larger one—that is, $p(A_1) = .95$ and $p(A_2) = .95$—the success probability for the engine function becomes

$$p(A) = .95 + .95 - (.95)^2 = 1.9 - .9025 = .9975,$$

which is a great improvement over the original .95. Together with the other components, this engine setup raises the probability of success for the entire mission from .774 to

$$p(S) = .9975 (.95)^4 = .8125.$$

Indeed, one can do better than that! Consider parallel configurations for all of the subsystems, as shown in panel (b) of Figure B. Assuming probabilities of success of .95 for all the components, and applying the same reasoning, one can raise the probabilities of success for all the other subsystems to .9975 as well.

Thus, the probability of success for the entire system to sustain the three astronauts, shown in panel (b) of Figure B, becomes

$$p(S) = (.9975)^5 = .9876,$$

which is a dramatic improvement over the original .774. Further improvements of the same type can also be made, of course—for example, by the substitution of two engines for A_1 and of two more engines for A_2. Nor is this idle speculation: During the Apollo 13 launch, the center engine of the second stage of the Saturn rocket failed, yet a satisfactory earth orbit was achieved with the remaining *four* rocket motors.

And note: the same principle of reliability control applies in the case of other and possibly less complex systems, be they automobiles, household appliances, power supplies, or telephones. Indeed, auto manufacturers introduced the five-year unconditional guarantee after examining the reliability of the affected components and determining the overall reliability of their cars in the fashion indicated here.

Application 2: Probability in Court*

On many an occasion, probability theory has had its day in court. Two famous cases stand out.

The Trujillo Case. In 1946, a police officer was shot to death. Joseph Trujillo was charged with the murder. The prosecutor argued that 11 matching fibers were found that apparently had been transferred between the victim's and the defendant's clothing. The probability for one accidental fiber match was given as .1; hence, that of 11 matches was calculated by the special multiplication law as $(.1)^{11}$ or a tiny .00000000001. More than that! The prosecutor also noted five matching characteristics between the bullet that had killed the victim and the pistol that belonged to the defendant.

1. Both were of .38 caliber, a probability given as .2.
2. Rifling was right-handed, a probability given as .5.
3. Both pistol and bullet showed five lands and grooves, a probability given as .2.
4. Both pistol and bullet showed the chamber out of line with the barrel, a probability given as .2.
5. The pistol was not, and the bullet had not come from, an automatic revolver, a probability given as .5.

Again, the prosecutor multiplied together the estimated probabilities to show the small joint probability of all these events and, hence, the tiny likelihood that someone else could have been the killer.

Trujillo was convicted and executed. No thought, apparently, was given to whether the stated events were independent (and the special multiplication law was even applicable). Some 18 years later, however, a similar case had quite a different outcome.

---

*Adapted from Darrell Huff, *How to Take a Chance* (New York: Norton, 1959), pp. 138–40; *Time,* January 8, 1965, p. 42, and April 26, 1968, p. 41; and W. Fairley and Frederick Mosteller, "A Conversation About Collins," *University of Chicago Law Review,* 1974.

The Collins Case. In 1964, an elderly woman was mugged in an alley. Shortly afterward, a witness saw a blonde girl, her ponytail flying, run out of the alley, get into a yellow car driven by a bearded black man with a moustache, and speed away. Eventually, the police arrested Janet and Malcolm Collins, a married couple who fit the physical description given above and who also owned a yellow Lincoln.

Once again, the prosecutor invoked the laws of probability. He suggested to the jury probability values:

1. for being a blonde of $p = .33$.
2. for wearing a ponytail of $p = .10$.
3. for being an interracial couple of $p = .001$.
4. for owning a yellow car of $p = .10$.
5. for wearing a moustache of $p = .25$.
6. for being a black man wearing a beard of $p = .10$.

He multiplied the numbers to produce a joint probability of .00000008 that the Collinses could have been duplicated on the morning of the crime. The jury, overwhelmed by the logic of it all, found them guilty beyond a reasonable doubt. Malcolm Collins received a sentence of "one year to life"; Janet Collins received a sentence of "not less than one year."

Yet, a few years later, the California Supreme Court overturned the verdict. There was no evidence, it said, that the individual probability values were even roughly accurate. Most important, they were not shown to be independent of one another, as they must be to satisfy the probability law that was applied. The fact that the black man had a beard, said the court, was not independent of the fact that the couple was interracial, for example. Indeed, the court used the very technique applied in Analytical Example 5.3, "The Miracle of the Matching Birthdays," to show that there was a 41 percent chance that at least one other couple in the area might have had the admittedly unusual characteristics listed above.

Application 3: The Paradox of the Chevalier de Méré*

A good example of what happens when the special law of multiplication is confused with the special law of addition comes to us from 17th-century France. A nobleman, Antoine Gombauld, also known as Chevalier de Méré, thought he had found a surefire way of winning in games of dice. For a single roll of a die, he reasoned (correctly), the probability of getting a 6 to come up was 1/6. Hence, he argued (incorrectly), the probability of at least one 6 to come up was 2/6 for two rolls of the die, 3/6 for three rolls (making for even odds), and 4/6 for four rolls (turning the odds in favor of this event). Thus, he was prepared to gamble that he would get at least one 6 in four rolls of a die; as he saw it, he would win two wagers out of three in the long run. Gamble he did, and, amazingly, he became quite rich.

Then, for reasons unknown, the Chevalier changed his strategy, betting that he could get at least one 12 in 24 rolls of two dice. His reasoning was the same. As he saw

---

*Adapted from Darrell Huff, *How to Take a Chance* (New York: Norton, 1959), pp. 63–67.

it, the probability of getting a 12 in a single roll of two dice was 1/36. So far, he was right, as panels (b) and (c) of Figure 5.2 attest. Hence, he concluded (incorrectly), the probability of getting a 12 in 24 such rolls was 24/36 or, once more, two out of three. Yet this time, his luck seemed to have run out; he lost more often than he won.

Perplexed, he turned for help to the mathematician and philosopher Blaise Pascal (see Biography 5.3) who, in turn, consulted with a friend, Pierre de Fermat (see Biography 5.2). Their joint inquiry into this *problem of points,* as it is called, gave birth to the theory of probability and quickly solved the Chevalier's paradox. The gambler's application of the special addition law for mutually exclusive events (see Box 5.H), argued the two, was a mistake. The events in question were not mutually exclusive at all: one could conceivably get a 6 on every one of four rolls of a die; one could get a 12 on every one of 24 rolls of two dice as well. Indeed, by the logic of our gambling friend, the probability of getting a 6 on *six* rolls of a die was 6/6; that is, such an event would occur with certainty!

The issue, argued Pascal and de Fermat, called for the use of the special *multiplication* law for independent events (see Box 5.J), but they found it easier first to calculate the probability for the gambler's failure rather than his success: the probability of *not* getting a 6 on a single roll, they figured, was 5/6. Hence, the probability of not getting a 6 on four such rolls was

$$\frac{5}{6} \cdot \frac{5}{6} \cdot \frac{5}{6} \cdot \frac{5}{6} = \frac{625}{1296}.$$

Thus, the first strategy had a probability of *success* of

$$1 - \frac{625}{1296} = \frac{671}{1296}$$

or about .52. This was less than the gambler's imagined 2/3, but high enough to make him rich.

By the same reasoning, the probability of *not* getting a 12 on a single roll of two dice was 35/36. Hence the probability of not getting a 12 on 24 such rolls was $(35/36)^{24}$ or about .509. Thus, the second strategy had a probability of success of $1 - (\approx .509)$, or about .491. No wonder the gambler's luck ran out.

Probability Laws and Tree Diagrams

Both addition and multiplication laws can be illustrated with the help of tree diagrams, such as those in Figure 5.9.

Both panels (a) and (b) illustrate the survey summarized in Table 5.2 but do so in slightly different ways. The first fork in panel (a) pictures the possibilities of a customer making or not making a purchase and lists the associated unconditional probabilities of these events along the two branches. Because a customer must do one or the other, the special additon law applies; the sum of .3 and .7 equals 1, implying certainty that one *or* the other of these events will occur. If a purchase is made, two mutually exclusive and collectively exhaustive possibilities exist once again at P: the customer does or does not remember seeing the ad. The conditional probabilities (*given* that a purchase

Figure 5.9 Probability Laws and Tree Diagrams

The two equivalent versions of the general multiplication law that helps us find the joint probability of two events by multiplying the unconditional probability of one event with the conditional probability of the other event, given the first event, are illustrated in panels (a) and (b). The special addition law is seen at work, too, although indirectly: the sum of probabilities at each fork and the sums of joint probabilities always equal 1, a requirement for mutually exclusive and collectively exhaustive events. Panel (c), finally, illustrates an application of the special multiplication law to the determination of joint probabilities for three independent events.

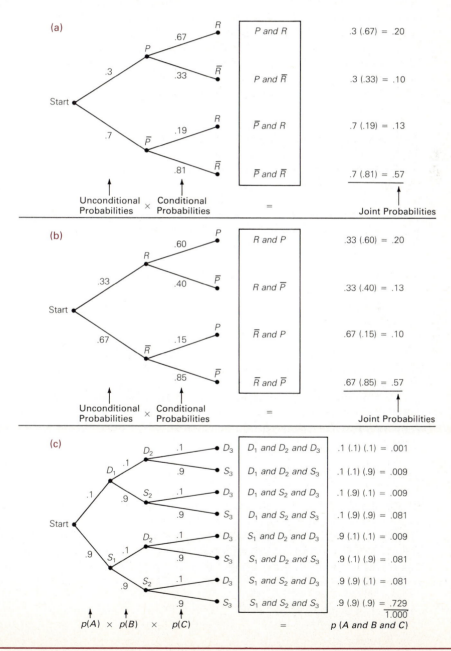

has been made) equal .67 and .33 (as was noted in an earlier section) and, once more, add to 1. If no purchase is made, the same two possibilities exist at $\overline{P}$, but the conditional probabilities are .19 and .81; once more, they add to 1. If the numbers along any path to the ultimate outcome are multiplied, joint probabilities are found, just as the general multiplication law suggests. Note that the joint probabilities of the four possible outcomes again add to 1 and that each equals the value previously calculated (and given in Table 5.3).

Panel (b) reaches the identical conclusion, but it does so in a different way, by employing the alternative version of the multiplication law. The choice between remembering and not remembering the ad is considered first; that between purchase and no purchase is considered second. The same joint probabilities result (but note that they are listed in a different order).

Note also how easy it is to recognize all the events in question as *dependent* ones. The unconditional probability of purchase, given as .3 in panel (a), differs from the conditional probabilities of purchase, given remembering (or not remembering) shown as .6 (or .15) in panel (b). So purchase and remembering (or not remembering) are dependent events. Similarly, the unconditional probability of not remembering, given as .67 in panel (b), differs from the conditional probabilities of not remembering, given purchase (or no purchase) shown as .33 (or .81) in panel (a). Not remembering and purchase (or no purchase) are dependent events as well.

Panel (c) pictures the case of quality inspection *with replacement* in which 3 items are picked from a lot of 100 of which 10 percent are defective. Thus, the panel simultaneously illustrates a case of independent events and a case of more than two events. The fact of independence is visible in the tree diagram because the probabilities at every fork remain the same (.1 for picking a defective item; .9 for picking a satisfactory one). Thus $p(A)$ symbolizes the unconditional probability of finding a defective or satisfactory item on the first trial; $p(B)$ symbolizes the corresponding conditional probabilities on the second trial, given the result of the first trial, but these probabilities equal the unconditional ones. Similarly, $p(C)$ symbolizes the conditional probabilities on the third trial, given the results of the first and second trials, but these probabilities, too, equal the unconditional ones, because the three trials are independent events when replacement occurs. Note how the tree diagram shows clearly why the 8 outcomes of the completed experiments are not equally likely. While the joint probability of finding three defective items in a row is a tiny .001, that of finding three satisfactory items in a row is a respectable .729.

REVISING PROBABILITIES: BAYES'S THEOREM*

It was noted earlier that *subjective* probability figures, although commonly used, are also rather controversial. In this section, we consider how such values can be confirmed or revised in the light of empirical evidence. The idea goes back to an 18th-century British clergyman, the Rev. Thomas Bayes (Biography 17.1), who stumbled onto probability theory when searching for a mathematical proof of the existence of

*Optional section.

God. Bayes pictured people as possessing, prior to any empirical investigation, an initial subjective estimate of the likelihood of an event; he called this value a **prior probability.** He then imagined an investigation to take place and to yield certain results. He developed what is now called **Bayes's theorem:** a formula for revising an initial subjective probability value on the basis of results obtained by an empirical investigation. The end result is a new probability value, a prior probability modified on the basis of new information and called a **posterior probability.** The sequence of probability revision can be summarized in the following sketch:

| Prior Probability | + | New Information | + | Application of Bayes's Theorem | → | Posterior Probability |

If we denote the probability of an event as $p(E)$, the probability of its complement as $p(\overline{E})$, and the result of an investigation as R, Bayes suggested that the posterior probability of event E could be calculated by the formula in Box 5.K.

5.K Bayes's Theorem

$$p(E|R) = \frac{p(E) \cdot p(R|E)}{p(E) \cdot p(R|E) + p(\overline{E}) \cdot p(R|\overline{E})}$$

Two things should be noted. First, the prior probability of the event, $p(E)$, is an unconditional one. The posterior probability, $p(E|R)$, is a conditional one, given the empirical result, R. Second, the entire formula can be rewritten to take the form of the expression given in Box 5.L. As a quick glance at the general multiplication law (Box 5.I) confirms, the numerator in Bayes's formula is nothing else but the joint probability of the given event and the empirical result, or $p(E \textit{ and } R)$. The denominator, therefore, is the *sum* of the joint probability of the given event and the observed result, or $p(E \textit{ and } R)$, and the joint probability of the given event's complement and the observed result, or $p(\overline{E} \textit{ and } R)$. This sum, as Table 5.3 confirms, is nothing else but the unconditional probability of the result, or $p(R)$. Thus, Bayes's theorem can also be written as in Box 5.L.

5.L Bayes's Theorem Rewritten

$$p(E|R) = \frac{p(E \textit{ and } R)}{p(R)}$$

The Bayesian posterior probability of an event is the ratio of the joint probability of this event and the empirical result to the unconditional probability of this result. The following examples illustrate the Bayesian approach.

Practice Problem 1

The Crooked Die

Consider being faced with three dice, one of which is known to be crooked. The crooked die, let us suppose, contains a lead weight that makes a 6 appear on half the tosses; yet the dice all look alike. If you know all of the above and now select one of the dice, what is the probability that it is the crooked one? Let us follow the revision sequence sketched above.

Step 1. Prior Probabilities. Letting C stand for "crooked" and $\overline{C}$ for "not crooked," the *prior* probabilities that you subjectively and arbitrarily select (because you are considering one die out of three) are the following:

$$p(C) = 1/3;$$
$$p(\overline{C}) = 2/3.$$

Step 2. New Information. Now comes the empirical investigation: you roll the die, and its face shows a 6. We have our result, and we can state the two conditional probabilities of this result, given each of the two possible events, easily enough:

$$p(6|C) = 1/2 \text{ (information given before)};$$
$$p(6|\overline{C}) = 1/6 \text{ (classical probability value)}.$$

Step 3. Applying the Theorem. We now apply Bayes's theorem from Box 5.K:

$$p(C|6) = \frac{p(C)p(6|C)}{p(C)p(6|C) + p(\overline{C})p(6|\overline{C})}$$

$$= \frac{\dfrac{1}{3} \cdot \dfrac{1}{2}}{\left(\dfrac{1}{3} \cdot \dfrac{1}{2}\right) + \left(\dfrac{2}{3} \cdot \dfrac{1}{6}\right)}$$

$$= \frac{\dfrac{1}{6}}{\dfrac{1}{6} + \dfrac{2}{18}}$$

$$= \frac{\dfrac{1}{6}}{\dfrac{5}{18}} = \frac{3}{5}.$$

Step 4. Posterior Probability. In light of the new information, the probability that the selected die is the crooked one is revised upward from the prior value of 1/3 (when nothing was known about the particular die) to a posterior value of 3/5 (when it was known that this die showed a 6 when rolled once). In light of the evidence, crookedness is more likely, but it is not certain, of course. (The 6 face could have been rolled by any die.) ▌▌▌

||| Practice Problem 2

Drilling for Oil

Consider an oil company about to drill for oil. The company geologists might state subjective probabilities for finding oil at the given site as follows:

$$p(\text{oil}) = .5$$
$$p(\text{no oil}) = .5$$

They drill for 500 feet, find no oil yet, but sample the soil. From past experience, they know the conditional probabilities of finding this type of soil, given oil or no oil:

$$p(\text{this type soil} \mid \text{oil}) = .2$$
$$p(\text{this type soil} \mid \text{no oil}) = .8$$

What is the posterior probability for finding oil, according to Bayes?

$$p(\text{oil} \mid \text{this type soil}) = \frac{.5(.2)}{[.5(.2)] + [.5(.8)]}$$
$$= \frac{.1}{.1 + .4} = \frac{.1}{.5} = .2$$

The new information reduces the probability of an oil find from .5 to .2. |||

||| Practice Problem 3

Screening Job Applicants

An employment agency administers a placement test to an equal number of men and women. Thus, the prior probability is .5 that any randomly selected test paper was written by a man.

The tests are graded. Some 60 percent of the tests score a C, and 30 percent of the tests score a C *and* are written by a man.

Given this new information, what is the likelihood that a randomly selected paper with a score of C was written by a man? The alternative version of Bayes's theorem can easily be applied:

$$p(\text{man} \mid \text{given score C}) = \frac{p(\text{man } and \text{ score C})}{p(\text{score C})}$$
$$= \frac{.3}{.6} = .5$$

The posterior probability that a C paper was written by a man is .5, which equals the prior probability of .5 for any paper. There is no reason for revising the prior probability value. |||

5.1 The ESP Mystery

In 1882, the Society for Psychical Research was founded in London to study paranormal (odd and unexplained) kinds of events. Mind reading, fortune-telling, and similar forms of alleged extrasensory perception (ESP) have in particular been subjected to experimental investigation. Is the phenomenon real?

Consider a typical card-guessing experiment. One person randomly deals 10 cards in one room, concentrating on their color (red or black). Another person, who knows that half the cards are red and half black, sits in another room and building. At the very time when the cards are dealt (watches have been synchronized), this person writes down the order of colors dealt. If, for example, 8 of the 10 cards are correctly identified, does this support the claim of ESP?

Probability theory can help. In how many ways could a person who was *guessing* assign red to 5 of the guesses (and black to 5)? This is the number of combinations of 10 items, taken 5 at a time, or

$$C_5^{10} = \frac{10!}{5!(10-5)!} = \frac{10 \cdot 9 \cdot 8 \cdot 7 \cdot 6}{5 \cdot 4 \cdot 3 \cdot 2 \cdot 1}$$
$$= \frac{30,240}{120} = 252.$$

If these 252 outcomes are equally likely, what is the probability of identifying correctly 8 of the 10 cards? The number of ways to identify correctly 4 of the 5 red cards is $C_4^5 = 5$; and for each of these ways, 4 black cards can be identified in $C_4^5 = 5$ ways as well. So the number of ways to identify 8 cards is $5 \times 5 = 25$. Hence, the probability of identifying correctly 8 of the cards is 25/252 or about .1. There is almost a 10 percent chance of getting the observed result by pure luck.

Many other experiments, however, have produced remarkable odds in favor of the existence of ESP. In one experiment, 558 correct identifications were made of 1,850 cards showing 5 types of symbols (cross, star, circle, square, wave). Guessing might have produced 370 correct identifications; the probability of achieving the actual result by accident was determined to be $1/10^{22}$—that is, once in 10 thousand billion billion times.

Sources: Adapted from J. Leroy Folks, *Ideas of Statistics* (New York: Wiley, 1981), pp. 86 and 87; and Darrell Huff, *How to Take a Chance* (New York: Norton, 1959), pp. 125–29.

5.2 Cosimo's Counting Problem

The mix-up of permutations with combinations can cause no end of troubles. Consider the case of some 17th-century Italian gamblers, such as His Serene Highness Cosimo II of Tuscany, who faced a paradox. When rolling three dice at a time, they argued, one could get the sum of 9 in six different ways; there were also six ways of getting the sum of 10. The possibilities they had in mind are illustrated in columns 1 and 3 of Table A, where 6–2–1, for example, represents getting a 6 on one die, a 2 on another die, and a 1 on another die yet.

There should exist, the gamblers figured, the same probability for getting the sum of 9 as for getting the sum of 10 (because six ways of getting such a sum exist in either case), but the gamblers had no such luck. Experience showed a 9 to come up less often than a 10. The paradox was taken to Galileo, First and Extraordinary Mathematician of the University of Pisa.

To appreciate his answer, we only need remind ourselves of the difference between permutations and combinations. The gamblers' selections of numbers, given in columns 1 and 3, are *combinations* of faces of dice that add to 9 or 10, respectively. Yet, Galileo noted, three dice often can produce a given combination in different ways. A single combination can have several *permutations*. There exist, for example, six ways for achieving the outcome 6–2–1, three ways of achieving 4–4–1, but only one way of achieving 3–3–3, as Table B attests.

The number of different ways to roll each one of the triplets of Table A, columns 1 and 3, are shown, respectively, in columns 2 and 4. This number of permutations equals 3! where 3 different components make up the combination, but 2! plus 1 (or 1! plus 2) where the combination consists of 3 components of which one (or two) are identical. As the two totals in

Table A

Possible Triplets for the Sum of 9 (1)	Number of Different Ways to Roll Triplet (2)	Possible Triplets for the Sum of 10 (3)	Number of Different Ways to Roll Triplet (4)
6-2-1	6	6-3-1	6
5-3-1	6	6-2-2	3
5-2-2	3	5-4-1	6
4-4-1	3	5-3-2	6
4-3-2	6	4-4-2	3
3-3-3	1	4-3-3	3
	Total: 25		Total: 27

Table B

Permutations of Combination 6-2-1			Permutations of Combination 4-4-1			Permutations of Combination 3-3-3		
1st Die	2nd Die	3rd Die	1st Die	2nd Die	3rd Die	1st Die	2nd Die	3rd Die
6	2	1	4	4	1	3	3	3
6	1	2	4	1	4			
2	6	1	1	4	4			
2	1	6						
1	6	2						
1	2	6						

Table A indicate, there exist 25 ways of rolling a sum of 9 with 3 dice, but 27 ways of rolling a sum of 10. On the other hand, Galileo noted, there exist $6 \cdot 6 \cdot 6 = 216$ possible triplets: 6 possible faces on the first die, each linked with 6 possible faces on the second die, and each of these linked with 6 possible faces on the third. Thus, the probability when rolling three dice of getting a sum of 9 equals 25/216, but that of rolling a sum of 10 is 27/216, a discrepancy that is consistent with the gamblers' experience.

Source: Adapted from F. N. David, *Games, Gods, and Gambling* (New York: Hafner, 1962), pp. 64–66.

5.3 The Miracle of the Matching Birthdays

Imagine yourself in a group of people and consider the probability that the birthdays (day and month, but not year) of at least two people in the group are exactly the same. If you are at all typical, you will view such a match as a highly unlikely event, unless, of course, the group is extremely large and contains, perhaps, one person for every day of the year. Let us be less hasty and instead consider the chances by using the laws of probability. We will assume that all birthdays are equally likely, except February 29, which we will equate with March 1. It is eas-iest to attack the problem backwards, by finding the probability of the complementary event that there is *no* match of birthdays within a group of n persons. We imagine ourselves asking one person at a time to reveal the birthday and then to compare the date so stated with all those previously revealed. Clearly, the probability of no match for the first person is

$$p(\overline{M}_1) = \frac{365}{365} = 1$$

Such certainty exists because there are 365 possible no-match dates when no previous date has been called (which accounts for the numerator), and there are, of course, 365 days in the year (which explains the denominator).

Yet the (conditional) probability for the second person's no-match, given the first person's no-match, is different because there are then only 364 possible no-match dates left (the date revealed by the first person might be a match). Thus,

$$p(\overline{M}_2|\overline{M}_1) = \frac{364}{365}$$

And so it goes, until, for the last and nth person in the group, the no-match probability equals

$$p(\overline{M}_n|\overline{M}_1 \text{ and } \overline{M}_2 \text{ and } \dots \overline{M}_{n-1})$$
$$= \frac{365 - (n - 1)}{365}$$

Combining our results with the help of the general multiplication law, we have

$$p(\text{no matches}) = \frac{365}{365} \times \frac{364}{365} \times \frac{363}{365}$$
$$\times \dots \times \frac{365 - n + 1}{365}$$

At this point, we can also state the complement of this expression and, thus, get the probability we really care about:

$$p(\text{at least one match}) = 1 - p(\text{no match})$$

What size of group is needed to make the probability of at least one match exceed .5? Amazingly, the answer is 23, as one can easily figure out by plugging alternative values for n into the above expression. Some of these values are shown in the accompanying table.

Optional extension: As the table shows, the probability of at least one match is considerably higher than what most people think. In particular, that probability for a group of n is not $n/365$, as is often thought. Consider a group of $n = 5$. As the previous analysis suggests,

$p(\text{at least one match in a group of 5})$
$$= 1 - \left[\frac{365}{365} \times \frac{364}{365} \times \frac{363}{365} \times \frac{362}{365} \times \frac{361}{365}\right]$$

This expression can be rewritten as

$p(\text{at least one match in a group of 5})$
$$= 1 - \left[\frac{365!}{(365 - 5)!} \times \frac{1}{365^5}\right]$$

which shows the probability of no match that is given in the brackets as equal to the set of all permutations of 365 days taken 5 at a time, divided by the sample space of 365^5 which gives all the possible comparisons of each birthday with every other birthday. Symbolically, we have

$p(\text{at least one match in a group of } n)$
$$= 1 - \left[\frac{365!}{(365 - n)! \, 365^n}\right]$$

which is a far cry from $n/365$.

Group Size, n	$p(\text{no matches})$	$p(\text{at least one match})$
5	.973	.027
10	.883	.117
15	.747	.253
20	.589	.411
23	.493	.507
30	.294	.706
40	.109	.891
50	.030	.970
60	.006	.994

5.1 Abraham de Moivre

Abraham de Moivre (1667–1754) was born at Vitry, France, where his father was a surgeon. De Moivre studied mathematics and physics in Paris, but in 1685, after the Edict of Nantes was revoked, he was imprisoned for being a Protestant. When released three years later, he emigrated to England to escape religious persecution. He never returned to France and never published anything in French.

By all accounts, he was a mathematical genius, and he was in constant touch (at the Royal Society) with the leading thinkers of his day, including Isaac Newton who became a close friend. Yet de Moivre never succeeded in obtaining a university appointment. He eked out a living by tutoring the sons of nobility and by advising gamblers and speculators. This unwelcome fate was posterity's gain, for his successful solution of the problems he met in his consulting practice led to his writing of two great books. His text on probability, *The Doctrine of Chances,* emanated from an article first published in Latin in 1711 and was published posthumously in its final and third edition in 1756. It is notable (among many other contributions) for the origin of the general laws of addition and multiplication of probabilities (discussed in the present chapter of this book), for the origin of the binomial distribution law (discussed in Chapter 6), and for the origin of the formula for the normal curve (discussed in Chapter 7), which de Moivre discovered in 1733. De Moivre's other book, *A Treatise of Annuities on Lives,* was published in 1752. It was highly original and laid foundations for the mathematics of life insurance.

De Moivre took great pains to free the science of probability from its connection with gambling and also to establish a connection between probability and theology. He says in *The Doctrine of Chances:*[1]

> And thus in all cases it will be found, that although Chance produces irregularities, still the Odds will be infinitely great, that in process of Time, those Irregularities will bear no proportion to the recurrency of that Order which naturally results from Original Design. . . . Again, as it is thus demonstrable that there are, in the constitution of things, certain Laws according to which Events happen, it is no less evident from Observation, that these Laws serve to wise, useful and beneficent purposes, to preserve the stedfast Order of the Universe, to propagate the several Species of Beings, and furnish to the sentient Kind such degrees of happiness as are suited to their State. . . . Yet there are Writers, of a Class indeed very different from that of *James Bernoulli,* who insinuate as if the *Doctrine of Probabilities* could have no place in any serious Enquiry; and that studies of this kind, trivial and easy as they be, rather disqualify a man for reasoning on every other subject. Let the Reader chuse.

[1]Abraham de Moivre, *The Doctrine of Chances,* 3rd ed. (London: 1756; reprinted New York: Chelsea, 1967), pp. 251–54.

Source: International Encyclopedia of Statistics, vol. 1 (New York: Free Press, 1978), pp. 601–4.

5.2 Pierre de Fermat

Pierre de Fermat (1601–1665) was born at Beaumont-de-Lomagne, France, where his father was a leather merchant. Even though he practiced law, when the courts went into recess he studied literature and mathematics. His knowledge of the chief European languages and the literature of Europe was wide; he even wrote verses in Latin, French, and Spanish. His main contribution, though, was to mathematics—in particular, the theory of probability. Many of his original ideas have been preserved in a vast correspondence with scientists all over Europe. Among the many correspondents was Blaise Pascal (Biography 5.3) whose name is always linked with Fermat as one of the joint discoverers of the laws of probability. Application 3: "The Paradox of the Chevalier de Méré," shows, perhaps, the most famous instance of this collaboration.

Source: F. N. David, *Games, Gods, and Gambling* (New York: Hafner, 1962), Chapters 8, 9, and Appendix 4.

5.3 Blaise Pascal

Blaise Pascal (1623–1662) was born at Clermont-Ferrand, France. He was a youngster of precocious mental ability and, from an early age, his father devoted himself almost entirely to developing the boy's reasoning powers. At age 14, the elder Pascal introduced Blaise into Mersenne's famous academy where musicians, mathematicians, and natural scientists met regularly. By age 16, Blaise wrote his famous *Essaie pour les Coniques,* which drew acclaim from many for the boy's astonishing mathematical powers. From projective geometry, he turned to mechanical computation, and two years later he invented a calculating machine. Still at a young age, he made a number of lasting contributions to mathematics (some of which are preserved in his correspondence with de Fermat—see Biography 5.2).

Unfortunately, Pascal's intellectual work was cut short by poor health and his desire to devote himself to prayer and self-denial. After his father's death, Pascal vacillated between a powerful entanglement with Jansenism, an anti-Jesuit religious cult, and a dissolute life (in which he quickly lost the moderate fortune he inherited but also met the Chevalier de Méré noted in Application 3).

Today, strangely enough, Pascal is best known for Pascal's Triangle, although, contrary to his own claims in his *Traité du Triangle Arithmétique,* he was surely not its originator. The triangle represents an ancient system of quickly answering questions such as these: In a family of 10 children, what is the likelihood that exactly 3 will be girls? When tossing a coin 10 times, what is the probability that exactly 3 heads will appear?

When spinning the roulette wheel 10 times (and not counting the zero) what is the chance for getting red precisely 3 times?

The answer, of course, is the same each time, and it can be found with the help of a tree diagram that notes all the possible $2^{10} = 1,024$ outcomes and allows one to count the ones that meet the specifications. Yet less laboriously, one can create the triangle in question and shown in the accompanying figure.

To begin with, note how the triangle is constructed, without trying to fathom its meaning. To construct it, one would begin by writing down two 1's in the first row, side by side. Next, in the second row, below to the left and below to the right, one would put two more 1's and would then fill in the gap between with the *sum* of the two numbers above the gap. These numbers in row 1 are 1's, so one would write 2. Next, one would put 1's at the extremes of the third line and would then fill in the two gaps with 3's (once more the *sum* of the numbers above each gap, which are 2 and 1). And so it goes. By the time one would reach the 10th line, one would have constructed Pascal's triangle as printed here. But note: there is no limit to the number of rows one could construct. Can you see why an 11th row would contain entries of 1, 11 (the sum of 1 + 10), 55 (the sum of 10 + 45), 165, 330, 462, 462, 330, 165, 55, 11, and 1?

What does it all mean? First, as the accompanying table lists, the *number of each row* represents the number of trials of some experiment with two possible types of outcome per trial. Row 1 represents *one* trial, such as tossing a coin once. Row 2 represents *two* trials, such as tossing a coin twice. Row 10 represents *ten* trials, such as tossing a coin ten times.

```
                1   1
              1   2   1
            1   3   3   1
          1   4   6   4   1
        1   5  10  10   5   1
      1   6  15  20  15   6   1
    1   7  21  35  35  21   7   1
  1   8  28  56  70  56  28   8   1
1   9  36  84 126 126  84  36   9   1
1  10  45 120 210 252 210 120  45  10   1
```

Row Number = Number of Trials, n	Number of Possible Outcomes, 2^n
1	2
2	4
3	8
4	16
5	32
6	64
7	128
8	256
9	512
10	1,024

Second, the *sum of the actual entries in each row* represents the number of possible outcomes. The numbers in row 1 sum to 2; there are 2 possible outcomes of one trial (such as heads or tails, if a coin is tossed once). The numbers in row 2 sum to 4; there are 4 possible outcomes of two trials (such as HH, HT, TH, and TT if a coin is tossed twice). And the sum of entries in row 10 tells us that there are 1,024 possible outcomes in 10 trials (as when a coin is tossed ten times).

Third, *each individual number in a given row* represents the possible number of times some particular outcome (such as x heads) will occur. The first number in a given row tells us how often n particular outcomes (such as n heads) will occur in n trials; the second number tells us how often $n - 1$ particular outcomes (such as $n - 1$ heads) will occur in n trials, and so on. In row 1, for example, the first 1 might denote 1 possibility of getting $n = 1$ heads in 1 trial; the second 1 might denote 1 possibility of getting $n - 1 = 0$ heads in 1 trial. In row 2, similarly, the first 1 might denote 1 possibility of getting $n = 2$ heads in 2 trials; the 2 might denote 2 possibilities of getting $n - 1 = 1$ head in 2 trials, and the last 1 might denote 1 possibility of getting $n - 2 = 0$ heads in 2 trials. Given the 4 possible outcomes, these numbers imply, of course, probabilities of 1/4, 2/4, and 1/4, respectively. Thus row 10 gives, for example, the probabilities for a sequence of 10 children, 10 tosses of a coin, 10 spins of the wheel at roulette, or 10 of any series of

equal chances. Note how the numbers in row 10 add to 1,024. These are all the *possible* outcomes. We can read the numbers in the row, from left to right. The first one is a 1. There is a probability of 1 in 1,024, it tells us, of having 10 girls out of 10 children (or 10 heads in 10 tosses or 10 reds in 10 spins of the wheel).

We turn to the second number, a 10. There is a probability of 10 in 1,024, it says, of having 9 girls out of 10 children (or 9 heads or 9 reds). And there is a probability of 45 in 1,024, the third number says, of having 8 girls out of 10 (or 8 heads or 8 reds). And so it goes until the final number in row 10 gives us the probability of 0 girls (or heads or reds) out of 10: it is 1 in 1,024.

Thus, we have our answer: there are 120 chances in 1,024 for each of the 3-in-10 events listed.

Note: Can you see that row 2 tells us the same thing as the tree diagram in panel (a) of Figure 5.2? There are four possibilities, it says, the sum of the numbers in line 2. There is 1 chance in 4 of getting 2 heads, there are 2 chances in 4 of getting 1 head, there is 1 chance in 4 of getting 0 heads. Pascal's triangle, it turns out, is another way (and a quicker one) of getting some of the results a tree diagram might provide.

Sources: F. N. David, *Games, Gods, and Gambling* (New York: Hafner, 1962), Chapters 8, 9, and Appendix 4; Darrell Huff, *How to Take a Chance* (New York: Norton, 1959), pp. 170–73; *Dictionary of Scientific Biography,* vol. 10 (New York: Scribner's, 1974), pp. 330–42.

Summary

1. The theory of probability helps us figure the likelihood of specific occurrences. Among the important concepts that the theory employs are those of the random experiment, the sample space, and the random event. Any activity that will result in one and only one of several well-defined outcomes, but that does not allow us to tell in advance which of these will prevail in any particular instance, is called a *random experiment.* Any one of a random experiment's possible outcomes, the occurrence of which rules out the occurrence of all the alternative outcomes, is called a *basic outcome;* a listing of all basic outcomes constitutes the *sample space.* Any subset of the sample space is a *random event;* it can be simple (containing one basic outcome) or composite (containing more than one). It is crucial to distinguish between mutually exclusive, collectively exhaustive, complementary, and compatible events as well as to distinguish the union from the intersection of two events.

2. Two types of random experiments exist: those that can be repeated over and over again and those that are unique. Accordingly, there are two numerical measures of chance that estimate the likelihood that a particular event will occur. Such a measure for a repeatable random experiment is called an *objective* probability; a corresponding value for a unique random experiment is called a *subjective* probability. The objective probability can be determined theoretically (by conducting repeated thought experiments) as the ratio of the number of equally likely basic outcomes favorable to an event A to the possible number of such outcomes. An objective probability can also be determined empirically (by conducting repeated actual experiments) as the ratio of the number of times an event A did occur in the past to the maximum number of times event A could have occurred. In either case, the resultant ratio must lie between 0 and 1. The same is true for a subjective probability

value that, in contrast, is derived neither by logical deduction nor from repeated experiments but reflects solely a personal degree of belief in the likelihood of an occurrence.

3. When sample spaces are extremely large, special counting techniques are needed to determine the numbers of favorable and possible outcomes of random experiments (and, thus, probability values). Such counting techniques employ the concepts of factorials, permutations, and combinations.

4. Deriving probability values for single events is one thing; manipulating these values to determine the combined probability for the union or intersection of several events is another. The procedure for computing the probability for the occurrence of a *union* of two or more events is summarized by the *addition law*. A general version applies to all types of events; a special law applies to mutually exclusive events only. The procedure for computing the

probability for the occurrence of an *intersection* of two or more events is summarized by the *multiplication law*. A general version applies to all types of events; a special law applies to independent events only. A full understanding of the general multiplication law requires an understanding of the concepts of unconditional, conditional, and joint probability; any application of the special law presupposes our ability to distinguish dependent from independent events. There are a number of famous applications of the probability laws of addition and multiplication, and these laws can be illustrated with the help of tree diagrams.

5. An important extension of the traditional calculus of probability is provided by *Bayes's theorem,* a formula for revising a subjective *prior* probability value on the basis of results obtained by an empirical investigation and for, thus, obtaining a *posterior* probability value.

Key Terms

addition law
basic outcome
Bayes's theorem
collectively exhaustive events
combinations
compatible events
complementary events
composite events
conditional probability
dependent events
disjoint events
elementary events
factorial

factorial product
incompatible events
independent events
intersection of two events
joint probability
law of large numbers
marginal probabilities
multiplication law
mutually exclusive events
objective probability
outcome space
permutations
posterior probability

prior probability
probability space
random event
random experiment
sample points
sample space
simple event
subjective probability
theory of probability
tree diagram
unconditional probability
union of two events
Venn diagrams

Questions and Problems

The computer programs noted at the end of this chapter can be used to work many of the subsequent problems.

1. a. Determine the number of basic outcomes on the *American-style* roulette wheel shown in Figure A. Are they univariate?

b. Determine the number of basic outcomes when tossing three coins once. Are they univariate? Find your answer with the help of a tree diagram.

2. With the help of a tree diagram, determine the sample space for a random experiment that takes 3 balls from an urn containing 1 red ball, 1 green ball, and 1 white ball, and do so *with* replacement.

Figure A

3. In the four sample spaces given in Figures B, C, D, and E, show the favorable outcomes for the events noted below by shading the appropriate subsets of the sample spaces. (i) Shade the subset of Figure B that corresponds to getting a sum of 5 or 6 when rolling two dice once. (ii) Shade the subset of Figure C that corresponds to getting a sum of 2 when rolling two dice once. (iii) Shade the subset of Figure D that corresponds to drawing a black king from a deck. (iv) Shade the subset of Figure E that corresponds to drawing a black card or an ace from a deck.

4. Classify each of the following random events as simple or composite: (i) Getting a 3 when rolling a single die once. (ii) Getting an odd number when rolling a single die once. (iii) Getting a sum of 5 or 6 when rolling two dice once. (iv) Getting a sum of 2 when rolling two dice once. (v) Getting a 7 when drawing one card from a deck.

5. Classify each of the following sets of events in two ways: first, as mutually exclusive or compatible and, second, as collectively exhaustive or not collectively exhaustive: (i) Money supply growth of 5 percent per year; money supply growth of 9

Figure B

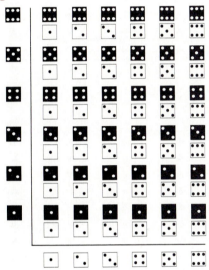

Figure C

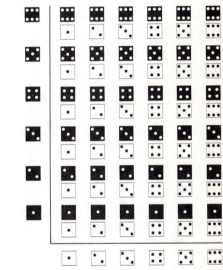

Figure D

	Clubs	Diamonds	Hearts	Spades
King	♣ K	♦ K	♥ K	♠ K
Queen	♣ Q	♦ Q	♥ Q	♠ Q
Jack	♣ J	♦ J	♥ J	♠ J
10	♣ 10	♦ 10	♥ 10	♠ 10
9	♣ 9	♦ 9	♥ 9	♠ 9
8	♣ 8	♦ 8	♥ 8	♠ 8
7	♣ 7	♦ 7	♥ 7	♠ 7
6	♣ 6	♦ 6	♥ 6	♠ 6
5	♣ 5	♦ 5	♥ 5	♠ 5
4	♣ 4	♦ 4	♥ 4	♠ 4
3	♣ 3	♦ 3	♥ 3	♠ 3
Deuce	♣ 2	♦ 2	♥ 2	♠ 2
Ace	♣ A	♦ A	♥ A	♠ A

Figure E

	Clubs	Diamonds	Hearts	Spades
King	♣ K	♦ K	♥ K	♠ K
Queen	♣ Q	♦ Q	♥ Q	♠ Q
Jack	♣ J	♦ J	♥ J	♠ J
10	♣ 10	♦ 10	♥ 10	♠ 10
9	♣ 9	♦ 9	♥ 9	♠ 9
8	♣ 8	♦ 8	♥ 8	♠ 8
7	♣ 7	♦ 7	♥ 7	♠ 7
6	♣ 6	♦ 6	♥ 6	♠ 6
5	♣ 5	♦ 5	♥ 5	♠ 5
4	♣ 4	♦ 4	♥ 4	♠ 4
3	♣ 3	♦ 3	♥ 3	♠ 3
Deuce	♣ 2	♦ 2	♥ 2	♠ 2
Ace	♣ A	♦ A	♥ A	♠ A

percent per year. (ii) Sales under $5 million per year; sales between $5 million and under $10 million per year; sales of $10 million per year or more. (iii) Profits between $1 and $7 million per year; profit of $5 million per year. (iv) Defective parts below 7 percent of output; defective parts above 2 percent of output; defective parts equal to 5 percent of output. (v) Real GNP growth of 3 percent per year; inflation of 7 percent per year; unemployment of 8 percent of the labor force.

6. Consider the sample space $S = \{1, 2, 3, 4, 5, 6, 7, 8, 9, 10\}$, as well as these random events:

$A = \{2, 4, 6, 8, 10\}$ $D = \{3, 4, 5, 6, 7\}$

$B = \{1, 2, 9, 10\}$ $E = \{7, 8, 9\}$

$C = \{1, 3, 5, 7, 9\}$ $F = \{2, 3, 4, 5\}$

Are the following sets of events mutually exclusive, collectively exhaustive, both, or neither?

A, B	B, C	C, E	A, D, F
A, C	B, D	C, F	B, D, E
A, D	B, E	D, E	C, E, F
A, E	B, F	D, F	
A, F	C, D	E, F	

7. State the complements of each of the following events: (i) Finding defective items in a group of items. (ii) Inflation above 10 percent per year. (iii) Real GNP growth of 3 percent per year or more. (iv) Drawing an ace from a deck of cards. (v) Finding no defective items in a group of items.

8. Identify each of the following descriptions as either the union or the intersection of events: (i) Drawing the king of hearts from a deck of cards. (ii) Drawing a black king from a deck of cards. (iii) Drawing a black card or an ace from a deck of cards. (iv) Drawing an ace or a deuce from a deck of cards. (v) Drawing a red face card from a deck of cards.

9. A questionnaire is sent to 500 homes. If we define encountering a brick house as event A, encountering a house over 40 years of age as event B, and encountering an all-electric house as event C, are the following unions, intersections, or complements of events: A and B; $\overline{A}$; $\overline{B}$ and C; A or C; B or $\overline{C}$.

10. In Figures F and G, depict your answers to 8 (iv) and 8 (v) graphically.

Figure F

	Clubs	Diamonds	Hearts	Spades
King	♣ K	♦ K	♥ K	♠ K
Queen	♣ Q	♦ Q	♥ Q	♠ Q
Jack	♣ J	♦ J	♥ J	♠ J
10	♣ 10	♦ 10	♥ 10	♠ 10
9	♣ 9	♦ 9	♥ 9	♠ 9
8	♣ 8	♦ 8	♥ 8	♠ 8
7	♣ 7	♦ 7	♥ 7	♠ 7
6	♣ 6	♦ 6	♥ 6	♠ 6
5	♣ 5	♦ 5	♥ 5	♠ 5
4	♣ 4	♦ 4	♥ 4	♠ 4
3	♣ 3	♦ 3	♥ 3	♠ 3
Deuce	♣ 2	♦ 2	♥ 2	♠ 2
Ace	♣ A	♦ A	♥ A	♠ A

Figure G

	Clubs	Diamonds	Hearts	Spades
King	♣ K	♦ K	♥ K	♠ K
Queen	♣ Q	♦ Q	♥ Q	♠ Q
Jack	♣ J	♦ J	♥ J	♠ J
10	♣ 10	♦ 10	♥ 10	♠ 10
9	♣ 9	♦ 9	♥ 9	♠ 9
8	♣ 8	♦ 8	♥ 8	♠ 8
7	♣ 7	♦ 7	♥ 7	♠ 7
6	♣ 6	♦ 6	♥ 6	♠ 6
5	♣ 5	♦ 5	♥ 5	♠ 5
4	♣ 4	♦ 4	♥ 4	♠ 4
3	♣ 3	♦ 3	♥ 3	♠ 3
Deuce	♣ 2	♦ 2	♥ 2	♠ 2
Ace	♣ A	♦ A	♥ A	♠ A

11. Classify each of the following as an objective or subjective probability statement (and if the former, as a classical or empirical one): (i) There is one chance in two for a recession this fall. (ii) There is 1 chance in 100 that we will meet a car coming up the other side of this hill. (iii) The probability is 1/52 that one picks the king of spades when taking one card from a deck. (iv) There is no chance for a meltdown of this nuclear plant. (v) The odds are 20 to 1 in favor of sales going up because of our product's new name. (vi) Past experience indicates

that the odds against engine-mount failure on this type of airplane are a million to one. (vii) There is one chance in three that the next U.S. President will be a Democrat. (viii) In 600 tosses of a die, you can expect to get a 3 face 100 times. (ix) The probability is .0001 that radicals will get hold of nuclear weapons. (x) The probability of picking an even-denomination card from a deck equals 5/13.

12. Calculate the following (objective) probabilities:
  a. Getting red when spinning the American roulette wheel (Figure A).
  b. Getting two heads and one tail in the random experiment described in problem 1b.
  c. Getting two red balls in the random experiment described in problem 2.
  d. For each of the four events described in problem 3.
  e. For each of the five events described in problem 4.
  f. For each of the five events described in problem 8.

13. Comment on the following bet someone suggests to you: "Let's toss two dimes at a time. Three outcomes are possible: 2 heads, 2 tails, or one of each. If two heads show or two tails, you win; if there is one tail and one head, I win. Because you have two chances out of three to win and I have only one, I will pay you 50¢ whenever you win, but you must pay me $1 whenever I win. In the long run, neither one of us will win or lose."

14. Comment on this reasoning by a sales clerk: "From past experience, I know that 1 customer out of 20 buys my product. I have just had 19 customers who didn't buy. The next one is certain to buy."

15. a. Consider Figure A-13 at the back of the book. If the data still pertain, what is the probability that a randomly selected car renter is an Avis customer?
  b. Consider Table A.11 at the back of the book. What is the probability that a randomly selected can of peas will have a net weight in the 15.5 to under 15.7 oz. range?

16. Select a single letter at random from the English alphabet. What is the probability of it being a vowel? What if you had selected this letter from among the alphabet's first 10 letters?

17. Consider Table 4B that accompanies Chapter 4's problem 8. If an order is randomly selected, what is the probability of its value being under $100? Between $100 and under $500? Above $500?

18. Compute factorials for the following numbers: 3, 11, 19, 25.

19. For each of the following sets, compute the number of permutations of x at a time out of n distinct items, with no repetitions allowed.

n	x
2	1
6	4
15	2
25	7
25	32

20. Create a table entitled "Number of Permutations (x at a time out of n distinct items)" that lists values of n from 0 to 10 in the first column, values of x from 0 to 10 in the first row, and the values of P_x^n in the body of the table.

21. A ship has five flags each of green, red, and yellow. If it uses all 15 flags for each, how many different messages can it send? What if it had ten flags, of which two each were green and red, while three each were blue and yellow?

22. For each of the following sets, compute the number of combinations of x at a time out of n distinct items, with no repetitions allowed.

n	x
2	2
7	3
13	3
22	10
30	35

23. Create a table entitled "Number of Combinations (x at a time out of n distinct items; no repetitions allowed)" that lists values of n from 0 to 50 in the first column, values of x from 0 to 10 in the first row, and the values of C_x^n in the body of the table.

24. In the Massachusetts Megabuck Lottery, six numbers out of 1 to 36 must be picked (in any order, with no repetitions allowed). How many different tickets would one have to buy in order to be *certain* to win? Given the fact that the jackpot usually varies between $3 million and $10 million, and a ticket costs only $1, why doesn't anyone do it?

25. If a restaurant serves 3 salads, 5 entrees, 4 vegetables, 6 drinks, and 7 desserts, in how many different ways can one get 2 entrees, 2 vegetables, and 2 desserts? If a complete dinner includes two of everything, how many different dinner combinations are being offered?

26. How many different combinations of 3 cards can be taken from a deck of 52 cards? How many permutations? Show your calculations.
27. How many different committees of 3 males and 2 females can be formed from a group of 10 males and 10 females?
28. If each participant in a 10-player tournament must play every other participant, how many plays must be made?
29. If 13 cards are taken from a deck, what is the probability that they include the king of spades and the king of hearts?
30. If 5 cards are taken from a deck, what is the probability of getting a *royal* flush (a sequence from 10 to king plus ace of the same suit)? Of getting *any* straight flush (a sequence of 5 cards from the same suit, starting with any number between 2 and 10)?
31. If you think it applies, use the general or special addition law to determine each of the following probabilities.
    a. The probability of getting a sum of 5 *or* 6 when rolling two dice.
    b. The probability of getting a black card *or* an ace from a deck of cards.
    c. The probability of inflation *or* recession, if that of inflation is .8, that of recession is .2, and that of both at the same time is .1.
    d. The probability for a city experiencing, before the end of this century, a natural disaster such as an earthquake, a flood, or a tornado, if the separate probabilities are .1, .4, and .7, respectively.
32. A market survey shows that 60 percent of consumers like Brand A, 40 percent like Brand B, 15 percent like both. What is the probability of someone chosen at random liking A *or* B?

33. A market survey yields the results given in Table 5.A. What is the probability of someone chosen at random liking A *or* B *or* C:

Table 5.A   Percent of Consumers Who Like Brand

A	20
B	10
C	5
A and B	8
A and C	2
B and C	7
All three	5

34. From the information in Table 5.B, construct a joint probability table and clearly identify all the joint and marginal probabilities.
35. Calculate all the conditional probabilities associated with problem 34.
36. Calculate the six conditional probabilities associated with the joint probabilities of .10, .13, and .57 that are given in Table 5.3.
37. From the information given in Tables 5.C to 5.F, calculate the following probabilities and identify them as conditional, joint, or marginal: (i) The probability of an inspector tagging a satisfactory item as unacceptable. (ii) The probability of a female guest ordering dish A. (iii) The probability of a random person being a nonsmoker with heart disease. (iv) The probability of a person with heart disease being a nonsmoker. (v) The probability of a smoker being free of heart disease. (vi) The probability of a loan being under $500 and also in excess of 30 days old. (vii) The probability of a loan being above $500 in size.

Table 5.B   Number of Workers Whose . . .

Score on Hiring Test Was . . .	Performance on the Job During First Year Was . . .		Total
	Good, G	Bad, B	
High, H	240	60	300
Low, L	10	90	100
Total	250	150	400

38. Still using Tables 5.C to 5.F, calculate the following probabilities and identify them as conditional, joint, or marginal: (i) The probability of an item being defective. (ii) The probability of a defective item being tagged acceptable. (iii) The probability of a B order having come from a male. (iv) The probability of a B order. (v) The probability of heart disease. (vi) The probability of a loan being above $500 and also in excess of 30 days old.

39. Apply the general multiplication law, according to which $p(A \text{ and } B) = p(A) \cdot p(B|A)$, to the intersection of the two events given in the first row and

Table 5.C   Number of Items . . .

Inspected and . . .	Tagged as Acceptable, A	Tagged as Unacceptable, U	Total
In Fact Satisfactory, S	600	30	630
In Fact Defective, D	35	35	70
Total	635	65	700

Table 5.D   Number of Restaurant Patrons Who . . .

	Ordered Dish A	Ordered Dish B	Total
Were Male, M	20	180	200
Were Female, F	90	10	100
Total	110	190	300

Table 5.E   Number of Persons Who . . .

	Had Heart Disease, H	Were Free of Heart Disease, F	Total
Were Smokers, S	300	200	500
Were Nonsmokers, N	100	300	400
Total	400	500	900

Table 5.F   Number of Loans . . .

	30 Days Old or Less, L	In Excess of 30 Days Old, E	Total
Of up to $500 Inclusive, U	700	1,000	1,700
Above $500, A	1,400	2,000	3,400
Total	2,100	3,000	5,100

second column of each of Tables 5.B to 5.F. Confirm your results by direct calculation from the table data.

40. Rework problem 39, but with respect to the intersection of the two events given in the second row and first column of Tables 5.B to 5.F.

41. An analysis of a recent labor-union vote on a new contract shows this: In the Northeast, 5,000 of 20,000 members voted yes; in the Southeast, 3,000 of 7,000 members voted no; in the Northwest, 4,000 of 10,000 members voted yes; and in the Southwest, 13,000 of 21,000 members voted yes. (There were no abstentions.) Calculate these probabilities: (a) that a randomly chosen member of this union was from the Northeast and voted no; (b) that a randomly chosen no-voter was from the Southwest; (c) that a randomly chosen member from the Northwest or Southwest voted yes; (d) that a randomly chosen member voted yes. In each case, identify the nature of the probabilities involved.

42. If you discovered any joint probabilities among the four cases noted in problem 41, confirm your answer with the help of an appropriate law from the calculus of probability.

43. Given $p(A) = .55$, $p(B) = .44$, $p(C|A) = .6$, $p(D|A) = .4$, $p(C|B) = .25$, $p(D|B) = .75$, compute $p(A \text{ and } C)$, $p(A \text{ and } D)$, $p(B \text{ and } C)$, $p(B \text{ and } D)$.

44. Given $p(A \text{ and } B \text{ and } C) = .014$, $p(B|A) = .7$, and $p(C|A \text{ and } B) = .1$, compute $p(A)$.

45. Consider Tables 5.C to 5.F and, using an appropriate formula from the calculus of probability determine $p(S \text{ or } A)$, $p(M \text{ or } A)$, $p(N \text{ or } H)$, $p(U \text{ or } L)$.

46. Once again find the answers to problem 45, but this time do so without a formula, simply by inspecting the table entries in question.

47. Are the events noted in problem 39 dependent or independent? Why?

48. Are the events noted in problem 40 dependent or independent? Why?

49. Comment on the following statement: "A World War II bomber pilot had a 2-percent chance of being shot down on a given mission. Thus, a pilot asked to do 50 missions would be shot down with certainty."

50. If the probability is .9 that a given helicopter remains operable in a three-hour desert sandstorm, what is the probability of successfully carrying out a 6-helicopter mission (such as the rescue attempt of Americans in Iran in 1980) if only 6 helicopters are available and such a storm occurs?

51. Comment on the following: "Incoming missiles have to pass five defense barriers. The probability of a missile being shot down is .20 at any one barrier; hence, the probability of a given missile getting through all five of the barriers is zero."

52. Let events A and B be independent. If $p(A) = .4$ and $p(B) = .5$, what is $p(A \text{ or } B)$, $p(A \text{ and } B)$, $p(\overline{A})$, $p(\overline{A} \text{ and } \overline{B})$, and $p(A|B)$?

53. Given a vaccination, the probability of getting the flu is .2, but without a vaccination, it is .4. Some 40 percent of the population is vaccinated.
a. If a random selection of a person is made, what is the person's probability of getting the flu?
b. How many out of 1 million will probably get the flu?
c. Are not getting the flu and being vaccinated independent events?

54. On the basis of the data given in Table 5.C, draw two tree diagrams analogous to panels a and b of text Figure 5.9.

55. On the basis of the data given in Table 5.F, draw two tree diagrams analogous to panels a and b of text Figure 5.9.

56. In what important way do the two sets of illustrations produced in answering problems 54 and 55 differ from one another?

57. A new movie is released. The producing company judges the probability of a smashing success at $p = .3$. It also knows that a certain reviewer has liked 75 percent of all movies that became greatly successful and has disliked 90 percent of all movies that later turned out to be ghastly failures. If the company followed the Bayesian approach, how would it revise its probability-of-success figure if it learned that the reviewer praised the movie? What if the reviewer criticized it severely?

58. The local weather forecaster gives an 80-percent chance of thunderstorms. A pilot uses this forecast as a prior probability. However, the pilot also "reads" the degree of arthritic pain experienced; it is severe on 80 percent of all days with thunderstorms and on 10 percent of all days without them. Right now, the pain is severe. What is the pilot's posterior probability of thunderstorms?

59. A wine producer has designed a new and distinctive bottle in the hope of increasing sales. The manager views the probability of success as .5 but also orders a survey of customers. The manager knows that when consumers are enthusiastic and sales are about to rise, the type of survey about to be taken will so indicate 90 percent of the time but that in 10 percent of the cases, the survey will say the opposite. When consumers are unimpressed and sales prospects look dim, the survey will so indicate 60 percent of the time, but in 40 percent of the cases, it will say the opposite. The survey is taken and shows great consumer enthusiasm about the bottle. What is the manager's posterior probability of success?

60. A personnel manager knows that 50 percent of the workers who are hired without a screening test perform satisfactorily on the job. However, among workers who take the test and later do well on the job 90 percent pass it, while among those who take the test and later do badly on the job, 90 percent fail it. What is the manager's posterior probability of satisfactory job performance if a randomly chosen applicant (a) passes the test and (b) fails the test?

Selected Readings

David, F. N. *Games, Gods, and Gambling.* New York: Hafner, 1962. On the origin and history of probability from the earliest times to the Newtonian era.

Feller, William. *An Introduction to Probability Theory and Its Applications.* New York: Wiley, 1968–1971, 2 volumes. A classic text.

Heron House Editors. *The Odds on Virtually Everything.* New York: Putnam's, 1980. A popular book that uses empirical data to determine probabilities for practically anything, anywhere, any time.

Huff, Darrell, *How to Take a Chance.* New York: W. W. Norton, 1959. An amusing but well-informed book on probability and its many applications.

Kerrick, J. E. *An Experimental Introduction to the Theory of Probability.* Copenhagen: Jorgensen, 1946. Reports on the experimental determination of probability, including a lengthy coin-tossing experiment (that yielded a probability of .507 for heads after the 10,000th toss).

Laplace, Pierre Simon, Marquis de. *A Philosophical Essay on Probabilities.* New York: Dover, 1951.

Mises, Richard von. *Probability, Statistics and Truth.* London: Allen and Unwin, 1961. A classic statement of the objectivist view of probability theory that defines probability as the relative frequency of the observed attribute that would be found if the observations were indefinitely continued.

Mumford, A. G. "A Note on the Uniformity Assumption in the Birthday Problem." *The American Statistician,* August 1977, p. 119. An extension of the matching birthday problem discussed in Analytical Example 5.3 that shows that the probability of at least one match is increased if all birthdays are not equally likely.

Savage, Leonard J. *The Foundations of Statistics.* New York: Dover, 1972. A classic statement arguing that subjective probability alone is essential to decision making and rejecting the von Mises view noted above.

Schlaifer, Robert. *Probability and Statistics for Business Decisions.* New York: McGraw-Hill, 1959. One of the first books emphasizing the Bayesian approach to decision making that blends subjective probabilities with objective ones.

Todhunter, Isaac. *A History of the Mathematical Theory of Probability.* New York: Chelsea, 1949. Deals with the development of the theory from the time of Pascal to the time of Laplace.

Computer Programs

The DAYSAL-2 personal-computer diskettes that accompany this book contain two programs of interest:

6. Factorials, Permutations, Combinations lets you make the calculations called for by Boxes 5.C–5.F, as well as others that involve permutations or combinations which do allow repetition among the x items picked at a time.

18. Cointoss, if you have graphics capability, lets you perform the kind of experiment illustrated in Figure 5.7.

CHAPTER SIX

Discrete Probability Distributions

In an earlier chapter, we defined *variables* as characteristics possessed by persons or objects, called elementary units, in which a statistician is interested. Such variables, we noted, can be *qualitative* or *quantitative*. Qualitative variables differ in kind rather than in degree among elementary units and are normally not described numerically. Quantitative variables, on the other hand, differ in degree rather than in kind among elementary units, and these variables are expressed numerically. Consider conducting a random experiment such as tossing three coins or taking a sample of five employees of a firm. If one were interested only in qualitative characteristics, such as the types of faces showing after the coins had been tossed or the races of the selected employees, one might record the experimental outcomes in words, such as "head, tail, head" or "black, white, black, white, white." If, on the other hand, one were interested in quantitative characteristics such as the numbers (or proportions) of heads showing in different trials of the experiment or the numbers (or proportions) of black employees found in different samples, one would record the outcomes numerically; perhaps, as 0, 1, 2, or 3 heads in a toss, as 0, 1, 2, 3, 4, or 5 blacks in a sample, or as the proportions such outcomes represent among all possible outcomes (1/3 heads, 2/5 blacks, and so on). This chapter and the next will deal exclusively with such quantitative variables. The present chapter, moreover, will be concerned only with *discrete* quantitative variables that can assume values only at specific points on a scale of values, with inevitable gaps between. The possible observations about such variables are countable and (with some exceptions, such as the Poisson variable to be discussed later in the chapter) are finite. (The number of heads showing when coins are tossed or the number of black employees in firms are examples of such discrete quantitative variables.) The next chapter will focus on *continuous* quantitative variables that can assume values at all points on a scale of values, with no breaks between possible values. The possible observations about such variables are not countable and are infinite in number. (Unlike the numbers of black employees, their ages, heights, or weights provide examples of such continuous quantitative variables.)

BASIC CONCEPTS

Before proceeding, we must become familiar with two important new concepts—namely, *random variable* and *probability distribution*.

The Random Variable

It is customary to refer to any quantitative variable the numerical value of which is determined by a random experiment (and, thus, by chance) as **random variable,** or **chance variable.** The number of heads observed when tossing three coins is such a variable; chance determines whether the number is 0, 1, 2, or 3. The number of points showing on the top face of a die that is rolled once is a random variable, too; so is the difference or *point spread* observed between the top faces of two dice that are rolled at the same time. The serial number of a lottery ticket about to be drawn from among 100,000 tickets in a bowl is a random variable as well.

As is usual in the field of statistics, this new concept is, however, applicable far beyond the world of games and gambling. Imagine yourself about to draw a random sample from among a list of employees of a firm in order to determine, say, their mean salary (or the proportion of blacks on the payroll). Before the sample is taken, the sample mean (or proportion) can also be viewed as a random variable because chance will determine which of many possible samples will actually be selected and, therefore, which of many possible sample means or proportions will actually be observed. Innumerable other variables that are of interest to business executives and economists are random variables, too. These range widely from tomorrow's closing price of a given stock to the number of customers arriving during the next 10 minutes at the supermarket checkout counter to the future rate of return on an investment undertaken today.

The Probability Distribution

Each of the possible numerical values of a random variable will occur, however, with a certain probability; a systematic listing of each possible value of a random variable with the associated likelihood of its occurrence is called the random variable's **probability distribution.** Table 6.1 shows two such distributions for discrete random variables. Part A gives the possible values, *x,* of the point spread when two dice are rolled at the same time, along with the associated probability, *p,* that the actual value of the random variable, *R,* emerging from such a random experiment equals any given possible value, *x.* The stated probabilities can be confirmed by referring to panel (b) of Figure 5.2 on page 139. Naturally, the sum of the separate probabilities of the six mutually exclusive and collectively exhaustive events equals 1. Part B shows another discrete probability distribution (the data are assumed); this distribution concerns the

Table 6.1   Two Discrete Probability Distributions

As these examples indicate, a probability distribution is similar to a relative frequency distribution. A probability distribution provides a probability value, p, for each value, x, of a random variable, R.

| (A) Random Variable, R, = Point Spread from Rolling Two Dice | | (B) Random Variable, R, = Hourly Credit-Verification Requests | |
Possible Values, x	Probability, $p(R = x)$	Possible Values, x	Probability, $p(R = x)$
0	6/36	0	.05
1	10/36	1	.10
2	8/36	2	.15
3	6/36	3	.17
4	4/36	4	.19
5	2/36	5	.09
	$\Sigma p = (36/36) = 1.00$	6	.08
		7	.06
		8	.06
		9	.02
		10	.02
		11	.01
			$\Sigma p = 1.00$

hourly credit-verification requests received by a credit bureau that never receives more than 11 such requests per hour.[1] Once again, the separate probabilities of all the mutually exclusive and collectively exhaustive events sum to 1.

The information given in Table 6.1 can be presented in graphical form as well, as Figure 6.1 demonstrates. Panels (a) and (b) depict the data of Table 6.1 directly. Panels (c) and (d), on the other hand, indicate how any regular probability distribution that shows probabilities for individual values of a random variable can be converted into a **cumulative probability distribution** that shows the probabilities of a random variable being less than or equal to any given possible value. Nothing more complicated is involved than adding the probability of any given value of the random variable to the probabilities associated with all the preceding values.

A final note: Often it is impossible to express even a discrete probability distribution in a table or a graph because too many possible values of the random variable exist. When that happens, the distribution can be expressed only symbolically. Take the case of drawing one lottery ticket from among 100,000 tickets (or of sampling one item for every 100,000 items sitting in a warehouse or moving along a production line.) It would be tedious and unnecessary to construct a table with 100,000 rows, stating that

---

[1]Throughout this chapter, we follow the simplified summation notation that was first explained on pp. 94 of Chapter 4. Without exception, whenever we encounter the Σ sign, we sum *all* the available values of the variables following that sign.

Figure 6.1   **Regular and Cumulative Probability Distributions of Discrete Random Variables**

Stick diagrams, such as those in panels (a) and (b), are graphical devices that illustrate discrete probability distributions, such as those given in Table 6.1. As comparisons of panel (a) with (c) and of panel (b) with (d) indicate, any regular probability distribution can easily be converted into a cumulative one by adding the probability of any given value of the random variable to that of all the preceding values. While the probability of a point spread of 2 equals 8/36, for example, that of a point spread of 2 or less equals 6/36 + 10/36 + 8/36 = 24/36.

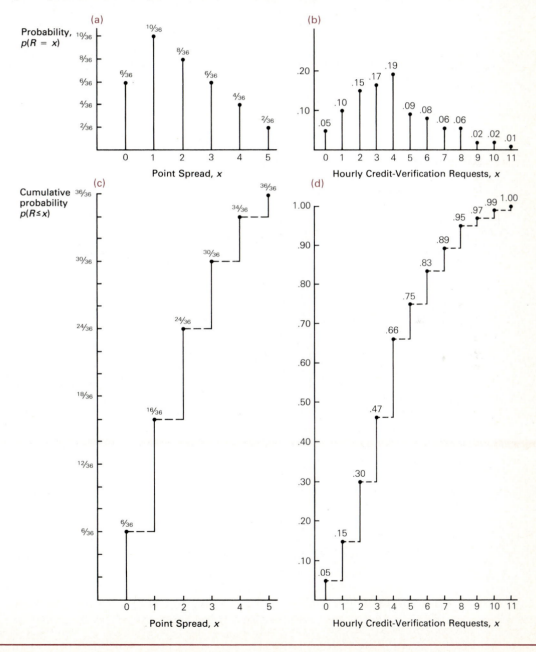

$$p(R = \text{serial number 1}) = \frac{1}{100,000}$$

$$p(R = \text{serial number 2}) = \frac{1}{100,000}$$

$$\vdots \qquad \qquad \vdots$$

$$p(R = \text{serial number 100,000}) = \frac{1}{100,000}$$

It is much easier simply to designate the actual number selected by R and possible serial numbers by x and to write

$$p(R = x) = \frac{1}{100,000}, \text{ where } x = 1, 2, \ldots, 100,000.$$

We will employ this notation throughout. It tells us succinctly that the probability is 1/100,000 for the random variable R to be equal to any specific possible number, x, where the latter can be any integer between 1 and 100,000.

SUMMARY MEASURES FOR THE PROBABILITY DISTRIBUTION

A gambler contemplating the information given in Part A of Table 6.1 [and graphed in panel (a) of Figure 6.1] might wonder what the *average* point spread would be for a large number of rolls. The manager of our credit bureau contemplating the information given in Part B of Table 6.1 [and graphed in panel (b) of Figure 6.1] might similarly wonder about the average hourly credit-verification requests over the long haul. For purposes such as these, it is useful to calculate summary measures. A probability distribution can most easily be summarized (and compared with other such distributions) by using measures of location and spread, like those introduced in Chapter 4. Consider calculating a random variable's arithmetic mean, variance, and standard deviation.

The Arithmetic Mean of a Random Variable

The most important measure of location for a probability distribution—the (weighted) arithmetic mean of the random variable's possible values—is calculated by weighting each of these possible values by its associated probability. This mean of the random variable R is symbolized by μ_R and is also called the **expected value, E_R,** of the random variable because it is the value of the random variable one can expect to find on the average by numerous repetitions of the random experiment that generates the variable's actual value.

Consider, for instance, Part A of Table 6.1. If the experiment of rolling two dice were repeated over and over again, one could expect a point spread of 0 in 6 out of 36 cases, a point spread of 1 in 10 out of 36 cases, and so on. The (weighted) average point spread for a large number of rolls can be calculated as shown in Table 6.2.

The same procedure can be applied to calculate the expected number of hourly credit-verification requests from the data in Part B of Table 6.1. The calculation is shown in Table 6.3.

Table 6.2   Expected Value of Point Spread from Rolling Two Dice

Calculating the expected value of a random variable is equivalent to calculating the weighted arithmetic mean from grouped data, as in Table 4.1 on page 96. Our present x corresponds to the earlier class midpoint, X, and our present p corresponds to the earlier class frequency, f. Thus, the earlier formula (equivalent to Box 4.C), $\mu = \dfrac{\Sigma f \cdot X}{\Sigma f}$, becomes $\mu = \dfrac{\Sigma p \cdot x}{\Sigma p}$, but (as the second column shows) $\Sigma p = 1$ for a probability distribution, making μ simply equal to the sum of the last column, or $\Sigma p \cdot x$.

Possible Value of Point Spread, x	Probability, $p(R = x)$	Weighted Value, $p(R = x) \cdot x$	Calculation
0	6/36	0	
1	10/36	10/36	
2	8/36	16/36	
3	6/36	18/36	
4	4/36	16/36	$\mu_R = E_R = \dfrac{70}{36} = 1.9444$
5	2/36	10/36	
	$\Sigma p = (36/36) = 1$	$\Sigma p \cdot x = (70/36)$	

Source: Table 6.1.

Table 6.3   Expected Number of Hourly Credit-Verification Requests

Possible Number of Requests, x	Probability, $p(R = x)$	Weighted Value, $p(R = x) \cdot x$
0	.05	0
1	.10	.10
2	.15	.30
3	.17	.51
4	.19	.76
5	.09	.45
6	.08	.48
7	.06	.42
8	.06	.48
9	.02	.18
10	.02	.20
11	.01	.11
	1.00	$\mu_R = 3.99$

Source: Table 6.1.

The Variance of a Random Variable

The variance of a random variable, denoted by σ_R^2, is also calculated by using probability weights. As the variance formula demands (see Box 4.J on page 111), these weights are, of course, attached to the squared deviations of the random variable's possible values from its expected value. This procedure is illustrated for our two examples in Tables 6.4 and 6.5.

Table 6.4   Variance of Point Spread from Rolling Two Dice

Calculating the variance of a random variable is equivalent to calculating the variance from grouped data, by the formula given in Box 4.J on p. 111: $\sigma^2 = \dfrac{\Sigma f(X - \mu)^2}{N}$. However, as was true when calculating the arithmetic mean (Table 6.2), the probability values, p, now replace the earlier class frequencies, f, while the sum of these weights (second column below), which corresponds to the earlier $N = \Sigma f$, now equals 1. Thus, the variance simply equals the sum of the last column. Note: the same result can be reached by utilizing the corresponding shortcut method given in Box 4.N on p. 114. Can you show this to be true?

Possible Value of Point Spread, x	Probability, $p(R = x)$	Deviation, $x - \mu_R$	Squared Deviation, $(x - \mu_R)^2$	Weighted Value, $p(R = x) \cdot (x - \mu_R)^2$
0	(6/36) = .1667	−1.9444	3.7807	.6302
1	(10/36) = .2778	−.9444	.8919	.2478
2	(8/36) = .2222	.0556	.0031	.0007
3	(6/36) = .1667	1.0556	1.1143	.1858
4	(4/36) = .1111	2.0556	4.2255	.4695
5	(2/36) = .0556	3.0556	9.3367	.5191
	$\Sigma p = 1.0000$			$\sigma_R^2 = 2.0531$

Source: Tables 6.1 and 6.2.

Table 6.5   Variance of Number of Hourly Credit-Verification Requests

Possible Number of Requests, x	Probability, $p(R = x)$	Deviation, $x - \mu_R$	Squared Deviation, $(x - \mu_R)^2$	Weighted Value, $p(R = x) \cdot (x - \mu_R)^2$
0	.05	−3.99	15.9201	.7960
1	.10	−2.99	8.9401	.8940
2	.15	−1.99	3.9601	.5940
3	.17	−.99	.9801	.1666
4	.19	.01	.0001	.0000
5	.09	1.01	1.0201	.0918
6	.08	2.01	4.0401	.3232
7	.06	3.01	9.0601	.5436
8	.06	4.01	16.0801	.9648
9	.02	5.01	25.1001	.5020
10	.02	6.01	36.1201	.7224
11	.01	7.01	49.1401	.4914
	1.00			$\sigma_R^2 = 6.0898$

Source: Tables 6.1 and 6.3

The Standard Deviation of a Random Variable

As always, the standard deviation equals the square root of the variance. Thus, in our two examples, $\sigma_R = 1.4329$ for case (A) and $\sigma_R = 2.4678$ for case (B).

The formulas for the (weighted) arithmetic mean, the variance, and the standard deviation of a random variable, R, are summarized in Box 6.A.

6.A Summary Measures for the Probability Distribution of a Random Variable, R

$$\mu_R = E_R = \Sigma p(R = x) \cdot x$$
$$\sigma_R^2 = \Sigma p(R = x) \cdot (x - \mu_R)^2$$
$$\sigma_R = \sqrt{\sigma_R^2}$$

where μ_R is the arithmetic mean, σ_R^2 is the variance, and σ_R is the standard deviation of the probability distribution of random variable, R, while $p(R = x)$ is the probability of the random variable equaling x and x is an observed value of that variable.

THE BINOMIAL PROBABILITY DISTRIBUTION

Different types of random experiments give rise to different types of probability distributions; the **binomial probability distribution** is the most important one for discrete variables. It shows the probabilities associated with possible values of a discrete random variable that are generated by a type of experiment called a *Bernoulli process.*

The Bernoulli Process

A **Bernoulli process** (named after James Bernoulli, discussed in Biography 6.2 at the end of the chapter) is a sequence of n identical trials of a random experiment such that each trial (a) produces one of two possible complementary outcomes that are conventionally called *success* and *failure* and (b) is independent of any other trial so that the probability of success or of failure is constant from trial to trial. The number of successes achieved in a Bernoulli process is the **binomial random variable.**

A sequence of coin tosses provides a perfect example. Each toss necessarily produces one of two possible outcomes: heads or tails. Quite arbitrarily, one of these outcomes is called a "success," the other one a "failure"; it does not matter which is called which. No matter how many tosses are made, the probability of success (for example, of getting a head) is precisely 1/2; that of failure (for example, of getting a tail) remains a constant 1/2 as well. The same situation arises in a series of die tosses, provided one classifies the outcomes as "odd" and "even" numbers of dots. Each toss must lead to one of these two outcomes, and each of these outcomes has an unchanging probability of 3/6. But note: the Bernoulli process requires that the success and failure probabilities be constant from one trial to the next; it does not require these two probabilities to be *equal* to each other, as happens to be the case in the examples just cited. A constant success probability of .1 and a constant failure probability of .9 satisfies the definition as well.

Interestingly, the conditions defining the Bernoulli process are met by a multitude of random experiments all of which lead to only one of two outcomes. Consider how a new baby must be male or female, how a new product must be liked or disliked, how a new drug must be effective or ineffective. Consider how a salesperson will sell or not sell, how a loan will be granted or denied, how a bid will be won or lost, how the quality of a product will be acceptable or unacceptable, how an account will be correct or in error, how an exam will be passed or failed, how a patient will be alive or dead, how an organ will be infected or not infected . . . the list goes on. But note: Whenever the random experiment involves sampling, successive elementary units must be selected *after replacement* of the units previously sampled in order to meet condition (b) and, thus, keep the probabilities of success and failure from changing as the sequence of trials proceeds. Consider the case of acceptance sampling. If there are 10 defective items in a lot of 100, and a sample of 20 items is to be taken, the probability of finding a defective item is 10/100 for the first item sampled. If it is set aside rather than replaced, the corresponding probability is 10/99 or 9/99 for the second item sampled (depending on whether the first item taken was satisfactory or defective). And the probability continues to change for the next items selected until it might lie between 10/81 and 0/81 for the 20th item (depending on how many defectives were found in the meantime). Only by replacing each item after inspection and remixing the lot can the probability in question be kept unchanged (and can the requirement of the Bernoulli process be strictly met). For practical purposes, sampling without replacement can, however, often be viewed as closely resembling a Bernoulli process, provided the sample is small relative to the population and constitutes, say, 5 percent or less of the population. In such a case, the probabilities involved change from sampling one unit to the next, but the changes are negligible. (If one took a sample of only 2 out of 100 or only 20 out of 1,000—and did so without replacement—the second item's probability of being defective could hardly be distinguished from that of the first item, as the above numbers indicate.)

||| Practice Problem 1

Tossing Three Coins

To illustrate how a binomial probability distribution is generated, let us first consider a random experiment of tossing three coins, while defining the binomial random variable of interest as the number of heads appearing. Panel (a) of Figure 6.2 uses a probability tree diagram of the type employed in Chapter 5 to determine the possible outcomes of such an experiment. The 8 possible outcomes are shown in the box, wherein H and T refer to head or tail showing and the subscripts, 1, 2, and 3, refer to the first, second, and third coins, respectively. The random variable of interest (number of heads) is given in the next column; the probability of each of these events is calculated by the special multiplication law using the probabilities given along the branches of the tree diagram. Since we are not interested in which particular coin shows a head, but only in the total number of heads, we can combine all the outcomes with the same total and derive the following binomial probability distribution:

$$
\begin{aligned}
p(R = 3 \text{ heads}) &= .125 \\
p(R = 2 \text{ heads}) &= .375 \\
p(R = 1 \text{ head}) &= .375 \\
p(R = 0 \text{ heads}) &= \underline{.125} \\
& 1.000
\end{aligned}
$$

Figure 6.2 Bernoulli Trials

These two diagrams illustrate the Bernoulli process with an example in panel (a) concerning the tossing of 3 coins and an example in panel (b) concerning the sampling of 3 items to determine quality. In each case, one of two outcomes must occur: a head or a tail must appear in one case; the sampled item must be defective or satisfactory in the other. In both examples, each trial is independent of all others, and the probabilities involved are constant throughout the experiment. As a comparison of panels (a) and (b) shows, the probabilities of success or failure need not, however, be equal to each other.

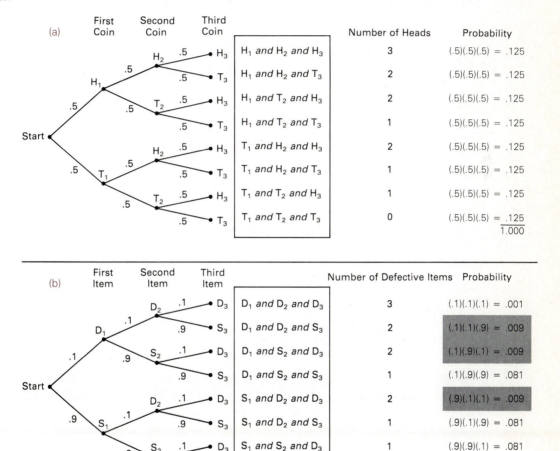

Note, for example, how 2 heads can occur in 3 different ways: via the H_1, H_2, T_3 route (with a probability of .125) *or* via the H_1, T_2, H_3 route (with a like probability) *or* via the T_1, H_2, H_3 route (once again with a like probability). Because these three ways of achieving the 2-heads outcome are mutually exclusive events, we can apply the special addition law (Box 5.H on p. 158) and say that the probability of getting 2 heads equals .125 + .125 + .125 = .375. |||

||| Practice Problem 2

Sampling Three Items for Quality

In order to confirm that the Bernoulli process requires the constancy rather than the equality of probabilities from trial to trial, let us reconsider the case of acceptance sampling first illustrated in panel (c) of Figure 5.9 on p. 171 and now reproduced in panel (b) of Figure 6.2. It is assumed that 3 items are picked from a lot of 100 items of which 10 percent are defective. If each sampled item is replaced before the next one is picked (which we assume), the strict conditions of the Bernoulli process are met. The probability of "success" (finding a defective item, say) remains a constant .1 from one trial to the next, and the probability of "failure" (finding a satisfactory item) equals a constant .9. As the box in panel (b) tells us, once again 8 possible outcomes exist, but, unlike in the coin-tossing experiment, these outcomes are not equally likely. Once again, however, several outcomes can be combined because they involve the same number of defective items, and when this is done (and the special addition law is applied), the following binomial probability distribution is found:

$$P(R = 3 \text{ defectives}) = .001$$
$$p(R = 2 \text{ defectives}) = .027$$
$$p(R = 1 \text{ defective}) = .243$$
$$p(R = 0 \text{ defectives}) = \underline{.729}$$
$$1.000 \quad |||$$

The Binomial Formula

Luckily, we need not always draw complicated tree diagrams to find the types of probabilities just calculated. They can also be determined more directly with the help of the formula given in Box 6.B.

6.B The Binomial Formula

$$p(R = x|n,\pi) = c_x^n \cdot \pi^x \cdot (1 - \pi)^{n-x},$$

where $x = 0, 1, 2, \ldots, n$ is the number of successes in n trials, $c_x^n = \dfrac{n!}{x! (n - x)!} = $ the binomial coefficient, and π is the probability of success in any one trial.

This **binomial formula** gives us the probability of x successes in n trials of a random experiment that satisfies the conditions of a Bernoulli process. Although the formula looks complicated, it is fairly easy to interpret:

1. The left-hand side of the equation symbolizes the probability of x successes, given the occurrence of n trials and given a probability of success of π in each of these trials. (As the box indicates, x cannot exceed n: one cannot succeed more often than one tries.)

2. The first term on the right-hand side, or c_x^n, is called the **binomial coefficient.** It tells us how many permutations of x successes (and, therefore, $n - x$ failures) can be achieved in n trials. The answer is indicated in Box 6.B as

$$c_x^n = \frac{n!}{x!\,(n - x)!}$$

It tells us, for example, how many sequences exist that contain $x = 2$ successess in $n = 3$ trials. The answer is 3 sequences:

$$c_2^3 = \frac{3!}{2!(3 - 2)!} = \frac{3 \cdot 2 \cdot 1}{2 \cdot 1(1)} = 3.$$

Indeed, as another look at Figure 6.2 confirms, three instances of 2 successes in 3 trials appear in the column labeled "Number of Heads" and also in the column labeled "Number of Defective Items." The kind of information given in these columns, and laboriously derived in Figure 6.2, can thus be gained instantly by calculating the binomial coefficient.

Pascal's triangle, given in Biography 5.3 on page 180, is another device for generating binomial coefficients. Consider how each row stands for a different value of n and how each successive entry in a given row stands for the number of outcomes with $0, 1, 2, \ldots, n$ successes. Thus, the third row represents $n = 3$ trials and the numbers therein tell us to expect 1 outcome with 0 successes, 3 outcomes with 1 success, 3 outcomes with 2 successes, and 1 outcome with 3 successes—a total of 8 possible outcomes. This result corresponds precisely to the answers given in the "Number of Heads" and "Number of Defective Items" columns of Figure 6.2.[2]

3. The next term in the formula, or π^x, represents the probability of success, π, in any one trial of the experiment, such as that of getting a head (.5) or finding a defective item (.1), raised to the xth power, where x is the number of successes.[3]

---

[2]The binomial coefficient, denoted by a lowercase c, is often confused with the combinatorial symbol, denoted by a capital C, but fortunately it does not matter because both are equal to the same ratio. Compare Box 6.B with Box 5.F. The binomial coefficient, however, does not measure combinations. Rather, it measures *permutations* involving *nondistinct* items (Box 5.E). It so happens that in a Bernoulli process any n trials contain precisely *two* subsets of like items (namely, x successes and $n - x$ failures); hence, the Box 5.E formula simplifies to $P_{x, n - x}^n = \dfrac{n!}{x!(n - x)!}$ and, thus, gives precisely the same result (now simply denoted as c_x^n) that one obtains when counting *combinations* of x items out of n distinct items (usually denoted by C_x^n). A look at Figure 6.2 can quickly establish the truth of the foregoing. Note, for example, that there are $c_2^3 = 3$ sequences of 2 heads and 1 tail: HHT, HTH, and THH. Because order matters in the case of permutations, these are 3 *permutations*. Because order does not matter in the case of combinations, these results represent a single *combination*. Then why does the combinatorial formula also give the correct answer? It is pure accident, and one cannot see this in Figure 6.2 because the combinatorial $C_2^3 = 3$ tells us a totally different thing, such as the number of combinations of 2 items that one can take out of 3 *distinct* items. If those 3 items are *A, B, C,* the combinations in question are $AB = BA$, $AC = CA$, and $BC = CB$.

[3]The symbol π is commonly used because the probability in question in fact equals a population proportion; for example, the ratio of defective items to all items in the population. Thus, π must not be confused here with the physical constant that is the ratio of the circumference to the diameter of the circle.

This term provides us with the joint probability of x successes, calculated by using the special multiplication law for independent events (Box 5.J on page 165). In each of the three shaded rows of panel (b) in Figure 6.2 (which correspond to the three outcomes with 2 successes), π^x appears as $(.1)(.1)$, which is, of course, the same thing as $\pi^x = (.1)^2$.

4. The last term in the formula, or $(1 - \pi)^{n-x}$, represents the probability of failure, $1 - \pi$, in any one trial of the experiment, such as that of getting a tail $(.5)$ or finding a satisfactory item $(.9)$, taken to the $(n - x)$th power, where $n - x$ is the number of failures. This term provides us with the joint probability of $n - x$ failures, once more calculated using the special multiplication law. In each of the three shaded rows of panel (b) in Figure 6.2 (which correspond to the three outcomes with 2 successes and, therefore, 1 failure), $(1 - \pi)^{n-x}$ appears as $(.9)^{3-2} = (.9)^1 = .9$.

5. When the binomial formula *combines* the last two terms and calculates $\pi^x \cdot (1 - \pi)^{n-x}$, it is in fact calculating (once more by using the special multiplication law for independent events) the joint probability of x successes *and* $n - x$ failures in a specific sequence of experimental outcomes. Thus, in panel (b) of Figure 6.2, the joint probability of a sequence of 2 successes *and* 1 failure equals $(.1)^2 (.9)^{3-2} = (.1)(.1)(.9) = .009$, precisely as each one of the shaded rows indicates.

6. Because, however, there exist $c_2^3 = 3$ different ways of getting the type of sequence just discussed (2 successes and 1 failure in 3 trials) and because these 3 outcomes are mutually exclusive events, the special addition law (Box 5.H on page 158) can be applied in order to calculate the overall probability of getting 2 successes in 3 trials of this experiment *somehow*—that is, regardless of whether this result occurs in the fashion shown in row 2 or 3 or 5 of panel (b), Figure 6.2. This overall probability equals $.009$ (first shaded row) + $.009$ (second shaded row) + $.009$ (third shaded row) = $.027$ (which is the very result for the probability of 2 defective items that was calculated above). Instead of adding $\pi^x (1 - \pi)^{n-x} = .009$ three times, one can, however, simply multiply the term by 3, which is precisely what the binomial coefficient, c_x^n, in the binomial formula accomplishes.

We can test the accuracy of the binomial formula by pretending to seek with its help some of the answers already provided in Figure 6.2.

||| Practice Problem 3

Coin Tossing

What is the probability of getting 3 heads when tossing 3 coins?

$$p(R = 3 | n = 3, \pi = .5) = c_3^3 \cdot (.5)^3 (1 - .5)^{3-3}$$

$$= \frac{3!}{3!(3 - 3)!} (.125)(.5)^0 = 1(.125)\,1 = .125.$$

This result is precisely the same as that found in row 1 of panel (a). |||

||| Practice Problem 4

Defective Items

What instead is the probability of finding 0 defective items when sampling 3 items under the conditions of panel (b)?

$$p(R = 0 | n = 3, \pi = .1) = c_0^3 (.1)^0 (1 - .1)^{3-0}$$

$$= \frac{3!}{0!(3 - 0)!} (1)(.9)^3 = 1(1)(.9)^3 = .729$$

Once more, the result is confirmed by that found in the last line of panel (b). |||

Note: While the binomial formula tells us directly the probability of *x* successes in *n* trials, it tells us indirectly the probability of the complementary event. Thus, the probability of *not* getting 3 heads in the Practice Problem 3 example equals $1 - .125 = .875$; that of finding *some* defective items in the Practice Problem 4 example equals $1 - .729 = .271$.

||| Practice Problem 5

The Production Process

To show how an entire binomial probability distribution can rapidly be determined by using the binomial formula (without resort to the type of tree diagrams seen in Figure 6.2), consider another problem: a manufacturer uses a production process that produces 20 percent defective items. A defect in any one item is, however, independent of possible defects in other items. (The problem is *not* a matter of some machine becoming occasionally maladjusted and turning out a whole series of defective items until a correction is made. If that were the case, the appearance of one defective item would significantly affect the probability of finding another such item.) If under these conditions 5 units are produced, what is the probability of 0, 1, 2, 3, 4, and 5 of them being defective? The answers are quickly found by a systematic application of the binomial formula, as Table 6.6 attests. |||

The Binomial Probability Distribution Family

A look at the binomial formula in Box 6.B above clearly indicates that the probability for any given number of successes varies with two factors: the number of trials, *n,* and the probability of success in any one trial, π. Each different combination of *n* and π thus produces a different binomial probability distribution, even though the same formula is applied to derive the distribution. It is customary to refer to any specific binomial probability distribution, such as that given in Table 6.6, as *a member of the binomial distribution family,* it being understood that other members can be found by varying the values of *n* or π. Figure 6.3, on page 205, depicts nine members of this family, each of them determined by the procedure illustrated in Table 6.6. Indeed, the distribution calculated in Table 6.6 is graphed in panel (d). Three things should be noted.

Table 6.6   Determining a Binomial Probability Distribution for $n = 5$ and $\pi = .2$ Using the Binomial Formula

The entries in the first column together with those in the last column represent a binomial probability distribution. Its mathematical determination, using the three terms in the binomial formula, is illustrated with the help of the three middle columns. The second rows tells us, for example, that there exist 5 different ways of achieving 1 success (along with 4 failures) in the experiment in question and that the probability of each of these sequences equals (.2) multiplied by (.8)(.8)(.8)(.8). Thus, the probability of getting any one of these 5 sequences is five times the former product, or .4096.

Number of "Successes" (defective items found), x	$C_x^n = \dfrac{n!}{x!\,(n-x)!}$	π^x	$(1-\pi)^{n-x}$	$p(R = x \mid n, \pi)$
0	$\dfrac{5!}{0!(5-0)!} = 1$	$(.2)^0 = 1$	$(.8)^{5-0} = .32768$	$1(1)(.32768) = .32768$
1	$\dfrac{5!}{1!(5-1)!} = 5$	$(.2)^1 = .2$	$(.8)^{5-1} = .4096$	$5(.2)(.4096) = .4096$
2	$\dfrac{5!}{2!(5-2)!} = 10$	$(.2)^2 = .04$	$(.8)^{5-2} = .512$	$10(.04)(.512) = .2048$
3	$\dfrac{5!}{3!(5-3)!} = 10$	$(.2)^3 = .008$	$(.8)^{5-3} = .64$	$10(.008)(.64) = .0512$
4	$\dfrac{5!}{4!(5-4)!} = 5$	$(.2)^4 = .0016$	$(.8)^{5-4} = .8$	$5(.0016)(.8) = .0064$
5	$\dfrac{5!}{5!(5-5)!} = 1$	$(.2)^5 = .00032$	$(.8)^{5-5} = 1$	$1(.00032)(1) = .00032$

$$1.00000$$

First, given any number of trials, n, the probability of a high proportion of successes rises with the probability of achieving success in a given trial. Consider, for example, panels (a) through (c), all of which involve two trials of a random experiment. The probability of achieving two successes in two trials rises from .04 to .25 and to .64 as the probability of success in a single trial, π, rises from .2 to .5 and then to .8. Similar observations can be made for panels (d) through (f) and panels (g) through (i), respectively.

Second, any binomial probability distribution with a success probability of $\pi = .5$ is perfectly symmetrical regardless of the value of n, as panels (b), (e), and (h) indicate. On the other hand, any such distribution with $\pi < .5$ or $\pi > .5$ is skewed to the right or left, respectively. This is illustrated by panels (a), (d), and (g) on the one hand and panels (c), (f), and (i) on the other. Although Figure 6.3 does not show it, the skewness gets more pronounced the closer π gets to 0 or 1, respectively.

Third, given any value of $\pi \neq .5$ and, therefore, the presence of skewness, the skewness becomes less pronounced as n rises. For high values of n, all such binomial

Figure 6.3 Nine Members of the Binomial-Probability-Distribution Family

In accordance with the binomial formula, each different combination of n (number of trials) and π (success probability in a given trial) gives birth to a different member of the binomial probability distribution family. Each of the nine distributions shown here has been calculated by the procedure illustrated in Table 6.6, with the results rounded to two decimals.

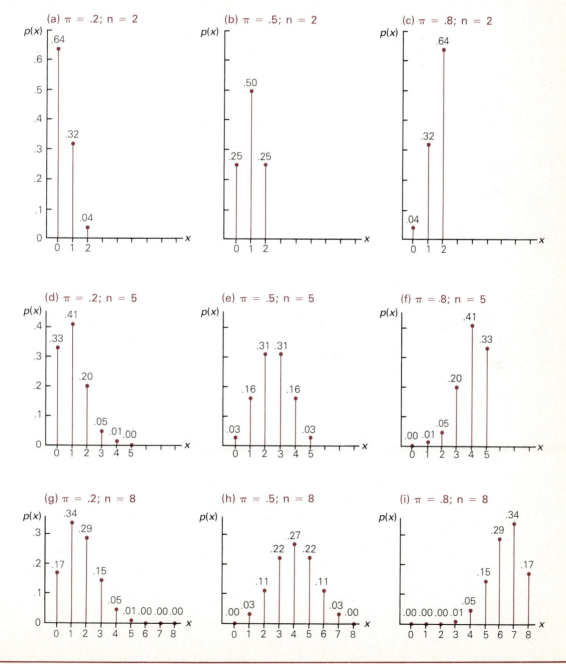

probability distributions approach symmetry, as a progressive comparison of panel (a) with (d) and (g) or of panel (c) with (f) and (i) indicates. As will be noted in the next chapter, this tendency of the binomial probability distribution to take on a bell-shaped form as n increases allows one to approximate such distributions by a normal probability distribution.

Binomial Probability Tables

Undoubtedly, the use of the binomial formula for determining binomial probabilities is less cumbersome than drawing tree diagrams. However, even this mathematical approach can become awkward and can require much time and effort—especially when large factorials are involved and the formula requires us to raise numbers to large powers. Under such circumstances, it would be convenient to have the probabilities already computed for us and listed in tables similar to those of square roots and logarithms. Luckily, such **binomial probability tables** do exist; they list binomial probabilities for various combinations of possible values of n and π. Two types of binomial probability tables exist; one for individual values of x (such as Appendix Table C) and another for cumulative values of x (such as Appendix Table D).

Binomial probabilities for individual values of x. Probabilities for achieving x numbers of successes in a Bernoulli process, given various combinations of n and π, are listed in Appendix Table C. Each value in the body of the table was calculated by the formula in Box 6.B. This particular table is set up for values of n ranging from 1 to 20 and for selected probabilities of success in any one trial, π, between .05 and .50. The table can, however, still be used to determine binomial probabilities for π in excess of .50, as will be shown presently.

To become familiar with Appendix Table C consider the following experiments, always assuming that they meet the conditions of a Bernoulli process.

Practice Problem 6

Producing Defective Items

a. A manufacturer uses a production process that produces 20 percent defective items. If 5 units are produced, what is the probability of none of them being defective? Of 3 units being defective? Of all 5 being defective? Looking at Appendix Table C in the section for $n = 5$ and under the column for $\pi = .20$, the desired probabilities can be found instantly: the probability for $x = 0$ is seen to be .3277, that for $x = 3$ is .0512, and that for $x = 5$ is .0003. Note: These answers correspond precisely to those calculated directly in the corresponding rows of Table 6.6.

b. Pursing (a) further, what is the probability of at most 1 unit being defective? Of 2 or 3 units being defective? Of at least 3 units being defective? The special addition law for mutually exclusive events can be applied to the numbers given in Appendix Table C. The probability of at most 1 unit being defective equals the sum of the separate probabilities of 0 or 1 unit being defective; that is, $.3277 + .4096 = .7373$. By similar reasoning, the probability of 2 or 3 units being defective is found as .2048 + .0512 = .256, while that of at least 3 units being defective equals the sum of the separate probabilities of 3 and 4 and 5 units being defective; that is, .0512 + .0064 + .0003 = .0579. |||

||| Practice Problem 7

Punching Data Cards

a. Now consider a case in which the probability of success in any one trial, π, exceeds .50, the highest value found in Appendix Table C. Take a data processor who punches cards correctly 95 percent of the time. If 20 cards are punched, what is the probability that precisely 16 cards are correct?

The Appendix Table C section with $n = 20$ seems to be of no help because the row containing $x = 16$ contains no value for $\pi = .95$. Yet the desired answer can still be found by rephrasing the question in terms of failure rather than success. Our data processor fails to punch cards correctly 5 percent of the time. If we now read π as the probability of failure on any one trial, we can find, in the $\pi = .05$ column, a probability of getting precisely 4 incorrect cards as .0133. This implies a probability of getting 16 correct cards of .0133 as well.

b. Finally, pursing (a) further, what is the probability of getting at least 4 incorrect cards? Of getting at most 3 incorrect cards? Here the answer requires once again the addition of various probabilities. Getting at least 4 incorrect cards involves adding the probabilities of 4 incorrect cards, of 5 incorrect cards, of 6 incorrect cards, . . . , and, finally, of 20 incorrect cards. As the $\pi = .05$ column in the $n = 20$ section of Appendix Table C tells us, this amounts to .0133 + .0022 + .0003 = .0158. Similarly, the probability of getting at most 3 incorrect cards is found as the sum of the probabilities of getting 0, 1, 2, and 3 incorrect cards; that is, .3585 + .3774 + .1887 + .0596 = .9842. It is no accident that our two answers sum to 1; the events described are complementary. |||

Binomial probabilities for cumulative values of x. Practice Problem 7b illustrates why it has been found desirable to construct tables of binomial probabilities for cumulative values of x as well as for individual values of x: while it is easy enough to calculate cumulative probabilities when n is small (we had to sum only four separate probability values in Problem 7b), this process can become extremely tedious for large values of n. Appendix Table D is such a binomial probability table for cumulative values of x. It is set up for values of n ranging from 1 to 20, as well as for $n = 50$ and $n = 100$. Once again, selected probability-of-success values between .05 and .50 are given. Unlike in Appendix Table C, the values in the body of Appendix Table D now give the probabilities not of x successes, but of x *or fewer* successes.

||| Practice Problem 8

Consumers' Product Preferences

Consider using Appendix Table D to answer questions about a population of consumers 30 percent of whom favor and 70 percent of whom dislike a new product. If 100 persons are sampled, what are the following probabilities?

a. What is the probability of finding 8 or fewer consumers who favor the product? As a look at the $\pi = .30$ column in the $n = 100$ section of Appendix Table D tells us, the probability is 0.

b. What is the probability of finding precisely 40 consumers who favor the product? The answer, in the $\pi = .30$ column and the $n = 100$ section of Appendix Table D, is the difference between the probabilities of finding 40 or fewer such consumers (.9875) and finding 39 or fewer such consumers (.9790), or .0085.

c. What is the probability of finding fewer than 20 consumers who favor the product? The answer is equivalent to finding 19 or fewer such consumers, or .0089.

d. What is the probability of finding 26 such consumers or more? The answer is equivalent to 1 minus the probability of finding 25 or fewer such consumers, or $1 - .1631 = .8369$.

e. What is the probability of finding more than 35 such consumers? The answer is equivalent to 1 minus the probability of finding 35 or fewer such consumers, or $1 - .8839 = .1161$.

f. What is the probability of finding between 20 and 40 such consumers, inclusive? The answer is equivalent to the probability of finding 40 or fewer such consumers (.9875) minus that of finding 19 or fewer of them (.0089), or .9786. **|||**

It should be noted that a cumulative probability table provides no information that could not also be gathered from the equivalent section of an ordinary probability table; the only (and not inconsiderable) advantage of the cumulative table is that it frees the user from the need to add or subtract large numbers of probability values. Analytical Examples 6.1, "Budgeting Research and Development," and 6.2, "Acceptance Sampling Plans," provide two cases in point.

Summary Measures for the Binomial Probability Distribution

Earlier in this chapter (see Box 6.A), we noted that the arithmetic mean or expected value of *any* random variable, R, can be calculated by a simplified formula, $E_R = \Sigma p \cdot x$. For a *binomial* random variable, an even simpler approach can be used.

The expected value. As Box 6.C (below) states, the expected value of a binomial random variable always equals the number of trials of the experiment multiplied by

---

6.C Summary Measures for the Binomial Probability Distribution of Random Variable, R.

$$\mu_R = E_R = n \cdot \pi$$
$$\sigma_R^2 = n \cdot \pi(1 - \pi)$$
$$\sigma_R = \sqrt{\sigma_R^2}$$

where μ_R is the arithmetic mean, σ_R^2 is the variance, and σ_R is the standard deviation of the probability distribution of binomial random variable, R, while n is the number of trials and π is the probability of success in 1 trial.

Skewness: zero for $\pi = .5$; positive for $\pi < .5$; negative for $\pi > .5$

the probability of success in any one trial; that is, $E_R = n \cdot \pi$. This equality is shown in Table 6.7, wherein the expected value of a binomial random variable is calculated in both ways, using the binomial probability distribution of Table 6.6 as a starting point.

The variance and standard deviation. We noted earlier (see Box 6.A) that the variance of *any* random variable, R, can be calculated by a simplified formula as well:

$$\sigma_R^2 = \Sigma p(x - \mu_R)^2 \tag{6.1}$$

—the standard deviation being, of course, the square root of this expression. For a *binomial* random variable an even simpler formula exists: the variance of a binomial random variable always equals the number of trials of the experiment multipled by the probability of success in any one trial, multiplied by the probability of failure in any one trial; that is,

$$\sigma_R^2 = n \cdot \pi(1 - \pi). \text{ (See Box 6.C.)} \tag{6.2}$$

The equivalence of equations 6.1 and 6.2 is demonstrated clearly in Table 6.8, wherein the variance of a binomial random variable is calculated both ways, again using the binomial probability distribution of Table 6.6 as a starting point.

Table 6.7 **Determining the Expected Value of a Binomial Random Variable Given $n = 5$ and $\pi = .2$**

The expected value of a binomial random variable can be calculated using the formula in Box 6.A, or it can be calculated more simply using Box 6.C as the number of trials of the experiment, n, multiplied by the probability of success in any one trial, π.

Number of Successes, x	Probability, $p(R = x)$	Weighted Value, $p(R = x) \cdot x$
0	.32768	0
1	.4096	.4096
2	.2048	.4096
3	.0512	.1536
4	.0064	.0256
5	.00032	.0016
		$E_R = \Sigma p \cdot x = 1.0000$

Alternative Calculation: $E_R = n \cdot \pi = 5(.2) = 1$

Source: Table 6.6

Table 6.8   Determining the Variance of a Binomial Random Variable Given $n = 5$ and $\pi = .2$

The variance of a binomial random variable can be calculated using the formula in Box 6.A, or it can be calculated more simply using Box 6.C as the number of trials of the experiment, n, multiplied by the probability of success in any one trial, π, multiplied by the probability of failure in any one trial, $(1 - \pi)$. The standard deviation is, of course, the square root of this expression; in this case $\sigma_R = \sqrt{.8} = .8944$.

Number of Successes, x	Probability, $p(R = x)$	Deviation, $x - \mu_R$	Squared Deviation, $(x - \mu_R)^2$	Weighted Value, $p(R = x)(x - \mu_R)^2$
0	.32768	-1	1	.32768
1	.4096	0	0	0
2	.2048	1	1	.2048
3	.0512	2	4	.2048
4	.0064	3	9	.0576
5	.00032	4	16	.00512
				$\sigma_R^2 = .8$

Alternative Calculation: $\sigma_R^2 = n \cdot \pi(1 - \pi) = 5(.2)(.8) = .8$

Source: Table 6.6 and Table 6.7.

Skewness. *   As was noted in connection with the discussion of Figure 6.3, the probability-of-success value for any one trial, π, can in effect be used as an indicator of skewness. When $\pi = .5$, any binomial probability distribution is symmetrical; that is, it exhibits zero skewness. When $\pi < .5$, the distribution is positively skewed (to the right); when $\pi > .5$, it is negatively skewed (to the left). This relationship can, in fact, be seen most clearly by studying the entries in Appendix Table C for different values of π.

Chapter 6 of the *Student Workbook* that accompanies this text contains a number of mathematical derivations concerning the binomial probability distribution.

THE HYPERGEOMETRIC PROBABILITY DISTRIBUTION†

As the previous section has shown, the binomial probability distribution helps us find the probability of x successes in a Bernoulli process. As we noted, when the random experiment in question involves sampling, the binomial probability distribution provides a precise answer *only if* sampling occurs *with replacement* so that the probability of success in any one trial remains constant throughout the experiment. Ordinarily, however, sampling is performed without replacement, often as a matter of conve-

---

*Requires prior study of optional Chapter 4 section, "Skewness."

†Optional section.

nience: one unit is selected from a population and set aside; then another is selected, and another. This procedure is easier than inspecting each unit, making a record of its characteristics, returning it to the population, and remixing the population before the next unit is selected. Sometimes sampling without replacement is unavoidable. Consider cases wherein sampling destroys the unit in question (as in the testing of flash bulbs, for instance). In such circumstances one could not return sampled items to the population, even if one wished to do so. The **hypergeometric probability distribution** provides probabilities associated with possible values of a discrete random variable precisely in these situations in which these values are generated by *sampling without replacement* and in which the probability of success, therefore, changes from one trial to the next.

||| Practice Problem 9

Selecting Job Applicants

Consider a personnel manager who has on file the names of $N = 10$ job applicants, of whom $S = 6$ are female; hence, $N - S = 4$ are male. If the manager were to select at random $n = 4$ names from the file, and were to do so without replacement (taking one name at a time and setting it aside), what kind of probability distribution would describe a random variable, R, such as the number of females selected?

As in the previous section, a tree diagram can help us find the answer, and such a diagram appears as Figure 6.4. Consider yourself at the starting point, the leftmost box in the diagram, representing the original content of the file, 6 female and 4 male applicants. The first selection made might be a female (F_1) or a male (M_1), and the respective probabilities of selection are 6/10 and 4/10, reflecting the makeup of the population. If a female is selected, we have moved along the color line in Figure 6.4 to a new box, containing the names of 5 females and 4 males. The conditional probabilities that the second selection will be a female (F_2) or a male (M_2), given F_1 (that a female was selected on the first try), are 5/9 and 4/9, respectively, as indicated once more along the forks of the tree. If now a male is selected on the second try, we move on further, perhaps along the colored line to the one of the 16 possible outcomes that is shown, along with its probability, in the colored box. Note, however, that there exist three other ways (shaded) of selecting 3 females by this procedure; thus, the probability of selecting 3 females equals not 480/5040, but four times this value, or 1920/5040. By similar reasoning, the entire hypergeometric probability distribution for selecting female applicants can be derived:

$$p(R = 4 \text{ females}) = 1 \times \frac{360}{5040} = \frac{360}{5040}$$

$$p(R = 3 \text{ females}) = 4 \times \frac{480}{5040} = \frac{1920}{5040}$$

$$p(R = 2 \text{ females}) = 6 \times \frac{360}{5040} = \frac{2160}{5040}$$

$$p(R = 1 \text{ female}) = 4 \times \frac{144}{5040} = \frac{576}{5040}$$

$$p(R = 0 \text{ females}) = 1 \times \frac{24}{5040} = \frac{24}{5040}$$

$$\text{Total} = \frac{5040}{5040} \text{ |||}$$

The Hypergeometric Formula

Fortunately, as was true of the binomial probability distribution, there exists a less cumbersome way of determining the hypergeometric one. This approach uses the formula given in Box 6.D.

6.D The Hypergeometric Formula

$$p(R = x|n, N, S) = \frac{c_x^S \cdot c_{n-x}^{N-S}}{c_n^N}$$

where x is the number of "successes" in a sample sized n, taken from a population sized S, that contains S units with the "success" characteristic. Clearly, $x = 0, 1, 2, \ldots, n$ or N (whichever is smaller); $n < N$; $S < N$.

The **hypergeometric formula** gives the probability of x successes when a random sample of n is drawn without replacement from a population of N within which S units exist with the characteristic denoting success. The number of successes achieved under these circumstances is the **hypergeometric random variable.** Once again, the formula, complicated as it seems, is easy to interpret:

1. The left-hand side of the equation symbolizes the probability of x successes, given that a random sample of size n is drawn without replacement from a population sized N that contains S units with whatever characteristic has been labeled "success." (As the note indicates, x cannot exceed n or S: one cannot succeed more often than one tries; nor can one succeed more often than is possible, which depends on the number of units, S, in the population with the success characteristic.)
2. The term c_x^S represents the number of ways in which one can get x successes out of S possible successes, while c_{n-x}^{N-S} is the associated number of ways in which $n - x$ failures can be selected from among $N - S$ possible failures. Since each of the former can be combined with each of the latter, the product of the two terms gives the possible number of outcomes with exactly x successes (and, therefore, $n - x$ failures).
3. The term c_n^N, finally, gives the total number of ways of selecting a sample of n (whether "successful" or not) from a population of N. Hence, the numerator described above, divided by c_n^N, gives the probability of x successes.

We can test the accuracy of the hypergeometric formula by pretending to seek with its help some of the answers already provided above.

Figure 6.4   Tree Diagram for Sampling Without Replacement

This tree diagram illustrates sampling of $n = 4$ persons without replacement from a group of $N = 10$, containing 6 females and 4 males. The probabilities of selecting a female or a male change from one trial to the next, depending on the prior selections made, as is indicated along the forks of the tree. If selecting a female is called a "success," the probabilities associated with achieving various degrees of success (that is, with selecting various numbers of them) can be determined from the information generated here.

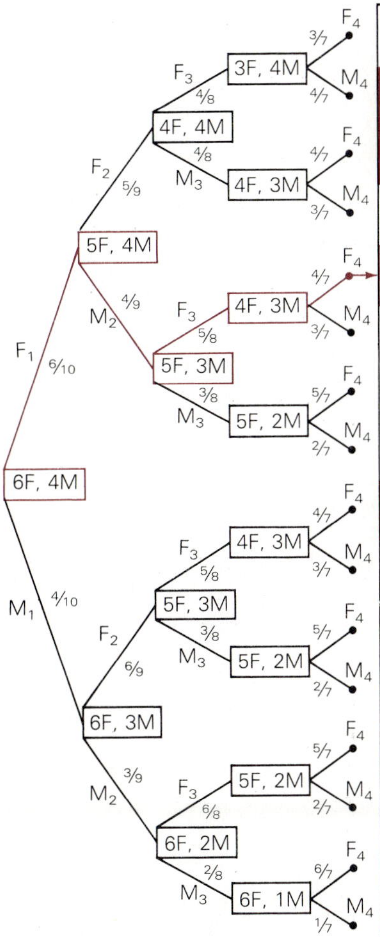

Outcome	Number of Females	Probability
$F_1\ F_2\ F_3\ F_4$	4	$\left(\frac{6}{10}\right)\left(\frac{5}{9}\right)\left(\frac{4}{8}\right)\left(\frac{3}{7}\right) = \frac{360}{5040}$
$F_1\ F_2\ F_3\ M_4$	3	$\left(\frac{6}{10}\right)\left(\frac{5}{9}\right)\left(\frac{4}{8}\right)\left(\frac{4}{7}\right) = \frac{480}{5040}$
$F_2\ F_2\ M_3\ F_4$	3	$\left(\frac{6}{10}\right)\left(\frac{5}{9}\right)\left(\frac{4}{8}\right)\left(\frac{4}{7}\right) = \frac{480}{5040}$
$F_1\ F_2\ M_3\ M_4$	2	$\left(\frac{6}{10}\right)\left(\frac{5}{9}\right)\left(\frac{4}{8}\right)\left(\frac{3}{7}\right) = \frac{360}{5040}$
$F_1\ M_2\ F_3\ F_4$	3	$\left(\frac{6}{10}\right)\left(\frac{4}{9}\right)\left(\frac{5}{8}\right)\left(\frac{4}{7}\right) = \frac{480}{5040}$
$F_1\ M_2\ F_3\ M_4$	2	$\left(\frac{6}{10}\right)\left(\frac{4}{9}\right)\left(\frac{5}{8}\right)\left(\frac{3}{7}\right) = \frac{360}{5040}$
$F_1\ M_2\ M_3\ F_4$	2	$\left(\frac{6}{10}\right)\left(\frac{4}{9}\right)\left(\frac{3}{8}\right)\left(\frac{5}{7}\right) = \frac{360}{5040}$
$F_1\ M_2\ M_3\ M_4$	1	$\left(\frac{6}{10}\right)\left(\frac{4}{9}\right)\left(\frac{3}{8}\right)\left(\frac{2}{7}\right) = \frac{144}{5040}$
$M_1\ F_2\ F_3\ F_4$	3	$\left(\frac{4}{10}\right)\left(\frac{6}{9}\right)\left(\frac{5}{8}\right)\left(\frac{4}{7}\right) = \frac{480}{5040}$
$M_1\ F_2\ F_3\ M_4$	2	$\left(\frac{4}{10}\right)\left(\frac{6}{9}\right)\left(\frac{5}{8}\right)\left(\frac{3}{7}\right) = \frac{360}{5040}$
$M_1\ F_2\ M_3\ F_4$	2	$\left(\frac{4}{10}\right)\left(\frac{6}{9}\right)\left(\frac{3}{8}\right)\left(\frac{5}{7}\right) = \frac{360}{5040}$
$M_1\ F_2\ M_3\ M_4$	1	$\left(\frac{4}{10}\right)\left(\frac{6}{9}\right)\left(\frac{3}{8}\right)\left(\frac{2}{7}\right) = \frac{144}{5040}$
$M_1\ M_2\ F_3\ F_4$	2	$\left(\frac{4}{10}\right)\left(\frac{3}{9}\right)\left(\frac{6}{8}\right)\left(\frac{5}{7}\right) = \frac{360}{5040}$
$M_1\ M_2\ F_3\ M_4$	1	$\left(\frac{4}{10}\right)\left(\frac{3}{9}\right)\left(\frac{6}{8}\right)\left(\frac{2}{7}\right) = \frac{144}{5040}$
$M_1\ M_2\ M_3\ F_4$	1	$\left(\frac{4}{10}\right)\left(\frac{3}{9}\right)\left(\frac{2}{8}\right)\left(\frac{6}{7}\right) = \frac{144}{5040}$
$M_1\ M_2\ M_3\ M_4$	0	$\left(\frac{4}{10}\right)\left(\frac{3}{9}\right)\left(\frac{2}{8}\right)\left(\frac{1}{7}\right) = \frac{24}{5040}$
		$\frac{5040}{5040}$

Tree forks (reading from $F_1\ \frac{6}{10}$ and $M_1\ \frac{4}{10}$): 6F, 4M; 5F, 4M; $F_2\ \frac{5}{9}$; $M_2\ \frac{4}{9}$; 4F, 4M; 5F, 3M; $F_3\ \frac{4}{8}$; $M_3\ \frac{3}{8}$; 3F, 4M; 4F, 3M; 5F, 2M; 6F, 3M; 5F, 3M; 6F, 2M; 6F, 1M; with final forks F_4 and M_4 at rates such as $\frac{3}{7}$, $\frac{4}{7}$, $\frac{5}{7}$, $\frac{2}{7}$, $\frac{6}{7}$, $\frac{1}{7}$.

||| **Practice Problem 10**

Selecting Female Applicants

a. What is the probability of selecting 4 females in the sampling case depicted in Figure 6.4? (The calculations for this problem are on the next page.)

$$p(R = 4 | n = 4, N = 10, S = 6) = \frac{c_4^6 \cdot c_{4-4}^{10-6}}{c_4^{10}}$$

$$= \frac{\dfrac{6!}{4!2!} \cdot \dfrac{4!}{0!4!}}{\dfrac{10!}{4!6!}}$$

$$= \frac{6!}{4!2!} \cdot \frac{4!6!}{10!}$$

$$= \frac{6 \cdot 5 \cdot 4 \cdot 3}{10 \cdot 9 \cdot 8 \cdot 7}$$

$$= \frac{360}{5040}$$

The result is precisely the one given in the top row of Figure 6.4.

b. What is the probability of selecting 3 females in the case shown in Figure 6.4?

$$p(R = 3 | n = 4, N = 10, S = 6) = \frac{c_3^6 \cdot c_{4-3}^{10-6}}{c_4^{10}}$$

$$= \frac{\dfrac{6!}{3!3!} \cdot \dfrac{4!}{1!3!}}{\dfrac{10!}{4!6!}}$$

$$= \frac{6!}{3!3!} \cdot \frac{4!}{1!3!} \cdot \frac{4!6!}{10!}$$

$$= \frac{6 \cdot 5 \cdot 4 \cdot 4 \cdot 4}{10 \cdot 9 \cdot 8 \cdot 7}$$

$$= \frac{1920}{5040}$$

Once more, the result corresponds precisely to that calculated before. ▌▌▌

▌▌▌ Practice Problem 11

Selecting Committee Members

To show how an entire hypergeometric probability distribution can rapidly be determined by using the hypergeometric formula (and without resort to the type of tree diagram seen in Figure 6.4), consider this situation: A committee of $n = 5$ members is to be formed randomly from among $N = 50$ members of a labor union local in which $S = 40$ workers are electricians. If, as is typical under such circumstances, the selection is made without replacement, what are the probabilities for $x = 0, 1, 2, 3, 4,$ and 5 committee members to be electricians? The answers are quickly found by a systematic application of the hypergeometric formula, as Table 6.9 shows. ▌▌▌

Table 6.9   Determining a Hypergeometric Probability Distribution for $n = 5$, $N = 50$, and $S = 40$ Using the Hypergeometric Formula

The entries in the first column together with those in the last column represent a hypergeometric probability distribution. Its mathematical determination, using the hypergeometric formula, is illustrated with the help of the three middle columns. The last row tells us, for example, that there exist $\dfrac{40!}{5!35!}$ different ways of getting 5 successes out of 40 possible ones, while there are $\dfrac{10!}{0!10!}$ associated ways of getting 0 failures from among 10 possible failures, as well as $\dfrac{50!}{5!45!}$ different ways of selecting a sample of 5 from a population of 50. The probability of getting a sequence of 5 successes and 0 failures under the specified conditions then equals the product of the first two expressions, divided by the third expression, or .3106.

Number of "Successes" (electricians selected), x	$c_x^S = \dfrac{S!}{x!(S-x)!}$	$c_{n-x}^{N-S} = \dfrac{(N-S)!}{(n-x)![(N-S)-(n-x)]!}$	$c_n^N = \dfrac{N!}{n!(N-n)!}$	$p(R = x\|n, N, S)$
0	$\dfrac{40!}{0!\,40!}$	$\dfrac{10!}{5!\,5!}$	$\dfrac{50!}{5!\,45!}$	$\dfrac{40!}{0!40!}\ \dfrac{10!}{5!5!}\ \dfrac{5!45!}{50!} = .0001$
1	$\dfrac{40!}{1!\,39!}$	$\dfrac{10!}{4!\,6!}$	$\dfrac{50!}{5!\,45!}$	$\dfrac{40!}{1!39}\ \dfrac{10!}{4!6!}\ \dfrac{5!45!}{50!} = .0040$
2	$\dfrac{40!}{2!\,38!}$	$\dfrac{10!}{3!\,7!}$	$\dfrac{50!}{5!\,45!}$	$\dfrac{40!}{2!38!}\ \dfrac{10!}{3!7!}\ \dfrac{5!45!}{50!} = .0442$
3	$\dfrac{40!}{3!\,37!}$	$\dfrac{10!}{2!\,8!}$	$\dfrac{50!}{5!\,45!}$	$\dfrac{40!}{3!37!}\ \dfrac{!10}{2!8!}\ \dfrac{5!45!}{50!} = .2098$
4	$\dfrac{40!}{4!\,36!}$	$\dfrac{10!}{1!\,9!}$	$\dfrac{50!}{5!\,45!}$	$\dfrac{40!}{4!36!}\ \dfrac{10!}{1!9!}\ \dfrac{5!45!}{50!} = .4313$
5	$\dfrac{40!}{5!\,35!}$	$\dfrac{10!}{0!\,10!}$	$\dfrac{50!}{5!\,45!}$	$\dfrac{40!}{5!35!}\ \dfrac{10!\,5!45!}{0!10!\ \ 50!} = .3106$

| | | | | 1.0000 |

Summary Measures for the Hypergeometric Probability Distribution

As was true for the binomial random variable, the mean, variance, and standard deviation of the hypergeometric random variable can be calculated either by the general formula derived earlier in this chapter (Box 6.A) or by a simpler version.

The expected value. As Box 6.E states, the expected value of a hypergeometric random variable always equals the number of trials of the experiment (in this case, the sample size) multiplied by the probability of success on the *first* trial, π, which, of course, equals the ratio of units with the success characteristic in the population, S, to population size, N. Thus, $E_R = n \cdot \pi = n(S/N)$. This equality is illustrated in Table

6.10, wherein the expected value of a hypergeometric random variable is calculated in both ways, using the hypergeometric distribution of Table 6.9 as a starting point.

6.E Summary Measures for the Hypergeometric Probability Distribution of Random Variable, R

$$\mu_R = E_R = n \cdot \pi = n(S/N)$$

$$\sigma_R^2 = n \cdot \pi(1 - \pi)\left(\frac{N - n}{N - 1}\right)$$

$$\sigma_R = \sqrt{\sigma_R^2}$$

where μ_R is the arithmetic mean, σ_R^2 is the variance, and σ_2^R is the standard deviation of hypergeometric random variable, R, while n is sample size, N is population size, S is the number of population units with the "success" characteristic, and π is the probability of success in the first trial, or S/N.

The variance and standard deviation. While the simplified formula for calculating the mean of a hypergeometric random variable equals that for a binomial random variable, the simplified formulas for calculating the variance (and, therefore, the standard deviation) differ for the two types of random variables. The variance of a hypergeometric random variable always equals the number of trials of the experiment (in this case, the sample size), n, multiplied by the probability of success on the *first* trial, $\pi = (S/N)$, multiplied by the probability of failure on the first trial, $(1 - \pi)$, multiplied by a so-called *finite population correction factor*, $(N - n)/(N - 1)$. It is the latter factor alone

Table 6.10 Determining the Expected Value of a Hypergeometric Random Variable, Given $n = 5$, $N = 50$, and $S = 40$

The expected value of a hypergeometric random variable can be calculated using the formula in Box 6.A, or it can be calculated more simply—using the Box 6.E formula—as the number of trials of the experiment (that is, the sample size), n, multiplied by the probability of success on the *first* trial, $\pi = (S/N)$.

Number of Successes, x	Probability, $p(R = x)$	Weighted Value, $p(R = x) \cdot x$
0	.0001	0
1	.0040	.0040
2	.0442	.0884
3	.2098	.6294
4	.4313	1.7252
5	.3106	1.5530

$$E_R = \Sigma px = 4.0000$$

Alternative Calculation: $E_R = n \cdot \pi = 5\left(\frac{40}{50}\right) = 5(.8) = 4$

Source: Table 6.9.

Table 6.11   Determining the Variance of a Hypergeometric Random Variable, Given $n = 5$, $N = 50$, and $S = 40$

The variance of a hypergeometric random variable can be calculated using the formula in Box 6.A, or it can be calculated more simply using the formula in Box 6.E—that is, as the number of trials of the experiment (in this case, the sample size), n, multiplied by the probability of success on the *first* trial, $\pi = (S/N)$, multiplied by the probability of failure on the first trial, $(1 - \pi)$, multiplied by the finite population correction factor, $(N - n)/(N - 1)$. The standard deviation is, of course, the square root of this expression; in this case, $\sigma_R = \sqrt{.7347} = .8571$.

Number of Successes, x	Probability, $p(R = x)$	Deviation, $x - \mu_R$	Squared Deviation, $(x - \mu_R)^2$	Weighted Value, $p(R = x) \cdot (x - \mu_R)^2$
0	.0001	−4	16	.0016
1	.0040	−3	9	.0360
2	.0442	−2	4	.1768
3	.2098	−1	1	.2098
4	.4313	0	0	0
5	.3106	1	1	.3106

$$\sigma_R^2 = .7348$$

Alternative Calculation: $\sigma_R^2 = n \cdot \pi (1 - \pi)\left(\dfrac{N - n}{N - 1}\right) = 5(.8)(.2)\left(\dfrac{50 - 5}{50 - 1}\right) = .7347$

Source: Table 6.9 and Table 6.10.

that distinguishes the variance formulas for the binomial and hypergeometric random variables. As always, the corresponding formula for the standard deviation is the square root of the variance (see Box 6.E).

The calculation of the variance by the general formula of Box 6.A and by the simplified one of Box 6.E is shown in Table 6.11, again with the help of the hypergeometric probability distribution of Table 6.9.

Approximating the Hypergeometric by the Binomial Probability Distribution

Sometimes it is tempting to represent one thing (such as a hypergeometric probability distribution) by something else (such as a binomial probability distribution) even though the results are not precisely correct. This is especially true when acquiring the former takes much time and effort (involves lengthy calculations), while obtaining the latter is easy (involves a quick look at ready-made tables). Using the binomial probability tables as a quick approximation of the hypergeometric probability distribution is particularly justified when the sample size, n, is small relative to the population size, N (a situation that arises when the sample size is, say, 5 percent or less of a small population or when the population size approaches infinity). In that case, the differences in the two probability distributions become negligible, and even their variances (and standard deviations) tend to become equal (because the finite population correction factor tends to become equal to 1).

Table 6.12 **Exact and Approximate Probability Distributions for Sampling Without Replacement, Given $n = 5$, $N = 50$, and $S = 40$**

While the hypergeometric probability distribution and its summary measures are strictly correct, the binomial probability distribution often approximates these values closely, the more so the smaller is the ratio of n/N.

Number of Successes, x	*Exact* Hypergeometric Probability, $p(R = x\|n,N,S)$	*Approximate* Binomial Probability, $p(R = x\|n,\pi)$
0	.0001	.0003
1	.0040	.0064
2	.0442	.0512
3	.2098	.2048
4	.4313	.4096
5	.3106	.3277
	$E_R = 4$ $\sigma_R^2 = .7347$ $\sigma_R = .8571$	$E_R = 4$ $\sigma_R^2 = .8$ $\sigma_R = .8944$

Source: Table 6.9 and Appendix Table C.

Table 6.12 illustrates the rather small differences between the two types of probability distributions for the example discussed in Tables 6.9–6.11. The binomial distribution has been taken from Appendix Table C for values of $n = 5$ and $\pi = (S/N) = .8$ (interpreting the x values in the table as failures with π of .2 so that the probability value for $x = 0$ failures stands for that of $x = 5$ successes, and so on).

THE POISSON PROBABILITY DISTRIBUTION*

In the previous two sections of this chapter, we have met two major types of discrete probability distributions, one for binomial random variables and another one for hypergeometric random variables. Both of these random variables were defined as *numbers of successes* and in both instances these were reaped *within a fixed number of trials* of some random experiment. Early in the nineteenth century, Siméon Poisson (see Biography 6.1 at the end of this chapter) noted the existence of many processes that generate random variables that cannot be described by either of the above distributions because only a single type of outcome or "event" is occurring, and it is doing so not within a confined number of trials, but along a continuum of time or space. Accordingly, he defined what is now called a **Poisson random variable** as the number of occurrences of a specified event within a specified time or space. He also defined the process generating the values of this variable.

_____

*Optional section.

The Poisson Process

A **Poisson process** is the occurrence of a series of events of a given type in a random (and, hence, unpredictable) pattern over time or space such that (1) the number of occurrences within a specified time or space can equal any integer between zero and infinity, (2) the number of occurrences in one unit of time or space is independent of that in any other such unit, and (3) the probability of an occurrence is the same in all such units. The kinds of events Poisson had in mind are encountered in all areas of life, as the next two sections abundantly show.

Events occurring randomly over time. A large class of events satisfying the definition of the Poisson process involves arrivals of "customers" at a "service facility," the two terms in quotation marks being defined in the widest possible sense. Consider the arrival of people demanding service at the bank, the barber shop, or the bus stop; at the checkout counter of a supermarket, the doctor's office, or the elevator door; at the hospital emergency room, the post office, or the restaurant. Consider the arrival of cars at car washes, entrances to major roads, or gas stations; at parking lots, traffic lights, or toll stations. Think of the arrival of planes at airports, of ships at ports, of trucks at terminals. Consider the arrival of telephone calls at the switchboard requesting services from ambulances, fire fighters, police officers. Contemplate the arrival of claims at the insurance company, of parcels at the post office, of orders at the warehouse, of semi-finished products at the next stage of production, of refund requests at the complaint department, of suicide victims at the morgue. In all these cases and a million more, the conditions of the Poisson process may well apply. Note: within a given span of time and for any one of these types of events, one might count any number of occurrences, yet one can never count the number of complementary events or nonoccurrences. Unlike in the Bernoulli process, wherein one might count successes (heads) as well as failures (tails), in the Poisson process, one can count only one thing (such as the people arriving at the bank), but not its complement (such as the people *not* arriving at the bank).

Arrivals at service facilities, however, are not the only types of Poisson events occurring randomly over time. The Poisson process has been found to describe well a great variety of other events, including occurrences of accidents at manufacturing plants, breakdowns of equipment at hospitals, failures of components in satellites, and even vacancies on the U.S. Supreme Court.

Events spread randomly over space. Examples of Poisson events are just as abundant in the medium of space as in that of time, regardless of whether "space" is viewed as lines, areas, or volumes. View space as a line and consider such events as the appearance of defects in a roll of coated wire, of leaks in a pipeline, of misspelled names in a telephone book, of potholes on a road, of typesetting errors in a book. View space as an area and contemplate the occurrence of births (or bomb bursts) in a city, of blisters in a painted wall, of bubbles in a plate-glass window. Think of crimes in a town, of defects in a carpet, of farmhouses in a county, of lightning-caused fires in a forest; of mines, meteorites, or weeds in a field. Consider stars in a photograph; schools of fish or submarines in the ocean. Finally, view space as volume and envision events such as finding bacteria in a gel, raisins in a loaf, or weed seeds in a packet of flower seeds.

Conclusions. Possible examples of the Poisson process at work are truly infinite in number. In the areas of business and economics, major applications involve waiting-line or queuing problems, inventory policy, and quality control. A bank or restaurant, for example, must know something about the probability distribution of customer arrivals. During a given hour, might there be 2 customers or 20 or 299? Should there be 5 employees ready to serve them, or would 25 be a better bet? Clearly, a trade-off is involved between possible idle time of employees (if there are too many of them) and possible waiting time of customers (if there are too many of *them*). Paying idle employees is costly, but losing angry customers who don't like to wait is costly, too.

Having the right amount of inventory in the warehouse presents a similar problem. If inventories in, say, an automotive-parts department are huge relative to requests for parts, all customers can always be satisfied without waiting, but inventory holding costs are high. Like the employees above, much of the inventory is "idle" much of the time. When, however, inventories are extremely low, many angry customers who don't like waiting will be lost. Having an inventory that is too "busy" is costly as well.

In a sense, both of the examples just given can be viewed as involving quality control (of various types of services). Quality control of physical products—be they carpets, heart-lung machines, plate-glass windows, or taxicabs—can, thus, be seen as an analogous issue. A manufacturer surely would not want to produce a heart-lung machine that fails during open-heart surgery, but costs can again be raised in two ways: by building in lots of redundant capability (akin to having mostly idle bank tellers or parts inventories) and by not having enough capability and losing customers (akin to having people switch away from crowded banks and unreliable parts suppliers). Once more, a knowledge of the relevant probability distribution is extremely helpful.

The Poisson Formula

The probabilities associated with alternative values of the Poisson random variable can be determined with the help of the formula given in Box 6.F.

6.F The Poisson Formula

$$p(R = x|\mu) = \frac{e^{-\mu} \cdot \mu^x}{x!}, \text{ where } x = 0, 1, 2, \ldots, \infty$$

where x is the number of occurrences when μ is the mean number of them within the examined units of time or space, while $e = 2.71828$.

The **Poisson formula** gives the probability, within a specified time or space, of x occurrences of a specified event that satisfy the conditions of a Poisson process. As on previous occasions, the formula is less forbidding than appears to be the case at first sight:

1. The left-hand side of the equation symbolizes the probability of x occurrences of the event in question within a specified time or space under examination, given that the expected or mean number of such occurrences in this time or space equals μ.

2. The constant e equals 2.71828; it is the base of the natural logarithms.[4] This constant is raised to the power of $-\mu$, which makes the expression equal to $1/e^{\mu}$. Appendix Table E, "Exponential Functions," provides ready-made calculations of this expression for selected values of μ.

3. The expression μ can itself be viewed as the product of two magnitudes: the **Poisson process rate,** or mean number of occurrences of the event in question *per unit* of time or space (symbolized by the lowercase Greek lambda, λ), and the total number of units of time or space examined, t. Thus, $\mu = \lambda \cdot t$. The expression λ might, for instance, measure the arrival of cars at a toll booth at a rate of 2 per minute (or the making of typesetting errors at a rate of 1 per page). The expression t would then be defined analogously, in this case as the total number of minutes studied, say, 60 (or the total number of pages examined, say, 100). Thus, $\mu = \lambda \cdot t$ would equal 120 cars per 60 minutes (or 100 errors per 100 pages).

Practice Problem 12

Customers at the Bank

Consider a bank manager who must decide how many tellers should be available during the Friday afternoon rush when the customers arrive at a rate of 5 customers per minute but in a pattern described by the Poisson process. Such a manager might wish to know the probability distribution for the numbers of customer arrivals within a 30-second period.

Clearly, $\lambda = 5$ customers per minute; $t = .5$ minutes; hence, $\mu = 2.5$ customers per half-minute. Table 6.13 shows the derivation of the desired probability distribution. When examining Table 6.13, two things should be noted. First, the probabilities of 11 or 12 customer arrivals in a 30-second period are not zero, but tiny positive numbers rounded to zero. For example, $p(R = 12) = .00001021$. This fact reminds us that the Poisson process sets no upper limit to the number of possible occurrences in the specified time or space. In this example, there exists even a positive tiny probability for the arrival of 500 customers in a 30-second period.

---

[4]The constant e is equal to $\left(1 + \dfrac{1}{n}\right)^{n}$ as n approaches infinity. For example:

$$\left(1 + \frac{1}{1}\right)^{1} = 2.00000$$

$$\left(1 + \frac{1}{2}\right)^{2} = 2.25000$$

$$\left(1 + \frac{1}{3}\right)^{3} = 2.37037$$

$$\left(1 + \frac{1}{4}\right)^{4} = 2.44141$$

$$\lim_{n \to \infty} \left(1 + \frac{1}{n}\right)^{n} = 2.71828$$

Table 6.13 Determining a Poisson Probability Distribution for $\lambda = 5$, $t = .5$ (hence, $\mu = 2.5$) Using the Poisson Formula

The entries in the first column together with those in the last column represent a Poisson probability distribution. Its mathematical determination, using the three terms in the Poisson formula, is illustrated with the help of the three middle columns.

Number of Occurrences of Specified Event (customer arrival), x	$e^{-\mu}$ (from Appendix Table E)	μ^x	$x!$	$p(R = x\|\mu)$
0		$2.5^0 = 1$	$0! = 1$	.0821
1		$2.5^1 = 2.5$	$1! = 1$	.2052
2		$2.5^2 = 6.25$	$2! = 2$	.2565
3		$2.5^3 = 15.625$	$3! = 6$	.2138
4		$2.5^4 = 39.0625$	$4! = 24$	.1336
5	.082085	$2.5^5 = 97.6563$	$5! = 120$	.0668
6		$2.5^6 = 244.1406$	$6! = 720$	.0278
7		$2.5^7 = 610.3516$	$7! = 5,040$	.0099
8		$2.5^8 = 1,525.8788$	$8! = 40,320$	.0031
9		$2.5^9 = 3,814.697$	$9! = 362,880$	.0009
10		$2.5^{10} = 9,536.7425$	$10! = 3,628,800$	.0002
11		$2.5^{11} = 23,841.856$	$11! = 39,916,800$	.0000
12		$2.5^{12} = 59,604.64$	$12! = 479,001,600$	.0000
				.9999

Second, one must be careful not to assume that the value of λ holds over an extended duration. The Poisson process rate frequently varies with the time of day, the day of the week, the season, and more. Customers surely arrive at the bank at a different rate during Friday afternoons than on Tuesday mornings. The same holds for the arrival of cars at toll booths and for all the other examples. ▮▮▮

▌▌▌ Practice Problem 13

Searching for Atlantic Salmon

Consider a fishing company on the coast of New England. It is operating a search plane to find schools of salmon that are randomly located in the North Atlantic, there being, on the average, 1 school per 100,000 square miles of sea. On a given day, the plane can fly 1,000 miles, effectively searching a lateral distance of 5 miles on either side of its path. (a) What is the probability of finding at least 1 school of salmon during 3 days of searching? (b) How many days of search are needed before the probability of finding at least 1 school reaches .95?

To answer question (a), we realize that $\lambda = .00001$ school per square mile and that the plane can search a total area of $t = 1,000(10)(3) = 30,000$ square miles in 3 days. Thus, the expected number of schools to be found in 3 days is $\mu = \lambda t = .3$. We also note that the probability of finding at least 1 school equals 1 minus that of finding no school:

$$p(R \geq 1) = 1 - p(R = 0) = 1 - \frac{e^{-.3} (.3)^0}{0!}$$

$$= 1 - e^{-.3} = 1 - .740818 = .259.$$

To answer question (b), we note that $p(R \geq 1)$ is supposed to equal .95. Hence,

$$.95 = 1 - p(R = 0) = 1 - \frac{e^{-\mu} \cdot \mu^0}{0!} = 1 - e^{-\mu}.$$

Thus,

$$e^{-\mu} = 1 - .95 = .05.$$

Appendix Table E shows this value of $e^{-\mu}$ to be associated with $\mu = 3$. Hence, we know that $\lambda \cdot t$ must equal 3. Given $\lambda = .00001$, t must equal $(3/.00001) = 300,000$. To achieve the desired result, 300,000 square miles of ocean must be searched, and this search, at a rate of 10,000 square miles per day, would take 30 days. |||

Poisson Probability Tables

As was shown above to be true for the binomial distribution, a whole family of Poisson probability distributions can be discovered. In this case, the members of the family are distinguished from one another by different values of μ. Once again, there exist ready-made **Poisson probability tables** that list probabilities of x occurrences in a Poisson process for various values of μ. Appendix Table F gives Poisson probabilities for *individual* values of x and for selected values of μ, ranging from .1 to 20. Appendix Table G gives Poisson probabilities for *cumulative* values of x and for the same range of values of μ.

To become familiar with these tables, consider the following examples of Poisson processes, using in each case first Appendix Table F and then Appendix Table G to solve the problem. (These problems will show why in some cases a table for individual probabilities is preferable, while in other cases the result is more quickly obtained from the table for cumulative probabilities.)

||| Practice Problem 14

Plane Arrivals at an Airport

a. Planes arrive at an airport at the rate of 2 per minute during a Friday evening. What are the probabilities that 6, 9, and 20 planes arrive between 8 P.M. and 8:10 P.M.?

In this case, $\lambda = 2, t = 10$, and $\mu = 20$. Therefore, Appendix Table F shows the answers directly in the $\mu = 20$ column as .0002, .0029, and .0888, respectively. Appendix Table G gives the answers only indirectly as $.0003 - .0001 = .0002; .0050 - .0021 = .0029;$ and $.5591 - .4703 = .0888$. In each case, the probability of one number or fewer ones must be reduced by the probability of the next lower number or fewer ones in order to find the desired probability of the first-stated number of occurrences.

b. In the same situation, what are the probabilities of at most 18 planes arriving? From Appendix Table F, we must sum the separate probabilities of 0, 1, 2, . . ., 18 planes arriving (still in the $\mu = 20$ column), and this sum is $.0000 + .0000 + .0000 + . . . + .0844 = .3814$. Appendix Table G, on the other hand, gives the same result directly in the $x = 18$ row and $\mu = 20$ column. |||

||| Practice Problem 15

Leaks in a Pipeline

a. In a pipeline, 2 leaks occur per 100 miles. What is the probability of finding between 7 and 9 leaks in a 500-mile stretch? In this case, $\lambda = 2, t = 5, \mu = 10$. Therefore, Appendix Table F shows the answer in the $\mu = 10$ column as the sum of $.0901 + .1126 + .1251 = .3278$. Appendix Table G shows the answer as the difference between the probabilities of 9 or fewer leaks (.4579) and 6 or fewer leaks (.1301)—that is, as .3278 as well.

b. Considering again the situation in (a), what is the probability of finding at least 20 leaks in that 500-mile stretch? Appendix Table F suggests the answer as the sum of $.0019 + .0009 + .0004 + .0002 + .0001 = .0035$. Appendix Table G gives the answer as the difference between 1 and the probability of finding 19 or fewer leaks (.9965)—that is, the answer is .0035 as well. |||

Close-Ups 6.1, "Probability Applied to Anti-Aircraft Fire," and 6.2, "Supplying Spare Parts to Polaris Submarines," provide other examples yet of the manifold uses of Poisson probabilities.

Summary Measures for the Poisson Probability Distribution

As was noted in Box 6.A, simplified formulas let us calculate summary measures for random variables, but in the case of the Poisson variable an even simpler approach exists. Both the mean and variance of the Poisson random variable are equal to $\lambda \cdot t$, as will be shown presently.

The expected value. The expected value of a Poisson random variable can be calculated in accordance with the formula given in Box 6.A, as Table 6.14 shows with the help of the Poisson distribution derived in Table 6.13. However, such complicated calculation is never necessary because, for a Poisson random variable, E_R always equals $\lambda \cdot t$. (See Box 6.G.)

Table 6.14 Determining the Expected Value of a Poisson Random Variable, Given $\lambda = 5$ and $t = .5$

The expected value of a Poisson random variable can be calculated using the formula in Box 6.A, or it can be calculated more simply using Box 6.G, as $\mu_R = \lambda \cdot t$; that is, as the product of the Poisson process rate, λ, and the total number of units of time or space examined, t.

Number of Occurences, x	Probability, $p(R = x)$	Weighted Value, $p(R = x) \cdot x$
0	.0821	0
1	.2052	.2052
2	.2565	.5130
3	.2138	.6414
4	.1336	.5344
5	.0668	.3340
6	.0278	.1668
7	.0099	.0693
8	.0031	.0248
9	.0009	.0081
10	.0002	.0020
11	.0000	.0000
12	.0000	.0000

$$E_R = \Sigma px = 2.499$$

Alternative Calculation: $E_R = \lambda \cdot t = 5(.5) = 2.5$. (The difference is due to rounding.)

Source: Table 6.13.

6.G Summary Measures for the Poisson Probability Distribution of Random Variable, *R*

$$\mu_R = E_R = \lambda \cdot t$$

$$\sigma_R^2 = \lambda t$$

$$\sigma_R = \sqrt{\sigma_R^2}$$

where μ_R is the arithmetic mean, σ_R^2 is the variance, and σ_R is the standard deviation of Poisson random variable *R*, while λ is the Poisson process rate and t is the total number of units of time or space examined.

Skewness: always positive.

The variance and standard deviation. Table 6.15 illustrates how the variance of our Poisson random variable can be derived by the formula in Box 6.A but can also be found simply by multiplying λ and t. (See Box 6.G.)

Table 6.15   Determining the Variance of a Poisson Random Variable, Given $\lambda = 5$ and $t = .5$

The variance of a Poisson random variable can be calculated using the formula in Box 6A, or it can be calculated more simply using Box 6.G as $\sigma_R^2 = \lambda \cdot t$; that is, as the product of the Poisson process rate, λ, and the total number of units of time or space examined, t. The standard deviation is, of course, the square root of this expression; in this case, $\sigma_R = \sqrt{2.5} = 1.5811$.

Number of Occurrences, x	Probability, $p(R = x)$	Deviation, $x - \mu_R$	Squared Deviation, $(x - \mu_R)^2$	Weighted Value, $p(R = x) \cdot (x - \mu_R)^2$
0	.0821	−2.5	6.25	.5131
1	.2052	−1.5	2.25	.4617
2	.2565	−.5	.25	.0641
3	.2138	.5	.25	.0535
4	.1336	1.5	2.25	.3006
5	.0668	2.5	6.25	.4175
6	.0278	3.5	12.25	.3406
7	.0099	4.5	20.25	.2005
8	.0031	5.5	30.25	.0938
9	.0009	6.5	42.25	.0380
10	.0002	7.5	56.25	.0113
11	.0000	8.5	72.25	.0000
12	.0000	9.5	90.25	.0000

$$\sigma_R^2 = 2.4947$$

Alternative Calculation: $\sigma_R^2 = \lambda \cdot t = 5(.5) = 2.5$. (The difference is due to rounding.)

Source: Tables 6.13 and 6.14.

Skewness. * All Poisson probability distributions are skewed to the right. This positive skewness can be observed by studying the values in Appendix Table F; it is also the reason why the Poisson probability distribution has been called the probability distribution of *rare* events (the probabilities tend to be high for small numbers of occurrences).

Approximating the Binomial by the Poisson Probability Distribution

We noted earlier in this chapter why the binomial probability distribution under some circumstances can be used as an approximation of the hypergeometric probability distribution. Now we must note other circumstances in which it is possible, in turn, to use the Poisson probability distribution as an approximation of the binomial one. This use is particularly justified in the case of a Bernoulli process in which the number of trials, *n,* is large and the probability of success in any one trial, π, is small. Under such circumstances, the two distributions are very similar. This similarity is not surprising because Poisson derived his formula, given in Box 6.F on page 220, from the binomial formula, given in Box 6.B, precisely by assuming that *n* approaches infinity while π approaches zero in such a fashion that the expected value of the binomial random

---

*Requires prior study of optional Chapter 4 section, "Skewness."

variable, $\mu_R = n \cdot \pi$, remains fixed.[5] Because binomial probabilities are usually not available in tabulated form for large values of n, the similarity of the Poisson distribution is then a particularly happy circumstance.

||| Practice Problem 16

Inspecting a Week's Production

As an illustration, consider the case of a quality inspector who knows that the week's production stored in a warehouse comes from a production process producing 5 percent defective items. The inspector wants to know the probability of finding no defective items in a sample of 20, if sampling is done with replacement. Using Appendix Table C for binomial probabilities, the precise answer is found (in the $n = 20$ and $\pi = .05$ section) as $p(R = 0) = .3585$. Yet an approximate answer could have been found from Appendix Table F for Poisson probabilities (in the section for $\mu = n \cdot \pi = 1$) as $p(R = 0) = .3679$. These values are given in the first line of Table 6.16, along with others for larger values of x. Clearly, the approximation is quite good.

Finally, consider the possibility of our inspector taking a sample not of 20, but of 200, all else being equal. The binomial probability table only goes to $n = 20$ and cannot be used, but Appendix Table F (in the section for $\mu = n \cdot \pi = 10$) instantly gives the probability for any given x, such as $p(R = 5) = .0378$. (The binomial *formula* in this case gives the precise value as .0339, but only after a complex set of calculations.) |||

Table 6.16   Exact and Approximate Probability Distributions for Sampling with Replacement, Given $n = 20$ and $\pi = .05$ (hence, $\mu = 1$)

While the binomial probability distribution and its summary measures are strictly correct, the Poisson probability distribution often approximates these values closely, the more so the larger is n and the smaller is π.

Number of Successes, x	Exact Binomial Probability, $p(R = x \mid n, \pi)$	Approximate Poisson Probability, $p(R = x \mid \mu)$
0	.3585	.3679
1	.3774	.3679
2	.1887	.1839
3	.0596	.0613
4	.0133	.0153
5	.0022	.0031
6	.0003	.0005
7	.0000	.0001
	$E_R = 1$ $\sigma_R^2 = .95$ $\sigma_R = .97$	$E_R = 1$ $\sigma_R^2 = 1$ $\sigma_R = 1$

Source: Appendix Tables C and F.

---

[5]The mathematical derivation of the Poisson formula from the binomial formula is presented in J. Leroy Folks, *Ideas of Statistics* (New York: Wiley, 1981), pp. 118–19.

6.1 Budgeting Research and Development

Firms and governments throughout the world spend large sums of money on basic or applied research and on product development. In the 1950s, an approach called *parallel-path strategy* was developed by the Rand Corporation to help achieve research-and-development (R&D) goals at minimum cost. This strategy involves n teams working independently of one another trying to achieve some goal, such as the development of a new product. Clearly, each team needs an annual budget of, say, $10 million and even then has a less than perfect chance of, say, merely 5 percent of succeeding. The more teams that are put to work, the greater is the chance of at least one of them having a success, but the greater must the overall budget be. This is where probability theory comes in; it helps answer questions such as: how much money must be spent by a sponsoring firm or government agency if it wants to have a 90 percent chance of at least one success? Because the activity in question involves a Bernoulli process (one of two outcomes, independent teams), the binomial probability distribution can be used to establish the probability of at least one success by n teams.

$$p(R \geq 1) = 1 - p(R = 0)$$
$$= 1 - c_0^n \cdot \pi^0 \cdot (1 - \pi)^{n-0}$$
$$= 1 - (1 - \pi)^n.$$

Given $\pi = .05$, we get $p(R \geq 1) = 1 - .95^n$

If the agency were content with only a 90 percent chance of success, this expression would become

$$p(R \geq 1) = 1 - .95^n = .90,$$

hence,

$$.95^n = .10.$$

The value of n implied by this expression is about 45, implying (at $10 million a team) an annual budget of $450 million. The accompanying table shows other budgets required for other acceptable probabilities of achieving at least one success, given a .05 success probability of any one team.

Note how certainty of at least one success would require an infinite budget, but quite feasible budgets can produce near-certainty. (In 1981, in spite of a recession, General Motors spent $2,250 million on R&D, which was the highest total spent by any U.S. company. Next came Ford with $1,718 million, AT&T with $1,686 million and IBM with $1,612 million. Firms such as Boeing, General Electric, United Technologies, DuPont, Exxon, Eastman Kodak, Xerox, and ITT each spent between $500 million and $1 billion.)

Desired Probability of at Least 1 Success	Number of R&D Teams Needed	Budget per Team	Total Required Budget (millions of dollars per year)
.50	14	10	140
.60	18	10	180
.70	24	10	240
.80	32	10	320
.90	45	10	450
.95	59	10	590
.99	90	10	900
1.00	∞	10	∞

Sources: Richard R. Nelson, "Uncertainty, Learning, and the Economics of Parallel Research and Development Efforts, "*Review of Economics and Statistics,* November 1961, pp. 351–64; Moshe Ben Horim and Haim Levy, *Statistics* (New York: Random House, 1981), pp. 246–48; *Business Week,* July 5, 1982, p. 54.

6.2 Acceptance Sampling Plans

Many firms and government agencies regularly receive shipments of goods from outside suppliers, and they want to accept these shipments only if they contain a reasonably low percentage of defective items. Thus, they take random samples of each shipment to determine the percentage of defective items. To make a decision on whether to accept or reject a shipment, however, they must have a *sampling plan;* that is, they must decide (a) on the desirable sample size and (b) on the maximum number of defective units they are willing to find, while still accepting the shipment. The binomial probability distribution can be extremely helpful here.

Consider the U.S. Air Force purchasing, say, electronic navigation equipment. If it sampled 5 such items in each shipment of 100 units and followed the self-imposed rule of accepting the entire shipment only if 0 defective units were found, what would it in fact be doing? The binomial formula provides a quick answer: With $n = 5$ and $x = 0$, the Air Force would under this sampling plan nevertheless be accepting shipments containing 5 percent defective items ($\pi = .05$) 77 percent of the time, it would be accepting shipments containing 10 percent defective items 59 percent of the time, and it would even accept shipments with 30 percent defective items 17 percent of the time.

(These numbers can be confirmed quickly in Appendix Table C, section $n = 5, x = 0$, for values of π of .05, .10, and .30.)

If such a situation were deemed undesirable, matters could be improved by increasing the sample size to, say, $n = 20$. A sample plan of $n = 20$ and $x = 0$, would cause acceptance of lots with 5, 10, or 30 percent defectives only 36, 12, and 0 percent of the time.

Naturally, sampling plans need not be based on $x = 0$; one can decide to accept lots that contain $x = 3$ defective items or fewer of them. As Appendix Table D shows, shipments with 5 percent defective items ($\pi = .05$) will be accepted 98 percent of the time when samples of 20 are taken and findings of 3 or fewer defective items are deemed acceptable. On the other hand, shipments with 10 or 30 percent defective items will then be accepted only 87 or 11 percent of the time.

Can you show why someone who wanted to accept shipments with 10 percent defectives at most 3 percent of the time, but who was willing to accept all shipments the samples of which contained up to 1 defective item, would have to choose a sample size of $n = 50$? (Hint: Note in Appendix Table D how the cumulative probability numbers in the $x \leq 1$ row and the $\pi = .10$ column steadily decline as n increases.)

6.1 Probability Applied to Anti-Aircraft Fire

During World War II, long before the days of guided missiles with homing devices, British scientists were considering possible improvements in the accuracy of anti-aircraft fire. The probability of crippling an enemy aircraft by firing a single shell and making a direct hit, they noted, was extremely low. This low probability was explained by two factors: positioning error (the shells often burst in the wrong place) and unfavorable fragmentation characteristics of the shell (even fragments of a shell that exploded some distance away from an aircraft often did little damage because the mass and velocity of the fragments hitting the aircraft provided insufficient penetrating power).

Ground tests that exploded shells in the midst of wooden targets showed that the mean number of perforations in the targets was 2 per unit area and could be represented by a Poisson distribution. As Appendix Table F (for $\mu = 2$) confirms, the probabilities for 0, 1, 2, 3, and more perforations equaled .14, .27, .27, .18, and so on. This information became part of a larger mathematical model that ultimately helped improve the design characteristics of shells: by varying the thickness of the shell casing and its explosive filling, the mass and velocity of fragments was controlled. Along with reductions in positioning errors, this improvement in shell design led to improved performance by the anti-aircraft batteries.

Source: E. S. Pearson, "Statistics and Probability Applied to Problems of Anti-Aircraft Fire in World War II," in Judith M. Tanur et al., eds., *Statistics: A Guide to the Unknown* (San Francisco: Holden-Day, 1972), pp. 407–15.

6.2 Supplying Spare Parts to Polaris Submarines

Polaris submarines go on missions that last about 60 days; during this time they must rely on their own supply of spare parts. At the end of each mission, they are met by a supply ship that replenishes spare parts used up during the mission. Defense Department analysts who are planning the inventory policy of the supply ships have found the *Poisson probability distribution* extremely helpful. The minimum number of units of any given spare part that must be on hand when the two ships meet equals the number that failed during the mission, and the occurrence of failures is well described by a Poisson process.

According to a Defense Department study of the first 61 patrols of these submarines, the value of μ var-

ied between 0 and 5, depending on the part. Thus, Appendix Table F quickly provides the implied probabilities: For a part with $\mu = 2$, the probability of 0 failures on a mission equals .1353, that of 1 failure equals .2707, and so on, while that of 7 or more failures is practically zero. The knowledge of this distribution has, thus, helped to promote an effective and yet economical inventory policy for submarine tenders.

Sources: Sheldon E. Haber and Rosedith Sitgreaves, "An Optimal Inventory Model for the Intermediate Echelon When Repair is Possible," *Management Science,* February 1975, pp. 638–48; Edwin Mansfield, *Statistics for Business and Economics* (New York: Norton, 1980), pp. 164–65.

BIOGRAPHIES

6.1 Siméon Poisson

Siméon Denis Poisson (1781–1840) was born at Pithiviers, France. He was sent to study at the famous École Polytechnique of Paris and performed so well that he was exempted from the final exams and remained there as a professor for 40 years. He also held a great variety of supplementary posts and wrote 300 papers on mathematics, astronomy, and physics. Most famous at the time was his *Traité de Mécanique* (1811), but today he is best known for his presentation in 1837 of the exponential limit of the binomial distribution. Although the same result had been reached in 1718 by de Moivre (see Biography 5.1), the probability distribution so derived is now called the *Poisson probability distribution.*

Important as it has turned out to be, Poisson's achievement remained, nevertheless, practically unknown for more than 60 years. Attention was drawn to it in 1898 in a paper by Ladislaus von Bortkiewicz.[1] He noted how deaths of Prussian soldiers, caused by horsekicks, could be described as a Poisson process and how their number could be estimated by Poisson's formula. When examining the experience of 10 army

corps over 20 years (a total of 200 observations), von Bortkiewicz found the absolute and relative frequency distribution as given in the first three columns of the table on the next page. That is, 109 of these records showed zero deaths, 65 showed one death, and so on. No army corps ever recorded more than four such deaths. Von Bortkiewicz calculated the weighted mean annual number of such deaths as

$$\frac{109(0) + 65(1) + 22(2) + 3(3) + 1(4)}{200}$$

$$= \frac{122}{200} = .61,$$

and he introduced this value of μ in the Poisson formula:

$$p(R = x | \mu = .61) = \frac{e^{-.61}(.61)^x}{x!}.$$

The result, given in the last column of the table, closely matched the empirical probabilities shown in the relative-frequency column. (Note: You can confirm the von Bortkiewicz results with the help of Appendix Table F, using a value of $\mu = .6$.) From then on, the Poisson probability distribution has remained in the limelight.

[1] *Das Gesetz der kleinen Zahlen* (Leipzig: Teubner, 1898).

Yearly Number of Deaths from Horsekicks, x	Absolute Frequency	Relative Frequency	Poisson Probabilities
0	109	.545	.544
1	65	.325	.331
2	22	.110	.101
3	3	.015	.021
4	1	.005	.003
	200	1.000	1.000

Source: International Encyclopedia of Statistics, vol. 2 (New York: The Free Press, 1978), pp. 704–706.

6.2 James Bernoulli

James Bernoulli (1654–1705), also known as Jacques or Jakob, was born in Basel, Switzerland, into a family that had fled Antwerp a few decades earlier because of its Protestant faith and that was to produce nine mathematicians of first rank within three generations. James himself was made to study theology and completed these studies, but, against his father's will, he soon turned to mathematics and astronomy. In travels through Holland, France, and England, he familiarized himself with the state of these sciences, and he returned with theories of his own concerning comets and gravity. Before long, he was appointed professor of mathematics at the University of Basel. Inspired by the work of Leibniz concerning the infinitesimal calculus, he used it to tackle an abundance of mathematical problems (especially in astronomy and mechanics) and to describe the properties of many important curves. Indeed, he became so fascinated by the logarithmic spiral that he wrote[1]:

Because our wonderful curve always in its changes remains constantly the same and identical in type, it can be regarded as the symbol of fortitude and constancy in adversity: or even of the resurrection of our flesh after various changes and at length after death itself. Indeed, if it were the habit to imitate Archimedes today, I would order this spiral to be inscribed on my tomb with the epitaph

[1] As cited by F. N. David, *Games, Gods, and Gambling* (New York: Hafner, 1962), pp. 138–39.

Eadem mutata resurgo (Although changed, I will resurrect).

Such was done. Yet, much more important than Bernoulli's description of the logarithmic spiral was his work on probability. Having been born in the very year in which de Fermat and Pascal (see Biographies 5.2 and 5.3) exchanged the famous letters that developed the fundamental principles of probability, James Bernoulli became the first to devote an entire book to the subject. His *Ars Conjectandi* (The Art of Conjecturing) was written in Latin and published posthumously in 1713. It contains, among other things, a systematic presentation of the theory of permutations and combinations, applications of probability theory to contemporary games (including the *jeu de paume,* a forerunner of tennis), and the development of the binomial probability distribution (discussed in this chapter). Bernoulli's great work also takes up a great many philosophical ideas concerning the very nature of probability (objective vs. subjective) and is interspersed with delightful examples of wit[2]:

Even as the finite encloses an infinite series
And in the unlimited limits appear,
So the soul of immensity dwells in minutia
And in narrowest limits no limits inhere.
What joy to discern the minute in infinity!
The vast to perceive in the small, what divinity!

[2] *Dictionary of Scientific Biography,* vol. 2 (New York: Scribner's, 1970), p. 50.

Source: International Encyclopedia of Statistics, vol. 1 (New York: The Free Press, 1978), pp. 18–19.

Summary

1. Any quantitative variable the numerical value of which is determined by a random experiment (and, thus, by chance) is called a *random variable*. A systematic listing of each possible value of a random variable with the associated likelihood of its occurrence is called the random variable's *probability distribution*. This chapter focuses on probability distributions of discrete random variables (that can assume values only at specific points on a scale of values, with inevitable gaps between values).

2. Any probability distribution can be summarized by calculating the random variable's arithmetic mean (or expected value), along with its variance and standard deviation. (Appropriate formulas are given in Box 6.A.)

3. The *binomial probability distribution* is the most important probability distribution for discrete variables. It shows the probabilities associated with possible values of a random variable that are generated by a *Bernoulli process*. Such a process is a sequence of n identical trials of a random experiment such that each trial (a) produces one of two possible complementary outcomes that are conventionally called *success* and *failure* and (b) is independent of any other trial so that the probability of success or of failure is constant from trial to trial. The number of successes achieved in a Bernoulii process is called the *binomial random variable;* the probabilities for different values of it can be calculated with the help of the *binomial formula* (Box 6.B). The formula shows that the probability for any given number of successes varies with the number of trials, n, and with the probability of success in any one trial, π. A different binomial probability distribution can, therefore, be derived for each combination of n and π; many such distributions have been tabulated in *binomial probability tables* for both individual and cumulative values of the binomial random variable. Simplified formulas exist for determining a binomial random variable's summary measures of location, spread, and skewness (Box 6.C).

4. The *hypergeometric probability distribution* provides probabilities associated with possible values of a discrete random variable in situations in which these values are generated by sampling and in which such sampling is done without replacement so that the probability of success changes from one trial to the next. The number of successes achieved when a random sample of n is drawn without replacement from a population of N (within which S units exist with the characteristic denoting success) is the *hypergeometric random variable*. Probabilities for different values of it can be calculated with the help of the *hypergeometric formula* (Box 6.D). Once more various simplified formulas can be derived for calculating summary measures for the hypergeometric random variable (Box 6.E). Binomial probability distributions closely resemble hypergeometric ones when the sample size is small relative to population size; under such circumstances, therefore, binomial probability tables can be used to make quick approximations of hypergeometric probability distributions.

5. In contrast to binomial and hypergeometric random variables, both of which are defined as numbers of successes reaped within a fixed number of trials of some random experiment, a *Poisson random variable* is the number of occurrences of a specified event within a specified time or space. In a *Poisson process,* a series of events of a given type occur in a random (and, hence, unpredictable) pattern over time or space such that (a) the Poisson random variable can equal any integer between zero and infinity, (b) the number of occurrences in one unit of time or space is independent of that in any other such unit, and (c) the probability of an occurrence is the same in all such units. The probabilities associated with alternative values of the Poisson random variable can be determined with the help of the *Poisson formula* (Box 6.F). A different Poisson probability distribution can be derived for each value of μ, the mean number of occurrences; many such distributions have been tabulated in *Poisson probability tables* for both individual and cumulative values of the Poisson random variable. Once again, simplified formulas exist for calculating summary measures for the Poisson random variable (Box 6.G). Under certain circumstances (namely, when the number of trials, n, in a Bernoulli process is large and the probability of success in any one trial, π, is small), the Poisson probability distribution can serve as an approximation of the binomial one.

Key Terms

Bernoulli process
binomial coefficient (c_x^n)
binomial formula
binomial probability distribution
binomial probability tables
binomial random variable
chance variable
cumulative probability distribution
expected value, E_R
hypergeometric formula

hypergeometric probability distribution
hypergeometric random variable
Poisson formula
Poisson probability tables
Poisson process
Poisson process rate
Poisson random variable
probability distribution
random variable

Questions and Problems

The computer programs noted at the end of the chapter can be used to work many of the subsequent problems.

1. Consider a canasta deck; it consists of four jokers plus two ordinary decks of 52 playing cards. Cards are given point values as follows: red 3 = 100 points; joker = 50 points; ace or deuce = 20 points; 8 through king = 10 points; 4 through 7 = 5 points; black 3 = 5 points. Determine the probability distribution for a random variable, R, that equals the point value of the first card drawn from a shuffled canasta deck.

2. For the random variable described in problem 1, determine
   a. the expected value.
   b. the variance
   c. the standard deviation.

3. Could the following represent probability distributions? If so, write out the distribution and determine its summary measures (mean, variance, standard deviation).

   a. $p(x) = \dfrac{x}{10}$, where x = 1, 2, 3, or 4.

   b. $p(x) = .7x$, where x = -1, 0, 1, or 2.

   c. $p(x) = \dfrac{x^2}{14}$, where x = 0, 1, 2, or 3.

   d. $p(x) = \dfrac{1}{x}$, where x = 1, 2 or 3.

   e. $p(x) = (10 - x)/50$, where x = -1, 0, 5, or 6.
   f. $p(x) = (10 - x)/40$, where x = 0, 1, 2, 3, or 4.

4. Determine the probability distribution for a random variable, R, that equals the number of points showing when a regular six-sided die is rolled once.

5. For the random variable described in problem 4, determine
   a. the expected value.
   b. the variance.
   c. the standard deviation.

6. Referring to Table 6.4, show why the sum of weighted values in the last column of the table equals the result yielded by using the shortcut method of determining variance given in Box 4.N on page 114.

7. The Kentfield Hardware store is about to place an order for the *expected number* of lawn mowers demanded, based on the probability distribution in Table 6.A. How many will be ordered? What is the variance and standard deviation of the number demanded?

Table 6.A

Number Demanded	Probability
0	.05
1	.10
2	.20
3	.30
4	.20
5	.10
6	.05

8. In a certain hour each Monday, the Metropolitan Transport Authority noted that the following data

applied. What would be the expected revenue if the Authority instituted a toll of $1 per car?

Table 6.B

Cars Crossing Throgs Neck Bridge	Absolute Frequency of Observation (Mondays)
100	20
120	35
140	16
150	7

9. A state auto inspection station collected the following data during the past year. If the observed pattern continued to hold, what would be the station's expected daily revenue if it instituted a $10 fee per car?

Table 6.C

Cars Coming For Inspection	Absolute Frequency of Observation (Days)
20	79
25	121
32	19
40	25
47	39
51	30

10. An office service company has collected the accompanying data. The manager wants to know
  a. the probability of fewer than 30 service calls on a given day.
  b. the expected company revenue per day if the pattern observed in the past continues and all service requests are answered (at a $50 fee per call).

Table 6.D

Requests For Servicing Broken-Down Xerox Machines	Absolute Frequency of Observation (Days)
5	276
9	59
13	30
25	36
31	38
52	21

11. The Massachusetts government has collected the following data over the past 50 years. If the pattern continues next year, what is the expected number of trees that will be defoliated by gypsy moths? What is the standard deviation?

Table 6.E

Number of Trees Defoliated by Gypsy Moths in Year	Probability
4,000 to under 7,000	.05
7,000 to under 10,000	.07
10,000 to under 13,000	.11
13,000 to under 16,000	.54
16,000 to under 19,000	.11
19,000 to under 22,000	.07
22,000 to under 25,000	.05

12. The Massachusetts sales tax is supposed to collect 5 cents on the dollar, but, on the cents part of any purchase, sales tax is collected according to the accompanying schedule. What is the percentage of tax actually collected? In your answer, make use of a probability distribution.

Table 6.F

Cents	Tax
1–29	1¢
30–49	2¢
50–69	3¢
70–89	4¢
90–99	5¢

13. An automobile manufacturer has fitted all cars with the identical pollution-control device (designed to meet government standards), yet experience shows that 5 percent of cars tested perform below government pollution standards. Assume that 20 cars coming off the assembly line during a given month are selected at random and that the condition of a Bernoulli process are met.
  a. What is the probability distribution for the numbers, x, of below-standard cars found in the sample?
  b. What is the probability that a government inspector who tests 20 cars a month in the above fashion will unjustly accuse the manufacturer of producing more than 5 percent of all cars below standard?

14. Determine the expected value, variance, and standard deviation of the random variable noted in problem 13.

15. a. A family wants ten children: five boys and five girls. Assuming there is an equal chance for getting a child of either sex on any one try, what is the probability of this happening? Make your determination in two ways: first with Pascal's triangle on p. 180, then with the help of Appendix Table C.

  b. Grover C. Jones of Peterson, W. Virginia, in fact had 15 children—all boys. What is the probability of this happening?

16. Determine the expected value, variance, and standard deviation of the random variable in problems 15a and 15b.

17. An employer has noted a 10-percent annual quit rate of the firm's employees. Assume that three employees are sampled at random and that the conditions of a Bernoulli process are met. With the help of a tree diagram, determine the probability distribution for the numbers, x, of employees in the sample who are likely to quit next year. Check your result in Appendix Table C.

18. Assuming that the conditions of a Bernoulli process are met, determine the following:

  a. A test is administered 10 times; the probability of getting the correct answer is .9. Use an appropriate formula to figure the likelihood of getting precisely 9 correct answers.

  b. What is the probability of having exactly 3 boys in a family of 10 children (assuming that boys and girls are equally likely and different births are independent events)? Find your answer with the help of a table and check it by using Pascal's Triangle on page 180.

19. A car-rental firm rents only compact and medium-sized cars; experience shows that three persons out of four prefer compacts. Considering the next 15 requests, what is the probability distribution (a) for the numbers of medium-sized cars requested? (b) For the numbers of compact cars requested? (You may assume that the conditions of a Bernoulli process apply.)

20. A manufacturer is sampling (with replacement) incoming shipments of 100 parts produced by other firms. (a) If (unbeknownst to the manufacturer) 5 percent of the parts are defective, what is the probability distribution for discovering defective units for a sample of 5 parts? (b) What is the probability distribution if in fact 40 percent of all units are defective and 5 of them are sampled? (c) What if 10 percent are defective and 3 units are sampled? (d) What if 60 percent are defective and 3 units are

sampled? (You may assume that the conditions of a Bernoulli process apply.)

21. A polling organization randomly samples 100 consumers from among the U.S. population about the new design of a product. If in fact 40 percent of all consumers favor the new design, what is the likelihood that among the sampled individuals the design will be favored by (i) 20 or fewer? (ii) Precisely 30? (iii) Fewer than 40? (iv) More than 45? (v) 50 or more? (vi) Between 30 and 40, inclusive? (You may assume that the conditions of a Bernoulli process apply.)

22. A new drug in nationwide use seems to be effective 40 percent of the time. What are the chances that a random sample of 100 patients using the drug will show success (i) in at most 20 patients? (ii) In exactly 30 patients? (iii) In at least 40 patients? (iv) In fewer than 50 patients? (v) In more than 50 patients? (You may assume that the conditions of a Bernoulli process apply.)

23. A new disease affects 20 percent of the population (or a new "bug" affects 20 percent of all units produced by a production process). If the conditions of the Bernoulli process are met, what is the likelihood that in a random sample of 5, (i) all are affected? (ii) None are affected? (iii) At least 1 is affected? (iv) At least 3 are affected? (v) Precisely 2 are affected? (vi) Between 2 and 4, inclusive, are affected?

24. A fair coin is tossed 16 times. What is the probability of getting (i) exactly 2 heads? (ii) Exactly 8 heads? (iii) Exactly 1 tail? (iv) Exactly 14 tails? [Compare your answer to (iv) with that given to (i) and explain.]

25. A lopsided coin (that gives heads a 70 percent chance of coming up) is tossed 50 times. What is the probability of getting (i) 44 or more heads? (ii) Exactly 40 heads? (iii) More than 37 heads? (iv) 31 or fewer heads? (v) Fewer than 28 heads? (vi) Between 30 and 40 heads, inclusive?

26. Twenty radar transponders are sitting on the bench at an avionics repair shop; of these, nine are inoperative, but this is not obvious to a casual observer. A thief steals five transponders. Assuming that the conditions of a Bernoulli process apply, determine the probability that the thief got (i) five good transponders, (ii) three bad transponders, (iii) only bad transponders, (iv) at least two good transponders.

27. What are the expected value, the variance, and the standard deviation of the binomial random variable (number of affected units) in problem 23? What is the skewness of the probability distribution?

28. A bottling company checks out empty returned bottles before cleaning and refilling them. Some 10 percent are chipped and discarded. If the conditions of a Bernoulli process apply and 18 bottles are sampled,
   a. What is the probability of 10 or fewer being chipped?
   b. What are the distribution's summary measures?

29. A market survey shows that 30 percent of all families own a Polaroid camera. If the conditions of a Bernoulli process apply and 100 families are sampled,
   a. What is the probability that 20 or fewer have such a camera?
   b. What is the probability that precisely 31 have a camera?
   c. What is the probability that 47 or more have a camera?
   d. What are the distribution's summary measures?

30. According to a well-known private firm, 90 percent of its parcels are delivered within 2 days. If the conditions of a Bernoulli process apply and 14 parcels (that were shipped at different times) are sampled,
   a. What is the probability that all 14 arrive within 2 days?
   b. What is the probability that none arrive within 2 days?
   c. What is the probability that precisely 8 arrive within 2 days?
   d. What are the summary measures for the probability distribution of a random variable that denotes *late* deliveries?

31. According to a well-known stockbroker, 95 percent of all purchase orders are executed within 15 minutes of request. If the conditions of a Bernoulli process apply and 7 such orders are sampled, determine these probabilities: (i) that all are executed within 15 minutes; (ii) that none are executed within 15 minutes; (iii) that at least 5 are so executed; (iv) that at most 6 are so executed.

32. A medical laboratory received a shipment of 16 microscopes of which (unbeknownst to the lab) 4 were defective. If a random sample of 3 were taken without replacement, what would be the chances of finding 0, 1, 2, 3, or 4 defective units? Determine the answer with a tree diagram.

33. Determine the answer to problem 32 by an appropriate formula.

34. Determine the summary measures of the problem 32 and 33 probability distribution.

35. Determine the probability distribution for an identical sample as in problem 32, but assume that 16,000 microscopes with 4,000 defective units had been shipped.

36. A random sample of 3 units is taken from a group of 10 in which 4 units are defective. With the help of the appropriate formulas, determine the probability distribution for the numbers of defective units found, along with summary measures, (i) if sampling occurs with replacement and (ii) if sampling occurs without replacement. (iii) Compare your result for (i) with that given in the appropriate probability table.

37. A polling organization samples (without replacement) 3 women out of a group of 10 (among whom 40 percent prefer brand A to brand B). Determine the sampling distribution of the proportion of those who favor brand A, along with appropriate summary measures.

38. Four women and five men apply for three job openings. The manager finds them "equally qualified" and, therefore, "draws names randomly out of a hat." He hires three men; the women sue. A judge asks you to determine the probability of this hiring having truly occurred by random choice.

39. At Christmas time, a furniture manufacturer has 20 unsold chairs, but 10 of them are damaged. The chairs are offered to the employees as gifts; only 10 requests come in. The manager alleges that she picked the gift chairs at random, but 9 of them turn out to be damaged. What is the probability that the selection was truly random?

40. The President appoints a special task force of 4 persons from a group of 5 Democrats and 5 Republicans. What is the probability that the selection was made randomly if (i) all appointees are Democrats, (ii) all appointees are Republicans, (iii) 2 appointees are Democrats?

41. Determine the summary measures for the probability distribution of problem 40.

42. An IRS officer has just completed 20 tax returns. Unbeknownst to him, 4 of these returns are in error. If a sample of 4 returns is now taken without replacement, what is the probability that none of these contain an error? That all of these contain an error?

43. Determine the summary measures for the probability distribution of problem 42.

44. A garden center has 100 elm trees for sale. Although it is not obvious, 35 trees have Dutch elm disease. If you buy 10 trees, what is the probability that all of them are infected? That half of them are? That all are healthy?

45. Determine the summary measures for the probability distribution of problem 44.

46. A city government is giving away wrappers for hot-water heaters to cut people's fuel bills. The mayor "randomly" selects 50 lucky recipients from among 56 Eastside and 64 Westside applicants. Of the recipients, 47 come from the Eastside. Do you believe that the distribution was "random?"

47. Draw a set of graphs for individual and cumulative values of x for those members of the Poisson probability distribution family with $\mu = .1$, 1.0, and 10.0.

48. Determine the summary measures for each of the problem 47 distributions.

49. Between 9:00 and 12:00 A.M. on Saturdays, customers arrive at a supermarket checkout counter at a rate of 50 per hour, but on Monday mornings they arrive at a rate of 2 per hour. Determine the probability distribution for up to 11 arrivals between 9:00 and 9:06 A.M. on either day by formula.

50. Rework problem 49 with the help of an appropriate table. Is the probability of any number of arrivals always higher during the Saturday peak than the Monday slump?

51. A Xerox machine fails to print on every page. On average, 1 percent of the pages are blank. If 100 copies are run off, what is the probability of none of them being blank? Of one being blank?

52. On the average, Amtrak repair crews have to replace three railroad ties per mile when they check the tracks. What is the probability of no needed replacements in the next mile? Of three or fewer? If each replacement costs $30, what is the expected cost for the next 10 miles?

53. Accidents at a chemical plant occur at a rate of 1.9 per month. What is the expected number of accidents in a year? What is the probability of no accident next month? Of fewer than 3 accidents?

54. Small businesses are going bankrupt at a rate of 9.7 per month. Determine the probability of at least 6 going bankrupt next month; of 24 businesses going bankrupt; of 5 or 6 going bankrupt.

55. In each of the following cases, determine whether the statement is *true* or *false* and explain why.
   a. If for a binomial probability distribution the mean is 4 and the variance is .8, the probability of success in any one trial equals. .5.
   b. The number of trials, n, that determines the probability distribution referred to in (a) equals 5.
   c. The entire probability distribution implied by (a), for 0–5 successes, respectively, equals .0003, .0064, .0512, .2048, .4096, and .3277.
   d. The mean number of hearts one would obtain when drawing 4 cards from a deck is the same regardless of whether each card received is set aside or is only recorded and returned to the deck which is then reshuffled before the next card is drawn.

56. In each of the following cases, determine whether the statement is *true* or *false* and explain why.
   a. A store permits returns of purchases within 10 days and knows from experience that the chance of return is 10 percent. If eighteen units are sold to different customers, the probability of having at most 5 items returned equals .9936.
   b. The probability of picking 2 blacks when selecting 5 jurors (without replacement) from a pool of 9 persons (of whom 4 are black) equals .6371.
   c. The probability of picking 8 males when selecting 10 persons (without replacement) from a pool of 1,000 (half of whom are male) equals .0439.
   d. If cars arrive at a gas station at a rate of 5 per minute, the probability of precisely 2 arrivals in 2 minutes equals .0023.

57. In each of the following cases, determine whether the statement is *true* or *false* and explain why.
   a. If potholes are found at the rate of 10 per 1,000 square yards in a stretch of road, the probability of finding at most 10 potholes in a 500-square-yard area equals .6160.
   b. If a typesetter makes errors at a rate of .5 error per page, the probability of finding at least 10 errors in 40 pages equals .995.
   c. If ships pass the Statue of Liberty at the rate of 2 per hour, a tourist looking for ships for 30 minutes has at most a 50 percent chance of seeing 1 or 2 ships.
   d. If crimes between 1:00 and 3:00 A.M. in a city district occur at a rate of 2 per hour, the chances for more than two crimes in any 15-minute period are .0144.

58. In each of the following cases, determine whether the statement is *true* or *false* and explain why.
   a. If a taxicab company has on the average 4 cars laid up for repairs on a given day, it needs 6 spare cabs in order to keep the probability of having a driver without a car below 3 percent.
   b. If people arrive at a restaurant at a rate of 15 groups per hour, the probability of more than 5 groups appearing in a 10-minute span equals .042.
   c. If 4 bubbles are found per 1,000 square feet of plate glass, the chances that a 20-by-5-foot window contains no bubbles are 32.97 percent.

d. If an auditor randomly samples without replacement 1,000 of 1 million electric bills (among which 1 percent are in error), the probability of 10 errors being found in the sample equals .1251.

59. In each of the following cases, determine whether the statement is *true* or *false* and explain why.

  a. Assume that a tornado will strike a given area at the rate of 5 times per century. According to the Poisson formula, the probability of at least one tornado per 50 years is .8120, but according to the binomial formula it is .8970.

  b. The probability of getting 4 heads in 6 tosses of a balanced coin is .2344.

  c. The probability is .4114 that fewer than 4 students among any 20 being picked at random from the very large number of U.S. entering freshmen will drop out of college before graduation, provided that 20 percent of all entering freshmen drop out.

d. The probability is .1511 that 9 acceptable items will be found in a sample of 10 picked at random from a very large annual output of a plant that produces 90 percent acceptable units.

60. In each of the following cases, determine whether the statement is *true* or *false* and explain why.

  a. The probability is .1310 that at least 30 correct answers will be given in a 50-question true/false test, provided the answers are given at random by a totally unprepared student.

  b. The probability is .2430 that oil will be found in precisely one of three holes being drilled (in different locations), although 90 out of 100 such holes drilled are dry.

  c. The probability is .0729 of finding precisely 2 books that have never been borrowed when taking a random sample of 5 books from a large library in which 10 percent of all books have never been borrowed.

Selected Readings

Feller, William. *An Introduction to Probability Theory and Its Applications,* 2 vols. New York: Wiley, 1950–1966. By far the best textbook on the theory and application of discrete probability distributions.

Gridgeman, N. T. "Probability and Sex." *The American Statistician.* June 1968, p. 29. Shows what the binomial distribution can tell us about the sexes of children in families, including the order of their appearance.

Mullet, Gary M. "Siméon Poisson and the National Hockey League." *The American Statistician,* February 1977, pp. 8–12. Shows that the Poisson distribution describes the number of goals scored for or against each of the teams that played in the National Hockey League during the 1973–74 season.

Owen, Donald B. *Handbook of Statistical Tables.* Reading, Mass.: Addison-Wesley, 1962. One of the most complete and useful volumes of statistical tables.

Wallis, W. Allen. "The Poisson Distribution and the Supreme Court." *Journal of the American Statistical Association,* June 1936, pp. 376–80. Shows that the Poisson distribution describes the number of vacancies on the U.S. Supreme Court.

Computer Programs

The DAYSAL-2 personal-computer diskettes that accompany this book contain two programs of interest to this chapter:

  7. *Probability Distributions,* lets you compute binomial, hypergeometric, and Poisson probabilities, along with the summary measures of the respective probability distributions.

  16. Expected Monetary Values, although designed for another purpose, can be used to compute the mean of probability distributions, such as that in Table 6.3.

CHAPTER SEVEN

Continuous Probability Distributions

Chapter 6 focused on quantitative random variables that are *discrete*—that can assume values only at specific points on a scale of values, with inevitable gaps between. As was noted, many everyday processes generate values of such variables, and the likelihood of occurrence of specific values can be gauged with the help of one or another of certain common probability distributions, such as the binomial, hypergeometric, or Poisson distribution. The present chapter extends the earlier discussion to quantitative random variables that are *continuous* in nature—that can assume values at all points on a scale of values, with no breaks between possible values. Consider characteristics measured in units of money, time, distance, or weight, to name just a few of the possibilities. The possible observations about such variables are infinite in number.

The profit per dollar of sales earned by a firm is a case in point. It might be -10 cents or $+23$ cents or even $+41.37895$ cents. The random experiment of operating a firm for a year can generate *any* value for this variable—positive, zero, or negative. Or contemplate the possible measurements concerning the lifetime of new appliances, of car batteries, of light bulbs, or of tires, the flight time of a plane, the completion time of a task, the reaction time after a stimulus. Consider distance measurements of the drilling depths of oil wells, of inches of rainfall, of the length of rods of steel or of ears of corn, of the thickness of tablets, of the height of weather satellites. Imagine measuring the weight of apples harvested per tree, of cereal boxes filled, of ingots of metal produced. In all these cases and a million more, the observations actually made will be among an infinite number of conceivable ones. As a result, one cannot *list* all the possible values of continuous random variables; nor can one list the associated probabilities for each one of the infinite number of conceivable values. When continuous random variables are involved, it is common practice instead to associate probabilities with *ranges* of values along the continuum of possible values that the random variable might take on.

THE PROBABILITY DENSITY FUNCTION

Panel (a) of Figure 7.1 represents one way of picturing the probability distribution of a continuous random variable. The graph is a relative frequency histogram of the type first introduced in Chapter 3. The continuous random variable in this example is the length of service, as of December 31 of a given year, of the 49,000 employees of a firm. For any one employee, this period could clearly be measured by any positive number: as 1 year, 5.2 years, or even as 7.13946 years. The possible values of this variable have been grouped on the horizontal axis in 5-year intervals. The height of any column standing on top of any one of these equal-sized intervals represents the relative frequency density, telling us simultaneously the proportion of all employees whose service length falls within the given range and the probability that any one employee, randomly chosen, will have a service length within that range. Naturally, the sum of these proportions (and of these probabilities) equals 1.

Thus, we can tell instantly that the probability equals .3 that a randomly chosen employee has been with the firm between 10 and 15 years: $p(R = 10$ to 15 years$) = .3$. Yet we cannot tell the probability of our random variable lying in the range from 11 to 13 years because that interval is completely contained within the 10- to 15-year range in our histogram. We can imagine, however, constructing from our data a series of alternative histograms with ever narrower years-of-service intervals. If we did, the

Figure 7.1   **A Continuous Probability Distribution**

The probability distribution of a continuous random variable can be represented by a relative frequency histogram, as in panel (a); it can also be approximated by a smoothed frequency curve, such as the color line in panel (b), which is called a *probability density function.*

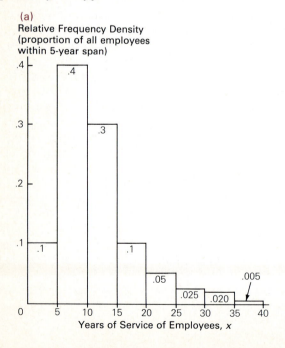

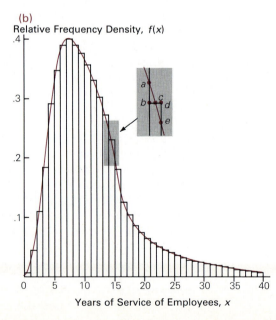

Figure 7.2   Two Probability Density Functions

The probability that a continuous random variable (such as years of employee service) takes on values within any given range of possible values (such as 11 to 13 years) is represented by the ratio of two areas: the area (here crosshatched) under the portion of the probability density function covering the range of interest and the entire area under the function (which is taken to equal 1). In this example, the probability that a randomly chosen employee has seen between 11 and 13 years of service is larger, therefore, in Firm A, than in Firm B.

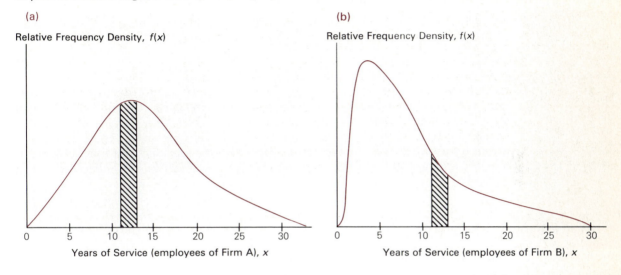

columns would get ever more numerous and also narrower until a line along their tops approached the smooth frequency curve given in panel (b). This curve is called a **probability density function.** It describes the probability distribution of a continuous random variable, R, and is denoted by $f(x)$, pronounced "f of x," where x represents all the possible values of this variable. Note that the total area under the probability density fucntion, just as that under the relative frequency histogram, equals 1. As the inset in panel (b) indicates, the smooth frequency curve adds some areas, such as *abc*, to the total histogram area, but it cuts off equivalent areas, such as *cde*. Given the probability density function, the probability that a continuous random variable takes on values within any given range is, thus, represented by the area under the portion of this curve covering this range.

Consider Figure 7.2, which shows probability density functions for the years of service of employees in two different firms, A and B. (The total area underneath either function equals 1). One can instantly tell that the probability of finding an employee with 11 to 13 years of service equals whatever proportion the shaded area constitutes of the total area underneath the curve; this probability is clearly larger in Firm A than in Firm B. The total area equals 1 in both cases, but the crosshatched area is larger in panel (a).

Note: in the case of continuous random variables, it is customary never to consider the probability of occurrence of a single value, such as 11.92374 years of service, to be anything but zero. This probability would have to equal the area above a *point* on the horizontal axis up to the probability density function, and, given the fact that a point has no width, this area above the point equals zero. Indeed, given an infinite number of possible values, the probability of one specific value is practically nil. When calculating cumulative probabilities for continuous random variables, it is, therefore, deemed irrelevant to make a distinction between the probability of the random variable being "smaller than x" and being "smaller than or equal to x":

$$p(R < x) = p(R \le x) \text{ because } p(R = x) = 0.$$

Similarly, no distinction is made between the probability of a continuous random variable being "larger than x" and being "larger than or equal to x":

$$p(R > x) = p(R \ge x) \text{ because } p(R = x) = 0.$$

Figure 7.2 pictures two rather differently shaped probability density functions; as one can expect, an infinite variety exists, each one describing the probabilities associated with different types of continuous random variables. Three types of probability density functions, however, are encountered more often than most. These are (1) the *normal* probability distribution, (2) the *exponential* probability distribution, and (3) the *uniform* probability distribution, and to these we turn in the remainder of this chapter.

THE NORMAL PROBABILITY DISTRIBUTION

The **normal probability distribution** is a probability density function that (a) is single-peaked above the random variable's mean, median, and mode (all of which are equal to one another), (b) is perfectly symmetrical about this central value (and, thus, said to be "bell-shaped"), and (c) is characterized by tails extending indefinitely in both directions from the center, approaching but never touching the horizontal axis (which implies a positive probability for finding values of the random variable within any range from minus infinity to plus infinity).[1] This probability distribution was first described in 1733 by Abraham de Moivre (Biography 5.1), and it was popularized by Adolphe Quetelet (Biography 2.1), who used it to discuss "the average man" (l'homme moyen), and by Carl Friedrich Gauss (see Biography 7.1 at the end of the chapter), who used it to describe errors of measurement in astronomy. Indeed, many measurements of natural phenomena—ranging from the heights, weights, or IQs of people to the distances, volumes, or speeds of heavenly bodies—have frequency distributions that closely resemble the normal distribution. As we shall see in Chapter 8, this fact makes the normal distribution very important for making inferences about unknown populations from known samples.

A Graphical Exposition

Figure 7.3 shows three members of the normal probability distribution family. They differ from one another only by the magnitudes of the mean, μ_R, and standard deviation, σ_R, of the random variable in question. Note that every one of the three probabil-

---

[1]Chapter 7 of the *Student Workbook* that accompanies this text considers the mathematical properties of the normal distribution in detail.

Figure 7.3   Three Members of the Normal Probability Distribution Family

Members of the normal probability distribution family, called *normal curves* for short, differ from one another only by the values of the distribution's mean (μ_R) and standard deviation (σ_R). The value of μ_R positions the distribution on the horizontal axis; that of σ_R determines its spread and, thus, its appearance as peaked (leptokurtic), when σ_R is small, or as flat (platykurtic), when σ_R is large. In every case, the normal probability density function has points of inflection precisely 1 standard deviation below and above the mean and then approaches, respectively, minus and plus infinity on the horizontal axis (which, naturally, the graph can only suggest but not show). As a practical matter, the height of every normal curve is near 0 within 3 standard deviations of the mean.

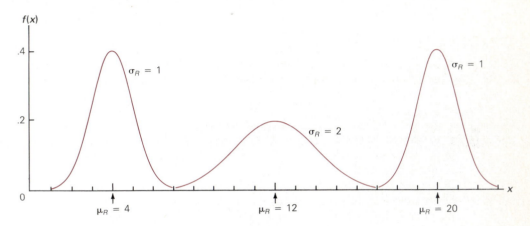

ity density functions shown in the graph exhibits the three characteristics listed before, although it is, of course, impractical in a graph to show how the tails of the functions extend indefinitely in both directions without ever touching the axes. Analytical Example 7.1, "Honest Weights and the Normal Curve of Error," provides additional illustrations.

The Formula

The precise height of the normal probability density function, $f(x)$, for different values, x, of the normal random variable, R, can be determined with the help of the formula given in Box 7.A. This rather formidable expression is the equation for the types of curves seen in Figure 7.3; as usual, it becomes less formidable upon closer inspection.

7.A The Normal Probability Density Function

$$f(x) = \frac{1}{\sigma_R \sqrt{2\pi}} \cdot e^{-\frac{1}{2}\left(\frac{x - \mu_R}{\sigma_R}\right)^2}$$

where $f(x)$ is a function of observed value, x, while μ_R is the mean and σ_R is the standard deviation of the probability distribution of random variable, R, and $\pi = 3.14159$, $e = 2.71828$.

The mean and standard deviation of the normal random variable's probability distribution appear as μ_R and σ_R, and they appear alongside two of the most famous constants in mathematics: π is in this case not the population proportion, but the ratio of the circumference to the diameter of the circle; it equals 3.14159. The constant e, already noted in the Poisson formula (Box 6.F and footnote 4 in Chapter 6), is the base of the natural logarithms; it equals 2.71828.

Thus, given μ_R and σ_R, one can calculate the height of the probability density function $f(x)$ for each x; by integrating that function over a specified range of x, one can calculate the type of crosshatched area shown in Figure 7.2 and, thus, determine the probability of finding values of the random variable within that range of x. In order to facilitate the determination of such probabilities, statisticians have found a way of converting normal curves with differing shapes and positions into a single *standard* normal curve and of then finding probabilities in appropriately prepared tables.

The Standard Normal Curve

The **standard normal curve** is a normal probability density function with a mean of 0 and a standard deviation of 1. As Figure 7.4 illustrates, any normal curve—whatever the value of μ_R and σ_R—can quickly be transformed into this standard curve by plotting the value of μ_R as 0 and expressing values above and below this mean in units of σ_R. The deviation from the mean, μ_R, of any given value, x, of a normally distributed random variable, when measured in units of standard deviations, σ_R, is called the **standard normal deviate** (see Box 7.B), and it is symbolized by z. (In education statistics, z is often referred to as the *standard score,* as was noted in Analytical Example 4.1).

7.B The Standard Normal Deviate

$$z = \frac{x - \mu_R}{\sigma_R}$$

where x is an observed value of random variable, R, while μ_R is the mean and σ_R is the standard deviation of its probability distribution.

Thus, if $\mu_R = 4$ and $\sigma_R = 1$, a value of $x = 2$ becomes $z = -2$, indicating (as Figure 7.4 shows) that this value of the random variable lies 2 standard deviations below the mean. Similarly, when $\mu_R = 12$ and $\sigma_R = 2$, a value of $x = 14$ becomes $z = +1$, indicating a distance of 1 standard deviation *above* the mean. By the same procedure (not shown in Figure 7.4), a value of $x = 22.5$, given $\mu_R = 20$ and $\sigma_R = 1$, would be plotted as $z = +2.5$ (but as $z = +5.25$, given $\mu_R = 12$ and $\sigma_R = 2$).

The Standard Normal Tables

When x values have been transformed into z values by setting $\mu_R = 0$ and $\sigma_R = 1$, the Box 7.A formula simplifies to that given in Box 7.C.

Figure 7.4   Constructing the Standard Normal Curve

The three different normal curves of Figure 7.3, reproduced here in the top panel, can be converted into an identical *standard* normal curve, shown in the bottom panel. Each actual value, *x*, of the normal random variable is translated into a standardized value, *z*, by first calculating the deviation of *x* from the mean and then expressing this deviation in terms of standard deviation units. This procedure is equivalent to setting μ equal to zero and treating the distance of 1σ as equal to one.

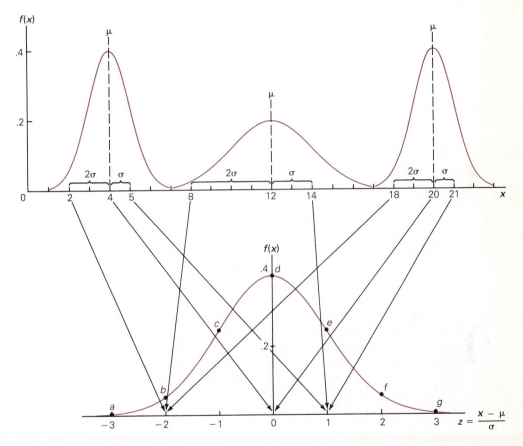

7.C The Standard Normal Probability Density Function

$$f(z) = \frac{1}{\sqrt{2\pi}} \cdot e^{-\frac{1}{2}z^2}$$

where $\pi = 3.14159$, $e = 2.71828$, $z = \dfrac{x - \mu_R}{\sigma_R}$ and $-\infty \le z \le +\infty$.

Based on this simplified formula, two types of tables can be prepared: tables of ordinates and tables of cumulative relative frequencies.[2] We will discuss the nature of these tables first and will then show how helpful they can be in typical business and economics applications.

Tables of ordinates. One set of these tables shows the *height* of the standard normal probability density function for different values of z. Thus, the height of the curve corresponding to z values of 0, 1, 2, 3, and 4 is shown in the accompanying table.

As was noted earlier, strictly speaking, the normal curve touches the horizontal axis only at minus and plus infinity, but for all practical purposes it can be considered to do so within 3 standard deviations of the mean (or when $z = 3$).

z	Height of Curve	Corresponding Points in Figure 7.4
0	.39894	*d*
1	.24197	*c* and *e*
2	.05399	*b* and *f*
3	.00443	*a* and *g*
4	.00013	—

Tables of cumulative relative frequencies. A second set of tables typically shows the *area* under the standard normal probability density function between $\mu = 0$ and alternative *positive* values of z. (Because of the symmetry of the normal curve, these tables can also be interpreted to show the noted area between $\mu = 0$ and alternative *negative* values of z.) This type of table, as we shall see presently, is of much greater importance than a table of ordinates, and it has been reproduced, for selected values of z, as Appendix Table H. Note that the value of z, up to the first decimal, is shown in the leftmost column of that table, while the second decimal of z is found in one of the 10 entries in the top row. By *adding together* any given row heading with any given column heading, one obtains a z value with two decimal places. Thus, a row heading of .5 and a column heading of .00 corresponds to a z value of .5 + .00 = .50. At the intersection of the .5 row and the .00 column, one finds, accordingly, the area under the standard normal curve between the mean of 0 and a z value of .50, and this area value equals .1915. This means that there is a probability of .1915 that the value of a normally distributed random variable lies within half a standard deviation above the mean. Given the curve's symmetry, there exists, of course, an equal probability of .1915 that such a value lies within half a standard deviation *below* the mean. Thus, there exists a probability of .1915 + .1915 = .383 that such a value lies somewhere between half a standard deviation below to half a standard deviation above the mean.

Moving along the $z = .5$ row to the column headed .01, .02, .03, and so on, we can find, similarly, the probability of any value of the random variable lying $z = .51$, .52, .53 standard deviations above the mean. The probabilities involved equal .1950, .1985, .2019, respectively. Hence, the probability of finding a value of a normally dis-

---

[2]These tables were first prepared in 1799 by Chrétien Kramp, a French mathematician and physicist (who also invented the factorial notation—n!—that was discussed in Chapter 5). The most recent tables were issued in 1953 by the U.S. National Bureau of Standards, as noted in Appendix Table H.

Figure 7.5 Finding Areas Under the Normal Curve

With the help of Appendix Table H, one can determine areas under the normal curve and, thus, the probability of finding values of a normally distributed random variable within specified numbers of standard deviations below or above the mean. Because the area under the normal curve between the mean and 1 standard deviation below it covers .3413 of the entire area under the curve, as represented by the shaded portion in panel (a), and because the same is true for the range of values between the mean and 1 standard deviation above it, as shown in panel (b), .3413 + .3413 = .6826 of the total area is found within the range of μ ± 1σ; thus .6826 is the probability of finding a value of the random variable within that range. Similarly, the probability is .9544 of finding a value of the normal random variable within μ ± 2σ.

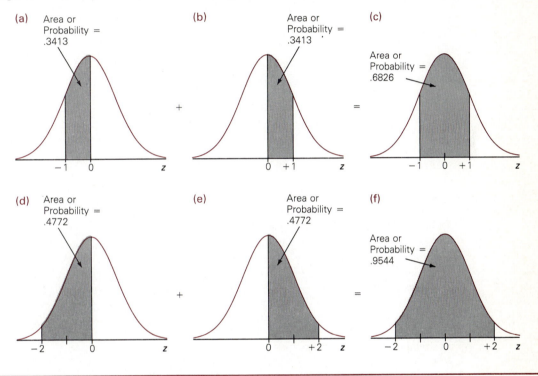

tributed random variable somewhere between .53 standard deviations below the mean and .53 standard deviations above it equals .2019 + .2019 = .4038.

Figure 7.5 shows, similarly, the likelihood of encountering values of such a random variable within 1 or 2 standard deviations below or above the mean. The graph clearly implies that the probability of finding a value of a normally distributed random variable that lies more than 2 standard deviations either below or above its mean equals a tiny 1 − .9544 = .0456. Yet the probability remains positive even for large values of z. (According to the calculations by the National Bureau of Standards, noted in Appendix Table H, the probability of finding a value that lies more than 3 standard deviations either below or above the mean equals 1 − .9972 = .0028; that for a value beyond 8 standard deviations above or below the mean equals a tiny .00000 00000 00001.)

Note: The graphical exposition in Figure 7.5 illustrates the manner in which more complicated questions can also be answered. For example, the probability of finding a value of this normal random variable lying between 2 standard deviations below the mean and 1 standard deviation above it can easily be found by combining panels (d) and (b) as .4772 + .3413 = .8185. Similarly, the probability of finding a value that lies more than 1 standard deviation below the mean *or* more than 2 standard deviations above it can be found by using panels (a) and (e) and combining the unshaded area in the left half of panel (a)—that is, .5000 − .3413 = .1587—with the unshaded area in the right half of panel (e)—that is, .5000 − .4772 = .0228; the result is .1815. (Naturally, the area under half the normal curve always equals .5000.) Analytical Example 7.2, "The Decision to Seed Hurricanes," provides a vivid illustration of why one may want to know normal-curve probabilities. Other examples are discussed in the next sections.

Practice Problem 1

Finding Probabilites Concerning the Lifetime of Aircraft Engines

Suppose that a manufacturer of aircraft engines knows their lifetimes to be a normally distributed random variable with a mean of 2,000 hours and a standard deviation of 100 hours. Here are some of the questions of interest to the manufacturer and to potential users (along with their answers).

a. What is the probability that a randomly chosen engine has a lifetime between 2,000 and 2,075 hours? The problem amounts to finding an area under the normal curve between the mean and a value above it. It helps to sketch the problem, as in Figure A, to convert x values to z values, and to find the answer in Appendix Table H. Because μ = 2,000 hours and σ = 100 hours, the x value of 2,075 hours equals a z value of 2,000 − 2,075 divided by 100, or of .75. The area between μ = 0 and z = .75 (shaded in Figure A) can be found in Appendix Table H at the intersection of the .7 row and .05 column as .2734. This number is also the probability sought.

Figure A

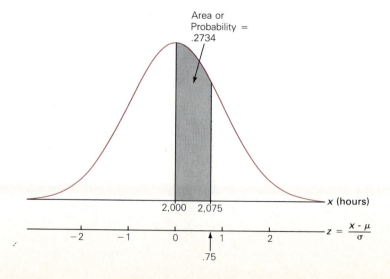

Area or Probability = .2734

2,000 2,075   x (hours)

−2   −1   0   1   2 $z = \dfrac{x - \mu}{\sigma}$

.75

b. What is the probability that a randomly chosen engine has a lifetime between 1,880 and 2,000 hours? The problem amounts to finding an area under the normal curve between the mean and a value below it. It is, therefore, solved by a procedure analogous to (a) above, as shown by Figure B. Because μ = 2,000 hours and σ = 100 hours, the x value of 1,880 hours equals a z value of −1.20. The area between μ = 0 and z = −1.20 (shaded in Figure B) can be found in Appendix Table H at the intersection of the 1.2 row and .00 column as .3849. This number is also the probability sought.

Figure B

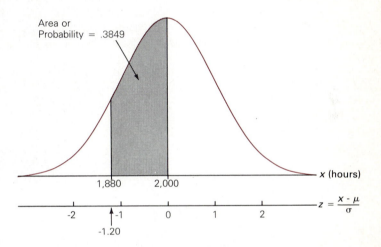

c. What is the probability that a randomly chosen engine has a lifetime between 1,950 and 2,150 hours? The problem amounts to finding an area under the normal curve overlapping the mean, as shown in Figure C. Because μ = 2,000 hours and σ = 100 hours, the x value of 1,950 hours (or 2,150 hours) equals a

Figure C

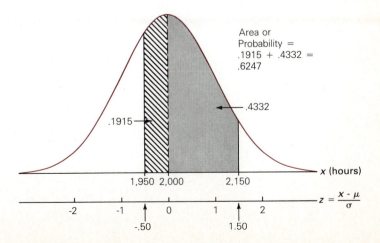

z value of $-.50$ (or $+1.50$). The area between $\mu = 0$ and $z = -.50$ (cross-hatched) and that between $\mu = 0$ and $z = +1.50$ (shaded) can be found in Appendix Table H at the intersection of the .5 row and .00 column (and of the 1.5 row and .00 column) as .1915 (and .4332). Thus, the probability sought equals the sum of the crosshatched and shaded areas, or $.1915 + .4332 = .6247$.

d. What is the probability that a randomly chosen engine has a lifetime above 2,170 hours? The problem amounts to finding an area under the upper tail of the normal curve, as shown in Figure D. Because $\mu = 2,000$ hours and $\sigma = 100$ hours, the x value of 2,170 hours equals a z value of $+1.70$. The entire area to the right of μ always equals .5, that between $\mu = 0$ and $z = +1.70$ (dotted in Figure D) can be found in Appendix Table H at the intersection of the 1.7 row and .00 column as .4554. Thus, the probability sought (which corresponds to the shaded upper-tail area) equals $.5000 - .4554$, or $.0446$.

Figure D

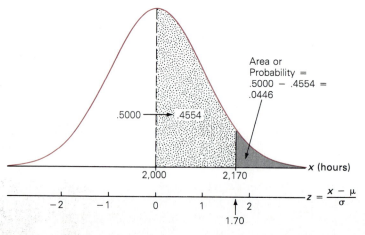

e. What is the probability that a randomly chosen engine has a lifetime below 1,840 hours? The problem amounts to finding an area under the lower tail of the

Figure E

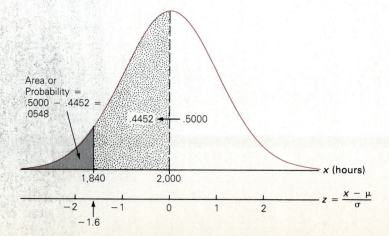

normal curve, as shown in Figure E. Because $\mu = 2,000$ hours and $\sigma = 100$ hours, the x value of 1,840 hours equals a z value of -1.60. The entire area to the left of μ always equals .5; the area between $\mu = 0$ and $z = -1.60$ (dotted in Figure E) can be found in Appendix Table H at the intersection of the 1.6 row and .00 column as .4452. Thus, the probability sought (which corresponds to the shaded lower-tail area) equals .5000 − .4452, or .0548.

f. What is the probability that a randomly chosen engine has a lifetime between 2,071 and 2,103 hours? The problem amounts to finding an area under the normal curve between two values both of which lie above the mean, as shown in Figure F. Because $\mu = 2,000$ hours and $\sigma = 100$ hours, the x values of 2,071 and 2,103 hours equal z values, respectively, of $+.71$ and $+1.03$. The area between $\mu = 0$ and $z = +1.03$ (the dotted plus the shaded areas in Figure F) can be found in Appendix Table H at the intersection of the 1.0 row and .03 column as .3485. Similarly, the area between $\mu = 0$ and $z = +.71$ (the dotted area in Figure F) can be found at the intersection of the .7 row and .01 column as .2612. Thus, the probability sought (which corresponds to the shaded area only) equals .3485 − .2612, or .0873.

Figure F

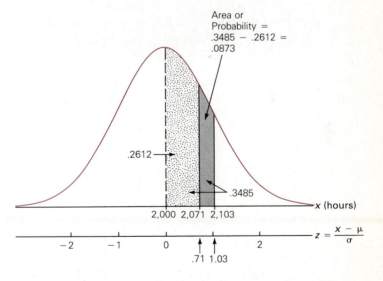

g. What is the probability that a randomly chosen engine has a lifetime between 1,849 and 1,923 hours? The problem amounts to finding an area under the normal curve between two values both of which lie below the mean, as shown in Figure G. Because $\mu = 2,000$ hours and $\sigma = 100$ hours, the x values of 1,849 and 1,923 hours equal z values, respectively, of -1.51 and $-.77$. The area between $\mu = 0$ and $z = -1.51$ (the dotted plus the shaded areas in Figure G) can be found in Appendix Table H at the intersection of the 1.5 row and .01 column as .4345. Similarly, the area between $\mu = 0$ and $z = -.77$ (the dotted area in Figure G) can be found at the intersection of the .7 row and .07 column as .2794. Thus, the probability sought (which corresponds to the shaded area only) equals .4345 − .2794, or .1551.

Figure G

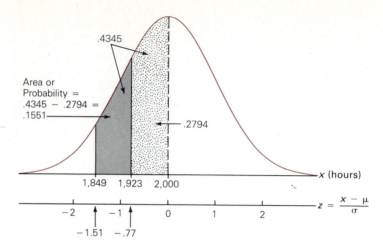

h. What is the probability that a randomly chosen engine has a lifetime of less than 2,075 hours? The problem amounts to finding an area under the normal curve to the left of a value that itself lies above the mean. Thus, the answer equals .5 (for the area left of the mean) plus whatever probability is associated with values between the mean and the given value of 2,075 hours. In this case, as a look at Figure A confirms, the answer equals .5000 + .2734, or .7734.

i. What is the probability that a randomly chosen engine has a lifetime of more than 1,880 hours? The problem amounts to finding an area under the normal curve to the right of a value that itself lies below the mean. Thus, the answer equals .5 (for the area right of the mean) plus whatever probability is associated with values between the mean and the given value of 1,880 hours. In this case, as a look at Figure B confirms, the answer equals .3849 + .5000, or .8849. |||

||| Practice Problem 2

Designing an Aircraft Cockpit

A manufacturer of aircraft (or, for that matter, of any product) is likely to be very much concerned about the ability of potential users to use the product. When 60 percent of all pilots cannot reach the rudder pedals when seated in the pilot's seat or find themselves unable to tune navigation equipment because it is nicely out of reach, something is wrong. Suppose a manufacturer knows that the lengths of pilots' legs and arms are normally distributed random variables with mean of 33 and 28 inches, respectively, and standard deviations of 2 inches in both cases.

a. If the manufacturer wants to design a cockpit such that precisely 90 percent of all pilots can reach the rudder pedals with their feet while seated (and can, thus, fly this particular airplane), what is the desired distance between seat and pedals? The question requires us to find the 10th percentile—that value of the random variable (length of pilots' legs) below which lies 10 percent of the area under the normal curve. Because the 10th percentile lies below the mean and 50 percent of the entire area under the normal curve lies below the mean, the 10th percentile corresponds to that negative value of z which places $50 - 10 = 40$ percent of the area between itself and the mean, as shown in Figure H. Accordingly, we can search the body of Appendix Table H for an entry of .4000, and such is found between a z of -1.28 (area: .3997) and a z of -1.29 (area: .4015). Interpolation yields a z value of -1.2817 for the 10th percentile. Just as x values can be converted into z values, so the reverse is also true; a z of -1.2817 corresponds to 30.44 inches. This means that 10 percent of all pilots cannot reach as far as 30.44 inches—which is, therefore, the distance sought because it assures that 90 percent of all pilots easily can reach the rudder pedals.

Figure H

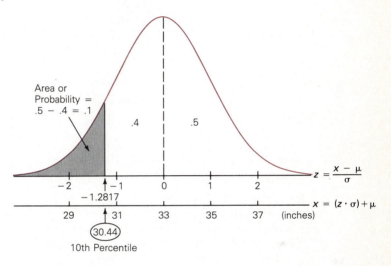

b. If it were true that precisely 60 percent of all pilots could not reach the navigation equipment when seated, what must have been the distance between the pilot's seat and the relevant knobs? The question requires us to find the 60th percentile—that value of the random variable (length of pilots' arms) below which lies 60 percent of the area under the normal curve. Because the 60th percentile lies above the mean, but 50 percent of the entire area under the normal curve lies below it, the 60th percentile corresponds to that positive value of z which places 10 percent of the area between itself and the mean, as shown in Figure I. Accordingly, we can search the body of Appendix Table H for an entry of .1000, and such is found between a z of .25 (area .0987) and a z of .26 (area .1026). Interpolation yields a z value of .2533 for the 60th percentile. Again, z values can be converted into x values; in this case, the 60th percentile corresponds to 28.51 inches. Such a distance would account for only 40 percent of all pilots being able to reach the navigation-equipment knobs.[3]

---

[3]The concerns noted in Practice Problem 2 are far from being merely academic exercises. In 1984, the Navy, for example, issued new flight-training standards that stipulated a pilot's maximum "sitting height" at 40.5 inches, maximum "buttock-to-leg length" at 48 inches, and maximum "functional reach" at 29 inches—all because of the design of new Navy fighter planes. The standards excluded 73 percent of all college-age women and 13 percent of all college-age men from naval flying. (*The New York Times*, November 25, 1984, p. 49.)

Figure I

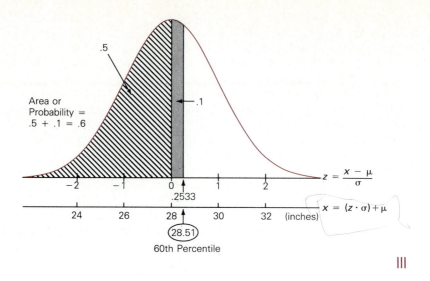

Approximating Discrete Probability Distributions by the Normal Probability Distribution*

Under certain circumstances, and in spite of the fact that it is a continuous one, the normal probability distribution can be used to approximate various discrete probability distributions. This use of the normal probability distribution is very convenient when the calculation of precise probabilities is a cumbersome procedure.

Approximating binomial probabilities. Binomial probability tables typically do not extend to large values of n (numbers of trials of a random experiment), or they may not cover a given value of π (the probability of success in one trial). Appendix Table C, for example, contains no section for $n = 50$, nor does the more detailed table from which it is derived. Appendix Table C, as well as its source, similarly contains no column for $\pi = .255$. If one were interested in establishing the probabilities for an experiment involving these numbers and did not have a computer, one could still do so, but one would have to go through the type of cumbersome calculations illustrated in Table 6.6 on page 204, using the binomial formula given in Box 6.B. Luckily, there are circumstances when the normal probability distribution gives almost identical results. Such results can be obtained whenever the value of n is large or whenever that of π is close to .5. It is easy to see why this similarity in results occurs by having another look at Figure 6.3 on page 205: regardless of the value of π, as n increases, a binomial probability distribution becomes less and less skewed (and, thus, looks more and more like

---

*Optional section. Assumes prior study of the Chapter 6 optional sections on the hypergeometric and Poisson probability distributions.

the symmetrical normal probability distribution). Furthermore, whenever $\pi = .5$, any binomial probability distribution is perfectly symmetrical regardless of the value of n.

It is, of course, a matter of arbitrary judgment which combination of n and π makes a binomial probability distribution "sufficiently" symmetrical to justify the use of the normal curve as an approximation to it. According to one widely used rule of thumb, the normal-curve approximation can be used without undue loss of accuracy whenever $n \cdot \pi$ *as well as* $n(1 - \pi)$ equals or exceeds 5.

Once the decision to substitute normal for binomial probabilities has been made, the relevant probabilities can be discovered quickly. Because the probability of any one x value, such as 11 successes in so many trials, is theoretically zero for a continuous distribution, one converts the discrete value of x into a small *range* of x by subtracting and adding half the distance between the discrete values to any one value of x. This so-called *continuity correction factor* turns a discrete 11 into a range of 10.5 to 11.5, for example, and it turns a discrete "11 or larger" into a range of "10.5 or more," while replacing a discrete "11 or fewer" by an "11.5 or less." This correction having been made, the new continuous x value is transformed into a z value by the standard formula, while noting (Box 6.C) that the binomial $\mu = n \cdot \pi$ and $\sigma = \sqrt{n \cdot \pi(1 - \pi)}$. Hence,

$$z = \frac{x - \mu}{\sigma}$$
$$= \frac{x - (n\pi)}{\sqrt{n \cdot \pi(1 - \pi)}}.$$

Given values of z, the desired probabilities can be found in Appendix Table H.

Practice Problem 3

Invoice Errors

Consider a company that has experienced 25.5 percent errors in its invoices. It takes a sample (with replacement) of 50 invoices. What is the probability that exactly 11 of the sample invoices are in error? One could calculate the precise probability by the binomial formula:

$$p(R = 11|50, .255) = \frac{50!}{11!(50 - 11)!} (.255)^{11}(.745)^{39} = .1144.$$

One can also note that $\mu = n \cdot \pi = 50(.255) = 12.75$, while $\sigma = \sqrt{n \cdot \pi(1 - \pi)} = \sqrt{50(.255)(.745)} = \sqrt{9.49875} = 3.082$. Accordingly, the problem can be pictured as shown in Figure J. The binomial value of 11 appears as the range from 10.5 to 11.5, and the corresponding z values equal $-.73$ and $-.41$. Accordingly, Appendix Table H gives the desired probability as the area (the dotted plus the shaded areas in Figure J) corresponding to $z = -.73$ (or .2673) minus the area (dotted) corresponding to $z = -.41$ (or .1591)—that is, as .1082. This value lies within 5.4 percent of the correct one.

Figure J

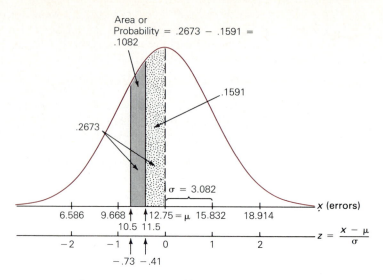

By the same procedure, one can quickly estimate the probability of finding, say 13 or fewer errors, as shown in Figure K. The binomial value of 13 or fewer appears as "13.5 or less." The corresponding z value equals $+ .24$. Accordingly, Appendix Table H gives the probability of the shaded area as .0948. The desired probability (cross-hatched plus shaded areas), therefore, is approximated by .5948.

Figure K

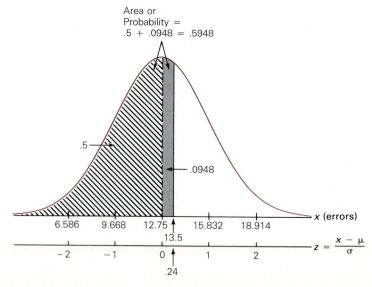

III

Approximating hypergeometric probabilities. Just as the normal probability distribution can be used to approximate the binomial one, so it can be used also to approximate the hypergeometric one. Once more the appropriate summary measures of μ and σ (this time calculated according to the formulas for the hypergeometric distribution in Box 6.E) can be applied to calculate z values, as in the previous examples on page 256. Once again, the approximations are best when n is large and $\pi = (S/N)$ is close to .5.

Approximating Poisson probabilities. As the value of μ in the Poisson probability distribution increases, the distribution becomes ever less skewed and approaches the shape of the normal one. Thus, for values of $\mu > 20$ (which cannot be found in Appendix Table F), the normal probability distribution can be used. Naturally, the appropriate summary measures of μ and σ must now be calculated according to the formulas in Box 6.G before z values are determined.

THE EXPONENTIAL PROBABILITY DISTRIBUTION*

Chapter 6 described how the Poisson probability distribution provides probabilities for x occurrences, in a segment of time or space, of a specified type of event. The applications, we noted, range widely: the Poisson formula and the tables derived from it give us the likelihood, for any specified period of time, of any specified number of people arriving at the bank, at the hospital emergency room, or at the airline ticket counter, or the likelihood of the arrival of specified numbers of cars at the car wash, of ships at the dock, of orders at the warehouse, of claims at the insurance company—the list goes on. The formula also gives us the likelihood of encountering, within any specified space, any specified number of events, be they defects on a roll of wire, leaks in a pipeline, potholes in a road, or errors on a page.

There are occasions, however, when one is less interested in the *number* of occurrences of such events than in the time elapsed or space encountered *between* any two such occurrences. Pilots in a holding pattern near their destination airport, for example, are less interested in the number of airplanes arriving at the final approach fix during a given half-hour period than in the time they have to wait before being cleared for the approach. Captains of fishing fleets, similarly, might be less interested in the number of schools of fish encountered in 10 days of searching than in the time it will take to find the first school or to find a second one once the first of them has been processed. As it turns out, whenever the *number* of occurrences of an event is determined by a Poisson process (and the associated probabilities, therefore, are described by the Poisson formula in Box 6.F), the likelihood of encountering specified *intervals* of time or space between consecutive occurrences can be described by the exponential probability distribution, as illustrated in the three panels of Figure 7.6.

---

*Optional section. Assumes prior study of the Chapter 6 optional section on the Poisson probability distribution.

Figure 7.6   Poisson Distribution vs. Exponential Distribution

The Poisson probability distribution associates probabilities with numbers of occurrences of some event, shown by the colored dots in panel (a), within specified intervals of time or space. The exponential distribution instead associates probabilities with the various gaps, shown by the values of x_1 to x_8 in panel (a), *between* the Poisson events. The frequency of occurrence of different-sized gaps can be depicted by a histogram, as in panel (b) or by a smooth frequency curve, as in panel (c). Inevitably, the mean size of these gaps equals the reciprocal, $1/\lambda$, of the Poisson process rate, λ. When, on the average, planes arrive at a rate of .38 per minute, the average gap between arriving planes is $1/.38$ or 2.63 minutes.

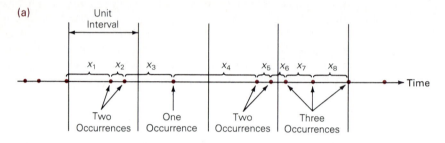

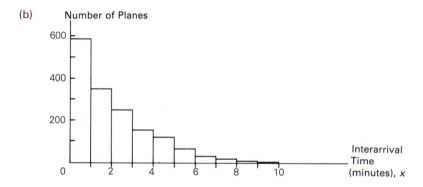

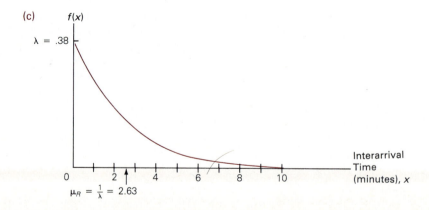

A Graphical Exposition

The horizontal line in panel (a) pictures a continuum of time during which a Poisson process, such as the arrival of planes at a final approach fix, is occurring; the random arrival of planes is indicated by the placement of the colored dots. The number of arrivals in each of four intervals of, say, 10 minutes can be determined by merely counting the dots within these intervals. Considering the four time intervals shown, the mean number of occurrences, λ, of this event per 10-minute span of time clearly equals

$$\frac{2 + 1 + 2 + 3}{4} = 2$$

or 2 planes per 10 minutes. Panel (a) also shows, however, the precise timing of the arrivals. Thus, the first plane that arrives within the 40-minute period between the first and last vertical line arrives at the end of the sixth minute, the second plane arrives at the end of the eighth minute, and so on, with the last plane arriving at the end of the 40th minute. The *interarrival times,* or the time gaps between the arrival times of the planes, have been indicated by the values of x_1, x_2, and so on, and they represent the **exponential random variable,** the uncertain time or space between any two successive events in a Poisson process.

For the 40-minute period just considered, the mean interarrival time equals $x_1 + x_2 + \ldots + x_8$, divided by 8, which comes to

$$\frac{6 + 2 + 7 + 12 + 2 + 2 + 4 + 5}{8} = \frac{40}{8}$$

or 5 minutes between planes. It is no accident that this number equals $1/\lambda$. If, on the average, $\lambda = 2$ planes arrive per 10 minute interval, then, on the average, 10 minutes elapse per 2 arriving planes (which is, of course, the same thing as 5 minutes per plane). Thus, the mean of the exponential random variable, μ_R, always equals $1/\lambda$, where λ is the Poisson process rate. (The standard deviation of the exponential random variable, σ_R, also equals $1/\lambda$, which differs from the case of the Poisson random variable, the *variance* of which equals its mean.)

Panel (b) of Figure 7.6 pictures the type of histogram we can expect to find if the arrival of planes is a Poisson process and if we record the frequency of various interarrival times over an extended period. After studying the arrivals of 1,603 planes, for example, we might note that the time between the arrivals of consecutive planes equaled 1 minute 586 times, 2 minutes 350 times, 3 minutes 250 times, and so on, and that a full 10 minutes elapsed between consecutive arrivals only 3 times. All these data are represented by the heights of the histogram columns.

Panel (c), finally, shows how a frequency curve can summarize the histogram data. The vertical intercept of this (and every) exponential probability density function equals the Poisson process rate, λ, here equal to .38 (meaning that planes were arriving at a rate of .38 per minute and implying a mean time between arrivals of 1/.38 or 2.63 minutes). Panel (c) shows the probability density function for the exponential random variable. As in this example, the **exponential probability distribution** is a probability density function for a continuous random variable that measures intervals of time or space between consecutive events in a Poisson process. Note that this distribution applies (a) only to positive values of x and (b) only to situations in which smaller values of x are more likely than larger ones. (The distribution is always positively skewed.) Like the right tail of the normal probability distribution, that of the exponential one never touches the horizontal axis, although it is impractical to show this graphically. This fact means that, theoretically, a gap, x, of *any* positive size might occur; the time between the arrival of one plane and the next one might be 400 hours or even infinity, for example, but such a possibility carries a very low probability indeed.

The Formula

Depending only on the size of λ, the Poisson process rate, many different members of the exponential probability distribution family can be created; each one looks similar to the probability density function given in panel (c) of Figure 7.6. (The larger is λ, the smaller is $\mu_R = \sigma_R$ and the less spread out is the distribution.) In all cases, however, the precise height of the exponential probability density function, $f(x)$, for different values, x, of the exponential random variable, R, can be determined with the help of the formula given in Box 7.D. Naturally, to make any sense, λ and x must be referring to the same units. If λ is expressed per minute, x must be measured in minutes (not seconds or hours); if x is to be measured in units of 100,000 hours, the value of λ likewise must represent occurrences per 100,000 hours. Beyond that, as in the Poisson formula and in the normal probability density function, the constant e appears; it equals 2.71828.

Note: Because the mean interval between events, μ_R, equals $1/\lambda$, one can, of course, substitute $1/\mu_R$ for λ in the exponential formula.

7.D The Exponential Probability Density Function

$$f(x) = \lambda \cdot e^{-\lambda x}, \text{ where } x > 0 \text{ and } \lambda > 0 \text{ and } e = 2.71828.$$

As was true in the case of the normal curve, the total area under the exponential probability density function equals 1, and various probabilities can be found by focusing on areas under this curve for different ranges of the exponential random variable. The area under the exponential probability density function to the *right* of a given value, x, of the exponential random variable is given by the formula in Box 7.E. (Recall that for continuous random variables the symbols $>$ and $\geq$ are equivalent.)

7.E Greater-Than Cumulative Exponential Probabilities

$$p(R > x) = e^{-\lambda x} = e^{-\frac{x}{\mu_R}}$$

On the other hand, the area to the *left* of a given value of the exponential random variable (that is, the area between zero and x) is represented by the formula in Box 7.F on the next page.

7.F Less-Than Cumulative Exponential Probabilities

$$p(R < x) = 1 - e^{-\lambda x} = 1 - e^{-\frac{x}{\mu_R}}$$

Boxes 7.E and 7.F imply that the area between two positive values of this random variable, x_1 and x_2, can be found by deducting from the total area (which equals 1) the area to the right of the upper limit of this range (the area to the right of x_2) as well as the area to the left of the lower limit of this range (the area to the left of x_1), as stated in Box 7.G.

7.G Combination Formula for Exponential Probabilities

$$p(x_1 < R < x_2) = 1 - [e^{-\lambda x_2} + (1 - e^{-\lambda x_1})] = e^{-\lambda x_1} - e^{-\lambda x_2},$$

where $1/\mu_R$ can be substituted for λ.

‖ Practice Problem 4

Plane Arrivals

Consider a Poisson process in which planes arrive at a fix at a rate of $\lambda = 4$ planes per hour. (This makes $\mu_R = 1/4$, meaning that the average gap between consecutive arrivals equals 1/4 hour.)

a. What is the probability for interarrival times in excess of 1/4 hour? The formula in Box 7.E applies; hence,

$$p(R > 1/4) = e^{-4(1/4)} = e^{-1}$$

Appendix Table E, utilized earlier in connection with the Poisson formula, can be used again by simply treating the column labeled μ as if it referred to our exponent, $\lambda \cdot x$, while reading the column labeled $e^{-\mu}$ as if it read $e^{-\lambda x}$. We find, accordingly, that $e^{-1} = .367879$, which is the probability sought.

The solution can also be pictured as shown in Figure L. Note that the formula in Box 7.D can be used to calculate, for example, the height of the exponential density function above various values of x, given $\lambda = 4$. Thus,

when $x = 0, f(x) = 4 \cdot e^{-4(0)} = 4 \cdot e^0 = 4$

when $x = .25, f(x) = 4 \cdot e^{-4(.25)} = 4 \cdot e^{-1} = \dfrac{4}{e^1} = \dfrac{4}{2.71828} = 1.47$

when $x = .50, f(x) = 4 \cdot e^{-4(.50)} = 4 \cdot e^{-2} = \dfrac{4}{e^2} = \dfrac{4}{7.38905} = .54$

when $x = .75, f(x) = 4 \cdot e^{-4(.75)} = 4 \cdot e^{-3} = \dfrac{4}{e^3} = \dfrac{4}{20.08550} = .20$

when $x = 1.00, f(x) = 4 \cdot e^{-4(1)} = 4 \cdot e^{-4} = \dfrac{4}{e^4} = \dfrac{4}{54.59800} = .07$

Figure L

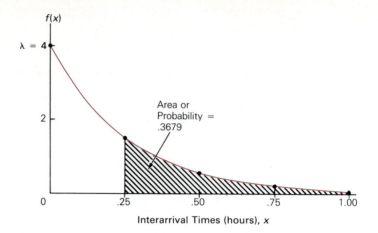

These values are given by the five fat dots in Figure L. The probability calculated above corresponds, in turn, to the shaded area of the graph.

b. What is the probability for interarrival times below 1/10 hour? The formula in Box 7.F applies; hence

$$p(R < 1/10) = 1 - e^{-4(1/10)}$$
$$= 1 - e^{-.4} = 1 - .670320,$$

according to Appendix Table E; and the probability sought is .32968.

This solution, too, can be found by the technique employed earlier, as shown in Figure M. In this case, the probability sought (crosshatched area) equals the total area under the curve (1.0000) minus the white area (of .6703).

Figure M

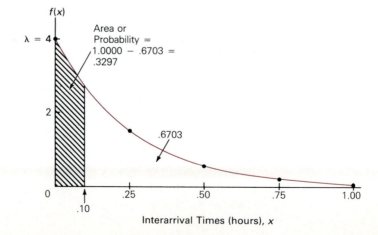

c. What is the probability for interarrival times between 1/10 hour and 1/4 hour (that is, between 6 and 25 minutes)? This time, the above-noted Box 7.G combination of the Box 7.E and 7.F formulas gives us

$$p(1/10 < R < 1/4) = e^{-4(1/10)} - e^{-4(1/4)} = e^{-.4} - e^{-1}$$
$$= .670320 - .367879 = .302441.$$

This solution is shown graphically in Figure N. Note that the desired probability equals the difference between the white area in Figure M and the crosshatched one in Figure L. It is no accident, of course, that the three crosshatched areas in Figures L, M, and N add precisely to 1.0000. It is inevitable that airplanes arrive in intervals of *either* more than .25 hours *or* less than .10 hours *or* something between.

Figure N

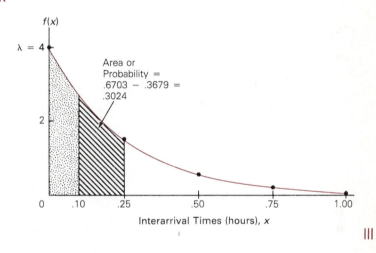

Area or
Probability =
.6703 − .3679 =
.3024

Interarrival Times (hours), *x*

Tables of Exponential Probabilities

The use of various types of tables that contain exponential probabilities already computed makes it unnecessary to apply any of the above formulas. Appendix Table I contains exponential probabilities for cumulative values of *x*, computed with the help of the formula in Box 7.F.

Practice Problem 5

Plane Arrivals Revisited

Note how the three answers to parts (a) to (c) of Practice Problem 4 can easily be found with the help of Appendix Table I.

a. The probability of interarrival times in excess of 1/4 hour equals 1 minus the probability (found in Appendix Table I) of interarrival times below 1/4 hours. Entering the table at $\lambda \cdot x = 4(1/4) = 1.0$ (and in the .00 column because the second decimal here is 0 as well), we find .6321. Deducting this value from 1, we get .3679, the same answer found in Practice Problem 4.

b. The probability of interarrival times below 1/10 hour can be read directly from Appendix Table I (because it shows areas under the curve between 0 and some positive value of x). Entering the table at $\lambda \cdot x = 4(1/10) = .4$ (and again in the .00 column), we find .3297, the same answer found in Practice Problem 4.

c. The probability of interarrival times between 1/10 and 1/4 hours must be read from Appendix Table I as the area between 0 and .25 minus the area between 0 and .10 (in Figure N, this area is the dotted plus the crosshatched areas minus the dotted area). Thus, we enter the table at $\lambda \cdot x_2 = 4(.25) = 1$ (and find .6321) and again at $\lambda \cdot x_1 = 4(.10) = .4$ (and find .3297). The difference, $.6321 - .3297$, equals .3024, the same answer found in Practice Problem 4.

Note: In cases where $\lambda \cdot x$ yields a number with two decimal points, the remaining columns of Appendix Table I come in handy, too. Thus, the probability associated with $\lambda = 2.5$ and $x \leq .5$ (making $\lambda \cdot x \leq 1.25$) can be found at the intersection of the 1.2 row and .05 column as .7135. |||

||| Practice Problem 6

Finding Probabilities for Times Between Failures

Manufacturers of all kinds of products—ranging from aircraft navigation equipment to automobile batteries, from simple light bulbs to nuclear power reactors, from home sound systems to satellite power cells—are keenly interested in the mean length of time that elapses before their product fails (or in the average time between a first failure and another one). Anybody about to offer a guarantee to customers, for example, will view this knowledge as crucial, indeed. Consider a manufacturer of aircraft distance-measuring equipment (DME). Experience might show that failures of DME units can be described by a Poisson process and that a unit fails on the average 4 times per 100,000 hours of use. The manufacturer may want to know:

a. the probability that a unit will operate without failure in excess of 50,000 hours,

b. the probability that a unit will operate without failure for up to 20,000 hours,

c. the required mean time between failures (often referred to as MTBF) before one can safely assume that 90 percent of all units will perform without failure in excess of 25,000 hours.

Answers: The value of λ equals 4 failures per 100,000 hours; thus, the times between failures must be measured in units of 100,000 hours as well.

a. $p(R > .5) = 1 - p(R < .5)$, and the latter expression is found in Appendix Table I for $\lambda \cdot x = 4(.5) = 2$ as .8647. Thus, the probability sought equals $1 - .8647 = .1353$.

b. $p(R < .2)$ can be found directly in Appendix Table I for $\lambda \cdot x = 4(.2) = .8$ as .5507.

c. The answer requires that $p(R > .25) = .9$, which is equivalent (according to Box 7.E) to $e^{-\lambda x} = .9$. From Appendix Table E, we can see than $e^{-\lambda x}$ equals .90 (rounded) when $\lambda \cdot x = .10$. Given $x = .25$, λ must equal $.10/.25$, or .4 (failures per 100,000 hours). Since the MTBF equals $1/\lambda$, the answer is $1/.4$, or 2.5 (units of 100,000 hours). Only when the mean time between failures has been raised from the present 25,000 hours (implied by 4 failures per 100,000 hours) to 250,000 hours can the manufacturer rely on 90 percent of all units performing without failure in excess of 25,000 hours.

The answer to (c) can now be checked with the help of Appendix Table I. If there were only .4 failures per 100,000 hours, we could find the probability of a time between failures in excess of 25,000 hours as in answer (a) above by consulting Appendix Table I for $\lambda \cdot x = .4 (.25) = .1$ and subtracting the value found (.0952) from 1. The result is .9048. In this case, the probability would be slightly above .9 that no failures would occur prior to 25,000 hours of use. |||

||| Practice Problem 7

Assuring the Reliability of an Airport Lighting System

An airport manager may justly worry about the possible loss, as a result of a power blackout, of crucial approach, runway, and taxiway lighting. A standby generator may exist, but it would also be subject to failure, having a mean time between failures (MTBF) of 100 hours. The manager may ask:

a. What is the probability of the standby generator failing during the next 12-hour blackout?
b. What is the probability of two such generators failing during such a circumstance, assuming that the two generators operate independently of one another?
c. If there were 5 such blackouts in a year, what are the chances that a single generator would work through them all?
d. How would those chances look with a second backup or even with a third one?

Answers: Since the MTBF of 100 hours equals $1/\lambda$, the value of $\lambda = .01$ (failure per hour).

a. The first question asks about $p(R \leq 12)$, which can be found directly in Appendix Table I for $\lambda \cdot x = .01(12) = .12$ as .1131.
b. The probability of two such independent generators failing would equal $(.1131)^2 = .0128$. (See Box 5.J.)
c. The binomial formula (Box 6.B) can be applied for $n = 5$ and for a probability of success of $\pi = 1 - .1131 = .8869$. Thus,

$$p(R = 5|n = 5, \pi = .8869) = \frac{5!}{5!\ 0!} (.8869)^5 (.1131)^0 = .5487.$$

The manager may not like this rather low probability.

d. Once more the binomial formula can be used, but (given 2 backups) with a probability of success of $\pi = 1 - .0128 = .9872$, in accordance with (b). Thus,

$$p(R = 5|n = 5, \pi = .9872) = \frac{5!}{5!\ 0!} (.9872)^5 (.0128)^0 = .9376.$$

For 3 backups, the probability of failure would be $(.1131)^3 = .0014$; hence, the probability of at least one of them working through all 5 blackouts would be

$$p(R = 5|n = 5, \pi = .0014) = \frac{5!}{5! \, 0!} (.9986)^5(.0014)^0$$
$$= .9930. \; \text{III}$$

Note: The same type of reasoning is, of course, being applied to a multitude of similar situations that call for a high degree of reliability: the communications of air-traffic-control centers, power supplies to hospital surgical units or satellites, and more.

THE UNIFORM PROBABILITY DISTRIBUTION*

Among continuous random variables, the **uniform random variable** undoubtedly is the simplest one; it has an equal chance of assuming any value within a specified range along a continuous scale. Accordingly, a **uniform probability distribution** is the probability density function for this type of random variable that is equally likely to take on any of the values within a given range. Consider the dial on a wheel of fortune that is equally likely to point to any one of various segments of the circle after being spun. Consider the arrival time of a bus, plane, or train that is equally likely to occur at any time in a 20-minute period. The same may be true about the ripening time of a crop, the random decimals generated by a computer, the daily sales of a wholesaler, the rate of inflation next year, next week's demand for electricity, or the time it takes to process a loan application. In all these cases and many more, the probability of any actual value may well be the same anywhere within a specified range.

A Graphical Exposition

Figure 7.7 pictures two members of the uniform probability distribution family.

The Formula

The precise height of the uniform probability density function, $f(x)$, for different values, x, of the uniform random variable, R, can be determined with the help of the formula given in Box 7.H.

7.H The Uniform Probability Density Function

$$f(x) = \frac{1}{b - a} \text{ if } a \leq x \leq b; \text{ otherwise } f(x) = 0$$

---

*Optional section.

Figure 7.7   Two Members of the Uniform Probability Distribution Family

A uniformly distributed random variable has an equal chance of assuming any value within a specified range, *ab,* along a continuous scale. When graphed, the probability-density function is seen to be a rectangle; this probability distribution, therefore, is also called the **rectangular probability distribution.**

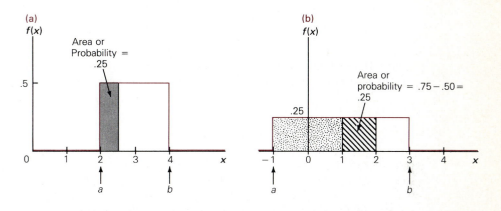

The uniform probability density function, this formula tells us, is a horizontal line segment of constant height,

$$\frac{1}{b-a}$$

over the interval, *ab.* Because values of the random variable below *a* and above *b* are impossible, $f(x) = 0$ outside the segment *ab.* Note, for example, in panel (a) of Figure 7.7, that the lowest possible value, *x*, of the random variable is $a = 2$, and that the highest possible value is $b = 4$. According to Box 7.H, the height of the probability density function within this range equals

$$f(x) = \frac{1}{b-a} = \frac{1}{4-2} = \frac{1}{2} = .5.$$

The height of the density function in panel (b), within the range of $a = -1$ and $b = 3$, is similarly calculated as

$$f(x) = \frac{1}{b-a} = \frac{1}{3-(-1)} = \frac{1}{4} = .25.$$

Note how, in both cases, $f(x) = 0$ outside the range *ab.* Note also that the total area under the probability function once again equals 1: the rectangle's base times height equals $2 \times (.5) = 1$ panel (a) and $4(.25) = 1$ in panel (b).

Probabilities. Naturally, probability is once again measured by the *area* above the interval of *x* values that is of interest. If one cumulates *x* values from left to right, one gets the formula in Box 7.I.

7.I Less-Than Cumulative Uniform Probabilities

$$p(R \leq x) = \frac{x - a}{b - a} \text{ if } a \leq x \leq b; \text{ otherwise } p(R \leq x) = 0$$

In panel (a) of Figure 7.7, for example, the probability of x lying between the lower limit of $a = 2$ on the one hand and a value of 2.5 on the other hand equals the shaded area:

$$p(R \leq 2.5) = \frac{x - a}{b - a}$$
$$= \frac{2.5 - 2}{4 - 2} = \frac{.5}{2} = .25.$$

(This probability implies, of course, a probability of $1 - .25 = .75$ for x between 2.5 and 4.)

In panel (b) of Figure 7.7, similarly, the probability of x lying between 1 and 2 equals the crosshatched area. This probability can be calculated as the dotted plus crosshatched area minus the dotted area. Hence:

$$p(1 \leq x \leq 2) = p(R \leq 2) - p(R \leq 1)$$
$$= \frac{2 - (-1)}{3 - (-1)} - \frac{1 - (-1)}{3 - (-1)}$$
$$= \frac{3}{4} - \frac{2}{4} = \frac{1}{4} = .25.$$

This probability is, of course, visually obvious because the crosshatched area takes up one quarter of the entire rectangle under the density function.

Summary measures. Because the uniform probability density function is such a simple one, summary measures can also be calculated in a simple fashion, as indicated by the formulas in Box 7.J.

7.J Summary Measures for the Uniform Probability Density Function

$$\mu_R = \frac{a + b}{2}$$

$$\sigma_R^2 = \frac{(b - a)^2}{12}$$

$$\sigma_R = \sqrt{\frac{(b - a)^2}{12}}$$

Thus, in panel (a) of Figure 7.7, the mean or expected value of the random variable is

$$\mu_R = \frac{a + b}{2} = \frac{2 + 4}{2} = 3,$$

which is visually quite obvious. The variance is

$$\sigma_R^2 = \frac{(b - a)^2}{12} = \frac{(4 - 2)^3}{12} = \frac{4}{12} = .25,$$

and the standard deviation is

$$\sigma_R = \sqrt{.25} = .5.$$

||| Practice Problem 8

Finding Probabilities for Making a Connecting Flight

Consider a flight scheduled to arrive in Keene, New Hampshire, at 1:30 P.M. but in fact equally likely to arrive at any time between 1:10 and 1:55 P.M. Someone might wish to know the probability of (a) the plane being on time or early and (b) being late for a connecting flight at 1:45 P.M.

The problem can be sketched as shown in Figure O, denoting the scheduled arrival time of 1:30 P.M. as 0, denoting the earliest possible arrival time of 1:10 P.M. as $a = -20$ minutes, and denoting the latest possible arrival time of 1:55 P.M. as $b = +25$ minutes. According to the formula in Box 7.H, the height of the probability density function in the *ab* range equals

$$\frac{1}{b - a} = \frac{1}{25 - (-20)} = \frac{1}{45}.$$

The probability of being on time or early, according to Box 7.I equals

$$p(R \leq 0) = \frac{x - a}{b - a} = \frac{0 - (-20)}{25 - (-20)} = \frac{20}{45} = .4444,$$

and this probability is shown by the crosshatched area. On the other hand, the probability of being late for the 1:45 P.M. flight is shown by the dotted area; it equals 1 minus the combined crosshatched and white areas under the density function:

$$p(R > 15) = 1 - p(R \leq 15) = 1 - \frac{15 - (-20)}{25 - (-20)} = 1 - \frac{35}{45} = \frac{10}{45} = .2222.$$

Note: In both cases, the probability can, of course, be calculated directly by multiplying the base of the relevant (crosshatched or dotted) rectangle (measuring 20 or 10, respectively) by its height (of 1/45). |||

Practice Problem 9

Probable Rates of Inflation

Consider an "almost informationless state" in which economists know only that next year's inflation will not be below 5 percent nor above 15 percent. If all values between 5 and 15 percent are deemed equally likely,

    a. What is the likelihood of inflation of 6 percent or less?
    b. What is the likelihood of inflation of more than 8.3 percent?
    c. What is the likelihood of inflation between 9.5 and 11.5 percent?
    d. What is the random variable's mean and standard deviation?

Figure O

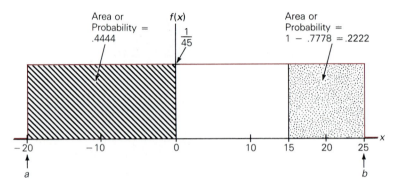

The problem can be sketched as shown in Figure P.

Figure P

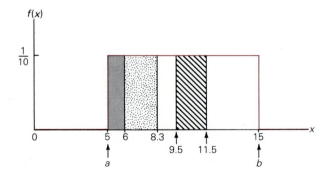

    a. The probability of 6 percent inflation or less equals the shaded area. Geometry (multiplying base times height) instantly reveals it to be equal to 1(1/10) = .1. Using the formula in Box 7.I, we get the same result:

$$p(R \leq 6) = \frac{x - a}{b - a} = \frac{6 - 5}{15 - 5} = \frac{1}{10} = .1.$$

b. The probability of more than 8.3 percent inflation equals the sum of the two white areas plus the crosshatched area. Geometry says the result is $(15 - 8.3)(1/10) = (6.7)(.1) = .67$. Because $p(R > x) = 1 - p(R \leq x)$, the formula in Box 7.I gives us

$$p(R > 8.3) = 1 - p(R \leq 8.3) = 1 - \frac{8.3 - 5}{15 - 5} = 1 - \frac{3.3}{10} = 1 - .33 = .67.$$

c. The probability of $9.5 - 11.5$ percent inflation equals the crosshatched area. Geometrically, this area equals $2(1/10) = .2$. According to the formula in Box 7.I.

$$p(9.5 \leq R \leq 11.5) = p(R \leq 11.5) - p(R \leq 9.5) = \frac{11.5 - 5}{15 - 5} - \frac{9.5 - 5}{15 - 5}$$

$$= \frac{6.5}{10} - \frac{4.5}{10} = \frac{2}{10} = .2.$$

d. According to the formulas in Box 7.J,

$$\mu_R = \frac{a + b}{2} = \frac{5 + 15}{2} = 10$$

$$\sigma_R = \sqrt{\frac{(b - a)^2}{12}} = \sqrt{\frac{(15 - 5)^2}{12}} = \sqrt{\frac{100}{12}} = \sqrt{8.3333} = 2.8868 \quad \text{|||}$$

Note: A more complex application (that combines the uniform with the exponential probability distribution) is Analytical Example 7.3, "Waiting Times at the Aircraft Repair Station: The Monte Carlo Approach."

||| ANALYTICAL EXAMPLES

7.1 Honest Weights and the Normal Curve of Error

Since time immemorial, merchants and their customers have had problems with accurate measurement. Whether they were weighing bags of spices, live sheep, or flasks of olive oil, whether they were measuring off lengths of cloth or lengths of pastureland, they were soon aware that each trial of measurement could easily produce a different result. Even today, with infinitely more refined measurement techniques, repeated measurement of the same thing tends to produce different results each time. Each result is contaminated by a different *random error*. (See Chapter 2 for a more detailed discussion of this concept.) These errors are as likely to be positive as negative; they are

also more likely to be small than large (that is, they are more likely to be close to zero than farther away from it). In short, errors of careful measurement can be described by the normal curve. This fact was clearly stated in 1809 by Gauss (Biography 7.1) and, some eight decades later, led Francis Galton (Biography 13.1) to exclaim[1]: "I know of scarce anything so apt to impress the imagination as the wonderful form of cosmic order expressed by the Law of Frequency of Error."

_____

[1]Francis Galton, *Natural Inheritance* (London: Macmillan, 1888).

Consider, for example, how it is decided that a pound of sugar you buy really is a pound. In 1875, a Treaty of the Meter was signed in Paris.[2] Among other things, the signatories agreed to define a certain object as the International Prototype Kilogram and to determine all other weights in relation to it. (The object in question is made of platinum-iridium and is held, under standard conditions of air pressure and temperature, at the International Bureau of Weights and Measures in Paris.) Each signatory nation owns a copy of the original Kilogram; that of the United States (number 20) is held at the National Bureau of Standards in Washington. With its help, similar weights in each state are calibrated, and these are used to check merchant scales periodically. In the end, the amount of sugar you buy is called 1 pound because its weight equals .4539237 of the original kilogram.

Yet even at the National Bureau of Standards, weighing the same object repeatedly may produce results such as these: 999.9231 grams, 999.9752 grams, 999.9603 grams, 1000.0598 grams, 1000.1003 grams,

999.9501 grams, 1000.0001 grams, and so on. Slight variations in the position of weights or imperceptible amounts of play in the balance mechanism will yield ever new results, but the probability distribution of these results may well fit a normal curve perfectly. If the distribution of results looks like Figure A, the object may well be deemed to weigh 1 kilogram; if the distribution looks like Figure B, the object may well be deemed to weigh less than 1 kilogram.

It is clear from the previous discussion why the normal curve has also been called the *normal curve of error*. If the mean result of many measurements is defined as the correct weight, the mean translates into an error of zero, while positive and negative errors (associated with results in excess of or below the mean) are normally distributed around the mean. Note: repeated comparisons of the U.S. Kilogram (K_{20}) with the Paris prototype suggest that the U.S. kilogram is 19 parts per billion lighter than the original.[3]

[2]See David Freedman, Robert Pisani, and Roger Purves, *Statistics* (New York: W. W. Norton, 1978), Chapter 6.

[3]Mort LaBrecque, "After 185 Years, Physicists Are Still Weighing the Kilogram," *Popular Science,* May 1984, pp. 96–99 and 170.

Figure A

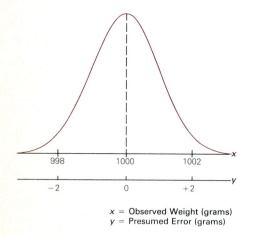

x = Observed Weight (grams)
y = Presumed Error (grams)

Figure B

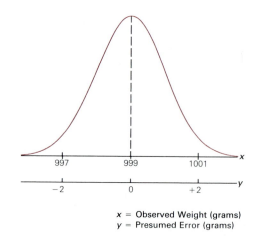

x = Observed Weight (grams)
y = Presumed Error (grams)

7.2 The Decision to Seed Hurricanes

In 1961, it was suggested that seeding hurricanes with silver iodide would mitigate their destructive force. The extent of that force was evidenced when Hurricane Betsy in 1965 and Hurricane Camille in 1969 each caused $1.5 billion in property damage. Also

in 1969, strong evidence for the effectiveness of the seeding procedure was obtained when Hurricane Debbie was seeded twice and reductions in peak wind speed of 31 and 15 percent, respectively, were observed as a result. The U.S. Department of Commerce subsequently sponsored a study of the behavior of hurricanes; it was found that in a "typical" hurricane of 100 miles per hour the probability distribution of various percentage changes in maximum sustained surface winds within 12 hours before landfall could be described by a normal curve with a mean of 0 and a standard deviation of 15.6 percent. Such a curve is shown in the accompanying figure; typical property damage totals are also indicated.

Thus, a hurricane the 100-mph peak sustained surface wind of which intensifies 32 percent within the 12 hours prior to landfall can be expected to destroy property well in excess of $300 million, but the damage will only be about $15 million if the winds decrease by 34 percent. (Note: The probabilities for the two events can be determined by using the normal probabilities table. In this example, +32 percent equals a z score of $32 - 0$, divided by 15.6, or 2.05; -34 percent equals a z score of -2.18. The associated probabilities, according to Appendix Table H, for $z \geq 2.05$ or $z \leq -2.18$ equal $.5 - .4798 = .0202$ and $.5 - .4854 = .0146$, respectively.) Clearly, the kind of information embodied in the accompanying normal

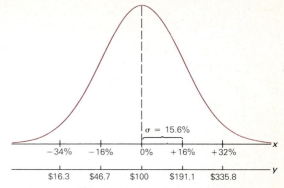

x = Percentage Change in Maximum Sustained Surface Wind Within 12 Hours Before Landfall, Given a 100mph Hurricane
y = Associated Property Damage (millions of dollars)

curve can become a crucial input in the decision to engage in, or refrain from engaging in, costly hurricane seeding.

Source: R. A. Howard, J. E. Matheson, D. W. North, "The Decision to Seed Hurricanes," *Science*, June 16, 1972, pp. 1191–1202.

7.3 The Monte Carlo Approach: Waiting Times at the Aircraft Repair Station

As we have seen throughout this chapter and the previous one, probability distributions can often be determined mathematically. There are, however, real-life processes that are so complex that they cannot be described by a completely determined set of equations. Under such circumstances, probabilities can still be found by the so-called **Monte Carlo method**, a procedure that simulates a real-life process in order to generate a body of data that is then viewed as a hypothetical "historical record" and from which the desired information is then derived. (The method was first employed during World War II when the behavior and interaction of elementary atomic particles was simulated as an aid to the design of atomic reactors.)

As an illustration of how the method works (and not necessarily as an example of a situation that is too complex to be handled mathematically), consider the following situation.

Scheduled airlines, as well as air-charter companies, typically operate their own facilities for the repair of their aircraft. It is not easy, however, to determine the optimum size of such a facility, because problems with mechanical or electrical equipment are apt to crop up at random, and their solution is likely to require varying amounts of time. If the capacity of the

service facility is very small, waiting lines may form, and lines of airplanes waiting for repair (rather than flying passengers or freight) spell loss of potential revenues. If the repair station's capacity is very large, on the other hand, waiting prior to servicing may never be required, yet the facility is bound to be idle often, and that is costly, too. It is easy to see why an airline that is about to open a new service facility (or that wants to review the capacity of an existing one) would want to know the probability distribution for various waiting times to which its planes will be subjected at the facility. The following only may be known: that the service facility can handle only one plane at a time, that the arrival of planes at the facility is a Poisson process (occurring at an average rate of, say, $\lambda = 1$ plane per week), that the interarrival times, therefore, can be described by an exponential probability density function (as in Figure A), and that servicing itself takes anywhere from .2 to .8 weeks, any interval within that range being equally likely (hence, servicing times can, in turn, be described by the uniform probability density function of Figure B).

The probabilities for various interarrival times (such as .1 week or less, .2 week or less, .3 week or less, and so on) can, of course, be established with the

Figure A

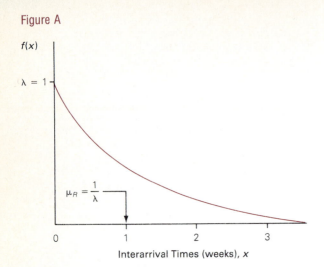

Interarrival Times (weeks), x

Figure B

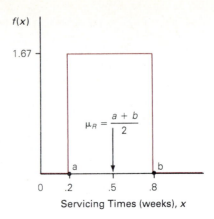

Servicing Times (weeks), x

help of the formula in Box 7.F or Appendix Table I. (Given $\lambda = 1$, the probabilities for interarrival times of $x = .1$ or less, .2 or less, and .3 or less, for example, can be found in the .00 column of Appendix Table I as .0952, .1813, and .2592, respectively.)

The probabilities for various servicing times (such as .3 week or less, .4 week or less, and so forth) can, similarly, be established with the help of the formula in Box 7.I. (Given $a = .2$ and $b = .8$, these probabilities equal .1667, .3333, and so on.) Finding probabilities for various *waiting times* (that involve waiting for service to start and waiting while service is rendered) is a more complex matter. A company statistician may use the Monte Carlo method to create an analogue of the situation:

a. Using a table of random numbers (Appendix Table A), the statistician might pick a large set of four-digit numbers as in column 1 of the table on page 275. (To keep things manageable, we pick only 20 numbers, using the first four digits of the last column of numbers in Appendix Table A.)

b. Dividing the numbers by 10,000, the statistician can turn them into (randomly selected) probabilities associated with (randomly occurring) interarrival times, as in column 2.

c. With the help of Appendix Table I, these probabilities can then be associated with values of interarrival times, x, as in column 3. In order for the probability distribution of interarrival times to be an exponential one (and for the arrivals themselves to be a Poisson process), a probability of .3355, for example, must

be associated with a value of λx and (given $\lambda = 1$) of x of .41. (The value of .3363 at the intersection of the .4 row and .01 column comes closest to .3355.) Similarly, a probability of .3817 must go with $x = .48$, and so on. The column 3 data represent a randomly selected exponentially distributed set of interarrival times.

d. From column 3, the actual arrival times in column 4 can be determined, starting with time 0. Thus, the first plane arrives for servicing after .41 week has elapsed following the opening of the facility; the second one arrives .48 week later, or .89 week after opening time; and so on. The last plane arrives 20.93 weeks after opening time. Given the exponentially distributed interarrival times, these times represent the simulated Poisson process of plane arrivals.

e. The probabilities of various servicing times can be established by a procedure analogous to the derivation of the column 2 data. In this case, the values in column 5 have been determined by using the last four digits of the last column of numbers in Appendix Table A, reading from the bottom up, and dividing by 10,000.

f. With the help of the formula in Box 7.I, these probabilities can be associated with values of servicing times, x, as in column 6. Since $p = (x - a)/(b - a)$, and since $a = .2$ and $b = .8$, it follows that $x = .6p + .2$, which is how all the values in column 6 have been derived from column 5. These column 6 data represent a randomly selected uniformly distributed set of servicing times.

g. Based on the previous data, the waiting times before service in column 7, the waiting times before and during service in column 8, and, hence, the departure times in column 9, can be simulated. The first plane, for example, arrives .41 week after opening and, being first, has a zero wait before service begins, as column 7 indicates. Its repairs take .74 week, as listed in column 6; hence, its total wait equals .74 week, as listed in column 8. The plane departs .41 + .74 = 1.15 weeks after the opening of the facility, as totaled in column 9. Plane 2 arrives .89 weeks after opening, but its servicing cannot begin until plane 1 has departed; thus, it has a preserve wait of 1.15 − .89 = .26 week. Its servicing takes only .40 week; hence, its total wait equals .26 + .40 = .66 week, and its departure occurs at .89 + .66 = 1.55 weeks.

The remaining entries in columns 7 through 9 have been determined similarly. (Note: The eighth plane once more has a zero wait before service because its arrival occurs after the departure of the seventh plane.)

Given the simulated "historical record" (and in an actual Monte Carlo application it would have to involve a vastly larger set of data), all kinds of interesting questions can be answered:

a. What is the probability that a newly arrived plane will have to wait before service is initiated? Our column 7 says 9 times out of 20, implying a probability of .45.
b. What is the mean time a plane will have to spend waiting before service is inititated? Our column 7 indicates .143 week.

Monte Carlo Study of Waiting Times at Repair Facility

Random Number (1)	Probability of Interarrival Time (2)	Interarrival Time (weeks since previous arrival) (3)	Arrival Time (weeks after time 0)	Probability of Servicing (time) (5)	Servicing Time (weeks) (6)	Waiting Time		Departure Time (weeks after time 0) (9) = (4) + (8)
						Before Service (weeks) (7)	Before and During Service (weeks) (8) = (7) + (6)	
3355	.3355	.41	.41	.9074	.74	0	.74	1.15
3817	.3817	.48	.89	.3256	.40	.26	.66	1.55
2404	.2404	.28	1.17	.3553	.41	.38	.79	1.96
3054	.3054	.36	1.53	.2133	.33	.43	.76	2.29
2265	.2265	.26	1.79	.0894	.25	.50	.75	2.54
3449	.3449	.42	2.21	.4989	.50	.33	..83	3.04
3063	.3063	.37	2.58	.8159	.69	.46	1.15	3.73
8860	.8860	2.17	4.75	.1170	.27	0	.27	5.02
1651	.1651	.18	4.93	.3796	.43	.09	.52	5.45
2927	.2927	.35	5.28	.2185	.33	.17	.50	5.78
2275	.2275	.26	5.54	.7629	.66	.24	.90	6.44
7825	.7825	1.53	7.07	.3157	.39	0	.39	7.46
6491	.6491	1.05	8.12	.7039	.62	0	.62	8.74
4808	.4808	.66	8.78	.1392	.28	0	.28	9.06
2449	.2449	.28	9.06	.5354	.52	0	.52	9.58
9951	.9951	5.03	14.09	.4795	.49	0	.49	14.58
9970	.9970	5.08	19.17	.0423	.23	0	.23	19.40
3990	.3990	.51	19.68	.0888	.25	0	.25	19.93
4093	.4093	.53	20.21	.1065	.26	0	.26	20.47
5119	.5119	.72	20.93	.1036	.26	0	.26	21.19

c. What is the mean time a plane will have to be grounded before or during service? Our column 8 shows .5585 week.

d. What is the mean time that it takes for a crew to service a plane? It is implied by the two previous answers as .5585 − .143 = .4155 week, which is also the mean of the column 6 data.

e. What are the probabilities that a plane in need of repairs is grounded (before or during repairs) for half a week or less and for more than .3 week? Our column 8 data tell us, respectively, 9/20 and 14/20.

But note: answers provided by a larger set of data may well be different. The theoretically precise answer to (a), for example, does not equal .45, but .50, the known average weekly arrival rate (λ = 1 plane per week), divided by the known average weekly number of planes, n, that can be serviced (and n = 2 because, as Figure B shows, the mean servicing time is .5 week). Thus (λ/n) = .5. The previous answers, therefore, should not be viewed as precise answers to the particular questions posed, but merely as illustrative of how answers can be derived from the Monte Carlo approach (answers that can be viewed with confidence provided the simulated "history" is long enough).

BIOGRAPHY

7.1 Carl Friedrich Gauss

Carl Friedrich Gauss (1777–1855) was born in Brunswick, Germany, where he grew up in humble circumstances (his father was a bricklayer and gardener; his mother had had no schooling and could not even write). Yet Gauss grew up to become one of the greatest mathematicians of all time. His precocity was extraordinary and probably unequaled. Before he was three, while watching his father's wage calculations, he detected an error and announced the correct result. At school, the speed and accuracy of his mental calculations astonished his beginning teacher in arithmetic; before long, Gauss mastered the best available textbooks and went beyond his teachers. Gauss's amazing powers came to the attention of the Duke of Brunswick who became his patron, sending him to the University of Göttingen. While a student there, Gauss discovered that a regular polygon of 17 sides is amenable to straightedge-and-compass construction; such a feat had eluded mathematicians for 2,000 years. Shortly thereafter, in 1801, Gauss published *Disquisitiones Arithmeticae,* a masterpiece that took the theory of numbers far beyond its earlier state and established Gauss as a mathematical genius of the first rank, equal to Archimedes and Newton. One great work was followed by another in 1809, the *Theoria Motus Corporum Coelestium,* in which Gauss developed methods for calculating the orbits of heavenly bodies. It was here that he made first use of

two of the most common tools of all sciences today, the theory of least squares (to be discussed in Chapter 13) and the normal curve (discussed in this chapter). Although Gauss acknowledged the priority of Laplace with respect to the normal curve,[1] it soon came to be known as the *Gaussean curve,* and it is still known as such in many countries today. (It was only in 1893 that Karl Pearson, Biography 10.1, introduced the term *normal* curve, and this is somewhat unfortunate because it creates the false impression that other-than-normal probability density functions are somehow abnormal or rare.) Gauss, apparently, was the first to view the bell-shaped curve as a model for depicting random error, as noted in Analytical Example 7.1. Writing in the context of his book on celestial mechanics, he said[2]:

The investigation of an orbit having, strictly speaking, the *maximum* probability, will depend upon a knowledge of the law according to which the probability of errors decreases as the errors increase in magnitude: but that depends upon so many vague and doubtful considerations—physiological included—which cannot be subjected to calculation, that it is scarcely, and indeed less than scarcely, possible to assign

[1]A biographical sketch of Laplace can be found in Chapter 5 of the *Student Workbook* that accompanies this text.
[2]Cited in Helen M. Walker, *Studies in the History of Statistical Method* (Baltimore: Williams and Wilkins, 1929), p. 23.

properly a law of this kind in any case of practical astronomy. Nevertheless, an investigation of the connection between this law and the most probable orbit, which we will undertake in its utmost generality, is not to be regarded as by any means a barren speculation. Let $\phi\Delta$ be the probability to be assigned to each error Δ. Now although we cannot precisely assign the form of this function, we can at least affirm that it should be a maximum for $\Delta = 0$, equal, generally, for equal opposite values of Δ, and should vanish, if, for Δ is taken the greatest error, or a value greater than the greatest error.

By age 30, Gauss was director of the Göttingen observatory, where he stayed for the rest of his life. He contributed mightily to astronomy but also to every branch of mathematics and physics. Hundreds of his publications testify to a man who combined in a most unusual way a pure mathematician's interest in abstract ideas and logical rigor, a theoretical physicist's interest in the creation of mathematical models of the physical world, an astronomer's talent for keen observation, and an experimentalist's skill in the application and invention of methods of measurement. (Among other things Gauss invented the heliotrope, the magnetometer, the photometer, and the telegraph.)

Summary

1. This chapter focuses on probability distributions of *continuous* random variables, which can assume values at all points on a scale of values. Naturally, one cannot list all the conceivable values of such variables and then associate their infinite number with an equally infinite number of probabilities. Instead, one associates probabilities of continuous random variables with various *ranges* of their values, and one measures these probabilities as appropriate areas under a smooth frequency curve that approximates a relative frequency histogram and is called a *probability density function*. An infinite variety of continuous probability distributions exist, although the normal, exponential, and uniform ones are most frequently encountered.

2. A *normal probability distribution* is a probability density function that is (a) single-peaked above the random variable's mean, median, and mode, (b) perfectly symmetrical about this central value, and (c) characterized by tails extending indefinitely in both directions from the center, approaching but never touching the horizontal axis. The height of the normal probability density function for different values of the normal random variable can be determined by formula (Box 7.A), and probabilities for specified ranges of values of this variable can be found by integrating that function over these ranges. It is common practice, however, to convert all members of the normal probability distribution family into a *standard normal curve* (with a mean of zero and a standard deviation of

1); probabilities for this curve are readily available in appropriately prepared tables (such as Appendix Table H). In certain circumstances, normal probabilities can be used to approximate various discrete probability distributions.

3. Whenever the number of occurrences of an event is determined by a Poisson process, the likelihood of encountering specified intervals of time or space between consecutive occurrences can be described by the *exponential probability distribution*. The height of the exponential probability density function can be determined by formula (Box 7.D); various probabilities can be found by integrating that function for different ranges of the exponential random variable. The exponential functions of Appendix Table E are helpful in finding quick answers from the formula. Other tables, such as Appendix Table I, provide ready-made probabilities for cumulative values of the exponential random variable.

4. The *uniform* or *rectangular probability distribution* is a probability density function for a random variable that is equally likely to take on any of the values within a given range. The (constant) height of the uniform probability density function can be determined by formula (Box 7.H). The formula is so simple, as is that (in Box 7.I) for calculating areas under the density function, that there is no need to tabulate uniform probabilities. Summary measures are also easy to calculate (Box 7.J), quite unlike those for other continuous random variables.

Key Terms

exponential probability distribution
exponential random variable
Monte Carlo method
normal probability distribution
probability density function

rectangular probability distribution
standard normal curve
standard normal deviate
uniform probability distribution
uniform random variable

Questions and Problems

The computer programs noted at the end of this chapter can be used to work many of the subsequent problems.

1. The time required by a bank teller to cash a check has a mean of 30 seconds and a standard deviation of 10 seconds. Find the times representing the tenth and seventy-fifth percentiles. (Assume that you are dealing with a normally distributed random variable.)

2. An architect wishes to design doors so that 95 percent of all people have at least a 1-inch clearance when passing through. If people have a mean height of 66 inches with a standard deviation of 4 inches, how high must the doors be? (Assume that people's heights are a normally distributed random variable.)

3. The Stanford-Binet IQ test has a mean of 100 and a standard deviation of 16. How likely is it that a randomly chosen person gets a score of at least 140? (Assume that IQs and test scores are normally distributed random variables.)

4. The output of workers in a plant is a normally distributed random variable. The plant's manager suggests that workers switch from hourly pay ($5.16 per hour) to a piece-work plan. If workers on the average produce 30 pieces per hour and the standard deviation is 5 pieces, which pay per piece would assure workers their present hourly pay at least 80 percent of the time? If the workers were paid 25 cents per piece, what percentage of the time would they earn between $4.50 and $5.50 per hour?

5. A company administers a test to all of its employees. The mean score is 500, and the standard deviation is 100. If the workers with the 25 percent highest scores are to be given special training, what is the lowest acceptable score for entrance into the training program? Furthermore, if the average time required to finish the test equals 60 minutes, with a standard deviation of 12 minutes, when should the exam be terminated so that 95

percent of the workers have completed all parts of the test? (Assume that test scores as well as finishing times are normally distributed random variables.)

6. A firm uses 2,000 light bulbs; their lifetime has a mean of 500 hours and a standard deviation of 50 hours. How often must the bulbs be replaced if all of them are to be replaced at once and at most 1 percent of them are to burn out between replacements? (Assume that the lifetime in question is a normally distributed random variable.)

7. The city miles-per-gallon (mpg) ratings of cars have a mean of 25.9 and a standard deviation of 2.45. If an automobile manufacturer wants to build a car the mpg ratings of which improve upon 99 percent of existing cars, what must the new car's mgp rating be? (Assume that the mpg rating is a normally distributed random variable.)

8. The shelf life of a battery has a mean of 525 days and a standard deviation of 50 days. If the battery has been on the shelf for 647 days, what is the probability of it being dead? (Assume that the shelf life of the battery is a normally distributed random variable.)

9. The time required to install a new aircraft engine is a normally distributed random variable with a mean of 20 hours and a standard deviation of 1 hour. What is the probability that the next installation takes
   a. between 20 and 21.5 hours?
   b. between 18 and 20 hours?
   c. between 19 and 22 hours?
   d. over 23 hours?
   e. at most 16.1 hours?
   f. between 21 and 22 hours?
   g. between 17 and 18 hours?
   h. at most 23.7 hours?
   i. more than 18.3 hours?

10. A manufacturer has deveoped a new type of automobile tire. The marketing department believes that the mileage guarantee offered to customers will be the crucial factor in winning consumer acceptance. Naturally, management wants to know the probability distribution of the lifetimes of these tires. Accordingly, 400 tires are tested. Engineers report that lifetimes are normally distributed with a mean of 36,000 miles and a standard deviation of 4,800 miles.
   a. How many of the tested tires had a lifetime between 31,200 and 40,800 miles?
   b. How many lasted at most 45,600 miles?
   c. How many lasted at least 26,400 miles?
   d. How many lasted between 26,400 and 31,200 miles?
   e. Within what limits were the lifetimes of the 20 percent worst tires?
   f. What mileage guarantee should the firm offer if it wants to make no more than 10 percent of all tires eligible?
11. The weekly number of checks cleared by a bank is a normally distributed random variable with a mean of 122,000 checks cleared and a standard deviation of 10,000 checks. In what proportion of weeks will the bank have to clear more than 140,000 checks?
12. The number of gallons of effluent processed at a sewage treatment plant is a normally distributed random variable with a mean of 50,000 gallons per day and a standard deviation of 7,000 gallons per day. The chief engineer wants to know: In what proportion of days will the plant be called upon to process in excess of 60,000 gallons (the plant's designed capacity)?
13. A state's turnpike authority finds its annual toll revenues to be normally distributed with a mean of $700,000 and a standard deviation of $50,000. The manager wants to know
   a. the probability that next year's operating expenses of $680,000 will be covered.
   b. the probability that these expenses will be covered in the next two years (assuming that the size of one year's revenue is independent of the size of the previous year's).
   c. the size of revenue in the best 25 percent of years.
14. In a recent year, the mathematics SAT scores were normally distributed with a mean of 500 for men and of 460 for women; both had a standard deviation of 100.

a. Estimate the percentages of men and women, respectively, with a score of 600 or more. Of 300 or less.
   b. Now you are to pick one man at random from the group and guess his score; if you are correct within 50 points, you get $100. What score do you pick if you want to win? What is your chance of winning?
15. Each year Amherst College accepts 700 students. From historical experience it is known that the number from this group who choose to come is a normally distributed random variable with a mean of 375 and a standard deviation of 40.
   a. Calculate the probability that more than 400 students will choose to come.
   b. Calculate the probability that between 350 and 400 will choose to come.
   c. Calculate the probability that fewer than 350 will choose to come.
16. College Board scores are normally distributed with standard deviations of 100 points for both left- and right-handed students. However, the mean math score for left-handed students is 525 whereas it is 500 for right-handed students. In the population as a whole, 10 percent of all students are left-handed. What fraction of students with math scores over 700 are left-handed?
17. A machine makes parts the lengths of which are normally distributed with a mean of 3 inches and a standard deviation of .15 inches. The acceptable range of lengths is 2.87 to 3.10 inches. What proportion of output is *not* acceptable?
18. A quality inspector tests the strength of aircraft control cables. Their breaking pressure is a normally distributed random variable with a mean of 500 lb. and a standard deviation of 50 lb. Some 100 cables ruptured under 475 lb. pressure or less. How many withstood pressure of 560 lb. or more?
19. A quality inspector tests the strength of window panes. Their breaking pressure is a normally distributed random variable with a mean of 80 mph wind velocity and a standard deviation of 7 mph. Some 50 panes broke under simulated wind velocities of 73 mph and less. How many withstood the hurricane velocity of 100 mph or more?
20. There are x gas stations in a state. Their incomes are normally distributed with a mean of $29,000 per year and a standard deviation of $5,100 per year. Some 189 stations earn between $26,000 and $31,000 per year. How many gas stations exist in the state?

21. There are x psychiatrists in a city. Their incomes are normally distributed with a mean of $89,000 per year and a standard deviation of $19,000 per year. Three psychiatrists earn more than $127,000 per year. How many psychiatrists are there in the entire city?

22. A time-and-motion study of 100,000 workers shows that they can accomplish a task in 35 seconds on the average. Seventy-four percent of the workers need more than 32 seconds. What percentage needs fewer than 36 seconds?

23. A study of 50,000 coal miners shows that the daily output per worker equals 17.3 tons on the average. Some 89.25 percent produce 18.1 tons or fewer. What percentage produces more than 16 tons?

24. An airline has 105 seats available per flight. If it sells 115 tickets, the actual arrivals are a normally distributed random variable with a mean of 100 and a standard deviation of 5. What is the probability that the number of passengers showing up exceeds the number of available seats?

25. A cattle feedlot finds that the daily weight gain of animals is 1.5 lb on the average with a standard deviation of .3 lb. If weight gain is a normally distributed random variable, what is the probability that a randomly chosen animal gains in a day
  a. less than 1 lb.
  b. less than 2 lb.
  c. more than 1.4 lb.
  d. between 1.6 and 1.7 lb.

26. The rents charged in a city are a normally distributed random variable with a mean of $551 per month and a standard deviation of $62. The mayor wants to know: What percentage of households would be eligible for rent subsidies if all households paying more than $600 were made eligible.

27. A personnel manager has discovered that the hours of sickleave taken by employees during a year are a normally distributed random variable with a mean of 52 hours and a standard deviation of 10 hours. She considers the hours taken by A (63) and B (93) highly unusual. Do you agree?

28. The weights of fish in Long Island Sound are normally distributed with a mean of 37 lb. and a standard deviation of 8 lb. One fisherman claims to have caught an 80-lb. fish. Do you believe it? Explain.

29. An efficiency expert knows that the time required by electricians to wire a certain type of house is a normally distributed random variable with a mean of 14 hours and a standard deviation of 1.3 hours. Evaluate the claims by two electricians.

  a. "I can wire a house like this in 11 hours."
  b. "It took me 17 hours of hard work to wire this house."

30. A city water department finds household water use to be a normally distributed random variable with a mean of 27 gallons and a standard deviation of 3 gallons per day.
  a. Find the probabilities that a randomly chosen household uses more than 25 gallons per day, between 20 and 25 gallons per day, fewer than 20 gallons per day.
  b. If the mayor wants to give a tax rebate to the 17-percent lowest water users, what should the gallons-per-day cutoff be?
  c. If the rebate changes the mean to 22 gallons and the standard deviation to 4 gallons per day, what should the 17-percent cutoff be?

31. The manager of a flower shop has a cash flow problem: 20 days of inflow are followed by 10 days of outflow, but the latter is a normally distributed random variable with a mean of $1,500 and a standard deviation of $400. If the manager wants only a 5-percent chance of running out of cash before the end of the outflow period, how much money should he have on hand at the beginning of that period?

32. The debt-asset ratios of all firms in an industry are normally distributed with a mean of 48 percent and a standard deviation of 9.9 percent. A particular firm's ratio is 37.2 percent. Do you find it unusual? Explain.

33. A bank fails. The FDIC finds it had *insider loans* (to directors and stockholders) of over $1.9 million. Given the fact that the size of such loans in the banking industry is a normally distributed random variable with a mean of $437,000 and a standard deviation of $129,000, is the situation unusual?

34. An investigation reveals that the local electric company has been overcharging customers. The overcharges are normally distributed with a mean of 700 kwh per month and a standard deviation of 200 kwh per month. What is the probability that a randomly chosen household was overcharged 800 kwh or more? A maximum of 500 kwh?

35. An unemployed worker sets out on a job search. The available jobs pay wages that are normally distributed with a mean of $400 per week and a standard deviation of $30 per week. What is the probability that the first job offered to the worker will pay $500 per week or more? $350 per week or less? Between $410 and $430 per week?

36. An insurance company finds that the life span of participants in its annuity program is a normally distributed random variable with a mean of 69 years and standard deviation of 8.1 years. What proportion of participants will get benefits beyond age 70? Age 75? Age 85?

37. The price-earnings ratio of all stocks is a normally distributed random variable with a mean of 9.6 and a standard deviation of 2.1. You suspect that some of the 10-percent lowest P/E ratio stocks are undervalued. You want to look at the companies involved. In order to make the selection, you should look at companies with a P/E ratio of x or less. What is x?

38. Since they are needed so often, find the z values corresponding to percentiles of 5, 10, 15, and so on up to 100 in increments of 5. Make your calculations to four decimals.

39. Experience shows that 25 percent of the people entering a store make a purchase.
  a. What is the probability distribution for purchases made for the next 20 customers? (Do *not* use binomial probability tables.)
  b. What is the probability that at most 3 of the next 20 customers will make a purchase?

40. In bridge tournaments players are assigned randomly to positions at the table (North, South, East, and West). Charles Goren once complained that he sat at the North position far more often than he should have. If Goren sat North in 250 out of 800 tournaments, was his complaint justified? What is the probability that such a happening could have occurred by chance? (Hint: Use the normal approximation to the binomial distribution.)

41. The absentee rate in a class of 500 statistics students is 10 percent. With the help of an appropriate approximation, compute the probability that in the next class
  a. more than 400 will attend.
  b. fewer than 450 will attend.
  c. between 413 and 463 will attend.

42. A firm ships thermometers in boxes of 1,000. Typically, 10 percent are broken upon arrival. Using an approximation, determine the probability that, in a randomly chosen box,
  a. 100 to 130 thermometers are broken.
  b. 75 thermometers are broken.
  c. 200 thermometers are broken.
  d. none are broken.

43. A census of telephone bills has revealed that the conversation times charged in 10 percent of the bills exceed the actual conversation times found elsewhere in the company records. A random sample of 100 bills is taken. Using an appropriate approximation, determine the probability of discovering fewer than 5 such errors.

44. Experience shows that in 40 percent of IRS audits taxpayers do not owe additional taxes. If we take a sample of 1,000 audited taxpayers, what is the probability that between 300 and 600 owe nothing? (Hint: Use an appropriate approximation.)

45. Consider the (exact) hypergeometric probability distribution given in Table 6.12 on page 218. What probabilities would one find by using the normal approximation?

46. During the rush hour, cars pass over a bridge at a rate of 20 per minute. Using the normal approximation to an appropriate probability distribution, determine the probability of exactly 10 cars crossing the bridge during a given minute. Then compare your result with the precise answer.

47. The time to service a car at a gas station is an exponential random variable with a mean of 2 minutes. Using formulas only, determine the probability that a newly arriving car will be serviced
  a. within 1 minute.
  b. within 4 minutes.
  c. within 2 to 6 minutes.
  d. only in 5 or more minutes.

48. The arrival of claims at an insurance company can be described as a Poisson process occurring at a rate of 2 claims per day. Using an appropriate table, determine the probability that the next claim will be made
  a. within 4 days.
  b. after the passage of 2 days.
  c. at some time between 3 and 5 days hence.

49. A hospital surgical unit cannot afford "ever" to be caught without electric power. Accordingly, the hospital administrator plans to purchase one or more standby generators to reduce the chances of such an occurrence to below 1 percent. The models available have a mean time between failures of 500 hours. Determine the probability of electric power loss in the surgical unit if
  a. one standby generator is bought and the city has a 10-hour blackout.
  b. two such generators are bought and the same event occurs.
  What are the chances of avoiding any power loss if
  c. one generator is bought, but there are 10 five-hour blackouts during a given year?
  d. two such generators are bought, and there are 10 such blackouts?

50. The length of life of a computer component is an exponential random variable with a mean of 7 years. If the warranty period is 5 years, what proportion of the components can the manufacturer expect to replace under the warranty? What should the warranty period be if the manufacturer doesn't want to be bothered with having to replace more than 10 percent of the components?

51. The time people have to wait in line at a fast-food outlet is an exponential random variable with a mean of 2 minutes. Compute the probability that a customer must wait
  a. more than 2 minutes.
  b. more than 3 minutes.
  c. less than 30 seconds.
  d. between 2 and 3 minutes.

52. A hospital administrator has noted that treatments of patients in the emergency room take 45 minutes on the average. Assuming that treatment time is an exponential random variable, determine
  a. the median treatment time.
  b. the probability that the next treatment exceeds 45 minutes.
  c. the probability that the next treatment exceeds 2 hours.
  d. the probability that the next treatment exceeds 45 minutes for each of the next 3 patients.

53. The manager of an aircraft avionics repair station has found that the repair time on radar transponders is an exponential random variable with a mean of 66 minutes. Determine the probability that the next repair takes
  a. at most 30 minutes.
  b. at most 66 minutes.
  c. at most 1.5 hours.
  d. between 2 and 3 hours.

54. An auto manufacturer claims that only 5.82 percent of its car radios have to be replaced under the firm's 5-year warranty. What must be the mean lifetime of the radio, assuming that the lifetime is an exponential random variable?

55. A washing-machine manufacturer makes two claims:

  a. that the lifetime of the firm's product is an exponential random variable;
  b. that only 4.88 percent of the firm's washing machines have to be replaced under the firm's generous 10-year unconditional warranty program.

Do you believe the claims? Explain.

56. A supervisor checks an employee's work every afternoon between 2:00 and 3:00 P.M., any arrival time within that range being equally likely. Graph the probability density function, and indicate
  a. the height of the function, its mean, and its standard deviation.

What is the probability that the supervisor arrives
  b. before 2:15 P.M.? **d.** between 2:41 and 2:43 P.M.?
  c. after 2:50 P.M.? **e.** precisely at 2:31:10 P.M.?

57. A computer generates random decimal values such that any value between zero and one is equally likely to be picked. Determine the probability that the next value chosen is
  a. less than .25.
  b. less than .79.
  c. greater than .79.
  d. between .25 and .79.

58. The waiting time for an elevator is uniformly distributed between zero and 3 minues.
  a. determine the probability of reaching the next floor in one minute or less if the trip itself takes 10 seconds.
  b. determine the distribution's summary measures.

59. The arrival of a bus is equally likely at any time during the next half hour. Determine
  a. your expected waiting time.
  b. the probability that you have to wait more than 25 minutes.
  c. the probability that you have to wait less than 5 minutes.
  d. the probability that you have to wait between 7 and 10 minutes.

60. If you wanted to draw the uniform probability distributions discussed in problems 57 to 59, what would the heights of the functions have to be? Prove that your answers are correct.

Selected Readings

Barlow, Richard, E., et al., eds. *Reliability and Fault Tree Analysis: Theoretical and Applied Aspects of System Reliability and Safety Assessment.* Philadelphia: Society for Industrial and Applied Mathematics, 1975. Papers of a conference on the application of probability models to safety problems. The technique called *fault tree analysis* provides an efficient way to compute the probability of failure of a complex system.

Cohen, Daniel. *Intelligence, What Is It?* New York: M. Evans, 1974. Traces the testing of intelligence from the ideas of Galton (Biography 13.1) in the 1890s to lawsuits about IQ tests in the 1970s.

Walker, Helen M. *Studies in the History of Statistical Method.* (Baltimore: Williams and Wilkins, 1929), Chapter 2. A superb discussion of the history of the normal curve.

Computer Program

The DAYSAL-2 personal-computer diskettes that accompany this book contain one program of interest to this chapter:

7 Probability Distributions, among other things, lets you find areas under the normal curve between any two z-values you specify.

PART THREE

Analytical Statistics: Basic Inference

CHAPTER EIGHT

Sampling Distributions and Estimation

As was noted in Chapter 2, decision makers frequently gain vital information about a population not by taking a full-fledged census, but merely by taking a sample. Consider a manufacturer who wants to know the average time workers require to complete a given task, the average amount of fuel needed to ship a truckload a given distance, or the average age of those who use a given product. Consider questions about a product's market share, about the percentage of acceptable units in a shipment of parts, or about the proportion of people watching a TV show, favoring a tax hike, or opposing a labor contract. In all these cases and many more, the statistician is interested in learning something about a statistical population. In the absence of a census, the desired knowledge about such *parameters* as the population mean (μ), the population standard deviation (σ), or the population proportion (π) can be gained only by drawing a random sample from the population, calculating such *statistics* as the sample mean ($\overline{X}$), the sample standard deviation (s), or the sample proportion (P), and making inferences about the parameters from the statistics.[1] The process of inferring the values of unknown population parameters from known sample statistics is called **estimation.**

ESTIMATORS AND ESTIMATES

The type of sample statistics that is used to make inferences about a given type of population parameter is called the **estimator** of that parameter. A later section of this chapter will consider major criteria for choosing an appropriate estimator—that is, one the actual numerical value of which is likely to be close to the unknown value of the parameter of interest. Obviously, the control of error is of major im-

---

[1]The initial discussion of *parameters* and *statistics* can be reviewed in the introductory section of Chapter 4.

portance when making estimates. For this reason, random sampling, rather than convenience or judgment sampling, is a necessity; this type of sampling helps eliminate *systematic* error or bias. In addition, as will also be shown in a later section, *random* or sampling error can be reduced (at a cost) by constructing samples that are sufficiently large. (For a review of the concepts of error, see the Chapter 2 section on "Errors in Survey Data.")

When the estimate of a population parameter is expressed as a single numerical value, it is referred to as a **point estimate.** Thus, the average time workers require to complete a given task may be estimated at 32 minutes; a product's market share may be estimated at 70 percent. These are point estimates. Any estimate, of course, can be wrong; users of estimates are always eager to know how wrong any given estimate might be. That is why it is often deemed desirable to provide a range of values within which the unknown population parameter presumably lies; such a range is called an **interval estimate.** As will be shown below, depending on the size of the interval, one can attach varying degrees of confidence to it and state with, say, 95-percent confidence that a product's market share lies between 64 and 76 percent or state with, say, 80-percent confidence that it lies between 67 and 73 percent. This type of statement calls attention to the fact that any estimate can be wrong. Thus, a procedure that yields interval esimates with 95-percent confidence is expected to be correct 95 percent of the time but wrong 5 percent of the time.

A final note: the process of estimation begins by sampling an already existing, clearly identifiable population. It must not be confused with *forecasting,* to be discussed in Chapter 15, which seeks to make statements about parameters of *future* populations—as about the median incomes of Americans in the year 2000, about the proportion of consumers likely to favor a new product that is still on the drawing board, or about the average daily demand for electric power five years hence.

THE CONCEPT OF THE SAMPLING DISTRIBUTION

To understand the nature of statistical estimation, one needs a firm grasp of the concept of the sampling distribution because it provides an important link between the single sample that is typically taken and the population about which inferences are to be made. As a basis for illustrating the concept, let us consider the two statistical populations given in Table 8.1. Merely in order to keep our example simple, we imagine that there exist only $N = 5$ business executives in an industry. Their salaries and sexes are given in the table. With this complete information in hand, we can, of course, calculate such population summary measures as the mean annual salary (μ), along with its variance (σ^2) and standard deviation (σ), as well as, say, the proportion of females (π_F) in the executive ranks. The summary calculations are indicated underneath the table. (The formulas involved were discussed in Chapter 4.) The mean annual salary equals $40,000 (with a standard deviation of $9,508); there are 40-percent females among the executives.

Suppose, however, that this census-type information were not available to us and couldn't easily be acquired. (In reality, there might be 30,000 executives in the industry, and contacting them all would be too costly or time-consuming.) Suppose we, therefore, decided to *estimate* the four population parameters by taking a simple random sample of the executives. As an example, let us take a sample of size $n = 3$, and let us

Table 8.1   Hypothetical Populations of Executive Salaries and Sexes

Executive	Annual Salary (thousands of dollars)	Sex
A	39	M
B	41	F
C	25	F
D	55	M
E	40	M

Summary Measures

a. Salary

$$\mu = \frac{\Sigma X}{N} = \frac{200}{5} = 40$$

$$\sigma^2 = \frac{\Sigma(X - \mu)^2}{N} = \frac{452}{5} = 90.4$$

$$\sigma = \sqrt{90.4} = 9.508$$

b. Sex

$$\pi_F = \frac{2}{5} = .40$$

do so without replacement. (Sampling without replacement is also common practice; once an elementary unit has been chosen, there is typically no chance that it will be chosen again.) The sample we pick might consist of executives B, D, and E, and we could instantly calculate the appropriate sample summary measures (once more using the applicable formulas from Chapter 4). Thus, the summary measures for salary would be the following:

$$\overline{X} = \frac{\Sigma X}{n} = \frac{136}{3} = 45.333 \text{ (thousands of dollars)}$$

$$s^2 = \frac{\Sigma(X - \overline{X})^2}{n - 1} = \frac{140.666}{2} = 70.333 \text{ (thousands of squared dollars)}$$

$$s = \sqrt{70.333} = 8.386 \text{ (thousands of dollars)}$$

The summary measure for sex would be the following:

$$P_F = \frac{1}{3} = .333$$

From our sample, we would estimate the mean annual salary at $45,333 (with a standard deviation of $8,386); and we would presume that 33.3 percent of the executives are female.

We might also, however, be bothered by a disturbing thought: if the random-selection procedure had produced a different sample of $n = 3$, we would surely have calculated different summary measures for the sample and would, thus, have made different point estimates of the corresponding population parameters. Indeed, what the point estimates might have been (if we had selected different samples) can easily be visualized with the help of Table 8.2. The table lists all the possible simple random

Table 8.2 Possible Samples of $n = 3$ Taken Without Replacement from a Population of $N = 5$, Along with Summary Measures of $\overline{X}$, s^2, s and P_F

| Sample Number | Units in Sample | Summary Measures | | | | Probability |
| | | (a) for Salary | | | (b) for Sex | |
		$\overline{X}$	s^2	s	P_F	p
(1)	ABC	35.000	76.000	8.718	.667	.1
(2)	ABD	45.000	76.000	8.718	.333	.1
(3)	ABE	40.000	1.000	1.000	.333	.1
(4)	ACD	39.667	225.333	15.011	.333	.1
(5)	ACE	34.667	70.333	8.386	.333	.1
(6)	ADE	44.667	80.333	8.963	0	.1
(7)	BCD	40.333	225.333	15.011	.667	.1
(8)	BCE	35.333	80.333	8.963	.667	.1
(9)	BDE	45.333	70.333	8.386	.333	.1
(10)	CDE	40.000	225.000	15.000	.333	.1

Source: Table 8.1.

samples of size $n = 3$ that we might have taken (without replacement) from among our population of size $N = 5$. (This ability to *list* all the possible samples is a lucky circumstance due to the utter simplicity of our example; it would not be practical to do so if we had considered a population of, say, 30,000 executives.) If we consider the order in which people enter a sample as irrelevant (so that a sample of B, D, E is treated as identical to a sample of E, D, B, and the like), we can apply the combinatorial formula of Box 5.F and determine that the possible number of samples of $n = 3$ out of a population of $N = 5$ equals

$$C_3^5 = \frac{5!}{3! \, (5-3)!} = \frac{5!}{3! \, 2!} = \frac{5 \cdot 4}{2} = 10$$

These 10 samples are listed in Table 8.2; four summary measures (calculated on the basis of the corresponding data in Table 8.1) are given for each one. Note that the sample evaluated above now appears as sample number (9).

When we contemplate Table 8.2, we realize that every sample summary measure (such as $\overline{X}$, s^2, s, and P_F) is in fact a random variable *prior* to the selection of an actual sample. Before we picked sample number (9), the sample mean, $\overline{X}$, for instance, could have turned out to be any one of the values in the $\overline{X}$ column of Table 8.2, depending entirely on which elementary units happened to be included in our sample. Because we were about to engage in simple random sampling, each of the 10 possible samples given in Table 8.2 was equally likely to be picked; hence, each of the 10 sets of associated summary measures was equally likely to become the set of sample statistics ultimately revealed to us. The probablities involved are given in the last column of Table 8.2. For any given value that is listed only once in the $\overline{X}$ column (or s^2 column or *s* column or P_F column), there existed, prior to our sample selection, a proability of .1 that it would become the sample statistic (because there was a probability of .1 that any one of the listed samples would actually be drawn). Similarly, the probability equaled two times .1, or .2, for any value listed twice, and so on.

This list of probabilities in the last column of Table 8.2 brings us to the concept of the **sampling distribution**—a probability distribution of a random variable that happens to be a summary measure based on sample data. The sampling distribution, thus, shows the likelihood of occurrence associated with all the possible values of a sample statistic (which values would be obtained when drawing all possible simple random samples of a given size from a population). This probability distribution, therefore, describes the manner in which sample characteristics vary from sample to sample. In Table 8.2, for example, the data in the $\overline{X}$ column, along with those in the probability column, represent the sampling distribution of the sample mean; similarly, the data in the s^2, the *s*, and the P_F columns, along with those in the probability column, represent the sampling distributions, respectively, of the sample variance, the sample standard deviation, and the sample proportion.

Note: to the extent that some values of the random variable occur more than once, the presentation of the sampling distribution can, of course, be simplified. The sampling distribution of the mean executive salary (or of the proportion of female executives) can, for example, be presented as shown in Table 8.3.

One way of interpreting Table 8.3 is: if simple random samples of $n = 3$ were to be selected, say, 1,000 times from the population of $N = 5$ given in Table 8.1, one could expect to calculate a $34,667 mean salary 100 times, a $40,000 mean salary 200 times, and so on. In addition, one could expect to find a 0 proportion of females 100 times, a proportion of .333 some 600 times, and a proportion of .667 in the remaining 300 times.

Table 8.3   Sampling Distributions of the Mean and of the Proportion

These sampling distributions, culled from Table 8.2, show that, prior to sample selection, there exists a probability of .1 that the sample statistic of the mean salary will turn out to be $34,667; the probability also equals .1 that $\overline{X}$ will be calculated as $35,000, and so on. Similarly, the probability is .1 that the proportion of females will be found to be 0, but .6 that it will be calculated as .333, and .3 that this sample statistic will equal .667.

$\overline{X}$ (thousands of dollars)	Probability	P_F	Probability
34.667	.1	0	.1
35.000	.1	.333	.6
35.333	.1	.667	.3
39.667	.1		1.0
40.000	.2		
40.333	.1		
44.667	.1		
45.000	.1		
45.333	.1		
	1.0		

Source: Table 8.2.

Table 8.4   Summary Measures of Two Sampling I

With the help of the formulas presented in Box 6.
marized as shown here.

(a) Summary Measures of the Sampling Distribution of the Mean, $\overline{X}$
$\mu_{\overline{X}} = 40$
$\sigma_{\overline{X}}^2 = 15.067$
$\sigma_{\overline{X}} = 3.882$

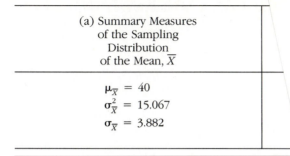

SUMMARY MEASURES OF THE SAMPLING DISTRIBUTION

Any sampling distribution can, in turn, be described by summary measures of its own. Given the data in the $\overline{X}$, s^2, s, and P_F columns of Table 8.2 (or given a consolidated version of these data as shown in Table 8.3), we can calculate their means or expected values, their variances, and their standard deviations. Such calculations have in fact been performed, with the help of the formulas in Box 6.A, for the two sampling distributions of $\overline{X}$ and P_F. The results are given in Table 8.4.

Thus, $\mu_{\overline{X}}$ (pronounced "mu sub X bar") is the mean of all the possible sample salary means (or the expected value of the sampling distribution of the mean); μ_P is, similarly, the mean of all the possible sample proportions of female executives. The expressions $\sigma_{\overline{X}}^2$ and σ_P^2 are the respective variances of the sampling distributions of the sample mean and of the sample proportion; their square roots (the standard deviations) are also referred to as the *standard errors* of the respective sample statistics. Thus, $\sigma_{\overline{X}}$ is the **standard error of the sample mean**, and σ_P is the **standard error of the sample proportion.**

A note of caution: it is important not to confuse $\sigma_{\overline{X}}$ with s. This mistake is easy to make because both are standard deviations. However, $\sigma_{\overline{X}}$ (the standard error of the sample mean from Table 8.4) measures the variability of *possible* $\overline{X}$ values that might be obtained—a variability a statistician is likely to contemplate in the planning stage of sampling. In contrast, s (the standard deviation of a particular sample result found in any one row of Table 8.2) measures the variability of X values around an *actual* $\overline{X}$ value observed; this variability is calculated at the final stage of sampling, after one of the many possible samples has actually been selected.

THE SAMPLING DISTRIBUTION VERSUS ITS PARENT POPULATION

The answer to one question is of crucial importance to statistical estimation—that is, the answer is crucial to inferring the values of unknown population parameters from known sample statistics: what is the relationship between a sampling distribution and its parent population? We can answer the question in two ways: by focusing on the general shape of the two frequency distributions (of possible sample statistics on the one hand and of population values on the other) or by focusing on the mathematical relationships between summary measures. We will do both.

...es of the Distributions and the Central Limit Theorem

The sampling distribution of the sample mean. Let us first consider the sampling distribution of the sample mean, $\overline{X}$, and how its shape relates to the associated population values. Two of the most important theorems in all of statistics describe this relationship.

> *Theorem 1:* If $\overline{X}$ is the mean of a random sample taken from a population and if the population values are normally distributed, the sampling distribution of $\overline{X}$ is also normally distributed, regardless of sample size.

The above discussion, although not precisely applicable, at least hints at the validity of this theorem. Note how the salary population values in Table 8.1 are symmetrically distributed around the population mean of $\mu = 40$; then notice how the sampling distribution of the mean in Table 8.3 is also symmetrically distributed around its mean of $\mu_{\overline{X}} = 40$. To be sure, a symmetrical distribution is not the same thing as a precisely normal distribution. Nevertheless, we can consider the Table 8.1 salary distribution *approximately* normal.[2]

Given a population of X values that is normally distributed in the strict sense, rather than only approximately so, and given an associated sampling distribution of $\overline{X}$ that is—in accordance with the above theorem—therefore also normal, it is easy to find the probability that the mean of any given random sample will be within any given interval. All one needs to do is transform the range of $\overline{X}$ values into the corresponding z values and use Appendix Table H to establish the appropriate areas under the standard normal curve. The following problems apply to a normally distributed sampling distribution of $\overline{X}$ in which $\mu_{\overline{X}} = 40$ and $\sigma_{\overline{X}} = 4.655$. (These values might have been derived like those in Table 8.4 but from a population, unlike that in Table 8.1, that was very large and normally distributed in the strict sense. Thus, Theorem 1 permits the stated assertion concerning the normal shape of the sampling distribution of $\overline{X}$.)

||| Practice Problem 1

A Sample Mean Below 45

What is the probability that a sample mean will lie below 45? The area in question lies below $z = \dfrac{45 - 40}{4.655} = 1.07$. Thus, the probability equals .8577, as shown in Figure A by the sum of the shaded plus crosshatched areas.

---

[2]First, 60 percent of these population values lie within 1 standard deviation of the mean (within the range of 30.492 to 49.508) as opposed to 68.26 percent in the case of the normal distribution. Second, 100 percent of the population values lie within 2 standard deviations of the mean (within the range of 20.984 and 59.016) as opposed to 95.44 percent in the case of the normal distribution. Third, the measure of skewness (Box 4.P) equals 0, precisely as in a normal distribution. Fourth, the measure of kurtosis (Box 4.Q) yields $K = 2.48$, while it equals 3 for a normal distribution. Similar statements can be made for the Table 8.3 sampling distribution of the mean.

Figure A

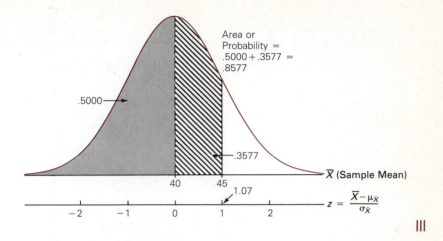

Area or
Probability =
.5000 + .3577 =
.8577

.5000

.3577

$\bar{X}$ (Sample Mean)

1.07

$z = \dfrac{\bar{X} - \mu_{\bar{X}}}{\sigma_{\bar{X}}}$

|||

||| Practice Problem 2

A Sample Mean Between 37 and 41.5

What is the probability that a sample mean will lie between 37 and 41.5? The area in question lies between $z = \dfrac{37 - 40}{4.655} = -.64$ and $z = \dfrac{41.5 - 40}{4.655} = .32$. Thus, the probability equals .2389 + .1255 = .3644 or the sum of the shaded plus crosshatched areas in Figure B.

Figure B

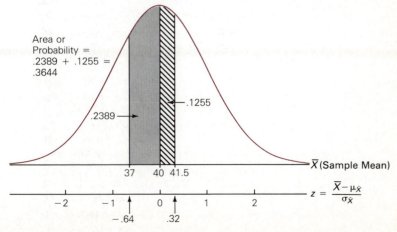

Area or
Probability =
.2389 + .1255 =
.3644

.1255

.2389

$\bar{X}$ (Sample Mean)

$z = \dfrac{\bar{X} - \mu_{\bar{X}}}{\sigma_{\bar{X}}}$

-.64

.32

|||

Practice Problem 3

A Sample Mean Above 43

What is the probability that a sample mean will exceed 43? The area in question lies above $z = \dfrac{43 - 40}{4.655} = .64$. Thus, the probability equals $.5000 - .2389 = .2611$ (the shaded area in Figure C).

Figure C

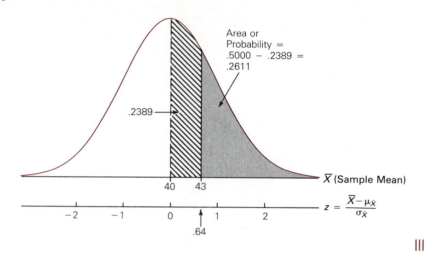

Theorem 2: If $\overline{X}$ is the mean of a random sample taken from a population and if the population values are *not* normally distributed, the sampling distribution of $\overline{X}$ nevertheless approaches a normal distribution as sample size increases—provided sample size, n, remains small relative to population size, N; this approximation is near-perfect for samples of $n \geq 30$ but $n < .05N$.

This theorem is called the **central limit theorem** and is, perhaps, the most important one in the entire field of statistical inference. As long as we take random samples that are sufficiently large absolutely ($n \geq 30$) but that are fairly small relative to population size ($n < .05N$), the theorem allows us to infer population parameters from sample statistics without knowing the shape of the population distribution (which is precisely the type of knowledge that is often unavailable). In addition, as the next section shows, the theorem can be adapted for use with discrete as well as continuous random variables.

Figure 8.1 pictures a variety of population distributions that are anything but normal, and it clearly shows the tendency toward normality in the associated sampling distributions as sample size, n, increases.

The sampling distribution of the sample proportion. Under specified circumstances, the sampling distribution of the sample proportion, P, can also be treated as if it were normally distributed. Recall from Chapter 6 that the proportion of successes in a Bernoulli process is a random variable having a binomial distribution. As sample size

Figure 8.1 Illustrations of the Central Limit Theorem

These graphs picture vividly the incredibly rapid tendency toward normality in sampling distributions as sample size increases, regardless of the shape of the associated population distributions—provided only that $n < .05N$. (The vertical dotted lines coincide with the means of the various distributions.)

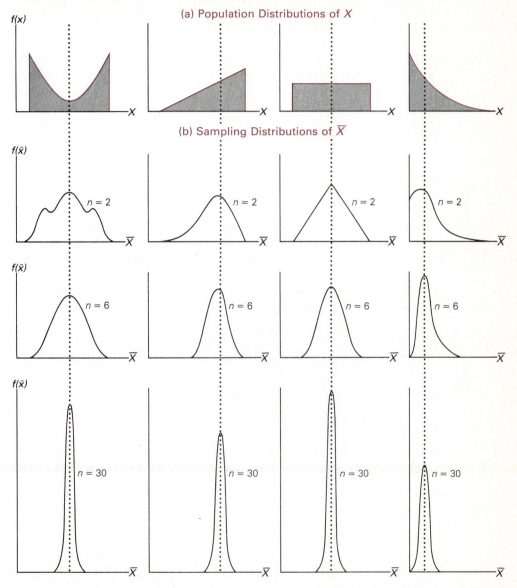

increases, the binomial distribution, too, approaches normality (see Figure 6.3 on page 205). Also recall the Chapter 7 section, "Approximating Discrete Probability Distributions by the Normal Probability Distribution." According to the rule of thumb there discussed, the binomial distribution is extremely close to the normal distribution whenever $n \cdot \pi$ as well as $n(1 - \pi) \geq 5$. In this expression, n stands for sample size, and π stands for the probability of success in any one trial (which is nothing else but the population proportion, π). In our above example, involving the proportion of females among executives, $n = 3$ and $\pi = .4$; thus, the rule of thumb is not satisfied [$n \cdot \pi = 1.2$ and $n(1 - \pi) = 1.8$]. Indeed, the above sampling distribution of the sample proportion (Table 8.3) is not normally distributed. If the rule of thumb is satisfied, however, one can determine probabilities for various ranges of values of the sample proportion by means of the standard-normal-curve table, using the procedure illustrated in the aforementioned Chapter 7 section, and calculating $z = (P - \mu_P)/\sigma_P$. Thus, we can state the following adaptation of the central limit theorem:

> *Theorem 2A:* If P is the proportion in a random sample taken from a population and if the population values are *not* normally distributed, the sampling distribution of P nevertheless approaches a normal distribution provided $n \cdot \pi$ as well as $n(1 - \pi) \geq 5$.

Mathematical Relationships Among Summary Measures

Concerning the sampling distribution of $\overline{X}$. The way in which the summary measures of the sampling distribution of the sample mean relate to those of the parent population is shown in Box 8.A. In addition, mathematical derivations are provided in Chapter 8 of the *Student Workbook* that accompanies this text.

8.A Summary Measures of the Sampling Distribution of $\overline{X}$

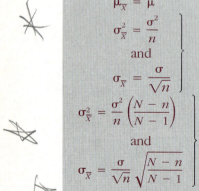

$$\mu_{\overline{X}} = \mu$$

$$\left. \begin{array}{c} \sigma_{\overline{X}}^2 = \dfrac{\sigma^2}{n} \\ \text{and} \\ \sigma_{\overline{X}} = \dfrac{\sigma}{\sqrt{n}} \end{array} \right\}$$ when selections of sample elements are statistically *independent* events; typically refered to as "the large-population case," because $n < .05N$.

$$\left. \begin{array}{c} \sigma_{\overline{X}}^2 = \dfrac{\sigma^2}{n}\left(\dfrac{N-n}{N-1}\right) \\ \text{and} \\ \sigma_{\overline{X}} = \dfrac{\sigma}{\sqrt{n}}\sqrt{\dfrac{N-n}{N-1}} \end{array} \right\}$$ when selections of sample elements are statistically *dependent* events; typically referred to as "the small-population case," because $n \geq .05N$.

where $\mu_{\overline{X}}$ is the mean, $\sigma_{\overline{X}}^2$ is the variance, and $\sigma_{\overline{X}}$ is the standard deviation of the sampling distribution of the sample mean, $\overline{X}$, while μ, σ^2, and σ are population mean, variance, and standard deviation, respectively, and N or n are population or sample size.

Notice that the mean of all possible sample means, $\mu_{\overline{X}}$, always equals the population mean, μ. This fact was evident in the above example (Tables 8.1 and 8.4) that showed $\mu = \mu_{\overline{X}} = 40$. This fact is also shown by the vertical dotted lines in Figure 8.1 that coincide with the means of the various distributions shown there.

The relationship between the variance of the sampling distribution, $\sigma_{\overline{X}}^2$ (or the standard deviation, $\sigma_{\overline{X}}$), on the one hand and the corresponding population measures is somewhat more complicated. The relationship also differs depending on whether the selections of sample elements are statistically independent or dependent events. As was noted in Chapter 5, two events are statistically independent if the probability of the occurrence of one is not affected by the occurrence of the other; in the case of statistical dependence, the occurrence of one event does affect the probability of the occurrence of the other. Now consider the event of selecting an element of the population for inclusion in a sample. Selecting one element does not in the least affect the probability of selecting any other element if the population is infinite in size or if sampling occurs with replacement. Even if sampling occurs without replacement, selecting one element does not affect the probability of selecting any other element noticeably as long as population size, N, is large relative to the sample size, n. (By tradition, the latter condition is considered fulfilled whenever $n < .05N$.) Statistical independence in sample selection is, therefore, frequently referred to as "the large-population case."

On the other hand, selecting an element of the population for inclusion in a sample *does* noticeably affect the probability of selecting any other element when sampling occurs without replacement and when the population is small relative to the sample ($n \geq .05\ N$). Statistical dependence in sample selection is, therefore, typically referred to as "the small-population case."

As the formulas in Box 8.A attest, as long as $n > 1$, the variance (or standard deviation) of the sampling distribution of $\overline{X}$ is smaller than the corresponding population summary measure. (Tables 8.1 and 8.4 once more provide examples: $\sigma^2 = 90.4$, but $\sigma_{\overline{X}}^2 = 15.067$; $\sigma = 9.508$, but $\sigma_{\overline{X}} = 3.882$.) Note, in addition, that the variance (or standard deviation) of the sampling distribution of $\overline{X}$ is larger the larger is the population variance (or standard deviation), given sample size, n. On the other hand, the larger the sample, given the population variance (or standard deviation), the smaller is the variance (or standard deviation) of the sampling distribution.

Note: the summary data of the sampling distribution of $\overline{X}$ in Table 8.4 were derived with the help of the general formulas first presented in Chapter 6. We can now confirm these results with the Box 8.A formulas (small-population case), applied to the population summary data of Table 8.1.

$$\mu_{\overline{X}} = \mu = 40$$

$$\sigma_{\overline{X}}^2 = \frac{\sigma^2}{n}\left(\frac{N-n}{N-1}\right) = \frac{90.4}{3}\left(\frac{5-3}{5-1}\right) = 30.133(.5) = 15.067$$

$$\sigma_{\overline{X}} = \frac{\sigma}{\sqrt{n}}\sqrt{\frac{N-n}{N-1}} = \frac{9.508}{\sqrt{3}}\sqrt{\frac{5-3}{5-1}} = 5.489(.707) = 3.881$$

Except for rounding error, the results are identical.

Concerning the sampling distribution of P. The way in which the summary measures of the sampling distribution of the sample proportion relate to that of the parent population (namely, π) is shown in Box 8.B.

8.B Summary Measures of the Sampling Distribution of *P*

$$\mu_P = \pi$$

$$\sigma_P^2 = \frac{\pi(1 - \pi)}{n}$$

$$\sigma_P = \sqrt{\frac{\pi(1 - \pi)}{n}}$$

when selections of sample elements are statistically *independent* events; typically referred to as "the large-population case," because $n < .05N$.

$$\sigma_P^2 = \frac{\pi(1 - \pi)}{n}\left(\frac{N - n}{N - 1}\right)$$

$$\sigma_P = \sqrt{\frac{\pi(1 - \pi)}{n}}\sqrt{\frac{N - n}{N - 1}}$$

when selections of sample elements are statistically *dependent* events; typically referred to as "the small-population case," because $n \geq .05N$.

where μ_P is the mean, σ_P^2 is the variance, and σ_P is the standard deviation of the sampling distribution of the sample proportion, *P*, while π is the population proportion and *N* or *n* are population or sample size.

Once more, we can confirm our earlier results (Table 8.4) that were derived independently. When applying the Box 8.B formulas (small-population case) to the only population summary value available (Table 8.1):

$$\mu_P = \pi = .40$$

$$\sigma_P^2 = \frac{\pi(1 - \pi)}{n}\left(\frac{N - n}{N - 1}\right) = \frac{.4(.6)}{3}\left(\frac{5 - 3}{5 - 1}\right) = .08(.5) = .04$$

$$\sigma_P = \sqrt{\frac{\pi(1 - \pi)}{n}}\sqrt{\frac{N - n}{N - 1}} = \sqrt{\frac{.4(.6)}{3}}\sqrt{\frac{5 - 3}{5 - 1}} = .2828(.707) = .20$$

THE CHOICE OF A GOOD ESTIMATOR

The linkage between understanding sampling distributions and engaging in statistical estimation becomes evident the moment one considers the question of what makes a good estimator. Three major criteria are commonly employed: unbiasedness, efficiency, and consistency.

The Criterion of Unbiasedness

A sample statistic that, on the average, across many samples, takes on a value equal to the population parameter that is being estimated with its help is called an **unbiased estimator.** On the other hand, if repeated random samplings from a given population produce a statistic with a mean or expected value that differs from the target to be estimated, bias is said to exist, as illustrated in panel (a) of Figure 8.2. The target, or true value of the population parameter, is designated by *T*. The sampling distribution of the statistic that is being used as an estimator, *E*, is also shown. In the upper graph

Figure 8.2 Criteria Employed in the Choice of Estimators

This set of graphs illustrates major criteria commonly used to define good estimators of unknown population parameters. Panel (a) shows that an estimator the expected value of which (μ_E) equals the true value of the population parameter *(T)* is *unbiased.* Panel (b) shows that, among several unbiased estimators, the one with the least variance is *efficient.* As panel (c) shows, an estimator that homes in on its target as sample size increases (so that bias and variance approach zero as sample size approaches infinity) is *consistent.*

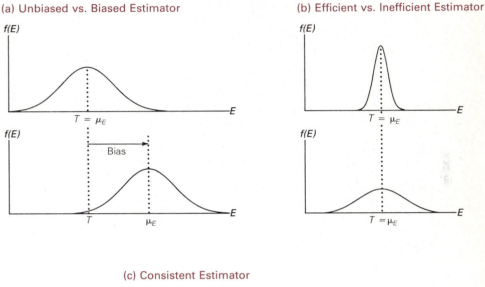

(a) Unbiased vs. Biased Estimator

(b) Efficient vs. Inefficient Estimator

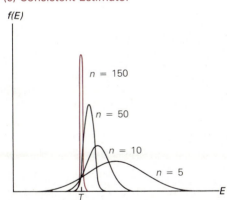

(c) Consistent Estimator

of panel (a), the mean of the sampling distribution, μ_E, equals the true parameter; the estimator, therefore, is said to be unbiased. In the lower graph of panel (a), the mean of the sampling distribution lies far above *T;* the estimator, therefore, contains upward bias. The extent of bias is always measured as $\mu_E - T$. (For an alternative discussion of the subject, see Figure 2.1 on page 25.)

We can now see why the sample mean, $\overline{X}$, is considered an unbiased estimate of the population mean, μ. As was noted in Box 8.A above, regardless of whether samling occurs from a large or small population, the mean of sample means, $\mu_{\overline{X}}$, always equals μ, and that equality satisfies the definition of unbiased estimator.

Similarly, because $\mu_P = \pi$ (Box 8.B), regardless of whether sampling occurs from a large or small population, the sample proportion, P, can always be viewed as an unbiased estimator of the population proportion, π.

At this point, we can also solve a riddle first noted in Chapter 4 (on page 111) when introducing the formula for the sample variance. Instead of being defined (in perfect analogy to the formula for the population variance) as the mean of the squared deviations of observed sample values from their mean, or as $\Sigma(X - \overline{X})^2/n$, the divisor was changed to $n - 1$ (see Box 4.I). The reason for this change in the divisor is this: the mean of the squared deviations of sample values from their mean turns out to be a *biased* estimator of the population variance; it produces an underestimate of it. This result is intuitively reasonable because a part cannot be more varied than the whole from which it comes, but it can easily be less diverse. Therefore, in order to estimate correctly the population variance from sample data, one may want to inflate the above expression somewhat, and this inflation is achieved by using $(n - 1)$ as a divisor and defining the sample variance as $s^2 = \Sigma(X - \overline{X})^2/(n - 1)$. It turns out that s^2, so defined, is an unbiased estimator of σ^2, *provided* that selections of sample elements are statistically independent events. In the small-population case, however, s^2 (even with the divisor of $n - 1$) is a biased estimator of σ^2. (In our small-population example, the expected value of s^2, when calculated from Table 8.2, is found to be 113, which is a far cry from the Table 8.1 population variance of 90.4).

The sample standard deviation, s, finally, is a biased estimator of the population standard deviation, σ, regardless of whether sampling involves the large- or small-population case. (Again, to use our above small-population example, the expected value of s, when calculated from Table 8.2, turns out to be 9.816, while Table 8.1 shows σ to equal 9.508.)

The Criterion of Efficiency

As a look at the upper graph in panel (a) of Figure 8.2 confirms, even though an estimate may be right on target *on the average,* it can, nevertheless, be off on many occasions. That is why a statistician naturally comes to think of another criterion for the merit of an estimator: all else being equal, one would surely prefer an estimator with a small variance (with a distribution that is highly concentrated near the mean) over one with a large degree of variability. Accordingly, among all the available unbiased estimators, the one with the smallest variance for a given sample size is considered to be the **efficient estimator.** This criterion is illustrated in panel (b) of Figure 8.2. Note that both estimators in panel (b) have a sampling distribution the mean of which hits the target precisely. Yet the statistic with the sampling distribution shown in the upper graph of panel (b) is the efficient one because of its lower variability around the mean. A good example is provided by the sample median and sample mean, both of which are unbiased estimators of the mean of a normally distributed population. Yet in the normally distributed sampling distributions shown, the variance of the sample median is 57 percent larger than that of the sample mean (hence, the variance of the sample mean is only 64 percent as large as that of the sample median); this difference makes $\overline{X}$ the more efficient estimator of the two. Sometimes this relative efficiency is expressed by the ratio of variances:

$$\text{Relative efficiency of a given statistic} = \frac{\text{variance of efficient statistic}}{\text{variance of given statistic}}$$

In the example just noted, this expression would amount to

$$\text{Relative efficiency of sample median} \atop \text{(as estimator of population mean)} = \frac{\text{variance of sample mean}}{\text{variance of sample median}}$$

$$= \frac{\sigma^2/n}{\left(\dfrac{\pi}{2}\right)\left(\dfrac{\sigma^2}{n}\right)}$$

$$= \frac{2}{\pi} = .64$$

(In this expression, π does refer to the well-known constant, not to the population proportion.) This relationship means that a sample median is considerably less likely than a sample mean to be a correct estimate of a population mean.

The Criterion of Consistency

If a large variance of the sampling distribution of the estimator seems undesirable, anything that can reduce the variance becomes of interest. When discussing the formulas in Box 8.A above, we noted one such possibility. At least so far as the sampling distribution of $\overline{X}$ is concerned, one can lower the variance by increasing sample size. In the large-population case, for example, given the population variance, σ^2, a large n reduces $\sigma^2_{\overline{x}}$ because the latter equals σ^2/n. Whenever the probability that a statistic will be close to the parameter being estimated gets ever larger (and, thus, approaches unity) as sample size increases, the statistic is called a **consistent estimator.** The sample mean and sample proportion are such consistent estimators of the corresponding population parameters. Panel (c) of Figure 8.2 illustrates a consistent estimator the sampling distribution of which increasingly concentrates on its target as sample size increases.

A Final Note: Mean Squared Error

More often than not, the search for a good estimator yields several prospects, none of which are perfect. One estimator may be unbiased but have a huge variance. Another may have a tiny variance but a huge bias. A third one may have some of both faults. This problem of choosing among imperfect estimators is illustrated in Figure 8.3. The estimator with distribution A has an expected value right on target but a large variance. The estimator with distribution C has a low variance but a large bias. The estimator with distribution B stands in the middle with respect to both criteria. Surely it would be foolish in this case to insist on no bias and to choose estimator A. Note how a trade-off could be made by switching from A to B, accepting some bias, but vastly reducing variability.

Statisticians do in fact choose the best estimator under such circumstances by calculating (and then minimizing) the **mean squared error,** which is the sum of an estimator's squared bias plus its variance, or $(\mu_E - T)^2 + \sigma^2_E$. In terms of Figure 8.3, this procedure picks the statistic with sampling distribution B as the best estimator.

Figure 8.3 Choosing Among Estimators

When all available estimators are imperfect, it is wise to choose the one with a combination of small bias and small variance (such as the one with distribution *B*).

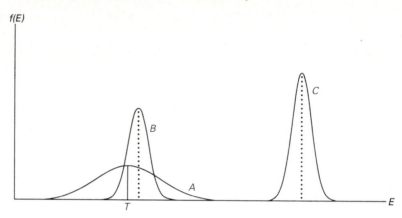

MAKING POINT ESTIMATES

Whenever the statistical population of interest is normally distributed or whenever the conditions of the central limit theorem are fulfilled [$n \geq 30$, but $n < .05N$, or $n \cdot \pi$ as well as $n(1 - \pi) \geq 5$], and regardless of whether sampling occurs from a large or a small population, sample mean and sample proportion are unbiased, efficient, and consistent estimators of the corresponding population parameters. In the large-population case, the same can be said about the sample variance. No wonder these statistics are routinely used as estimators under the stated circumstances. Indeed, because of a lack of better alternatives, statisticians employ the sample variance also as an estimator of the population variance in the small-population case (even though the estimate is then biased). For the same reason, and being equally aware of its bias under all circumstances, statisticians use the sample standard deviation as an estimator of that of the population. Let us consider two examples of point estimation made on this basis.

▌▌▌ Practice Problem 4

The Length of a Manuscript

Suppose a publisher wishes to estimate the number of words in a 935-page manuscript. Since nothing is known about the distribution of the words-per-page population, the central limit theorem can be invoked, and a random sample of 30 pages can be taken. The results are shown in Table 8.5.

The sample mean is 248.4 words per page; this sample mean becomes the point estimate of the population mean, μ. Hence, the length of the manuscript is estimated at 248.4(935) = 232,254 words. Indeed, if desired, the standard deviation of words per page for the entire manuscript can be estimated, too. Underneath the table, the sample

Table 8.5 Estimating the Length of a Manuscript by Means of Sampling

Number of Words on Sample Page X	X^2	Number of Words on Sample Page X	X^2
240	57,600	250	62,500
226	51,076	296	87,616
292	85,264	203	41,209
227	51,529	269	72,361
244	59,536	204	41,616
221	48,841	277	76,729
232	53,824	244	59,536
257	66,049	254	64,516
254	64,516	293	85,849
232	53,824	230	52,900
255	65,025	232	53,824
262	68,644	200	40,000
274	75,076	218	47,524
246	60,516	287	82,369
246	60,516	287	82,369
		$\Sigma X = 7,452$	$\Sigma X^2 = 1,872,754$

$$\overline{X} = \frac{\Sigma X}{n} = \frac{7,452}{30} = 248.4$$

$$s^2 = \frac{\Sigma X^2 - n\overline{X}^2}{n - 1} = \frac{1,872,754 - 30(61,702.56)}{29} = 747.49$$

variance, s^2, has been calculated (by means of the shortcut method in Box 4.M); it becomes the point estimate of the population variance, σ^2. Thus, the population standard deviation, σ, can be estimated at $\sqrt{747.49} = 27.34$ words. |||

||| Practice Problem 5

Profit Survey of Fast-Food Outlets

Suppose a franchiser of fast-food outlets wishes to gather information about the annual profits of current franchise holders. Past experience suggests that profits are normally distributed; a random sample of $n = 12$ outlets yields these profit data for the previous year:

$61,242	$10,404	$91,053	$48,912
$28,785	$76,326	$67,422	$40,056
$92,203	$96,803	$97,357	$92,233

a. What is a point estimate for the mean annual profit of all the franchised outlets? The sample mean equals

$$\overline{X} = \frac{\Sigma X}{n} = \frac{\$802,796}{12} = \$66,899.67.$$

This figure is the point estimate of the *population* mean as well because the sample mean, $\overline{X}$, is an unbiased, efficient, and consistent estimator of μ.

b. What is a point estimate for the standard deviation of the annual profit of all the franchised outlets?

$$s = \sqrt{\frac{\Sigma X^2 - n\overline{X}^2}{n-1}} = \sqrt{\frac{63,203,823,000 - 53,706,784,800}{11}}$$

$$= \sqrt{863,367,109} = \$29,383.10.$$

The sample's standard deviation serves as the estimate of the population's.

c. What is the proportion of profits, among all the franchisees, in excess of \$50,000 (below \$20,000)? Because these proportions in the sample are 8/12 and 1/12, respectively, and because the sample proportion, P, is an unbiased, efficient, and consistent estimator of π, these sample proportions are the estimates of the population proportions as well. **|||**

THE NATURE OF CONFIDENCE INTERVALS

Even though an unbiased point estimator will *on the average* take on a value equal to the parameter being estimated, any one estimate is unlikely to be on target. Yet, typically, only one estimate is made because only one sample is taken; hence, the typical point estimate is almost certain to lie below or above the true value of the parameter.

Take, for example, the sample mean, $\overline{X}$, as an estimator of the population mean, μ. As Table 8.2 so vividly demonstrates, only if sample number (3) or sample number (10) happens to be selected, will $\overline{X} = 40$, which is also the value of μ (see Table 8.1). In the other eight cases, $\overline{X} \neq \mu$, and the fact that $\mu_{\overline{X}} = \mu$ is then of little comfort, indeed. We must recognize that any parameter being estimated equals the statistic serving as a point estimate minus (negative or positive) sampling error, e. For instance,

$$\mu = \overline{X} - e.$$

In the previous example, if sample number 1 is selected, $\overline{X} = 35$, but a negative sampling error of $e = -5$ occurs; hence, $\mu = 35 - (-5) = 40$. If sample number 2 is selected, $\overline{X} = 45$, but a positive sampling error of $e = +5$ occurs; hence, $\mu = 45 - (+5) = 40$. And so it goes.

The inevitable uncertainty attached to any point estimate can be made explicit by presenting an interval estimate and stating, say, that $a \leq \mu \leq b$. Statisticians construct an interval within which the unknown population parameter presumably lies with the help of a point estimate (that serves as the interval center) and the standard error of the estimator (which, as we noted, is the standard deviation of the sampling distribution of the estimator). The limits of the interval, however, need not lie precisely 1 standard error below or above the point estimate that serves as the interval center. The limits can lie a fraction or a multiple of a standard error to either side of the point estimate. If we define a coefficient A as any (fractional or whole) positive value (such as .1, .5, 1, 2, and so on), we can make the following statement.

Upper and lower limits of interval estimate
= point estimate ± [A · (standard error of estimator)]

Consider, for example, estimating the population mean by using the sample mean. If sample number 4 in Table 8.2 had been selected, one could present $\overline{X} = 39.667$ as a point estimate of μ. Given $\sigma_{\overline{X}} = 3.882$ (Table 8.4), one could, however, also present various interval estimates for μ, each one being constructed such that μ falls within the range of $\overline{X} \pm (A \cdot \sigma_{\overline{X}})$, as follows:

1. If A is taken as .5, μ is estimated as lying between the limits of $39.667 \pm [.5(3.882)]$; hence, $37.726 \le \mu \le 41.608$.
2. If A is taken as 1.0, μ is estimated as lying between the limits of $39.667 \pm [1(3.882)]$; hence, $35.785 \le \mu \le 43.549$.
3. If A is taken as 2.0, μ is estimated as lying between the limits of $39.667 \pm [2(3.882)]$; hence, $31.903 \le \mu \le 47.431$.

And so it goes for other values of A. The fact that larger values of A produce wider intervals (less precise estimates) is no accident because the width of the interval has important implications for the degree of confidence with which one can state that the population parameter that is being estimated lies within the two limits, also called **confidence limits,** that define the interval. Let us see why.

Recall that a sampling distribution, such as that of the sample mean, is a normal distribution (a) when the underlying population values are normally distributed and (b) even when the population is not normal, provided only that the sample size is sufficiently large absolutely ($n \ge 30$), but fairly small relative to population size ($n < .05N$), to validate the central limit theorem. By choosing a proper sample size, a statistician, therefore, can *assure* that the sampling distribution of the estimator being used is approximately normal. As a result, the coefficient A in the previous expression can be treated as a z value, such that

Upper and lower limits of interval estimate
= point estimate ± [z · (standard error of estimator)].

Appendix Table H can then be employed to figure the areas under the normal curve lying within the ranges so calculated. Consider again the example employed in Practice Problems 1–3 earlier: a normally distributed sampling distribution of $\overline{X}$ with $\mu_{\overline{X}} = 40$ and $\sigma_{\overline{X}} = 4.655$. In order to make interval estimates of the population mean, we can view μ as falling in the range of $\overline{X} \pm (z \cdot \sigma_{\overline{X}})$. If we assume a random sample produces an $\overline{X}$ equal to 35, what can we conclude?

1. If $z = .5$, we can figure that 38.3 percent of all $\overline{X}$ values fall within the range of $\mu_{\overline{X}} \pm .5\sigma_{\overline{X}}$ because Appendix Table H tells us that the area under the standard normal curve between the center and $z = .5$ equals .1915; hence, the area between $z = -.5$ and $z = +.5$ equals $2(.1915) = .383$. Because the center of the sampling distribution lies at $\mu_{\overline{X}} = \mu$, we can be 38.3 percent confident that our method of interval construction will produce an interval that will actually contain μ. In fact, we can calculate an interval that reaches from 32.6725 to 37.3275; this interval does not contain $\mu = \mu_{\overline{X}} = 40$.

2. If $z = 1.0$, we can similarly figure that 68.26 percent of all $\overline{X}$ values fall within the range of $\mu_{\overline{X}} \pm 1\sigma_{\overline{X}}$ [because $2(.3413) = .6826$]. We can, thus, be 68.26 percent confident that our method of interval construction will produce an interval that will contain μ. In fact, we can calculate an interval that reaches from 30.345 to 39.655.

3. If $z = 2.0$, it is likely that 95.44 percent of all $\overline{X}$ values fall within the range of $\mu_{\overline{X}} \pm 2\sigma_{\overline{X}}$ [because $2(.4772) = .9544$]. We can, thus, be 95.44 percent confident that our method of interval construction will produce an interval that will contain μ. In fact, we can calculate an interval that reaches from 25.69 to 44.31.

The foregoing discussion shows why the intervals in which estimates are stated are called **confidence intervals.** Given sample size, the level of confidence attached to an estimate varies directly with the value of z and, thus, with the interval width. Notice how a smaller z value means greater precision of an estimate (a narrower interval) but also implies a smaller degree of confidence in the estimate, while a larger z value means less precision (a wider interval) but implies a greater degree of confidence. The **confidence level,** or the percentage of interval estimates (obtained from repeated random samples, each of size n, taken from a given population) that can be expected to contain the actual value of the parameter being estimated, can clearly be set by choosing the value of z. Anyone who wishes to make an estimate with 68.26 percent confidence, for example, must choose the z value corresponding to $(.6826/2) = .3413$ in Appendix Table H (hence, $z = 1$), for this will assure that the resultant interval will cover $2(.3413) = .6826$ of the area under the normal curve. Appendix Table J, "Critical Normal Deviate Values," contains z values for the most commonly used confidence levels; others can be derived from Appendix Table H in the manner just shown.

What is meant by confidence level must, however, be considered with care. The three above statements, for example, do *not* imply probabilities of .383, .6826, or .9544 that the unknown population mean lies within the limits of the respective intervals. (The unknown parameter either does or does not lie within any given interval. Period.) The statements do mean the following:

If all possible samples of a given size are selected from a given population and if all possible values of a given estimator are then calculated and a confidence interval of a given width is then constructed around each of these values, the percentage of resulting intervals that will contain the true population value is referred to as *the confidence level* of the interval.

This meaning of *confidence level* is illustrated in Figure 8.4. It depicts the sampling distribution of $\overline{X}$ that was first postulated in Practice Problems 1–3 and was imagined to have been derived from a very large and normally distributed population. Even though this sampling distribution is known to us (the readers of this textbook), it

Figure 8.4 Confidence Intervals

An interval estimate of a population mean (μ) can be constructed around a sample mean ($\overline{X}$) as $\overline{X} \pm z\sigma_{\overline{X}}$. Depending on the value of z, one can place different degrees of confidence in the method of estimation. In this example, $z = 1.28$, which implies that 80 percent of the $\overline{X}$ values (heavy dots) that might be found by repeated sampling from the same population lie within the range of $\mu_{\overline{X}} \pm 1.28\,\sigma_{\overline{X}}$. Because $\mu = \mu_{\overline{X}}$, this procedure also implies that 80 percent of the intervals that might be constructed will contain μ. Any interval constructed by this method is, therefore, called an "80-percent confidence interval." By choosing smaller or larger z values, one can, similarly, construct narrower or wider intervals with smaller or higher confidence levels. What is illustrated here for population mean and sample mean holds equally for other parameters and other estimators.

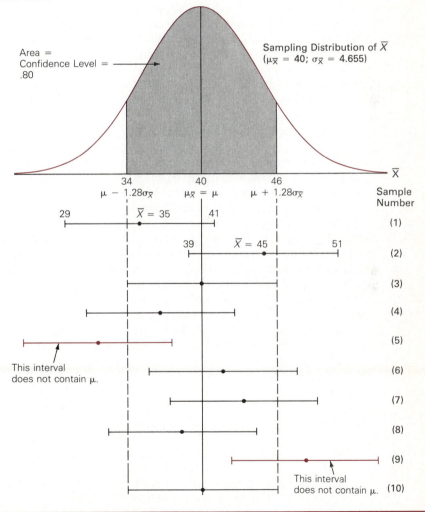

would not, of course, be known to the statistician about to estimate μ by taking a single random sample from a population. If this statistician wanted to estimate μ with, say, an 80-percent confidence level, that statistician would have to select a z value such that 80 percent of all possible $\overline{X}$ values fall within the range of $\pm z\sigma_{\overline{X}}$ from the unknown μ. As Appendix Table J indicates, that z value is 1.28. Now let us imagine for a moment that our statistician knew $\sigma_{\overline{x}} = 4.655$. Then the 80-percent confidence interval could be set as lying between the limits of $\mu \pm 1.28(4.655)$ or $\mu \pm 6$ (rounded). As we learned from our simplified example of Table 8.2, however, many different random samples of a given size can be taken from a given population. Our statistician might select any one of them. If from, perhaps, thousands of possible samples sample number 1 is selected, the statistician might calculate $\overline{X} = 35$. He or she would then use this point estimate of μ to make an interval estimate of μ as the range of 35 ± 6—that is, between 29 and 41, as shown in Figure 8.4. We can see (but the statistician could not) that the point estimate of 35 is wrong (μ actually equals 40; hence, a sampling error of -5 was made); we can also see (and, again, the statistician would not know) that the interval estimate of 29 to 41 does contain the true value of $\mu = 40$.

But what if the statistician had selected a different sample that we might call sample number (2)? Then $\overline{X}$ would probably have been different, perhaps equal to 45 (also, unbeknownst to the statistician, a wrong point estimate of μ), and the interval estimate would have ranged from 39 to 51 (again, unbeknownst to the statistician, containing the true value of $\mu = 40$). A few of the many other possible sample results are indicated by the unlabeled fat dots in Figure 8.4, along with the confidence intervals our statistician would calculate. Notice that all but two $\overline{X}$ values—samples (3) and (10) being the exceptions—would be incorrect point estimates of μ, yet fully 8 of the 10 interval estimates would, nevertheless, contain the true value of μ—the two exceptions here being the interval constructed after taking sample numbers (5) and (9). And therein lies the meaning of the term "80-percent confidence interval": Any one confidence interval is either dead right or dead wrong (it does or does not contain μ, but the statistician does not know it). Yet the statistician can have 80-percent confidence in *the method employed.* If repeated sampling occurred, 80 percent of the intervals would be right and 20 percent would be wrong, as this example illustrates.

Notice again: There is *not* an 80-percent probability that μ lies within the confidence limits of the first interval. Unbeknownst to the statistician, there is certainty that μ is contained in that interval (just as there is certainty that it is not bracketed by the fifth interval). All that the statistician knows is that of, say, 1000 intervals, each constructed with a value of $z = 1.28$, some 800 will contain the unknown parameter, but the statistician has no way of identifying these particular 800 intervals. A stated confidence level gives the probability *prior to actual sample selection* that the parameter being estimated will lie within the interval being constructed. (Once a specific sample has been taken and a specific interval has been constructed, the parameter is certain to lie or not to lie within that interval.)

MAKING INTERVAL ESTIMATES OF THE POPULATION MEAN AND PROPORTION: LARGE SAMPLES

Let us consider a series of cases, focusing first on those involving (a) large samples [when estimating means, $n \geq 30$; when estimating proportions, $n \cdot \pi$ and also $n(1 - \pi) \geq 5$] and (b) normally distributed sampling distributions (either because the population values are themselves normally distributed or because the central limit theorem applies). Confidence intervals for the population mean (μ) or proportion (π) can be constructed as shown in Box 8.C.

8.C Upper and Lower Limits of Confidence Intervals for the Population Mean, μ, and Proportion, π (using the normal distribution)

$$\mu = \overline{X} \pm (z\sigma_{\overline{x}}) \text{ and } \pi = P \pm (z\sigma_P)$$

where $\overline{X}$ is the sample mean and P is the sample proportion, while $\sigma_{\overline{x}}$ and σ_P are the standard deviations of their respective sampling distributions.

Assumptions: Sampling distribution of $\overline{X}$ or P is normal because (i) population distribution is normal *or* (ii) the central-limit theorem applies [for the mean, $n \geq 30$, $n < .05N$; for the proportion, $n \cdot \pi \geq 5$ and $n(1 - \pi) \geq 5$].

The actual sample taken will clearly yield a value for $\overline{X}$ or P, and the value of z (and hence, the desired confidence level) can be determined at will. Once values for $\overline{X}$ or P and z are determined, we are left with the necessity to determine $\sigma_{\overline{x}}$ or σ_P. If only one sample is taken (and that is the usual case), these values cannot be obtained in the fashion indicated in Table 8.4 because the respective sampling distributions are not available. Boxes 8.A and 8.B provide an alternative approach.

According to Box 8.A, $\sigma_{\overline{x}}$ can be calculated if σ, the population standard deviation, is known, and on occasion such is the case. (The statistician obviously does know the size of the sample taken; hence, $\sigma_{\overline{x}}$ can be calculated, given σ, in one of the two ways shown in Box 8.A, depending on whether the large- or small-population case applies.) This information provides us with all the ingredients necessary for solving Practice Problems 6 and 7 below.

Note: According to Box 8.B, σ_P can be calculated if π, the population proportion, is known, but if that were the case, there would be no need to estimate it. This approach, therefore, leads nowhere.

What if σ, like π, is not known, all else being the same? In that case, statisticians estimate these values by substituting, in the Box 8.A and 8.B formulas, the sample standard deviation, *s*, and the sample proportion, *P*, for σ and π, respectively. This substitution provides us with all the ingredients for solving Practice Problems 8 and 9.

Practice Problem 6

Estimating Mean Family Income ($n \geq 30$, $n < .05N$, σ known)

Consider a large city. From prior surveys, it seems reasonable to assume that the city population's family incomes have a standard deviation of $4,000. A 95-percent confidence interval for the city's mean family income is to be constructed with the help of a simple random sample (without replacement) of 100 families. The sample mean income turns out to be $19,763.

The sample size $n = 100$ clearly is less than $.05N$; thus, the large-population case (Box 8.A) applies, and

$$\sigma_{\overline{x}} = \frac{\sigma}{\sqrt{n}} = \frac{\$4,000}{\sqrt{100}} = \$400.$$

STANDARD ERROR

A 95-percent confidence level requires $z = 1.96$ (see Appendix Table J). Hence, the limits of the confidence interval are

$$\mu = \overline{X} \pm (z\sigma_{\overline{X}}) = \$19{,}763 \pm [1.96\,(\$400)] = \$19{,}763 \pm \$784.$$

Note: In order to simplify the wording, the expression, $x = a \pm b$, shall henceforth imply that x can take on any value in the interval from $(a - b)$ to $(a + b)$. Thus, the expression, $x = a \pm b$, shall be considered equivalent to $(a - b) \le x \le (a + b)$ and shall not be construed to imply that x can equal only $(a - b)$ or $(a + b)$. The expression, $\mu = \$19{,}763 \pm \$784$, therefore, implies that the 95-percent confidence interval goes from \$18,979 to \$20,547. Remember: This does not mean that there exists a 95-percent probability that the city population's mean income lies within this range. As Figure 8.4 showed, the true mean either does or does not lie within any given interval. But we can say that μ would on the average lie within 95 out of 100 such intervals, if we took repeated samples of $n = 100$ from the city population and repeatedly made interval estimates by this method. III

Practice Problem 7

Estimating Mean Miles per Gallon ($n \ge 30$, $n \ge .05N$, normal population, σ known)

Consider a rental car agency. It knows from experience that the standard deviation of the normally distributed miles-per-gallon population of its cars is 4 miles per gallon. A 99-percent confidence interval for the mean miles per gallon of the agency's fleet of 100 cars is to be constructed with the help of a simple random sample (without replacement) of 36 of its cars. The sample mean miles per gallon turns out to be 29.

The sample size $n = 36$ exceeds $.05N$; thus, the small-population case (Box 8.A) applies, and

$$\sigma_{\overline{X}} = \frac{\sigma}{\sqrt{n}}\sqrt{\frac{N - n}{N - 1}} = \frac{4\text{ mpg}}{\sqrt{36}}\sqrt{\frac{100 - 36}{100 - 1}} = \frac{4\text{ mpg}}{6}(.804) = .536\text{ mpg}.$$

A 99-percent confidence level requires $z = 2.57$. Hence,

$$\mu = \overline{X} \pm (z\sigma_{\overline{X}}) = 29\text{ mpg} \pm [2.57(.536\text{ mpg})] = 29\text{ mpg} \pm 1.378\text{ mpg}.$$

The 99-percent confidence interval goes from 27.6 miles per gallon to 30.4 miles per gallon. III

Practice Problem 8

Estimating Mean Weight per Animal ($n \ge 30$, $n < .05N$, σ unknown)

Consider a cattle feed lot. It is about to ship 800 head of cattle by train. A 90-percent confidence interval for the entire shipment's mean weight per animal is to be constructed with the help of a simple random sample (without replacement) of 30 animals. The sample mean weight per animal turns out to be 1,301 pounds, with a sample standard deviation of 290 pounds.

The sample size of $n = 30$ is less than $.05N$; thus, the large-population case (Box 8.A) applies, and (in the absence of σ) one can estimate (using s):

$$\sigma_{\overline{X}} \cong \frac{s}{\sqrt{n}} = \frac{290 \text{ lbs.}}{\sqrt{30}} = 52.95 \text{ lbs.}$$

A 90-percent confidence level requires $z = 1.64$. Hence,

$$\mu = \overline{X} \pm (z\sigma_{\overline{X}}) = 1{,}301 \text{ lbs.} \pm [1.64(52.95 \text{ lbs.})] = 1{,}301 \text{ lbs.} \pm 86.84 \text{ lbs.}$$

The 90-percent confidence interval goes from 1,214.14 pounds to 1,387.84 pounds. |||

||| Practice Problem 9

Estimating the Proportion of Buses Arriving on Schedule ($n \cdot \pi$ and $n(1 - \pi) \geq 5$, $n < .05N$)

Consider a bus company. Each month, thousands of its buses arrive at a certain depot. A 99.9-percent confidence interval for the proportion of all buses arriving on schedule is to be constructed with the help of a simple random sample (without replacement) of 49 buses. The sample proportion of correct arrivals is .64; the sample standard deviation is .2.

The sample size of $n = 49$ is less than $.05N$; thus, the large-population case (Box 8.B) applies, and (in the absence of π) one can estimate (using P):

$$\sigma_P \cong \sqrt{\frac{P(1 - P)}{n}} = \sqrt{\frac{.64(.36)}{49}} = .0686.$$

A 99.9-percent confidence level requires $z = 3.27$. Hence,

$$\pi = P \pm (z\sigma_P) = .64 \pm [3.27(.0686)] = .64 \pm .22.$$

The 99.9-percent confidence interval goes from .42 to .86.

Analytical Example 8.1, "Election Advice," is another application of this formula. |||

MAKING INTERVAL ESTIMATES OF THE POPULATION MEAN AND PROPORTION: SMALL SAMPLES

Now let us consider a series of cases involving small samples (when estimating means, $n < 30$; when estimating proportions, $n \cdot \pi$ or $n(1 - \pi) < 5$). Such cases are quite common for a variety of reasons, including: the high cost of sampling, the desire to get faster results, and the rarity of some phenomena (only two skyscrapers are produced per year, only one nuclear reactor accident occurs in a decade, and so on). Under such circumstances, it is not safe to invoke the central limit theorem. Yet it is still possible

for the sampling distribution to be normal, provided only that the underlying population values are normally distributed. If such is known to be the case, and if the population standard deviation is also known, a confidence interval for a population mean
can be estimated with the help of z values as in Practice Problems 6 and 7. One example is provided by Practice Problem 10. Practice Problem 11 is an example of the
opposite case in which it is known that the underlying population values are *not* normally distributed but in which the population standard deviation is given. In this case,
a confidence interval can still be estimated with the help of Chebyshev's theorem (see
Biography 4.1).

‖‖‖ Practice Problem 10

Estimating Mean Managerial Salaries ($n < 30$, $n \geq .05N$, population normal, σ known)

Consider a normally distributed population of 5 managerial salaries, but assume that
only the fact of this normality along with $\sigma = \$11{,}400$ are known. An 80-percent con-
fidence interval for the mean salary of all 5 managers is to be constructed with the
help of a simple random sample (without replacement) of 3 managers. The sample
mean turns out to be $\$48{,}333$.

The sample size of $n = 3$ exceeds $.05N$; thus, the small-population case (Box 8.A)
applies, and

$$\sigma_{\overline{X}} = \frac{\sigma}{\sqrt{n}} \sqrt{\frac{N - n}{N - 1}} = \frac{\$11{,}400}{\sqrt{3}} \sqrt{\frac{5 - 3}{5 - 1}} = \$4{,}654.$$

An 80-percent confidence level requires $z = 1.28$. Hence,

$$\mu = \overline{X} \pm (z\sigma_{\overline{X}}) = \$48{,}333 \pm [1.28(\$4{,}654)] = \$48{,}333 \pm \$5{,}957.$$

The 80-percent confidence interval goes from $\$42{,}376$ to $\$54{,}290$. ‖‖‖

‖‖‖ Practice Problem 11

Estimating Mean Waiting Times ($n < 30$, $n < .05N$, population not normal, σ known)

Consider a hotel manager who wants to know the mean time guests have to wait for
room service. The population of waiting times is known to be not normally distributed,
with a standard deviation of 11 minutes. A 95-percent confidence interval of the mean
waiting time for 1,000 orders is to be constructed with the help of a simple random
sample (without replacement) of 20 orders. The sample mean turns out to be 32 minutes.

The sample size of $n = 20$ is less than $.05N$; thus, the large-population case (Box
8.A) applies, and

$$\sigma_{\overline{X}} = \frac{\sigma}{\sqrt{n}} = \frac{11 \text{ minutes}}{\sqrt{20}} = 2.46 \text{ minutes.}$$

According to Chebyshev's theorem, at least $[1 - (1/k^2)]$ 100 percent of all observations fall within k standard deviations of the mean. For a 95-percent confidence level, we need

$$95 = \left(1 - \frac{1}{k^2}\right)100, \text{ or } .95 = 1 - \frac{1}{k^2}.$$

This confidence level requires $k = 4.472$. Hence,

$$\mu = \overline{X} \pm (k\sigma_{\overline{X}}) = 32 \text{ minutes} \pm [4.472(2.46 \text{ minutes})]$$
$$= 32 \text{ minutes} \pm 11 \text{ minutes}.$$

The 95-percent confidence interval goes from 21 minutes to 43 minutes. |||

Note: When $n < .05N$, the wide interval inevitably derived for any reasonable confidence level when using Chebyshev's theorem can easily be escaped by raising the sample size to 30 or above. This increase in sample size validates the central limit theorem and allows the kind of estimation procedure shown in Practice Problem 6 above. Thus, if the sample result in Practice Problem 11 had been obtained from 30 orders, we could estimate

$$\mu = \overline{X} \pm (z\sigma_{\overline{X}}) = 32 \text{ minutes} \pm [1.96(2.46 \text{ minutes})]$$
$$= 32 \text{ minutes} \pm 4.82 \text{ minutes}.$$

This escape route (of raising n) is also recommended when the population is not normally distributed and when σ is not known. Then the procedure described in Practice Problem 8 above can be used.

Student's *t* Distribution

What remains is the most important small-sample case of all: the presence of a (presumably) normally distributed population but the absence of any other information about this population. In this small-sample case (unlike in situations illustrated by Practice Problems 8–9), the standard errors of the mean and proportion, $\sigma_{\overline{X}}$ and σ_P, are only poorly estimated with the help of the sample standard deviations, s and P. As was first shown by William S. Gosset, who wrote under the name "Student" (see Biography 8.1 at the end of the chapter), under these circumstances, better interval estimates can be derived by using a probability density function somewhat different from the normal curve.

Gosset described a sampling distribution for a random variable, t, derived from a normally distributed population and defined in analogy to the standard normal deviate, z:

$$t = \frac{\overline{X} - \mu}{\dfrac{s}{\sqrt{n}}}$$

$$z = \frac{X - \mu}{\sigma}$$

Like the standard normal curve, Gosset's probability density function, now called **Student's *t* distribution,** is (a) single-peaked above the random variable's mean, median, and mode of zero, (b) perfectly symmetrical about this central value, and (c) characterized by tails extending indefinitely in both directions from the center, approaching but never touching the horizontal axis. The only difference is that the random variable is *t* rather than *z;* as a result, the distribution's variance does not equal 1 (as is true of the *z* distribution), but equals

$$\frac{n - 1}{(n - 1) - 2}.$$

This variance of *t* implies that a different *t* distribution exists for each sample size, *n,* and also that the *t* distribution approaches the *z* distribution as sample size increases. The *t* distribution for $n = \infty$ has a variance of 1 and is indistinguishable from the standard normal curve. All these characteristics of the *t* distribution are illustrated in Figure 8.5.

As Figure 8.5 shows, members of the *t*-distribution family are labeled not on the basis of sample size, *n,* but on the basis of **degrees of freedom,** $n - 1$, which equal the number of values that can be "freely chosen" when calculating a statistic. Consider the number of independent deviations, $X - \overline{X}$, of sample observations from the sample mean that enter the computation of the sample standard deviation, *s.* Once we have "freely chosen" $n - 1$ of these deviations, the remaining one is determined for us because the sum of all deviations necessarily equals 0. In a sample of $n = 3$, for example, we have only $n - 1 = 2$ degrees of freedom. Take sample number (1) in Table 8.2. The individual sample observations (for sample units A, B, and C) are 39, 41, and 25 (see Table 8.1). The sample mean, therefore, equals $\overline{X} = 35$. Proceeding to compute *s,* once we have determined two deviations—say, $39 - 35 = +4$ and $41 - 35 = +6$—the last one has effectively been determined for us by the fact that all three deviations must sum to 0. The last one can be found logically as $0 - (+4) - (+6) = -10$; indeed, this value is correct ($25 - 35 = -10$). For an earlier discussion of degrees of freedom (that does not mention the term) see the Chapter 5 section on "Factorials."

The *t* Distribution Table

Just as areas under the standard normal curve have been tabulated to make it easy for us to gauge various probabilities, so there exist tables for areas under various *t* distributions. Appendix Table K, "Student *t* Distributions," is a case in point. By tradition, the table shows the total area under a specified curve (defined by a given number of degrees of freedom) that lies *to the right of* a specified value of *t.* As is pictured in Appendix Table K, this upper-tail area is called α (the lowercase Greek letter *alpha*), and this *t* value is designated as t_α. Frequently, the applicable degrees of freedom (d.f.) are added to the α subscript, either in parentheses or following a comma: $t_{\alpha(\text{d.f.})}$ or $t_{\alpha,\text{d.f.}}$ As a look at Appendix Table K will confirm, for 10 degrees of freedom and an upper-tail area of $\alpha = .1$, $t_{.1(10)} = 1.372$. In other words, .1 of the area under the *t* curve appropriate for a sample of $n = 11$ is associated with $t > 1.372$. Because of the curve's

Figure 8.5 Student *t* Distributions

Different bell-shaped *t* distributions exist for different degrees of freedom, *d.f.,* defined as sample size minus one, *n* − 1. Unless the degrees of freedom are infinite (which makes the *t* distribution equal to the standard normal curve), the *t* distribution is flatter than the standard normal curve, and more of its area is found in the tails. Strictly speaking, each *t* distribution is a sampling distribution of the *t* statistic derived by taking a small sample from a normally distributed population. Nevertheless, in practice, statisticians use the *t* distribution to make small-sample inferences about all types of populations, provided only that these populations are not highly skewed.

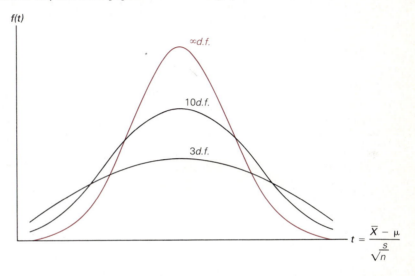

symmetry, .1 of the area under this curve is also associated with *t* < − 1.372. Consequently, .8 of the area under this curve is associated with *t* values between − 1.372 and +1.372; the probability for such *t* values equals .8. To construct an interval estimate with a confidence level of .8, or 80 percent, we must use *t* = 1.372.

By comparing this discussion to our earlier discussion, confidence intervals can be constructed with the help of Appendix Table K's *t* values, which are shown in Box 8.D below.

8.D Upper and Lower Limits of Confidence Interval for the Population Mean (using the *t* distribution)

$$\mu = \overline{X} \pm \left(t \, \frac{s}{\sqrt{n}} \right)$$

where $\overline{X}$ is the sample mean, *s* is the sample standard deviation, and *n* is sample size.

Assumption: *n* < 30, but population distribution is normal.

||| Practice Problem 12

Estimating the Mean Time Lost from Machinery Breakdowns ($n < 30$; $n < .05N$; population normal; σ unknown)

Consider a manufacturing plant. On many occasions during the course of a year, labor hours are lost because machinery breaks down. The manager (who knows from previous experience that the time-lost population is normally distributed) decides to keep track of the time lost for a week and calculates the mean time lost during 10 breakdowns as 820 hours, with a standard deviation of 510 hours. A 98-percent confidence interval is to be constructed for the mean time lost during all of the year's breakdowns.

Given a normal population, but an unknown σ and a small sample, the t distribution must be used. For sample size $n = 10$, there are 9 degrees of freedom, while the 98-percent confidence level corresponds to an upper-tail area of $\alpha = .01$. It is helpful to designate the corresponding t value with respect to α as well as with respect to the degrees of freedom—in this case as $t_{.01(9)}$. From Appendix Table K, we find $t_{.01(9)} = 2.821$. According to Box 8.D,

$$\mu = \overline{X} \pm \left(t \frac{s}{\sqrt{n}} \right) = 820 \text{ hours} \pm \left(2.821 \frac{510 \text{ hours}}{\sqrt{10}} \right)$$

$$= 820 \text{ hours} \pm 455 \text{ hours}.$$

The 98-percent confidence interval is disconcertingly wide; it goes from 365 hours to 1,275 hours. |||

MAKING INTERVAL ESTIMATES FOR THE DIFFERENCE BETWEEN TWO POPULATION MEANS

Oftentimes decision makers seek an estimate not of a single population mean, but of the difference between the means of two populations. Two methods of production may be available; on the average, which one requires the greater amount of labor time? Two drugs may be effective in reducing high blood pressure; which one, on the average, is more effective? Two types of raw material may be usable; which one, on the average, has the greater tensile strength? Which of two work rules is more conducive to raising average labor productivity? Which of two types of tires has the longer average life? To answer questions such as these, statisticians take samples from two populations, A and B, with means of μ_A and μ_B and estimate the difference, $\mu_A - \mu_B$. Once again, the proper estimating procedure depends on whether the samples are large or small, but the existence of two populations requires a further choice: whether to take independent or matched-pairs samples. We will consider all of these possibilities.

Large and Independent Samples

We first imagine taking two completely independent samples of sizes n_A and n_B from the two populations. These samples need not be of equal size; they will yield sample means $\overline{X}_A$ and $\overline{X}_B$; their difference, $d = \overline{X}_A - \overline{X}_B$, is an unbiased point estimator of $\mu_A - \mu_B$.

We could take such samples again and again, each time getting a different d. If we take large and independent samples of $n_A \geq 30$ and $n_B \geq 30$ (as we now assume), the sampling distribution of d will be normally distributed when $n < .05 N$ (because the central limit theorem then holds) and even when $n \geq .05 N$ provided the underlying populations are normally distributed. This sampling distribution will have a mean of $\mu_A - \mu_B$ and a standard deviation of

$$\sigma_d = \sqrt{\frac{\sigma_A^2}{n_A} + \frac{\sigma_B^2}{n_B}}$$

where σ_A^2 and σ_B^2 are the two population variances. Given large samples, we can also estimate the standard error of the difference between the means, σ_d, using the sample variances:

$$\sigma_d \cong s_d = \sqrt{\frac{s_A^2}{n_A} + \frac{s_B^2}{n_B}}$$

We can now adapt the earlier expression of a confidence interval for the population mean (Box 8.C) to establish a confidence interval for the difference between population means, $\mu_A - \mu_B$, as lying between the limits of $d \pm z\sigma_d$. These limits can also be written as shown in Box 8.E.

8.E. Upper and Lower Limits of Confidence Interval for the Difference Between Two Population Means (large and independent samples from large or normal populations)

$$\mu_A - \mu_B \cong (\bar{X}_A - \bar{X}_B) \pm \left[z \sqrt{\frac{s_A^2}{n_A} + \frac{s_B^2}{n_B}} \right]$$

where $\bar{X}$ is the sample mean, s is the sample standard deviation, and n is sample size.

Assumptions: $n_A \geq 30$ and $n_B \geq 30$ and sampling distribution of $\bar{X}_A - \bar{X}_B$ is normal because $n_A < .05N_A$ and $n_B < .05N_B$ *or* population distributions are normal.

Note: As will be shown, for *small* and independent samples, the same formula applies, except that t replaces z. In that case, $n_A < 30$ and $n_B < 30$, but the two populations are assumed to be normally distributed and to have equal variances.

Practice Problem 13

Comparing the Mean Lives of Two Types of Tires

Consider a tire manufacturer who wishes to estimate the difference between the mean lives of two types of tires, type A and type B, as a prelude to a major advertising campaign. A sample of 100 tires is taken from each production process, making $n_A = n_B = 100$. (As noted earlier, this equality of the two sample sizes is not a necessity.) The sample mean lifetimes equal $\bar{X}_A = 30,100$ miles and $\bar{X}_B = 25,200$ miles; the sample variances are $s_A^2 = 1,500,000$ miles squared and $s_B^2 = 2,400,000$ miles squared.

Clearly, the point estimate of the difference is $30,100 - 25,200 = 4,900$ miles, but the manufacturer desires a 99-percent confidence interval of the difference (requiring, as Appendix Table J indicates, a z value of 2.57). Using the Box 8.E formula,

$$\mu_A - \mu_B \cong 4,900 \pm \left[2.57 \sqrt{\frac{1,500,000}{100} + \frac{2,400,000}{100}} \right]$$

$$= 4,900 \pm (2.57 \sqrt{39,000}) = 4,900 \pm [2.57(197.48)]$$

$$= 4,900 \pm 507.52.$$

The 99-percent confidence interval goes from 4,392.48 miles to 5,407.52 miles. The tire from population A is clearly superior. |||

||| Practice Problem 14

The Mean Lives of Tires Revisited

What if $\overline{X}_B$ had equaled 29,800 miles, all else being the same? Then the point estimate would have been only 300 miles (suggesting that tire A is slightly better), and the 99-percent confidence interval would have been $\mu_A - \mu_B = 300 \pm 507.52$, ranging from -207.52 miles to 807.52 miles. The lower limit would indicate a possibility that tire A might have a mean lifetime of 207.52 miles *less* than tire B (making tire B better), while the upper limit would suggest a possible tire A mean lifetime of 807.52 miles *more* than tire B (making tire A better). In short, the test would be inconclusive. |||

Large Matched-Pairs Sample

Now let us turn to an alternative approach to making interval estimates for the difference between two means. Consider taking a sample after each elementary unit in population A has been matched with a "twin" from population B so that any sample observation about a unit in population A automatically yields an associated observation about a unit in population B. This procedure is referred to as taking a **matched-pairs sample.** Such matching is based on other factors besides the one under study that might also influence the characteristic being measured. For example, when testing the effectiveness of a new drug compared to a traditional one, each patient in an experimental group might be matched with a partner in a control group of the same age, weight, height, sex, occupation, medical history, lifestyle, and so on. (For an earlier discussion of this procedure, see the Chapter 2 section on "The Randomized Block Design.") The individual differences in response to the experimental stimulus between each pair are then used to estimate population differences.

The procedure can be summarized as follows: Let the sample observation for the ith pair equal X_{Ai} and X_{Bi}, depending on whether it refers to the partner from population A or B. Then the matched-pair difference equals $D_i = X_{Ai} - X_{Bi}$. From all the matched-pair differences involving n pairs, a mean and standard deviation can be calculated in the usual fashion as:

$$\overline{D} = \frac{\Sigma D_i}{n}$$

and

$$s_D = \sqrt{\frac{\Sigma D_i^2 - n\overline{D}^2}{n - 1}}$$

Now we can treat $\overline{D}$ and s_D equivalently to $\overline{X}$ and s in our earlier examples and, given our large sample ($n \geq 30$) and large population ($n < .05\ N$), we can assume that $\overline{D}$ is normally distributed with a mean of $\mu_A - \mu_B$ and a standard deviation of $s_D/\sqrt{n}$. Accordingly, we can establish the desired confidence interval in the fashion indicated in Box 8.F.

8.F Upper and Lower Limits of Confidence Interval for the Difference Between Two Population Means (large and matched-pairs sample from large or normal populations)

$$\mu_A - \mu_B = \overline{D} \pm \left(z\, \frac{s_D}{\sqrt{n}} \right)$$

where $\overline{D}$ is the mean of sample matched-pairs differences, s_D is the associated standard deviation, and n is the number of matched sample pairs.

Assumptions: $n \geq 30$ and sampling distribution of $\overline{D}$ is normal because $n < .05N$ or the populations are normally distributed.
Note: As will be shown, for a *small* and matched-pairs sample, the same formula applies, except that t replaces z. In that case, $n < 30$, but the populations are assumed to be normally distributed and to have equal variances.

Practice Problem 15

Comparing the Effectiveness of Two Drugs

Consider a drug manufacturer who wishes to compare the performance of a new drug for treating high blood pressure with that of one traditionally used, perhaps as a prelude to a marketing campaign. Sixty people are combined into 30 pairs of "near-twins." In each pair, one person receives the new drug (A), and the other person receives the old drug (B). The individual percentage reductions in blood pressure are recorded, as shown in Table 8.6.

The matched-pair difference, D_i, is calculated for each pair, along with the mean and standard deviation of the 30-pair sample. For a 95-percent confidence interval, we use a z value of 1.96; hence, according to Box 8.F:

$$\mu_A - \mu_B = 4.933 \pm \left[1.96\left(\frac{4.948}{\sqrt{30}} \right) \right] = 4.933 \pm 1.771.$$

Thus, the 95-percent confidence interval goes from 3.162 percent to 6.704 percent. We can say with 95 percent confidence that drug A reduces blood pressure between 3.2 and 6.7 percent more than drug B.

Table 8.6   Calculation of Matched-Pairs Differences of Effects on Patients Using New Drug (A) and Old Drug (B)

Pair i	Percentage Reduction in Blood Pressure, While Using Drug A, X_{Ai}	Drug B, X_{Bi}	Matched-Pair Difference, $D_i = X_{Ai} - X_{Bi}$	Squared Matched-Pair Difference, D_i^2
1	22	18	4	16
2	21	22	−1	1
3	26	19	7	49
4	23	22	1	1
5	21	22	−1	1
6	23	23	0	0
7	24	19	5	25
8	30	18	12	144
9	25	23	2	4
10	24	22	2	4
11	27	23	4	16
12	23	19	4	16
13	27	22	5	25
14	26	17	9	81
15	23	25	−2	4
16	25	17	8	64
17	21	19	2	4
18	23	23	0	0
19	25	15	10	100
20	28	25	3	9
21	28	16	12	144
22	30	18	12	144
23	29	16	13	169
24	28	20	8	64
25	21	20	1	1
26	30	18	12	144
27	24	25	−1	1
28	30	20	10	100
29	25	15	10	100
30	22	25	−3	9
			148	1,440

$$\overline{D} = \frac{\Sigma D_i}{n} = \frac{148}{30} = 4.933$$

$$s_D = \sqrt{\frac{\Sigma D_i^2 - n\overline{D}^2}{n-1}} = \sqrt{\frac{1{,}440 - 30(4.933)^2}{29}} = \sqrt{24.482} = 4.948 \quad \text{|||}$$

Small and Independent Samples

Let us now consider taking independent samples from two (probably normal) populations when both sample sizes are small ($n_A < 30$ and $n_B < 30$). As we learned earlier, in the absence of any precise knowledge about population variances, the Student t distribution must be used. Essentially, the same procedure can be used as in the case of large samples, except that t_α replaces z. (Strictly speaking, use of the t distribution in this case requires an equality of the two unknown population variances, and these are often estimated by a weighted average of the two sample variances, but we leave this refinement to the next chapter.)

‖‖ Practice Problem 16

Comparing the Effectiveness of Two Pesticides

Consider an agricultural experiment in which the effectiveness of a new pesticide compared to an old one is tested by measuring cotton production on 30 sample plots of land. Of the 30 plots, 15 use the new pesticide (A); the mean output is $\overline{X}_A = 1,575$ pounds per acre, with a standard deviation of $s_A = 250$ pounds. Another 15 plots use an old pesticide (B); the mean output is $\overline{X}_B = 1,450$ pounds per acre, with a standard deviation of $s_B = 300$ pounds.

Clearly, the point estimate of the difference between the population means is 125 pounds per acre, but suppose we desire to establish a 99-percent confidence interval for the difference between the population means. To find the appropriate t value, we consult Appendix Table K in the column for an upper-tail $\alpha = .005$ and in the row for 28 degrees of freedom, finding $t_{.005(28)} = 2.763$. [We use $(n_A - 1) + (n_B - 1)$ for the degrees of freedom.] Accordingly our 99-percent confidence interval equals

$$\mu_A - \mu_B = \overline{X}_A - \overline{X}_B \pm \left[t \sqrt{\frac{s_A^2}{n_A} + \frac{s_B^2}{n_B}} \right]$$

$$= 125 \pm \left[2.763 \sqrt{\frac{250^2}{15} + \frac{300^2}{15}} \right] = 125 \pm [2.763(100.830)]$$

$$= 125 \pm 278.59$$

The interval ranges from -153.59 pounds per acre to 403.59 pounds per acre, leaving open the possibility that the new pesticide is worse *or* better—hardly conclusive evidence in its favor. ‖‖‖

Small Matched-Pairs Sample

In the case of small matched-pairs sampling, the formula in Box 8.F is applied, with t_α again replacing z (and $n - 1$ degrees of freedom being employed).

Practice Problem 17

Comparing the Performance of Two Car Models

An auto manufacturer, annoyed by a rival's advertising claims, wishes to estimate the difference between the mean miles per gallon of two car models, A and B. A 98-percent confidence interval for the difference between the mean miles per gallon is desired. Ten pairs of drivers, matched according to driving skill, are observed; the mean and standard deviation of the differences between A's miles per gallon and B's miles per gallon are found to be $\overline{D} = 5$ miles per gallon, and $s_D = 2$ miles per gallon.

Adapting formula 8.F by replacing z with $t_{.01(9)} = 2.821$ (upper-tail area $\alpha = .01$; 9 degrees of freedom),

$$\mu_A - \mu_B = \overline{D} \pm \left(t \frac{s_D}{\sqrt{n}} \right) = 5 \text{ mpg} \pm \left[2.821 \left(\frac{2}{\sqrt{10}} \right) \text{mpg} \right] = 5 \text{ mpg} \pm 1.784 \text{ mpg}.$$

The 98-percent confidence interval reaches from 3.216 to 6.784 miles per gallon, which suggests that model A is, indeed, superior as far as gasoline mileage is concerned.

Note: Analytical Example 8.2, "Matched-Pairs Sampling in the Chemical Industry," provides another application of this procedure. |||

MAKING INTERVAL ESTIMATES FOR THE DIFFERENCE BETWEEN TWO POPULATION PROPORTIONS

Interval estimates not of a single population proportion, but of the difference between two such proportions are made by procedures similar to the ones described in the previous sections. We illustrate this fact with the case involving large and independent samples.

Imagine taking two large and independent samples of sizes $n_A = 250$ and $n_B = 350$ from two large populations of interest—say, cookies produced by process A and cookies produced by process B. Let the sample proportions of broken cookies equal $P_A = .08$ and $P_B = .10$; the difference, $d = P_A - P_B = -.02$, is the unbiased point estimate of the difference between the population proportions, $\pi_A - \pi_B$.

We can assume that the sampling distribution of d is normally distributed with a mean of $\pi_A - \pi_B$ and a standard deviation of

$$\sigma_d = \sqrt{\frac{\pi_A (1 - \pi_A)}{n_A} + \frac{\pi_B (1 - \pi_B)}{n_B}}$$

and that the latter can be approximated by

$$\sigma_d \cong s_d = \sqrt{\frac{P_A(1 - P_A)}{n_A} + \frac{P_B(1 - P_B)}{n_B}}$$

We can now adapt the earlier expression of a confidence interval for the population proportion (Box 8.C) to establish a confidence interval for the difference between population proportions, $\pi_A - \pi_B$ as lying between the limits of $d \pm z\,\sigma_d$. These limits can also be written as shown in Box 8.G.

> **8.G Upper and Lower Limits of Confidence Interval for the Difference Between Two Population Proportions (large and independent samples from large populations)**
>
> $$\pi_A - \pi_B \cong (P_A - P_B) \pm \left(z \sqrt{\frac{P_A\,(1 - P_A)}{n_A} + \frac{P_B\,(1 - P_B)}{n_B}} \right)$$
>
> Assumptions: $n_A \geq 30$ and $n_B \geq 30$ and sampling distribution of $P_A - P_B$ is normal because $n\,\pi \geq 5$ and $n\,(1 - \pi) \geq 5$ for both samples.

In our example and for a 95-percent confidence level, this interval comes to

$$\pi_A - \pi_B \cong -.02 \pm \left(1.96 \sqrt{\frac{.08(.92)}{250} + \frac{.10(.90)}{350}} \right) = -.02 \pm [1.96(.0235)]$$

$$= -.02 \pm .046.$$

The 95-percent confidence interval ranges from $-.066$ to $+.026$, suggesting that production process A may be better *or* worse than process B (it may yield a smaller *or* larger proportion of broken cookies). Caution is advised: a true difference of zero in the proportion of broken cookies lies within the interval.

THE OPTIMAL SAMPLE SIZE

One question inevitably arises whenever sampling is employed to estimate population parameters: how can one minimize the total cost of sampling? Figure 8.6 brings together the major elements that must be considered. We imagine a population of size $N = 1,000$ for which the cost of making each observation equals $40. Thus, the total cost of collecting data equals $40 for a sample of 1, $400 for a sample of 10, and so on, finally amounting to $40,000 for a census. For alternative sample sizes, these costs of collection are shown by line C_C. To keep down these costs, one clearly wants to keep sample size n as small as possible.

Another consideration, however, militates against this desire. As a look at the formulas in Boxes 8.C through 8.G can confirm, given a specified confidence level (and, hence, a specified value of z or t), any decrease in sample size, n, raises the standard error of the estimator (such as $\sigma_{\bar{x}}$ or σ_P in Box 8.C). Take, for example, the case where $\sigma_{\bar{x}} = (\sigma/\sqrt{n})$. Given a population standard deviation of $\sigma = 10$ and a sample of 25, $\sigma_{\bar{x}} = 2$; but if n is reduced to 16, $\sigma_{\bar{x}} = 2.5$. Unfortunately, any such increase in the standard error widens the confidence interval (that is, raises the sampling error), and any sampling error is costly, too. Decisions made on the basis of erroneous figures may, for example, reduce the profits of a firm. Line C_E in Figure 8.6 represents a hypothetical estimate of the costs of sampling errors (which are zero for a census but get increasingly larger the smaller is the sample size).

Figure 8.6   The Costs of Sampling

The optimal sample size (here $n = 460$) minimizes the total cost of sampling, C_c, and maximizes the savings (here $11,000) that can be achieved by sampling rather than by taking a census. The total cost of sampling equals the sum of the costs of collection, C_T, and of errors, C_E. Note how, at the optimal sample size, $C_C = 18,400$ (distance ac) plus $C_E = 10,600$ (distance ab) equals $C_T = 29,000$ (distance ad).

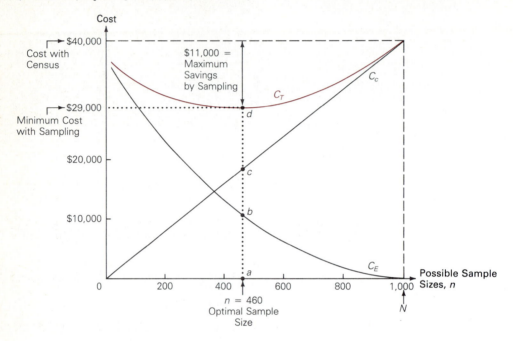

We can now add these two types of sampling costs for each possible sample size and determine the total cost of sampling, C_T. Note how this total cost is minimized at point *d*, implying an optimal sample size of $n = 460$.

Having clarified the principle involved, let us now consider how statisticians in practice go about determining the best sample size. We will first focus on situations when a mean is to be estimated, then on those involving the estimation of a proportion.

Determining the Best Sample Size When Estimating the Population Mean

Because it is impractical to draw the types of curves shown in Figure 8.6 (to determine, that is, C_C and C_E for all possible sample sizes), it is common practice to find the optimal sample size in a different way, which involves three simple steps.

Step 1. A **tolerable error level,** *e,* is set.

The tolerable error level equals the maximum amount a point estimate should, in the opinion of the statistician, extend above or below the parameter being estimated. If the parameter is μ and its point estimate is $\overline{X}$, the desired limits are

$$\overline{X} = \mu \pm e. \tag{1}$$

Take Practice Problem 6, which was concerned with estimating the mean family income in a large city. The statistician may want to have $e = \$500$, making sure that $\overline{X}$ lies within \$500 of the true value of μ.

It is one thing, however, to desire the kind of precision just decided upon; it is quite another actually to achieve it. In fact, as long as sampling is used, such perfection is out of the question. The statistician can, however, decide on an acceptable degree of reliability or confidence level, which brings us to the second step.

Step 2. An acceptable confidence level is set (and the appropriate z value is found).

Suppose the statistician decides, as was the case in Problem 6 earlier, to accept a 95-percent confidence level (which implies a z value of 1.96). Setting a confidence level of 95 percent means that the statistician will be satisfied if, in the long run, 95 percent of all $\overline{X}$ values (rather than an ideal 100 percent) lie within the interval of $\mu \pm e$ given in equation 1.

Step 3. The tolerable error is related to the actual standard error of the estimator.

As we have seen, the parameter μ can in fact be estimated with varying degrees of confidence, in accordance with the Box 8.C formula, $\mu = \overline{X} \pm z\sigma_{\overline{X}}$. This formula implies

$$\overline{X} = \mu \pm (z\sigma_{\overline{X}}). \tag{2}$$

Equations 1 and 2, in turn, imply that anyone who sets a tolerable error equal to e is in fact specifying that $z\sigma_{\overline{X}}$ be equal to e, as shown in equation 3.

$$e = z\sigma_{\overline{X}}. \tag{3}$$

When, as in Practice Problem 6, $\sigma_{\overline{X}} = (\sigma/\sqrt{n})$, equation 3 implies further that $e = z\dfrac{\sigma}{\sqrt{n}}$; hence, $\sqrt{n} = \dfrac{z \cdot \sigma}{e}$. Therefore, n can be determined by the formula in Box 8.H.

8.H Required Sample Size for Specified Tolerable Error and Confidence Level When Estimating the Population Mean

$$n = \left(\frac{z \cdot \sigma}{e}\right)^2$$

where n is sample size, z is the standard normal deviate appropriate for the desired confidence level, σ is the population standard deviation, and e is the tolerable error level.

The equation in Box 8.H is of great importance, for it shows that only one specific sample size will produce the tolerable error, e, at a specified confidence level (and, therefore, z value), given the population standard deviation σ (which the statistician is unable to influence).

For example, given a tolerable error of $e = \$500$, a standard normal deviate of $z = 1.96$, and a population standard deviation of \$4,000 (as in Practice Problem 6), only a sample size of 246 will make the actual sampling error equal to the tolerable error because

$$n = \left(\frac{1.96(4,000)}{500}\right)^2$$
$$= 245.86.$$

Note that the actual error in Practice Problem 6 was \$784; the difference between the actual error and the tolerable error can be explained precisely by the fact that the sample size was 100—too small to reduce the actual error to the tolerable one here specified. By the same token, if one were determined to reduce the tolerable error even further to within, say, \$100 of μ, while keeping the same confidence level, one would have to increase the sample size further to 6,147. Presumably, the user of the estimate could decide whether such increased precision was worth the added cost of data collection. Note: Analytical Example 8.3, "Reducing Sample-Collection Costs by Work Sampling," provides an interesting case study of how imaginative sampling procedures can reduce the cost of collecting sample data.

Determining the Best Sample Size When Estimating the Population Proportion

The determination of the best sample size for estimating π rather than μ follows an analogous route. First, the tolerable error, expressed as a decimal fraction, is specified as the maximum allowable deviation between the population proportion, π, and its estimator, the sample proportion, P. Thus, the desired limits are $P = \pi \pm e$. Second, an acceptable confidence level is set and, thus, a z value is determined. Finally, the tolerable error is related to the actual standard error of the estimator. Given the Box 8.C formula, $\pi = P \pm z\sigma_P$, which implies $P = \pi \pm z\sigma_P$, we note the requirement that $e = z\sigma_P$. When, as in Practice Problem 9,

$$\sigma_P = \sqrt{\frac{P(1 - P)}{n}},$$

this equality requires

$$e = z\sqrt{\frac{P(1 - P)}{n}},$$

or

$$\sqrt{n} = \frac{z\sqrt{P(1 - P)}}{e}.$$

Thus, n can be determined as shown in Box 8.I.

8.1 Required Sample Size for Specified Tolerable Error and Confidence Level When Estimating the Population Proportion

$$n = \frac{z^2 P(1 - P)}{e^2}$$

where n is sample size, z is the standard normal deviate appropriate for the desired confidence level, P is an estimate of the population proportion, and e is the tolerable error level.

Note: The value of P in this formula can be a preliminary estimate of the unknown π (since P is not known until after the sample is taken). Barring that, one simply sets P (and, therefore, $1 - P$) equal to .5 for purposes of establishing the sample size.

If someone, for example, had specified a tolerable error of .10 in Practice Problem 9, while keeping the confidence level at 99.9 percent (and, therefore, $z = 3.27$), while having no idea about the value of π, the sample size would have had to be set not at 49 buses (given in the problem) but at 268 buses, because

$$n = \frac{(3.27)^2(.5)(.5)}{(.10)^2} = 267.32,$$

which would be rounded to 268.

A FINAL NOTE ON BAYESIAN ESTIMATION*

The estimation techniques discussed so far in this chapter regard the parameter to be estimated as a constant, while the estimator is viewed as a random variable. (The reader may wish to review Figure 8.4 as an illustration of this fact.) **Bayesian estimation** differs from this approach in that it views the parameter itself as a random variable as well. This may be reasonable whenever statisticians, as a result of prior experience or *a priori* reasoning, hold strong opinions about what the value of the unknown parameter is likely to be. These opinions about likely values of population mean or proportion, for example, may be represented by a prior (subjective) probability distribution; unlike standard techniques, Bayesian estimation techniques take into account prior probabilities. (For an earlier discussion of prior probabilities see the Chapter 5 section on "Revising Probabilities: Bayes's Theorem.") For example, if a statistician's prior (subjective) probability distribution of the population mean is normal, the Bayesian estimate of the population mean is given by

$$\mu = \left(\frac{n\sigma_*^2}{n\sigma_*^2 + \sigma^2}\right)\overline{X} + \left(\frac{\sigma^2}{n\sigma_*^2 + \sigma^2}\right)\mu_*,$$

where n is the sample size, $\overline{X}$ is the sample mean, and σ^2 is the population variance (as usual), while σ_*^2 and μ_*, respectively, are the variance and the mean of the prior (subjective) probability distribution of the population mean.

---

*Optional section.

Thus, the Bayesian estimate of the population mean is a weighted average of the sample mean, $\overline{X}$, and the mean of the prior (subjective) probability distribution of the population mean, μ_*. It is, in short, a weighted average of what the sample actually finds ($\overline{X}$) and what the statistician believes to be true about the population (μ_*). Note that the formula gives more weight to $\overline{X}$ (and less to μ_*) the larger is the sample (the greater is n) and the greater is the statistician's uncertainty about the population mean (the greater is σ_*^2). The opposite is true for a small n and a small σ_*^2 because then $\overline{X}$ is less trustworthy, but the low variance makes μ_* more believable. This approach is eminently sensible.

As an example, let us reconsider the point estimate of $\mu = 248.4$ words per page derived in Practice Problem 4 about the length of a manuscript. Suppose that the publisher had held a strong prior opinion about the population mean and that the mean and standard deviation of the publisher's prior probability distribution had been, respectively, $\mu_* = 300$ and $\sigma_* = 5$ words per page. What would a Bayesian estimate have been? If the sample variance of $s^2 = 747.49$ squared words per page is used as an estimate of the missing population variance,

$$\mu = \left(\frac{30(5)^2}{30(5)^2 + 747.49}\right) 248.4 + \left(\frac{747.49}{30(5)^2 + 747.49}\right) 300$$
$$= .5008(248.4) + .4992(300)$$
$$= 274.2 \text{ words per page.}$$

The result, clearly, is a compromise between what the sample discovered and what the publisher believed all along.

What, however, if the publisher had been less certain and the value of σ_* had been 100? In that case,

$$\mu = \left(\frac{30(100)^2}{30(100)^2 + 747.49}\right) 248.4 + \left(\frac{747.49}{30(100)^2 + 747.49}\right) 300$$
$$= .9975(248.4) + .0025(300)$$
$$= 248.5 \text{ words per page.}$$

This result is practically the same as that obtained by the non-Bayesian technique because the large variance of the publisher's prior probability distribution causes little weight to be attached to the associated prior mean.

ANALYTICAL EXAMPLES

8.1 Election Advice

Those who advise presidential candidates continually poll the electorate to assess their candidate's chance of winning and to decide on where the greatest campaign effort is needed.

A poll of 1,000 voters in California once indicated a week before the election that 450 voters favored the Republican, while an identical poll in Louisiana showed only 250 voters similarly inclined. Should the

Republican have conceded both states to the Democrat and shifted the campaign elsewhere?

The point estimates of $P_C = .45$ and $P_L = .25$ might so indicate, but a 95-percent confidence interval for the true proportion was constructed just to make sure.

For California, the result was

$$\pi_C = P \pm (z\sigma_P) \cong$$

$$.45 \pm \left(1.96\sqrt{\frac{.45(.55)}{1000}}\right) = .45 \pm .03.$$

For Louisiana, the result was

$$\pi_L = P \pm (z\sigma_P) \cong$$

$$.25 \pm \left(1.96\sqrt{\frac{.25(.75)}{1000}}\right) = .25 \pm .03.$$

Clearly, there was practically no chance of winning Louisiana because, under the best of circumstances, a population proportion of .28 would have had to be increased to slightly over .50 in a mere week. Yet moving a population proportion that was possibly as high as .48 up to .50 was not equally inconceivable. The candidate was advised to campaign heavily in California.

8.2 Matched-Pairs Sampling in the Chemical Industry

Imperial Chemical Industries, a British firm, wanted to test the effect of a chlorinating agent on the abrasion resistance of a certain type of rubber. Ten pieces of rubber were cut in half; one of the two halves in each pair was chosen by the toss of a coin to be treated with the agent, the other one was not treated. Subsequently, the abrasion resistance of all the pieces was tested by a machine and the difference for each pair of halves was recorded, as indicated in the accompanying table.

The mean and standard deviation for the sample differences were computed as $\overline{D} = 1.27$ and $s_D = 1.1265$. The firm's statisticians then calculated a 98-per-

cent confidence interval for the mean difference (using $t_{.01}$ for 9 degrees of freedom):

$$\mu_A - \mu_B = \overline{D} \pm \left(t\frac{s_D}{\sqrt{n}}\right)$$

$$= 1.27 \pm \left[2.821\left(\frac{1.1265}{\sqrt{10}}\right)\right] = 1.27 \pm 1.005.$$

Thus, the mean difference in abrasion resistance of treated and untreated (otherwise identical) rubber pieces was an increase between .265 and 2.275, suggesting, with 98-percent confidence, a positive effect of such treatment.

Pair	Abrasion Resistance of Treated Minus that of Untreated Piece, D	Pair	Abrasion Resistance of Treated Minus that of Untreated Piece, D
1	2.6	6	1.2
2	3.1	7	2.2
3	−.2	8	1.1
4	1.7	9	−.2
5	.6	10	.6
			$\Sigma D = 12.7$

Source: Owen L. Davies, ed., *The Design and Analysis of Industrial Experiments* (London: Oliver and Boyd, 1960), p. 13.

8.3 Reducing Sample-Collection Costs by Work Sampling

Frequently, there exist several ways of collecting sample data, and they are unlikely to be equally costly. The management of one firm wanted to know the mean time its workers required to spray paint a house-

hold appliance. One manager assigned 30 individuals to keep a continuous watch over a 30-member random sample of the firm's 1,000-member work force. During a 40-hour period, every move the sampled workers made was duly recorded and measured. The observed workers, of course, disliked being watched, and critics feared that the very presence of observers may have changed the behavior of the observed so as to bias the results obtained. Nevertheless, at a cost of $200 per observed worker (and a lot of griping), the mean time required to complete the task was established for the workers in the sample at 46 minutes, with a standard deviation of 12 minutes. The firm's statisticians then computed the limits of a 95-percent confidence interval for the mean painting time by all workers as

$$\mu = \overline{X} \pm \left(z \frac{s}{\sqrt{n}} \right) = 46 \pm \left(1.96 \frac{12}{\sqrt{30}} \right)$$

$$= 46 \pm 4.29 \text{ minutes.}$$

This knowledge cost the firm 30($200) = $6,000.

In the meantime, another executive took a different approach. Two observers were hired for a week at $200 each. Without anyone's knowledge, these observers watched another sample of 30 workers, but each worker was observed only at randomly scattered *moments*. With the help of a random numbers table, it was determined, for example, that worker A would be observed on Monday at 9:05, 9:47, 11:12, and 13:32, on Tuesday at 8:31, 10:11, and so on. Only the type of activity undertaken at these moments was then recorded. (Thus, in 100 observations, worker A might have been found 81 times painting the appliance, 3 times being absent from the workbench, twice drinking coffee, twice chatting, and so on.) It was then assumed that the proportion of *time* spent by a worker

in the pursuit of a given activity during the entire week was identical to the relative frequency with which momentary observations of this worker engaged in this activity were made. (Thus, it would have been concluded that worker A spent 81/100 of the 40-hour week, or 32.4 hours = 1,944 minutes, actually painting the appliance. If worker A also painted 36 appliances, which would be easy enough to check, that particular worker's mean completion time could be figured as 54 minutes.) The mean painting time of all workers in the second sample turned out to be 47 minutes, with a standard deviation of 10 minutes—almost identical to the results noted above. Thus, the second method of collecting sample data, which is referred to as **work sampling,** was found to be a considerably cheaper way of getting identical information (costing, in our example, a mere $400 and completely avoiding the alienating effect of the overseer's presence).

Work sampling is being applied widely to set reasonable work standards and to improve labor productivity. The method allows auto manufacturers to know precisely how much time must be planned for installing a bumper or a steering wheel; it allows restaurants and hospitals to plan personnel assignments, respectively, for clearing tables, pouring coffee, and figuring bills or charting fevers, dispensing pills, and nursing patients. Indeed, the method has been used outside the workplace, as by city governments, intent on charting traffic flows. By observing anybody or anything often enough (for a second at a time) one can figure out quite easily the proportion of total time anybody or anything is spending on various activities.

BIOGRAPHY

8.1 William S. Gosset

William Sealy Gosset (1876–1937) was born in Canterbury, England, the latest descendant of an old Huguenot family that had left France after the revocation of the Edict of Nantes. He studied at Winchester, then at Oxford, where he focused on mathematics and natural sciences. Upon graduation, he joined Arthur Guinness and Son, a Dublin brewery, and he remained employed there throughout his life, ultimately becoming chief brewer at a new brewery in London.

Early on, Gosset saw a need for careful scientific analyses of a variety of processes, from barley production to yeast fermentation, all of which profoundly affected the quality of the brewery's final product, beer. His firm sent him to study under Karl Pearson (Biography 10.1) at the University College, London.

At the time, the theory of estimation based on large samples had been fully worked out, but Gosset noticed a void with respect to small-sample estimation theory. Small samples, however, were typical of Gosset's work at the brewery. So Gosset developed the theory himself. In a now famous 1908 paper, "The Probable Error of a Mean," he noted that s is an erratic estimator of σ when n is small; hence, customary measures of the precision of estimates were invalid for small samples and unknown σ. His paper presented the sampling distribution of a statistic now known as "Student's t" and introduced small-sample estimation by means of the t-distribution family. It is difficult to overestimate the importance of this achievement. It has proven fundamental to statistical inference as it exists today, not only in the realm of estimation, but also in hypothesis testing and the analysis of variance. Sir Ronald Fisher (Biography 9.1), who held great admiration for Gosset and shared with him a keen interest in agricultural experimentation, quite aptly called Gosset "the Faraday

of statistics," for Gosset had a similar ability to grasp general principles and develop them further by applying them to practical ends.

All but one of Gosset's numerous papers were published under the pseudonym "Student" to protect the advances in his firm's quality control from nosy competitors. For many years, an air of romanticism surrounded the appearance of "Student's" papers, and only a few individuals knew his real identity, even for some time after his death.

Source: Dictionary of Scientific Biography (New York: Scribner's, 1972), pp. 476–77; *International Encyclopedia of Statistics,* vol. 1 (New York: Free Press, 1978), pp. 409–13.

Summary

1. The process of inferring the values of unknown population parameters from known sample statistics is referred to as *estimation.* The type of sample statistic that serves the purpose of making inferences about a given type of population parameter is called the *estimator* of that parameter. The estimate of a population parameter can be expressed as a single numerical value (called a *point estimate*) or as a range of values within which the unknown parameter presumably lies (called an *interval estimate*).

2. The concept of the *sampling distribution* is crucial to an understanding of the nature of estimation. Every sample summary measure is in fact a random variable prior to the selection of an actual sample, and a sampling distribution is simply a probability distribution of a random variable that happens to be a summary measure based on sample data. It shows the likelihood of occurrence associated with all the possible values of a sample statistic (which values would be obtained when drawing all possible simple random samples of a given size from a population).

3. Any sampling distribution can, in turn, be described by summary measures of its own, such as mean, variance, and standard deviation. The standard deviation of a sampling distribution is also referred to as the *standard error* of the sample statistic in question.

4. The relationship between a sampling distribution and its parent population is of great importance; it can be described in a general way by focusing on the general shapes of the two frequency distributions or more precisely by focusing on mathemat-

ical relationships among their summary measures. Two important theorems relate the general shape of the sampling distribution of the mean to the population distribution:

   a. If $\overline{X}$ is the mean of a random sample taken from a population and if the population values are normally distributed, the sampling distribution of $\overline{X}$ is also normally distributed, regardless of sample size.

   b. If $\overline{X}$ is the mean of a random sample taken from a population and if the population values are *not* normally distributed, the sampling distribution of $\overline{X}$ nevertheless approaches a normal distribution as sample size increases provided sample size, n, remains small relative to population size, $N;$ this approximation is near-perfect for samples of $n \geq 30$, but $n < .05\ N$ (*central limit theorem*).

   Similar general relationships exist between other types of sampling distributions and their associated population distributions. More specific mathematical relationships are given in Boxes 8.A and 8.B.

5. Ideally, an estimate should be unbiased, efficient, and consistent. An *unbiased estimator* is a sample statistic that, on the average, across many samples, takes on a value equal to the population parameter that is being estimated with its help. An *efficient estimator* is that sample statistic, among all the available unbiased estimators, which has the smallest variance for a given sample size. A *consistent estimator* is a sample statistic such that the probability of its being close to the parameter being estimated gets ever larger (and, thus, approaches

unity) as sample size increases. Whenever several possible estimators exist, but none of them is perfect by the above criteria, statisticians choose the best one by calculating (and then minimizing) *mean squared error,* the sum of an estimator's squared bias plus its variance.

6. Whenever the statistical population of interest is normally distributed, or whenever the conditions of the central limit theorem are fulfilled, sample mean and sample proportion are unbiased, efficient, and consistent estimators of the corresponding population parameters. The same is true of the sample variance, provided sampling occurs under conditions of statistical independence—but not otherwise. The sample standard deviation is always a biased estimator of that of the population, but it is still used routinely for lack of a better alternative.

7. The typical point estimate, even if the estimator is unbiased, is almost certain to lie below or above the true value of the parameter being estimated. Statisticians, therefore, construct interval estimates centered on the point estimate and ranging below and above this value by A standard errors of the estimator, where A can equal any (fractional or whole) positive value. The width of the interval has important implications for the degree of confidence with which one can state that the population parameter lies within the two limits of the interval.

8. Confidence intervals for the mean or proportion can be constructed from large samples with the help of the normal distribution, and this procedure can be applied under a variety of conditions.

9. Confidence intervals for the mean or porportion can be constructed from small samples as well. Depending on circumstances, the procedure may rely on the normal distribution, may use Chebyshev's theorem, or may proceed from one of Student's t distributions. Tabular presentations of t values facilitate the use of these special types of sampling distributions.

10. Interval estimates can also be derived for the difference between two means, a parameter that frequently is of great interest to decision makers. The proper estimating procedure depends on whether the samples taken from the two populations are large or small but also on whether two independent samples are taken or a matched-pairs sample is taken.

11. Interval estimates can be constructed similarly for the difference between two proportions.

12. Anyone engaged in estimation is inevitably concerned about finding the optimal sample size that minimizes the total cost of sampling, or the sum of data-collection and error costs. Special procedures exist that can be employed for finding the best sample size.

13. While regular estimation procedures view the parameter being estimated as a constant, *Bayesian estimation* views it, just like the estimator itself, as a random variable. This approach may be reasonable whenever statisticians, as a result of prior experience or *a priori* reasoning, hold strong opinions about what the value of the unknown parameter is likely to be. The Bayesian estimate of the population mean, for example, is a weighted average of what the sample actually finds and what the statistician believes to be true about the population that is being sampled. Bayesian estimation, thus, makes use of *all* available information, not just of that which is revealed by sampling.

Key Terms

Bayesian estimation
central limit theorem
confidence intervals
confidence level
confidence limits
consistent estimator
degrees of freedom, *d.f.*
efficient estimator
estimation
estimator
interval estimate

matched-pairs sample
mean squared error
point estimate
sampling distribution
standard error of the sample mean, $\sigma_{\overline{X}}$
standard error of the sample proportion, σ_P
Student's t distribution
tolerable error level
unbiased estimator
work sampling

Questions and Problems

The computer programs noted at the end of this chapter can be used to work many of the subsequent problems.

1. Public-health officials in charge of the safety of swimming pools and beaches have determined that the population of index numbers (that define bacterial contamination per gallon of water) is normally distributed with a mean of 150 and a standard deviation of 50. If a random sample of 25 gallons is taken, what is the probability that the sample mean index will
   a. lie between 140 and 160?
   b. exceed 140?
   c. lie below 130?
   d. lie between 160 and 170?
   e. exceed 155?

2. The manufacturer of batteries for aircraft emergency-locator transmitters claims that the lifetime of these batteries is normally distributed with a mean of 30 months and a standard deviation of 3 months. An aircraft manufacturer checks out 50 batteries and discovers a sample mean of only 29 months. What is the probability that the manufacturer's claim is true?

3. Various samples are drawn from normally distributed (infinite) populations with means and variances as shown below. In each case, determine the mean and standard deviation of the sampling distribution of $\overline{X}$.
   a. $n = 10; \mu = 30; \sigma^2 = 9$
   b. $n = 15; \mu = 50; \sigma^2 = 4$
   c. $n = 30; \mu = 100; \sigma^2 = 100$
   d. $n = 100; \mu = 400; \sigma^2 = 64$

4. In each of the four cases given in problem 3, determine the probability of observing a sample mean of $(\mu_{\overline{x}} - 1)$ or less.

5. A bank reports (accurately) that the population of its demand deposit balances is normally distributed with a mean of $1,200 and a standard deviation of $250. An auditor refuses to certify the bank's claim and takes a random sample of 36 account balances. He will certify the bank's report only if the sample mean lies within $50 of the alleged population mean. What is the probability for such a finding?

6. A manufacturer of strapping tape claims that the lengths of tape on the firm's rolls have a mean of 90.10 feet and a standard deviation of .14 feet. If you take a random sample of the firm's output and $n = 49$, what is the nature of the sampling distri-

bution of the sample mean? What are the chances that your sample mean will be 90 feet or less?

7. Consider a group of persons with these ages: 34, 29, 22, 30, 25. If a simple random sample of 2 is to be taken *without replacement,* what is the sampling distribution for their mean age? What are the distribution's expected value, variance, and standard deviation?

8. Consider panel (b) of Figure 6.2 on page 199, which pictures the case of sampling 3 items *with replacement* from a large lot containing 10 percent defective items. Find the sampling distribution for the sample proportion of defectives; then determine its summary measures.

9. Consider Figure 6.4 on page 213 as an example of sampling *without replacement* from a small population ($n = 4, N = 10$, so $n \geq .05N$). Find the sampling distribution for the sample proportion of females; then determine its summary measures.

10. Construct *point* estimates in the following situations:
    a. At an automobile assembly plant, a simple random sample of 40 workers is taken. Of these, 21 are found to have been vaccinated against tetanus. Estimate the proportion of all the workers who are so vaccinated.
    b. The state fisheries department caught 1,000 fish in a lake, marked them, and returned them to the lake. A week later, it caught another 1,000 fish and found 200 of them marked. Estimate the proportion of all the fish in the lake so marked (and, hence, the size of the lake's fish population).
    c. The governor wants to finance a new prison by raising the state's sales tax from 3 percent to 4 percent. Of 900 residents randomly sampled, 810 oppose the idea. Estimate the proportion of all residents so opposed.

11. Construct *point* estimates in the following situations:
    a. A labor union is negotiating a new contract. The bargaining committee randomly samples the opinions of 100 workers of whom 61 favor the contract. Estimate the proportion of all workers in favor.
    b. After taking a random sample of 50 students, a professor discovers that 30 of them have no

idea of the meaning of the term *c.i.f.* Estimate the proportion of all the students equally ignorant.

c. A rental-car company takes a sample of three of its cars, finding annual repair costs of $510, $98, and $121. Estimate the annual repair cost per car for all the firm's cars.

12. A credit-card company has been billing its millions of customers on the last day of each month, receiving payment, on the average, 13 days later. It decides to experiment with the billing date, hoping to accelerate payments. A simple random sample of 500 customers is switched to a billing date in the middle of the month. The average length of time between billing and payment in the sample comes to 18.9 days, with a standard deviation of 5.3 days. Prepare a 98-percent confidence interval for the mean period between billing and payment for all customers if all of them were given the mid-month billing date.

13. The operator of a telephone exchange is aware that the population of telephone-call durations is normally distributed with a standard deviation of 4 minutes. A sample of 50 calls yields a mean duration of 9.1 minutes. Construct a 95-percent confidence interval for the mean duration of all calls.

14. In a given year, 1,000 apartments are vacated in a city; a random sample of 169 of them shows a mean repair cost of $171.32, with a standard deviation of $15.39. Construct a 90-percent confidence interval for the mean repair costs of all the vacated apartments.

15. The normally distributed population of stock investments by the 75 employees of a firm is known to have a standard deviation of $99. A random sample of 36 employees shows a mean investment of $736. Construct a 99.8-percent confidence interval for the mean investment of all the employees.

16. The personnel manager of a large corporation takes a random sample of $n = 100$ from $N = 600,000$ employees and determines their average annual sick leave at 9.8 days, with a sample standard deviation of 3.1 days. Determine a 90-percent confidence interval for the average annual sick leave of all employees.

17. The Internal Revenue Service is auditing the operators of some 13,000 private airports by taking a random sample of 100 of them. The IRS discovers an average error in reported taxable income of $14,750, with a sample standard deviation of $3,600. Determine a 95-percent confidence interval for the average error made by all the existing airports.

18. A real estate agent wants to determine a 99-percent confidence interval for the average home value in a large metropolitan area. A random sample of 36 homes shows a mean value of $156,900 and a sample standard deviation of $40,333. Make the determination.

19. There are 200 gas stations in a city; an economist takes a random sample of 50 of them. Their average gasoline price is $1.339 per gallon, with a sample standard deviation of 23.1¢ per gallon. Determine an 80-percent confidence interval for the average price citywide, while assuming that the population distribution is normal.

20. The state police has taken a random sample of 300 cars on the turnpike and has calculated an average speed of 59.13 mph, with a sample standard deviation of 3.13 mph. Determine a 98-percent confidence interval for the average speed of all cars.

21. An airline has randomly sampled 81 of its thousands of annual flights and has discovered a sample mean of 13.9 unoccupied seats, with a standard deviation of 5.2 seats. Determine a 99.8-percent confidence interval for the mean number of unoccupied seats per flight on all of the airline's flights.

22. On the New York Stock Exchange in 1985, the average price of a share of stock in a random sample of 32 stocks was $52.07, with a standard deviation of $17.09. Determine a 90-percent confidence interval for the average price of all stocks.

23. The manager of United Parcel Service has taken a random sample of the week's 30,195 packages of $n = 100$ and has determined the average weight at 5.75 lb., with a sample standard deviation of 2.01 lb. Determine a 97.5-percent confidence interval for the average weight of all the week's packages.

24. The manager of a canning factory wants a 94.26-percent confidence interval for the average amount of green peas put into cans by a filling machine. A sample of 94 cans is taken from the ongoing production process; sample mean and standard deviation are 15.98 oz. and .21 oz., respectively. Make the determination.

25. Just before the Presidential elections, a Gallup poll of 1,500 voters showed the proportions of voters favoring the candidates indicated in Table 8.A. Construct 95-percent confidence intervals for the population proportion favoring the Democratic candidate. (The actual election outcomes were: Kennedy .501, Johnson .613, Humphrey .497, McGovern .382.)

Table 8.A

	Democratic		Republican	
1960	Kennedy	.51	Nixon	.49
1964	Johnson	.64	Goldwater	.36
1968	Humphrey	.50	Nixon	.50
1972	McGovern	.38	Nixon	.62

26. In a random sample of 100 households, 59 are found to prefer brand X. Construct a 98-percent confidence interval for the proportion of all households so inclined.

27. In a random sample of 100 taken from 55,000 fatal-traffic-accident records, alcohol is the apparent cause 39 times. Construct a 95-percent confidence interval for the proportion of all accidents so caused.

28. A lumber company ships 1 million pine boards. A random sample of 50 boards shows 17 being excessively warped. Construct a 95-percent confidence interval for the proportion of all boards so warped.

29. A random sample of 30 police officers taken from a 100-person police force shows that 21 believe their chief is doing a good job. Construct a 99-percent confidence interval for the proportion of all officers so inclined.

30. A random sample of 300 floppy disks taken from an ongoing production process shows 12 defective. Construct a 95-percent confidence interval for the proportion of defective disks in the firm's entire output.

31. A home-improvement contractor takes a random sample of 50 homes in a city and finds that 19 of them have vinyl siding. Construct a 77.76-percent confidence interval for the proportion of all city homes with such siding.

32. A mayoral study commission takes a sample of 500 apartments in the city (from among a total of 5,000) and finds 50 to be vacant. Construct a 98-percent confidence interval for the citywide proportion of vacant apartments.

33. The Federal Aviation Administration takes a sample of 50 planes landing at an airport where a total of 150 planes are landing. In the sample, 13 planes have illegal transponders (that have not been inspected and certified within the past 24 months as required). Determine a 98.8-percent confidence interval for the proportion of illegal transponders on all 150 planes.

34. A personnel manager is investigating why 1,900 employees quit last year. A random sample of 500 shows "boredom" to be the answer 225 times. Determine a 68.76-percent confidence interval for the proportion of all former employees who left the firm because of boredom. What would a 95-percent confidence interval be?

35. A bank claims that no more than 2 percent of its monthly customer statements are in error. An auditor doesn't believe the claim, selects 100 accounts randomly from the bank's 15,233 accounts, and contacts the customers in question. Of these, 12 report an error in last month's statement and prove it. Construct a 98-percent confidence interval for the true proportion of erroneous bank statements that the bank mails out. What, precisely, is the *meaning* of this interval?

36. A manufacturer is aware that the lifetimes of batteries produced by the firm are normally distributed. A random sample of 10 batteries shows a mean lifetime of 6 hours with a standard deviation of 1 hour. Construct a 99-percent confidence interval for the mean lifetime of all batteries produced by the same process.

37. The daily gasoline consumption of a fleet of 35 taxis is known to be normally distributed with a population standard deviation of 3.7 gallons. One day, a random sample of 10 taxis is taken, and their mean gasoline consumption is found to have been 20 gallons. Construct a 96-percent confidence interval for the day's mean gasoline consumption by all the taxis.

38. An insurance company wants to estimate the average claim on its automobile collision policies. It believes the sizes of such claims to be normally distributed. It uses the last 21 claims as a sample and finds their average to equal $657, with a standard deviation of $310. Construct a 95-percent confidence interval for the average claim on all policies.

39. A pharmaceutical company deliberately infects 20 volunteers and then tests a new drug on them, finding a mean recovery time of 8 days, with a standard deviation of 2.5 days. Construct a 95-percent confidence interval for the mean recovery time of all potential users of this drug (i) on the assumption that recovery times are normally distributed and (ii) on the assumption that recovery times are definitely *not* normally distributed, but the population standard deviation equals 3 days.

40. A quality engineer is checking out the machinery that is supposed to put 20 oz. of liquid detergent into a container. A sample of 12 containers shows the mean amount dispensed to be 18.9 oz., the standard deviation being 3.1 oz. Construct a 90-

percent confidence interval for the mean amount dispensed by the machine, assuming that the amounts dispensed are normally distributed.

41. Military analysts believe that the top speeds of the 5,000 Soviet submarines are normally distributed around an unknown mean that is of great interest to them. They do, however, have information on a random sample of five submarines. Their average top speed is 52 nautical miles per hour, the standard deviation is 2.3 knots. Construct a 99-percent confidence interval for the average top speed of all Soviet submarines.

42. An accountant believes that the value of telephone services stolen by the theft of telephone-company credit cards is a normally distributed random variable, but knows nothing else about the population in question. However, a random sample of 15 thefts shows the mean amount stolen to be $532, with a standard deviation of $17.63. Construct a 98-percent confidence interval for the mean stolen amount for all such thefts.

43. Upon randomly sampling 25 of the firm's 900 employees, the personnel manager finds that 17 prefer the newly proposed plan of working only 4 days a week, but more hours each day. Construct a 99-percent confidence interval for the proportion of all employees so inclined.

44. The credit-card company noted in problem 12 takes a second random sample of 500 customers and switches their billing date to the twenty-fourth of the month. The average period between billing and payment in that sample comes to 7.5 days, with a standard deviation of 2.1 days. Construct a 99-percent confidence interval for the difference between the mean days until payment if all customers are switched to a billing date of the fifteenth or the twenty-fourth.

45. A Federal Aviation Administration flight-service specialist wants to determine the difference between the mean thunderstorm intensities (measured by peak gust velocities) for two regions of the country. The following data for a five-year period are obtained and used as a sample:

Region A: 290 occurrences; sample mean-59 miles per hour; sample standard deviation-20 miles per hour.

Region B: 333 occurrences; sample mean-71 miles per hour; sample standard deviation-30 miles per hour.

Construct a 99-percent confidence interval for the difference between the mean intensities.

46. The manager of a chain of retail stores wants to determine, with 95-percent confidence, whether an advertising campaign has had any effect. A sample of 50 days prior to the campaign shows mean sales per store of $3,952 (with a standard deviation of $296). A sample of 40 days after the campaign shows corresponding numbers of $4,102 and $306. Make the determination.

47. On a given day, the average price per share is $59.05 for 42 randomly selected stocks traded on the New York Stock Exchange (and the standard deviation is $7.92), while it is $70.08 for 32 stocks on the American Stock Exchange (and the standard deviation for that sample is $25.39). Construct a 90-percent confidence interval for the difference between the mean prices per share.

48. A Pentagon official wants to determine the gasoline savings if all U.S. Army cars switch from regular to radial tires. Some 150 cars are equipped with new tires; half of them are equipped with regular tires and half of them with radial tires. One car in each group, furthermore, is matched with one in the other group by make, model, age, region of the country, and department. After three months, the mean increase in mileage on cars with radial tires is found to be 5 miles per gallon, with a standard deviation of 3 miles per gallon. Construct a 98-percent confidence interval for the gasoline savings if a similar switchover is made on all Army cars.

49. A space agency wants to compare the magnification achieved by two types of cameras. One of them (A) is placed on one satellite, the other (B) on another satellite; subsequently, 15 photographs are taken by each satellite camera of the identical areas at almost identical moments, using identical films and exposure settings. Afterward, the number of houses in each picture are counted; the sample means and standard deviations are as follows: $\overline{X}_A = 139; s_A = 21; \overline{X}_B = 115; s_B = 52$. The relevant population distributions are normal; the populations have equal variances. Construct a 95-percent confidence interval for the difference between mean house counts for the two types of cameras.

50. A housing-construction-company executive is contacted by a glass manufacturer who claims that a revolutionary new shipping container can reduce the breakage of window panes dramatically and who graciously offers to replace the construction company's current supplier. At the same price, more usable panes will be received. The construction-company executive is intrigued but decides to test the claim. A dozen cases of 100 panes are ordered from the new would-be supplier, and the number of unbroken panes is counted; a similar count is made on the latest shipment of 18 cases received from the old supplier. The results of this sampling are shown in Table 8.B. Construct a 95-percent confidence interval for the difference between the mean numbers of unbroken panes in the populations of all possible cases delivered by the new and the old supplier.

Table 8.B **Number of Unbroken Panes in Cases . . .**

. . . From New Supplier		. . . From Old Supplier		
88	92	85	82	84
91	89	81	81	86
91	93	85	85	81
95	91	87	87	80
86	88	88	88	84
87	94	90	92	83

51. A restaurant owner wishes to estimate the difference between the mean daily sales of two restaurants on Main Street. A 98-percent confidence interval is desired. The two restaurants are matched up for 14 days, their sales being compared on Monday, then on Tuesday, and so on. The mean and standard deviation of the differences between A's daily sales and B's daily sales are found to be $\overline{D}$ = \$133 and s_D = \$41. Make the estimate, assuming a normal population distribution.

52. A polling organization wants to estimate the difference between the proportion of urban and rural residents favoring agricultural price-support programs. It draws a simple random sample of 500 urban residents (250 favor the programs) and of 500 rural residents (400 favor the programs). Construct a 95 percent confidence interval for the difference between the proportions of those favoring the program.

53. An airline wants to estimate the difference between the proportion of passengers who carry only hand luggage on its New York-to-Chicago versus its New York-to-Miami flights. Two random samples of 50 passengers each show 34 versus 11 such passengers. Construct a 98-percent confidence interval for the difference between the proportions of hand-luggage-only passengers.

54. A firm producing batteries has introduced two different quality-control systems, each on a different assembly line. One day, 400 batteries are randomly selected from the 10,000-unit output of line 1, and an equal number is pulled from the identical output of line 2. The proportions of defective items are, respectively, .06 and .04. Construct a 99-percent confidence interval for the difference in the proportion of defective items produced under the two quality-control systems.

55. Table 8.C shows the estimated unemployment rates for the whole civilian labor force and for two parts of manufacturing industry for July 1983. Assume these rates have been estimated on the basis of simple random samples of the indicated sizes.

Table 8.C

	Sample Size	Estimated Unemployment Rate (July 1983)
Total, 16 years and over	50,000	.094
By Industry		
Durable goods manufacturing	4,900	.116
Nondurable goods manufacturing	3,900	.092

Find a 98-percent confidence interval for the difference between π_D and π_N, the population unemployment rates in durable goods manufacturing and nondurable manufacturing.

56. There is an old rule of thumb: "Always sample about 10 percent of the statistical population of interest." Comment on the rule.

57. In each of the following situations, determine the required sample size.
  a. A firm is split into several divisions; each of these uses a central computer. Management is interested in the amount of time the computer is used by division A on the average day. Given that the standard deviation of the population of times equals .5 hour, that the tolerable error equals .1 hour, and that the desired reliability is 95 percent, what is the sample size needed to establish the mean daily computer time used by division A?
  b. A new product is to be marketed. Management wants to know the proportion of people who will like it enough to buy it. How big a sample must be interviewed to assure a 98-percent confidence level and a .05 tolerable error if preliminary estimates make a population proportion of .25 likely? What if no such preliminary estimate exists?
  c. A personnel manager wants to know the average time a staff member spends with job appli-

cants. Given that the population standard deviation is 5 minutes and that a 99-percent confidence interval is desired, along with a tolerable error of 2 minutes, what must the sample size be? What if a 95-percent confidence interval is deemed acceptable?

58. In each of the following situations, determine the required sample size, assuming normally distributed populations.

a. The population of starting salaries of engineers is known to have a standard deviation of $2,800. How big a sample is needed to estimate the mean starting salary with 95-percent confidence and a tolerable error of $100? Of $200? Of $400?

b. A pharmaceutical company wants to know the mean number of milligrams a machine puts into capsules. The population standard deviation is known to equal 2 milligrams, and the desired reliability is 95 percent. What must the sample size be for a tolerable error of .05 milligrams? Of .1 milligrams? Of .5 milligrams?

c. The population of sick leaves at a large corporation is known to have a standard deviation of 3.1 days. How big a sample is needed to estimate the mean number of employee sick days with 90-percent confidence and a tolerable error of .51 days?

59. In each of the following situations, determine the required sample size, assuming normally distributed populations.

a. The Environmental Protection Agency wants to determine the mean daily sulfur emissions from an industrial complex. How many air samples must be taken if the population standard deviation of pollution measurements is known to equal 2 parts per million (ppm), the tolerable error level is 1 ppm, and a 95-percent confidence level is desired?

b. An auto importer wants to estimate the average age of all registered cars in the Northeastern United States. The standard deviation of ages is 3 years, the desired confidence level is 99.9 percent, the tolerable error level is .1 year.

c. A bill is about to be brought before the state senate. A senator wants to know, with 95-percent confidence and a tolerable error of .01, the proportion of voters favoring the bill.

d. A bank auditor believes that 12 percent of a bank's 15,233 monthly customer statements are in error. A 98-percent confidence level is desired; the tolerable error level is .0757.

60. Take another look at Practice Problem 8 on page 310 in which a cattle shipment's mean weight per animal was estimated. The sample mean was $\overline{X} = 1,301$ pounds (and this was the point estimate of μ); the sample standard deviation was $s = 290$ pounds. What would a Bayesian point estimate have been if the shipper (from prior experience) had held a prior probability distribution with $\mu_* = 1,500$ pounds and $\sigma_* = 500$ pounds?

Selected Readings

Adams, William J. *The Life and Times of the Central Limit Theorem*. New York: Kaedmon, 1974.

Arthur Guinness Son & Co. *Letters from W. S. Gosset to R. A. Fisher: 1915–1936*. Dublin: Guinness, 1967. A stimulating collection of nearly 200 letters that are of great interest with respect to the history of statistical theory and practice.

Pearson, E. S. and John Wishart, eds. *"Student's" Collected Papers*. London: University College, 1943. A collection of Gosset's writings published between 1907 and 1938.

"Student." "The Probable Error of a Mean" *Biometrica* 6(1908):1–25. The crucial article on the *t* distribution.

Computer Programs

The DAYSAL-2 personal-computer diskettes that accompany this book contain three programs of interest to this chapter:

7. Probability Distributions, among other things, lets you find areas under the normal curve between any two z values you specify.

8. Interval Estimation lets you make interval estimates for a population mean, a population proportion, a difference between 2 population means—independent samples or matched pairs—and a difference between 2 population proportions. Estimates are made for large and small samples, using the normal distribution, the t-distribution, or Chebyshev's theorem, as appropriate.

16. Expected Monetary Values, although designed for another purpose, can be used to find the means of sampling distributions.

CHAPTER NINE

Hypothesis Testing: Classical Techniques

When people make decisions, they inevitably do so on the basis of beliefs they hold concerning the true state of the world. They carry in their minds a certain image of reality; they hold some things to be true, and others to be false, and they act accordingly. Thus, one government agency may ban cigarette ads because its officials have come to believe that cigarettes cause heart and lung disease; another one may refuse to license a new anticancer drug because no credible case has been made for its alleged effectiveness. A third agency yet may require the wearing of helmets on motorcycles because this precaution is believed to reduce fatality rates, and a fourth one may set out to stem the destruction of orchards by gypsy moths—not by means of the traditional spraying of pesticides, but by means of the introduction of intestinal (gypsy-moth) parasites, a procedure deemed much more effective in achieving the desired goal. Business executives, similarly, make crucial decisions every day because they hold certain beliefs: that a given type of filling machine puts at least one pound of detergent into a box, that a certain steel cable has a breaking strength of 5,000 pounds or more, that the average lifetime of a battery equals 100 hours, that a certain process yields shirt sleeves that are precisely 33 inches long (or capsules that contain precisely 100 milligrams of a drug), that the wages of construction workers in New York exceed those in Chicago by $8 per hour, that one type of fertilizer is more effective than another, that shipping company A has faster delivery times than company B, that the output of the east-coast plant contains fewer defective units than that of the west-coast plant, that one type of engine yields fewer exhaust emissions than another—the list goes on. In all these cases and a million more, people act on the basis of some belief about reality, a belief that first came into the world as a mere conjecture—perhaps a little more than an informed guess, a proposition tentatively advanced as possibly true, and therefore, called a **hypothesis.** Sooner or later, however, every hypothesis is confronted with evidence that either substantiates or refutes it, and in this way people's image of reality moves from much uncertainty to less.

Indeed, **hypothesis testing** (which is the subject of this chapter and the next two) is nothing else but a *systematic* approach to assessing beliefs about reality: it is confronting a belief (such as a tentative idea about the value of an unknown population parameter) with evidence (such as the value of a statistic computed from a random sample taken from the population in question) and then deciding, in light of this evidence, whether the initial belief (or hypothesis) can be maintained as reasonable or must be discarded as untenable. In this chapter, we will consider the testing of hypotheses about a population mean or proportion and about the difference between two such means or proportions. In every case, such testing involves four major steps.

STEP 1: FORMULATING TWO OPPOSING HYPOTHESES

The first step in hypothesis testing is always the formulation of two hypotheses that are mutually exclusive and also collectively exhaustive of the possible states of reality. Each of these complementary hypotheses is a proposition about a population parameter such that the truth of one hypothesis implies the falsity of the other. The first hypothesis in the set is called the **null hypothesis** and is symbolized by H_0; often, but not necessarily, it represents that proposition about an unknown population parameter which is tentatively assumed to be true. The second hypothesis is called the **alternative hypothesis** and is symbolized by H_A; often, but not necessarily, it is initially assumed to be false (but is later accepted as true if there is strong evidence against the null hypothesis). This procedure of formulating hypotheses is quite common and is similar to that in a criminal trial in which the defendant is assumed innocent until proven guilty. Being free of guilt is tentatively assumed to be true (a *null* hypothesis of *zero* guilt is made), but contrary evidence can reverse this judgment (an alternative hypothesis of guilt can be substituted).

Note: Which of two opposing propositions is called H_0 and which is called H_A is obviously quite arbitrary. Unlike defense lawyers, prosecutors, for example, view the assumption of innocence as something to be attacked and, ultimately, rejected in favor of a verdict of "guilty." In the same way, some researchers prefer to set up the null hypothesis in such a way that it *contradicts* what they suspect to be true. If they are right, the hypothesis-testing process leads to the acceptance of the alternative hypothesis, while the null hypothesis is rejected or "nullified" (which gives us another way yet to account for its name).

Hypotheses About a Population Mean

The opposing hypotheses about the value of a population mean, μ, are typically stated in one of three forms by reference to a specified value of the mean, μ_0.

Form 1	*Form 2*	*Form 3*
$H_0: \mu = \mu_0$	$H_0: \mu \geq \mu_0$	$H_0: \mu \leq \mu_0$
$H_A: \mu \neq \mu_0$	$H_A: \mu < \mu_0$	$H_A: \mu > \mu_0$

In Form 1, the null hypothesis is stated as an **exact hypothesis;** that is, as one that specifies a single value for the unknown parameter. The alternative hypothesis, on the other hand, is stated as an **inexact hypothesis,** or one that specifies a range of values for the unknown parameter. As examples, consider the null hypothesis that an industrial process yields metal parts with an average length, μ, of $\mu_0 = 5$ inches, drills holes with an average diameter of 1.2 inches, or makes shirt sleeves with an average length of 33 inches. The alternative hypothesis in each case suggests that the average is in fact larger or smaller than μ_0. Whenever an alternative hypothesis holds for deviations from the null hypothesis in either direction, it is called a **two-sided hypothesis.**

While the alternative hypothesis is always stated as an inexact one, the null hypothesis may be so stated (as in Forms 2 and 3), but even then it is customary to have it contain the equality sign. In Form 2, the null hypothesis claims that the parameter is greater than or equal to a specified value, while the alternative hypothesis claims it to be smaller than that value. As examples consider the null hypothesis that the average quantity of detergent, μ, put into a box by a filling machine equals or exceeds $\mu_0 = 1$ pound, that the average breaking strength of a certain type of cable is at least 5,000 pounds, or that the average lifetime of a certain type of battery is at least 100 hours. The alternative hypothesis in each case suggests that the average in fact falls short of μ_0. Whenever an alternative hypothesis holds for deviations from the null hypothesis in one direction only, it is called a **one-sided hypothesis.**

In Form 3, finally, the null hypothesis claims that the parameter is less than or equal to a specified value, while the alternative hypothesis claims it to be larger than that value. As examples, consider the null hypothesis that a shipping company's average delivery time, μ, equals or falls short of $\mu_0 = 3$ days, that the average drying time of a paint is at most 4 hours, or that an engine produces exhaust emissions of at most 17 parts per million. The (one-sided) alternative hypothesis in each case suggests that the average in fact exceeds μ_0.

Hypotheses About the Difference Between Two Population Means

The opposing hypotheses about the difference between two population means, μ_A and μ_B, are, similarly, stated in one of three forms:

Form I	*Form II*	*Form III*
$H_0: \mu_A = \mu_B$	$H_0: \mu_A \geq \mu_B$	$H_0: \mu_A \leq \mu_B$
$H_A: \mu_A \neq \mu_B$	$H_A: \mu_A < \mu_B$	$H_A: \mu_A > \mu_B$

Thus, the null hypothesis may claim (Form I) that two population means are the same (for example, that the average lifetimes of two types of tires are identical). Or it may claim (Form II) that one population mean equals or exceeds another (for example, that average construction-industry wages in New York are at least equal to, but possibly larger than, those in Chicago). Or it may claim (Form III) that one population mean equals or falls short of another (for example, that the average yield on acres using one type of fertilizer is at most equal to, but possibly less than, that on acres using a different type of fertilizer). In each case, the alternative hypothesis suggests the opposite.

Note: Often hypotheses about the difference between two population means are stated in the following ways that are mathematically equivalent to Forms I–III:

Form A	*Form B*	*Form C*
$H_0: \mu_A - \mu_B = 0$	$H_0: \mu_A - \mu_B \geq 0$	$H_0: \mu_A - \mu_B \leq 0$
$H_A: \mu_A - \mu_B \neq 0$	$H_A: \mu_A - \mu_B < 0$	$H_A: \mu_A - \mu_B > 0$

Here, the null hypothesis claims that the difference between the mean of population A and population B is precisely zero (Form A), is zero or positive (Form B), or is zero or negative (Form C). The alternative hypothesis sees this difference, respectively, as positive or negative (Form A), as negative (Form B), or as positive (Form C).

Hypotheses About a Population Proportion (or the Difference Between Two Such Proportions)

Hypotheses about a population proportion, π (such as the proportion of defective units put out by a production process), are formulated in a manner precisely analogous to those about a population mean. One simply substitutes π for μ in Forms 1 to 3 above, leaving all else unchanged, including subscripts.

Similarly, hypotheses about the difference between two population proportions, π_A and π_B (such as the proportions of cancer victims among smokers on the one hand and nonsmokers on the other), are formulated in analogy to those about the difference between population means. One simply substitutes π for μ in Forms I to III above (or in the equivalent Forms A to C), again leaving all else unchanged, subscripts included.

STEP 2: SELECTING A TEST STATISTIC

The second step in hypothesis testing is the selection of a **test statistic,** which is the type of statistic to be computed from a simple random sample taken from the population of interest and to be used for establishing the probable truth or falsity of the null hypothesis. Obvious choices are (1) the sample mean, $\overline{X}$, when the hypothesis test involves the population mean; (2) the difference between two sample means, $\overline{X}_A - \overline{X}_B$, when the test involves the difference between two population means; and (3) the sample proportion, P, or (4) the difference between two sample proportions, $P_A - P_B$, respectively, when the test is about the population proportion or the difference between two such proportions.

As we learned in Chapter 8, however, every sample statistic has a sampling distribution of its own, and such a distribution can often be approximated by the normal distribution for large samples [when $n \geq 30$ but $n < .05N$ or when $n \cdot \pi$ as well as $n(1 - \pi) \geq 5$] or by a Student t distribution for small samples (when the underlying population values are normally distributed or at least unimodal and fairly symmetrical). When a sample statistic's sampling distribution can be, thus, approximated, one can convert each of the four sample statistics noted above into a corresponding z value or t value by *dividing the difference between the sample statistic and the extreme value of the population parameter postulated in the null hypothesis by the standard error of the sample statistic.* Such z or t values are commonly used as test statistics, too (as is shown in Appendix 9A on page 394.)

STEP 3: DERIVING A DECISION RULE

Having formulated two opposing hypotheses and having selected the type of statistic with which to test them, the next step in hypothesis testing is the derivation of a **decision rule** that specifies in advance, for all possible values of the test statistic that might be computed from a sample, whether the null hypothesis should be accepted or whether it should be rejected in favor of the alternative hypothesis. At first thought, such a rule may seem superfluous. Isn't it bound to be obvious whether the computed value of the test statistic agrees with the null hypothesis or contradicts it? On reflection it becomes clear, however, why the matter is more complicated than might appear to be the case at first sight. While it is true enough that anyone can tell instantly whether the computed value of the test statistic does or does not agree with the null hypothesis, it is not obvious that a given divergence between the observed value and the hypothetical value automatically proves the null hypothesis to be false. What is proved by such a divergence is questionable because any sample statistic is a *random variable;* its value depends very much on the particular sample that happens to be selected from the population in question. Consider, for example, the population of metal parts produced by an industrial process. The average length of these parts may in fact be $\mu_0 = 5$ inches, as required by the users of these parts. A quality inspector who does not know this fact may formulate two hypotheses as follows:

H_0: $\mu = 5$
H_A: $\mu \neq 5$

Now let a large sample of $n \geq 30$ be taken from this infinite population. Even though *we* know H_0 to be true, the chance factor operating in the sampling process can, nevertheless, provide the inspector with many possible sample means, as shown in Figure 9.1. All kinds of sample means might be found: $\overline{X} = 5$, $\overline{X} = 4.9$, $\overline{X} = 5.2$, and even $\overline{X} = 5.32$; yet all these results (many of which are apparently contradicting the null hypothesis) are consistent with the null hypothesis being true! So what is our quality inspector to do?

One escapes the dilemma by recognizing that even though all kinds of $\overline{X}$ values might be discovered when a simple random sample is taken from a population for which the null hypothesis is true, not all of these values are equally probable. In Figure 9.1, for example, finding a sample mean between 5 and 5.1 is more likely than finding one between 5.1 and 5.2, and that is more likely still than finding one between 5.2 and

Figure 9.1   A Sampling Distribution of the Sample Mean

This graph shows the sampling distribution of the sample mean, $\overline{X}$, for a large simple random sample taken from a population of metal parts with a mean length of $\mu = 5$ inches. The sampling distribution is approximately normally distributed with a mean, $\mu_{\overline{X}}$, equal to the population mean of 5 and a standard deviation, $\sigma_{\overline{X}}$, of .1 inch. Thus, even though a null hypothesis of H_0: $\mu = 5$ is true, half of the possible sample statistics show $\overline{X} < 5$, and half show $\overline{X} > 5$. Thus, a finding of $\overline{X} \neq 5$ does not prove the correctness of the alternative hypothesis of H_A: $\mu \neq 5$.

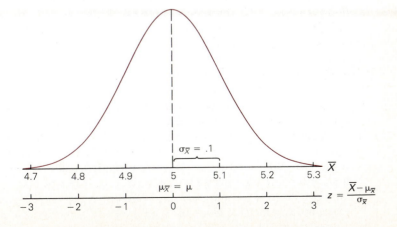

5.3. Given that $H_0: \mu = 5$ is true, finding a sample mean of 5.5 or larger, even though possible, would be a highly unusual result. Thus, one can proceed as follows.

If a particular sample result is very likely when H_0 is true (if, in our example, the probability of finding $\overline{X}$ in a given range is very high when H_0 is true), one ascribes any divergence between the observed and hypothesized values to chance factors operating during the sampling process, and one accepts the null hypothesis as true. If, on the other hand, a particular sample result is very unlikely to occur when H_0 is true (if, in our example, the probability of finding $\overline{X}$ in a given range is very low when H_0 is true), one ascribes a divergence between the observed and hypothesized values to nonchance factors (such as the fact that H_A is true), and one rejects the null hypothesis as false. In short, even seemingly contradictory sample results that are highly likely and nonsurprising when H_0 is true lead to the acceptance of the null hypothesis; others that are highly unusual and surprising when H_0 is true lead to the rejection of it.

The Level of Significance

The foregoing discussion of surprising and nonsurprising results makes a lot of sense but still leaves us with a crucial question: given that H_0 is true, when is the probability of observing a given sample statistic "high" (and the result, therefore, "nonsurprising") and when is that probability "low" (and the result "surprising")? The only possible answer is a *totally arbitrary* one. Whoever undertakes to test a hypothesis must make a subjective decision here; he or she must decide which value of the test statistic is unlikely enough when the null hypothesis is true that it can be viewed as sufficient evidence to reject the null hypothesis and embrace the alternative one. In the situation depicted in Figure 9.1, for example, one might decide to reject the null hypothesis for all sample results that differ by more than two standard errors of the sample mean (by more than $2 \sigma_{\overline{X}}$) from the mean of the sampling distribution, $\mu_{\overline{X}}$ (and, therefore, from the population mean, μ, if the null hypothesis is true). Thus, values of $\overline{X}$ below 4.8 and above 5.2 (two ranges that correspond to all the values beyond $\pm 2z$) would be regarded as occurring with a sufficiently low probability (if $H_0: \mu = 5$ were true) to call H_0 false. As we can see in Appendix Table H, the (combined) area under the normal curve beyond $\pm 2z$ equals $1 - [2(.4772)] = .0456$. Thus, the decision rule just stated treats all sample results that differ from the null hypothesis—and that could occur by chance at most 4.56 percent of the time if H_0 were true—as sufficiently strong evidence to reject H_0. Any such sample result that leads to the rejection of H_0 is called a **statistically significant result.** On the other hand, the above decision rule treats all sample results that differ from the null hypothesis—but that could occur by chance more than 4.56 percent of the time if H_0 were true—as evidence too weak to reject H_0. Any such sample result that leads to the continued acceptance of H_0 is called a **not statistically significant result.** Such a result does not prove H_0 to be true, but it doesn't disprove it either. It does, however, make H_0 somewhat more credible than it was before the sample was taken.

Indeed, among all the sample results that are possible when the null hypothesis is true, the (arbitrary) maximum proportion that is considered sufficiently unusual to reject the null hypothesis is called the **significance level,** or **size of a hypothesis test,** and it is symbolized by α (the lowercase Greek letter *alpha*). In the example, α equals .0456; more commonly used values are .10, .05, .025, and even lower ones. Thus, when the significance level, α, is set at .05, or 5 percent, only statistics that occur with a probability of .05 or less when H_0 is true are sufficient evidence to reject H_0.

Three Illustrations

The three panels of Figure 9.2 illustrate how easily a decision rule can be derived once the desired significance level of a hypothesis test has been specified. The three examples focus (a) on hypothesis tests about a population mean (stated in each of the three forms noted above) and (b) on the large-sample case of $n \geq 30$ but $n < .05N$ (so that the sampling distribution is represented by the normal curve). As will be shown later in this chapter, decision rules for other types of hypothesis tests (or for sampling distributions that are more accurately described by a t distribution) can easily be derived in an analogous fashion.

A two-tailed hypothesis test. Panel (a) pictures the case in which the opposing hypotheses take Form 1. The null hypothesis claims, for example, that the lengths of metal parts produced average 5 inches; the alternative hypothesis denies it. The alternative hypothesis, thus, is a two-sided one. There is concern about departures from the 5-inch norm in *both* directions; very low as well as very high values of the test statistic ($\overline{X}$ or z) are to be viewed as evidence against H_0. If the desired significance level is set at $\alpha = .05$, then values of the test statistic that are so far below the hypothesized mean as to occur with a probability of at most $(\alpha/2) = .025$ when H_0 is true are to be viewed as evidence against H_0, and the same is true for values so far above the hypothesized mean as to occur with the same low probability. Jointly, the statistically significant values of the test statistic, thus, have a probability of $\alpha = .05$ of occurring when H_0 is true. By consulting Appendix Table H, one can easily determine the lower value of the test statistic (here of $-z_{\alpha/2}$ or $-z_{.025}$) and the upper value of the test statistic (here $+z_{\alpha/2}$ or $+z_{.025}$) that marks the border between two regions of values—namely, those that are to signal the acceptance of H_0 (and, thus, lie in the **acceptance region**) and those that are to signal the rejection of H_0 (and, thus, lie in the **rejection region**). The value of z below (or above) which .025 of the area under the standard normal curve lies is also the value that places $.500 - .025 = .475$ of that area between itself and the mean; therefore, as Appendix Table H shows, $-z_{.025} = -1.96$, while $+z_{.025} = +1.96$. If we assume that $\sigma_{\overline{X}} = .1$ (as in Figure 9.1), the corresponding values of $\overline{X}$ equal $5 - [1.96(.1)] = 4.804$ and $5 + [1.96(.1)] = 5.196$. Note: the value of a test statistic that in this way divides all possible values into an acceptance and rejection region is called a **critical value,** a **cutoff point,** or an **acceptance number.** As panel (a) of Figure 9.2 clearly indicates, the decision rule emerging in this example is: "Accept H_0 for any sample mean between 4.804 and 5.196 inches (or for any z value between -1.96 and $+1.96$); reject H_0 (and accept H_A) for values of the test statistic outside these limits."

A lower-tailed hypothesis test. Panel (b) of Figure 9.2 pictures the case in which the opposing hypotheses take Form 2. The null hypothesis claims, for example, that the lifetime of a battery is at least 100 hours; the alternative hypothesis denies it. The alternative hypothesis is one-sided. There is concern about departures from the 100-hour norm only in the lower direction (nobody will complain if the batteries in fact last longer than 100 hours); only very low values of the test statistic are to be viewed as evidence against H_0. If the desired significance level is again set at $\alpha = .05$, values of the test statistic so far below the hypothesized mean as to occur with a maximum probability of $\alpha = .05$ when H_0 is true must be viewed as evidence against H_0. According to Appendix Table H, the critical z value of $-z_{\alpha}$ or $-z_{.05}$ equals -1.64 because

Figure 9.2   Two-Tailed vs. One-Tailed Hypothesis Tests

This set of graphs pictures sampling distributions of sample means (and their corresponding standard normal deviates) on the assumption that the stated null hypotheses are true. When the alternative hypothesis is two-sided, as in panel (a), any hypothesis test is said to be a **two-tailed test** because the null hypothesis is then rejected for values of the test statistic located in either tail of that statistic's sampling distribution. The rejection region is then split between the two tails of that distribution. When, on the other hand, the alternative hypothesis is one-sided, as in panels (b) and (c), any hypothesis test is said to be a **one-tailed test** because the null hypothesis is then rejected only for very low (or only for very high) values of the test statistic located entirely in one tail of the sampling distribution. The entire rejection region is then found in either the lower tail, making for a **lower-tailed test,** as in panel (b), or in the upper tail, making for an **upper-tailed test,** as in panel (c). Note that the critical values of the test statistics are encircled.

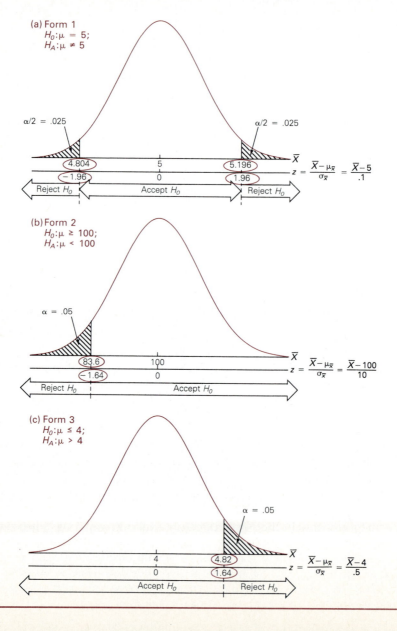

this value places $.50 - .05 = .45$ of the area under the standard normal curve between itself and the mean above it (and thus, leaves .05 of that area in the lower tail of the sampling distribution). If we assume that $\sigma_{\overline{x}} = 10$ hours, the corresponding critical value of $\overline{X}$ equals 83.6 hours. Thus, the decision rule is: "Accept H_0 for any sample mean of 83.6 hours or more (or for z values of -1.64 and above); reject H_0 (and accept H_A) for all lower values of the test statistic."

An upper-tailed hypothesis test. Panel (c) of Figure 9.2 pictures the case in which the opposing hypotheses take Form 3. The null hypothesis claims, for example, that the drying time of a paint is at most 4 hours; the alternative hypothesis denies it. The alternative hypothesis is again one-sided, but this time the story differs from that in panel (b). Now there is concern about departures from the 4-hour norm only in the *upper* direction (nobody will complain if the paint in fact dries faster than claimed); only very *high* values of the test statistic are to be viewed as evidence against H_0. If the desired significance level is again set at $\alpha = .05$, values of the test statistic so far above the hypothesized mean as to occur with a maximum probability of $\alpha = .05$ when H_0 is true must be viewed as evidence against H_0. This time, the critical z value equals $+1.64$; if we know that $\sigma_{\overline{x}} = .5$ hour (because, say, $\sigma = 3$ hours and sample size is $n = 36$), the corresponding critical value of $\overline{X}$ equals 4.82 hours. Thus, the decision rule is "Accept H_0 for any sample mean of 4.82 hours or less (or for z values of 1.64 and below); reject H_0 (and accept H_A) for all higher values of the test statistic." Note: Part (b) of Appendix Table J provides critical z values for two-tailed and one-tailed hypothesis tests and for the most commonly used significance levels.

Decision rules summarized. The formulas in Box 9.A summarize the decision rules for large-sample hypothesis tests about a mean, which we have discussed so far. In Box 9.A, $\overline{X}$ and z are the computed values of the test statistics; their critical values are printed in black.

9.A Decision Rules for Large-Sample Hypothesis Tests About a Mean*

1. Two-tailed test, $H_A: \mu \neq \mu_0$
   Accept H_0 if $\mu_0 - (z_{\alpha/2} \cdot \sigma_{\overline{x}}) \leq \overline{X} \leq \mu_0 + (z_{\alpha/2} \cdot \sigma_{\overline{x}})$
   or if $-z_{\alpha/2} \leq z \leq z_{\alpha/2}$

2. Lower-tailed test, $H_A: \mu < \mu_0$
   Accept H_0 if $\overline{X} \geq \mu_0 - (z_\alpha \cdot \sigma_{\overline{x}})$
   or if $z \geq -z_\alpha$

3. Upper-tailed test, $H_A: \mu > \mu_0$
   Accept H_0 if $\overline{X} \leq \mu_0 + (z_\alpha \cdot \sigma_{\overline{x}})$
   or if $z \leq z_\alpha$

*Assuming $n \geq 30$ and $n < .05N$ or population is normal.

The decision rules discussed above can easily be extended to the small-sample case, provided the underlying population values are normally distributed. One simply substitutes Student t values for z in the expressions of Box 9.A. Similarly, one can derive analogous decision rules for hypothesis tests about a proportion (by substituting π for μ and P for $\overline{X}$) or about the difference between means or proportions (by substituting differences where now individual values appear).

STEP 4: SELECTING A SAMPLE, COMPUTING THE TEST STATISTIC, AND CONFRONTING IT WITH THE DECISION RULE

The final step in hypothesis testing involves (a) the selection of a simple random sample of arbitrary size, n, from the population of interest, (b) the computation of the actual (as opposed to critical) value of the test statistic (selected in Step 2) and (c) its confrontation with the decision rule (derived in Step 3). Note: as a later section of this chapter will show, even though sample size is arbitrary, some sample sizes may be preferred to others—in order to control, for example, the occurrence of various types of errors (also discussed below) that are inherent in the hypothesis-testing process.

If, for example, a sample of metal parts proved to have an average length of $\overline{X} = 5.1$ inches ($z = 1.00$), the null hypothesis stated in panel (a) of Figure 9.2 would be upheld. Similarly, if a sample of batteries turned out to have an average lifetime of $\overline{X} = 90$ hours ($z = -1.00$), the null hypothesis stated in panel (b) of Figure 9.2 would be upheld. If, however, a sample of paint canisters showed an average drying time of $\overline{X} = 4.9$ hours ($z = 1.80$), the null hypothesis given in panel (c) of Figure 9.2 would be rejected.

But note: given $\alpha = .05$, the latter conclusion (that the paint's average drying time exceeded the mere 4 hours claimed for it) would be incorrect in 5 out of 100 times in which this type of statistical procedure was employed. After all, the unusual sample result cited does occur with a 5 percent probability when the panel (c) null hypothesis is true!

The Possibility of Error

The hypothesis-testing procedure outlined in this chapter so far can lead to one of four results that are qualitatively quite different. As is noted in Table 9.1, the result of the procedure can be correct, but it can also be erroneous.

When the null hypothesis is true. When the null hypothesis is in fact true (and those who engage in hypothesis testing, of course, do not know this, or they would have no reason to undertake the test), two possible consequences can flow from a hypothesis test: the null hypothesis may be accepted (a correct result), or it may be rejected (an erroneous outcome). The latter possibility—the rejection of a true null hypothesis—is referred to as a **type I error,** or as the **error of rejection.** Such an error can be costly, indeed. It may lead to the interruption of the production process in order to adjust machinery that needs no adjusting; it may lead a firm to switch to a new supplier of raw materials although the performance of the old one was satisfactory; it may lead a government agency to condemn a firm for ignoring pollution standards even though it has done no such thing. Committing a type I error is equivalent to condemning a defendant in a criminal trial even though he or she is innocent. But note: the logic of hypothesis testing *guarantees* that this type of error occurs sometimes. Picture in your mind a sampling distribution, such as Figure 9.1, that shows all the possible values of a test statistic, given that the null hypothesis is true. Then recall what happens when we select a significance level for a hypothesis test. We arbitrarily place some of these possible values in the rejection region; we interpret the occurrence of those values that have a low probability of occurring when H_0 is true as evidence that H_0 is false. Hence, in the long run, as we use a given procedure again and again, we are bound to reject

Table 9.1 Evaluating the Outcome of a Hypothesis Test

Depending on the true state of the world, the result reached in a hypothesis test can be correct—(a) and (d)—or erroneous—(b) and (c). The probabilities associated with these outcomes are given in the brackets.

True State of the World	Test Result	
	Null Hypothesis is Accepted	Null Hypothesis is Rejected
Null Hypothesis is in Fact True	(a) Correct result $[p = 1 - \alpha]$	(b) Type I error $[p = \alpha]$
Null Hypothesis is in Fact False	(c) Type II error $[p = \beta]$	(d) Correct result $[p = 1 - \beta]$

a true null hypothesis the proportion of times we have set as the significance level. In short, the probability of making a type I error—of rejecting a null hypothesis that is true—is equal to α, the significance level of the hypothesis test, which is why it is also referred to as the **α risk.**

$$p \text{ (rejecting } H_0 | H_0 \text{ is true)} = \alpha = \text{significance level of the hypothesis test}$$

It follows that the complementary probability of avoiding a type I error when the null hypothesis is true equals $1 - \alpha$, which is also called the **confidence level of the hypothesis test.**

$$p \text{ (accepting } H_0 | H_0 \text{ is true)} = 1 - \alpha = \text{confidence level of the hypothesis test}$$

When the null hypothesis is false. When the null hypothesis is in fact false, a hypothesis test once again leads to one of two possible consequences: the null hypothesis may be accepted (an erroneous outcome), or it may be rejected (a correct result). The former possibility—the acceptance of a false null hypothesis—is referred to as a **type II error,** or as the **error of acceptance.** Such an error, too, has its costs. It may lead to uninterrupted production at a time when the machinery needs to be adjusted badly (because this fact is not recognized); it may lead a firm to continue using a supplier whose performance is unsatisfactory (because this fact is not recognized); it may lead a government agency to do nothing about a firm's violation of pollution standards (because this fact is not recognized). Committing a type II error is equivalent to acquitting a defendant in a criminal trial even though he or she is guilty. And just as in the field of criminal justice it is deemed less serious to let a guilty person go than to convict an innocent one, it is common practice to phrase the null hypothesis in such a way that a type I error is more serious than a type II error.

The probability of committing a type II error is symbolized by β and is referred to as the **β risk.**

$$p \text{ (accepting } H_0 | H_0 \text{ is false)} = \beta$$

In turn, the complementary probability of avoiding a type II error when the null hypothesis is false equals $1 - \beta$, which is also called the **power of the hypothesis test.**

$$p \text{ (rejecting } H_0 | H_0 \text{ is false)} = 1 - \beta = \text{power of the hypothesis test}$$

The Trade-Off Between a Type I Error and a Type II Error

Whoever undertakes to test a hypothesis knows neither before sampling nor afterwards whether a type I or type II error has been committed. It is possible, however, to control the probability of having a given error occur. Unfortunately, given sample size, n, anything that reduces α automatically raises β; the reverse is also true, as can easily be seen with the help of Figure 9.3. Panel (a) of Figure 9.3 is a copy of panel (c) of Figure 9.2. It shows the sampling distribution of the mean drying time of paint when the null hypothesis ("average drying time equals at most 4 hours") is in fact true. If a significance level of $\alpha = .05$ is selected and if, as in the earlier example, $\sigma_{\overline{x}} = .5$, the critical values are $\overline{X} = 4.82$ hours and $z = 1.64$, respectively. The probability of making a type I error, thus, appears as the crosshatched area in panel (a); it equals the area under the upper tail of the sampling distribution prevailing if H_0 is true.

Panel (b) of Figure 9.3 shows the sampling distribution when the null hypothesis is in fact false (and the average drying time equals 5.8 hours). The critical value of $\overline{X} = 4.82$ (that was established on the basis of assuming H_0 to be true) now is shown to be lower than the mean of the true panel (b) sampling distribution; a corresponding z value is -1.96. To the left of this value (and, thus, in the acceptance region), we find .025 of the area under the standard normal curve. This crosshatched area in panel (b) gives us the probability of making a type II error. For any given sample, the selection of α, which leads to the decision rule, thus, ultimately determines β.

Two things should be noted about type I and type II errors:

1. A lower α (which would move the dashed borderline between the acceptance and rejection region to the right) would clearly raise β. A higher α (which would move that borderline left) would lower β. As we shall see in a later section of this chapter, only a larger sample size can reduce α and β at the same time (or can reduce one of the two without raising the other).

2. Unscrupulous investigators clearly could manipulate the size of α (and, thus, the position of the borderline between the acceptance and rejection regions) so as to accommodate *any* observed value of a test statistic. Someone eager to have a null hypothesis accepted could make α extremely small (at the cost of having β very large). Someone eager to have a null hypothesis rejected could make α extremely large (which would make β very small). Obviously, such behavior would make a mockery of scientific inquiry. The only way to avoid it is to select α *before* the investigator has seen the sample data. Step 3 must be performed before Step 4.

A HISTORICAL PERSPECTIVE

Although the construction of formal hypothesis tests is a 20th-century innovation in statistical theory, certain ideas underlying this theory have a long history, going back at least eight centuries, probably to the reign of England's King Henry II (1154–89). Close-Up 9.1, "The Trial of the Pyx," tells the story. The modern theory was developed

Figure 9.3   **The Trade-Off Between α and β**

Someone about to test a null hypothesis obviously does not know whether it is true or false and whether, therefore, the sampling distribution looks like that shown in panel (a) or like that in panel (b). By assuming the former and selecting a value for α, a decision rule is derived; this decision rule automatically determines the value of β if H_0 is in fact false. Given sample size, any decrease in α raises β (as the dashed vertical line moves right). The opposite is also true; any increase in α lowers β (as that line moves left). This result again is analogous to our experience with the criminal justice system: anything that lowers the probability, α, of convicting an innocent person (such as a rule against self-incrimination) also raises the probability, β, of acquitting a guilty one. Anything that lowers the probability, β, of acquitting a guilty person (such as a rule that allows split juries to convict) thereby raises the probability, α, of convicting an innocent one.

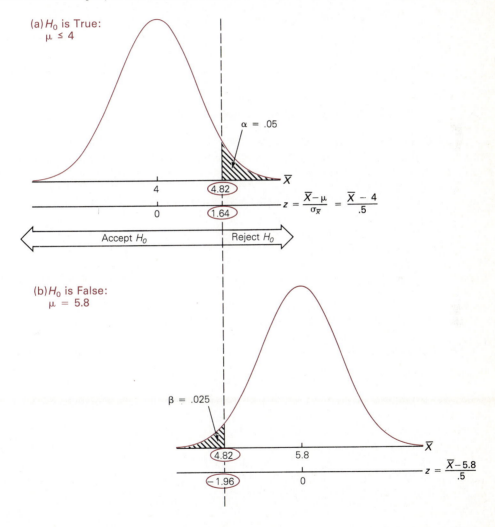

in England, too, but not without a lot of controversy, as is evidenced in Biographies 9.1 to 9.3 at the end of the chapter. The following sections show applications of the modern theory.

LARGE-SAMPLE TESTS

Let us now apply the four-stage procedure introduced above to typical testing situations. We begin by assuming that the samples to be taken are large ($n \geq 30$ but $n <$ $.05N$ for tests involving one or two means; $n \cdot \pi$ and $n(1 - \pi) \geq 5$ for tests involving one or two proportions); therefore, we can assume that the test statistic's sampling distribution is approximated by the normal curve.

Tests of a Population Mean

We first consider each of the three types of hypothesis tests of a single population mean.

||| Practice Problem 1

The Thickness of Aluminum Sheets (a two-tailed test)

An aircraft manufacturer needs aluminum sheets that are .01 inches thick—no more, no less. The firm's quality inspector is to test the quality of supplies accordingly by measuring thickness in a simple random sample of incoming sheets.

Step 1: Formulating two opposing hypotheses.

H_0: $\mu = .01$ inch
H_A: $\mu \neq .01$ inch

Step 2: Selecting a test statistic.[1] The inspector selects the standard normal deviate for the sample mean thickness of the sheets,

$$z = \frac{\overline{X} - \mu_0}{\sigma_{\overline{X}}}.$$

Step 3: Deriving a decision rule. The inspector selects a significance level of $\alpha = .05$, which (according to Appendix Table J) implies critical values of $\pm z_{\alpha/2} = \pm 1.96$ (this being a two-tailed test). Thus, the decision rule must be: "Accept H_0 if $-1.96 \leq z \leq +1.96$." The critical values are encircled in Figure A.

Step 4: Selecting a sample, computing the test statistic, and confronting it with the decision rule. After taking a sample of $n = 100$, the inspector finds a sample mean thickness of $\overline{X} = .009$ and a sample standard deviation of $s = .01$. Given an infinite population (the sheets come from an ongoing production process), the large-population case applies; hence $\sigma_{\overline{X}}$ is estimated as

$$\sigma_{\overline{X}} \cong \frac{s}{\sqrt{n}} = \frac{.01}{\sqrt{100}} = \frac{.01}{10} = .001.$$

---

[1]See Appendix 9A, "Commonly Used Test Statistics," at the end of this chapter for a discussion of how to determine the test statistic for a given type of hypothesis test.

Figure A

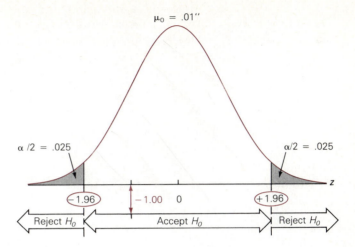

Accordingly, the computed value of the test statistic equals

$$z = \frac{\overline{X} - \mu_0}{\sigma_{\overline{X}}} = \frac{.009 - .01}{.001} = -1.00.$$

This value corresponds to the colored arrow in Figure A; it suggests that the null hypothesis should be *accepted*. At the 5 percent significance level, the sample result is not statistically significant. The observed divergence from the desired average .01-inch standard of thickness is more likely due to chance factors at work during the sampling process than due to a faulty production process that systematically puts out sheets with an average thickness other than .01 inch. **III**

||| Practice Problem 2

The Tensile Strength of Steel Rods (a lower-tailed test)

For purposes of mounting engines on aircraft, a manufacturer needs specialty steel rods with a tensile strength of at least 5,000 pounds. The firm's quality inspector is to test incoming supplies accordingly by figuring the tensile strength evidenced in a simple random sample of these rods. (This test is performed by noting the force at which the rods become distorted.)

Step 1: Formulating two opposing hypotheses.

H_0: $\mu \geq 5,000$ pounds
H_A: $\mu < 5,000$ pounds

Step 2: Selecting a test statistic. The inspector selects the standard normal deviate for the sample mean tensile strength of rods,

$$z = \frac{\overline{X} - \mu_0}{\sigma_{\overline{X}}}.$$

Step 3: Deriving a decision rule. The inspector selects a significance level of $\alpha = .01$, which (according to Appendix Table J) implies a critical value of $-z_\alpha = -2.33$ (this being a lower-tailed test). Thus, the decision rule must be: "Accept H_0 if $z \geq -2.33$." The critical value is encircled in Figure B.

Figure B

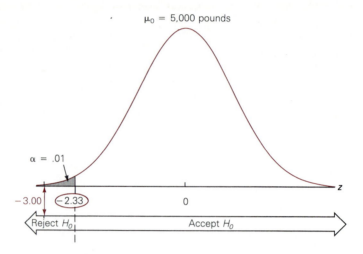

$\mu_0 = 5,000$ pounds

$\alpha = .01$

-3.00 (-2.33) 0 z

Reject H_0 Accept H_0

Step 4: Selecting a sample, computing the test statistic, and confronting it with the decision rule. After taking a sample of 64 rods, the inspector finds that the rods become distorted by a mean strength of $\overline{X} = 4,700$ pounds, the sample standard deviation being $s = 800$ pounds. Once again, the large-population case applies, so $\sigma_{\overline{X}}$ is estimated as

$$\sigma_{\overline{X}} \cong \frac{s}{\sqrt{n}} = \frac{800}{\sqrt{64}} = \frac{800}{8} = 100 \text{ pounds.}$$

Accordingly, the computed value of the test statistic equals

$$z = \frac{\overline{X} - \mu_0}{\sigma_{\overline{X}}} = \frac{4,700 - 5,000}{100} = -3.00.$$

This value corresponds to the colored arrow in Figure B; it suggests that the null hypothesis should be *rejected*. At the 1-percent significance level, the sample result is statistically significant. The observed divergence from the 5,000-pound standard is unlikely to be due to chance factors operating during sampling; it is more likely to be the result of a production process that puts out low-quality (and, therefore, unusable) rods. ▐▐▐

▐▐▐ Practice Problem 3

Additive in Aviation Fuel (an upper-tailed test)

An airline wants to test a manufacturer's claim that 1 ounce of additive will significantly increase the performance of aviation fuel. The airline's records show that the fuel exclusive of the additive has delivered an average of 6 minutes of flight time per gallon (mpg). Future flights are to be flown with the additive; a random sample is to be evaluated by the airline's statistician who has reason to assume that the minutes-per-gallon population will be normally distributed.

Step 1: Formulating two opposing hypotheses.

H_0: $\mu \leq 6$ minutes per gallon
H_A: $\mu > 6$ minutes per gallon

Step 2: Selecting a test statistic. The statistician selects the sample mean minutes per gallon, $\overline{X}$.

Step 3: Deriving a decision rule. The statistician selects a significance level of $\alpha = .10$, which (according to Appendix Table J) implies a critical value of $+z_\alpha = 1.28$ (this being an upper-tailed test). Thus, the decision rule must be: "Accept H_0 if $\overline{X} \leq \mu_0 + (1.28\ \sigma_{\overline{X}})$." Assuming the population standard deviation, σ, is unknown, the precise magnitude of this critical $\overline{X}$ cannot be determined until after the sample has been taken and $\sigma_{\overline{X}} = (\sigma/\sqrt{n})$ can be estimated as $s/\sqrt{n}$. The critical value of 6.1 given in Figure C is in this case found in Step 4.

Figure C

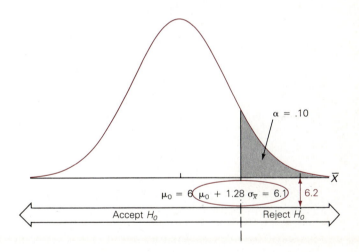

Step 4: Selecting a sample, computing the test statistic, and confronting it with the decision rule. After 100 flights have been flown with the additive, the statistician takes a sample of $n = 36$ and finds the mean minutes per gallon to be $\overline{X} = 6.2$, with a sample standard deviation of $s = .6$ minutes. Because the small-population case applies ($n \geq .05N$), the statistician estimates $\sigma_{\overline{X}}$ as

$$\sigma_{\overline{X}} \cong \frac{s}{\sqrt{n}} \sqrt{\frac{N-n}{N-1}} = \frac{.6}{\sqrt{36}} \sqrt{\frac{100-36}{100-1}} = .1(.804) = .0804 \text{ minutes per gallon.}$$

Accordingly, the critical value of $\overline{X}$ equals $6 + [1.28(.0804)] = 6.1$ minutes per gallon. This result suggests that the null hypothesis should be *rejected*. At the 10 percent significance level, the sample result is statistically significant. The additive does stretch the flight time squeezed out of a gallon. ▮▮▮

Note: This is a good place to observe that *statistical* significance is not necessarily the same thing as *practical* significance. The statistical significance of this test simply tells us that the divergence of the sample result (an average 6.2 minutes of flight time per gallon) from the previous experience (an average 6 minutes per gallon) is probably not merely the result of the sampling process, but quite likely attributable to the additive. Yet the practical significance of all this may be nil: it may be cheaper, for example, to buy the added 3.3 percent of flight time by paying for extra fuel than by paying for an expensive additive. (This example also highlights the fact that the term *statistically significant,* being a technical term, must not be confused with such everyday terms as *important* or *impressive.*)

Tests of a Population Proportion

Hypothesis tests concerning a single population proportion are conducted analogously to those about a mean; Practice Problem 4 is a two-tailed test and Practice Problem 5 is a lower-tailed test.[2]

||| Practice Problem 4

The Weights of Drug Dosages (a two-tailed test)

A hospital administrator needs to know whether it is true that 90 percent of the drug dosages prepared by a machine weigh precisely 100 milligrams. A random sample of dosages from among thousands prepared during a week is to be taken to evaluate the claim.

Step 1: Formulating two opposing hypotheses.

H_0: $\pi = .9$
H_A: $\pi \neq .9$

Step 2: Selecting a test statistic.[3] The evaluator selects the standard normal deviate for the sample proportion,

$$z = \frac{P - \pi_0}{\sigma_P}.$$

Step 3: Deriving a decision rule. A significance level of .05 is set; hence, critical values are $\pm z_{\alpha/2} = \pm 1.96$ (this being a two-tailed test). Thus, the decision rule must be: "Accept H_0 if $-1.96 \leq z \leq +1.96$." The critical values are encircled in Figure D.

Step 4: Selecting a sample, computing the test statistic, and confronting it with the decision rule. After taking a sample of $n = 100$, it is found that the sample proportion

---

[2]For an illustration of an upper-tailed test of a population proportion, see Problem 13 in Chapter 9 of the *Student Workbook* that accompanies this text.

[3]See Appendix 9A.

Figure D

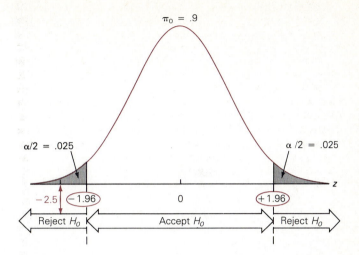

of precise 100-milligram dosages equals $P = .8$. For the large-population case, σ_P is, thus, estimated as

$$\sigma_P \cong \sqrt{\frac{P(1 - P)}{n}} = \sqrt{\frac{.8(.2)}{100}} = .04.$$

Accordingly, the computed value of the test statistic equals

$$z = \frac{P - \pi_0}{\sigma_P} = \frac{.8 - .9}{.04} = -2.5.$$

This value corresponds to the colored arrow in Figure D; it suggests that the null hypothesis should be *rejected.* At the 5-percent significance level, the sample result is statistically significant. The observed divergence between the hypothesized population proportion and the sample proportion is unlikely to be the result of pure chance. ‖

‖ Practice Problem 5

Brand-Name Recognition (a lower-tailed test)

A firm believes that at least 50 percent of shoppers entering a department store recognize its brand name. A random sample of shoppers is to test the claim.

Step 1: Formulating two opposing hypotheses.

H_0: $\pi \geq .5$
H_A: $\pi < .5$

Step 2: Selecting a test statistic. The evaluator selects the sample proportion, P.

Step 3: Deriving a decision rule. A significance level of .005 is set, which implies a critical value of $-z_\alpha = -2.57$. Thus, the decision rule must be: "Accept H_0 if $P \geq \pi_0 - (2.57\sigma_P)$." The precise magnitude of this critical P cannot be determined until after the sample has been taken and σ_P has been estimated. The critical value of .25 given in Figure E can be found only in Step 4.

Figure E

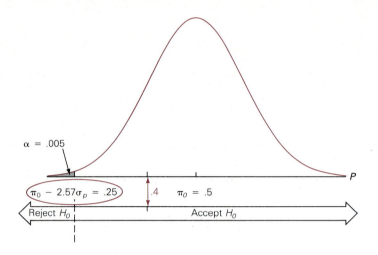

Step 4: Selecting a sample, computing the test statistic, and confronting it with the decision rule. After taking a sample of $n = 25$, it is found that 10 of the 25 shoppers recognize the brand name; thus, $P = .4$. Hence,

$$\sigma_P \cong \sqrt{\frac{P(1 - P)}{n}} = \sqrt{\frac{.4(.6)}{25}} = .098.$$

Accordingly, the critical value of P equals $.5 - [2.57(.098)] = .25$. This result suggests that the null hypothesis should be *accepted.* At the .5 percent significance level, the sample result is not statistically significant. It is quite consistent with at least 50 percent of *all* shoppers recognizing the brand name. ▮▮▮

Tests of the Difference Between Two Population Means: Independent Samples

We first consider hypothesis tests of the difference between means, while employing *independent samples* such that the elements making up the sample taken from population A are chosen independently of the elements making up the sample taken from population B.[4]

---

[4]Practice Problems 6 and 7 here illustrate a two-tailed test and an upper-tailed test; for an illustration of a lower-tailed independent-samples test of the difference between two population means, see Problem 6 or Problem 14 in Chapter 9 of the *Student Workbook* that accompanies this text.

||| Practice Problem 6

The Lifetimes of Aircraft Radios (a two-tailed test)

An airline wants to test the claim that the mean lifetimes of two types of aircraft radios are identical. It installs 800 radios of each type in its current fleet and later selects two simple random samples of radios of each type.

Step 1: Formulating two opposing hypotheses.

H_0: $\mu_A - \mu_B = 0$
H_A: $\mu_A - \mu_B \neq 0$

Step 2: Selecting a test statistic.[5] The firm's statistician selects

$$z = \frac{d}{\sigma_d} = \frac{\overline{X}_A - \overline{X}_B}{\sigma_{\overline{X}_A - \overline{X}_B}}.$$

Step 3: Deriving a decision rule. The statistician selects a significance level of $\alpha = .10$, which implies critical values of $\pm z_{\alpha/2} = \pm 1.64$ (this being a two-tailed test). Thus, the decision rule must be: "Accept H_0 if $-1.64 \leq z \leq + 1.64$." The critical values are encircled in Figure F.

Figure F

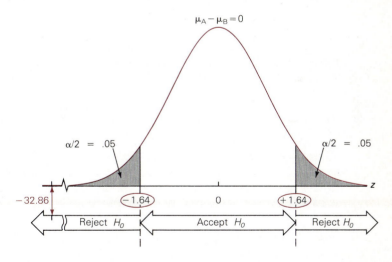

Step 4: Selecting the samples, computing the test statistic, and confronting it with the decision rule. After taking two samples of $n = 36$ each, the statistician finds mean lifetimes and sample standard deviations of $\overline{X}_A = 4,120$ hours, $s_A = 80$ hours and $\overline{X}_B = 4,910$ hours, $s_B = 120$ hours. The large-population case applies ($n \leq .05N$); σ_d is estimated as

$$\sigma_d \cong \sqrt{\frac{s_A^2}{n_A} + \frac{s_B^2}{n_B}} = \sqrt{177.78 + 400} = 24.04.$$

---

[5]See Appendix 9A.

Accordingly, the computed value of the test statistic equals

$$z = \frac{d}{\sigma_d} = \frac{4{,}120 - 4{,}910}{24.04} = -32.86.$$

This value corresponds to the colored arrow in Figure F; it suggests that the null hypothesis should be *rejected*. At the 10-percent significance level, the sample result is statistically significant. It is very likely that radio B has a longer lifetime than radio A. ▐▐▐

▐▐▐ Practice Problem 7

The Spraying of Fruit Trees (an upper-tailed test)

An orchardist wants to compare the mean yield of fruit trees sprayed with gypsy-moth parasites (A) with that of fruit trees sprayed with traditional pesticides (B). Some 250 trees are sampled in each of two large orchards that were given one treatment or the other.

Step 1: Formulating two opposing hypotheses.

H_0: $\mu_A - \mu_B \leq 0$
H_A: $\mu_A - \mu_B > 0$

Step 2: Selecting a test statistic. The orchardist selects the standard normal deviate,

$$z = \frac{d}{\sigma_d}.$$

Step 3: Deriving a decision rule. The orchardist selects a significance level of .005, which implies a critical value of $z_\alpha = 2.57$ (this being an upper-tailed test). Thus, the decision rule must be: "Accept H_0 if $z \leq 2.57$." The critical value is encircled in Figure G.

Figure G

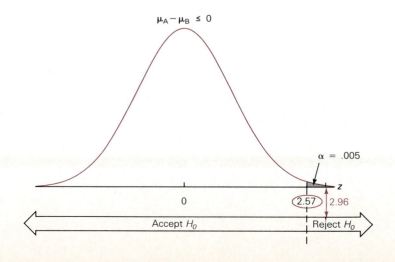

Step 4: Selecting the samples, computing the test statistic, and confronting it with the decision rule. After taking the samples, the orchardist finds mean fruit yields and sample standard deviations of $\overline{X}_A = 10.5$ bushels, $s_A = 8$ bushels and $\overline{X}_B = 8.9$ bushels, $s_B = 3$ bushels. The large-population case applies; σ_d is estimated as

$$\sigma_d \cong \sqrt{\frac{s_A^2}{n_A} + \frac{s_B^2}{n_B}} = \sqrt{.256 + .036} = .54.$$

Accordingly, the computed value of the test statistic equals

$$z = \frac{d}{\sigma_d} = \frac{10.5 - 8.9}{.54} = 2.96.$$

This value corresponds to the colored arrow in Figure G; it suggests that the null hypothesis should be *rejected*. At the .5-percent significance level, the sample result is statistically significant. Parasites are more effective than pesticides. |||

Tests of the Difference Between Two Population Means: Matched-Pairs Samples

We now turn to hypothesis tests of the difference between means, while employing *matched-pairs samples,* taken after each elementary unit in population A has been matched with a "twin" from population B.

||| Practice Problem 8

Smoking and Heart Disease (a two-tailed test)
A medical researcher wants to test whether there exists no connection between smoking and heart disease because the mean age at which heart disease is first detected is the same for smokers (A) and nonsmokers (B). Some 100 smokers are matched with 100 nonsmokers according to age, lifestyle, medical history, occupation, sex, and so on.

Step 1: Formulating two opposing hypotheses.

$H_0: \mu_A - \mu_B = 0$
$H_A: \mu_A - \mu_B \neq 0$

Step 2: Selecting a test statistic.[6] The researcher selects

$$z = \frac{\overline{D}}{s_D/\sqrt{n}}$$

Step 3: Deriving a decision rule. The researcher selects a significance level of $\alpha = .05$, which implies critical values of $\pm z_{\alpha/2} = \pm 1.96$ (this being a two-tailed test). Thus, the decision rule must be: "Accept H_0 if $-1.96 \leq z \leq +1.96$." The critical values are encircled in Figure H.

---

[6]See Appendix 9A.

Figure H

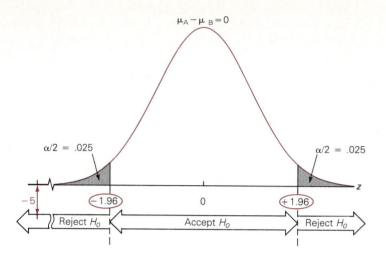

Step 4: Selecting the sample, computing the test statistic, and confronting it with the decision rule. After taking the sample, the researcher finds the mean age of the onset of heart disease to be 6 years earlier for smokers than for their. "twin" nonsmokers; thus, $\overline{D} = -6$ years. The sample standard deviation of $s_D = 12$ years. Thus, the computed value of the test statistic equals

$$z = \frac{\overline{D}}{s_D/\sqrt{n}} = \frac{-6}{12/\sqrt{100}} = -5.$$

This value corresponds to the colored arrow in Figure H; it suggests that the null hypothesis should be *rejected.* At the 5-percent significance level, the sample result is statistically significant. Because the test statistic is negative, we conclude, furthermore, that smoking *hastens* the onset of heart disease. (If z had been in the right-hand rejection region, it would indicate that smoking delays heart disease.) |||

||| Practice Problem 9

Fertilizing Strawberries (a lower-tailed test)

An agronomist who has matched adjacent plots of land is growing strawberries on 49 such pairs, always using nitrite-based fertilizer on one "twin" (A) and phosphate-based fertilizer on the other (B). The agronomist wants to determine whether (as claimed) nitrites lead to yields greater than or equal to phosphates.

Step 1: Formulating two opposing hypotheses.

$H_0: \mu_A - \mu_B \geq 0$
$H_A: \mu_A - \mu_B < 0$

Step 2: Selecting a test statistic. The agronomist selects

$$z = \frac{\overline{D}}{s_D/\sqrt{n}}.$$

Step 3: *Deriving a decision rule*. The researcher selects a significance level of $\alpha = .1$, which implies a critical value of $-z_\alpha = -1.28$ (this being a lower-tailed test). Thus, the decision rule must be: "Accept H_0 if $z \geq -1.28$." The critical value is encircled in Figure I.

Figure I

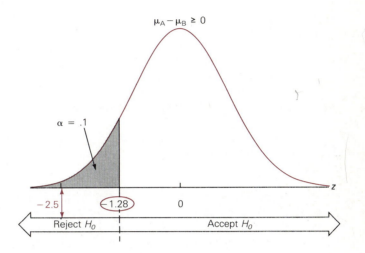

Step 4: *Selecting the sample, computing the test statistic, and confronting it with the decision rule*. After taking the sample, the agronomist finds the mean yield on A plots to be 50 quarts less than on B plots; thus, $\overline{D} = -50$ quarts. The sample standard deviation is $s_D = 140$ quarts. Thus, the computed value of the test statistic equals

$$z = \frac{\overline{D}}{s_D/\sqrt{n}} = \frac{-50}{140/\sqrt{49}} = -2.5.$$

This value corresponds to the colored arrow in Figure I; it suggests that the null hypothesis should be *rejected*. At the 10-percent significance level, the sample result is statistically significant. Phosphates, not nitrates, lead to a greater yield of strawberries. |||

Tests of the Difference Between Two Population Proportions: Independent Samples

We, finally, consider large-sample hypothesis tests of the difference between proportions, again employing independent samples.

||| Practice Problem 10

Sex and TV-Program Preference (a two-tailed test)

A television station wants to know whether there is any difference in the proportion of males (A) or females (B) who favor a given program over another during a certain time slot. Two simple random samples of 100 each are to be taken from among the city's adult population.

Step 1: Formulating two opposing hypotheses.

$H_0: \pi_A - \pi_B = 0$
$H_A: \pi_A - \pi_B \neq 0$

Step 2: Selecting a test statistic.[7] The station manager selects

$$z = \frac{d}{\sigma_d} = \frac{P_A - P_B}{\sigma_{P_A - P_B}}.$$

Step 3: Deriving a decision rule. The station manager selects a significance level of $\alpha = .05$, which implies critical values of $\pm z_{\alpha/2} = \pm 1.96$ (this being a two-tailed test). Thus, the decision rule must be: "Accept H_0 if $-1.96 \leq z \leq +1.96$." The critical values are encircled in Figure J.

Figure J

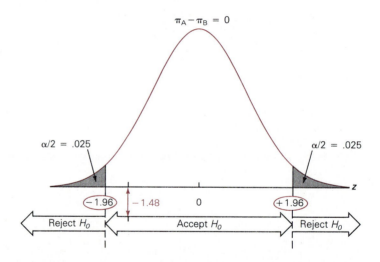

Step 4: Selecting the samples, computing the test statistic, and confronting it with the decision rule. After taking the samples, the manager finds proportions of $P_A = .6$ and $P_B = .7$ and then calculates the pooled estimator of the population proportion as

$$P_P = \frac{100(.6) + 100(.7)}{200} = .65.$$

Accordingly, σ_d is estimated as

$$\sigma_d \cong \sqrt{P_P(1 - P_P)\left(\frac{1}{n_A} + \frac{1}{n_B}\right)} = \sqrt{.65(.35)\left(\frac{1}{100} + \frac{1}{100}\right)} = .0675.$$

---

[7]See Appendix 9A.

Thus, the value of the test statistic is computed as

$$z = \frac{d}{\sigma_d} = \frac{.6 - .7}{.0675} = -1.48.$$

This value corresponds to the colored arrow in Figure J; it suggests that the null hypothesis should be *accepted*. At the 5-percent significance level, the sample result is not statistically significant. It is quite likely that the proportions of males and females are the same.

Note: Analytical Example 9.2, "Antitrust Pork Barrel," provides another instance of the use of this procedure, as do Close-Ups 9.2, "Death Month and Birth Month: An Unexpected Connection," and 9.3, "The Case of the Missing Women." III

||| Practice Problem 11

Enforcing Antipollution Standards (a lower-tailed test)

A government agency wants to know whether it is true that a larger proportion of plants in region A than in B are abiding by its antipollution standards. It samples 60 plants in region A and 30 plants in region B.

Step 1: Formulating two opposing hypotheses.

$$H_0: \pi_A - \pi_B \geq 0$$
$$H_A: \pi_A - \pi_B < 0$$

Step 2: Selecting a test statistic. The agency director selects the standard normal deviate,

$$z = \frac{d}{\sigma_d}.$$

Step 3: Deriving a decision rule. The agency director selects a significance level of .05, which implies a critical value of $-z_\alpha = -1.64$ (this being a lower-tailed test). Thus, the decision rule must be: "Accept H_0 if $z \geq -1.64$." The critical value is encircled in Figure K.

Step 4: Selecting the samples, computing the test statistic, and confronting it with the decision rule. After selecting the samples from among thousands of plants in each region, the director finds the proportions of those meeting the standards as $P_A = .7$ and $P_B = .5$. Thus, the pooled estimator of the population proportion is

$$P_P = \frac{60(.7) + 30(.5)}{90} = .63.$$

Accordingly, σ_d is estimated as

$$\sigma_d \cong \sqrt{P_P(1 - P_P)\left(\frac{1}{n_A} + \frac{1}{n_B}\right)} = \sqrt{.63(.37)\left(\frac{1}{60} + \frac{1}{30}\right)} = .108$$

Figure K

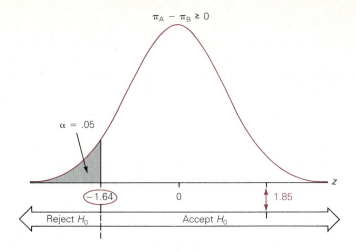

Thus, the value of the test statistic is computed as

$$z = \frac{d}{\sigma_d} = \frac{.7 - .5}{.108} = 1.85.$$

This value corresponds to the colored arrow in Figure K; it suggests that the null hypothesis should be *accepted*. At the 5-percent significance level, the sample result is not statistically significant. It seems that the proportion of plants complying with government standards is, indeed, larger in region A than in region B. ▌▌▌

SMALL-SAMPLE TESTS

We now turn to cases where samples are small ($n < 30$), but the test statistic's sampling distribution can be approximated by a t distribution (because the underlying population values are approximately normally distributed and because, when two populations are involved, their standard deviations, though unknown, are known to be equal). Having discussed large-sample tests in such detail, we can restrict ourselves here to fewer examples, and we can also omit the graphical illustrations. (All graphs would be analogous to those given in the previous section, except that t distributions for specified degrees of freedom would replace the normal curves.)

Tests of a Population Mean

We first consider two tests concerning a single population mean.[8]

---

[8]Practice Problems 12 and 13 illustrate a two-tailed test and a lower-tailed test; for an illustration of an upper-tailed small-sample test of a population mean, see Problem 9 or Problem 15 in Chapter 9 of the *Student Workbook* that accompanies this text.

|||| Practice Problem 12

The Length of Long-Distance Calls (a two-tailed test)

A telephone company executive wants to know whether it is still true (as in the past) that the average length of evening long-distance telephone calls equals 18.1 minutes. A simple random sample of 25 of an evening's calls is to be used to find the answer.

Step 1: Formulating two opposing hypotheses.

H_0: μ = 18.1 minutes
H_A: $\mu \neq$ 18.1 minutes

Step 2: Selecting a test statistic. The company statistician selects

$$t = \frac{\overline{X} - \mu_0}{s/\sqrt{n}},$$

given that the population of telephone-call durations is likely to be normally distributed and that the large-population case applies ($n < .05N$ because of the large number of calls made).

Step 3: Deriving a decision rule. The statistician selects a significance level of $\alpha = .05$, which (according to Appendix Table K) implies critical values of $\pm t_{\alpha/2} = \pm t_{.025,24} = \pm 2.064$ (this being a two-tailed test and there being $n - 1 = 24$ degrees of freedom). Thus, the decision rule must be: "Accept H_0 if $-2.064 \leq t \leq +2.064$."

Step 4: Selecting a sample, computing the test statistic, and confronting it with the decision rule. After taking the sample, the statistician finds a sample mean duration of calls of $\overline{X} = 17.2$ minutes and a sample standard deviation of $s = 4$ minutes. Thus, the value of the test statistic is computed as

$$t = \frac{\overline{X} - \mu_0}{s/\sqrt{n}} = \frac{17.2 - 18.1}{4/\sqrt{25}} = -1.125.$$

Accordingly, the null hypothesis should be *accepted*. At the 5-percent level of significance, the sample result (a somewhat smaller duration of the calls than noted in the past) is not statistically significant. The observed divergence is likely due to chance factors. |||

|||| Practice Problem 13

The Weight of Raisin Boxes (a lower-tailed test)

A government agency has received a consumer complaint that boxes allegedly containing 15 ounces of raisins actually contain less. A simple random sample of 10 such boxes is to be selected from various stores to clarify the issue.

Step 1: Formulating two opposing hypotheses.

H_0: $\mu \geq$ 15 ounces
H_A: $\mu <$ 15 ounces

Step 2: Selecting a test statistic. The government statistician selects

$$t = \frac{\overline{X} - \mu_0}{s/\sqrt{n}},$$

given the likelihood of a normal distribution of the population of net weights and the large-population case ($n < .05N$ because of the large number of such boxes being produced and sold).

Step 3: Deriving a decision rule. The government desires a significance level of $\alpha = .01$, which (according to Appendix Table K) implies a critical value of $-t_\alpha = -t_{.01,9} = -2.821$ (this being a lower-tailed test and there being $n - 1 = 9$ degrees of freedom). Thus, the decision rule must be: "Accept H_0 if $t \geq -2.821$."

Step 4: Selecting a sample, computing the test statistic, and confronting it with the decision rule. After taking the sample, the statistician finds a sample mean weight of $\overline{X} = 13.5$ ounces and a sample standard deviation of $s = 1$ ounce. Thus, the value of the test statistic is computed as

$$t = \frac{\overline{X} - \mu_0}{s/\sqrt{n}} = \frac{13.5 - 15}{1/\sqrt{10}} = -4.743.$$

Accordingly, the null hypothesis should be *rejected.* At the 1-percent level of significance, the sample result in satistically significant. The observed divergence of the mean sample weight from the advertised weight is unlikely to be the result of chance factors operating in the sampling process; it is more likely due to the fact that the population of raisin boxes being marketed has a mean net weight below 15 ounces, as the alternative hypothesis claims. |||

Tests of the Difference Between Two Population Means: Independent Samples

We will consider two examples of hypothesis tests concerning two means that involve independent samples.

||| Practice Problem 14

The Cavity-Fighting Ability of Two Toothpastes (a two-tailed test)

The American Dental Association wants to determine whether there is a difference in the cavity-fighting ability of two toothpastes, A and B. A simple random sample of 21 users of each type is taken, and the mean number of cavities over a decade is counted.

Step 1: Formulating two opposing hypotheses.

$H_0: \mu_A - \mu_B = 0$
$H_A: \mu_A - \mu_B \neq 0$

Step 2: Selecting a test statistic. The investigator selects

$$t = \frac{\overline{X}_A - \overline{X}_B}{\sqrt{s_P^2\left(\dfrac{1}{n_A} + \dfrac{1}{n_B}\right)}},$$

given that the two populations of numbers of cavities are likely to be normally distributed and that the population variances are likely to be the same.[9]

Step 3: Deriving a decision rule. The investigator selects a significance level of $\alpha = .01$, which (according to Appendix Table K) implies critical values of $\pm t_{\alpha/2} = \pm t_{.005,40} = \pm 2.704$ (this being a two-tailed test and there being $n_A + n_B - 2 = 40$ degrees of freedom). Thus, the decision rule must be: "Accept H_0 if $-2.704 \leq t \leq +2.704$."

Step 4: Selecting the samples, computing the test statistic, and confronting it with the decision rule. After taking the samples, the investigator finds sample mean numbers of cavities (and associated sample standard deviations) of $\overline{X}_A = 27$ and $\overline{X}_B = 23$ (with $s_A = 6$ and $s_B = 2$). Thus, the pooled variance is

$$s_P^2 = \frac{(n_A - 1)s_A^2 + (n_B - 1)s_B^2}{n_A + n_B - 2} = \frac{20(6)^2 + 20(2)^2}{40} = 20,$$

while the value of the test statistic is computed as

$$t = \frac{\overline{X}_A - \overline{X}_B}{\sqrt{s_P^2\left(\dfrac{1}{n_A} + \dfrac{1}{n_B}\right)}} = \frac{27 - 23}{\sqrt{20\left(\dfrac{1}{21} + \dfrac{1}{21}\right)}} = 2.898.$$

Accordingly, the null hypothesis should be *rejected.* At the 1-percent significance level, the sample result (fewer cavities with toothpaste B) is statistically significant. The observed divergence of the sample difference from the hypothesized zero population difference is unlikely to be due to chance; toothpaste B does a better job.

Note: Analytical Example 9.1, "The Never-Ending Search for New and Better Drugs," provides another instance of the use of this procedure. |||

||| Practice Problem 15

Annual Income from Leasing Arrangements (an upper-tailed test)

A property manager of thousands of apartments wants to test the difference in net annual income between two types of leasing arrangements: arrangement A is to charge a lower rent but to require that tenants make repairs; arrangement B is to charge a

---

[9]As noted in Appendix 9A, s_P^2 is an estimate of these equal population variances derived by pooling the information contained in the two sample variances.

higher rent and to provide that the landlord makes repairs. Two simple random samples of $n_A = 15$ and $n_B = 12$ leases are to be used to test the claim that A produces at most the same mean net annual income as B.

Step 1: Formulating two opposing hypotheses.

$H_0: \mu_A - \mu_B \leq 0$
$H_A: \mu_A - \mu_B > 0$

Step 2: Selecting a test statistic. The manager selects *t*, as in the previous example.

Step 3: Deriving a decision rule. The manager selects a significance level of $\alpha = .025$, which (according to Appendix Table K) implies a critical value of $t_\alpha = t_{.025,25} = 2.060$ (this being an upper-tailed test and there being $n_A + n_B - 2 = 25$ degrees of freedom). Thus, the decision rule must be: "Accept H_0 if $t \leq 2.060$."

Step 4: Selecting the samples, computing the test statistic, and confronting it with the decision rule. After taking the samples, the manager finds sample mean net annual incomes (and associated sample standard deviations) of $\overline{X}_A = \$1,532.50$ and $\overline{X}_B = \$1,489.20$ (with $s_A = \$400$ and $s_B = \$100$). Thus, the pooled variance is

$$s_P^2 = \frac{(n_A - 1)s_A^2 + (n_B - 1)s_B^2}{n_A + n_B - 2}$$
$$= \frac{14(400)^2 + 11(100)^2}{25} = 94,000,$$

while the value of the test statistic is computed as

$$t = \frac{\overline{X}_A - \overline{X}_B}{\sqrt{s_P^2\left(\dfrac{1}{n_A} + \dfrac{1}{n_B}\right)}} = \frac{1,532.50 - 1,489.20}{\sqrt{94,000\left(\dfrac{1}{15} + \dfrac{1}{12}\right)}} = .365.$$

Accordingly, the null hypothesis should be *accepted.* At the 2.5-percent significance level, the sample result (more income from plan A) is not statistically significant. The manager is justified in the continuing belief that plan B is at least as good as plan A in creating net income for landlords.

Note: Analytical Example 9.3, "Do Nursing Homes Discriminate Against the Poor?" provides another example of this procedure. ▌▌▌

Tests of the Difference Between Two Population Means: Matched-Pairs Sampling

We consider, finally, a hypothesis test concerning two means that makes use of matched-pairs sampling.[10]

---

[10]Practice Problem 16 illustrates a two-tailed test; for an illustration of an upper-tailed, matched-pairs-sample test, see Problem 12 or Problem 16 in Chapter 9 of the *Student Workbook* that accompanies this text.

||| Practice Problem 16

Setting Retail Prices (a two-tailed test)

A manufacturer markets two models of a product, A and B. The manufacturer has suggested to retailers nationwide that the retail prices of the two models should be kept the same. The manufacturer wishes to test whether this suggestion is being followed by sampling 10 retail outlets, all of which sell both models.

Step 1: Formulating two opposing hypotheses.

H_0: $\mu_A - \mu_B = 0$
H_A: $\mu_A - \mu_B \neq 0$

Step 2: Selecting a test statistic. The statistician selects

$$t = \frac{\overline{D}}{s_D/\sqrt{n}}$$

given that each retail outlet in the sample provides two price observations (and, thus, the desired difference) and that the population of price differences in all of the nation's retail outlets is likely to be normally distributed.

Step 3: Deriving a decision rule. The statistician selects a significance level of $\alpha = .05$, which (according to Appendix Table K) implies critical values of $\pm t_{\alpha/2} = \pm t_{.025,9} = \pm 2.262$ (this being a two-tailed test and there being $n - 1 = 9$ degrees of freedom). Thus, the decision rule must be: "Accept H_0 if $-2.262 \leq t \leq +2.262$."

Step 4: Selecting the sample, computing the test statistic, and confronting it with the decision rule. After taking the sample, the company statistician finds an average difference in model A vs. B prices of $\overline{D} = \$18.23$ (along with a standard deviation of $s_D = \$1.25$). Thus, the computed value of the test statistic is

$$t = \frac{\overline{D}}{s_D/\sqrt{n}} = \frac{18.23}{1.25/\sqrt{10}} = 46.12.$$

Accordingly, the null hypothesis should be *rejected*. At the 5 percent significance level, the sample result (that a large price difference is being maintained) is statistically significant. Model A is being sold for considerably more than model B; the observed sample difference is not a fluke inherent in the sampling process. |||

ERROR PROBABILITIES AND OPTIMAL SAMPLE SIZE*

The traditional hypothesis-testing procedure, as noted above, leaves open the possibility of rejecting a null hypothesis even though it is true (a type I error) or of accepting a null hypothesis even though it is false (a type II error). Table 9.1 summarized the discussion about the possibility of error; it also designated the probabilities of committing the two types of error as α or β, respectively.

---

*Optional section.

Figure 9.3 illustrated, in turn, how the typical hypothesis test that proceeds from a given sample size, n, and then arbitrarily selects a maximum allowable value of α thereby also specifies (though indirectly) a maximum value of β. Given n, we noted, any increase in α lowers β, and any decrease in α raises β. Yet, given sample size, we must now add, there exists more than one possible β for any specified maximum value of α. The sampling distributions in Figure 9.3, for example, were based on an upper-tailed test of a single population mean (the drying time of paint, measured in hours), a test that took the form of:

H_0: $\mu \leq 4$
H_A: $\mu > 4$

Given a sample of $n = 36$ and given a maximum α of .05, the value of β in Figure 9.3 turned out to be .025, but only because we imagined H_0 to be false *in a specific way*, such that $\mu = 5.8$. Clearly, however, there exists a different value of β for each possible population value for which H_0 is false; and, in the example just recalled, H_0 can be false for an infinite number of values; namely, all those exceeding $\mu = 4$. All else being equal, picture the sampling distribution in panel (b) of Figure 9.3 as moving to the left, being centered on, say, $\mu = 5.4$ or $\mu = 5.0$ or even $\mu = 4.1$. Can you see how the crosshatched area representing β would increase to engulf, ultimately, more than half of the area under the panel (b) normal curve? Then picture the sampling distribution in panel (b) moving to the *right*, becoming centered on, say, $\mu = 6.2$ or $\mu = 6.6$. Can you see how the crosshatched area representing β would decrease in size? The error probabilities associated with a hypothesis test can be determined systematically, of course, as is shown in Table 9.2. The first column shows selected possible values of the population mean when the null hypothesis is false; the corresponding values of β and $1 - \beta$ are derived in the remainder of the table. Given $n = 36$ (and, therefore, the general shape of the sampling distributions shown in Figure 9.3) and given a maximum $\alpha = .05$ (and, therefore, a critical value of the sample mean of $\overline{X}_C = 4.82$), a true population mean of 5.8, for example, is associated with a critical

Table 9.2   Beta and Power Values

Given (a) sample size (here $n = 36$), (b) a maximum specified type I error probability (here $\alpha = .05$), and (c) a stated null hypothesis (here H_0: $\mu \leq 4$), different type II error probabilities (β) and, therefore, different power values ($1 - \beta$) are associated with different possible population parameters that contradict the null hypothesis.

Selected Possible Values of Population Mean, μ When H_0 is False	$z_\beta = \dfrac{\overline{X}_C - \mu}{\sigma_{\overline{X}}}$	Probability of Type II Error, β	Power of Test, $1 - \beta$
4.1	1.44	$.5000 + .4251 = .9251$	.0749
4.6	.44	$.5000 + .1700 = .6700$	.3300
5.0	$-.36$	$.5000 - .1406 = .3594$	.6406
5.4	-1.16	$.5000 - .3770 = .1230$	.8770
5.8	-1.96	$.5000 - .4750 = .0250$	.9750
6.2	-2.76	$.5000 - .4971 = .0029$	.9971

z_β of -1.96 and with a β risk of $.5000 - .4750 = .025$ [as Appendix Table H tells us and as was noted in panel (b) of Figure 9.3]. Accordingly, the power of the test to detect that the null hypothesis is false when $\mu = 5.8$ is .975—as shown in the boxed row of Table 9.2.

Imagine instead a true μ not of 5.8, but of 4.1 (*first* row of Table 9.2). Graphically, this μ would move the center of panel (b) of Figure 9.3 to the *left* of the critical $\overline{X}_C = 4.82$. The associated z_β would equal 1.44, and the crosshatched area representing β would expand to more than half of the panel (b) normal curve area, equaling $.5000 + .4251 = .9251$. In contrast, imagine a true μ of 6.2 (*last* row of Table 9.2). Graphically, this μ would move the center of panel (b) of Figure 9.3 further to the right of the critical $\overline{X}_C = 4.82$. The associated z_β would become -2.76, and the crosshatched area representing β would shrink to $.5000 - .4971 = .0029$.

The data so derived in Table 9.2 are traditionally pictured with the help of one of two curves.

The Operating-Characteristic Curve (Upper-Tailed Test)

The **operating-characteristic curve,** or **OC curve,** of a hypothesis test graphs, for all possible values of a population parameter that contradict the null hypothesis, the probability, β, of erroneously accepting H_0 (of committing a type II error), given sample size and a specified maximum α risk. The color line in panel (a) of Figure 9.4 provides an example; it is based on the first and third columns of Table 9.2. Note how, in this example, the null hypothesis is in fact false for all values of $\mu > 4$. Given sample size and maximum α (also visible in the graph), we see how β gets smaller and smaller (and, therefore, $1 - \beta$ gets larger and larger), the more the true population parameter lies above the extreme value (here $\mu = 4$) specified in the null hypothesis. The progressive decline in β as μ rises that is shown here is equivalent to the progressive shrinking of the crosshatched area one would observe in panel (b) of Figure 9.3, if one initially centered the bell-shaped curve of panel (b) at $\mu = 4$ and then moved it to the right toward $\mu = 5.8$ and beyond. Note how the *height* of the OC curve at $\mu = 5.8$ corresponds to the crosshatched panel (b) *area* in Figure 9.3.

The Power Curve (Upper-Tailed Test)

The **power curve** of a hypothesis test graphs, for all possible values of a population parameter that contradict the null hypothesis, the probability, $1 - \beta$, of correctly rejecting H_0, given sample size and a specified maximum α risk. The color line in panel (b) of Figure 9.4 provides an example; it is based on the first and last columns of Table 9.2. Note how, in this example, the null hypothesis is in fact false for all values of $\mu > 4$. Given sample size and maximum α (also visible in the graph), we see how the power of the test $(1 - \beta)$ gets larger and larger (and, therefore, β gets smaller and smaller) the more the true population parameter lies above the extreme value (here $\mu = 4$) specified in the null hypothesis. It is obvious that the two curves are mirror images of one another; they give us the same probability information in two different ways. Note: because a false null hypothesis must either be accepted or rejected, the probability of doing the former plus that of doing the latter must necessarily add to 1 for any given value of μ. When $\mu = 4.6$, for example, β of .67 plus $(1 - \beta)$ of .33 sum to 1.

Figure 9.4   Operating-Characteristic (OC) and Power Curves for an Upper-Tailed Hypothesis Test

Panel (a) shows the operating-characteristic curve of a hypothesis test (here concerning a single population mean). It graphs, for all possible values of the population parameter that contradict the null hypothesis, the probability, β, of erroneously accepting H_0, given sample size and a specified maximum α risk. Panel (b) shows the power curve of the same hypothesis test. It graphs, for all possible values of the population parameter that contradict the null hypothesis, the probability, $1 - \beta$, of correctly rejecting H_0, again given sample size and maximum α.

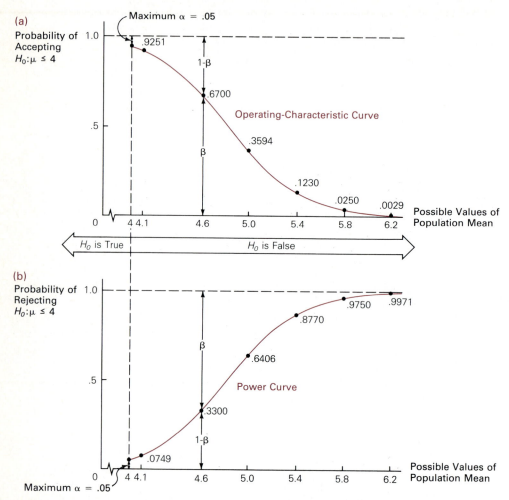

Other Types of OC Curves and Power Curves

The two types of curves just discussed look somewhat different for lower-tailed and two-tailed hypothesis tests—as illustrated in Figure 9.5. The information conveyed is, however, the same in all cases. Note: in all cases, given sample size, a change in maximum allowable α changes all the β values in the opposite direction and changes all the power values in the same direction. An increase in maximum α, for example, pushes the entire OC curve down or pushes the entire power curve up.

Figure 9.5   OC Curves and Power Curves for Lower-Tailed and Two-Tailed Hypothesis Tests

The OC curves and power curves for lower-tailed and two-tailed hypothesis tests (shown here) look different from those for an upper-tailed test (shown in Figure 9.4), but they convey the same type of probability information.

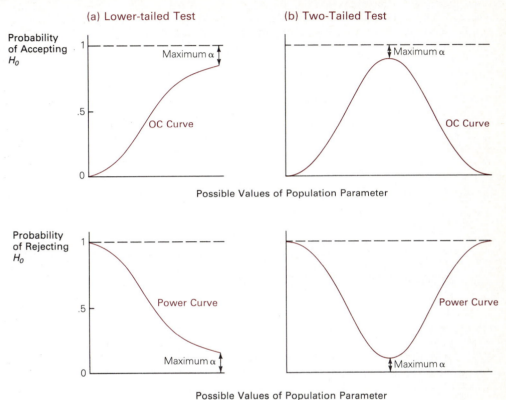

The Effect of Sample Size

We are now ready to note an important point: when sample size is increased, both α and β error probabilities can be reduced at the same time. This simultaneous reduction in both error probabilities is ultimately due to the fact that large sample size reduces the spread of the sampling distribution and, thus, the chance of extreme sampling error. Consider, for example, the hypothesis test underlying Table 9.2 (and Figure 9.3). It involved a sample of $n = 36$ and a value of $\sigma_{\bar{x}} = .5$. Because $\sigma_{\bar{x}} = (\sigma/\sqrt{n})$, the implied value of σ was 3. All else being equal, an increase in sample size to $n = 100$ would lower $\sigma_{\bar{x}}$ to $(3/\sqrt{100}) = .3$. Accordingly, the critical value $\bar{X}_C = 4.82$ would change to 4.492, and this change, along with that of $\sigma_{\bar{x}}$ from .5 to .3, would change all the β and power values in Table 9.2, as shown in Table 9.3. In short, for a given maximum α, lower β values can be achieved, but it is also possible to lower α along with β.

Table 9.3   Beta and Power Values After Increase in Sample Size

This table repeats the calculations given in Table 9.2 with only a single change: an increase in sample size from $n = 36$ to $n = 100$. As a result, given $\sigma = 3$, $\sigma_{\overline{x}}$ changes from .5 to .3 and $\overline{X}_C$ from 4.82 to 4.492. Hence, each β value declines, and the corresponding power of the test rises.

Selected Possible Values of Population Mean, μ, When H_0 is False	$z_\beta = \dfrac{\overline{X}_C - \mu}{\sigma_{\overline{x}}}$	Probability of Type II Error, β	Power of Test, $1 - \beta$
4.1	1.31	.5000 + .4049 = .9049	.0951
4.6	− .36	.5000 − .1416 = .3584	.6416
5.0	−1.69	.5000 − .4545 = .0455	.9545
5.4	−3.03	.5000 − .4988 = .0012	.9988
5.8	−4.36	.5000 − .49999 = .0000*	.9999
6.2	−5.69	.5000 − .49999 = .0000*	.9999

*The value is tiny, but positive.

The Optimum

Knowing that any combination of maximum α and maximum β can in fact be achieved (for example, by choosing the desired α and then adjusting sample size to get the desired β) allows the user of a hypothesis test a great deal of choice. Clearly, there are benefits to be reaped both from lowering maximum α *and* lowering maximum β. A drug manufacturer, for instance, who erroneously rejects new and effective drugs as ineffective is forgoing future profits (and, therefore, wants to reduce α as much as possible). A drug manufacturer who erroneously accepts new and ineffective drugs as effective is inviting future losses (and, therefore, wants to reduce β as much as possible). Yet the simultaneous reduction of α and β requires larger samples, and the cost of such sampling also lowers profits—perhaps at the rate of $1,000 for each extra patient on whom a new drug is tested. Somewhere, therefore, a balance must be found where the marginal benefit of raising sample size (namely, higher profit due to lowered α and β risks) just equals the marginal cost of doing so (namely, lower profit because of higher sampling costs).

Because of the need to balance marginal benefit with marginal cost, every user of a hypothesis test will ultimately specify some positive value for the maximum allowable α as well as β, figure the implied sample size, and then consider whether it would be more beneficial to raise or lower sample size (and, therefore, accept different error probabilities).

||| Practice Problem 17

The Drying Time of Paint

Consider the paint manufacturer, noted in Figure 9.3, who wanted to test the hypothesis that the company's paint had a maximum drying time of 4 hours. The manufacturer might initially specify the maximum tolerable α as .05 (as in Figure 9.3). If H_0: $\mu \leq 4$ is in fact true, there should be at most a 5 percent chance for the test result to be wrong (for the company to reject its perfectly valid claim). Yet the manufacturer might also wish to specify a maximum tolerable β of .025 (again as in Figure 9.3) if the company's paint had in fact a drying time of 5.8 hours. If, that is, H_A: $\mu = 5.8$ is in fact true, there should be at most a 2.5 percent chance for the test result to be wrong (for the company to persist with an invalid claim).

If these were the initial instructions, the company's statistician could make a sketch similar to Figure 9.3, noting that the rejection region for one sampling distribution (centered on $\mu = 4$) must have an upper tail area of $\alpha = .05$ and, hence, a critical z value of $z_\alpha = +1.64$, while the *acceptance* region for the other sampling distribution (centered on $\mu = 5.8$) must have a lower tail area of $\beta = .025$ and, hence, a critical z value of $z_\beta = -1.96$. The statistician could then formulate two equations, one for the critical value of the sample mean, $\overline{X}_C$, assuming $\mu = 4$, and the other for the same critical value of the sample mean, assuming $\mu = 5.8$. These equations would read:

If $\mu = \mu_0 = 4$,

$$z_\alpha = \frac{\overline{X}_C - \mu_0}{\sigma_{\overline{X}}} \text{ and } \overline{X}_C = \mu_0 + z_\alpha \sigma_{\overline{X}}. \tag{1}$$

If $\mu = \mu_1 = 5.8$,

$$z_\beta = \frac{\overline{X}_C - \mu_1}{\sigma_{\overline{X}}} \text{ and } \overline{X}_C = \mu_1 + z_\beta \sigma_{\overline{X}}. \tag{2}$$

Because the critical value of the sample mean is the same whether the hypothesis is in fact true or false (as Figure 9.3 confirms), the statistician could equate the two equations and determine:

$$\mu_0 + (z_\alpha \, \sigma_{\overline{X}}) = \mu_1 + (z_\beta \, \sigma_{\overline{X}}) \tag{3}$$

$$(z_\alpha - z_\beta) \, \sigma_{\overline{X}} = \mu_1 - \mu_0 \tag{4}$$

$$\sigma_{\overline{X}} = \frac{\sigma}{\sqrt{n}} = \frac{\mu_1 - \mu_0}{z_\alpha - z_\beta} \tag{5}$$

$$\frac{\sigma^2}{n} = \frac{\mu_1 - \mu_0}{z_\alpha - z_\beta} \tag{6}$$

$$\sqrt{n} = \frac{\sigma(z_\alpha - z_\beta)}{\mu_1 - \mu_0} \tag{7}$$

$$n = \left[\frac{\sigma(z_\alpha - z_\beta)}{\mu_1 - \mu_0} \right]^2 \tag{8}$$

By using a preliminary estimate of the population standard deviation (of, say, $\sigma = 3$), the statistician could use equation (8) to derive:

$$n = \left[\frac{3[1.64 - (-1.96)]}{5.8 - 4} \right]^2 = 36.$$

Thus, the sample size necessary for achieving the maximum error probabilities specified above equals 36. The user of the test must decide whether this is optimal or whether it is worthwhile to "buy" a lower α or β with a higher n (or to buy a lower n with a higher α or β). ▮

CRITICISMS OF HYPOTHESIS TESTING

The testing procedures discussed in this chapter are widely used, yet there is considerable controversy about their usefulness. Without trying to present an exhaustive list, we shall consider four of the criticisms that are commonly encountered.

Serious Violations of Assumptions

Some critics charge that many hypothesis tests are performed even though the assumptions that would validate the procedure used are not met. There might be a difference, for example, between the statistical (target) population about which inferences are to be made and the population that is actually being sampled. (This difference is likely because a perfect frame is usually unavailable for any nontrivial population.) In addition, an intended simple random sample can in fact turn out to be a nonprobability sample. (This result is likely because high nonresponse rates are common.) Or, perhaps, the populations being sampled may not be normally distributed, or they may not have equal variances (requirements for using the t statistic in two-sample tests). In situations such as these, the testing procedure will grind out nonsense.

Neglect of Power Considerations

It is quite common, critics charge, to neglect power—the probability that a given test will (quite properly) reject the null hypothesis when it is false. For example, if that probability [graphed in panel (b) of Figure 9.4] is very low for alternatives close to the extreme value given in the null hypothesis, then many a sample result that leads to the acceptance of H_0 is also very likely if H_A is true. Thus, the conclusion ("accept H_0") seems hardly warranted.

Rigid Interpretation of the "Acceptance" Criterion

While it is meaningful to *reject* a null hypothesis as a result of uncovering contrary evidence that cannot reasonably be attributed to chance, it is not meaningful, critics claim, to *accept* a null hypothesis in the absence of such contrary evidence. Yet these words are commonly used. It would be much wiser, critics urge, to talk of "failure to reject" rather than "acceptance of" a null hypothesis. All the things we know ultimately are things that haven't yet been proven false and are things concerning which a wait-and-see attitude is most appropriate. All that we know is at best *tentatively* accepted. Any rigid interpretation of the acceptance criterion (as if it assured us of the truth with *certainty*), critics conclude, is to be avoided at all cost.

Note: this criticism is clearly semantic and can be met by agreeing to interpret "acceptance" loosely, as a token word for "failure to reject," except, perhaps, in cases where the probability β of a Type II error is known to be extremely small. Admittedly, it is often difficult to know this probability.

Nonpublication of Nonsignificant Results

Results of hypothesis tests that are statistically nonsignificant (and, thus, nonsurprising) are unlikely to be published. This situation has serious consequences, critics note. Consider a null hypothesis that is in fact true (unbeknownst to researchers) and that is tested independently by many, always at the $\alpha = .05$ level of significance. Each of these investigators has only 5 chances out of 100 of (incorrectly) finding statistical significance (and falsely rejecting H_0). Yet the chance that *at least one* among the many will reach such a result is much higher. If there are 10 such investigators, the probability that at least one will falsely reject H_0 equals $1 - (.95)^{10} = .401$—and that person's (wrong) results *will* be published!

ANALYTICAL EXAMPLES

9.1 The Never-Ending Search for New and Better Drugs

Pharmaceutical companies are engaged in a never-ending search for new and better drugs. There exists a virtually infinite supply of substances that might possibly be effective against one disease or another. In the United States, perhaps as many as 175,000 compounds are evaluated in research labs each year; only 20 of these eventually make it to the drug store. Sometimes success comes fairly soon (the remarkable antibiotic Aureomycin was found after screening "merely" 4,000 soil samples); at other times, success eludes researchers for decades. The search for an anticancer drug is a case in point. In all cases, however, the hypothesis-testing procedures introduced in this chapter are put to work.

In one actual case, researchers implanted cancer cells in 9 mice, then treated 3 mice with a new chemical compound to test whether it would retard the growth of cancer tumors. After a certain period, all the tumors were removed and weighed. It was expected that the compound, if effective, would have caused the tumor weights of the treated mice (experimental group) to be less than those of the untreated ones (control group). The results are shown in the accompanying table.

Was the .53-gram reduction in the average tumor weight of the experimental group significant enough to warrant further tests with the compound? A two-sided hypothesis test employing a known standard error of the difference between the two population means of $\sigma_d = .35$ grams and a significance level of $\alpha = .001$ would conclude in the negative.

Step 1:

$$H_0: \mu_A - \mu_B = 0$$
$$H_A: \mu_A - \mu_B \neq 0$$

Tumor Weights (grams)	
Experimental Group	Control Group
.96	1.29
1.59	1.60
1.14	2.27
	1.31
	1.88
	2.21
$\overline{X}_A = 1.23$	$\overline{X}_B = 1.76$

Step 2:

$$t = \frac{\overline{X}_A - \overline{X}_B}{\sqrt{\sigma_d^2\left(\dfrac{1}{n_A} + \dfrac{1}{n_B}\right)}}$$

Step 3: Critical values for 7 degrees of freedom:

$$\pm t_{\alpha/2} = \pm t_{.0005(7)} = \pm 5.408.$$

Decision rule: "Accept H_0 if $-5.408 \leq t \leq +5.408$."

Step 4:

$$t = \frac{1.23 - 1.76}{.247} = -2.15$$

Source: Charles W. Dunnett, "Drug Screening: The Never-Ending Search for New and Better Drugs," in Judith M. Tanur et al., eds., *Statistics: A Guide to the Unknown* (San Francisco: Holden-Day, 1972), pp. 23–33.

9.2 Antitrust Pork Barrel

It has long been asserted that federal regulatory agencies are significantly impaired in their tasks by their dependence on Congress. This impairment is said to occur because each member of Congress inevitably seeks to further the provincial interests of citizens in the home district, the welfare of which may depend disproportionately on a few key industries. When those industries are threatened in any way by regulatory agencies, members of Congress who have power over these agencies will naturally deflect their actions. A group of researchers decided to test whether the case-bringing activity of the Federal Trade Commission (FTC) during the 1961–79 period was biased in favor of firms that operated in the districts of members of those congressional committees (three in the Senate and five in the House) that have important budgetary and oversight powers with respect to the FTC.

Although the FTC is empowered to initiate antitrust investigations on its own, almost 90 percent of its investigations are begun as a result of cases brought to it by the public. Ultimately, these investigations lead to a consent decree (by which the accused party promises to reform) or to a formal complaint; the latter, in turn, leads to dismissal for insufficient evidence of a violation or to a cease-and-desist order. The investigators formulated the null hypothesis that there was a zero difference between (a) the proportion of dismissals to cases brought affecting districts with representatives on FTC-relevant committees (π_A) and (b) this proportion in districts without such representatives (π_B). Accordingly, the alternative hypothesis supported the pork-barrel thesis:

$$H_0: \pi_A - \pi_B = 0$$
$$H_A: \pi_A - \pi_B \neq 0$$

Congressional Committee	Ratio of Dismissals to Cases Brought (1961–69)		z
	Within Committee Members' Districts, P_A	Outside Committee Members' Districts, P_B	
(1) Senate Committee on Interior and Insular Affairs	$\frac{17}{285} = .0596$	$\frac{148}{2,190} = .0675$	$-.50$
(2) Senate Committee on Commerce, Science, and Transportation	$\frac{32}{570} = .0561$	$\frac{133}{1,905} = .0698$	-1.14
(3) Senate Subcommittee on Antitrust and Monopoly of the Senate Judiciary Committee	$\frac{60}{638} = .0940$	$\frac{105}{1,837} = .0572$	3.22
(4) House Subcommittee on Independent Offices of the House Appropriations Committee	$\frac{14}{87} = .1609$	$\frac{151}{2,388} = .0632$	3.59
(5) All Five Relevant House Subcommittees	$\frac{84}{1,104} = .0761$	$\frac{81}{1,371} = .0591$	1.69

Using

$$z = \frac{P_A - P_B}{\sigma_d}$$

as a test statistic and choosing the 10 percent significance level, the critical values for accepting H_0 were $\pm z_{\alpha/2} = \pm z_{.05} = \pm 1.64$. Selected sample results and computed values of z are shown in the table on the preceding page.

There was no reason to reject the null hypothesis in cases 1 and 2 but plenty of reason to do so for cases 3–5 because of the statistically significant z values. After reviewing similar data for the 1970–79 period (as well as for the ratio of dismissals to *formal complaints*), the investigators concluded that there was considerable support for the thesis that members of certain congressional committees that have important oversight and budgetary powers with respect to the FTC do deflect commission decisions in favor of firms in their home districts.

Source: Roger L. Faith, Donald R. Leavens, Robert D. Tollison, "Antitrust Pork Barrel," *Journal of Law and Economics,* October 1982, pp. 329–42. Table copyright © 1982 by the University of Chicago. All rights reserved. Reprinted by permission of the University of Chicago Press.

9.3 Do Nursing Homes Discriminate Against the Poor?

For-profit nursing homes set their own fees for private patients but are dictated lower fees, called *reimbursements,* for patients supported by government agencies. Provided that the cost of care is the same for both types of patients, it seems reasonable to assume that such nursing homes will give preference in admissions to those patients (such as private ones) who bring in the greatest revenue and will discriminate in admissions against other patients (such as the poor who are supported by Medicaid) who bring in less revenue. A group of researchers investigated the 1969 practices of 18 southern California nursing homes to test this proposition. Of these, $n_A = 11$ facilities (group A) claimed not to discriminate in admissions, the other $n_B = 7$ facilities (group B) clearly preferred non-Medicaid patients. The mean percentage of Medicaid-supported patient days (and the standard deviations) were found to be $\overline{X}_A = 75.4$ ($s_A = 16.3$) and $\overline{X}_B = 40.4$ ($s_B = 30.8$). A hypothesis test was performed as follows:

Step 1:

$$H_0: \mu_A - \mu_B \leq 0$$
$$H_A: \mu_A - \mu_B > 0$$

Step 2:

$$t = \frac{\overline{X}_A - \overline{X}_B}{\sqrt{s_P^2 \left(\frac{1}{n_A} + \frac{1}{n_B}\right)}}$$

Step 3: $\alpha = .05$; given 16 degrees of freedom, $t_\alpha = t_{.05(16)} = 1.746$. The decision rule is: "Accept H_0 if $t \leq 1.746$."

Step 4:

$$s_P^2 = \frac{(n_A - 1)s_A^2 + (n_B - 1)s_B^2}{n_A + n_B - 2}$$
$$= \frac{(10)(16.3)^2 + (6)(30.8)^2}{16} = 521.8$$

$$t = \frac{\overline{X}_A - \overline{X}_B}{\sqrt{s_P^2 \left(\frac{1}{n_A} + \frac{1}{n_B}\right)}}$$
$$= \frac{75.4 - 40.4}{\sqrt{521.8 \left(\frac{1}{11} + \frac{1}{7}\right)}}$$
$$= \frac{35}{11.04} = 3.17$$

Accordingly, H_0 was rejected. At the 5 percent level of significance, the alternative hypothesis (that group A facilities, on the average, had more Medicaid-supported patient days than group B) was accepted.

Source: John S. Greenlees, John M. Marshall, Donald E. Yett, "Nursing Home Admission Policies Under Reimbursement," *The Bell Journal of Economics,* Spring 1982, pp. 93–106.

‖‖‖ CLOSE-UPS

9.1 The Trial of the Pyx

The trial of the Pyx is an ancient ceremony at the Royal Mint in London. The avowed purpose of the ceremony is to ascertain whether the coinage issued by the mint meets the Crown's specifications. During a period of time, coins are taken at random from those minted and placed in a locked box, called *the Pyx*. At irregular intervals, perhaps as frequently as once each year but usually less often, a trial of the Pyx is declared, and a jury of members of the Worshipful Company of Goldsmiths assembles. At this time, the Pyx is opened, and the contents are counted, weighed, and assayed. The results are then compared with a set standard that includes an allowable tolerance above and below a given number. After a successful trial, a banquet is held in celebration.

From the statistician's point of view, the trial is a marvelous example of a hypothesis test: it involves the drawing of a random sample, a null hypothesis to be tested (the standard), a two-sided alternative (the

Crown is worried about excessive deviations either above or below the standard), a test statistic (the total weight of the coins in the box), and a critical region (the allowable tolerance).

To be sure, there is also much to criticize from the point of view of modern theory: the way the sampled coins are selected has always been ill defined (leaving open the possibility of serious bias). The effect of sample size on the probability of a type I error (the sample size traditionally chosen makes the probability of type I error practically zero) was long unrecognized, as was the fact that the power of the test to detect a false null hypothesis is extremely low for parameter values near the one specified in the null hypothesis.

Source: Stephen M. Stigler, "Eight Centuries of Sampling Inspection: The Trial of the Pyx," *Journal of the American Statistical Association,* September 1977, pp. 493–500.

9.2 Death Month and Birth Month: An Unexpected Connection

In literature and in movies one sometimes comes across a deathbed scene in which a dying person holds onto life until some special event has occurred: a son returning from the war, a daughter getting married, a religious holiday. One researcher wanted to know whether such feats occur in real life. He set out to examine the *death-dip hypothesis* versus the *independence hypothesis.* According to the death-dip hypothesis, if we compare the date of death with the date of birth for a large number of people, we find fewer deaths than expected just before the birthday (and more than expected afterwards) because dying people postpone death in order to witness their birthdays. According to the independence hypothesis, no such effect takes place: 1/12 of all deaths occur in people's

birth month or in any of the preceding and following months of the year.

Four samples of "notable" Americans (who were likely to receive more than usual attention on their birthdays) were used (involving the period 1897–1960 and a total of 1,251 individuals). In each sample, the death-dip hypothesis was clearly upheld! Just before the birth month, there were 17 percent fewer deaths than expected under the independence hypothesis; just afterwards, there were 11 percent more. Following traditional hypothesis-testing procedures, this finding could not reasonably be ascribed to chance.

Source: David P. Phillips, "Deathday and Birthday: An Unexpected Connection," in Judith M. Tanur et al., eds., *Statistics: A Guide to the Unknown* (San Francisco: Holden-Day, 1972), pp. 52–65.

9.3 The Case of the Missing Women

In 1968, the pediatrician-author Dr. Benjamin Spock was tried in the U.S. District Court in Boston for conspiracy to violate the Selective Service Act by encouraging resistance to the Vietnam War. In that trial, the defense challenged the jury-selection method:

more than half of all eligible jurors in Boston were women, but there were no women in the Spock jury. Spock would have wanted some women on the jury because so many mothers had raised their children "according to Dr. Spock" and might have been sympathic to him; also because polls showed more women than men opposed to the war.

A hypothesis test was performed to determine whether the absence of women from the Spock jury was an accident or whether it was the result of systematic discrimination. Investigation showed that juries of Dr. Spock's trial judge had an average of 14.6 percent women, while those of his six judicial colleagues had an average of 29 percent (in both cases, the variance was small). The hypothesis test concluded that such a difference (or a larger one) could have occurred by chance only once in 10^{18} times.

Source: Hans Zeisel and Harry Kalven, Jr., "Parking Tickets and Missing Women: Statistics and the Law," in Judith M. Tanur et al., eds., *Statistics: A Guide to the Unknown* (San Francisco: Holden-Day, 1972), pp. 104–5.

BIOGRAPHIES

9.1 Ronald A. Fisher

Ronald Aylmer Fisher (1890–1962) was born in London, England, and studied mathematics and physics at Cambridge University. He spent his early years indecisively moving about—working at an investment house, on a farm in Canada, as a teacher in British public schools. He developed an interest in biometrics, however, and this interest led him, in 1919, to join the staff of the world-famous agricultural experiment station at Rothamsted. There he was charged with sorting and reassessing a 66-year accumulation of data on field trials and weather records—and in the process, he became one of the century's leading statisticians. Early on, he published the epochal *Statistical Methods for Research Workers* (1925), a book that was destined to go through 14 editions in many languages and to become the "bible" for researchers throughout the world. This book was followed by two equally influential works, *The Genetical Theory of Natural Selection* (1930), a book that reconciled Darwinian evolution with Mendelian genetics, and *The Design of Experiments* (1935); these books established Fisher as a top-ranking geneticist as well as statistician. Indeed, prior to his move to Australia late in life, Fisher held long-term positions as professor of eugenics, first at University College, London, and later at Cambridge University.

The works just cited, however, hardly begin to tell the story of Fisher's productivity. Over the course of half a century, he published one paper every two months, and most of them broke new ground! Thus, it is difficult to decide which of his many contributions are the most praiseworthy, and it is surely impossible in a few paragraphs to show how thoroughly the land of statistics is crisscrossed with the footprints of this prolific scholar. He is the pioneer in experimentation who revolutionized the field with randomized blocks, Latin squares, factorial designs, and confounding. (See Chapter 2 of this text and of the accompanying *Student Workbook* for a discussion of these concepts.) Then he is the man who pushed ahead the theory of estimation (and introduced concepts of unbiasedness, consistency, efficiency, and more) and who made correlation, regression, and analysis of variance (and covariance) what they are today. It was Fisher who erected upon the work of William S. Gosset (Biography 8.1) a comprehensive theory of hypothesis testing based on small samples.

It is not surprising that honors and prizes were heaped upon Fisher throughout his life (he was even knighted in 1952); it is unfortunate that Fisher also became involved in an extraordinarily long-lasting and acrimonious battle with other statisticians concerning the nature of estimation and hypothesis testing. Fisher's views will be summarized here; those of his opponents will be outlined in Biographies 9.2 and 9.3. In his theory of estimation, Fisher introduced the concept of the **fiducial interval,** a range of values that can be *trusted,* with a specified probability, to contain the value of some parameter. Such an interval might take the form of $\mu = \overline{X} \pm (z\sigma_{\overline{X}})$; Fisher would establish this range *after* a sample had been taken and $\overline{X}$ had been calculated. Given the specific sample result, he would, thus, make a probability statement about the parameter in the form of the fiducial interval and would tell us that we could trust the parameter to lie within that range (the Latin *fiducia* means *trust*)—with a specified probability, of course, depending on the value of z. (This view is generally rejected by statisticians today, a fact already implied by the Chapter 8 discussion of *confidence intervals* and to be explained further in Biographies 9.2 and 9.3.)

In subtle but important ways, Fisher's view of hypothesis testing also differed from that of other statisticians (and from that found in the body of Chapter 9). Fisher viewed hypothesis testing as a form of data analysis and opinion formation engaged in by someone in a research situation. This person would

1. hypothesize the value of some parameter,
2. select a test statistic (such as z or t) the distribution of which was completely known if the hypothesized parameter value was true,
3. take a random sample from the population of interest,
4. calculate the value of the test statistic,
5. calculate the probability, also called the **p value,** or **observed significance level,** that a value of the test statistic as extreme as or more extreme than the one actually observed could have occurred by chance, assuming that the hypothesized parameter value is true, and
6. use the following arbitrary cutoff points to judge the plausibility of the hypothesized parameter value:
   a. If the p value is $\leq .01$, the person would render a strong judgment against the hypothesized parameter value.
   b. If the p value is $\leq .05$ but $> .01$, the person would render a weak judgment against the hypothesized parameter value.
   c. If the p value is $> .05$, the person would judge in favor of the hypothesized parameter value.

Figure 9.1 in the chapter can be used to illustrate the Fisher approach. The hypothesized value might be $\mu = 5$; the test statistic might be z, and its distribution would be the normal curve shown. The random sample might yield $\overline{X} = 5.1$, from which the value of the test statistic would be calculated as $z = 1$. Thus, Fisher would figure the p value (from Appendix Table H) as $.5000 - .3413 = .1587$, and he would form the opinion that the hypothesized parameter value was reasonable.

This Fisherian view of hypothesis testing, too, is generally rejected by statisticians today, a fact already implied by this chapter's discussion and to be explained further in Biographies 9.2 and 9.3. Nevertheless, Fisher's work greatly helped the development of the modern theory of hypothesis testing, and his scientific achievements are in any case so varied and so crucial that nothing can dim their luster or reduce his ranking as one of the greatest scientists of this century.

Sources: Dictionary of Scientific Biography, vol. 5 (New York: Scribner's, 1972), pp. 7–11; *International Encyclopedia of Statistics,* vol. 1 (New York: Free Press, 1978), pp. 352–58. For more information on Fisher, see *Encyclopedia of Statistical Sciences* (New York: Wiley-Interscience, 1982–86), vol. 3, pp. 103–10.

9.2 Jerzy Neyman

Jerzy Neyman (1894–1981) was born in Benderey, Bessarabia. He studied at the University of Warsaw and, in the 1920s, achieved world renown by extending sampling theory to sampling from finite populations without replacement and carrying out complex stratified sampling schemes for the Polish government. In the 1930s, he joined the faculty at University College, London, where Ronald A. Fisher (Biography 9.1) and Egon S. Pearson (Biography 9.3) had just filled two positions created to replace the one long held by Karl Pearson (Biography 10.1). Neyman, who soon was to move on to a long and distinguished career at the University of California,

Berkeley, shared many interests with Fisher—agricultural experimentation, weather-modification experiments, genetics, astronomy, medical diagnosis—yet the two ended up on opposites sides when the famous and bitter controversy arose concerning the nature of estimation and hypothesis testing. For years, however, Neyman collaborated fruitfully with Egon S. Pearson; jointly the two men were immortalized by the Neyman-Pearson theory of estimation and hypothesis testing that is now generally accepted.

Neyman and Pearson introduced the concept of *confidence intervals* into the theory of estimation at about the same time that Fisher wrote about *fiducial intervals,* and for a time the two concepts lived amiably side by side, appearing to be two names for the same thing. Eventually, however, it became clear that these were different concepts, indeed. A 95-percent fiducial interval à la Fisher, for example, would claim a 95-percent probability that a given parameter lay within the interval constructed around a sample statistic already calculated. Unlike Fisher, Neyman-Pearson would establish the interval *before* the sample was taken and before any statistic was calculated. A 95-percent confidence interval à la Neyman-Pearson only claims that use of their formula in the long run produces intervals such that 95 out of 100 of them contain the parameter, while any actual interval, constructed *after* sampling, was *certain* either to contain or to exclude the parameter. This important concept of confidence levels has already been discussed in Chapter 8 in connection with Figure 8.4 which can be found on page 307).

For additional information on Neyman, see *Encyclopedia of Statistical Sciences* (New York: Wiley-Interscience, 1982–86), vol. 6, pp. 215–23.

9.3 Egon S. Pearson

Egon Sharpe Pearson (1895–1980) was born in London, England, the son of Karl Pearson (Biography 10.1). Egon was educated at Cambridge University and closely followed in his father's footsteps. Early on, he joined his father's department at University College, London; in 1933, when his father resigned, he took over one of the two new positions created as replacements, the other one going to R. A. Fisher (Biography 9.1). In this position, and as editor of *Biometrika,* he contributed importantly to statistics (he himself published some 133 papers); above all, he is known, along with Jerzy Neyman (Biography 9.2), as the developer of the modern theory of hypothesis testing (as it is found in this chapter). The Neyman-Pearson approach differed considerably from Fisher's and this difference gave rise to a lifelong and bitter controversy. Unlike Fisher who viewed hypothesis testing as a procedure by which a *researcher* could *form an opinion* about some population parameter, Neyman-Pearson viewed it as a means by which a *decision maker* operating under uncertainty could *make a clear choice* between two alternatives, while at the same time controlling the chances for error (and minimizing costs associated therewith).

While Fisher hypothesized the value of only one parameter, Neyman-Pearson explicitly formulated *two* rival hypotheses, H_0 and H_A—a procedure suggested to them by William S. Gosset (Biography 8.1). This first step contrasts sharply with Fisher's total neglect of an alternative hypothesis. (If, under the Fisher approach, the parameter in question is judged unlikely to be true, what is the truth?) In addition, Neyman-Pearson introduced the formal acceptance/rejection rule employed throughout this chapter, as well as the notion of the two error types (with probabilities α and β), and they investigated the costs of committing these errors. (Fisher, who moved in the world of the research scientist, perhaps understandably had shown no concern about such costs. Indeed, there seems to be a qualitative difference between the cost to, say, a drug manufacturer of a wrong *decision* and the cost to, say, a genetic researcher of a wrong *opinion*.) Finally, Neyman-Pearson introduced the concept of the power of a hypothesis test and noted how the cost of making observations (which depends on sample size) can be traded against the costs of type I or type II errors.

If you would like to find additional information on Pearson, consult the *Encyclopedia of Statistical Sciences* (New York: Wiley-Interscience, 1982–86), vol. 6, pp. 650–53.

Summary

1. When people make decisions, they inevitably do so on the basis of beliefs they hold concerning the true state of the world. Every one of these beliefs originates as a *hypothesis,* a proposition tentatively advanced as possibly true. *Hypothesis testing* is a systematic approach to assessing beliefs about reality; it is confronting a belief with evidence and deciding whether that belief can be maintained as reasonable or must be discarded as untenable. Four major steps are involved.

2. Step 1 is the formulation of two opposing hypotheses, called the *null hypothesis* (H_0) and the *alternative hypothesis* (H_A). They are mutually exclusive and also collectively exhaustive of the pos-

sible states of reality. These hypotheses can be stated in various forms. While the null hypothesis may be stated either as an exact or inexact one, the alternative hypothesis is always stated as an inexact one (two-sided or one-sided).

3. Step 2 is the selection of a *test statistic*—the statistic to be computed from a simple random sample taken from the population of interest and to be used for establishing the probable truth or falsity of the null hypothesis.

4. In Step 3, a *decision rule* is derived that specifies in advance for all possible values of a test statistic that might be computed from a sample whether the null hypothesis should be accepted or whether it should be rejected in favor of the alternative hypothesis. Such a rule is needed because the test statistic is a random variable; simply as a result of sampling error, it can take on all kinds of values apparently contradicting the null hypothesis even when H_0 is true. One escapes the dilemma by accepting the null hypothesis for all sample results (even seemingly contradictory ones) that are highly likely and nonsurprising when H_0 is true and by rejecting it for results that are highly unusual and surprising when H_0 is true. The cutoff point between values of the test statistic that are "surprising" (that have a "low" probability of occurring when H_0 is true) and those that are "not surprising" (that have a "high" probability of occurring when H_0 is true) must necessarily be an arbitrary one. "Surprising" results that lead to the rejection of the null hypothesis are also called *statistically significant* results. "Nonsurprising" results that lead to the acceptance of the null hypothesis are called *not statistically significant* results. Among all the sample results that are possible when the null hypothesis is true, the (arbitrary) maximum proportion, α, that is considered sufficiently unusual to reject the null hypothesis is called the significance level of the test.

When the null hypothesis is rejected for values of the test statistic located in either tail of that statistic's sampling distribution (because the alternative hypothesis is two-sided), the hypothesis test is called a *two-tailed test.* When, instead, the null hypothesis is rejected only for very low (or very high) values of the test statistic located entirely in one tail of that statistic's sampling distribution (because the alternative hypothesis is one sided), the hypothesis test is called a *one-tailed test.* Such a test can be *lower-tailed* or *upper-tailed,* depending on the location of the rejection region.

5. In Step 4, a simple random sample is selected, the actual value of the test statistic is computed, and it is confronted with the decision rule. The null hypothesis, accordingly, is accepted or rejected. We must note, however, that either outcome can be a correct or an erroneous interpretation of the true state of reality. The erroneous rejection of a null hypothesis that is in fact true is called a *type I error;* it occurs with a probability of α. The erroneous acceptance of a null hypothesis that is in fact false is called a *type II error;* it occurs with a probability of β. Given sample size, $n,$ anything that reduces α automatically raises β. The two complementary probabilities, $1 - \alpha$ (with respect to α) and $1 - \beta$ (with respect to β) are, respectively, referred to as the *confidence level* and the *power* of the hypothesis test.

6. The development of the theory of hypothesis testing began at least eight centuries ago, probably in the reign of England's King Henry II.

7. The modern theory can be applied to both large-sample and small-sample situations.

8. The relationship between type I and type II errors can be depicted systematically with the help of the *operating-characteristic curve* that shows, for all possible values of a population parameter that contradict the null hypothesis, the probability, β, of erroneously accepting the null hypothesis (of committing a type II error), given sample size and a specified maximum α risk (of committing a type I error). An alternative to this presentation is the *power curve* that shows, for all possible values of a population parameter that contradict the null hypothesis, the probability, $1 - \beta$, of correctly rejecting it, again given sample size and maximum α. An inspection of these relationships reveals that it is possible to set maximum α as low as desired and then, by increasing sample size, to lower β to any desired level as well. The user of a hypothesis test can, therefore, find an optimal sample size that balances the marginal cost of increasing sample size (the rising cost of making additional observations) with the marginal benefit of doing so (the lowered cost of committing type I or type II errors).

9. Modern hypothesis-testing procedures are still subject to considerable controversy. Critics are concerned about serious violations of assumptions, the neglect of power considerations, a rigid interpretation of the "acceptance" criterion, the non-publication of nonsignificant results, and other concepts of significance.

Key Terms

α risk
acceptance number
acceptance region
alternative hypothesis
β risk
confidence level of the hypothesis test
critical value
cutoff point
decision rule
error of acceptance
error of rejection
exact hypothesis
fiducial interval
hypothesis
hypothesis testing
inexact hypothesis
lower-tailed test
not statistically significant result
null hypothesis

observed significance level
one-sided hypothesis
one-tailed test
operating-characteristic curve (OC curve)
pooled variance (in Appendix 9A)
power curve
power of the hypothesis test
p value
rejection region
significance level of a hypothesis test
size of a hypothesis test
statistically significant result
test statistic
two-sided hypothesis
two-tailed test
type I error
type II error
upper-tailed test

Questions and Problems

The computer programs noted at the end of this chapter can be used to work many of the subsequent problems.

1. Reconsider Practice Problem 1 in this chapter on p. 352. What would have been the result of that test if a significance level of $\alpha = .001$ had been set? Of $\alpha = .25$?

2. Reconsider Practice Problem 2 on p. 353. What would have been the result of that test if the sample size had been 500 rods? Or 16 rods?

3. The buyer of shirts for a department store wants to test whether shirts with sleeve labels of "33 inches" really meet that specification. A random sample of $n = 100$ from 10,000 incoming shirts is to be taken; the desired significance level is $\alpha = .025$. The sample shows a mean length of $\overline{X} = 34$ inches, with a standard deviation of $s = 2$ inches. Make the test.

4. An economist wants to test whether the average salary of aircraft mechanics really is $600 per month as has been alleged. A random sample of $n = 100$ from the nation's 29,952 mechanics is taken; the desired significance level is $\alpha = .05$. The sample shows a mean salary of $\overline{X} = \$657$, with a standard deviation of $s = \$22$. Make the test.

5. In order to rule on a potential federal grant, a government official must confirm that the average family income in an Illinois county is $12,357, as alleged. A random sample of $n = 36$ families is taken; the desired significance level is $\alpha = .001$. The sample shows a mean income of $13,950, with a standard deviation of $3,972. To prove the validity of the figures make the test.

6. The manager of a firm wants to test the advertising claim of a competitor: "There are 50 chunks of beef in every can of Eric's Homemade Stew." A random sample of 50 cans of Eric's stew is acquired; a hypothesis test at the $\alpha = .025$ level of significance is to be performed. The sample shows a mean of 48 chunks of beef per can, with a standard deviation of 2 chunks. Make the test.

7. The manager of an airport has the distinct impression that the monthly maintenance cost of planes used in agricultural work does not equal the projected $500 per plane. A sample of 32 planes is taken; a hypothesis test at the $\alpha = .02$ level of significance is to be performed on the assumption that the population of cost figures is normally distributed. The sample shows costs of $592 per plane, with a standard deviation of $101. Make the test.

8. The mayor wants to know whether mean apartment rents in the town really are now higher than last year's $375 per month. A random sample of 50 of the town's 450 apartments is taken; the desired significance level is $\alpha = .01$. The sample shows current mean rents of $\overline{X} = \$399$ per month; it is assumed that the current population standard deviation equals last year's: $\sigma = \$25$. Make the test, assuming a normally distributed population.

9. The manager of a retail business wonders whether people other than customers are using the store's parking lot. She wants to test, at a significance level of $\alpha = .05$, whether cars in the lot are typically parked for over an hour. A random sample of 50 cars shows a mean parking time of 101 minutes, with a standard deviation of 21 minutes.

10. The Environmental Protection Agency is allowing a plant to dump its effluent into a river—as long as the effluent contains a maximum of 4 parts per million (ppm) of a certain toxic substance. During the course of one week, the EPA randomly samples the effluent and finds, in $n = 64$ samples, an average of $\overline{X} = 4.2$ ppm of toxic substance, with a standard deviation of $s = 1$ ppm. Is the plant in violation? Conduct an appropriate hypothesis test at $\alpha = .025$.

11. The Federal Aviation Administration believes that the mean number of takeoffs and landings at American airports last year was at most 50 per day. Make an appropriate hypothesis test of this belief at the $\alpha = .001$ level of significance, while using these sample data: $n = 100$, $\overline{X} = 71$, $s = 30$. (There are 13,000 airports in the United States.)

12. Once again consider problems 3–6. In each case, compute a *confidence interval* for the mean in question, always at the $100 (1 - \alpha)$ percent level of confidence.

13. Once again consider problems 7–10. In each case, compute a *confidence interval* for the mean in question, always at the $100 (1 - \alpha)$ percent level of confidence.

14. The buyer of tomatoes for a ketchup producer wants to test whether it is true that 80 percent of the tomatoes being received deserve the label "superior." A random sample of $n = 100$ from 5,000 incoming tomatoes is to be taken; the desired significance level is $\alpha = .05$. The sample shows 72 "superior" tomatoes. Make the test.

15. An advertiser wants to test a magazine publisher's claim that 25 percent of the magazine's readers are college students. A random sample of $n = 200$ from 2.1 million subscribers shows that 38 are col-

lege students. Perform an appropriate hypothesis test at the $\alpha = .01$ level of significance.

16. The manager of a firm wants to test the accountant's claim that the firm's cash-flow problem is due to the fact that at least 80 percent of accounts receivable are more than three months old. A random sample of $n = 50$ from 10,000 accounts is to be taken; the desired significance level is $\alpha = .001$. The sample shows 30 such delinquent accounts. Make the test.

17. A Bible salesman has been told that at least 50 percent of all families in Atlanta attend religious services regularly. He makes a random sample of 25 families and finds that 10 of them attend regularly.
   a. Does the finding support the claim at the $\alpha = .05$ level of significance? Explain.
   b. What would your answer be if the same result had been obtained from a random sample of $n = 100$? Explain.

18. A presidential candidate has decided to enter primaries in those states in which at least 20 percent of the voters support the candidate. Random samples of $n = 200$ in four states show the following numbers of supporters: New Hampshire 25, Florida 44, Illinois 31, California 17. Prepare hypothesis tests at the $\alpha = .05$ level of significance to determine where the candidate should enter.

19. An aircraft manufacturer finds that 12 percent of engine crankcases shrink too much when cooling down after being cast. A new process is proposed and tried: of 250 castings, some 27 are equally bad. At the $\alpha = .05$ level of significance, conduct a hypothesis test to find out whether the new process produces crankcases that are, at worst, as bad as, but possibly better than, the old process.

20. The manufacturer of Lycoming aircraft engines knows from long experience that 10 percent of these engines develop problems by the time they reach 2,000 flight hours. The manufacturer believes, however, that a new series of these engines will do better. Accordingly, a random sample of 225 of such engines is selected from the 9,000 owners of them. Of the sampled owners, 198 had used the engine for at least 2,000 hours and 22 of these had had problems. Perform an appropriate hypothesis test at the $\alpha = .05$ level of significance and determine whether the manufacturer's faith in the new engine series is justified.

21. Just prior to the election of President Kennedy, a political analyst believed that 53 percent of the voters would vote for Kennedy. Yet a Gallup poll of 1,500 voters showed only 765 (that is, 51 percent)

so inclined. Conduct a hypothesis test of the analyst's belief at the $\alpha = .05$ level of significance. Then compare your result with that for problem 25 in Chapter 8.

22. Reconsider Chapter 8's Practice Problem 13 on p. 317. In it, a tire manufacturer wishes to estimate the difference between the mean lives of two types of tires. A 99-percent confidence interval of the difference in tire A and tire B lifetimes was computed as 4,900 miles ± 507.52 (as reaching, that is, from 4,392.48 miles to 5,407.52 miles). Perform a hypothesis test on the same example and discover the relationship between estimation and hypothesis testing.

23. An economist wants to test the claim that the wages of construction workers in New York (A) are different from those in Chicago (B). Random samples of workers, $n_A = 100$ and $n_B = 80$, are to be taken; the desired significance level is $\alpha = .025$. The samples show average weekly wages of $\overline{X}_A = \$500$ and $\overline{X}_B = \$410$, with sample standard deviations of $s_A = \$200$ and $s_B = \$100$. Make the test, assuming normally distributed populations with equal variances.

24. A firm's manager wants to know whether the delivery time of raw materials is truly longer with one shipping company (A) than another (B). Random samples of $n_A = 50$ and $n_B = 30$ shipments from among the thousands received in a month are to be taken; the desired significance level is .10. The samples show mean delivery times of $\overline{X}_A = 14$ days and $\overline{X}_B = 12$ days, with sample standard deviations of $s_A = 1$ day and $s_B = .4$ day. Make the test.

25. A government administrator wants to know whether there is any difference in the quality of training given to new air traffic controllers at two different training centers. Exams are administered to two random samples of recent trainees. At center A, the sample is $n_A = 36$, the mean score is $\overline{X}_A = 93$, the standard deviation is $s_A = 8$. At center B, the corresponding data are $n_B = 49$, $\overline{X}_B = 87$, $s_B = 3$. Conduct a hypothesis test at the $\alpha = .005$ level of significance, assuming normally distributed populations with equal variances.

26. A psychologist wants to test whether age influences IQ. A random sample of 50 middle-aged individuals whose IQ score at age 20 is available is to be taken; the desired significance level is .001. The sample shows a mean difference between the current and earlier score of 12 points, with a sample standard deviation of 8 points. Make the test.

27. The owner of a sawmill is about to purchase one of two tree lots, but wonders which lot has the larger trees. Two random samples of $n_A = n_B = 100$ are taken. The average diameter of trees in lot A is $\overline{X}_A = 48.25$ inches with a standard deviation of $s_A = 18.2$ inches. The corresponding data for the sampled trees in lot B are $\overline{X}_B = 39.95$ inches and $s_B = 3.2$ inches. Conduct a hypothesis test to find out whether lot A trees are larger than lot B trees, using a significance level of $\alpha = .05$.

28. The IRS commissioner wants to test whether the quality of work done in two regional offices is the same. Random samples of $n_A = 300$ and $n_B = 200$ tax returns are taken from each office and checked for accuracy; the desired significance level is .01. The samples show 39 errors in office A, 20 in office B. Conduct the test.

29. The study described in Analytical Example 9.2, "Antitrust Pork Barrel," also showed the following for the House Subcommittee on Monopolies and Commercial Law of the House Judiciary Committee: during the 1970–79 period, 18 of 25 formal complaints brought by the FTC were dismissed when a committee member's district was affected; 46 of 171 such complaints were dismissed when no committee member's district was involved. Test the null hypothesis of "no difference" in these proportions at the 10-percent level of significance.

30. The manager of a pizza parlor is testing the proposition that more of the phone orders received during the day (A) than during the night (B) are in fact being picked up. In a three-month test, 250 each of day and night phone orders are sampled; the desired significance level is $\alpha = .05$. It turns out that 240 of the day orders but only 220 of the night orders were picked up. Make the test.

31. An economist is testing the proposition that a larger percentage of firms in industry A than in industry B is making advertising expenditures in excess of 5 percent of sales revenue. A random sample of 200 firms from each industry reveals percentages of 18 and 19, respectively. Conduct an appropriate hypothesis test at a significance level of $\alpha = .01$.

32. An advertising agency wishes to test the claim that a larger percentage of adult women watches *Dallas* than *The Guiding Light* on TV. One random sample of 150 women shows 102 watching *Dallas;* another random sample of 250 women shows 182 watching *The Guiding Light.* Conduct an appropriate hypothesis test at a significance level of $\alpha = .10$.

33. A quality inspector wants to test the claim that the proportion of acceptable electronics components delivered by a foreign supplier (A) is at best equal to that coming from a domestic supplier (B). Random samples of $n_A = 100$ and $n_B = 150$ are to be taken from incoming shipments; the desired significance level is $\alpha = .005$. The samples show 90 good components coming from A and 105 from B. Make the test.

34. A manufacturer has been receiving complaints from customers about their orders arriving fully 12 days after being shipped. The manufacturer selects 20 of the following week's orders at random and ships them differently. A statistician is to test whether the new procedure is better or worse and to do so at a significance level of $\alpha = .05$. The mean delivery time in the sample turns out to be $\overline{X} = 9$ days, with a sample standard deviation of $s = 3$ days. Make the test.

35. A researcher wants to determine whether rats (and maybe even people) can live normal lives entirely with artificial blood. At a significance level of .001, a test is to be conducted on 16 rats that would normally live another 5 months. The mean remaining lifetime of the tested rats, whose natural blood was replaced with artificial blood, turns out to be $\overline{X} = 4.1$ months, with a sample standard deviation of $s = 1.6$ months. Make the test.

36. The process of producing steel is designed to use up 6 quarts of hydrochloric acid per ton of steel. Yet a random sample of 9 different tons of steel shows a mean usage of 9 quarts per ton, with a standard deviation of 1 quart. At a significance level of $\alpha = .05$, conduct an appropriate hypothesis test to determine whether the production process needs recalibration.

37. The mean operating temperature of an aircraft engine during level cruise is supposed to be 190° F. One particular engine was tested at randomly selected times. The mean temperature on these 25 tests was 193° F, with a standard deviation of 3° F. Should the pilot begin to worry about the engine running hot? Conduct an appropriate hypothesis test at a significance level of $\alpha = .05$.

38. A manager wishes to test the tensile strength of yarn that is to be used in the firm's new machines and that must be at least 25 lb. A random sample of 16 spools are taken from a number of incoming shipments; the average tensile strength is 23 lb., with a standard deviation of .2 lb. Conduct an appropriate hypothesis test at a significance level of

$\alpha = .10$ and tell the manager whether the yarn is suitable.

39. A manager wishes to test the weight of crankcases intended for aircraft engines that must be no heavier than 105 lb. A random sample of 20 crankcases is taken from the day's output at the foundry; their average weight is 107 lb., with a standard deviation of .1 lb. Conduct an appropriate hypothesis test at a significance level of $\alpha = .01$ and tell the manager whether the crankcases are suitable.

40. A developer believes that the labor hours required to build a new home average 1,000 at most. A random sample of 11 new homes is taken, the mean is found to be $\overline{X} = 1,230$ hours, with a standard deviation of $s = 470$ hours. Conduct an appropriate hypothesis test at a significance level of $\alpha = .025$ to see whether the developer is correct.

41. Reconsider problem 41 of Chapter 8 on page 336. Suppose an analyst believed the average top speed of Soviet submarines to be at least 60 knots. Conduct an appropriate hypothesis test at the $\alpha = .01$ level of significance.

42. An advertising agency wants to test whether beer drinkers rate alike two brands that are presented in unmarked cans. Two independent samples of 16 each are used; the desired significance level is $\alpha = .05$. Scoring on a scale from 0 to 100, the two sample mean scores are 65 and 79, with sample standard deviations of 20 and 3. Make the test assuming that the two populations in question are normally distributed and have equal variances.

43. The manager noted in problem 38 now wishes to test whether the tensile strength of yarn is alike for two suppliers. In addition to the sample noted earlier ($n_A = 16$, $\overline{X}_A = 23$ lb., $s_A = .2$ lb.), a second sample of yarns coming from another firm is taken and the results are: $n_B = 26$, $\overline{X}_B = 24$ lb., $s_B = 2.3$ lb. Conduct an appropriate hypothesis test at the $\alpha = .10$ level of significance, while assuming that the two populations of tensile strengths are normally distributed and have equal variances.

44. A student wants to test, at the $\alpha = .02$ level of significance, whether the final exam scores given by two different instructors, A and B, are alike. A random sample of each class is taken: $n_A = 10$, $\overline{X}_A = 88.2$, $s_A = 8$; $n_B = 15$, $\overline{X}_B = 82.7$, $s_B = 2$. Conduct the test, assuming the two populations of scores to be normally distributed with equal variances.

45. A business executive wants to test the claim that the firm's sales personnel will make the same average daily sales whether they are salaried or working for a commission. Accordingly, a random sample of 16 salespeople is to be drawn from the New York office and put on salary, while another such sample from the Chicago office is to be put on commission. (Individuals in the first sample are to be matched with "twins" in the second sample according to past sales performance.) The test is to be conducted at the $\alpha = .05$ significance level. After three months, the average difference in the mean daily sales of the salaried minus the other "twin" is found to equal $\overline{D} = \$100$, with $s_D = \$59$. Make the test, assuming that the underlying population of differences is normally distributed.

46. A production engineer wants to test the assertion that workers using method A will on average complete a job in the same time as they would by using method B. Six workers are selected at random, and each is made to do a given job first in one way and then in the other way (although half use A first and the other half use B first). A significance level of $\alpha = .01$ is desired. The results of the experiment are shown in the accompanying table. Make the test, assuming that the underlying population of differences is normally distributed.

Table 9.A

Worker	Completion Time (minutes)	
	Method A	Method B
1	10.0	9.8
2	11.1	11.0
3	9.8	8.2
4	10.0	9.5
5	10.3	10.6
6	10.5	10.2

47. A farmer wants to test whether fertilizer really makes a difference in the yield of strawberry plants. Accordingly, 4 acres of strawberry plants are heavily fertilized; 4 similar acres nearby get no fertilizer at all. (Each acre is matched with a nearby "twin.") At the end of the season, the average difference of yield from fertilized acres minus yield from "twin" unfertilized acres is found to equal $\overline{D} = -57$ pints, with $s_D = 12$ pints. Perform an appropriate hypothesis test at the $\alpha = .05$ level of significance, assuming that the underlying population of differences is normally distributed.

48. A government official wants to determine whether it is true that the average motorcycle-accident fatality rate (average ratio of fatal accidents to all accidents) is lower in states requiring drivers to wear helmets. The record is reviewed: the official finds that 5 states had helmet laws (A) during the entire past decade, while 6 states were without such laws (B) throughout the period. A test at the $\alpha = .05$ significance level is to be conducted, using the observed mean fatality rates of $\overline{X}_A = .1021$ and $\overline{X}_B = .2133$, along with standard deviations of $s_A = .0918$ and $s_B = .0547$. Make the test, assuming that the underlying populations are normally distributed and have equal variances and that for each state only one figure is available that refers to the entire decade in question.

49. A researcher hypothesizes that occasional coffee drinkers (A), when given a cup of coffee just before bedtime, will take longer to fall asleep than habitual coffee drinkers (B) under the same circumstances. A test is to be conducted on $n_A = 9$ occasional and $n_B = 16$ habitual coffee drinkers; a significance level of $\alpha = .10$ is desired. The average times needed to fall asleep turn out to be $\overline{X}_A = 56$ minutes and $\overline{X}_B = 29$ minutes, with sample standard deviations of $s_A = 25$ minutes and $s_B = 22$ minutes. Make the test, assuming that the underlying populations of times are normally distributed and have equal variances.

50. Given "H_0: Tire A is better than tire B," consider the following four situations. In each case, indicate whether the decision is correct or false and, if it is false, what type of error is being made. (i) Tire A is in fact better, and A is used. (ii) Tire A is in fact better, but B is used. (iii) Tire B is in fact better, and B is used. (iv) Tire B is better, but A is used.

51. An auto manufacturer is testing the proposition that a new car model provides at least 25 miles per gallon (mpg). A random sample of 36 cars is to be used, the significance level is to be set at $\alpha = .025$, and the population standard deviation is known to be $\sigma = 5$ miles per gallon.
  a. If the mean sample mpg is to be used as the test statistic, what is the decision rule?
  b. What is the probability of a type II error if the actual mileage equals 22? 24? 26?
  c. Repeat the calculations for β if α is initially set at .10.
  d. What must sample size be if maximum α is supposed to be .025, while β is supposed to equal .10 when the mileage in fact equals 24?

52. Given the sample data of Table 9.B, perform the indicated hypothesis tests, assuming each N is infinite and the desired $\alpha = .05$.

Table 9.B

Case	$\overline{X}$	s	n	H_0
a.	13.7	2	39	$\mu = 14$
b.	207	7.1	150	$\mu = 210$
c.	93	10	42	$\mu \geq 95$
d.	13	3	300	$\mu \geq 18$

53. Determine the significance levels of the following hypothesis tests.
   a. Two-tailed tests: $z_{\alpha/2} = 1.96$, $z_{\alpha/2} = .85$, $z_{\alpha/2} = .15$
   b. One-tailed tests: $z_{\alpha} = 1.96$, $z_{\alpha} = 1.25$, $z_{\alpha} = .15$
54. Expand Appendix Table J, Part B, to include entries
   a. for two-tailed tests and $\alpha = .08, .06, .02$.
   b. for one-tailed tests and $\alpha = .09, .08, .02$.
55. If you were the manager of a firm making a hypothesis test and the rejection of H_0 would mean that the firm would have to be liquidated, would you favor a large or a small value for α? Explain.
56. As was noted in Biography 9.1 (R. A. Fisher), many statisticians like to report a *p-value* for each hypothesis test. On the assumption that the null hypothesis is true, this p-value measures the probability of getting a test statistic as extreme as or more extreme than the one actually computed from the sample data. For a believer in H_0, a large

p-value signals "no surprise" (and acceptance of H_0), but the smaller the p-value is, the greater is the surprise (and the more do sample data point to the rejection of H_0 and the acceptance of H_A). Thus, in Figure 9.A, the crosshatched area represents the (arbitrarily chosen) maximum for probability α (the probability of making the Type I error of falsely rejecting a null hypothesis that is true), while the crosshatched plus dotted area represents the p-value corresponding to an actually observed z-value of $-.82$ in the this lower-tailed test. Compute similar p-values for the results in
   a. Practice Problems 2, 5, 9, and 11.
   b. end-of-chapter problems 8, 9, 16–18, 24, 27, 30–32, and 52c and d.
Hint: All of these are lower-tailed tests.
57. a. Similar to Figure 9.A, draw a graph depicting the difference between α and p-value for a two-tailed hypothesis test.
   b. Then compute the p-values for Practice Problems 1, 4, 6, 8, and 10; for Analytical Example 9.1 (five parts); and for end-of-chapter problems 3–7, 14, 15, 21–23, 25, 26, 28, 29, and 52a and b (all of which are two-tailed tests).
Hint: Part a *is* a difficult problem; if you can't do it, consult the answer at the back of the book, then do part b.
58. a. Similar to Figure 9.A, draw a graph depicting the difference between α value and p-value for an upper-tailed hypothesis test.
   b. Then compute p-values for Practice Problems 3 and 7 and for end-of-chapter problems 10, 11, 19, 20, and 33 (all of which are upper-tailed tests).

Figure 9.A

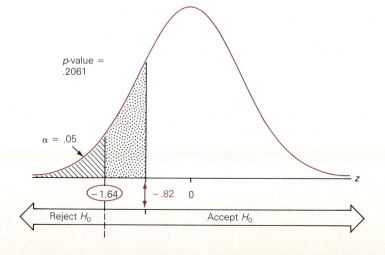

59. The *p*-values for hypothesis tests using a *t* statistic are computed in a fashion analogous to that used in the preceding three problems. Naturally, Appendix Table K rather than Appendix Table H must be consulted. In addition, there is not enough detail in any given degrees-of-freedom row; therefore, *p*-values are typically reported as a range. Compute *p*-values for Practice Problems 13 and 15, for Analytical Example 9.3, on page 381, and for end-of-chapter chapter problems 38–41 on page 390 and 48–49 on page 391.

60. Compute *p*-values for the following two-tailed hypothesis tests (all of which used the Student *t* distribution): Practice Problems 12, 14, and 16; Analytical Example 9.1; end-of-chapter problems 34–37 and 42–47.

Selected Readings

Bartlett, Maurice S. "Egon Sharpe Pearson, 1895–1980." *Biometrika,* April 1981, pp. 1–12. Contains a listing of 133 papers published by E. S. Pearson.

Bennett, J. H. *Collected Papers of R. A. Fisher.* Adelaide, Australia: The University of Adelaide, 1971–74. This five-volume set of more than 300 articles is a monument to Fisher's genius.

Fisher, R. A. "On the Mathematical Foundations of Theoretical Statistics." *Philosophical Transactions of the Royal Society,* 1922, pp. 309–68. An early landmark in the Fisherian revolution in statistics.

Fisher, R. A. *Statistical Methods for Research Workers,* 14th ed. New York: Hafner Press, 1970. A classic work on statistical estimation, hypothesis testing, analysis of variance, correlation, and more.

Fisher, R. A. *The Design of Experiments,* 9th ed. New York: Hafner Press, 1971. A classic work on all aspects of experimentation, including randomized blocks, Latin squares, factorial designs, confounding, hypothesis testing, and more.

Fisher, R. A. *Statistical Methods and Scientific Inference,* 3rd ed. New York: Hafner, 1973.

McCloskey, Donald N. "The Loss Function Has Been Mislaid: The Rhetoric of Significance Tests," *The American Economic Review,* May 1985, pp. 201–205. An important and delightfully written warning about the common misuse of significance tests.

Menges, G. "Inference and Decision." *Selecta Statistica Canadiana,* vol. 1. New York: Wiley, 1973. Contrasts the Fisher view of statistics with the Neyman-Pearson view.

Morrison, D., and R. E. Henkel. *The Significance Test Controversy.* Chicago: Aldine, 1970. Raises serious questions about the usefulness and sense of employing significance tests in social-science research.

Neyman, Jerzy, and E. S. Pearson. "On the Use and Interpretation of Certain Test Criteria for Purposes of Statistical Inference." *Biometrika,* 1928, pp. 175–240 and 263–94.

Neyman, Jerzy, and E. S. Pearson. "On the Problem of the Most Efficient Tests of Statistical Hypotheses." *Philosophical Transactions of the Royal Society,* 1933, pp. 289–337. This and the preceding entry are two crucial articles developing the modern theory of hypothesis testing.

Popper, Karl R. *The Logic of Scientific Discovery.* New York: Harper, 1965. This book presents a detailed discussion of the formulation of hypotheses, their testing through observation, and their role in the development of scientific theories. It was first published in 1935.

Savage, Leonard J. *The Foundations of Statistics,* rev. ed. New York: Dover, 1972. Contains a sharp attack on hypothesis testing.

Computer Programs

The DAYSAL-2 personal-computer diskettes that accompany this book contain one program of interest to this chapter:

9. Classical Hypothesis Tests lets you make large- and small-sample tests of the population mean or proportion or of the difference between two such means or proportions. You can select two-tailed, lower-tailed, or upper-tailed tests, and you can choose among numerous significance levels when you use this program to work problems.

Commonly Used Test Statistics

STATISTICS COMMONLY USED FOR TESTS ABOUT A POPULATION MEAN

Hypothesis tests about a population mean either employ the sample mean, $\overline{X}$, directly for judging the probable truth or falsity of the null hypothesis, or they make use of the corresponding values (implied by Boxes 8.C and 8.D) of

$$z = \frac{\overline{X} - \mu_0}{\sigma_{\overline{X}}}$$

for samples of $n \geq 30$ (given $n < .05N$ or a normal population), and

$$t = \frac{\overline{X} - \mu_0}{s_{\overline{X}}}$$

for samples of $n < 30$, given the population is normal.

In these formulations, μ_0 is the value specified in the null hypothesis. Furthermore (as was noted in Box 8.A), in the large-population case (when the population being sampled is infinite or when sampling occurs with replacement or when there is no replacement, but $n < .05N$),

$$\sigma_{\overline{X}} = \left(\frac{\sigma}{\sqrt{n}}\right).$$

In the small-population case (when sampling occurs without replacement and $n \geq .05N$),

$$\sigma_{\overline{X}} = \left(\frac{\sigma}{\sqrt{n}}\right)\sqrt{\frac{N - n}{N - 1}}.$$

In either case, σ can be estimated by s. Similarly,

$$s_{\bar{X}} = \frac{s}{\sqrt{n}} \text{ or } \left(\frac{s}{\sqrt{n}}\right)\sqrt{\frac{N-n}{N-1}},$$

respectively.

STATISTICS COMMONLY USED FOR TESTS ABOUT THE DIFFERENCE BETWEEN TWO POPULATION MEANS

Hypothesis tests about the difference between two population means can involve the taking of two independent samples or of a matched-pairs sample.

Independent Samples

In the case of independent samples, hypothesis tests either employ the difference between the sample means, $d = \bar{X}_A - \bar{X}_B$, directly, or they make use of the corresponding values (implied by Box 8.E) of

$$z = \frac{d - (\mu_A - \mu_B)}{\sigma_d}$$
$$= \frac{(\bar{X}_A - \bar{X}_B) - (\mu_A - \mu_B)}{\sigma_{\bar{X}_A - \bar{X}_B}}$$

for samples of $n \geq 30$ (given $n < .05N$ or a normal population), and

$$t = \frac{d - (\mu_A - \mu_B)}{s_d}$$
$$= \frac{(\bar{X}_A - \bar{X}_B) - (\mu_A - \mu_B)}{s_{\bar{X}_A - \bar{X}_B}}$$

for samples of $n < 30$, given that the two populations are normal and their variances are equal.

Three things about the values of z and t should be noted:

1. The expression, $\mu_A - \mu_B$, has an extreme value of 0 in all three formulations of the null hypothesis (Forms A to C in Chapter 9); hence, this expression can be deleted from the above formulas.
2. For the large-population case, the standard error of the difference between sample means equals

$$\sigma_d = \sigma_{\bar{X}_A - \bar{X}_B} = \sqrt{\frac{\sigma_A^2}{n_A} + \frac{\sigma_B^2}{n_B}} \cong \sqrt{\frac{s_A^2}{n_A} + \frac{s_B^2}{n_B}}.$$

3. Because the t distribution assumes equal population variances, the information given by the two sample variances (one pertaining to each population) is usually combined to get the best available estimate of the (identical) variance in each population. This **pooled variance,** s_P^2, is the weighted average of the two sample variances, where the weights equal the degrees of freedom associated with each variance:

$$s_P^2 = \frac{(n_A - 1)s_A^2 + (n_B - 1)s_B^2}{n_A + n_B - 2}.$$

Thus, the t value turns into

$$t = \frac{\overline{X}_A - \overline{X}_B}{\sqrt{\dfrac{s_A^2}{n_A} + \dfrac{s_B^2}{n_B}}} = \frac{\overline{X}_A - \overline{X}_B}{\sqrt{s_P^2\left(\dfrac{1}{n_A} + \dfrac{1}{n_B}\right)}}.$$

Matched-Pairs Sample

In the case of matched-pairs sampling, hypothesis tests either employ the average sample difference, $\overline{D} = (\Sigma D_i/n)$, directly, or they make use of the corresponding value (implied by Box 8.F) of

$$z = \frac{\overline{D} - (\mu_A - \mu_B)}{s_D/\sqrt{n}}$$

for samples of $n \geq 30$, given that the two populations are normal and their variances are equal. (For $n < 30$, t replaces z.)

Once again note that the expression $\mu_A - \mu_B$ equals 0 in this case, while (as was noted in Chapter 8)

$$s_D = \sqrt{\frac{\Sigma D_i^2 - n\overline{D}^2}{n - 1}}.$$

STATISTICS COMMONLY USED FOR TESTS ABOUT A POPULATION PROPORTION

Hypothesis tests about a population proportion either employ the sample proportion, P, directly or they make use of the corresponding value (implied by Box 8.C) of

$$z = \frac{P - \pi_0}{\sigma_P},$$

provided the sampling distribution of the proportion can be assumed to be normal because both $n\pi$ and $n(1 - \pi) \geq 5$.

Note: In this formulation, π_0 is the value specified in the null hypothesis; furthermore (as was noted in Box 8.B), in the large-population case,

$$\sigma_P = \sqrt{\frac{\pi(1 - \pi)}{n}} \cong \sqrt{\frac{P(1 - P)}{n}}.$$

In the small-population case,

$$\sigma_P = \sqrt{\frac{\pi(1 - \pi)}{n} \left(\frac{N - n}{N - 1}\right)} \cong \sqrt{\frac{P(1 - P)}{n} \left(\frac{N - n}{N - 1}\right)}.$$

STATISTICS COMMONLY USED FOR TESTS ABOUT THE DIFFERENCE BETWEEN TWO POPULATION PROPORTIONS

The procedure for testing the difference between two proportions is analogous to that concerning the difference between means. Consider the case of two large and independent samples (for which the assumption of normality is warranted): one can either employ the difference between the sample proportions, $d = P_A - P_B$, directly, or make use of the corresponding value (implied by Box 8.G) of

$$z = \frac{d - (\pi_A - \pi_B)}{\sigma_d} = \frac{(P_A - P_B) - (\pi_A - \pi_B)}{\sigma_{P_A - P_B}}$$

Note: Once again, the expression $\pi_A - \pi_B$ can be deleted because it has an extreme value of 0 in each of the common formulations of the null hypothesis (corresponding to Forms A to C in Chapter 9). For the large-population case, the standard error of the difference between sample proportions equals

$$\sigma_d = \sigma_{P_A - P_B} = \sqrt{\frac{\pi_A(1 - \pi_A)}{n_A} + \frac{\pi_B(1 - \pi_B)}{n_B}} \cong \sqrt{\frac{P_A(1 - P_A)}{n_A} + \frac{P_B(1 - P_B)}{n_B}}.$$

In the latter expression, a pooled estimator of the population proportion,

$$P_P = \frac{n_A P_A + n_B P_B}{n_A + n_B},$$

is usually substituted, so that

$$\sigma_{P_A - P_B} = \sqrt{P_P(1 - P_P)\left(\frac{1}{n_A} + \frac{1}{n_B}\right)}.$$

CHAPTER TEN

Hypothesis Testing: The Chi-Square Technique

Chapters 8 and 9 have amply demonstrated the importance of the normal probability distribution and of the *t* distribution for purposes of making estimates of unknown population parameters or testing hypotheses about them. There are occasions, however, when these distributions cannot be used; the *chi-square technique,* to be discussed in this chapter, is often employed in such circumstances. We will consider four major applications of this technique: (1) testing the alleged independence of two qualitative population variables, (2) making inferences about more than two population proportions, (3) making inferences about a population variance, and (4) conducting goodness-of-fit tests to assess the plausibility that sample data come from a population the elements of which conform to a specified type of probability distribution.

TESTING THE ALLEGED INDEPENDENCE OF TWO QUALITATIVE VARIABLES

Qualitative variables, as noted in Chapter 2, are characteristics (like type of business or job title) that are usually not expressed numerically because they differ in kind rather than in degree among the elementary units of a statistical population. There are many occasions when it is important to know whether two such variables are *statistically independent* of one another (when, therefore, the probability of the occurrence of one variable is unaffected by the occurrence of the other) or whether these variables, on the contrary, are *statistically dependent* (when, therefore, the probability of the occurrence of one *is* affected by the occurrence of the other). Consider how important it might be to know whether there exists a relationship of independence or dependence between heart attacks and lifestyle, between the occurrence of disease and having been vaccinated, between equipment failure and prior years of service, between the percentage of defective units produced and the time of day of their production, between the causes of airplane accidents and the extent of pilot qualification, between the educational level (or age or sex or geographic location) of consumers and their brand preferences, between the severity of auto accidents and the location of their occurrence, between people's major in college and their type of employment later on, between attitudes toward a piece of legislation and people's political affiliation, between sex and type of job held—the list is endless.

Statistical Independence Reviewed

Table 10.1 is designed to help us review the nature of statistical independence. We imagine that a census has revealed that 100 pilots are residing in a city and that they have been classified by sex (male, female) and by the type of license held (private, commercial, airline-transport). We note that the proportion of males in the entire pilot population (80/100) is the same as that among private pilots only (56/70) or among commercial pilots only (16/20) or among airline-transport pilots only (8/10). Similarly, the proportion of females in the entire pilot population (20/100) is identical with that in any of the three subcategories. Because sex and type of license held are, thus, statistically independent, the special multiplication law for independent events (Box 5.J on page 165) can be applied to any of the probabilities involved. For example, the probability of finding a male private pilot is simply the product of the probabilities of finding a male pilot and of finding a private pilot:

$$p(\text{male } and \text{ private pilot}) = p(\text{male pilot}) \cdot p(\text{private pilot})$$

$$\frac{56}{100} = \frac{80}{100} \cdot \frac{70}{100}.$$

Similarly, the probablity of finding a female airline-transport pilot is the product of the probabilities of finding a female pilot and of finding an airline-transport pilot:

$$p(\text{female } and \text{ airline-transport pilot}) = p(\text{female pilot}) \cdot p(\text{airline-transport pilot})$$

$$\frac{2}{100} = \frac{20}{100} \cdot \frac{10}{100}.$$

Table 10.1   Pilots Residing in City, Classified by Sex and by Type of License

This table exemplifies the statistical independence of two qualitative population variables, A and B (here sex and type of pilot's license held). When such independence exists, the proportion of units having one attribute (such as female sex) is the same in the total population as in any part of the population having the other attribute (such as commercial-pilot-license holder). Hence, the frequency for any particular pair of attributes can be found by multiplying the respective frequencies for the two individual attributes and dividing the product by the total number of units observed. Note how

$$\frac{20 \text{ female pilots} \times 20 \text{ commercial pilots}}{100 \text{ pilots}} = 4 \text{ female commercial pilots.}$$

(A) Sex	(B) Type of License			Total
	Private	Commercial	Airline-Transport	
Male	56	16	8	80
Female	14	4	2	20
Total	70	20	10	100

The foregoing discussion suggests an easy way of testing *census* data for statistical independence:

1. Set up a **contingency table,** such as Table 10.1, that classifies data according to two or more categories associated with each of two qualitative variables that may or may not depend on each other statistically. (Such a table, thus, shows all possible combinations or *contingencies,* which accounts for its name.)
2. For each cell of the table, check whether the joint probability (such as 56/100 for male *and* private pilot) equals the product of the corresponding unconditional probabilities (such as 80/100 for male pilot times 70/100 for private pilot). If the answer is yes, the characteristics in question are independent; if it is no, they are dependent.

Typically, however, population frequencies are unknown and only sample data are available. In that case, the type of test just suggested will not work because many a joint probability may then fail to equal the product of the relevant unconditional probabilities merely as a result of sampling error and not because of any statistical dependence between the two qualitative variables of interest in the population. It is in such circumstances that the chi-square technique finds its first application.

A Contingency-Table Test

Suppose the Federal Aviation Administration wanted to determine whether there existed any link between the official causes of aviation accidents and the type of license held by the involved pilots-in-command. The experience of the past five years, summarized in Table 10.2, might be viewed as a sample of all possible accidents. The observed frequencies given in the various cells of this table can subsequently be compared with the frequencies one would expect if the two variables in question (accident

Table 10.2   Five-Year Sample of Nonmilitary Aviation Accidents, Classified by Cause and by Type of Pilot's License: Observed Frequencies

This contingency table shows observed frequencies (typically symbolized by f_o) for various combinations of aviation-accident causes (A) and pilot's-license types (B). Setting up such a table is the first step in testing for independence between the two variables, A and B.

(A) Official Accident Cause	(B) License Held by Pilot-in-Command		Total
	Student or Private	Commercial or Airline-Transport	
(1) Thunderstorm	50	20	70
(2) Icing	40	10	50
(3) Equipment failure	25	6	31
(4) Controller error	25	5	30
(5) Pilot incapacitation	80	6	86
(6) Other	30	3	33
Total	250	50	300

Table 10.3   Five-Year Sample of Nonmilitary Aviation Accidents, Classified by Cause (A) and by Type of Pilot's License (B): Expected Frequencies, Assuming Independence Between A and B.

This contingency table shows expected frequencies (typically symbolized by f_e) for various combinations of aviation-accident causes (A) and pilot's license types (B) under the assumption of independence. Setting up such a table is the second step in testing for independence between the two variables, A and B.

(A) Official Accident Cause	(B) License Held by Pilot-in-Command		Total
	Student or Private	Commercial or Airline-Transport	
(1) Thunderstorms	[70(250)/300] = 58.33	[70(50)/300] = 11.67	70
(2) Icing	[50(250)/300] = 41.67	[50(50)/300] = 8.33	50
(3) Equipment failure	[31(250)/300] = 25.83	[31(50)/300] = 5.17	31
(4) Controller error	[30(250)/300] = 25.00	[30(50)/300] = 5.00	30
(5) Pilot incapacitation	[86(250)/300] = 71.67	[86(50)/300] = 14.33	86
(6) Other	[33(250)/300] = 27.50	[33(50)/300] = 5.50	33
Total	250	50	300

cause and type of license held) were statistically independent of one another. Such expected frequencies are shown in Table 10.3. On the assumption of independence between variables A and B, each of the expected frequencies in Table 10.3 has been calculated as

$$\text{expected frequency}, f_e = \frac{\text{row total} \times \text{column total}}{\text{sample size}}.$$

This procedure corresponds to that described in the caption of Table 10.1.

For purposes of conducting the test of independence, it is convenient to show both the observed and expected frequencies in the same table, which is accomplished in Table 10.4. This table, in turn, can be used to calculate the **chi-square statistic,** which was introduced in 1900 by Karl Pearson (see Biography 10.1 at the end of the chapter). The chi-square statistic is admirably suited to assess the issue of independence or dependence between qualitative variables. Contrary to the common practice of denoting population parameters by Greek letters and sample statistics by Roman letters, this statistic is denoted by the lowercase Greek letter χ, which is pronounced "ki" as in "kite." According to one commonly used formula (that is shown in Box 10.A), chi square (χ^2) equals the sum of all the ratios that can be constructed by taking the difference between each observed and expected frequency, squaring it, and dividing this squared deviation by the expected frequency.

10.A A Chi-Square Statistic

$$\chi^2 = \sum \frac{(f_o - f_e)^2}{f_e}.$$

where f_o is an observed frequency and f_e is the associated expected frequency.

Table 10.4   Five-Year Sample of Nonmilitary Aviation Accidents, Classified by Cause (A) and by Type of Pilot's License (B): Observed Frequencies vs. Expected Frequencies, Assuming Independence Between A and B.

This contingency table combines the information contained in Tables 10.2 and 10.3. Frequencies actually observed are shown in the upper-left portion of each cell; frequencies expected on the assumption of independence between variables A and B are shown in the lower-right portion.

(A) Official Accident Cause	(B) License Held by Pilot-in-Command		Total
	Student or Private (1)	Commercial or Airline-Transport (2)	
(1) Thunderstorm	50 / 58.33	20 / 11.67	70
(2) Icing	40 / 41.67	10 / 8.33	50
(3) Equipment failure	25 / 25.83	6 / 5.17	31
(4) Controller error	25 / 25.00	5 / 5.00	30
(5) Pilot incapacitation	80 / 71.67	6 / 14.33	86
(6) Other	30 / 27.50	3 / 5.50	33
Total	250	50	300

For our example, the value of χ^2 can be calculated as 14.88, which is shown in Table 10.5 at the top of the facing page.

Note: The preceding χ^2 formula makes intuitive sense. We are interested in assessing the differences between the observed and expected; hence, it is reasonable to calculate the difference between f_o and f_e for each cell of the contingency table. Yet we are unable to reach an overall assessment by then summing these differences because such a sum necessarily equals zero, as Table 10.5 shows. For this reason, the deviations are squared. That procedure has the added advantage of magnifying large deviations (which are probably the result of some kind of dependence between the variables studied) relative to small deviations (which are more likely the result of sampling error). Consider, in Table 10.5, the deviation of 8.33 relative to that of .83 (a tenfold difference in magnitude). When squared, the numbers become 69.39 and .69 (a 100-fold difference in magnitude). Why, finally, does the formula divide each squared dif-

Table 10.5 The Calculation of Chi Square

This table shows the calculation of the chi-square statistic from the data embodied in Table 10.4. Note: because the χ^2 formula calls for the squaring of deviations between the observed and expected, χ^2 can never be negative but can take on values only of zero or above.

Row, Column	Observed Frequency, f_o	Expected Frequency (given independence), f_e	Deviation, $f_o - f_e$	Squared Deviation, $(f_o - f_e)^2$	Standardized Squared Deviation, $\dfrac{(f_o - f_e)^2}{f_e}$
(1, 1)	50	58.33	−8.33	69.39	1.19
(1, 2)	20	11.67	8.33	69.39	5.95
(2, 1)	40	41.67	−1.67	2.79	.07
(2, 2)	10	8.33	1.67	2.79	.33
(3, 1)	25	25.83	− .83	.69	.03
(3, 2)	6	5.17	.83	.69	.13
(4, 1)	25	25.00	0	0	0
(4, 2)	5	5.00	0	0	0
(5, 1)	80	71.67	8.33	69.39	.97
(5, 2)	6	14.33	−8.33	69.39	4.84
(6, 1)	30	27.50	2.50	6.25	.23
(6, 2)	3	5.50	−2.50	6.25	1.14
Total	300	300	0		$\chi^2 = 14.88$

ference by the expected frequency? This division converts absolute squared deviations into relative ones and puts all the cells on an equal footing. Consider the first two rows of Table 10.5. The absolute value of the deviation (and, therefore, of its square) is the same in each case, but a deviation of 8.33 from an expected value of 58.33 is surely less significant than an identical deviation from an expected value of 11.67; this fact is evidenced by the different entries in the last column (1.19 versus 5.95).

Also note: If we had taken a different sample, all the observed frequencies would have been different, and a different value of χ^2 would have been calculated as well. Thus, it becomes crucial to know something about the sampling distribution of this statistic. That knowledge, in turn, will enable us to test the hypothesis of independence between any two qualitative variables that are of interest to us and to do so while specifying a maximum acceptable α risk (of rejecting a hypothesis of independence even though it is true) which, given sample size, implies a specific β risk (of accepting a hypothesis of independence even though it is false).

Sampling distributions of chi square. The nature of the sampling distribution of chi square depends on the number of degrees of freedom associated with the problem under investigation. This number equals the number of expected-frequency values one is free to set before the constraints of the problem dictate the remaining values. Consider Table 10.4. Given the column and row totals (which are the constraints of that problem), we are free to set only any 5 of the 12 expected-frequency values. Once we have calculated, say, the first five f_e values in column (1)—once we have calculated, that is, 58.33, 41.67, 25.83, 25.00, and 71.67—all the remaining f_e values are automatically determined by the constraints. Because the first column sums to 250, the last f_e value of column (1) must equal 27.50. Given the six column (1) f_e values and the row totals, the remaining column (2) f_e values, in turn, must necessarily be precisely what the table shows them to be. Thus, the problem involves 5 degrees of freedom. In a contingency-table test, this number can most easily be found as shown in Box 10.B. In our example, this number, of course, is

$$(6 - 1) \cdot (2 - 1) = 5.$$

10.B Number of Degrees of Freedom (d.f.) in a Contingency-Table Test

$$d.f. = (\text{number of rows} - 1) \cdot (\text{number of columns} - 1)$$

Because the χ^2 random variable is a sum of squares and if certain conditions are met, the sampling distribution of the χ^2 statistic for n degrees of freedom can be represented by the probability distribution of the sum of the squares of n independent standard normal random variables and is then referred to as a **chi-square distribution** (with n degrees of freedom). The conditions just referred to are (1) a sample sufficiently large so that each expected frequency equals 5 or more (in a contingency-table test, this result can often be assured by combining two or more rows or columns, as will be shown below) and (2) an underlying population that is normally distributed.

To understand what a chi-square distribution is, consider a population of values, x, that are normally distributed. The corresponding standard normal deviates, $z = [(x - \mu)/\sigma]$, would be normally distributed as well. Yet the squares of these deviates, z^2, would not be normally distributed. Their distribution would be continuous and unimodal, but strongly skewed to the right, with values ranging from zero to positive infinity. Under the conditions just cited, this distribution would equal the sampling distribution of χ^2 with 1 degree of freedom.

Now consider two independent standard normal random variables, z_a and z_b. If we squared each and added the squares, the distribution of this expression ($z_a^2 + z_b^2$) would be similar to the skewed distribution just discussed and, again under the cited conditions, would equal the sampling distribution of χ^2 with 2 degrees of freedom. In the same way, we can imagine finding the probability distributions for the sum of 4, 10, 20, or any other number of squared z values; in each case, these distributions would correspond (under the cited conditions) to the sampling distribution of χ^2 with 4, 10, 20, or any other number of degrees of freedom. Figure 10.1 shows such chi-square distributions for 2, 4, 10, and 20 degrees of freedom. What is not shown in the graph is that the expected value or mean of every chi-square distribution equals the number of degrees of freedom, while the variance equals twice that number, as shown in Box 10.C on page 406.

Figure 10.1 Four Members of the Chi-Square Distribution Family

A different chi-square distribution exists for each possible number of degrees of freedom *(d.f.)*. Because the χ^2 random variable is a sum of squares, negative values are impossible; all values, therefore, range between zero and positive infinity. Even though all chi-square distributions are, thus, skewed to the right, the extent of skewness decreases with increasing numbers of degrees of freedom. Chi-square distributions with 30 degrees of freedom or more very closely approach the normal distribution, a fact that is already evident by comparing panels (a) through (d).

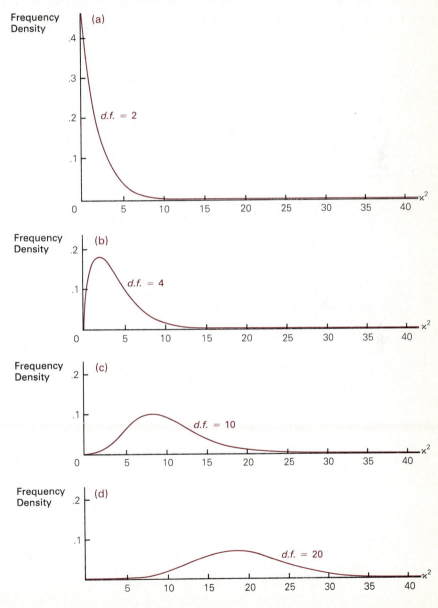

10.C Summary Measures of the Chi-Square Distribution

$$E_{\chi^2} = \mu_{\chi^2} = d.f.$$
$$\sigma^2_{\chi^2} = 2\ d.f.$$
$$\sigma_{\chi^2} = \sqrt{2\ d.f.}$$

where μ_{χ^2} is the mean, $\sigma^2_{\chi^2}$ is the variance, and σ_{χ^2} is the standard deviation of the chi-square distribution, while d.f. are the associated degrees of freedom.

The chi-square table. For the convenience of conducting χ^2 tests, critical values of χ^2 that correspond to specified upper-tail areas of the distribution and to specified numbers of degrees of freedom have been tabulated, as in Appendix Table L. Consider our example that involves 5 degrees of freedom. As the fifth row of Appendix Table L shows, .99 of the area under the appropriate chi-square distribution lies to the right of $\chi^2 = .554$; .98 of the area lies to the right of $\chi^2 = .752$, and so on, until .001 of the area lies to the right of $\chi^2 = 20.517$. If we cared to specify a significance level of $\alpha = .05$, the critical chi-square value would be $\chi^2_\alpha = 11.070$. This value can also be designated as $\chi^2_{.05,5}$ to remind us instantly of the significance level ($\alpha = .05$) and the degrees of freedom (5 *d.f.*). Our choice of $\alpha = .05$ implies that, due to sampling error, only 5 percent of the χ^2 values we would compute by taking repeated samples from a population for which the hypothesis of independence was true would exceed our critical value of 11.07. In our example, the computed value of χ^2 equals 14.88 (see Table 10.5). Thus, we are justified in suspecting that more than sampling error is involved and in *rejecting* the hypothesis of independence between accident cause and type of pilot's license held. But remember: our willingness to live with an α risk of .05 implies a willingness to make such a rejection erroneously at most 5 percent of the time when using this procedure over and over again.

Summary of procedure. We can summarize the procedure just introduced by showing how it relates to the hypothesis-testing steps introduced in Chapter 9.

Step 1: Formulating two opposing hypotheses.

H_0: The two variables (accident cause and type of pilot's license) are independent.
H_A: The two variables are dependent.

Step 2: Selecting a test statistic. We select

$$\chi^2 = \sum \frac{(f_o - f_e)^2}{f_e}.$$

Step 3: Deriving a decision rule. We set a significance level of $\alpha = .05$; hence (for 5 degrees of freedom), the critical value of χ^2_α is $\chi^2_{.05,5} = 11.070$ (Appendix Table L). Thus, the decision rule must be: "Accept H_0 if $\chi^2 \leq 11.070$." Note: χ^2 tests of independence are always upper-tailed because a perfect fit between f_o and f_e makes $\chi^2 = 0$. Because deviations between f_o and f_e are being squared, the greater the deviations are (positive or negative), the greater becomes the computed value of χ^2. The decision rule is illustrated in Figure A; the critical value is encircled.

Figure A

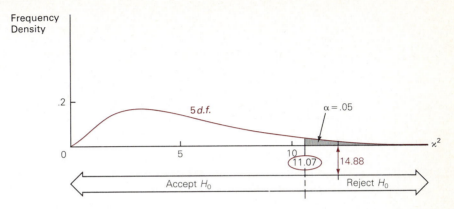

Step 4: Selecting a sample, computing the test statistic, and confronting it with the decision rule. After taking the sample, as noted above (Table 10.2), an actual χ^2 of 14.88 is calculated (Table 10.5). This value corresponds to the colored arrow in Figure A; it suggests that the null hypothesis should be *rejected.* At the 5-percent significance level, it appears that the accident cause and the type of pilot's license are statistically *dependent* variables. The following section considers another example of a χ^2 test for independence.

 Practice Problem 1

Sex and Brand Preferences

An advertising agency wants to know whether the sex of consumers is independent of their preferences for 4 brands of coffee. (The answer will determine whether different ads must be created for men's and women's magazines.) A test of independence is to be conducted at the 5-percent significance level using a simple random sample of 100 persons.

Step 1: Formulating two opposing hypotheses.

H_0: Sex and brand preference are independent variables.
H_A: Sex and brand preference are dependent variables.

Step 2: Selecting a test statistic:

$$\chi^2 = \sum \frac{(f_o - f_e)^2}{f_e}.$$

Step 3: Deriving a decision rule. Given $\alpha = .05$ and 3 degrees of freedom (2 sexes and 4 brands make for a 2-row, 4-column contingency table and, hence, 1×3 degrees of freedom), the critical $\chi^2_{.05,3} = 7.815$ (Appendix Table L), making the decision rule: "Accept H_0 if $\chi^2 \leq 7.815$."

Step 4: Selecting a sample, computing the test statistic, and confronting it with the decision rule. The sample reveals, we assume, the frequencies given in Table A. Thus,

Table A

Sex	Brand Preference				Total
	A	B	C	D	
(1) Male	18	25	15	2	60
(2) Female	32	5	1	2	40
Total	50	30	16	4	100

Table B

Row, Column	Observed Frequency f_o	Expected Frequency (if H_0 is true), f_e	$f_o - f_e$	$(f_o - f_e)^2$	$\dfrac{(f_o - f_e)^2}{f_e}$
(1, A)	18	30.0	−12	144	4.80
(1, B)	25	18.0	7	49	2.72
(1, C)	15 }17	9.6 }12			2.08
(1, D)	2	2.4	5	25	
(2, A)	32	20.0	12	144	7.20
(2, B)	5	12.0	−7	49	4.08
(2, C)	1 } 3	6.4 } 8			3.13
(2, D)	2	1.6	−5	25	
Total	100	100	0		$\chi^2 = 24.01$

χ^2 is calculated as shown in Table B. Note that some categories have been combined in order to avoid expected frequencies of less than 5 (in accordance with the rule noted earlier on page 404). The result suggests that the null hypothesis should be *rejected*. Sex and coffee brand preference are statistically dependent variables.

Note: Analytical Example 10.1, "Leukemia and the Atomic Bomb," provides another example of the use of this technique. |||

MAKING INFERENCES ABOUT MORE THAN TWO POPULATION PROPORTIONS

As the previous examples have shown, when expected frequencies are being calculated for the various cells of a contingency table, one effectively assumes that certain *proportions* (such as those of males among all pilots or of females among all consumers) are the same for all the categories of some variable (such as type of pilot's license held or preferred brand of product). It is not surprising, therefore, that the same technique that allows us to test for the independence of two qualitative population variables can also be used to test whether a number of population proportions are equal to each other or are equal to any predetermined set of values. The next two sections provide illustrations.

Case 1

Are Population Proportions Equal to Each Other?

Consider the process of assembling television sets. Management may be interested in testing the hypothesis that the proportion of defective units produced (which has been .05 in the past) would be the same for each of 6 possible assembly-line speeds. The company statistician may be asked to perform a test at the 1-percent significance level, taking 6 samples of 100 TV sets each, while different assembly-line speeds are being maintained. The statistical procedure may well be the following four steps.

Step 1: Formulating two opposing hypotheses.

H_0: The population proportion of defectives is the same for each of 6 assembly-line speeds; that is, $\pi_1 = \pi_2 = \pi_3 = \pi_4 = \pi_5 = \pi_6 = .05$.
H_A: The population proportion of defectives is not the same for each of 6 assembly-line speeds; that is, at least one of the equalities in H_0 does not hold.

Step 2: Selecting a test statistic:

$$\chi^2 = \sum \frac{(f_e - f_e)^2}{f_e}.$$

Step 3: Deriving a decision rule. Given $\alpha = .01$ and 5 degrees of freedom (2 possible product qualities and 6 possible assembly-line speeds make for 1×5 *d.f.*), the critical $\chi^2_{.01,5} = 15.086$ (Appendix Table L), making the decision rule: "Accept H_0 if $\chi^2 \leq 15.086$."

Step 4: Selecting the samples, computing the test statistic, and confronting it with the decision rule. The samples reveal, we assume, the observed frequencies given in the upper-left corners of the cells in combined contingency Table C, while the expected frequencies (assuming H_0 is true) are given in the lower-right corners. Accordingly, the value of the χ^2 statistic is calculated as shown in Table D. As a result, H_0 should be *accepted* at the 1-percent level of significance. The population proportion of defectives is the same for each assembly-line speed tested. |||

Table C

Product Quality	Assembly-Line Speed (units per hour)						Total
	A = 60	B = 70	C = 80	D = 90	E = 100	F = 110	
(1) Defective	6 / 5	4 / 5	5 / 5	5 / 5	6 / 5	4 / 5	30
(2) Acceptable	94 / 95	96 / 95	95 / 95	95 / 95	94 / 95	96 / 95	570
Total	100	100	100	100	100	100	600

Table D

Row, Column	Observed Frequency, f_o	Expected Frequency (if H_0 is true), f_e	$f_o - f_e$	$(f_o - f_e)^2$	$\dfrac{(f_o - f_e)^2}{f_e}$
(1, A)	6	5	1	1	.20
(1, B)	4	5	−1	1	.20
(1, C)	5	5	0	0	0
(1, D)	5	5	0	0	0
(1, E)	6	5	1	1	.20
(1, F)	4	5	−1	1	.20
(2, A)	94	95	−1	1	.01
(2, B)	96	95	1	1	.01
(2, C)	95	95	0	0	0
(2, D)	95	95	0	0	0
(2, E)	94	95	−1	1	.01
(2, F)	96	95	1	1	.01
Total	600	600	0		$\chi^2 = .84$

Case 2

Are Population Proportions Equal to a Given Set of (Unequal) Values?

Consider three firms, A, B, and C, that currently hold shares of 10, 40, and 50 percent of the market for a particular product. The smallest firm (A) wants to change the design of its product and wonders whether the subsequent market shares will be the same as before. A statistical test is to be conducted at the 5-percent significance level after asking a random sample of 300 consumers to indicate preferences among the newly designed product of A and the traditional products of B and C.

Step 1: Formulating two opposing hypotheses.

H_0: The population proportions (market shares) will be the same as in the past; that is, $\pi_A = .10$, $\pi_B = .40$, $\pi_C = .50$.

H_A: The population proportions will not be the same as in the past; that is, they will differ from those given in H_0.

Step 2: Selecting a test statistic:

$$\chi^2 = \sum \frac{(f_o - f_e)^2}{f_e}.$$

Step 3: Deriving a decision rule. Given $\alpha = .05$ and 2 degrees of freedom, the critical $\chi^2_{.05,2} = 5.991$ (Appendix Table L), making the decision rule: "Accept H_0 if $\chi^2 \leq 5.991$."

Step 4: Selecting the sample, computing the test statistic, and confronting it with the decision rule. The sample reveals, we assume, the observed frequencies shown in Table E; the expected frequencies are calculated on the assumption of the null hypothesis

Table E

| | Number Preferring Product | | | | |
Product	Observed Frequency, f_o	Expected Frequency (if H_0 is true), f_e	$f_o - f_e$	$(f_o - f_e)^2$	$\dfrac{(f_o - f_e)^2}{f_e}$
new A	90	$300(.10) = 30$	60	3600	120.0
old B	90	$300(.40) = 120$	-30	900	7.5
old C	120	$300(.50) = 150$	-30	900	6.0
Total	300		300	0		$\chi^2 = 133.5$

being true. Accordingly, H_0 should be *rejected.* The market shares are unlikely to be the same as in the past.

Note: Another example of this procedure is provided by Analytical Example 10.2, "Were Mendel's Data Fudged?" ▐▐▐

MAKING INFERENCES ABOUT A POPULATION VARIANCE

We noted in Chapter 8 that the sample variance, s^2, is an unbiased estimator of the population variance, σ^2, provided that the selections of sample elements are statistically independent events. Prior to the selection of an actual sample, the sample variance, like every sample summary measure, is a random variable. As it turns out, probabilities concerning the sample variance can be established with the help of χ^2 distributions because the sample variance, s^2, which equals $(\Sigma X^2 - n\overline{X}^2)/n - 1$ (see Box 4.M on p. 113), can easily be converted into a chi-square random variable with $n - 1$ degrees of freedom, as shown in Box 10.D.

10.D Alternative Chi-Square Statistic

$$\chi^2 = \frac{s^2(n - 1)}{\sigma^2}.$$

where s^2 is the sample variance, n is sample size, and σ^2 is the population variance.

Probability Intervals for the Sample Variance

Given the Box 10.D expression of χ^2, one can quickly establish a probability interval for s^2 with the help of Appendix Table L. Consider, for example, a pharmaceutical firm that produces tablets the weights of which are normally distributed with a population standard deviation of $\sigma = 2$ milligrams (hence, the population variance is $\sigma^2 = 4$). If a simple random sample of 25 tablets is taken from this population, what is the probability of finding a sample variance between 3 and 5? For the lower limit of $s_L^2 = 3$, we get a lower χ^2 value of

$$\chi_L^2 = \frac{3(24)}{4} = 18.$$

For the upper limit of $s_U^2 = 5$, we get an upper χ^2 value of

$$\chi_U^2 = \frac{5(24)}{4} = 30.$$

The probability sought, accordingly, equals the probability of finding χ^2 values within the range of 18 to 30. An inspection of the appropriate row of Appendix Table L (for 24 degrees of freedom) reveals that an upper-tail area of slightly more than .8 lies to the right of $\chi^2 = 18$, while an upper-tail area of slightly less than .2 lies to the right of $\chi^2 = 30$. Thus, an area of about $.8 - .2 = .6$ lies between these values; this value of about .6 is also the probability of finding a sample variance between 3 and 5 in our example.

The same procedure can also be used to determine the limits below and above which lie specified percentages (say, 5 percent each) of all possible s^2 values. For our example (with 24 degrees of freedom), the lower χ^2 value would then be $\chi_{.95,24}^2 = 13.848$, and the upper χ^2 value would be $\chi_{.05,24}^2 = 36.415$. Accordingly, the lower s^2 limit would be

$$s_L^2 = \frac{\sigma^2 \cdot \chi_L^2}{n-1} = \frac{4(13.848)}{24} = 2.31$$

while the upper s^2 limit would be

$$s_U^2 = \frac{\sigma^2 \cdot \chi_U^2}{n-1} = \frac{4(36.415)}{24} = 6.07.$$

Thus, 5 percent of all s^2 values would lie below 2.31, another 5 percent would lie above 6.07, and 90 percent would lie between these values. The general formula for establishing a probability interval for the sample variance is shown in Box 10.E.

10.E Probability Interval for the Sample Variance, s^2

$$\frac{\sigma^2 \cdot \chi_L^2}{n-1} \le s^2 \le \frac{\sigma^2 \cdot \chi_U^2}{n-1}$$

where σ^2 is the population variance, n is sample size, and χ_L^2 and χ_U^2 are lower and upper chi-square values.

Confidence Intervals for the Population Variance

Unlike in the example just employed, the population variance, σ^2, is typically unknown and the problem is to make an estimate of it. In that case, the procedure used above can easily be adjusted to yield a confidence interval estimate of σ^2. By appropriately rewriting the expression found in Box 10.E, we get that in Box 10.F.

10.F Confidence Interval for the Population Variance, σ^2

$$\frac{s^2(n-1)}{\chi_U^2} \leq \sigma^2 \leq \frac{s^2(n-1)}{\chi_L^2}$$

where χ_U^2 = value of χ^2 variable with $n-1$ degrees of freedom such that larger values have a probability of $\alpha/2$; χ_L^2 = value of χ^2 variable with $n-1$ degrees of freedom such that smaller values have a probability of $\alpha/2$; and $\alpha = 1 -$ confidence level.

Consider again our pharmaceutical firm, but this time imagine that it wants to *estimate* the variance in the population of tablet weights. A sample of 30 tablets might yield a sample variance, $s^2 = 3$. What would a 90-percent confidence interval of σ^2 be?

Given 29 degrees of freedom, and $\alpha = .10$, Appendix Table L yields $\chi_U^2 = \chi_{.05,29}^2 = 42.557$ and $\chi_L^2 = \chi_{.95,29}^2 = 17.708$. Accordingly, the interval sought is

$$\frac{3(29)}{42.557} \leq \sigma^2 \leq \frac{3(29)}{17.708}$$

or

$$2.04 \leq \sigma^2 \leq 4.91.$$

With 90-percent confidence, the population variance can be said to lie between 2.04 and 4.91 milligrams squared; hence, the population standard deviation is between 1.43 and 2.22 milligrams, the square roots of these values.

Testing Hypotheses About the Population Variance

It is easy to adapt the foregoing technique further in order to test various hypotheses about the population variance. These hypotheses can be one-sided or two-sided and, in the former case, lower-tailed or upper-tailed (just as noted in Chapter 9 with respect to other population parameters). Three examples will illustrate the possibilities.

Practice Problem 2

A Bank's Waiting-Line Policy (a lower-tailed test)

The manager of a bank is thinking of introducing a "single-line" policy (that directs all customers to enter a single waiting line in the order of their arrival and that, in turn, "feeds" them to different tellers as they become available). Although such a policy does not change the average time customers must wait, the manager prefers it because it decreases waiting-time variability. The manager's critics, however, claim that this variability will be at least as great as for a policy of multiple independent lines (which in the past had a standard deviation of $\sigma_0 = 8$ minutes per customer). A hypothesis test at the 2-percent significance level is to settle the issue. It is to be based on the experience of a random sample of 30 customers subjected to the new policy.

Step 1: Formulating two opposing hypotheses.

$H_0: \sigma^2 \geq 64$
$H_A: \sigma^2 < 64$

Note: Because σ_0 has equaled 8 in the past, the hypothesized value of the population variance, σ_0^2, becomes 64.

Step 2: Selecting a test statistic. The bank's statistician selects

$$\chi^2 = \frac{s^2(n-1)}{\sigma_0^2}.$$

Step 3: Deriving a decision rule. Given a desired significance level of $\alpha = .02$ and 29 degrees of freedom, Appendix Table L suggests a critical value of $\chi^2_{.98,29} = 15.574$ (this being a lower-tailed test). Thus, the decision rule must be: "Accept H_0 if $\chi^2 \geq 15.574$." The critical value is encircled in Figure B.

Figure B

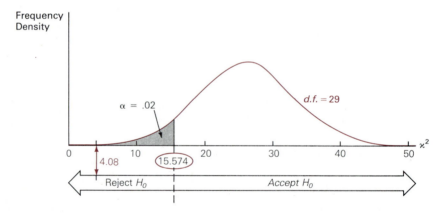

Step 4: Selecting a sample, computing the test statistic, and confronting it with the decision rule. After taking a sample of 30 customers, the statistician finds the sample single-line waiting times to have a standard deviation of $s = 3$ minutes per customer. Accordingly, the computed value of the test statistic equals:

$$\chi^2 = \frac{s^2(n-1)}{\sigma_0^2} = \frac{3^2(30-1)}{64} = 4.08.$$

This value corresponds to the colored arrow in Figure B; it suggests that the null hypothesis should be *rejected*. At the 2-percent significance level, the sample result is statistically significant. The observed divergence from the hypothesized value of $\sigma_0 = 8$ minutes is unlikely to be the result of chance factors operating during sampling; it is more likely to be the result of the manager being right: a single line feeding to many tellers does reduce waiting-time variability. **III**

||| Practice Problem 3

Altimeter Readouts (an upper-tailed test)

A manufacturer of precision instruments is bringing out a new model of a radar altimeter (an instrument that measures aircraft height above the ever-changing ground rather than above mean sea level). The standard deviation of altitude readouts on the old model was $\sigma_0 = 5$ feet. A hypothesis test at the 5 percent significance level is to be conducted concerning management's claim that the readout variability on the new model is equal to or less than that on the old model. The test involves getting 20 readings from a known height above the ground.

Step 1: Formulating two opposing hypotheses.

H_0: $\sigma^2 \leq 25$
H_A: $\sigma^2 > 25$

Step 2: Selecting a test statistic. The company statistician selects

$$\chi^2 = \frac{s^2(n - 1)}{\sigma_0^2}.$$

Step 3: Deriving a decision rule. Given a desired significance level of $\alpha = .05$ and 19 degrees of freedom, Appendix Table L suggests a critical value of $\chi^2_{.05,19} = 30.144$ (this being an upper-tailed test). Thus, the decision rule must be: "Accept H_0 if $\chi^2 \leq 30.144$." The critical value is encircled in Figure C.

Figure C

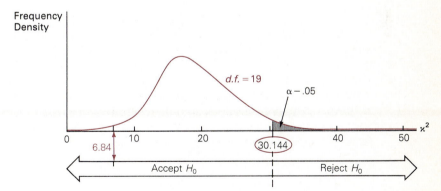

Step 4: Selecting a sample, computing the test statistic, and confronting it with the decision rule. After taking the 20 readings, an engineer finds them to have a standard deviation of $s = 3$ feet. Accordingly, the computed value of the test statistic equals:

$$\chi^2 = \frac{s^2(n - 1)}{\sigma_0^2} = \frac{3^2(20 - 1)}{25} = 6.84.$$

This value corresponds to the colored arrow in Figure C; it suggests that the null hypothesis should be *accepted*. At the 5-percent significance level, the sample result is not statistically significant. The new model is equal to or better than the old one, as claimed. |||

||| Practice Problem 4

The Diameters of Fuel-Tank Lids (a two-tailed test)

An aircraft manufacturer is concerned about variability in the diameters of lids used to seal the fuel tanks that are located inside aircraft wings. Only a narrow range of diameters is acceptable. Lids that fit too tightly prevent air from entering the tanks as the fuel is being used, creating a vacuum and, ultimately, causing collapse of the wing structure. Lids that fit too loosely can allow fuel to be sucked out of the tank during flight, which is equally undesirable from the point of view of flight safety. A test at the 2-percent significance level is to be conducted with a random sample of 20 fuel-tank lids to see whether the population variance of lid diameters equals .0001 inches squared, as specified by engineers.

Step 1: Formulating two opposing hypotheses.

H_0: σ^2 = .0001
H_A: $\sigma^2 \neq$ .0001

Step 2: Selecting a test statistic. The company statistician selects

$$\chi^2 = \frac{s^2(n-1)}{\sigma_0^2}.$$

Step 3: Deriving a decision rule. Given a desired significance level of α = .02 and a two-tailed test, a lower and upper critical χ^2 value must be established (so that .01 of the area under the χ^2 distribution lies below and above each of these values). According to Appendix Table L (and for 19 degrees of freedom) these values are $\chi^2_{.99,19}$ = 7.633 and $\chi^2_{.01,19}$ = 36.191. Thus, the decision rule must be: "Accept H_0 if 7.633 $\leq \chi^2$ $\leq$ 36.191." The critical values are encircled in Figure D.

Figure D

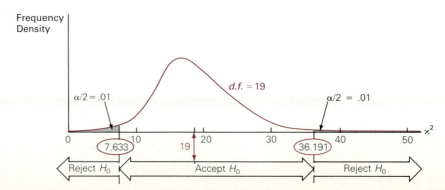

Step 4: Selecting a sample, computing the test statistic, and confronting it with the decision rule. After measuring the 20 diameters, an engineer finds them to have a standard deviation of $s = .01$ inches. Accordingly, the computed value of the test statistic equals:

$$\chi^2 = \frac{s^2(n-1)}{\sigma_0^2} = \frac{(.01)^2(19)}{.0001} = 19.$$

This value corresponds to the colored arrow in Figure D; it suggests that the null hypothesis should be *accepted*. At the 2-percent significance level, the sample result is not statistically significant. The lids do meet specifications. ‖‖

CONDUCTING GOODNESS-OF-FIT TESTS

A **goodness-of-fit test** is a statistical test to determine the likelihood that sample data have been generated from a population that conforms to a specified type of probability distribution. For this purpose, the test compares the entire shapes of two (discrete or continuous) probability distributions: one describing known sample data and the other one describing hypothetical population data. The aim of the test might be limited to identifying only the family to which the underlying distribution belongs, or it might go further, seeking even to identify a particular member of that family. Accordingly, the null hypothesis to be tested might be rather general, such as "The sample data come from a population that is normally distributed." It can also be more specific: "The sample data come from a normally distributed population with a mean of 100 and a standard deviation of 10." In every case, the alternative hypothesis claims that "the sample data come from some other kind of population."

Note: In this type of test, a null hypothesis can clearly be false in many different ways. In our example, the sample data might be coming from all sorts of populations that are not normally distributed or even from one that is normally distributed but has population parameters that differ from those specified. As a result, it is difficult to calculate the β risk (of accepting a null hypothesis that is false) for a goodness-of-fit test, unless one first specifies in what particular way the null hypothesis is false. The best approach to protecting against a type II error, therefore, lies in taking a fairly large sample.

It is easy to see why one might want to conduct a goodness-of-fit test: a knowledge of the underlying population distribution is crucial whenever the statistical procedures used rely on sample data. Someone about to construct a table of random numbers, for example, would want to be sure that the population of numbers from which table entries were being selected was uniformly distributed. Someone who wanted to build a queuing model might want to be sure that the underlying population values were Poisson-distributed. Someone engaged in small-sample hypothesis testing with the help of the *t* distribution would grind out nonsense unless the underlying population values were normally distributed. In the following sections, we will consider a number of examples of how observed frequency distributions based on sample data can be used to make inferences about the distribution of the underlying population data. Given a single sample, a perfect match between the two is highly unlikely, of course; but on the average, when many samples are taken, we expect that the sample data will reveal to us the nature of the population distribution.

Goodness of Fit to the Binomial Distribution

The following example illustrates the procedure of testing whether sample data come from a binomially distributed population.

||| Practice Problem 5

Housing-Code Violations

A city inspector investigates the compliance of landlords with six conditions specified in the housing code. A simple random sample of 200 apartments reveals the data of Table F. The inspector wants to conduct a hypothesis test at the 5-percent significance level to determine whether the sample comes from a population in which the number of actual violations per apartment (out of six possible violations) is a binomially distributed random variable. Accordingly, the inspector proceeds with Steps 1 to 4.

Table F

Number of Possible Violations per Apartment	Observed Frequency, f_o
0	31
1	51
2	70
3	32
4	9
5	5
6	2
Total	200

Step 1: Formulating two opposing hypotheses.

H_0: The number of violations per apartment in the population of all city apartments is binomially distributed with a probability of success in any one trial of $\pi = .3$. [Note: the inspector estimates the parameter, π, by noting that the mean of a binomial random variable is $n \cdot \pi$ (see Box 6.C on p. 208) and that the mean number of violations per apartment in the sample equals

$$n \cdot \pi = \frac{0(31) + 1(51) + 2(70) + 3(32) + 4(9) + 5(5) + 6(2)}{200} = 1.8$$

Hence, π can be estimated as $1.8/6 = .3$]

H_A: The number of violations per apartment in the population of all city apartments is not correctly described by H_0.

Step 2: Selecting a test statistic. The inspector selects

$$\chi^2 = \sum \frac{(f_o - f_e)^2}{f_e}.$$

Step 3: Deriving a decision rule. The inspector first establishes in Table G the expected frequencies for the various numbers of possible violations per apartment, assuming H_0 is true. This process involves finding the binomial probabilities for $n = 6$ and $\pi = .3$ from Appendix Table C and multiplying each by the total number of observations (of 200). Subsequently, if expected frequencies in any class are less than 5 (which is true for 5 and 6 violations per apartment in this example), adjacent classes are combined as needed to yield minimum expected frequencies of 5 in every class (a requirement for using the χ^2 test that was first noted on page 404). The number of degrees of freedom must be established next.

Table G

Number of Possible Violations per Apartment	Binomial Probability, p, for $n = 6$ and $\pi = .3$	Expected Frequency (if H_0 is true), $f_e = 200p$
0	.1176	23.52
1	.3025	60.50
2	.3241	64.82
3	.1852	37.04
4	.0595	11.90 ⎤
5	.0102	2.04 ⎬ 14.04
6	.0007	.14 ⎦
Total		200*

*allowing for rounding error.

‖‖‖

In goodness-of-fit tests, the number of degrees of freedom equals the number of classes, k, for which observed and expected frequencies are to be compared (possibly after some adjacent classes have been combined to assure that all expected frequencies are at least equal to 5) minus 1 minus the number of population parameters that are being estimated (see Box 10.G).

10.G Number of Degrees of Freedom (d.f.) in a Goodness-of-Fit Test

$d.f.$ = number of classes (adjusted) − 1 − number of estimated parameters

The negative 1 in the Box 10.G expression reflects the fact that once all but one of the expected frequencies have been calculated, the last one is predetermined by the fact that the sum of expected frequencies must equal sample size (that is, the sum of observed frequencies). The further reduction of the number of degrees of freedom by 1 for each estimated parameter provides more precise results, as more advanced texts show.

||| Practice Problem 5 *(continued)*

In our example, 5 classes remain after 3 of 7 classes are combined (note the bracket in Table G), while 1 parameter (π) must be estimated. Hence, there are $5 - 1 - 1 = 3$ degrees of freedom. Given a desired significance level of $\alpha = .05$ and 3 degrees of freedom, Appendix Table L suggests a critical value of $\chi^2_{.05,3} = 7.815$. Thus, the decision rule must be: "Accept H_0 if $\chi^2 \leq 7.815$."

Step 4: Computing the test statistic and confronting it with the decision rule. The computation of the test statistic is given in Table H. In view of the value of χ^2 computed in Table H, the null hypothesis should be *accepted.* Any discrepancy between the observed set of frequencies and the set of frequencies expected from a binomially distributed population is most likely the result of sampling error.

Table H

Number of Possible Violations	Observed Frequency f_o	Expected Frequency (if H_0 is true), f_e	$f_o - f_e$	$(f_o - f_e)^2$	$\dfrac{(f_o - f_e)^2}{f_e}$
0	31	23.52	7.48	55.950	2.379
1	51	60.50	-9.50	90.250	1.492
2	70	64.82	5.18	26.832	.414
3	32	37.04	-5.04	25.402	.686
4–6	16	14.08	1.92	3.686	.262
Total	200	200	0		$\chi^2 = 5.233$

|||

Goodness of Fit to the Poisson Distribution*

The following example shows how one can test whether sample data come from a Poisson-distributed population.

||| Practice Problem 6

Outpatient Arrivals

To construct employee work schedules a hospital administrator wants to determine whether the number of hourly arrivals in the outpatient department can be described by a Poisson distribution. A test at the 1-percent level of significance is to be conducted; a simple random sample of 50 hours reveals the data of Table I.

Step 1: Formulating two opposing hypotheses.

H_0: The number of hourly arrivals in the outpatient department is Poisson-distributed with a mean number of arrivals of $\mu = 3.8$. (Note: this value is assumed on the basis of the hospital administrator's prior experience.)

H_A: The number of hourly arrivals in the outpatient department is not correctly described by H_0.

---

*Optional section. Requires prior study of the Chapter 6 optional section on the Poisson probability distribution.

Table I

Number of Outpatient Arrivals per Hour	Observed Frequency, f_o
0	0
1	1
2	5
3	8
4	15
5	9
6	7
7	3
8 or more	2
Total	50

Step 2: Selecting a test statistic. The administrator selects

$$\chi^2 = \sum \frac{(f_o - f_e)^2}{f_e}.$$

Step 3. Deriving a decision rule. The administrator establishes in Table J the expected frequencies for the various numbers of hourly arrivals, assuming H_0 is true. This process involves finding the Poisson probabilities for $\mu = 3.8$ from Appendix Table F and multiplying each by the total number of observations (of 50). The number of degrees of freedom equals $6 - 1 - 0 = 5$. Together with $\alpha = .01$, this *d.f.* implies a critical value of $\chi^2_{.01,5} = 15.086$ (Appendix Table L). Thus, the decision rule must be: "Accept H_0 if $\chi^2 \leq 15.086$."

Table J

Number of Outpatient Arrivals per Hour	Poisson Probability, p, for $\mu = 3.8$	Expected Frequency (if H_0 is true), $f_e = 50p$	
0	.0224	1.12	
1	.0850	4.25	5.37
2	.1615	8.075	
3	.2046	10.23	
4	.1944	9.72	
5	.1477	7.385	
6	.0936	4.68	
7	.0508	2.54	9.22
8 or more	.0400	2.00	
Total	1.0000	50	

Step 4: Computing the test statistic and confronting it with the decision rule. In view of the χ^2 value computed in Table K, the null hypothesis should be *accepted.* The number of hourly arrivals can be viewed as a Poisson process with a mean of 3.8.

Table K

Number of Outpatient Arrivals per Hour	Observed Frequency, f_o	Expected Frequency, (if H_0 is true), f_e	$f_o - f_e$	$(f_o - f_e)^2$	$\dfrac{(f_o - f_e)^2}{f_e}$
0–1	1	5.37	−4.37	19.097	3.556
2	5	8.075	−3.075	9.456	1.171
3	8	10.23	−2.23	4.973	.486
4	15	9.72	5.28	27.878	2.868
5	9	7.385	1.615	2.608	.353
6 or more	12	9.22	2.78	7.728	.838
Total	50	50	0		$\chi^2 = 9.272$

Note: Analytical Example 10.3, "The Distribution of Word Frequencies," provides another example of a goodness-of-fit test involving the Poisson distribution. |||

Goodness of Fit to the Normal Distribution

The following example illustrates how one can test whether sample data come from a normally distributed population.

||| Practice Problem 7

Trading in the Futures Market

A financial analyst wishes to determine whether the daily volume of futures contracts traded at U.S. commodities exchanges is still normally distributed with a mean of 50 million contracts and a standard deviation of 10 million contracts, as indicated by a study conducted a decade ago. A test at the 2 percent level of significance is desired. The analyst collects the data of Table L during the next 90 business days. Subsequently, the analyst proceeds with Steps 1 to 4.

Step 1: Formulating two opposing hypotheses.

H_0: The daily volume of futures contracts traded is normally distributed with a mean of 50 million contracts and a standard deviation of 10 million contracts.

H_A: The daily volume of futures contracts traded is not correctly described by H_0.

Step 2: Selecting a test statistic. The analyst selects

$$\chi^2 = \sum \frac{(f_o - f_e)^2}{f_e}.$$

Table L

Number of Futures Contracts Traded per Day (millions)	Observed Frequency, f_o
under 10	5
10–under 20	9
20–under 30	15
30–under 40	23
40–under 50	20
50–under 60	8
60–under 70	6
70–under 80	3
80 and above	1
Total	90

Step 3: Deriving a decision rule. The analyst establishes in Table M the expected frequencies for the various classes of daily volumes, assuming H_0 is true. This process involves finding areas under the normal curve for the various classes in Table L with the help of Appendix Table H and then multiplying each of these probabilities by the total number of observations (of 90). To bring each expected frequency up to a minimum of 5 (a requirement first noted on page 404), the analyst combines several classes, as shown, leaving $k = 4$ classes for the χ^2 test. Since no parameters were estimated,

Table M

Number of Futures Contracts Traded per Day (millions): upper class limit, x	Normal Deviate, $z = \dfrac{x - \mu}{\sigma} = \dfrac{x - 50}{10}$	Area Under Standard Normal Curve Left of x	Area of Class Interval, p	Expected Frequency (if H_0 is true), $f_e = 90p$
10	−4.0	.0000	.0000	0 ⎫
20	−3.0	.0014	.0014	.126 ⎬ 14.283
30	−2.0	.0228	.0214	1.926
40	−1.0	.1587	.1359	12.231 ⎭
50	0	.5000	.3413	30.717
60	1.0	.8413	.3413	30.717
70	2.0	.9772	.1359	12.231 ⎫
80	3.0	.9986	.0214	1.926 ⎬ 14.283
∞	∞	1.0000	.0014	.126 ⎭
Total			1.0000	90

there are $4 - 1 - 0 = 3$ degrees of freedom, which implies (for $\alpha = .02$) a critical value of $\chi^2_{.02,3} = 9.837$ (Appendix Table L). Thus, the decision rule must be: "Accept H_0 if $\chi^2 \leq 9.837$."

Step 4: Computing the test statistic and confronting it with the decision rule. In view of the χ^2 value computed in Table N, H_0 should be *rejected*. This decision does not necessarily mean, of course, that the daily numbers of futures contracts traded are not normally distributed, but it does mean (at the 2-percent level of significance) that the distribution, if it is normal, does not conform to the historical $\mu = 50$ and $\sigma = 10$. [For a further discussion of this issue see problem 53 in the "Questions and Problems" section at the end of this chapter and its solution at the back of this book.] |||

Table N

Number of Futures Contracts Traded per Day (millions)	Observed Frequency, f_o	Expected Frequency (if H_0 is true), f_e	$f_o - f_e$	$\dfrac{(f_o - f_e)^2}{f_e}$
under 40	52	14.283	37.717	99.599
40–under 50	20	30.717	−10.717	3.739
50–under 60	8	30.717	−22.717	16.801
60 and above	10	14.283	−4.283	1.284
Total	90	90	0	$\chi^2 = 121.423$

GOODNESS OF FIT TO THE UNIFORM DISTRIBUTION*

Perhaps the simplest of all the goodness-of-fit tests involves the hypothesis that sample data come from a uniformly distributed population because in that case, as the next example shows, the expected frequencies in all categories are the same.

||| Practice Problem 8

Customers at the Dry Cleaner

The manager of a dry-cleaning establishment wants to know whether requests for service are spread evenly over the six business days of the week. (Employee work schedules are to be set accordingly.) A test at the 5-percent significance level is desired; records for the past three months reveal the data in the first two columns of Table O. Accordingly, the manager proceeds with Steps 1 to 4.

Step 1: Formulating two opposing hypotheses.

H_0: The numbers of service requests are uniformly distributed over the six business days of the week.

H_A: The numbers of service requests are not uniformly distributed over the six business days of the week.

---

*Optional section. Requires prior study of the Chapter 7 optional section on the uniform probability distribution.

Table O

Day of the Week When Service Was Requested	Observed Frequency, f_o	Expected Frequency (for uniform distribution), f_e	$f_o - f_e$	$\dfrac{(f_o - f_e)^2}{f_e}$
Monday	1,251	1,836	-585	186.397
Tuesday	1,830	1,836	-6	.020
Wednesday	1,675	1,836	-161	14.118
Thursday	1,450	1,836	-386	81.153
Friday	1,905	1,836	69	2.593
Saturday	2,905	1,836	1,069	622.419
Total	11,016	11,016	0	$\chi^2 = 906.7$

Step 2: Selecting a test statistic. The manager selects

$$\chi^2 = \sum \frac{(f_o - f_e)^2}{f_e}.$$

Step 3: Deriving a decision rule. Under the null hypothesis, the manager expects the total of 11,016 service requests to be equally distributed among the six business days; this expectation is reflected in the third column of Table O. There being six classes with expected frequencies of at least 5, and because no parameters are being estimated, the degrees of freedom equal $6 - 1 - 0 = 5$, and the critical $\chi^2_{.05,5} = 11.070$ (Appendix Table L). Thus, the decision rule must be: "Accept H_0 if $\chi^2 \leq 11.070$."

Step 4: Computing the test statistic and confronting it with the decision rule. The computation of χ^2 is shown in the last two columns of Table O. In view of this χ^2 value, H_0 should be *rejected.* In other words, the numbers of service requests are not uniformly distributed over the six business days of the week.

Note: Analytical Example 10.4, "Testing Random Digits for Randomness," provides another illustration of this type of situation. |||

GOODNESS OF FIT TO ANY SPECIFIED DISTRIBUTION

Goodness-of-fit tests can be performed with respect to any specified distribution, not just with respect to members of some well-known distribution family. This fact is illustrated by the following final example on the television viewing habits of people in 1950 and 1980.

||| Practice Problem 9

Television Viewing Habits

A TV network polled 100 viewers in 1950 and 1980 and noted viewing habits as indicated in Table P. It now wants to test, at the 5-percent level of significance, whether 1980 viewing habits of the population at large differ from 1950. (The answer may affect its ability to sell advertising time to various sponsors.)

Table P

Hours of TV Watched per Week	Observed Frequency, f_o	
	1950	1980
0–under 3	3	9
3–under 5	4	3
5–under 7	10	9
7–under 15	20	21
15–under 25	45	44
25 or more	18	14
Total	100	100

Step 1: Formulating two opposing hypotheses.

H_0: 1980 viewing habits equal those of 1950.
H_A: 1980 viewing habits differ from those of 1950.

Step 2: Selecting a test statistic. The network manager selects

$$\chi^2 = \sum \frac{(f_o - f_e)^2}{f_e}.$$

Step 3: Deriving a decision rule. Under the null hypothesis, the manager expects the frequencies of 1950 to prevail also in 1980; thus, the data in the 1950 column become the expected frequencies for 1980, but the first two classes must be combined to yield a minimum value of 5. Thus, there are 5 remaining classes and $5 - 1 - 0 = 4$ degrees of freedom, making the critical $\chi^2_{.05,4} = 9.488$ (Appendix Table L). Thus, the decision rule is: "Accept H_0 if $\chi^2 \leq 9.488$."

Step 4: Computing the test statistic and confronting it with the decision rule. The computation of χ^2 is shown in Table Q. In view of this χ^2 value, H_0 should be *accepted.* 1980 viewing habits equal those of 1950.

Table Q

Hours of TV Watched per Week	1980 Observed Frequency, f_o (1980 values)	1980 Expected Frequency (if H_0 is true) f_e (1950 values)	$f_o - f_e$	$\frac{(f_o - f_e)^2}{f_e}$
0–under 5	12	7	5	3.571
5–under 7	9	10	−1	.100
7–under 15	21	20	1	.050
15–under 25	44	45	−1	.022
25 and more	14	18	−4	.889
Total	100	100	0	$\chi^2 = 4.632$

III

10.1 Leukemia and the Atomic Bomb

Was the high incidence of leukemia among survivors of the atomic bomb at Hiroshima attributable to radiation? One study provided the data given in the accompanying Table A. A χ^2 test of the null hypothesis that bomb radiation had no such effect, undertaken at the 1-percent significance level, yields the results given in the accompanying Table B. Compared with the critical $\chi^2_{.01,2} = 9.210$, the computed value of the χ^2 statistic provides strong evidence *against* the null hypothesis.

Table A

1945 Radiation Dosage (rads)	Percent of Population Affected	1950–58 Leukemia Cases
0–20	75.56	12
21–80	13.41	5
81 and over	11.03	34
Total	100.00	51

Table B

Radiation Dosage (rads)	Observed Frequency, f_o	Expected Frequency (if H_o is true), f_e	$f_o - f_e$	$\dfrac{(f_o - f_e)^2}{f_e}$
0–20	12	$51(.7556) = 38.54$	−26.54	18.276
21–80	5	$51(.1341) = 6.84$	−1.84	.495
81 and over	34	$51(.1103) = 5.62$	28.37	142.959
Total	51	51	0	$\chi^2 = 161.73$

Source: A. Bertrand Brill, Masanobu Tomonaga, Robert M. Heyssel, "Leukemia in Man Following Exposure to Ionizing Radiation," *Annals of Internal Medicine* 56, 4 (April 1962):599.

10.2 Were Mendel's Data Fudged?

In the 1860s, the Austrian monk Gregor Mendel wrote an important paper about inheritance. He postulated the existence of entities, now called *genes,* that determine how various characteristics are passed on from one generation to the next. For example, according to one of his theories (now known as Mendel's Second Law of Inheritance), the cross-fertilization of two pure strains of pea plants—one producing only round yellow seeds, the other one only wrinkled green seeds—would produce a first generation of hybrids with nothing but round yellow seeds. Yet a mating of these hybrids with one another would yield plants with round yellow, round green, wrinkled yellow, as well as wrinkled green seeds, and would produce these in definite proportions of 9:3:3:1. Mendel performed numerous experiments to back up his theories. One experiment involved the just-mentioned law and yielded the data for the second-generation hybrids given in the accompanying Table A.

It is easy to perform a χ^2 test of the null hypothesis that this sample of 556 observations has come from a population obeying Mendel's Second Law. The theo-

Table A

Pea Type	Observed Frequency, f_o
Round yellow	315
Round green	108
Wrinkled yellow	101
Wrinkled green	32
Total	556

retical proportions of 9:3:3:1 translate into the expected frequencies given in the accompanying Table B, which also shows the calculation of χ^2. If one were to perform a χ^2 test at the 5-percent level of significance (there being 3 degrees of freedom), the critical value of $\chi^2_{.05,3}$ would be 7.815. Thus, the null hypothesis would be *accepted;* the experimental data would clearly vindicate Mendel's Second Law.

But note the tiny value of χ^2 calculated in Table B. The fit between expected and observed frequencies is almost perfect. The great statistician Ronald A. Fisher (Biography 9.1) thought this was too good to be true, especially after noting that such unusually close agree-

ment between the observed and the expected could be found in *all* of Mendel's experiments! Fisher, therefore, made an experiment of his own. For each of Mendel's experiments, he computed the χ^2 value. Then he pooled the results by adding all the χ^2 values and also adding all the numbers of degrees of freedom. (This is a legitimate procedure as long as all the experiments are independent of one another. Thus, if one experiment yields $\chi^2 = .47$ and has 3 degrees of freedom, while another one yields $\chi^2 = .61$ and has 5 degrees of freedom, the two have a pooled χ^2 of $.47 + .61 = 1.08$ with $3 + 5 = 8$ degrees of freedom.) For all of Mendel's data, Fisher found a pooled χ^2 value of 41.6 with 84 degrees of freedom. More importantly, such a low pooled χ^2 value or a lower one would occur, Fisher found, only 7 times in 100,000 experiments. That is, if 100,000 scientists each repeated Mendel's experiments, only 7 of them would manage to match Mendel's own low χ^2; 99,993 would get results with a larger χ^2. Thus, Mendel was either very, very lucky indeed or his data were fudged. It is hard to escape the latter conclusion. (Indeed, Fisher believed that Mendel's gardening assistant, who was overly eager to please his master, is to blame.)

Table B

Pea Type	Observed Frequency, f_o	Expected Frequency (if H_0 is true), f_e	$f_o - f_e$	$\dfrac{(f_o - f_e)^2}{f_e}$
Round yellow	315	312.75	2.25	.016
Round green	108	104.25	3.75	.135
Wrinkled yellow	101	104.25	−3.25	.101
Wrinkled green	32	34.75	−2.75	.218
Total	556	556	0	$\chi^2 = .47$

Source: J. H. Bennett, ed., *Experiments in Plant Hybridisation* (Edinburgh: Oliver and Boyd, 1965), especially pp. 23 and 78. This book reprints Mendel's original article, along with commentary by Ronald A. Fisher.

10.3 The Distribution of Word Frequencies

In the past, many attempts have been made to represent word-frequency counts by some kind of well-known statistical distribution. None of the attempts was particularly successful, however, when applied to a variety of data over the entire length of the observed word distributions. More recently, a compound type of Poisson distribution, on the other hand,

proved to be an excellent fit of some 20 observed distributions quoted in the literature. This finding is not really surprising because whenever the different words in a text are arranged in the order of their frequency of usage, a word-frequency distribution is generated that has an extraordinarily long upper tail and closely resembles the Poisson distribution. A sample of

2,048 English nouns in Macaulay's *Essay on Bacon,* for example, exhibits the distribution given in the accompanying Table A.

The accompanying Table B presents some of the results of goodness-of-fit tests of this and other observed distributions to a Poisson-type distribution. Recall the discussion of p values in Biography 9.1, page 383. Each of these numbers indicates the probability

that a value of the test statistic as extreme as or more extreme than the one observed could have occurred by chance, given that the null hypothesis is true (here, H_0: The observed word distribution fits a Poisson distribution.). According to Fisher (as noted in Biography 9.1), a p value greater than .05 strongly favors the null hypothesis, which is certainly the case in the examples cited here.

Table A

Frequency of Word Usage	Number of Words Observed, f_o	Frequency of Word Usage	Number of Words Observed, f_o
1	990	10	17
2	367	11	24
3	173	12	19
4	112	13	10
5	72	14	10
6	47	15	13
7	41	16–20	31
8	31	21–30	31
9	34	≥ 31	26
		Total	2,048

Table B

Author	Title	Word Types	Sample Size (words)	Degrees of Freedom	Observed χ^2	p Value
Macaulay	*Essay on Bacon*	nouns	2,048	15	10.626	.832
Bunyan	*Life and Death of Mr. Badman*	nouns	1,030	18	16.998	.523
St. John	*Gospel of St. John*	nouns	353	12	5.909	.920
Shakespeare	*As You Like It*	nouns	1,241	15	16.817	.330
—	*De Imitatione Christi*	adjectives	529	17	12.711	.755
St. Paul	*Gospel of St. Paul*	all	2,648	22	23.996	.347
Corneille	*Illusion Comique*	all	1,906	18	16.169	.581

Source: H. S. Sichel, "On a Distribution Law for Word Frequencies," *Journal of the American Statistical Association,* September 1975, pp. 542–47.

10.4 Testing Random Digits for Randomness

The Rand Corporation's famous table of a million random digits (from which Appendix Table A has been excerpted) has been subjected to a great variety of χ^2 tests to determine whether the table is free from serious bias. (It is.) In one of these tests, the million digits were divided into 20 blocks of 50,000 digits each, and it was then noted how frequently each possible digit between 0 and 9 appeared within each block. For block number 1, the observed frequencies—along with the expected ones, given perfect randomness—are shown in the accompanying table. The calculation of the χ^2 statistic is also shown. Given 9 degrees of freedom and a significance level of 1 percent, the critical $\chi^2_{.01,9} = 21.666$. Therefore, the null hypothesis of perfect randomness is quite acceptable.

Digit	Observed Frequency, f_o	Expected Frequency (given uniform distribution), f_e	$f_o - f_e$	$\dfrac{(f_o - f_e)^2}{f_e}$
0	4,923	5,000	−77	1.186
1	5,013	5,000	13	.034
2	4,916	5,000	−84	1.411
3	4,951	5,000	−49	.480
4	5,109	5,000	109	2.376
5	4,993	5,000	−7	.010
6	5,055	5,000	55	.605
7	5,080	5,000	80	1.280
8	4,986	5,000	−14	.039
9	4,974	5,000	−26	.135
Total	50,000	50,000	0	$\chi^2 = 7.556$

Source: The Rand Corporation, *A Million Random Digits With 100,000 Normal Deviates* (Glencoe, Ill.: The Free Press, 1955), p. xiii.

||| BIOGRAPHY

10.1 Karl Pearson

Karl Pearson (1857–1936) was born in London, England, the son of a successful trial lawyer. While attending University College, London, and then the universities of Heidelberg, Berlin, and Cambridge (where he received a law degree), he exhibited a phenomenal range of interests, moving from mathematics and physics to philosophy and religion to history and law to German folklore, socialism, and Darwinism! Much of this study had little to do with those things for which Pearson is now remembered, but early in his life he was simply overwhelmed with all there is to know and noted that "not one subject in the universe is unworthy of study."

Pearson's interest in analytical statistics was kindled only in the late 1880s after he had become a professor of applied mathematics and mechanics at University College. (Later, he was named the first Galton professor of eugenics there.) His 1892 book, *The Grammar of Science,* illustrates his growing conviction that analytical statistics lies at the foundation of all knowledge. Going beyond Adolphe Quetelet (Biography 2.1) who believed that almost all phenomena can be described by the normal distribution (provided only that the number of cases examined is large enough), Pearson derived a system of generalized frequency curves that recognized the importance of asymmetrical ones. (In that connection, he noted that true variability among individuals was a concept very different from chance variation among errors made while estimating a single value. He introduced the term *standard deviation* and the symbol σ for the former and called the latter *probable error.*) Having established ways of fitting all kinds of curves to all kinds of observations, Pearson searched for a criterion that could measure the goodness of fit. Thus, in a famous paper published in 1900, he introduced chi square.

When Quetelet and his followers wanted to demonstrate the closeness of agreement between the frequencies in a distribution of observed data and the frequencies calculated on the assumption of a normal distribution, they merely printed the two series side by side, and that was that! Readers could look at these and reach their own conclusions. They had no measure of discrepancy between the observed and the expected. By introducing chi square, Pearson provided such a measure, and he worked out its distribution as well. As this chapter shows, chi square has turned out to be an enormously useful statistic, and it now occupies a major position in statistical theory.

Note: Unbeknownst to Pearson, a German—Friedrich Helmert—had discovered a chi-square distribution in 1875, when studying the sampling distribution of the sample variance, while sampling from a normal population. Pearson, of course, discovered it in a different context—namely, that of a goodness-of-fit problem—and he later extended its application to the analysis of frequencies in contingency tables. Pearson himself was unclear, however, on the proper number of degrees of freedom to use; he always used $k - 1$. As was later shown by Ronald A. Fisher (Biography 9.1), a more accurate result is achieved when this number is reduced by one for each estimated parameter.

Pearson also is responsible for developing considerably further the idea of correlation introduced by Francis Galton (Biography 13.1). He generalized Galton's conclusions and methods, derived the formula now called "Pearson's product moment" (discussed in Chapter 13), derived a simple routine for the computation of regression equations, and much more.

Most important, perhaps, is the fact that Pearson aroused the scientific world from a state of total indifference to statistical studies and convinced thousands in all fields of the necessity to gather and analyze data. He showed statistics to be a general method that was applicable to all sciences. Undoubtedly, there was no

science for which Pearson himself demonstrated this fact more conclusively than biology. In 1900, Pearson became the cofounder (with Galton and Weldon) of *Biometrika*, a journal devoted to the statistical study of biological problems. Pearson edited the journal until his death and made it into the world's leading medium for the discussion of statistical theory and practice. The first issue of the journal carried a picture of a statue of Charles Darwin with the words, *"Ignoramus, in hoc signo laboremus"* (We are ignorant; so let us work). These words pretty much sum up Pearson's own philosophy of life. In the pursuit of that vast universe of knowledge that so overwhelmed him in his youth, he published hundreds of articles, and he facilitated the work of other statisticians by creating (and publishing in *Biometrika*) the types of tables found in the appendix of this book (the existence of which we all now take for granted). In his relentless pursuit of the truth, with the help of newly developing statistical techniques, Pearson also became embroiled in many controversies, often bitter and prolonged. As a young man, Pearson was something of a crusader, battling for such causes as socialism, the emancipation of women, and the ethics of free thought. This spirit was still evident later in his life when, as head of the Eugenics Laboratory, he battled the medical profession on such issues as the causes of tuberculosis or the effects of alcoholism on future generations. (Using statistical analysis, Pearson showed tuberculosis more related to hereditary factors than to environmental ones, which made the widespread use of sanatoria look foolish. He also showed that, contrary to common opinion, alcoholic parents were not producing mentally or physically deficient children.)

Sources: Dictionary of Scientific Biography, vol. 10 (New York: Scribner's, 1974), pp. 447–73; *International Encyclopedia of Statistics,* vol. 2 (New York: Free Press, 1978), pp. 691–98.

Summary

1. The chi-square technique can often be employed for purposes of estimation or hypothesis testing when use of the normal or *t* distributions is inadmissible. The first major application of the technique involves tests of the alleged statistical independence of two qualitative population variables. Typically, sample data are classified according to two or more categories associated with each of these variables, and the *chi-square statistic* is constructed on the basis of observed frequencies and frequencies expected on the assumption of statistical independence. Different sampling distributions of the chi-square statistic exist for different numbers of degrees of freedom; critical values of χ^2 that correspond to specified upper-tail areas of such distributions have been tabulated in the chi-square table (Appendix Table L). This table can be used in tests of statistical independence.

2. A second major application of the χ^2 technique is in the making of inferences about more than two population proportions. This technique can be used to test the equality or inequality of proportions.

3. Third, the χ^2 technique allows us to make inferences about a population variance. Confidence intervals for a population variance (and also probability intervals for a sample variance) can be established with the help of an appropriate χ^2 dis-

tribution. Hypotheses about a population variance can also be tested with the help of the χ^2 statistic.

4. Fourth, the χ^2 technique allows us to conduct goodness-of-fit tests to assess the plausibility that sample data come from a population the elements of which conform to a specified type of probability distribution. This technique can be used to test the goodness of fit of sample data to the binomial distribution, to the Poisson distribution, to the normal distribution, to the uniform distribution, and more.

Key Terms

chi-square distribution
chi-square statistic

contingency table
goodness-of-fit test

Questions and Problems

The computer program noted at the end of this chapter can be used to work many of the subsequent problems.

1. Using the sample data of Table 10.A, conduct an appropriate hypothesis test of independence between the two stated variables (A and B) and do so at the 2-percent level of significance.

2. Using the sample data of Table 10.B, conduct an appropriate hypothesis test of independence between the two stated variables (A and B) and do so at the 10-percent level of significance.

3. Using the sample data of Table 10.C, conduct an appropriate hypothesis test of independence between the two stated variables and do so at the 1-percent level of significance.

4. Using the sample data of Table 10.D, conduct an appropriate hypothesis test of independence between the two stated variables and do so at the 5-percent level of significance.

5. In each of the following situations, determine whether a null hypothesis of independence between the stated variables should be accepted on the basis of a χ^2 test.

a. The variables are sex of viewers and their most preferred of 10 TV shows. Desired $\alpha = .001$; a sample yields $\chi^2 = 41.72$.

b. The variables are marital status (single, married) and type of car owned (foreign, domestic). Desired $\alpha = .01$; a sample yields $\chi^2 = 2.39$.

c. The variables are type of soap used (4 choices) and most preferred TV show (10 choices). Desired $\alpha = .02$; a sample yields $\chi^2 = 137.02$.

d. The variables are college major (12 choices) and type of employment (3 choices). Desired $\alpha = .05$; a sample yields $\chi^2 = 55.19$.

Table 10.A Auto Accidents (Illinois sample, 1986)

| (A) Severity of Accident | (B) Location of Accident | | | |
	Freeway (1)	Rural Road (2)	City Road (3)	Total
(1) Property Damage	10	20	20	50
(2) Injury	10	10	5	25
(3) Fatality	10	10	5	25
Total	30	40	30	100

Table 10.B Heart-Attack Victims (New York sample, 1986)

(A) Age	(B) Sex		Total
	Male	Female	
< 30	6	4	10
30–60	38	42	80
> 60	6	4	10
Total	50	50	100

Table 10.C Raw-Material Units (Peoria plant sample, 1986)

Supplier	Type of Defect						Total
	A	B	C	D	E	F	
(1) Jones	5	0	3	12	50	30	100
(2) Smith	18	32	42	68	0	40	200
(3) Green	27	38	35	70	20	110	300
Total	50	70	80	150	70	180	600

Table 10.D U.S. Economic Performance (1887–1986)

Unemployment	Inflation			Total
	(A) Abated	(B) Unchanged	(C) Accelerated	
(1) Lower	5	5	10	20
(2) Unchanged	5	35	20	60
(3) Higher	20	0	0	20
Total	30	40	30	100

6. In each of the following situations, determine whether a null hypothesis of independence between the stated variables should be accepted on the basis of a χ^2 test.

a. The variables are income level (5 levels) and political affiliation (3 choices). Desired $\alpha = .10$; a sample yields $\chi^2 = 105.00$.

b. The variables are social class (3 classes) and newspaper read (5 choices). Desired $\alpha = .01$; a sample yields $\chi^2 = 3.52$.

c. The variables are sex and age at heart attack (3 classes). Desired $\alpha = .001$; a sample yields $\chi^2 = 1$ [as in problem 2, which students will find

on page 432, where α was .10 and H_0 was accepted].

d. The variables are women's preferred clothing stores (3 choices) and women's ages (3 classes). Desired $\alpha = .05$; a sample yields $\chi^2 = 7.78$.

7. Using the sample data of Table 10.E, conduct an appropriate hypothesis test of independence between the two stated variables and do so at the 2-percent level of significance.

8. Using the sample data of Table 10.F, conduct an appropriate hypothesis test of independence between the two stated variables and do so at the 2-percent level of significance.

Table 10.E   American Stock Exchange (1986 sample)

Stock Price per Share	Dividend Yield		
	0 to under 3%	3 to under 10%	10% and more
(1) $0–25	27	52	12
(2) $25.01–$75	38	46	3
(3) $75.01 and more	19	3	40

Table 10.F   Patients at General Hospital (Boston, 1986 sample)

Maternity Benefits Provided by Insurance	Days New Mothers Stayed in Maternity Ward					
	0–2	3–4	5–6	7	8	9 and more
(1) Poor	4	10	6	5	1	0
(2) Fair	2	27	38	17	6	5
(3) Excellent	0	38	59	20	15	7

9. Using the sample data of Table 10.G, conduct an appropriate hypothesis test of independence between the two stated variables and do so at the 5-percent level of significance.

Table 10.G   Magazine Subscribers (U.S. sample, 1986)

Awareness of Scott Oxygen-Systems Ad	Magazine		
	The Pilot	Flying	Aviation Consumer
Don't Remember	100	125	132
Remember Vaguely	52	65	32
Remember Well	15	43	136

10. Using the sample data of Table 10.H, conduct an appropriate hypothesis test of independence between the stated variables and do so at the .1-percent level of significance.

Table 10.H   Shoppers (Hartford suburbs, 1986 sample)

Distance of Residence	Preferred Shopping Center			
	A	B	C	D
0 to under 5 miles	40	55	27	18
5 to under 15 miles	128	72	12	3
15 miles and more	178	12	6	3

11. Using the sample data of Table 10.I, conduct an appropriate hypothesis test of independence between the stated variables and do so at the 10-percent level of significance.

Table 10.I   Atlanta Residents (1986 sample)

Type of Chicken Consumed	Regularly Watch "Falconcrest"	
	Yes	No
Purdue Brand	255	127
Regular Unbranded	301	609
None	62	146

12. Using the sample data of Table 10.J, conduct an appropriate hypothesis test of independence between the stated variables and do so at the 5-percent level of significance.

Table 10.J   Individuals With Auto Insurance (U.S. sample, 1986)

Action	Age			
	Under 25	25–35	36–55	Over 55
Filed Claim	86	92	121	67
No Claim	92	121	603	139

13. The Massachusetts Department of Motor Vehicles found that 163 of the last 550 accidents were serious and 49 of these involved drivers wearing seat belts. Yet 222 of the drivers in nonserious accidents wore belts. Conduct an appropriate hypothesis test of independence between the severity of accidents and the wearing of seat belts and do so at the 1-percent level of significance.

14. Using the sample data of Table 10.K, conduct an appropriate hypothesis test of independence between the stated variables and do so at the 10-percent level of significance.

Table 10.K   Number of Sales (U.S. market, 1986)

Type of Sales Letter Used	Response	
	Sale	No Sale
A	17	83
B	33	67
C	51	49
D	23	77

15. Using the sample data of Table 10.L, conduct an appropriate hypothesis test of independence be-

tween the stated variables and do so at the 1-percent level of significance.

16. For problems 7 through 11, calculate the expected value, variance, and standard deviation of the χ^2 random variable.

17. For problems 12 through 15, calculate the expected value, variance, and standard deviation of the χ^2 random variable.

18. An economist wants to test (at the 2-percent significance level) whether the proportion of firms planning to increase investment spending next year is the same in each of four industries as it is overall. A sample of 200 firms reveals the frequencies listed in Table 10.M. Make the test.

19. A manager wants to determine (at the 10-percent significance level) whether the proportion of defective pie crusts produced is the same for each of four oven temperatures as it is overall. A sample of 200 crusts reveals the frequencies listed in Table 10.N on page 436. Make the test.

20. The manager of a travel bureau wants to determine (at the 5-percent significance level) whether the proportion of potential customers who prefer the Massachusetts historical tour to the Bay State nature tour is the same for four ethnic groups as it is overall. A sample reveals the frequencies listed in Table 10.O on page 436. Make the test.

Table 10.L   Number of New Policies Sold (U.S. market, 1986)

Life-Insurance Contract	Region			
	Northeast	Midwest	South	West
Whole Life	69	72	23	56
Level Term	23	29	38	77
Decreasing Term	105	93	105	19

Table 10.M   Number of Firms (U.S. sample, 1986)

Investment Plans	Industry				Total
	A	B	C	D	
(1) Increased Spending	10	20	30	40	100
(2) Unchanged or Decreased Spending	40	30	20	10	100
Total	50	50	50	50	200

Table 10.N   Number of Pie Crusts (January 1986 sample)

Product Quality	Oven Temperature				Total
	A = 300°	B = 350°	C = 400°	D = 450°	
(1) Defective	10	0	20	40	70
(2) Perfect	40	50	30	10	130
Total	50	50	50	50	200

Table 10.O   Potential Travelers (U.S. sample, 1986)

	Ethnic Group			
	A = French	B = English	C = Spanish	D = German
(1) Massachusetts Historical	57	33	29	76
(2) Bay State Nature	103	105	132	77

21. Someone flips a penny, a nickel, and a dime, each 50 times. The number of heads are 20, 25, and 30, respectively. Make a test on the fairness of the coins and do so at the 5-percent level of significance. (Hint: Test whether the sample data are consistent with an equality of proportions, of presumably .5, for the population of all possible coin flips.)

22. A sales manager wants to determine, at the 10-percent level of significance, whether the proportion of people who own the Polaroid SX-70 camera is the same in all parts of the country. A sample shows that the camera is owned by 10 of 30 Northerners, by 20 of 50 Easterners, by 30 of 60 Southerners, and by 30 of 90 Westerners. Make the test.

23. An engineer wants to determine, at the 1-percent level of significance, whether the proportion of defective pieces produced is the same for two machines. A sample shows that machine A produces 24 defectives out of 400, and machine B produces 42 out of 600. Make the test.

24. A car dealer wants to determine, at the 2-percent level of significance, whether the proportion of people under 30 who own cars is the same for 3 makes of cars. A sample shows that 27 of 120 car A owners are under 30; the corresponding numbers are 56 of 203 car B owners, and 13 of 31 car C owners. Make the test.

25. An insurance industry analyst wants to determine, at the .1-percent level of significance, whether the proportion of people who renew their policies only during the last week of the grace period is the same for 4 companies. A sample shows that 17 out of 52 customers of company A fall into this category; the corresponding numbers for company B are 12 out of 19, for company C they are 30 out of 51, and for company D they are 22 out of 25. Make the test.

26. A personnel manager wants to determine whether the taking of sick days by workers is explained by the workers' ages. The last 100 sick days taken are viewed as a random sample; they were distributed as in row 1 of Table 10.P. The percentage of the firm's labor force in the various age groups is given in row 2. A test at the 5-percent level of significance is desired. Make the test, showing your calculations.

27. A lawyer wants to know whether the educational level of jurors chosen at the county court reflects the makeup of the county population that is eligible for jury duty. The last 100 jurors chosen are viewed as a random sample; their makeup is given in row 1 of Table 10.Q. Row 2 shows countywide percentages. A test at the 10-percent level of significance is desired. Make the test, showing your calculations. (What if a 1-percent level of significance had been chosen?)

Table 10.P

Distribution of	Age Groups				Total
	Under 25 (A)	25–35 (B)	36–50 (C)	51 and Over (D)	
(1) 100 Sick Days (number)	50	15	15	20	100
(2) Firm's Labor Force	20%	30%	30%	20%	100%

Table 10.Q

Distribution of	Educational Level				Total
	Elementary School (A)	High School (B)	Some College (C)	College Degree (D)	
(1) 100 Jurors (number)	12	64	12	12	100
(2) County Population Eligible for Jury Duty	20%	50%	10%	20%	100%

28. A lawyer wants to know whether the available professional and managerial jobs are distributed among the races in accordance with the racial make-up of the population (which is 80-percent white, 15-percent black, 5-percent others). A random sample of 1,000 professional/managerial jobs shows whites to hold 792, blacks 98, others the rest. Make the test at the 5-percent level of significance.

29. A random sample of 500 deaths in Massachusetts shows that 43 percent died within 3 months after their birthday, 29 percent died in the next 3 months after that, but only 9 percent died in the 3 months preceding their birthday. Conduct an appropriate hypothesis test of the apparent link between time of death and birthday and do so at the 1-percent level of significance.

30. A pharmaceutical firm is bringing out a new sleeping pill; it is tested on a random sample of 20 people. The sample standard deviation of the time to sleep turns out to be 11 minutes. An 80-percent confidence interval is desired for the population variance of the time to sleep. Construct it.

31. The manager of a chain of muffler-repair shops takes a random sample of 30 days from last year's sales records. The sample shows mean sales of 51.7 mufflers per day, with a standard deviation of 11.3 mufflers. Construct a 96-percent confidence interval for the population variance of daily muffler sales.

32. A product's diameter must have a maximum variance of .006 squared inches. The production manager takes a random sample of 25 units from the week's output. The sample shows a standard deviation of diameters of .07 inches. Construct a 90-percent confidence interval for the population variance of diameters and, thus, determine whether the production process turns out a product meeting specifications.

33. An aircraft altimeter must have readouts with a maximum variance of 25 squared feet. An inspector takes a random sample of 10 new altimeters, the sample variance is 26 squared feet. Construct a 98-percent confidence interval for the population variance of readouts and, thus, determine whether the altimeters meet specifications.

34. A machine that fills cereal boxes is supposed to do so with a maximum standard deviation of .01 ounces. An inspector takes a random sample of 15 boxes that were filled during the day; the sample standard deviation is .013 ounces. Construct a 96-percent confidence interval for the population standard deviation and, thus, determine whether the machine meets specifications.

35. In each of the following cases, indicate whether the hypothesis concerning the population variance is lower-tailed, upper-tailed, or two-sided; then determine the critical and actual χ^2 values and show whether the null hypothesis should be accepted or rejected.

   a. H_0: $\sigma^2 \geq 64$; $n = 26$, $\alpha = .01$, $s^2 = 10$.
   b. H_0: $\sigma^2 \geq .64$; $n = 15$, $\alpha = .02$, $s^2 = .1$.
   c. H_0: $\sigma^2 \leq 100$; $n = 5$, $\alpha = .001$, $s^2 = 110$.

36. For many years, the Federal Aviation Administration has been giving a pilot's exam with a mean score of 85 and a variance of 64. It is introducing a completely new exam and wants to know whether the population of exam scores for the new exam has the same variance. A hypothesis test at the 2-percent significance level is desired. When a sample of 15 persons takes the exam, the sample variance equals 36. Make the test.

37. The manager of a firm has found the variance in the firm's daily net cash flow to be very important for proper cash management. In the past, this figure was $\sigma^2 = 100,000$ dollars squared. The manager has reason to suspect, however, that this figure might have changed drastically. A hypothesis test at the 4-percent level of significance is to settle this issue. Table 10.R shows the net-cash-flow figures over a recent two-week period that are to be used in the test. Make the test.

Table 10.R

Day	Net Cash Flow (thousands of dollars)
1	−17
2	+22
3	+ 5
4	0
5	+ 8
6	− 3
7	+13
8	+20
9	+25
10	−41
11	0
12	− 4
13	+ 8
14	+13

38. A manufacturer claims that the light emitted by a cathode ray tube used in computer terminals is so uniform that the light intensity variance is .0001 squared units per pixel. A user takes a random sample of 14 pixels (dots on the screen) and measures the light intensity, finding a sample standard deviation of .0094 units per pixel. Can the manufacturer's claim be accepted? Make a test at the 4-percent level of significance.

39. A builder claims that the compressive strength of his concrete has a variance of at least 80 squared pounds per square inch. A skeptic takes a random sample of 25 batches of concrete; the sample standard deviation is 8 psi. Make a hypothesis test at the 5-percent level of significance.

40. A tape manufacturer is criticized because the tape strength has a variance of at least 52 squared pounds per square inch. The manufacturer takes a random sample of 30 rolls of tape produced in a week, the sample standard deviation is 5.3 psi. Make a hypothesis test at the 1-percent level of significance.

41. An educator believes that the variance in college students' IQ scores is at least 200 squared scores. A random sample of 20 students shows a sample variance of 198 squared scores. Conduct a hypothesis test at the 10-percent level of significance.

42. The manager of a commuter airline claims that the variance in airplane arrival times is at most 225 squared minutes. A random sample of 15 arrivals during a given month reveals a sample standard deviation of 16.3 minutes. Conduct a hypothesis test at the 10-percent level of significance.

43. A gas station manager claims that the readout error variance on the station's pumps is at most .01 squared gallons. A random sample of 30 readouts shows a sample standard deviation of .24 gallons. Conduct a hypothesis test at the 2-percent level of significance.

44. A supplier of biological laboratories claims that the variance of the weight of rats shipped by her is at most 4 squared grams. An experimenter takes a random sample of 7 rats received; the standard deviation of their weight is 6 grams. Conduct a hypothesis test of the supplier's claim and do so at the 5-percent level of significance.

45. The pilot of a four-engine plane knows the probability to be $\pi = .5$ that an engine requires additional oil after a 7-hour flight. The experience of the last 200 flights is summarized in Table 10.S.

The pilot wonders whether these sample data are consistent with an underlying binomial distribution of population data. A test at the 2-percent significance level is desired. Make the test.

Table 10.S

Number of Engines Requiring Oil After 7-Hour Flight	Observed Frequency, f_o
0	15
1	44
2	75
3	56
4	10
Total	200

46. A quality inspector wishes to test the null hypothesis that the number of defective items found in a box of 3 is a binomially distributed random variable with a probability of success in any one trial of $\pi = .2$. A test at the 10-percent level of significance is desired; a sample of 200 boxes yields the results given in Table 10.T. Make the test.

Table 10.T

Number of Defectives Found in a Box of 3	Observed Frequency, f_o
0	46
1	60
2	58
3	36
Total	200

47. A hundred high-school seniors have each applied to three colleges. Of these students, 24 received no acceptances, 36 received one acceptance, 26 received two acceptances, and the rest received three acceptances. A guidance counselor wants to test, at the 5-percent level of significance, whether the number of acceptances is a binomially distributed random variable with a probability of success in any one trial of $\pi = .25$. Make the test.

48. A quality inspector has made a random sample of 50 batches of work and has recorded how many batches contain how many errors. Zero errors were found in only 1 batch, 1 error occurred four times, 2 errors were found in 20 batches, 3 errors in 10, 4 errors in 8, 5 errors in 7. At the 2-percent level of significance, make a test of H_0: The number of errors per batch is a binomially distributed random variable.

49. The state department of transportation believes that the population of yearly automobile accidents per driver has a Poisson distribution. The data given in Table 10.U on a year's performance of the state's 112,956 drivers are to be used for a hypothesis test of this proposition at the 10-percent significance level. Make the test.

Table 10.U

Number of Yearly Auto Accidents per Driver	Observed Frequency, f_o
0	103,628
1	7,389
2	1,910
3	29
Total	112,956

50. An airline flight instructor wonders whether the hourly number of errors made by pilots operating a flight simulator can be described by a Poisson distribution. Records are kept for 30 training hours and are given in Table 10.V. A test at the 1-percent level of significance is desired. Make the test.

Table 10.V

Number of Errors per Hour	Observed Frequency, f_o
0	3
1	8
2	5
3	7
4	2
5	1
6	2
7	1
8	0
9	0
10	1
Total	30

51. A hotel manager wonders whether the number of cancellations per day can be described by a Poisson distribution. Using the data of Table 10.W, make the appropriate test and do so at the 2-percent level of significance.

Table 10.W

Number of Cancellations per Day	Observed Frequency, f_o
0	22
1	30
2	22
3	10
4	5
5	2
6	0
7	1
8	1
9	0
10 and more	3
Total	96

52. The director of a computer center wonders whether the number of computer access requests per hour can be described by a Poisson distribution. Using the data of Table 10.X, make the appropriate test and do so at the 5-percent level of significance.

Table 10.X

Number of Access Requests per Hour	Observed Frequency, f_o
0	55
1	61
2	50
3	32
4	18
5	9
6	5
7	2
8	1
9 and more	0
Total	233

53. Reconsider Practice Problem 7 about trading in the futures market (on page 265). How would you determine whether the data fit a normal distribution, if the population parameters $\mu = 50$ and $\sigma = 10$ were *not* given to you? Make the test and show your computations in detail.

54. An economist wants to determine (at the 5-percent level of significance) whether the annual incomes of the nation's lawyers can be described by a normal curve with $\mu = \$45,000$ and $\sigma = \$10,000$. A simple random sample of 100 lawyers yields the results given in Table 10.Y. Make the test.

Table 10.Y

Annual Incomes of Lawyers (thousands of dollars)	Observed Frequency, f_o
under 20	2
20—under 30	5
30—under 40	10
40—under 50	60
50—under 60	10
60—under 70	5
70—under 80	3
80—under 90	0
90 and over	5
Total	100

55. An advertising agency asks a random sample of 500 people to taste 5 brands of coffee and records the brand most preferred. It wishes to test, at the 10-percent level of significance, whether the preferences of coffee drinkers in general are uniformly distributed among the brands. The sample results are listed in Table 10.Z. Make the test.

Table 10.Z

Most Preferred Brand	Observed Frequency, f_o
A	91
B	109
C	85
D	100
E	115
Total	500

56. The city's fire chief wants to know whether the occurrences of fires are uniformly distributed among the days of the week. To perform a test at a 1-percent level of significance, the record of the last 700 fires given in Table 10.AA is to be used. Make the test.

Table 10.AA

Day Fire Occurred	Observed Frequency, f_o
Monday	85
Tuesday	129
Wednesday	72
Thursday	91
Friday	123
Saturday	135
Sunday	65
Total	700

57. In order to test whether a certain die is loaded, a man makes 120 rolls and records the frequencies with which faces of 1 through 6 show up. The frequencies are, respectively, 5, 36, 8, 9, 7, and 55. Make a hypothesis test, at the 5-percent level of significance, to determine whether, in all possible rolls of the die, the six possible faces are uniformly distributed.

58. A mayoral candidate wants to test why firms are leaving the city. She finds that 53 firms left for "greener pastures" during the past three years; of

these, 13 left because of too-high taxes, 9 because of too-high wages, 24 because of too-high transportation costs. Make a hypothesis test, at the 2-percent level of significance, that no one reason for leaving is more important than any other.

59. A polling organization recorded the opinions given in Table 10.BB at the two indicated times. It wishes to test, at the 2-percent significance level, whether public opinion at the more recent date equals that at the earlier date. Make the test.

60. A medical doctor checks the record of 200 patients who have taken a certain drug and makes the observations given in Table 10.CC. A test at the .1-percent level of significance is to be conducted to see whether these data could possibly come from a larger population that does conform with the manufacturer's claim, in which 50 percent of users improve, 30 percent show no change, 15 percent show minor deterioration, and 5 percent show major deterioration. Make the test.

Table 10.CC

Effect of Drug on Patients	Observed Frequency, f_o
Improvement	90
No change	40
Minor negative effect	30
Major negative effect	40
Total	200

Table 10.BB

Opinion Expressed About the President's Performance	Observed Frequency, f_o	
	1 Year After Inauguration	3 Years After Inauguration
Does superb job	50	68
Does good job	25	33
Does fair job	15	22
Does lousy job	10	27
Total	100	150

Selected Readings

Pearson, Karl. "On the Criterion That a Given System of Deviations From the Probable in the Case of a Correlated System of Variables is Such That It Can Be Reasonably Supposed to Have Arisen From Random Sampling." *Philosophical Magazine,* 5th series (1900), pp. 157–75. Presents the chi-square test of goodness of fit.

Computer Programs

The DAYSAL-2 personal-computer diskettes that accompany this book contain one program of interest to this chapter:

10. Chi-Square Tests lets you make (a) contingency-table tests to check on the alleged independence of 2 qualitative variables, (b) tests for more than 2 population proportions, (c) tests about the population variance, and (d) goodness-of-fit tests to check whether an observed distribution fits a specified theoretical distribution.

CHAPTER ELEVEN

Hypothesis Testing: Nonparametric Techniques

The sampling distributions of many test statistics employed in hypothesis tests are based on strict assumptions concerning the underlying populations that are being sampled. We have encountered many examples of linking test statistics with such assumptions in the previous two chapters. Consider the use of Student's t statistic (from Chapter 9) in an independent-sample test of the difference between two population means. The validity of the entire test rests on an assumption about the shape of the population distributions (namely, that they are normal) and on an assumption about the values of certain population parameters (namely, that the two population variances are equal). Whenever a hypothesis test is in this way based on certain specific assumptions about the probability distribution of population values or the sizes of population parameters, it is referred to as a **parametric test,** and the test statistic used is called a **parametric statistic.** Often, however, it is difficult to fulfull such assumptions, especially if they are demanding and narrow. Not surprisingly, therefore, statisticians are keenly interested in how sensitive their tests are to deviations from basic assumptions. Some tests show a great deal of **robustness;** their degree of sensitivity to errors in assumptions is low. Any application of the test to situations in which the assumptions are violated then causes the test to grind out nonsense. In such cases, it is desirable to find inferential procedures that are free from restrictive assumptions about the population that is being sampled. Such methods of inference, which make no assumptions whatsoever about the nature of underlying population distributions or their parameters, are referred to as **nonparametric tests.** Their test statistics are called **nonparametric statistics;** the χ^2 statistic employed in Chapter 10's test for independence is a case in point. Nonparametric tests (which are the exclusive subject matter of the present chapter) are often also referred to as **distribution-free tests,** because no assumption about the nature of the *population* distribution is being made—but this term can easily contribute to confusion, given the fact that these tests make very definite assumptions about the *sampling* distributions of their test statistics.

An obvious advantage of a nonparametric test is one's ability to employ it safely in situations where nothing at all is known about the population from which sample data are being drawn (and where the use of a parametric test would carry with it the risk of possibly violating crucial assumptions). Another advantage is the fact that such a test can be made with nominal and ordinal data, while parametric tests require data of a higher order. (For a review of these concepts see the Chapter 2 section, "Types of Data.")

Nonparametric tests have disadvantages, too. For example, they are less powerful than parametric tests under the same circumstances: given sample size and a specified significance level (that is, a specified type I error probability), the probability of a type II error is larger for a nonparametric test. In addition, use of a nonparametric test, when a parametric one could be employed, can be less efficient; a nonparametric test often tends to ignore available sample information (for example, by focusing only on the direction rather than also on the sizes of observed differences).

Innumerable nonparametric tests exist; an entire book could easily be devoted to them (and several such books are noted in the "Selected Readings" at the end of this chapter). In the following sections, eight of the more popular tests are introduced.

THE WILCOXON RANK-SUM TEST

The **Wilcoxon rank-sum test** is a nonparametric test based on two independent simple random samples and is designed to determine whether two statistical populations of continuous values are identical to or different from one another. The test is named after Frank Wilcoxon (see Biography 11.1 at the end of the chapter), who proposed it in 1945, and is also called the **W test.**

The Wilcoxon rank-sum test uses a test statistic, symbolized by W, that is derived by pooling the data contained in two independent samples (the sizes of which can be called n_A and n_B), ranking the combined data from the smallest value (to be called 1) to the largest (equal to $n_A + n_B$), recreating the original two samples with the rank data, summing the ranks in each sample, and, finally, designating *either one* of these rank sums (typically that of sample A) as the test statistic. See Box 11.A.

11.A The Wilcoxon Rank-Sum Test Statistic

$$W = \text{rank sum of sample A}$$

Note: In this book, W is defined as the rank sum of sample A, but, in principle, it could also be the rank sum of sample B.

All this is most easily illustrated by considering the results of two samples, as shown in Table 11.1. The two samples show randomly selected average monthly checking-account balances in two hypothetical banks, bank A and bank B. These data will be used below to solve Practice Problem 1. The combined sample data can be ranked, as shown in Table 11.2 on page 446, by writing down the pooled sample data in order of their magnitude, as in column 1, and then assigning, as in column 3, a value of 1 to the lowest number, a value of 2 to the next one, and so on, until the highest number contained in the sample is given a rank of $n_A + n_B$ (in this case, of $10 + 12 = 22$).

Table 11.1 Average Monthly Checking-Account Balances in Two Banks: Sample Data

The average monthly checking-account balances found in two independent samples might equal the values shown here. As this example indicates, the two samples need not be of equal size (here $n_A = 10$, but $n_B = 12$).

Bank A		Bank B	
Number of Sampled Account	Average Balance Last Month (dollars)	Number of Sampled Account	Average Balance Last Month (dollars)
33552	201	15107	3,362
38174	950	12985	129
24041	1,209	97616	201
30547	367	64241	1,579
22651	792	08592	485
34492	804	81036	2,639
30631	42	98059	79
88602	950	44951	92
16510	505	23078	3,010
29278	4,099	92793	3,159
		80756	3,412
		52799	2,910

As Table 11.2 shows, by noting in column 2 the sample that contained the original column 1 data, the pooled rank data in column 3 can easily be separated into sample A ranks and sample B ranks in columns 4 and 5, and a sum of ranks can then be calculated for each sample. Either one of these rank sums can be used as the test statistic. (For Practice Problem 1 below, we will arbitrarily choose the rank sum of sample A for that purpose.)

Note: Because this type of test deals with continuous data, values of equal size in the original sample data—column 1—are highly unlikely in theory, but in practice such ties do occur, mainly as a result of rounding. In our example, average balances of $201 and of $950 each occur twice; in the absence of rounding to whole dollars, this duplication probably would not have occurred. *When assigning ranks—column 3—tied values are each given the mean of the next ranks to be assigned.* Note how the two values of $201 are made to share ranks 5 and 6 and are both ranked as 5.5. Similarly the two valves of $950 are made to share ranks 12 and 13 and are both ranked as 12.5. This procedure is crucial when the tied values belong to different samples, as is the case here for the two balances of $201. Depending on whether one rank (say, 5) was assigned to the sample A figure and the next rank (say, 6) to the sample B figure or whether rank 6 went to the A figure and 5 to the B figure, one would produce a different sum of ranks for the two samples. The averaging procedure avoids such arbitrariness. For the same reason, the procedure, although employed here, is not crucial (and an arbitrary assignment of ranks to tied values would be acceptable) when the tied values belong to the same sample, as is true for the two balances of $950. Note how the two rank sums remain the same whether the two $950 values are ranked 12.5 each or whether they are ranked 12 and 13.

Table 11.2　Finding the Wilcoxon Rank Sums

After pooling the original sample data, in column 1, and ranking them, in column 3, the rank data can again be separated by noting the sample that contained the original data, in column 2. The separation of rank data, in columns 4 and 5, in turn, allows the calculation of a sum of ranks for each sample. One of these Wilcoxon rank sums, typically that of sample A, is used as the test statistic. In this case, the computed value of the rank sum of sample A is $W = 104.5$.

Average Balance Last Month (dollars) (1)	Sample of Origin (2)	Rank (3)	Sample A Ranks (4)	Sample B Ranks (5)
42	A	1	1	
79	B	2		2
92	B	3		3
129	B	4		4
201	A	5.5	5.5	
201	B	5.5		5.5
367	A	7	7	
485	B	8		8
505	A	9	9	
792	A	10	10	
804	A	11	11	
950	A	12.5	12.5	
950	A	12.5	12.5	
1,209	A	14	14	
1,579	B	15		15
2,639	B	16		16
2,910	B	17		17
3,010	B	18		18
3,159	B	19		19
3,362	B	20		20
3,412	B	21		21
4,099	A	22	22	
		Rank Sums:	$W = 104.5$	148.5

The Sampling Distribution of W

Knowing how to calculate the observed value of our rank-sum test statistic is one thing; knowing how to interpret it is another. Wilcoxon argued that if the null hypothesis of identical populations were *not true,* the value of W (the rank sum of sample A) would either be very small or very large (depending on whether the population A data and, thus, the original sample A data were generally smaller or larger than the population B data and, thus, the original sample B data). In our example, the smallest possible value of W equals $1 + 2 + 3 + \ldots + 10 = 55$ (and that value would occur if every single dollar balance in sample A were smaller than every single one in sample B). The largest possible value of W on the other hand, equals $13 + 14 + 15 + \ldots + 22 = 175$ (and that value would occur if every single dollar balance in sample A were larger than every single one in sample B). If, in contrast, the null hypothesis of identical populations were *true,* one would expect sample A to produce about as many low ranks as high ranks, making the expected value of W the mean of the minimum and maximum values just noted, or $(55 + 175) \div 2 = 115$. Indeed, argued Wilcoxon, if the sample sizes were at least equal to 10 each and if the identical-population hypothesis were true, the entire sampling distribution of W could be approximated by the normal curve and would have a mean and standard deviation as given in Box 11.B.

11.B Expected Value and Standard Deviation of the Sampling Distribution of W (given H_0: The sampled populations are identical.)

$$\mu_W = \frac{n_A \cdot (n_A + n_B + 1)}{2}$$

$$\sigma_W = \sqrt{\frac{n_A \cdot n_B(n_A + n_B + 1)}{12}}$$

Note: If W were defined as the rank sum of sample B, one would have to interchange subscripts A and B in the μ_W formula.

Note how using the first formula from Box 11.B produces the same expected value of W as calculated above by averaging the smallest possible value of W (55) and the largest possible value of W (175):

$$\mu_W = \frac{10(10 + 12 + 1)}{2} = 115.$$

Box 11.C gives the formula for calculating the normal deviate of W which can also be used as a test statistic instead of W itself.

11.C Normal Deviate for the Wilcoxon Rank-Sum Test

$$z = \frac{W - \mu_W}{\sigma_W}$$

Assumption: $n_A \geq 10$, $n_B \geq 10$

Applications

The following Practice Problems illustrate the procedure for using the Wilcoxon rank-sum test. The first problem uses the data in Tables 11.1 and 11.2.[1]

---

[1]Problem 3 at the end of this chapter, along with its solution at the back of the book, is an example of an upper-tailed Wilcoxon rank-sum test, in contrast to the two-tailed and lower-tailed tests illustrated here.

Practice Problem 1

Comparing Average Account Balances in Two Banks (a two-tailed test)

Consider the manager of a bank holding company who wants to know quickly whether the average monthly balance in customers' checking accounts is the same in two banks, A and B. Two small simple random samples are to be taken; one from the accounts in bank A and another one from those in bank B. A test at the 5-percent significance level is to be conducted. A statistician might proceed with Steps 1 to 4.

Step 1: Formulating two opposing hypotheses.

H_0: The average balance is identical in the two banks.
H_A: The average balance differs in the two banks.

Step 2: Selecting a test statistic. Conceivably, the statistician could conduct a parametric test, using the t statistic as in the Chapter 9 section that introduced small-sample tests of the difference between two population means (pp. 368–371). But suppose our statistician was far from certain that the requirements for such a test (that the two populations of average monthly bank balances were normally distributed and also had identical variances) were met. In that case, the statistician could turn to the Wilcoxon rank-sum test and use as the test statistic:

W (the arbitrarily chosen rank sum of sample A) or
z (the normal deviate of W given in Box 11.C).

For illustrative purposes, we shall employ both W and z in this first problem.
Step 3: Deriving a decision rule. Given a desired significance level of $\alpha = .05$, and this being a two-tailed test, we find in Appendix Table J critical normal deviate values of $\pm z_{\alpha/2} = \pm 1.96$. Thus, the decision rule must be either: (a) "Accept H_0 if $\mu_W - (1.96\sigma_W) \leq W \leq \mu_W + (1.96\sigma_W)$" or (b) "Accept H_0 if $-1.96 \leq z \leq +1.96$." With the help of the formulas in Box 11.B, the critical rank sums given in decision rule (a) can, of course, quickly be calculated as 85.28 and 144.72 (given that, in this example, $\mu_W = 115$ and $\sigma_W = 15.16575$). The critical values are encircled in Figure A.

Figure A

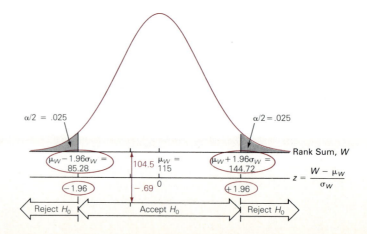

$\alpha/2 = .025$ $\alpha/2 = .025$

Rank Sum, W

$\mu_W - 1.96\sigma_W = 85.28$ 104.5 $\mu_W = 115$ $\mu_W + 1.96\sigma_W = 144.72$

$z = \dfrac{W - \mu_W}{\sigma_W}$

-1.96 $-.69$ 0 $+1.96$

Reject H_0 Accept H_0 Reject H_0

Step 4: Selecting the samples, computing the test statistic, and confronting it with the decision rule. At this point, the statistician would select the data we have already previewed in Table 11.1 and would calculate the observed value of W or its normal deviate as

$$W = 104.5$$

(as noted in Table 11.2) or

$$z = \frac{104.5 - 115}{15.16575} = -.69$$

(in accordance with Box 11.C). These values correspond to those marked by the colored arrow in Figure A; they suggest that the null hypothesis should be *accepted*. Although the sample A rank sum is smaller than the value to be expected if H_0 is true, the sample result is not statistically significant. At the 5-percent significance level, it is quite likely that the populations of average monthly balances in the two banks are identical and that sampling error alone accounts for the deviation of W from its expected value.

Note: What if W had been defined as the rank sum not of sample A, but of sample B? Then the subscripts in the Box 11.B formulas would have to be interchanged, making $\mu_W = 138$, while leaving σ_W unchanged at 15.16575. Accordingly, the critical W values would have been 108.28 to 167.72, still corresponding to $z = \pm 1.96$. Given the computed value of the sample B rank sum of 148.5 (Table 11.2), H_0 would also have been accepted. This rank-sum value would have corresponded to

$$z = \frac{148.5 - 138}{15.16575} = +.69,$$

suggesting that sample B values were *larger* than expected if H_0 is true, but not large enough to rule out sampling error as the sole cause. ‖‖

‖‖ Practice Problem 2

Comparing the Effectiveness of Two Fertilizers (a lower-tailed test)

We now imagine a farm manager who wants to test a manufacturer's claim that cheaper fertilizer A is at least as effective as more expensive fertilizer B. Twenty identical plots of strawberries are randomly selected such that half are fertilized with A and half with B; the yields are to be recorded, and a statistical test at the 5-percent significance level is to be conducted.

Step 1: Formulating two opposing hypotheses.

H_0: Fertilizer A is at least as effective as B. (Yields on A plots exceed or equal those on B plots.)

H_A: Fertilizer A is less effective than B. (Yields on A plots are lower than those on B plots.)

Step 2: Selecting a test statistic. The manager selects the normal deviate for the Wilcoxon rank-sum test from Box 11.C:

$$z = \frac{W - \mu_W}{\sigma_W}.$$

Step 3: Deriving a decision rule. Given a desired significance level of $\alpha = .05$ and this being a lower-tailed test, Appendix Table J suggests a critical normal deviate of $z_\alpha = -1.64$. Thus, the decision rule must be: "Accept H_0 if $z \geq -1.64$." The critical value is encircled in Figure B.

Figure B

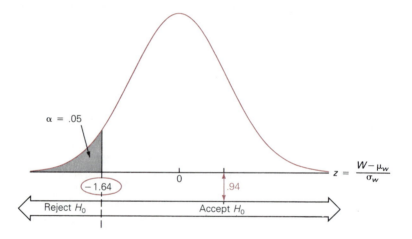

Step 4: Selecting the samples, computing the test statistic, and confronting it with the decision rule. We assume that the experiment on yields from two types of fertilizer yields the results shown in the first four columns of Table A and that the value of W is calculated accordingly, as shown in the last four columns. Applying the Box 11.B formulas:

$$\mu_W = \frac{10(10 + 10 + 1)}{2} = 105$$

and

$$\sigma_W = \sqrt{\frac{10 \cdot 10(10 + 10 + 1)}{12}} = 13.2288.$$

Hence, we have an observed value of

$$z = \frac{117.5 - 105}{13.2288} = .94.$$

This value is marked by the colored arrow in Figure B; accordingly, H_o should be *accepted.* Fertilizer A is at least as effective as fertilizer B.

Table A

Fertilizer A		Fertilizer B					
Plot Number	Quarts of Strawberries	Plot Number	Quarts of Strawberries	Ordering of Original Data	Sample of Origin	Ranks	Sample A Ranks
3	91	1	79	79	B	1	
4	97	2	90	80	A	3	3
7	85	5	80	80	B	3	
9	88	6	95	80	B	3	
11	86	8	83	81	A	5.5	5.5
12	93	10	93	81	A	5.5	5.5
13	80	14	85	82	B	7	
15	81	16	80	83	B	8	
18	81	17	82	84	B	9	
20	90	19	84	85	A	10.5	10.5
				85	B	10.5	
				86	A	12	12
				88	A	13	13
				90	A	14.5	14.5
				90	B	14.5	
				91	A	16	16
				93	A	17.5	17.5
				93	B	17.5	
				95	B	19	
				97	A	20	20

$W = 117.5$

III

THE MANN-WHITNEY TEST

The **Mann-Whitney test** is a nonparametric test that is equivalent to the Wilcoxon rank-sum test. Both tests serve the identical purpose of determining whether two statistical populations of continuous values are identical to or different from one another. The test discussed in this section is named after Henry B. Mann and D. R. Whitney; it is often referred to as the **U test** because U is the symbol employed for its test statistic.

The statistic, U, is defined as the difference between the largest possible value of W and its actual value (where W, as in the previous section, equals the sum of sample A ranks). Box 11.D gives the formula for U.

11.D Mann-Whitney Test Statistic

$$U = \left[(n_A \cdot n_B) + \frac{n_A(n_A + 1)}{2} \right] - W$$

Note: W is defined as the rank sum of sample A. If it were defined as the rank sum of sample B, one would have to interchange subscripts A and B in this formula.

Note how the expression in the square brackets in Box 11.D, when applied to the data in Table 11.2, does indeed produce the largest possible value of W calculated earlier:

$$(10 \cdot 12) + \frac{10(10 + 1)}{2} = 175.$$

The Sampling Distribution of U

When the sizes of the two independent samples are equal to at least 10 each and the identical-population hypothesis is true, the sampling distribution of U can also be approximated by the normal curve (as is true for W); the expected value and standard deviation of the sampling distribution of U are given in Box 11.E. Accordingly, the normal deviate of U can be calculated as in Box 11.F.

11.E Expected Value and Standard Deviation of the Sampling Distribution of U (given H_0: The sampled populations are identical.)

$$\mu_U = \frac{n_A \cdot n_B}{2}$$

$$\sigma_U = \sqrt{\frac{n_A \cdot n_B(n_A + n_B + 1)}{12}} = \sigma_W$$

11.F Normal Deviate for Mann-Whitney Test

$$z = \frac{U - \mu_U}{\sigma_U}$$

Assumption: $n_A \geq 10, n_B \geq 10$

Application

Consider applying the U test to Practice Problem 1 concerning average account balances in two banks.

Practice Problem 3

Average Account Balances Revisited

Step 1 of the test would be identical; in Step 2, we now imagine, the test statistic, U, is selected.

The decision rule derived in Step 3 would become either (a) "Accept H_0 if $\mu_U -$ $(1.96\sigma_U) \leq U \leq \mu_U + (1.96\sigma_U)$" or (b) "Accept H_0 if $-1.96 \leq z \leq +1.96$." With the help of the Box 11.E formulas, the critical U values could quickly be calculated as:

$$\mu_U = \frac{10 \cdot 12}{2} = 60$$

and as the equation that follows:

$$\sigma_U = \sigma_W = 15.16575.$$

Hence, the critical values of decision rule (a) are 30.28 and 89.72, respectively.

In Step 4, given $W = 104.5$ (from Table 11.2), one would find $U = 175 - 104.5 = 70.5$ or its normal deviate of

$$z = \frac{70.5 - 60}{15.16575} = .69.$$

Given the decision rule, either figure would indicate that H_0 should be *accepted,* just as the W test did. A note of caution, however, is in order. The earlier W test (Practice Problem 1) produced a z value of $-.69$ when W was defined as the sum of sample A ranks. Yet the U test here, which *utilizes,* in the Box 11.D formula, a value of W defined as the sum of sample A ranks, produces a z value of $+.69$, just as a W test would if W were the sum of sample B ranks (see the last paragraph of Practice Problem 1). In a two-tailed test, all this makes little difference. If H_0 is accepted for $z = -.69$ (as in our W test), it is equally accepted for $z = +.69$ (as in our U test). In a one-tailed test, on the other hand, there might be a difference. A test statistic of -2.14 might spell accep-tance of H_0; one of $+2.14$, rejection. In *one-tailed U* tests, therefore, correct test results are found only if the decision rule is confronted with the *negative* of whatever z value emerges from the Box 11.F formula. |||

THE SIGN TEST

The **sign test** is yet another nonparametric test designed to determine whether two statistical populations are identical to or different from one another. In fact, the history of nonparametric statistics dates back to the invention of this test by J. Arbuthnott in 1710. Unlike the W test and the U test, this test, however, is not based on two indepen-dent samples, but on a matched-pairs sample. In addition, and unlike the matched-pairs tests discussed in Chapter 9, the sign test does not consider the absolute sizes of dif-ferences between the sample pairs but notes only the direction of these differences, initially expressed by a zero (0), a plus sign ($+$), or a minus sign ($-$). Subsequently, matched pairs with a zero difference are omitted from the analysis, while the number of plus signs ($+$) among the remaining n matched pairs is used as the test statistic, S.

11.G The Sign Test Statistic
S = Number of plus signs ($+$) among the n matched pairs with nonzero differences

The sign test is the simplest of all the nonparametric tests, but is also considerably less discriminating than a t test. Whenever quantitative measures of matched-pairs dif-ferences are available and when the population of all such potential differences can be assumed to have a normal distribution, the more powerful t test is preferred; the sign test then is a definite second choice. However, when quantitative measures cannot be acquired or when the above assumption of normality does not hold, the t test cannot be employed and the sign test is most useful.

The Sampling Distribution of S

If the null hypothesis of no differences is true, the sampling distribution of S is binomial, with the probability of "success" (finding a plus sign) equaling $\pi = .5$ in any one of the n trials. Accordingly, the mean and standard deviation of the sampling distribution of S can be derived from Box 6.C on page 208, as now shown in Box 11.H.

11.H Expected Value and Standard Deviation of the Sampling Distribution of S (given H_0: There are no population differences.)

$$\mu_S = n(.5)$$
$$\sigma_S = \sqrt{n(.5)(.5)}$$

Yet, as was noted in Chapter 7, when both $n \cdot \pi$ and $n(1 - \pi) \geq 5$, the normal distribution becomes a good approximation of the binomial distribution. Given $\pi = 1 - \pi = .5$, this condition is fulfilled whenever $n \geq 10$. Accordingly, the normal deviate of S, given in Box 11.I, can replace S as the test statistic when $n \geq 10$.

11.I Normal Deviate for Sign Test

$$z = \frac{S - \mu_S}{\sigma_S}$$

Assumption: $n \geq 10$

Applications

The following Practice Problems demonstrate the usefulness of the sign test.[2]

⫼ Practice Problem 4

Comparing the Quality of Two House Paints (a two-tailed test)
Consider the manufacturer of house paint who wants to determine whether the company's paint (A) is of the same quality as that of a competitor (B). Each of 100 wooden

---

[2]Problem 19 at the end of this chapter, along with its solution at the back of the book, is an example of a lower-tailed sign test, in contrast to the two-tailed and upper-tailed tests illustrated here.

planks is painted half with A and half with B; accordingly, the two halves of each individual plank constitute a matched pair. The planks are then placed outdoors. Eventually, for each plank, a plus, minus, or zero is recorded, depending on whether the portion painted with A, compared to that painted with B, lasts a longer time (+), a shorter time (−), or the same time (0) before the first signs of cracking or peeling appear. A test at the 5-percent level of significance is desired. A company statistician may proceed with Steps 1 to 4.

Step 1: Formulating two opposing hypotheses.

H_0: There are no differences in quality.
H_A: There are differences in quality.

Step 2: Selecting a test statistic. Because the test procedure does not call for any recording of actual time differences (such as the A portion of a plank lasting 9 months longer than, 3 months less than, or the same time as the B portion) and also because there exists no knowledge about the probability distribution of all such potential time differences, the statistician employs the sign test. Accordingly, the test statistic can be the following:

S, the number of plus signs (+) among the *n* matched pairs remaining after pairs with a zero difference are omitted from the analysis, or
z, the normal deviate of *S,* as defined in Box 11.I.

For illustrative purposes, we shall employ both *S* and *z.*

Step 3: Deriving a decision rule. Given a desired significance level of $\alpha = .05$, and this being a two-tailed test, we find in Appendix Table J critical normal deviate values of $\pm z_{\alpha/2} = \pm 1.96$. Thus, the decision rule must be either (a) "Accept H_0 if $\mu_S - (1.96\sigma_S) \leq S \leq \mu_S + (1.96\sigma_S)$" or (b) "Accept H_0 if $-1.96 \leq z \leq +1.96$." The critical values are encircled in Figure C.

Figure C

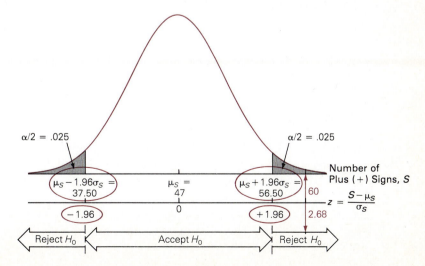

Step 4: Selecting the sample, computing the test statistic, and confronting it with the decision rule. At this point, the statistician would consider the data provided by the experiment and find, say, a plus ($+$) 60 times, a minus ($-$) 34 times, and a zero (0) 6 times, making $n = 60 + 34 = 94$. This information would allow the calculation, with the Box 11.H formulas, of the encircled critical values of S, as shown in Figure C. Because $\mu_S = 47$ and $\sigma_S = 4.8477$, $\mu_S - (1.96\sigma_S) = 37.50$, and $\mu_S + (1.96\sigma_S) = 56.50$.

On the other hand, given the actual value of $S = 60$, the above summary measures imply a normal deviate of

$$z = \frac{60 - 47}{4.8477} = 2.68.$$

These actual values are marked by the colored arrow in Figure C; they suggest that the null hypothesis should be *rejected*. At the 5-percent level, the sample result is statistically significant. It is unlikely that the observed better performance of paint A is a result of sampling error; paint A is of better quality. |||

||| Practice Problem 5

Comparing the Effectiveness of Two Types of Gasoline (an upper-tailed test)

Consider a government agency that wants to test whether it is true that the mileage achieved with unleaded gasoline (A) is equal to or less than that achieved with leaded gasoline (B). Fifteen car models are chosen; a given driver drives each model over the identical route—first with unleaded, then with leaded gas. A test at the 2.5-percent level of significance is to be conducted on the basis of the observed differences in miles per gallon.

Step 1: Formulating two opposing hypotheses.

H_0: Miles per gallon with unleaded gasoline are at most equal to those with leaded gasoline.

H_A: Miles per gallon with unleaded gasoline exceed those with leaded gasoline.

Step 2: Selecting a test statistic. The agency statistician knows nothing about the probability distribution of the population of miles-per-gallon differences and selects the normal deviate of S as the test statistic:

$$z = \frac{S - \mu_S}{\sigma_S}.$$

Step 3: Deriving a decision rule. Given a desired significance level of $\alpha = .025$, and this being a one-tailed test, Appendix Table J suggests a critical normal deviate value of $z_\alpha = 1.96$. Thus, the decision rule must be: "Accept H_0 if $z \leq +1.96$." The critical value is encircled in Figure D.

Figure D

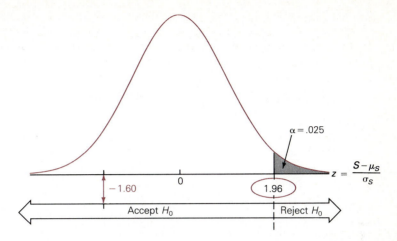

$$z = \frac{S - \mu_S}{\sigma_S}$$

$\alpha = .025$

-1.60

0

1.96

Accept H_0 Reject H_0

Step 4: Selecting the sample, computing the test statistic, and confronting it with the decision rule. After conducting the experiment on miles per gallon achieved with two gasoline types with the 15 cars, the agency records, we assume, the results given in Table B. Accordingly, $n = 14$ (the tie is eliminated), $S = 4$, $\mu_S = 7$, $\sigma_S = 1.8708$, and

$$z = \frac{4 - 7}{1.8708} = -1.60.$$

Table B

Car Model	Miles Per Gallon		Sign of Difference A − B
	Unleaded Gas (A)	Leaded Gas (B)	
1	22.1	23.7	−
2	15.7	16.1	−
3	18.2	19.0	−
4	19.0	19.0	0
5	25.7	24.2	+
6	19.2	22.0	−
7	11.9	12.1	−
8	28.0	30.2	−
9	35.0	37.8	−
10	27.0	30.0	−
11	19.7	19.0	+
12	16.3	15.1	+
13	21.2	23.4	−
14	22.2	20.7	+
15	27.8	28.1	−

This z value is marked by the colored arrow in Figure D; it suggests that the null hypothesis should be *accepted*. ▐

An Important Variation: A Test Concerning the Population Median

The sign test lends itself admirably to answering another kind of question—namely, whether a sample comes from a population with a specified median. In this case, the null hypothesis asserts "H_0: The sample comes from a population with a median of x."

Practice Problem 6

The Median Lifetime of Tires

Consider a tire manufacturer who claims that the firm's tires have a median life of 25,000 miles. If the claim is true, one would expect that half of a random sample of the firm's tires last a longer time ($+$), while half last a shorter time ($-$). What if 100 tires are tested and 40 are found to last a longer time ($S = 40$), while 60 are found to last a shorter time? Can this result be attributed to sampling error or does it suggest that the tires do not come from a population with a median life of 25,000 miles (but from one with a lower median)? A two-tailed test at the 5-percent level of significance calls for this decision rule: "Accept H_0 (a median of 25,000 miles) if $-1.96 \leq z \leq +1.96$." The value of the test statistic can, in turn, be calculated as follows: Given $n = 100$, $S = 40$, $\mu_S = 50$, $\sigma_S = 5$,

$$z = \frac{40 - 50}{5} = -2.00.$$

Accordingly, the null hypothesis should be *rejected*. The median life of this type of tire is less than 25,000 miles. |||

THE WILCOXON SIGNED-RANK TEST

The **Wilcoxon signed-rank test** is a more powerful alternative to the sign test (see Box 11.J). Like the sign test, it also seeks to determine whether two statistical populations are identical to or different from one another, and it also employs matched-pairs sampling. However, it considers not only the direction, but also the magnitude of differences between the matched sample pairs. As a result, the test uses more of the information contained in sample data; for a given sample size, it is more efficient. Thus, if the magnitude of differences is known (as in Practice Problem 5), it pays to use this more discriminating test. After collecting data for each sample pair, the procedure involves these steps:

1. Calculating the difference between the sample data for each matched pair. (Pairs with zero differences are eliminated from the test at this stage, and sample size, *n,* is reduced accordingly to the remaining number of pairs with nonzero differences.)
2. Ranking the absolute values of the (nonzero) differences from the smallest (rank = 1) to the largest (rank = n).
3. Attaching to each rank the sign of the original difference corresponding to it.
4. Calculating the sum of these signed ranks and designating it as the test statistic, T.

11.J The Wilcoxon Signed-Rank Test Statistic

T = sum of signed ranks among the n matched pairs with nonzero differences

The Sampling Distribution of T

If the null hypothesis (that the two populations are identical) is true, half of the matched-pair differences will be positive, and half will be negative. In addition, positive or negative differences of a given magnitude will be equally likely. (A -3, for example, will then have an equal probability as a $+3$.) As a result, the sum of signed ranks, T, has an expected value of zero. Given sufficient sample size ($n \geq 10$), the entire sampling distribution of T can again be approximated by the normal curve; the mean and standard deviation are given in Box 11.K, and the normal deviate of T can be calculated as in Box 11.L.

11.K Expected Value and Standard Deviation of the Sampling Distribution of T (given H_0: There are no population differences.)

$$\mu_T = 0$$

$$\sigma_T = \sqrt{\frac{n(n + 1)(2n + 1)}{6}}$$

11.L Normal Deviate for Wilcoxon Signed-Rank Test

$$z = \frac{T - \mu_T}{\sigma_T} = \frac{T}{\sigma_T}$$

Assumption: $n \geq 10$

Application

We can most easily learn about the nature of this test by reworking Practice Problem 5.

Practice Problem 7

The Effectiveness of Two Types of Gasoline Revisited

To rework Practice Problem 5 using the Wilcoxon signed-rank test, we formulate the identical hypotheses in Step 1, but now select the normal deviate of T (from Box 11.L) as the test statistic in Step 2. The decision rule (Step 3) is identical; the graphical portrayal in Figure E is almost identical to that in Practice Problem 5's Figure D. Step 4

Figure E

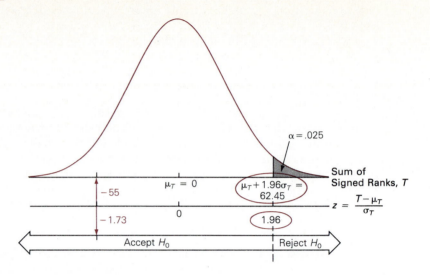

brings out the major differences between the earlier sign test and the signed-rank test now employed. The necessary calculations, based on Table B, are given in Table C, which reveals $T = -55$.

Table C

Car Model	Miles Per Gallon		Difference $(A - B)$	Absolute Value of Nonzero Difference	Rank of Absolute Nonzero Difference	Signed Rank
	Unleaded Gas (A)	Leaded Gas (B)				
1	22.1	23.7	−1.6	1.6	9	−9
2	15.7	16.1	−.4	.4	3	−3
3	18.2	19.0	−.8	.8	5	−5
4	19.0	19.0	0	—	—	—
5	25.7	24.2	+1.5	1.5	7.5	+7.5
6	19.2	22.0	−2.8	2.8	12.5	−12.5
7	11.9	12.1	−.2	.2	1	−1
8	28.0	30.2	−2.2	2.2	10.5	−10.5
9	35.0	37.8	−2.8	2.8	12.5	−12.5
10	27.0	30.0	−3.0	3.0	14	−14
11	19.7	19.0	+.7	.7	4	+4
12	16.3	15.1	+1.2	1.2	6	+6
13	21.2	23.4	−2.2	2.2	10.5	−10.5
14	22.2	20.7	+1.5	1.5	7.5	+7.5
15	27.8	28.1	−.3	.3	2	−2

$$T = -55$$

Because one pair (with zero difference) is eliminated, $n = 14$, and (according to Box 11.K) $\sigma_T = 31.86$. Hence, the critical value of T in Figure E can be calculated as 62.45, and the normal deviate of the observed T value can be calculated as $z = -1.73$ (using Box 11.L). As the position of the colored arrow in Figure E indicates, H_0, accordingly, should be *accepted.* The earlier sign test, of course, reached the same conclusion, but note: it can easily happen that the more discriminating signed-rank test (because it uses more information) reverses the judgment of the sign test (that may well discard available information). Whenever either test could be used, the signed-rank test, therefore, is the one to apply. |||

THE NUMBER-OF-RUNS TEST

The occurrence of an unbroken succession of like observations in a sequence containing potentially different types of observations is referred to as a **run.** Consider tossing a coin 10 times and recording the appearance of heads (H) or tails (T). The result may be as given in sequence 1.

Sequence 1: H̲ H̲ T̅ T̅ T̅ H̲ H̲ T̅ H̲ H̲

As the horizontal lines below or above the letters indicate, this sequence contains 3 runs of heads and 2 runs of tails; as the example also indicates, a run may be as short as a single observation that is preceded or followed by a different type of observation (or that begins or ends a sequence). If the observations are generated by a random process, some sequences, such as sequence 1, will instantly appear more probable than others, such as sequence 2 or 3 below.

Sequence 2: T̅ T̅ T̅ T̅ T̅ H̲ H̲ H̲ H̲ H̲

Sequence 3: H̲ T̅ H̲ T̅ H̲ T̅ H̲ T̅ H̲ T̅

There are many situations in which it is important to know whether the elements of a sequence appear in random order; the **number-of-runs test** is a procedure designed to discover the presence or absence of such randomness. Whenever sample data can be separated into two categories, such as heads or tails (including such symbolic "heads" or "tails" as fatal vs. nonfatal accidents, income below $3,000 vs. income of $3,000 and above, defective vs. nondefective parts, and so on), they can be represented by a string of letters (H and T) appearing in the chronological order in which the data were collected or in the order in which the events that they represent occurred. The test to be discussed in this section is designed to determine whether occurrences describe a random pattern or whether instead there exists some form of nonrandomness, such as *serial dependency* in which the order of occurrences is affected by previous events. (Note how, in sequence 3, each T is preceded by an H—which suggests something other than randomness. Sequence 2, similarly, does not exactly conjure up the image of a random process at work.)

The runs test begins with the null hypothesis that the elements of the sequence appear randomly. The test is two-sided: if the null hypothesis is true, too few runs, as in sequence 2, are considered just as unlikely as too many, as in sequence 3. The number of runs of a given category (such as "heads") is typically used as the test statistic and is denoted by R_H. For the three sequences, R_H equals 3, 1, and 5, respectively. (Note: It would also be possible to define the test statistic as R_T, the number of runs of some other category, such as "tails," or simply as R, the total number of all types of runs.)

11.M The Number-of-Runs Test Statistic

$$R = \text{number of runs in category H}$$

or

$$R = \text{number of runs in category T}$$

or

$$R = \text{total number of runs in all types of categories}$$

The Sampling Distribution of R_H

The sampling distribution of R_H, too, can be approximated by the normal distribution under certain conditions. If n_H and n_T are defined, respectively, as the number of "heads" and "tails" in the sample (not to be confused with the number of runs), the summary measures of the sampling distribution of R_H are given in Box 11.N. Accordingly, the normal deviate of R_H can be calculated as in Box 11.O.

11.N Expected Value and Standard Deviation of the Sampling Distribution of R_H (given H_0: Sampling is random.)

$$\mu_{R_H} = \frac{n_H(n_T + 1)}{n_H + n_T}$$

$$\sigma_{R_H} = \sqrt{\frac{n_H(n_T + 1)(n_H - 1)}{(n_H + n_T)^2}\left(\frac{n_T}{n_H + n_T - 1}\right)}$$

11.O Normal Deviate for the Number-of-Runs Test

$$z = \frac{R_H - \mu_{R_H}}{\sigma_{R_H}}$$

Application

We can quickly apply the test to the very examples embodied in sequences 1–3, realizing, of course, that *heads* and *tails* can symbolize all kinds of things besides the actual sides of coins observed in a coin-tossing experiment.

||| Practice Problem 8

Sequences of "Heads" and "Tails"

In each of these cases, the first three steps of the hypothesis test are the same.

Step 1: Formulating two opposing hypotheses.

H_0: The different types of sample elements in the sequence are distributed randomly.

H_A: The different types of sample elements in the sequence are not distributed randomly.

Step 2: Selecting a test statistic. We select the normal deviate for the number-of-runs test from Box 11.O.

Step 3: Deriving a decision rule. We choose a significance level of, say, $\alpha = .05$. This being a two-tailed test, we find in Appendix Table J critical normal deviate values of $\pm z_{\alpha/2} = \pm 1.96$. Thus, the decision rule is: "Accept H_0 if $-1.96 \leq z \leq +1.96$." The critical values are encircled in Figure F.

Figure F

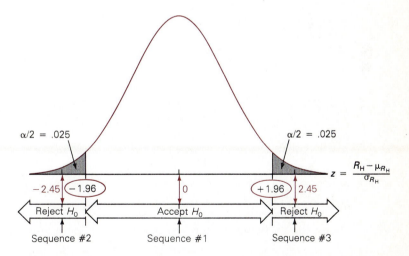

Step 4: Inspecting the sample, computing the test statistic, and confronting it with the decision rule. This final step in our procedure differs, of course, depending on which of the three sequences is being tested for randomness. [The observed normal deviates calculated below for sequences 1–3 are marked by the three colored arrows in Figure F.]

Sequence 1: $n_H = 6, n_T = 4, R_H = 3$

$$\mu_{R_H} = \frac{6(4 + 1)}{6 + 4} = 3$$

$$\sigma_{R_H} = \sqrt{\frac{6(4 + 1)(6 - 1)}{(6 + 4)^2} \left(\frac{4}{6 + 4 - 1}\right)} = .8165$$

$$z = \frac{3 - 3}{.8165} = 0.$$

Accordingly, H_0 should be *accepted* in the case of sequence 1.

Sequence 2: $n_H = 5, n_T = 5, R_H = 1$

$$\mu_{R_H} = \frac{5(5 + 1)}{5 + 5} = 3$$

$$\sigma_{R_H} = \sqrt{\frac{5(5 + 1)(5 - 1)}{(5 + 5)^2} \left(\frac{5}{5 + 5 - 1}\right)} = .8165$$

$$z = \frac{1 - 3}{.8165} = -2.45.$$

Accordingly, H_0 should be *rejected* in the case of sequence 2. There are too few runs to make randomness believable.

Sequence 3: $n_H = 5, n_T = 5, R_H = 5$

$$\mu_{R_H} = \frac{5(5 + 1)}{5 + 5} = 3$$

$$\sigma_{R_H} = \sqrt{\frac{5(5 + 1)(5 - 1)}{(5 + 5)^2} \left(\frac{5}{5 + 5 - 1}\right)} = .8165$$

$$z = \frac{5 - 3}{.8165} = 2.45.$$

Accordingly, H_0 should be *rejected* in the case of sequence 3. There are too many runs to make randomness believable.

Note: Analytical Example 11.1, "Examining the 1971 Draft Lottery," vividly demonstrates the usefulness of this test in practical applications. ▮▮▮

THE KRUSKAL-WALLIS TEST

The **Kruskal-Wallis test** is an extension of the Wilcoxon rank-sum test from two to more than two statistical populations. The purpose of the test, named after W. H. Kruskal and W. A. Wallis, remains the same: to determine whether the populations of interest are identical to or different from one another; the Kruskal-Wallis test accomplishes this goal by drawing independent simple random samples from more than two populations and analyzing these samples. As in the W test, the observations contained in the various samples are pooled and ranked, and a rank sum is computed for each original sample. A test statistic, K, is computed on that basis, as given in Box 11.P.

11.P The Kruskal-Wallis Test Statistic

$$K = \left[\frac{12}{n(n+1)} \left(\sum \frac{W_i^2}{n_i} \right) \right] - [3(n+1)]$$

In Box 11.P, n is the number of observations in all samples; W_i is the rank sum of an individual sample; n_i is the number of observations in that sample.

A precise sampling distribution of K can be determined from the knowledge that the rank assigned to any particular observation has an equal chance of being any number between 1 and n, regardless of the sample to which the observation belongs, provided the null hypothesis of identical populations is true. Furthermore, if each of these x samples contains more than 5 observations, the sampling distribution of K can be approximated by a χ^2 distribution with $x - 1$ degrees of freedom. Practice Problem 9 illustrates how the Kruskal-Wallis test is conducted.

Practice Problem 9

Comparing the Effectiveness of Four Advertisements

Consider an advertising agency that has been running one of four distinctly different advertisements, A–D, in each of four cities. It wishes to test, at the 5-percent level of significance, whether these ads are equally effective in stimulating sales. For this purpose, monthly extra sales figures (compared to a year ago) are collected for half a year, as shown in the "Extra Units Sold" columns of Table D.

Table D

Month	Advertisement A Extra Units Sold	Rank of Sales	Advertisement B Extra Units Sold	Rank of Sales	Advertisement C Extra Units Sold	Rank of Sales	Advertisement D Extra Units Sold	Rank of Sales
May	80	18.5	25	4	97	24	78	16
June	66	13	62	10	15	2	75	15
July	66	13	22	3	43	7	85	20
August	80	18.5	87	21	27	5	94	23
September	50	8	63	11	42	6	52	9
October	79	17	66	13	12	1	93	22
	$W_A = 88$		$W_B = 62$		$W_C = 45$		$W_D = 105$	
	$\dfrac{W_A^2}{n_A} = 1{,}290.67$		$\dfrac{W_B^2}{n_B} = 640.67$		$\dfrac{W_C^2}{n_C} = 337.5$		$\dfrac{W_D^2}{n_D} = 1{,}837.5$	

Step 1: Formulating two opposing hypotheses.

H_0: There are no differences in the effectiveness of the ads. (The four populations of extra units sold are identical.)

H_A: There are differences in the effectiveness of the ads.

Step 2: Selecting a test statistic. The agency selects the K statistic (from Box 11.P).

Step 3: Deriving a decision rule. Because there are $x = 4$ samples, we have $x - 1 = 3$ degrees of freedom. According to Appendix Table L, the critical value of $\chi^2_{.05,3} = 7.815$. Thus, the decision rule is: "Accept H_0 if $K \leq 7.815$." The critical value is encircled in Figure G.

Figure G

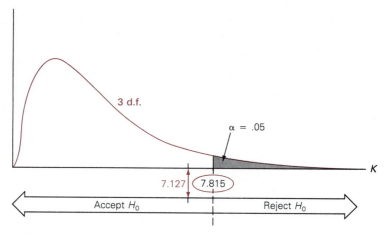

Step 4: Inspecting the samples, computing the test statistic, and confronting it with the decision rule. In the "Rank of Sales" columns of Table D, ranks for the pooled sample data have been inserted, and the separate sample rank sums, W_A to W_D, have been calculated. Each of these values has, in turn, been squared and divided by the corresponding sample size, n_A to n_D, as noted underneath the table. Thus, the observed value of the test statistic (which makes use of the *sum* of these values) is

$$K = \left[\frac{12}{24(24 + 1)}(4,106.34) \right] - [3(24 + 1)] = 7.1268.$$

Accordingly, H_0 should be *accepted.* There are no differences in the effectiveness of the ads. |||

THE KOLMOGOROV-SMIRNOV ONE-SAMPLE TEST

The **Kolmogorov-Smirnov one-sample test** is an alternative to the chi-square test for goodness of fit. It is named after A. N. Kolmogorov (who proposed it in 1933) and N. V. Smirnov (who later tabulated critical values and also extended the test to the two-sample case). As was shown in the previous chapter, the χ^2 test requires that each expected frequency equal at least 5; this condition can often be achieved only by taking large samples (which is costly) or by combining classes with low frequencies (which results in a loss of valuable information). The Kolmogorov-Smirnov test does not impose lower limits on class frequencies and, thus, avoids this difficulty. The test compares a *cumulative* relative frequency distribution derived from sample data with a theoretical one that pertains to a specified type of population from which, the null hypothesis claims, the sample was drawn. The test statistic, *D,* is the (absolute value of the) maximum deviation between the observed cumulative relative frequency distribution (F_o) and the expected cumulative relative frequency distribution (F_e), as noted in Box 11.Q. Depending on the probability that such a deviation would occur if the sample data really came from the specified type of population, the null hypothesis should be accepted or rejected. How the Kolmogorov-Smirnov test works is most easily seen by example.

11.Q The Kolmogorov-Smirnov One-Sample Test Statistic

$$D = max|F_o - F_e|$$

Goodness of Fit to the Binomial Distribution

We can conduct a Kolmogorov-Smirnov test (rather than a χ^2 test) on Chapter 10's Practice Problem 5 (page 418) concerning housing-code violations in order to determine whether the data fit a binomial distribution.

||| Practice Problem 10

Housing-Code Violations Revisited

Step 1: Formulating two opposing hypotheses.

H_0: The number of violations per apartment in the population of all city apartments is binomially distributed with a probability of success in any one trial of $\pi = .3$.

H_A: The number of violations per apartment in the population of all city apartments is not correctly described by H_0.

Step 2: Selecting a test statistic. We select the Kolmogorov-Smirnov test statistic,

$$D = max|F_o - F_e|,$$

where F_o is a cumulative relative frequency actually observed and F_e is a cumulative relative frequency expected, provided the null hypothesis is true.

Step 3: Deriving a decision rule. Given a desired significance level of $\alpha = .05$ (as in the Chapter 10 computation) and a sample size of $n = 200$ observations (from Table F in Chapter 10), we establish a critical value of D_α by consulting Appendix Table M. It suggests that $D_{.05}$ should be calculated as $1.36/\sqrt{n}$; thus, it equals $1.36/\sqrt{200}$, or .09617. Therefore, when H_0 is true, the probability is .05 that an observed value of D equals or exceeds .09617. Accordingly, the decision rule one would want to use in this case must be: "Accept H_0 if $D \leq .09617$."

Step 4: Inspecting the sample, computing the test statistic, and confronting it with the decision rule. The accompanying Table E helps us see how the value of D is computed. The entries in columns 1, 2, and 5 have been taken from Tables F and G of Chapter 10; note how the current test does not require the combining of data in the expected-frequency column 5. The entries in the remaining columns are computed from the original data in a fashion indicated by the column headings: absolute values in columns 2 and 5 are converted into relative values in columns 3 and 6 by means of division by sample size ($n = 200$), and these relative values are then cumulated in columns 4 and 7. After compiling the absolute deviations between the cumulative values, the maximum deviation—and, thus, the size of the test statistic—immediately emerges as $D = .0374$. Accordingly, H_0 should be *accepted* (exactly as in Chapter 10's χ^2 test).

Table E

Number of Possible Violations per Apartment (1)	Observed Frequencies			Expected Frequencies (if H_0 is true)			Absolute Deviations, $|F_o - F_e|$ (8)
	Absolute Values, f_o (2)	Relative Values, $f_o/200$ (3)	Cumulative Relative Values, F_o (4)	Absolute Values, f_e (5)	Relative Values, $f_e/200$ (6)	Cumulative Relative Values, F_e (7)	
0	31	.155	.155	23.52	.1176	.1176	.0374 $= D$
1	51	.255	.410	60.50	.3025	.4201	.0101
2	70	.350	.760	64.82	.3241	.7442	.0158
3	32	.160	.920	37.04	.1852	.9294	.0094
4	9	.045	.965	11.90	.0595	.9889	.0239
5	5	.025	.990	2.04	.0102	.9991	.0091
6	2	.010	1.000	.14	.0007	1.0000	0
	200	1.000		200	1.0000		

Note: Detail may not always add to totals due to rounding.

Note: The Kolmogorov-Smirnov test has one major disadvantage. It does not allow for the estimation of population parameters from sample data; these must be specified in advance of testing. (In the null hypothesis above, $\pi = .3$ must be based on information other than that contained in the sample, such as previous experience.) In contrast, a χ^2 test allows such estimation from a sample, provided the degrees of freedom are appropriately reduced. **|||**

Goodness of Fit to the Poisson Distribution*

Now consider conducting a Kolmogorov-Smirnov test on Chapter 10's Practice Problem 6 (page 420) concerning outpatient arrivals in order to test whether the data fit a Poisson distribution.

---

*Requires prior study of the Chapter 6 section on the Poisson probability distribution.

Practice Problem 11

Outpatient Arrivals Revisited

Step 1 can take the same form as in the earlier χ^2 test; Step 2 now selects the D statistic. In Step 3, we establish a critical value of $D = .21068$ with the help of Appendix Table M, using $\alpha = .01$ (as before) and noting that $n = 50$ (from Table I of Chapter 10). Hence, the decision rule is: "Accept H_0 if $D \leq .21068$." In Step 4, we compute the value of D with the help of the accompanying Table F. The entries in columns 1, 2, and 5 come from Tables I and J of Chapter 10; the remaining ones are computed as in this chapter's Practice Problem 10. Given $D = .1935$, H_0 should be *accepted* (as in Chapter 10's χ^2 test).

Table F

Number of Outpatient Arrivals per Hour (1)	Observed Frequencies			Expected Frequencies (if H_0 is true)			Absolute Deviations, $\lvert F_o - F_e \rvert$ (8)
	Absolute Values, f_o (2)	Relative Values, $f_o/50$ (3)	Cumulative Relative Values, F_o (4)	Absolute Values, f_e (5)	Relative Values, $f_e/50$ (6)	Cumulative Relative Values, F_e (7)	
0	0	0	0	1.12	.0224	.0224	.0224
1	1	.02	.02	4.25	.0850	.1074	.0874
2	5	.10	.12	8.075	.1615	.2689	.1489
3	8	.16	.28	10.23	.2046	.4735	.1935 = D
4	15	.30	.58	9.72	.1944	.6679	.0879
5	9	.18	.76	7.385	.1477	.8156	.0556
6	7	.14	.90	4.68	.0936	.9092	.0092
7	3	.06	.96	2.54	.0508	.9600	0
8 or more	2	.04	1.00	2.00	.0400	1.0000	0
	50	1.00		50	1.0000		

Goodness of Fit to the Normal Distribution

Finally, consider making a Kolmogorov-Smirnov test on Chapter 10's Practice Problem 7 (page 422) about trading in the futures market in order to test whether the data fit a normal distribution.

Practice Problem 12

Trading in the Futures Market Revisited

Step 1 remains unchanged; Step 2 selects the D statistic. In Step 3, we establish a critical value of $D = .14117$ with the help of Appendix Table M, using $\alpha = .025$ (the closest available value to the earlier .02) and noting that $n = 90$ (from Table L of Chapter 10).

Hence, the decision rule is: "Accept H_0 if $D \leq .14117$." In Step 4, we compute the value of D with the help of the accompanying Table G. The entries in columns 1, 2, and 5 come from Table L and M of Chapter 10; the remaining ones are computed as in Practice Problems 10 and 11. Given $D = .4192$, H_0 should be *rejected* (again, as in Chapter 10's χ^2 test).

Table G

Number of Futures Contracts Traded per Day (millions) (1)	Observed Frequencies			Expected Frequencies (if H_0 is true)			Absolute Deviations, $\lvert F_o - F_e \rvert$ (8)
	Absolute Values, f_o (2)	Relative Values, $f_o/90$ (3)	Cumulative Relative Values, F_o (4)	Absolute Values, f_e (5)	Relative Values, $f_e/90$ (6)	Cumulative Relative Values, F_e (7)	
under 10	5	.0556	.0556	0	0	0	.0556
10–under 20	9	.1000	.1556	.126	.0014	.0014	.1542
20–under 30	15	.1667	.3223	1.926	.0214	.0228	.2995
30–under 40	23	.2556	.5779	12.231	.1359	.1587	.4192 = D
40–under 50	20	.2222	.8001	30.717	.3413	.5000	.3001
50–under 60	8	.0889	.8890	30.717	.3413	.8413	.0477
60–under 70	6	.0667	.9557	12.231	.1359	.9772	.0215
70–under 80	3	.0333	.9890	1.926	.0214	.9986	.0096
80 and above	1	.0111	1.0000	.126	.0014	1.0000	0
	90	1.0000		90	1.0000		

Note: The bad fit of the observed distribution (F_o values) to the expected one (F_e values) in Table G is made visually obvious in Figure 11.1, which also shows the value of the test statistic graphically. ▮

SPEARMAN'S RANK-CORRELATION TEST

In this last section, we consider a popular nonparametric test introduced by C. E. Spearman (see Biography 11.2 at the end of the chapter). The **Spearman rank-correlation test** measures the degree of association between two variables for which only rank-order data are available.) (In Chapter 13, we will turn to other correlation measures that focus on the association between two variables for which quantitative data are available.) The test statistic produced by this test is called the **Spearman rank-correlation coefficient** and is symbolized by ρ (the lower-case Greek *rho,* pronounced "rō").

If one defines x_i as the rank of an individual sample element with respect to one variable, y_i as the rank of that element with respect to another variable, d_i as the difference between these rankings (such that $d_i = x_i - y_i$), and n as the total number of elements being ranked in these two ways, Spearman's measure is defined as in Box 11.R.

Figure 11.1 The Kolmogorov-Smirnov Maximum Deviation Test for Goodness of Fit

This graph is based on columns 4 and 7 of Table G and, thus, provides a visual picture of what the Kolmogorov-Smirnov test accomplishes. In this case (see Practice Problem 12), it determines that the maximum deviation, *D*, between the observed and expected distributions is too large to be attributed to sampling error alone. Thus, the null hypothesis of Practice Problem 12 (that the sample has come from a population with the expected distribution shown here) should be rejected.

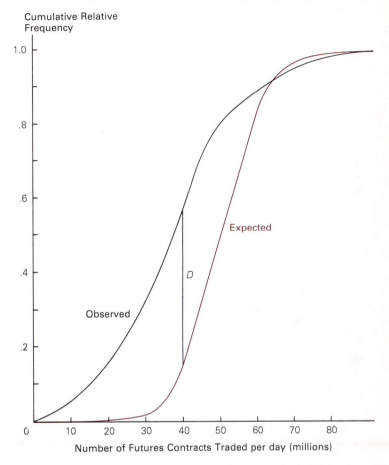

11.R Spearman's Rank-Correlation Coefficient

$$\rho = 1 - \frac{6\Sigma d_i^2}{n(n^2 - 1)}$$

This statistic can take on values between -1 and $+1$. Negative values near -1 indicate that between the two variables of interest a monotonically decreasing relationship exists (such that decreases in one variable are associated with decreases in the other). Positive values near $+1$ indicate that a monotonically increasing relationship exists between the two variables (such that increases in one are associated with increases in the other).

The Sampling Distribution of ρ

Whenever $n \geq 10$, and if the two variables of interest are not correlated (if their rankings are independent of one another), the sampling distribution of ρ can be approximated by the normal distribution with the summary measures given in Box 11.S. Accordingly, the normal deviate of ρ can be calculated as in Box 11.T.

11.S Expected Value and Standard Deviation of the Sampling Distribution of ρ (given H_0: There exists no rank correlation.)

$$\mu_\rho = 0$$

$$\sigma_\rho = \sqrt{\frac{1}{n-1}}$$

11.T Normal Deviate for Spearman's Rank-Correlation Coefficient

$$z = \frac{\rho - \mu_\rho}{\sigma_\rho} = \frac{\rho}{\sigma_\rho}$$

Assumption: $n \geq 10$

Application

Consider a test to determine, at the 5-percent level of significance, whether there exists a significant rank correlation between the speed and price of airplanes. Such a test would proceed with Steps 1 to 4 as in Practice Problem 13.

Practice Problem 13

Speed vs. Price of Airplanes

Step 1: Formulating two opposing hypotheses.

H_0: The speed and price rankings are not correlated.
H_A: The speed and price rankings are correlated.

Step 2: Selecting a test statistic. We select the normal deviate for Spearman's rank-correlation coefficient (from Box 11.T).

Step 3: Deriving a decision rule. Given a significance level of $\alpha = .05$ and this being a two-tailed test (either a negative or a positive correlation might exist), we find in Appendix Table J critical normal deviate values of $\pm z_{\alpha/2} = \pm 1.96$. Thus, the decision rule is: "Accept H_0 if $-1.96 \leq z \leq +1.96$." The critical values are encircled in Figure H.

Figure H

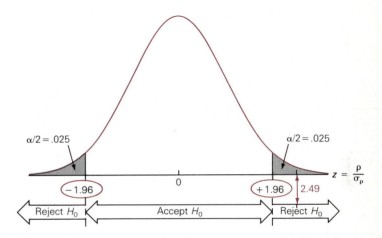

$\alpha/2 = .025$

$\alpha/2 = .025$

-1.96

0

$+1.96$ $\quad$ 2.49

$$z = \frac{\rho}{\sigma_\rho}$$

Reject H_0 $\qquad$ Accept H_0 $\qquad$ Reject H_0

Step 4: Selecting the sample, computing the test statistic, and confronting it with the decision rule. Suppose our investigation proceeds by taking a sample of airplane types, which reveals the rank data given in columns 1–3 of Table H.

Table H

Airplane Model (1)	Ranking According to Speed, x_i (2)	Ranking According to Price, y_i (3)	Rank Difference, $d_i = x_i - y_i$ (4)	d_i^2 (5)
A	1	2	-1	1
B	2	1	$+1$	1
C	3	5	-2	4
D	4	4	0	0
E	5	3	$+2$	4
F	6	9	-3	9
G	7	7	0	0
H	8	8	0	0
I	9	6	$+3$	9
J	10	10	0	0
				$\Sigma d_i^2 = 28$

$$\rho = 1 - \frac{6(28)}{10(10^2 - 1)} = +.83$$

Then the value of ρ can be calculated at +.83, as shown in the remainder of the table. Furthermore, $n = 10$; hence,

$$\sigma_\rho = \sqrt{\frac{1}{10 - 1}} = .\overline{33}.$$

Accordingly,

$$z = \frac{.83}{.\overline{33}} = 2.49,$$

as indicated by the colored arrow in Figure H. Thus, H_0 should be *rejected*; there exists a significant degree of (positive) rank correlation between speed and price. Analytical Example 11.2, "Searching for Leviathan," provides another illustration of the use of this test. ▐▐▐

11.1 Examining the 1971 Draft Lottery

Close-Up 2.1, "The 1970 Draft Lottery Fiasco," described how that year's determination of draft-priority birth dates was anything but random. Was the 1971 lottery any better? We can apply the *number-of-runs test* to the lottery results to see for ourselves. The result of the 1971 lottery is given in the table on the next page.

If the lottery was truly random, one would expect each month to contain an equal number of draft-priority numbers that are "dangerously low" and "safely high." Because half the number of days in 1971 is 182.5, we might classify numbers between 1 and 182 as "dangerous" (more likely to lead to the draft) and those between 183 and 365 as "safe" (less likely to lead to the draft). The "dangerous" numbers have been identified in the table by colored dots. Let us apply the number-of-runs test to answer the question posed in the previous paragraph.

Step 1:

H_0: "Dangerous" and "safe" numbers are randomly distributed over the dates of the year.
H_A: The randomness asserted in H_0 does not exist.

Step 2: We select the normal deviate for the number-of-runs test (from Box 11.O).

Step 3: We select a significance level of $\alpha = .05$; hence, the decision rule is: "Accept H_0 if $-1.96 \le z \le +1.96$."

Step 4: We inspect our data, and, calling "dangerous" numbers "heads" and "safe" numbers "tails," we find: $n_H = 182$, $n_T = 183$, $R_H = 94$. (The latter number is found by counting the runs of colored dots, going down successive columns of the table). Thus, according to Box 11.N,

$$\mu_{R_H} = \frac{182(183 + 1)}{365} = 91.75;$$

$$\sigma_{R_H} = \sqrt{\frac{182(183 + 1)(182 - 1)}{(182 + 183)^2}\left(\frac{183}{182 + 183 - 1}\right)}$$
$$= 4.7826.$$

Therefore, according to Box 11.O,

$$z = \frac{94 - 91.75}{4.7826} = .47.$$

Birthday	1971 Draft Priority Numbers											
	Jan.	Feb.	Mar.	Apr.	May	Jun.	Jul.	Aug.	Sep.	Oct.	Nov.	Dec.
1	133•	335	14•	224	179•	65•	104•	326	283	306	243	347
2	195	354	77•	216	96•	304	322	102•	161•	191	205	321
3	336	186	207	297	171•	135•	30•	279	183	134•	294	110•
4	99•	94•	117•	37•	240	42•	59•	300	231	266	39•	305
5	33•	97•	299	124•	301	233	287	64•	295	166•	286	27•
6	285	16•	296	312	268	153•	164•	251	21•	78•	245	198
7	159•	25•	141•	142•	29•	169•	365	263	265	131•	72•	162•
8	116•	127•	79•	267	105•	7•	106•	49•	108•	45•	119•	323
9	53•	187	278	223	357	352	1•	125•	313	302	176•	114•
10	101•	46•	150•	165•	146•	76•	158•	359	130•	160•	63•	204
11	144•	227	317	178•	293	355	174•	230	288	84•	123•	73•
12	152•	262	24•	89•	210	51•	257	320	314	70•	255	19•
13	330	13•	241	143•	353	342	349	58•	238	92•	272	151•
14	71•	260	12•	202	40•	363	156•	103•	247	115•	11•	348
15	75•	201	157•	182•	344	276	273	270	291	310	362	87•
16	136•	334	258	31•	175•	229	284	329	139•	34•	197	41•
17	54•	345	220	264	212	289	341	343	200	290	6•	315
18	185	337	319	138•	180•	214	90•	109•	333	340	280	208
19	188	331	189	62•	155•	163•	316	83•	228	74•	252	249
20	211	20•	170•	118•	242	43•	120•	69•	261	196	98•	218
21	129•	213	246	8•	225	113•	356	50•	68•	5•	35•	181•
22	132•	271	269	256	199	307	282	250	88•	36•	253	194
23	48•	351	281	292	222	44•	172•	10•	206	339	193	219
24	177•	226	203	244	22•	236	360	274	237	149•	81•	2•
25	57•	325	298	328	26•	327	3•	364	107•	17•	23•	361
26	140•	86•	121•	137•	148•	308	47•	91•	93•	184	52•	80•
27	173•	66•	254	235	122•	55•	85•	232	338	318	168•	239
28	346	234	95•	82•	9•	215	190	248	309	28•	324	128•
29	277		147•	111•	61•	154•	4•	32•	303	259	100•	145•
30	112•		56•	358	209	217	15•	167•	18•	332	67•	192
31	60•		38•		350		221	275		311		126•

We should *accept* H_0. The 1971 draft-lottery results were random.

Note: Problem 57 in the "Questions and Problems" section at the end of this chapter (the answer to which appears near the end of this book) calls for a rank-correlation test of the 1970 draft lottery.

Source: The New York Times, July 2, 1970, p. 12.

11.2 Searching for Leviathan

Two economists, Geoffrey Brennan and James Buchanan, have put forth a striking and controversial view of the economy's public sector. As they see it, just as monopoly in the private sector tends to harm people by decreasing output and raising prices (compared to competitive conditions), so a monolithic government (Leviathan) tends to harm its citizenry through the maximization of the tax revenues that it extracts from the economy. Yet Leviathan can be constrained! The authors predict that "total government intrusion into the economy should be smaller, *ceteris paribus,* the greater the extent to which taxes and expenditures are decentralized."[1]

One economist, Wallace E. Oates, recently set out to test the Leviathan thesis, one implication of which is that the size of the economy's public sector should vary inversely with the extent of fiscal decentralization. (If making more fiscal decisions locally limits the capacity of government to tax and spend, less tax revenue and less government spending should go hand in hand with more decentralization.) However, in order not to prejudge the issue of whether decentralization constrains or expands the public sector, a two-tailed test was designed. Its null hypothesis was

H_0: The size of the public sector and the extent of government decentralization are not correlated.

Three different sets of data were explored, a U.S. sample and two country samples.

The U.S. sample. Data were assembled for the 48 contiguous states: The size of each state's state-local public sector, G, was measured by state-local tax receipts as a fraction of state personal income. The degree of government decentralization within each state was then measured in three alternative ways: first, as R, the state government's share in state-local government revenue; second, as E, the state government's share in state-local government expenditures; third, as L, the absolute number of local government units in the state. If the Leviathan thesis were true, one would expect a positive correlation between G and R and between G and E, but a negative correlation between G and L. One would expect H_0 to be rejected in all three cases. Table A shows the test results. At a significance level of $\alpha = .05$, H_0 was *accepted.*

Table A: Sample of 48 U.S. States

Variables	Spearman's rank correlation coefficient, ρ	z value
G vs. R	−.22	−1.50
G vs. E	−.25	−1.73
G vs. L	−.06	−.41

[1]Geoffrey Brennan and James Buchanan, *The Power to Tax: Analytical Foundations of a Fiscal Constitution* (Cambridge: Cambridge University Press, 1980), p. 185.

Sample of 18 industrialized countries. Data were available for 18 industrialized countries. The size of each county's public sector, G, was now measured by total government tax receipts as a fraction of gross domestic product. The degree of government decentralization within each country was measured, first, as R, the central government's share of all government revenues; second, as E, the central government's share of all government expenditures. Once, again, if the Leviathan thesis were true, one would expect a positive correlation between both G and R and G and E and a rejection of H_0. Table B shows the test results; H_0 was *accepted.*

Table B: Sample of 18 Industrialized Countries

Variables	Spearman's rank correlation coefficient, ρ	z value
G vs. R	−.02	−.08
G vs. E	−.15	−.61

Sample of 25 developing countries. Data were also available for 25 developing countries. Using the same variables as for the industrialized countries, the test results are given in Table C; once again H_0 was *accepted.*

Table C: Sample of 25 Developing Countries

Variables	Spearman's rank correlation coefficient, ρ	z value
G vs. R	.20	.98
G vs. E	.12	.58

Because Spearman's rank correlation test is a crude test that fails to hold constant the possible influence of other variables on the size of the public sector, further tests were performed (see problem 59b, Chapter 14). The conclusion, however, remained the same: "Perhaps, after all, Leviathan is a mythical beast."

Source: Wallace E. Oates, "Searching for Leviathan: An Empirical Study," *The American Economic Review,* September 1985, pp. 748–57.

BIOGRAPHIES

11.1 Frank Wilcoxon

Frank Wilcoxon (1892–1965) was born in Glengarriffe Castle, near Cork, Ireland, to wealthy American parents. He was soon brought to the United States where he attended Pennsylvania Military College, Rutgers, and Cornell. After receiving his doctorate as a physical chemist, Wilcoxon joined the Boyce Thompson Institute for Plant Research and began to study the use of copper compounds as fungicides. While doing so, he became part of a group, along with W. J. Youden, that studied the newly published *Statistical Methods for Research Workers* by Ronald A. Fisher (Biography 9.1). In the process (and either in spite or because of his lifelong preoccupation with biochemistry, plant pathology, and entomology), he became a significant member of that small group of twentieth-century pioneers who developed new statistical methodology. In a now famous 1945 paper, he presented the rank-sum test and the signed-rank test now named after him. The basic idea of replacing actual sample data by their ranks, which seems so utterly simple in retrospect, proved to be inspirational to the further development of the entire field of nonparametric statistics. The elegant simplicity of these tests led to their widespread adoption; the fact that Wilcoxon was not even aware (in 1945) of all the advantages of his new methods does not dim the luster of his contribution. (These advantages—not all of them discussed in this chapter—include the ease and rapidity of calculation, the availability of exact significance levels without the restrictive normality assumption, the relative insensitivity to outlying sample observations, the invariance under certain monotonic transformations of the data, the applicability to situations where the data are ordinal, the excellent power properties for wide classes of alternative distributions, and the availability of distribution-free confidence intervals for the location parameters of interest.) In addition, Wilcoxon contributed mightily to other aspects of statistics, in particular biological assay methods and sequential analysis (discussed in Chapter 17 of this book).

Source: International Encyclopedia of Statistics, vol. 2 (New York: The Free Press, 1978), pp. 1245–50.

11.2 Charles E. Spearman

Charles Edward Spearman (1863–1945) was born in London, to parents of some eminence. He became an officer in the regular army because, he said, this offered him more leisure and freedom to study than did other professions. He served as an officer in the Burmese and Boer War, then resigned to study psychology. Rather late in life, he took a position as a professor of psychology at University College, London.

Above all, he is now known for two major contributions: (1) his work on factor analysis and (2) the development of a rational basis for determining the concept of general intelligence and for validating intelligence testing. These interests also forced him to study statistical methods and led to the development of the rank-correlation coefficient with which his name is now associated eponymously.

In a nutshell, Spearman's theory of intelligence states that any cognitive performance is a function of two factors: a general ability common to most cognitive performances and an ability specific to a given test. He showed that it was possible to determine this general factor objectively. This possibility gave to intelligence testing a theoretical basis that was missing from the empirical approaches of men like Binet and Wechsler whose work ultimately led, as Spearman foresaw, to the cynical view that "intelligence is what intelligence tests measure."

Spearman's contributions have been further developed and have led to multiple factor analysis and multivariate experimental design (assessing the simultaneous effects of many variables); indeed, they have revolutionized many fields besides psychology.

Source: International Encyclopedia of Statistics, vol. 2 (New York: The Free Press, 1978), pp. 1036–39.

Summary

1. Whenever a hypothesis test is based on certain specific assumptions about the probability distribution of population values or about the sizes of population parameters, it is referred to as a *parametric test*. In contrast, any test that makes no such assumptions is called a *nonparametric test*. When nothing is known about the statistical populations from which samples are being drawn, nonparametric tests alone can be employed.
2. Innumerable nonparametric tests exist; this chapter focuses on eight of the more popular ones.
   a. The Wilcoxon rank-sum test (or *W* test) is based on two independent simple random samples and is designed to determine whether two statistical populations are identical to or different from one another.
   b. The Mann-Whitney test (or *U* test) is equivalent to the Wilcoxon rank-sum test.
   c. The sign test is based on a matched-pairs sample; otherwise, it seeks to make the same determination as the *W* test and the *U* test, but it can also be employed to test whether a sample comes from a population with a specified median.
   d. The Wilcoxon signed-rank test is similar to the sign test but uses more of the information provided in the sample.
   e. The number-of-runs test determines whether the elements of a sequence appear in random order.
   f. The Kruskal-Wallis test extends the *W* test from two to more than two populations.
   g. The Kolmogorov-Smirnov one-sample test is an alternative to the χ^2 test for goodness of fit.
   h. Spearman's rank-correlation test measures the degree of association between two variables for which only rank-order data are available.

Key Terms

distribution-free tests
Kolmogorov-Smirnov one-sample test
Kruskal-Wallis test
Mann-Whitney test (*U* test)
nonparametric statistics
nonparametric tests
number-of-runs test
parametric statistic

parametric test
robustness
run
sign test
Spearman rank-correlation coefficient (ρ)
Spearman rank-correlation test
Wilcoxon rank-sum test (*W* test)
Wilcoxon signed-rank test

Questions and Problems

The computer program noted at the end of this chapter can be used to work many of the subsequent problems.

1. Given the information in Table 11.A, which was gathered in two random samples of recent engineering graduates, test at the 10-percent significance level whether starting salaries are the same for the graduates of the two institutions. Show your calculations.
2. Given the sample information in Table 11.B, a pharmaceutical researcher wants to test, at the 5-percent significance level, whether drug A really provides at least as much relief as drug B. Make the test, showing your calculations.
3. A personnel manager wants to test the claim, at the 1-percent significance level, that output per worker under an hourly-pay plan (A) is at best equal to that under a piece-work plan (B). A four-week experiment provides the information shown in Table 11.C. Make the test, showing your calculations.

Table 11.A

Student	University A Monthly Starting Salary (dollars)	Student	University B Monthly Starting Salary (dollars)
Abrams	1,200	Adler	1,390
Bak	1,500	Bacon	1,750
Cotter	2,000	Conrad	1,250
Derin	1,850	Heber	2,350
Ford	1,900	Kott	2,100
McRae	2,200	Noke	2,050
Topa	2,300	Ryan	1,710
Torok	1,450	Trane	1,490
Upton	1,600	Vogel	1,410
Weber	1,800	Weise	2,200

Table 11.B

Hours of Relief Experienced by 25 Patients

Drug A	Drug B
3.0	3.8
5.9	8.3
4.7	7.9
3.3	6.5
3.8	4.5
4.2	6.1
4.5	5.9
6.1	7.6
8.0	4.2
7.1	3.3
	8.1
	8.2
	7.0
	5.0
	6.9

Table 11.C

Daily Output per Worker (Units)

Under Plan A		Under Plan B	
552	492	751	517
274	631	252	707
241	602	917	908
547	510	981	937
651	278	492	199

4. Given the sample information in Table 11.D, test at the .1-percent level of significance whether tenants stay about the same length of time in an apartment complex regardless of whether they were screened by realtor A or B.

Table 11.D

Length of Tenancy (months)

Realtor A			Realtor B	
5	15	6	22	31
13	27	17	27	30
36	31	34	26	5
8	49	19	28	6
12	3		24	8

5. Given the sample information in Table 11.E, test at the .5-percent level of significance whether model A cars really last as long as or longer than model B cars, as has been alleged.

Table 11.E

Odometer Reading at Scrappage (thousands of miles)

Model A			Model B		
86.1	99.3	77.9	100.3	66.3	99.9
92.7	100.7	98.9	110.3	75.7	95.6
80.3	110.5	99.9	125.7	87.9	98.7
85.7	77.8		111.6	92.0	
90.6	62.0		85.4	87.9	

6. Given the sample information in Table 11.F, test at the 1-percent level of significance whether Japanese labor turnover rates are really at most equal to, but probably lower than U.S. rates.

Table 11.F

Electronics Industry Labor Turnover (percent per year)

Japanese Firms			U.S. Firms		
1.9	2.1	3.1	3.1	2.1	2.2
0.3	2.3	0.5	3.6	3.6	2.3
1.2	1.8	0.7	4.1	2.9	
1.4	1.7	0.9	5.7	3.3	
1.6	3.0		1.9	4.6	

7. Given the sample information in Table 11.G, test at the 2.5-percent level of significance whether house assessments in the two neighborhoods are about the same.

Table 11.G

Assessed Value of Houses (thousands of dollars)			
Neighborhood A		Neighborhood B	
60.7	120.7	150.8	88.7
63.7	130.9	145.2	77.5
75.6	88.2	167.5	
82.9		178.3	
88.7		52.3	
77.3		66.8	
92.7		99.3	
89.7		77.5	
52.9		103.0	
99.5		43.6	

8. Given the sample information in Table 11.H, test at the 5-percent level of significance whether scores on an identical exam are really as high or higher under scheme A as under scheme B.

Table 11.H

Exam Scores			
A (Questions Ordered from Easy to Hard)		B (Questions Ordered from Hard to Easy)	
50	62	55	95
63	89	61	73
98	95	68	78
75	79	72	82
84	89	75	89
92		85	
95		86	
88		88	
74		79	
73		90	

9. Reconsider problem 3, along with its solution at the back of the book. Perform a U test with the same data.

10. Given the sample information in Table 11.I, test at the 10-percent level of significance whether now-bankrupt firms lived about equally long in the two industries. Make a U test.

Table 11.I

Lifetimes of Now-Bankrupt Firms (years)				
Industry A			Industry B	
2.3	16.5	2.5	30.6	5.8
3.7	9.2	3.1	25.5	0.6
10.6	0.6	1.0	3.3	0.2
8.9	0.9		7.0	0.3
27.0	1.3		6.7	8.9

11. Given the sample information in Table 11.J, test at the 5-percent level of significance whether the test scores on the civil-service exam are really as high or higher for high-school dropouts as for high-school graduates. Make a U test.

Table 11.J

Test Scores on Civil-Service Exam					
High-school Dropouts			High-school Graduates		
66	72	91	88	89	69
82	73		83	91	71
93	77		84	62	75
50	69		85	69	75
50	85		86	75	

12. Given the sample information in Table 11.K, test at the 2.5-percent level of significance whether the daily sales at store A are really at most equal to, but probably lower than at store B. Make a U test.

Table 11.K

Daily Sales at Two Computer Stores (in millions of bytes)			
Store A		Store B	
64	704	128	320
192	192	256	640
384	0	192	896
576	384	0	1,088
896	128	64	0

13. Given the sample information in Table 11.L, test at the 1-percent level of significance whether planes purchase about equal amounts of fuel at the two airports. Make a U test.

Table 11.L

Airplane Fuel Purchases (in gallons)

City Airport		Suburban Airport		
25	348	14	75	89
49	17	420	99	97
191	5	39	187	103
15	29	56	231	121
297	99	65	123	

14. Given the sample information in Table 11.M, test at the 5-percent level of significance whether the weekly number of customer complaints is at least as high under the new manager as under the old one. Make a *U* test.

Table 11.M

Number of Customer Complaints per Week

New Manager			Old Manager		
2	3	7	12	17	15
1	2	1	15	6	14
7	2	4	19	0	17
0	5	5	21	9	12
12	6	9	41	12	1

15. Given the sample information in Table 11.N, test at the 5-percent level of significance whether the career earnings of baseball players are at most equal to those of golfers. Make a *U* test.

Table 11.N

Career Earnings (in hundreds of thousands of dollars)

Baseball Players			Golfers		
942	742	4,123	462	652	6,652
768	651		593	1,980	192
6,592	5,600		982	2,520	357
3,091	3,092		1,125	890	
13,100	16,500		229	2,541	

16. Problems 1 through 8 were solved by means of a *W* test. In each case, compute the value of U, μ_U, σ_U, and z as one would in preparation for a *U* test.

17. Problems 10 through 15 were solved by means of a *U* test. In each case, compute the values of W, μ_W, σ_W, and z as one would in preparation for a *W* test.

18. An advertising agency wants to know whether people in general are indifferent between two detergents, A and B. A simple random sample of 30 households is selected; each uses A or B for one month, then the other brand for another month. In the end, 19 households prefer A (+); 2 cannot decide; the rest prefer B (−). Make an appropriate hypothesis test at the 5-percent level of significance.

19. The Federal Aviation Administration matches (into 30 pairs) similar individuals who are about to take its air-traffic-controller course. One member of each pair is subjected to a new curriculum (A); the other member is subjected to an old one (B). Then both take the same test, and the difference in scores is recorded. It is hypothesized that the new curriculum will produce equal or better scores. In fact, there are 20 better scores (+) and no ties. Make an appropriate hypothesis test at the 5-percent level of significance.

20. A plant manager believes that there are in general fewer accidents during the day shift than during the night shift. The difference in the number of accidents is recorded for 100 days, more accidents during the day than night are recorded as (+); fewer during the day are recorded as (−). There are 8 ties and 50 pluses. Make an appropriate hypothesis test at the 5-percent level of significance.

21. The manager of a chain of department stores wants to know whether sales of children's clothing are promoted more effectively by making periodic loudspeaker announcements or by visual advertising throughout the stores. A simple random sample of 12 stores is selected. In each, sales are recorded during one week of loudspeaker announcements (A), then during a second week of visual displays (B). In the end, sales were the same in one store, higher (+) in 7 stores under A, lower (−) in 4 stores under A. Make an appropriate hypothesis test at the 2.5-percent level of significance.

22. A marketing manager wants to know whether people truly prefer an identical candy bar when wrapped in colorful paper rather than plain paper, as has been alleged. A random sample of 30 individuals is asked to taste this bar after being presented in colorful paper (A) and again after being presented in the plain wrapper (B). While 2 persons cannot decide, 22 individuals prefer A (+), 6 individuals prefer B (−). Make an appropriate hypothesis test at the 5-percent level of significance.

23. The manager of a multiple-listing service believes that realtor A is slower to sell houses than realtor B. Accordingly, the next 100 listings are given to both simultaneously; the sale by A obviously implies that B would have taken longer to sell the house; the sale by B implies the same about A. In the end, there are 82 sales by A (+) and 16 sales by B (−). Make an appropriate hypothesis test at the 5-percent level of significance.

24. At a time when the median price of a single-family home was alleged to be $80,000 nationally, a government agency investigated the median prices in the 100 largest cities. Of these, 30 exceeded the above figure (+), and 70 were below it (−). Did the 100 largest cities come from a population with an $80,000 median price? Make an appropriate hypothesis test at the 5-percent level of significance.

25. At a time when the median life of an aircraft spark-plug was being advertised as 800 service hours, the FAA checked the logbooks of 50 types of planes using this type of plug. For 12 airplane types, the figure was exceeded (+), but the median service hours were lower (−) for the other 38 airplane types. Was the advertising claim correct? Make an appropriate hypothesis test at the 5-percent level of significance.

26. An advertising agency wants to know whether a national ad campaign has been successful in boosting sales. A simple random sample of 10 urban areas reveals the information in Table 11.O at the bottom of the page (and it is assumed that observed differences are not attributable to anything but the ad campaign). Make an appropriate hypothesis test at the 1-percent level of significance, showing your calculations.

27. The Food and Drug Administration wants to test whether a traditional preservative produces a shorter shelf life for meats than a newly developed one. A random sample of meats from 12 suppliers reveals the information in Table 11.P. Make an appropriate hypothesis test at the 1-percent level of significance, showing your calculations.

Table 11.P

Supplier	Shelf Life of Meats (days)	
	With Old Preservative	With New Preservative
A	4.2	4.8
B	6.1	6.3
C	5.3	5.0
D	3.9	4.5
E	4.5	4.2
F	6.0	5.5
G	5.8	5.9
H	4.9	4.9
I	5.3	5.0
J	5.5	5.5
K	5.2	5.5
L	4.8	4.0

28. A marketing manager wants to test whether the sales of a particular product differ depending on whether it is displayed at eye-level or above and below eye-level on supermarket shelves. A random sample of 20 stores is taken; they are grouped by two's according to similarity of weekly sales, and the data of Table 11.Q are collected. Make an appropriate hypothesis test at the 5-percent level of significance.

Table 11.Q

Store Pair	A Week's Sales (dollars)	
	Product at Eye-Level	Product Not at Eye-Level
A	$ 52	$ 17
B	79	29
C	26	28
D	197	120
E	220	190
F	92	120
G	105	85
H	69	70
I	97	70
J	53	23

Table 11.O

City	Monthly Sales (thousands of dollars)	
	Before Campaign	After Campaign
Atlanta	100	110
Boston	120	130
Cincinnati	27	25
Detroit	35	30
Houston	51	70
Los Angeles	75	82
Sacramento	12	19
Seattle	80	75
Toledo	60	58
Washington	34	64

29. An economist wants to test whether the salaries of economists differ in academic and nonacademic employment. A random sample of 30 economists is taken; they are grouped into 15 pairs in accordance with years of experience and type of employment. The resultant data are given in Table 11.R. Make an appropriate hypothesis test at the 2.5-percent level of significance.

Table 11.R

Economist Pair	Annual Salary (thousands of dollars)	
	Academic Employment	Nonacademic Employment
A	$52.7	$81.3
B	33.8	75.0
C	21.5	32.9
D	25.7	38.6
E	32.1	42.6
F	40.0	41.2
G	45.0	38.6
H	35.4	38.9
I	42.6	52.1
J	31.9	42.7
K	50.0	48.0
L	56.0	62.7
M	52.0	66.6
N	39.0	75.0
O	27.8	42.0

30. An economist wants to test whether antitrust enforcement has become more or less vigorous during the 1980s. A sample of 10 firms is selected from *Fortune's 500* list; the data of Table 11.S are collected. Make an appropriate hypothesis test at the 5-percent level of significance.

Table 11.S

Firm	Number of Antitrust Indictments	
	1975–1979	1980–1984
A	5	7
B	6	8
C	2	3
D	7	9
E	10	9
F	15	20
G	9	18
H	12	7
I	8	12
J	3	7

31. The Environmental Protection Agency wants to test, at the 5-percent significance level, whether the acidity level in a lake varies randomly over time. The acidity is measured on 60 days, its average level is computed, and each day's measurement is then designated to be above average (A) or below average (B) with the results shown in Table 11.T. Make an appropriate hypothesis test.

Table 11.T

Day	Acidity	Day	Acidity	Day	Acidity	Day	Acidity
1	B	16	B	31	B	46	A
2	B	17	B	32	B	47	A
3	B	18	B	33	B	48	A
4	A	19	B	34	B	49	A
5	A	20	B	35	B	50	A
6	A	21	B	36	B	51	B
7	A	22	A	37	B	52	B
8	A	23	A	38	B	53	A
9	A	24	A	39	B	54	A
10	A	25	A	40	A	55	A
11	A	26	A	41	A	56	B
12	A	27	A	42	A	57	B
13	B	28	A	43	A	58	A
14	B	29	A	44	A	59	A
15	B	30	A	45	A	60	A

32. A stock-market analyst wants to test the *random-walk hypothesis,* according to which stock prices move at random over time with no discernable pattern whatsoever. A test at the 1-percent significance level is to be conducted, based on the information given in Table 11.U about a given stock. *Hint:* First, calculate the average closing price and classify each day according to whether its price deviated positively (P) or negatively (N) from that average. Then conduct an appropriate hypothesis test.

33. A medical-school admissions board stands accused of manipulating admissions on a daily basis according to a secret quota system based on sex. The board denies the charge and claims that sex is determined randomly for each succeeding admission because sex is never even considered during the admission process. The information given in Table 11.V is acquired by investigators for 60 successive admissions; a test at the 2.5-percent level of significance is desired (M = male, F = female, and data are to be read in successive rows). Make an appropriate hypothesis test. Also ask yourself this: Would your answer be different if a 5-percent significance level had been specified?

34. Corporate managers who consider the price-per-share of their firm's stock too high (in that it discourages trading) might engineer a *stock split,* issuing, say, 4 new shares for each old share. Then a price per share of $300 will drop to $75. Some analysts have argued, however, that there always is a reason for the high price to begin with (such as a recognized growth potential of the firm) and that this reason is bound to reassert itself after the split, driving the new and lower price up to ever-higher levels, and making such stock a "sure winner." Test this theory at the 5-percent level of significance, using the following data. They show how the price of a stock moved on the 50 days following a split: + + + + − − − + − + − − − + + + + + − − − − − − − − − + + + + − + − + + + − − − + + − − − − + + + +

35. An ice-hockey coach wants to test, at the 5-percent level of significance, whether his team's wins and losses occur at random. Make the test, using the following data on the last 30 games: *W W W W W L L L L L L W W W W W L L L W W L L W W L L W W L* .

Table 11.U

Day	Closing Price (dollars)	Day	Closing Price (dollars)	Day	Closing Price (dollars)
1	41	9	23	17	23
2	43	10	39	18	42
3	37	11	61	19	57
4	51	12	48	20	51
5	63	13	48	21	47
6	39	14	60	22	41
7	43	15	33	23	39
8	58	16	39	24	50

Table 11.V

Order of Admission	Sex of Admission									
1–10	M	M	M	M	M	F	F	M	M	M
11–20	M	M	F	F	F	M	M	M	M	M
21–30	F	F	M	F	M	M	M	M	M	F
31–40	F	F	M	M	M	M	M	M	M	F
41–50	F	F	M	F	M	M	M	M	M	M
51–60	M	M	F	F	M	M	F	F	M	M

36. A production engineer is closely watching a machine that is supposed to fill a box with 20 ounces of cereal. She has weighed the contents of the last 40 boxes filled and recorded underfilling *(U)* and overfilling *(O)* in the order in which they occurred. At the 2.5-percent level of significance, make an appropriate test to determine whether underfilling and overfilling occurs at random, using these data: *U U O O U U O O U U O O U U O O U U O O U U O O U U O O U U O O O O O O U U U U* .

37. Using an appropriate hypothesis test, determine, at the 10-percent level of significance, whether the following sequence is random: 1 9 9 1 1 9 9 9 1 1 1 1 1 9 9 9 1 9 9 1 1 9 9 9 9 9 9.

38. An airline wants to test, at the 1-percent level of significance, whether there is any difference in the times to failure of three types of radar transponders. Random samples of new units installed in the airline's fleet yield the data in Table 11.W. Make the test, showing your calculations.

Table 11.W

Times to Failure (months)		
Type A	Type B	Type C
48.1	40.3	13.8
26.5	43.7	20.4
41.8	50.1	19.1
38.1	40.4	21.8
36.3	36.0	24.7
40.7	28.8	26.8
26.6		19.8
28.8		

39. A government agency wants to determine, at the 5-percent level of significance, whether the training given to air-traffic controllers is equally effective at four locations. The same test is administered everywhere to the most recent class; the scores are given in Table 11.X. Make the test, showing your calculations.

40. A corporate executive wants to test, at the 2-percent level of significance, whether the percentage of overdue customer accounts differs from one region of the country to another. Random samples of 5–7 sales outlets in regions A–D reveal the information given in Table 11.Y. Make the test, showing your calculations.

Table 11.X

Test Scores at Four Locations			
A	B	C	D
96	65	60	95
82	74	73	93
88	72	85	90
70	66	61	88
90	79	79	91
91	82	85	87
87		88	90

Table 11.Y

Percentage of Accounts Overdue in Four Regions			
A	B	C	D
10.1	7.2	20.7	4.1
7.2	8.5	19.0	3.0
8.5	9.1	18.0	2.7
9.0	6.3	13.7	8.3
7.0	5.9	21.2	7.1
5.2	7.3		1.0
			1.5

41. A builder wants to know whether the appreciation of house prices in Illinois differs with the quality of houses. Using a 2-percent level of significance and the sample data of Table 11.Z, make an appropriate hypothesis test.

Table 11.Z

Annual Percentage Price Increase of Houses of Three Qualities		
A	B	C
8.9	6.1	1.2
7.2	5.0	2.6
5.6	3.1	3.0
13.6	4.2	1.9
11.0	3.9	2.0

42. A farmer wants to know whether crop yield differs with the type of pest control employed. Using a 1-percent level of significance and the sample data of Table 11.AA, make an appropriate hypothesis test.

Table 11.AA

Bushels per Acre with Three Types of Pest Control

A	B	C
50.6	60.1	40.6
48.0	55.2	38.9
39.9	32.0	29.7
52.6	54.0	42.6
42.7	42.0	27.0
48.8	56.1	25.0
		31.3

43. An economist wants to know whether economists' salaries differ with type of employer. Using a 10-percent level of significance and the sample data of Table 11.BB, make an appropriate hypothesis test.

Table 11.BB

Economists' Annual Salaries (in thousands of dollars)

Academia	Business	Government
25.6	42.9	52.0
28.7	29.9	50.0
29.3	53.6	48.7
33.7	69.0	42.0
35.6	42.3	39.0
58.7		37.9
60.0		

44. A flight instructor wants to know whether pilots' reaction time to a certain flight situation differs with the amount of sleep recently experienced. Using a 2-percent level of significance and the sample data of Table 11.CC, make an appropriate hypothesis test.

Table 11.CC

Pilots' Reaction Time to Flight Emergency (seconds)

A. No Sleep for 5 Hours	B. No Sleep for 10 Hours	C. No Sleep for 20 Hours
3.2	8.2	10.1
2.9	9.0	8.9
1.9	6.3	9.7
4.3	5.0	7.3
5.0	4.7	12.6
2.7	6.9	15.3
		14.1

45. The manager of a national car-rental company wants to know whether the company's service differs at various airport locations. Past customers have mailed in questionnaires about the cars' mechanical condition, cleanliness, timeliness of service, and more. The answers have been coded as scores, as in the Table 11.DD sample. Make an appropriate hypothesis test at the 5-percent level of significance.

Table 11.DD

Scores Given By Rental-Car Customers

A. New York La Guardia	B. New York Kennedy	C. Washington National
100	98	61
88	66	59
65	85	63
79	92	72
86	96	85
92	72	99

46. A telephone-company manager wants to know whether the time customers take to pay their bills differs in four districts. Using a 5-percent significance level and the sample data of Table 11.EE, make an appropriate hypothesis test.

47. An oil company has been drilling 3 wells each at 500 sites around the world; the number of prom-

Table 11.EE

Time Taken to Pay Telephone Bills (days)			
District A	District B	District C	District D
3	6	20	17
17	6	22	20
5	7	29	33
8	9	33	25
9	10	60	9
22	29	3	

ising wells per site was 0 in 250 tries, 1 in 198 tries, 2 in 47 tries, and 3 in 5 tries. The company wants to conduct a hypothesis test at the 1-percent significance level that the number of promising wells per site (out of 3 drillings) is a binomially distributed random variable with a probability of success in any one trial of $\pi = .2$ (as suggested by past experience). Make the test, showing your calculations.

48. Apply the *Kolmogorov-Smirnov one-sample test* to Chapter 10's problem 45, but use $\alpha = .025$. Show your calculations.

49. Apply the *Kolmogorov-Smirnov one-sample test* to Chapter 10's problem 49. Show your calculations.

50. Apply the *Kolmogorov-Smirnov one-sample test* to Chapter 10's problem 51. Show your calculations.

51. Apply the *Kolmogorov-Smirnov one-sample test* to Chapter 10's problem 54. Show your calculations.

52. Apply the *Kolmogorov-Smirnov one-sample test* to Chapter 10's problem 55. Show your calculations.

53. Apply the *Kolmogorov-Smirnov one-sample test* to Chapter 10's Practice Problem 8: Customers at the Dry Cleaner (on page 424). Show your calculations.

54. These are the ranks of exam scores received by 10 students:

Student:	A	B	C	D	E	F	G	H	I	J
Midterm:	1	2	3	4	5	6	7	8	9	10
Final:	10	9	8	7	6	5	4	3	2	1

Determine whether a significant degree of *rank correlation* exists, using $\alpha = .10$ as the desired significance level.

55. These are the ranks of evaluation scores given to 10 salespeople:

Salesperson:	A	B	C	D	E	F	G	H	I	J
When hired:	1	2	3	4	5	6	7	8	9	10
After 2 years:	1	2	3	4	5	6	7	8	9	10

Determine whether a significant degree of *rank correlation* exists, using $\alpha = .10$ as the desired significance level.

56. These are the ranks of preference scores given to 10 TV shows:

TV show:	A	B	C	D	E	F	G	H	I	J
Men:	1	2	3	4	5	6	7	8	9	10
Women:	10	1	9	2	8	3	7	4	6	5

Determine whether a significant degree of *rank correlation* exists, using $\alpha = .10$ as the desired significance level.

57. The average draft-priority numbers picked during the 1970 draft lottery (discussed in Close-Up 2.1), are given in Table 11.FF. Determine whether a significant degree of *rank correlation* exists, using $\alpha = .10$ as the desired significance level.

58. These are the evaluation scores given by two critics to 10 paintings at an exhibition:

Painting:	A	B	C	D	E	F	G	H	I	J
Critic A:	1	2	3	4	5	6	7	8	9	10
Critic B:	3	7	1	5	8	2	6	10	4	9

Determine whether a significant degree of *rank correlation* exists, using $\alpha = .05$ as the desired significance level.

59. For each of the sampling situations, specify an appropriate nonparametric hypothesis test.
  a. After running military vehicles with ordinary gasoline for half a year and then with gasohol,

Table 11.FF

Month	Average Number Picked	Month	Average Number Picked
January	201.2	July	181.5
February	203.0	August	173.5
March	225.8	September	157.3
April	203.7	October	182.5
May	208.0	November	148.7
June	195.7	December	121.5

the Pentagon wants to determine the relative performance of the two types of fuel.

b. An advertising agency asks a group of men to shave with one shaver for a while and then to shave with another (although the order is varied from man to man). It wants to determine which shaver is perceived to give a closer shave.

c. An economist wants to determine whether the past movements of silver futures prices have been random over time.

d. An economist wants to determine whether the median house price in an area equals an alleged number.

e. A researcher wants to determine which of two drugs cures a disease faster.

60. For each of the following sampling situations, specify an appropriate nonparametric hypothesis test.

a. A union official wants to determine the preferences of union members between two proposed labor contracts.

b. After running 25 military vehicles with regular gasoline for half a year, then with unleaded gasoline, then with gasohol, and then with a new

top-secret type of fuel, the Pentagon wants to determine the relative performance of the four types of fuel.

c. An executive wants to know whether brand preference and price are related.

d. A quality inspector wants to determine whether the appearance of defective items in a continuous production process is random.

e. An economist wants to know whether riskiness and salaries of different jobs are related.

f. An executive wants to compare the delivery times of steel coming from two mills; he knows that the times are available for the last 50 shipments received.

g. An executive wants to know whether output is higher without or with piped-in music at the workplace. Data are available for the performance of 15 employees.

h. An executive wants to compare the delivery times of steel coming from seven mills; times are available for the last 50 shipments received.

i. An executive wants to study the relationship between newspaper advertising and sales results; data are available for the last 17 weeks.

Selected Readings

Gibbons, J. *Nonparametric Methods for Quantitative Analysis.* New York: Holt, Rinehart and Winston, 1976.

Lehmann, E. L. *Nonparametrics: Statistical Methods Based on Ranks.* San Francisco: Holden-Day, 1975.

Noether, Gottfried E. *Introduction to Statistics: A Nonparametric Approach,* 2nd ed. Boston: Houghton Mifflin, 1976.

Savage, I. Richard. *Bibliography of Nonparametric Statistics.* Cambridge, Mass.: Harvard University Press, 1962.

Siegel, Sidney. *Nonparametric Statistics for the Behavioral Sciences.* New York: McGraw-Hill, 1956. Gives detailed information on applying many nonparametric procedures.

Solterer, J. "A Sequence of Historical Random Events: Do Jesuits Die in Three's?" *Journal of the American Statistical Association,* December 1941, pp. 477–84. Examines the deaths of 597 Jesuit priests in the United States between 1900 and 1939 by means of a runs test to assess the folklore that holds that accidents or tragedies occur in triplets. (They do.)

Spearman, Charles E. "The Proof and Measurement of Association Between Two Things." *American Journal of Psychology* 15(1904):72–101. The original article on rank correlation.

Wilcoxon, Frank. "Individual Comparisons by Ranking Methods." *Biometrics Bulletin* 1, 6(1945):80–83. The pathbreaking article in which the two-sample rank-sum statistic as well as the paired-sample signed-rank statistic first appeared.

Computer Programs

The DAYSAL-2 personal-computer diskettes that accompany this book contain one program of interest to this chapter:

11. Nonparametric Hypothesis Tests lets you perform two-tailed, lower-tailed, or upper-tailed Wilcoxon rank-sum tests and Mann-Whitney tests, as well as the Spearman rank-correlation test.

PART FOUR

Analytical Statistics: Advanced Inference

CHAPTER TWELVE

Analysis of Variance

We learned in Chapter 9 that independent random samples taken from two appropriate quantitative populations can help us answer questions such as these: Are the mean lifetimes of two types of aircraft radios the same? Is the mean yield of fruit trees sprayed with gypsy-moth parasites larger than that of trees sprayed with traditional pesticides? Is the mean number of cavities associated with the use of toothpaste 1 smaller than that associated with the use of toothpaste 2? We also learned in that chapter that the normal probability distribution (in the case of large samples) and Student's t distribution (in the case of small samples) are ideally suited to help us perform any desired hypothesis test about the comparative magnitudes of two population means. In this chapter, we will take this type of analysis one step further. We will imagine ourselves wanting to compare the means not merely of two, but of more than two quantitative populations, and we will learn that such extended comparisons are not performed very well by the earlier procedures.

The desire to compare the means of more than two statistical populations is common indeed. Consider the farmer who wants to know whether there is any difference in the average crop yield associated with the use of five types of fertilizer (or with five alternative quantities of a given type of fertilizer that might be applied, or even with five alternative times at which a given quantity of a given type might be applied). Consider the industrial manager who wants to know whether the average lifetime (or delivery time) of components is the same regardless of which one of seven outside contractors supplies them or the manager who wants to test whether the average number of units assembled in a day differs among seven possible production methods (or workers or machines) that may be employed. Consider the seller who wonders whether average sales vary among six alternative window displays, TV ads, types of packaging, or price levels. Think of the teacher who ponders whether the average scores of students would differ among four teaching methods or even whether a dozen alternative styles or colors of textbook print might be associated with different degrees of comprehension. Finally, consider the government agency that might wish to know whether average housing prices or average pollution levels or average price levels in general differ among the 50 states. In all these cases and a million more, sample evidence can be employed to make inferences about the population means of interest. Yet tempting as it may be, stringing together the results of several two-sample tests is not wise.

To see why it is not wise to combine the results of several two-sample tests for a multiple-sample test, imagine that we wanted to compare the mean numbers of cavities associated with five types of toothpaste, 1 through 5, and that we hypothesized an equality of these means. If our samples were small, we could perform a series of t tests, comparing the mean associated with the use of 1 vs. 2, then 1 vs. 3, 1 vs. 4, and 1 vs. 5, then 2 vs. 3, 2 vs. 4, 2 vs. 5, and, finally, 3 vs. 4, 3 vs. 5, and 4 vs. 5. We would be performing a total of 10 separate tests of the type introduced in Chapter 9. Yet if each of these tests were performed at the $\alpha = .05$ level of significance, the maximum probability of making a type I error in the entire series of tests would not be 5 percent, but considerably larger (more than 40 percent). If the null hypothesis of equal means were in fact true, the chances of falsely rejecting equality in at least one of the 10 pairwise comparisons outlined above can be found in Appendix Table C for binomial probabilities. For $n = 10$ and $\pi = .05$, for probability of $x \geq 1$ equals $1 - .5987 = .4013$. This probability figure is hardly acceptable. This chapter is devoted to an alternative procedure that avoids compounding the probability of a type I error when making comparisons of many means.

The **analysis of variance,** frequently referred to by the contraction **ANOVA,** is a statistical technique specially designed to test whether the means of more than two quantitative populations are equal. (It can, however, also be used to test the equality of only two means and then yields the same result as the normal-distribution or t-distribution tests of Chapter 9.) The technique involves taking an independent simple random sample from each of several populations of interest (such as the users of toothpastes 1, 2, 3, and so on) and then analyzing the data. Like the t test, the ANOVA test assumes that the sampled populations are normally distributed and have identical variances. (The latter assumption is often referred to as that of **homoscedasticity,** an awkward word created by the Greek *homo* = "same" and *skedastikos* = "scatter.") It turns out that the analysis-of-variance test is quite robust with respect to the normality assumption (even moderate departures from this assumption do not change the results much), but any violation of the equal-variances assumption seriously affects the validity of the test. This fact is not surprising because the basis of the entire test is the development, from the sample data, of two independent estimates of what is assumed to be the common variance, σ^2, of the populations of interest. A first estimate of σ^2 is based on the variation *among the sample means.* This estimate is denoted by s_A^2; it is an unbiased estimate of σ^2 only if the population means are in fact equal. A second estimate of σ^2, based on the variation *of individual sample observations within each sample,* is denoted by s_W^2 and is a weighted average of the individual sample variances—which (as we noted in Chapter 8) always provide an unbiased estimate of σ^2. It can easily be seen that the ratio, s_A^2/s_W^2, will be close to 1 if, and only if, the population means are equal to each other. On the other hand, the more the value of this ratio diverges from 1 (and, in principle, this ratio of squares can take on any value between zero and positive infinity), the greater is the probability that the population means are not equal to each other. Thus, an analysis of *variances* helps us test hypotheses about the equality of *means* (which accounts for the seemingly inappropriate name of the test). Figure 12.1 presents these ideas in another way.

Figure 12.1 Basic ANOVA Concepts

In each panel, the density functions shown refer to three populations, consisting, say, of the numbers of cavities observed, X, among all the users of toothpastes 1, 2, and 3, respectively. The populations are normally distributed and have identical variances, σ^2. The positions of the density functions, however, differ in the three panels. In panel a, the population means (μ_1, μ_2, and μ_3) are wide apart. In panel b, they are closer together. In panel c, they are identical; hence, the three density functions have merged into one. If a sample of five were taken from each of these populations, the individual sample observations might have values corresponding to the positions of the small circles underneath each graph. The corresponding sample means are given by the colored squares. It is immediately clear that the common population variance ($\sigma^2 = \sigma_1^2 = \sigma_2^2 = \sigma_3^2$) would be estimated well by the dispersion among the five individual observations (small circles) within each sample as well as by the average of these sample variances, s_W^2. It is also clear that this common population variance would not be estimated well (but would be vastly exaggerated) by the dispersion among the three sample means (colored squares) in panel a and, to only a slightly lesser extent, in panel b. Only in panel c, wherein the colored squares are not any more dispersed than any one set of small circles, could one derive, from the dispersions among the sample means, an unbiased estimate, s_A^2, of σ^2. *Conclusion:* Whenever the estimate, s_A^2 of the common population variance that is derived from the dispersion *among* sample means is considerably larger than such an estimate, s_W^2, that is based on the dispersion *within* samples, the populations have unequal means, as panels a and b show. Whenever the two estimates are about the same, as in c, the populations have equal means.

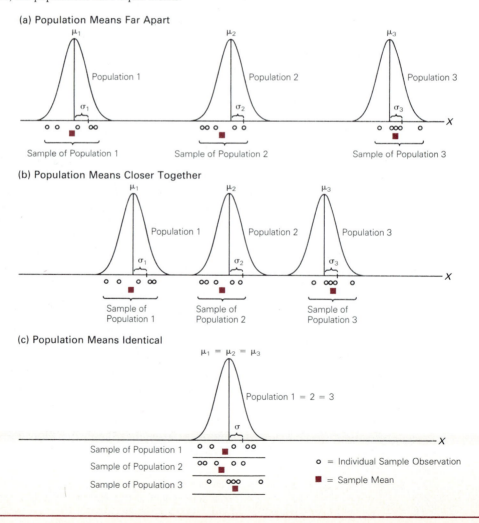

ALTERNATIVE VERSIONS OF ANOVA

In the following sections, we will discuss two major versions of the analysis of variance. These versions of ANOVA can most easily be distinguished by briefly recalling the Chapter 2 section, "Experiments." As we then noted, in one way or another, all experiments involve the assignment of different "treatments" (such as different types of toothpaste) to different experimental units (such as people) in order to see whether effects differ. (All the situations noted above that involve comparisons of means are experiments in this sense.) Yet it is all too easy to mix up or confound the effects of the treatments (such as types of toothpaste) with the effects of other extraneous factors that are also operating (such as the users' ethnic backgrounds and, hence, dietary habits). That is why good experimental designs carefully control extraneous factors, either by randomization or by blocking (also discussed in Chapter 2).

Extraneous factors are controlled by *randomization* when they are allowed to operate but when the experimenter assures, by virtue of the random assignment of experimental units to treatments, that each treatment has an equal chance of being enhanced or handicapped by these factors. (Randomization is achieved, for example, when each toothpaste is assigned to a mixed group of people whose composition with respect to ethnic background is roughly identical.) Extraneous factors are controlled by *blocking* when their effect is eliminated because the experimenter forms blocks of experimental units within each of which all units are as alike as possible with respect to these factors. (Blocking occurs, for example, when each toothpaste is tested with persons of Italian background but also with persons of German background.)

From these considerations, we noted in Chapter 2, a number of experimental designs have emerged. The *completely randomized design* uses randomization as a control device. It creates one treatment group for each treatment (such as toothpastes 1, 2, and 3) and assigns each experimental unit to one of these groups by a random process (such that, for example, a randomly chosen third of all participants ends up using toothpaste 1, another such third uses toothpaste 2, and the last third uses toothpaste 3). In the end, only one factor (the treatment) is considered as possibly affecting the variable of interest (such as the number of cavities observed). The analysis of this type of problem, taken up in the next section, is referred to as **one-factor ANOVA,** or **one-way ANOVA.**

Another favorite experimental plan, the *randomized block design,* uses blocking as an additional control device. It divides the available experimental units into distinctly different, but internally homogeneous blocks (each of which contains as many units as there are treatments or some multiple thereof) and then randomly matches each treatment with one or more units within any given block. Within any one block (among persons of Italian background, say) only the treatment (such as the type of toothpaste used) is believed to have an effect (as on the number of cavities observed), but it is now also possible to observe potential differences for any given treatment (such as toothpaste 1) among blocks. (For example, different numbers of cavities may be observed when toothpaste 1 is used by spice-loving Italians than when it is used by sweets-loving Germans.) The analysis of this type of problem is referred to as **two-factor ANOVA,** or **two-way ANOVA** (the treatments and the blocks).[1]

---

[1]More complicated experimental designs, such as those discussed in Chapter 2 of the *Student Workbook* that accompanies this text, give rise, in turn, to a more complex analysis of variance, such as the *three-factor ANOVA,* or *three-way ANOVA,* that is introduced in chapter 12 of the *Student Workbook*.

ONE-WAY ANOVA

It is easiest to introduce the technique of one-way ANOVA with the help of an example. Consider a 10-year study in which a sample of 15 people has been observed while using toothpaste 1, 2, or 3, respectively. Let us assume that five of the participants have been randomly assigned to each of the treatments and that the study has provided the data given in Table 12.1. (This study is the subject of Practice Problem 1 at the end of this section.)

A statistician about to analyze these data would formulate opposing hypotheses such as:

H_0: The mean number of cavities for all users of toothpaste 1 is the same as that for all users of toothpaste 2 or 3; that is, $\mu_1 = \mu_2 = \mu_3$.

H_A: At least one of the population means is different from the others. (At least one of the equalities doesn't hold.)

The analyst would proceed by studying the variation in the sample data listed in the $r = 5$ rows and $c = 3$ columns of our table. This variation has two components.

Variation Among Columns: Explained by Treatments

The variation among the $c = 3$ sample means ($\overline{X}_1$, $\overline{X}_2$, and $\overline{X}_3$), which summarize the data associated with each of the treatments, is referred to as **explained variation,** or **treatments variation,** because it is attributable not to chance, but to inherent differences among the treatment populations. As noted above, the measurement of this variation constitutes a first estimate, s_A^2, of the population variance, σ^2; this estimate is based on the following considerations.

Given the assumed normality of the sampled populations, the sampling distribution of each sample mean, $\overline{X}_j$ (where j represents treatments and can equal 1, 2, or 3 in our case) will be normally distributed and will (as a quick review of Box 8.A on page 296 can confirm) itself have a variance of

$$\sigma_{\overline{X}_j}^2 = (\sigma_j^2/n_j). \tag{1}$$

Given the assumed equality of population variances ($\sigma_1^2 = \sigma_2^2 = \sigma_3^2$) and equal sample sizes ($n_1 = n_2 = n_3$), the value of $\sigma_{\overline{X}_j}^2$ is the same for each j; hence, we can drop the j-subscripts in equation **1.** Furthermore, n, the number of observations in each sample (which equals 5 in our case) also equals r, the number of rows in Table 12.1; hence, the common population variance sought equals

$$\sigma^2 = n\sigma_{\overline{X}}^2 = r\sigma_{\overline{X}}^2. \tag{2}$$

The variance of the sampling distribution of $\overline{X}$, or $\sigma_{\overline{X}}^2$, can itself be estimated by the variance of the individual sample means, $\overline{X}_j$, about their mean, which is called the **grand mean, $\overline{\overline{X}}$** (pronounced "X double bar" and calculated, as Table 12.1 shows, as the arithmetic mean of the sample means). In accordance with the sample variance

Table 12.1 The Completely Randomized Design: Numbers of Cavities Observed
During 10-Year Period

Observation, i	Treatment, j (type of toothpaste used)		
	1	2	3
1	19	20	18
2	15	25	12
3	22	22	16
4	17	19	17
5	19	23	15
Totals:	92	109	78
Sample Means	$\overline{X}_1 = \dfrac{92}{5} = 18.4$	$\overline{X}_2 = \dfrac{109}{5} = 21.8$	$\overline{X}_3 = \dfrac{78}{5} = 15.6$
Grand Mean	$\overline{\overline{X}} = \dfrac{18.4 + 21.8 + 15.6}{3} = 18.6$		

formula (in Box 4.I on page 110), this estimate of the variance of the sampling distribution of $\overline{X}$ can be calculated as

$$\sigma_{\overline{X}}^2 \cong s_{\overline{X}}^2 = \frac{\Sigma(\overline{X}_j - \overline{\overline{X}})^2}{c - 1}. \tag{3}$$

(Note: The individual sample observations, called X in Box 4.I, are in equation 3 now themselves sample means, $\overline{X}_j$, while the earlier sample mean, $\overline{X}$, is in equation 3 the mean of these means, $\overline{\overline{X}}$. In addition, the number of individual observations, called n in Box 4.I, is now equal to c, the number of columns, each of which provides us with one sample mean.)

From equations **2** and **3**, it follows that

$$\sigma^2 = r\sigma_{\overline{X}}^2 \cong \frac{r\Sigma(\overline{X}_j - \overline{\overline{X}})^2}{c - 1}. \tag{4}$$

The numerator of the ratio in equation 4—the sum of the squared deviations between each sample mean and the grand mean, multiplied by the number of observations made for each treatment—is referred to as the **treatments sum of squares**, or **TSS**. In our example, *TSS* can be calculated as

$$
\begin{aligned}
TSS &= r\Sigma(\overline{X}_j - \overline{\overline{X}})^2 \\
&= 5[(18.4 - 18.6)^2 + (21.8 - 18.6)^2 + (15.6 - 18.6)^2] \\
&= 96.4.
\end{aligned}
$$

The denominator of the ratio in equation 4 equals the degrees of freedom associated with the estimation of σ^2 with the help of the variation among sample means. Thus,

$$d.f. = c - 1 = 3 - 1 = 2$$

in our case. (This number of degrees of freedom indicates that two of the three sample means are free to vary, given the grand mean.)

The equation 4 ratio as a whole, finally, is also called the **treatments mean square (TMS),** or **explained variance,** and is identical to the population variance estimate symbolized by s_A^2 above (where the subscript A was used to remind us that this estimate of σ^2 is based on the observed variation *among* samples).

$$\sigma^2 \cong \frac{r\Sigma(\overline{X}_j - \overline{\overline{X}})^2}{c - 1} = TMS = s_A^2 \tag{5}$$

In our case, $TMS = s_A^2 = (96.4/2) = 48.2$.

Variation Within Columns: Due to Error

The variation of the sample data within each of the $c = 3$ columns (or samples) about the respective sample means is referred to as **unexplained variation,** or **residual variation,** or simply as (experimental or sampling) **error;** it is attributed to chance. As noted above, the measurement of this variation constitutes a second estimate, s_W^2, of the population variance, σ^2; this estimate is based on the following considerations.

From each of the j samples in Table 12.1, a (probably different) sample variance can be derived by using the familiar Box 4.I formula (page 110) such that the variance of sample j equals

$$s_j^2 = \frac{\Sigma(X_{ij} - \overline{X}_j)^2}{n_j - 1}, \tag{6}$$

where X_{ij} is the sample observation in row i and column j, $\overline{X}_j$ is the mean of sample j, and n_j is the number of observations in sample j (which also equals r, the number of rows in Table 12.1).

To get a single estimate of σ^2, one can, in turn, take the weighted average of the j sample variances; in our case, wherein all the samples are of equal size, taking this weighted average requires only the summing of the $c = 3$ sample variances and division by c. Thus, the second estimate of the population variance can be calculated as

$$\sigma^2 \cong \frac{\Sigma\Sigma(X_{ij} - \overline{X}_j)^2}{(r - 1) \cdot c}. \tag{7}$$

The numerator of the ratio in equation 7—the sum of each sample's sum of squared deviations of individual observations from that sample's mean—is referred to as the **error sum of squares,** or **ESS.** Its calculation for our example, as given in Table 12.2, is

$$ESS = \Sigma\Sigma(X_{ij} - \overline{X}_j)^2 = 27.2 + 22.8 + 21.2 = 71.2.$$

Table 12.2   Calculation of the Error Sum of Squares

Observation, i	Sample 1 $(X_{i1} - \bar{X}_1)^2$	Sample 2 $(X_{i2} - \bar{X}_2)^2$	Sample 3 $(X_{i3} - \bar{X}_3)^2$
1	$(19 - 18.4)^2 = \quad .36$	$(20 - 21.8)^2 = \quad 3.24$	$(18 - 15.6)^2 = \quad 5.76$
2	$(15 - 18.4)^2 = 11.56$	$(25 - 21.8)^2 = 10.24$	$(12 - 15.6)^2 = 12.96$
3	$(22 - 18.4)^2 = 12.96$	$(22 - 21.8)^2 = \quad .04$	$(16 - 15.6)^2 = \quad .16$
4	$(17 - 18.4)^2 = \quad 1.96$	$(19 - 21.8)^2 = \quad 7.84$	$(17 - 15.6)^2 = \quad 1.96$
5	$(19 - 18.4)^2 = \quad .36$	$(23 - 21.8)^2 = \quad 1.44$	$(15 - 15.6)^2 = \quad .36$
Total	27.20	22.80	21.20

$$ESS = 27.2 + 22.8 + 21.2 = 71.2$$

The denominator of the ratio in equation 7 equals the degrees of freedom associated with the estimation of σ^2 with the help of the variation within the samples. Thus,

$$d.f. = (r - 1) \cdot c = (5 - 1) \cdot 3 = 12$$

in our case. [This number of degrees of freedom indicates that, given a sample mean, $\bar{X}_j$, only $r - 1 = 4$ of the 5 observations in the sample are free to vary; there being $c = 3$ samples, however, the total number of values free to vary is $4(3) = 12$.]

The ratio in equation 7 as a whole, finally, is called the **error mean square (EMS)**, or **unexplained variance,** and is identical to the population variance estimate symbolized by s_W^2 above (where the subscript W was used to remind us that this estimate of σ^2 is based on the observed variation *within* samples).

$$\sigma^2 \cong \frac{\Sigma\Sigma(X_{ij} - \bar{X}_j)^2}{(r - 1)c} = EMS = s_W^2 \tag{8}$$

In our case, $EMS = s_W^2 = (71.2/12) = 5.93$.

The ANOVA Table

Our computations so far are conveniently summarized in Table 12.3. This **ANOVA table** shows, for each source of variation, the sum of squares, the degrees of freedom, and the *ratio* of the sum of squares to the degrees of freedom. This ratio is called the mean square and is the desired estimate of the population variance, as equations **5** and **8** have just shown. The last row of the table also introduces the concept of the **total sum of squares,** or *Total SS,* which can be calculated independently by summing the squared deviations of each individual sample observation (regardless of the sample to which it belongs) from the mean of all observations, $\bar{\bar{X}}$. Thus, the data of Table 12.1 can be used to calculate

$$Total\ SS = (19 - 18.6)^2 + (15 - 18.6)^2 + (22 - 18.6)^2 + \ldots + (15 - 18.6)^2$$
$$= 167.6.$$

Table 12.3   The One-Way ANOVA Table

Source of Variation	Sum of Squares (1)	Degrees of Freedom (2)	Mean Square (3) = (1) ÷ (2)	Test Statistic (4)
Treatments	$TSS = 96.4$	$c - 1 = 2$	$TMS = \dfrac{96.4}{2} = 48.2$	$F = \dfrac{TMS}{EMS} = \dfrac{48.2}{5.93} = 8.13$
Error	$ESS = 71.2$	$(r - 1)c = 12$	$EMS = \dfrac{71.2}{12} = 5.93$	
Total	$Total\ SS = 167.6$	$rc - 1 = 14$		

In one-way ANOVA, this *Total SS* value is, however, also equal to the sum of the treatments sum of squares and the error sum of squares:

$$Total\ SS = TSS + ESS = 96.4 + 71.2 = 167.6.$$

Table 12.3 also shows that the total degrees of freedom equal the sum of those associated with the two separate sources of variation:

$$\text{total degrees of freedom} = (c - 1) + (r - 1)c = c - 1 + rc - c = rc - 1.$$

This sum is $5(3) - 1 = 14$ in our case and equals the total number of observations (15) minus 1. (This *d.f.* reflects the fact that, given the grand mean, all but one of the $rc = 15$ values in Table 12.1 are free to vary.)

The next section, finally, discusses the meaning of the last column of our ANOVA table and reveals the result of our hypothesis test.

The *F* Distribution

When discussing basic ANOVA concepts earlier in this chapter, we noted that the ratio of the two independent estimates of the common population variance—then denoted by s_A^2/s_W^2 and now found to equal *TMS/EMS*—would be close to 1 whenever the null hypothesis of equal population means was true. Indeed, this ratio is used as the ANOVA test statistic and is denoted by *F,* in honor of Ronald A. Fisher (Biography 9.1) who developed its probability distribution in the 1920s.[2] Thus, the *F* statistic is given in Box 12.A.

The probability distribution of *F* helps us decide whether any given divergence of *F* from 1 (note that $F = 8.13$ in Table 12.3) is significant enough to warrant the rejection of the null hypothesis of equal population means. Appendix Table N provides

---

[2]Strictly speaking, he developed a Z distribution that can be transformed into the F distribution by the fact that $F = e^{2Z}$, where $e = 2.71828$. One of the earliest tabulations of F distributions appeared in George W. Snedecor, *Calculation and Interpretation of Analysis of Variance and Covariance* (Ames, Iowa: Collegiate Press, 1934), which also explains why the statistic is sometimes referred to as *Snedecor's F*. Biography 14.1 tells more of Snedecor's contributions.

12.A The *F* Statistic

$$F = \frac{s_A^2}{s_W^2} = \frac{TMS}{EMS} = \frac{\text{explained variance}}{\text{unexplained variance}}$$

where s_A^2 is the variation *among* sample means, s_W^2 is the variation of individual sample observations *within* samples, *TMS* is treatments mean square, and *EMS* is error mean square.

critical values of F_α for significance levels of $\alpha = .05, .025,$ and $.01$. Note: in contrast to the *t* and χ^2 statistics, the **F statistic** (being a ratio of two variances) is associated not with a single number of degrees of freedom, but with a pair of them. One of these (equal to $c - 1 = 2$ in our example) goes with the numerator of the ratio; the other one (equal to $(r - 1)c = 12$ in our example) goes with the denominator. As Appendix Table N shows, with 2 numerator degrees of freedom and 12 denominator degrees of freedom, the critical value of $F_{.05}$ is 3.89, of $F_{.025}$ is 5.10, and of $F_{.01}$ is 6.93. This means there is only a 5-percent, 2.5-percent, or 1-percent chance, respectively, that a computed value of *F* will exceed the indicated magnitudes when the null hypothesis is true.

Thus, regardless of which of the three significance levels our analyst had selected, the actual value of $F = 8.13$ would exceed the critical value, and H_0 would be *rejected*.

Figure 12.2 pictures a number of **F distributions** for selected pairs of degrees of freedom. In each case, the first *d.f.* number that appears within the parentheses refers

Figure 12.2   *F* Distributions

A different *F*-distribution curve can be drawn for each possible pair of numerator and denominator degrees of freedom that is associated with the two variance estimates making up the *F* ratio. All the members of this continuous-probability-distribution family are positively skewed between values of zero and positive infinity, but *F* distributions tend towards normality as both numbers of degrees of freedom become large. Indeed, the normal, *t*, and χ^2 distributions are special cases of the *F* distribution family. (Note: The numerator degrees of freedom and the denominator degrees of freedom happen to be equal to one another in each of the three examples given here. There is no logical necessity for this, as Table 12.3 and numerous other examples in this chapter illustrate.)

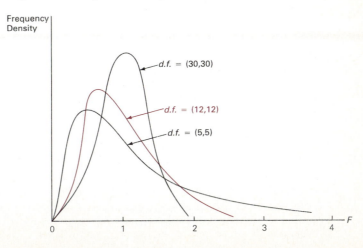

to the numerator; the second one refers to the denominator of the F ratio. Extreme caution is advised: an F distribution with 2 numerator degrees of freedom and 12 denominator degrees of freedom must not be confused with one that has 12 numerator degrees of freedom and 2 denominator degrees of freedom. The shapes of these distributions are entirely different. This fact becomes evident when inspecting Appendix Table N. Note, for example, how the critical value of $F_{.05}$ equals 3.89 for a numerator/denominator degrees of freedom pair of 2, 12 but equals 19.41 for a pair of 12, 2.

||| Practice Problem 1

The Effect of Three Toothpastes on Cavity Development

Figure A pictures the F distribution applicable to our experiment (which seeks to determine whether 3 toothpastes are equally effective by randomly assigning 5 different participants to use each toothpaste). The critical value of $F_\alpha = 3.89$ for a test at the 5-percent significance level is encircled. Thus, if an experiment of this type were to be repeated over and over again, while the null hypothesis of equal population means was true, the computed value of F, shown by the colored arrow for our example, would exceed 3.89 only 5 percent of the time. This result leads us to reject the null hypothesis and to conclude the three toothpastes do have different effects.

Figure A

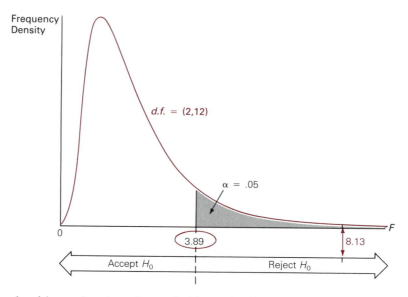

What should one do, given the probable truth of the alternative hypothesis that the three toothpastes are *not* equally effective? One possible reaction is recommending use of toothpaste 3 because sample 3 yielded the lowest mean number of cavities (Table 12.1). Yet the sample results are still clouded by sampling error. It is conceivable that toothpaste 1 is better than 3 but that an unfortunate sample was drawn from population 1 and a lucky one from population 3. Thus, further analysis is in order, and we will examine the possibilities in the last section of this chapter. |||

TWO-WAY ANOVA

There are times when it might be desirable to consider the effects of a second factor on the variable of interest. If there is reason to believe, for example, that dietary habits associated with ethnic background affect the number of cavities independently of and in addition to the type of toothpaste used, one can explain in this way some of the previously unexplained variation *within* the samples taken from the different toothpaste-user populations, can correspondingly reduce the error sum of squares, and can ultimately derive a more discriminating *F* statistic.

Two basic types of two-way ANOVA will be introduced in this part of the chapter. In a first section, we will assume that the two factors under consideration (such as type of toothpaste used and ethnic background) do not interact with each other. For example, if use of toothpaste 1 instead of 2 has a significant cavity-reducing effect among people in general, we assume this to be equally true for all ethnic groups, regardless of their (possibly differing) initial rate of cavity formation. Similarly, if being German and, therefore, sweets-loving has a significant cavity-increasing effect in general, we assume this to be equally true regardless of the type of toothpaste used. As a result, the joint effect of the two factors on the number of cavities simply equals the sum of the factors' separate effects. The factor effects are *additive*: if using toothpaste 1 instead of 2 lowers cavities by 5 a year (regardless of ethnic group), while eating sweets raises them by 8 a year (regardless of toothpaste type), doing both will raise them by 3 a year $(-5 + 8 = 3)$.

In a section section, we will show that two factors can also interact with each other, and when that happens their joint effect does not equal the sum of their separate individual effects. (Such would be the case, for example, if the use of toothpaste 2 instead of 1 had a cavity-increasing effect when considered by itself, if eating lots of sweets had the same bad effect when considered by itself, while doing both together miraculously *reduced* cavities.) As we shall see, this type of situation (unlikely to occur in the toothpaste-diet example, but not unusual in other circumstances) calls for a more complex type of analysis.

Two Factors, No Interaction

The toothpaste study discussed in the previous section was based on the completely randomized design, which is supposed to control for extraneous factors. Yet someone who was convinced that dietary habits, determined by ethnic background, exert a strong influence on the number of cavities observed might well feel uncomfortable about that experimental design. What if the random process that assigns experimental units to treatments just happened to assign lots of sweets-loving Germans to toothpaste 2 and lots of sweets-disdaining Britishers to toothpaste 3? Then it would *appear* that toothpaste 3 was more effective than 2 (as the data in Table 12.1 possibly suggest), while this result was in fact the consequence not of differences in the toothpastes used, but in the dietary habits of their users. To avoid such a possible pitfall, one can employ the randomized block design: One can first split the experimental units into homogeneous blocks such that the different blocks reflect the second factor (ethnic grouping in our case) that is believed to influence the results. Having created blocks, one can then assign units within each block randomly to each of the treatments. As a result, each block will be represented equally in each treatment. The possible results of such a study are given in Table 12.4.

Table 12.4 The Randomized Block Design (No Interaction): Number of Cavities Observed During 10-Year Period

Block, i (ethnic group)	Treatment, j (type of toothpaste used)			Row Totals	Row Means
	1	2	3		
(A) British	15	12	16	43	$\overline{X}_A = (43/3) = 14.33$
(B) French	19	19	19	57	$\overline{X}_B = (57/3) = 19.00$
(C) German	25	23	22	70	$\overline{X}_C = (70/3) = 23.33$
(D) Italian	22	20	18	60	$\overline{X}_D = (60/3) = 20.00$
(E) Spanish	17	15	17	49	$\overline{X}_E = (49/3) = 16.33$
Column Totals	98	89	92		
Column Means	$\overline{X}_1 = \dfrac{98}{5} = 19.6$	$\overline{X}_2 = \dfrac{89}{5} = 17.8$	$\overline{X}_3 = \dfrac{92}{5} = 18.4$		
Grand Mean	$\overline{\overline{X}} = \dfrac{19.6 + 17.8 + 18.4}{3} = 18.6$ or $\overline{\overline{X}} = \dfrac{14.33 + 19 + 23.33 + 20 + 16.33}{5} = 18.6$				

The analysis of the data proceeds in perfect analogy to that of the previous section, the only difference being that the effects of two factors (type of toothpaste used and ethnic group) are being assessed. The major hypothesis, however, remains the same as in the one-way ANOVA test:

H_0: The mean number of cavities for all users of toothpaste 1 is the same as that for all users of toothpaste 2 or 3; that is, $\mu_1 = \mu_2 = \mu_3$.

H_A: At least one of these population means is different from the others. (At least one of the equalities doesn't hold.)

However, it is now possible to test simultaneously a second hypothesis:

H_0: The mean number of cavities for all the members of ethnic group A is the same as that for all the members of ethnic group B, C, D, and E, respectively; that is, $\mu_A = \mu_B = \mu_C = \mu_D = \mu_E$.

H_A: At least one of these population means is different from the others. (At least one of the equalities doesn't hold.)

Ordinarily, this second hypothesis about blocks is of secondary interest because the only purpose of blocking is to allow us to produce a more reliable test concerning treatments. The total variation in the data is again broken into explained variation and unexplained variation (or error), but explained variation now has two components: treatments variation and blocks variation. (Table 12.5 summarizes the results of this two-way ANOVA test with no interaction.)

Variation among columns: explained by treatments. Treatments variation is calculated precisely as in the previous section: as the number of *rows* multiplied by the sum of the squared deviations of the *column* means from the grand mean. Thus,

$$TSS = r\Sigma(\overline{X}_j - \overline{\overline{X}})^2 = 5[(19.6 - 18.6)^2 + (17.8 - 18.6)^2 + (18.4 - 18.6)^2] = 8.4.$$

Also as before, the applicable degrees of freedom equal $c - 1 = 2$. Thus, the treatments mean square is $TMS = (8.4/2) = 4.2$.

Table 12.5 A Two-Way ANOVA Table Without Interaction

Source of Variation	Sum of Squares (1)	Degrees of Freedom (2)	Mean Square (3) = (1) ÷ (2)	Test Statistic (4)
Treatments	$TSS =$ 8.4	$c - 1 = 2$	$TMS = \dfrac{8.4}{2} = 4.2$	$F_T = \dfrac{TMS}{EMS} = \dfrac{4.2}{1.95}$ $= 2.15$
Blocks	$BSS = 143.6$	$r - 1 = 4$	$BMS = \dfrac{143.6}{4} = 35.9$	$F_B = \dfrac{BMS}{EMS} = \dfrac{35.9}{1.95}$ $= 18.41$
Error	$ESS =$ 15.6	$(r - 1)(c - 1) = 8$	$EMS = \dfrac{15.6}{8} = 1.95$	
Total	$Total\ SS = 167.6$	$(rc) - 1 = 14$		

Variation among rows: explained by blocks. The variation among the $r = 5$ block means ($\overline{X}_A$, $\overline{X}_B$, $\overline{X}_C$, $\overline{X}_D$, and $\overline{X}_E$) that summarize the data associated with each of the blocks (or ethnic groups in our example) is referred to as **blocks variation.** In perfect analogy to treatments variation, it is calculated as the number of *columns* multiplied by the sum of the squared deviations of the *row* means from the grand mean. Thus, the **blocks sum of squares (BSS),** is

$$BSS = c\Sigma(\overline{X}_i - \overline{\overline{X}})^2 = 3[(14.33 - 18.6)^2 + (19 - 18.6)^2 + (23.33 - 18.6)^2 + (20 - 18.6)^2 + (16.33 - 18.6)^2] = 143.6.$$

This time, the applicable degrees of freedom are $r - 1 = 4$. Thus, the **blocks mean square (BMS)** $= (143.6/4) = 35.9$.

Variation due to error. In the two-way ANOVA problem, the unexplained variation due to error is calculated as the difference between the total sum of squares on the one hand and the explained variation on the other; the explained variation, of course, is the sum of the treatments sum of squares and the blocks sum of squares. Thus,

$$Total\ SS - TSS - BSS = ESS.$$

In our example, the total sum of squares can be calculated from Table 12.4 as

$$Total\ SS = (15 - 18.6)^2 + (19 - 18.6)^2 + \ldots + (17 - 18.6)^2 = 167.6.$$

Thus, $ESS = 167.6 - 8.4 - 143.6 = 15.6$. Similarly, the degrees of freedom applicable to the total sum of squares, minus those associated with the treatments and blocks sums of squares, equal those associated with the error sum of squares.

$$d.f._{Total\ SS} - d.f._{TSS} - d.f._{BSS} = d.f._{ESS}$$
$$[(rc) - 1] - (c - 1) - (r - 1) = (r - 1)(c - 1).$$

In our case, $(r - 1)(c - 1) = 8$; hence, the error mean square is $EMS = (15.6/8) = 1.95$.

||| Practice Problem 2

The Effects of Toothpastes and Ethnic Background on Cavity Development

Assuming that we want to test our two null hypotheses (about toothpastes and ethnic background) at a significance level of 5 percent, we can quickly reach a conclusion. The critical values for $F_{.05}$ from Appendix Table N are $F_{.05} = 4.46$ for treatments with $d.f. = (2, 8)$, and $F_{.05} = 3.84$ for blocks with $d.f. = (4, 8)$. This critical F value for treatments exceeds the computed value of F_T; this F value for blocks falls short of the computed value of F_B (see Table 12.5). Thus, we should *accept* the null hypothesis about treatments but should *reject* the null hypothesis about blocks: the mean number of cavities is the same for the three toothpastes but differs among ethnic groups. As usual, the critical values are encircled in Figure B, and the computed values are designated by the colored arrows.

Figure B

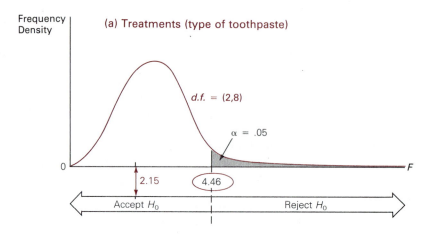

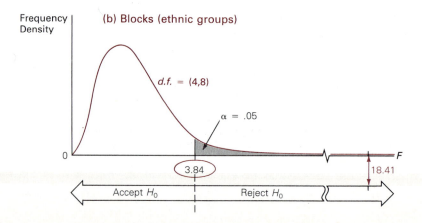

Analytical Example 12.1, "The Fairness of Earnings Differentials," also illustrates the procedure for a two-way ANOVA test with no interaction. |||

Two Factors, With Interaction

The analysis of variance for two factors that do interact with each other—and jointly produce effects that do not exist when each operates alone—is only slightly more complicated. Consider the manager (whose problem is solved in Practice Problem 3 below) who wants to test whether there is any difference in the average output produced by 3 machines. He suspects that the answer may depend on which of 4 crews operates the machines and that there may even be interaction such that one crew may be more productive with one machine than another crew, while that crew may be less productive than another crew with another machine (rather than one crew being consistently better or worse than another, regardless of the type of machine). Such a situation calls for an experiment with the randomized block design, but in order to test interaction, at least two observations are required for each combination of treatments (machines) and blocks (crews). An appropriate experiment (in which precisely two observations are made for each of these combinations) may yield the results given in Table 12.6.

Table 12.6 The Randomized Block Design (with Interaction): Number of Units Produced During Six Monthly Experiments

Block, i (crew)	Treatment, j (type of machine) 1	Treatment, j (type of machine) 2	Treatment, j (type of machine) 3	Block Totals	Block Means
(A)	$\left.{50 \atop 48}\right\}$ Cell total $= 98$ Cell mean, $\overline{X}_{A1} = 49$	$\left.{42 \atop 46}\right\}$ Cell total $= 88$ Cell mean, $\overline{X}_{A2} = 44$	$\left.{43 \atop 45}\right\}$ Cell total $= 88$ Cell mean, $\overline{X}_{A3} = 44$	274	$\overline{X}_A = \dfrac{274}{6}$ $= 45.67$
(B)	$\left.{56 \atop 58}\right\}$ Cell total $= 114$ Cell mean, $\overline{X}_{B1} = 57$	$\left.{38 \atop 32}\right\}$ Cell total $= 70$ Cell mean, $\overline{X}_{B2} = 35$	$\left.{40 \atop 38}\right\}$ Cell total $= 78$ Cell mean, $\overline{X}_{B3} = 39$	262	$\overline{X}_B = \dfrac{262}{6}$ $= 43.67$
(C)	$\left.{51 \atop 55}\right\}$ Cell total $= 106$ Cell mean, $\overline{X}_{C1} = 53$	$\left.{39 \atop 33}\right\}$ Cell total $= 72$ Cell mean, $\overline{X}_{C2} = 36$	$\left.{42 \atop 40}\right\}$ Cell total $= 82$ Cell mean, $\overline{X}_{C3} = 41$	260	$\overline{X}_C = \dfrac{260}{6}$ $= 43.33$
(D)	$\left.{40 \atop 38}\right\}$ Cell total $= 78$ Cell mean, $\overline{X}_{D1} = 39$	$\left.{47 \atop 51}\right\}$ Cell total $= 98$ Cell mean, $\overline{X}_{D2} = 49$	$\left.{45 \atop 39}\right\}$ Cell total $= 84$ Cell mean, $\overline{X}_{D3} = 42$	260	$\overline{X}_D = \dfrac{260}{6}$ $= 43.33$
Column Totals	396	328	332		
Column Means	$\overline{X}_1 = \dfrac{396}{8} = 49.5$	$\overline{X}_2 = \dfrac{328}{8} = 41$	$\overline{X}_3 = \dfrac{332}{8} = 41.5$		

Grand Mean

$$\overline{\overline{X}} = \frac{49.5 + 41 + 41.5}{3} = 44 \text{ or } \frac{45.67 + 43.67 + 43.33 + 43.33}{4} = 44 \text{ or}$$

$$\frac{(49 + 44 + 44) + (57 + 35 + 39) + (53 + 36 + 41) + (39 + 49 + 42)}{12} = 44$$

This design allows the analyst to test three null hypotheses simultaneously.

1. Testing this first hypothesis may help the manager decide, perhaps, which type of machine to buy in the future:

H_0: The average number of units produced is the same with each type of machine; that is, $\mu_1 = \mu_2 = \mu_3$.

H_A: At least one of these population means is different from the others. (At least one equality doesn't hold.)

2. Testing this second hypothesis may help the manager decide, perhaps, on wages, promotions, or firings:

H_0: The average number of units produced is the same with each crew; that is, $\mu_A = \mu_B = \mu_C$.

H_A: At least one of these population means is different from the others. (At least one equality doesn't hold.)

3. Testing this third hypothesis may help the manager decide, perhaps, on the best assignment of particular crews to various machines:

H_0: Machines and crews do not interact with respect to the average number produced.

H_A: There is interaction betweeen machines and crews.

Once again, the total variation in the data is broken down into explained and unexplained categories, but this time the explained variation has *three* components: treatments variation, blocks variation, and interaction variation. (Table 12.7 summarizes this two-way ANOVA test with interaction.)

Table 12.7   A Two-Way ANOVA Table With Interaction

Source of Variation	Sum of Squares (1)	Degrees of Freedom (2)	Mean Square (3) = (1) ÷ (2)	Test Statistic (4)
Treatments	$TSS = 364$	$c - 1 = 2$	$TMS = \dfrac{364}{2} = 182$	$F_T = \dfrac{TMS}{EMS} = \dfrac{182}{7.5}$ $= 24.27$
Blocks	$BSS = 22.67$	$r - 1 = 3$	$BMS = \dfrac{22.67}{3} = 7.56$	$F_B = \dfrac{BMS}{EMS} = \dfrac{7.56}{7.5}$ $= 1.01$
Interactions	$ISS = 629.33$	$(c - 1)(r - 1) = 6$	$IMS = \dfrac{629.33}{6} = 104.89$	$F_I = \dfrac{IMS}{EMS} = \dfrac{104.89}{7.5}$ $= 13.99$
Error	$ESS = 90$	$rc(k - 1) = 12$	$EMS = \dfrac{90}{12} = 7.5$	
Total	$Total\ SS = 1106$	$(rck) - 1 = 23$		

Variation among columns: explained by treatments. Treatments variation is calculated almost precisely as above, the only difference being that the above *TSS* formula must be multiplied by *k,* the number of observations per cell. Thus, treatments variation equals *k* times the number of *rows* (or blocks) multiplied by the sum of the squared deviations of the *column* (or treatment) means from the grand mean:

$$TSS = kr\Sigma(\overline{X}_j - \overline{\overline{X}})^2 = 2(4)[(49.5 - 44)^2 + (41 - 44)^2 + (41.5 - 44)^2] = 364.$$

The applicable degrees of freedom equal the number of treatments columns minus 1, as before: $c - 1 = 2$. Thus, the treatments mean square is $TMS = (364/2) = 182$.

Variation among rows: explained by blocks. Blocks variation, too, is calculated in analogy to the way *BSS* was calculated above: as *k* times the number of *columns* (or treatments) multiplied by the sum of the squared deviations of the row (or block) means from the grand mean. Thus,

$$BSS = kc\Sigma(\overline{X}_i - \overline{\overline{X}})^2 = 2(3)[(45.67 - 44)^2 + (43.67 - 44)^2$$
$$+ (43.33 - 44)^2 + (43.33 - 44)^2] = 22.67.$$

The applicable degrees of freedom equal the number of block rows minus 1, as before: $r - 1 = 3$. Thus, the blocks mean square is $BMS = (22.67/3) = 7.56$.

Variation explained by interaction. Interaction is considered to be present if the deviation of the mean observation for a given treatment/block combination (such as $\overline{X}_{A1}$ = 49 in the first cell of Table 12.6) from the grand mean (such as $\overline{\overline{X}} = 44$) does not equal the combined deviations of the corresponding block mean from the grand mean (such as $\overline{X}_A - \overline{\overline{X}} = 45.67 - 44$) and the corresponding treatment mean from the grand mean (such as $\overline{X}_1 - \overline{\overline{X}} = 49.5 - 44$). In short, interaction is measured as

$$I = (\overline{X}_{ij} - \overline{\overline{X}}) - [(\overline{X}_i - \overline{\overline{X}}) + (\overline{X}_j - \overline{\overline{X}})].$$

This expression can be simplified to

$$I = \overline{X}_{ij} - \overline{X}_i - \overline{X}_j + \overline{\overline{X}}.$$

The **interactions sum of squares (ISS)** equals the number of observations per cell, *k,* times the sum of the squared interactions:

$$ISS = k\Sigma(\overline{X}_{ij} - \overline{X}_i - \overline{X}_j + \overline{\overline{X}})^2.$$

In our example, *ISS* can be calculated as

$$\begin{aligned}
ISS = 2[&(49 - 45.67 - 49.5 + 44)^2 + (44 - 45.67 - 41 + 44)^2 \\
&+ (44 - 45.67 - 41.5 + 44)^2 + (57 - 43.67 - 49.5 + 44)^2 \\
&+ (35 - 43.67 - 41 + 44)^2 + (39 - 43.67 - 41.5 + 44)^2 \\
&+ (53 - 43.33 - 49.5 + 44)^2 + (36 - 43.33 - 41 + 44)^2 \\
&+ (41 - 43.33 - 41.5 + 44)^2 + (39 - 43.33 - 49.5 + 44)^2 \\
&+ (49 - 43.33 - 41 + 44)^2 + (42 - 43.33 - 41.5 + 44)^2] \\
= {}& 629.33.
\end{aligned}$$

The applicable degrees of freedom equal $(c - 1)(r - 1)$ where c is the number of treatment columns and r is the number of block rows; then $(c - 1)(r - 1) = 6$. Thus, the **interactions mean square,** or *IMS,* is $(629.33/6) = 104.89$.

Variation due to error. The unexplained variation is again calculated as the difference between the total sum of squares and the explained variation, but explained variation now is composed of the treatments sum of squares, the blocks sum of squares, and the interactions sum of squares:

$$\text{Total SS} - \text{TSS} - \text{BSS} - \text{ISS} = \text{ESS}.$$

In our example, the total sum of squares can be calculated from Table 12.6 as

$$\text{Total SS} = (50 - 44)^2 + (48 - 44)^2 + \ldots + (39 - 44)^2 = 1,106.$$

Thus, $\text{ESS} = 1,106 - 364 - 22.67 - 629.33 = 90$. Similarly, the degrees of freedom applicable to the total sum of squares [now equal to $(rck) - 1$] minus those associated with the treatments, blocks, and interactions sums of squares equal those associated with the error sum of squares:

$$d.f._{\text{Total SS}} - d.f._{\text{TSS}} - d.f._{\text{BSS}} - d.f._{\text{ISS}} = d.f._{\text{ESS}}$$
$$[(rck) - 1] - (c - 1) - (r - 1) - [(c - 1)(r - 1)] = rc(k - 1).$$

In our case, $rc(k - 1) = 12$; hence, the error mean square is $\text{EMS} = (90/12) = 7.5$.

Practice Problem 3

The Effect of Crews, Machines, and Machine/Crew Interaction on Output

Assuming that the manager described above wants to test the above three null hypotheses (about the effect of machines, crews, and machine/crew interaction on output) at a significance level of 1 percent, we can quickly reach a conclusion. The critical Appendix Table N values are $F_{.01} = 6.93$ for treatments with $d.f. = (2, 12)$; $F_{.01} = 5.95$ for blocks with $d.f. = (3, 12)$; and $F_{.01} = 4.82$ for interactions with $d.f. = (6, 12)$. Given the corresponding computed values of F_T, F_B, and F_I in column 4 of Table 12.7, the manager should (1) reject H_0 with respect to treatments, (2) accept H_0 with respect to blocks, and (3) reject H_0 with respect to interactions. The manager concludes that: (a) the average number of units produced is not the same with each type of machine but (b) is the same with each crew, yet (c) there is interaction between machine types and crews. Can you show that a test at the 5-percent level of significance would have reached the same conclusions? |||

DISCRIMINATING AMONG DIFFERENT POPULATION MEANS

Whenever the analysis of variance leads to the rejection of the null hypothesis of equal population means and, thus, suggests that there are differences among them, the analyst inevitably asks *which* of the means differ. Consider the *one-way* ANOVA problem above concerning the effectiveness of three different brands of toothpaste. It concluded that they differed, but one would surely want to know whether $\mu_1 = \mu_2$ while μ_3 is different, or whether $\mu_1 = \mu_3$ while μ_2 is different, or whether, perhaps, $\mu_2 = \mu_3$ while μ_1 is different. Yet, given sampling error, one would be hesitant to make that determination solely on the basis of the known sample means. In this final section of the chapter, we discuss three different approaches to discriminating among three or more population means, some of which are known to be different from the others.

Establishing Confidence Intervals for Individual Population Means

The first approach makes use of the technique, introduced in Chapter 8, of estimating an individual mean. As noted in Box 8.D on page 315, the limits of a confidence interval for a population mean can be estimated, from a small sample using the t distribution, as

$$\mu = \overline{X} \pm \left(t \frac{s}{\sqrt{n}} \right)$$

In this expression, $s/\sqrt{n}$ serves as an estimate of the unknown population variance, σ^2, but in the ANOVA situation (in which several samples are taken from different populations that have identical variances, σ^2) a better estimate of σ^2 can be obtained by pooling the sample results and using the unexplained variance (or error mean square, EMS) in place of the single sample variance, s^2. If we let n equal sample size (the number of observations made about any treatment, j, and, thus, used to calculate any one sample mean), we get the formula given in Box 12.B.

12.B Limits of a Confidence Interval of a Population Mean for jth Treatment

$$\mu_j = \overline{X}_j \pm \left(t \sqrt{\frac{EMS}{n}} \right)$$

where $\overline{X}_j$ is the sample mean for jth treatment, EMS is the error mean square, and n is sample size, while t is the t-statistic (Appendix Table K) associated with a chosen confidence level and the degrees of freedom of EMS.

In this formula, μ_j and $\overline{X}_j$ refer to the population and sample mean, respectively, for the jth treatment. To find the value of t for any given confidence level—say, 95 percent—we must employ the degrees of freedom associated with EMS—for example, $(r - 1)c$ in the case of one-way ANOVA.

In our earlier example (Table 12.3), $EMS = 5.93$; $(r - 1)c = 12$; accordingly, the value of $t_{.025,12}$ can be found in Appendix Table K as 2.179. Thus, the following confidence intervals can be calculated from the sample means of Table 12.1.

For the population mean of toothpaste 1, the 95-percent confidence-interval limits are the following:

$$\mu_1 = 18.4 \pm \left(2.179 \sqrt{\frac{5.93}{5}} \right)$$

$$\mu_1 = 18.4 \pm 2.37$$

Therefore, the confidence interval is: $16.03 \le \mu_1 \le 20.77$.

For the population mean of toothpaste 2, the 95-percent confidence-interval limits are

$$\mu_2 = 21.8 \pm 2.37.$$

Therefore, the confidence interval is: $19.43 \leq \mu_2 \leq 24.17$.

For the population mean of toothpaste 3, the 95-percent confidence-interval limits are

$$\mu_3 = 15.6 \pm 2.37.$$

Therefore, the confidence interval is: $13.23 \leq \mu_3 \leq 17.97$.

This analysis suggests that μ_1 and μ_2 may well be the same (the confidence intervals overlap) and that μ_1 and μ_3 are the same, but μ_2 and μ_3 clearly differ (the confidence intervals do not overlap).

Establishing Confidence Intervals for the Difference Between Two Population Means

A second approach also uses Chapter 8 material but focuses not on one mean at a time, but on the difference between two means. For this purpose, we refer to Box 8.E on page 317, according to which the limits of a confidence interval for the difference between two population means can be estimated from small and independent samples as

$$\mu_1 - \mu_2 \cong (\overline{X}_1 - \overline{X}_2) \pm \left(t\sqrt{\frac{s_1^2}{n_1} + \frac{s_2^2}{n_2}} \right).$$

The square root can be replaced by $\sqrt{2EMS/n}$ because the common population variance is estimated by the error mean square and because variability is additive for the two samples. The result is the formula in Box 12.C.

12.C Limits of a Confidence Interval of the Difference Between Two Population Means

$$\mu_1 - \mu_2 \cong (\overline{X}_1 - \overline{X}_2) \pm \left(t\sqrt{\frac{2EMS}{n}} \right)$$

where the $\overline{X}$s are the sample means, EMS is the error mean square, n is sample size, while t is the t-statistic (Appendix Table K) associated with a chosen confidence level and the degrees of freedom of EMS.

The value of t is again found as in the previous section; hence, the 95-percent confidence intervals can be calculated as follows.

For the difference between the population means of toothpastes 1 and 2, the 95-percent confidence-interval limits are

$$\mu_1 - \mu_2 \cong (18.4 - 21.8) \pm \left(2.179\sqrt{\frac{2(5.93)}{5}}\right) = -3.4 \pm 3.36.$$

Therefore, the confidence interval is: $-6.76 \le (\mu_1 - \mu_2) \le -.04$. The fact that the interval does not overlap zero and lies entirely below it suggests that a null hypothesis, H_0: $\mu_1 = \mu_2$, should be rejected at the 5-percent significance level (that toothpaste 1 produces *fewer* cavities than toothpaste 2).

For the difference between the population means of toothpastes 1 and 3, the 95-percent confidence-interval limits are

$$\mu_1 - \mu_3 \cong (18.4 - 15.6) \pm 3.36 = 2.8 \pm 3.36.$$

Therefore, the confidence interval is: $-.56 \le (\mu_1 - \mu_3) \le 6.16$. Because this interval does overlap zero, the null hypothesis, H_0: $\mu_1 = \mu_3$, can be accepted at the 5-percent significance level.

For the difference between the population means of toothpastes 2 and 3, the 95-percent confidence-interval limits are

$$\mu_2 - \mu_3 \cong (21.8 - 15.6) \pm 3.36 = 6.2 \pm 3.36.$$

Therefore, the confidence interval is: $2.84 \le (\mu_2 - \mu_3) \le 9.56$. The fact that this interval does not overlap zero and lies entirely above it suggests that the null hypothesis, H_0: $\mu_2 = \mu_3$, should be rejected at the 5-percent significance level (that toothpaste 2 produces *more* cavities than toothpaste 3). Caution is advised: The stated confidence level of 95 percent applies to each test individually, but not to the series of all three. We cannot state with a 95-percent degree of confidence that toothpaste 1 is better than toothpaste 2, which is worse than toothpaste 3, which is as good as toothpaste 1. The chance of making at least one erroneous rejection in the *series* of null hypotheses exceeds 5 percent (for reasons indicated in the third paragraph of this chapter on page 493).

Tukey's *HSD* Test

Another approach yet is **Tukey's *HSD* test** that seeks out "honestly significant differences" among paired sample means. Imagine that an analysis of variance has rejected equality among all the population means of interest, as in our one-way example (Table 12.3). We call k the number of such means to be compared in pairwise fashion, such as the $k = 3$ treatment means in our earlier example. Thus, the number of comparisons equals

$$\frac{k(k - 1)}{2}$$

which, as we have already noted, also equals 3 in our case. According to Tukey, the difference between any two population means is statistically significant at level α if the absolute difference between the corresponding sample means equals or exceeds his honestly significant difference, *HSD,* which is calculated according to the formula given in Box 12.D.

12.D Tukey's *HSD*

$$HSD = q_\alpha \sqrt{\frac{EMS}{n}}$$

where *EMS* is the error mean square, n is sample size, and q_α is found in Appendix Table O.

A key element in the formula is a number denoted by q_α that varies with the number of means to be compared, k, with the error degrees of freedom, and with the desired significance level, α. Appendix Table O provides values of q_α for different values of k, for various error degrees of freedom, and for α levels of .05 and .01. The ANOVA test itself, of course, provides the value of the error mean square, *EMS,* that also appears in the formula, and n equals the number of observations used to calculate the sample mean (which corresponds to r, the number of rows, in Table 12.1).

For our example, with $k = 3$ means, error degrees of freedom $= 12$, and $\alpha = .05$, Appendix Table O yields $q_\alpha = 3.77$. Given $r = 5$ observations per sample and $EMS = 5.93$,

$$HSD = 3.77 \sqrt{\frac{5.93}{5}} = 4.11.$$

Now consider the (absolute) differences between our sample means from Table 12.1:

$$\overline{X}_1 \text{ vs. } \overline{X}_2: 3.4$$
$$\overline{X}_1 \text{ vs. } \overline{X}_3: 2.8$$
$$\overline{X}_2 \text{ vs. } \overline{X}_3: 6.2$$

Only the last number equals or exceeds $HSD = 4.11$; the test, therefore, concludes that population means μ_2 and μ_3 are different.

Practice Problem 4

Practice Problem 3 Revisited

What if the manager in Practice Problem 3 wanted to discriminate among the reportedly different treatment means in Table 12.7 at the $\alpha = .01$ level? He or she would note that in Table 12.6 there were $n = 8$ observations for $k = 3$ treatments and that in Table 12.7 $EMS = 7.5$ with 12 *d.f.* Hence, $q_\alpha = 5.05$ (Appendix Table O), while

$$HSD = 5.05 \sqrt{\frac{7.5}{8}} = 4.89.$$

Then the manager would note the absolute differences among the Table 12.6 column means:

$$\overline{X}_1 \text{ vs. } \overline{X}_2 \text{: } 8.5$$
$$\overline{X}_1 \text{ vs. } \overline{X}_3 \text{: } 8.0$$
$$\overline{X}_2 \text{ vs. } \overline{X}_3 \text{: } .5$$

The manager would conclude that treatment population means μ_1 and μ_2, as well as μ_1 and μ_3 differed, while μ_2 and μ_3 did not (because the former two sample means, but not the latter, equaled or exceeded $HSD = 4.89$). When called upon to make the choice, the manager would, therefore, buy machine 1 in preference to machine 2 and would buy machine 1 in preference to machine 3 but would be indifferent about buying machine 2 or machine 3. |||

||| ANALYTICAL EXAMPLE

12.1 The Fairness of Earnings Differentials

In a 1976 experiment, 100 skilled blue-collar workers, who were employed by a large industrial corporation, were asked to evaluate various pay plans involving different combinations of wages for blue-collar workers and their immediate supervisory personnel. The aim was to see what types of earnings differentials (either among different blue-collar workers or between blue-collar workers and their supervisors) produce perceptions of injustice and, therefore, fuel labor-management disputes and give rise to costly labor turnover and strikes. A perceived-injustice score was recorded twice for each of the situations that correspond to the four cells of the accompanying table: (a) for workers who gave priority to comparisons of their own individual pay with that of like fellow workers (upper-left entry in each cell), and (b) for workers who gave priority to comparisons of their own group's pay with that of the supervisory group (lower-right entry in each cell). "Injustice" is represented by a score of less than 4; "justice" is represented by a score of 4 or more.

A two-way analysis of variance was performed on each scoring. The researcher rejected the null hypothesis of equal column mean scores (identical perceived injustice regardless of the pay differential between groups) at the 5-percent level of significance for both (upper-left and lower-right) sets of scores. (The computed F values were 5.31 and 4.19, respectively, the

critical $F_{.05}$ being less than 4.00.) Thus, regardless of how the workers determined their own measure of fairness (by looking at their relative position vis-à-vis other workers or vis-à-vis supervisors), any feeling of injustice was always accentuated by large as compared to small differentials between group pay levels. (Compare 4.19 and 5.10 with 3.85 and 3.71 or compare 3.81 and 4.52 with 3.40 and 3.38.)

Pay Differentials Within Blue-Collar Group	Pay Differentials Between Blue-Collar Workers and Supervisory Personnel	
	Small	Large
Small	4.19 — 3.81	3.85 — 3.40
Large	5.10 — 4.52	3.71 — 3.38

Source: Joanne Martin, "The Fairness of Earnings Differentials: An Experimental Study of the Perceptions of Blue-Collar Workers," *The Journal of Human Resources,* Winter 1982, pp. 110–22.

Summary

1. The *analysis of variance,* frequently referred to by the contraction *ANOVA,* is a statistical technique specially designed to test whether the means of more than two quantitative populations are equal. An independent simple random sample is taken from each of several populations (that are assumed to be normally distributed and to have identical variances). From the sample data, in turn, two independent estimates are derived of σ^2, the assumed common variance of the populations of interest. A first estimate, s_A^2, is based on the variation *among* the sample means; it is an unbiased estimate of σ^2 only if the population means happen to be equal. A second estimate, s_W^2, is based on the variation of individual sample observations *within* each sample; this weighted average of individual sample variances is always an unbiased estimate of σ^2. Therefore, the ratio s_A^2/s_W^2 is close to 1 if the population means are equal to each other; the more it diverges from 1, the greater is the probability that the population means are not equal to each other. Thus, an analysis of *variance* helps us test hypotheses about *means.*

2. Alternative versions of ANOVA reflect alternative designs of experiments. *One-way ANOVA* is performed with data derived from experiments based on the completely randomized design that controls extraneous factors by creating one treatment group for each treatment and assigning each experimental unit to one of these groups by a random process; thus, only one factor (the treatment) is considered as possibly affecting the variable of interest. The one-way ANOVA table summarizes the one-way analysis of variance and contains the ANOVA test statistic, *F*. Because *F* equals the ratio of s_A^2/s_W^2, the probability distribution of *F* (first developed by Ronald A. Fisher) helps us decide whether any given divergence of *F* from 1 is significant enough to warrant rejection of a null hypothesis of equal population means.

3. *Two-way ANOVA* is performed with data derived from experiments based on the randomized block design that controls extraneous factors by dividing the available experimental units into distinctly different, but internally homogeneous blocks and then randomly matching each treatment with one or more units within any given block; thus, two factors (treatments and blocks) are considered as possibly affecting the variable of interest. Two-way ANOVA procedures differ depending on whether the two factors of interest (a) are not interacting or (b) are interacting.

4. Whenever the analysis of variance leads to the rejection of the null hypothesis of equal population means, further analysis can be conducted to determine how these means differ. This further analysis can involve the establishment of confidence intervals for individual population means (or for the difference between them); it can also involve the application of Tukey's *HSD* test.

Key Terms

analysis of variance (ANOVA)
ANOVA table
blocks mean square *(BMS)*
blocks sum of squares *(BSS)*
blocks variation
error
error mean square *(EMS)*
error sum of squares *(ESS)*
explained variance
explained variation
F distributions
F statistic
grand mean
homoscedasticity

interactions mean square *(IMS)*
interactions sum of squares *(ISS)*
one-factor ANOVA
one-way ANOVA
residual variation
total sum of squares *(Total SS)*
treatments mean square *(TMS)*
treatments sum of squares *(TSS)*
treatments variation
Tukey's *HSD* test
two-factor ANOVA
two-way ANOVA
unexplained variance
unexplained variation

Questions and Problems

The computer program noted at the end of the chapter can be used to work most of the subsequent problems.

1. A manager wants to test, at the 2.5-percent level of significance, whether the mean delivery time of components supplied by five outside contractors is the same. A random sample of past records reveals the delivery times (in days after orders were placed) given in Table 12.A. Perform the test desired by the manager, showing your computations in detail, along with an ANOVA table.

Table 12.A   Delivery Times of 5 Contractors (days)

Observation	Supplier 1	2	3	4	5
1	3	4	2	7	2
2	3	1	4	1	3
3	2	3	3	1	4
4	5	5	1	2	5
5	3	2	2	3	4

2. A government agency wants to test, at the 1-percent level of significance, whether the average cost of a given market basket of goods is the same in three cities. A random sample of 8 stores in each of the cities reveals the data given in Table 12.B. Perform the test desired by the agency, showing your computations in detail, along with an ANOVA table.

Table 12.B   The Average Cost of a Given Market Basket

Store	Boston (1)	Chicago (2)	Los Angeles (3)
1	$73.64	$67.05	$72.20
2	69.27	77.50	53.20
3	63.88	75.30	52.22
4	77.50	68.15	55.55
5	74.11	73.11	56.80
6	67.82	69.72	54.72
7	68.28	76.00	68.39
8	70.03	71.85	51.20

3. An economist wants to test, at the 2.5-percent level of significance, whether mean housing prices are the same regardless of which of 3 air-pollution levels typically prevails. A random sample of house purchases in 3 areas yields the price data (in thousands of dollars) given in Table 12.C. Perform the test desired by the economist, showing your computations in detail, along with an ANOVA table.

Table 12.C   Mean Housing Prices (thousands of dollars)

Observation	Air Pollution Level Low	Moderate	High
1	120	61	40
2	68	59	55
3	40	110	73
4	95	75	45
5	83	80	64

4. An advertising manager wants to test whether average sales are the same regardless of which of four available displays is used to advertise a product. Independent random samples of stores using displays A–D, respectively, show the following monthly sales (in dollars).
A: 102, 113, 98, 123, 141, 152, 173, 129
B: 53, 66, 88, 79, 41, 52, 71, 50
C: 172, 189, 193, 205, 252, 98, 87, 79
D: 130, 131, 133, 141, 129, 118, 106, 100
Perform the desired test at the 5-percent level of significance, showing your ANOVA table. What assumptions are you making?

5. An office manager wants to test whether output per worker is the same regardless of which of three degrees of office-crowding prevails. Independent random samples from three offices show the following data (in typed pages per day).
Office A (no crowding): 56, 63, 71, 46, 61, 66
Office B (some crowding): 32, 41, 43, 46, 38, 49
Office C (severe crowding): 28, 32, 30, 26, 25, 31
Perform the desired test at the 2.5-percent level of significance, showing your ANOVA table. What conditions must hold in order for your results to be valid?

6. A corporate executive wants to test whether the average total cost of producing lawnmowers is the same in four parts of the country. Independent random samples of seven firms in each of the four regions show the following data (in dollars per mower).

Region A: 88, 92, 85, 80, 78, 81, 80
Region B: 92, 96, 98, 91, 88, 97, 106
Region C: 57, 88, 102, 99, 62, 49, 59
Region D: 76, 79, 80, 81, 88, 59, 93

Perform the desired test at the 1-percent level of significance, showing your ANOVA table. What conditions must hold in order for your results to be valid?

7. An educator wants to test whether the average reading score of children is the same regardless of the typeface used in a text. The identical book is printed in three different ways and used in three different schools. Subsequent independent random samples in the three schools show the following data (reading scores).

Typeface A: 88, 92, 66, 55, 73, 82, 86
Typeface B: 75, 56, 88, 93, 89, 76, 66
Typeface C: 56, 63, 78, 82, 82, 88, 96

Perform the desired test at the 5-percent level of significance, showing your ANOVA table. What conditions must hold in order for your results to be valid?

8. An economist wants to test whether the average total cost of producing lawnmowers is, on average, the same for small, medium-sized, and large firms. Independent random samples of 10 firms in each category show the following data (in dollars per mower).

A. Small Firms: 101, 99, 89, 98, 105, 110, 82, 107, 105, 98
B. Medium-sized Firms: 73, 82, 88, 86, 110, 57, 69, 83, 66, 77
C. Large Firms: 49, 52, 66, 51, 53, 57, 69, 76, 81, 60

Perform the desired test at the 2.5-percent level of significance, showing your ANOVA table. What conditions must hold in order for your results to be valid?

9. The manufacturer of dog food wants to test whether dogs, on average, like three new foods equally well. Each of the food is fed to six different dogs at their regular mealtime; their consumption is recorded as follows (in ounces).

Food A: 16, 18, 22, 10, 19, 23
Food B: 3, 6, 18, 29, 36, 48
Food C: 12, 22, 32, 28, 10, 6

Perform the desired test at the 1-percent level of significance, showing your ANOVA table. What conditions must hold in order for your results to be valid?

10. An airline manager wants to test whether airplanes, on average, get the same minutes of flight time per gallon (mpg) with three different brands of gas. An appropriate experiment yields the following data (in mpg).

A. 100 octane, low-lead: 50, 59, 51, 48, 43, 63
B. 130 octane, low-lead: 59, 63, 66, 71, 65, 73
C. 180 octane, no-lead: 63, 76, 81, 79, 80, 85

Perform the desired test at the 5-percent level of significance, showing your ANOVA table. What conditions must hold in order for your results to be valid?

11. A police chief wants to test whether, on average, the incidence of crime is the same in four city districts. The number of police calls received in the last ten days is used as a sample:

District A: 11, 17, 9, 3, 20, 18, 15, 13, 8, 15
District B: 8, 15, 7, 0, 12, 16, 10, 11, 6, 10
District C: 6, 10, 6, 1, 6, 8, 8, 8, 3, 7
District D: 29, 33, 12, 13, 33, 22, 17, 15, 8, 20

Perform the desired test at the 2.5-percent level of significance, showing your ANOVA table. What conditions must hold in order for your results to be valid?

12. A doctor wants to test whether dieting men, on average, lose the same weight during a week regardless of which of three diets are involved. The following sample data are available (in pounds of weight loss per week).

Diet A: 5.1, 6.7, 7.0, 5.3, 9.1, 4.3
Diet B: 6.1, 7.3, 8.1, 9.1, 10.3, 12.1
Diet C: 7.3, 8.9, 9.4, 12.0, 15.7, 9.5

Perform the desired test at the 1-percent level of significance, showing your ANOVA table. What conditions must hold in order for your results to be valid?

13. An economist wants to test whether the debt-to-equity ratio is, on average, the same for firms in four industries. Independent random samples of six firms in each industry show the following results:

Industry A: .1, .2, .3, 0, .1, .2
Industry B: .3, .2, .4, .5, .3, .1
Industry C: .1, .2, .3, .1, 0, 0
Industry D: .9, .8, .1, 1.2, 1.3, .4

Perform the desired test at the 5-percent level of significance, showing your ANOVA table. What conditions must hold in order for your results to be valid?

14. An analyst wants to test whether the price per share, on average, differs among the New York Stock Exchange, the American Stock Exchange, and

the over-the-counter market. Independent random samples of eight stocks from each market yield the following (in dollars per share).

A. NYSE: 45, 56, 82, 49, 53, 61, 48, 51
B. ASE: 17, 19, 27, 22, 31, 41, 15, 16
C. OTC: 30, 19, 82, 49, 31, 19, 16, 51

Perform the desired test at the 2.5-percent level of significance, showing your ANOVA table. What conditions must hold in order for your results to be valid?

15. An agricultural researcher wants to test, at the 2.5-percent level of significance, whether the mean yield associated with 5 types of fertilizer and with 3 types of soil is the same. No interaction is assumed. A (randomized block design) experiment reveals the yields (in pints of strawberries per season per unit of land) given in Table 12.D. Perform the test desired by the researcher, showing your computations in detail, along with an ANOVA table.

Table 12.D Strawberry Yields (pints)

Soil Quality, i	Fertilizer, j				
	1	2	3	4	5
A	15	13	8	17	9
B	10	12	15	15	11
C	14	9	6	14	12

16. A manager wants to test, at the 5-percent level of significance, whether the mean amount of delinquent debt repaid is the same for 4 types of collection methods (a friendly letter, a nasty letter, a telephone call, a personal visit) and 3 categories of amount overdue. Table 12.E shows the sample data that are available (in dollars paid per overdue account). Perform the test desired by the manager, showing your computations in detail, along with an ANOVA table. Assume no interaction.

Table 12.E Dollars Paid Per Overdue Account

Amount Overdue, i	Collection Method, j			
	Friendly Letter, F	Nasty Letter, N	Telephone Call, T	Personal Visit, P
(A) under $500	10	30	10	200
(B) $500 to under $1,000	100	300	50	500
(C) $1,000 and more	500	1,000	100	1,200

17. A manager wants to test, at the 2.5-percent level of significance, whether the average quality of a paint job is the same for 5 types of paint and 5 types of surface to which it is applied. Table 12.F shows the sample data that are available (in quality scores). Perform the desired test, showing your ANOVA table. Assume no interaction.

Table 12.F Quality of Paint Jobs (scores)

Surface, i	Paint Type, j				
	1	2	3	4	5
A (Pine)	80	86	89	93	100
B (Oak)	90	94	95	97	100
C (Cedar)	73	69	88	92	96
D (Plastic)	60	73	72	79	86
E (Metal)	100	85	73	84	61

18. The manager of a fast-food chain wants to test, at the 1-percent level of significance, whether the average monthly sales are the same for 5 types of outlet and 6 months of the year. Table 12.G shows the sample data that are available (in thousands of dollars). Perform the desired test, showing your ANOVA table. Assume no interaction.

19. An engineer wants to test, at the 5-percent level of significance, whether the tensile strength of steel rods is, on average, the same for 4 production temperatures and 3 production pressures. Table 12.H shows the sample data that are available (in pounds per square inch). Perform the desired test, showing your ANOVA table. Assume no interaction.

20. An economist wants to test, at the 2.5-percent level of significance, whether the amount of taxes figured by 3 tax preparers for 5 types of returns is, on average, the same. Table 12.I shows the sample data that are available (in dollars per return). Perform the desired test, showing your ANOVA table. Assume no interaction.

Table 12.G Monthly Sales (thousands of dollars)

Month, i	Outlet, j				
	1	2	3	4	5
A. July	15.1	8.9	22.3	5.9	9.2
B. August	20.7	23.6	56.6	25.0	19.2
C. September	13.2	13.2	27.8	25.0	19.2
D. October	12.8	10.9	25.8	16.7	10.3
E. November	8.7	9.9	25.9	10.2	9.3
F. December	5.2	9.6	13.0	2.1	5.0

Table 12.H Tensile Strength of Steel Rods (psi)

Pressure During Production, i	Production Temperatures, j			
	(1) Low	(2) Medium	(3) High	(4) Extreme
A. Low	600	700	800	900
B. Medium	700	800	900	1,000
C. High	800	900	1,000	500

Table 12.I Taxes Due Figured By Tax Preparers (dollars)

Type of Return, i	Preparer, j		
	1	2	3
A	250.97	147.31	0
B	1,900.00	1,782.57	1,300.28
C	2,878.00	2,878.00	1,300.99
D	9,822.87	9,822.87	7,929.25
E	22,357.00	39,852.00	18,999.53

21. An economist wants to test, at the 1-percent level of significance, whether the price of a 1-minute advertisement is, on average, the same for 4 geographic areas and 3 sizes of radio stations. Table 12.J shows the sample data that are available (in dollars per minute). Perform the desired test, showing your ANOVA table. Assume no interaction.

22. A mayor wants to test, at the 5-percent level of significance, whether property-tax assessments are, on average, the same for 4 assessors and 5 types of property. Table 12.K shows the sample data that are available (in thousands of dollars of assessed value). Perform the desired test, showing your ANOVA table. Assume no interaction.

23. The Environmental Protection Agency wants to test, at the 2.5-percent level of significance, whether, on average, sulfur dioxide levels in the air over Massachusetts are found to be the same by 3 alternative detection methods and 4 laboratories that analyze air samples. Table 12.L shows the sample data that are available (in parts per million reported). Perform the desired test, showing your ANOVA table. Assume no interaction.

Table 12.J Price of 1-Minute Advertisement (dollars)

Size of Radio Station, i	Geographic Area, j			
	North	East	South	West
A. 1,000 watts	100	150	90	150
B. 5,000 watts	200	250	120	290
C. 10,000 watts	500	600	220	850

Table 12.K Property-Tax Assessments (thousands of dollars)

Property Type, i	Assessor, j			
	1	2	3	4
A	20.5	19.8	21.0	22.5
B	30.7	39.8	43.0	29.9
C	89.0	92.0	87.0	73.5
D	100.7	110.8	120.0	105.6
E	23.2	37.9	120.0	67.3

Table 12.L Sulfur Dioxide Reported in Massachusetts Air (ppm)

Laboratory, i	Detection Method, j		
	1	2	3
A	3.1	5.6	11.7
B	3.1	6.5	7.1
C	5.9	11.6	23.0
D	1.0	6.7	2.3

Table 12.M Annual Percentage Price Increases

Size of Firm, i	Industry Concentration Level, j		
	Low = 1	Medium = 2	High = 3
Small = A	12	5	2
	10	6	5
	9	9	7
	9	7	4
Medium = B	8	6	5
	8	9	9
	9	11	7
	10	8	8
Large = C	3	8	12
	5	10	15
	6	12	20
	4	11	16

24. An economist wants to test, at the 1-percent level of significance, whether average annual price increases associated with 3 levels of industry concentration and 3 sizes of firm are the same. Interaction between the two factors is suspected. The pertinent data on annual percentage price increases have been collected in Table 12.M. Perform the test desired by the economist, showing your computations in detail, along with an ANOVA table.

25. An engineer wants to test, at the 5-percent level of significance, whether the average tensile strength of plastic sheets is the same for 3 temperatures and 3 pressures applied during production. Interaction between the two factors is suspected. Given the sample data of Table 12.N, perform the test, showing your ANOVA table.

Table 12.N   Tensile Strength of Plastic Sheets (*psi*)

Pressure, i	Temperature, j					
	Low = 1		Medium = 2		High = 3	
Low = A	18 19 21	23 16 20	18 19 21	23 17 22	27 29 35	39 40 39
Medium = B	27 29 35	39 40 42	39 40 41	42 42 42	18 19 20	15 14 17
High = C	35 38 42	48 50 56	18 17 14	12 13 10	9 7 8	6 5 4

Table 12.O   Strength of Plywood (*psi*)

Type of Wood, i	Type of Glue, j					
	1		2		3	
Birch	30 36	33 35	40 46	43 45	50 75	56 63
Fir	40 45	48 47	56 63	61 70	71 82	79 87
Larch	41 47	49 56	69 75	79 86	22 36	29 31
Pine	72 80	75 86	50 51	49 47	22 36	31 27

26. An engineer wants to test, at the 2.5-percent level of significance, whether the average strength of plywood boards is the same for 3 types of glue and 4 types of wood. Interaction between the two factors is suspected. Given the sample data of Table 12.O, perform the test, showing your ANOVA table.

27. The captain of a fishing fleet wants to test, at the 1-percent level of significance, whether the average weight of fish caught is the same for 5 species and 3 fishing areas. Interaction between the two factors is suspected. Given the sample data of Table 12.P, perform the test, showing your ANOVA table.

Table 12.P   Weight of Fish (pounds)

Fishing Area, i	Species, j									
	Cod		Flounder		Haddock		Perch		Whiting	
Casco Bay	.9 1.1	.3 2.0	2.6 3.1	2.9 5.0	8.9 10.5	9.3 5.1	5.4 2.9	4.0 3.6	5.4 4.3	6.6 3.3
Georges Bank	2.0 6.1	2.3 6.0	6.9 7.0	5.1 4.9	10.5 9.3	11.2 7.5	1.7 7.7	1.9 5.0	3.3 9.0	6.9 10.3
Long Island Sound	.9 1.1	.3 1.2	1.2 3.7	3.1 6.1	6.1 1.7	2.3 1.9	1.2 2.1	3.6 7.0	3.0 2.6	2.1 .9

Table 12.Q   Approval Scores

Sex of Listener, i	Theme Song, j								
	1			2			3		
Male	50	60	63	75	75	81	99	100	100
Female	99	100	100	75	75	81	50	50	0

28. The manager of a new radio station wants to test, at the 5-percent level of significance, whether the average approval score for a station theme song is the same for 3 alternative songs and the 2 sexes. Interaction between the two factors is suspected. Given the sample data of Table 12.Q, perform the test, showing your ANOVA table.

29. A psychologist wants to test, at the 2.5-percent level of significance, whether the average job satisfaction score is the same for 3 types of work and 4 types of employment experience. Interaction between the two factors is suspected. Given the sample data of Table 12.R, perform the test, showing your ANOVA table.

30. An economist wants to test, at the 1-percent level of significance, whether the average price per pound of soap is the same for 3 brands and 5 supermarkets. Interaction between the two factors is suspected. Given the sample data of Table 12.S, perform the test, showing your ANOVA table.

31. A supervisor wants to test, at the 5-percent level of significance, whether the average productivity score is the same for 3 types of work and the two sexes. Interaction between the two factors is sus-

Table 12.R Job Satisfaction Scores

Experience on the Job, i	Nature of Work, j					
	Unstructured		Some Structure		Highly Structured	
A. Less than 5 years	50	60	60	70	70	60
B. 5–under 10 years	70	80	70	80	80	80
C. 10 years or more	89	92	92	88	100	92

pected. Given the sample data of Table 12.T, perform the test, showing your ANOVA table.

32. An economist wants to test, at the 2.5-percent level of significance, whether the average charge per minute of telephone conversation originating in Boston is the same for 3 telephone companies and 5 destination cities. Interaction between the two

Table 12.S Prices of Soap (dollars per pound)

Supermarket, i	Brand, j					
	1		2		3	
A	5.60	7.20	10.25	11.50	2.20	1.70
B	5.90	8.33	7.90	12.00	2.90	3.10
C	6.71	7.90	14.00	17.00	2.90	3.10
D	8.50	8.50	10.00	11.00	1.70	1.29
E	9.10	5.90	11.75	12.01	1.56	6.00

Table 12.T Productivity Scores

Employee Sex, i	Nature of Work, j								
	Relaxed			Some Stress			Highly Stressful		
Male	88	92	61	66	76	85	66	76	88
Female	90	99	100	85	76	60	75	89	93

factors is suspected. Given the sample data of Table 12.U, perform the test, showing your ANOVA table.

Table 12.U Charge per Minute of Telephoning (dollars)

Cities Called, i	AT&T		GTE		Sprint	
Atlanta	.29	.35	.15	.12	.10	.09
Dallas	.40	.32	.23	.23	.19	.21
Fresno	.40	.40	.53	.61	.59	.61
Lincoln	.18	.19	.60	.42	.33	.29
Roanoke	.18	.20	.33	.39	.46	.58

(Column group header: Company, j)

33. A farm manager wants to test, at the 1-percent level of significance, whether the average yield per acre of wheat is the same for 3 degrees of fertilization and 3 types of wheat. Interaction between the two factors is suspected. Given the data of Table 12.V, perform the test, showing your ANOVA table.

34. An economist wants to test, at the 5-percent level of significance, whether the average number of strikes is the same in 4 industries and 4 years. Interaction between the two factors is suspected. Given the sample data of Table 12.W, perform the test, showing your ANOVA table.

35. An orchardist wants to test, at the 2.5-percent level of significance, whether the average yield of apple trees is the same for 3 types of tree and 4 types of weather. Interaction between the two factors is suspected. Given the sample data of Table 12.X, perform the test, showing your ANOVA table.

36. A shopper wants to test, at the 1-percent level of significance, whether the average price of electric dryers is the same in 4 department stores and for 3 brands. Interaction between the two factors is suspected. Given the sample data of Table 12.Y (that were collected at different times of the year), perform the test, showing your ANOVA table.

37. An economist wants to test, at the 5-percent level of significance, whether the average price of a di-

Table 12.V Wheat Yields (bushels per acre)

Type of Wheat, i	Low			Medium			High		
A	24	29	31	66	70	71	87	85	81
B	36	39	43	29	23	24	15	12	10
C	40	43	45	77	87	80	39	29	33

(Column group header: Fertilizer Amount, j)

Table 12.W The Incidence of Strikes (number per firm per year)

Year, i	1			2			3			4		
1983	0	2	3	1	2	0	4	3	6	1	0	2
1984	0	1	0	0	0	1	1	0	0	1	0	0
1985	2	2	1	2	1	3	3	5	2	1	1	1
1986	7	6	8	5	2	2	10	3	6	0	0	1

(Column group header: Industry, j)

Table 12.X Apple Tree Yields (pounds per tree)

Time of Last Winter Frost, i	Type of Tree, j								
	Golden Delicious			Macintosh			Northern Spy		
January	189	192	210	291	287	245	348	352	370
February	210	223	231	286	279	236	320	338	341
March	231	245	269	241	242	250	356	361	300
April	272	287	291	260	240	280	327	333	333

Table 12.Y Prices of Electric Dryers (dollars)

Brand, i	Department Store, j								
	1			2			3		
A	129	150	135	150	150	150	200	150	100
B	220	220	220	200	200	200	300	310	315
C	150	160	175	250	250	250	150	250	170

Table 12.Z The Price of a Divorce (lawyer's fee in dollars)

Residence, i	State, j								
	Alabama			California			Wyoming		
Rural	333	401	510	792	890	642	200	200	210
Urban	792	853	980	1,560	1,710	1,900	510	600	850

vorce is the same in 3 states and for cities and rural areas. Interaction between the two factors is suspected. Given the sample data of Table 12.Z, perform the test, showing your ANOVA table.

38. Reconsider Table 12.5 and note that in Practice Problem 2 the null hypothesis about the equality of block means is rejected. Calculate 98-percent confidence intervals for each of these population means.

39. Reconsider problem 7; calculate a 95-percent confidence interval for each of the three treatment population means.

40. Reconsider problem 2; calculate a 99-percent confidence interval for each of the three treatment population means.

41. Reconsider problem 9; calculate a 99-percent confidence interval for each of the three treatment population means.

42. Reconsider problem 4; calculate a 95-percent confidence interval for each of the four treatment population means.

43. Reconsider problem 13; calculate a 95-percent confidence interval for each of the four treatment population means.

44. Reconsider problem 6; calculate a 99-percent confidence interval for each of the four treatment population means.

45. Reconsider problem 19; calculate a 95-percent confidence interval for each of the four treatment and three block population means.

46. Reconsider problem 16; calculate a 95-percent confidence interval for each of the four treatment and three block population means.

47. Reconsider problem 21; calculate a 99-percent confidence interval for each of the four treatment and three block population means.

48. Reconsider problem 18; calculate a 99-percent confidence interval for each of the five treatment and six block population means.

49. Reconsider problem 2 and calculate 95-percent confidence intervals for the difference between each of the three pairs of population means.

50. Reconsider problem 24; calculate a 99-percent confidence interval for the difference between each of the three pairs of treatment population means and between each of the three pairs of block population means.

51. Reconsider problem 25; calculate a 95-percent confidence interval for the difference between each of the three pairs of treatment population means and between each of the three pairs of block population means.

52. Reconsider problem 28; calculate a 95-percent confidence interval for the difference between each of three pairs of treatment population means and between two block population means.

53. Reconsider problem 27; calculate a 99-percent confidence interval for the difference between each of the three pairs of block population means.

54. Reconsider problem 30; calculate a 99-percent confidence interval for the difference between each of the three pairs of treatment population means.

55. Reconsider problem 31; calculate a 95-percent confidence interval for the difference between each of the three pairs of treatment population means and between the two block population means.

56. As in problem 49, once again reconsider problem 2, but this time employ Tukey's HSD test at $\alpha = .05$ to assess the difference between each of the three pairs of population means.

57. Reconsider Practice Problem 3. Note that the F test rejected the null hypothesis of equality of interactions means. Apply Tukey's HSD test at $\alpha = .01$ to discriminate among these means.

58. Employ Tukey's HSD test at $\alpha = .05$ to assess the difference between each of the three pairs of treatment population means in problem 10.

59. Employ Tukey's HSD test at $\alpha = .01$ to assess the difference between each of the three pairs of treatment population means and each of the three pairs of block population means in problem 33 on page 522.

60. Employ Tukey's HSD test at $\alpha = .01$ to assess the difference between each of the three pairs of treatment population means and each of the three pairs of block population means in problem 36.

Selected Readings

Anderson, V. L., and R. A. McLean. *Design of Experiments: A Realistic Approach*. New York: Marcel Dekker, 1974.

Cochran, W. G., and G. M. Cox. *Experimental Designs,* 2nd ed. New York: Wiley, 1957.

Davies, Owen L., ed. *The Design and Analysis of Industrial Experiments,* 2nd ed. New York: Hafner, 1956.

Fisher, Ronald A. *Statistical Methods for Research Workers,* 14th ed. New York: Hafner, 1970.

Fisher, Ronald A., and W. A. Mackenzie. "Studies in Crop Variation, II. The Manurial Response of Different Potato Varieties." *Journal of Agricultural Science,* 1923, pp. 311–20. The seminal article that introduced the analysis of variance.

Neter, John, and William Wasserman. *Applied Linear Statistical Models*. Homewood, Ill.: Irwin, 1974.

Scheffé, H. *The Analysis of Variance*. New York: Wiley, 1959. A classic and fairly advanced book on the subject.

Computer Programs

The DAYSAL-2 personal-computer diskettes that accompany this book contain one program of interest to this chapter:

12. Analysis of Variance lets you perform one-way ANOVA, two-way ANOVA without and with interaction, as well as three-way ANOVA, which is introduced in the *Student Workbook*. You can also develop confidence intervals for all the means—and the differences between them—that are inevitably computed during ANOVA. Numerous confidence levels are available.

CHAPTER THIRTEEN

Simple Regression and Correlation

In many situations that are of interest to business administrators or to economists or to the practitioners of almost any other field of human endeavor, the value of one variable is associated with the value of another variable in some systematic way. Consider how quantity demanded (or supplied) varies with price, how output varies with input, cost varies with output, saving varies with income, income varies with education, interest rate varies with money supply, investment varies with interest rate, unemployment varies with inflation. Consider the link between the frequency of repairs and the age of equipment, between advertising expenditures and sales, R&D spending and corporate profits, corporate profits and stock prices, entry barriers and economic concentration, grade-point averages of graduates and their starting salaries, pre-employment test scores and subsequent job performance, assembly-line speeds and defective units produced, the unemployment rate at election time and the percentage of votes captured by an incumbent president. The list goes on, but one thing is instantly clear: an awareness of such linkages between variables can be of great usefulness to researchers and decision makers because it enables them to predict, from a knowledge of the value of one variable, the value of the other variable. Regression and correlation analysis aid such an endeavor by establishing (a) what types of linkages between variables exist and (b) how strong these linkages are. A word of caution must, however, be sounded at the outset: the techniques to be described in this chapter and the next are designed to determine the existence and strength of associations between variables, *but they are unable to prove anything about possible cause-and-effect relationships.* If high R&D spending, for example, were shown to be related to high corporate profit, this relationship would not prove that high R&D spending was *causing* high corporate profit. In fact, the opposite might be true (high profit might give rise to high R&D spending), or neither variable might be causally related to the other, despite their association. Nevertheless, a knowledge of such an association would be very helpful in that it would allow one to estimate, say, an unknown R&D spending figure from a known value of corporate profit. The knowledge of such an association would also be useful as a first step in a broader investigation of cause-and-effect relations. Indeed, the discovery of a systematic association between two variables frequently provides the initial impetus to further studies about causation. Close-Up 13.3, "Smoking and Health," provides a case in point.

The first major section of this chapter introduces **regression analysis,** a statistical method the central focus of which is the establishment of an *equation* that allows the unknown value of one variable to be estimated from the known value of one or more other variables. When a single variable is used to estimate the value of an unknown variable (a procedure that might be employed with respect to all of the two-variable examples given above), the method is referred to as **simple regression analysis;** thus, the term *simple* is not to be interpreted as a synonym for *easy.* When, on the other hand, several variables are used to estimate the value of an unknown variable (a procedure that will be introduced in the next chapter), the method is referred to as **multiple regression analysis.** The origin of the term *regression* itself is described in Biography 13.1 (at the end of the chapter), which introduces Sir Francis Galton who first developed the quantitative expressions that show how two or more variables are related and that are now referred to as *regression equations.*

The second major section of this chapter introduces **simple correlation analysis,** a statistical method the central focus of which is the establishment of an *index* that provides, in a single number, an instant measure of the strength of association between two variables. Depending on the size of this quantitative measure, one can tell how closely two variables move together and, therefore, how reliably one can estimate one with the help of the other.

SIMPLE REGRESSION ANALYSIS

It is easiest to introduce the nature of regression analysis with the help of the extended example introduced in Practice Problem 1 below. Before we turn to it, however, it is necessary to understand a number of basic concepts.

Basic Concepts

To begin with, we must become familiar with the names typically given to the variables under study and with the nature of the relationship between them.

Independent versus dependent variable. In regression analysis, the variable the value of which is assumed known (and that is being used to explain or predict the value of the other variable) is symbolized by X and is referred to as the **independent, explanatory,** or **predictor variable** and, on occasion, even as the **regressor.** In contrast, the variable the value of which is assumed unknown (and that is being explained or predicted on the basis of its relationship with the other variable) is symbolized by Y and is referred to as the **dependent, explained,** or **predicted variable** and, on occasion, even as the **regressand.**

Deterministic versus stochastic relationships. The relationship between any two variables, X and Y, can be one of two types. It can be a precise, exact, or **deterministic relationship,** such that the value of Y is uniquely determined when ever the value of X is specified. It can, in contrast, be an imprecise, inexact, or **stochastic relationship,** such that many possible values of Y can be associated with any one value of X. The former type of relationship abounds in the physical sciences; the latter one (which alone will concern us in this chapter and the next) is common in the social sciences.

Panel (a) of Figure 13.1 illustrates a deterministic relationship that is described by the equation $Y = a + bX$. In this particular example, Y equals degrees Fahrenheit and X equals degrees Celsius, while a and b are constants that represent, respectively, the vertical intercept and the slope of the straight line. Note that for any given value of X only one possible value of Y exists. Thus, 0° Celsius corresponds to 32° Fahrenheit, 50° C corresponds to 122° F, and 100° C corresponds to 212° F. Just like the three fat dots shown in the graph, every other conceivable pair of X and Y values lies precisely on the straight line that represents the equation $Y = 32 + 1.8X$.

Panel (b) of Figure 13.1, in contrast, illustrates a stochastic relationship that is described by the equation $\hat{Y}_X = a + bX$. In this particular example, $\hat{Y}_X$ (pronounced "Y hat sub X") equals the value of Y (sales) that is predicted from a knowledge of X (advertising expenditures), while a and b once again represent, respectively, the vertical intercept and the slope of the straight line. This time, however, it is possible to observe different values of Y for any given X. Points A and B, for example, represent those two pairs of observations, among the 25 pairs plotted in the graph, that involve advertising expenditures of $X = \$1,000$ per month. Yet sales were not the same in these two instances: sales equaled 130 units in one case (point A) and 200 units in the other (point B). Thus, anyone who would use the equation of the straight line in panel (b) to estimate the volume of sales associated with \$1,000 of advertising expenditures would get an imprecise answer of $\hat{Y}_X = 100 + .05(1,000) = 150$ (which equals the height of the straight line at $X = 1,000$). Neither one of the actual observations of Y at $X = 1,000$, however, equals $\hat{Y}_X$. Instead, both can be found as

$$Y = \hat{Y}_X + e = a + bX + e,$$

where e is an error term that can take on negative, zero, or positive values and that reflects the net effect on Y of numerous other variables besides X that are excluded from the analysis. (Observation A, for example, might refer to the experience of one firm, while observation B refers to that of another firm that finds itself in very different circumstances even though it is also spending \$1,000 on advertising.) For observation A, the value of $e = -20$; for observation B, $e = +50$. As will be shown below, regression analysis seeks to establish a line such as the color **regression line** in panel (b) that summarizes the stochastic relationship between X and Y while also minimizing the errors made when the equation of that line is employed to estimate Y from X.

Direct versus inverse relationships. The relationship between any two variables, X and Y, can be classified in still another way. If the values of the dependent variable, Y, *increase* with larger values of the independent variable, X, theirs is said to be a **direct relationship**—in which case, b, the slope of the regression line, is positive. If instead the values of the dependent variable, Y, *decrease* with larger values of the independent variable, X, theirs is an **inverse relationship**—in which case, the value of b is negative. Panels (a) and (b) of Figure 13.2 graphically illustrate the difference between a direct and an inverse relationship. The straight lines of panels (a) and (b) might summarize, respectively, observations about the familiar macroeconomic consumption

Figure 13.1 Deterministic versus Stochastic Relationship

In a deterministic relationship, shown in panel (a), the value of Y is uniquely determined whenever the value of X is specified. The equation of the straight line, $Y = 32 + 1.8X$, predicts Y with precision for any given X. In a stochastic relationship, shown in panel (b), many possible values of Y can be associated with any one value of X. The equation of the straight line, $\hat{Y}_X = 100 + .05X$, therefore, predicts Y only roughly for any given X. This line in panel (b) is a *regression line*.

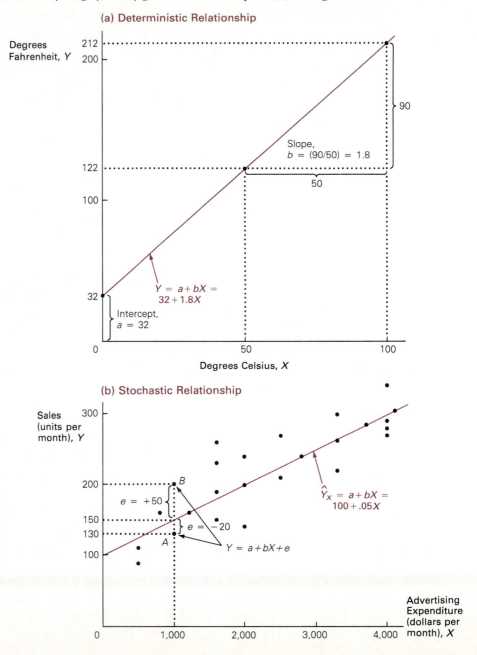

Figure 13.2   Alternative Relationships Between *X* and *Y*

Each dot in these graphs represents a hypothetical pair of observations about an independent variable, *X,* and a dependent variable, *Y.* The color lines summarize the nature of their relationship. It can be linear, as in panels (a) and (b), or curvilinear, as in panels (c) through (h). In addition, the relationship can be direct, as in panels (a), (e), and (g); it can be inverse, as in panels (b) and (f); or it can be a combination of both, as in panels (c), (d), and (h). Naturally, the values of the constants, such as *a, b,* and *c,* differ from one equation to the next.

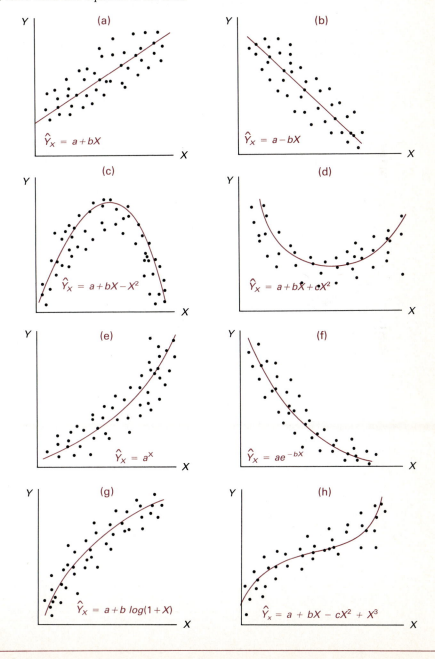

function or the equally familiar microeconomic demand line. Yet relationships between our two variables need not necessarily be *linear*.

As the remaining panels of Figure 13.2 indicate, the association between two variables is often better described as *curvilinear*. Yet, because it is easier to explain, and also because such an analysis can easily be extended to the curvilinear case (as the final section of this chapter will show), the bulk of this chapter will focus on the analysis of linear relationships.

||| Practice Problem 1

A Preliminary Graphical Test of the Association Between Education and Income

Suppose someone desired to investigate the possible association between the extent of formal education attained by the residents of a country and their annual incomes. Because it would be too cumbersome to gather relevant data for every adult member of society, a simple random sample might be taken. To minimize our burden of calculation, we assume that only 20 individuals are included in the sample and that the result is that given in Table 13.1. Typically, however, it is difficult to tell much about the nature of the relationship between the variables of interest from such tabular data. This difficulty is all the more present when the sample data are more extensive than in our

Table 13.1 Sample Data on Education and Income

Individual	Education (years), X	Income (dollars per year), Y
A	2	5,012
B	4	9,680
C	8	28,432
D	8	8,774
E	8	21,008
F	10	26,565
G	12	25,428
H	12	23,113
I	12	22,500
J	12	19,456
K	12	21,690
L	13	24,750
M	14	30,100
N	14	24,798
O	15	28,532
P	15	26,000
Q	16	38,908
R	16	22,050
S	17	33,060
T	21	48,276

Figure 13.3   The Scatter Diagram

The 20 fat dots found in this scatter diagram represent the data pairs of Table 13.1 and help us visualize the relationship of interest to us. The diagram suggests the existence of a direct relationship between *X* and *Y*: more education seems to be associated with higher income. This fact may well be summarized best by a positively sloping straight line.

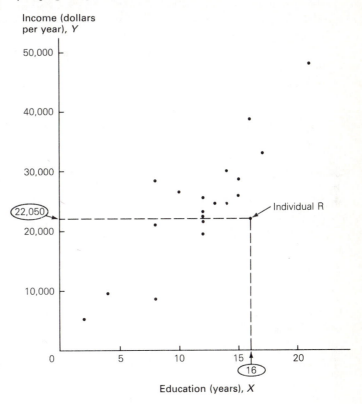

example. The need to make at least a preliminary judgment about what type of regression line might be an appropriate summary of the data (and the panels of Figure 13.2 indicate just some of the many possibilities), it is customary to plot the available data pairs as dots in a **scatter diagram** in such a way that values of the independent variable are measured along the horizontal axis and associated values of the dependent variable along the vertical axis. The data pairs of Table 13.1, for example, appear as the 20 fat dots of Figure 13.3. Note how the information provided by individual R (16 years of education and an annual income of $22,050) has been highlighted in the graph. The scatter of dots provides an important clue: A direct relationship seems to exist between education *(X)* and income *(Y)*; this association may well be summarized best by a (positively sloping) straight line of the type found in panel (a) of Figure 13.2. (The situation just described in this Practice Problem will be referred to throughout this chapter and in many of the following Practice Problems.) |||

Finding the Regression Line: The Use of Eyesight

One conceivable way of establishing a regression line (with the help of which one might predict the value of Y for any given value of X) involves nothing more complicated than using one's eyesight, a ruler, and a pen to draw a line in the scatter diagram that seems to "fit" the scattered dots. The procedure has been used in Figure 13.4. Having drawn the line, one can, of course, easily determine its equation from the graph. In this case, the vertical intercept of the line is seen to equal $a = \$10,000$, while its slope can be determined by placing a triangle underneath the line and calculating the ratio of its vertical side to its horizontal side (which comes to $b = \$1,333.33$ in this case).

This regression line, therefore, suggests that a person's annual income can be estimated as \$10,000 plus an additional \$1,333.33 for each year of education. Naturally, this is only true on average, but not for any particular individual. Thus, the income of

Figure 13.4   A Regression Line Drawn by Sight

The regression line shown here has been drawn by sight with a view toward making it somehow "fit" the scatter of dots that were first seen in Figure 13.3. Given this line, one would estimate a person's annual income as \$10,000 plus an additional \$1,333.33 for each year of education.

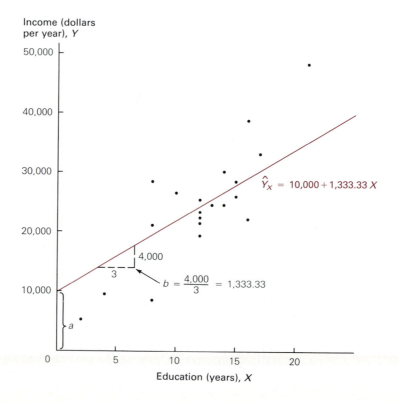

individual R (which we know to be $22,050) would be overestimated from a knowledge of this person's 16-year education and this regression line as $\hat{Y}_{16} = 10,000 + 1,333.33(16) = 31,333.33$. Similarly, the income of individual T (which Table 13.1 shows to be $48,276) would be underestimated from a knowledge of this person's 21-year education as $\hat{Y}_{21} = 10,000 + 1,333.33(21) = 38,000$.

Given the nature of the problem (our desire to summarize a set of scattered dots by a straight line), such overestimation or underestimation is, of course, inevitable. Yet there is another problem with the above procedure, and this problem is the reason why statisticians reject regression lines drawn with the help of eyesight alone as unacceptably crude. The problem is that such a procedure cannot be duplicated, except by sheer accident. Different people, faced with an identical scatter of dots (such as that in Figure 13.3), would undoubtedly use different judgment and draw regression lines with different intercepts and slopes. Which of these different lines could we trust to be the best summary of the data plots? On what basis, furthermore, could we establish confidence intervals for the estimates we would make with the help of such a regression line? Because satisfactory answers to these questions do not exist, statisticians establish regression lines (and their associated equations) in a different way.

Finding the Regression Line: The Method of Least Squares

The commonly used approach to fitting a line to the sample data displayed in the scatter diagram is called the **method of least squares.** The line is derived in such a way that the sum of the squares of the vertical deviations between the line and the individual data plots is minimized. As we noted above, these vertical deviations represent the errors associated with using the regression line to predict Y with the help of X. Figure 13.5 illustrates the nature of the squares the sum of which is being minimized.[1]

Why do statisticians use this particular method of establishing a regression line? Their desire to minimize errors is understood easily enough, but why minimize the *squares* of the errors? One reason is that a summation of unsquared deviations can easily yield a number close to zero (and, thus, give the impression of low error) even when errors are common and huge—simply because positive errors (arising when observed data points lie above the line) and negative errors (arising when observed data points lie below the line) tend to cancel each other. Another reason will be noted later in this section (see footnote 2).

The regression line calculated from sample data by the method of least squares is called the **sample regression line,** or **estimated regression line;** the values of a and b found in its equation, $\hat{Y}_X = a + bX$, are referred to as the **estimated regression coefficients.** By minimizing $\Sigma(Y - \hat{Y}_X)^2 = \Sigma(Y - a - bX)^2$, these coefficients can be calculated as shown in Box 13.A, which appears in the middle of the following page.

---

[1]A supplemental section in Chapter 13 of the *Student Workbook* that accompanies this text shows the mathematics of the procedure and derives the formulas given in Box 13.A.

Figure 13.5 The Method of Least Squares

This scatter diagram displays eight data plots, only two of which lie precisely on the regression line drawn. The six remaining plots diverge from the line by varying distances, as shown by the solid vertical line segments. The method of least squares establishes the regression line, among all the conceivable ones, that minimizes the sum of the squares of such vertical deviations or errors. These squares are represented by the boxes in the graph; it is their combined area that the line minimizes. For any given data set, there exists only one line that minimizes this error sum of squares, $ESS = \Sigma(Y - \hat{Y}_X)^2$, where Y is a point such as A (an individual value of the dependent variable, Y, that is actually observed for a given value of the independent variable, X), while $\hat{Y}_X$ is a point such as B (the value of Y that is estimated for the given value of X with the help of the regression line).

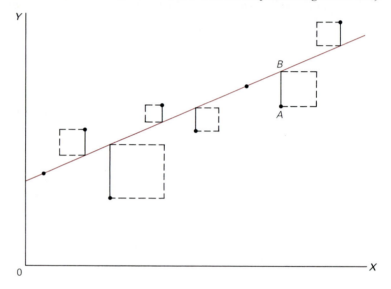

13.A Estimated Regression Coefficients, Using Method of Least Squares

$$b = \frac{\Sigma XY - n\overline{X}\,\overline{Y}}{\Sigma X^2 - n\overline{X}^2} \qquad\qquad a = \overline{Y} - b\overline{X}$$

where b is the slope and a is the intercept of the estimated regression line, while X's are observed values of the independent variable ($\overline{X}$ being their mean) and Y's are the associated observed values of the dependent variable ($\overline{Y}$ being *their* mean), and n is sample size.

||| Practice Problem 2

Using the Least-Squares Method to Determine the Estimated Regression Line for Education and Income

The calculation of the sample regression coefficients, using the formulas in Box 13.A, is performed in Table 13.2 with the help of our sample data given in Table 13.1. Note that the last three columns of Table 13.2 are not needed for the calculation of our

Table 13.2 Calculating the Estimated Regression Line by the Method of Least Squares

Individual	X	Y	XY	X^2	$\hat{Y}_X$	$Y - \hat{Y}_X$	Y^2
A	2	5,012	10,024	4	5,839.23	−827.23	25,120,144
B	4	9,680	38,720	16	9,534.23	145.77	93,702,400
C	8	28,432	227,456	64	16,924.23	11,507.77	808,378,624
D	8	8,774	70,192	64	16,924.23	−8,150.23	76,983,076
E	8	21,008	168,064	64	16,924.23	4,083.77	441,336,064
F	10	26,565	265,650	100	20,619.23	5,945.77	705,699,225
G	12	25,428	305,136	144	24,314.23	1,113.77	646,583,184
H	12	23,113	277,356	144	24,314.23	−1,201.23	534,210,769
I	12	22,500	270,000	144	24,314.23	−1,814.23	506,250,000
J	12	19,456	233,472	144	24,314.23	−4,858.23	378,535,936
K	12	21,690	260,280	144	24,314.23	−2,624.23	470,456,100
L	13	24,750	321,750	169	26,161.73	−1,411.73	612,562,500
M	14	30,100	421,400	196	28,009.23	2,090.77	906,010,000
N	14	24,798	347,172	196	28,009.23	−3,211.23	614,940,804
O	15	28,532	427,980	225	29,856.73	−1,324.73	814,075,024
P	15	26,000	390,000	225	29,856.73	−3,856.73	676,000,000
Q	16	38,908	622,528	256	31,704.23	7,203.77	1,513,832,464
R	16	22,050	352,800	256	31,704.23	−9,654.23	486,202,500
S	17	33,060	562,020	289	33,551.73	−491.73	1,092,963,600
T	21	48,276	1,013,796	441	40,941.73	7,334.27	2,330,572,176
	$\Sigma X =$ 241	$\Sigma Y =$ 488,132	$\Sigma XY =$ 6,585,796	$\Sigma X^2 =$ 3,285		$\Sigma(Y - \hat{Y}_X) = 0$	$\Sigma Y^2 =$ 13,734,414,590

$$\bar{X} = \frac{\Sigma X}{n} = \frac{241}{20} = 12.05 \qquad \bar{Y} = \frac{\Sigma Y}{n} = \frac{488,132}{20} = 24,406.60$$

$$b = \frac{\Sigma XY - n\bar{X}\,\bar{Y}}{\Sigma X^2 - n\bar{X}^2} = \frac{6,585,796 - 20(12.05)(24,406.60)}{3,285 - 20(145.2025)} = \frac{703,805.40}{380.95} = 1,847.50$$

$$a = \bar{Y} - b\bar{X} = 24,406.60 - 1,847.50(12.05) = 2,144.23$$

regression coefficients (these columns will be used later) and that coefficient b must be calculated before a because b appears in the formula for a. Thus, our estimated regression line comes to $\hat{Y}_X = 2,144.23 + 1,847.50\,X$, which is quite different from the one drawn freehand in Figure 13.4. This least-squares line is graphed in Figure 13.6, along with the original scatter of dots. Note how one can estimate, with the help of this line, the annual income of someone with 5 years or 20 years of education (and no such persons were included in the original sample) as

$$\hat{Y}_5 = 2,144.23 + 1,847.50(5) = 11,381.73, \text{ and}$$
$$\hat{Y}_{20} = 2,144.23 + 1,847.50(20) = 39,094.23.$$

These numbers represent the height of the colored line above the X values of 5 and 20, respectively.

Figure 13.6   A Regression Line Derived by the Method of Least Squares

The regression line shown here has been fitted by the method of least squares to the scatter of dots first seen in Figure 13.3. Given this line, one would estimate a person's annual income as $2,144.23 plus an additional $1,847.50 for each year of education. Two features of this and every least-squares regression line should be noted. First, the line goes through point $(\overline{X}, \overline{Y})$—the respective means of the X and Y observations in the sample, here equal to 12.05 and 24,406.60. Second, the sum of the vertical deviations of Ys from the regression line is zero: $\Sigma(Y - \hat{Y}_X) = 0$ (a fact that is illustrated in the right-hand side of Table 13.2 and serves as an excellent check on the accuracy of calculations). In these two senses, the line can be said to "go through the center" of the scatter diagram.

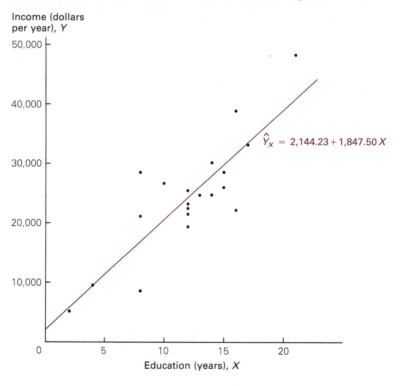

Caution: Most people recognize that it would be unwise to generalize the relation-ship summarized in our regression line to completely different times and places. What is true for 1985 may not be true for 1995; what is true for the United States may not be true for France. Fewer people realize, however, that it is equally unwise to use a regression line to make estimates of Y for values of X that lie beyond the range of the data (here below 2 or above 21 years of education, respectively) from which the esti-mated regression line has been derived. Such extrapolation can easily lead to absurd results (such as the association of certain low incomes with negative years of education or of certain high levels of education with near-infinite incomes). Such absurd results could be obtained because the apparently linear relationship that we see in our scatter diagram may well fail to hold for smaller or larger values of X (and even if it does hold, additional data would probably lead us to estimate a much flatter or steeper regression line with quite a different intercept). In addition, we must become aware of another matter yet that concerns the quality, even within the range of observed X values, of the least-squares regression line itself. To that matter we now turn.

The Estimated versus the True Regression Line

Let us return briefly to our above example (Practice Problems 1 and 2) and recall that we set out to discover the nature of a possible association between the extent of formal education attained by the residents of a country and their annual incomes. Does the estimated regression line of Figure 13.6 provide an answer to our original quest? It certainly appears to do just that. It establishes the existence of a direct relationship between the two variables and also specifies, by means of the equation of the regression line, the mathematical details of that relationship. But now consider this sobering thought: the regression line that appears in Figure 13.6 was fitted to data (first given in Table 13.1) that were obtained not by taking a full-fledged census of the population of interest (of all the adult residents of the country), but by taking a tiny simple random sample of them. What if we were to take another such sample from the same population? Surely, we would acquire a different set of data, would end up with a different scatter diagram, and would derive a different estimated regression line! We could repeat the process; for every new sample, we would estimate yet another regression line. Which one of these could be trusted to reveal the true relationship (if there is a relationship) that exists between X and Y in the population as a whole and that we would find if we took a census of the entire population of adult individuals, recorded data on education and income for every one of them, and derived a least-squares regression line on that basis?

Imagine a scatter diagram for the entire adult population of our country. For each and every value of X (and not just for some of them), it would record not merely a few values of Y (as in Figure 13.6), but a great multitude of them. The least-squares regression line that we might derive from such census data is called the **population regression line,** or **true regression line** (in contrast to the *sample* or *estimated regression line* that we have met above). Such a line might be described by the equation $E(Y) = \alpha + \beta X$, where $E(Y)$ is the expected value of Y for any given X. In this expression, α and β are the **true regression coefficients,** the vertical intercept and the slope, respectively, of the true regression line. These population parameters are the values that are really of interest to us rather than the *sample* or *estimated regression coefficients, a* and *b,* which, as a result of sampling error, may seriously distort the true underlying relationship between X and Y. (Conceivably, the value of b might be 1,847.50, as in Figure 13.6, while β is zero, indicating that, unlike in this particular sample, there exists no relationship whatsoever between X and Y *in the population as a whole.*)

As it turns out, statisticians have devised ways of making inferences concerning the true regression line from sample data, but the validity of these inferences rests on some or all of the following assumptions:

Assumption 1. It is assumed that the values of X are known without error and that the different sample observations about Y that are associated with any given X are statistically independent of each other. (The fact that an observed value of Y is low, for example, is no reason for another Y observation to be low as well.)

Assumption 2. It is assumed that every population of Y values (a different one of which is associated with every possible value of X) is normally distributed. Because each of these normal curves of Y values is associated with a specific value of X, each curve is referred to as a **conditional probability distribution of Y** and is said to have a **conditional mean of Y ($\mu_{Y \cdot X}$)** and a **conditional standard deviation of Y ($\sigma_{Y \cdot X}$).** This assumption is most easily understood with the help of Figure 13.7. The graph is three-dimensional, containing the now familiar Y vs. X plane but also measuring, vertically above this plane the probability density of Y values.

Imagine plotting, for specified values of $X = 5$, $X = 10$, and $X = 15$, all the associated values of Y found in the entire population. Given $X = 5$, the conditional mean of Y values might equal distance 0b; it is designated as $\mu_{Y \cdot 5}$. Given $X = 10$, the conditional mean of Y values might equal distance 0c; it is designated as $\mu_{Y \cdot 10}$. Finally, given $X = 15$, the conditional mean of Y values might equal distance 0d; it is designated as $\mu_{Y \cdot 15}$. In each case, many individual Y values would be smaller or larger than this mean, of course, but the second assumption of linear regression analysis asserts that the frequency curve of Y values for any given X is a normal curve centered on $\mu_{Y \cdot X}$, such as each of the three solid color lines shown in Figure 13.7.

Assumption 3. It is assumed that all conditional probability distributions of Y (a different one of which is associated with every possible value of X) have the same conditional standard deviation of Y. This value, $\sigma_{Y \cdot X}$, is also referred to as the **population standard error of the estimate** (of Y, given X). This third assumption, thus, is the familiar one of homoscedasticity; we have met it in previous chapters, but here it postulates that the scatter of observed Y values above and below each conditional mean is the same. (Note how the normal curves shown in Figure 13.7 look alike even though they are centered on different means.)

Assumption 4. It is assumed that all the conditional means, $\mu_{Y \cdot X}$, lie on a straight line that is the true regression line and is described by the equation

$$E(Y) = \mu_{Y \cdot X} = \alpha + \beta X.$$

The dashed color line in the Y vs. X plane of Figure 13.7 is such a line. Its Y-intercept equals α = distance 0a, and its slope equals β.

In the following sections, we will show how a number of important inferences can be made about population data from sample data when all these assumptions are fulfilled and sometimes even when only some of them are fulfilled.

Estimating the Average Value of Y, Given X

Practice Problem 3

Estimating the Average Value of Income, Given Education

Suppose we wanted to estimate the value of a conditional mean, $\mu_{Y \cdot X}$, such as $\mu_{Y \cdot 10}$, the mean income of all those people in the population from which the Table 13.1 sample data were drawn who had 10 years of education. Just as the sample mean, $\overline{X}$, has been

Figure 13.7 Three Conditional Probability Distributions of *Y*

The solid color lines shown here represent the three conditional probability distributions of *Y* that are associated with *X* values of 5, 10, and 15, respectively. Each of these frequency curves rises above the *Y* vs. *X* plane into the third dimension; each is also a normal curve clustering around a central value, $\mu_{Y \cdot X}$; hence, the probability of finding *Y* values declines progressively and equally the farther one moves along the dashed lines that parallel the *Y* axis either above or below any one of the three means. The dashed color line that connects all the conditional means in the *Y* vs. *X* plane is the true regression line.

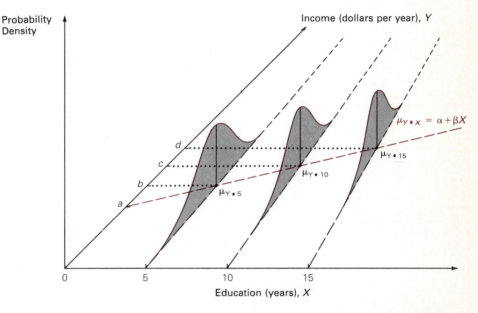

used in an earlier chapter to estimate the population mean, μ, we can now use $\hat{Y}_{10}$, a point on the sample regression line, as an estimate of $\mu_{Y \cdot 10}$, a point on the population regression line. Our least-squares sample regression line, depicted in Figure 13.6, is described by the equation,

$$\hat{Y}_X = a + bX = 2{,}144.23 + 1{,}847.50X.$$

Hence, $\hat{Y}_{10} = 2{,}144.23 + 1{,}847.50(10) = 20{,}619.23.$ ▐

This number turns out to be the best *point estimate* of $\mu_{Y \cdot 10}$ that is available.[2] Yet we may wish for more than that; we may wish to acknowledge explicitly the existence

---

[2]The reason that $\hat{Y}_X$ is the best point estimate of $\mu_{Y \cdot X}$ is: The method of least squares produces estimated regression coefficients, *a* and *b,* that are unbiased, efficient, and consistent estimators of the corresponding true regression coefficients, α and β. The concepts of unbiasedness, efficiency, and consistency were first introduced in Chapter 8 and may be reviewed there. In the present context, if all possible random samples of a given size were taken from the population of interest and if a least-squares regression equation were calculated for each of these, the average value of the estimated coefficient *a* would equal the true coefficient α, while the average value of *b* would equal β (unbiasedness). In addition, the coefficients estimated by the method of least squares have the smallest variance for a given sample size among all available unbiased estimators (efficiency), and they get progressively closer to α and β as sample size is increased (consistency).

of sampling error. If we had drawn a different sample, we would have estimated a different regression equation (perhaps one stating that $\hat{Y}_X = 1{,}800.50 + 2{,}001.73X$) and, thus, would have made a different point estimate of $\mu_{Y\cdot 10}$ (in this case equal to $\hat{Y}_{10} = 21{,}817.80$). This degree of uncertainty about the value of $\mu_{Y\cdot X}$ can be made explicit by presenting not a point estimate, but an interval estimate. In the context of regression analysis, such a confidence interval within which a population parameter that is being estimated presumably lies is typically referred to as a **prediction interval.** We will discuss its determination separately for the small-sample case ($n < 30$) and the large-sample case ($n \geq 30$).

Establishing a prediction interval for $\mu_{Y\cdot X}$ *from a small sample* (n < 30). We noted in Chapter 8 that the limits of a confidence interval for the population mean can be established in the small-sample case, using the t distribution, as

$$\mu = \overline{X} \pm (t\sigma_{\overline{X}})$$

and that $\sigma_{\overline{X}}$ can be estimated by $s/\sqrt{n}$ (Box 8.D on page 315). In regression analysis, the variable being estimated is Y rather than X; hence, an expression analogous to the previous one is

$$\mu_{Y\cdot X} = \hat{Y}_X \pm t\sigma_{\hat{Y}_X}.$$

The term, $\sigma_{\hat{Y}_X}$, is the standard error of $\hat{Y}_X$; it measures the variability (caused by the selection of different samples) of all the possible $\hat{Y}_X$ values (of which the above values of 20,169.23 and 21,817.80 are just two examples) at the X value for which the prediction interval of the conditional mean is to be established. This variability has two components. One of these, the variability in the mean of observed Y values, depends on the conditional standard deviation, $\sigma_{Y\cdot X}$, of the relevant population of Ys, as well as on sample size, n. (In terms of Figure 13.7, this variability in the mean of observed Y values depends on the shape of the frequency curve at the relevant X and on the number of sample observations made.) A second source of variability in $\hat{Y}_X$, however, is the distance of the specified X from $\overline{X}$, as illustrated in Figure 13.8. The graph features, in color, the (unknown) true regression line and, in black, estimated regression lines fitted to four different samples drawn from the same population. Note how these lines diverge increasingly the farther the specified value of X lies from $\overline{X}$. As a result, the estimated values of $\hat{Y}_X$ become more varied, too.

As a matter of fact, the value of $\sigma_{\hat{Y}_X}$ cannot be calculated directly because all possible values of $\hat{Y}_X$ are never determined. In practice the standard error of $\hat{Y}_X$ is estimated as

$$\sigma_{\hat{Y}_X} \cong s_{Y\cdot X}\sqrt{\frac{1}{n} + \frac{(X - \overline{X})^2}{\Sigma X^2 - n\overline{X}^2}}$$

and this complicated expression reflects both of the components of $\hat{Y}_X$ variability that have just been noted. In this expression $s_{Y\cdot X}$ is the **sample standard error of the**

Figure 13.8 One Source of Variability in $\hat{Y}_x$

This graph features a true regression line (in color) with the equation $\mu_{Y \cdot X} = \alpha + \beta X$. It also shows estimated regression lines (in black) that might have been fitted to four different samples and are described by the equation $\hat{Y}_X = a + bX$ (where a and b, of course, differ from one line to the next). Note how the computed values of $\hat{Y}_{10}$ (the four fat dots above $X = 10$) are closer together than the computed values of $\hat{Y}_{21}$ (the four fat dots above $X = 21$) even though they are derived from identical samples. This illustrates one source in the variability of $\hat{Y}_X$: the distance of the specified X from $\overline{X}$.

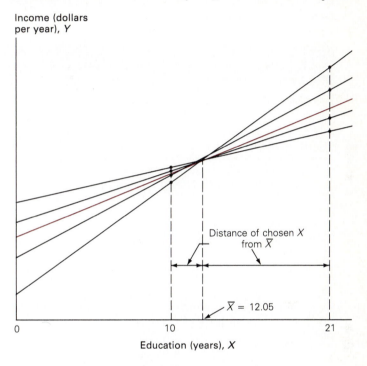

Income (dollars per year), Y

Distance of chosen X from $\overline{X}$

$\overline{X} = 12.05$

0   10   21

Education (years), X

---

estimate about the regression line that (as the subscript reminds us) gives estimates of Y for any specified X. This standard error equals

$$s_{Y \cdot X} = \sqrt{\frac{\Sigma(Y - \hat{Y})^2}{n - 2}}$$

which is mathematically equivalent to

$$s_{Y \cdot X} = \sqrt{\frac{\Sigma Y^2 - a\Sigma Y - b\Sigma XY}{n - 2}}$$

and measures the dispersion of the values of Y observed in a sample around the least-squares regression line estimated from this sample.

Note: The divisor, $n - 2$, is used in the formula because this makes $s_{Y \cdot X}^2$ an unbiased estimator of the variance of Y values around the true regression line. The reduction of n by 2 indicates the loss of 2 degrees of freedom because the coefficients a and b (that allow the calculation of $\hat{Y}_X$) are being calculated from the same data as $s_{Y \cdot X}$. The nature of this measure of dispersion is illustrated with the help of Figure 13.9.

The formula given in Box 13.B, finally, puts together the various strands of our discussion and becomes the basis for calculating a prediction interval for the conditional mean in our example.

13.B The Limits of the Prediction Interval for the Average Value of Y, Given X (small-sample case)

$$\mu_{Y \cdot X} = \hat{Y}_X \pm \left(t_{\alpha/2} \cdot s_{Y \cdot X} \sqrt{\frac{1}{n} + \frac{(X - \overline{X})^2}{\Sigma X^2 - n\overline{X}^2}} \right)$$

where Xs are observed values of the independent variable ($\overline{X}$ being their mean), while $\hat{Y}_X$ is the estimated value of dependent variable Y, $s_{Y \cdot X}$ is the sample standard error of the estimate of Y, n is sample size, and the t-statistic is found in Appendix Table K for $n - 2$ degrees of freedom.

Practice Problem 4

Establishing a Prediction Interval for the Average Value of Income, Given Education

As we return to our example from Practice Problems 1–3, we must recall having earlier determined a point estimate of $\mu_{Y \cdot 10}$ as $\hat{Y}_X = 20{,}619.23$. To calculate, say, a 95-percent prediction interval for this conditional mean, we must now find the appropriate value of $t_{.025}$ from Appendix Table K. Table 13.1 shows $n = 20$; hence, we have $n - 2 = 20 - 2 = 18$ degrees of freedom; $t_{.025,18} = 2.101$.

Next we can calculate, with the help of Table 13.2,

$$s_{Y \cdot X} = \sqrt{\frac{\Sigma Y^2 - a\Sigma Y - b\Sigma XY}{n - 2}}$$

$$= \sqrt{\frac{13{,}734{,}414{,}590 - 2{,}144.23(488{,}132) - 1{,}847.50(6{,}585{,}796)}{18}}$$

$$= \sqrt{\frac{520{,}489{,}201.64}{18}}$$

$$= \sqrt{28{,}916{,}066.76} = 5{,}377.37.$$

Next, we can calculate:

$$\sqrt{\frac{1}{n} + \frac{(X - \overline{X})^2}{\Sigma X^2 - n\overline{X}^2}} = \sqrt{\frac{1}{20} + \frac{(10 - 12.05)^2}{3{,}285 - 20(12.05)^2}}$$

$$= \sqrt{.05 + .011} = .247.$$

Figure 13.9   The Sample Standard Error of the Estimate About the Regression Line

This graph copies the scatter diagram and the estimated regression line of Figure 13.6. It also illustrates (on the right-hand side) the kind of dispersion that is being measured by the sample standard error of the estimate about the regression line. Note how the dashed arrows (that parallel the color estimated regression line) project points A and C (the two extreme deviations of observed Y values above and below the regression line) onto the vertical line at points B and D. All other observed Y values clearly diverge from the regression line to a lesser extent (and would be projected to various points on the vertical line lying between B and D). The sample standard error of the estimate about the regression line, $s_{Y \cdot X}$, measures this type of dispersion of observed Y values around the estimated regression line. But caution is advised: this sample standard error of the estimate about the regression line should not be confused with the sample standard deviation of observed Y values, s_Y, that measures the dispersion of these values around their sample mean (as shown on the left-hand side of this graph).

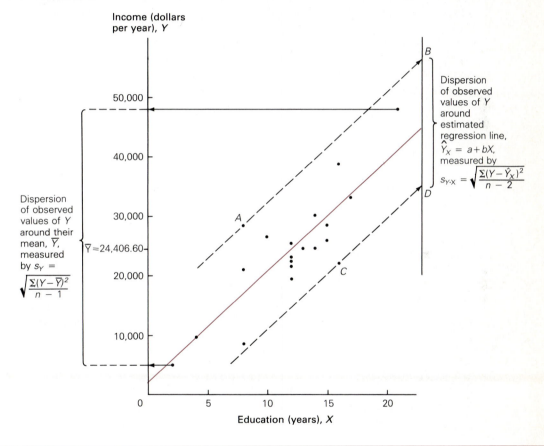

Thus, the limits of our prediction interval are

$$\mu_{Y \cdot 10} = 20{,}619.23 \pm [2.101(5{,}377.37).247]$$
$$= 20{,}619.23 \pm 2{,}790.57,$$

and the 95-percent prediction interval is $17{,}828.66 \leq \mu_{Y \cdot 10} \leq 23{,}409.80$. We can conclude with 95-percent confidence that the average income of all the people within our population who have 10 years of education lies between \$17,828.66 and \$23,409.80. But note: the true conditional mean could lie outside the confidence limits just computed. If we follow the above procedure of sampling and computation again and again, however, it will yield ever new, but similar intervals, and they will contain the true conditional mean 95 percent of the time. ▐▐▐

Figure 13.10 shows graphically the prediction interval we have just computed. It also shows that if one calculated similar 95-percent prediction intervals for all other values of X and connected the limits of these intervals graphically, one could envelope the entire estimated regression line in a 95-percent **confidence band** (the width of which would be narrowest for $X = \overline{X}$ and would become ever broader the farther the X values were from their mean).

Establishing a prediction interval for $\mu_{Y \cdot X}$ from a large sample ($n \geq 30$). We also noted in Chapter 8 (in Box 8.C on page 309) that the limits of a confidence interval for the population mean can be established in the large-sample case using the normal distribution as

$$\mu = \overline{X} \pm (z\sigma_{\overline{X}})$$

and that $\sigma_{\overline{X}}$ can be estimated by $s/\sqrt{n}$. In regression analysis, the analogous expression is

$$\mu_{Y \cdot X} = \hat{Y}_X \pm (z\sigma_{\hat{Y}_X})$$

and $s_{Y \cdot X}/\sqrt{n}$ can be used to estimate $\sigma_{\hat{Y}_X}$ in the large-sample case. Thus, the result is the formula in Box 13.C,

---

13.C The Limits of the Prediction Interval for the Average Value of *Y*, Given *X* (large-sample case)

$$\mu_{Y \cdot X} = \hat{Y}_X \pm \left(z_{\alpha/2} \cdot \frac{s_{Y \cdot X}}{\sqrt{n}} \right)$$

where $\hat{Y}_X$ is the estimated value of Y, the dependent variable, $s_{Y \cdot X}$ is the sample standard error of the estimate of Y, n is sample size, and z is the standard normal deviate from Appendix Table J.

▐▐▐ Practice Problem 5

Establishing a Prediction Interval for the Average Wheat Output, Given Fertilizer Input

Suppose that a sample of 100 acres has revealed the association between pounds of fertilizer input per acre *(X)* and bushels of wheat output per acre *(Y)* that is summarized by the estimated regression line: $\hat{Y}_X = 100 + .1X$. Let $s_{Y \cdot X} = 10$ bushels and

Figure 13.10 A 95-Percent Confidence Band for the Conditional Mean

The color line in this graph is the least-squares estimated regression line of $\hat{Y}_X = 2,144.23 + 1,847.50X$ that was fitted to the sample data of Table 13.1 and first shown in Figure 13.6. Segment *AC* represents the 95-percent prediction interval for $\mu_{Y \cdot 10}$ that reaches from \$17,828.66 (point *A*) past the point estimate of \$20,619.23 (point *B*) to \$23,409.80 (point *C*). Similar intervals can easily be calculated for other values of *X*. For example, they reach from \$6,582.21 to \$16,181.25 for $X = 5$, from \$21,880.33 to \$26,932.89 for $\bar{X} = 12.05$, and from \$33,844.59 to \$44,343.87 for $X = 20$. Thus, it is possible to envelope the entire estimated regression line in a 95-percent (nonlinear) confidence band and be that confident that any given conditional mean will lie within its limits.

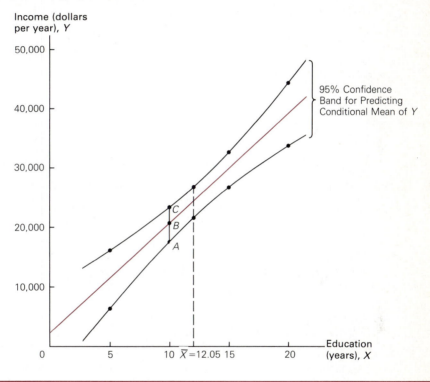

assume that a 95-percent prediction interval is desired for the conditional mean of *Y*, given $X = 100$ pounds. Clearly, the point estimate is $\mu_{Y \cdot 100} = 100 + .1(100) = 110$ bushels per acre. The appropriate value of *z* can be found in Appendix Table J as 1.96. Thus, the interval limits are:

$$\mu_{Y \cdot 100} = 110 \pm \left[1.96 \left(\frac{10}{\sqrt{100}} \right) \right]$$
$$= 110 \pm 1.96 \text{ bushels,}$$

and the interval is $108.04 \leq \mu_{Y \cdot 100} \leq 111.96$ bushels of wheat. **III**

Predicting an Individual Value of *Y*, Given *X*

Suppose now that we wished to predict, at a given *X*, not the average value of *Y*, or $\mu_{Y \cdot X}$, but an individual value, $I_{Y \cdot X}$, such as the next one we are likely to encounter in the process of sampling. In our example from Practice Problem 1, we may wish to predict the income of the next member of our population who has an education of, say, 10 years.

The point estimate for an individual *Y* is the same as for the average *Y*; in our case it equals once again $I_{Y \cdot X} = \hat{Y}_{10} = 20,619.23$. The prediction interval is once again determined differently for small and large samples.

Establishing a prediction interval for $I_{Y \cdot X}$ *from a small sample (n < 30).* The formula given in Box 13.B has to be amended only slightly, as shown in Box 13.D. The only difference is the "+ 1" appearing in the square root; its presence acknowledges the existence of a third component yet in the variability of $\hat{Y}_X$: the dispersion, illustrated in Figure 13.9, of individual *Y* values around the estimated regression line.

13.D The Limits of the Prediction Interval for an Individual Value of *Y*, Given *X* (small-sample case)

$$I_{Y \cdot X} = \hat{Y}_X \pm \left(t_{\alpha/2} \cdot s_{Y \cdot X} \sqrt{\frac{1}{n} + \frac{(X - \overline{X})^2}{\Sigma X^2 - n\overline{X}^2} + 1} \right)$$

where *X*s are observed values of the independent variable ($\overline{X}$ being their mean), while $\hat{Y}_X$ is the estimated value of dependent variable *Y*, $s_{Y \cdot X}$ is the sample standard error of the estimate of *Y*, *n* is sample size, and the *t*-statistic is found in Appendix Table K for $n - 2$ degrees of freedom.

⫼ Practice Problem 6

Establishing a Prediction Interval for an Individual Value of Income, Given Education

Using the Box 13.D formula and the values previously calculated, we can quickly establish the limits of a 95-percent prediction interval for an individual value of *Y*, given an *X* of 10 years, as

$$I_{Y \cdot 10} = 20,619.23 \pm [2.101(5,377.37)\sqrt{.05 + .011 + 1}]$$
$$= 20,619.23 \pm 11,636.79.$$

Thus, the interval is: $\$8,982.44 \le I_{Y \cdot 10} \le \$32,256.02$. In other words, we can be 95-percent confident that the next person encountered in sampling who has 10 years of education will have an income that lies within this range. ⫼

This interval, it should be noted, is considerably wider than that previously calculated for $\mu_{Y \cdot 10}$. Indeed, as Figure 13.11 shows, the entire confidence band for individual values of *Y*, given *X*, lies outside that for average values of *Y*.

Figure 13.11   A 95-Percent Confidence Band for the Individual Y, Given X

The color line in this graph is the least-squares estimated regression line of $\hat{Y}_X = 2{,}144.23 + 1{,}847.50X$ that was fitted to the sample data of Table 13.1 and first shown in Figure 13.6. Segment *AC* represents the 95-percent prediction interval for $I_{Y \cdot 10}$ that reaches from \$8,982.44 (point *A*) past the point estimate of \$20,619.23 (point *B*) to \$32,256.02 (point *C*). Similar intervals for individual *Y* values can easily be calculated given other specified values of *X*; in all cases, they are wider than those for $\mu_{Y \cdot X}$. Note how the solid confidence band for predicting individual values of *Y* lies beyond the dashed confidence band (copied from Figure 13.10) for average values of *Y*.

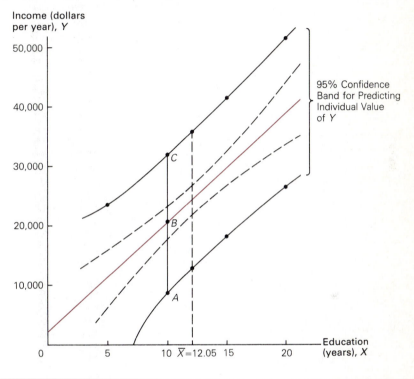

Establishing a prediction interval for $I_{Y \cdot X}$ from a large sample (n ≥ 30). In this case, the formula given in Box 13.C has to be amended only slightly, as shown in Box 13.E.

13.E The Limits of the Prediction Interval for an Individual Value of Y, Given X (large-sample case)

$$I_{Y \cdot X} = \hat{Y}_X \pm (z_{\alpha/2} \cdot s_{Y \cdot X})$$

where $\hat{Y}_X$ is the estimated value of *Y*, the dependent variable, $s_{Y \cdot X}$ is the sample standard error of the estimate of *Y*, and *z* is the standard normal deviate from Appendix Table J.

||| Practice Problem 7

Establishing a Prediction Interval for an Individual Value of Wheat Output, Given Fertilizer Input

Consider applying the Box 13.E formula to the earlier fertilizer-and-wheat example in Practice Problem 5, the goal now being to predict with 95-percent confidence the output of the next acre (rather than the average output of many acres) on which $X = 100$ pounds of fertilizer was applied. In that case, the interval limits are:

$$I_{Y \cdot 100} = 110 \pm [1.96(10)]$$
$$= 110 \pm 19.6 \text{ bushels,}$$

and the interval is $90.4 \leq I_{Y \cdot 100} \leq 129.6$ bushels of wheat. Once again the interval is considerably wider than that for $\mu_{Y \cdot 100}$. |||

Making Inferences About the True Regression Coefficients

Often the making of inferences about α and β, the coefficients of the true regression line, is even more important to researchers and decision makers than the prediction, for any specified value of the independent variable, X, of an associated average or individual value of the dependent variable, Y. As was noted above, when the method of least squares is employed, the estimated regression coefficients, a and b, are unbiased, efficient, and consistent estimators of α and β; yet, here, too, it is often desirable to establish confidence intervals. The establishment of confidence intervals is particularly common with respect to β, the slope of the true regression line, because there is always the danger that a positive or negative b is a fluke resulting from the selection of a highly improbable sample and that β is in fact zero (meaning there is no association whatsoever between X and Y in the population as a whole).

Establishing a confidence interval for β. In our small-sample case, the limits of a confidence interval for β can be established using the t distribution as

$$\beta = b \pm (t\sigma_b)$$

where σ_b is the standard error of slope b and

$$\sigma_b = \frac{\sigma_{Y \cdot X}}{\sqrt{\Sigma X^2 - n\overline{X}^2}}$$

but where σ_b can be estimated by s_b and

$$s_b = \frac{s_{Y \cdot X}}{\sqrt{\Sigma X^2 - n\overline{X}^2}}.$$

Box 13.F gives the formula for the limits of the confidence interval for β.

13.F The Limits of the Confidence Interval for β (small-sample case)

$$\beta = b \pm \left(t_{\alpha/2} \cdot \frac{s_{Y \cdot X}}{\sqrt{\Sigma X^2 - n\overline{X}^2}} \right)$$

where b is the estimated (slope) regression coefficient, Xs are observed values of the independent variable ($\overline{X}$ being their mean), while $s_{Y \cdot X}$ is the sample standard error of the estimate of dependent variable Y, n is sample size, and the t-statistic is found in Appendix Table K for $n - 2$ degrees of freedom.

Note: For large samples, the normal deviate $z_{\alpha/2}$ replaces $t_{\alpha/2}$.

||| Practice Problem 8

Establishing a Confidence Interval for β in the Case of Education and Income

In Table 13.2, we calculated $b = 1,847.50$. As noted in Practice Problem 4, the value of t for a 95-percent confidence interval and $n - 2 = 18$ degrees of freedom is $t_{.025,18} = 2.101$. Also in Practice Problem 4, we found $s_{Y \cdot X} = 5,377.37$. Thus, our interval limits are

$$\beta = 1,847.50 \pm \left(2.101 \, \frac{5,377.37}{\sqrt{3,285 - 20(12.05)^2}} \right)$$

$$= 1,847.50 \pm [2.101(275.51)]$$

$$= 1,847.50 \pm 578.84.$$

Thus, the interval is $\$1,268.66 \leq \beta \leq \$2,426.34$, and we can be 95-percent confident that the true value of β lies within this range. In other words, for the country's population as a whole, each additional year of education yields between $\$1,269$ and $\$2,426$ of additional annual income.

Note again: β may lie outside this range, but if we follow the above procedure of sampling and computation a large number of times, it will yield ever new but similar intervals, and they will contain the true β 95 percent of the time.

Note also: The confidence interval just calculated is distressingly imprecise, but nothing can increase the degree of precision except an increase in sample size. In that case, of course, the t distribution approaches the normal one, and the normal deviate, z, can replace t in our Box 13.F formula. |||

Establishing a confidence interval for α. In our small-sample case, the limits of a confidence interval for α can be established in a similar fashion as that for β.

$$\alpha = a \pm (t\sigma_a)$$

but σ_a—the standard error of intercept a—can be estimated by s_a and

$$s_a = s_b \sqrt{\frac{\Sigma X^2}{n}}.$$

Thus, we can calculate the confidence interval for α using the formula in Box 13.G.

13.G The Limits of the Confidence Interval for α (small-sample case)

$$\alpha = a \pm \left(t_{\alpha/2} \frac{s_{Y \cdot X}}{\sqrt{\Sigma X^2 - n\bar{X}^2}} \sqrt{\frac{\Sigma X^2}{n}} \right)$$

where a is the estimated (intercept) regression coefficient, Xs are observed values of the independent variable ($\bar{X}$ being their mean), while $s_{Y \cdot X}$ is the sample standard error of the estimate of dependent variable Y, n is sample size, and the t-statistic is found in Appendix Table K for $n - 2$ degrees of freedom.

   Note: For large samples, the normal deviate $z_{\alpha/2}$ replaces $t_{\alpha/2}$.

Practice Problem 9

Establishing a Confidence Interval for α in the Case of Education and Income

Using once again the calculations of Table 13.2 and those of Practice Problem 8, we get the following 95-percent confidence-interval limits for our example about education and income:

$$\alpha = 2,144.23 \pm \left(2.101(275.51) \sqrt{\frac{3,285}{20}} \right)$$
$$= 2,144.23 \pm 7,418.50$$

Thus, the interval is $-5,274.27 \leq \alpha \leq 9,562.73$. In other words, we can be 95-percent confident that the vertical intercept of the true regression line between education and income lies within this income range. |||

Testing hypotheses about β. A commonly used alternative to establishing a confidence interval for β is testing a hypothesis about it. Sometimes the key question is not the precise value of β, but merely whether the value is zero. If it is zero—no matter what the value of X—the value of $\mu_{Y \cdot X}$ equals α, the true regression line is parallel to the X axis, and regression analysis is of no use in making predictions of Y from X. The usual four steps of hypothesis testing are involved.

Practice Problem 10

A Two-Tailed Test of β in the Case of Education and Income

Step 1: The set of opposing hypotheses typically is that for a two-tailed test:

H_0: $\beta = 0$ (because Y is not associated with X in the population)
H_A: $\beta \neq 0$ (because Y is associated with X in the population).

Step 2: For any small-sample case, such as our limited data set from Practice Problem 1, the Student t-statistic is used as the test statistic. Given Box 13.F above, and setting β equal to zero,

$$t = \frac{b}{\dfrac{s_{Y \cdot X}}{\sqrt{\Sigma X^2 - n\overline{X}^2}}}$$

Step 3: Given once again a desired confidence level of 95 percent (that is, a significance level of $\alpha = .05$, where α clearly does *not* equal the vertical intercept of the true regression line), Appendix Table K implies (for $n - 2 = 18$ degrees of freedom) critical values of $\pm t_{\alpha/2} = \pm t_{.025,18} = \pm 2.101$. Thus, the decision rule must be: "Accept H_0 if $-2.101 \le t \le +2.101$."

Step 4: In our example, the actual value of t is computed as

$$t = \frac{1,847.50}{\dfrac{5,377.37}{\sqrt{3,285 - 20(12.05)^2}}} = \frac{1,847.50}{275.51} = 6.71.$$

Accordingly, the null hypothesis should be *rejected.* At the 5-percent level of significance, the sample result (a positive slope b) is statistically significant. The observed divergence of b from the zero value of β hypothesized in the null hypothesis is unlikely to be caused merely by chance factors operating during sampling. There is good reason to embrace the alternative hypothesis: more education does mean more income. Note: This conclusion is also evident by the nature of the 95-percent confidence interval for β calculated in Practice Problem 8; that interval does not contain $\beta = 0$ as it would have to if $H_0: \beta = 0$ were true. |||

||| Practice Problem 11

An Upper-Tailed Test of β in the Case of Education and Income

When one is fairly certain that Y varies directly with X, an upper-tailed test can be used instead of a two-tailed test.

Step 1: The set of opposing hypotheses becomes:

$H_0: \beta \le 0$
$H_A: \beta > 0$

Step 2: The test statistic remains the same.

Step 3: If we again use $\alpha = .05$ as a significance level, the critical value is $t_\alpha = t_{.05,18} = 1.734$ (again from Appendix Table K and for 18 degrees of freedom), making the decision rule: "Accept H_0 if $t \le 1.734$."

Step 4: Given the computed value of $t = 6.71$, H_0 should be *rejected.* Thus, this test also indicates that the slope of the true regression line is greater than zero. In the population as a whole, more education does mean more income. Note: When Y is inversely related to X, an analogous *lower-tailed test* can be employed, with opposing claims of $H_0: \beta \ge 0$ and $H_A: \beta < 0$.

Figure 13.12   Different Degrees of Correlation

The three panels of this graph illustrate different degrees of correlation. Depending on whether the scatter of data points around a regression line is nonexistent, considerable, or moderate, correlation between the variables is said to be (a) perfect, (b) zero, or (c) weak.

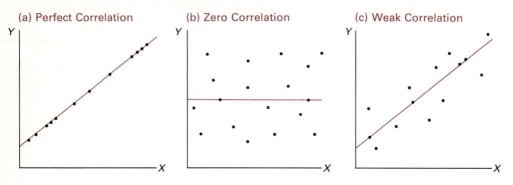

Analytical Examples 13.1, "Evaluating a New Food Product," and 13.2, "Of Bulls, Bears, and Beauty" provide typical illustrations of the use of simple regression analysis in business and economics. ▐▐▐

SIMPLE CORRELATION ANALYSIS

While simple regression analysis establishes a precise equation that links two variables, simple correlation analysis seeks to establish a general index about their strength of association. Consider the three panels of Figure 13.12. When all the points in a scatter diagram lie precisely on the estimated regression line, as in panel (a), the index to be discussed in this section will show the variables to be *perfectly correlated* (and in that case, Y can always be predicted with precision from a knowledge of the regression line and any value of X). When, on the other hand, the points in a scatter diagram are so widely scattered as to make X completely worthless as a predictor of Y, as in panel (b), the index will show the variables to be *completely uncorrelated*. When, finally, many scattered points diverge from the regression line, but not without suggesting some definite association between the variables, such as the direct association in panel (c), the index will show the variables to be *imperfectly correlated*.

Several different indexes of association between quantitative variables are used by statisticians, and we will discuss them in order of importance.[3]

The Coefficient of Determination

The most important measure of how well the estimated regression line fits the sample data on which it is based is the **sample coefficient of determination**—which equals the proportion of the total variation in the values of the dependent variable, Y, that can

---

[3]Chapter 11 contains a discussion of Spearman's rank-correlation coefficient, which is the most important measure of the degree of association between *qualitative* data.

Figure 13.13 Total, Explained, and Unexplained Deviation in *Y*

The total deviation (here *AC*) of any observed value of *Y* from the mean of all such observations ($\bar{Y}$) always equals the sum of the explained deviation (here *BC*) and the unexplained deviation (here *AB*).

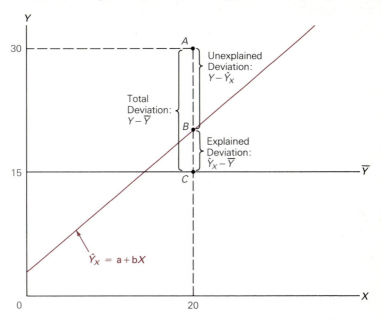

be explained by means of the association of *Y* with *X* as measured by the estimated regression line. Imagine that Figure 13.13 is a scatter diagram with all the dots removed except point *A* (which corresponds to the data pair of $X = 20, Y = 30$). Assume that the color line represents the estimated regression line calculated from the complete data set and that the average value of *Y* is $\bar{Y} = 15$, as illustrated by the horizontal line. The difference between any particular observation of the dependent variable, such as $Y = 30$ when $X = 20$ (point *A*) and the mean of the dependent variable, such as $\bar{Y} = 15$ (point *C*), is the **total deviation of *Y*** (distance *AC*).

There are, of course, many such deviations in a scatter diagram, one between each individual observation of *Y* (most of which are not shown here) and the common mean of all these *Y*s. Now imagine squaring all these total deviations and summing them. The resulting sum of squares of total deviations, $\Sigma(Y - \bar{Y})^2$, is called the **total variation of *Y*,** or the **total sum of squares *(Total SS)*.**

Next focus on the difference between the regression estimate of our observed *Y*, or $\hat{Y}_{20}$ (point *B*) and the average *Y* (point *C*); this difference is the **explained deviation of *Y*** (distance *BC*). We can see where this name originates: the positive deviation, *BC*, of $\hat{Y}_X$ from $\bar{Y}$ is explained by the fact that *Y* rises with *X* and the value of $X = 20$ is above average. (If we had chosen $X = 5$ for our example, we would explain a similar negative deviation in an analogous fashion—namely, by $X = 5$ being below average.) Now imagine squaring all these explained deviations in the scatter diagram and summing them. The resulting sum of squares of explained deviations, $\Sigma(\hat{Y}_X - \bar{Y})^2$, is the **explained variation of *Y*,** or the **regression sum of squares *(RSS)*.**

Finally, focus on the difference between our particular observation of Y when $X = 20$ (point A) and the corresponding regression estimate, $\hat{Y}_{20}$ (point B); this difference is the **unexplained deviation of Y** (distance AB). Again, we can see the reason for this name: the fact that Y rises with X as shown by the regression line explains why our observed Y (point A) lies above $\overline{Y}$ to the extent of BC, but it does not explain the additional deviation of AB. Now imagine squaring all these unexplained deviations in the scatter diagram and summing them. The resulting sum of squares of unexplained deviations, $\Sigma(Y - \hat{Y}_X)^2$, is the **unexplained variation of Y,** or the **error sum of squares (ESS).** (The method of least squares, of course, guarantees that this error component is minimized.)

Just as the total deviation is the sum of explained and unexplained deviations, so the relationships given in Box 13.H hold as well.

13.H Components of Total Variation

$$\begin{array}{ccc} \text{Total} & = & \text{Explained} & + & \text{Unexplained} \\ \text{Variation} & & \text{Variation} & & \text{Variation} \end{array}$$

or

$$\begin{array}{ccc} \text{Total Sum} & = & \text{Regression Sum} & + & \text{Error Sum} \\ \text{of Squares} & & \text{of Squares} & & \text{of Squares} \\ \textit{(Total SS)} & & \textit{(RSS)} & & \textit{(ESS)} \end{array}$$

or

$$\Sigma(Y - \overline{Y})^2 = \Sigma(\hat{Y}_X - \overline{Y})^2 + \Sigma(Y - \hat{Y}_X)^2$$

where Ys are observed values of the dependent variable ($\overline{Y}$ being their mean) and $\hat{Y}_X$ is the estimated value of Y, given a value of independent variable X.

Given the relationships in Box 13.H, the meaning of the sample coefficient of determination, usually symbolized by r^2, is quite clear—as shown in Box 13.I.

13.I Sample Coefficient of Determination

$$r^2 = \frac{\text{explained variation}}{\text{total variation}} = \frac{RSS}{Total\ SS} = \frac{\Sigma(\hat{Y}_X - \overline{Y})^2}{\Sigma(Y - \overline{Y})^2}$$

Note: Because *Total SS* $-$ *ESS* $=$ *RSS*, we can also write

$$r^2 = \frac{Total\ SS - ESS}{Total\ SS} = 1 - \frac{ESS}{Total\ SS} = 1 - \frac{\Sigma(Y - \hat{Y}_X)^2}{\Sigma(Y - \overline{Y})^2}.$$

where Ys are observed values of the dependent variable ($\overline{Y}$ being their mean) and $\hat{Y}_X$ is the estimated value of Y, given a value of independent variable X.

When X and Y are perfectly correlated, as in panel (a) of Figure 13.12, every $Y = \hat{Y}_X$; hence, $\Sigma(Y - \hat{Y}_X)^2 = 0$, and $r^2 = 1 - 0 = 1$. When instead X and Y are completely uncorrelated, the regression line has a slope of zero, as in panel (b) of Figure 13.12. Then $\hat{Y}_X = \bar{Y}$, the last fraction in Box 13.I equals one, and $r^2 = 1 - 1 = 0$. Thus, the value of r^2 always lies between 1 (perfect correlation) and 0 (no correlation). The formula in Box 13.J is one of several mathematical equivalents to the formulas in Box 13.I and is easier and less cumbersome to employ.

13.J Sample Coefficient of Determination: Alternative Formulation

$$r^2 = \frac{a\Sigma Y + b\Sigma XY - n\bar{Y}^2}{\Sigma Y^2 - n\bar{Y}^2}$$

where Xs are observed values of the independent variable, Ys are observed associated values of the dependent variable ($\bar{Y}$ being their mean), while n is sample size, and a and b are the estimated (intercept and slope) regression coefficients, respectively.

||| Practice Problem 12

Calculating the Sample Coefficient of Determination (r^2) for Education and Income

To calculate r^2 for our example, we could apply the Box 13.I formula to the data of Table 13.2, but it would be less cumbersome to apply the formula in Box 13.J, using values already derived during regression analysis. Thus, the sample coefficient of determination can be calculated as

$$r^2 = \frac{2{,}144.23(488{,}132) + 1{,}847.5(6{,}585{,}796) - 20(24{,}406.6)^2}{13{,}734{,}414{,}590 - 20(24{,}406.6)^2} = .71.$$

In other words, 71 percent of the total variation in Y (income) observed in Figure 13.6 can be explained by the association of Y (income) with X (education) as estimated by the regression line shown there. Naturally, statisticians are always pleased with a high r^2 because it indicates that they have explained a large proportion of the variation in the variable of interest.

Note: The sample coefficient of determination varies from sample to sample; it is only an estimate of the underlying **population coefficient of determination (ρ^2).** The symbol ρ^2 is the lowercase Greek rho, squared.

Analytical Examples 13.4, "The Inflation Rate Versus Its Predictability," and 13.5, "Predicting Money Supply Changes," illustrate how research results are routinely assessed with the help of the coefficient of determination. |||

The Coefficient of Correlation

The square root of the sample coefficient of determination, $\sqrt{r^2}$, is a common alternative index of the degree of association between two quantitative variables. This measure is called the **sample coefficient of correlation (r)** and is a point estimator of the **population coefficient of correlation (ρ).** Just like r^2, the coefficient r takes on absolute values between 0 and 1, but it is common to take advantage of the fact that the square root of any number can be positive or negative and to place a plus sign or a minus sign in front of r to denote, respectively, a direct relationship or an inverse relationship between X and Y. Thus, $r = 0$ denotes no correlation, while $r = +1$ or $r = -1$ denotes, respectively, perfect correlation between directly related variables or perfect correlation between inversely related variables. In our example, therefore, the $r^2 = .71$ translates into $r = \sqrt{.71} = +.84$ because the relationship between education and income is a direct one. Note: An alternative formula for calculating r without prior regression analysis is given in Box 13.K.

13.K Pearson's Sample Coefficient of Correlation

$$r = \frac{\Sigma XY - n\overline{X}\overline{Y}}{\sqrt{(\Sigma X^2 - n\overline{X}^2)(\Sigma Y^2 - n\overline{Y}^2)}}$$

where Xs are observed values of the independent variable ($\overline{X}$ being their mean), Ys are observed associated values of the dependent variable ($\overline{Y}$ being *their* mean), and n is sample size.

‖‖ Practice Problem 13

Calculating Pearson's Sample Coefficient of Correlation (r) for Education and Income

When applied to our Table 13.2 data, the formula in Box 13.K yields

$$r = \frac{6,585,796 - 20(12.05)(24,406.6)}{\sqrt{[3,285 - 20(12.05)^2][13,734,414,590 - 20(24,406.6)^2]}}$$

$$= \frac{703,805.4}{\sqrt{(380.95)(1,820,772,118.8)}} = \frac{703,805.4}{836,434.7} = .84,$$

and this value confirms the earlier calculation. Income is strongly correlated with education. ‖‖

Finally, three notes of caution must be sounded. First, the correlation coefficient discussed here is capable of detecting only linear associations between variables. It is quite possible to calculate $r = 0$ from a set of data such as that given in panel (c) of Figure 13.2—thus, falsely suggesting no association between X and Y. In fact, there is no *linear* association, but quite a pronounced curvilinear one.

Second, statisticians prefer to use r^2 rather than r as a measure of association because fairly high absolute values of r (such as .70) can give the false impression that there exists a strong association between Y and X (but in this case, $r^2 = .49$; thus, the association of Y with X explains less than half of the variation of Y). Only at the extremes, when $r = \pm 1$ or when $r = 0$ (and, therefore, $r = r^2$) does the value of r directly convey what proportion of the variation of Y is explained by X. Analytical Example 13.3, "Predicting Changes in the GNP," illustrates the use of these coefficients; so do Close-Ups 13.1, "TV Commercials and the Demand for Water," and 13.2, "The Uneven Burden of International Trade Restraints."

Third, we must reemphasize that the correlation coefficient measures only the strength of a statistical relationship between two variables, but the discovery of a strong statistical relationship between X and Y need not imply a causal one. Perhaps X does cause Y; perhaps Y is causing X; perhaps each is part cause, part effect of the other. Perhaps the two variables move together by pure chance. Indeed, the existence of a high positive or negative correlation between two variables that have no logical connection with one another is common and is called **nonsense correlation** or **spurious correlation.** By pure chance one might discover, for example, a correlation coefficient of $r = +.98$ between the numbers of ministers and illegitimate pregnancies in the 50 states or a correlation coefficient of $r = -.85$ between the numbers of whales caught in the world's oceans and Congressional expenditures on stationery. Sometimes, of course, such statistical associations reflect the operation of some third factor (such as population differences) that affects both of the variables (producing simultaneously more ministers and also more illegitimate pregnancies in states where population is larger). Analytical Example 13.6 "Snowfall and Unemployment," provides a vivid illustration of the problem.

Other Coefficients

On occasion, one is apt to meet two other indexes of association, but they are merely mirror images of the two just discussed. Thus, the **sample coefficient of nondetermination (k^2)** equals the proportion of total variation in the values of the dependent variable, Y, that is *not* explained by means of the association of Y with X as measured by the estimated regression line. The formula for k^2 is given in Box 13.L.

13.L Sample Coefficient of Nondetermination

$$k^2 = \frac{\text{unexplained variation}}{\text{total variation}} = \frac{ESS}{Total\ SS} = \frac{\Sigma(Y - \hat{Y}_X)^2}{\Sigma(Y - \overline{Y})^2} = 1 - r^2$$

||| Practice Problem 14

Calculating the Sample Coefficient of Nondetermination (k^2) for Education and Income
In our example from Practice Problem 12, k^2 comes to $1 - .71 = .29$, indicating that 29 percent of the total variation in Y (income) observed in Figure 13.6 remains unexplained by the association of Y (income) with X (education) as estimated by the regression line shown there. |||

This measure, in turn, gives rise to the **sample coefficient of alienation *(k)*,** which is used even less frequently. This square root of k^2 is an abstract measure of the lack of correlation (equal to $\sqrt{.29} = .54$ in our example).

TWO EXTENSIONS*

In this final section, we will consider two extensions of the foregoing analysis. The first one is a new application of the previous chapter's analysis of variance; the second one is a discussion of curvilinear regression.

Testing β With Analysis of Variance

In Practice Problem 10, we tested the hypothesis that $\beta = 0$. The analysis of variance, introduced in Chapter 12, provides an alternative way of testing this hypothesis. The ANOVA procedure yields the same result as the t test employed above, but it has one major advantage: it can be applied to multiple regression problems as well.

The ANOVA test makes use of the three sums of squares discussed previously:

1. the regression sum of squares, $RSS = \Sigma(\hat{Y}_X - \overline{Y})^2$,
2. the error sum of squares, $ESS = \Sigma(Y - \hat{Y}_X)^2$, and
3. the total sum of squares, $Total\ SS = \Sigma(Y - \overline{Y})^2$.

||| Practice Problem 15

An ANOVA Test of β in the Case of Education and Income

The sums of squares can be calculated from the data of Table 13.2 and the formulas already noted, but they can also be derived from mathematically equivalent expressions that yield the following for our example:

$$
\begin{aligned}
RSS = \Sigma(\hat{Y}_X - \overline{Y})^2 &= a\Sigma Y + b\Sigma XY - n\overline{Y}^2 \\
&= 2{,}144.23(488{,}132) + 1{,}847.5(6{,}585{,}796) - 20(24{,}406.6)^2 \\
&= 1{,}046{,}667{,}278.36 + 12{,}167{,}258{,}110 - 11{,}913{,}642{,}471.2 \\
&= 1{,}300{,}282{,}917.16. \\
ESS = \Sigma(Y - \hat{Y}_X)^2 &= \Sigma Y^2 - a\Sigma Y - b\Sigma XY \\
&= 520{,}489{,}201.64
\end{aligned}
$$

(as already noted in the calculation of $s_{Y \cdot X}$ in Practice Problem 4).

$$
\begin{aligned}
Total\ SS = \Sigma(Y - \overline{Y})^2 &= \Sigma Y^2 - n \cdot \overline{Y}^2 \\
&= 1{,}820{,}772{,}118.80
\end{aligned}
$$

(as already noted in the calculation of r using the Box 13.K formula in Practice Problem 13). These sums of squares have also been entered in column 1 of Table 13.3. The appropriate degrees of freedom, which appear in column 2, equal the number of independent variables, m, for the regression sum of squares; they equal sample size, n, minus 2 for the error sum of squares (there are n errors and 2 constraints—namely,

---

*Optional section.

Table 13.3   The ANOVA Table

Source of Variation	Sum of Squares (1)	Degrees of Freedom (2)	Mean Square (3) = (1) ÷ (2)	Test Statistic (4)
Regression (variation explained by X)	RSS = 1,300,282,917.16	$m = 1$	RMS = 1,300,282,917.16	$F = \dfrac{RMS}{EMS} = 44.97$
Error (unexplained variation)	ESS = 520,489,201.64	$n - 2 = 18$	EMS = 28,916,066.76	
Total	*Total SS* = 1,820,772,118.80	$n - 1 = 19$		

the estimated regression coefficients, a and b); they equal $n - 1$ for the total sum of squares [there are n deviations subject to the constraint that $\Sigma(Y - \overline{Y}) = 0$].

From this information, the regression mean square and error mean square of column 3 can be calculated; the error mean square has, in fact, been derived in Practice Problem 4 in connection with the calculation of $s_{Y \cdot X}$. These mean squares are sample variances that can be used to calculate F, as shown in column 4.

Suppose we want to test H_0: $\beta = 0$ at the 5-percent level of significance. We can find the critical value of F in Appendix Table N for 1 numerator $d.f.$ and 18 denominator $d.f.$ as $F_{.05} = 4.41$. Thus, there is only a 5-percent chance that the computed value of F exceeds 4.41 if H_0 is true. Accordingly, as in our t test in Practice Problem 10, we should *reject* H_0. We conclude that $\beta \neq 0$; X (education) does help explain the variation in Y (income) in the population as a whole. ‖‖

Curvilinear Regression

So far, we have focused entirely on linear regression, but we should note that the technique can be adapted to analyze data that clearly reveal a curvilinear association between two variables of interest.

Approximating curvilinear by linear regression lines. One approach, illustrated in Figure 13.14, involves fitting piecewise linear regression lines to the data, with each of these lines being restricted to specified intervals of X values. This approach is particularly good when no simple curve can be found that summarizes the data very well.

Data transformation. A second approach to applying linear regression analysis to curvilinear data involves transforming, prior to analysis, the original values of one or both of the variables and then using the transformed values to compute the regression equation. For example, all the values of Y might be squared or replaced by their reciprocals or perhaps even by their logarithms. (The statement says "perhaps" because there are exceptions; for example, there is no logarithm for zero.) Such systematic conversion of all the numerical values of some variable is **data transformation.** If, as a result, the sample standard error of the estimate ($s_{Y \cdot X}$) declines or the sample coefficient of determination (r^2) rises, the procedure is considered a success.

Figure 13.14   Approximating Curvilinear by Linear Regression

A curvilinear association between two variables can sometimes be approximated well by fitting several linear regression lines to the data and restricting the use of each line to specified ranges of X values. In the example given here, one linear regression line estimates the values of Y for X values between a and b, a second linear regression line estimates Y for X values between b and c, and a third linear regression line estimates Y for X values between c and d.

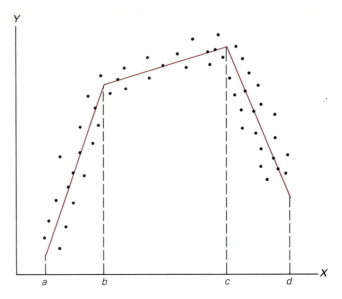

Practice Problem 16

The Relationship Between Output and Marginal Cost

Consider the data in columns 1 and 2 of Table 13.4. When plotted in a scatter diagram, they strongly suggest a curvilinear association between output and marginal cost, such as the one shown by the color line in panel (a) of Figure 13.15. But now consider transforming the Y values to logarithms, as in column 3. When plotted on semilog paper, as in panel (b) of our graph, the column 1 vs. column 3 data plots approximately a straight line. Given this linear relationship between log Y and X, the procedure employed earlier in this chapter can be applied to the column 1 and column 3 data. For this purpose, the remaining data in Table 13.4 have been computed. With their help and the formulas of Box 13.A (appropriately adjusted to our transformed data), we can also compute the following:

$$\overline{X} = \frac{\Sigma X}{n} = \frac{225}{16} = 14.0625$$

$$\overline{\log Y} = \frac{\Sigma \log Y}{n} = \frac{37.2596}{16} = 2.3287$$

Table 13.4 Sample Data on Output and Marginal Cost

Output (thousands of units per year), X (1)	Marginal Cost of Production (dollars per unit), Y (2)	Marginal Cost (transformed to logarithms), log Y (3)	X log Y (4)	X^2 (5)	$(\log Y)^2$ (6)
5	20	1.3010	6.5050	25	1.6926
7	60	1.7782	12.4474	49	3.1620
9	60	1.7782	16.0038	81	3.1620
9	100	2.0000	18.0000	81	4.0000
11	120	2.0792	22.8712	121	4.3231
12	120	2.0792	24.9504	144	4.3231
12	180	2.2553	27.0636	144	5.0864
14	240	2.3802	33.3228	196	5.6654
16	280	2.4472	39.1552	256	5.9888
16	360	2.5563	40.9008	256	6.5347
18	360	2.5563	46.0134	324	6.5347
18	480	2.6812	48.2616	324	7.1888
19	520	2.7160	51.6040	361	7.3767
19	640	2.8062	53.3178	361	7.8748
20	700	2.8451	56.9020	400	8.0946
20	1,000	3.0000	60.0000	400	9.0000
$\Sigma X = 225$		$\Sigma \log Y =$ 37.2596	$\Sigma(X \log Y) =$ 557.319	$\Sigma X^2 =$ 3,523	$\Sigma(\log Y)^2 =$ 90.0077

$$\log b = \frac{\Sigma(X \log Y) - n\,\overline{X}\,\overline{\log Y}}{\Sigma X^2 - n\,\overline{X}^2} = \frac{557.319 - 16(14.0625)(2.3287)}{3,523 - 16(14.0625)^2}$$

$$= \frac{33.3559}{358.9376} = .0929$$

$$\log a = \overline{\log Y} - \log b\,\overline{X} = 2.3287 - .0929(14.0625) = 1.0223$$

$$\log \hat{Y}_X = \log a + \log b\,X = 1.0223 + .0929\,X$$

As with any other regression equation, we can now estimate any desired Y for any given X. Let us take two cases: What, for example, is Y when X = 10 or when X = 20?

$$\log \hat{Y}_{10} = 1.0223 + .0929(10) = 1.9513, \text{ and } \hat{Y}_{10} = \text{antilog } (1.9513) = 89.4.$$

$$\log \hat{Y}_{20} = 1.0223 + .0929(20) = 2.8803, \text{ and } \hat{Y}_{20} = \text{antilog } (2.8803) = 759.17.$$

All four of these results make sense with respect to the relevant panels in Figure 13.15.

Figure 13.15   The Effect of Data Transformation

A curvilinear relationship between two variables appearing on an arithmetic scale, as in panel (a), may be transformed into a linear one on a semilog scale, as in panel (b).

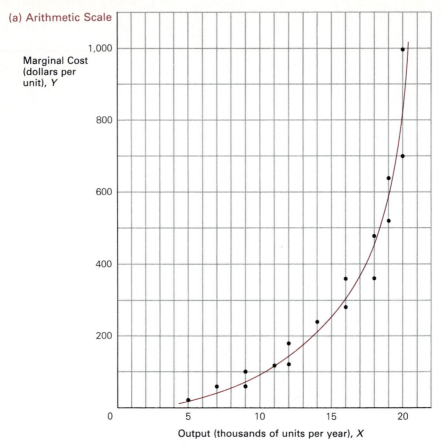

(a) Arithmetic Scale

Marginal Cost (dollars per unit), Y

Output (thousands of units per year), X

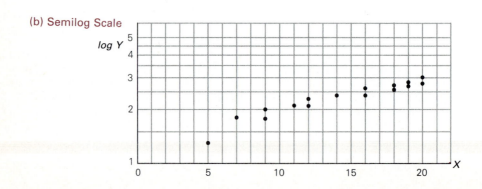

(b) Semilog Scale

log Y

Note: The regression equation highlighted above can also be written in antilog form as $\hat{Y}_X = 10.5268(1.2386)^X$, a form suggested in panel (e) of Figure 13.2. This way of writing the relationship also tells us that marginal cost rises by 23.86 percent for every 1,000-unit increase in output.

We can also calculate the coefficient of correlation with the help of our transformed data and an appropriately adjusted Box 13.K formula:

$$r = \frac{\Sigma(X \log Y) - n \bar{X} (\overline{\log Y})}{\sqrt{(\Sigma X^2 - n \bar{X}^2)[\Sigma(\log Y)^2 - n \overline{\log Y^2}]}}$$

$$= \frac{557.319 - 16(14.0625)(2.3287)}{\sqrt{[3,523 - 16(14.0625)^2][90.0077 - 16(2.3287)^2]}}$$

$$= \frac{33.3559}{\sqrt{(358.9376)(3.2422)}} = \frac{33.3559}{\sqrt{1163.7474}} = \frac{33.3559}{34.1137} = .9778$$

Thus, the coefficient of determination is $r^2 = .956$, which suggests that the above regression equation is capable of explaining more than 95 percent of the variation in Y by means of the association of Y with X. The level of marginal cost is almost perfectly explained by the level of output. |||

||| ANALYTICAL EXAMPLES

13.1 Evaluating a New Food Product

Americans are said to be eager for on-the-run meals, provided only that they are also palatable and nutritious. The General Foods Corporation decided to bring out such a product, called H. In order to test the palatability and nutritive quality of H, it arranged for a 28-day rat-feeding experiment (because rats have metabolic processes similar to people), and it used regression techniques to evaluate the experimental data to draw conclusions about marketing such a product.

In order to provide randomization, 30 rats were divided into 10 blocks of 3 rats each such that the rats within any one block had about the same initial weight. One member of each trio was then assigned, at random, to a diet of either liquid H or solid H or to a control diet of casein. Each of the three diets contained about 9 percent protein, and the rats could eat all they wanted. Ultimately, the rats' final and initial weights were compared.

Panel (a) of the figure on page 566 shows a scatter diagram of the results; panel (b) shows the regression line for the 3 dietary groups, relating 28-day weight gain, Y, to protein intake, X.

The equations were: for liquid H, $\hat{Y}_X = -4.124 + 3.72 X$; for solid H, $\hat{Y}_X = -8.478 + 3.66 X$; for casein control, $\hat{Y}_X = -3.644 + 2.91 X$.

The company statisticians concluded (a) from the position of the H lines to the right of the casein line that rats ate more H than casein because H was more palatable (at least to rats) and (b) from the higher slope of the H lines that the protein efficiency of H was superior to that of the control diet (protein contained in H resulted in greater weight gain). The difference between the slopes of the H lines (3.72 vs. 3.66) was not statistically significant; the difference between the H-line slopes and the slope of the casein line (2.91) was statistically significant.

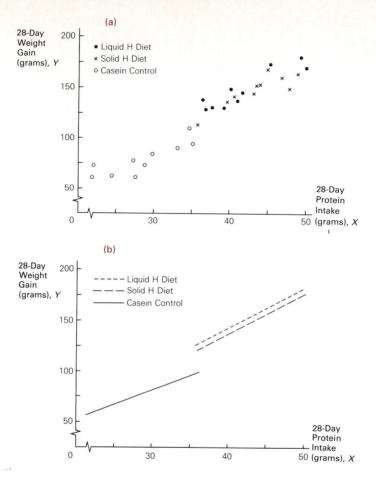

Source: Elisabeth Street and Mavis B. Carroll, "Preliminary Evaluation of a New Food Product," in Judith M. Tanur et al., eds., *Statistics: A Guide to the Unknown* (San Francisco: Holden-Day, 1972), pp. 220–28.

13.2 Of Bulls, Bears, and Beauty

Stock market analysts sometimes characterize the riskiness of a particular stock (such as TWA or IBM or Procter & Gamble) with the help of a sample regression line that relates the stock's rate of return to the average rate of return on all stocks, as in Figure A.

Depending on whether the slope, β, of the true regression line is believed to exceed, equal, or fall short of unity, the stock is referred to as an *aggressive, marketlike,* or *defensive* stock. In a bull market (when the average rate of return on all stocks is rising), an aggressive stock gains more than the average; a defensive stock gains less. In a bear market (when the average rate of return on all stocks is falling), an aggres-

sive stock loses more than the average; a defensive stock loses less. Clearly, an investor is well advised to own aggressive stocks in a bull market, but defensive stocks in a bear market. Thus, stock market analysts are always engaging in "beta analysis."

Recently, an economist derived 1971–1984 rates-of-return on particular types of *paintings* from components of the well-known art-market price index produced by Sotheby's London auction house and compared these returns on particular assets with the average rate-of-return on a market portfolio of bonds, real estate, and stocks. As Figure B shows, differences in riskiness similar to these connected with particular stocks emerged.

Figure A

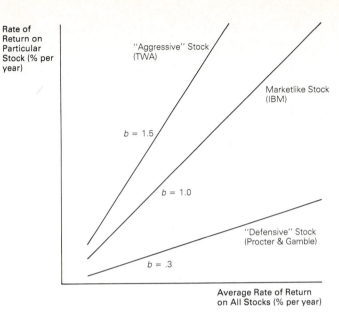

Figure B

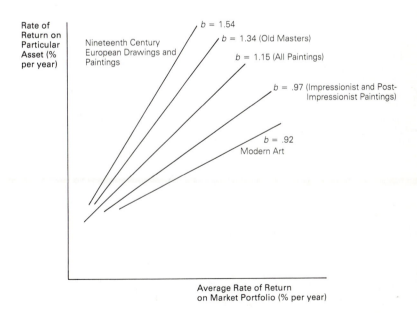

Sources: Charles L. Olson and Mario J. Picconi, *Statistics for Business Decision Making* (Glenview, IL: Scott, Foresman and Company, 1983), p. 496, and Michael F. Bryan, "Beauty and the Bulls: The Investment Characteristics of Paintings," *Federal Reserve Bank of Cleveland Economic Review,* I, 1985, pp. 2–10.

13.3 Predicting Changes in the GNP

There are two major types of policies that governments can pursue in order to affect the Gross National Product (GNP). Monetary policy works through changes in the money supply, ΔM; and fiscal policy works through changes in government expenditures (ΔG) or in tax collections (ΔT). One economist wanted to test whether a knowledge of ΔM or ΔG is better for predicting an associated change in GNP, ΔY.

Using U.S. data about quarterly changes between 1953 and 1969, the accompanying scatter diagrams were produced. Coefficients of determination and correlation were calculated from the data; they are shown along with the two graphs. It was concluded that ΔY can be predicted more accurately from ΔM than from ΔG—but note: even so only 44 percent of the variation in ΔY is explained by ΔM. Caution: This is shown by $r^2 = .44$ in panel (a), not by $r = .44$ in panel (b).

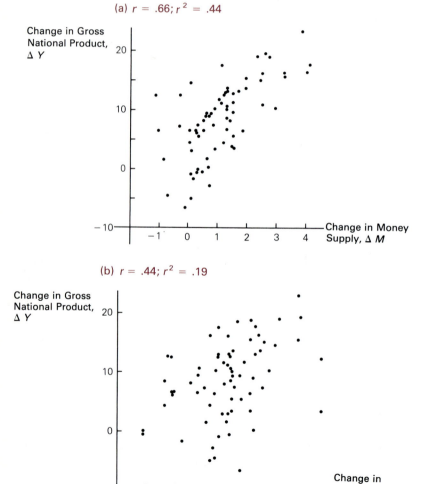

(a) $r = .66; r^2 = .44$

(b) $r = .44; r^2 = .19$

Source: Leonall C. Andersen, "Statistics for Public Financial Policy," in Judith M. Tanur et al., eds., *Statistics: A Guide to the Unknown* (San Francisco: Holden-Day, 1972), pp. 321–25.

13.4 The Inflation Rate Versus Its Predictability

Two researchers suspected the existence of a direct relationship between the level of inflation and its variability (and, thus, an inverse relation between inflation and its predictability). Using a sample of 40 economies during the 1970s, they recorded each country's average rate of inflation, $X = \bar{I}$ (which ranged from 4.89 percent in Germany to 59.25 percent in Uruguay) as well as each country's standard deviation of this inflation rate, $Y = SDI$. The following regression equation was estimated from these data:

$$SDI = a + b\bar{I} = -.119 + .507\,\bar{I}.$$

The coefficient of determination was $r^2 = .84$, and a similar coefficient for ρ^2 was confirmed. Thus, it was confirmed that higher levels of inflation are strongly and positively related to inflation variability (and, thus, reduce inflation predictability).

Source: R. W. Hafer and Gail Heyne-Hafer, "The Relationship Between Inflation and Its Variability: International Evidence from the 1970s," *Journal of Macroeconomics,* Fall 1981, pp. 571–77.

13.5 Predicting Money Supply Changes

Participants in the money market continually monitor Federal Reserve behavior because this helps them predict changes in the money supply ($M1$) and, thus, in interest rates. Two economists produced regressions relating the actual change in the money supply, $\Delta AM1$, to the expected change, $\Delta EM1$ (as measured by the median figure found in a regular survey taken by Money Market Services). Their regressions took the form

$$\Delta AM1 = a + b\Delta EM1 + e.$$

The accompanying table shows some of their results. (The values in parentheses are t values, the ratios of the estimated regression coefficients to their standard errors, that are helpful in making judgments about the true regression coefficients—see Practice Problems 10 and 11.)

Sample Period	a	b	$s_{Y \cdot X}$	r^2
9/29/77 to 10/4/79 (103 observations)	−.13 (−2.64)	1.16 (9.91)	.42	.49
10/11/79 to 10/1/82 (150 observations)	.05 (1.06)	1.14 (8.12)	.54	.30
10/2/82 to 1/27/84 (68 observations)	.05 (1.04)	1.12 (8.44)	.37	.51
2/3/84 to 12/20/84 (46 observations)	−.14 (−3.07)	1.48 (11.69)	.28	.75

Source: William T. Gavin and Nicholas V. Karamouzis, "The Reserve Market and the Information Content of $M1$ Announcements," *Federal Reserve Bank of Cleveland Economic Review,* I, 1985, pp. 11–28.

13.6 Snowfall and Unemployment

A splendid example of spurious correlation is provided by the data in the accompanying table. A scatter diagram reveals a strong positive association between the two variables; their relationship can be summarized by the least-squares regression line, $\hat{Y}_X = .3764 + .0955X$, which is also shown in the accompanying graph. The coefficient of determination equals $r^2 = .967$, implying that 96.7 percent of the variation in U.S. unemployment between 1973 and 1982 can be explained by variations in snowfall at Amherst, Massachusetts. Yet, unlike the association between smoking and health (Close-Up 13.3), this association is totally unreasonable. Nobody would be prepared to argue that the strong association between U.S. unemployment and the extent of snowfall in a small New England town is anything but pure coincidence.

Year	Snowfall at Amherst, Massachusetts (inches)	U.S. Unemployment Rate (percent)
1973	45	4.9
1974	59	5.6
1975	82	8.5
1976	80	7.7
1977	71	7.1
1978	60	6.1
1979	55	5.8
1980	69	7.1
1981	79	7.6
1982	95	9.7

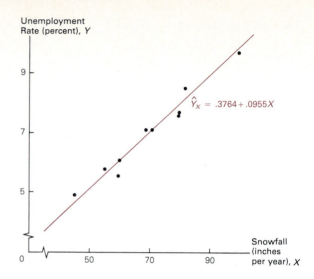

$$\hat{Y}_X = .3764 + .0955X$$

Sources: Economic Report of the President (Washington, D.C.: U.S. Government Printing Office, 1983), p. 199, and data collected by the author.

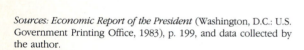

CLOSE-UPS

13.1 TV Commercials and the Demand for Water

Some years ago, the manager of a Long Island water district was perplexed: during the evening hours, an otherwise steady demand for water was punctuated by short bursts of enormous increases in demand. Following a hunch, he recorded the amount of water demanded during specific time periods and also the precise timing of commercials during popular TV shows playing at the time. It turned out that the greatest increase in the demand for water occurred during the commercial breaks of the program ("I Love Lucy") then holding the top spot in the Nielsen ratings, with correspondingly smaller surges in demand occurring during the commercials of the next nine shows among the top ten. Apparently, commercials were sending people to the bathroom in droves; the correlation was almost perfect.

Source: Richard P. Runyon, *How Numbers Lie: A Consumer's Guide to the Fine Art of Numerical Deception* (Lexington, Mass.: Lewis Publishing Co., 1981), pp. 125–26.

13.2 The Uneven Burden of International Trade Restraints

More and more U.S. industries have recently requested governmental protection from international competition. When granted, such protection helps those who supply resources to the industries involved (such as factory owners and workers), but it harms consumers who must pay higher prices. In 1984, for example, American consumers spent a minimum of $16 billion more just because of the protective measures designed to help the domestic automobile, clothing, sugar, and steel industries. However, this burden was not shared equally. The higher prices caused by protection were equivalent to a 23-percent income-tax surcharge for families earning less than $10,000 a year; they were equivalent to a 3-percent income-tax surcharge for families earning over $60,000 a year. The correlation between harm so measured and size of family income was strong and negative: − .83.

Source: Based on Susan Hickok, "The Consumer Cost of U.S. Trade Restraints," *Federal Reserve Bank of New York Quarterly Review,* Summer 1985, pp. 1–12.

13.3 Smoking and Health

The long-standing controversy on smoking and health provides an excellent example of how regression and correlation analysis has been employed to discover possible *causal* connections between variables yet has been unable to provide definite proof. After hundreds of years of tobacco use, governments throughout the world now condemn the practice as a serious health hazard, claiming that smoking tobacco causes cancer, heart attacks, and many other deleterious effects. (In the early 1980s, some 340,000 deaths per year were blamed on tobacco smoking in the United States alone.) Such conclusions have been based on scatter diagrams, showing a strong positive association between different countries' per capita cigarette consumption and lung-cancer deaths. They have been based on life expectancy data such as those in the accompanying graph, showing that only 45 percent of heavy smokers, but 65 percent of nonsmokers survive to age 60 (and similarly, for all other ages). They have been based on prospective studies that select large numbers of people who either smoke or don't smoke and whose health is then monitored until death. (One such study has shown heavy smokers to die from lung cancer at a rate of 1.66 per 1,000 per year, while the corresponding rate for nonsmokers was .07. The corresponding numbers for deaths from heart attacks were 5.99 and 4.22.)

Yet despite the mounting evidence against smoking, there are many who remain unconvinced, citing such problems as selection bias in the studies noted above. (Sir Ronald Fisher—Biography 9.1—was among these dissenters.)

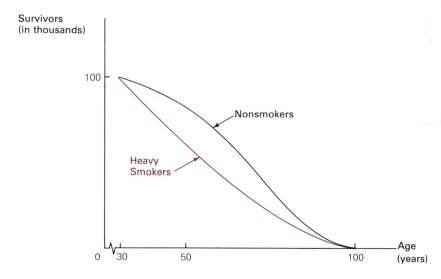

Sources: R. Doll, "Etiology of Lung Cancer," *Advances in Cancer Research* 3(1955):1–50; R. A. Fisher, *Smoking: The Cancer Controversy* (Edinburgh: Oliver and Boyd, 1959); Report of the U.S. Surgeon General, *Smoking and Health,* 1964; B. W. Brown, Jr., "Statistics, Scientific Method, and Smoking," in Judith M. Tanur et al., eds., *Statistics: A Guide to the Unknown* (San Francisco: Holden-Day, 1972), pp. 40–51.

13.1 Francis Galton

Francis Galton (1822–1911) was born in Birmingham, England. Intellectually precocious, he turned to studying mathematics, and then medicine, at an unusually early age. Yet he never finished his studies. When his father died and left him a fortune at age 22, he took to travelling: down the Danube to the Black Sea, to Egypt, and on to then-unexplored parts of southwest Africa. Upon his return, he settled in London for the rest of his life (and promptly won the Royal Geographic Society's gold medal for his earlier exploits). He never held any academic or professional posts, but he was driven by an endless curiosity about humanity and nature, and he never ceased to roam every nook and cranny of the world of thought. A rich flow of original ideas yielded 16 books and more than 200 articles and, eventually, brought him knighthood shortly before his death.

Consider just these few examples of his accomplishments: He was a meteorologist who systematically charted weather patterns and in the process discovered and named the anticyclone. He was a psychologist who measured sensory acuity and character traits and sowed the seeds of mental testing. Like his cousin, Charles Darwin, he was a biologist and as such established fingerprinting as an infallible means of human identification. (His taxonomy of prints is used to this day.) Using data on notable families, he also studied the inheritance of talent (artistic, athletic, scholarly) and concluded, in such works as *Hereditary Genius* (1869) and *Natural Inheritance* (1889), that inheritance played a significant role in transmitting talent, even after making due allowance for environmental factors. Indeed, he became convinced that a "eugenics" program (he coined the word) "to foster talent and healthiness and suppress stupidity and sickliness" was crucial for the promotion of a high-quality society. He, therefore, endowed the Galton chair of eugenics at University College, London; the first occupant of that chair, Karl Pearson (Biography 10.1), extended and refined much of Galton's work—particularly Galton's work in statistics.

Galton had a passion for collecting numerical data and exhibited an astonishing cleverness in analyzing them. Thus, he became the originator of the modern regression and correlation techniques discussed in this chapter (that were foreshadowed by Carl Friedrich Gauss—Biography 7.1). In 1875, Galton experimented with sweet peas to determine the law of inheritance of size. He selected 7 groups of different sizes, persuaded friends in different parts of England to plant 10 seeds from each group, and eventually compared the sizes of parental seeds with those of their offspring. When he plotted the results, as in the accompanying figure, he discovered that instead of each offspring being just

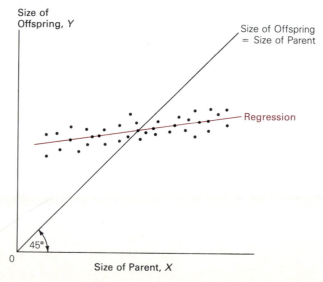

like its parent (which would put his pairwise data plots precisely on the 45° line in our graph), dwarf peas produced larger offspring, while giant peas produced smaller offspring (note how the actual data plots diverge from the 45° line).[1] Galton called this phenomenon "reversion" (toward some average ancestral type) and later "regression toward mediocrity." Thus, Galton argued, for purposes of predicting the transmission of natural characteristics from one generation to the next, the 45° line in the accompanying figure was inferior to another or *regression* line going through the scatter of dots. Nowadays, this tendency of those members of any population who, with respect to a given characteristic, are in an extreme position (below or above the population mean) at one time to be in a less extreme position at a later time (either personally or by means of their offspring) is called the **regression effect.** It is found to apply widely and not only to the sizes of sweet peas from one generation to the next. Thus, as Galton also discovered, on average, extremely short fathers tend to have somewhat taller

sons; extremely tall fathers tend to have somewhat shorter sons. The subset of all students with extremely poor grades on exam 1 tends to have better grades (closer to the average of all students) on exam 2, while the subset of all students with extremely good grades on exam 1 tends to have worse grades (again closer to the average of all students) on exam 2. And firms with the worst profit picture in year 1, on average, are not among the worst in year 2, while those with the highest profit in year 1, on average, are not in the top group in year 2.

But caution is advised: it is one thing merely to observe this regression effect; it is quite another to succumb to the **regression fallacy** and incorrectly attribute the regression effect (as Galton did) to the operation of some important unseen factor (such as his "tendency toward mediocrity"). The real explanation is that extreme values (whether of size, grades, or profit) often occur by pure chance; the odds are that they do not occur twice in a row.

[1]Francis Galton, "Typical Laws of Heredity," *Proceedings of the Royal Institute* 8(1877):282–301.

Sources: Dictionary of Scientific Biography, vol. 5 (New York: Scribner's, 1972), pp. 265–67; *International Encyclopedia of Statistics,* vol. 1 (New York: Free Press, 1978), pp. 359–64.

Summary

1. In many situations that are of interest to researchers or decision makers, the value of one variable is associated with the value of another variable in some systematic way. Regression and correlation analysis is concerned with such linkages. The central focus of *regression analysis* is the establishment of an equation that allows the unknown value of one variable to be estimated from the known value of one or more other variables. (A single variable is used to do the estimating in *simple* regression analysis; several variables are used in *multiple* regression analysis.) The central focus of *simple correlation analysis* is the establishment of an index that provides, in a single number, an instant picture of the strength of association between two variables.

2. An understanding of regression analysis requires an understanding of the basic concepts of independent vs. dependent variables, deterministic vs. stochastic relationships, and direct vs. inverse relationships. The technique of simple regression analysis for two variables that are related in linear fashion begins with the plotting of sample data in a scatter diagram. In principle, regression lines can

be fitted to the scatter diagram (a) by eyesight or (b) by the method of least squares, but (b) is preferred to (a) because the result of (b) alone is reliably reproducible and usable for making a variety of inferences. An *estimated regression line* shows the association between two variables present in a sample, while the *true regression line* shows the association present in the population. Certain basic assumptions must hold before one can make valid inferences about the true regression line with the help of the estimated regression line. Inferences can be made concerning the average value of Y, given X (for both the small-sample and large-sample case), concerning an individual value of Y, given X (for both the small-sample and large-sample case), and concerning the true regression coefficients (using confidence intervals or hypothesis tests).

3. Simple correlation analysis provides a variety of general indices measuring the strength of linear association between two variables. These indexes include (a) the sample coefficient of determination, r^2 (which equals the proportion of the total variation in the values of Y that can be explained by means of the association of Y with X as measured by the estimated regression line), (b) the sample coefficient of correlation, r, and (c) the sample coefficient of nondetermination (k^2) and the sample coefficient of alienation *(k)*, which are employed less frequently.

4. The analysis of variance can play a role in regression analysis (such as in testing hypotheses concerning β, the slope of the true regression line). Curvilinear regression analysis can be accomplished by using several linear regression lines to approximate curvilinear ones and by transforming data.

Key Terms

conditional mean of Y ($\mu_{Y \cdot X}$)
conditional probability distribution of Y
conditional standard deviation of Y ($\sigma_{Y \cdot X}$)
confidence band
data transformation
dependent variable
deterministic relationship
direct relationship
error sum of squares *(ESS)*
estimated regression coefficients
estimated regression line
explained deviation of Y
explained variable
explained variation of Y
explanatory variable
independent variable
inverse relationship
method of least squares
multiple regression analysis
nonsense correlation
population coefficient of correlation (ρ)
population coefficient of determination (ρ^2)
population regression line
population standard error of the estimate
predicted variable
prediction interval

predictor variable
regressand
regression analysis
regression effect
regression fallacy
regression line
regression sum of squares *(RSS)*
regressor
sample coefficient of alienation *(k)*
sample coefficient of correlation *(r)*
sample coefficient of determination (r^2)
sample coefficient of nondetermination (k^2)
sample regression line
sample standard error of the estimate ($s_{Y \cdot X}$)
scatter diagram
simple correlation analysis
simple regression analysis
spurious correlation
stochastic relationship
total deviation of Y
total sum of squares *(Total SS)*
total variation of Y
true regression coefficients
true regression line
unexplained deviation of Y
unexplained variation of Y

Questions and Problems

The computer program noted at the end of the chapter can be used to work most of the subsequent problems.

1. The following has been hypothesized about U.S. presidential elections: the percentage of the popular vote, Y, that goes to the candidate put up by the party of the incumbent president depends on the change in the unemployment rate during the election year, X, in such a way that any increase in unemployment detracts from, and any decrease in unemployment adds to, the vote for the incumbent-party candidate. On the basis of the data provided in Table 13.A, test this hypothesis by (a) preparing a scatter diagram with axes like those in Figure A, (b) estimating a regression line with the equation, $\hat{Y}_X = a + bX$, and (c) drawing this line in your diagram. Do your results make sense?

Table 13.A

Election Year	Party of Incumbent President[a]	Change in Election-Year Unemployment Rate, X	Percentage of Popular Vote Captured by Party of Incumbent President, Y
1892	1	−2.4	48.3
1896	0	+0.7	47.8
1900	1	−1.5	53.2
1904	1	+1.5	60.0
1908	1	+5.2	54.5
1912	1	−2.1	54.7
1916	0	−3.4	51.7
1920	0	+3.8	36.1
1924	1	+2.6	54.3
1928	1	+0.9	58.8
1932	1	+7.7	40.9
1936	0	−3.2	62.5
1940	0	−2.6	55.0
1944	0	−0.7	53.8
1948	0	−0.1	52.4
1952	0	−0.3	44.6
1956	1	−0.3	57.8
1960	1	0.0	49.9
1964	0	−0.5	61.3
1968	0	−0.2	49.6
1972	1	−0.3	61.8
1976	1	−0.8	49.4
1980	0	+1.3	48.4
1984	1	−2.1	59.0

[a]Democrats = 0; Republicans = 1.

Figure A

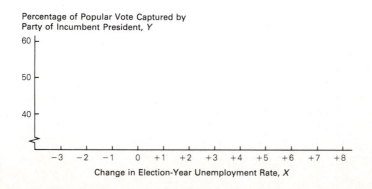

Percentage of Popular Vote Captured by Party of Incumbent President, Y

Change in Election-Year Unemployment Rate, X

2. An avionics manufacturer wants to establish the relationship between the monthly sales of long-range navigation units, Y, and the number of ads placed in monthly magazines, X. Given the data of Table 13.B, determine an estimated regression equation.

Table 13.B

Loran Units Sold, Y	Number of Ads, X	Loran Units Sold, Y	Number of Ads, X
7	20	18	187
4	36	9	70
8	36	11	81
7	50	12	111
18	200	16	200
19	195	16	180
9	140	14	167
9	126	11	172
13	195	12	160
17	158	6	50
7	77	9	96
7	100	12	122
16	155	15	130
10	98	11	40

3. The manager of the city water department wants to establish the relationship between monthly household water consumption, Y, and household size, X. Given the data of Table 13.C, determine an estimated regression equation.

Table 13.C

Gallons of Water Used, Y	Household Size, X
650	2
1,200	7
1,300	9
430	4
1,400	12
900	6
1,800	9
640	3
793	3
925	2

4. A realtor wants to establish the relationship between the number of weeks homes are on the market prior to sale, Y, and the asking price, X. Given the data of Table 13.D, determine an estimated regression equation.

Table 13.D

Weeks to Sale, Y	Asking Price (thousands), X	Weeks to Sale, Y	Asking Price (thousands), X
6.5	20	8.6	99
6.8	80	10.6	99
7.0	100	15.0	125
8.6	99	15.0	130
12.1	125	19.0	180
9.0	140	12.5	120
9.5	110	27.0	200

5. A marketing manager wants to establish the relationship between the number of cereal boxes sold, Y, and the shelf space devoted to them, X. Given the data of Table 13.E, determine an estimated regression equation.

Table 13.E

Number of Boxes Sold, Y	Feet of Shelf Space, X	Number of Boxes Sold, Y	Feet of Shelf Space, X
145	3	125	7
151	6	190	5
235	9	210	10
120	5	118	8
272	13	100	7
300	15	390	12
110	2	210	6

6. The manager of an insurance company wants to establish the relationship between people's life insurance, Y, and their salaries, X. Given the data of Table 13.F, determine an estimated regression equation.

Table 13.F

Life Insurance in Force, Y (thousands of dollars)	Annual Salary, X (thousands of dollars)
50	10
80	29
100	30
130	31
150	36
150	40
200	40
300	29
300	50
350	127

7. The manager of a railroad wants to establish the relationship between fuel costs per ton mile, Y, and the number of cars on a freight train, X. Given the data of Table 13.G, determine an estimated regression equation.

Table 13.G

Fuel Costs per Ton-Mile, Y (cents)	Cars on Train, X (number)
16.3	20
15.0	30
14.3	36
13.0	40
15.1	49
12.0	51
12.0	60
12.1	63
14.7	70
15.2	70
12.1	81
11.0	97
9.2	100
10.1	120
6.2	152
4.8	156

8. The operator of a small airport wants to establish the relationship between annual maintenance costs of airplanes, Y, and the airplanes' age, X. Given the data of Table 13.H, determine an estimated regression equation.

Table 13.H

Annual Maintenance Cost, Y (dollars per plane)	Age of Plane, X (years)
562	1
622	1
1,922	1
3,502	2
222	3
2,050	5
751	5
1,200	6
512	8
723	10
1,980	12
2,050	12
751	15
3,171	20
622	21

9. An economist wants to establish the relationship between the unemployment rate, Y, and the Treasury bill rate, X. Given the data of Table 13.I, determine an estimated regression equation.

Table 13.I

Unemployment Rate, Y (percent of labor force)	Treasury Bill Rate, X (percent per year)
6.7	9.7
7.3	9.8
8.9	7.6
9.1	6.1
7.2	10.2
5.2	12.7
6.9	14.3
6.9	7.9
7.1	8.9

10. A farmer wants to establish the relationship between the per-acre yield of corn, Y, and the average July temperature in Iowa, X. Given the data of Table 13.J, determine an estimated regression equation.

Table 13.J

Corn Yield per Acre, Y (bushels)	Average July Temperature, X (degrees Fahrenheit)
115	91
119	95
126	101
91	88
107	92
89	89
111	95
90	87

11. An economist wants to establish the relationship between the price per gallon of gasoline, Y, and the price per barrel of crude oil, X. Given the data of Table 13.K, determine an estimated regression equation.

Table 13.K

Price per Gallon of Gasoline, Y (dollars)	Price per Barrel of Crude Oil, X (dollars)
.39	3.50
.42	4.29
.56	5.66
.79	18.21
.82	19.33
.97	29.10
1.11	36.12
1.22	36.05
1.36	42.12
1.22	30.50

12. A marketing manager wants to establish the relationship between the second-year sales of sales representatives, Y, and their first-year sales, X.

Given the data of Table 13.L, determine an estimated regression equation.

Table 13.L

Second-Year Sales, Y (units)	First-Year Sales, X (units)
69	170
75	133
86	86
111	161
129	112
133	133
134	136
136	82
140	60
152	152
161	83
170	97

13. A marketing manager wants to establish the relationship between sales, Y, and the price of a similar product produced by a competitor, X. Given the data of Table 13.M, determine an estimated regression equation.

Table 13.M

Sales, Y (units)	Competitor's Price, X (dollars per unit)
520	13
550	13
600	15
610	15
620	16
724	21
680	21
300	14
962	40
270	12

14. An insurance company executive wants to establish the relationship between fire damage, Y, and distance of the victims' property from the fire station, X. Given the data of Table 13.N, determine an estimated regression equation.

Table 13.N

Fire Damage, Y (thousands of dollars)	Distance to Fire Station, X (miles)
12.6	1.1
18.9	.9
33.0	2.0
4.1	.7
150.0	1.2
77.0	.8
6.1	3.1
9.2	3.2
10.2	7.0
300.0	1.3

Table 13.P

New Car Sales, Y (millions per year)	Old Car Stock, X (millions at mid-year)
9.3	68.9
8.4	80.4
8.6	95.2
10.7	104.7
9.0	104.6
8.5	105.8
8.0	106.9

15. An advertising executive wants to establish the relationship between the firm's market share, Y, and its annual spending on television ads, X. Given the data of Table 13.O, determine an estimated regression equation.

Table 13.O

Market Share, Y (percent)	Annual Spending on TV ads, X (thousands of dollars)
8.8	23
12.7	33
13.8	36
15.0	39
17.2	42
28.6	69
30.8	88
41.2	127
53.0	130

Table 13.Q

Electricity Demand, Y (million megawatt-hours)	Natural-Gas Price, X (dollars per unit)
173	33
161	36
142	22
193	41
211	38
219	42
260	50
250	51
181	30
129	25

Table 13.R

Airplanes Sold, Y (number per year)	Price per Plane, X (thousands of dollars)
451	49
437	63
411	89
389	111
359	141
350	150
345	155
331	169
267	233
207	293
87	413
85	415

16. An economist wants to establish the relationship between annual sales of new cars, Y, and the stocks of old cars, X. Given the data of Table 13.P, determine an estimated regression equation.

17. An economist wants to establish the relationship between the demand for electricity, Y, and the price of natural gas, X. Given the data of Table 13.Q, determine an estimated regression equation.

18. An economist wants to establish the relationship between the number of airplanes sold per year, Y, and their price, X. Given the data of Table 13.R, determine an estimated regression equation.

19. An economist wants to establish the relationship between U.S. personal consumption expenditures, Y, and gross national product, X. Given the data of Table 13.S, determine an estimated regression equation.

Table 13.S

Personal Consumption Expenditures, Y (billions of dollars)	Gross National Product, X (billions of dollars)
452.00	737.20
461.40	756.60
482.00	800.30
500.50	832.50
528.00	876.40
557.50	929.30
585.70	984.80
602.70	1011.40
634.40	1058.10
657.90	1087.60
672.10	1085.60
696.80	1122.40
737.10	1185.90
767.90	1254.30
762.80	1246.30
779.40	1231.60
823.10	1298.20
864.30	1369.70
903.20	1438.60
927.60	1479.40
931.80	1475.00
956.80	1513.80
970.20	1485.40
1011.40	1534.80

20. An economist wants to establish the relationship between U.S. gross private domestic investment, Y, and Moody's Aaa corporate bond yields, X. Given the data of Table 13.T, determine an estimated regression equation.

21. Reconsider problem 1. Make a *point estimate* for the voting percentage captured by the incumbent-party candidate if election-year unemployment goes from 2 to 7 percent (hence, the change is $+5$ points); then figure a *98-percent prediction interval* for the voting percentage received, on the average, by candidates who find themselves in this

Table 13.T

Gross Private Domestic Investment, Y (billions of dollars)	Moody's Aaa Corporate Bond Yields, X (percent per annum)
75.90	4.41
74.80	4.35
85.40	4.33
90.90	4.26
97.40	4.40
113.50	4.49
125.70	5.13
122.80	5.51
133.30	6.18
149.30	7.03
144.20	8.04
166.40	7.39
195.00	7.21
229.80	7.44
228.70	8.57
206.10	8.83
257.90	8.43
324.10	8.02
386.60	8.73
423.00	9.63
401.90	11.94
474.90	14.17
414.50	13.79
471.30	12.04

predicament. Would your answer differ for a particular incumbent-party candidate you might be advising? What would it be?

22. Reconsider problem 2. Make a *point estimate* for monthly sales if there are 150 ads. Then compute *95-percent prediction intervals* for $\mu_{Y \cdot 150}$ and $I_{Y \cdot 150}$.

23. Reconsider problem 3. Make a *point estimate* of household water consumption if household size is 5 persons. Then compute *90-percent prediction intervals* for $\mu_{Y \cdot 5}$ and $I_{Y \cdot 5}$.

24. Reconsider problem 4. Make a *point estimate* of weeks to sale if the asking price is $138,000. Then compute *95-percent prediction intervals* for $\mu_{Y \cdot 138}$ and $I_{Y \cdot 138}$.

25. Reconsider problem 5. Make a *point estimate* of boxes sold if 11 feet of shelf space are available. Then compute *98-percent prediction intervals* for $\mu_{Y \cdot 11}$ and $I_{Y \cdot 11}$.

26. Reconsider problem 6. Make a *point estimate* of life insurance in force if annual salary is $80,000. Then compute *99-percent prediction intervals* for $\mu_{Y \cdot 80}$ and $I_{Y \cdot 80}$.

27. Reconsider problem 7. Make a *point estimate* of fuel costs per ton-mile if there are 140 cars on the train. Then compute *99.5-percent prediction intervals* for $\mu_{Y \cdot 140}$ and $I_{Y \cdot 140}$.

28. Reconsider problem 8. Make a *point estimate* of annual maintenance cost if the plane is 7 years old. Then compute *99.8-percent prediction intervals* for $\mu_{Y \cdot 7}$ and $I_{Y \cdot 7}$.

29. Reconsider problem 9. Make a *point estimate* of the unemployment rate when the Treasury Bill rate is 14 percent. Then compute *99.9-percent prediction intervals* for $\mu_{Y \cdot 14}$ and $I_{Y \cdot 14}$. Also ask yourself this: Does the observed relationship between Y and X make sense?

30. Reconsider problem 10. Make a *point estimate* of corn yield if the average July temperature is an even 90°F. Then compute *90-percent prediction intervals* for $\mu_{Y \cdot 90}$ and $I_{Y \cdot 90}$.

31. Reconsider problem 11. Make a *point estimate* of the gasoline price if the crude oil price is $25. Then compute *95-percent prediction intervals* for $\mu_{Y \cdot 25}$ and $I_{Y \cdot 25}$.

32. Reconsider problem 12. Make a *point estimate* of the second-year sales if first-year sales are 100 units. Then compute *98-percent prediction intervals* for $\mu_{Y \cdot 100}$ and $I_{Y \cdot 100}$.

33. Reconsider problem 1. Determine *95-percent confidence intervals* for β and α. Show your computations.

34. Determine *95-percent confidence intervals* for β and α in problems 2, 4, and 6.

35. Determine *95-percent confidence intervals* for β and α in problems 3, 5, and 7.

36. Determine *95-percent confidence intervals* for β and α in problems 8, 10, and 12.

37. Determine *95-percent confidence intervals* for β and α in problems 9, 11, and 13.

38. Determine *95-percent confidence intervals* for β and α in problems 14, 16, and 18.

39. Reconsider problem 1. Make a *two-tailed hypothesis test* (H_0: $\beta = 0$), using a confidence level of 95 percent. Show your computations.

40. Using a confidence level of 95 percent, make *two-tailed hypothesis tests* of H_0: $\beta = 0$ for problems 2, 4, 6, 8, and 10.

41. Using a confidence level of 95 percent, make *two-tailed hypothesis tests* of H_0: $\beta = 0$ for problems 3, 5, 7, 9, and 11.

42. Using a confidence level of 95 percent, make *two-tailed hypothesis tests* of H_0: $\beta = 0$ for problems 12, 14, 16, 18, and 20.

43. Using a confidence level of 95 percent, make *two-tailed hypothesis tests* of H_0: $\beta = 0$ for problems 13, 15, 17, and 19.

44. Calculate the sample coefficients of determination and of correlation for problems 2, 4, 6, 8, and 10. For problem 2, explain verbally what r^2 means.

45. Calculate the sample coefficients of determination and of correlation for problems 1, 3, 5, 7, and 9. For problem 1, explain verbally what r^2 means.

46. Calculate the sample coefficients of determination and of correlation for problems 12, 14, 16, 18, and 20.

47. Calculate the sample coefficients of determination and of correlation for problems 11, 13, 15, 17, and 19.

48. At the 5-percent level of significance, make an ANOVA test of H_0: $\beta = 0$ for
   a. problem 2.
   b. problem 4.

49. At the 5-percent level of significance, make an ANOVA test of H_0: $\beta = 0$ for
   a. problem 1.
   b. problem 3.

50. At the 5-percent level of significance, make an ANOVA test of H_0: $\beta = 0$ for
   a. problem 6.
   b. problem 8.

51. At the 5-percent level of significance, make an ANOVA test of H_0: $\beta = 0$ for
   a. problem 5.
   b. problem 7.

52. At the 5-percent level of significance, make an ANOVA test of H_0: $\beta = 0$ for
   a. problem 10.
   b. problem 12.

53. At the 5-percent level of significance, make an ANOVA test of H_0: $\beta = 0$ for
   a. problem 9.
   b. problem 11.
   c. problem 13.

54. At the 5-percent level of significance, make an ANOVA test of H_0: $\beta = 0$ for
   a. problem 14.
   b. problem 16.

55. At the 5-percent level of significance, make an ANOVA test of H_0: $\beta = 0$ for
   a. problem 15.
   b. problem 17.
   c. problem 19.

56. At the 5-percent level of significance, make an ANOVA test of H_0: $\beta = 0$ for
  a. problem 18.
  b. problem 20.
57. A personnel manager wants to establish the relationship between the weekly number of grievances filed by workers, Y, and the average weekly humidity index, X. Given the data of Table 13.U, determine an estimated linear regression equation. Also determine the values of $s_{Y \cdot X}$, r^2, and F (in an ANOVA test of H_0: $\beta = 0$).
58. Once again consider problem 57. Suppose the personnel manager's boss believes that the personnel manager (a) should have drawn a scatter diagram, and (b) should have taken a hint from that diagram and produced a regression of Y versus X^2 "for a much better fit." Check the boss's claims.
59. Analytical Example 13.6, "Snowfall and Unemployment," presents an estimated regression line as well as a coefficient of determination. Make the calculations to check their accuracy. Show your computations.
60. Consider the accompanying Figure B. Which line (A, B, C, or D) is the regression line for the four colored dots? Explain.

Table 13.U

Grievances Filed, Y (number per week)	Average Weekly Humidity Index, X
11	.7
9	1.4
11	2.1
13	2.7
22	3.7
18	4.2
22	5.0
41	5.6
62	6.9
50	7.3
88	7.7
68	7.8
109	8.1
125	8.4
142	8.5
96	8.6
112	8.8
132	9.0

Figure B

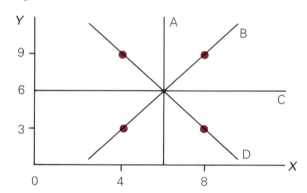

Selected Readings

Bibby, J. "The General Linear Model—A Cautionary Tale." In C. A. O'Muircheartaigh and Clive Payne, eds. *The Analysis of Survey Data,* vol. 2: *Model Fitting.* New York: Wiley, 1977, pp. 35–79. Argues that regression analysis is *fragile* (that violations of assumptions make the results of such analysis useless).

Bohrnstedt, G. W. and T. M. Carter. "Robustness in Regression Analysis." In H. Costner, ed. *Sociological Methodology.* San Francisco: Jossey-Bass, 1971. An excellent discussion of whether violations of assumptions render the results of regression analysis useless or leave them intact.

Folks, J. Leroy. *Ideas of Statistics.* New York: Wiley, 1981. Chapters 20 and 21 provide fascinating insights into Galton's discovery of regression and correlation; Chapter 22 sheds light on Legendre's principle of least squares.

Galton, Francis. "Regression Towards Mediocrity in Hereditary Stature." *Journal of the Anthropological Institute,* 1885.

Kerlinger, F. N. and E. J. Pedhazur. *Multiple Regression in Behavioral Research.* New York: Holt, Rinehart, and Winston, 1973. Argues that regression analysis is *robust* (that violations of assumptions are not serious).

Neter, J., W. Wasserman, and M. H. Kutner. *Applied Linear Regression Models.* Homewood, Il.: Richard D. Irwin, 1983. A good text.

Walker, Helen M. *Studies in the History of Statistical Method.* Baltimore: Williams and Wilkins, 1929. An excellent history of regression and correlation theory in chap. 5.

Wonnacott, Thomas H. and Ronald J. Wonnacott. *Regression: A Second Course in Statistics.* New York: Wiley, 1981. A superb text devoted to regression and correlation analysis at a more advanced level.

Computer Programs

The DAYSAL-2 personal-computer diskettes that accompany this book contain one program of interest to this chapter:

13. Regression and Correlation Analysis lets you enter any 2 sets of data and view a scatter diagram of their association, along with all summary measures of central tendency and of dispersion. You can perform simple regression analysis that provides not only regression coefficients, standard errors, and *t*-values, but also a correlation matrix, an ANOVA table, confidence intervals for α and β coefficients, hypothesis tests about them, and prediction intervals, at numerous confidence levels, for average and individual *Y*-values associated with each of the *X*-values originally entered. You can also take advantage of a "live regression" feature: You specify any *X*-value . . . and the computer provides an instant point estimate of *Y*, along with limits of prediction intervals for the average and individual *Y* at your chosen confidence level. You can use data files A–P that contain numerous data on the U.S. economy.

CHAPTER FOURTEEN

Multiple Regression and Correlation

Simple regression analysis, we learned in Chapter 13, helps us predict the value of one variable from a knowledge of that of another variable, while simple correlation analysis helps us measure the overall strength of association between two such variables. **Multiple regression analysis,** in contrast, is a technique that employs several independent variables to predict the value of a dependent variable; hence, each of these predictor variables explains part of the total variation of the dependent variable. **Multiple correlation analysis,** accordingly, measures the overall strength of association among all these variables.

There are plenty of good reasons for utilizing the more complex techniques to be discussed in this chapter; foremost among them is the fact that the world is a complicated network of interdependencies, which is rarely captured very well by models involving two variables only. Typically, the value of any one variable is influenced not only by one other variable, but by two, three, or even a multitude of other variables. People's income, for example, may indeed, depend on their education (as noted in Chapter 13), but it can also depend on their job experience, perhaps on their age, and, in the presence of discriminatory practices, even on their sex or race. The demand for a good may well depend on the good's price but, in addition, may vary with the income and taste of consumers, the prices of substitutes or complements, and a host of other factors. A firm's sales may, indeed, vary with advertising expenditures but also with the number of competitors it has, with the size of the community in which it is located, and even with the local unemployment rate. A farmer's output per acre may be a function of fertilizer input but will surely also vary with the quantity of labor used, with the amounts of pesticide used, and even with the extent of rainfall.

Examples such as these could be multiplied without end, but their message is already clear: simple regression and correlation analysis is rarely adequate for exploring relationships, even between two variables of interest. Such simple analysis always leaves open the possibility of finding a fuller explanation of the dependent variable by considering more than one independent variable. (A substantial amount of the variation in Y = income that is not explained by X_1 = education may, for example, be explained by X_2 = job experience, X_3 = age, or even X_4 = sex.) In addition, the precise effect of a particular independent variable (such as education) on a dependent variable of interest (such as income) may be seriously distorted by the results of simple regression and correlation analysis because the influence of other independent variables (such as job experience, age, or sex) has been ignored. The sample results noted in Table 13.1 on p. 532, for example, are ambiguous if high or low income can also be caused by high or low job experience and not only by high or low education. If the sampled high-income individuals just happened to have above-average job experience, while the low-income individuals had below-average job experience, a simple regression that ignored job experience but used these data would exaggerate both the income-raising effect of high education and the income-depressing effect of low education. In other words, if only a single independent variable is used in the regression equation, but several such variables are in fact operating in the real-life situation that this equation attempts to describe, the influence of all the ignored independent variables is absorbed into the coefficient attached to that single variable—which makes the coefficient misleading. The problem noted here is once again that of controlling for extraneous factors.

Especially in the fields of business and economics, the influence of X on Y can rarely be established by *experimental control,* the literal control of all aspects of the environment in a laboratory wherein the effect of X on Y might be observed, other things being equal. As we shall see, however, multiple regression analysis, unlike simple regression analysis, allows us to exercise *statistical control* and to determine the influence of any X on Y for specified constant values of other variables that might affect Y as well. (In business and economics texts, an "other things being equal" or *ceteris paribus* clause is frequently inserted in sentences to warn the reader that a given relationship between two variables was found while exercising statistical control and might not be evident when other variables fail to remain constant.)

LINEAR MULTIPLE REGRESSION: TWO EXPLANATORY VARIABLES

The techniques of multiple regression are straightforward extensions of those of simple regression. Let us consider the case in which one dependent variable, Y, is related, in linear fashion, to two independent variables, X_1 and X_2. (Y might be income, X_1 might be education, and X_2 might be job experience.) A first goal of the analysis of such a case is the establishment of an *estimated multiple-regression equation,* such as

$$\hat{Y} = a + b_1 X_1 + b_2 X_2.$$

This equation gives us the estimated value, $\hat{Y}$, of the dependent variable for any specified pair of values of the independent variables. In contrast to our experience in Chapter 13, there now exist three estimated regression coefficients, a, b_1, and b_2. Their meaning is easy to comprehend: a is the estimated value of Y when $X_1 = X_2 = 0$; b_1 gives us the change in Y (also referred to as the *partial* change or *net* change in Y) associated with a unit change in X_1 when X_2 is held constant; while b_2, similarly, equals the change in Y associated with a unit change in X_2 when X_1 is held constant. The values of b_1 and b_2 are called the **estimated partial-regression coefficients;** they are, in fact, the partial derivatives of Y with respect to either X_1 or X_2. [Thus, b_1 = $(\delta Y/\delta X_1)$ and b_2 = $(\delta Y/\delta X_2)$ where δ, the lowercase Greek *delta,* stands for "the change in."] But note: Unlike the last chapter's two-variable simple regression equation (that related Y to X), our three-variable multiple regression equation (that relates Y to X_1 and X_2) does not correspond to a line in a two-dimensional scatter diagram (such as Figure 13.6 on page 538); it corresponds to a plane in three-dimensional space.

The Regression Plane

In our three-variable case, three observations are made for each sample unit: one for the value of Y, one for X_1, and one for X_2. These observations, too, can be depicted in a scatter diagram, but the scatter diagram has to be three-dimensional, as Figure 14.1 indicates. In this particular graph, the value of Y associated with any sample unit is measured vertically from the origin, 0; the value of X_1 is measured toward the right, and the value of X_2 is measured toward the left. Thus, a sample observation of Y = distance 0A, X_1 = distance 0B, and X_2 = distance 0C appears as the color sample point, labeled P.

Now imagine that there were lots of sample observations, such as all the colored dots suspended in the three-dimensional space of Figure 14.2. The three-variable multiple regression technique estimates the above-noted multiple regression equation in such a way that all the estimates derived from it fall on a surface, such as shaded area *ABCD* in our graph, that is called the **regression plane** and that is positioned among the sample points in such a way as to minimize the sum of the squared vertical devia-

Figure 14.1 **Sample Point in Three-Dimensional Scatter Diagram**

When three observations are associated with each sample unit (such as Y = distance 0A, X_1 = distance 0B, and X_2 = distance 0C), any set of these observations can be depicted like sample point P, as a dot suspended in three-dimensional space.

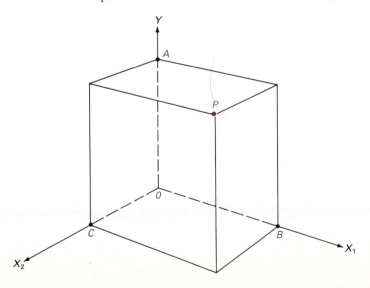

Figure 14.2 The Regression Plane

Three-variable linear multiple-regression analysis estimates an equation ($\hat{Y} = a + b_1X_1 + b_2X_2$) in such a way that all the estimates of Y made with its help (and depicted by crosses in this graph) fall on a surface, such as *ABCD,* that is called the *regression plane* and that is positioned among the sample points (the color dots) in such a way as to minimize the sum of the squared vertical deviations between these sample points and their associated estimates. Note how point P from Figure 14.1 reappears here at a position above the regression plane and is projected by a vertical line to the cross lying on the plane directly below it. Similarly, a point such as Q is suspended below the plane and is projected by a vertical line to the cross lying on the plane directly above it.

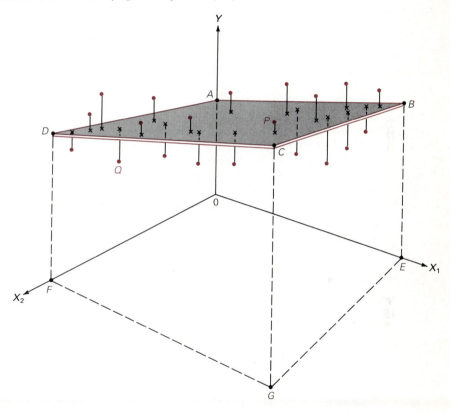

tions between these sample points and their associated estimates. In Figure 14.2, the observed sample points (or the actual values of Y for any given combination of X_1 and X_2) are represented by the colored dots; the associated estimates, $\hat{Y}$, are represented by the black crosses lying on the shaded plane. These crosses, of course, are positioned immediately below or above the colored dots, depending on whether the latter are suspended above or below the regression plane.

We can also note the values of the regression coefficients in Figure 14.2. The value of a estimates Y for $X_1 = X_2 = 0$; hence, it is still the Y intercept, equal to distance $0A$ in our graph. The value of b_1 is the slope of the regression plane when holding X_2 constant (at any desired level). Imagine cutting the regression plane parallel to the X_1 axis at $X_2 = 0$. The cut would trace line AB in the YX_1 plane and show the value of Y

rising from $0A$ at $X_1 = 0$ to EB at $X_1 = 0E$. The slope of line AB, relative to $0E$, equals b_1. Imagine instead cutting the regression plane parallel to the X_1 axis at $X_2 = 0F$. The cut would trace line DC and show the value of Y rising from FD at $X_1 = 0$ to GC at $X_1 = 0E$. The slope of line DC, relative to FG, also equals b_1. Thus, b_1 always indicates how Y changes with X_1, while *not* changing X_2.

Finally, b_2 is the slope of the regression plane when holding X_1 constant (again at any desired level). Imagine cutting the regression plane parallel to the X_2 axis at $X_1 = 0$. That cut would trace line AD in the YX_2 plane and show the value of Y rising (even though slightly) from $0A$ at $X_2 = 0$ to FD at $X_2 = 0F$. The slope of line AD, relative to $0F$, equals b_2. Imagine instead cutting the regression plane parallel to the X_2 axis at $X_1 = 0E$. The cut would trace line BC and show the value of Y rising from EB at $X_2 = 0$ to GC at $X_2 = 0F$. The slope of line BC, relative to EG, also equals b_2. Thus, b_2 always indicates how Y changes with X_2, while not changing X_1.

Least Squares Revisited

As we have just noted, the regression plane is positioned so as to minimize the sum of squared errors, or $\Sigma(Y - \hat{Y})^2$. In Figure 14.2, the heights of the color sample points above the X_1X_2 plane represent the values of Y; the heights of the crosses on the regression plane immediately below or above the color dots represent the associated values of $\hat{Y}$ that are estimated from the equation given above. The vertical differences, $Y - \hat{Y}$, are the deviations or errors, of course; the sum of their squares is to be minimized. Ronald A. Fisher (Biography 9.1), George Snedecor (see Biography 14.1 at the end of this chapter) and others have shown that when the following three so-called *normal equations* are satisfied, the sum of the squares of the deviations is minimized.[1]

$$\Sigma Y = na + b_1\Sigma X_1 + b_2\Sigma X_2$$
$$\Sigma X_1 Y = a\Sigma X_1 + b_1\Sigma X_1^2 + b_2\Sigma X_1 X_2$$
$$\Sigma X_2 Y = a\Sigma X_2 + b_1\Sigma X_1 X_2 + b_2\Sigma X_2^2.$$

From these equations one can, in turn, derive the expressions given in Box 14.A.

14.A Estimated Regression Coefficients (two independent variables)

$$b_1 = \frac{[\Sigma X_2^2 - n\overline{X}_2^2][\Sigma X_1 Y - n\overline{X}_1\overline{Y}] - [(\Sigma X_1 X_2 - n\overline{X}_1\overline{X}_2)(\Sigma X_2 Y - n\overline{X}_2\overline{Y})]}{[\Sigma X_1^2 - n\overline{X}_1^2][\Sigma X_2^2 - n\overline{X}_2^2] - [\Sigma X_1 X_2 - n\overline{X}_1\overline{X}_2]^2}$$

$$b_2 = \frac{[\Sigma X_1^2 - n\overline{X}_1^2][\Sigma X_2 Y - n\overline{X}_2\overline{Y}] - [(\Sigma X_1 X_2 - n\overline{X}_1\overline{X}_2)(\Sigma X_1 Y - n\overline{X}_1\overline{Y})]}{[\Sigma X_1^2 - n\overline{X}_1^2][\Sigma X_2^2 - n\overline{X}_2^2] - [\Sigma X_1 X_2 - n\overline{X}_1\overline{X}_2]^2}$$

$$a = \overline{Y} - b_1\overline{X}_1 - b_2\overline{X}_2$$

where X_1 and X_2 are observed values of the two independent variables ($\overline{X}_1$ and $\overline{X}_2$ being their means), Ys are associated observed values of the independent variable ($\overline{Y}$ being their mean), and n is sample size.

---

[1]A mathematical proof is given in the "Supplementary Topics" section of Chapter 14 of the *Student Workbook* that accompanies this text.

Table 14.1 Sample Data on Education, Job Experience, and Income

Individual	Education (years), X_1	Job Experience (years), X_2	Income (thousands of dollars per year), Y
A	2	9	5.0
B	4	18	9.7
C	8	21	28.4
D	8	12	8.8
E	8	14	21.0
F	10	16	26.6
G	12	16	25.4
H	12	9	23.1
I	12	18	22.5
J	12	5	19.5
K	12	7	21.7
L	13	9	24.8
M	14	12	30.1
N	14	17	24.8
O	15	19	28.5
P	15	6	26.0
Q	16	17	38.9
R	16	1	22.1
S	17	10	33.1
T	21	17	48.3

Practice Problem 1

Regressing Income on Education and Job Experience

Consider the hypothetical study of the link between income and education, first introduced in Chapter 13's Practice Problem 1, but now suppose that income is believed to be linked to years of job experience as well. Conceivable sample data, based on Table 13.1 on page 532, are presented in Table 14.1 (but Y is now measured in *thousands* of dollars per year, instead of *just plain dollars,* as in Table 13.1).

Given the formulas of Box 14.A, the Table 14.1 data must be subjected to a number of intermediate calculations before the multiple regression equation can be determined. These calculations are shown in Table 14.2, along with a variety of others that will be utilized later in this chapter. From these calculations, we can determine

$$b_1 = \frac{[(566.55)(704.285)] - [(-49.65)(271.905)]}{[(380.95)(566.55)] - (-49.65)^2} = 1.933;$$

$$b_2 = \frac{[(380.95)(271.905)] - [(-49.65)(704.285)]}{[(380.95)(566.55)] - (-49.65)^2} = .649;$$

$$a = 24.415 - 1.933(12.05) - .649(12.65) = -7.088.$$

When inserted into the normal equations given above, these values will be found to satisfy the equations (except for rounding errors). Consequently, the estimated regression equation is

$$\hat{Y} = -7.088 + 1.933X_1 + .649X_2.$$

The values of the estimated partial regression coefficients can be interpreted easily enough: $b_1 = 1.933$ implies that an estimated increase of $1,933 of income is associated with each additional year in education, if job experience is held constant at any level, while $b_2 = .649$ implies that an estimated increase of $649 of income is associated with each additional year of job experience, if education is similarly held constant. Not much should be made of $a = -7.088$ (and the apparent implication that a person's income would be *negative* $7,088 with zero education and zero job experience). As noted in Chapter 13, such extrapolation beyond the range of sample data is not appropriate. Yet the equation can be used to predict the income of someone whose education and job experience does fall within the observed ranges of these variables. Thus, one could

Table 14.2   Intermediate Calculations for Multiple Regression

Individual	X_1	X_2	Y	X_1Y	X_2Y	X_1X_2	X_1^2	X_2^2	Y^2
A	2	9	5.0	10.0	45.0	18	4	81	25.00
B	4	18	9.7	38.8	174.6	72	16	324	94.09
C	8	21	28.4	227.2	596.4	168	64	441	806.56
D	8	12	8.8	70.4	105.6	96	64	144	77.44
E	8	14	21.0	168.0	294.0	112	64	196	441.00
F	10	16	26.6	266.0	425.6	160	100	256	707.56
G	12	16	25.4	304.8	406.4	192	144	256	645.16
H	12	9	23.1	277.2	207.9	108	144	81	533.61
I	12	18	22.5	270.0	405.0	216	144	324	506.25
J	12	5	19.5	234.0	97.5	60	144	25	380.25
K	12	7	21.7	260.4	151.9	84	144	49	470.89
L	13	9	24.8	322.4	223.2	117	169	81	615.04
M	14	12	30.1	421.4	361.2	168	196	144	906.01
N	14	17	24.8	347.2	421.6	238	196	289	615.04
O	15	19	28.5	427.5	541.5	285	225	361	812.25
P	15	6	26.0	390.0	156.0	90	225	36	676.00
Q	16	17	38.9	622.4	661.3	272	256	289	1,513.21
R	16	1	22.1	353.6	22.1	16	256	1	488.41
S	17	10	33.1	562.7	331.0	170	289	100	1,095.61
T	21	17	48.3	1,014.3	821.1	357	441	289	2,332.89
	241 $= \Sigma X_1$	253 $= \Sigma X_2$	488.3 $= \Sigma Y$	6,588.3 $= \Sigma X_1Y$	6,448.9 $= \Sigma X_2Y$	2,999 $= \Sigma X_1X_2$	3,285 $= \Sigma X_1^2$	3,767 $= \Sigma X_2^2$	13,742.27 $= \Sigma Y^2$

$$\bar{X}_1 = \frac{\Sigma X_1}{n} = \frac{241}{20} = 12.05 \qquad \bar{X}_2 = \frac{\Sigma X_2}{n} = \frac{253}{20} = 12.65 \qquad \bar{Y} = \frac{\Sigma Y}{n} = \frac{488.3}{20} = 24.415$$

predict the income of someone with 11 years of education and 20 years of job experience as $27,155 because

$$\hat{Y} = -7.088 + 1.933(11) + .649(20) = 27.155.$$

Similarly, one can predict the income of someone with 8 years of education and 2 years of job experience as $9,674 because then

$$\hat{Y} = -7.088 + 1.933(8) + .649(2) = 9.674.$$

In fact, such a person is included in our sample (individual D) and is earning about $8,800, as Table 14.1 confirms.

Note: The simple regression equation of Figure 13.6 on page 538 predicts D's income as $\hat{Y}_8 = 2{,}144.23 + 1{,}847.5(8) = \$16{,}924.23$—that is, with a considerably larger error, which brings us to the next topic. |||

The Sample Standard Error of the Estimate of Y

We can once again evaluate the quality of our estimated regression equation by computing the *sample standard error of the estimate of Y* which in this case measures the dispersion of observed Y values about the regression *plane*. Because two independent variables are being used to estimate Y, the standard error is designated not as $s_{Y \cdot X}$ (as in Chapter 13), but as $s_{Y \cdot X_1 X_2}$, or, more simply, as $s_{Y \cdot 12}$. In addition, because 3 degrees of freedom are being lost when estimating the 3 regression coefficients, the divisor in the formula is adjusted accordingly from $n - 2$ (noted in Chapter 13) to $n - 3$ (which makes $s_{Y \cdot 12}^2$ an unbiased estimator of the variance of Y about the regression plane). Thus,

$$s_{Y \cdot 12} = \sqrt{\frac{\Sigma(Y - \hat{Y})^2}{n - 3}},$$

which is mathematically equivalent to

$$s_{Y \cdot 12} = \sqrt{\frac{\Sigma Y^2 - a\Sigma Y - b_1\Sigma X_1 Y - b_2\Sigma X_2 Y}{n - 3}}.$$

||| Practice Problem 2

Calculating the Sample Standard Error of the Estimate of Income

In our example, the sample standard error of the estimate of Y can be calculated, using the previous formula, as

$$s_{Y \cdot 12} = \sqrt{\frac{13{,}742.27 - (-7.088)(488.3) - 1.933(6{,}588.3) - .649(6{,}448.9)}{17}}$$

$$= \sqrt{\frac{282.82}{17}} = \sqrt{16.64} = 4.079.$$

Compare this standard error of $4,079 with that of $5,377, calculated in Chapter 13's Practice Problem 4 for the simple regression relating income to education alone. It is obvious that our multiple regression provides a better explanation of the data. Much less of the variation in Y is left unexplained here. Therefore, predictions of Y with the help of our multiple regression equation are better than those made with the help of our simple regression equation. ▐▐▐

Making Inferences

Provided that certain conditions can be assumed to hold, we can once again use our regression equation to make a variety of inferences. The necessary assumptions are analogous to the four assumptions described in Chapter 13 (pages 539 to 540); we review them here, along with one addition:

Assumption 1: It is assumed that the values of X_1 and X_2 are known without error and that the different sample observations about Y that are associated with any X_1X_2 combination are statistically independent of each other.

Assumption 2: It is assumed that every population of Y values (a different one of which is associated with every possible X_1X_2 combination) is normally distributed, with a conditional mean of $\mu_{Y \cdot X_1X_2}$ (also denoted as $\mu_{Y \cdot 12}$) and a conditional standard deviation of $\sigma_{Y \cdot X_1X_2}$ (also denoted as $\sigma_{Y \cdot 12}$).

Assumption 3: It is assumed that all conditional probability distributions of Y (a different one of which is associated with every possible X_1X_2 combination) have the same conditional standard deviation, $\sigma_{Y \cdot 12}$ (homoscedasticity exists).

Assumption 4: It is assumed that all the conditional means, $\mu_{Y \cdot 12}$, lie on a surface (similar to *ABCD* in Figure 14.2) that is the *true regression plane* and is described by the equation $E(Y) = \mu_{Y \cdot 12} = \alpha + \beta_1X_1 + \beta_2X_2$. (The estimated regression coefficients a, b_1, and b_2 are, of course, estimators of α, β_1, and β_2). As in the simple regression case, this assumption is still referred to as the assumption of linearity (even though the line has become a plane).

Assumption 5: It is assumed that there exists no exact linear relationship between different independent variables. (If different Xs are perfectly linearly correlated, one cannot calculate the parameter estimates because two or more of the normal equations are then not independent. If different Xs are imperfectly but highly correlated, one can calculate the regression coefficients, but one cannot isolate the individual effects on Y of each of these highly linearly correlated variables.) As the scatter diagram of Figure 14.3 indicates, there certainly exists no such perfect linear correlation between the independent variables in our example.

Estimating the average value of Y, *given* X_1 *and* X_2. Suppose we wanted to estimate the value of a conditional mean, $\mu_{Y \cdot X_1X_2}$, such as $\mu_{Y \cdot 11,20}$ (the average income of those members of our population with 11 years' education and 20 years' job experience). The best *point estimate* equals 27.155 (that is, $27,155), as noted above. For a small sample, such as ours, a prediction interval is computed by using the formula in Box

Figure 14.3   The Relation of X_1 to X_2

This scatter diagram illustrates the fulfillment of Assumption 5 noted in the text: There exists no perfect linear correlation between the values of the two independent variables listed in Table 14.1. (In fact, the simple correlation coefficient between X_1 and X_2 is $r_{1 \cdot 2} = -.11$.)

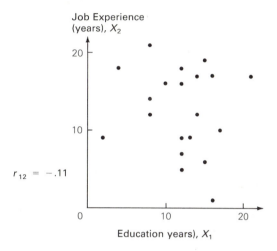

14.B, which involves matrix notation. Matrix algebra, however, is too complex for the intended level of this text; therefore, we cannot use Box 14.B directly. It has, however, been built into the DAYSAL-2 computer program that accompanies the text, and Practice Problem 3 can be solved by computer.[2]

14.B The Limits of a Prediction Interval for the Average Value of Y, Given X_1 and X_2 (small-sample case, $n < 30$)

$$\mu_{Y \cdot 12} = \hat{Y} \pm (t_{\alpha/2} \cdot s_{Y \cdot 12} \cdot \sqrt{M})$$

where $\hat{Y}$ is the estimated value of dependent variable Y, $s_{Y \cdot 12}$ is the sample standard error of the estimate of Y, and M is the product of three matrices.

Note: Given sample size n and m independent variables, the value of $t_{\alpha/2}$ is found in Appendix Table K for $n - m - 1$ degrees of freedom. In the large-sample case ($n \geq 30$), the normal deviate, z, replaces t.

---

[2]Without the use of matrix algebra or computers, the prediction interval for the average Y can still be *approximated* by substituting, for the parenthetical expression in Box 14.B, the value $\left(t_{\alpha/2} \cdot \dfrac{s_{Y \cdot 12}}{\sqrt{n}} \right)$.

| | Practice Problem 3

Estimating Average Income, Given Education and Job Experience

Assuming that a 95-percent prediction interval is desired, the interval limits in our example can be calculated for 11 years' education and 20 years' job experience, using the DAYSAL-2 computer program, as:

$$23.8753 \leq \mu_{Y \cdot 11,20} \leq 30.44026.$$

We can be 95-percent confident that the height of the regression plane at $X_1 = 11$ and $X_2 = 20$ lies within these limits (that, on the average, people with 11 years of education and 20 years of work experience have annual incomes between \$23,875.30 and \$30,440.26). | | |

Predicting an individual value of Y, given X_1 and X_2. What if we had wanted to predict $I_{Y \cdot X_1 X_2}$ such as $I_{Y \cdot 11,20}$ (the particular income of the next member of our population encountered in sampling who falls into the 11 years' education and 20 years' job-experience category)? The best point estimate still equals 27.155, as noted in the previous section. A small-sample prediction interval is computed by using the formula in Box 14.C, which once again employs matrix algebra.[3]

14.C The Limits of a Prediction Interval for the Individual Value of Y, Given X_1 and X_2 (small-sample case, $n < 30$)

$$I_{Y \cdot 12} = \hat{Y} \pm (t_{\alpha/2} \cdot s_{Y \cdot 12} \sqrt{1 + M})$$

where $\hat{Y}$ is the estimated value of dependent variable Y, $s_{Y \cdot 12}$ is the sample standard error of the estimate of Y, and M is the product of three matrices.

Note: Given sample size n and m independent variables, the value of $t_{\alpha/2}$ is found in Appendix Table K for $n - m - 1$ degrees of freedom. In the large-sample case ($n \geq 30$), the normal deviate, z, replaces t.

| | Practice Problem 4

Estimating Individual Income, Given Education and Job Experience

Assuming that a 95-percent prediction interval is desired, the interval limits in our example can be calculated for 11 years' education and 20 years' job experience, using the DAYSAL-2 computer program, as:

$$17.95563 \leq I_{Y \cdot 11,20} \leq 36.35993.$$

---

[3]Without the use of matrix algebra or computers, the prediction interval for the individual Y can still be *approximated* by substituting, for the parenthetical expression in Box 14.C, the value $\left(t_{\alpha/2} \cdot s_{Y \cdot 12} \sqrt{\dfrac{n + 1}{n}} \right)$.

We can be 95-percent confident that the next person with the characteristics of $X_1 = 11$ and $X_2 = 20$ whom we encounter in sampling has an annual income between $17,955.63 and $36,359.93. |||

Establishing confidence intervals for β_1 and β_2. In multiple regression, no less than in simple regression, it is crucial to determine whether a nonzero value of any sample coefficient, b (which estimates the corresponding population coefficient, β), is statistically significant or is instead the result of a sampling error that is perfectly consistent with a β value of zero. Limits of confidence intervals for β values can be established using the t distribution, in a fashion analogous to that noted in Chapter 13, as $\beta_1 = b_1 \pm t\sigma_{b_1}$ and $\beta_2 = b_2 \pm t\sigma_{b_2}$. The two standard errors, σ_{b_1} and σ_{b_2}, must again be estimated by s_{b_1} and s_{b_2}, which yields the formulas of Box 14.D.

14.D The Limits of Confidence Intervals for β_1 and β_2 (small-sample case, $n < 30$)

$$\beta_1 \cong b_1 \pm (t_{\alpha/2} \cdot s_{b_1})$$

$$= b_1 \pm \left(t_{\alpha/2} \cdot s_{Y \cdot 12} \sqrt{\frac{\Sigma X_2^2 - n\overline{X}_2^2}{[\Sigma X_1^2 - n\overline{X}_1^2][\Sigma X_2^2 - n\overline{X}_2^2] - [\Sigma X_1 X_2 - n\overline{X}_1\overline{X}_2]^2}} \right)$$

$$\beta_2 \cong b_2 \pm (t_{\alpha/2} \cdot s_{b_2})$$

$$= b_2 \pm \left(t_{\alpha/2} \cdot s_{Y \cdot 12} \sqrt{\frac{\Sigma X_1^2 - n\overline{X}_1^2}{[\Sigma X_1^2 - n\overline{X}_1^2][\Sigma X_2^2 - n\overline{X}_2^2] - [\Sigma X_1 X_2 - n\overline{X}_1\overline{X}_2]^2}} \right)$$

where b_1 and b_2 are estimated partial regression coefficients (and s_{b_1} and s_{b_2} are their standard errors), Xs are observed values of independent variables ($\overline{X}$s being their means), while $s_{Y \cdot 12}$ is the sample standard error of the estimate of Y.

Note: Given sample size n and m independent variables, the value of $t_{\alpha/2}$ is found in Appendix Table K for $n - m - 1$ degrees of freedom.

||| Practice Problem 5

Establishing Confidence Intervals for β_1 and β_2 in the Regression of Income on Education and Job Experience

Given the data of Table 14.2 and the above-noted values of $b_1 = 1.933$, $b_2 = .649$, $s_{Y \cdot 12} = 4.079$, and $t_{\alpha/2} = 2.11$ (for a 95-percent confidence level), we can use the Box 14.D formulas to calculate the following confidence-interval limits:

$$\beta_1 = 1.933 \pm \left[2.11(4.079) \sqrt{\frac{3,767 - 20(12.65)^2}{[3,285 - 20(12.05)^2][3,767 - 20(12.65)^2] - [2,999 - 20(12.05)(12.65)]^2}} \right]$$

$$= 1.933 \pm \left[2.11(4.079) \sqrt{\frac{566.55}{(380.95)(566.55) - (-49.65)^2}} \right]$$

$$= 1.933 \pm [2.11(.210)] = 1.933 \pm .443$$

$$\beta_2 = .649 \pm \left[2.11(4.079) \sqrt{\frac{3,285 - 20(12.05)^2}{[3,285 - 20(12.05)^2][3,767 - 20(12.65)^2] - [2,999 - 20(12.05)(12.65)]^2}} \right]$$

$$= .649 \pm \left[2.11(4.079) \sqrt{\frac{380.95}{(380.95)(566.55) - (-49.65)^2}} \right]$$

$$= .649 \pm [2.11(.172)] = .649 \pm .363$$

Thus, we can be 95-percent confident that the values of β_1 and β_2 are not zero but lie within these intervals:

$$1.49 \leq \beta_1 \leq 2.376$$
$$.286 \leq \beta_2 \leq 1.012 \quad \text{III}$$

t values. Frequently, the information derived in the previous section is provided by a ***t* value,** or ***t* ratio,** which is nothing else but the ratio of an estimated partial regression coefficient to its standard error.

||| **Practice Problem 6**

Using *t* Values for Practice Problem 5
For b_1:

$$t_{b_1} = \frac{b_1}{s_{b_1}}$$
$$= \frac{1.933}{.210}$$
$$= 9.205$$

and for b_2:

$$t_{b_2} = \frac{b_2}{s_{b_2}}$$
$$= \frac{.649}{.172}$$
$$= 3.773.$$

Because these t values exceed the critical t value of 2.11 established for a 95-percent confidence level and $n - 3 = 17$ degrees of freedom, one concludes that both b_1 and b_2 are statistically significant at that level. Typically, the computed t values appear in parentheses underneath the coefficients of the regression equation, as shown:

$$\hat{Y} = -7.088 + 1.933 X_1 + .649 X_2$$
$$\qquad\qquad (9.205) \qquad (3.773) \quad \text{III}$$

Note: For almost any sample size, the critical t value for a 95-percent confidence level approximates 2. As the appropriate column of Appendix Table K shows, the critical t equals 2.228 for $n = 10$ and 1.96 for $n = \infty$. This convenient fact leads to the following *rule of thumb:*

> If the absolute value of the computed t value is greater than 2 (we ignore a possible minus sign), the parameter estimate, b, is significant at the 95-percent level of confidence.

An examination of the parenthetical values underneath the above equation, with this rule of thumb in mind, instantly reveals the significance of the coefficients.

p *values.* There exists a third alternative yet for noting the degree of significance of an estimated regression coefficient. As noted in Biography 9.1, one can calculate a ***p* value** that gives the probability that a test statistic (such as a t value) as extreme as or more extreme than the one actually observed could have occurred by chance (merely as a result of sampling error), assuming some hypothesis about a parameter is true (assuming, for example, that the true regression coefficient, β, equals zero).

||| Practice Problem 7

Using p Values for Practice Problem 5

In our example, what is the probability of finding t values at least as large as those given in the previous section if the corresponding β values are zero? The 17 degrees-of-freedom row in Appendix Table K is of little help with respect to t_{b_1} since the largest t value listed there equals 3.965. A computer calculation, however, shows the tiny probabilities involved:

$$p = .000 \text{ for } t_{b_1} \geq 9.205 \text{ if } \beta_1 = 0 \text{ and}$$

$$p = .001 \text{ for } t_{b_2} \geq 3.773 \text{ if } \beta_2 = 0. \text{ |||}$$

*Testing the overall significance of the regression.** A test of the overall significance of the regression, rather than a test of the significance of individual coefficients, amounts to testing the hypothesis that *all* of the true regression coefficients are zero and that, therefore, *none* of the independent variables helps explain the variation of the dependent one.

||| Practice Problem 8

An Overall Significance Test of the Regression of Income on Education and Job Experience

In our example containing two independent variables (education and job experience) the opposing hypotheses are the following:

H_0: $\beta_1 = \beta_2 = 0$
H_A: At least one $\beta \neq 0$.

---

*Requires prior study of the Chapter 13 section, "Testing β with Analysis of Variance."

Table 14.3 The ANOVA Table

Source of Variation	Sum of Squares (1)	Degrees of Freedom (2)	Mean Square (3) = (1) ÷ (2)	Test Statistic (4)
Variation Explained by Regression	RSS = 1,537.61	m = 2	RMS = 768.81	$F = \dfrac{RMS}{EMS}$ = 46.20
Unexplained Variation	ESS = 282.82	$n - m - 1 =$ $n - 3 = 17$	EMS = 16.64	
Total	*Total SS* = 1,820.43	$n - 1 = 19$		

Once again, the test can be performed most easily by constructing an ANOVA table, such as Table 14.3. Column 1 contains the regression sum of squares, the error sum of squares, and the total sum of squares:

$$RSS = \Sigma(\hat{Y} - \overline{Y})^2 = a\Sigma Y + b_1\Sigma X_1 Y + b_2\Sigma X_2 Y - n\overline{Y}^2$$

$$= -7.088(488.3) + 1.933(6,588.3) + .649(6,448.9) - 20(24.415)^2$$

$$= 1,537.61.$$

$$ESS = \Sigma(Y - \hat{Y})^2 = \Sigma Y^2 - a\Sigma Y - b_1\Sigma X_1 Y - b_2\Sigma X_2 Y$$

$$= 13,742.27 - (-7.088)(488.3) - 1.933(6,588.3) - .649(6,448.9)$$

$$= 282.82.$$

$$\textit{Total SS} = \Sigma(Y - \overline{Y})^2 = \Sigma Y^2 - n\overline{Y}^2 = 13,742.27 - 20(24.415)^2$$

$$= 1,820.43.$$

Column 2 lists the appropriate degrees of freedom: equal to the number of independent variables, m, for the regression sum of squares; equal to sample size, n, minus 3 for the error sum of squares (there are n errors and 3 constraints—namely, the estimated regression coefficients, a, b_1, and b_2); and equal to $n - 1$ for the total sum of squares [there are n deviations subject to the single constraint that $\Sigma(Y - \overline{Y}) = 0$].

From this information, the regression and error mean squares of column 3 can be calculated (the latter has, in fact, been derived earlier in connection with the calculation in Practice Problem 2 of $s_{Y\cdot 12}$, which equals $\sqrt{EMS}$). Column 4, finally, contains the F ratio.

If we wish to test our hypothesis at the 5-percent level of significance, Appendix Table N provides a critical value of $F_{.05}$ (for 2 numerator and 17 denominator degrees of freedom) of 3.59. Thus, we should *reject* H_0 resoundingly. Far from being of no help, our regression is highly significant, indeed; a knowledge of X_1 and X_2 (years of education and job experience) does explain almost all of the variation of Y (income) in the population from which we drew our sample. ▌▌▌

MULTIPLE CORRELATION ANALYSIS

The concept of correlation can also be extended easily from an analysis dealing with one independent variable to one dealing with several of them. The aim is once again to calculate a single value which, in this case, is to describe the overall strength of association between a dependent variable and two or more independent ones. To distinguish between the coefficients of simple and multiple correlation, capital letters are used to designate the coefficients of multiple correlation.

The Coefficient of Multiple Determination

The **sample coefficient of multiple determination** is denoted by R^2 and equals the proportion of the total variation in the values of the dependent variable, Y, that is explained by the multiple regression of Y on X_1, X_2, and possibly additional independent variables (X_3, X_4, and so on). This coefficient describes, therefore, how well the regression *plane* fits the data; its definition is given in Box 14.E.

14.E Sample Coefficient of Multiple Determination

$$R^2 = \frac{\text{explained variation}}{\text{total variation}}$$

$$= \frac{RSS}{Total\ SS}$$

$$= \frac{\Sigma(\hat{Y} - \bar{Y})^2}{\Sigma(Y - \bar{Y})^2}$$

where Ys are observed values of the dependent variable ($\bar{Y}$ being their mean), while $\hat{Y}$s are estimated values of Ys.

As in the case of simple regression, R^2 can take on values between 0 and 1. The closer R^2 is to 0, the worse is the fit of the regression plane to the data; the closer it is to 1, the better is the fit.

The value of R^2 can, however, also be calculated directly from a knowledge of the regression coefficients, as shown in Box 14.F.

14.F Sample Coefficient of Multiple Determination: Alternative Formulation

$$R^2 = \frac{a\Sigma Y + b_1\Sigma X_1 Y + b_2\Sigma X_2 Y - n\bar{Y}^2}{\Sigma Y^2 - n\bar{Y}^2}$$

where a and b's are estimated partial regression coefficients, Xs are observed values of the independent variables, Ys are associated observed values of the dependent variable ($\bar{Y}$ being their mean), and n is sample size.

Calculating the Sample Coefficient of Multiple Determination for the Regression of Income on Education and Job Experience

Using the values in Table 14.3 and the Box 14.E formula, we can immediately calculate

$$R_{Y\cdot12}^2 = \frac{1,537.61}{1,820.43} = .84$$

for our example (which is an improvement over the value of $r_{Y\cdot1}^2 = .71$ calculated in Practice Problem 12 of Chapter 13 that related income only to education).

Using the Box 14.F formula, R^2 can be calculated alternatively for our example as

$$R_{Y\cdot12}^2 = \frac{-7.088(488.3) + 1.933(6,588.3) + .649(6,448.9) - 20(24.415)^2}{13,742.27 - 20(24.415)^2} = .84,$$

which confirms the calculation.

Note: We need not test the statistical significance of our sample R^2 for fear that the corresponding population coefficient, ρ^2 (lowercase Greek rho squared), might be zero. Such a test is not needed because of the large F ratio noted in Table 14.3. Our analysis of variance, by rejecting the null hypothesis of H_0: $\beta_1 = \beta_2 = 0$, implicitly rejected a similar hypothesis of H_0: $\rho^2 = 0$. ▐

The Coefficient of Multiple Correlation

The **sample coefficient of multiple correlation** is denoted by R and equals the square root of R^2. Its sign is always considered positive. In our example, $R_{Y\cdot12} = \sqrt{.84} = .92$. This value, too, has risen substantially from that of $r_{Y\cdot1} = .84$, found in Chapter 13's Practice Problem 13.

The Adjusted Coefficient of Multiple Determination

As a look at Box 14.E confirms, the calculation of the sample coefficient of multiple determination does not involve any adjustment for degrees of freedom. As a result, there is a tendency for R^2 to be "too large," or to overestimate the true or population coefficient. This bias can be removed by calculating an **adjusted sample coefficient of multiple determination, $\overline{R}^2$,** using the formula in Box 14.G, wherein n equals sample size and m is the number of independent variables in the regression.[4]

14.G Adjusted Sample Coefficient of Multiple Determination

$$\overline{R}^2 = 1 - \left[\frac{ESS}{Total\ SS} \left(\frac{n-1}{n-m-1} \right) \right]$$

---

[4]The formula in Box 14.G can easily be seen as an adjustment of that in Box 14.E by noting that *RSS/Total SS* is the same as $1 - (ESS/Total\ SS)$.

An alternative, but mathematically equivalent formula for $\bar{R}^2$ is given in Box 14.H, wherein s_Y^2 is the sample variance of the dependent variable, Y, around its mean, and $s_{Y \cdot 12 \ldots m}^2$ is the variance of Y around the estimated regression line or plane.

14.H Adjusted Sample Coefficient of Multiple Determination: Alternative Formulation

$$\bar{R}^2 = \frac{s_Y^2 - s_{Y \cdot 12 \ldots m}^2}{s_Y^2}$$

||| Practice Problem 10

Adjusting the Sample Coefficient of Multiple Determination for the Regression of Income on Education and Job Experience

Using the Table 14.3 data and Box 14.G, the adjusted coefficient in our example can be calculated as:

$$\bar{R}_{Y \cdot 12}^2 = 1 - \left[\frac{282.82}{1,820.43} \left(\frac{19}{17} \right) \right] = 1 - .17 = .83.$$

Alternatively, however, in order to use the Box 14.H formula, we can use the Table 14.2 data to calculate

$$s_Y^2 = \frac{\Sigma Y^2 - n\bar{Y}^2}{n - 1} = \frac{13,742.27 - 20(24.415)^2}{19} = \frac{1,820.43}{19} = 95.81.$$

We have already calculated $s_{Y \cdot 12} = \sqrt{\dfrac{282.82}{17}}$ in Practice Problem 2 of this chapter; hence,

$$s_{Y \cdot 12}^2 = \frac{282.82}{17} = 16.64$$

and

$$\bar{R}_{Y \cdot 12}^2 = \frac{95.81 - 16.64}{95.81} = .83. \quad |||$$

Two things should be noted:

1. While an unadjusted coefficient of determination always increases when a relevant and previously ignored independent variable is added to a regression model (note the increase in our example from $r_{Y \cdot 1}^2 = .71$ to $R_{Y \cdot 12}^2 = .84$), $\bar{R}^2$ can actually decrease and will do so when the addition of a new independent variable does not reduce the error sum of squares *(ESS)* enough to offset the loss of

one degree of freedom that is occasioned by the increase in m. (In that case, in an ANOVA table such as Table 14.3, a lower ESS can nevertheless raise EMS and lower F.) If that happens, the addition of the new independent variable was not meaningful.

2. The formulas given in Boxes 14.G and 14.H can also be used to calculate adjusted coefficients of simple determination. For example, the coefficient of $r_{Y\cdot1}^2 = .71$ calculated in Practice Problem 12 of Chapter 13 can be adjusted to

$$\bar{r}_{Y\cdot1}^2 = 1 - \left[\frac{520,489,201.64}{1,820,772,118.80} \left(\frac{19}{18} \right) \right] = .70$$

by using the data of Table 13.3 and the Box 14.G formula.

Partial Correlation

Our regression of income (Y) on education (X_1) in Chapter 13 produced, as we just recalled, an unadjusted simple coefficient of determination of $r_{Y\cdot1}^2 = .71$. Earlier in the current chapter, the addition to our model of a second explanatory variable (job experience, X_2) yielded a regression equation with considerably greater predictive power and produced an unadjusted multiple coefficient of determination of $R_{Y\cdot12}^2 = .84$. It is common practice to quantify the improvement achieved by such higher dimensional regression analysis. The result is an index of association between the dependent variable and just *one* of the independent variables, given a prior accounting of the effects of one or more other independent variables. The index is calculated as the proportional decrease in the previously unexplained variation (or increase in the previously explained variation) resulting from the higher dimensional regression. Box 14.I defines the new measure, called the **sample coefficient of partial determination,** for the case in which a regression with one independent variable is replaced by one with two such variables.[5]

14.I Sample Coefficient of Partial Determination

$$r_{Y2\cdot1}^2 = \frac{\text{reduction in previously unexplained variation in } Y}{\text{previously unexplained variation in } Y} = \frac{R_{Y\cdot12}^2 - r_{Y\cdot1}^2}{1 - r_{Y\cdot1}^2}$$

Note the subscripts in the Box 14.I formula: The coefficient of simple determination (calculated for a regression of Y on X_1 only) is denoted by $r_{Y\cdot1}^2$; the subscript tells us that the effect of variables other than X_1 (such as X_2, X_3, etc.) is totally being ignored. The coefficient of multiple determination (calculated for a regression of Y on X_1 and X_2) is denoted by $R_{Y\cdot12}^2$; here the subscript tells us that the mutual effect of X_1

---

[5]The coefficient is defined analogously for additional moves to ever higher dimensional regressions. Thus, going from two to three independent variables produces

$$r_{Y3\cdot12}^2 = \frac{R_{Y\cdot123}^2 - R_{Y\cdot12}^2}{1 - R_{Y\cdot12}^2}.$$

and X_2 is being considered. Finally, the measure of partial correlation that shows the correlation of Y on X_2, when X_1 is present but constant, is denoted by $r^2_{Y2 \cdot 1}$; the subscript reminds us that the (additional) effect of X_2 on Y is being considered, given that the effect of X_1 on Y has already been accounted for. Such coefficients of partial determination, like the simple ones, range from -1 to $+1$ in value; they take on a sign corresponding to that of the associated parameter estimate (b_2 in this case).

> ### ||| Practice Problem 11
>
> #### Calculating the Sample Coefficient of Partial Determination to Show the Correlation of Income With Job Experience, When the Effect of Education is Present but Constant
>
> For our example, we can use the Box 14.I formula to calculate:
>
> $$r^2_{Y2 \cdot 1} = \frac{.84 - .71}{1 - .71} = .45.$$
>
> This result implies that 45 percent of the variation in Y (income) that was left unexplained by the simple regression of Y on X_1 (education) in Chapter 13 has been explained by the addition here in Chapter 14 of X_2 (job experience) as an explanatory variable. |||

Indeed, we can confirm this result independently by comparing the error sum of squares found in the ANOVA tables. The value of *ESS* changed from 520.49 (million) in Table 13.3 (on p. 561) to 282.82 in Table 14.3, which is a 45-percent reduction (except for rounding error). It is easy to see why the sample coefficient of partial determination, or its square root, the **sample coefficient of partial correlation,** is often used to evaluate the merit of potentially adding an explanatory variable to a regression analysis. In our case, the addition of X_2 is worthwhile because X_2 is quite effective in sharpening the analysis.

LINEAR MULTIPLE REGRESSION AND CORRELATION WITH THREE OR MORE EXPLANATORY VARIABLES (COMPUTER APPLICATIONS)

When simple regression analysis turns into multiple regression analysis, hand calculations become quite burdensome. We have seen how cumbersome the calculations are even for the relatively simple case of two independent variables; they are all the more cumbersome when there are three or more independent variables. The principles involved remain the same as the number of variables rises, but the calculations, luckily, can nowadays be performed by computer.

When three independent variables are involved, the computer finds an estimated least-squares regression equation of the form,

$$\hat{Y} = a + b_1 X_1 + b_2 X_2 + b_3 X_3,$$

and this procedure can be extended without end when additional predictor variables are included in the analysis (by adding $b_4 X_4$, $b_5 X_5$, and so on to the equation).

The moment the analysis involves four variables (one dependent and three independent ones), the data can no longer be pictured in a scatter diagram. The human mind can envision two dimensions (note Figure 13.3 on p. 533) and even three dimensions (note Figure 14.2), but it balks at four dimensions and more. As mathematicians put it, the estimation of a least-squares regression equation with three or more independent variables is analogous to the fitting of a **regression hyperplane** to the sample data. Such a plane exists in four or higher-dimensional space, but no one has ever seen it!

Higher-dimensional regression analysis still includes the calculation of the standard error of the estimate, but it is now symbolized by $s_{Y \cdot 123}$, by $s_{Y \cdot 1234}$, and so on, depending on how many independent variables are being considered. The calculation of these standard errors is again a straightforward extension of that noted above. While

$$s_{Y \cdot 12} = \sqrt{\frac{\Sigma Y^2 - a\Sigma Y - b_1\Sigma X_1 Y - b_2\Sigma X_2 Y}{n - 3}},$$

the calculation of $s_{Y \cdot 123}$ adds $-b_3\Sigma X_3 Y$ to the numerator and changes the denominator to $n - 4$. And so it goes.

Prediction intervals, too, are calculated as noted above (in Box 14.B or 14.C, for example), except that the degrees of freedom for $t_{\alpha/2}$ change as they do for the standard error, from $n - 3$ to $n - 4$, and so on (always being equal to $n - m - 1$, where n is sample size and m is the number of independent variables). The use of the computer saves an enormous amount of time, especially when sample size and the number of independent variables are large, but it does more than that. It also produces a greater level of accuracy by (a) preventing a cascade of errors once a single one has been made and (b) being able to handle a large number of significant figures.

Consider the data of Table 14.4. They were fed into a computer, which promptly produced the accompanying printout.

"Canned" computer programs are widely available; all of them, when fed sample data, quickly determine and print out sample regression coefficients ($a, b_1, b_2,$ etc.) along with their standard errors ($s_a, s_{b_1}, s_{b_2},$ etc.) and t values (which, of course are the ratios of each estimated regression coefficient to its standard error, or $a/s_a, b_1/s_{b_1}, b_2/s_{b_2},$ etc.) Some programs print p values instead of t ratios. Most of them also provide the standard error of the estimate of Y, the ANOVA table, the sample coefficient of determination, and more.[6]

The following sections describe how to interpret and use the computer results.

The Estimated Multiple Regression Equation

We can quickly formulate our desired multiple regression equation as

$$\hat{Y} = \underset{(-1.265)}{-2.983} + \underset{(15.817)}{2.099\, X_1} + \underset{(\ 8.151)}{1.197\, X_2} - \underset{(-5.381)}{.3107\, X_3}$$

---

[6]See, for example, the DAYSAL-2 floppy disks that accompany this text and are designed to be used with personal computers.

Table 14.4 Sample Data on Income, Using Education, Job Experience, and Age as Independent Variables

Individual	Income (thousands of dollars per year), Y	Education (years), X_1	Job Experience (years), X_2	Age (years), X_3
A	5.0	2	9	29
B	9.7	4	18	50
C	28.4	8	21	41
D	8.8	8	12	55
E	21.0	8	14	34
F	26.6	10	16	36
G	25.4	12	16	61
H	23.1	12	9	29
I	22.5	12	18	64
J	19.5	12	5	30
K	21.7	12	7	28
L	24.8	13	9	29
M	30.1	14	12	35
N	24.8	14	17	59
O	28.5	15	19	65
P	26.0	15	6	30
Q	38.9	16	17	40
R	22.1	16	1	23
S	33.1	17	10	58
T	48.3	21	17	44

```
Problem title : Textbook Table 14.4
The Number of Paired Observations, n ....  = 20
Dependent Variable .......................  = Y
Independent Variables ...................  = X1 X2 X3

Regression Results

Variable        Coefficient    Std.Error    t-Value

Intercept        -2.98295      2.35732     -1.26540
X1 (Education)    2.09943      0.13274     15.81660
X2 (Job Exp.)     1.19746      0.14690      8.15145
X3 (Age)         -0.31067      0.05773     -5.38144

The ANOVA Table

Regression sum of squares (RSS) .........  = 1720
Regression degrees of freedom (m) .......  = 3
Regression mean square (RMS) ............  = 573.3333

Error sum of squares (ESS) ..............  = 100.4275
Error degrees of freedom (n-m-1) ........  = 16
Error mean square (EMS) .................  = 6.27672

Total sum of squares (SS) ...............  = 1820.428
Total degrees of freedom (n-1) ..........  = 19

Sample standard error of estimate of Y... = 2.505338
F statistic (RMS/EMS) ...................  = 91.34282

Sample coeff. of multiple determination,
  R² = (RSS/SS) .................................  = .9448331
Adjusted sample coefficient of multiple
  determination, R̄² ............................  = .9344892
Sample coefficient of nondetermination
  (ESS/SS) ......................................  = .055167
```

The interpretation of this equation is straightforward:

1. A b_1 of 2.099 means that each extra year of education (X_1) raises estimated annual income by $2,099, if the values of the other two variables (X_2 and X_3) are held constant.
2. A b_2 of 1.197 means that each extra year of work experience (X_2) raises estimated annual income by $1,197, if the values of the other two variables (X_1 and X_3) are held constant.
3. A b_3 of $-.3107$ means that each extra year of age (X_3) *lowers* estimated annual income by $310.70, if the values of the other two variables (X_1 and X_2) are held constant.

The corresponding t values (in parentheses below the coefficients) all pass the rule-of-thumb test of 95-percent significance noted earlier in the chapter. As noted previously (and for the reasons noted), we need not concern ourselves with the intercept of -2.983, nor with its t value that does not pass the 95-percent significance test.

A note of caution for those about to study the results of other people's regression analyses: Instead of t values, some publications print standard errors or even p values underneath the estimated coefficients. It is mandatory to read the fine print, lest utter confusion prevail.

The Standard Error of the Estimate of Y

Note that the standard error of the estimate of Y equals $s_{Y \cdot 123} = 2.505$. Compare this standard error of the estimate of Y with $s_{Y \cdot 12} = 4.079$ from this chapter's Practice Problem 2 or with $s_{Y \cdot X} = 5.377$ from Chapter 13's Practice Problem 4. Clearly, there has been further improvement in our predictive ability.

Predicting the Average or an Individual Value of Y

Using a computer program like DAYSAL-2, we can quickly produce prediction intervals for any average or individual Y. Suppose that we wanted such predictions for $X_1 = 11$ years of education, $X_2 = 20$ years of job experience, and $X_3 = 45$ years of age. In both cases, the point estimate comes to:

$$\hat{Y} = -2.983 + 2.099(11) + 1.197(20) - .3107(45) = 30.0645.$$

We predict, that is, that a person with 11 years of education, 20 years of job experience, and an age of 45 will have an annual income of $30,064.50. Yet the limits of a 95-percent prediction interval for the *average* person with these characteristics equal:

$$27.74799 \leq \mu_{Y \cdot 123} \leq 32.41188.$$

We can be 95-percent confident that people with the noted characteristics will, on the average, have annual incomes between $27,747.99 and $32,411.88. Similarly, the limits of a 95-percent prediction interval for the *next* person encountered with these characteristics equal:

$$24.27924 \leq I_{Y \cdot 123} \leq 35.88063.$$

We can be 95-percent confident that such a person's annual income will lie somewhere between $24,279.24 and $35,880.63.

Confidence Intervals for the True Regression Coefficients

Using the Box 14.D formula, we can also produce confidence intervals for any one of the true regression coefficients from our computer printout. Choosing, once again, 95-percent intervals, the limits are:

$$\alpha = a \pm (t\, s_a) = -2.983 \pm [2.12(2.357)] = -2.983 \pm 4.997$$
$$\beta_1 = b_1 \pm (t\, s_{b_1}) = 2.099 \pm [2.12(.133)] = 2.099 \pm .282$$
$$\beta_2 = b_2 \pm (t\, s_{b_2}) = 1.197 \pm [2.12(.147)] = 1.197 \pm .312$$
$$\beta_3 = b_3 \pm (t\, s_{b_3}) = -.3107 \pm [2.12(.0577)] = -.3107 \pm .1223$$

Thus, we conclude:

$$-7.98 \leq \alpha \leq 2.014$$
$$1.817 \leq \beta_1 \leq 2.381$$
$$.885 \leq \beta_2 \leq 1.509$$
$$-.433 \leq \beta_3 \leq -.1884$$

Only the true intercept, α (which concerns us little), might conceivably be zero.

Hypothesis Tests About True Regression Coefficients

We can use an alternative approach as well to test whether any one of our parameters might equal zero; we can look at the associated t value. If the parameter is zero, the t value follows a t distribution with $n - m - 1$ degrees of freedom (16 in our case); thus, our critical t value is 2.12 (as noted before). All of the printed t values (except that for the intercept) have absolute value in excess of 2.12; hence, we can be (at least) 95-percent confident that the associated parameters β_1, β_2, and β_3 (but not the intercept, α) are not zero.

Testing the Overall Significance of the Regression

Our computer printout provides all the information needed to create an ANOVA table like Table 14.3. It also gives the F value directly. The likelihood of having gotten such a large F by pure chance is tiny, indeed.

The Sample Coefficient of Multiple Determination

The value of $R^2_{Y \cdot 123}$ is given directly as .9448. We can also check it with the help of the Box 14.E formula: the printout indicates that $RSS = 1{,}720$ and $ESS = 100.4275$. Hence, *Total SS* $= 1{,}820.4275$ and

$$R^2_{Y \cdot 123} = \frac{1{,}720}{1{,}820.4275} = .9448.$$

The *adjusted* sample coefficient of multiple determination, of course, must equal (according to Box 14.G):

$$\overline{R}^2_{Y \cdot 123} = 1 - \left[\frac{100.4275}{1{,}820.4275} \left(\frac{19}{16} \right) \right] = .9345.$$

Many computer programs print this adjusted R-squared as well.

Examples

Close-Up 14.1, "Revenue Sharing and the Flypaper Effect," as well as Analytical Examples 14.1, "Advertising and the Demand for Electricity," 14.2, "Explaining the Changing Prices of Farm Land," and 14.3, "Testing the Crowding-Out Effect," illustrate how the linear-multiple-regression technique is used to answer innumerable questions.

EXTENSIONS*

In this section, we will consider three major extensions of our discussion so far: (a) stepwise multiple-regression procedures, (b) the use of dummy variables, and (c) the analysis of covariance.

Stepwise Multiple Regression

Many computers enable us to perform with ease a complicated procedure that, in the view of some statisticians, should not be performed at all. This procedure is called **stepwise multiple regression** and, as the name implies, it develops a regression equation in carefully delineated steps. The following discussion will explain why some statisticians disapprove of the procedure but is included, nevertheless, because it is being used by many other statisticians. (In the view of this author, a textbook should make students aware of controversies rather than keep students ignorant of them.) The two versions of stepwise multiple regression are the *forward-selection method* and the *backward-elimination method.*

The forward-selection method. The **step-up,** or **forward-selection, method** starts with *no* independent X variables in a model designed to explain the behavior of the dependent variable Y and then adds one X at a time. In each successive step, an explanatory variable is selected for inclusion in the analysis from a list of potential ones because it provides the greatest decrease in the hitherto unexplained variation in Y (has the highest coefficient of partial determination and, thus, promises to make the greatest marginal contribution to explaining Y). At the first stage, the computer performs simple regressions separately for each of a number of independent variables, chooses the most promising of these (the one that produces the highest regression sum of squares and, therefore, the lowest error sum of squares), and prints these results only. At the second stage, the computer performs multiple regressions separately for each combination of the independent variable selected in the first stage with one of the remaining independent variables. Once again the combination that reduces the unexplained variation of Y the most is chosen. This process continues until all potential independent variables have been included in the analysis or until further reductions in the unexplained variation of Y prove impossible. In this way, statisticians can determine which among a long list of potential independent variables provide the best explanation of the behavior of some dependent variable.

Caution is advised: The list of potential independent variables to be included in a regression analysis must not be drawn up indiscriminately. It is important to have good *a priori* reasons (a meaningful nonstatistical explanation) for potentially including each of the independent variables in the analysis. Given the ease with which modern com-

---

*Optional section.

puter technology allows us to perform stepwise multiple-regression analysis, it is possible to continue the search for an ever better-fitting regression equation until some set of independent variables is found that explains almost all the variation in the dependent variable *even if these independent variables have no effect whatsoever on the dependent one.* As already mentioned in Chapter 13, some variables are bound to be highly correlated with others by pure chance (recall Analytical Example 13.6, "Snowfall and Unemployment"), or two variables could be influenced by some third factor without being causally related to each other. (The number of ministers and the volume of whiskey consumption may be higher in one state than in another, for example, simply because that state has a larger population and not because ministers do most of the drinking.) When such is the case, the selection of a coincidentally correlated variable (snowfall or number of ministers) as the independent variable may well "explain" perfectly the dependent one (unemployment or whiskey consumption), yet the relationship is nonsensical because there is no logical connection between the variables. Nevertheless, if one uses the computer to hunt long enough, one is likely to find such a nonsensical, yet well-fitting regression equation. Only a good dose of plain common sense can ultimately prevent such foolishness. Indeed, some statisticians would rather not employ this procedure at all. They argue that one should specify on theoretical grounds, and *in advance of data selection and computer analysis,* which independent variables are logically capable of explaining the behavior of the dependent variable. *Following* such a one-time specification of the model, data should be selected and analyzed, the model should be accepted or rejected, and that should be the end of it! Statisticians who hold this view reject the forward-selection method and the backward-elimination method to be discussed below as unscientific "data mining."

The backward-elimination method. The **step-down,** or **backward-elimination, method** starts by including *all* potential independent X variables in a regression model and then eliminates one X at a time. In step 1, a multiple regression equation containing all independent variables under consideration is estimated. In each successive step, an explanatory variable is removed from the analysis because its coefficient has that t value, among all insignificant ones, that is closest to zero (and, therefore, has that p value, among all those in excess of some designated level of significance, that is the largest). Another regression equation is then estimated with the remaining independent variables and, once again, the process of elimination occurs. This process continues until all the independent variables remaining in the analysis have coefficients that are significantly different from zero.

Caution: The step-up and step-down methods cannot be relied upon to produce the same end result. It is possible that the regression model ultimately chosen contains a different subset of the original list of Xs in one case than in the other.

Practice Problem 12

Applying Stepwise Multiple-Regression Analysis to an Explanation of Income (forward-selection method)

Suppose we wanted to apply the forward-selection method to the data of Table 14.4, arguing on *a priori* grounds that it is reasonable to assume that income (Y) might be explained by education (X_1), by job experience (X_2), or by age (X_3). The results of a stepwise regression analysis can be summarized by Table 14.5.

Table 14.5 Stepwise Regression Analysis: The Influence of Successively Higher Dimensional Regression Equations on the Proportion of Explained Variation in Y

	Independent Variables Included (1)	Standard Error of the Estimate of Y (2)	Proportion of Variation in Y Explained by Regression (3)	Proportional Reduction in Previously Unexplained Variation in Y Resulting from Additional X (4)
Step 1	none	$s_Y = 9.788$	0	—
Step 2	X_1	$s_{Y \cdot X_1} = 5.366$	$r^2_{Y \cdot 1} = .7152$	$r^2_{Y \cdot 1} = .7152$
Step 3	X_2	$s_{Y \cdot 12} = 4.074$	$R^2_{Y \cdot 12} = .8450$	$r^2_{Y2 \cdot 1} = .4558$
Step 4	X_3	$s_{Y \cdot 123} = 2.505$	$R^2_{Y \cdot 123} = .9448$	$r^2_{Y3 \cdot 12} = .6439$

Step 1: If we perform no regression analysis whatsoever, we can best estimate Y by its mean, $\bar{Y} = 24.415$; the standard error of the estimate of Y equals

$$s_Y = \sqrt{\frac{\Sigma Y^2 - n\bar{Y}^2}{n - 1}} = \sqrt{\frac{13,742.27 - 20(24.415)^2}{19}} = 9.788.$$

Step 2: If we instruct the computer to regress Y on X_1, then on X_2, and then on X_3, it selects the regression of Y on X_1 as the best alternative. The standard error of the estimate of Y (now around the regression line rather than its mean) is $s_{Y \cdot X_1} = 5.366$. (This result, like others in this section, differs slightly from our Chapter 13 result because of rounding.) The computer gives a regression sum of squares of 1,302.055 and a total sum of squares of 1,820.426 (which also equals the numerator of the ratio under the above square root); thus, (1,302.055/1,820.426)100 = 71.52 percent of the variation in Y has been explained by the regression of Y on X_1, and this fact is reflected in the $r^2_{Y \cdot 1}$ value listed in column 3.

Step 3: If we now instruct the computer to regress Y on X_1 and X_2 and then on X_1 and X_3, it selects the regression of Y on X_1 and X_2 as the best alternative. The standard error of the estimate of Y (now around the regression plane rather than line) is $s_{Y \cdot 12} = 4.074$. This time the computer gives a regression sum of squares of 1,538.226; the total sum of squares, of course, is the same as before. Hence (1,538.226/1,820.426)100 = 84.50 percent of the variation in Y has been explained by the regression of Y on X_1 and X_2, and this fact is reflected in the $R^2_{Y \cdot 12}$ value listed in column 3. Note: The previously unexplained variation in Y (the Step 2 *ESS*) was 518.4. The new unexplained variation in Y (the Step 3 *ESS*) is 282.2. Thus, the unexplained variation in Y has been reduced by 236.2, or 45.56 percent. This reduction in unexplained variation shows up in a calculation, performed using the formula in Box 14.I, of the sample coefficient of partial determination:

$$r^2_{Y2 \cdot 1} = \frac{.8450 - .7152}{1 - .7152} = .4558,$$

as shown in column 4. There is a slight rounding error.

Step 4: We finally instruct the computer to regress Y on X_1, X_2, and the remaining independent variable, X_3. The standard error of the estimate of Y (now around the regression hyperplane) drops further to $s_{Y \cdot 123} = 2.505$. This time, the computer gives a regression sum of squares of 1,720; hence, $(1,720/1,820.426)100 = 94.48$ percent of the variation in Y has been explained by the regression of Y on X_1, X_2, and X_3; this fact is reflected by the value of $R^2_{Y \cdot 123}$ listed in column 3. Note: The previously unexplained variation in Y (the Step 3 *ESS*) was 282.2. The new unexplained variation in Y (the Step 4 *ESS*) is 100.4. Thus, the unexplained variation in Y has been reduced by 181.8, or by 64.42 percent. This reduction in the unexplained variation in Y shows up in a calculation, performed using the formula in Box 14.I, of the sample coefficient of partial determination:

$$r^2_{Y3 \cdot 12} = \frac{.9448 - .8450}{1 - .8450} = .6439.$$

(Again, there is a negligible rounding error.)

Conclusion: The end result is precisely the regression of Y on X_1, X_2, and X_3 that was introduced in the above section, "The Estimated Multiple Regression Equation." This is a coincidence. It is not a coincidence, however, that the forward-selection method always adds independent variables in such an order that each addition makes the greatest marginal contribution to the explanation of the behavior of Y. ▌▌▌

Dummy Variables

The type of regression analysis we have conducted so far requires that all variables be quantitative, but often one has reason to believe that the dependent variable under study is influenced in some important way by an independent variable that is qualitative in nature—one that cannot be measured numerically (as so many thousands of dollars of income or so many years of education) but can only be described categorically (as male/female; black/white; home owner/renter; urban/rural; war time/peace time; recession/prosperity; Catholic/Protestant/Jewish). Under such circumstances, all is not lost; one can incorporate any qualitative variable into regression analysis by creating one or more **binary, indicator,** or **dummy variables,** which are variables that can take on two values only, namely 0 or 1, which are used to indicate the absence or presence of a particular qualitative characteristic. If the qualitative variable of interest has two categories only (is dichotomous), a single dummy variable will do; one might code "male" as 0 and "female" as 1, for instance. If the variable has more than two categories, several dummy variables are needed, as will be shown below.

A qualitative variable with two categories. Consider, for example, the sample data contained in columns 1–4 of Table 14.6. If one has reason to believe that income is influenced not only by education, but also by sex, one does not have to capitulate before the fact that the column 4 data are qualitative. One can create the column 5 *nominal* data and develop a multiple regression equation of the form

$$\hat{Y} = a + b_1 X_1 + b_2 D_1$$

where D_1 represents the dummy variable. One need not confine oneself to the simple regression equation pictured in last chapter's Figure 13.6 (on p. 538).

Our new multiple regression equation predicts the income of any male as $\hat{Y}_M = a + b_1X_1$ (because D_1 then equals 0) and that of any female as $\hat{Y}_F = a + b_1X_1 + b_2 = (a + b_2) + b_1X_1$ (because D_1 then equals 1).

Using the data of Table 14.6, one can estimate the multiple regression equation to be

$$\hat{Y} = 5.376 + 1.786\,X_1 - 7.103\,D_1$$
$$(1.884)\quad(8.410)\quad(-3.655)$$

Once more t values are given in parentheses; $R^2 = .84$. This equation could be graphed as a regression plane similar to Figure 14.2, except that the slope of Y vs. X_2 (now called D_1) would be negative. However, only two values of D_1 are possible, $D_1 = 0$ and $D_1 = 1$, and if the regression plane is cut parallel to the YX_1 plane at these two values of D_1 and the cut is then projected onto the YX_1 plane, the result is the graph in Figure 14.4.

Note that our dummy variable can be viewed like a switch; it is "on" when the income of females is being estimated (the lower broken regression line then applies); it is "off" when the income of males is being estimated (and the upper solid regression line applies).

Table 14.6   Sample Data on Income, Using Education and Sex as Independent Variables

Individual (1)	Income (thousands of dollars per year), Y (2)	Education (years), X_1 (3)	Sex (4)	Sex Dummy (male = 0, female = 1), D_1 (5)
A	5.0	2	male	0
B	9.7	4	female	1
C	28.4	8	male	0
D	8.8	8	female	1
E	21.0	8	male	0
F	26.6	10	male	0
G	25.4	12	male	0
H	23.1	12	male	0
I	22.5	12	male	0
J	19.5	12	female	1
K	21.7	12	female	1
L	24.8	13	male	0
M	30.1	14	male	0
N	24.8	14	female	1
O	28.5	15	male	0
P	26.0	15	female	1
Q	38.9	16	male	0
R	22.1	16	female	1
S	33.1	17	male	0
T	48.3	21	male	0

Figure 14.4   Scatter Diagram and Multiple Regression with Dummy Variable

The scatter of points in this graph depicts the data of columns (2)–(4) of Table 14.6. The data are precisely the same as in Figure 13.6 (on p. 538), but those pertaining to females have been highlighted by the color crosses. This time, furthermore, a multiple regression equation has been estimated, including sex as a dummy variable: $\hat{Y} = 5.376 + 1.786 \, X_1 - 7.103 \, D_1$. However, income estimates for males can be made by the upper line ($D_1 = 0$); income for females can be estimated by the lower line ($D_1 = 1$). Thus, a female's income is estimated at \$7,103 less than that of a comparably educated male. The standard error of the estimate of Y is considerably lower than for Chapter 13's simple regression: $s_{Y \cdot X_1 D_1}$ equals 4.132 here, $s_{Y \cdot X_1}$ equaled 5.377 earlier.

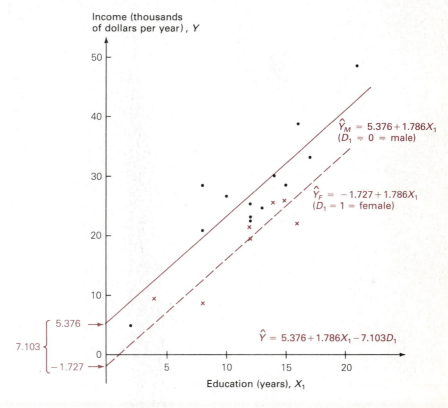

Note also that the accuracy of the estimates improves considerably: the standard error of the estimate of Y changes from \$5,377 in Chapter 13's simple regression of Practice Problem 4 to \$4,132 in this multiple one; the coefficient of determination changes from $r^2 = .71$ (Chapter 13's Practice Problem 12) to $R^2 = .84$.

A qualitative variable with more than two categories. Even qualitative variables with multiple categories (such as first/second/third/fourth quarter of the year, Catholic/Protestant/Jewish, or extremely interested/somewhat interested/not interested) can be incorporated into a multiple regression equation using the dummy-variables technique. In that case, given k categories, $k - 1$ dummy variables are used. Suppose, for example, one suspects that income, Y, can be explained solely by religion and that there are three religions. One can then specify a multiple regression equation of the form $\hat{Y} = a + b_1D_1 + b_2D_2$, where both of the Ds are dummy variables, such that $D_1 = 1$ if Catholic and $D_1 = 0$ if not Catholic; $D_2 = 1$ if Protestant and $D_2 = 0$ if not Protestant. If someone is Jewish, that person's income is estimated as $\hat{Y} = a$ because then D_1 as well as D_2 are zero. There is, however, more than convenience involved in proceeding in this way. If a third dummy variable were included ($D_3 = 1$ if Jewish, $D_3 = 0$ if not Jewish), the parameters of the expanded equation (then including "$+ b_3D_3$" as well) could not be uniquely determined because Assumption 5 (noted on page 592) would be violated. The third dummy, D_3, would be an exact linear function of the other two, D_1 and D_2. Knowing the values of D_1 and D_2 for any person, one would always know the value for D_3 as well. Assuming that only the three noted religious affiliations exist, someone who is not-Catholic and not-Protestant, for example, necessarily is Jewish. When Assumption 5 is violated, proper estimation is impossible. (We will return to this matter in the later section on multicollinearity.)

Note: The use of dummy independent variables is nicely illustrated in Analytical Examples 14.4, "Do Safety Caps on Aspirin Bottles Help?" and 14.5, "Does Photocopying Harm Authors and Publishers?"

A dummy dependent variable. A dummy variable can even take the place of the dependent variable. Consider the hypothesis that poverty is a function of race and sex. One might collect sample data on the subject and code the results using *three* dummy variables such that $D_1 = 1$ means "poverty," $D_1 = 0$ means "the absence of poverty"; $D_2 = 1$ means "black," $D_2 = 0$ means "white"; and $D_3 = 1$ means "female," and $D_3 = 0$ means "male." The multiple regression equation might read

$$\hat{D}_1 = .06 + .25\, D_2 + .50\, D_3,$$

or

$$\text{Poverty} = .06 + .25\ \text{Race} + .50\ \text{Sex}.$$

This equation would give us the *probability* of being poor—equal to .06 for a white male and $.06 + .25 + .50 = .81$ for a black female, for example. However, extreme caution is advised. Such *linear probability models* violate some of the assumptions of the analysis (for example, the assumption concerning homoscedasticity), and there is no guarantee that the estimated probability number lies between 0 and 1. (One can deal with these problems, but we must leave that resolution to more advanced texts.)

Analysis of Covariance

A combination of (a) regression analysis that incorporates quantitative and dummy variables with (b) an analysis of variance is referred to as an **analysis of covariance,** frequently abbreviated as ANOCOVA. We shall consider two examples.

Example 1: Inclusion of dummy yields two regression lines with different intercepts, same slope. Suppose we are worried that the apparent *shift* in the estimated regression line when sex differences are considered (a shift that is depicted in Figure 14.4) is entirely due to sampling error and that the observed sexual difference in the income/education relationship does not exist in the population as a whole. We can make a test in one of two ways.

||| Practice Problem 13

Analyzing the Effect of Sex on Income Using a *t* Test

First, we can look at the *t* value for the coefficient ($b_2 = -7.103$) of the dummy variable (D_1). That value equals the ratio of that coefficient to its standard error, or

$$t_{b_2} = \frac{b_2}{s_{b_2}} = \frac{|7.103|}{1.943} = |3.66|.$$

The critical value of *t* for a one-tail area $\alpha = .025$ (corresponding to a 95-percent confidence level) can be gleaned for our $n - 3 = 17$ degrees of freedom from Appendix Table K as $t_{.025,17} = 2.11$. Thus, the probability of β_2 (the true difference between the sexes in the population as a whole) being zero, yet finding a β_2 estimate of $b_2 = |7.103|$ in a random sample of the population, is considerably less than 2.5 percent. In fact, the *p* value corresponding to our computed *t* value equals .001 (as the 17 degrees-of-freedom row of Appendix Table K also reveals). We, therefore, should *reject* the hypothesis, H_0: $\beta_2 = 0$, and conclude that the shift observed in the sample data is real rather than apparent. |||

||| Practice Problem 14

Analyzing the Effect of Sex on Income Using an *F* Test

Using the Table 14.6 data, a regression of income on education alone yields the equation $\hat{Y}_X = 2.137 + 1.849 X_1$ (which differs from the equivalent Chapter 13 regression only due to rounding). The error sum of squares can be computed as $ESS_{Y \text{ vs } X_1} = 518.4$ with $n - 2 = 18$ degrees of freedom (which again differs from the Table 13.3 result only due to rounding). This chapter's regression of income on education and sex, in contrast, produces an $ESS_{Y \text{ vs } X_1 D_1} = 290.3$, with $n - 3 = 17$ degrees of freedom. If β_2, the coefficient of the dummy variable in the population, really were zero, the error variances in our two regression estimates should be the same; thus, the ratio,

$$F = \frac{\dfrac{ESS_{Y \text{ vs } X_1} - ESS_{Y \text{ vs } X_1 D_1}}{18 - 17}}{\dfrac{ESS_{Y \text{ vs } X_1 D_1}}{17}} = \frac{\dfrac{518.4 - 290.3}{1}}{\dfrac{290.3}{17}} = 13.36,$$

would come from an *F* distribution with 1 and 17 degrees of freedom. As Appendix Table N shows, the critical value of $F_{.025}^{1,17} = 6.04$, which (when compared with 13.36) again leads us to *reject* H_0: $\beta_2 = 0$. Sex *does* have an effect on income.

Note: The *t* and *F* distributions are related in such a way that *t* (for *n* degrees of freedom) when squared equals *F* for (1 and *n* degrees of freedom) when both are computed for the same set of data. This relationship affords a check on our calculations in Practice Problems 13 and 14: $t^2 = (3.66)^2 = 13.40$, which, except for rounding error, equals our computed value of *F*. |||

Example 2: Inclusion of dummy yields two regression lines with different intercepts as well as slopes. Let us now consider the possibility that the relationship between our two quantitative variables (income and education) is entirely different for the two categories (males and females) contained in our qualitative variable. Suspecting that an estimated regression that ignores sex ($\hat{Y}_X = 2.137 + 1.849X_1$) is grossly misleading, we might estimate two separate regressions from the Table 14.6 data—one for males only, the other one for females only. The results follow (with t values in parentheses).

$$\text{For males:}\quad \hat{Y}_M = 3.905 + 1.906\,X_1 \qquad r^2 = .8144$$
$$\phantom{\text{For males:}\quad}(1.084)\ \ (6.948)$$

$$\text{For females:}\ \hat{Y}_F = 1.751 + 1.486\,X_1 \qquad r^2 = .819$$
$$\phantom{\text{For females:}\ }(\ .459)\ \ (4.756)$$

These regression lines are shown in Figure 14.5, together with the original scatter diagram. The lines certainly appear to be different, but once again an obvious question arises: How do we know that the apparent sexual differences are not simply due to sampling error?

Figure 14.5   Scatter Diagram with Separate Simple Regressions for Males and Females

Separate regressions of income against education for males only (solid black) and females only (color) display different intercepts as well as slopes. A regression line based on the combined data, first graphed in Figure 13.6 on p. 538, is also shown for comparison (dashed black).

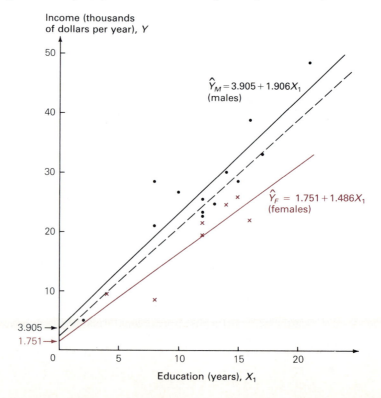

||| Practice Problem 15

Analyzing the Effect of Sex on the Income/Education Relationship Using an *F* Test
We can conduct a test with this null hypothesis:

H_0: The income/education relationship in the population as a whole is the same for men as for women.

We can focus on the observed improvement of the fit, noting that the single-group error sum of squares was $ESS_S = 518.4$ (with $n - 2 = 18$ degrees of freedom), while the *combined* error sums of squares of the separate regressions equals:

$$ESS_C = ESS_M + ESS_F = 224.1 + 52.5 = 276.6$$

(with $n - 2 - 2 = 16$ degrees of freedom). If the null hypothesis were true, the ratio $(ESS_S - ESS_C)/ESS_C$ should be near zero and the ratio,

$$F = \frac{\dfrac{ESS_S - ESS_C}{18 - 16}}{\dfrac{ESS_C}{16}} = \frac{\dfrac{518.4 - 276.6}{2}}{\dfrac{276.6}{16}} = 6.99,$$

would come from an *F* distribution with 2 and 16 degrees of freedom. As Appendix Table N shows, the critical value of $F_{.025}^{2,16} = 4.69$, which is considerably smaller than the computed 6.99. Thus, we should *reject* H_0: "the income/education relationship in the population as a whole is the same for men and women" and conclude that it is not the same. |||

||| Practice Problem 16

Analyzing the Effect of Sex on the Income/Education Relationship Using an Alternative Test
With the help of dummy variables, the problem of whether two separate estimated regressions reflect the same true relationship can also be converted into a different one: whether several coefficients in a single regression are equal to zero. If we estimate the following single regression equation (which is equivalent to the two separate ones),

$$\hat{Y} = a + b_1 X_1 + b_2 D_1 + b_3 X_1 D_1$$

we find that

$$\hat{Y} = 3.905 + 1.906 \, X_1 - 2.153 \, D_1 - .4202 \, X_1 D_1$$
$$\phantom{\hat{Y} = } (1.177) \ (7.542) \quad (-.364) \quad (-.8871)$$

and that

$$R^2 = .848.$$

As proof that this result is equivalent to the above separate equations for men and women, note that this equation reduces as follows.

For men (when $D_1 = 0$),

$$\hat{Y}_M = a + b_1 X_1$$
$$= 3.905 + 1.906\, X_1$$

For women (when $D_1 = 1$),

$$\hat{Y}_F = a + b_1 X_1 + b_2 + b_3 X_1$$
$$= (a + b_2) + (b_1 + b_3)\, X_1$$
$$= (3.905 - 2.153) + (1.906 - .4202)\, X_1$$
$$= 1.752 + 1.486\, X_1.$$

These results are the same as those for our separate regressions. Even our new equation's error sum of squares (of 276.6) is the same as ESS_C calculated earlier.

Asking whether the income/education relationship is the same for both men and women is now equivalent to asking whether both b_2 and b_3 are zero (a kind of question typically asked in an analysis of variance). To test the hypothesis, "$H_0: b_2 = b_3 = 0$," we compare the improvement in fit between the income/education regression ignoring sex and our latest regression. The test statistic is

$$F = \frac{\dfrac{ESS_{Y \text{ vs } X_1} - ESS_{Y \text{ vs } X_1, D_1,\ X_1 D_1}}{18 - 16}}{\dfrac{ESS_{Y \text{ vs } X_1,\ D_1,\ X_1 D_1}}{16}} = \frac{\dfrac{518.4 - 276.6}{2}}{\dfrac{276.6}{16}} = 6.99.$$

Because the test is equivalent to that in Practice Problem 15, the result and the conclusion are the same.

Note: As a look at some of the Selected Readings at the end of this chapter shows, there is a lot more to the analysis of covariance than this brief introduction can indicate, which must be left to more advanced texts. ▮

DISCOVERING POSSIBLE VIOLATIONS OF ASSUMPTIONS

In this final section of the chapter, we consider how one might discover violations of the crucial assumptions noted earlier (p. 592) that underlie all of regression analysis. Such a check is of obvious importance because serious violations of the assumptions can call into question all the conclusions drawn from a regression analysis. One basic technique is the **analysis of residuals,** a careful study of the differences, or errors, $e = Y - \hat{Y}$, between actual observations, Y, concerning the dependent variable and the associated estimates, $\hat{Y}$, made by the regression equation.

Table 14.7, for example, lists these residuals for the multiple regression equation derived from Table 14.4 and discussed earlier. Many computer programs print out a table of residuals such as this one. Consider what a study of such residuals can tell us.

Table 14.7   Table of Residuals

Individual	Actual Observation, Y	Regression Estimate, $\hat{Y} = -2.983 + 2.099X_1 + 1.197X_2 - .3107X_3$	Residual or Error, $e = Y - \hat{Y}$
A	5.0	2.978	2.022
B	9.7	11.424	−1.724
C	28.4	26.207	2.193
D	8.8	11.085	−2.285
E	21.0	20.003	.997
F	26.6	25.974	.626
G	25.4	22.404	2.996
H	23.1	23.968	−.868
I	22.5	23.866	−1.366
J	19.5	18.869	.631
K	21.7	21.884	−.184
L	24.8	26.067	−1.267
M	30.1	29.893	.207
N	24.8	28.421	−3.621
O	28.5	31.050	−2.550
P	26.0	26.363	−.363
Q	38.9	38.522	.378
R	22.1	24.652	−2.552
S	33.1	26.649	6.451
T	48.3	47.774	.526

Note: The sum of the last column, $\Sigma(Y - \hat{Y})$, should equal zero but equals .248 in this case due to rounding error.

Statistical Independence, Serial Correlation, and the Durbin-Watson Test

The first assumption of regression analysis noted above deals with the statistical independence of different sample observations concerning Y, the dependent variable. If this assumption is fulfilled, the associated residuals will be statistically independent as well. In many business and economic applications, however, regression analysis involves **time-series data,** data that pertain to a given population observed during different past periods of time. More often than not, such time-series data exhibit **autocorrelation,** or **serial correlation,** which is said to exist whenever the value of Y (and the associated error term, e) that is observed at one time, t, is correlated with that observed at an earlier time, such as $t - 1$ or $t - 2$. If e_t is correlated with e_{t-1}, one speaks of **first-order serial correlation;** if e_t is correlated with e_{t-2}, one speaks of **second-order serial correlation;** and so on for higher-order types of such correlation. In addition, if one large value (Y above its mean and $e > 0$) follows another, or if one small value follows another, one speaks of **positive serial correlation.** In contrast, if a small value (Y below its mean and $e < 0$) follows a large one, or if a large value follows a small one, one speaks of **negative serial correlation.** An example of positive serial correlation might be the behavior of the Dow-Jones index of stock prices. If stock prices are high by historical standards today, it is quite likely that they will be high tomorrow as well, and this pattern may continue for weeks. After a sudden break, the opposite may occur: average prices that are low by historical standards may be followed by average prices that are low as well. Positive serial correlation is illustrated in panel (a) of Figure 14.6. Panel (b) illustrates the opposite case of negative serial correlation (that might depict the behavior of hog prices following the famous cobweb cycle discussed in economics texts: low prices discourage hog breeding and lead to a shortage and high prices; they encourage breeding and lead to a surplus and low prices . . .)

The problem. Why is serial correlation a problem? In its presence, the least-squares method will still produce estimated regression coefficients (a, b_1, b_2, etc.) that are unbiased estimators of the true regression coefficients (α, β_1, β_2, etc.). However, the standard errors of these estimators (s_a, s_{b_1}, etc.) will be seriously underestimated. As a result, all kinds of inferences will be totally wrong: confidence intervals will be more precise than is warranted; t values, the coefficient of determination, and the F statistic will all be vastly exaggerated; hypothesis tests will report statistical significance of parameters where there is none. Nor is this a complete list of problems associated with serial correlation!

The Durbin-Watson test. Several techniques for detecting the presence of serial correlation exist. These include the use of Chapter 11's *number-of-runs test* (to count runs of positive or negative *errors*) and the calculation of a special *serial-correlation coefficient.* Most common, however, is the use of the **Durbin-Watson test,** which is based on the statistic given in Box 14.J.

14.J The Durbin-Watson Test Statistic (for first-order serial correlation)

$$D\text{-}W \text{ or } d = \frac{\sum_{t=2}^{n} (e_t - e_{t-1})^2}{\sum_{t=1}^{n} e_t^2}$$

The inventors of this test (after whom it is named) have provided tables, such as Appendix Table P, that show, for specified significance levels (α), sample sizes (n), and numbers of independent variables in a regression (m), lower and upper limits of d (denoted by d_L and d_U, respectively) with the help of which the null hypothesis, "H_0: No serial correlation exists," can be tested.

As Figure 14.7 shows, values of d can range from 0 to 4; serial correlation is absent for values in the vicinity of 2.

Figure 14.6 Serial Correlation

Serial correlation between residuals (that are depicted by the color dots) may be positive, as in panel (a), or negative, as in panel (b). In the first case, one large residual typically follows another, or one small one follows another; in the second case, large and small residuals continually alternate over time.

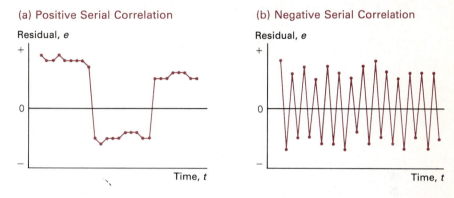

Three types of tests can be made, depending on how the alternative hypothesis is formulated:

1. The null hypothesis can be pitted in a *one-tailed test* against "H_A: Positive serial correlation exists (e_t is directly related to e_{t-1})." In that case, one *rejects* H_0 if $d < d_L$, accepts H_0 if $d > d_U$, and *cannot decide* if $d_L \leq d \leq d_U$.
2. The null hypothesis can be pitted in a *one-tailed test* against "H_A: Negative serial correlation exists (e_t is inversely related to e_{t-1})." In that case, one *rejects* H_0 if $d > 4 - d_L$, *accepts* H_0 if $d < 4 - d_U$, and *cannot decide* if $4 - d_U \leq d \leq 4 - d_L$.
3. The above null hypothesis can be pitted in a *two-tailed test* against "H_A: Positive *or* negative serial correlation exists" (and in this case, the significance levels given in Appendix Table P double: if a level of $\alpha = .05$ is desired, the $\alpha = .025$ section of Appendix Table P must be consulted). Now one *rejects* H_0 if $d < d_L$ or $d > 4 - d_L$, *accepts* H_0 if $d_U < d < 4 - d_U$, and *cannot decide* if $d_L \leq d \leq d_U$ or $4 - d_U \leq d \leq 4 - d_L$.

Figure 14.7 Critical Values of the Durbin-Watson Test Statistic

These critical values of d define the presence or absence of positive or negative first-order autocorrelation in the Durbin-Watson test.

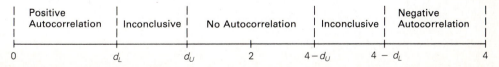

||| Practice Problem 17

Determining the Presence of Serial Correlation for a Demand Function

In order to illustrate the procedure, suppose that the residuals given in the second column of Table 14.8 came from a time-series regression of a good's quantity demanded (Y) against the good's own price (X_1), the price of a substitute (X_2), and consumer income (X_3). We can calculate the Durbin-Watson test statistic using the Box 14.J formula as

$$d = \frac{19.2733}{24.7864} = .78.$$

Suspecting positive serial correlation, we might select the first version of an alternative hypothesis given above. Choosing $\alpha = .05$, and given $n = 15$ and $m = 3$, we find, in Appendix Table P, $d_L = .82$ and $d_U = 1.75$. Accordingly, we *reject* H_0 (no serial correlation) and *accept* H_A (positive serial correlation) with 95-percent confidence.

A final note: Techniques for *dealing with* autocorrelation (once it has been discovered) exist but go beyond the intended scope of this book. They include the creation of *general difference equation models* (noted in the "Selected Readings" section at the end of this chapter) that transform regressions of Y on X to regressions of some function of Y on some function of X.

Table 14.8   Calculation of the Durbin-Watson Test Statistic

Year	e	$e_t - e_{t-1}$	$(e_t - e_{t-1})^2$	e_t^2
1970	1.30	—	—	1.6900
1971	1.01	$-.29$	.0841	1.0201
1972	.90	$-.11$	.0121	.8100
1973	.20	$-.70$	.4900	.0400
1974	-3.41	-3.61	13.0321	11.6281
1975	-1.37	2.04	4.1616	1.8769
1976	-1.25	.12	.0144	1.5625
1977	-1.12	.13	.0169	1.2544
1978	$-.10$	.04	.0016	.0100
1979	$-.01$	.09	.0081	.0001
1980	.14	.15	.0225	.0196
1981	.29	.15	.0025	.0841
1982	.37	.08	.0064	.1369
1983	1.56	1.19	1.4161	2.4336
1984	1.49	$-.07$	.0049	2.2201
	0		19.2733	24.7864
			$= \sum\limits_{t=2}^{n} (e_t - e_{t-1})^2$	$= \sum\limits_{t=1}^{n} e_t^2$

|||

Normality

The second assumption (concerning normality) is equally crucial if the inferences made in the course of regression analysis are to be valid. A plot of residuals against their associated regression estimates is often used to test this assumption. The procedure is complex; in order to understand the procedure, it is helpful to review Figure 13.7 (on p. 541) and to recall that in the *simple* regression case depicted there a normally distributed Y population that was assumed to exist at each possible value of what was then the single independent variable, X. Analogously, in the *multiple* regression case, a normally distributed Y population is assumed to exist for each possible combination of, say, three independent variables, such as X_1, X_2, and X_3. If the assumption of normality holds, a large number of actual Y observations (that would be associated with any given combination of X_1, X_2, X_3 and, therefore, with a single regression estimate, $\hat{Y}$, of the dependent variable, Y) would be associated with an equally large number of *normally distributed* residuals or errors, $e = Y - \hat{Y}$. If we graphed e against $\hat{Y}$, as in Figure 14.8, we would find a vertical line of many dots at *each* value of $\hat{Y}$. Some actual Y observations for a given X_1, X_2, X_3 (and, therefore, $\hat{Y}$) would equal $\hat{Y}$; the corresponding errors would be zero; the dots representing the e vs. $\hat{Y}$ combination would lie on the solid horizontal line in the center of the graph. Half of the remaining Y observations at a given $\hat{Y}$ would exceed $\hat{Y}$; such positive errors would show up as dots above the horizontal center line. The other half of the remaining Y observations at a given $\hat{Y}$ would fall short of $\hat{Y}$; such negative errors would show up as dots below the horizontal center line. More than that! Roughly 68 percent of the dots at a given $\hat{Y}$ would lie within plus or minus one standard error (shown by the dashed horizontal lines in Figure 14.8) of the zero mean error, while roughly 95 percent of these dots would lie within plus or minus two such standard errors. (Recall from Appendix Table H that $2(.3413) = .6826$ of the area under the standard normal curve falls within $\pm 1z$ of the mean, while $2(.4772) = .9544$ of the area falls within $\pm 2z$ of the mean.) In our case, we do not have enough observations to make this test. The color dots in Figure 14.8 represent the Table 14.7 data we do have; together with the remaining dots, they illustrate the kind of pattern one expects to see if the normality assumption holds.

Homoscedasticity

The third assumption (concerning equal variances of all the Y populations) can be tested by a residuals plot like Figure 14.8 as well. One notes whether the spread of the dots around the $e = 0$ line is about the same at all levels of $\hat{Y}$. If not, the homoscedasticity assumption is probably violated, and the analysis is said to suffer from the problem of **heteroscedasticity** (unequal variances associated with the different Y populations defined by different values of the independent variables). Figure 14.9 illustrates a clear-cut case of heteroscedasticity that might be discovered by a study of residuals. If such a problem is present, the least-squares parameter estimates are still unbiased and consistent, but they are inefficient (they have larger than minimum variances). The use of inefficient parameter estimates leads again, as in the case of autocorrelation, to biased confidence intervals (that are unjustifiably precise) and also to incorrect hypothesis tests.

Here, too, methods exist that cope with the problem, as the "Selected Readings" to this chapter note. (They also point to some alternative ways of detecting a lack of homoscedasticity.)

Figure 14.8   Testing for Normality

If one had large numbers of data of the type given in Table 14.7, plotted residuals, *e*, against associated regression estimates, $\hat{Y}$, and observed the kind of pattern shown in this graph, one could be satisfied that the normality assumption underlying regression analysis was fulfilled.

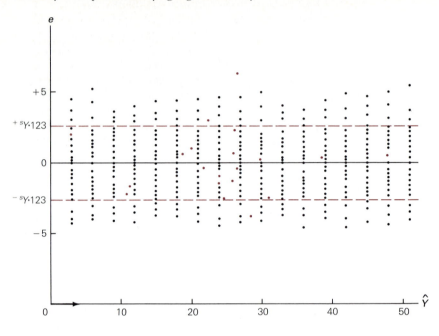

Figure 14.9   The Heteroscedastic Fan

A residuals plot such as this one strongly suggests the presence of heteroscedasticity. In this particular case, variances increase with larger values of $\hat{Y}$.

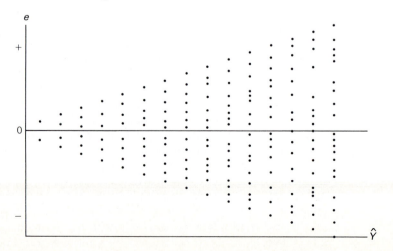

Linearity

The fourth assumption (of linearity) can also be tested with the help of a residuals plot, such as Figure 14.8. If the linearity assumption holds, one expects the dots to lie in a broad band around the horizontal $e = 0$ line; anything else is highly suspicious and suggests possible curvilinear relationships among the variables (which, in turn, suggests fitting the data to a nonlinear equation or transforming the data in a fashion suggested by Chapter 13's section on "Curvilinear Regression").

Avoiding Multicollinearity

We have already noted the reason for the fifth assumption. When two or more independent variables in a regression model are *perfectly* correlated, one cannot calculate the parameter estimates (because no unique solution of the normal equations exists); when the independent variables are *imperfectly but highly* correlated, it is next to impossible to isolate their individual effects on the dependent variable (because the regression coefficients then cannot be estimated accurately). This problem of high correlation between or among independent variables is referred to as the problem of **multicollinearity.**

While it is easy enough to detect *perfect* multicollinearity (one simply cannot find the estimated regression coefficients), it is often difficult to detect imperfect but high multicollinearity. Some telltale symptoms are the following:

1. R^2 is large, yet the estimated regression coefficients have huge standard errors and are statistically insignificant. (Multicollinearity causes the estimated coefficients to vary substantially from sample to sample; this fact raises their standard errors; hence, the ratio $(b/s_b) = t$ is unlikely to be greater than 2, or statistically significant.)
2. The estimated regression coefficients change greatly in value as independent variables are dropped from or added to the regression equation.
3. The magnitudes of the estimated regression coefficients are unreasonable. (They are unexpectedly large or small.)
4. The signs of the estimated regression coefficients are nonsensical. (They are negative when theory suggests positive signs and vice versa.)

A simple graphical test of multicollinearity has already been suggested by Figure 14.3; an obvious arithmetic one is the construction of a **correlation matrix,** a table showing simple correlation coefficients between each possible pair of variables included in a regression model. Many computer programs print out simple correlation matrices of the form given in Table 14.9, which contains the actual coefficients for the income vs. education, job experience, and age model used earlier in this chapter. Such coefficients can also be calculated by hand, of course, using the Box 13.K formula that measures the tendency of paired sample observations to vary concurrently, which tendency is called the **sample covariance.**

Naturally, each variable is correlated perfectly with itself; hence, we find four entries equal to 1.00 in the main diagonal of the table. In any case, it is the boxed entries that are of crucial interest here. A coefficient of .80 or larger between any pair of independent variables is considered suggestive of multicollinearity (which is not the case here). But note: The simple correlation matrix by itself is not sufficient to detect

Table 14.9 A Simple Correlation Matrix

Variables	Income, Y	Education, X_1	Job Experience, X_2	Age, X_3
Income, Y	$r_{Y \cdot Y} = 1.00$	$r_{Y \cdot 1} = .85$	$r_{Y \cdot 2} = .27$	$r_{Y \cdot 3} = .10$
Education, X_1	—	$r_{1 \cdot 1} = 1.00$	$r_{1 \cdot 2} = -.11$	$r_{1 \cdot 3} = .10$
Job Experience, X_2	—	—	$r_{2 \cdot 2} = 1.00$	$r_{2 \cdot 3} = .68$
Age, X_3	—	—	—	$r_{3 \cdot 3} = 1.00$

the problem because this matrix fails to take into account the possible correlation between any one independent variable with *all* the others as a group. Therefore, it is customary to regress each of the independent variables against all the others and to note whether any of the resultant R^2 values is near 1. (If that is the case, high multicollinearity is present.) If we follow this procedure (using again the data of Table 14.4), we find the following:

$$\hat{X}_1 = 11.11 - .2613 \, X_2 + .1011 \, X_3 \qquad R^2_{1 \cdot 23} = .0648$$
$$\hat{X}_2 = 3.777 - .2133 \, X_1 + .2725 \, X_3 \qquad R^2_{2 \cdot 13} = .4866$$
$$\hat{X}_3 = 13.24 + .5344 \, X_1 + 1.764 \, X_2 \qquad R^2_{3 \cdot 12} = .4857$$

Clearly, we do not have a multicollinearity problem.

A variety of options exist for dealing with serious cases of multicollinearity. The easiest approach is to drop one of the highly collinear variables from the regression, but this option amounts to the willful commission of **specification error,** which is the exclusion (in this case) of an explanatory variable that theory suggests should be included in a regression or the inclusion (in other cases) of an explanatory variable that theory suggests to be irrelevant. Other approaches include collecting more data, using *a priori* information about parameters (gathered in a previous study, perhaps), or transforming data.

As an example of data transformation, consider the following: In a regression relating a good's quantity demanded to the good's price and people's money income, the two independent variables may be highly correlated in a period of inflation. (Increases in the price of this good then go hand in hand with increases in prices in general and, therefore, with increases in money income). The transformation of money-income data to real-income data, however, may solve the problem.

APPLICATIONS: READING THE JOURNALS

The business and economics literature is filled with studies that are based on the multiple-regression techniques introduced in this chapter. Any reader who has mastered this introduction, therefore, will be able to understand most of what this literature has to offer. Consider the Analytical Examples contained in this chapter. Consider the four Applications discussed in this section. Consider, finally, the "Supplementary Topics" section of Chapter 14 of the *Student Workbook* that accompanies this text.

Application 1: Whom Do Regulators Serve? The Case of the Education Industry*

Economists and business administrators alike have long been fascinated by the regulatory process. In particular, they have wondered whether government regulators of an industry primarily serve the consumers of that industry's product (as is invariably alleged by the regulators) or whether these regulators really are "captured" by and serve, at the expense of consumers, those who supply resources to that industry. Recently that question was put to the test with respect to the primary and secondary public-education industry in the United States.

Most states have a state board of education and a state superintendent who regulate public education by prescribing curricula, teacher qualifications, textbook selections, the length of the school day and year, and more. Officially, the regulation serves to create "equal opportunities" for the consumers of education, but could it be that it really serves the interests of teachers and school administrators?

A study of 47 states during the 1969–1970 school year revealed one basic difference among the states: in some of them, the board of education and superintendent were *elected;* in others, they were *appointed.* It was hypothesized that the elected regulators would be more responsive to the wishes of consumers (who must elect them) and that the appointed regulators would be more responsive to the wishes of education "producers." A number of multiple-regression equations were fitted to the data, with the following three results, among others (*t* values appear in parentheses underneath the estimated regression coefficients, while * and ** indicate significance at the .05 or .01 level, respectively):

$$1. \quad \hat{Y} = 35.062 - 2.141\,D_1 - 1.967\,D_2 - .064\,X_1 - .096\,X_2 + .223\,X_3$$
$$\phantom{1. \quad \hat{Y} = 35.062}(1.718)^* \quad (1.610) \quad (.261) \quad (2.824)^{**} \quad (4.738)^{**}$$
$$+ .003\,X_4 - 2.727\,X_5 + .005\,X_6$$
$$(2.394)^* \quad (2.892)^{**} \quad (3.333)^{**}$$
$$R^2 = .746$$

In this first case, the *dependent variable, Y,* is the ratio of state private to public school enrollment; it is used as an indirect measure of the consumer's benefit from public education (the larger is the ratio, the smaller is that benefit). The *independent variables* are as follows:

D_1 = dummy variable, = 1 if state board members are elected, = 0 if appointed;
D_2 = dummy variable, = 1 if state superintendent is elected, = 0 if appointed;
X_1 = length of term (in years) of state board members;
X_2 = percentage of state educators unionized;
X_3 = percentage of state population Catholic;
X_4 = median income of state population;
X_5 = median number of years of schooling of state's adult population;
X_6 = the number of school districts in state.

Partial interpretation. The election of state board members (D_1 = 1) subtracts 2.141 from the estimated private- to public-school enrollment ratio, all else being

---

Source: Eugenia Froedge Toma, "Institutional Structures, Regulation, and Producer Gains in the Education Industry," *Journal of Law and Economics* (April 1983), pp. 103–16.

equal. This result suggests *greater consumer benefits from elected than from appointed members* ($D_1 = 0$). The election of the state superintendent ($D_2 = 1$) subtracts 1.967 from $\hat{Y}$, with the same implication.

2. $\hat{Y} = -589.575 - 100.385\,D_1 - 47.724\,D_2 + 23.560\,X_1 - 2.006\,X_2 - 1.087\,X_3$
$\qquad\qquad\quad (2.355)^* \qquad (1.195) \qquad (2.772)^{**} \quad (2.016)^* \qquad (.552)$
$\qquad\quad + .221\,X_4 + 57.755\,X_5 - .064\,X_6$
$\qquad\quad\;\; (5.073)^{**} \qquad (1.835)^* \quad (1.175)$

$\quad R^2 = .741$

In this second case, the *dependent variable, Y,* is the per-pupil expenditure; it is used as a measure of benefit to producers (the larger is the figure, the greater is the benefit to educators and administrators). The *independent variables* are the same as those listed above for the first case.

Partial interpretation. The election of state board members ($D_1 = 1$) or of the state superintendent ($D_2 = 1$) reduces the per-pupil expenditure by $100 or $48, respectively. These results suggest *smaller producer benefits from elected than from appointed members* ($D_1 = D_2 = 0$). The longer is the term of these officials (X_1), the greater is the per-pupil expenditure (producer benefits).

3. Other equations used "number of administrators per pupil," "pupils per teacher" (work load), and "average teacher salaries" as dependent variables. Each time, the results were the same: all else being equal, appointed regulators tended to favor producers by giving them larger budgets, or more administrative staff, or lower work loads, or larger salaries (all the while providing lower benefits to consumers) than elected ones.

"This may explain," concluded the author, "why professional educators continuously advocate removing policymakers from the 'political' sphere. What better way is there to assure benefits for themselves? Educational testing scores suggest that they may be the only ones who are benefiting."

Application 2: Housing Prices and Proposition 13*

On June 7, 1978, California voters approved a statewide property-tax-limitation initiative, known as Proposition 13. This initiative led to a substantial *and differential* reduction in property taxes among localities, and one would expect corresponding impacts on housing prices.

A multiple regression equation was estimated from data on the San Francisco Bay Area, relating the change in post-Proposition 13 mean house prices, *Y,* to a number of independent variables (*t* values appear in parentheses):

$\hat{Y} = -.171 + 7.275\,X_1 + .5468\,X_2 + .00073\,X_3 + .0638\;X_4$
$\qquad\;\; (3.21) \quad (2.97) \qquad (2.32) \qquad (1.34) \qquad (3.26)$
$\qquad\qquad\quad - .0043\,X_5 + .857\,X_6$
$\qquad\qquad\quad\;\; (2.24) \qquad (1.80)$

$\quad \bar{R}^2 = .89.$

---

*Source: Kenneth T. Rosen, "The Impact of Proposition 13 on House Prices in Northern California: A Test of the Interjurisdictional Capitalization Hypothesis," *Journal of Political Economy* (February 1982), pp. 191–200.

The *independent variables* in the above equation were defined as follows:

X_1 = post-Proposition 13 decrease in property-tax bill on mean house;
X_2 = mean square footage of house;
X_3 = median income;
X_4 = mean age of house;
X_5 = transportation time to San Francisco;
X_6 = housing-quality index.

Partial interpretation. As the coefficient of X_1 indicates, each $1 decrease in relative property taxes increased relative property values by about $7. Thus, "the results of this regression provide strong confirmation that the differential interjurisdictional tax reductions of Proposition 13 were partially capitalized in the year following the effective date of the statewide initiative. The capitalization rate implied by this equation is about 7, which is precisely the magnitude that one would expect given an interest rate of about 12–15 percent."

[Anyone who expects an eternal annual return of $1 (for example, as a result of a $1 property-tax reduction) will find that a one-time receipt of $7 is precisely equivalent to this future flow of returns, provided the applicable interest rate is (1/7)100, or 14.29 percent. This is so because $7, when invested at this rate, will generate a $1 return annually, forever. For a more complete discussion of capitalization, the reader may wish to consult Heinz Kohler, *Intermediate Microeconomics: Theory and Applications,* 2nd ed. (Glenview, Ill.: Scott, Foresman and Co., 1986), chapter 15.]

Application 3: The Price of Heroin and the Incidence of Crime*

It has often been claimed that there is an important relationship between drug abuse and crime. This theory was tested recently with the help of monthly data from New York City in the early 1970s. A number of multiple regression equations were estimated, relating the incidence of various crimes (Y) to various independent variables, such as the per-gram retail price of heroin (X_1), the average temperature (X_2), and a time trend (X_3). The table on page 630 summarizes selected results (*t* values appear in parentheses).

Partial interpretation. The data in each row can be used to formulate a regression equation, such as, for murder,

$$\hat{Y} = 51.66 + 1.45\,X_1 + .04\,X_2 + .05\,X_3.$$

Focusing on the b_1 column, we invariably find higher heroin prices being associated with higher numbers of crimes. All else being equal, a $1-per-gram increase in the retail price of heroin, for instance, was associated with and possibly even caused one additional murder, fifty-nine additional robberies, and forty-six additional auto thefts per month.

Source: George F. Brown, Jr. and Lester P. Silverman, "The Retail Price of Heroin: Estimation and Applications," *Journal of the American Statistical Association* (September 1974), pp. 595–606.

Crime	a = Constant	b_1 = Coefficient of X_1	b_2 = Coefficient of X_2	b_3 = Coefficient of X_3	R^2
Murder	51.66	1.45 (2.89)	.04 (.22)	.05 (.07)	.523
Rape	113.26	.07 (.10)	.81 (3.03)	3.84 (4.11)	.670
Robbery	6,351.3	58.67 (2.24)	−14.18 (−1.40)	−87.84 (−2.48)	.314
Aggravated Assault	1,188.3	9.88 (1.73)	19.74 (8.94)	9.52 (1.23)	.813
Burglary	13,251.0	61.19 (1.68)	25.11 (1.79)	−213.05 (−4.33)	.605
Larceny over $50	9,220.7	15.13 (.51)	49.83 (4.33)	−202.09 (−5.02)	.806
Auto Theft	6,235.3	46.07 (2.38)	26.58 (3.55)	−157.92 (−6.04)	.771
Total[a]	36,412.0	192.48 (1.83)	107.93 (2.66)	−647.50 (−4.56)	.655

[a]Total includes above plus manslaughter, simple assault, and larceny under $50.

Application 4: An Economic Interpretation of Congressional Voting*

Many analysts have adduced a variety of explanations for the alleged fact that congressmen and congresswomen often do *not* vote for the interests of the majority of their constituents. Others, on the contrary, believe that profound changes in congressional voting patterns during the twentieth century can be traced precisely to corresponding changes in the economic interests of constituents. Recently, the controversy was subjected to an empirical test.

One multiple-regression equation regressed SPEND/TAX (the ratio of federal government expenditures in a state to the federal tax burden on the state—a proxy for the state's benefit from federal programs) against *HHINC* (median household income in a state), *MFG* (the percent of a state's nonagricultural labor force in manufacturing), and *URB* (the percent of a state's population in urban areas). The result (with t values in parentheses) had an $R^2 = .58$.

$$SPEND/TAX = -11.8 \ HHINC - 1.22MFG + .20 \ URB$$
$$(-5.5) \qquad\qquad (-3.8) \qquad\quad (.8)$$

*Source: Sam Peltzman, "An Economic Interpretation of the History of Congressional Voting in the Twentieth Century," *The American Economic Review,* September 1985, pp. 656–75.

Conclusion: The federal budget has tended to redistribute benefits away from states with high incomes and large manufacturing sectors.

A second multiple-regression equation described voting patterns in the Senate. It regressed *ADA* (the senators' ratings by the pro-spending Americans for Democratic Action) against *HHINC, MFG, URB, METRO* (the percent of state population in standard metropolitan statistical areas) and *DEMS* (the number of Democratic senators from the state divided by two). The result (with *t* values in parentheses) had an $R^2 = .51$.

$$ADA = .65\,HHINC + .88\,MFG + .03\,URB - .13\,METRO + 29.5\,DEMS$$
$$(3.4) \qquad\quad (2.4) \qquad (.1) \qquad (-.5) \qquad\quad (3.9)$$

Conclusion: The characteristics clearly *negatively* correlated with net spending benefits (income and manufacturing) have been *positively* correlated with voting for larger federal spending. There has been a perverse connection between the interests of constituents and the votes of their senators.

Note: Additional analysis suggests a fascinating trend. As the South has become relatively better off economically and the North relatively worse off, the "price" of liberal votes (hurting constituents at home through the redistribution effect noted earlier) has risen in the South, but fallen in the North. This explains ever fewer liberal votes from Southern and ever more liberal votes from Northern members of Congress and, ultimately, the gain in "market share" by Republicans in the South and Democrats in the North.

ANALYTICAL EXAMPLE

14.1 Advertising and the Demand for Electricity

Following the 1973 oil embargo, a variety of measures were taken in the United States to decrease the growth of demand for electric power. One of these measures involved advertising to conserve energy. Yet questions arose: Where was the *incentive* for profit-seeking power companies to urge their customers to use less electricity? What was the *effect* of such advertising; was demand in fact sensitive to it?

Recently these questions were addressed. One of several multiple regressions related *Q* (the quantity of kilowatt hours sold per residential customer) to P_e (the constant-dollar average price of electricity), to P_{ng} (the constant-dollar average price of natural gas), to P_{fo} (the constant-dollar average price of fuel oil), to *PCI* (the constant-dollar per capita income in a given company's home state), to *H* (heating degree days in the state), to *C* (cooling degree days in the state), to *A* (constant-dollar advertising expenditures), and *EX* (a measure of excess capacity), all data expressed in natural logarithms. The result (with *t* values in parentheses) had an $R^2 = .522$.

$$\ln Q = 4.98 \;\; - .772 \ln P_e - .054 \ln P_{ng}$$
$$(2.15) \;\; (-11.58) \;\; (-.58)$$

$$+ .096 \ln P_{fo} + .136 \ln PCI + .395 \ln H$$
$$(.72) \qquad\quad (.48) \qquad\quad (1.31)$$

$$+ .220 \ln C - .089 \ln A + .003 \ln EX$$
$$(1.13) \qquad (-3.02) \qquad (2.53)$$

Given the highly significant coefficient of ln *A*, the authors concluded that conservational advertising by electric power companies did promote the conservation of electricity in the postembargo era. (It was also shown that regulated firms did have an incentive to engage in demand-reducing advertising so long as the market demand curve intersected the regulated price in the region of increasing average costs.)

Source: David L. Kaserman and John W. Mayo, "Advertising and the Residential Demand for Electricity," *Journal of Business,* October 1985, pp. 399–408.

14.2 Explaining the Changing Prices of Farm Land

One economist recently set out to explain the rise and fall of land prices. Using 1955–81 U.S. data from the 48 contiguous states, a multiple regression was estimated that related PFL (the percentage change in the price of farm land) to $3YRDEF$ (a 3-year moving average of the percentage change in the GNP deflator[a]—a proxy for expected inflation), to $3YRNR$ (a 3-year moving average of real farm net returns[b]—a proxy for the expected growth in such returns), and to RTB (the percentage change in the difference between the nominal Treasury Bill rate and the expected rate of inflation—a proxy for earnings available on alter-

native investments). The resulting regression (with t values in parentheses) had an $\bar{R}^2 = .49$.

$$PFL = 2.981$$
$$(2.18)$$

$$+ 1.398 \ \ 3Y\dot{R}DEF + .196 \ \ 3Y\dot{R}NR + .001 \ R\dot{T}B$$
$$(4.54) \qquad\qquad (2.65) \qquad\qquad (.15)$$

Thus the percentage change in the price of farmland was explained (in part) by the expected rate of inflation and the expected growth in real farm net returns. Given the highly significant coefficient of $3YRDEF$, farmland was a perfect hedge against inflation.

[a]For a discussion of moving averages, see Chapter 15; for a discussion of the GNP deflator, see Chapter 16.
[b]Defined as sales revenue plus government payments minus variable costs, all deflated by the GNP deflator.

Source: Michael T. Belongia, "Factors Behind the Rise and Fall of Farmland Prices: A Preliminary Assessment," *Federal Reserve Bank of St. Louis Review,* August-September 1985, pp. 18–24.

14.3 Testing the Crowding-Out Effect

Economists have long argued that massive government deficits, such as the $200 billion U.S. federal deficit of 1985, will "crowd out" private demand. The mechanism is fairly simple: The massive government borrowing drives up interest rates on government securities, this entices funds to flow from the private to the public sector, this drives up interest rates on private borrowing, which discourages private consumption and investment demand.

Recently, the proposition was tested with the help of 1970–82 U.S. data. A multiple regression was estimated that related PR (the rate of change of the prime interest rate) to inflation, INF (the rate of change of the consumer price index), to MS (the rate of change of the M1 money stock), and to TB (the rate of change of the interest rate on 3-month Treasury Bills). The resulting regression (with t values in parentheses) had an $\bar{R}^2 = .64$.

$$\dot{PR} = 3.336 + .3256 \ I\dot{N}F - 3.248 \ \dot{M}S + .68911 \ \dot{T}B$$
$$(1.35) \qquad (-2.85) \qquad (8.75)$$

As the highly significant coefficient of $\dot{TB}$ indicates, government borrowing has had a considerable impact on private-sector interest rates. To the extent that private spending is sensitive to interest rates, a very strong argument on behalf of the crowding-out thesis can be made.

Source: Richard J. Cebula, "New Evidence on Financial Crowding Out," *Public Choice,* 46, 1985, pp. 305–9. But note: Oftentimes, small changes in model specification lead to dramatically different results; this is demonstrated in Dennis Placone, Holley Ulbrich, and Myles Wallace, "The Crowding Out Debate: It's Over When It's Over And It Isn't Over Yet," *Journal of Post-Keynesian Economics,* Fall 1985, pp. 91–96.

14.4 Do Safety Caps on Aspirin Bottles Help?

Every year, some 30,000 Americans are killed while using consumer products in the home. Have government regulators of health and safety managed to make a significant difference? The question was recently investigated with respect to the Consumer Product Safety Commission (CPSC) that has taken up work in the 1970s.

One multiple-regression equation using 1949–81 data related the year t home accident rate, HAR_t, to a number of independent variables, including the lagged value of that rate, HAR_{t-1}, real per capita consumption, $RPCC$, a dummy variable, $CPSC$, denoting the existence (1973–81 = 1) or absence (1949–72 = 0) of the Commission, the percentage of children under 5,

%UNDER5, and more. HAR_{t-1} was included because any one year's accident rate is affected not only by newly purchased products but also by the continued use of a large stock of preexisting products that produced last year's accident rate. $RPCC$ was included because rising consumer affluence can be expected to give rise to a greater demand for safety which will reduce the accident rate. Finally, %UNDER5 was included because this demographic group is particularly vulnerable to death by fire, ingestion, poisoning, etc. The equation (with t values in parentheses) had an $R^2 = .97$.

$$HAR_t = 17.79 + .548\ HAR_{t-1} - .002\ RPCC$$
$$(1.79)\quad (315)\qquad\quad (-2.00)$$

$$-\ .333\ CPSC - 53.80\%UNDER5 \ldots \text{and so on.}$$
$$(-.90)\qquad\quad (-1.28)$$

While the effects on the current home accident rate of HAR_{t-1} and $RPCC$ were positive and negative, respectively (as expected), and were significant (given the t-values), the effect of the $CPSC$ was negative (as one would hope), but not much and certainly not significant. Taken at face value, the coefficient of $-.333$ suggested that—all else being equal—the $CPSC$ reduced the 1981 home accident rate, for example, from 9.5 deaths to 9.2 deaths per 100,000.

The same study also related PDR_t (the year t aspirin poisoning death rate of children under 5) to the same variable's lagged value, PDR_{t-1} to real per capita consumption, $RPCC$, to SAFETYCAPS (the fraction of aspirin sold with safety caps), and to $PROD$ (the per capita production of aspirin tablets, which has shown a declining trend as a result of the use of substitutes, such as Tylenol). The equation (with t values in parentheses) had an $R^2 = .95$.

$$PDR_t = 9.500 + .483\ PDR_{t-1} - .002\ RPCC$$
$$(2.17)\quad (2.15)\qquad\quad (-2.00)$$

$$+\ .100\ SAFETYCAPS - .032\ PROD$$
$$(.10)\qquad\qquad\quad (-.31)$$

The SAFETYCAPS coefficient was not statistically significant. There was no downward shift in the poisoning rate attributable to safety caps! This can be explained in large part by a wholly unjustified lulling effect on consumers who thought all was well with the advent of the caps. In fact, an ever increasing percentage of poisonings has come from safety cap bottles (73 percent in 1978); fully half of all poisonings came from bottles with the cap left off, because consumers were tired of grappling with the troublesome caps.

(A similar effect has been noted with respect to auto fatality rates and seat belts. Drivers with seat belts have exercised less care in driving, thus offsetting the seat-belt safety effect.)

Source: W. Kip Viscusi, "Consumer Behavior and the Safety Effects of Product Safety Regulation," *Journal of Law and Economics,* October 1985, pp. 527–53.

14.5 Does Photocopying Harm Authors and Publishers?

Creators and owners of intellectual properties have become increasingly alarmed by new technologies, such as the Xerox machine, that make it easy to copy these properties. Yet some have argued that unauthorized copying of intellectual properties need not be harmful, and may actually be beneficial, because authors and publishers can *indirectly* appropriate revenues from users who are not original purchasers.

The issue was addressed recently by one economist who argued that journals, for example, are most heavily photocopied in libraries and that publishers were getting revenue for this service by *price discrimination* (charging libraries considerably more than individuals). Data on institutional and individual subscription prices for 80 economics journals in 1959 and 1982 were used. A multiple-regression equation was estimated relating P_{LIB}/P_{IND} (the ratio of library price to individual price) to CIT (the number of citations per page received by each journal in 1981 to articles written between 1975 and 1979—a proxy for popularity and, hence, photocopying activity), to D_{PUB} (a dummy variable $= 1$, if the publisher was a commercial firm presumably interested in profit maximization), and to D_{AGE} (a dummy variable $= 1$, if the journal was in existence prior to 1959, the year the Xerox machine was born). The result (with t values in parentheses) had an $\overline{R}^2 = .17$.

$$\frac{P_{LIB}}{P_{IND}} = 1.38 + .0071\ CIT + .578\ D_{PUB} - .160\ D_{AGE}$$
$$(2.14)\qquad (3.36)\qquad (-1.01)$$

The coefficient of CIT is of greatest interest, it has the expected sign (more citations and, presumably, photocopying does lead to a higher ratio of library to individual price), and is highly significant ($t = 2.14$). In addition, the coefficient and *t* value of D*PUB* shows that pricing by commercial publishers is significantly more discriminatory than that of noncommercial publishers. There is also other indirect evidence: In 1959, only 3 of 38 economics journals then in existence price-discriminated between institutions and individuals. In 1983, 59 out of 80 journals did. On the other hand, the ratio of expenditures on periodicals to expenditures on books for American academic libraries was .41 in 1959; it was .88 in 1981.

Source: S. J. Liebowitz, "Copying and Indirect Appropriability: Photocopying of Journals," *Journal of Political Economy,* October 1985, pp. 945–57.

||| CLOSE-UP

14.1 Revenue-Sharing and the Flypaper Effect

It has often been claimed that a disproportionate amount of intergovernmental aid dispensed by higher levels of government is subsequently used by the recipient lower levels of government to increase spending rather than reduce taxes. Revenue-sharing money seems to stick to the public sector like flies stick to flypaper! A ready explanation exists: Bureaucrats have an insatiable appetite for ever-increasing budgets, because bigger budgets (unlike lower taxes) enhance the personal prestige, income, and power of the bureaucrats. When they get additional funds, they will always find some new project on which to spend them. Recently, the flypaper effect was tested with the help of data pertaining to 105 small city governments in Michigan. A multiple regression ($\bar{R}^2 = .633$) was estimated relating governmental capital expenditures (on construction, equipment, and land) to 9 independent variables, one of which was the proportion of the total income of the median voter (private plus revenue-sharing funds) that was provided by revenue sharing. The estimated coefficient was positive (more revenue-sharing funds, more government spending), it was significant (with a *t*-value of 2.087), and its size suggested that a large flypaper effect was occurring.

Source: Paul Gary Wyckoff, "A Bureaucratic Theory of Flypaper Effects," Federal Reserve Bank of Cleveland *Working Paper 8501,* June 1985.

||| BIOGRAPHY

14.1 George Snedecor

George Waddell Snedecor (1882–1974) was born in Memphis, Tennessee. He studied mathematics and physics at the Universities of Alabama and Michigan, then became a professor of mathematics at Iowa State University. While there, he offered the first statistics course and began a famous collaboration with Henry A. Wallace, then editor of *Wallace's Farmer* in Des Moines, but later vice-president of the United States. They were interested in agricultural research, jointly organized a seminar to study multiple regression, and did pioneering work in the use of punch cards and punchcard machines. Jointly, they published *Correlation and Machine Calculation* in 1925, and they established a Mathematical Statistical Service in 1927 and a now famous Statistical Laboratory in 1933.

In 1931, Snedecor had invited Ronald A. Fisher (Biography 9.1) to Ames and their meeting set in motion developments that produced many of the techniques that are discussed in this chapter. Among the visible signs of this development was the founding, at Iowa State, of the first Department of Statistics in the United States, Snedecor's presidency of the American Statistical Association, and the publication by Snedecor of two famous works: *Calculation and Interpretation of Analysis of Variance and Covariance* (1934) and *Statistical Methods* (1937). The latter work, ultimately co-authored with William G. Cochran, went through seven editions before the death of both authors and sold more than 125,000 copies.

Source: O. Kempthorne, "George W. Snedecor," *International Statistical Review* 42(1974):319–21.

Summary

1. *Multiple-regression analysis* is a technique that uses several independent variables (rather than a single one) to estimate the value of a dependent variable; *multiple-correlation analysis* measures the strength of association among all these variables. Unlike simple regression analysis, multiple-regression analysis allows us to exercise statistical control over extraneous factors and to determine the influence of any X on Y for specified constant values of other variables that might affect Y.

2. The techniques of multiple regression are straightforward extensions of those of simple regression. In the presence of two explanatory variables, they yield an estimated multiple-regression equation of the form, $\hat{Y} = a + b_1X_1 + b_2X_2$. The bs are the *estimated partial-regression coefficients;* they give the partial change in Y that is associated with a unit change in one of the independent variables when the other one is held constant. The above three-variable multiple-regression equation (involving Y, X_1, and X_2) can be pictured as a *regression plane* in a three-dimensional scatter diagram. In a manner analogous to that for simple regression, the corresponding equation can be estimated, the standard error of the estimate of Y can be calculated, and various inferences (that are valid only if five crucial assumptions are fulfilled) can be made. Such inferences include the estimation of an average (or of an individual) value of Y for a given X_1 and X_2 and the establishment of confidence intervals for the βs. In making inferences, t values and p values can be used, and the overall significance of a multiple regression can also be tested by means of an analysis of variance. (This analysis of variance is the testing of the hypothesis that *all* the true regression coefficients are zero and, therefore, that *none* of the independent variables helps explain the variation of the dependent one.)

3. *The sample coefficient of multiple determination,* denoted by R^2, is the most important measure of how well an estimated regression plane (or hyperplane) fits the sample data on which it is based. It equals the proportion of the total variation in the values of the dependent variable, Y, that is explained by the multiple regression of Y on X_1, on X_2, and possibly on additional independent variables (X_3, X_4, and so on). The square root of R^2 is sometimes used as an alternative; it is called the *sample coefficient of multiple correlation*. Because R^2 overestimates ρ^2, the corresponding population coefficient, an *adjusted sample coefficient of multiple determi-* nation, $\overline{R}^2$—which is not so biased—is often calculated. Another important measure employed in multiple correlation analysis, paradoxical as it may sound, is the *sample coefficient of partial determination,* which is an index of association between the dependent variable and just *one* of the independent variables, given a prior accounting of the effects of one or more other independent variables. The index measures the proportional decrease in the previously unexplained variation in Y resulting from the addition of another independent variable to the regression model; thus, it can be used to evaluate the potential merit of adding an explanatory variable to the analysis.

4. While the principles remain the same, the calculations involved in multiple-regression analysis become extremely burdensome once the analysis encompasses more than two explanatory variables. Luckily, the modern computer comes to the rescue.

5. Several extensions of basic multiple-regression analysis include the following:

   a. *Stepwise multiple-regression* is a procedure that develops a multiple regression equation in carefully delineated steps, by means of either the forward-selection method or the backward-elimination method. The *forward-selection method* starts with no independent X variables in a model designed to explain the behavior of the dependent Y variable and then adds one X at a time—always the one that provides the greatest decrease in the hitherto unexplained variation in Y. The *backward-elimination method,* in contrast, starts with all potential independent X variables in a model designed to explain the behavior of the dependent Y variable and then eliminates one X at a time—always the one the coefficient of which has that t value, among all the insignificant ones, that is closest to zero.

   b. *Dummy variables,* which can take on only two values—0 or 1—can be used to incorporate qualitative variables in regression analysis.

   c. *Analysis of covariance* is a combination of (i) regression analysis that incorporates quantitative and dummy variables with (ii) an analysis of variance. Two examples are discussed: one in which inclusion of a dummy yields two regression lines with different intercepts but the same slope and another in which the inclusion of a dummy yields two regression lines with different intercepts as well as different slopes.

6. Regression analysis must not be applied blindly. It is important to consider whether the assumptions underlying that analysis are at least roughly valid, lest the inferences made by means of regression analysis become invalid. One common problem, equivalent to a violation of Assumption 1 (concerning the statistical independence of observations) is the presence of *serial correlation,* a situation in which the value of Y (and the associated error term, e) that is observed at one time, t, is correlated with that observed at an earlier time, such as $t - 1$ or $t - 2$. When such is the case, the standard errors of the estimated regression coefficients are being seriously underestimated, which, in turn, makes all kinds of inferences totally wrong. The most common technique of detection is the *Durbin-Watson test.* Assumption 2 (concerning normality), Assumption 3 (concerning homoscedasticity), and Assumption 4 (concerning linearity) can be tested by means of a plot of residuals against their associated regression estimates. Assumption 5 requires us to avoid *multicollinearity,* a high correlation between or among independent variables. The problem can be detected graphically or by studying a variety of correlation coefficients.

7. The business and economics literature is filled with studies that are based on the multiple-regression techniques introduced in this chapter.

Key Terms

adjusted sample coefficient of multiple determination ($\bar{R}^2$)
analysis of covariance (ANOCOVA)
analysis of residuals
autocorrelation
backward-elimination method
binary variables
correlation matrix
dummy variables
Durbin-Watson test
estimated partial-regression coefficient
first-order serial correlation
forward-selection method
heteroscedasticity
indicator variables
multicollinearity
multiple-correlation analysis
multiple-regression analysis
negative serial correlation

positive serial correlation
p value
regression hyperplane
regression plane
sample coefficient of multiple correlation (R)
sample coefficient of multiple determination (R^2)
sample coefficient of partial correlation
sample coefficient of partial determination
sample covariance
second-order serial correlation
serial correlation
specification error
step-down method
step-up method
stepwise multiple regression
time-series data
t ratio
t value

Questions and Problems

The computer program noted at the end of this chapter can be used to work many of the subsequent problems.

1. Reconsider Chapter 13's problem 2. Using the data of Table 13.B and the data on the interest rate charged credit customers, X_2, given below, determine a multiple-regression equation, including t-values, R^2, $\bar{R}^2$, the ANOVA table, and the critical $F_{.05}$.
X_2 (percent per year): 17, 18, 17, 17, 14, 14, 17, 17, 16, 16, 19, 19, 14, 15, 13, 16, 16, 16, 14, 14, 14, 15, 15, 18, 18, 17, 16, 17 (for successive columns).

2. Reconsider Chapter 13's problem 3. Using the data of Table 13.C and the additional data on household income, X_2, given below, determine a multiple regression equation, including t-values, R^2, $\bar{R}^2$, the ANOVA table, and the critical $F_{.05}$.
X_2 (annual household income, thousands of dollars): 19.2, 22.7, 33.0, 17.2, 9.6, 8.2, 11.9, 27.3, 33.0, 14.9.

3. Reconsider Chapter 13's problem 4. Using the data of Table 13.D and the additional data on the selling realtor's sex, D_1, given below, determine a multiple-regression equation, including t-values, R^2, $\overline{R}^2$, the ANOVA table, and the critical $F_{.05}$.
D_1 ($= 0$ if male; $= 1$ if female): 1, 1, 1, 1, 0, 1, 1, 1, 0, 0, 0, 0, 0, 0 (for successive columns).

4. Reconsider Chapter 13's problem 5. Using the data of Table 13.E and the additional data on the price of cereal, X_2, given below, determine a multiple-regression equation, including t-values, R^2, $\overline{R}^2$, the ANOVA table, and the critical $F_{.05}$.
X_2 (dollars per box): 2.20, 2.22, 1.50, 2.20, 1.60, 1.40, 3.00, 3.00, 1.70, 1.70, 2.90, 2.90, 1.70, 2.30 (for successive columns).

5. Reconsider Chapter 13's problem 6. Using the data of Table 13.F and the additional data on age, X_2, given below, determine a multiple regression-equation, including t-values, R^2, $\overline{R}^2$, the ANOVA table, and the critical $F_{.05}$.
X_2 (age in years): 60, 33, 49, 42, 43, 39, 48, 29, 27, 30.

6. Reconsider Chapter 13's problem 7. Using the data of Table 13.G and the additional data on terrain, D_1, given below, determine a multiple regression equation, including t-values, R^2, $\overline{R}^2$, the ANOVA table, and the critical $F_{.05}$.
D_1 ($= 0$ if flat; $= 1$ if mountainous): 1, 1, 1, 1, 1, 0, 0, 0, 1, 1, 0, 0, 0, 0, 0, 0.

7. Reconsider Chapter 13's problem 8. Using the data of Table 13.H and the additional data on hours flown, X_2, given below, determine a multiple regression equation, including t-values, R^2, $\overline{R}^2$, the ANOVA table, and the critical $F_{.05}$.
X_2 (annual hours flown): 56, 62, 190, 350, 22, 200, 75, 120, 53, 72, 198, 205, 118, 299, 21.

8. Reconsider Chapter 13's problem 9. Using the data of Table 13.I and the additional data on the tax rate, X_2, given below, determine a multiple regression equation, including t-values, R^2, $\overline{R}^2$, the ANOVA table, and the critical $F_{.05}$.
X_2 (average federal tax rate, percent): 21, 22, 23, 25, 18, 19, 18, 19, 20.

9. Reconsider Chapter 13's problem 10. Using the data of Table 13.J and the additional data on fertilizer input, X_2, given below, determine a multiple regression equation, including t-values, R^2, $\overline{R}^2$, the ANOVA table, and the critical $F_{.05}$.
X_2 (pounds of fertilizer per acre): 200, 200, 220, 180, 190, 180, 190, 180.

10. Reconsider Chapter 13's problem 11. Using the data of Table 13.K and the additional data on real

per capita income, X_2 given below, determine a multiple regression equation, including t-values, R^2, $\overline{R}^2$, the ANOVA table, and the critical $F_{.05}$.
X_2 (per capita income in thousands of 1972 dollars): 3.39, 3.62, 3.86, 4.32, 4.67, 5.08, 5.48, 5.97, 6.62, 7.33.

11. Reconsider Chapter 13's problem 12. Using the data of Table 13.L and the additional data on sales territory, D_1, given below, determine a multiple regression equation, including t-values, R^2, $\overline{R}^2$, the ANOVA table, and the critical $F_{.05}$.
D_1 (sales territory, $= 0$ if North; $= 1$ if South): 0, 0, 0, 0, 1, 0, 0, 1, 1, 0, 1, 1.

12. Reconsider Chapter 13's problem 13. Using the data of Table 13.M and the additional data on the price of a complementary good, X_2, given below, determine a multiple regression equation, including t-values, R^2, $\overline{R}^2$, the ANOVA table, and the critical $F_{.05}$.
X_2 (price of complementary good, dollars per unit): 50, 45, 40, 39, 38, 25, 31, 70, 20, 70 (for successive columns).

13. Reconsider Chapter 13's problem 14. Using the data of Table 13.N and the additional data on property type, D_1, given below, determine a multiple regression equation, including t-values, R^2, $\overline{R}^2$, the ANOVA table, and the critical $F_{.05}$.
D_1 (property type, $= 0$ if commercial, $= 1$ if residential): 1, 1, 0, 1, 0, 0, 1, 1, 1, 0.

14. Reconsider Chapter 13's problem 15. Using the data of Table 13.O and the additional data on the number of competitors, X_2, given below, determine a multiple regression equation, including t-values, R^2, $\overline{R}^2$, the ANOVA table, and the critical $F_{.05}$.
X_2 (number of competitors): 61, 58, 57, 60, 49, 33, 30, 25, 18.

15. Reconsider Chapter 13's problem 16. Using the data of Table 13.P and the additional data on import quotas, D_1, given below, determine a multiple regression equation, including t-values, R^2, $\overline{R}^2$, the ANOVA table, and the critical $F_{.05}$.
D_1 (import quotas, $0 =$ not in effect; $1 =$ in effect): 0, 0, 0, 1, 1, 1, 1.

16. Reconsider Chapter 13's problem 17. Using the data of Table 13.Q and the additional data on electricity price, X_2, given below, determine a multiple regression equation, including t-values, R^2, $\overline{R}^2$, the ANOVA table, and the critical $F_{.05}$.
X_2 (electricity price, cents per kilowatt hour): 2.1, 2.3, 2.5, 2.6, 2.8, 2.9, 2.9, 4.6, 5.3, 7.9.

17. Reconsider Chapter 13's problem 18. Using the data of Table 13.R and the additional data on

aircraft type, D_1, given below, determine a multiple regression equation, including t-values, R^2, $\overline{R}^2$, the ANOVA table, and the critical $F_{.05}$.
D_1 (aircraft type, 0 = single-engine, 1 = multiengine): 0, 0, 0, 0, 1, 1, 1, 1, 1, 1, 1, 1, 1.

18. Reconsider Chapter 13's problem 57. Using the data of Table 13.U and the additional data on type of plant, D_1, determine a multiple regression equation, including t-values, R^2, $\overline{R}^2$, the ANOVA table, and the critical $F_{.05}$.
D_1 (type of plant, 0 = old; 1 = new): 0, 0, 0, 0, 0, 0, 0, 0, 0, 0, 1, 1, 1, 1, 1, 1, 1, 1.

19. The manager of an art gallery wants to determine the relationship between the auction price of paintings, Y, the number of bidders, X_1, and the type of painting, D_1. Given the data of Table 14.A, determine a multiple regression equation, including t-values, R^2, $\overline{R}^2$, the ANOVA table, and the critical $F_{.025}$.

20. A publisher wants to determine the relationship between the number of book copies sold, Y, the number of pages in the book, X_1, the age of the author, X_2, the advertising expenditure made, X_3, and the number of similar books on the market, X_4. Given the data of Table 14.B, determine a multiple regression equation, including t-values, R^2, $\overline{R}^2$, the ANOVA table, and the critical $F_{.025}$.

21. A textbook publisher wants to determine the relationship between the number of book copies sold, Y, the number of pages in the book, X_1, the number of analytical examples, X_2, the number of end-of-chapter problems, X_3, the ratio of favorable to unfavorable prepublication reviews, X_4, and the nature of the support package, D_1. Given the data of Table 14.C, determine a multiple-regression

Table 14.A

Auction Price, Y (thousands of dollars)	Number of Bidders, X_1	Type of Painting, D_1 (0 = Modern Art; 1 = Old Masters)
150	12	0
147	7	0
299	8	1
500	3	0
840	7	1
77	12	0
22	9	1
179	15	1
200	33	0
20	3	0

equation, including t-values, R^2, $\overline{R}^2$, the ANOVA table, and the critical $F_{.025}$.

22. An engineer wants to determine whether there is a relationship between the tensile strength of plastic sheets, Y, and the respective quantities of ingredient 1, X_1, and 2, X_2. Given the data of Table 14.D, determine a multiple regression equation, including t-values, R^2, $\overline{R}^2$, the ANOVA table, and the critical $F_{.01}$.

23. A marketing manager wants to determine the relationship between average monthly purchases with the company's credit card, Y, the cardholder's family income, X_1, and the cardholder's family size, X_2. Given the data of Table 14.E, determine a multiple regression equation, including t-values, R^2, $\overline{R}^2$, the ANOVA table, and the critical $F_{.01}$.

Table 14.B

Copies Sold, Y (thousands)	Pages, X_1 (number)	Author's Age, X_2 (years)	Advertising Expense, X_3 (thousands of dollars)	Competing Books, X_4 (number)
5	352	22	5	7
13	490	29	9	3
23	250	38	5	20
36	670	67	7	39
57	805	70	8	61
72	390	52	6	10
98	510	50	12	7
102	711	31	22	2
122	905	29	39	5
190	472	75	30	0

Table 14.C

Copies Sold, Y (thousands)	Pages, X_1 (number)	Analytical Examples, X_2 (number)	Problems, X_3 (number)	Favorable to Unfavorable Reviews, X_4 (ratio)	Support Package, D_1 (Study Guide, Intstructor's Manual, Computer Disk, 0 = unavailable, 1 = available)
7	390	0	0	5.1	0
11	472	0	0	4.6	0
14	512	0	0	6.0	0
17	408	15	100	2.0	0
29	610	15	100	3.1	1
39	830	30	900	.2	1
52	310	15	500	.1	1
6	910	92	900	10.3	0
3	211	0	10	5.0	0
2	172	0	0	12.0	0

Table 14.D

Tensile Strength, Y (psi)	Ingredient 1, X_1 (units)	Ingredient 2, X_2 (units)
31.7	12	88
52.0	13	144
45.6	14	126
29.1	15	81
49.3	16	136
37.0	17	102
48.0	18	133

Table 14.E

Average Purchases, Y (dollars per month)	Family Income, X_1 (thousands of dollars per month)	Family Size, X_2 (number of persons)
22.50	6.9	1
33.77	3.1	2
52.41	2.6	3
60.00	3.7	4
68.72	2.4	1
75.53	3.9	2
82.99	4.7	3
91.07	.8	4
98.99	.6	2
107.23	1.7	1

Table 14.F

Per Capita Government Expenditure in County, Y (dollars per year)	Per Capita Income in County, X_1 (thousands of dollars per year)	Population Density in County, X_2 (persons per square mile)
92.70	3.2	52
95.60	4.3	39
110.00	6.2	48
115.29	8.7	27
120.37	9.1	93
130.40	15.6	117
140.82	20.3	985
190.67	25.6	1,033
233.12	29.7	10,611
293.15	30.1	22

24. A politician wants to determine the relationship between per capita government expenditure, Y, in a state's counties, the county per capita income, X_1, and the county population density, X_2. Given the data of Table 14.F, determine a multiple regression equation, including t-values, R^2, $\bar{R}^2$, the ANOVA table, and the critical $F_{.01}$.

25. Compute the *sample standard error of the estimate of Y* for problems 1 through 8.

26. Compute the *sample standard error of the estimate of Y* for problems 9 through 16.

27. Compute the *sample standard error of the estimate of Y* for problems 17 through 24.

28. Reconsider problem 24. Given $X_1 = 15.6$ and $X_2 = 117$, compute a *95-percent prediction interval* for
   a. the average value of Y.
   b. the individual value of Y.

29. Reconsider problem 23. Given $X_1 = 3.9$ and $X_2 = 2$, compute a *95-percent prediction interval* for
   a. the average value of Y.
   b. the individual value of Y.

30. Reconsider problem 22. Given $X_1 = 17$ and $X_2 = 102$, compute a *95-percent prediction interval* for
   a. the average value of Y.
   b. the individual value of Y.

31. Reconsider problem 21. Given $X_1 = 830$, $X_2 = 30$, $X_3 = 900$, $X_4 = .2$, and $D_1 = 1$, compute a *95-percent prediction interval* for
   a. the average value of Y.
   b. the individual value of Y.

32. Reconsider problem 20. Given $X_1 = 390$, $X_2 = 52$, $X_3 = 6$, and $X_4 = 10$, compute a *95-percent prediction interval* for
   a. the average value of Y.
   b. the individual value of Y.

33. Reconsider problem 19. Given $X_1 = 12$ and $D_1 = 0$, compute a *95-percent prediction interval* for
   a. the average value of Y.
   b. the individual value of Y.

34. Reconsider your answer to problem 18. Use it to compute *95-percent confidence intervals for* β_1 *and* β_2.

35. Reconsider your answer to problem 17. Use it to compute *95-percent confidence intervals for* β_1 *and* β_2.

36. Reconsider your answer to problem 16. Use it to compute *95-percent confidence intervals for* β_1 *and* β_2.

37. Reconsider your answer to problem 15. Use it to compute *95-percent confidence intervals for* β_1 *and* β_2.

38. Given $r_{Y\cdot1}^2 = .89$, $r_{Y\cdot2}^2 = .10$, $R_{Y\cdot12}^2 = .95$, what is $r_{Y1\cdot2}^2$ and $r_{Y2\cdot1}^2$?

39. Given $r_{Y\cdot1}^2 = .715$, $r_{Y\cdot2}^2 = .072$, $r_{Y\cdot3}^2 = .011$, and $R_{Y\cdot12}^2 = .845$, what is $r_{Y1\cdot2}^2$ and $r_{Y2\cdot1}^2$?

40. In Step 2 of Practice Problem 12, it is claimed that "if we instruct the computer to regress Y on X_1, then on X_2, and then on X_3, it selects the regression of Y on X_1 as the best alternative." Prove it, after explaining what "best alternative" means.

41. In Step 3 of Practice Problem 12, it is claimed that "if we now instruct the computer to regress Y on X_1 and X_2 and then on X_1 and X_3, it selects the regression of Y on X_1 and X_2 as the best alternative." Prove it, after explaining what "best alternative" means.

42. Write down the three regression equations mentioned in problem 40, including t-values and r^2.

43. Write down the two regression equations mentioned in problem 41, including t-values and R^2.

44. The text section, "Dummy Variables," contains the following multiple-regression equation relating income, Y, to education (X_1) and sex (D_1)

$$\hat{Y} = 5.376 + 1.786 X_1 - 7.103 D_1$$
$$(1.884) \quad (8.410) \quad (-3.655)$$

and notes that

$$R^2 = .84.$$

Note: Numbers in parentheses are t values.
   a. With the help of the Table 14.6 data, check the accuracy of the three estimated regression coefficients. Show your computations.
   b. Determine $s_{Y\cdot X_1 D_1}$, the standard error of the estimate of Y.
   c. Make a point estimate of the annual income of a woman with 10 years' education.
   d. Determine a 99-percent confidence interval for the annual income of the average woman in the population who has 10 years' education.
   e. Determine such a confidence interval for the next woman sampled who has 10 years' education.
   f. Determine 98-percent confidence intervals for β_1 (the population coefficient of X_1) and β_2 (the population coefficient of D_1). What is the meaning of the β_2 interval?
   g. The computer printout for the above regression equation indicates that $ESS = 290.267$ and $F = 44.808$ with 2 and 17 degrees of freedom. Set up an ANOVA table like Table 14.3. Is the regression significant overall?
   h. Use the ANOVA table to determine $R_{Y\cdot X_1 D_1}^2$.

45. Reconsider the section, "Analysis of Covariance." From the information given therein, and the fact that the total sum of squares is 290.297, set up an ANOVA table for the females-only regression, $\hat{Y}_F = 1.751 + 1.486 X_1$. Determine the overall significance of that regression at the $\alpha = .025$ level, and check the value of r^2 given in the text.

46. Given a regression of a dependent variable on 4 independent ones and 25 observations, the Durbin-Watson statistic equals 2.95. Determine, at the $\alpha = .05$ level of significance, whether there is positive serial correlation.

47. Given the information in problem 46, determine, at the $\alpha = .01$ level of significance, whether there is negative serial correlation.

48. An economist collects data for a ten-year period on a good's quantity demanded, Y, in millions of units, on the good's price (X_1), on the price of a substitute good (X_2), and on consumer income (X_3). The following puzzling result appears in the computer (*standard errors* in parentheses):

$$\hat{Y} = 47 - .6\,X_1 + .1\,X_2 + .2\,X_3$$
$$\quad\;\; (.0000)\;(.0000)\;(.0000)$$
$$s_{Y\cdot123} = .6839\,(10)^{-6}$$
$$ESS = .2807\,(10)^{-11}$$
$$R^2_{Y\cdot123} = 1.0000$$
$$F = \infty$$

Can you figure it out?

49. Study problems 1–12, along with your solutions. Do you detect any signs of multicollinearity? Explain.

50. Study problem 13–24, along with your solutions. Do you detect any signs of multicollinearity? Explain.

51. A local tax assessor wants to develop a regression model that can explain the value of houses, Y, measured in thousands of dollars, by the age of houses (X_1), by their square footage (X_2), by the number of bathrooms (X_3), by the absence (0) or presence (1) of an attached garage (D_1), and by the absence (0) or presence (1) of a view (D_2). A random sample of 15 houses is used to gather observations, with these results (*standard errors* in parentheses):

$$\hat{Y} = 64.42 - .5415\,X_1 + .00845\,X_2$$
$$\quad\;\; (40.54)\quad (.5060)\qquad (.02166)$$

$$+ .76\,X_3 - 5.22\,D_1 + 12.16\,D_2$$
$$\;\; (12.01)\quad (18.77)\qquad (18.15)$$

The error sum of squares is 8,337; the total sum of squares is 9,773.
   a. Comment on the significance of the regression coefficients.
   b. Comment on the overall significance of this regression.

52. A firm operating airport facilities throughout the nation wants to develop a regression model that can explain its profit, Y, measured in thousands of dollars per year, by its annual sales of aircraft repair and maintenance services (X_1), by its annual sales of avionics equipment (X_2), and by its annual

sales of hotel and restaurant services (X_3). A random sample of 10 of its facilities is used to gather observations, with these results (*standard errors* in parentheses):

$$\hat{Y} = -2.33 - .0338\,X_1 + .0923\,X_2 + 3.864\,X_3$$
$$\quad\;\; (14.12)\;\;(.1514)\qquad (.0153)\qquad (2.381)$$

The error sum of squares is 866.6; the total sum of squares is 6,326.2.
   a. Comment on the significance of the regression coefficients.
   b. Comment on the overall significance of this regression.
   c. Do you see any evidence of a possible violation of crucial assumptions?

53. Use the U.S. data (in billions of current dollars) in Table 14.G to work the following problems.

Table 14.G

Year	Personal Consumption Expenditures, C	Disposable Personal Income, DPI
1939	67.0	70.0
1940	71.0	75.3
1941	80.8	92.2
1942	88.6	116.6
1943	99.4	133.0
1944	108.2	145.6
1945	119.5	149.1
1946	143.8	158.9
1947	161.7	168.7
1948	174.7	188.0
1949	178.1	187.9
1950	192.0	206.6

Source: Economic Report of the President (Washington, D.C.: U.S. Government Printing Office, 1982), p. 261.

   a. Estimate a simple regression equation of the form $\hat{C} = a + b\,DPI$.
   b. Interpret the coefficient of DPI.
   c. Graph this aggregate consumption function in a scatter diagram, showing war years (1942–45) as crosses, peace years as dots.
   d. Because people's consumption behavior during the war years was obviously different from that during peace years (probably as a result of rationing, moral suasion, dislocation, and the unavailability of goods), estimate two separate

consumption functions, for war years only and for peace years only. Graph them as well and interpret your results.

e. Compute the standard error of the estimate of C for the three equations.

f. Given the following information from a computer printout, test the overall significance of your three regressions by computing F; also compute r^2. For *all years*, $ESS = 1{,}496.547$ and *Total SS* $= 21{,}986.24$; for *war years, ESS* $= 40.02113$ and *Total SS* $= 516.1975$; for *peace years, ESS* $= 76.09928$ and *Total SS* $= 19{,}115.82$.

54. In each case listed, interpret the meaning of the estimated regression coefficients; then assess the overall significance of the regression.

a. One economist ranked U.S. utilities by assets, selected the 25 largest companies, and regressed their 1983 net income, Y (in millions of dollars) against assets, X_1 (in billions of dollars) and employees, X_2 (numbers). The result (t values in parentheses):

$$\hat{Y} = 371.01330 + 8.88493\,X_1 - .00137\,X_2$$
$$\qquad\;\;(3.57713)\quad\;(.57669)\quad(-.60151)$$
$$r_{X_1X_2} = .99497 \qquad R^2_{Y \cdot 12} = .01826$$
$$RSS = 18{,}821.51 \qquad ESS = 1{,}011{,}779$$

b. One economist selected the top 29 companies on the *Fortune 500* list and regressed the 1984 stock trading activity, Y (the percent of the company's outstanding shares traded during the average week) against the company's net profit margin, X_1 (percent) and debt/equity ratio, X_2 (percent). The result (t values in parentheses):

$$\hat{Y} = 1.92869 \quad - .07573\,X_1 - \qquad .00312\,X_2$$
$$\quad\;(7.25846)(-2.41312)\qquad\quad(-1.15834)$$
$$r_{X_1X_2} = .00362 \qquad R^2_{Y \cdot 12} = .21652$$
$$RSS = 2.442634 \qquad ESS = 8.838742$$

55. a. The study noted in Close-Up 14.1, "Revenue Sharing and the Flypaper Effect," provides the following estimated regression coefficients (and t-values): $a = -12.325\,(-1.439)$, $b_1 = -.704$ (-1.434), $b_2 = -1.354\,(-2.719)$, $b_3 = 1.972$ (1.873), $b_4 = 1.079\,(2.087)$, $b_5 = .627\,(3.464)$, $b_6 = -.0000835\,(-.747)$, $b_7 = -.008704$ $(-.665)$, $b_8 = -.00949\,(-.678)$, and $b_9 = -.05642\,(-1.424)$. Compute the coefficients' standard errors.

b. Reconsider Analytical Example 14.1, "Advertising and the Demand for Electricity." What per-

centage of the variation in ln Q is explained by the regression?

56. a. Reconsider Analytical Example 14.2, "Explaining the Changing Prices of Farm Land." What percentage of the variation in *PFL* is explained by the regression? All else being equal, what is the effect of a 2-percentage point increase in expected inflation?

b. Reconsider Analytical Example 14.3, "Testing the Crowding-Out Effect." The source of the regression equation shown there also notes that $F = 26.69$ and $DW = 1.86$. Explain.

57. a. Reconsider Analytical Example 14.4. Let the poisoning death rate of children under 5 equal 5 per 100,000 when 10 percent of aspirin tablets are sold in bottles with safety caps. According to the equation as it stands (and ignoring the issue of significance of its coefficients), what would happen to PDR_t if the percentage changed from 10 to 80?

b. Reconsider Analytical Example 14.5, "Does Photocopying Harm Authors and Publishers?" Let the ratio of library price to individual price be 10/1 when the citations per page equals .5. What would happen to P_{LIB}/P_{IND} if the journal's popularity rose such that .5 became 1.0?

58. Consider the auto-theft multiple-regression equation given in Application 3, "The Price of Heroin and the Incidence of Crime": $\hat{Y} = 6{,}235.3 + 46.07$ $X_1 + 26.58\,X_2 - 157.92\,X_3$.

a. Interpret the coefficients of X_1 and X_2.

b. For $X_1 = \$30$, $X_2 = 100°F$, and $X_3 = 50$, make a point estimate of the number of auto thefts.

59. a. What will happen to the *SPEND/TAX* ratio in Application 4, "An Economic Interpretation of Congressional Voting" if the percent of a state's population in urban areas rises from 50 to 75?

b. Analytical Example 11.2, "Searching for Leviathan," discussed tests to find out whether the size of the economy's public sector is reduced by fiscal decentralization and increased by centralization (see pages 475–476). The sample data for the 48 U.S. states were also used to derive three multiple regression equations:

$$G = -2.9 + .0001Y - .003P - .002U$$
$$\quad + .01I - .006R$$

$$G = -2.8 + .0001Y - .0002P - .002U$$
$$\quad + .008I - .004E$$

$$G = -3.0 + .0001Y - .008P - .002U$$
$$\quad + .007I - .0002L$$

In these equations, *G, R, E,* and *L* are defined as in Analytical Example 11.2; *Y* is state per capita personal income, *P* is state population, *U* is the percentage of a state's population residing in urban areas, and *I* is intergovernmental grants as a percentage of state-local government revenues. Clearly, these equations were estimated in an attempt to control statistically for the influence of other key variables besides *R, E,* or *L* on the size of the public sector, *G.* If the relevant coefficients *were* statistically significant (and they were not), would these equations support the Leviathan thesis?

60. Reread the discussion of paintings in Analytical Example 13.2, "Of Bulls, Bears, and Beauty" (on page 566). The values of *b* given in Figure B were derived from a series of regression equations, one of which (on modern art) was (with *t*-values in parentheses):

$$R_P = .061 + .92\,R_M + 2.70\,I - .37\,D + e$$
$$(3.10) \quad (2.64) \quad (3.11)\ (-4.87)$$

$$R^2 = .80$$
$$DW = 1.45$$
$$F = 13.02$$

In this equation, R_P is the rate of return on paintings in excess of a risk-free rate of return; R_M is the rate of return on the market portfolio in excess of a risk-free rate of return; *I* is the rate of inflation in excess of the expected rate, and *D* is a dummy variable measuring the special (art-price depressing) effects of a 1975 British tax on art and U.S. legalization of gold holdings.

a. Assess the significance of the regression coefficients and of the regression as a whole.

b. Assess the meaning of the coefficient of *I*.

Selected Readings

Allen, D. M., and F. B. Cady. *Analyzing Experimental Data by Regression.* Belmont, Calif.: Wadsworth, 1982.

Chatterjee, S., and B. Price. *Regression Analysis by Example.* New York: Wiley, 1977.

Daniel, C., F. S. Wood, and J. W. Gorman. *Fitting Equations to Data: Computer Analysis of Multifactor Data,* 2nd ed., New York: Wiley, 1980.

Draper, N., and H. Smith. *Applied Regression Analysis,* 2nd ed., New York: Wiley, 1981.

Durbin, J. R., and G. S. Watson. "Testing for Serial Correlation in Least Squares Regression," Parts 1–3. *Biometrika* (1950), pp. 409–28; (1951), pp. 159–78; (1971), pp. 1–20.

Goldfeld, S. M., and R. E. Quandt. "Some Tests for Homoscedasticity." *Journal of the American Statistical Association* (1965), pp. 539–47.

Katz, David A. *Econometric Theory and Applications.* Englewood Cliffs, N.J.: Prentice-Hall, 1982. Chapter 4 discusses how to correct for autocorrelation, heteroscedasticity, and multicollinearity.

Mendenhall, William, and James T. McClave. *A Second Course in Business Statistics: Regression Analysis.* San Francisco: Dellen, 1981. Chapter 5 discusses matrix algebra and its application to multiple regression (as, for example, in the formulas of Box 14.B and 14.C).

Park, R. E. "Estimation with Heteroscedastic Error Terms." *Econometrica* (October 1966), p. 888.

Paul, Chris W., II. "Competition in the Medical Profession: An Application of the Economic Theory of Regulation." *Southern Economic Journal* (January 1982), pp. 559–69. A study similar to Application 1, focusing on selecting licensing boards in the medical industry.

Computer Programs

The DAYSAL-2 personal-computer diskettes that accompany this book contain one program of interest to this chapter:

13. Regression and Correlation Analysis can perform multiple-regression analysis that provides not only regression coefficients, standard errors, and *t*-values, but also a correlation matrix, an ANOVA table, confidence intervals for α and β coefficients, hypothesis tests about them, and prediction intervals, at numerous confidence levels, for average and individual *Y*-values associated with each set of *X*-values originally entered.

You can view a scatter diagram of the association between any two variables, along with all the summary measures of central tendency and of dispersion. You can take advantage of a "live regression" feature, specifying any set of *X*-values, while the computer provides an instant point estimate of *Y*, along with limits of prediction intervals for the average and individual *Y* at your chosen confidence level. You can select follow-up programs of backwards elimination, make Durbin-Watson tests, or multicollinearity tests. You can use data files *Q–T* that contain data on the U.S. economy.

PART FIVE

Special Topics for Business and Economics

CHAPTER FIFTEEN

Time Series and Forecasting

Every organization must plan for the future; businesses and governments are no exception. Implicitly or explicitly, all such planning involves the making of predictions, and such predictions are inevitably linked to experiences gained in the past. An electric power producer, for example, may study how demand has grown in the past decades, project this growth into the future, and on that basis plan the construction of new generating capacity. A clothing manufacturer, likewise, may use past sales data as a guide to what can be expected and place new orders for raw materials and equipment accordingly. A government agency may study the past behavior of national income to estimate its future course and, thus, the tax revenues likely to be available.

The business and economic data about the past that are available to decision makers usually fall into one of two categories; most of them are either cross-section or time-series data. **Cross-section data** are numerical values pertaining to units of different populations that have been observed simultaneously at the same point in time (in the case of a *stock variable*) or during the same period of time (in the case of a *flow variable*). Examples are listings of the closing prices of different securites traded at the New York Stock Exchange on June 15, 1987 or listings of the profits of the different U.S.-based multinationals during the (Jan. 1–Dec. 31) 1981 period (as in Table 3.1 on page 50). Such cross-section data will *not* concern us in this chapter. We will focus instead on **time-series data,** or numerical values pertaining to units of a given population that have been observed repeatedly at different points in time (in the case of a stock variable) or during different periods of time (in the case of a flow variable). Examples are listings of the closing prices of a given security at the end of each of 30 successive business days or listings of the profits of a given multinational company during each of the last 15 years. It is precisely the latter type of chronological sequence of numerical data (each one of which is associated with a different moment or period of time) that becomes the raw material for **time-series analysis,** a statistical procedure that employs time-series data, usually for the purpose of explaining past or forecasting future events.

THE COMPOSITION OF A TIME SERIES

Several models seek to describe the typical time series. The most popular one, no doubt, is the **classical time-series model;** it attempts to explain the pattern observed in an actual time series by the presence of four components: trend, cyclical, seasonal, and irregular components.[1]

The Trend

The **trend component** of a time series is denoted by T and is evidenced as a relatively smooth, progressively upward or downward movement of the variable of interest, Y, over an extended period of time. The trend is viewed as the consequence of long-range gradual change in such factors as population size or composition, technology, or consumer preferences, and is typically computed from data that cover a minimum of 20 years.

As a result of this trend factor, a firm's sales of avionics equipment, for example, may be rising steadily at a rate of 50 units per year. (*Avionics equipment* is electronic instruments used by aircraft, such as communication and navigation units, distance-measuring units, radar transponders, and more.) Figure 15.1 illustrates this trend component in the firm's sales (with trend sales, T_t, measured vertically against time, t, measured horizontally). Note how anyone in 1981 with a knowledge of initial sales (point a) would, on the basis of this trend line, have predicted quarterly sales of 300 units four years hence—that is, in the winter of 1985 (point b). Yet this may not be what actually happened. We also have to take into account the cyclical component in the firm's sales.

Cyclical Fluctuations

The **cyclical component** of a time series is denoted by C and is evidenced by wide up-and-down swings of the variable of interest around the trend, with the swings lasting from one to several years each and typically differing in length and amplitude from one cycle to the next. Such irregular, but recurring, swings are seen to reflect the endless ebb and flow of economic activity in general, the eternal *business cycle* of boom and bust, that is bound to affect any particular variable in the economy.

Returning to our example, consider the influence on avionics sales of alternating sequences of boom and bust in the economy in general. If such business cycles occurred, avionics sales might lie above the trend in good years and below it in bad

---

[1]As will be shown below, all four components are likely to be present in data that pertain to periods shorter than a year, such as weekly, monthly, or quarterly data. When data pertaining to periods of a year or longer are being used, however, the seasonal component disappears from the model.

Figure 15.1   Trend Component in Avionics Sales

The line in this graph illustrates the trend component in a firm's sales that has been determined, we assume, on the basis of data pertaining to at least 20 years, but only some of which are shown here. Taking into account the trend only, sales of 100 units in the winter of 1981 (point *a*) rise by 50 units per year—for example, to 450 units in the winter of 1988 (point *c*). Note: *W, S, S,* and *F* refer to the winter, spring, summer, and fall quarters of the various years.

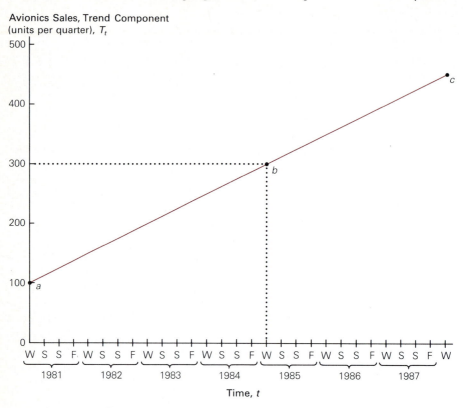

years, as depicted in panel (a) of Figure 15.2. (The cyclical component is measured as a proportion of the trend, which is now depicted as the horizontal line.) Thus, "booms" raise avionics sales to 1.3 times the expected in the winter of 1981 (point *a*), and again to well above the trend in late 1983 (point *c*) and in the winter of 1988 (point *f*). Yet, by the same token, economic "busts" lower sales below their trend, to 82 and 61 percent of the trend, respectively, in mid-1982 and mid-1985 (points *b* and *e*). By combining the information in panel (a) with that given in Figure 15.1, we can produce the trend/cyclical time series of sales given in panel (b). Note how the trend sales for the winter of 1985 (300 units according to point *b* of either Figure 15.1 or Figure 15.2) do not materialize. Panel (a) of Figure 15.2 shows how the (assumed) business cycle trough at that time depresses sales to 64 percent of their trend level (point *d*); thus, trend and cycle together produce sales of 300(.64) = 192 units only, as represented by point *c* in panel (b) of Figure 15.2. Thus, the combined trend/cyclical component of

Figure 15.2 Recognizing a Cyclical Component in Avionics Sales

The influence of the business cycle on a particular time series, shown in panel (a), will make the movement of the series over time diverge from the trend. Compare the color line in panel (b) with the black line representing the trend.

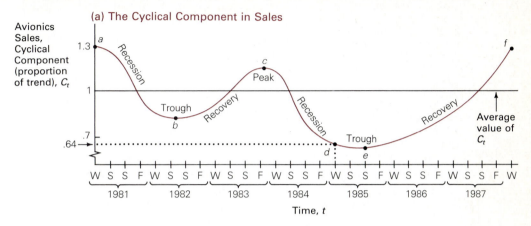

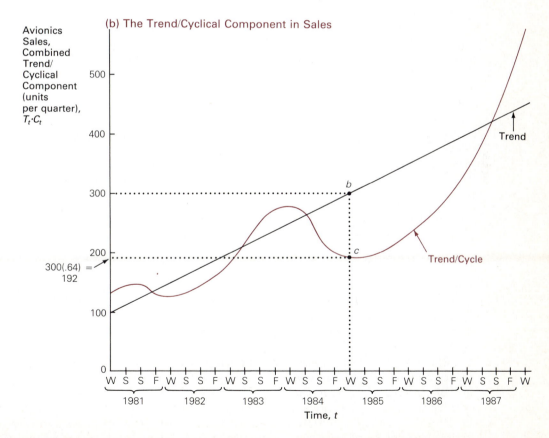

our time series can always be found by multiplying T_t (expressed in physical units sold) by C_t (expressed as a proportion of T_t). Yet trend and cycle do not tell the whole story; seasonal fluctuations play a role as well.

Seasonal Fluctuations

The **seasonal component** of a time series is denoted by S and is evidenced by narrow up-and-down swings of the variable of interest around the trend/cyclical components, with the swings predictably repeating each other within periods of one year or less. Such regularly recurring intrayear variations often reflect the influence of the weather and the calendar on economic activity. Consider how the sales of coal, gas, and oil (or of heavy clothing and snow-removal equipment) rise predictably each winter but fall during other seasons. Consider how the sales of air conditioners, gardening supplies, and swimming pools rise during the spring and summer and stagnate in the fall and winter. Consider the booming activity of tax-preparation firms just before April 15, of resorts after schools let out, of caterers of June weddings, or of sellers of Christmas-tree ornaments before December 24.

Once again we can depict the story graphically, as in Figure 15.3. Panel (a) shows (assumed) seasonal fluctuations in avionics sales, this time around the longer term trend/cyclical component of the time series. Note how winter sales always are 80 percent of average quarterly sales, while spring and summer sales rise to 100 and 120 percent, respectively, with fall sales returning to the average. The color line in panel (b) shows the actual time series, taking into account trend, cycle, and seasonal factors; the trend-only sales and the trend/cycle components are shown in black for comparison. Note how neither the 300-unit trend sales for the winter of 1985 (point b) nor the 192-unit trend/cycle sales (point c) materialize. The assumed seasonal low at that time, represented by point a in panel (a) of Figure 15.3, depresses sales to 80 percent of their trend/cyclical level; thus, trend, cycle, and season together produce sales of $300(.64)(.8) = 154$ units, rounded, which is represented by point d in panel (b) of Figure 15.3. Thus, the combined trend/cycle/seasonal component of our time series can always be found by multiplying T_t (expressed in physical units sold) by C_t (expressed as a proportion of T_t) and by S_t (expressed as a proportion of $T_t \cdot C_t$). Still, there is more to the story than trends, cycles, and seasons, as we will see in the next section.

Irregular Variations

The **irregular component** of a time series is denoted by I and is evidenced by random movements of the variable of interest around the trend/cyclical/seasonal components. Such movements are viewed as arising from completely unpredictable and probably nonrecurring chance events, such as fads in fashion, strikes, natural disasters, or wars.

The effects of such unsystematic influences on our avionics sales might be summarized by panel (a) of Figure 15.4 on page 652 (it being assumed that these factors average to zero in the long run). Note how the winter 1985 sales are only 95 percent of what would be expected as a result of trend, cyclical, and seasonal factors, perhaps because of a strike in the industry (point a). The color line in panel (b) shows the actual time series, taking into account all the components considered in the classical model. This line differs from the color line in panel (b) of Figure 15.3 by the inclusion of the irregular factors. Note how, for the winter of 1985, neither the 300-unit trend

Figure 15.3 Recognizing a Seasonal Component in Avionics Sales

The influence of seasonal factors on a particular time series, shown in panel (a), will make the movement of the series over time diverge from the path that trend alone or trend plus cyclical factors alone would suggest. Compare the color line in panel (b) with the solid and dashed black lines that represent, respectively, trend sales alone or trend/cyclical sales alone.

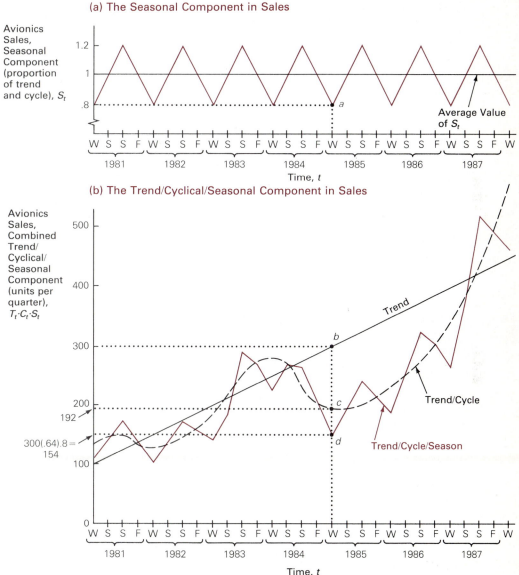

(a) The Seasonal Component in Sales

Avionics Sales, Seasonal Component (proportion of trend and cycle), S_t

(b) The Trend/Cyclical/Seasonal Component in Sales

Avionics Sales, Combined Trend/ Cyclical/ Seasonal Component (units per quarter), $T_t \cdot C_t \cdot S_t$

sales materialize (point b), nor the 192-unit trend/cycle sales (point c), nor even the 154-unit trend/cycle/seasonal sales (point d). Actual sales are 146 units (point e). The winter of 1985 is a bad time for avionics sales, indeed: The recession depresses sales to 64 percent below the trend (from b to c), the winter season, as usual, depresses

Figure 15.4   Recognizing an Irregular Component in Avionics Sales

The influence of irregular factors on a particular time series, shown in panel (a), will make the movement of the series over time diverge from the path that trend, cyclical, and seasonal factors alone would suggest [and that is depicted by the color line in panel (b) of Figure 15.3]. The color line shown here shows the actual avionics sales the company statistician would record.

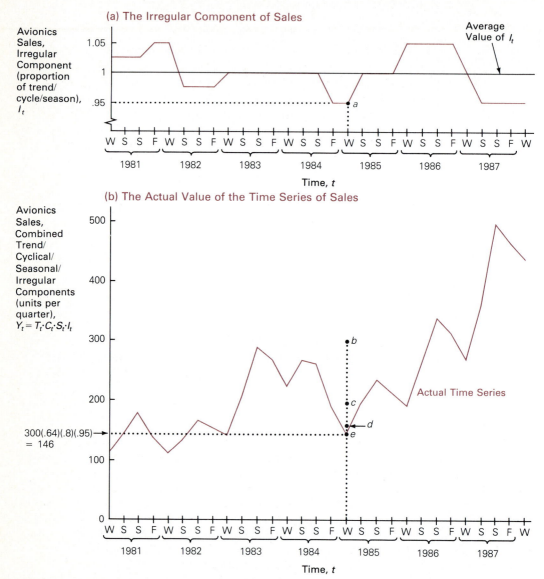

(a) The Irregular Component of Sales

(b) The Actual Value of the Time Series of Sales

sales to 80 percent of that (from c to d), and the strike reduces them further to 95 percent of that (from d to e). Thus, the classical time-series model asserts, the actual value of a time series, Y_t, at time t, can always be found by multiplying its trend value, T_t (expressed in physical units), by C_t (expressed as a proportion of T_t), by S_t (expressed as a proportion of $T_t \cdot C_t$), and by I_t (expressed as a proportion of $T_t \cdot C_t \cdot S_t$). Indeed, for this reason, the classical time-series model discussed here is referred to as a **multiplicative time-series model.**

15.A The Classical Multiplicative Time-Series Model

$$Y_t = T_t \cdot C_t \cdot S_t \cdot I_t$$

Note: The multiplicative model is well suited for a great variety of business and economic data and, therefore, has been widely accepted as the standard for analyzing time series. Yet, on occasion, time series are better represented by alternative models that allow for the possibility of components interacting in additive fashion or in a mixture of additive and multiplicative ways. In the **additive time-series model,**

$$Y_t = T_t + C_t + S_t + I_t,$$

and all the components are expressed in the same physical units. Examples of **mixed time-series models** are

$$Y_t = T_t \cdot C_t + S_t \cdot I_t$$

and

$$Y_t = T_t \cdot C_t \cdot I_t + S_t.$$

Analytical Example 15.1, "The Role of Time Series in Social Experiments," provides one example of the usefulness of time-series analysis; others can be found in the next section.

FORECASTING: EXTRAPOLATING TRENDS

Business and economic **forecasting** is making statements about one or more unknown, uncertain, and, typically, future events. As this section and those that follow will show, many such statements are derived with the help of time-series data. Indeed, the trend component of a time series is often considered the most valuable forecasting tool, especially for long-term projections. If one has available a trend line, similar to that given in Figure 15.1, that summarizes the historical movement of a time series over an extended period of time, it is tempting to predict the future by *extrapolation,* by extending that line beyond the range of past and current data. In the beginning of his or her analysis, however, such a trend line is not available to the forecaster.

The forecaster has available some actual time series, such as the quarterly data embodied in the color line of panel (b) of Figure 15.4, which clearly reflect a mixture of trend, cyclical, seasonal, and irregular influences. To the extent that forecasts are to be made by means of a trend line, however, *annual* data are typically used: they automatically exclude all the seasonal fluctuations (that, by definition, arise within years only) and many irregular fluctuations as well. The first two columns of Table 15.1 might

Table 15.1 The Avionics Sales of Butler Aviation and the Moving-Average Method

Year, t (1)	Units Sold, Y_t (2)	3-Year Moving Average, (3)	8-Year Moving Average Unadjusted (4)	8-Year Moving Average Adjusted (5)
1968	330	—		—
1969	241	257	—	—
1970	200	313	—	—
1971	499	340	—	—
1972	322	440	387	392
1973	500	474	396	437
1974	601	501	478	503
1975	401	468	528	546
1976	401	567	564	586
1977	899	633	607	611
1978	598	761	615	613
1979	787	684	610	643
1980	666	674	675	709
1981	569	599	743	736
1982	561	682	729	762
1983	915	808	794	844
1984	947	884	894	—
1985	791	951	—	—
1986	1,114	1,165	—	—
1987	1,591	—	—	—

represent such an annual time series available to a would-be forecaster. (In order to simplify calculations, we restrict the example to 20 years; in real-life cases, trend cal-culations are preferably performed with much longer time series.) To find a trend line for these data, the forecaster has a variety of methods available.

Moving Averages

One way of portraying a trend is the construction of a **moving-averages series,** which is a series of numbers obtained by successively averaging overlapping groups of two or more consecutive values in a time series and replacing the central value in each group by the group's average. The average is "moving" because an ever-new average is calculated by adding a more recent time-series value to the group and dropping the oldest one.

||| Practice Problem 1

Constructing a 3-Year Moving-Averages Series for the Avionics Sales of Butler Aviation

Consider constructing a 3-year moving average for the time series given in columns 1 and 2 of Table 15.1. The first three numbers (330, 241, and 200) are averaged first, and their average (257) is used to replace the central value of the group (241). This explains the first numerical entry in column 3. Then a new average is calculated, this time after adding to the group of three the 1971 time series value (499), while dropping the 1968 value (330). The (rounded) average of the new group (241, 200, and 499) equals 313, and it replaces the central value of the new group (200), which accounts for the second numerical entry in column 3. And so it goes. |||

Now let us picture what has been accomplished. We graph the original time series, along with the moving-averages series, in Figure 15.5. The moving-average line displays considerably fewer fluctuations, which explains why any statistical procedure that dampens (or averages out) fluctuations in a time series is referred to as a **smoothing technique.** Having used the technique, one is free to imagine that the line of moving averages, such as the color line in Figure 15.5, reflects the systematic movement of the underlying series over time, such as its trend. But caution is advised: Our original series of annual data, by definition, is free of seasonal fluctuations because such fluctuations are intrayear changes. Our series, perhaps, is free of most irregular fluctuations as well because such fluctuations are often short-lived. Given these facts, if our annual data series contained a *regular* 3-year cycle, the averaging out of these recurring cycles by our construction of a 3-year moving-averages series would make sense. Yet cycles are not that regular; their periodicity could vary from 3 years in the 1950s and 1960s to 8 years in the 1970s or 1980s, and in that case we could not conclude that the Figure 15.5 color line has accurately captured the underlying trend. Nor would the result be any more certain if we constructed an 8-year moving-averages series instead, but when we do, we learn two additional lessons.

As we construct moving averages for even-numbered groups of time-series data (rather than odd-numbered ones), the procedure gets more complicated (which is the first lesson). As we construct moving averages for larger groups of data, the resulting series becomes even smoother (which is the second lesson). We can illustrate both of these effects by constructing an 8-year moving-averages series from our data, as shown in the last columns of Table 15.1.

Figure 15.5   A 20-Year Time Series of Avionics Sales and 3-Year Moving Averages

The original 20-year time series of avionics sales—given in columns 1 and 2 of Table 15.1 and plotted here as the black line—shows considerably more fluctuations than a 3-year moving-averages series constructed with its help (color line). For this reason, the construction of a moving-averages series is referred to as a *smoothing technique*. The color line may be interpreted as a trend with the help of which values pertaining to future years (shaded) might be forecast. (Note the projection of the color line by the forecaster to some point such as *a* or *b*.)

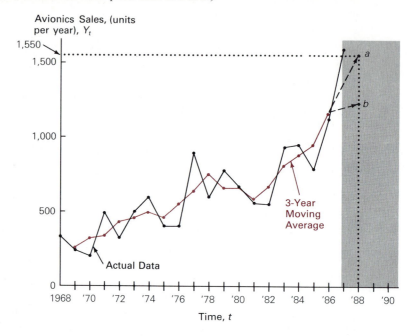

Practice Problem 2

Constructing an 8-Year Moving-Averages Series for the Avionics Sales of Butler Aviation

As we average the first group of 8 data—the 1968–75 values of column 2—we derive an average of 387, but placing it in the center of the group of 1968–75 values puts it right in the middle between 1971 and 1972. To which year does the average belong? The assignment of time-series values to proper times is crucial indeed.[2] In our present case, the calculated average really belongs to a year that includes the second half of 1971 and the first half of 1972; therefore, the entry of 387 in column 4 is printed *between* the other rows of our table. The same is true of all the other, similarly derived entries in that column. The entry of 396, for example (which is the average of the 1969–76 values), belongs to the second half of 1972 and to the first half of 1973. To find the moving average associated with a particular (whole) calendar year, one aver-

---

[2]This fact is vividly illustrated by the Close-Up, "The Importance of Proper Timing" that is contained in the "Supplementary Topics" section of Chapter 15 of the *Student Workbook* that accompanies this text.

ages adjacent averages. Thus, the averaging of 387 (which half belongs to the first half of 1972) and of 396 (which half belongs to the second half of 1972) gives us 392, and that number can be regarded as the 8-year moving average properly belonging to 1972. The remaining entries in column 5 of Table 15.1 have been similarly derived. |||

Figure 15.6 is a graph of the original time series and the 8-year moving-averages series. When compared with Figure 15.5, it clearly shows how the longer-term moving-averages series produces an even smoother picture of long-term change.

Disadvantages of moving averages. Although it is certainly possible to use a moving-averages series as an approximation of the underlying trend component of a time series, there are disadvantages. For one thing, many a moving-averages series still is quite irregular in appearance; the 3-year series of Figure 15.5 exemplifies this fact. Worse yet, it is possible for a moving-averages series to exhibit strong cyclical fluctuations even though there are none whatsoever in the original time-series data. Analytical Example 15.2 describes this so-called *Slutsky-Yule effect.* Indeed, a moving-averages series can produce all kinds of false impressions about the underlying data. To name a few additional ones, the series can lie consistently above or below the original data

Figure 15.6 A 20-Year Time Series of Avionics Sales and 8-Year Moving Averages

The original 20-year time series of avionics sales (given in Table 15.1 and plotted here as the black line) shows considerably more fluctuations than an 8-year moving-averages series constructed with its help (color line). Note that the 8-year series, in turn, is even smoother than the 3-year series depicted in Figure 15.5. Unfortunately, it is also shorter; as a result, it is all the more risky to "cast forward," by means of the dashed lines, the past movement of the (color) "trend" into the (shaded) future.

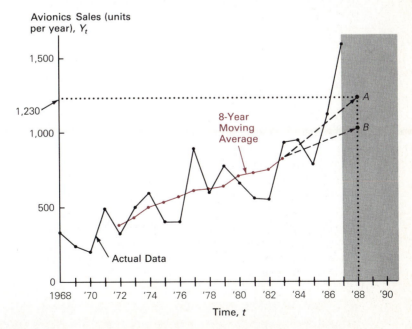

(namely, when they are growing or declining exponentially), it will anticipate or prolong changes in the original data (and thus show a different timing of turning points), it will be extremely sensitive to unusually large or small values in the times series (as any average is bound to be). Worst of all, the very procedure of calculating moving averages makes it impossible to construct this "trend line" for the earliest and latest years of one's time series (because the average centered on any one year must be calculated with the help of a number of preceding and following values). Note how Table 15.1 contains no entry in the first and last rows of column 3, nor in the first and last *four* rows of column 5. Accordingly, the moving-averages lines in Figures 15.5 and 15.6 are considerably shorter than the 20-year period for which the original time-series data are available. This fact leads directly, of course, to the most serious disadvantage of all, and to that we now turn.

Forecasting with moving averages. Suppose we wanted to forecast, in 1987, avionics sales for 1988 or later years. How would we do so with the help of our 3-year moving-averages series (that ends in 1986) or with the help of our 8-year moving-averages series (that ends in 1983)? Would we dare project our color "trend line" straight into the (shaded) future region depicted in our graphs? Note how such a projection of the 3-year moving-averages series brings us to point *a* in Figure 15.5, leading us to forecast 1988 trend sales of 1,550 units, if we project the last segment of the color line by means of the upper dashed arrow. (This procedure is nothing more complicated than extending the last segment of the color line, while leaving its slope unchanged.) Yet we project sales, perhaps, to point *b*, if we take into account the generally lower slope of the entire color line. On the other hand, a projection of the last segment of the 8-year moving-averages series brings us to point *A* in Figure 15.6, leading us to forecast 1988 trend sales of 1,230 units, while a projection of the general slope of that line leads us, perhaps, to point *B*. Even assuming we had good reason to choose one of these "trend lines" over the other, how reliable would such a forecast be? It would be excellent if one could assume that the behavior of the "trend" observed in the past will not change in the near future, but such is never true with certainty. Close-Up 15.1, "The Dangers of Extrapolation," issues an important warning on the subject.

Exponential Smoothing

Much of the uncertainty of making a forecast that is connected with the moving-averages method is eliminated by another and much more popular forecasting procedure, called **exponential smoothing.** It operates like a thermostat and produces self-correcting forecasts by means of a built-in adjustment mechanism that corrects for earlier forecasting errors. We will consider two versions of the procedure.

Single-parameter exponential smoothing. The simplest approach to exponential smoothing produces the next period's forecast value, F_{t+1}, directly from the current actual value of the time series, Y_t, and the current forecast value, F_t, by using a **smoothing constant,** α, as shown in Box 15.B.

15.B Single-Parameter Exponential Smoothing

$$F_{t+1} = \alpha Y_t + (1 - \alpha)F_t.$$

The parameter, α, is a value betwen 0 and 1 that is chosen by forecasters to indicate the weight they want to attach to the most recent value of the time series. As long as only one such parameter is being employed, the procedure is called **single-parameter exponential smoothing.**

||| Practice Problem 3

Predicting Future Avionics Sales Using Single-Parameter Exponential Smoothing

Consider yourself in 1968 ($t = 1$) knowing only that current avionics sales equal $Y_1 = 330$ units, that current sales had been forecast as $F_1 = 320$ units, and that the current sales for 1968, thus, exceed the forecast by 10 units, as in row 1 of Table 15.2. If you choose $\alpha = .20$ as your smoothing constant, you can use the Box 15.B equation to forecast the 1969 value as

$$F_2 = \alpha\, Y_1 + (1 - \alpha)F_1 = .20(330) + (1 - .20)320 = 322.$$

Table 15.2 Forecasting Avionics Sales by Single-Parameter Exponential Smoothing

Year, t (1)	Actual Sales, Y_t (2)	$\alpha = .20$		$\alpha = .50$	
		Forecast Sales, F_t (3)	Forecasting Error, $Y_t - F_t$ (4) = (2) − (3)	Forecast Sales, F_t (5)	Forecasting Error, $Y_t - F_t$ (6) = (2) − (5)
1968	330	320	10	320	10
1969	241	322	−81	325	−84
1970	200	306	−106	283	−83
1971	499	285	214	242	257
1972	322	328	−6	371	−49
1973	500	327	173	347	153
1974	601	362	239	424	177
1975	401	410	−9	513	−112
1976	401	408	−7	457	−56
1977	899	407	492	429	470
1978	598	505	93	664	−66
1979	787	524	263	631	156
1980	666	577	89	709	−43
1981	569	595	−26	688	−119
1982	561	590	−29	629	−68
1983	915	584	331	595	320
1984	947	650	297	755	192
1985	791	709	82	851	−60
1986	1,114	725	389	821	293
1987	1,591	803	788	968	623
1988	—	961	—	1,280	—

This value is the second entry in column 3. Note that the previous error of $+10$ has raised the subsequent forecast by α times the error, or by $+2$. (Thus, the next forecast is being adjusted by a fraction of and in the direction of the previous forecast error). As in Table 15.1, you might find actual 1969 ($t = 2$) sales of 241, however. Accordingly, you would record an error of -81 and make a new forecast for 1970 of

$$F_3 = \alpha\, Y_2 + (1 - \alpha)F_2 = .20(241) + (1 - .20)322 = 305.8.$$

Once again, you would have adjusted the previous forecast (of 322) by α times the previous error (of -81), or by -16.2 in this case. The remainder of the table can be filled out in similar fashion; all entries have been rounded. Note: What if no forecast had existed for 1968? Then the actual 1968 value of 330 would have been used as the 1969 forecast, and the equation would have been employed thereafter. ▮

Fine-tuning the smoothing constant. As an inspection of column 4 of Table 15.2 indicates, many of our forecasts were fairly wide off the mark; that is, they differed significantly from actual sales. Whenever this happens, forecasters adjust the value of α, increasing it when they want to give more weight to current experience (and less to the past) or decreasing it when they want to do the opposite. By trial and error, they can in this way select a smoothing constant that minimizes the forecasting error for the particular time series with which they are working. (Typically, a very stable time series calls for the use of a large α; an extremely volatile one calls for a smaller α.) Columns 5 and 6 of Table 15.2 indicate the forecasting record that would have been achieved with $\alpha = .5$, and Figure 15.7 shows graphically how actual and forecast values would have diverged under the two alternative procedures.

Advantages of exponential smoothing. The popularity of the exponential-smoothing method arises from two facts: (1) the calculations are simple, and (2) next to no data storage is required. To appreciate the latter, consider that only one number (namely F_{t+1}, the most recent forecast made) has to be remembered, and only until a new forecast has been made, because any given forecast embodies the entire time series in it. As we have noted,

$$F_{t+1} = \alpha Y_t + (1 - \alpha)F_t$$

but, in turn,

$$F_t = \alpha(Y_{t-1}) + (1 - \alpha)F_{t-1}.$$

Thus,

$$\begin{aligned} F_{t+1} &= \alpha Y_t + (1 - \alpha)[\alpha(Y_{t-1}) + (1 - \alpha)F_{t-1}] \\ &= \alpha Y_t + \alpha(1 - \alpha)Y_{t-1} + (1 - \alpha)^2 F_{t-1}. \end{aligned}$$

If we continued this breakdown for ever earlier periods, we would discover all the preceding Y values (Y_t, Y_{t-1}, Y_{t-2}, Y_{t-3}, and so on) hidden in the most recent forecast, and they would appear with successive weights of α, $\alpha(1 - \alpha)$, $\alpha(1 - \alpha)^2$, $\alpha(1 - \alpha)^3$, . . . , and so on. These weights decrease *exponentially* (the more distant the value of the time series, the less its forecasting weight), which explains the name of this technique.

To sum up, the exponential-smoothing technique produces a weighted average of all past time-series values (with weights decreasing exponentially as one goes back in

Figure 15.7 Avionics Sales, Actual and Single-Parameter Exponentially Smoothed

This graph indicates how an actual time series might be smoothed exponentially—in this case, by employing a smoothing constant of $\alpha = .2$ or $\alpha = .5$, respectively. The method can also be employed to make forecasts—for example, of *A* or *B* in 1988.

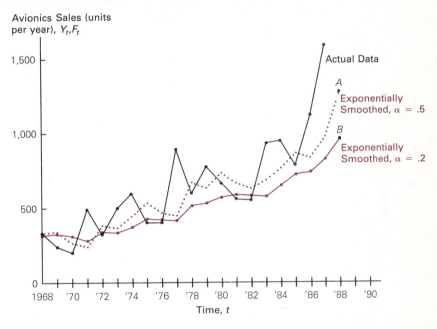

time), and the average so constructed serves as a forecast for the next period. The method is cheap and fast and is often employed when multiple forecasts for a large set of items are required—as, perhaps, in a firm's inventory-control system, which might require demand forecasts for tens of thousands of items.

Two-parameter exponential smoothing. Whenever there is a strong upward trend in a time series, single-parameter exponential smoothing can be relied upon to produce generally too-low forecasts, as is evident from an inspection of Figure 15.7. **Two-parameter exponential smoothing** is a technique that might eliminate this problem by explicitly taking into account the influence of the trend. The technique employs three equations: (1) to produce a smoothed time-series value, Y_t^*, for the current time period, (2) to produce a smoothed trend-component value, T_t^*, for the current time period, and (3) to produce a forecast, F_{t+1}, for the next time period. These equations are given in Box 15.C.

15.C Two-Parameter Exponential Smoothing

$$Y_t^* = \alpha Y_t + (1 - \alpha)(Y_{t-1}^* + T_{t-1}^*) \qquad (1)$$
$$T_t^* = \gamma(Y_t^* - Y_{t-1}^*) + (1 - \gamma)T_{t-1}^* \qquad (2)$$
$$F_{t+1} = Y_t^* + T_t^* \qquad (3)$$

In the equations in Box 15.C, the parameter γ (the lowercase Greek *gamma*) is the **trend smoothing constant,** also chosen to lie between 0 and 1, while the current trend is defined as the difference between the current and preceding smoothed value of the series, $Y_t^* - Y_{t-1}^*$.

||| Practice Problem 4

Predicting Future Avionics Sales Using Two-Parameter Exponential Smoothing

We can apply the technique to our familiar example, now using Table 15.3. Columns 1 and 2 are copies of Table 15.2; the column 3 entries have been calculated using equation (1) of Box 15.C, arbitrarily using a smoothing constant of $\alpha = .20$ and a value of $Y_2^* = Y_1 = 330$. The column (4) entries have been calculated using equation (2) of Box 15.C, arbitrarily using a trend smoothing constant of $\gamma = .40$ and an initial smoothed trend value of $T_2^* = Y_2 - Y_1 = -89$. All of the above implies (assumed)

Table 15.3   Forecasting Avionics Sales by Two-Parameter Exponential Smoothing ($\alpha = .20$, $\gamma = .40$)

Year, t (1)	Actual Sales, Y_t (2)	Equation (1): Smoothed Sales, Y_t^* (3)	Equation (2): Smoothed Trend, T_t^* (4)	Equation (3): Forecast Sales, F_t (5)	Forecasting Error, $Y_t - F_t$ (6) = (2) − (5)
1968	330	—	—	—	—
1969	241	330	−89	—	—
1970	200	233	−92	241	−41
1971	499	213	−63	141	358
1972	322	184	−49	150	172
1973	500	208	−20	135	365
1974	601	271	13	188	413
1975	401	307	22	284	117
1976	401	343	28	329	72
1977	899	477	70	371	528
1978	598	557	74	547	51
1979	787	662	86	631	156
1980	666	732	80	748	−82
1981	569	763	60	812	−243
1982	561	771	39	823	−262
1983	915	831	47	810	105
1984	947	892	53	878	69
1985	791	914	41	945	−154
1986	1,114	987	54	955	159
1987	1,591	1,151	98	1,041	550
1988	—	—	—	1,249	—

Figure 15.8 Avionics Sales, Actual and Two-Parameter Exponentially Smoothed

This graph indicates how an actual time series might be smoothed exponentially—in this case, by employing two parameters ($\alpha = .2$, $\gamma = .4$). The method can also be employed to make forecasts—as of 1,249 units (point *A*) in 1988.

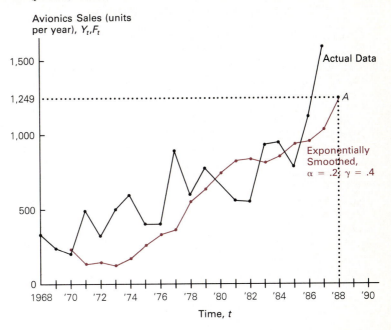

initial values of $F_3 = 241$ and $Y_3 - F_3 = -41$. Now we employ our equations to forecast 1971 sales. Thus, in 1970 ($t = 3$),

$$Y_3^* = .20(200) + (.80)(330 - 89) = 232.8,$$

and

$$T_3^* = .40(233 - 330) + (.60)(-89) = -92.2.$$

Thus, our 1971 forecast is

$$F_4 = 232.8 - 92.2 = 140.6.$$

The remaining entries in Table 15.3 have been similarly calculated (but always rounded to whole numbers). We can also picture our forecasting record, as in Figure 15.8. As the graph shows, and as column (6) of Table 15.3 confirms, that record is not too good and certainly makes us wonder whether we should trust the 1988 forecast of 1,249 units (point *A* in the graph). The use of different smoothing constants might improve that record, however, and this outcome could be determined by trial and error. |||

Other exponential-smoothing procedures. A great variety of other exponential-smoothing procedures exist. These include **three-parameter exponential smoothing** (a method that adds seasonal smoothing to trend smoothing), the adjustment of the smoothing constant, α, from one period to the next, and even the consideration of nonlinear relationships between values.

Figure 15.9 Alternative Trend Lines Suggested by Scatter Diagrams

The black dots in these four panels depict alternative scatters of some variable, Y_t, over time, t. The color lines represent trends that seem to fit the data best. Panel (a) might represent the steady *absolute* growth in a firm's sales (at a rate of 50 units per year). Panel (b) might depict a steady *percentage* growth of sales (at a rate of 2 percent per year), and panel (c) might show a steady percentage decline in sales (at a rate of 3 percent a year). Panel (d) shows a mixture, with growth at an increasing rate followed by growth at a decreasing rate.

Panel (b) depicts well the trend in such time series as the U.S. population, the U.S. GNP, the number of passengers carried by U.S. airlines, the U.S. per capita consumption of synthetic fibers, or that of frozen vegetables. Panel (c) depicts well the trend in such time series as the number of passengers carried by U.S. railroads, the U.S. per capita consumption of cotton, or that of fresh vegetables. Panel (d) might be a so-called Gompertz or logistic (Pearl-Reed) growth curve; it depicts the production and sales behavior of many a new product that experiences spectacular growth, followed by market saturation.

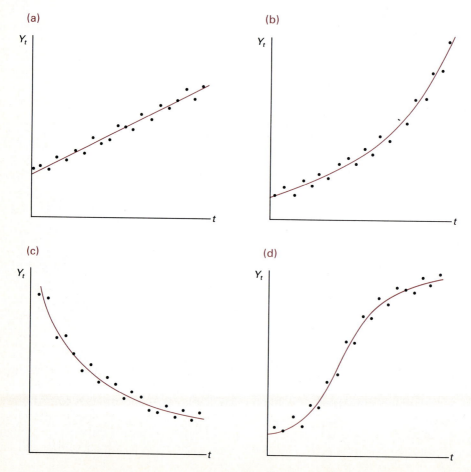

Fitting Regression Lines

It is also possible to estimate a trend, and then to make forecasts with its help, by fitting a least-squares regression line to the available time-series data. This approach is dangerous business, however, because the crucial assumptions underlying that procedure (and noted in Chapters 13 and 14) may well not be applicable. In that case, unbiased regression coefficients can, perhaps, still be estimated, but the making of all kinds of valid inferences ceases to be possible. Yet such inferences would be particularly desirable in a forecasting situation: Precarious as it is, the projection of the dependent variable (for example, of avionics sales), by means of a regression line, to time periods beyond the range of data from which that line was calculated is in this case the whole point of estimating the regression line in the first place. Thus, one would want to have reliable confidence intervals for any estimated future values at the very time that such intervals are unlikely to be available. Nevertheless, because the procedure is frequently employed, let us consider how regression analysis might be used in the service of forecasting.

In all cases, good judgment suggests first drawing a scatter diagram of the variable of interest against time. Such a diagram will suggest a basic type of trend line that is likely to fit the data best; Figure 15.9 shows several possibilities. The black dots in Figure 15.10, in turn, constitute a scatter diagram for our avionics-sales data; we will illustrate the fitting of a linear regression line to these data presently.

Figure 15.10 Scatter Diagram of Avionics Sales and Fitted Regression Line

This scatter diagram is based on the data in columns 1 and 2 of Table 15.4; the solid color line was fitted by the method of least squares. With its help, one would forecast 1990 trend sales of 1,234 units (point *a*).

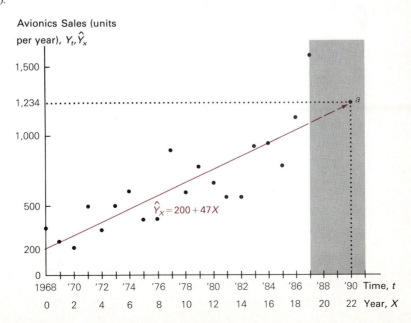

Letting a Linear Regression Line Approximate the Trend in Avionics Sales

The values pertaining to our example are given in columns 1 and 2 of Table 15.4. Typically, the time values (that serve as the independent variable) are transformed or coded as shown in column 3, and a regression line such as the following is then estimated:

$$\hat{Y}_X = a + bX$$

In this equation, $\hat{Y}_X$ is not equal to the actual time-series value (in our case, actual sales, Y_t) but is the computed or trend value. The calculations proceed according to the Box

Table 15.4 Fitting a Trend Line by Linear Regression

Year, t (1)	Units Sold, Y_t (2)	Time, X (3)	XY (4)	X^2 (5)	$\hat{Y}_X$ (6)	$Y - \hat{Y}_X$ (7)
1968	330	0	0	0	200	130
1969	241	1	241	1	247	−6
1970	200	2	400	4	294	−94
1971	499	3	1,497	9	341	158
1972	322	4	1,288	16	388	−66
1973	500	5	2,500	25	435	65
1974	601	6	3,606	36	482	119
1975	401	7	2,807	49	529	−128
1976	401	8	3,208	64	576	−175
1977	899	9	8,091	81	623	276
1978	598	10	5,980	100	670	−72
1979	787	11	8,657	121	717	70
1980	666	12	7,992	144	764	−98
1981	569	13	7,397	169	811	−242
1982	561	14	7,854	196	858	−297
1983	915	15	13,725	225	905	10
1984	947	16	15,152	256	952	−5
1985	791	17	13,447	289	999	−208
1986	1,114	18	20,052	324	1,046	68
1987	1,591	19	30,229	361	1,093	498
	12,933 = ΣY	190 = ΣX	154,123 = ΣXY	2,470 = ΣX^2		$\Sigma(Y - \hat{Y}_X) = 0$ (except for rounding)

$$\bar{Y} = \frac{\Sigma Y}{n} = \frac{12,933}{20} = 646.65$$

$$\bar{X} = \frac{\Sigma X}{n} = \frac{190}{20} = 9.5$$

13.A formulas given on p. 536; the preliminaries are indicated in columns 4–7 of Table 15.4. Hence,

$$b = \frac{\Sigma XY - n\overline{X}\,\overline{Y}}{\Sigma X^2 - n\overline{X}^2} = \frac{154{,}123 - 20(9.5)646.65}{2{,}470 - 20(9.5)^2} = 47 \text{ (rounded)};$$

$$a = \overline{Y} - b\overline{X} = 646.65 - 47(9.5) = 200 \text{ (rounded)}.$$

Therefore, our estimated trend line can be described as:

$$\hat{Y}_X = 200 + 47X,$$

where X is measured in years and $X = 0$ is located at mid-1968. This line is the solid color line in Figure 15.10; with its help, one might estimate 1990 trend sales as 1,234 units (point a). ▐▐▐

From annual to quarterly data. Most trend lines are fitted to annual data, but it may be desirable to have quarterly or even monthly trend values. These can be easily determined by modifying our above regression equation. Suppose our manager required quarterly information on sales (at annual rates). In that case, we would need X values measured in quarters from a zero base, and each of them would be centered in the middle of a quarter. If $X = 0$ were placed in the middle of the first or winter quarter of 1968 (that is, at mid-February, 1968), the base would shift backward in time by 1.5 quarters (the original base was mid-1968, at the end of the second or spring quarter). Our earlier calculation showed sales rising annually by $b = 47$ units; hence, they are rising by $b/4$ or 11.75 units per quarter, which would be the slope of our new regression line for quarterly data. Hence, the new line's intercept would have to equal 200 (or mid-1968 trend sales) minus 1.5(11.75), or 182.375. Thus, the transformed regression line (measuring sales quarterly at annual rates) would be

$$\hat{Y}_X = 182.375 + 11.75X,$$

where X is measured in quarters and $X = 0$ is located at the middle of the first or winter quarter of 1968. Note how this equation accurately predicts the 200-unit annual trend sales of 1968 that constitutes the intercept of the annual-data regression line in Figure 15.10:

$$\hat{Y}_{1.5} = 182.375 + 11.75(1.5) = 200.$$

Fitting a curvilinear regression line.* Let us now consider a time series with a scatter diagram that looks like panel (b) or (c) rather than panel (a) of Figure 15.9 and that, therefore, requires the fitting of a curvilinear rather than a linear regression line. Such an increasing exponential curve would have an equation of $\hat{Y}_X = a \cdot b^X$, while a decreasing exponential curve would be described by $\hat{Y}_X = a \cdot b^{-X}$, where a and b are

---

*Requires prior study of the Chapter 13 optional section on curvilinear regression.

Table 15.5 Domestic Revenue Passengers Carried by Major U.S. Airlines

Year, t	X	Passengers (millions), Y_t	Year, t	X	Passengers (millions), Y_t
1955	0	34	1970	15	123
1956	1	38	1971	16	124
1957	2	40	1972	17	137
1958	3	40	1973	18	145
1959	4	44	1974	19	148
1960	5	45	1975	20	147
1961	6	45	1976	21	160
1962	7	47	1977	22	172
1963	8	53	1978	23	196
1964	9	61	1979	24	212
1965	10	70	1980	25	222
1966	11	79	1981	26	205
1967	12	97			
1968	13	119			
1969	14	126			

Source: Moody's Transportation Manual (New York: Moody's Investors Service, 1982), p. a 39.

positive constants. As was noted in Chapter 13's section, "Curvilinear Regression," the determination of such a regression line is easy with the help of an appropriate data transformation. For purposes of illustration, let us consider the actual data given in Table 15.5.

||| Practice Problem 6

Letting a Curvilinear Regression Line Approximate the Trend in Passengers Carried by Major U.S. Airlines

A scatter diagram of the data in Table 15.5 looks like the black dots in panel (a) of Figure 15.11 and suggests that the first of the two equations just noted might be the proper one to summarize the data. If we take the logarithms of both sides, the equation becomes

$$\log \hat{Y}_X = \log a + X \log b,$$

which is the equation of our familiar straight regression line with $\log \hat{Y}_X$ on the vertical axis, $\log a$ as the intercept, and $\log b$ as the slope. If we fit the Table 15.5 data accordingly, we find this result:

$$a = 33.07 \text{ and } \log a = 1.5194$$
$$b = 1.081 \text{ and } \log b = .0338.$$

Figure 15.11 Exponential Trend Line and Its Logarithmic Transformation

A scatter diagram of the Table 15.5 data, shown in panel (a), suggests an exponential trend line as the best fit, and such a line is shown there. The same information, however, is embodied in panel (b), which contains a scatter diagram of the dependent variable, logarithmically transformed, against time, along with a straight-line logarithmic trend. Note how estimating, say, year 15 trend sales yields the same result (point A or B), regardless of which of the two trend equations is used.

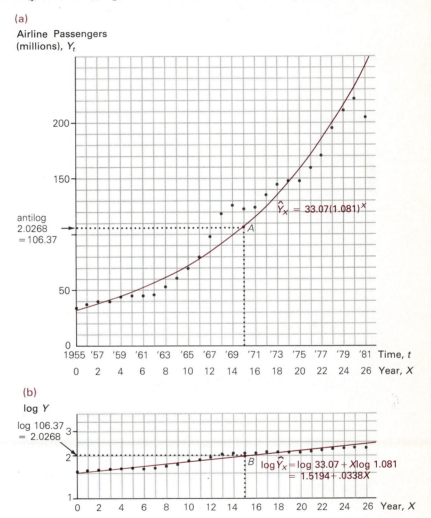

Thus, we can write our equation in one of two ways:

$$\hat{Y}_X = 33.07(1.081)^X \tag{1}$$

$$\log \hat{Y}_X = \log 33.07 + X \log 1.081 = 1.5194 + .0338\,X \tag{2}$$

where X is measured in years and $X = 0$ is located at mid-1955. Equation (1) has been graphed as the color line in panel (a) of Figure 15.11. Note how it estimates 1970 passengers as $\hat{Y}_{15} = 33.07(1.081)^{15} = 106.37$ million (point A), which equals antilog 2.0268. Note also that the value of b indicates an annual passenger-growth rate of 8.1 percent.

Equation (2) has been graphed as the color line in panel (b) of Figure 15.11, which also contains a scatter diagram of log Y values against X. This equation estimates 1970 passengers as log $\hat{Y}_{15} = 1.5194 + .0338(15) = 2.0264$ (point B) which is log 106.27. Except for rounding, this is the same result as that found above (and has been so recorded in the graph). |||

||| Practice Problem 7

Using a Curvilinear Regression to Predict Future Passenger Levels

It is easy to use either equation for forecasting, say, 1988 ($t = 33$) passenger levels. According to equation 1,

$$\hat{Y}_{33} = 33.07(1.081)^{33} = 432.2 \text{ (million).}$$

According to equation 2,

$$\log \hat{Y}_{33} = 1.5194 + .0338(33) = 2.6348, \text{ and antilog } 2.6348 = 431.3 \text{ (million).}$$

Once again, it is only due to rounding error that the answers differ.

Analytical Example 15.3, "Fitting Trends with Logarithms," gives other instances of the use of this technique. |||

Dealing with serial correlation. We have already noted that time-series data are *unlikely* to be a string of random numbers, each independent of the others. More likely than not, they are serially correlated. In fact, an inspection of (either panel of) Figure 15.11 should make us suspicious. Note how the first three data points lie above their regression line estimates; thus, we have three positive errors (of $Y_X - \hat{Y}_X$) in a row. This string is followed by strings of eight negative errors, nine positive errors, and seven negative errors, which certainly suggests the presence of positive serial correlation. Indeed, we can apply what we learned in Chapter 14 and perform the Durbin-Watson test with respect to the data in Table 15.5. If we test the hypothesis H_0: "No serial correlation exists" against the alternative H_A: "Positive-serial correlation exists," we find $d = .44$. If we then consult Appendix Table P for $\alpha = .05$, we find that H_0 should be resoundingly rejected.

One common approach to dealing with serial correlation is computing regressions for time series not with the help of the original data, but on the basis of their first differences (in this case, not on the basis of annual passenger numbers, but of year-to-year changes in these numbers). If such first differences are used, first-order serial correlation disappears, and better forecasts can often be made.

FORECASTING: THE USE OF SEASONAL INDEXES

On many occasions, business executives and economists are interested in isolating not the trend, but the seasonal component of a time series. If first-quarter (or winter) sales are typically low, a manager who is aware of this fact will not worry when January sales figures start to slip. If December food prices are typically high, an economist who is aware of this fact will not falsely conclude that inflation is accelerating when the month's price index begins to climb. We will consider two methods of isolating the seasonal component in a time series.

The Ratio-to-Moving-Average Method

Consider the quarterly time series given in columns 1 and 2 of Table 15.6. We will show how its seasonal component can be isolated by the so-called **ratio-to-moving-average-method.** This method involves four steps:

1. First, a 4-quarter moving-average series is constructed by the method introduced earlier in this chapter (see Table 15.1). This series appears in column 4 of Table 15.6 and is viewed as containing only the trend and cyclical component ($T_t \cdot C_t$) of the original time series (the shorter than 4-quarter-long seasonal and irregular components having been averaged out).

2. Second, the original time-series data of $Y_t = T_t \cdot C_t \cdot S_t \cdot I_t$, in column 2 of Table 15.6, are then divided by the associated 4-quarter moving average of $T_t \cdot C_t$, in column 4, which isolates $S_t \cdot I_t$, given in column 5. Note how, in the summer of 1981, actual sales were $Y_t = 177$, but the average quarterly sales for the 4 quarters centered in mid-summer 1981 (and presumably reflecting trend and cyclical factors only) were $T_t \cdot C_t = 143$; at this point, the *ratio* of actual sales *to* the associated *moving-average* sales was 1.24 (which accounts also for the strange name of this procedure). Thus, actual summer 1981 sales were 1.24 times (or 124 percent of) the level one would expect from trend and cyclical factors alone; the divergence must be attributed to seasonal and irregular factors.

3. Third, the irregular component is removed by grouping, as in Table 15.7, all the available actual-to-moving-average sales ratios—from column 5 of Table 15.6—by season and finding an average ratio for each season. That average (which might be the arithmetic mean or the median of the entries found in any given column) is believed to show the "typical" seasonal influence only, free of the irregular component.

4. Fourth, the quarterly seasonal time-series components, S_t, that were calculated in step 3 and that appear in row A of Table 15.7 are expected to average to 1 over the course of a year; thus, their sum must equal 4. If it does not, an appropriate adjustment is made. In our case, the sum equals 3.925, hence, each S_t is multiplied by 4/3.925; the result is given in row B. These so-called **seasonal indexes** indicate, for example, that seasonal factors alone typically depress winter sales to 80 percent but raise summer sales to 120 percent of the level that could be accounted for by trend, cycle, and irregular factors (while spring and fall sales are fully explained by those factors and, thus, equal the quarterly average).

Table 15.6   Avionics Sales of Butler Aviation and the Construction of Seasonal Indexes by the Ratio-to-Moving-Average Method

Time, t (1)	Actual Units Sold, $Y_t =$ $T_t \cdot C_t \cdot S_t \cdot I_t$ (2)	4-Quarter Moving Average, $T_t \cdot C_t$ Unadjusted (3)	4-Quarter Moving Average, $T_t \cdot C_t$ Adjusted (4)	Ratio of Actual to Moving-Average Sales, $S_t \cdot I_t$ (5) = (2)/(4)	Seasonal Index, S_t (6)	Deseasonalized Units Sold, $T_t \cdot C_t \cdot I_t$ (7) = (2)/(6)
1981 Winter	107	—	—	—	.80	133.75
Spring	146	—	—	—	1.00	146
		142				
Summer	177		143	1.24	1.20	147.5
		143				
Fall	139		141	.99	1.00	139
		139				
1982 Winter	108		138	.78	.80	135
		137				
Spring	130		139	.94	1.00	130
		140				
Summer	169		145	1.17	1.20	140.8
		149				
Fall	154		159	.97	1.00	154
		169				
1983 Winter	144		185	.78	.80	180
		200				
Spring	208		215	.97	1.00	208
		229				
Summer	292		239	1.22	1.20	243.3
		249				
Fall	271		257	1.05	1.00	271
		264				
1984 Winter	224		261	.86	.80	280
		257				
Spring	268		247	1.09	1.00	268
		237				
Summer	264		227	1.16	1.20	220
		217				
Fall	191		208	.92	1.00	191
		199				

Deseasonalizing data. Any original time series can be divided by the kind of seasonal indices found in row B of Table 15.7 to produce a **deseasonalized,** or **seasonally adjusted, time series.** The seasonal adjustment of the data on avionics sales is shown in columns 6 and 7 of Table 15.6. Because the division of $Y_t = T_t \cdot C_t \cdot S_t \cdot I_t$ by S_t produces $T_t \cdot C_t \cdot I_t$, a deseasonalized time series is one that contains all influences except the seasonal factor. Thus, the manager of Butler Aviation who sadly contemplates low winter 1981 avionics sales of 107 units—column 1 of Table 15.6—can take comfort in the fact that underlying nonseasonal factors are producing sales of 133.75 units per quarter—column 7 of Table 15.6—but usual seasonal factors are depressing this num-

Table 15.6 *Continued*

Time, t (1)	Actual Units Sold, $Y_t = T_t \cdot C_t \cdot S_t \cdot I_t$ (2)	4-Quarter Moving Average, $T_t \cdot C_t$ — Unadjusted (3)	4-Quarter Moving Average, $T_t \cdot C_t$ — Adjusted (4)	Ratio of Actual to Moving-Average Sales, $S_t \cdot I_t$ (5) = (2)/(4)	Seasonal Index, S_t (6)	Deseasonalized Units Sold, $T_t \cdot C_t \cdot I_t$ (7) = (2)/(6)
1985 Winter	146		196	.74	.80	182.5
		192				
Spring	194		195	.99	1.00	194
		198				
Summer	238		204	1.17	1.20	198.3
		210				
Fall	213		219	.97	1.00	213
		227				
1986 Winter	194		240	.81	.80	242.5
		253				
Spring	263		266	.99	1.00	263
		279				
Summer	340		288	1.18	1.20	283.3
		297				
Fall	317		310	1.02	1.00	317
		322				
1987 Winter	269		342	.79	.80	336.25
		361				
Spring	361		380	.95	1.00	361
		398				
Summer	495		419	1.18	1.20	412.5
		440				
Fall	466		—	—	1.00	466
		—				
1988 Winter	438		—	—	.80	547.5

Table 15.7 Calculating Seasonal Indexes from Groupings of Actual-to-Moving-Average Ratios ($S_t \cdot I_t$ data)

Year	Winter	Spring	Summer	Fall
1981	—	—	1.24	.99
1982	.78	.94	1.17	.97
1983	.78	.97	1.22	1.05
1984	.86	1.09	1.16	.92
1985	.74	.99	1.17	.97
1986	.81	.99	1.18	1.02
1987	.79	.95	1.18	—
(A) Average Ratio (median of column data) = Unadjusted S_t	.785	.98	1.18	.98
(B) Adjusted S_t = Seasonal Index	.80	1.00	1.20	1.00

Figure 15.12   Avionics Sales, Unadjusted vs. Deseasonalized Data

This graph is based on columns 1, 2, and 7 of Table 15.6 and vividly shows the difference between time series that do and those that do not contain a seasonal component. The fluctuations of the black line (that represents $Y_t = T_t \cdot C_t \cdot S_t \cdot I_t$) about the color line (that represents Y_t/S_t *or* $T_t \cdot C_t \cdot I_t$) are accounted for by seasonal factors alone (S_t).

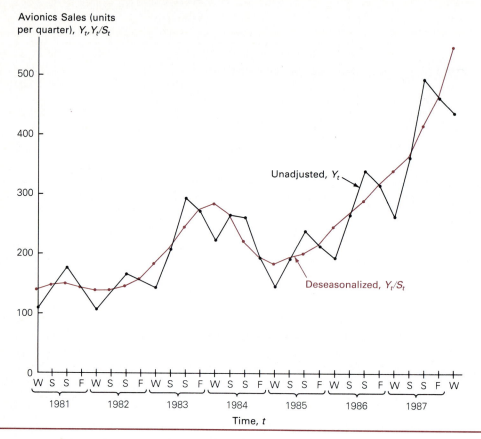

ber. By the same token, that manager should not be overly elated by the brisk summer 1981 sales of 177 units; the seasonal factor explains almost 17 percent of this, and the underlying nonseasonal factors are producing sales of only 147.5 units per quarter. Figure 15.12 indicates visually what a different picture a decision maker can get from deseasonalized as opposed to original, unadjusted data.

Note: The ratio-to-moving-average method was introduced in 1922 by Frederick R. Macaulay at the National Bureau of Economic Research, was adopted under the leadership of Julius Shiskin by the U.S. Bureau of the Census in 1954 and is now commonly utilized worldwide.[3] As a result, the method has found its way into many computer

---

[3] A biography of Julius Shiskin appears in the "Supplementary Topics" section of Chapter 15 of the *Student Workbook* that accompanies this text.

programs. Although the above example has focused on quarterly data, many business and economic applications require monthly data, and computers are often programmed to deseasonalize monthly data in an analogous fashion.

Making forecasts. Seasonal indices are ideally suited for making short-term forecasts—for example, of quarterly sales figures at annual rates.

Practice Problem 8

Predicting Future Avionics Sales Using Seasonal Indices
Consider our quarterly trend equation,

$$\hat{Y}_X = 182.375 + 11.75\,X$$

where $X = 0$ was located at the middle of the winter quarter of 1968. On the basis of this equation alone (and rounding to whole numbers), management would forecast 1990 and 1991 quarterly trend sales (at annual rates) as noted in column 4 of Table 15.8. Yet applying the seasonal indices of Table 15.7, now shown in column 3, the actual forecast would appear as in column 5.

We can also test the accuracy of the technique by pretending to forecast some of the data that are already available to us, such as the quarterly 1984 data (pertaining to quarters $X = 64 - 67$) in column 2 of Table 15.6. If we do, quarterly trend sales (at annual rates) of 934, 946, 958, and 970 are calculated (which numbers average reassuringly to the 952-unit sales figure that the annual trend equation of Figure 15.10 predicts). Yet forecasts taking seasonality into account would equal .8, 1.0, 1.2, and 1.0 times these figures, respectively, or 747, 946, 1150, and 970 (rounded). Since these are sales at annual rates achieved in the four quarters, the actual sales *per quarter* would be forecast as one quarter of these figures, or as 187, 237, 288, and 243. (From Table 15.6 we know actual sales, however, to have been 224, 268, 264, and 191.)

Table 15.8   Making Quarterly Forecasts of Butler Avionics Sales ($X = 0$ at midwinter 1968)

Time			Quarterly Sales	
t (1)	X (2)	Seasonal Index (3)	Trend, $\hat{Y}_X$ (4)	Forecast (5) = (3) · (4)
1990 Winter	88	.80	1216	973
Spring	89	1.00	1228	1228
Summer	90	1.20	1240	1488
Fall	91	1.00	1252	1252
1991 Winter	92	.80	1263	1010
Spring	93	1.00	1275	1275
Summer	94	1.20	1287	1544
Fall	95	1.00	1299	1299

III

The Dummy-Variable Method*

On occasion, a multiple regression equation is estimated in order to take the seasonal factor into account. For a quarterly time series such as the one given in Table 15.6, the equation might take the form,

$$\hat{Y}_t = a + b_1 t + c_1 Q_1 + c_2 Q_2 + c_3 Q_3$$

where Q_1, Q_2, and Q_3 are dummies equal to 1 if t is the first, second, or third quarter, respectively; they are equal to 0 otherwise. Note that for a fourth-quarter observation, $\hat{Y}_t = a + b_1 t$, while for a first (or second or third) quarter observation it equals $\hat{Y}_t = a + b_1 t + c_1$ (or c_2 or c_3). Thus, c_1, c_2, or c_3, respectively, give the difference in the expected value of the time series between the 4th quarter on the one hand and the 1st, 2nd, or 3rd quarter on the other, and these coefficients can be used to represent the seasonal variation.

Yet caution is advised: In this model, the seasonal effects are *added* to the linear trend of $a + b_2 t$; thus, the model is different from the classical multiplicative one. In addition, any performance of such a multiple regression in the presence of a likely violation of crucial assumptions can lead to absurd results.

FORECASTING: ECONOMETRIC MODELS

A considerably more sophisticated forecasting technique than the ones introduced above is embodied in that blend of economic theory and statistical method that is called the **econometric model.** This type of model is strongly identified with regression analysis; it may consist of a single equation or a system of equations estimated from past time-series data and used to show the effect of various independent variables (that are known or are themselves being estimated) on various dependent variables (such as the business and economic variables of interest to a forecaster).

Single-Equation Models

In one single-equation model, for instance, the quantity of onions supplied at one time, Y_t, was related to price and production cost in the previous period, P_{t-1} and C_{t-1}, with all the data being logarithmically transformed[4]:

$$\log Y_t = .134 + .0123(t - 1924) + .324 \log P_{t-1} - .512 \log C_{t-1}.$$

Note the reasonableness of the various signs of the coefficients: there was a positive trend in quantity, a higher price coaxed out a larger quantity, but higher production costs led to a lower quantity.

Another model linked the (male) military volunteer enlistment rate, Y, to the ratio of average civilian to average military pay, P, the civilian unemployment rate, U, a draft-

---

*Requires prior study of the Chapter 14 optional section on dummy variables.

[4]Daniel B. Suits and S. Koizumi, "The Dynamics of the Onion Market," *Journal of Farm Economics* (May 1956), pp. 475–84.

threat variable, D (the probability of not being drafted), and a trend, T (a proxy for systematic changes in "taste" with respect to military service)[5]:

$$Y = 1.340 - .355\,P - .200\,U - .283\,D - .014\,T.$$

In this case, all coefficients except that of U were statistically significant, and the signs of the significant ones also made sense: higher civilian pay, a higher probability of not being drafted, and a persistent trend in tastes away from military pursuits were all seen to *discourage* voluntary enlistments.

Multiple-Equation Models

Many econometric models—such as the Federal Reserve Board–Massachusetts Institute of Technology Model, the Chase Econometrics Model, or the Wharton Econometric Model—are, however, considerably more complex. They express relations among variables by means of a number of simultaneous equations such that the variables found in different equations interact. The Wharton Model contains more than 200 equations that are used jointly to forecast the level and composition of the GNP, the general price level, unemployment, and more. The household sector of the model, for instance, explains consumer expenditures on such bases as prices and disposable income and does so separately for automobiles, other durables, nondurable goods, and services. Various investment equations explain business spending on plant and equipment (as a function of interest rates, capacity utilization, and output growth), housing construction, and inventory change. The forecasts produced by equations such as these, in turn, enter other equations. Thus, the level of the GNP is explained as a function of consumer expenditures, investment expenditures, government expenditures, and foreign trade; the level of employment is explained by means of a production function, relating various resource inputs to the GNP; unemployment is explained by labor force and employment data, the rate of wage increase by unemployment and inflation, the rate of inflation by capacity utilization, interest rates by the money supply, and much more.

Econometric models have made excellent forecasts at times. At other times, however, their record has been spotty. (See Analytical Example 15.4, "How Accurate are Economic Forecasts?") One thing is certain: the mechanical application of such models does not assure forecasting success; the most successful forecasts have often involved the selective introduction into such models of additional pieces of information, such as those gained from sources discussed in the next section.

FORECASTING: BAROMETRIC INDICATORS

Many forecasters make use of qualitative or **barometric indicators** of business and economic conditions. These include (1) the results of anticipatory surveys and (2) leading economic indicators.

---

[5]Colin Ash, Bernard Udis, Robert F. McNown, "Enlistments in the All-Volunteer Force: A Military Personnel Supply Model and Its Forecasts," *The American Economic Review* (March 1983), pp. 145–55.

Anticipatory Surveys

A number of agencies make forecasts by the eminently sensible method of *asking* the economic actors involved about their intentions. Several **anticipatory surveys** of households, businesses, or governments are conducted on a regular basis to determine economic intentions. The Survey Research Center of the University of Michigan, for instance, develops quarterly data on consumer spending plans. The U.S. Department of Commerce, together with the Securities and Exchange Commission, produces surveys on business investment plans; so do the McGraw-Hill Publishing Company, *Fortune* magazine, and others. The oldest of these surveys, the Railroad Shippers' Forecast, has provided data on quarterly anticipated carloadings, differentiated by commodities, ever since 1927.

Business-Cycle Indicators

Many economic time series have been found to behave in a certain predictable fashion near the turning points of the business cycle, which makes them extremely useful for forecasting purposes. Originally, such series were collected and studied systematically

Table 15.9   Short List of Business-Cycle Indicators

Leading Indicators

   Average work week, production workers, manufacturing
   Lay-off rate per 100 employees, manufacturing
   New orders for consumer goods and materials, 1972 dollars
   Vendor performance, percent of companies receiving slower deliveries
   Net business formation, index: 1967 = 100
   Contracts and orders for plant and equipment, 1972 dollars
   New building permits, private housing units, index: 1967 = 100
   Change in inventories on hand and on order, smoothed series, 1972 dollars
   Percent change in sensitive crude-materials prices, smoothed series
   Stock prices, 500 common stocks, index: 1941–43 = 10
   Percent change in total liquid assets, smoothed series
   Money supply (M2), 1972 dollars

Coincident Indicators

   Employees on nonagricultural payrolls
   Personal income less transfer payments, 1972 dollars
   Industrial production, total, index: 1967 = 100
   Manufacturing and trade sales, 1972 dollars

Lagging Indicators

   Average duration of unemployment
   Manufacturing and trade inventories, 1972 dollars
   Labor cost per unit of output, manufacturing, index: 1967 = 100
   Average prime rate charged by banks
   Commercial and industrial loans outstanding
   Ratio of consumer installment credit to personal income

Source: U.S. Department of Commerce.

at the National Bureau of Economic Research (by such researchers as Wesley C. Mitchell, Arthur F. Burns, Geoffrey H. Moore, and Julius Shiskin), and the bureau still follows the behavior of many hundreds of monthly and quarterly series. More recently, the Bureau of Economic Analysis of the U.S. Department of Commerce has begun publishing its monthly *Business Conditions Digest* that provides several hundred time series of interest to economic analysts and forecasters.

The National Bureau's detailed examination, often for 100 years or more, of the behavior of time series led to an important discovery: some economic time series, now called **leading economic indicators,** anticipate business cycle turns; they turn down just before the GNP reaches its cyclical peak and turn up just before it reaches its trough. Other series, called **coincident economic indicators,** precisely coincide with the business cycle; they turn down at the GNP peak and up at its trough. Still other series, called **lagging economic indicators,** lag behind the cyclical turns; they turn down only after the GNP has begun its cyclical decline, while turning up only after the GNP has begun to rise. Table 15.9 gives the so-called "short list" of business-cycle indicators. That list, however, is not totally reliable and, therefore, subject to revision. (In 1952 and again in 1962, the leading indicators turned down, but the economy did not follow.) In addition, there are times when some leading indicators point to a change, while others do not. To deal with that situation, a **diffusion index** is published; it is a summary of a dozen leading economic indicators and equals the percentage of these indicators that have increased in a given month. Thus, the index lies between 0 and 100; if 8 of these indicators are higher this month than last month, the diffusion index is (8/12)100, or 67 (percent).

ANALYTICAL EXAMPLES

15.1 The Role of Time Series in Social Experiments

On many occasions, throughout every year, some kind of social experiment is being initiated. It may involve a tax cut designed to put an end to a recession or a cut in the money supply to stop inflation; it may involve a price cut or a change in product design to increase sales; it may involve the assignment of more police to a city district to deter crime or to the freeways to put an end to drunken driving. Sooner or later, one question naturally arises: how effective has the program been?

Unfortunately, these types of social experiments can rarely be conducted like laboratory experiments. Typically, one cannot assign human subjects randomly to a control group or to an experimental one, supply a "treatment" (tax cut, tight money, lower price, new product design, more police patrols) to the latter group only, and ultimately attribute observed differences between the groups to the treatment. In the realm of social policy, such controlled experiments happen but rarely. More likely than not, for example, *all* citizens get a tax cut at the same time, and later on it is difficult to know whether any subsequent behavior change was the result of the tax cut or would have happened anyway for all kinds of other reasons. Nevertheless, a comparison of time-series data from before and after the introduction of a new social policy is usually employed to evaluate the effectiveness of that policy. Clearly, this is tricky business.

Consider the case of Connecticut's crackdown on speeding that began on December 23, 1955. Traffic deaths in 1955 had been 324; in 1956, they dropped to 284, as shown in panel (a) of the figure on the next page. The governor attributed the saving of 40 lives to the crackdown. Yet he implicitly assumed that there would have been no change in traffic deaths in the absence of his program, and he was surely wrong on

(a)

Traffic Fatalities

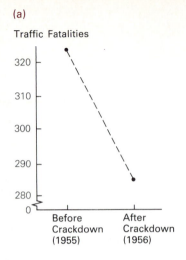

Before
Crackdown
(1955)

After
Crackdown
(1956)

(b)

Traffic Fatalities

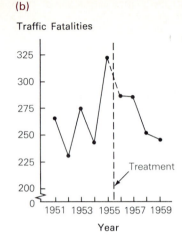

Treatment

Year

that, as panel (b) indicates. Throughout the 1950s, Connecticut traffic deaths had been zigzagging up and down. The 1955 figure had been an all-time high; thus, chances were high that the number would drop in the next year even without any crackdown (once again, the regression effect!).

Thus, an evaluation of social quasi-experiments by means of time-series analysis must carefully consider the observed effect—panel (a)—*in the context of the general trend*—panel (b). Any single rise or fall in a variable that is part of a zigzagging time series can hardly be attributed to the policy with which that rise or fall happens to have been associated. The same is

true for any rise embedded in a generally upward trend or any decline that is part of a generally downward trend. On the other hand, a rise, at the time of a policy change, in a long-standing downward trend is quite another matter; so is a decline at such a time in a long-term upward trend. Thus, time-series analysis is far from useless when it comes to evaluating social experiments. It just needs to be applied with caution.

Source: Donald T. Campbell, "Measuring the Effects of Social Innovations by Means of Time Series," in Judith M. Tanur et al., eds., *Statistics: A Guide to the Unknown* (San Francisco: Holden-Day, 1972), pp. 120–29.

15.2 The Slutsky-Yule Effect

While the method of moving averages will produce a fairly reliable estimate of a time-series value at any particular time, it also produces a misleading *pattern* of points over an extended period of time. It *creates* periodicities where there are none in the original data. This **Slutsky-Yule effect** can be illustrated most effectively by noting how a time series that consists of nothing but random numbers (and, therefore, contains no oscillations) can give birth to an oscillatory moving-averages series that shows clearly visible, yet artificially created and, therefore, misleading waves.

Consider the hypothetical time series plotted in panel (a) of the figure on the next page. It is nothing else but a plot of the 50 digits found in row 1 of Appendix Table A, "Random Numbers," or, what time-series analysts refer to as patternless "white noise." Now imagine taking 10-period moving averages of this

series and plotting these, as in panel (b). The oscillations are obvious, yet it would be wrong to look for any causal link here (such as the influence of strikes or holidays on our time series).

In fact, the reason for the waves is fairly simple. Because the original numbers vary randomly, the sum of any ten of these (and the average of that group) is bound to fluctuate. When that sum is unusually high, there are lots of large numbers in the group; chances are lower that a larger number will be added next than that a smaller number will be added next. But a large number will be dropped, so the sum (and average) probably decline after a peak. The reverse is true when we start with an unusually small sum (and average); thus, the two probably rise after a trough. If all this reminds you of the *regression effect,* discussed in Biography 13.1 of Francis Galton, it should.

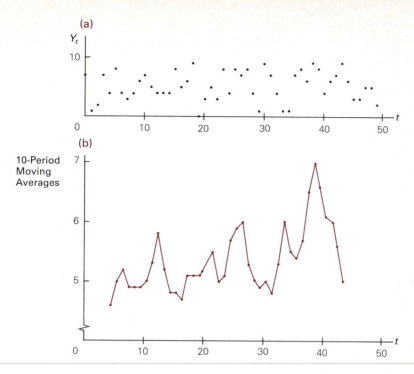

Source: Based on Eugen E. Slutsky, "The Summation of Random Causes as the Source of Cyclic Processes," *Econometrica* 5(1937):105–46; G. Udny Yule, "On a Method of Investigating Periodicities in Disturbed Series, with Special Reference to Wolfer's Sunspot Numbers," Royal Society of London, *Philosophical Transactions* (1927), pp. 267–98.

15.3 Fitting Trends with Logarithms

The commonality of fitting curvilinear trend lines, such as the ones in Figure 15.9, with the help of logarithms was pointed out two decades ago when two economists examined a number of famous ratios used in economics, including the consumption/income ratio, the capital/output ratio, the labor-income/total-income ratio, and more.

For example, a scatter diagram of the real consumption to real net national product ratio (C/Y) for the 1900–1953 period looked very much like panel (b) of Figure 15.9 and was well represented by

$$\log \frac{C}{Y} = -.03933 + .00054\, t$$

(based on semiannual U.S. data, centered at 1/1/1927) or

$$\frac{C}{Y} = .9134(1.00129)^{t}.$$

Thus, the ratio was rising exponentially at a semiannual rate of .129 percent.

On the other hand, a scatter diagram of the real capital stock to real net national product ratio (K/Y) for the same period looked very much like panel (c) of Figure 15.9 and was well represented by

$$\log \frac{K}{Y} = .54699 - .0015\, t$$

(based on semiannual U.S. data, centered at 1/1/1927) or

$$\frac{K}{Y} = 3.523(1.0033)^{-t}.$$

Thus, the ratio was falling exponentially at a semiannual rate of .33 percent.

Source: Lawrence R. Klein and Richard F. Kosobud, "Some Econometrics of Growth: Great Ratios of Economics," *Quarterly Journal of Economics* (May 1961), pp. 173–98.

15.4 How Accurate Are Economic Forecasts?

In few fields has the concentration of the best techniques and the best brains been as high as in short-term macroeconomic forecasting for the U.S. economy. Yet the track record of the best known economic forecasters since 1970 has been mixed at best. They missed the 1974 recession; they missed the 1978–79 acceleration of inflation; they underestimated first the rapid recovery from the 1980 recession and then the severity of the 1982 recession.

Many business cases illustrate a similar phenomenon. Take the oil industry, for example. The figure below shows one major oil company's forecast of oil demand, made in 1978. This company allocates more resources to analyzing the future than do most companies and is well respected for its professionalism. Yet note how far outside the shaded forecast demand range the reality of the mid-1980s (color line) proved to be.

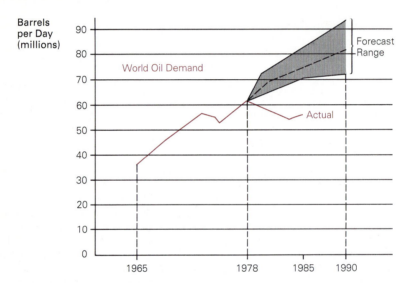

Source: Stephen K. McNees and John Ries, "The Track Record of Macroeconomic Forecasts," *New England Economic Review,* November-December 1983, p. 5; "Wrong When it Hurts Most," *Harvard Business Review,* September-October 1985, p. 75.

CLOSE-UP

15.1 The Dangers of Extrapolation

Even when the past data of a time series exhibit a clearly visible trend, forecasting the future on that basis (or, for that matter, making inferences about a more distant and unknown past) is dangerous business. Thus, President Lincoln, upon observing the 1790–1860 trend in U.S. population, predicted a U.S. population of 252 million by 1930. In fact, the actual 1930 population was 123 million. On the other hand, a 1938 presidential commission doubted that the U.S.

population would ever reach 140 million, yet that figure was reached and surpassed by 1945.

No one, perhaps, has ever put the matter more clearly than Mark Twain when he wrote these lines in 1874 in *Life on the Mississippi:*

In the space of one hundred and seventy-six years the Lower Mississippi has shortened itself two hundred and forty-two miles. That is an average of a trifle over one mile and a third per year. Therefore, any calm person, who is not blind or idiotic, can see that in the Old Oölitic Silurian Period, just a million years ago next November, the Lower Mississippi River was upward of one million three hundred thousand miles long, and stuck out over the Gulf of Mexico like a fishing-rod. And by the same token any person can see that seven hundred and forty-two years from now the Lower Mississippi will be only a mile and

three-quarters long, and Cairo and New Orleans will have joined their streets together, and be plodding comfortably along under a single mayor and a mutual board of aldermen. There is something fascinating about science. One gets such wholesale returns of conjecture out of such a trifling investment of fact.

Source: Darrell Huff, *How to Lie with Statistics* (New York: Norton, 1954), p. 142.

Summary

1. All economic organizations must plan for the future; inevitably, this planning requires the making of predictions that, in turn, are linked to experiences gained in the past. Most business and economic data are either *cross-section data* (which come from different populations observed at the same time) or *time-series data* (which come from the same population observed at different times). *Time-series analysis* employs time-series data for the purpose of explaining past or forecasting future events.

2. The *classical time-series model* attempts to explain the pattern observed in an actual time series by the presence of (1) trend, (2) cyclical, (3) seasonal, and (4) irregular components. The trend component, T, is a relatively smooth, progressively upward or downward movement of the variable of interest over an extended period of time. The cyclical component, C, is evidenced by wide up-and-down swings of the variable of interest around the trend, with the swings lasting from one to several years each and typically differing in length and amplitude from one cycle to the next. The seasonal component, S, is evidenced by narrow up-and-down swings of the variable of interest around the trend/cyclical components, with the swings predictably repeating each other within periods of one year or less. The irregular component, I, finally, adds random movements to all of the above movement. Because the classical model expresses the actual time-series value, Y, as the product of the four components just noted (making $Y = T \cdot C \cdot S \cdot I$), this model is also referred to as a *multiplicative time-series model*.

3. Business and economic forecasting is making statements about unknown, uncertain, and, typically, future events. Many such statements are derived with the help of time-series data. The trend component is often considered the most valuable forecasting tool, especially for long-term projections. The trend can be estimated in a variety of ways, including the construction of a *moving-averages series* (by successively averaging overlapping groups of two or more consecutive values in a time series and replacing the central value in each group by the group's average), the employment of single or multiple-parameter *exponential-smoothing techniques* (that produce self-correcting forecasts in the form of weighted averages of all past time-series values, with weights decreasing exponentially as one goes back in time), and the fitting of least-squares (linear or curvilinear) regression lines (which, however, can be problematic because of serial correlation).

4. On many occasions, especially when short-term forecasts are desired, business executives and economists are interested in isolating not the trend, but the seasonal component of a time series. Two favorite approaches are the *ratio-to-moving-average method* (that leads to the computation of seasonal indexes) and the computation of a multiple-regression equation in which dummy variables represent the seasonal factor.

5. A more complex method of business and economic forecasting is the construction of *econometric models* that typically consist of a system of regression equations estimated from past time-series data that show the effect of various independent variables on various dependent variables.

6. Many forecasters make use of qualitative *barometric indicators;* these include the results of *anticipatory surveys* (for example, of consumer spending plans) and a variety of *leading economic indicators* (time series that have been found to anticipate business-cycle turning points).

Key Terms

additive time-series model
anticipatory surveys
barometric indicators
classical time-series model
coincident economic indicators
cross-section data
cyclical component
deseasonalized time series
diffusion index
econometric model
exponential smoothing
forecasting
irregular component
lagging economic indicators
leading economic indicators
mixed time-series models

moving-averages series
multiplicative time-series model
ratio-to-moving-average method
seasonal component
seasonal indexes
seasonally adjusted time series
single-parameter exponential smoothing
Slutsky-Yule effect
smoothing constant
smoothing technique
three-parameter exponential smoothing
time-series analysis
time-series data
trend component
trend smoothing constant
two-parameter exponential smoothing

Questions and Problems

The computer programs noted at the end of this chapter can be used to work many of the subsequent problems.

1. An analyst has determined the values given in Table 15.A for the sales of a company. Assuming that the classical multiplicative time-series model applies, find the missing values. Explain how you found them.
2. The following time series represents the U.S. gross federal debt at the end of 1939–86 (in billions of dollars): 48.2 — 50.7 — 57.5 — 79.2 — 142.6 — 204.1 — 260.1 — 271.0 — 257.1 — 252.0 — 252.6 — 256.9 — 255.3 — 259.1 — 266.0 — 270.8 — 274.4 — 272.8 — 272.4 — 279.7 — 287.8 — 290.9 — 292.9 — 303.3 — 310.8 — 316.8 — 323.2 — 329.5 — 341.3 — 369.8 — 367.1 — 382.6 — 409.5 — 437.3 — 468.4 — 486.2 — 544.1 — 631.9 — 709.1 — 780.4 — 833.8 — 914.3 — 1,003.9 — 1,147.0 — 1,381.9 — 1,576.7 — 1,841.1 — 2,074.2.
   a. Estimate the trend in the form of a 5-year moving-averages series.

b. On the basis of the slope of the trend line between the last two numbers estimated in (a), forecast the U.S. gross federal debt at the end of 1989.
3. Using the data of problem 2,
   a. Estimate the trend in the form of a 10-year moving-averages series.
   b. On the basis of the slope of the trend line between the last two numbers estimated in (a), forecast the U.S. gross federal debt at the end of 1989.
4. The following time series represents new U.S. housing units started in 1959–84 (in thousands of units): 1,554 — 1,296 — 1,365 — 1,493 — 1,635 — 1,561 — 1,510 — 1,196 — 1,322 — 1,545 — 1,500 — 1,469 — 2,085 — 2,379 — 2,058 — 1,353 — 1,171 — 1,548 — 2,002 — 2,036 — 1,760 — 1,313 — 1,100 — 1,072 — 1,713 — 1,751.

Table 15.A

Quarter, t	Actual Sales, Y_t	Trend Component, T_t	Cyclical Component, C_t	Seasonal Component, S_t	Irregular Component, I_t
Winter	$93,060	$50,000	.99	2.00	?
Spring	?	55,000	.95	.50	1.02
Summer	22,344	60,000	.95	?	.98
Fall	69,998.50	65,000	?	1.10	1.10

a. Estimate the trend in the form of a 3-year moving-averages series.

b. On the basis of the slope of the trend line between the last two numbers estimated in (a), forecast new U.S. housing starts in 1989.

5. Using the data of problem 4,

a. Estimate the trend in the form of a 6-year moving-averages series.

b. On the basis of the slope of the trend line between the last two numbers estimated in (a), forecast new U.S. housing starts in 1989.

6. The following time series represents the service industry portion of the U.S. gross national product from 1947–83 (in billions of 1972 dollars): 55.9 — 57.5 — 57.6 — 59.7 — 60.8 — 61.6 — 62.7 — 62.9 — 67.6 — 70.9 — 74.1 — 76.2 — 80.8 — 83.5 — 86.6 — 90.3 — 94.0 — 98.8 — 103.1 — 109.0 — 115.0 — 118.8 — 124.0 — 126.7 — 128.4 — 136.5 — 144.8 — 147.9 — 148.5 — 154.7 — 164.3 — 174.2 — 183.0 — 189.1 — 197.6 — 200.2 — 206.8.

a. Estimate the trend in the form of a 7-year moving-averages series.

b. On the basis of the slope of the trend line between the last two numbers estimated in (a), forecast the service industry portion of U.S. real GNP in 1989.

7. Using the data of problem 6,

a. Estimate the trend in the form of a 14-year moving-averages series.

b. On the basis of the slope of the trend line between the last two numbers estimated in (a), forecast the service industry portion of U.S. real GNP in 1989.

8. The following time series represents the mining industry portion of the U.S. gross national product from 1947–83 (in billions of 1972 dollars): 6.8 — 9.4 — 8.1 — 9.3 — 10.2 — 10.1 — 10.6 — 10.9 — 12.4 — 13.4 — 13.5 — 12.4 — 12.3 — 12.6 — 12.7 — 12.8 — 13.1 — 13.4 — 13.5 — 14.2 — 14.6 — 15.3 — 16.1 — 17.6 — 17.4 — 19.0 — 21.7 — 32.2 — 38.8 — 43.0 — 47.4 — 52.0 — 66.8 — 96.0 — 132.3 — 125.1 — 112.4.

a. Estimate the trend in the form of a 9-year moving-averages series.

b. On the basis of the slope of the trend line between the last two numbers estimated in (a), forecast the mining industry portion of U.S. real GNP in 1989.

9. Using the data of problem 8,

a. Estimate the trend in the form of a 4-year moving-averages series.

b. On the basis of the slope of the trend line between the last two numbers estimated in (a), forecast the mining industry portion of U.S. real GNP in 1989.

10. The following time series represents U.S. federal government transfer payments from fiscal year 1960 to fiscal year 1986 (in billions of dollars): 20.6 — 23.6 — 25.1 — 26.5 — 27.4 — 28.4 — 31.8 — 37.2 — 42.7 — 48.7 — 55.0 — 67.7 — 76.1 — 87.2 — 101.8 — 131.4 — 153.8 — 166.6 — 178.7 — 197.8 — 234.6 — 273.7 — 304.5 — 338.3 — 340.7 — 361.0 — 377.6.

a. Estimate the trend in the form of a 3-year moving-averages series.

b. On the basis of the slope of the trend line between the last two numbers estimated in (a), forecast the U. S. federal government transfers in fiscal 1989.

11. Using the data of problem 10,

a. Estimate the trend in the form of a 4-year moving-averages series.

b. On the basis of the slope of the trend line between the last two numbers estimated in (a), forecast the U.S. federal government transfers in fiscal 1989.

12. Reconsider the data of problem 2. After equating the first actual value (48.2) with the 1939 forecast value, employ the method of single-parameter exponential smoothing with $\alpha = .2$ to develop a 1987 forecast. Show your computations.

13. Reconsider the data of problem 2. After equating the first actual value (48.2) with the 1939 forecast value, employ the method of single-parameter exponential smoothing with $\alpha = .8$ to develop a 1987 forecast. Which is better to use, $\alpha = .2$ (in problem 12) or $\alpha = .8$ (in problem 13)? Explain.

14. Reconsider the data of problem 4. After equating the first actual value (1,554) with the 1959 forecast value, employ the method of single-parameter exponential smoothing with $\alpha = .1$ to develop a 1985 forecast. Show your computations.

15. Reconsider the data of problem 4. After equating the first actual value (1,554) with the 1959 forecast value, employ the method of single-parameter exponential smoothing with $\alpha = .9$ to develop a 1985 forecast. Which is better to use, $\alpha = .1$ (as in problem 14) or $\alpha = .9$ (as in problem 15)? Explain.

16. Reconsider the data of problem 6. After equating the first actual value (55.9) with the 1947 forecast value, employ the method of single-parameter exponential smoothing with $\alpha = .3$ to develop a 1984 forecast. Show your computations.

17. Reconsider the data of problem 6. After equating the first actual value (55.9) with the 1947 forecast value, employ the method of single-parameter exponential smoothing with $\alpha = .7$ to develop a 1984 forecast. Which is better, $\alpha = .3$ (as in problem 16) or $\alpha = .7$ (as in problem 17)? Explain.

18. Reconsider the data of problem 8. After equating the first actual value (6.8) with the 1947 forecast value, employ the method of single-parameter exponential smoothing with $\alpha = .4$ to develop a 1984 forecast. Show your computations.

19. Reconsider the data of problem 8. After equating the first actual value (6.8) with the 1947 forecast value, employ the method of single-parameter exponential smoothing with $\alpha = .5$ to develop a 1984 forecast. Which is better, $\alpha = .4$ (as in problem 18) or $\alpha = .5$ (as in problem 19)? Explain.

20. Reconsider the data of problem 10. After equating the first actual value (20.6) with the 1960 forecast value, employ the method of single-parameter exponential smoothing with $\alpha = .15$ to develop a 1987 forecast. Show your computations.

21. Reconsider the data of problem 10. After equating the first actual value (20.6) with the 1960 forecast value, employ the method of single-parameter exponential smoothing with $\alpha = .8$ to develop a 1987 forecast. Which is better, $\alpha = .15$ (as in problem 20) or $\alpha = .8$ (as in problem 21)? Explain.

22. Reconsider the data of problem 2. Using two-parameter exponential smoothing ($\alpha = .2, \gamma = .3$), prepare a 1987 forecast. Show the entire series.

23. Reconsider the data of problem 4. Using two-parameter exponential smoothing ($\alpha = .1, \gamma = .4$), prepare a 1985 forecast. Show the entire series.

24. Reconsider the data of problem 6. Using two-parameter exponential smoothing ($\alpha = .3, \gamma = .6$), prepare a 1984 forecast. Show the entire series.

25. Reconsider the data of problem 8. Using two-parameter exponential smoothing ($\alpha = .4, \gamma = .8$), prepare a 1984 forecast. Show the entire series.

26. Reconsider the data of problem 10. Using two-parameter exponential smoothing ($\alpha = .15, \gamma = .9$), prepare a 1987 forecast. Show the entire series.

27. Have another look at Practice Problem 5 in the text.
    a. Test the overall significance of the regression with the help of an ANOVA table (using a critical $F_{.05}$).
    b. Compute the value of r^2.

28. Have another look at Practice Problem 5 in the text. For the regression shown there,
    a. compute the sample standard error of the estimate of Y.
    b. determine 95-percent confidence intervals for α and β.
    c. make 2-tailed tests of the hypotheses $H_0: \alpha = 0$ and $H_0: \beta = 0$.

29. Consider the data given in Table 15.B that pertain to the U.S. economy. Find the trend by fitting a linear or curvilinear regression line, whichever seems appropriate. Then make a forecast for 1988. (Normally, a trend would be determined only from a considerably larger data set of at least 20 years, but this is only an exercise. We assume that earlier and later data are unavailable to keep the exercise simple.)

30. Given the regression determined in problem 29,
    a. test the overall significance with the help of an ANOVA table (using a critical $F_{.05}$).
    b. determine the value of r^2.
    c. determine 95-percent confidence intervals for α and β.

31. Reconsider the data of problem 2. Estimate a linear regression line; then make a forecast for 1989.

32. Given the regression determined in problem 31,
    a. test the overall significance with the help of an ANOVA table (using a critical $F_{.05}$).
    b. determine the value of r^2.

Table 15.B

Year, t	Personal Consumption Expenditures (billions of 1972 dollars) C_t	Year, t	Personal Consumption Expenditures (billions of 1972 dollars) C_t
1970	672	1977	864
1971	697	1978	903
1972	737	1979	928
1973	768	1980	931
1974	763	1981	948
1975	779	1982	957
1976	823		

33. Given the regression determined in problem 31,
   a. compute the sample standard error of the estimate of Y.
   b. determine 95-percent confidence intervals for α and β.
   c. make 2-tailed tests of the hypotheses H_0: $\alpha = 0$ and H_0: $\beta = 0$.

34. Reconsider the data of problem 4. Estimate a linear regression line; then make a forecast for 1989.

35. Given the regression determined in problem 34,
   a. test the overall significance with the help of an ANOVA table (using a critical $F_{.05}$).
   b. determine 95-percent confidence intervals for α and β.

36. Reconsider the data of problem 6. Estimate a linear regression line; then make a forecast for 1989, along with a 95-percent prediction interval for the individual Y value.

37. Given the regression determined in problem 36,
   a. determine 95-percent confidence intervals for α and β.
   b. test the overall significance with the help of an ANOVA table (using a critical $F_{.05}$).

38. Reconsider the data of problem 8. Estimate a linear regression line; then make a forecast for 1989.

39. Given the regression determined in problem 38,
   a. test the overall significance with the help of an ANOVA table (using a critical $F_{.05}$).
   b. make 2-tailed tests of the hypotheses H_0: $\alpha = 0$ and H_0: $\beta = 0$.

40. Given the regression determined in problem 38,
   a. compute the value of r^2.
   b. determine 95-percent confidence intervals for α and β.

41. Reconsider the data of problem 10. Estimate a linear regression line; then make a forecast for 1989.

42. Test the overall significance of the regression determined in problem 41 with the help of an ANOVA table (using a critical $F_{.05}$).

43. The sales of airplane parts by Sporty's Pilot Shop from 1965–1986 (in thousands of dollars) were: 98 — 101 — 106 — 106 — 108 — 88 — 101 — 111 — 115 — 121 — 119 — 118 — 98 — 101 — 122 — 123 — 127 — 138 — 141 — 119 — 123 — 130. With the help of a linear regression, forecast sales in 1990.

44. Test the overall significance of the regression determined in problem 43 with the help of an ANOVA table (using a critical $F_{.025}$). What is the value of r^2?

45. The following data show U.S. pig iron production from 1975–84 (in millions of net tons): 79.9 — 86.9 — 81.3 — 87.7 — 87.0 — 68.7 — 73.6 — 43.3 — 48.7 — 51.9.
   a. Forecast the 1988 value using a linear regression.
   b. Test the regression's overall significance using an ANOVA table (using a critical $F_{.05}$).

46. The following data show hourly labor costs in the U.S. iron and steel industry from 1975–84 (in dollars per hour): 10.59 — 11.74 — 13.04 — 14.30 — 15.92 — 18.45 — 20.16 — 23.78 — 22.21 — 21.30.
   a. Forecast the 1988 value with the help of a linear regression, along with a 95-percent prediction interval for I_Y.
   b. Test the regression's overall significance with the help of an ANOVA table (using a critical $F_{.05}$).
   c. Determine 95-percent confidence intervals for α and β.

47. Figure 15.10 gives an annual trend equation. Convert it to a monthly equation (that gives month-to-month trend sales at annual rates).

48. Consider the annual trend equation determined in problem 36. Convert it to a monthly equation.

49. Consider the annual trend equation determined in problem 43. Convert it to a quarterly equation.

50. Consider the annual trend equation determined in problem 38. Convert it to a quarterly equation.

51. An analyst has available the values given in Table 15.C for the sales of a company and also knows the trend to be $T_X = 200,000 + 50,000\,X$ (with $X = 0$ at mid-1976). Assuming that the classical multiplicative time-series model applies, find the cyclical component.

Table 15.C

Year, t	Actual Sales, Y_t
1976	200,000
1977	300,000
1978	400,000
1979	400,000
1980	390,000
1981	370,000
1982	350,000
1983	330,000
1984	300,000
1985	250,000

52. Practice Problem 6 in this chapter derives this equation from the Table 15.5 data:

$$\log \hat{Y}_X = \log 33.07 + X \log 1.081.$$

Confirm the calculation. (*Hint:* If necessary, review the Chapter 13 section, "Curvilinear Regression.")

53. Panel (a) of Figure 15.11 (which is based on Table 15.5) shows an exponential regression line, $\hat{Y}_X = 33.07(1.081)^X$. Subsequently, the text claims that the underlying data were serially correlated. Perform the appropriate Durbin-Watson test, using the Box 14.J formula (p. 620) and $\alpha = .05$.

54. Perform a test for serial correlation on the problem 43 data, showing your computations. Use $\alpha = .05$ and a two-tailed test.

55. Perform a test for serial correlation on the problem 45 data, showing your computations. Use $\alpha = .05$ and a two-tailed test.

56. Perform a test for serial correlation on the problem 46 data, showing your computations. Use $\alpha = .05$ and a two-tailed test.

57. Table 15.D gives average weekly carloadings of all commodities carried by major U.S. railroads. Determine a set of monthly seasonal indices. (Normally, these would be determined from a considerably larger set of data, but this is only an exercise.)

Table 15.D Average Weekly Carloadings (thousands)

Month	Year		
	1979	1980	1981
January	392	426	410
February	407	439	437
March	584	457	456
April	478	442	382
May	494	435	385
June	611	450	424
July	446	387	411
August	476	434	439
September	476	436	435
October	488	454	441
November	459	439	403
December	399	410	360

Source: Moody's Transportation Manual (New York: Moody's Investors Service, 1982), p. a 12.

58. Assume that the monthly trend equation determined in problem 47 as well as the seasonal indices determined in problem 57 apply to U.S. railroads. Make a month-to-month forecast for 1988 carloadings that takes into account both trend and seasonality.

59. Table 15.E shows the monthly demand for blood at a blood bank. Determine a set of monthly seasonal indices. (Normally, these would be determined from a considerably larger set of data, but this is only an exercise.)

Table 15.E Monthly Blood Demand (pints)

Month	Year		
	1985	1986	1987
January	600	855	1,220
February	618	882	1,260
March	637	908	1,294
April	641	920	1,310
May	652	930	1,350
June	680	980	1,400
July	716	1,020	1,456
August	741	1,060	1,570
September	771	1,100	1,592
October	782	1,116	1,599
November	806	1,150	1,639
December	831	1,183	1,680

60. Table 15.F shows the monthly demand for electricity at a power station. Determine a set of monthly seasonal indices. (Normally, these would be determined from a considerably larger set of data, but this is only an exercise.) Show your computations.

Table 15.F Monthly Electricity Demand (megawatts)

Month	Year		
	1986	1987	1988
January	65	73	83
February	66	74	83
March	66	75	85
April	65	72	81
May	63	70	78
June	60	69	75
July	69	77	88
August	70	79	88
September	70	80	93
October	65	78	90
November	66	76	87
December	73	81	85

Selected Readings[6]

Armstrong, J. Scott. *Long-Range Forecasting: From Crystal Ball to Computer.* New York: Wiley, 1978. An excellent, wide-ranging discussion with lots of examples.

Brown, Robert G. *Smoothing, Forecasting, and Prediction of Discrete Time Series.* Englewood Cliffs, N.J.: Prentice-Hall, 1963. Shows, among other things, certain desirable theoretical properties of exponential smoothing.

Christ, Carl F. "Judging the Performance of Econometric Models of the U.S. Economy." *International Economic Review* 16(1975):54–74.

Granger, Clive W. J. *Forecasting in Business and Economics.* New York: Academic Press, 1980. A fairly advanced treatment of many aspects of forecasting, including the use of time-series models, econometric models, and more.

Johnston, J. *Econometric Methods.* New York: McGraw-Hill, 1972. A superb text.

Makridakis, Spyros G. "A Survey of Time Series Analysis." *International Statistical Review* 44(1976): 29–70.

Makridakis, Spyros, and Steven C. Wheelwright. *Forecasting: Methods and Applications.* New York: Wiley, 1978.

Moore, Geoffrey H., and Julius Shiskin. "Early Warning Signals for the Economy." In Judith M. Tanur et al., eds. *Statistics: A Guide to the Unknown.* San Francisco: Holden-Day, 1972, pp. 310–20. A discussion of the history, nature, and reliability of leading, coincident, and lagging business-cycle indicators.

Morgenstern, Oskar. *On the Accuracy of Economic Observations,* 2nd ed. Princeton, N.J.: Princeton University Press, 1963. A must reading for anyone working with business and economic statistics.

Shiskin, Julius, and Harry Eisenpress. "Seasonal Adjustments by Electronic Computer Methods." Technical Paper #12. New York: National Bureau of Economic Research, 1958.

Spencer, Milton H., Colin G. Clark, and Peter W. Hoguet. *Business and Economic Forecasting: An Econometric Approach.* Homewood, Ill.: Richard D. Irwin, 1965.

Wonnacott, Thomas H. and Ronald J. Wonnacott *Regression: A Second Course in Statistics.* New York: Wiley, 1981. Chapters 6 and 7 contain a more advanced discussion of time-series forecasting.

---

[6]Additional readings on spectrum analysis and ARIMA models are given in Chapter 15 of the *Student Workbook* that accompanies this text; these advanced techniques of time-series analysis are briefly described in the *Workbook* as well.

Computer Programs

The DAYSAL-2 personal computer diskettes that accompany this book contain three programs of interest to this chapter:

13. Regression and Correlation Analysis is described in detail at the end of Chapters 13 and 14.

14. Time Series and Forecasting lets you create moving-averages series, prepare forecasts by single-parameter or two-parameter exponential smoothing. The latter two programs also allow you to test a wide range of alpha values or of alpha-gamma combinations for the best results. Data files U-Z can be used for all of these options.

19. General Graphics, lets you draw time-series line graphs on graphs of your own design.

CHAPTER SIXTEEN

Index Numbers

On one occasion or another, almost everyone is interested in comparing economic conditions over time or space. Consider the consumer who wants to know whether and to what extent the cost of living now is higher or lower than in the past or how it differs now between Boston and Detroit. Consider the worker who wants to contrast current wages with wages 10 years ago or wages now and here with wages now and elsewhere. Consider the business executive who looks at the prices currently paid or received vis-à-vis those paid or received in the past or who wishes to compare the growth of sales in one region with that in another. Consider, finally, the government official (about to run for reelection, perhaps) who ponders the differences between the general price level now and four years ago, between the physical volume of national output now and then, or even between the levels of national output here and abroad. In all these cases and a million more, a certain kind of yardstick can greatly facilitate the desired comparison. Such a yardstick is provided by a set of **index numbers,** which are numbers that measure the magnitude of a variable at one time or place relative to the magnitude of the same variable at another time or place. The variable in question can be an individual price, quantity, or value figure; more often, however, entire sets of these prices, quantities, or values are being compared. When single items rather than sets of them are being compared, index construction is simple, indeed. When, perhaps, the prices or quantities of hundreds or even thousands of goods are being compared, things quickly become more complicated. This chapter discusses the construction of index numbers in a systematic way and also introduces the major types of business and economic index numbers that are regularly published and used in the United States.

We begin our discussion by considering a variety of methods that are commonly used to construct **price indexes,** which are numbers that measure the level of either a single price or of a set of prices at one time or place relative to another time or place. Later in the chapter, we will turn to **quantity indexes,** which are analogously defined as numbers that measure the magnitude of a single quantity or of a set of quantities at one time or place relative to another time or place.

SIMPLE PRICE INDEXES

The **simple price index,** or **price-relatives index,** is a number that compares the price of a single item at one time or place with the price of the same item at another time or place.

An Intertemporal Index

Table 16.1 shows the computation of an *intertemporal* price index. Columns 1 and 2 contain data about the average price of Grade-A eggs in Boston, Massachusetts, for a number of years. Each column 2 entry might be viewed as an average of the daily egg prices observed in a sample of Boston stores during the year. Accordingly, one can calculate a simple (intertemporal) price index by relating the price of each period to that of an *arbitrarily chosen* **base period,** or **reference period.** If we choose 1980 as the base period against which all comparisons are made, we can calculate the ratios of the current-year price (P_t) to the base-year price (P_0), as shown in column 3. For 1981, for instance, this ratio, which is also called a **price relative,** equals \$.88/\$.80, or 1.10, which is the second entry in column 3. In the same manner, all other entries in column 3 can be found by dividing the given year's price by the base-year price; naturally this procedure always makes the price relative for the base year itself equal to 1.00 (in our case, as a result of dividing \$.80 by \$.80). By tradition, the base-year index is, however, expressed as 100 (which can be thought of as a pure number or as 100 percent). Following this tradition, we can quickly convert all the column 3 price relatives into the simple price indexes of column 4. Thus, a formula for any simple (intertemporal) price index can be stated as in Box 16.A, where $I_{t,\,0}$ is the index for period t with period 0 base.

16.A Simple (intertemporal) Price Index

$$I_{t,0} = \frac{P_t}{P_0} \cdot 100$$

where P_t is the price in period t and P_0 is the price in base period 0.

Table 16.1 Constructing a Simple Price Index

Year (1)	Average Price of Grade-A Eggs in Boston (dollars per dozen) (2)	Current Price Relative to 1980 Price (P_t/P_0) (3)	Simple Price Index (1980 = 100) (4) = (3) × 100
1980	.80	1.00	100
1981	.88	1.10	110
1982	.96	1.20	120
1983	1.08	1.35	135
1984	1.12	1.40	140
1985	1.20	1.50	150

> ### Practice Problem 1
>
> #### Calculating a Simple Price Index for Grade-A Eggs in Boston in 1984
>
> Using calendar-year subscripts, this formula, for instance, gives us the 1984 index, with 1980 base, as
>
> $$I_{84,80} = \frac{P_{84}}{P_{80}} \cdot 100 = \frac{\$1.12}{\$.80} \cdot 100 = 140.$$
>
> The result, obviously, means that the 1984 Boston average price of Grade-A eggs equaled 140 percent of (or was 40 percent higher than) the corresponding 1980 price. III

An Interspatial Index

A simple *interspatial* price index can be calculated in an analogous manner. Imagine that column 1 of Table 16.1, instead of listing various years, contained the names of various cities, such as Boston, Detroit, New York, and so on, while column 2 listed the average egg prices in these cities for a *given* year, such as 1984. In that case, a base *location* (such as Boston) might be chosen and the price in each city, P_X, would be related to that at the base (P_b). For the Detroit-to-Boston comparison, the price relative might equal $(P_X/P_b) = (\$.88/\$.80) = 1.10$, and the corresponding simple (interspatial) price index would equal 110. This index number would indicate that the 1984 Detroit price of eggs was, on the average, 10 percent above the 1984 Boston price. Because this kind of analogy works for all other price indexes as well, we will henceforth focus our attention on intertemporal indexes entirely, but the reader should keep in mind the possibility of constructing interspatial price indexes using the same procedures.

COMPOSITE PRICE INDEXES

Typically, economic decision makers are concerned with more complex comparisons than the ones discussed so far. Consider their need to compare a whole set of prices prevailing at a given place at one time with a corresponding set of prices prevailing at the same place at a different time. Table 16.2 helps us visualize the problem. Given the indicated price quotations, it is easy enough to calculate simple price indexes separately for each of the food items listed in the table, and to note the degree to which some of them have risen and others have fallen, while others still haven't changed at all. What, however, has happened over time to the Boston price of breakfast food *in general?* Or how do such Boston prices in general compare to those in Detroit? A number of possible answers to these questions can be found by calculating a **composite price index,** a number that compares a set of prices for a variety of items at one time or place with such a set for the same items at another time or place. We will discuss a number of approaches to calculating composite price indexes always focusing on intertemporal comparisons.

An Unweighted Average of Simple Price Indexes

One way to construct a composite price index that combines the possibly divergent price movements of different items in a single number is to compute an *unweighted average* of the separate simple price indexes for all the items in question.

Table 16.2 Constructing a Composite Price Index by Averaging Simple Price Indexes: Unweighted Method

| Item (1) | Average Price in Boston (dollars per indicated unit) | | Simple Price Index, $\frac{P_{85}}{P_{84}} \cdot 100$ |
	in 1984, P_{84} (2)	in 1985, P_{85} (3)	(4) = [(3) ÷ (2)] · 100
Large eggs (dozen)	1.00	1.21	121
Whole milk (half gallon)	1.40	1.54	110
Wheat bread (loaf)	.88	.88	100
Margarine (pound)	1.80	1.71	95
Grape jelly (18 ounces)	1.50	2.01	134
Corn flakes (12 ounces)	.93	.93	100
			660

||| Practice Problem 2

Calculating an Unweighted Average of Simple Price Indexes for Boston Breakfast Food in 1985

For our example, such an unweighted average of price-relative indexes can quickly be calculated by summing the data of column 4 in Table 16.2 and dividing by the number of items, n, in that column:

$$I_{85,84} = \frac{\Sigma \left(\frac{P_{85}}{P_{84}} \cdot 100 \right)}{n} = \frac{660}{6} = 110.$$

This result indicates that Boston breakfast-food items in 1985 were, in general, 10 percent more expensive than in 1984. |||

This approach is stated more formally in Box 16.B.

16.B The Composite (intertemporal) Price Index as an Unweighted Average of Simple Price Indexes

$$I_{t,0} = \frac{\Sigma \left(\frac{P_t}{P_0} \cdot 100 \right)}{n}$$

where P_t is the price in period t and P_0 is the price in base period 0, while n is the number of indexes being averaged.

Indeed, this method of price indexation was the first one employed historically: In 1764, Giovanni R. Carli, an Italian nobleman, set out to measure the effect on European prices of the influx of precious metals following the discovery of America. He applied the Box 16.B formula to the prices of grain, oil, and wine in 1500 and 1750.

It is obvious, however, that this method of constructing a composite price index embodies one major flaw: every item is treated as if it had the same importance as every other item. This may, in fact, not be the case. Eggs and milk, for example, may be more important in people's diets than bread and margarine. This flaw brings us then to an alternative method.

A Weighted Average of Simple Price Indexes

In order to account for the possibly different importance of the various items entering the price index, one can weight each item by some appropriate measure of "importance." In the case of price indexes, the typical measure used is the set of quantities (Q) or values ($P \cdot Q = V$) associated with the set of prices for which the index is being constructed. This measure, however, sounds more simple than it is. Suppose we wanted to weight our simple price indexes by associated *value* figures (that is, according to the different amounts of money the typical Boston family was spending annually on the different goods listed in Table 16.2). Even in our simple example, an obvious question immediately arises: which set of values shall one use for weighting purposes, the values of the various foods purchased by people in 1984 (the base year) or the possibly different values purchased in 1985 (the current year) or even the "more typical" values purchased in some other, arbitrarily chosen year (an allegedly "normal" year)? As we shall illustrate presently, each of these possibilities (and there are others still) might lead to a different result!

Column 2 of Table 16.3 lists the simple price indexes calculated earlier; columns 3–5 show the amounts of money spent by an average Boston family on the six food items during 1984 (our base year), 1985 (our current year), and 1970 (an allegedly more "normal" year). Depending on which one of these value sets is used for weight-

Table 16.3 Constructing a Composite Price Index by Averaging Simple Price Indexes, Using Value Weights

Item (1)	Simple Price Index $\frac{P_{85}}{P_{84}} \cdot 100$ (2)	Value (dollars spent by average Boston family in year)			Weighted Simple Price Index		
		in 1984, V_{84} (3)	in 1985, V_{85} (4)	in 1970, V_{70} (5)	1984 Weights (6) = (2) · (3)	1985 Weights (7) = (2) · (4)	1970 Weights (8) = (2) · (5)
Large eggs	121	50	48.4	100	6,050	5,856.4	12,100
Whole milk	110	140	138.6	200	15,400	15,246.0	22,000
Wheat bread	100	352	396.0	200	35,200	39,600.0	20,000
Margarine	95	90	102.6	20	8,550	9,747.0	1,900
Grape jelly	134	150	180.9	50	20,100	24,240.6	6,700
Corn flakes	100	279	558.0	100	27,900	55,800.0	10,000
		1,061	1,424.5	670	113,200	150,490.0	72,700

ing purposes, a different 1985 price index (with 1984 base) emerges. In all cases, the index is calculated by the formula given in Box 16.C.

16.C The Composite (intertemporal) Price Index as a Value-Weighted Average of Simple Price Indexes

$$I_{t,0} = \frac{\Sigma\left(\frac{P_t}{P_0} \cdot 100\right) \cdot V}{\Sigma V}$$

where P_t is the price in period t, P_0 is the price in base period 0, and V can refer to values in period 0, in period t, or in any other period.

||| Practice Problem 3

Calculating the Value-Weighted Average of Simple Price Indexes for Boston Breakfast Food in 1985

a. If we use 1984 (base-year) value weights, our index equals

$$I_{85,84} = \frac{\Sigma\left(\frac{P_{85}}{P_{84}} \cdot 100\right) \cdot V_{84}}{\Sigma V_{84}} = \frac{113,200}{1,061} = 106.7.$$

This result suggests that Boston breakfast-food items in 1985 were, in general, about 6.7 percent more expensive than in 1984.

b. If we use 1985 (current-year) value weights, our index equals

$$I_{85,84} = \frac{\Sigma\left(\frac{P_{85}}{P_{84}} \cdot 100\right) \cdot V_{85}}{\Sigma V_{85}} = \frac{150,490}{1,424.5} = 105.6,$$

which suggests a 5.6 percent higher price level.

c. If we, finally, use 1970 (arbitrary) value weights, our index is

$$I_{85,84} = \frac{\Sigma\left(\frac{P_{85}}{P_{84}} \cdot 100\right) \cdot V_{70}}{\Sigma V_{70}} = \frac{72,700}{670} = 108.5,$$

which suggests 8.5 percent higher prices. |||

Note that none of the three results from Practice Problem 3 matches the 10-percent higher price index calculated by the unweighted-averaging method of index construction in Practice Problem 2, but which of the three results obtained by the supposedly superior weighted-averaging method is the correct one? An answer to that question does not exist because, when comparing a 1984 price set with a 1985 price set, it is equally logical to use 1984 value weights as 1985 value weights, and there may even be good reasons to use 1970 value weights if, say, consumer spending patterns in the mid-1980s were in some way abnormal and distorted (as a result, perhaps, of such unusual conditions as a war, a depression, or a hyperinflation). The problem of being able to construct a different index for a given phenomenon (such as price changes in Boston between 1984 and 1985), depending on which one of several *equally logical* sets of weights is being employed is called the **index-number problem.** The problem occurs whenever the relative magnitudes of the weights contained in one set (such as our 1984 values) differ from the relative magnitudes of the weights contained in another set (such as our 1985 or 1970 values). Unfortunately, this difference is usually present. (One can show, however, that the index-number problem disappears when the relative magnitudes of weights contained in alternative sets of weights are the same. Thus, the same composite price index would have been computed in our example if, compared to the 1984 value weights, say, *all* 1985 value weights had been precisely 10 percent higher and, say, *all* 1970 value weights had been precisely 40 percent lower.)

An Unweighted Aggregative Price Index

An altogether different approach to constructing a composite price index does not involve the prior computation of individual price relatives at all. In its simplest form, this approach is to construct an **unweighted aggregative price index** by simply summing all the prices of the current year (for which the index is desired) and relating that sum to a similar sum of all the base-year prices. The formula is given in Box 16.D.

16.D The Composite (intertemporal) Price Index as an Unweighted Aggregative Index

$$I_{t,0} = \frac{\Sigma P_t}{\Sigma P_0} \cdot 100$$

where P_t is the price in period t and P_0 in the price is base period 0.

||| Practice Problem 4

Calculating an Unweighted Aggregative Index for Boston Breakfast-Food Prices in 1985

Part A of Table 16.4 presents once again our earlier data from columns 1–3 of Table 16.2. By applying our Box 16.D formula, we can compute a 1985 price index, with 1984 base, of

$$I_{85,84} = \frac{\Sigma P_{85}}{\Sigma P_{84}} \cdot 100 = \frac{8.28}{7.51} \cdot 100 = 110.3,$$

which suggests a 10.3 percent increase in breakfast-food prices.

Table 16.4 Constructing a Composite Price Index as an Unweighted Aggregative Index

| | Average Price in Boston (dollars per indicated unit) | | | | | |
| | Part (A) | | | Part (B) | | |
Item	Unit	P_{84}	P_{85}	Unit	P^*_{84}	P^*_{85}
Large eggs	dozen	1.00	1.21	1,000	83.33	100.83
Whole milk	half gallon	1.40	1.54	5 gallons	14.00	15.40
Wheat bread	loaf	.88	.88	loaf	.88	.88
Margarine	pound	1.80	1.71	pound	1.80	1.71
Grape jelly	18 ounces	1.50	2.01	pound	1.33	1.79
Corn flakes	12 ounces	.93	.93	pound	1.24	1.24
		7.51	8.28		102.58	121.85

Now watch what happens if we present the *identical* price data but express prices with respect to different quantity units, as we do in Part B of Table 16.4. (Note, for example, that a $1 price per dozen eggs is the same thing as a $83.33 price per 1,000 eggs.) If we apply our Box 16.D formula in Part B, we can compute an entirely different index of

$$I_{85,84} = \frac{\Sigma P^*_{85}}{\Sigma P^*_{84}} \cdot 100 = \frac{121.85}{102.58} \cdot 100 = 118.8.$$

This result suggests that breakfast-food prices have risen not by 10.3 but by 18.8 percent! **|||**

Thus, an unweighted aggregative price index is a poor index because its magnitude is affected by the physical units in which goods happen to be priced. In addition, like the Box 16.B unweighted average of simple price indexes, this index, too, treats all goods as equally important—which is unlikely to be true. Both problems are avoided by introducing *quantity* weights into the Box 16.D formula.

Weighted Aggregative Price Indexes: Laspeyres, Paasche, and Others

We shall consider a number of **weighted aggregative price indexes.** Each of them computes a sum of quantity-weighted prices for the current year and then relates that sum to a similar sum for the base year. The indexes differ from one another with respect to the way in which the quantity weights are chosen.

The Laspeyres price index. The most commonly used price index nowadays is probably the weighted aggregative index introduced in 1864 by Étienne Laspeyres. The **Laspeyres price index** is determined by introducing base-year quantity weights (Q_0) to the numerator and denominator of the Box 16.D formula, which produces the new formula given in Box 16.E.

16.E The Composite (intertemporal) Price Index as the Weighted Aggregative Index of Laspeyres

$$I_{t,0} = \frac{\Sigma P_t \cdot Q_0}{\Sigma P_0 \cdot Q_0} \cdot 100$$

where P_t is the price in period t, P_0 is the price and Q_0 is the quantity in base period 0.

It is easy to see how this simple weighting device instantly eliminates both of the objections noted earlier about the unweighted aggregative index. First, the fact that each price is multiplied by quantity makes the choice of quantity units unproblematical because these units cancel. Thus, a price of $1 *per dozen* eggs, when multiplied by a quantity of, say, 100 *dozens,* simply yields $100 (the "dozens" have disappeared). An equivalent price of $83.33 *per thousand* eggs, when multiplied by the equivalent quantity of 1.2 *thousand* eggs, also yields $100 (the "thousand" has disappeared). Second, the use of differentiated quantity weights eliminates the equal treatment of all goods.

‖ Practice Problem 5

Calculating the Weighted Aggregative Laspeyres Index for Boston Breakfast-Food Prices in 1985

We can apply the Laspeyres formula to our familiar data, now rearranged in columns 1–4 of Table 16.5. The 1984 (base-year) quantity data appearing here are implied by column 3 of Table 16.3 and column 2 of Table 16.2; the remainder of Table 16.5 has been calculated to give us the values needed for the Laspeyres formula. Thus,

$$I_{85,84} = \frac{\Sigma P_{85} \cdot Q_{84}}{\Sigma P_{84} \cdot Q_{84}} \cdot 100 = \frac{1{,}132}{1{,}061} \cdot 100 = 106.7.$$

Table 16.5   Constructing a Composite Price Index: The Weighted Aggregative Index of Laspeyres

Item (1)	Average Price in Boston (dollars per indicated unit)		Quantity (indicated units bought by average Boston family in base year), Q_{84} (4)	Prices Weighted by Base-Year Quantities (dollars),	
	in 1984, P_{84} (2)	in 1985, P_{85} (3)		$P_{84} \cdot Q_{84}$ (5) = (2) × (4)	$P_{85} \cdot Q_{84}$ (6) = (3) × (4)
Large eggs (dozen)	1.00	1.21	50	50	60.5
Whole milk (half gallon)	1.40	1.54	100	140	154.0
Wheat bread (loaf)	.88	.88	400	352	352.0
Margarine (pound)	1.80	1.71	50	90	85.5
Grape jelly (18 ounces)	1.50	2.01	100	150	201.0
Corn flakes (12 ounces)	.93	.93	300	279	279.0
				1,061	1,132.0

According to the Laspeyres formula, Boston breakfast-food items in 1985 were, in general, 6.7 percent more expensive than in 1984. ||||

Note: It is no accident that the result in Practice Problem 5 equals the value-weighted average of simple price indexes calculated with 1984 weights in part (a) of Practice Problem 3. If base-year weights (V_0) are used in the Box 16.C formula, the formulas of Box 16.C and 16.E are identical:

$$I_{t,0} = \underbrace{\frac{\Sigma \left(\frac{P_t}{P_0} \cdot 100 \right) V_0}{\Sigma V_0} = \frac{\Sigma \left(\frac{P_t}{P_0} \cdot 100 \right) P_0 \cdot Q_0}{\Sigma P_0 \cdot Q_0} = \frac{\Sigma P_t \cdot Q_0}{\Sigma P_0 \cdot Q_0} \cdot 100}$$

Box 16.C
formula with
base-year
value weights

Box 16.E
Laspeyres
formula

The Paasche price index. An alternative type of weighted aggregative index was introduced in 1874 by Hermann Paasche. The **Paasche price index** is found by replacing the base-year quantity weights of Laspeyres by current-year quantity weights (Q_t), as is now indicated in Box 16.F.

16.F The Composite (intertemporal) Price Index as the Weighted Aggregative Index of Paasche

$$I_{t,0} = \frac{\Sigma P_t \cdot Q_t}{\Sigma P_0 \cdot Q_t} \cdot 100$$

where P_t is the price and Q_t is the quantity in base period t, while P_0 is the price in period 0.

|||| Practice Problem 6

Calculating the Weighted Aggregative Paasche Index for Boston Breakfast-Food Prices in 1985

We apply the Paasche formula to our familiar data, now rearranged in columns 1–4 of Table 16.6. The 1985 (current-year) quantity data appearing here are implied by column 4 of Table 16.3 and column 3 of Table 16.2; the remainder of Table 16.6 has been calculated to give us the values needed for the Paasche formula. Thus,

$$I_{85,84} = \frac{\Sigma P_{85} \cdot Q_{85}}{\Sigma P_{84} \cdot Q_{85}} \cdot 100 = \frac{1{,}424.5}{1{,}363} \cdot 100 = 104.5.$$

According to the Paasche formula, Boston breakfast-food items in 1985 were, in general, 4.5 percent more expensive than in 1984.

Table 16.6   Constructing a Composite Price Index: The Weighted Aggregative Index of Paasche

Item (1)	Average Price in Boston (dollars per indicated unit)		Quantity (indicated units bought by average Boston family in current year), Q_{85} (4)	Prices Weighted by Current-Year Quantities (dollars),	
	in 1984, P_{84} (2)	in 1985, P_{85} (3)		$P_{84} \cdot Q_{85}$ (5) = (2) × (4)	$P_{85} \cdot Q_{85}$ (6) = (3) × (4)
Large eggs (dozen)	1.00	1.21	40	40	48.4
Whole milk (half gallon)	1.40	1.54	90	126	138.6
Wheat bread (loaf)	.88	.88	450	396	396.0
Margarine (pound)	1.80	1.71	60	108	102.6
Grape jelly (18 ounces)	1.50	2.01	90	135	180.9
Corn flakes (12 ounces)	.93	.93	600	558	558.0
				1,363	1,424.5

III

Clearly, this result differs from that of the Laspeyres formula, but that is to be expected. Once again, we have encountered the index-number problem! Neither result is more correct or faulty than the other one because the two formulas measure two different things. Laspeyres tells us how much more or less money people would have had to spend in 1985, compared to 1984, if they had tried to buy in both years the physical quantities that they in fact bought in 1984. (The answer is "6.7 percent more"; thus, it is concluded that prices on the average rose by 6.7 percent.) Paasche tells us how much more or less money people would have had to spend in 1985, compared to 1984, if they had tried to buy in both years the physical quantities that they in fact bought in 1985. (The answer is "4.5 percent more"; thus, it is concluded that prices, on the average, rose by 4.5 percent.) The truth is, of course, that both of these stories are *pure fiction;* each serves only the purpose of constructing two value figures—$\Sigma P_t \cdot Q_0$ and $\Sigma P_0 \cdot Q_0$ in the case of Laspeyres and $\Sigma P_t \cdot Q_t$ and $\Sigma P_0 \cdot Q_t$ in the case of Paasche—(1) that conveniently contain the same quantity component, (2) that can, therefore, differ from one another only as a result of intertemporal differences in prices, and (3) that can, thus, be used to estimate price change. Yet, in truth, people bought 1984 quantities in 1984 and different 1985 quantities in 1985.

Indeed, the discrepancy between the two types of indexes is likely to get larger as the distance between the base period and the current one increases. Thus, the difference between Laspeyres and Paasche indexes that attempted to measure price changes, say, between 1985 and 1885 would surely be vast. First, people's tastes often change drastically over such long periods; the discovery of cholesterol in eggs and whole milk, for example, might make modern health-conscious consumers buy relatively lower quantities than their ancestors did a century earlier. The discrepancy would be large, second, because consumers change their buying patterns in response to changes in relative prices: a century of improvements in agricultural productivity, for example, might drastically reduce the relative prices of wheat and corn products and make consumers buy relatively larger quantities of these even in the absence of any changes in tastes. As a result of these and other factors, an 1885 quantity set (which Laspeyres would use) would hardly resemble the 1985 quantity set (which Paasche would use), and the indexes would differ from one another accordingly.

Other weighted aggregative price indexes. There is literally no end to the possibilities of constructing different types of weighted aggregative price indexes once one realizes that all kinds of other weighting schemes can be used as well; two of the more common ones are the *typical-year aggregative price index* and the *Edgeworth price index.*

The **typical-year aggregative price index** is constructed in the manner of the Laspeyres and Paasche indexes, except that neither a base-year nor a current-year quantity set is employed; instead, quantity weights (Q_a) of any other arbitrarily selected but "typical" year (or even the average quantities of several years) are substituted, as indicated in Box 16.G.

16.G The Composite (intertemporal) Price Index as the Typical-Year Aggregative Index

$$I_{t,0} = \frac{\Sigma P_t \cdot Q_a}{\Sigma P_0 \cdot Q_a} \cdot 100$$

where P_t is the price in period t, P_0 is the price in base period 0, and Q_a is the quantity of the "typical" year.

Sometimes this index is also referred to as the "fixed-weight" aggregative index, but that name is a most unfortunate one; the other weighted indexes discussed here have fixed weights, too; they just refer to a different kind of year.

Another version of a weighted aggregative price index was suggested by Alfred Marshall (1842–1924) and popularized by Francis Y. Edgeworth (1845–1926), both renowned British economists. The **Edgeworth price index,** as it is now called, is defined in Box 16.H; it combines features of the Laspeyres and Paasche indexes by using the sum of base-year and current-year quantities as weights.

16.H The Composite (intertemporal) Price Index as the Edgeworth Index

$$I_{t,0} = \frac{\Sigma P_t(Q_0 + Q_t)}{\Sigma P_0(Q_0 + Q_t)} \cdot 100$$

where P_t is the price and Q_t is the quantity in period t, while P_0 and Q_0 are price and quantity is base period 0.

> ||| Practice Problem 7

Calculating the Edgeworth Index for Boston Breakfast-Food Prices in 1985

For our example, the Edgeworth index can be calculated as

$$I_{85,84} = \frac{\Sigma P_{85}(Q_{84} + Q_{85})}{\Sigma P_{84}(Q_{84} + Q_{85})} \cdot 100 = \frac{2,556.5}{2,424} \cdot 100 = 105.5.$$

Thus, the result lies between the Laspeyres and Paasche results (of 106.7 and 104.5, respectively). |||

FISHER'S QUALITY TESTS AND HIS IDEAL INDEX

With so many indexes from which to choose, the index maker naturally looks for guidance to facilitate the inevitable choice. In a now classic work on *The Making of Index Numbers,* first published in 1922, the American economist Irving Fisher (see Biography 16.1 at the end of the chapter) set out to provide general principles according to which superior index numbers could be distinguished from inferior ones. Although he did not consider them equally important, we shall consider three of his index-number quality tests: the time-reversal test, the factor-reversal test, and the circularity test. We shall also consider the ideal index-number formula suggested by Fisher himself.

The Time-Reversal Test

A price-index formula is said to pass Fisher's **time-reversal test** if, in a comparison of prices at two dates, the same result is obtained regardless of which of the two dates is chosen as a base. For example, if the formula indicates that 1985 prices are twice as high as 1975 prices when 1975 is used as a base, then it should also indicate that 1975 prices are half as high as 1985 prices when 1985 is used as a base. Denoting the price index by I^P and the two dates by 0 and 1, the test requires that

$$\frac{I_{1,0}^P}{100} = \frac{100}{I_{0,1}^P}$$

or that

$$\frac{I_{1,0}^P}{100} \cdot \frac{I_{0,1}^P}{100} = 1.$$

The Factor-Reversal Test

The **factor-reversal test** presupposes that the weights used in a price-index formula are quantity weights. This being so, the formula is said to pass the test if division of the price index by 100, and subsequent multiplication by a corresponding quantity index (derived by interchanging the *P*s and *Q*s in the formula) produces a value equal to an independently derived value index. Denoting the price index by I^P and the quantity index by I^Q, the test requires that

$$\frac{I_{0,1}^P}{100} \cdot I_{0,1}^Q = \frac{\Sigma P_1 Q_1}{\Sigma P_0 Q_0} \cdot 100.$$

The Circularity Test

A price-index formula is said to pass Fisher's **circularity test** if the price index of year 1 with year 0 base, when multiplied by the price index of year 2 with year 1 base and divided by 100, equals the independently calculated price index of year 2 with year 0 base. For example, if the formula indicates that prices have doubled between 1975 and 1980 and then have doubled again between 1980 and 1985, it should also indicate that prices have quadrupled between 1975 and 1985 when the prices of the latter two years are compared directly. Thus, the test requires that

$$\frac{I_{1,0}^P \cdot I_{2,1}^P}{100} = I_{2,0}^P.$$

Testing the Laspeyres and Paasche Indexes

All of Fisher's quality tests make eminent sense. Surely, if A is twice as big as B, an examination of B should reveal it as being half as big as A (the time-reversal test). Surely, if A times B equals C, the aggregation of lots of As times the aggregation of lots of Bs should equal the aggregation of lots of Cs (the factor-reversal test). Surely, if B is twice A and C is twice B, C should be found to equal four times A (the circularity test). Yet, surprisingly, many index numbers in common use fail to meet these tests—including the Laspeyres and Paasche indexes, as the following calculations illustrate.

‖ Practice Problem 8

Testing the Laspeyres and Paasche Indexes of Boston Breakfast-Food Prices by Fisher's Criteria

a. *The time-reversal test:* **Laspeyres.** We already found in Practice Problem 5, using the Laspeyres index, that

$$I_{85,84}^{P \text{ (Laspeyres)}} = \frac{\Sigma P_{85} \cdot Q_{84}}{\Sigma P_{84} \cdot Q_{84}} \cdot 100 = \frac{1{,}132}{1{,}061} \cdot 100 = 106.7.$$

By using calculations already made in Table 16.6, we can compute a corresponding Laspeyres price index for 1984 with 1985 base as

$$I_{84,85}^{P \text{ (Laspeyres)}} = \frac{\Sigma P_{84} \cdot Q_{85}}{\Sigma P_{85} \cdot Q_{85}} \cdot 100 = \frac{1{,}363}{1{,}424.5} \cdot 100 = 95.7.$$

The Laspeyres index does not pass the time-reversal test because

$$\frac{106.7}{100} \cdot \frac{95.7}{100} = 1.021 \neq 1.$$

b. *The time-reversal test:* **Paasche.** We already found in Practice Problem 6, using the Paasche index, that

$$I^{P\ (\text{Paasche})}_{85,84} = \frac{\Sigma P_{85} \cdot Q_{85}}{\Sigma P_{84} \cdot Q_{85}} \cdot 100 = 104.5.$$

By using calculations already made in Table 16.5, we can compute a corresponding Paasche price index for 1984 with 1985 base as

$$I^{P\ (\text{Paasche})}_{84,85} = \frac{\Sigma P_{84} \cdot Q_{84}}{\Sigma P_{85} \cdot Q_{84}} \cdot 100 = \frac{1,061}{1,132} \cdot 100 = 93.7.$$

The Paasche index does not pass the time-reversal test because

$$\frac{104.5}{100} \cdot \frac{93.7}{100} = .979 \neq 1.$$

c. *The factor-reversal test:* **Laspeyres.** We already found in Practice Problem 5, using the Laspeyres index, that

$$I^{P\ (\text{Laspeyres})}_{85,84} = \frac{\Sigma P_{85} \cdot Q_{84}}{\Sigma P_{84} \cdot Q_{84}} \cdot 100 = \frac{1,132}{1,061} \cdot 100 = 106.7.$$

The corresponding Laspeyres *quantity* index is found by interchanging the price and quantity entries in the formula and using the calculations already available in Tables 16.5 and 16.6:

$$I^{Q\ (\text{Laspeyres})}_{85,84} = \frac{\Sigma Q_{85} \cdot P_{84}}{\Sigma Q_{84} \cdot P_{84}} \cdot 100 = \frac{1,363}{1,061} \cdot 100 = 128.5.$$

(This result indicates that 1985 quantities were, on average, 28.5 percent above 1984 quantities, provided both sets are aggregated with the help of 1984 prices.) Using Table 16.5 and 16.6 calculations, the corresponding value index is found to equal

$$I^{V}_{85,84} = \frac{\Sigma P_{85} \cdot Q_{85}}{\Sigma P_{84} \cdot Q_{84}} \cdot 100 = \frac{1,424.5}{1,061} \cdot 100 = 134.3.$$

(This result indicates that 1985 actual-dollar expenditures exceeded 1984 expenditures by 34.3 percent.) The Laspeyres index does not pass the factor-reversal test because

$$\frac{106.7}{100} (128.5) = 137.1 \neq 134.3.$$

d. *The factor-reversal test:* **Paasche.** We already found in Practice Problem 6, using the Paasche index, that

$$I_{85,84}^{P \text{ (Paasche)}} = \frac{\Sigma P_{85} \cdot Q_{85}}{\Sigma P_{84} \cdot Q_{85}} \cdot 100 = 104.5.$$

The corresponding Paasche *quantity* index is found by interchanging the price and quantity entries in the formula and using the calculations already available in Tables 16.5 and 16.6:

$$I_{85,84}^{Q \text{ (Paasche)}} = \frac{\Sigma Q_{85} \cdot P_{85}}{\Sigma Q_{84} \cdot P_{85}} \cdot 100 = \frac{1,424.5}{1,132} \cdot 100 = 125.8.$$

(This result indicates that 1985 quantities were, on average, 25.8 percent above 1984 quantities, provided both sets are aggregated with the help of 1985 prices.) The value index has already been found in part (c) to equal 134.3. The Paasche index does not pass the factor-reversal test because

$$\frac{104.5}{100} (125.8) = 131.5 \neq 134.3.$$

e. *The circularity test:* **Laspeyres.** To perform this test, let us assume the following additional data concerning the food items listed in Tables 16.5 and 16.6:

P_{86}:	2.00	1.75	1.00	2.00	2.50	1.00
Q_{86}:	45	100	400	80	120	500

We also denote 1984, 1985, and 1986 as year 0, 1, and 2, respectively. We already found in Practice Problem 5, using the Laspeyres index, that

$$I_{85,84}^{P \text{ (Laspeyres)}} = \frac{\Sigma P_{85} \cdot Q_{84}}{\Sigma P_{84} \cdot Q_{84}} \cdot 100 = \frac{1,132}{1,061} \cdot 100 = 106.7.$$

Using the new data and those in Tables 16.5 and 16.6, we can compute corresponding Laspeyres indexes of

$$I_{86,85}^{P \text{ (Laspeyres)}} = \frac{\Sigma P_{86} \cdot Q_{85}}{\Sigma P_{85} \cdot Q_{85}} \cdot 100 = \frac{1,632.5}{1,424.5} \cdot 100 = 114.6,$$

and

$$I_{86,84}^{P \text{ (Laspeyres)}} = \frac{\Sigma P_{86} \cdot Q_{84}}{\Sigma P_{84} \cdot Q_{84}} \cdot 100 = \frac{1,325}{1,061} \cdot 100 = 124.9.$$

The Laspeyres index does not pass the circularity test because

$$\frac{106.7}{100}(114.6) = 122.3 \neq 124.9.$$

f. *The circularity test:* **Paasche.** We assume the same data as in (e). We already found in Practice Problem 6, using the Paasche index, that

$$I_{85,84}^{P \text{ (Paasche)}} = \frac{\Sigma P_{85} \cdot Q_{85}}{\Sigma P_{84} \cdot Q_{85}} \cdot 100 = 104.5.$$

Using the new data and those of Table 16.5, we can compute corresponding Paasche indexes of

$$I_{86,85}^{P \text{ (Paasche)}} = \frac{\Sigma P_{86} \cdot Q_{86}}{\Sigma P_{85} \cdot Q_{86}} \cdot 100 = \frac{1,625}{1,403.45} \cdot 100 = 115.8,$$

and

$$I_{86,84}^{P \text{ (Paasche)}} = \frac{\Sigma P_{86} \cdot Q_{86}}{\Sigma P_{84} \cdot Q_{86}} \cdot 100 = \frac{1,625}{1,326} \cdot 100 = 122.5.$$

The Paasche index does not pass the circularity test because

$$\frac{104.5}{100}(115.8) = 121.0 \neq 122.5. \quad \text{|||}$$

The Ideal Index

Fisher himself did not place much importance on the circularity test (although most statisticians find it very appealing), and he suggested as ideal an index of his own that passes only the time-reversal and factor-reversal tests. **Fisher's ideal index** can be calculated for prices or quantities and equals the geometric mean of the Laspeyres and Paasche indexes, as stated in Box 16.I.[1]

16.I Fisher's Ideal Index

$$I_{t,0}^{\text{Fisher}} = \sqrt{I_{t,0}^{\text{Laspeyres}} \cdot I_{t,0}^{\text{Paasche}}}$$

||| Practice Problem 9

Calculating Fisher's Ideal Index for Boston Breakfast-Food Prices in 1985

Consider the Fisher index for 1985 prices, based on 1984, using once more our example of Boston breakfast-food prices. Given a Laspeyres price index of $I_{85,84}^{P \text{ (Laspeyres)}} = 106.7$ and a Paasche price index of $I_{85,84}^{P \text{ (Paasche)}} = 104.5$, the Fisher price index equals

$$I_{85,84}^{P \text{ (Fisher)}} = \sqrt{106.7(104.5)} = 105.6. \quad \text{|||}$$

---

[1]The geometric mean of two positive values, *a* and *b,* equals the square root of their product, $\sqrt{a \cdot b}$. This mean is discussed in the "Supplementary Topics" section of Chapter 4 of the *Student Workbook* that accompanies this text.

||| Practice Problem 10

Testing Fisher's Ideal Index for Boston Breakfast-Food Prices by Fisher's Criteria

We can quickly perform the three tests with the help of the data already available.

a. *The time-reversal test:* **Fisher.** We found in parts (a) and (b) of Practice Problem 8 that the Laspeyres and Paasche price indexes for 1984 with a 1985 base were, respectively,

$$I_{84,85}^{P \text{ (Laspeyres)}} = 95.7 \text{ and } I_{84,85}^{P \text{ (Paasche)}} = 93.7.$$

Hence, the Fisher price index for 1984 with 1985 as a base is

$$I_{84,85}^{P \text{ (Fisher)}} = \sqrt{95.7(93.7)} = 94.7.$$

The Fisher ideal index passes the time-reversal test because

$$\frac{105.6}{100}\left(\frac{94.7}{100}\right) = 1.00$$

b. *The factor-reversal test:* **Fisher.** We already know from Practice Problem 9 that the Fisher ideal price index for 1985, with 1984 base, is

$$I_{85,84}^{P \text{ (Fisher)}} = 105.6.$$

The corresponding Fisher *quantity* index equals

$$I_{85,84}^{Q \text{ (Fisher)}} = \sqrt{I_{85,84}^{Q \text{ (Laspeyres)}} \cdot I_{85,84}^{Q \text{ (Paasche)}}} = \sqrt{128.5(125.8)} = 127.1.$$

The Fisher index passes the factor-reversal test because

$$\frac{105.6}{100}(127.1) = 134.2$$

which, except for rounding error, equals the value index calculated in part (c) of Practice Problem 8.

c. *The circularity test:* **Fisher.** We already know from Practice Problem 9 that the Fisher ideal price index for 1985, with 1984 base, is

$$I_{85,84}^{P \text{ (Fisher)}} = 105.6.$$

From the above data in part (e) of Practice Problem 8, we can also compute

$$I_{86,85}^{P \text{ (Fisher)}} = \sqrt{I_{86,85}^{P \text{ (Laspeyres)}} \cdot I_{86,85}^{P \text{ (Paasche)}}} = \sqrt{114.6(115.8)} = 115.2,$$

and

$$I_{86,84}^{P \text{ (Fisher)}} = \sqrt{I_{86,84}^{P \text{ (Laspeyres)}} \cdot I_{86,84}^{P \text{ (Paasche)}}} = \sqrt{124.9(122.5)} = 123.7.$$

The Fisher index does *not* pass the circularity test because

$$\frac{105.6(115.2)}{100} = 121.7 \neq 123.7. \quad \blacksquare\blacksquare\blacksquare$$

CHAIN INDEXES

All the price-index formulas discussed so far (Boxes 16.A–16.I) have involved comparisons between two dates only (a current period and a base period). In many cases, when price indexes are constructed repeatedly (month after month, quarter after quarter, or year after year), the base is continually moved forward such that each period's prices are related to those of the immediately preceding period. Thus, one might get (as we did above) a Laspeyres price index for 1985, *with 1984 base,* of 106.7, followed a year hence by an index for 1986, *with 1985 base,* of 114.6, and so on indefinitely. Such series of index numbers, each of which uses the immediately preceding period as a base, rather than a common base, are called **chain indexes.**

A series of such numbers can, however, be converted to a common base if longer-term comparisons are desired, yet caution is advised. If the index-number formula in question does not meet the circularity test, one cannot multiply, say, $I^P_{85,84} = 106.7$ by $I^P_{86,85} = 114.6$, divide by 100, and expect to have found $I^P_{86,84} = 124.9$. [Recall part (e) of Practice Problem 8, "The circularity test: **Laspeyres.**"] Still, chain indexes *can* be converted into a series with a common base. The usual procedure employs the formula in Box 16.J, wherein Q_a is an arbitrary set of quantity weights that must be applied to all indexes alike.

16.J Converting Chain Price Indexes to a Common Base

$$I^P_{t,0} = I^P_{t-1,0} \frac{\Sigma P_t \cdot Q_a}{\Sigma P_{t-1} \cdot Q_a}$$

where P_t is the price in period t, P_{t-1} is the price in preceding period $t - 1$, and Q_a is the quantity in arbitrary period a.

‖‖‖ Practice Problem 11

Calculating a Series of Index Numbers with a Common Base for Boston Breakfast-Food Prices

If we let Q_a equal the 1984 quantity set given in Table 16.5 and use the calculations already made above in Practice Problem 8, we can find the correct 1986 Laspeyres price index with 1984 base as

$$I^P_{86,84} = I^P_{85,84} \frac{\Sigma P_{86} \cdot Q_{84}}{\Sigma P_{85} \cdot Q_{84}} = 106.7 \left(\frac{1,325}{1,132}\right) = 124.9,$$

which is the number calculated independently above. Thus, our short series of 1984, 1985, and 1986 price indexes *with a common 1984 base* is 100, 106.7, and 124.9. ‖‖‖

MANIPULATING INDEX-NUMBER TIME SERIES

On occasion, it is desirable to manipulate index-number times series in various ways. Three of the most common procedures are *shifting, splicing,* and *combining.*

Shifting the Base

Imagine the availability of an index-number time series with a common base, such as the one given in column 2 of Table 16.7. (This series is in fact the official U.S. *consumer price index* to be discussed below.) Over time, users of such an index may be less and less able to relate to the base period because it moves further and further away from the realm of their personal experience. For example, to someone who was not an adult in 1967 (and who has no concept of the meaning of "1967 purchasing power"), the

Table 16.7 Shifting the Base of an Index-Number Time Series

Year (1)	Original Price Index (1967 = 100) (2)	Price Index with Shifted Base (1980 = 100) (3)
1960	88.7	35.9
1961	89.6	36.3
1962	90.6	36.7
1963	91.7	37.2
1964	92.9	37.6
1965	94.5	38.3
1966	97.2	39.4
1967	100.0	40.5
1968	104.2	42.2
1969	109.8	44.5
1970	116.3	47.1
1971	121.3	49.1
1972	125.3	50.8
1973	133.1	53.9
1974	147.7	59.8
1975	161.2	65.3
1976	170.5	69.1
1977	181.5	73.5
1978	195.4	79.2
1979	217.4	88.1
1980	246.8	100.0
1981	272.4	110.4
1982	289.1	117.1
1983	298.4	120.9
1984	311.1	126.1
1985	322.2	130.6
1986	328.4	133.1

fact that 1983 prices were about three times as high as 1967 prices may not mean very much. Under such circumstances, it may be desirable to shift the base of the index series to a more recent date, such as 1980.

Another reason for wanting to shift the base is that one may want to compare one index-number series (such as a series of U.S. price indexes based on 1967) with another series that has a different base (such as a series of U.S. *quantity* indexes or *French* price indexes based on 1980). Such side-by-side comparisons are eased considerably by having the same base periods.

The desired conversion of one index-number series into another one with a new base, which is called **base shifting,** is accomplished by successively dividing each value of the original series by the original index number (encircled in Table 16.7) for the year that is to become the new base and multiplying each result by 100. The calculations are shown in column 3 of Table 16.7. Note how the 1960 value of 88.7, when divided by the encircled 1980 value of 246.8 and multiplied by 100, yields the first entry of 35.9 in column 3. Changing the 88.7 figure to 35.9 means, of course, that 1960 prices (that had been listed as 88.7 percent of 1967 prices) are now expressed as 35.9 percent of 1980 prices. All other column 3 entries are similarly calculated, and the 1980 value turns into the new base value of 100.

Note how much easier it might be for a younger person to relate to the new base. The fact that 1983 prices were 121 percent of 1980 prices may well conjure up a more concrete image than the equivalent statement that 1983 prices were 298 percent of 1967 prices. Note also that the two index-number series provide the same year-to-year percentage changes. Column 2 tells us, for example, that prices between 1980 and 1981 rose by $[(272.4/246.8) \cdot 100] - 100 = 10.4$ percent; column 3 tells us the same.

Splicing Two Short Series into a Longer Series

Sometimes two (or more) index-number time series are available that cover different ranges of time. Columns 1–3 of Table 16.8 provide a hypothetical example that represents, however, a fairly common experience. The publication of one index-number series may be abandoned (because a new formula is being applied, because a major change in weights has been made, or because new goods are being included); a new, revised series may be published instead. A user of the index in question may, however, require a series that covers a range of time that includes years covered by both series. As long as no better alternatives are available, the user may have to resort to **splicing** the two short series together (that is, to uniting them into a single and longer series). Such splicing is not very satisfactory because the two series clearly measure different things, but it can be done as long as the two series contain at least one overlapping period. Our example shows how one might retain the values of the new series (the column 4 entries below the horizontal line), while making the old series conform to the new one by shifting its base to the base of the new series. This base shifting requires, in our case, the division of each old-series index-number by the encircled 150.6 and multiplication of the result by 100.

Combining Two Specialized Series into a More Comprehensive Series

Sometimes two (or more) index-number time series are available that cover different items rather than different ranges of time. Columns 1–3 of Table 16.9 provide an example. The user of such indexes (a car dealer, perhaps) may wish to combine them

Table 16.8 Splicing Two Short Index-Number Series Into a Longer Series

Year (1)	Old Price Index (1970 = 100) (2)	Revised Price Index (1978 = 100) (3)	Spliced Price Index (1978 = 100) (4)
1970	100.0	—	66.4
1971	105.2	—	69.9
1972	111.3	—	73.9
1973	117.9	—	78.3
1974	129.1	—	85.7
1975	136.8	—	90.8
1976	142.4	—	94.6
1977	146.5	—	97.3
1978	(150.6)	100.0	100.0
1979	—	102.3	102.3
1980	—	104.7	104.7
1981	—	106.4	106.4
1982	—	112.6	112.6
1983	—	122.5	122.5
1984	—	131.9	131.9
1985	—	137.3	137.3
1986	—	142.9	142.9

into a more comprehensive measure (for new *and* used cars, for example). Such combining of two specialized series can be accomplished by computing a weighted average of the separate indexes for each year. If our car dealer's sales, for example, included 70 percent new cars and 30 percent used cars, the published national indexes of any one year might be combined accordingly, as is shown in column 4. Each entry in column 4 equals .7 times the column 2 index plus .3 times the column 3 index. The dealer's personal experience may then be compared with the national one now given in column 4.

Table 16.9 Combining Two Specialized Index-Number Series into a More Comprehensive Series

Year (1)	Price Index for New Cars (1978 = 100) (2)	Price Index for Used Cars (1978 = 100) (3)	Combined Price Index for New and Used Cars (1978 = 100) (4) = [.7 × (2)] + [.3 × (3)]
1978	100.0	100.0	100.0
1979	107.9	107.8	107.9
1980	116.6	111.6	115.1
1981	123.7	137.7	127.9
1982	128.5	158.9	137.6
1983	133.4	183.3	148.4
1984	140.1	210.8	161.3
1985	147.1	242.4	175.7
1986	154.4	278.8	191.7

MAJOR U.S. PRICE INDEXES

Having discussed the procedures that might be employed in index-number construction, it is time to consider some of the actual price indexes regularly published in the United States. These include the consumer price index, the producer price index, various stock price indexes, and a variety of other price indexes, such as implicit price deflators.

The Consumer Price Index

The **consumer price index,** usually referred to as the **CPI,** is published monthly by the Bureau of Labor Statistics of the U.S. Department of Labor and is readily available in such government publications as the *Monthly Labor Review,* the *Survey of Current Business,* and the *Economic Report of the President.* The index measures the average change in U.S. consumer prices over time. It is a typical-year aggregative chain index that is converted to a common base. (Note formulas 16.G and 16.J.) Column 2 of Table 16.7 above contains some of the more recent data; they also appear in panel (a) of Figure 16.1. Note that this index is based on 1967 = 100, but the base year is changed about once per decade. (In more recent times, it had been set at 1957–59 and 1947–49.)

How is the CPI produced? It is obviously impossible to monitor *all* the transactions U.S. consumers make; therefore, the data going into the making of the index are derived from a complex series of samples. In the 1980s, price data are collected in 85 urban areas across the country, ranging from New York City (the prices of which are given a weight of 9.585 out of 100) to Anchorage, Alaska (the prices of which are given a weight of .102 only because fewer people pay them). In addition, price data are collected only for a fixed *market basket* of about 400 commodities and services. The content of this basket was established by a 1972–73 Consumer Expenditure Survey during which some 20,000 randomly selected families kept two-week diaries of everything they bought. As one can expect, the basket does not contain van Gogh paintings, but it does include major items of food, clothing, shelter, fuel, medical care, and more. Table 16.10 on page 714 shows the relative importance of the major groups of commodities and services that are being priced for the CPI.

The 400 specific items included in the fixed market basket are being priced in a variety of ways. Mail questionnaires establish a few prices, such as transportation costs, public-utility rates, and newspaper prices. Most prices, however, are found by personal visits of bureau representatives who visit regularly 24,000 retail stores and service establishments, 18,000 tenants (for rent), and another 18,000 housing units (for property taxes). The list of those to be visited was also established by sampling (based on such frames as the Census of Retail Trade). Bureau representatives attempt to price identical quantities of identical items at each visit, and, for that purpose, they follow detailed instructions. (A dress may be described not only by size, but also by width in hem and seams, seams pressed open, taping on inside of hem, type of stitching, number of loose threads, and the presence of thread belt loops!) When all is said and done, price data for any given location are combined into a weighted average (a popular store's price counting more than one from a store with few customers) and then into a weighted average of all sampled locations and, ultimately, all sampled goods.

Figure 16.1 The Recent Behavior of Selected U.S. Prices

These four panels picture the movements of four major price-index time series that are regularly published in the United States.

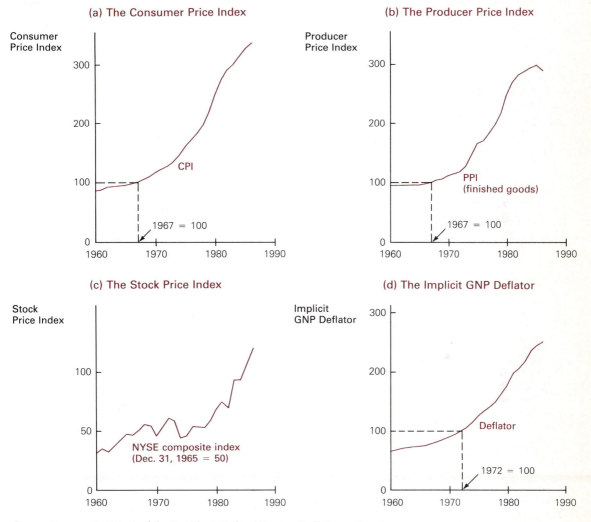

(a) The Consumer Price Index

(b) The Producer Price Index

(c) The Stock Price Index

(d) The Implicit GNP Deflator

Source: Economic Report of the President, Federal Reserve Bulletin, various issues.

In the process, a variety of consumer price indexes emerge. There are two overall indexes, one for urban wage and clerical workers only (the CPI-W), another one for all urban consumers (the CPI-U). The CPI-W excludes not only the prices paid by rural Americans, but also those paid by some urban ones, such as professional, managerial, and technical workers, the self-employed, short-term workers, the retired, and the (voluntarily or involuntarily) unemployed. Thus, it applies to only about 50 percent of the

Table 16.10   The December 1977 Relative Importance of Various Groups of Commodities and Services Priced for the CPI (all urban consumers)

Major Group	Relative Importance	Major Group	Relative Importance
All items	100.000	Apparel and upkeep	5.800
Food and beverages	18.813	Apparel commodities	5.137
Food	17.718	Mens' and boys' apparel	1.646
Food at home	12.235	Womens' and girls' apparel	2.044
Cereals and bakery products	1.530	Infants' and toddlers' apparel	.127
Meats, poultry, fish, and eggs	3.943	Footwear	.716
Dairy products	1.654	Other apparel commodities	.604
Fruits and vegetables	1.759	Apparel services	.662
Sugar and sweets	.435	Transportation	18.027
Fats and oils	.360	Private transportation	16.930
Nonalcoholic beverages	1.513	New cars	4.039
Other prepared food	1.041	Used cars	3.020
Food away from home	5.483	Gasoline	4.205
Alcoholic beverages	1.095	Maintenance and repair	1.516
Housing	43.911	Other private transportation	4.149
Shelter	29.181	Commodities	.733
Rent, residential	5.624	Services	3.416
Other rental costs	.711	Public transportation	1.097
Homeownership	22.846	Medical care	4.969
Home purchase	9.967	Medical-care commodities	.859
Financing, taxes, and insurance	9.211	Medical-care services	4.110
Maintenance and repairs	3.668	Professional services	2.007
Services	2.800	Other medical-care services	2.103
Commodities	.868	Entertainment	4.085
Fuel and other utilities	6.516	Entertainment commodities	2.423
Fuels	4.289	Entertainment services	1.662
Fuel oil, coal, and bottled gas	.897	Other goods and services	4.394
Gas (piped) and electricity	3.391	Tobacco products	1.202
Other utilities and public services	2.227	Personal care	1.752
Household furnishings and operation	8.215	Toilet goods, personal-care appliances	.791
House furnishings	4.602	Personal-care services	.961
Housekeeping supplies	1.559	Personal and educational expenses	1.441
Housekeeping services	2.053	School books and supplies	.189
		Personal and educational services	1.252

Source: Bureau of Labor Statistics, U.S. Department of Labor, *The Consumer Price Index: Concepts and Content Over the Years,* Report 517, May 1978 (revised), p. 22.

U.S. noninstitutionalized civilian population. The CPI-U, in contrast, covers 80 percent of this population; it is the index usually cited. In addition to the two overall indexes, separate indexes are published for many categories of goods and localities, and there are also seasonally unadjusted and seasonally adjusted CPI series.[2]

---

[2]An Analytical Example, "Consumer Price Indexes for Specific Population Groups," which can be found in Chapter 16 of the *Student Workbook* that accompanies this text, reports on a private effort to expand the CPI information further.

Criticizing the CPI is child's play, but there is no better index of the prices U.S. consumers pay. Among the criticisms are these:

1. The fixed market basket of 400 goods is always out of date, people in the 1980s are in fact buying different quantities and qualities of goods than the 1972–73 survey found; indeed, they are buying all kinds of goods that didn't even exist at the earlier time.
2. Bureau agents may not be finding the prices people actually pay. The agents may "shop" on Tuesdays, while most people buy goods on Saturdays, taking advantage of cheaper weekend specials. The agents may miss the presence of secret kickbacks, side payments, and quantity discounts. The agents will persist in pricing at a fixed sample of stores, while real shoppers are abandoning these for other, cheaper, or newer stores that may not be in the sample at all.
3. Bureau agents may miss changes in quality. The nickel candy bar that shrinks in size has surely risen in price, while the $30 tire that gives more mileage has surely fallen in price. Similarly, equal-priced medical care that can cure formerly incurable diseases is really cheaper, while equal-priced city bus rides that are dirtier, slower, and more crowded are surely more expensive.
4. The CPI is still irrelevant for the 20 percent of rural Americans whose goods are not being priced at all.

This list is by no means an exhaustive one.

The Producer Price Index

Like the CPI, the **producer price index,** formerly called the *wholesale price index,* is also a monthly product of the Bureau of Labor Statistics. It can be found in the same publications as the CPI. The index measures average price changes over time in U.S. primary (nonretail) markets. It is a value-weighted average of simple price indexes (see the Box 16.C formula); the weights are 1972 shipments. (A major revision of the index is to be completed by the late 1980s). Each month, bureau agents price several thousand domestically produced and imported products in U.S. nonretail markets. Ultimately, a number of indexes emerge: by stage of processing (for finished goods, intermediate materials, and crude materials) and by major commodity groups (for farm products, chemicals, machinery, and many more). Panel (b) of Figure 16.1 shows the recent behavior of the producer price index for finished goods (which is regarded as a leading indicator of the future behavior of consumer prices).

Stock Price Indexes

A variety of common-stock indexes are published regularly by the New York Stock Exchange (NYSE), by Dow Jones and Co., and by Standard & Poor's Corporation. All of them are computed from average daily closing prices, but coverages and base periods vary. The NYSE composite index—the recent behavior of which is shown in panel (c) of Figure 16.1—is based on December 31, 1965 = 50 and covers all of the more than 1,500 issues listed on the New York Stock Exchange; other NYSE indexes cover industrial, transportation, utilities, or finance stocks only. Standard & Poor's composite or "500 index" is based on 1941–43 = 10 and, as the name suggests, covers 500 stocks. The famous Dow-Jones Industrial Average, finally, is simply a weighted average of 30 stock prices; it is not even expressed as a percentage of any base-period average.

Implicit Price Deflators

The government publishes a multitude of economic time series (for example, in the annual *Economic Report of the President* or the Commerce Department's *Survey of Current Business*) that are expressed both in current dollars and constant (typically 1972) dollars. Such series include the gross national product and many of its expenditure subdivisions (consumption, investment, government, exports, imports), its industry subdivisions (such as agriculture, mining, manufacturing, services), and its product subdivisions (such as commodities, structures, and even autos). Given any current-dollar figure and a corresponding constant-dollar figure, one can divide the former by the latter, multiply by 100, and, thus, produce a so-called **implicit price deflator,** which is the price index for the item and period in question relative to the base period of the constant-dollar series.

Consider columns 1–3 of Table 16.11. They give data for the U.S. GNP at current and 1972 prices. The successive division of each column 1 figure by the associated column 2 figure, and multiplication by 100, yields the price-index series of column 4 that measures the change, relative to 1972, in the general price level of all the goods entering the GNP. Panel (d) of Figure 16.1 shows these and earlier data graphically.

MAJOR USES OF PRICE INDEXES

Price indexes are regularly put to use by all kinds of people; this section will describe some of these possible uses.

Table 16.11   Finding the Implicit GNP Deflator

Year (1)	GNP at Current Prices (billions of dollars) (2)	GNP at 1972 Prices (billions of dollars) (3)	Implicit GNP Deflator (4) = [(2) ÷ (3)] · 100
1972	1,185.9	1,185.9	100.0
1973	1,326.4	1,254.3	105.7
1974	1,434.2	1,246.3	115.1
1975	1,549.2	1,231.6	125.8
1976	1,718.0	1,298.2	132.3
1977	1,918.3	1,369.7	140.1
1978	2,163.9	1,438.6	150.4
1979	2,417.8	1,479.4	163.4
1980	2,631.7	1,475.0	178.4
1981	2,954.1	1,513.8	195.1
1982	3,073.0	1,485.4	206.9
1983	3,309.5	1,534.8	215.6
1984	3,690.9	1,587.6	232.5
1985	3,903.9	1,624.2	240.4
1986	4,109.3	1,669.1	246.2

Source: Economic Report of the President, various issues.

Evaluating Economic Policy

The most obvious use for price indexes is to help in evaluating the success or failure of economic policy. If a Chairman of the Federal Reserve Board or a President of the United States proclaims in 1970 that "inflation can and must be stopped during this decade," one can wait 10 years, look at the price-index series pictured in panels (a), (b), or (d) of Figure 16.1, and pronounce the policy a failure. (Given the fact that any price index can never be more than a rough estimate of the underlying phenomenon, one should be less hasty, however, about drawing conclusions from minute changes in the index from one month to the next. Nevertheless, newspapers never fail to make headlines when the CPI quivers by a decimal point.)

Deflating Current-Dollar Time Series

A less obvious, but important use of price indexes is their role as deflators of economic time series that are expressed in current dollars. Such **deflation** is a process that removes the effect of price changes from a current-dollar time series and, thus, restates the series in terms of *constant dollars* (dollars of constant purchasing power). The members of a labor union, for example, might be pondering the column 2 data of Table 16.12 just published by the U.S. government. Yet they might fail to be impressed by the 2.7-fold increase in their incomes since 1970. They might divide each entry in that column by the column 3 consumer price index (and multiply by 100) and, thus, compute the column 4 time series of their *real* income (that actually stayed pretty much the same).

Table 16.12 Deflating a Time Series with the Help of Index Numbers

Year (1)	Average Weekly Earnings in Manufacturing (current dollars) (2)	Consumer Price Index (1967 = 100) (3)	Average Weekly Earnings in Manufacturing (1967 dollars) (4)
1970	$133.33	116.3	$114.64
1971	142.44	121.3	117.43
1972	154.71	125.3	123.47
1973	166.46	133.1	125.06
1974	176.80	147.7	119.70
1975	190.79	161.2	118.36
1976	209.32	170.5	122.77
1977	228.90	181.5	126.12
1978	249.27	195.4	127.57
1979	269.34	217.4	123.89
1980	288.62	246.8	116.94
1981	318.00	272.4	116.74
1982	330.26	289.1	114.24
1983	354.08	298.4	118.66
1984	374.03	311.1	120.23
1985	385.97	322.2	119.79
1986	396.01	328.4	120.59

Source: Economic Report of the President, January 1987, pp. 293 and 307.

In the same way, a business executive might deflate current-dollar time series about company sales, production, inventory, or profit by some appropriate price index (a producer price index, perhaps) to discover the underlying change in physical volume or real purchasing power. Similarly, economists invariably deflate current-dollar series of national output, consumption, money supply, and a host of other variables because a knowledge of real change (undistorted by price movements) is more important to them.

Escalators

In recent years ever more legal contracts have come to contain so-called **escalator clauses,** according to which nominal monetary payments are raised periodically and automatically in accordance with increases in the consumer price index in order to preserve the real purchasing power of these payments. Such clauses are part of the labor contracts of millions of union members (who seek to get nominal pay raises of, say, 6.9 percent when the CPI rises by 6.9 percent so that the purchasing power of their pay remains unchanged). Similar clauses can be found in divorce settlements (with respect to alimony payments), in insurance contracts (with respect to premiums and benefits), and in long-term leases on commercial land and buildings (with respect to rents). Even the rents of some Chicago penthouse apartments have recently been tied to the index, as have been pensions, Social Security benefits, and certain welfare payments.

But note: Pay increases in strict accordance with the CPI are not necessarily required to keep an income recipient's welfare from being eroded by inflation. For one thing, people do not consume all of their income; some of it is taxed away or saved. Thus, a 6.9 percent increase in the CPI, together with a much smaller increase in money income, could easily be sufficient to raise the amount of money used for consumption by 6.9 percent. More important, inflation is never uniform, with all prices moving in the same direction and by the same percentage. While the CPI rises by 6.9 percent, some individual prices fall, others stay unchanged, and others rise. Consumers can and do adjust their purchases to these changes in *relative* prices: they buy more of relatively cheaper goods and less of relatively more expensive goods; they certainly do not continue to buy a fixed basket of goods forever. Thus, consumers who, following a rise in the CPI, are given enough extra dollars to be able to buy the very set of goods that they bought at the relative prices prevailing in the past, but who will in fact substitute a different set of goods because of the changed relative prices prevailing in the present, will in fact not be equally well off but will be better off. Hence, it is possible for consumers to keep a *constant* level of welfare (or utility), while spending *even on consumption goods* an added percentage that falls short of the rise in the CPI. Economists, in contrast to statisticians, therefore, have long been interested in establishing a *constant-utility index* that might indicate the extra amount of money people would require (say, 3.1 percent) to keep their utility constant (rather than to acquire an unchanged set of goods) in the face of a given increase (of, say, 6.9 percent) in the consumer price index.[3]

---

[3]For a brief discussion of the economic theory involved, see Heinz Kohler, *Intermediate Microeconomics: Theory and Applications,* 1st ed. (Glenview, Ill.: Scott, Foresman and Company, 1982), pp. 68–72. For further readings see Melville J. Ulmer, *The Economic Theory of Cost of Living Index Numbers* (New York: Columbia University Press, 1950).

A MAJOR U.S. QUANTITY INDEX: THE INDEX OF INDUSTRIAL PRODUCTION

While the major part of this chapter has been concerned with price indexes, quantity indexes are no less important. We need not, however, discuss at great length the theory of constructing them because all the above formulas in Boxes 16.A through 16.J apply fully if one interchanges the roles of P and Q. For example, a Laspeyres (intertemporal) *quantity* index equals

$$I_{t,0} = \frac{\Sigma Q_t \cdot P_0}{\Sigma Q_0 \cdot P_0} \cdot 100,$$

which is nothing else but a transformation of the Box 16.E formula and indicates that the index is constructed by aggregating time t and time 0 quantity sets using time 0 (or base-year) price weights. All other index formulas can be similarly transformed.

The most important quantity index used in the United States, perhaps, is the monthly (Federal Reserve) **index of industrial production,** published in the *Federal Reserve Bulletin.* It is designed to measure changes over time in the physical volume of industrial goods produced in the United States. The index uses *value added per unit of output* as weights. In addition to the overall index, separate indexes are computed for major industry divisions (such as manufacturing, mining, and utilities) and also for major market groupings (such as final products, intermediate products, and materials). Further subdivisions are available (final products, for example, are divided into automotive products, home goods, consumer staples, and equipment). Figure 16.2 shows

Figure 16.2 The Federal Reserve Index of Industrial Production

The physical volume of U.S. industrial production has grown significantly since World War II but not without the interruption of six major recessions (arrows).

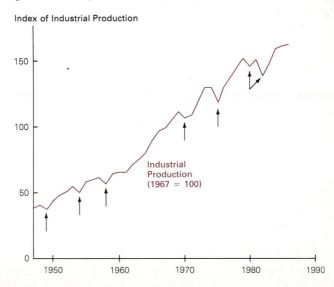

Source: Economic Report of the President, various issues.

the movement of the overall index since 1947; note the clear impact of the six postwar recessions.

Note: Just like price indexes, quantity indexes, too, can be plagued by the index-number problem. Analytical Examples 16.1, "The Index-Number Problem: Measuring Soviet Economic Growth," and 16.2, "The Index-Number Problem: Comparing U.S. with Soviet Real GNP," testify to this fact.

ANALYTICAL EXAMPLES

16.1 The Index-Number Problem: Measuring Soviet Economic Growth

The index-number problem emerges whenever an attempt is made to compare, over time or space, heterogeneous aggregates of prices or output. The sections of this chapter on "A Weighted Average of Simple Price Indexes" and "Weighted Aggregative Price Indexes" noted the difficulties associated with comparing price levels over time; consider the similar difficulty of comparing output levels over time. Heterogeneous outputs of apples, blast furnaces, medical care, and steel, for example, cannot be compared directly; they have to be added together using the common denominator of money. To yield meaningful results, the same price weights must be applied to the lists of quantities produced in two different periods, but one can use the prices of one year just as easily as those of another. If (as is likely) relative prices come to differ over time, different results can emerge.

Raymond P. Powell attempted to measure the growth of Soviet GNP from 1928–37, a time during which the Soviets industrialized their economy. Measured in 1928 prices, the average annual growth rate was 11.9 percent, but measured in 1937 prices, it was 6.2 percent.[1] Neither answer is more correct than the other. The hypothetical example in the accompanying table highlights the problem. Evaluate each year's physical output at earlier 1928 prices (as a Laspeyres *quantity* index would) and aggregate output is seen to have risen from $(100 \cdot 1) + (50 \cdot 1) = 150$ rubles to $(90 \cdot 1) + (300 \cdot 1) = 390$ rubles—at an average rate of more than 11 percent per year. Now evaluate each year's physical output at later 1937 prices (as a Paasche quantity index would) and aggregate output can be seen to have risen from $(100 \cdot 2) + (50 \cdot 0.5) = 225$ rubles to $(90 \cdot 2) + (300 \cdot 0.5) = 330$ rubles—at an average rate of less than 5 percent per year. There is no logical way to escape the problem.

[1] Raymond P. Powell, "Economic Growth in the U.S.S.R.," *Scientific American,* December 1968, pp. 17–23. Reprinted in Heinz Kohler, *Readings in Economics,* 2nd ed. (New York: Holt, Rinehart, & Winston, 1969), pp. 629–39.

Type of Output	1928		1937	
	Quantity	Price	Quantity	Price
Food	100	1	90	2
Machinery	50	1	300	0.5

Source: Heinz Kohler, *Intermediate Microeconomics: Theory and Applications,* 1st ed. (Glenview, Ill.: Scott, Foresman and Company, 1982), p. 71.

16.2 The Index-Number Problem: Comparing U.S. with Soviet Real GNP

The index-number problem is just as serious with respect to interspatial comparisons as it is with respect to intertemporal comparisons. One favorite type of interspatial comparison (that is continually being made by economists and politicians all over the world) involves the real GNPs of different countries. Thus, the author of this book once noted that, *measured in U.S. prices,* the Soviet real GNP has grown 10-fold between 1928 and 1975, while its U.S. counterpart has grown only 4-fold. As a result, the Soviet real GNP (that was equal to 25 percent of U.S. real GNP in 1928) rose to 40 percent of U.S. real GNP by 1955 and to 61 percent by 1975, making the Soviet Union the world's second largest economic power. The hypothetical ex-

ample in the table once again highlights the problem with this kind of statement.

Evaluate each country's physical output at U.S. prices, and Soviet output is seen to equal $(44 \cdot 1) + (200 \cdot 1) = 244$ dollars as opposed to U.S. output of $(200 \cdot 1) + (200 \cdot 1) = 400$ dollars; hence, Soviet output is 61 percent of U.S. output. Now evaluate each country's physical output at Soviet prices (which is just as logical) and Soviet output is seen to equal $(44 \cdot 1) + (200 \cdot 100) = 20,044$ rubles as opposed to U.S. output of $(200 \cdot 1) + (200 \cdot 100) = 20,200$ rubles; therefore, the Soviet output is 99 percent of the U.S. output. Both statements cannot be true at the same time.

Type of Output	1975 United States		1975 Soviet Union	
	Quantity	Price (dollars)	Quantity	Price (rubles)
Food	200	1	44	1
Machinery	200	1	200	100

Source: Heinz Kohler, *Scarcity and Freedom: An Introduction to Economics* (Lexington, Mass.: D.C. Heath, 1977), p. 449.

▌▌▌ BIOGRAPHY

16.1 Irving Fisher

Irving Fisher (1867–1947) was born at Saugerties, New York, the son of a Congregational minister. As did his father, Fisher studied at Yale. Mathematics was his favorite subject. He won first prize in a math contest even as a freshman; his doctoral dissertation, *Mathematical Investigations in the Theory of Value and Prices* (1892), was a landmark in the development of mathematical economics. It won immediate praise from no lesser figures than Francis Y. Edgeworth and Vilfredo Pareto, two renowned economists. Some 55 years later, Ragnar Frisch (eventual winner of the 1969 Nobel Prize in Economic Science) would say about Fisher: "He has been anywhere from a decade to two generations ahead of his time . . . it will be hard to find any single

work that has been more influential than Fisher's dissertation."[1] No wonder that Fisher was a full professor of political economy at Yale within seven years of graduation. He stayed there during his entire career.

Fisher's main contributions lie in the theory of utility and consumer choice, the theory of interest and capital, and the theory of statistics (index numbers, distributed lags). These contributions are reflected in such works as *The Nature of Capital and Income* (1906), *The Theory of Interest* (1907), *The Purchasing Power of Money* (1911)—a great pioneering venture in econometrics—and *The Making of Index Numbers* (1922).

[1]Ragnar Frisch, "Irving Fisher at Eighty," *Econometrica,* April 1947, pp. 71–72.

In *The Making of Index Numbers,* Fisher tested many of the formulas introduced in this chapter by the criteria also discussed here. The few formulas that got superlative ratings (such as the Edgeworth index and Fisher's own ideal index) include quantity (or price) weights from *both* of the time periods or geographic areas involved in the price (or quantity) comparison that is being made. The inclusion of two sets of weights tends to make such index construction costly and is the major reason why Fisher's own index is rarely used. (One exception: The Bureau of Foreign Commerce of the U.S. Department of Commerce calculates monthly quantity indexes of exports and imports using Fisher's formula.)

The works cited above established Fisher's reputation as the country's greatest scientific economist. As such, he served as president of the American Economic Association and was a founder and the first president of the Econometric Society. He played a major role in the establishment of the Cowles Foundation (now at Yale) as a means to nurture mathematical and quantitative research in economics.

Source: Heinz Kohler, *Intermediate Microeconomics: Theory and Applications,* 1rst ed. (Glenview, Ill.: Scott, Foresman and Company, 1982), pp. 264–65.

Summary

1. *Index numbers* measure the magnitude of a variable at one time or place relative to the magnitude of the same variable at another time or place. The variable in question can be an individual price, quantity, or value figure; more often, entire sets of these prices, quantities, or values are being compared. Thus, *price indexes* measure the level of either a single price or of a set of prices at one time or place relative to another time or place, while *quantity indexes* measure the magnitude of either a single quantity or of a set of quantities at one time or place relative to another time or place.

2. A *simple price index* compares the price of a single item at one time or place with the price of the same item at another time or place. The period to which a given price is related is called the *base period;* its index is expressed as 100.

3. A *composite price index* compares a set of prices for a variety of items at one time or place with such a set for the same items at another time or place. Such an index can be constructed in many ways: as an unweighted average of simple price indexes, a weighted average of simple price indexes, an unweighted aggregative price index, or a weighted aggregative price index. (Versions of the weighted aggregative price index, such as the Laspeyres price index, the Paasche price index, the typical-year aggregative price index, or the Edgeworth price index, are distinguished by the way in which the weights are chosen.) Weighted indexes are usually considered superior to unweighted ones in that all prices (or quantities) are not treated as being of equal importance, but weighting gives rise to the *index-number problem,* which arises when one is able to construct a different index for a given phenomenon depending on which one of several, equally logical sets of weights is being employed.

4. Irving Fisher suggested a number of quality tests designed to distinguish superior from inferior index numbers, including the time-reversal test, the factor-reversal test, and the circularity test. A price-index formula, for instance, is said to pass the *time-reversal test* if, in a comparison of prices at two dates, the same result is obtained regardless of which of the two dates is chosen as a base. Similarly, a (quantity-weighted) price-index formula is said to pass the *factor-reversal test* if division of the price index by 100, and subsequent multiplication by a corresponding quantity index, produces a value equal to an independently derived value index. A price-index formula is said to pass the *circularity test* if the price index of year 1 with year 0 base, when multiplied by the price index of year 2 with year 1 base and divided by 100, equals the independently calculated price index of year 2 with year 0 base. The Laspeyres and Paasche indexes, although commonly used, fail all of these tests. Fisher's ideal index, which equals the geometric mean of the Laspeyres and Paasche indexes, passes the time-reversal test and the factor-reversal test.

5. Series of index numbers each of which uses the immediately preceding period as a base, rather than using a common base, are called *chain indexes.* They can be converted into a single series with a common base to facilitate longer-term comparisons. Such conversion must, however, proceed with caution when the index-number formula in question fails to pass the circularity test.

6. Index-number time series can be manipulated in a variety of ways—for example, by shifting the base, by splicing two short series into a longer one, and by combining two specialized series into a more comprehensive one.

7. Among the major price indexes regularly published and used in the United States are the consumer price index, the producer price index, various stock price indexes, and implicit price deflators.

8. Price indexes serve a number of purposes. They are used to evaluate economic policy, to deflate current-dollar time series, and to facilitate computations required by escalator clauses of legal contracts. Escalator clauses that raise nominal income payments in strict accordance with increases in the consumer price index (to keep real income unchanged) probably *raise* the recipients' welfare.

9. Quantity indexes are no less important than price indexes, but they can easily be constructed by reversing the roles of P and Q in price-index formulas. The Federal Reserve index of industrial production is one of the more important quantity indexes used in the United States.

Key Terms

base period
base shifting
chain indexes
circularity test
composite price index
consumer price index (CPI)
deflation
Edgeworth price index
escalator clauses
factor-reversal test
Fisher's ideal index
implicit price deflator
index-number problem
index numbers
index of industrial production

Laspeyres price index
Paasche price index
price indexes
price relative
price-relatives index
producer price index
quantity indexes
reference period
simple price index
splicing
time-reversal test
typical-year aggregative price index
unweighted aggregative price index
weighted aggregative price indexes

Questions and Problems

The computer program noted at the end of the chapter can be used to work many of the subsequent problems.

1. Consider the hypothetical sample data concerning automobile operating costs contained in Table 16.A. Show your calculations in each case:
   a. Calculate separate 1984 simple price indexes for the four items, using 1985 as a base.
   b. Calculate the unweighted average of the simple price indexes computed in problem 1a. What does the index mean?
   c. Calculate the value-weighted average of the simple price indexes computed in problem 1a, using 1984 values. What does the index mean? Why is it so different from that obtained in problem 1b?

2. Once more consider the data of Table 16.A.
   a. Calculate separate 1985 simple price indexes for the four items, using 1984 as a base.

   b. Calculate the unweighted average of the simple price indexes computed in problem 2a.
   c. Calculate the value-weighted average of the simple price indexes computed in problem 2a, using 1985 values.

3. Consider the hypothetical sample data concerning interest rates charged by banks that are given in Table 16.B.
   a. Calculate separate simple interest-price indexes for the four types of loans, using 1986 as a base.
   b. Calculate the unweighted average of the simple price indexes computed in problem 3a.
   c. Calculate the value-weighted average of the simple price indexes computed in problem 3a, using 1986 values.

Table 16.A

Item	Average Price in Boston (dollars per indicated unit)		Value (dollars spent by average Boston family in year)	
	in 1984	in 1985	in 1984	in 1985
Gasoline (gallons)	1.40	1.30	1,484	1,690
Oil (quarts)	1.00	1.20	20	30
Repairs (hours)	20.00	25.00	200	500
Tires (each)	30.00	30.00	120	150

Table 16.B

Loan	Effective Annual Interest Rate (percent)		Value of Loans Outstanding (millions of dollars)	
	1986	1987	1986	1987
Auto	8.9	11.2	120	136
Credit Card	16.4	17.0	240	300
Mortgage	14.1	12.1	690	710
Signature	18.9	18.6	57	40

4. Once more consider the data of Table 16.B.
  a. Calculate separate simple interest-price indexes for the four types of loans, using 1987 as a base.
  b. Calculate the unweighted average of the simple price indexes computed in problem 4a.
  c. Calculate the value-weighted average of the simple price indexes computed in problem 4a, using 1987 values.

5. Once more consider the data of Table 16.A. Compute the 1985 unweighted aggregative price index, based on 1984. Show your computations.

6. Once more consider the data of Table 16.A. Compute the 1984 unweighted aggregative price index, based on 1985.

7. Once more consider the data of Table 16.B. Compute the 1987 unweighted aggregative price index, based on 1986.

8. Once more consider the data of Table 16.B. Compute the 1986 unweighted aggregative price index, based on 1987.

9. Reconsider the data of Table 16.A. Calculate a 1985 Laspeyres price index, based on 1984. Show your computations.

10. Consider the data of Table 16.C that pertain to an aircraft parts supplier. Compute a 1987 Laspeyres price index, based on 1986.

11. From the data of Table 16.C, compute a 1986 Laspeyres price index, based on 1985.

12. From the data of Table 16.C, compute a 1985 Laspeyres price index, based on 1984.

13. From the data of Table 16.C, compute a 1987 Laspeyres price index, based on 1984.

14. Reconsider the data of Table 16.A. Calculate a 1985 Paasche price index, based on 1984. Show your computations.

15. From the data of Table 16.C, compute a 1987 Paasche price index, based on 1986.

16. From the data of Table 16.C, compute a 1986 Paasche price index, based on 1985.

17. From the data of Table 16.C, compute a 1985 Paasche price index, based on 1984.

18. From the data of Table 16.C, compute a 1987 Paasche price index, based on 1984.

19. Reconsider the data of Table 16.A. Calculate a 1985 typical-year price index, based on 1984 and using *average* 1985 and 1984 quantities. Show your computations.

20. From the data of Table 16.C, compute a 1987 typical-year price index, based on 1986 and using *average* 1987 and 1986 quantities.

21. From the data of Table 16.A, compute a 1985 Edgeworth price index, based on 1984.

Table 16.C

Item	Price (dollars per unit)				Quantity (units sold)			
	1984	1985	1986	1987	1984	1985	1986	1987
Oxygen System	521.00	495.75	488.10	460.00	10	8	7	12
Flight-Plan Computer	695.00	395.00	300.00	257.95	15	27	33	69
Transceiver	999.00	695.00	500.00	495.00	5	7	9	15
Flight Case	125.00	139.50	141.00	152.00	20	20	20	23
Life Raft	999.00	786.00	820.00	629.00	1	3	7	5
Xenon Strobe	120.00	65.00	65.00	65.00	7	17	29	8
Tire (Nosewheel)	182.00	163.00	175.00	170.00	622	721	852	499
Altimeter	533.00	182.00	250.00	197.00	173	120	89	158
Battery Charger	35.00	30.70	29.00	33.50	39	63	85	60
Transponder	980.00	935.00	722.00	621.00	120	115	299	980

Table 16.D

Item	Output (thousands of bushels)				Price (dollars per bushel)			
	1979	1980	1981	1982	1979	1980	1981	1982
Almonds	12	13	15	9	20	19	15	26
Apricots	4	5	8	6	10	10	9	12
Cherries	20	15	16	25	25	28	30	15
Peaches	6	6	6	7	15	12	12	12

22. From the data of Table 16.C, compute a 1987 Edgeworth price index, based on 1986.

23. From the data of Table 16.C, compute a 1986 Edgeworth price index, based on 1985.

24. From the data of Table 16.C, compute a 1985 Edgeworth price index, based on 1984.

25. From the data of Table 16.C, compute a 1987 Edgeworth price index, based on 1984.

26. From the data of Table 16.A, compute Fisher's ideal price index for 1985, based on 1984. Show your computations.

27. From the data of Table 16.C, compute Fisher's ideal price index for 1987, based on 1986.

28. From the data of Table 16.C, compute Fisher's ideal price index for 1986, based on 1985.

29. From the data of Table 16.C, compute Fisher's ideal price index for 1985, based on 1984.

30. From the data of Table 16.C, compute Fisher's ideal price index for 1987, based on 1984.

31. The data of Table 16.D were gathered by the manager of an orchard. Always rounding to whole numbers, calculate Laspeyres quantity indexes for the orchard's output from 1980–82 with 1979 = 100 as a base. Show your computations.

32. Using the data of Table 16.C, calculate a Laspeyres quantity index for 1987, based on 1986.

33. Using the data of Table 16.C, calculate a Laspeyres quantity index for 1986, based on 1985.

34. Using the data of Table 16.C, calculate a Laspeyres quantity index for 1987, based on 1985.

35. Consider the data of Table 16.D. Always rounding to whole numbers, calculate Paasche quantity indexes for the orchard's output from 1980–82 with 1979 = 100 as a base. Show your computations. Do you know why so many people prefer the Laspeyres index (problem 31) over the Paasche index?

36. Using the data of Table 16.C, calculate a Paasche quantity index for 1987, based on 1986.

37. Using the data of Table 16.C, calculate a Paasche quantity index for 1986, based on 1985.

38. Using the data of Table 16.C, calculate a Paasche quantity index for 1987, based on 1985.

39. Consider the data of Table 16.D. Always rounding to whole numbers, calculate Fisher's ideal quantity

indexes for the orchard's output from 1980–82 with 1979 = 100 as a base. Show your computations.

40. Using the data of Table 16.C, calculate Fisher's ideal quantity index for 1987, based on 1986.

41. Using the data of Table 16.C, calculate Fisher's ideal quantity index for 1986, based on 1985.

42. Using the data of Table 16.C, calculate Fisher's ideal quantity index for 1987, based on 1985.

43. It has been claimed that a correct value index can always be found for any given intertemporal comparison by multiplying an appropriate Laspeyres price index with a corresponding Paasche quantity index or by multiplying an appropriate Paasche price index with a corresponding Laspeyres quantity index (and in each case dividing by 100). Check out the claim with the help of the Box 16.E and Box 16.F formulas.

44. Test the claim noted in problem 43 with the help of the actual 1985 indexes, based on 1984, that were calculated in Practice Problem 8.

45. Show your understanding of the *index-number problem* by recomputing your answer to problem 1c using 1985 values and commenting on the result. Show your computations.

46. Show your understanding of the *index-number problem* by proving the text statement on page 696 that "the same composite price index would have been computed in our example if, compared to the 1984 value weights, say, *all* 1985 value weights had been precisely 10 percent higher and, say, *all* 1970 value weights had been precisely 40 percent lower."

47. Show your understanding of the *index-number problem* by indicating what change in Soviet 1928

prices, in Analytical Example 16.1, would eliminate the index-number problem encountered there.

48. Show your understanding of the *index-number problem* by indicating what change in Analytical Example 16.2 would eliminate the index-number problem encountered there.

49. Consider the unweighted aggregative price index of problem 5. Do you think it meets the three Fisher tests? (Assume 1986 dollar prices of 1, 1, 26, and 39.) Explain your answer.

50. Consider your answer to problem 10. Does that index meet Fisher's time reversal test? Explain.

51. Consider your answer to problem 15. Does that index meet Fisher's time reversal test? Explain.

52. Consider your answer to problem 27. Does that index meet Fisher's time reversal test? Explain.

53. Consider your answers to problems 10 and 32. Does that index meet Fisher's factor-reversal test? Explain.

54. Consider your answers to problems 15 and 36. Does that index meet Fisher's factor-reversal test? Explain.

55. Consider your answers to problems 27 and 40. Does that index meet Fisher's factor-reversal test? Explain.

56. Consider your answers to problems 10–13. Does that index meet Fisher's circularity test?

57. Consider your answers to problems 15–18. Does that index meet Fisher's circularity test?

58. Consider your answers to problems 27–30. Does that index meet Fisher's circularity test?

59. Consider the index-number series in Table 16.E to solve the following problems (in all cases, round to whole numbers).

Table 16.E

	Part (A)			Part (B)	
Year	Old Index (1955 = 100)	Revised Index (1979 = 100)	Place	U.S. State Department 1984 Cost-of-Living Index	
1976	293	—	Washington, D.C.	100	
1977	301	—	London	110	
1978	311	—	Paris	139	
1979	317	100	Bonn	120	
1980	—	105	Moscow	80	
1981	—	111	Nairobi	60	
1982	—	97			
1983	—	95			
1984	—	103			

a. Splice the two series in Part A together into a combined series with 1979 = 100 as a base.

b. Shift the base of your spliced series to 1982 = 100.

c. Shift the base of the Part B series from Washington, D.C. = 100 to Moscow = 100.

60. *Deflate* the data in problems (a) and (b) as needed, using the appropriate CPI from column 2 of Table 16.7.

a. In real terms, which was the best Table 16.F salary offer given to a graduating student? Which was the worst offer?

Table 16.F

Year	Salary
1978	$17,000
1979	18,000
1980	19,000
1981	20,000
1982	27,000

b. Seen in a 1982 newspaper: "For Sale 1960 Cadillac. $100 (1960 dollars)." What was the price?

Selected Readings

Cagan, Phillip, and Geoffrey H. Moore. *The Consumer Price Index: Issues and Alternatives.* Washington, D.C.: American Enterprise Institute, 1981. Reviews the history and uses of the CPI, examines its limitations and deficiencies, and recommends feasible improvements.

Fisher, Irving. *The Making of Index Numbers: A Study of Their Varieties, Tests, and Reliability,* 3rd ed. rev. Boston: Houghton Mifflin, 1927. A classic work on the subject that examines many alternative formulas for computing index numbers.

Maunder, W. F., ed. *Bibliography of Index Numbers.* London: Athlone Press, 1970. An exhaustive listing from 1707 to 1968.

Theil, Henri. "Best Linear Index Numbers of Prices and Quantities." *Econometrica* (April 1960), pp. 464–80. A more advanced development of the theory of chain indexes. (Additional discussion in Kloek, T., and G. M. De Wit. "Best Linear and Best Linear Unbiased Index Numbers." *Econometrica* (October 1961), pp. 602–16.)

Ulmer, Melville J. *The Economic Theory of Cost of Living Index Numbers.* New York: Columbia University Press, 1950. An imaginative approach to establishing the upper and lower limit of a constant-utility index.

U.S. Department of Labor, Bureau of Labor Statistics. *BLS Handbook of Methods,* 2 vols., Bulletin 2134–1 and 2134–2. Washington, D.C.: U.S. Government Printing Office, December 1982. Presents detailed explanations of how the BLS obtains and prepares the economic data it publishes. Volume I contains this information for all BLS programs except the consumer price index (Chapter 7 deals with the producer price index). Volume II deals with the CPI.

U.S. Department of Labor, Bureau of Labor Statistics. *The Consumer Price Index: History and Techniques.* Bulletin 1517. Washington, D.C.: U.S. Government Printing Office, no date. A detailed discussion of procedures from the 1890s to the mid-1960s.

U.S. Department of Labor, Bureau of Labor Statistics. *The Consumer Price Index: Concepts and Content over the Years.* Report 517, May 1978 (revised).

Wilkerson, Marvin. "Sampling Error in the Consumer Price Index." *Journal of the American Statistical Association* (September 1967), pp. 899–914. A discussion of the inevitable errors contained in an index number produced by a highly complex network of samples.

Computer Programs

The DAYSAL-2 personal-computer diskettes that accompany this book contain one program of interest to this chapter:

15. Index Numbers lets you construct simple, unweighted price indexes, as well as composite, weighted price or quantity indexes. The program computes the indexes of Laspeyres, of Paasche, Edgeworth, and Fisher. It also makes time-reversal and factor-reversal tests of the computed indexes.

CHAPTER SEVENTEEN

Decision Theory

Only rarely do the actions of business and economic decision makers lead to unique consequences that are known with certainty in advance. In countless situations, such decision makers face a serious problem: They are called upon to choose among two or more alternative courses of action at a time when the associated consequences cannot be foreseen with certainty because they depend on the nature of some future event over which the decision maker has no control. Under such circumstances, it is difficult to make a rational decision. Consider the farmer who must choose now among plantings of alternative crops but knows that some of these crops will flourish if the season turns out to be cool and dry, while others will do so only if it is hot and wet. Consider the manufacturer who must decide now whether to introduce a new production process or to continue using the old one but who also knows that the profit consequences of either decision will differ greatly, depending on whether future oil prices (or future wages) are low or high. Consider the marketing manager whose advice about introducing a new product (or modifying the style, packaging, or labeling of an old product) is desired now but who knows the consequences of any of these actions to be dependent on as-yet-unknown levels of future demand. Examples such as these can be multiplied without end because the ultimate outcome of any choice is likely to be affected by chance. Think of the real-estate developer who must make a commitment now to a small- or medium-sized or large-scale project; the consulting firm that must decide now to install a small or a large computer; the oil company that must drill now or sell its rights; the record producer who must initiate a major sales campaign now or abort the production of a record altogether; the independent TV producer who can submit the pilot of a new program to a network now or sell the rights to someone else; the job applicant who must decide now to work for a fixed salary or on a commission basis. In all these cases and many more, the wisdom or foolishness of any present decision will ultimately be determined by (uncontrollable) future events: by whether the demand for housing (or consulting services) turns out to be high or low, by whether oil is found at the drilling site or the test hole is dry, by whether the new record (or TV show) becomes a hit or is a failure, by whether the new employee is successful at making sales or not. In this final chapter of the book, we will consider a variety of methods, collectively referred to as **decision theory,** that can be employed in the systematic analysis and solution of decision-making problems that arise because uncertainty exists about future events over which the decision maker has no control but which are bound to influence the ultimate outcome of a decision.

Note: Some writers, such as the economist Frank Knight (1885–1972), have introduced a fine distinction between decision making under *uncertainty* and decision making under *risk.* In either case, the ultimate outcome of a present decision is affected by future random events. In the case of decision making under uncertainty, however, the probabilities associated with all of these events, or even the nature of these events, are unknown at the time of decision making; in the case of decision making under risk, both the nature and the probabilities of these events are known in advance. Like most writers, we make no such distinction here and refer to all situations in which the ultimate outcome of a decision maker's choice depends on chance as **decision making under uncertainty.**

BASIC CONCEPTS

Whenever decisions are being made in the context of uncertainty, certain identical elements are present; it is helpful to focus on these elements systematically when searching for a solution to a decision-making problem.

Actions, Events, Payoffs

First, each decision maker has specific decision alternatives available. In the language of decision theory, the decision maker must make a choice among various mutually exclusive **actions,** a complete list of which is symbolized by $A_1, A_2, \ldots, A_n$. A business executive, for example, might be called upon to manufacture and market a new product by taking one of two actions: A_1 = constructing a small plant; A_2 = constructing a large plant.

Second, future occurrences, not under the control of the decision maker, will affect the outcome of any present action taken by the decision maker. These occurrences are commonly referred to as **states of nature** or, more simply, as **events;** a collectively exhaustive list of mutually exclusive events is symbolized by $E_1, E_2, \ldots, E_n$. The production and marketing of a new product, for example, might be linked with one of two events: E_1 = an environment of weak demand; E_2 = an environment of strong demand. As we shall see, probability values about such events may or may not be employed by a decision maker at the time when a choice among actions is taken.

Third, positive or negative net benefits are associated with each possible action/event combination. These net benefits are the joint outcome of choice and chance and are commonly referred to as **payoffs.** Payoffs can be measured in any kind of unit appropriate to the problem at hand: in dollars, in time, or even in utility, as we will see. The payoffs from constructing a small plant, for example, might be annual profits of $8 million if demand is weak or $5 million if demand is strong (the smaller payoff resulting from the need, perhaps, to then run the plant above its designed capacity if demand is strong, which leads to extremely high unit costs of production). The corresponding payoffs from a large plant might be − $2 million or + $12 million instead (the smaller payoff again, perhaps, occurring because the running of a large plant *below* its capacity also involves unusually high unit costs of production).

The Payoff Table

A decision-making situation in the context of uncertainty can be summarized in a variety of ways; one popular summary is the **payoff table,** a tabular listing of the payoffs associated with all possible combinations of actions and events. Table 17.1 is such a tabular summary for our example. Each row corresponds to one of the two possible actions; each column corresponds to one of the two possible events. The four cells of the table show the associated annual profits (in millions of dollars).

Note: In this payoff table neither action is unambiguously superior to the other in the sense that it produces payoffs that are as good as or better than those of the alternative action no matter which event occurs. If such an unambiguously superior action exists, it is called a **dominant action,** and its existence turns the alternative into an **inadmissible action** that need not be considered further because of its obvious inferiority. In our example, A_1 would be dominant if, all else being equal, the A_1/E_2 cell contained an entry of 12 instead of 5; if A_1 were dominant, A_2 would be inadmissible. Under those circumstances, the decision maker could choose A_1 without hesitation and would reap a profit of \$8 million if E_1 occurred or of \$12 million if E_2 occurred. The choice of A_2 would lead to worse or at best equally good results, as the A_2 row of Table 17.1 indicates.

The Decision Tree

Another popular summary of a decision-making situation is the **decision tree.** This graphical device is similar to the tree diagram encountered in Chapter 5; it illustrates a decision problem by showing, in chronological order from left to right, every potential action, event, and payoff. The decision tree diagram is particularly useful when a decision problem involves a sequence of many decisions that extend over a long stretch of time.

It is customary in decision-tree diagrams to denote any point of *choice* (at which the decision maker is in control) by a square symbol from which branches representing the possible actions of the decision maker (and, therefore, called **action branches** or **decision branches**) emanate toward the right. Such an **action point** (also variously referred to as a **decision point, decision node,** or **decision fork**) is illustrated in panel (a) of Figure 17.1, along with the relevant branches. In contrast, any

Table 17.1   A Payoff Table

This table shows the annual profits (in millions of dollars) associated with four possible action/event combinations.

	Events	
Actions	E_1 = weak demand for new product	E_2 = strong demand for new product
A_1 = constructing a small plant	8	5
A_2 = constructing a large plant	-2	12

Figure 17.1   A Decision Tree

As this illustration shows, the possible payoffs associated with a decision made under uncertainty result from a mixture of choice (square symbol) and chance (circular symbols). Note that the information contained in this diagram is precisely the same as that found in the payoff table (Table 17.1).

(a) Action Point and Branches

(b) Event Point and Branches

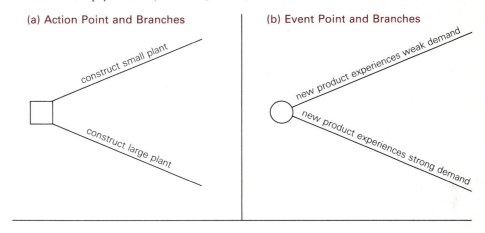

(c) Actions, Events, and Payoffs

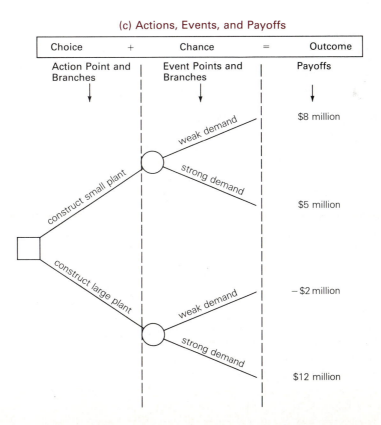

point of *chance* (at which the decision maker exercises no control, but "nature" is in charge) is denoted symbolically by a circle from which branches representing the possible events confronting the decision maker (and, therefore, called **event branches** or **state-of-nature branches**) emanate toward the right. Such an **event point** (also variously referred to as a **state-of-nature point, state-of-nature node,** or **state-of-nature fork**) is illustrated in panel (b) of Figure 17.1, again with the associated branches. Panel (c), finally, shows the entire decision tree that summarizes our example.

Summarizing a decision problem, either with the help of a payoff table or a decision tree, is helpful, but it is only the first step toward a solution. A best plan of action still remains to be chosen; the following sections consider some of the criteria that might be employed to find it.

DECISION MAKING WITHOUT PROBABILITIES

Consider a situation, such as that summarized by Table 17.1 or Figure 17.1, in which a decision must be made in the face of uncertainty, but nothing is known about the likelihood of occurrence of those alternative future events that are certain to affect the eventual outcome of the present decision. Imagine that the decision maker does not even care to guess what the event probabilities might be. In such a case, it is common practice to employ one of three decision criteria: *maximin* (or *minimax*), *maximax* (or *minimin*), or *minimax regret*.

Maximin (or Minimax)

The first of the nonprobabilistic criteria embodies a decidedly conservative approach to decision making. It guarantees that the decision maker does no worse than achieve the best among the poorest possible outcomes. This criterion takes one of two forms, depending on whether the decision maker aims to maximize benefit or minimize cost. If maximization of benefit is the objective (as in our example above), the criterion is called **maximin** because the decision maker is to find the lowest possible (or minimum) benefit associated with each possible action, identify the highest (or maximum) benefit among these minima, and then choose the action associated with this maximum of minima (which explains the name given to this criterion). Table 17.2 illustrates how the maximin criterion would be applied to our example.

If minimization of cost is the objective (as when a firm must choose, perhaps, between the installation of alternative antipollution devices), the analogous criterion is often referred to as **minimax** because the decision maker is to find the highest possible (or maximum) cost associated with each possible action, identify the lowest (or minimum) cost among these maxima, and then choose the action associated with this minimum of maxima (which explains the name given to the criterion in this instance). Table 17.3 illustrates how the minimax criterion would be applied.

Note that the maximin (or minimax) criterion is ideally suited to the born pessimist. If you always expect the worst, if you always suspect that "nature" or chance will work against you, then you cannot do better than employ this criterion. It will give you nothing worse than the best among all the worst things that can happen: the highest of the lowest-possible profits in Table 17.2 and the lowest of the highest-possible costs in Table 17.3.

Table 17.2 Maximizing Benefit: The Maximin Criterion

This table shows the annual profits (in millions of dollars) associated with four possible action/event combinations. The lowest of the possible profits associated with each action is shown in the last column; the maximin criterion suggests taking the best of these (encircled), or action A_1. Under the circumstances, the firm can do no worse than earn a profit of $5 million. It might even earn $8 million, if E_1 rather than E_2 occurs.

Actions	Events		Row Minimum
	E_1 = weak demand for new product	E_2 = strong demand for new product	
A_1 = constructing a small plant	8	5	⑤←Maximin
A_2 = constructing a large plant	−2	12	−2

Maximax (or Minimin)

The second among the nonprobabilistic criteria goes to the other extreme and seeks to achieve the best of the best possible outcomes. This criterion, too, comes in two forms. If maximization of benefit is the objective, the criterion is called **maximax** because the decision maker is to find the highest possible (or maximum) benefit associated with each possible action, identify the highest (or maximum) benefit among these maxima, and then choose the action associated with this maximum of maxima

Table 17.3 Minimizing Cost: The Minimax Criterion

This table shows the annual costs (in millions of dollars) associated with four possible action/event combinations. The highest of the possible costs associated with each action is shown in the last column; the minimax criterion suggests taking the best of these (encircled), or action A_1. Under the circumstances, the firm can do no worse than incur a cost of $8 million. It might even pay as little as $5 million, if E_1 rather than E_2 occurs.

Actions	Events		Row Maximum
	E_1 = low level of demand and production	E_2 = high level of demand and production	
A_1 = install small antipollution system	5	8	⑧←Minimax
A_2 = install large antipollution system	10	4	10

Table 17.4 Maximizing Benefit: The Maximax Criterion

This table shows the annual profits (in millions of dollars) associated with four possible action/event combinations. The highest of the possible profits associated with each action is shown in the last column; the maximax criterion suggests taking the best of these (encircled), or action A_2. Under the circumstances, the firm might earn a profit of $12 million, but it might also have a loss of $2 million, if E_1 rather than E_2 occurs.

	Events		
Actions	E_1 = weak demand for new product	E_2 = strong demand for new product	Row Maximum
A_1 = constructing a small plant	8	5	8
A_2 = constructing a large plant	-2	12	12 ← Maximax

(as the name suggests). Table 17.4 illustrates how the maximax criterion would be applied to our first example.

If minimization of cost is the objective (as in our second example), the analogous criterion is often referred to as **minimin** because the decision maker is to find the lowest possible (or minimum) cost associated with each possible action, identify the lowest (or minimum) cost among these minima, and then choose the action associated with this minimum of minima (as the name suggests). Table 17.5 illustrates how the minimin criterion would be applied to our second example.

Note that the maximax (or minimin) criterion is ideally suited to the born optimist. If you always expect the best, if you are always convinced that "nature" or chance will

Table 17.5 Minimizing Cost: The Minimin Criterion

This table shows the annual costs (in millions of dollars) associated with four possible action/event combinations. The lowest of the possible costs associated with each action is shown in the last column; the minimin criterion suggests taking the best of these (encircled), or action A_2. Under the circumstances, the firm might incur costs as low as $4 million, but it might also have costs of $10 million, if E_1 rather than E_2 occurs.

	Events		
Actions	E_1 = low level of demand and production	E_2 = high level of demand and production	Row Minimum
A_1 = install small antipollution system	5	8	5
A_2 = install large antipollution system	10	4	④←Minimin

be on your side, then you might go after the best of the best in the fashion just shown: the highest of the highest-possible profits in Table 17.4 and the lowest of the lowest-possible costs in Table 17.5.

Minimax Regret

The third one among the nonprobabilistic criteria is just a bit more complicated. To understand it, we must first define the concept of **opportunity loss (OL)** or **regret.** When the decision maker aims to maximize benefit, the opportunity loss equals the difference between (a) the optimal payoff for a given event (the highest benefit in a given event column of our tables) and (b) the actual payoff achieved as a result of taking a specified action and the subsequent occurrence of that event. Consider, for example, the E_1 column of Table 17.4. If E_1 occurs, the optimal payoff is \$8 million. If the decision maker has previously chosen A_1, the actual payoff will equal \$8 million as well; hence, the opportunity loss will equal \$8 million $-$ \$8 million $= 0$. The decision maker has no reason to regret anything because given the occurrence of E_1, the best possible action had, in fact, been taken. If, on the other hand, the decision maker has previously chosen A_2, the actual payoff will equal $-$ \$2 million; hence, the opportunity loss will equal \$8 million $- (-$ \$2 million$) =$ \$10 million. The decision maker has plenty of reason for regret because given the occurrence of E_1, if only A_1 had been chosen instead of A_2, the decision maker would be better off by \$10 million (having a gain of \$8 million instead of a loss of \$2 million). Similar opportunity-loss values can be calculated for the E_2 column, of course, as is shown in Table 17.6 (which is also referred to as a **regret table**). The application of the criterion of **minimax regret** is also illustrated in that table. According to this criterion, the decision maker finds the highest possible (or maximum) regret value associated with each possible action, identifies the lowest (or minimum) value among these maxima, and then chooses the action associated with this minimum of maximum regret values. (Note the encircled figure in Table 17.6.)

Table 17.6 Maximizing Benefit: The Minimax-Regret Criterion

This table, based on Table 17.4, shows the opportunity loss or regret values (in millions of dollars) associated with four possible action/event combinations. The highest of the possible regret values associated with each action is shown in the last column; the minimax-regret criterion suggests taking the least painful of these (encircled), or action A_1. Under the circumstances, the decision maker might come to regret the lost opportunity of earning \$7 million in extra profit, but the regret might also be zero (if E_1 rather than E_2 occurs) and it could not be worse than \$7 million (as it would be if A_2 were chosen and E_1 were to occur).

	Events		
Actions	E_1 = weak demand for new product	E_2 = strong demand for new product	Row Maximum
A_1 = constructing a small plant	$8 - 8 = 0$	$12 - 5 = 7$	⑦ ←Minimax Regret
A_2 = constructing a large plant	$8 - (-2) = 10$	$12 - 12 = 0$	10

The minimax regret criterion is applied in the same way to cost-minimization problems, but the regret table itself is computed in a slightly different way. When the decision maker aims to minimize cost, the opportunity loss equals the difference between (a) the actual cost incurred as a result of taking a specified action and the subsequent occurrence of an event and (b) the minimum cost achievable for that event (the lowest cost in a given event column of our tables). Table 17.7 gives the solution for our example.

Criticisms

When we survey the above solutions to our decision-making problems, we discover that the results differ depending on the criterion that is applied. In our benefit-maximizing example, the maximin criterion suggests A_1 (the construction of a small plant), the maximax criterion suggests A_2 (the construction of a large plant), and the minimax regret criterion suggests A_1 as well. Similarly, different results were obtained in the cost-minimizing case. Such differences, however, should not surprise us; they merely reflect the underlying differences in decision-making philosophies that stress, respectively, being careful, being daring, or minimizing future regret.

There exist, however, certain other problems with the above criteria that might make us think twice about applying them.

Undue sensitivity to irrelevant factors. The results obtained by the use of these criteria are easily affected by irrelevant factors. Consider Table 17.2 and imagine a government offering a fixed subsidy of $8 million, *regardless of the firm's action,* if demand should be weak. One would think that this incentive should not affect the firm's action, but under the maximin criterion it does: The E_1 column entries change from 8 and -2 to 16 and 6, the row minima change from 5 and -2 to 5 and 6, and the maximin action changes from A_1 to A_2.

Consider the same subsidy in the context of Table 17.3. The E_1 column entries (being costs) change from 5 and 10 to -3 and 2, the row maxima change from 8 and 10 to 8 and 4, and the minimax action changes from A_1 to A_2.

Although a fixed subsidy would not change the minimax-regret action, some other irrelevant factor easily might. Imagine that Table 17.2 contained a third possible action, A_3 = buying an established plant, with payoffs (in millions of dollars) of -5 and $+20$, respectively, for E_1 and E_2. The existence of this third alternative would change the Table 17.6 entries in the E_1 column from 0 and 10 to 0, 10, and 13, and it would change the E_2 column from 7 and 0 to 15, 8, and 0. Hence, the row maxima would become 15, 10, and 13, and the minimax-regret action would change from A_1 to A_2. But why should the mere existence of another (and rejected) alternative change the choice between the remaining alternatives?

Undue reliance on extreme values. Another factor that has worried decision makers is the fact that the above criteria ignore all payoff values except certain extreme ones found in a given row. Consider again Table 17.2 and imagine that the 12 changed to 12,000. Despite the drastic change (and the sudden possibility of making a profit of $12 *billion* by constructing a large plant) the maximin criterion would continue to counsel that action A_1 be taken (that a small plant be built in order to avoid a possible $2 million loss).

Table 17.7   Minimizing Cost: The Minimax-Regret Criterion

This table, based on Table 17.5, shows the opportunity loss or regret values (in millions of dollars) associated with four possible action/event combinations. The highest of the possible regret values associated with each action is shown in the last column; the minimax-regret criterion suggests taking the least painful of these (encircled), or action A_1. Under the circumstances, the decision maker might come to regret the lost opportunity of reducing costs by $4 million more, but the regret might also be zero (if E_1 rather than E_2 occurs) and it could not be worse than $4 million (as it would be if A_2 were chosen and E_1 were to occur).

	Events		
Actions	E_1 = low level of demand and production	E_2 = high level of demand and production	Row Maximum
A_1 = install small antipollution system	$5 - 5 = 0$	$8 - 4 = 4$	④ ←Minimax Regret
A_2 = install large antipollution system	$10 - 5 = 5$	$4 - 4 = 0$	5

Lack of consideration for the probabilities of events. All of the criteria ignore the probabilities of events, but, critics argue, even guesses could often be extremely helpful. If, in the example just cited (Table 17.2 with the 12 changed to 12,000), E_1 had a low probability and E_2 had a high one, it would surely be foolish to apply the maximin criterion and forgo an almost certain $12 billion gain in order to avoid a highly unlikely $2 million loss. Indeed, many decision makers prefer the criteria discussed in the next sections because they do take probabilities into account.

DECISION MAKING WITH PROBABILITIES: PRIOR ANALYSIS

Many times a decision maker can develop fairly good estimates for the probabilities of alternative future events. Decision criteria that make use of such estimates include the *maximum-likelihood criterion,* the *maximization* (or minimization) *of expected monetary value* (depending on whether the objective is the maximization of some benefit or the minimization of some cost), the *minimization of expected opportunity loss* or regret value, and the *maximization of expected utility.* In this section, we will discuss the use of these criteria in **prior analysis,** when decision making is based solely on whatever information on event probabilities is available prior to the gathering of *new* experimental or sample evidence about them. As noted in the Chapter 5 section, "Revising Probabilities: Bayes's Theorem," such probabilities are called *prior probabilities.* They are usually purely subjective (and, thus, representative of nothing else but a personal degree of belief in the likelihood of some event), but they can also be objective (and, thus, indicative of the relative frequency with which the event in question is

Table 17.8   Maximizing Benefit: The Maximum-Likelihood Criterion

This table shows the annual profits (in millions of dollars) associated with four possible action/event combinations. Given the assumed prior probabilities of the events [$p(E_1) = .7$ and $p(E_2) = .3$], the maximum-likelihood criterion leads a decision maker to focus entirely on the shaded E_1 column (because .7 is greater than .3) and to select the optimal result in that column (encircled). Thus, A_1 is the action chosen by those who employ the maximum-likelihood criterion.

Actions	Events	
	E_1 = weak demand for new product $p(E_1) = .7$	E_2 = strong demand for new product $p(E_2) = .3$
A_1 = constructing a small plant	(8)	5
A_2 = constructing a large plant	−2	12

bound to occur if logic or past experience can be used as a guide). Another Chapter 5 section, "The Meaning of Probability," can be used to review the concepts of subjective and objective probability.

Maximum Likelihood

A decision maker using the **maximum-likelihood criterion** simply ignores all the events that might occur except the most likely one and selects the action that produces the optimal result (maximum benefit or minimum cost) associated with this most likely event. Table 17.8, which is an adaptation of Table 17.1, illustrates the procedure.

Critics argue that the use of this criterion amounts to "playing ostrich" because so much that might happen is being ignored. In this example, the best of all possible outcomes (12) is never even considered; in other cases, the preoccupation with the event of the highest probability might lead the decision maker to focus on an event with a probability of .1, while ignoring other events with a combined probability of .9 (but individual probabilities of less than .1). This result is considered most unsatisfactory by critics of this criterion.

Proponents of this criterion, on the other hand, argue that people often do behave in precisely this fashion. They look at nothing but the most likely event, as evidenced, for example, in the famous *cobweb cycles* in agricultural and labor markets. Thus, farmers assume that the *most likely* future price of hogs is the one equal to the present price; hence, a low price in period 1 (because it is used as an indicator of a low price in period 2) discourages the raising of hogs in period 1 and leads to a shortage and an unexpectedly *high* price in period 2. This high price in period 2, in turn, encourages the raising of hogs in period 2 and leads to a surplus and an unexpectedly low price in period 3. And so it goes, low and high prices following one another eternally.[1]

---

[1]For a more detailed discussion of cobweb cycles, see Heinz Kohler, *Intermediate Microeconomics: Theory and Applications,* 2nd. ed. (Glenview, Ill.: Scott, Foresman and Company, 1986), pp. 196–201 and 302–3.

Table 17.9 Maximizing Benefit: The Expected-Monetary-Value Criterion

This table shows the annual profits (in millions of dollars) associated with four possible action/event combinations; it also shows the expected monetary value of each possible action, based on the assumed prior probabilities of the events $[p(E_1) = .7$ and $p(E_2) = .3]$. Because the objective is to maximize benefit, the *largest EMV* is optimal; thus, A_1 is the action chosen by those who employ the expected-monetary-value criterion.

Actions	Events		Expected Monetary Value, *EMV*
	E_1 = weak demand for new product $p(E_1) = .7$	E_2 = strong demand for new product $p(E_2) = .3$	
A_1 = constructing a small plant	8	5	$8(.7) + 5(.3) = 7.1 \leftarrow$ optimum
A_2 = constructing a large plant	-2	12	$-2(.7) + 12(.3) = 2.2$

Expected Monetary Value

A decision maker using the **expected-monetary-value criterion** determines an expected monetary value for each possible action and then selects the action with the optimal expected monetary value (the largest, if the objective is to maximize some benefit; the smallest if the objective is to minimize some cost). The **expected monetary value (EMV)** of an action equals the sum of the weighted payoffs associated with that action, the weights being the probabilities of the alternative events that produce the various possible payoffs. Tables 17.9 and 17.10, which are based on our earlier examples, illustrate the procedure.

Table 17.10 Minimizing Cost: The Expected-Monetary-Value Criterion

This table shows the annual costs (in millions of dollars) associated with four possible action/event combinations; it also shows the expected monetary value of each possible action, based on the assumed prior probabilities of the events $[p(E_1) = .7$ and $p(E_2) = .3]$. Because the objective is to minimize cost, the *smallest EMV* is optimal; thus, A_1 is the action chosen by those who employ the *EMV* criterion.

Actions	Events		Expected Monetary Value, *EMV*
	E_1 = low level of demand and production $p(E_1) = .7$	E_2 = high level of demand and production $p(E_2) = .3$	
A_1 = install small antipollution system	5	8	$5(.7) + 8(.3) = 5.9 \leftarrow$ optimum
A_2 = install large antipollution system	10	4	$10(.7) + 4(.3) = 8.2$

Figure 17.2   A Decision Tree and the Expected-Monetary-Value Criterion

The application of the expected-monetary-value criterion is often depicted by means of a decision tree. Event probabilities are attached to each event branch and, by a process termed *backward induction,* the ultimate payoff values appearing at the tips of the branches are translated, using these probabilities, into expected monetary values at each fork of the tree (here, $7.1 million at event point *b,* $2.2 million at event point *c,* and, ultimately, $7.1 million at action point *a*). When maximizing benefit, a decision maker employing the *EMV* criterion will take the action leading to the highest *EMV* (in this case, the decision maker will follow the branch from action point *a* to event point *b*). By refusing to follow the alternative action branch (from *a* to *c*) that leads to the lower *EMV,* the benefit-maximizing decision maker is said to "cut off" or "prune" the nonoptimal path through the tree. (Note the red line suggesting a cut by the pruning shears.) This pruning leaves intact only the optimal path from *a* to *b* and beyond and makes the expected monetary value of the decision maker's action equal to $7.1 million as well. Caution: Branch pruning can take place only at action points, never at event points, because only the former are controlled by the decision maker.

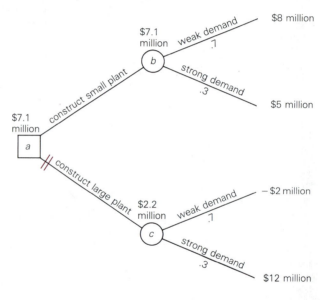

Note how a decision maker who, when confronted with the type of situation given in Table 17.9, consistently chooses A₁ will, on average, gain $7.1 million per year because $8 million will be gained 70 percent of the time and $5 million 30 percent of the time. In contrast, a consistent choice of nonoptimal A₂ would, in the long run, yield a gain of only $2.2 million per year because $2 million would be lost 70 percent of the time and $12 million would be gained 30 percent of the time in this type of situation. Similarly, a decision maker who, when confronted with the type of situation given in Table 17.10, consistently chooses A₁ will, on average, incur costs of only $5.9 million per year, while the nonoptimal choice of A₂ would, in the long run, produce costs of $8.2 million per year.

Note also that the probabilities used in these examples are assumed to be the best guesses of the decision maker. It may well happen, however, that a decision maker has no idea about event probabilities whatsoever. In that case, many decision theorists recommend that equal probabilities be assigned to all events. Given a list of collectively exhaustive and mutually exclusive events, the probability assigned to each then equals

1 divided by the number of such events. This procedure is variously referred to as the **equal-likelihood criterion,** the **criterion of insufficient reason, Bayes's postulate,** or (because he popularized it, even though it was first suggested by Thomas Bayes) the **Laplace criterion.**

The use of the expected-monetary-value criterion can also be depicted graphically. Consider the decision tree of Figure 17.2. It is based on Figure 17.1 and, as the caption indicates, the decision-tree analysis leads to the same decision as Table 17.9. Because the graphical process of finding the optimal action involves starting at the terminal (ultimate-payoff) points of the tree branches (at the "top" or right side of the tree) and then working backwards along the branches to find expected monetary values for each fork, this type of graphical solution of a decison problem is called **backward induction.**

Expected Opportunity Loss

A decision maker using the **expected-opportunity-loss criterion** determines an expected opportunity loss (or expected regret value) for each possible action and then selects the action with the smallest of these values. This procedure *always* yields the same result as the expected-monetary-value criterion. The **expected opportunity loss *(EOL),*** or **expected regret value,** of an action equals the sum of the weighted opportunity-loss values associated with that action—the weights being the probabilities of the alternative events that produce the various possible opportunity losses. Tables 17.11 and 17.12, which are based on our familiar examples, illustrate the procedure.

Note how a decision maker who, when confronted with the type of situation given in Table 17.11, consistently chooses A_1 will, on average, incur an opportunity loss of \$2.1 million per year because that loss will be zero 70 percent of the time and equal to \$7 million 30 percent of the time. In contrast, a consistent choice of nonoptimal A_2 would, in the long run, produce an opportunity loss of \$7 million per year because such a loss would equal \$10 million 70 percent of the time and be zero 30 percent of the time. A similar analysis can be made with respect to Table 17.12.

Table 17.11   Maximizing Benefit: The Expected-Opportunity-Loss Criterion

This table, based on Table 17.4, shows the opportunity-loss or regret values (in millions of dollars) associated with four possible action/event combinations; it also shows the expected opportunity loss for each possible action, based on the assumed prior probabilities of the events $[p(E_1) = .7$ and $p(E_2) = .3]$. The smallest *EOL* is optimal; thus, A_1 is the action chosen by those who employ the expected-opportunity-loss criterion. It is no accident that this result is identical with that given by the expected-monetary-value criterion (Table 17.9).

	Events		
Actions	E_1 = weak demand for new product $p(E_1) = .7$	E_2 = strong demand for new product $p(E_2) = .3$	Expected Opportunity Loss, *EOL*
A_1 = constructing a small plant	$8 - 8 = 0$	$12 - 5 = 7$	$0(.7) + 7(.3) = 2.1$ ←optimum
A_2 = constructing a large plant	$8 - (-2) = 10$	$12 - 12 = 0$	$10(.7) + 0(.3) = 7$

Table 17.12　Minimizing Cost: The Expected-Opportunity-Loss Criterion

This table, based on Table 17.5, shows the opportunity loss or regret values (in millions of dollars) associated with four possible action/event combinations; it also shows the expected opportunity loss for each possible action, based on the assumed prior probabilities of the events [$p(E_1)$ = .7 and $p(E_2)$ = .3]. The smallest *EOL* is optimal; thus, A_1 is the action chosen by those who employ the expected-opportunity-loss criterion. It is no accident that this result is identical with that given by the expected-monetary-value criterion (Table 17.10).

Actions	Events		Expected Opportunity Loss, *EOL*
	E_1 = low level of demand and production $p(E_1)$ = .7	E_2 = high level of demand and production $p(E_2)$ = .3	
A_1 = install small antipollution system	$5 - 5 = 0$	$8 - 4 = 4$	$0(.7) + 4(.3) = \boxed{1.2} \leftarrow$ optimum
A_2 = install large antipollution system	$10 - 5 = 5$	$4 - 4 = 0$	$5(.7) + 0(.3) = 3.5$

Expected Utility

All of the decision-making criteria discussed so far have employed *monetary* outcome measures. Yet critics have argued that using money to measure outcomes is a mistake, that people will invariably take those actions that maximize their welfare or *utility,* and that actions that maximize monetary benefit or minimize monetary cost may well not coincide with those that maximize utility. The kind of situation depicted in Table 17.13 can be used to illustrate what these critics have in mind.

　Would the typical decision maker really be indifferent between actions A_1 (investing $100 million in vineyards) and A_2 (investing $100 million in the auto industry) just because their expected monetary values are equal? Critics argue that the decision maker would not be indifferent because A_2, unlike A_1, involves *risk.* Choosing A_2 over A_1 is equivalent to taking a gamble, to giving up a sure thing (a $15 million return from action A_1 regardless of which event occurs) for an uncertain thing (*either* a $25 million return if E_1 occurs *or* a $5 million return if E_2 occurs). Although such a gamble would be a **fair gamble** (in which the expected monetary value of what is given up— that is, the $15 million certain to be received from action A_1—is precisely equal to the expected monetary value of what is received—that is, the $15 million received, on the average, from action A_2), most people would not take the gamble. If critics are right, most people would be even less inclined to take an **unfair gamble** in which the expected monetary value of what is given up exceeds the expected monetary value of what is received. (By changing all the entries of 15 in the A_1 row of Table 17.13 to 16, for example, action A_2 would be turned into an unfair gamble.) Indeed, many people, critics assert, would even reject a **more-than-fair gamble** in which the expected monetary value of what is given up is less than the expected monetary value of what is received. (By changing all the entries of 15 in the A_1 row of Table 17.13 to 14, for example, action A_2 would be turned into a more-than-fair gamble.)

Table 17.13   Doubts About the Expected-Monetary-Value Criterion

This table shows the annual profits (in millions of dollars) associated with four possible action/event combinations; it also shows the expected monetary value of each possible action, based on the assumed prior probabilities of the events [$p(E_1)$ = $p(E_2)$ = .5]. According to the *EMV* criterion, a decision maker would be indifferent between the two actions, but critics argue that most people would decisively opt for A_1.

Actions	Events		Expected Monetary Value, *EMV*
	E_1 = oil price rises moderately $p(E_1)$ = .5	E_2 = oil price rises sharply $p(E_2)$ = .5	
A_1 = investing $100 million in vineyards	15	15	15(.5) + 15(.5) = 15 ← optimum?
A_2 = investing $100 million in the auto industry	25	5	25(.5) + 5(.5) = 15

The St. Petersburg paradox. The apparent paradox of people refusing to take fair gambles was first solved some 250 years ago. A Swiss mathematician, Daniel Bernoulli (1700–1782) studied gamblers at the casinos of St. Petersburg, and he considered the game described below between two persons, A and B:

A fair coin is tossed until heads appears; if heads appears on the first toss, A pays B $1; if heads appears for the first time on the second toss, A pays B $2; if heads appears first on the third toss, A pays B $4; and so on, with A always paying 2^{n-1} at the *n*th toss if heads appears.

If playing this game is to be a fair gamble, what fee, Bernoulli asked, should B be willing to pay A for the privilege of playing this game? Because the player of a fair game is never asked to pay more than the expected monetary value of gain, this value can be calculated easily. Consider Table 17.14. Given a probability of .5 for heads to appear on the first toss, the expected monetary value of gain is $1(.5) = $.50 if the game ends after the first toss. Given a probability of $(.5)^2$ for heads to appear first on the second toss, the expected monetary value of gain is $2(.5)^2 = $.50 if the game ends after the second toss. And so it goes. Thus, the expected monetary value of gain during the entire game is the sum of the expected monetary values of all possible outcomes, or the infinite series $.50 + $.50 + . . . = ∞. Yet people are clearly not willing to pay such an infinite sum of money for the privilege of playing this game (even though such a payment would make the gamble precisely fair).

One probably could solve the paradox by postulating that gamblers cannot possibly be convinced to take this game seriously. How could they be so gullible as to believe that payoff would actually be made should they be lucky enough to win a large sum? (If heads did not appear until the 42nd toss, for example, the required payoff would equal more than the entire U.S. gross national product!) Bernoulli, however, had a different idea. He argued that people making decisions under uncertainty were not

Table 17.14   The St. Petersburg Game

The St. Petersburg game has an expected monetary value of gain of infinity (equal to the sum of the last column of this table).

Number of Toss (1)	Payoff if Heads First Appears at Given Toss (2)	Probability of Heads First Appearing at Given Toss (3)	Expected Monetary Value = Payoff × Probability (4) = (2) × (3)
1	$1	.5	$.50
2	2	$(.5)^2$	.50
3	4	$(.5)^3$	.50
4	8	$(.5)^4$	.50
5	16	$(.5)^5$	.50
.	.	.	.
.	.	.	.
.	.	.	.

attempting to maximize expected *monetary* values, but maximized expected *utilities* instead. He thought that the **total utility of money** (the overall welfare people derived from the possession of money) was rising the more money people had, while the **marginal utility of money** (the increase in total utility that was associated with a one dollar increase in the quantity of money, for example) was declining the more money people had. Any person starting with $500, for example, would, therefore, place a smaller subjective value on gaining an extra sum than on losing an equal amount. Any game with an equal probability of gaining and losing a given amount was, therefore, fair in monetary terms, but unfair in utility terms. The game's expected utility was negative. No wonder people refused to play it!

Figure 17.3 contains a hypothetical person's **utility-of-money function**—the relationship, that is, between alternative amounts of money the person might possess and the different utility totals associated with these amounts. It is assumed, in accordance with Bernoulli's postulate, that the total utility of money rises with greater amounts of it, while its marginal utility (shown by the slope of the total utility curve) declines.

In Figure 17.3, the total utility of money, U, is related to the amount of money, $, by the equation $U = \sqrt{\$}$. Thus, a person with $500 is assumed to receive $\sqrt{500}$, or 22.36 *utils* (units of utility) from it (point A). The same person would, however, receive only 5.92 extra utils from an added $300 (when moving from A to C) but would lose 8.22 utils by a loss of $300 (when moving from A to B). Given this type of utility function, the St. Petersburg paradox is easily solved. While the expected monetary value of the game equals infinity, its expected utility equals

$$EU = \sqrt{1}\,(.5) + \sqrt{2}\,(.5)^2 + \sqrt{4}\,(.5)^3 + \sqrt{8}\,(.5)^4 + \sqrt{16}\,(.5)^5 + \ldots$$
$$= 1(1/2) + \sqrt{2}\,(1/4) + 2(1/8) + 2\sqrt{2}\,(1/16) + 4(1/32) + \ldots$$
$$= 1/2 + \sqrt{2}\,(1/4) + 1/4 + \sqrt{2}\,(1/8) + 1/8 + \ldots$$
$$= 1 + \sqrt{2}\,(1/2) = 1.707 \text{ utils.}$$

The equation $U = \sqrt{\$}$ implies $U^2 = \$$; hence an expected utility of 1.707 translates into a dollar equivalent of $(1.707)^2 = \$2.91$. This $2.91, then, is the far-less-than-infinite

Figure 17.3   The Utility of Money

This graph illustrates a person's utility function that is characterized by declining marginal utility of money. The total utility of money, U, here is related to the amount of money, $, by the equation $U = \sqrt{\$}$. Other relationships are, of course, also possible.

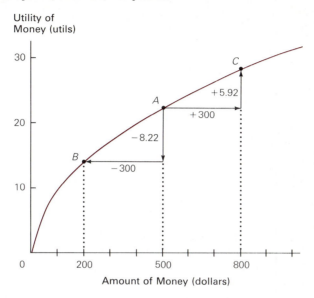

amount that a person who maximized expected utility, and possessed the $U = \sqrt{\$}$ utility function, would pay for the privilege of playing the St. Petersburg game.

Note: Someone else who also maximized expected utility, but possessed a different utility function, would act quite differently. This difference brings us to an important conclusion: Even if people, as is likely, maximize the expected utility rather than the expected monetary value of an action when facing uncertainty, we cannot predict their behavior without a knowledge of their utility functions. Indeed, the shape of these functions reveals important information about people's attitudes concerning the spread of possible outcomes of their action around the action's expected value. The extent of such spread (as from 15 down to 5 and up to 25 for action A_2 in Table 17.13) in fact measures the *risk* of an action, and people can view risk in one of three ways. They can be averse to it; they can be neutral toward it; they can seek it out.

Risk aversion. Imagine for a moment that the investor whose choices were pictured in Table 17.13 possesses the type of utility function postulated by Bernoulli. This type of utility function is illustrated in panel (a) of Figure 17.4. Action A_1, as we noted earlier, would bring the investor a profit of $15,000 a year, regardless of what happens to the price of oil. This amount of money is associated, according to point B on the utility function, with a total utility of $0b$. Action A_2, on the other hand, would bring the investor, with equal probability, a profit of $5,000 a year *or* of $25,000 a year. As we can see from points A and C, respectively, the associated utilities equal $0a$ and $0c$. The *expected* utility from an equally weighted $5,000 or $25,000 a year, however, equals the sum of half of $0a$ plus half of $0c$, and this sum is shown in the graph by utility $0d$. It

Figure 17.4   Attitudes Toward Risk

This set of graphs illustrates three basic attitudes toward risk: risk aversion in panel (a), risk neutrality in panel (b), and risk seeking in panel (c). These attitudes correspond, respectively, to declining, constant, and increasing marginal utilities of money.

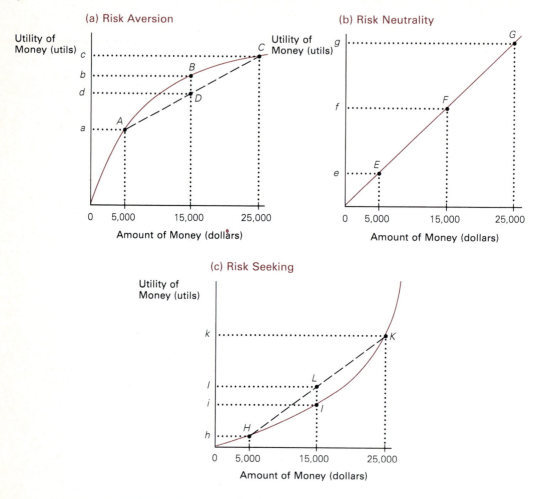

corresponds to point D, located on the dashed line connecting A and C and above the $15,000 expected monetary value of the $5,000 or $25,000 gamble. (Note: The expected utility of receiving $5,000 with a probability of .2 or $25,000 with a probability of .8 could similarly be read off of dashed line AC, but at a point above the $5,000(.2) + $25,000(.8) = $21,000 expected monetary value of this different gamble.)

Whenever a person in this way considers the utility (as at point B) of a certain prospect of money to be higher than the expected utility (as at point D) of an uncertain prospect of equal expected monetary value, the person is said to hold an attitude of **risk aversion.** This attitude is always present when a person's marginal utility of money (shown here by the slope of utility function $0ABC$) declines with larger amounts of money.

Risk aversion is, indeed, quite common. Consider how people do all kinds of things, small and large, to *escape* gambles: they place person-to-person calls instead of station-to-station calls; at airports, banks, and post offices, they prefer single lines feeding to many clerks to the chance of getting into a slow or fast line; they diversify their assets and do not "place all eggs in one basket"; they buy plenty of insurance; they reject fair gambles, as our risk-averse investor would by taking action A_1 (to gain utility $0b$) rather than action A_2 (that has an equal expected monetary value but a smaller expected utility $0d$).

Risk neutrality. Let us imagine instead that our investor's utility function is illustrated by panel (b) of Figure 17.4. In this case, the utility of action A_1, with its certain payoff of $15,000 a year, would equal $0f$, corresponding to point F. Action A_2 would bring, according to points E and G, utility of $0e$ or $0g$, and with equal probability. The expected utility of this gamble, however, would also equal $0f$. (A straight line between E and G leads to point F above the $15,000 expected monetary value of the $5,000 or $25,000 gamble.)

Whenever a person in this way considers the utility (as at point F) of a certain prospect of money to be equal to the expected utility (as at point F) of an uncertain prospect of equal expected monetary value, the person is said to hold an attitude of **risk neutrality.** This attitude is always present when a person's marginal utility of money (shown here by the slope of utility function $0EFG$) remains constant with larger amounts of money.

Risk neutrality is not very common, but it is precisely the implicit assumption made by those who advocate the use of the expected-monetary-value criterion. If all people were risk neutral, our investor would be truly indifferent between actions A_1 and A_2 because, as we saw, their expected monetary values are equal. By the same token, a risk-neutral person would be willing to pay $100 for the privilege of taking each of the following fair gambles, for the expected monetary value of each of these is also $100:

    a. a 99-percent chance of getting $101.01 and a 1-percent chance of getting nothing;
    b. a 50-percent chance of getting $101 and a 50-percent chance of getting $99;
    c. a 1-percent chance of getting $10,000 and a 99-percent chance of getting nothing;
    d. a 1-percent chance of getting $1 million, a 1-percent chance of losing $990,000, and a 98-percent chance of getting nothing.

After thinking about it, you might pay the price for (a) and (b) or something close to it, but do you have the sweepstakes mentality to go after (c)? Are you ready for the Russian roulette of (d)?

Risk seeking. Now imagine that our investor's utility function is illustrated by panel (c) of Figure 17.4. In this case, the utility of action A_1, with its certain payoff of $15,000 a year, would equal $0i$, corresponding to point I. Action A_2 would bring, according to points H and K, utility of $0h$ or $0k$, and with equal probability. By the now familiar procedure, we can establish the expected utility of this gamble as $0l$, corresponding to point L on dashed line HK.

Whenever a person in this way considers the utility (as at point I) of a certain prospect of money to be lower than the expected utility (as at point L) of an uncertain prospect of equal expected monetary value, the person is said to hold an attitude of

risk seeking. This attitude is always present when a person's marginal utility of money (shown here by the slope of utility function 0*HIK*) rises with larger amounts of money.

Risk seeking, like risk neutrality, is not very common. If all people were risk seekers, our investor would, of course, prefer action A_2 to A_1, and all of us would constantly seek out and accept huge riches-or-ruin gambles. In fact, of course, most of us have the opposite inclination and seek to buy insurance.

Conclusion. An evaluation of decision outcomes in terms of utility rather than money may well be a superior approach. If the investor depicted in Table 17.13, for example, were risk-averse and had Bernoulli's utility function, the decision problem could be laid out as in Table 17.15. This approach would explain why the investor would prefer A_1 to A_2, even though A_1 and A_2 have equal expected monetary values.

A decision maker using the **expected-utility criterion,** thus, determines the expected utility for each possible action and then selects the action that maximizes expected utility. The **expected utility *(EU)*** of an action equals the sum of the weighted utilities associated with that action—the weights being the probabilities of the alternative events that produce the various possible utility payoffs.

Note: Modern theorists have advanced utility theory far beyond Bernoulli and have devised a variety of ingenious methods of identifying a decision maker's utility function. Readers interested in pursuing the matter may begin by consulting the classic work of von Neumann and Morgenstern listed in the "Selected Readings" section at the end of this chapter.

Note also: Real-world examples of the use of decision trees to maximize benefit or minimize cost (and of making a decision on the basis of expected monetary values alone or by considering expected utilities instead) are given in Analytical Examples 17.1, "Decision Analysis Comes of Age," and 17.2, "The Decision to Seed Hurricanes— A Second Look."

Table 17.15   Maximizing Expected Utility

This table, based on Table 17.13, shows the utility associated with four possible action/event combinations for a risk-averse person with a utility function of $U = \sqrt{\$}$. The table also shows the expected utility of each possible action, based on the assumed prior probabilities of the events [$p(E_1) = p(E_2) = .5$]. Because the objective is to maximize expected utility, the largest *EU* is optimal; thus, A_1 is the action chosen by those who employ the expected-utility criterion.

	Events		
Actions	E_1 = oil price rises moderately $p(E_1) = .5$	E_2 = oil price rises sharply $p(E_2) = .5$	Expected Utility, *EU*
A_1 = investing $100 million in vineyards	$\sqrt{15} = 3.87$	$\sqrt{15} = 3.87$	$3.87(.5) + 3.87(.5) = \boxed{3.87}$ ←optimum
A_2 = investing $100 million in the auto industry	$\sqrt{25} = 5$	$\sqrt{5} = 2.24$	$5(.5) + 2.24(.5) = 3.62$

DECISION MAKING WITH PROBABILITIES: POSTERIOR ANALYSIS

In the previous section, we discussed various forms of decision making that made use of prior probabilities concerning events over which a decision maker has no control. **Posterior analysis** is a form of decision making under uncertainty that starts out with prior probabilities, proceeds to gather additional experimental or sample evidence about event probabilities, and then uses this new evidence to transform the set of prior probabilities, by means of Bayes's theorem, into a revised set of posterior probabilities with the help of which a final decision is reached. The procedure of revising probabilities, first discussed in the Chapter 5 section, "Revising Probabilities: Bayes's Theorem," can be sketched as follows:

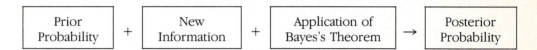

Gathering New Information

Consider our above example concerning a decision on constructing either a small or a large plant to manufacture and sell a new product (Table 17.9). The decision maker's prior probabilities for the two possible events (E_1 = weak demand and E_2 = strong demand) equaled $p(E_1) = .7$ and $p(E_2) = .3$; they might have been based on nothing but a hunch. Instead of proceeding with the analysis on that basis (as illustrated in Tables 17.9 and 17.11, or in Figure 17.2), the decision maker might decide to gather new information about the event probabilities. The required research can take many different forms. In this particular case, a consumer survey might be conducted. In other instances, such as those noted at the beginning of this chapter, other approaches might be more appropriate: obtaining a long-range weather forecast, perhaps, or a forecast of future prices and wages, sampling inputs bought or output produced for quality, building a pilot plant to test a new production process, conducting a seismic test (that can indicate the likely presence of oil deposits), taking an aptitude test (that can indicate future job performance), or even paying a bribe (that can provide inside information on the likely reaction of a TV network to a new pilot that might be offered to it). Unfortunately, none of this new information will be perfect. Even the best of aptitude tests will flunk some job applicants who later turn out to be superb, while passing others who later turn out to be incompetent. Even the best of seismic tests can deny the presence of oil in a field that is already producing oil, while confirming the presence of oil at a site that is later proven to be dry. Thus, the decision maker who wants to know about weak or strong demand for the new product will still have to contend with uncertainty, even if a consumer survey is conducted and a fancy market-research report is received. Indeed, the market-research firm's past record may be that indicated by Table 17.16 on the following page. As the caption explains, although the firm's record is not perfect, its "batting average" is quite impressive; one can place a fairly high degree of confidence in the research firm's reports. When it predicts weak demand, it is correct 90 percent of the time; when it predicts strong demand, it is correct 80 percent of the time.

Table 17.16 **The Past Record of a Market-Research Firm**

This table shows the probabilities that a firm's market research will discover the true state of affairs about the demand for a new product. Based on past performance, the conditional probability is .9 that the firm's research result indicates weak demand when demand is in fact weak, and it is .1 that the result indicates strong demand when demand is in fact weak. Similarly, the conditional probability is .2 that the firm's research result indicates weak demand when demand is in fact strong, and it is .8 that the result indicates strong demand when demand is in fact strong.

Market-Research Result	True State of Affairs	
	E_1 = weak demand	E_2 = strong demand
R_1 = demand will be weak	$p(R_1\|E_1) = .9$	$p(R_1\|E_2) = .2$
R_2 = demand will be strong	$p(R_2\|E_1) = .1$	$p(R_2\|E_2) = .8$

Applying Bayes's Theorem

Let us now suppose that our decision maker does consult the research firm and is given a report concerning the likely intensity of future demand. What happens next depends on the content of the report.

A report of weak demand. If the report indicates weak demand (the research result is R_1), the data contained in the first row of Table 17.16 come into play as *Bayes's theorem* (see Box 5.K on page 173 or Box 5.L on page 173) is applied. According to the Box 5.K formula, the posterior probability of an event is computed as

$$p(E|R) = \frac{p(E)\,p(R|E)}{p(E)\,p(R|E) + p(\bar{E})\,p(R|\bar{E})}$$

where E denotes the event, $\bar{E}$ its complement, and R the research result.

In our case, therefore (using the above-noted prior probabilities and the data of Table 17.16),

$$p(E_1|R_1) = \frac{p(E_1)\,p(R_1|E_1)}{p(E_1)\,p(R_1|E_1) + p(E_2)\,p(R_1|E_2)} = \frac{.7(.9)}{.7(.9) + .3(.2)} = .91,$$

and

$$p(E_2|R_1) = \frac{p(E_2)\,p(R_1|E_2)}{p(E_2)\,p(R_1|E_2) + p(E_1)\,p(R_1|E_1)} = \frac{.3(.2)}{.3(.2) + .7(.9)} = .09.$$

An alternative calculation is provided by the Box 5.L formula as

$$p(E|R) = \frac{p(E\ and\ R)}{p(R)}.$$

In our case, according to Box 5.I on page 162, the joint probability of E_1 *and* R_1 equals $p(E_1)p(R_1|E_1)$, or $.7(.9) = .63$, while the joint probability of E_2 *and* R_1 equals $p(E_2)p(R_1|E_2)$, or $.3(.2) = .06$. These being mutually exclusive events, we can calculate the unconditional probability of our result, $p(R_1)$ as $.63 + .06 = .69$. Therefore,

$$p(E_1|R_1) = \frac{p(E_1 \text{ } and \text{ } R_1)}{p(R_1)} = \frac{.63}{.69} = .91,$$

and

$$p(E_2|R_1) = \frac{p(E_2 \text{ } and \text{ } R_1)}{p(R_1)} = \frac{.06}{.69} = .09.$$

These results confirm those already derived.

A report of strong demand. What if the research firm's report had predicted a strong demand for the new product (the research result had been R_2)? In that case, of course, an analogous set of calculations would have been performed:

$$p(E_1|R_2) = \frac{p(E_1)\,p(R_2|E_1)}{p(E_1)\,p(R_2|E_1) + p(E_2)\,p(R_2|E_2)} = \frac{.7(.1)}{.7(.1) + .3(.8)} = .23;$$

$$p(E_2|R_2) = \frac{p(E_2)\,p(R_2|E_2)}{p(E_2)\,p(R_2|E_2) + p(E_1)\,p(R_2|E_1)} = \frac{.3(.8)}{.3(.8) + .7(.1)} = .77.$$

Finding the Optimal Strategy

If our decision maker were risk-neutral and cared to maximize expected *monetary* value (which we will assume), an optimal strategy could now be mapped out with the help of these posterior probabilities. An **optimal strategy** is a complete plan specifying the actions to be taken at each available action point, if the expected (monetary or utility) payoff is to be the best one available. The procedure is illustrated in Figure 17.5. It corresponds to the example discussed so far in which a decision maker, at action point *a,* has decided to make a consumer survey in order to gauge the likely intensity of future demand. This decision, inevitably, takes us to event point *b.* At event point *b,* the survey will predict either weak or strong future demand, taking us to either action point *c* or action point *d.* Once located at either *c* or *d,* the decision maker must then act and construct either a small plant or a large one. As a result, one of the event points labeled *f* through *i* is reached, and chance will decide whether demand is in fact weak or strong. The right-hand column shows the payoffs, which are equal to those given throughout this chapter (starting with Table 17.1) but are reduced by an assumed $100,000 cost of taking the survey.

We can find the optimal strategy by using the process of backward induction. Consider event point *f.* Given the conditional probabilities of $p(E_1|R_1) = .91$ and $p(E_2|R_1) = .09$ that were calculated above, the decision maker can calculate the expected monetary value at point *f* as $\$7.9(.91) + \$4.9(.09) = \$7.63$ (million). This number is indicated in color above event point *f.*

Figure 17.5   Posterior Decision-Tree Analysis

This decision tree summarizes the development of an optimal strategy, while making use of posterior probabilities. (Note: All dollar amounts are in millions.)

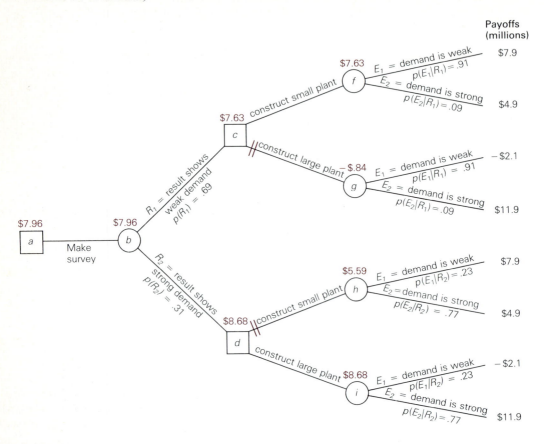

Now consider event point g. Given the same conditional probabilities as at f, the decision maker can calculate the expected monetary value at point g as $-\$2.1(.91) + \$11.9(.09) = -\$.84$ (million). This number is indicated in color above event point g. Surely, a decision maker at action point c would want to move toward f rather than g; thus, we can prune the branch going from c to g (note the color cut) and carry the color *EMV* figure at f to c.

We now turn to event point h. The probabilities are now different as we assume having received a survey report indicating strong demand (R_2). Given the conditional probabilities calculated above, the decision maker can calculate the expected monetary value at point h as $\$7.9(.23) + \$4.9(.77) = \$5.59$ (million). This number is indicated in color above point h.

Now consider event point i. Given the same conditional probabilities as at point h, the decision maker can calculate the expected monetary value at point i as $-\$2.1(.23) + \$11.9(.77) = \$8.68$ (million). This number is indicated in color above event point i. Surely, a decision maker at action point d would want to move towards i rather than h;

thus, we can prune the branch going from *d* to *h* (note the color cut) and carry the color *EMV* figure at *i* to *d*.

We must, finally, consider event point *b*. As we noted earlier, the unconditional probability of getting survey result R_1 equals .69. There being only two possible survey results, the unconditional probability of getting survey result R_2 can, thus, be found quickly as $p(R_2) = 1 - p(R_1) = 1 - .69 = .31$. This conclusion can, however, also be checked by adding the joint probability of E_1 *and* R_2 [equal to $p(E_1) \cdot p(R_2|E_1) = .7(.1) = .07$] to the joint probability of E_2 *and* R_2 [equal to $p(E_2) \cdot p(R_2|E_2) = .3(.8) = .24$]. As we can see, the same result is obtained: $p(R_2) = .07 + .24 = .31$. Thus, we can calculate the expected monetary value at event point *b* as $\$7.63(.69) + \$8.68(.31) = \$7.96$ (million), a figure that can be placed above *b* as well as *a*.

The grand conclusion is at hand. The optimal strategy (that carries an expected monetary value of \$7.96 million) is the following:
Given that a survey is made,

    a. if the result shows weak demand, to construct a small plant (at *c*) and let chance (at *f*) determine the reward: of \$7.9 million if the survey result was correct, of \$4.9 million if it was incorrect.

    b. if the result shows strong demand, to construct a large plant (at *d*) and let chance (at *i*) determine the reward: of −\$2.1 million if the survey result was incorrect, of \$11.9 million if it was correct.

Note: This course would be the best strategy, given the information available at the time. Obviously, this information can change over time and, if it is not too late, the decision can be changed accordingly. Information can change as a result of some deliberate action taken by the decision maker (such as the making of a second, independent survey) or as a result of some unpredicted event (such as a scandal that seriously hurts the firm's image). In the former case, new posterior probabilities can be calculated; in the latter case, the decision maker may wish to revise the prior probabilities as well. To illustrate the former contingency, imagine that a second (independent) survey were to confirm (rather than contradict) a report of weak demand obtained by a first one. The conditional probability of E_1, given the first R_1 (which was shown to be .91) would then change to become the conditional probability of E_1, given the first R_1 *and* a second R_1^*, and would be calculated as

$$p(E_1|R_1 \text{ } and \text{ } R_1^*) = \frac{p(E_1) \, p(R_1|E_1) \, p(R_1^*|E_1)}{p(E_1) \, p(R_1|E_1) \, p(R_1^*|E_1) + p(E_2) \, p(R_1|E_2) \, p(R_1^*|E_2)}$$

$$= \frac{.7(.9)(.9)}{.7(.9)(.9) + .3(.2)(.2)} = .98.$$

As more and more information comes in, therefore, the posterior probability depends less and less on the prior one. In this case, the prior .7 changed to .91 after the first survey, and we now have seen it change to .98 after the second. Caution: Evidence can, of course, go either way; hence, a posterior probability can decline below the prior one as well as rise above it.

Because the crucial insights of the Rev. Thomas Bayes have become such an integral part of modern decision theory, he can rightly be called the "father of decision theory" (see Biography 17.1 at the end of this chapter).

THE VALUE OF INFORMATION

In the previous section, we made an implicit assumption: that it was worthwhile to the decision maker to spend $100,000 on a survey of consumers. In fact, this assumption has been borne out, as we can see when comparing the expected monetary values of the optimal strategies that have emerged, respectively, from prior analysis ($7.10 million in Figure 17.2) and posterior analysis ($7.96 million in Figure 17.5). Yet this result was a lucky circumstance; it is time to consider the issue systematically.

The Expected Value of Perfect Information

We must note, first of all, that there exists a maximum amount, called the **expected value of perfect information,** or *EVPI,* that any decision maker can be expected to pay for obtaining perfect information about future events and, thus, for eliminating uncertainty completely. When the objective is the maximization of some benefit, this maximum equals the difference between the expected benefit when the optimal decision is made with perfect information and the expected benefit when that decision is made under uncertainty. (When the objective is the minimization of some cost, this maximum equals the difference between the expected cost when the optimal decision is made under uncertainty and the expected cost when that decision is made with perfect information.) Depending on whether the expected value of the optimal action under uncertainty is based on prior or posterior probabilities, one can also distinguish a **prior *EVPI*** and a **posterior *EVPI.***

Prior EVPI. Consider the case illustrated in Table 17.9 and Figure 17.2. The expected value of that decision under uncertainty equals $7.1 million. If the decision maker, however, were offered perfect information by someone about the guaranteed occurrence of either E_1 or E_2, things would be different. If told that E_1 will occur with certainty, the decision maker would instantly take action A_1 and gain $8 million. If told that E_2 will occur with certainty, the decision maker would instantly take action A_2 and gain $12 million. There is a problem, however. The purveyor of the perfect information may not reveal it until *after* being paid; otherwise, the decision maker would have the information and could use it while paying nothing for it. To determine the expected value of a decision with perfect information, the decision maker must guess, before possessing the information, about the likelihood of receiving a report that E_1 will occur, $p(R_1)$, and of receiving a report that E_2 will occur, $p(R_2)$. Suppose that such prior probabilities, like $p(E_1)$ and $p(E_2)$ earlier, are also set to equal .7 and .3, respectively. Then the expected value of the decision with perfect information is

$$\$8(.7) + \$12(.3) = \$9.2 \text{ (million)}.$$

Given an expected value of the decision under uncertainty of $7.1 million, the

$$\text{prior } EVPI = \$9.2 - \$7.1 = \$2.1 \text{ million.}$$

This amount is the maximum the decision maker would pay to the holder of perfect information for revealing it. Note that this figure also equals the expected opportunity loss of the optimal decision under uncertainty (Table 17.11). This equality of prior *EVPI* and *EOL* is always the case and explains why some decision makers prefer the expected-opportunity-loss criterion (Table 17.11) over the expected-monetary-value criterion (Table 17.9); the former automatically indicates the *EVPI*. (Nevertheless, because the computations are less complicated, most decision makers use the *EMV* criterion that, of course, always points to the same optimal action.)

   Posterior EVPI. Now consider the case illustrated in Figure 17.5. The expected value of that decision under uncertainty equals $7.96 million because this decision maker has the advantage of additional information (the survey result). Accordingly, we can determine a

$$\text{posterior } EVPI = \$9.20 - \$7.96 = \$1.24 \text{ million.}$$

After having already paid $100,000 for a survey of consumers, this decision maker would pay a purveyor of perfect information only this smaller maximum amount. Analytical Example 17.3," The Value of Perfect Information: The Case of the U.S. Cattle Industry," provides an interesting real-world application.

The Expected Value of Sample Information

In fact, no one will ever be able to provide *perfect* information about future events. Decision makers, therefore, are much more interested in another concept, the **expected value of sample information,** or *EVSI,* which equals the maximum amount any decision maker can be expected to pay for obtaining admittedly imperfect information about future events and, thus, for reducing (rather than eliminating) uncertainty. When the objective is the maximization of some benefit, this maximum equals the difference between the expected benefit when an optimal decision is made with the sample information (and before paying for it) and the expected benefit when that decision is made without it. (When the objective is the minimization of some cost, this maximum equals the difference between the expected cost when the optimal decision is made without the sample information and the expected cost when that decision is made with it, but again before paying for it.)

   Consider our example. The optimal decision with the survey information (Figure 17.5) is expected to yield $7.96 million only *after* paying an assumed $100,000 for conducting the survey. Hence, the expected gross benefit with survey information equals $8.06 million. The optimal decision without the survey information (Figure 17.2) is expected to yield $7.10 million. Hence, the

$$EVSI = \$8.06 - \$7.10 = \$.96 \text{ million.}$$

Our decision maker would pay a maximum of $960,000 for the survey. Since $100,000 was paid in fact, a shrewd survey firm could have extracted an additional $860,000 for its (imperfect) information.

The Efficiency of Sample Information

We just noted how our decision maker, before obtaining any sample information, would place a maximum value of $2.1 million on perfect information. At the same time, the expected value of (inevitably imperfect) sample information was shown to equal

\$.96 million. The ratio of the latter to the former, multiplied by 100, is called the **efficiency of sample information,** or **ESI.** Thus,

$$ESI = \frac{EVSI}{(\text{prior}) \, EVPI} \cdot 100.$$

In our example, the *ESI* can be calculated as:

$$ESI = \frac{\$.96 \text{ million}}{\$2.10 \text{ million}} \cdot 100 = 45.7.$$

This *ESI* value indicates that the consumer survey provided only 45.7 percent of what perfect information would have been worth. Some decision makers employ this type of efficiency rating to decide whether they should look for some other type of information. Presumably, they seek other information when the rating is low, but decide that the information is as good as one can hope to obtain when the rating is high. Naturally, what is "low" and what is "high" is a debatable point.

PREPOSTERIOR ANALYSIS

When decision makers are deciding the worth of obtaining additional (experimental or sample) information before proceeding to prior or posterior analysis, they are said to be engaged in **preposterior analysis.** This analysis can be simple or complicated.

When the new (and inevitably imperfect) information about future events is being offered to them at a price exceeding the (prior) expected value of perfect information, a decision can instantly be made in the negative; it is then better to stick with prior analysis. In the above example, for instance, it would clearly not be worthwhile to acquire (assuredly imperfect) information at a price of \$2.1 million or more because that sum equals the prior *EVPI.*

When the new (and inevitably imperfect) information about future events is being offered at a price below the prior *EVPI,* however, it becomes necessary to determine how that price compares with the expected value, *EVSI,* of the sample information that is being offered. On that basis, one can decide whether to buy the information or proceed without it. In our above example, the information was offered, we assumed, for \$100,000, while being worth, we calculated, \$960,000; hence, the decision to make the survey (a decision that was treated as a foregone conclusion in Figure 17.5) was validated in retrospect. It is customary, however, to make the decision about possibly acquiring additional information in a more systematic way, either with the help of a decision tree (that now includes the choice between posterior and prior analysis) or in tabular form. The decision-tree approach is referred to as **extensive-form analysis;** the tabular approach is **normal-form analysis.**

Extensive-Form Analysis

Figure 17.6 provides an example of the extensive form of preposterior analysis; it identifies the optimal strategy by means of backward induction on a decision tree. The lower branches of this particular tree—those emanating from action point *e*—correspond precisely to Figure 17.2 and illustrate prior analysis. The upper branches—those

Figure 17.6 Preposterior Decision-Tree Analysis

This decision tree summarizes the development of an optimal strategy that includes the initial choice between engaging in prior analysis only or engaging in posterior analysis. Thus, the tree combines Figures 17.2 and 17.5. (Note: All dollar amounts are in millions.)

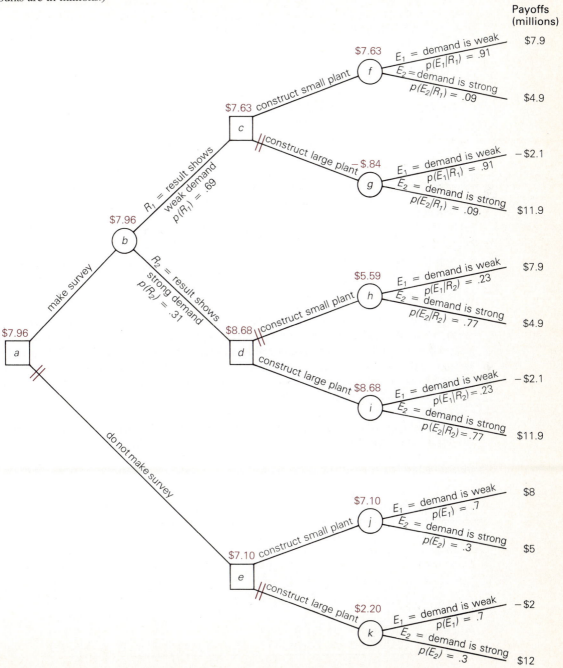

Payoffs (millions)

$7.63 construct small plant → f

$7.63 → c

$7.9
$E_1 = $ demand is weak
$p(E_1|R_1) = .91$

$4.9
$E_2 = $ demand is strong
$p(E_2|R_1) = .09$

construct large plant $-\$.84$ → g

$-\$2.1
$E_1 = $ demand is weak
$p(E_1|R_1) = .91$

$11.9
$E_2 = $ demand is strong
$p(E_2|R_1) = .09$

$R_1 = $ result shows weak demand
$p(R_1) = .69$

$7.96 → b

make survey

$7.96 → a

$R_2 = $ result shows strong demand
$p(R_2) = .31$

$8.68 construct small plant → h

$5.59 → h
$7.9
$E_1 = $ demand is weak
$p(E_1|R_2) = .23$

$4.9
$E_2 = $ demand is strong
$p(E_2|R_2) = .77$

$8.68 → d

construct large plant $8.68 → i

$-\$2.1
$E_1 = $ demand is weak
$p(E_1|R_2) = .23$

$11.9
$E_2 = $ demand is strong
$p(E_2|R_2) = .77$

do not make survey

$7.10 construct small plant → j

$7.10 → j
$8
$E_1 = $ demand is weak
$p(E_1) = .7$

$5
$E_2 = $ demand is strong
$p(E_2) = .3$

$7.10 → e

construct large plant $2.20 → k

$-\$2
$E_1 = $ demand is weak
$p(E_1) = .7$

$12
$E_2 = $ demand is strong
$p(E_2) = .3$

emanating from action point a—correspond to Figure 17.5 and illustrate posterior analysis. The tree as a whole now also depicts the initial choice (at a) of making or not making a survey. Because the former route leads to b (and an expected monetary value, *after* paying the $100,000 survey cost, of $7.96 million), while the latter route leads to e (and a lower expected monetary value of $7.10 million), the decision maker would prune the branch leading from a to e and beyond, as indicated here.

Thus, the optimal strategy is the following:

a. Do make a survey.
b. If the result shows weak demand, construct a small plant (and let chance at f determine the reward).
c. If the result shows strong demand, construct a large plant (and let chance at i determine the reward).

Note: The difference between the expected monetary values at b and e, or $860,000, represents the *additional* amount (on top of the actual $100,000 survey cost) that the decision maker could pay for this information before "making a survey" and "not making a survey" would become a matter of indifference. Thus, the *EVSI* equals $960,000, as noted above. Therefore, one could show that the "make survey" branch, rather than the "do not make survey" branch, would be pruned from this decision tree if the survey cost exceeded this sum and equaled, say, $1 million, rather than $100,000.

Normal-Form Analysis

The normal form of preposterior analysis always leads to the same conclusion as the extensive form, but uses a different route; it is the systematic computation of an expected payoff value for every conceivable strategy and the subsequent selection of the strategy with the best of these values as the optimal one. As we can see by inspecting Figure 17.6, our example includes 6 potential strategies, which we can label S_1 through S_6. They are the following:

S_1: making a survey; given a report of weak demand, constructing a small plant and, given a report of strong demand, constructing a small plant as well.
S_2: making a survey; given a report of weak demand, constructing a small plant and, given a report of strong demand, constructing a large plant.

Table 17.17   Calculation of Expected Payoff for Strategy S_1

First-Stage Event Probability (1)	Second-Stage Event Probability (2)	Joint Probability (3) = (1) × (2)	Payoff (millions of dollars) (4)	Joint Probability × Payoff (millions of dollars) (5) = (3) × (4)
$p(R_1) = .69$	$p(E_1\|R_1) = .91$	.69(.91) = .63	$7.9	.63(7.9) = $4.98
$p(R_1) = .69$	$p(E_2\|R_1) = .09$	.69(.09) = .06	4.9	.06(4.9) = .29
$p(R_2) = .31$	$p(E_1\|R_2) = .23$	.31(.23) = .07	7.9	.07(7.9) = .55
$p(R_2) = .31$	$p(E_2\|R_2) = .77$	.31(.77) = .24	4.9	.24(4.9) = 1.18
		1.00		Expected Payoff → $7.00

Table 17.18 Calculation of Expected Payoff for Strategy S_6

Event Probability (1)	Payoff (millions of dollars) (2)	Probability × Payoff (millions of dollars) (3) = (1) × (2)
$p(E_1) = .7$	$-\$2$	$.7(-\$2) = -\$1.4$
$p(E_2) = .3$	$\$12$	$.3(\$12) = \quad 3.6$

Expected Payoff → $\$2.2$

S_3: making a survey; given a report of weak demand, constructing a large plant and, given a report of strong demand, constructing a small plant.

S_4: making a survey; given a report of weak demand, constructing a large plant and, given a report of strong demand, constructing a large plant as well.

S_5: making no survey; constructing a small plant.

S_6: making no survey; constructing a large plant.

Normal-form analysis requires that for each possible strategy an expected payoff value be calculated, as is shown in Tables 17.17 and 17.18 for strategies S_1 and S_6, respectively.

Observe the logic of Table 17.17. The first two rows indicate that the probability of a report of weak demand is .69 and that the subsequent construction of a small plant (which is what S_1 requires in that case) can have two consequences: (a) that demand is in fact weak (which carries a probability of .91) and that a payoff of $7.9 million is received or (b) that demand is in fact strong (which carries a probability of .09) and that a payoff of $4.9 million is received. The last two rows indicate, similarly, that the probability of a report of strong demand is .31 and that the subsequent construction of a small plant (which is what S_1 requires in that case as well) can also have two consequences: (a) that demand is in fact weak (which carries a probability of .23) and that a payoff of $7.9 million is received or (b) that demand is in fact strong (which carries a probability of .77) and that a payoff of $4.9 million is received. The rest of the table is self-explanatory. So is the much simpler Table 17.18, which gives the normal-form analysis of strategy S_6.

Table 17.19 shows our results for S_1 and S_6 and those of similar calculations for the other strategies as well. Given that S_2 has the highest expected value (encircled), S_2 is the optimal strategy. This result corresponds precisely to the result of our extensive-form analysis (that is, to the portion of our Figure 17.6 decision tree that has not been pruned away). Note: Except for rounding error, the expected value of the optimal strategy has been found to be the same in the two forms of analysis.

Table 17.19 Normal-Form Analysis: Expected Payoffs (in millions of dollars) of Six Strategies

S_1: $7.00	S_4: $2.10
S_2: 7.98	S_5: 7.10
S_3: 1.12	S_6: 2.20

Pros and Cons

It is easy to see why most people prefer the extensive form of analysis: it is so much easier computationally. The pruning of the decision tree leaves only the optimal strategy behind; the expected values of all the nonoptimal strategies need not be computed at all.[2]

There are occasions, however, when the manifold and detailed calculations produced by the normal-form analysis come in handy. The most important of these occasions arises when a decision maker wants to determine how sensitive the selection of the optimal strategy is to the initial choice of prior probabilities [$p(E_1) = .7$ and $p(E_2) = .3$ in our example]. Many times, a decision maker may feel uncomfortable with probabilities based on subjective feelings and may, therefore, want to test whether slight or moderate changes in these values would change the ultimate result. In our example, for instance, there are three leading contenders for optimal strategy: S_1, S_2, and S_5 (as Table 17.19 shows). What would have happened, a decision maker might wonder, if the prior probabilities had been $p(E_1) = p(E_2) = .5$ instead of $p(E_1) = .7$ and $p(E_2) = .3$? Would S_1 have become the optimal strategy? Or S_5? Or even S_4? To answer such questions, the systematic and admittedly tedious calculations of the normal-form analysis might be repeated with the new probability values.

SEQUENTIAL SAMPLING

A logical extension of preposterior analysis has emerged from the pioneering work of Abraham Wald (see Biography 17.2 at the end of this chapter). Wald recognized that if a decision maker (as in Figure 17.6) can make the choice of either gathering information or doing entirely without it before deciding upon a course of action, that decision maker surely also has the option, after getting information, to seek still more information, and still more after that, and so on, before a final decision is made. In principle, the number of information-gathering stages could be infinite; the need to make a decision, of course, dictates an end to that process long before that. From this thought emerged the technique of **sequential sampling,** in which the sample size is not fixed in advance, but a decision is made, after each bit of information is received, either to continue sampling or to put a stop to it.

Wald argued that much of traditional sampling was wasteful in that people were forever scrutinizing complete samples of fixed size even when an examination of the first few units in the sample already indicated the decision to be made. Wald showed how the sequential-sampling technique could replace traditional hypothesis testing. After each observation, a test statistic could be computed, and once it reached predetermined limits, the null hypothesis could be accepted or rejected. Depending on what the observations turned out to be, the statistician might make the decision to accept or to reject after only a few observations or after many, but the crucial advantage of the procedure was that, on average, it would require a smaller number of observations

---

[2]These values *could* be computed easily enough, however, even from a decision tree. Consider Figure 17.6. Find the weighted average of the *EMV*s at *f* and *h*, the weights being $p(R_1) = .69$ and $p(R_2) = .31$, and you have the *EMV* of strategy S_1. The similarly weighted average of the *EMV*s at *f* and *i* equals the *EMV* of S_2; of the *EMV*s at *g* and *h* equals the *EMV* of S_3, and of the *EMV*s at *g* and *i* equals the *EMV* of S_4. The *EMV*s of S_5 and S_6 are, of course, given directly at points *j* and *k*.

than a fixed-size sample required, while attaining the same probability of making a Type I or Type II error. Clearly, this advantage is a crucial one, especially in the costly case of destructive sampling.

Tables are now available for carrying out various sequential-sampling plans. In the realm of industrial quality control, for example, such a table might state the set of alternative hypotheses (such as "H_0: the proportion of defectives is .01 or less" and "H_A: the proportion of defectives is greater than .01") and the probability of error (such as "the Type I error probability of rejecting H_0 although it is true = .05" and "the Type II error probability of accepting H_0 although it is false = .10") and then give appropriate instructions depending on the number of items sampled so far (such as "accept H_0 if the number of defectives is 5 or below," "reject H_0 if the number of defectives is 20 or more," and "inspect another unit").

THE GREAT CONTROVERSY: CLASSICAL VS. BAYESIAN STATISTICS

A great controversy erupted among statisticians in the early 1960s, and it continues unabated to this day. As a result of the development of decision theory by Wald and others, more and more statisticians began to replace traditional and so-called "classical" methods of analysis (that relied on *objective probabilities* exclusively and attempted to *make inferences*) by Bayesian methods (that incorporated personal judgment in the form of *subjective probabilities* and attempted to *make decisions* in the face of uncertainty). The quarrel between the two camps has turned out to be a heated one and is reminiscent of (and not entirely unrelated to) the Neyman-Pearson vs. Fisher controversy noted in Biographies 9.1–9.3. The essence of the controversy can, perhaps, be seen most easily by an example.

An Example

Suppose there are two urns, I and II, each containing 6 balls. The contents of the urns are as follows:

> Urn I: 5 white balls and 1 black ball.
> Urn II: 1 white ball and 5 black balls.

You are told all of the above, you are given one of the urns (but not told which), and you are asked to draw a ball from it at random. It turns out to be white. You are asked to determine the probability that the urn from which you drew was urn I. (Note: The wider applications of this general example should be obvious. The urns might represent two shipments of raw materials or the outputs of two plants; the white and black balls might represent satisfactory and defective items, respectively.)

It is easy enough to state the answer in a general way. The probability that you drew from urn I, given that a white ball was drawn, or $p(\text{I}|\text{W})$, equals the ratio of the joint probability of encountering urn I and a white ball, or $p(\text{I } and \text{ W})$, to the unconditional probability of drawing a white ball, $p(\text{W})$. Thus,

$$p(\text{I}|\text{W}) = \frac{p(\text{I } and \text{ W})}{p(\text{W})}. \tag{1}$$

According to the general multiplication law (Box 5.I on page 162, $p(I \; and \; W) = p(I) \cdot p(W|I)$; hence,

$$p(I|W) = \frac{p(I) \, p(W|I)}{p(W)}. \tag{2}$$

The probability of drawing a white ball can occur in two mutually exclusive ways, by means of urn I or urn II:

$$p(W) = p(I \; and \; W) + p(II \; and \; W), \tag{3}$$

and this probability, in turn, equals

$$p(W) = p(I) \, p(W|I) + p(II) \, p(W|II). \tag{4}$$

Combining equations 2 and 4,

$$p(I|W) = \frac{p(I) \, p(W|I)}{p(I) \, p(W|I) + p(II) \, p(W|II)} \tag{5}$$

Equation 5 is our answer stated in general terms. Do we know enough to be specific? We know that $p(W|I) = (5/6)$ because urn I contains 6 balls, 5 of which are white. We also know that $p(W|II) = (1/6)$ because urn II contains 6 balls, 1 of which is white. But we know nothing else. The best we can say is:

$$p(I|W) = \frac{p(I)(5/6)}{p(I)(5/6) + p(II)(1/6)}. \tag{6}$$

Classical statisticians, who like to rely on objective probabilities exclusively, would stop right here and despair of finding an answer unless someone supplies values for $p(I)$ and $p(II)$—the unconditional probabilities of being given urn I or urn II.

Bayesian statisticians would not give up as easily. Under most circumstances, they would argue, people hold some degree of belief concerning all kinds of probabilities, and if they are totally ignorant on the matter (as, perhaps, in this case), they can always invoke Bayes's postulate and assume that $p(I) = p(II) = (1/2)$. In that case, the answer is

$$p(I|W) = \frac{(1/2)(5/6)}{(1/2)(5/6) + (1/2)(1/6)} = \frac{5}{6}. \tag{7}$$

Note that the answer has risen, as a result of picking a white ball, from a prior 1/2 to a posterior 5/6. Note also that a similar increase in probability would have occurred if the original subjective probabilities had been different, such as $p(I) = 1/3$ and $p(II) = 2/3$. Then the answer would have been

$$p(I|W) = \frac{(1/3)(5/6)}{(1/3)(5/6) + (2/3)(1/6)} = \frac{5}{7}. \tag{8}$$

The Unresolved Issue

Classical statisticians are apt to reject all techniques that make use of subjective probabilities as nonsensical. They would, therefore, relegate to the ash heap most of the decision theory presented in this chapter. (The legitimacy of subjective probability values, however, is not the only issue. The cardinal measurement of utilities is seen as an equally weak link in the Bayesian analysis, for example.)

The Bayesians are far from speechless, however. They argue that real-life people act precisely as modern decision theory suggests. They hold beliefs about various probabilities; they obtain new information (if it is not too costly); they revise their beliefs; they finally act. Indeed, say the Bayesians, if one rigidly stuck to the use of objective probabilities only, many problems (especially in business and economics) could not be solved at all. Moreover, even classical methods are riddled with *subjective* judgment; the claim of exclusive use of objective data is patently false. Consider, the Bayesians say, how the statistician's subjective judgment determines how a hypothesis is formulated, what kind of probability distribution is employed, what the maximum error probability is to be (almost without thought, the significance level is set at $\alpha = .05$ or $.01$), and what kinds of data are to be collected.

More likely than not, the controversy will be with us for some time to come.

||| ANALYTICAL EXAMPLES

17.1 Decision Analysis Comes of Age

As recently as the early 1970s, decision analysis was still an experimental management technique. The idea that a choice facing a decision maker can be expressed as a mathematical function of probability and utility numbers (that measured, respectively, the decision maker's uncertainties and value judgments) and that the best action was the one with the highest expected utility had just begun to move out of business schools and into practical application in the business world. By the early 1980s, however, decision making with the help of quantitative models that incorporate personal judgment had gained acceptance in many large corporations and government departments.

For example, in 1974, the AIL Division of Cutler-Hammer, Inc. was offered the opportunity to acquire the defense-market rights to a new flight-safety-system patent. The inventor claimed he had a strong patent position as well as technical superiority, but the market for the product depended on legislative action and was very uncertain. Management had to decide fast whether to purchase the rights to the new patent or to reject the offer; it used standard decision-tree techniques. Its initial analysis is illustrated in the accompanying diagram.

The immediate choice (at action point a) was the purchase of a six-month option on the patent rights or rejection of the offer. Six months hence, it was judged, there was (at event point b) a 71-percent chance of exercising the option and a 29-percent chance of not doing so. Three years down the road (at event point c), there was a 15-percent chance of getting, and an 85-percent chance of not getting, a first defense contract. Five years down the road (at event point d), finally, there was a 25-percent chance of getting a second defense contract and a 75-percent chance of being able to sublicense. Thus, there were five possible outcomes. Using the technique of backward induction, the expected monetary values given in color were computed, and the tree was pruned as shown. Purchasing the option was shown to be worth $100,425; rejecting the offer was worth $0.

Upon further analysis, however, management discovered a third possible action at a—rejecting the offer and seeking a sublicense later on. That approach

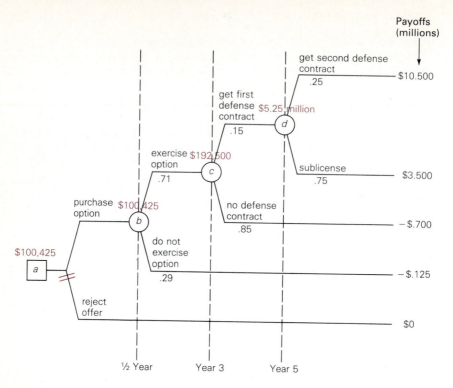

Payoffs
(millions)

get second defense
contract
.25 — $10.500

get first
defense $5.25 million
contract
.15

d

exercise $192,500
option
.71

c

sublicense
.75 — $3.500

purchase $100,425
option

b

no defense
contract
.85 — $.700

$100,425

a

do not
exercise
option
.29 — $.125

reject
offer — $0

½ Year   Year 3   Year 5

was shown to yield an expected value of $49,800, about half of the $100,425 expected monetary value of the option-purchase route. Nevertheless, being risk-averse, management chose the newly discovered third alternative: its analysis showed the *EMV* of $49,800 to be composed of a 94-percent chance of zero gain or loss and a 6-percent chance of gaining $830,000. On the other hand, as we can figure from the diagram, the *EMV* of $100,425 is provided by a riskier route composed of a 29-percent chance of losing $125,000, a 60-percent chance of losing $700,000, and an 11-percent chance of a positive return with an expectation of $5.25 million. Thus, management effectively maximized expected utility rather than monetary value.

Many other companies have also used the decision-tree technique. Ford has determined in this way whether to produce its own tires and whether to stop

producing convertibles; Honeywell used it to decide whether to pursue certain new weapons programs; Pillsbury used the technique to determine whether to switch from a box to a bag for a certain grocery product (and whether it was worthwhile to make a market test of the issue); Southern Railway employed it to choose whether to electrify part of its system; Gulf Oil used it to decide whether to explore certain sites; ITT used decision trees in deciding whether to make certain capital investments.

Source: Jacob W. Ulvila and Rex V. Brown, "Decision Analysis Comes of Age," *Harvard Business Review,* September-October 1982, pp. 130–41. For a fascinating discussion of the use of decision trees in legal disputes, see also Samuel E. Bodily, "When should you go to court?" *Harvard Business Review,* May-June 1981, pp. 103–13.

17.2 The Decision to Seed Hurricanes—A Second Look

Analytical Example 7.2 on page 272 analyzed the decision to seed hurricanes with the help of the normal curve. Decision-tree analysis, however, can be em-

ployed with equal success and was, in fact, used in that decision. The accompanying diagram is a copy of the government's analysis.

The diagram is almost self-explanatory. The government, at action point *a,* can decide to seed or not to seed. Subsequently, at event points *b* or *c,* changes in wind speed will occur, with varying probabilities. The ultimate "payoff" is here measured in millions of dollars of property damage. Naturally, the government wants to minimize this cost; the branch from *a* to *c* is pruned, and the decision is to go ahead and seed.

Can you figure out how the expected monetary values at *b* and *c* were calculated?

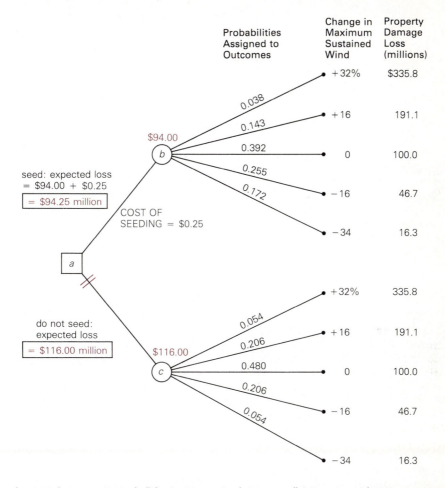

Probabilities Assigned to Outcomes	Change in Maximum Sustained Wind	Property Damage Loss (millions)
0.038	+32%	$335.8
0.143	+16	191.1
0.392	0	100.0
0.255	−16	46.7
0.172	−34	16.3

b — $94.00

seed: expected loss = $94.00 + $0.25 = $94.25 million

COST OF SEEDING = $0.25

a

do not seed: expected loss = $116.00 million

c — $116.00

0.054	+32%	335.8
0.206	+16	191.1
0.480	0	100.0
0.206	−16	46.7
0.054	−34	16.3

Source: R. A. Howard, J. E. Matheson, D. W. North, "The Decision to Seed Hurricanes," *Science,* June 16, 1972, pp. 1191–1202. (This article is well worth reading, especially for its expanded discussion of the value of gathering additional information.)

17.3 The Value of Perfect Information: The Case of the U.S. Cattle Industry

Two economists recently studied the willingness of risk-averse decision makers to pay for price information under uncertainty. They focussed on the U.S. cattle industry; the table on the next page shows some of their results. To an individual producer, the value of knowing the realized cattle price (column 1) beforehand ranged, in constant dollars, from 8 cents per hundredweight (cwt) to a $4.24 per cwt (column 2), with a mean value of 46 cents per cwt for the period as a whole.

Period	Realized Price ($ per cwt) (1)	Value of Perfect Information ($ per cwt) (2)
1978 Jan./Feb.	20.82	.59
Mar./Apr.	23.92	1.38
May/June	26.15	1.57
July/Aug.	25.20	1.63
Sept./Oct.	26.34	.80
Nov./Dec.	26.30	.42
1979 Jan./Feb.	30.33	1.13
Mar./Apr.	34.23	4.24
May/June	32.21	2.55
July/Aug.	29.03	.83
Sept./Oct.	29.28	.29
Nov./Dec.	28.05	.11
1980 Jan./Feb.	27.85	.08
Mar./Apr.	25.92	.23
May/June	24.73	.80
July/Aug.	25.65	.36
Sept./Oct.	24.55	.12

Multiplying the column 2 values by mean bimonthly production provides, in turn, a maximum estimate of the value of perfect price information to the entire industry: $296 million per period. This figure is a *maximum* for this reason: While the output response of an individual producer to the receipt of perfect information can be expected to have a negligible impact on market price, the output response by the entire industry (if everyone got the information) would change price markedly and depress the information value per unit of output.

Source: Terry Roe and Frances Antonovitz, "A Producer's Willingness to Pay for Information Under Price Uncertainty: Theory and Applications," *Southern Economic Journal,* October 1985, pp. 382–91. See also the more advanced discussion by the same authors in "A Theoretical and Empirical Approach to the Value of Information in Risky Markets," *The Review of Economics and Statistics,* February 1986, pp. 105–14.

BIOGRAPHIES

17.1 Thomas Bayes

Thomas Bayes (1702–61) was born in London, England, as the eldest son of the Rev. Joshua Bayes, one of the first nonconformist ministers to be publicly ordained in England. The younger Bayes was privately educated (as was the usual practice among nonconformists at the time) in languages, literature, and science. Quite possibly, he was tutored in mathematics by Abraham de Moivre (Biography 5.1). Like his father,

Thomas Bayes also became a Presbyterian minister and spent 30 years in that capacity. In that period, he published (nonmathematical) treatises under the pseudonym of John Noon, and they got him elected a Fellow of the Royal Society. Bayes's mathematical work is contained in two papers published posthumously due to the efforts of a friend, Richard Price. One of the papers, "An Essay Towards Solving a Problem in the

Doctrine of Chances" reversed de Moivre's focus of reasoning from the population to the sample and dealt with inferences from the sample to the population. The paper put forth a number of mathematical propositions; as "Proposition 9," it presented what is now known as *Bayes's theorem* (first discussed in Chapter 5 of this book). The essay is, perhaps, one of the least understood but most famous and controversial contributions ever made in the history of science. Many call it a masterpiece of mathematical elegance. In the words of Ronald Fisher (Biography 9.1), Bayes's "mathematical contributions . . . show him to have

been in the first rank of independent thinkers."[1] Certainly, Bayes has become, through his famous theorem, the father of modern decision theory. For this reason, the optimal strategy selected in expected-payoff-optimizing decision making problems is universally referred to as **Bayes's strategy.**

[1]Ronald Fisher, *Statistical Methods and Scientific Inference,* 2nd ed., rev. (New York: Hafner, 1959), p. 8.

Sources: Biometrika, December 1958, pp. 293–315 (includes a reproduction of the famous essay); *International Encyclopedia of Statistics,* vol. 1 (New York: Free Press, 1978), pp. 7–9.

17.2 Abraham Wald

Abraham Wald (1902–50) was born in Klausenburg, Hungary (which later became Cluj, Rumania). He was privately educated and self-taught (a consequence of complications arising from his family's Jewish orthodoxy). In the 1930s, he moved to Vienna, worked at the Austrian Institute for Business Cycle Research, and, at the time of the Nazi annexation, accepted an invitation by the Cowles Commission to work in the United States. That move surely saved him from death in a gas chamber— a fate that befell all but one member of his numerous family. Eventually, Wald taught statistics at Columbia University (his students remember him as a superb teacher of unrivalled clarity and precision), but his career was cut short by an airplane crash during a lecture tour of India.

Above all else, Wald's name will forever be linked with statistical decision theory and sequential analysis. In one of his earliest and finest papers, he introduced many fundamental concepts of decision theory, along with a mathematical structure for single-sample decision making that was sufficiently general to embrace estimation, hypothesis testing, and even the design of experiments. As he states in the Preface to his monumental *Statistical Decision Functions* (1950): "A major advance beyond previous results is the treatment of the design of experimentation as a part of the general decision problem." This tendency of Wald's (and of later decision theorists) to view all of statistics as a science of decision making under uncertainty (and the user of statistical data as someone in a decision-making

situation who seeks to make the one of several possible decisions that has the best expected payoff) invoked the ire of Ronald Fisher (Biography 9.1). Fisher thought that decision theory was appropriate, perhaps, for industry but certainly not for scientific work (in which inferences about the truth were to be made and not where payoffs were to be achieved through optimal decisions). Yet Wald's approach fit in nicely with the Neyman-Pearson theory of hypothesis testing (see Biographies 9.2 and 9.3).

Although he had forerunners, Wald was the first to formulate mathematically and solve generally the problem of sequential tests of statistical hypotheses. If evidence in sequentially unfolding data is sharply one-sided, Wald asked, why not stop an inquiry early? Why insist on observing and analyzing all the data contained in a single sample of predetermined fixed size? This line of work ultimately gave rise to a classic work, *Sequential Analysis* (1947).

Nor is this all. During his short life, Wald made major contributions in other fields as well. (He published more than 90 books and papers!) These include topology, measure and set theory, lattice theory, econometrics (the seasonal correction of time series, formulas for approximating economic index numbers), mathematical economics (indifference surfaces, proofs of the existence and uniqueness of solutions to the Walrasian system of equations), and much more.

Sources: Oskar Morgenstern, "Abraham Wald, 1902–1950," *Econometrica,* October 1951, pp. 361–67; *International Encyclopedia of Statistics,* vol. 2 (New York: Free Press, 1978), pp. 1235–38.

Summary

1. *Decision theory* consists of a variety of methods that can be employed in the systematic analysis and solution of decision-making problems that arise because uncertainty exists about future events over which the decision maker has no control, but which are bound to influence the ultimate outcome of a decision.

2. Certain identical elements are present in every decision problem involving uncertainty. They include the mutually exclusive decision alternatives, or *actions,* available to the decision maker, the mutually exclusive future occurrences, or *events,* that will affect the outcome of any present action taken but that are not under the control of a decision maker, and the positive or negative net benefits, or *payoffs,* that are associated with each possible action/event combination and that are, thus, the joint outcome of choice and chance.

   A *payoff table* summarizes a decision-making situation in the context of uncertainty; it consists of a tabular listing of the payoffs associated with all possible combinations of actions and events. A *decision tree* provides an alternative summary; it shows graphically, and in chronological order from left to right, every potential action, event, and payoff. In such a tree diagram, any point of choice (at which the decision maker is in control) is denoted by a square symbol and is called an *action point* (from which *action branches* representing the possible actions emanate). In contrast, any point of chance (at which the decision maker exercises no control but at which "nature" is in charge) is denoted by a circle and is called an *event point* (from which *event branches* representing the possible events emanate).

3. When nothing is known about the likelihood of those alternative future events that are certain to affect the eventual outcome of a present decision, and if the decision maker does not even care to guess what the event probabilities might be, one of three decision criteria is commonly employed: maximin (or minimax), maximax (or minimin), or minimax regret.

   a. According to the *maximin criterion,* a decision maker who seeks to maximize some benefit finds the minimum benefit associated with each possible action, identifies the maximum among these minima, and chooses the action associated with this maximum of minima. (The same criterion is called the *minimax criterion* when a decision maker seeks to minimize some cost; such a decision maker finds the maximum cost associated with each possible action, identifies the minimum among these maxima, and chooses the action associated with this minimum of maxima.)

   b. According to the *maximax criterion,* a decision maker who seeks to maximize some benefit finds the maximum benefit associated with each possible action, identifies the maximum among these maxima, and chooses the action associated with this maximum of maxima. (The same criterion is called the *minimin criterion* when a decision maker seeks to minimize some cost; such a decision maker finds the minimum cost associated with each possible action, identifies the minimum among these minima, and chooses the action associated with this minimum of minima.)

   c. According to the *minimax-regret criterion,* a decision maker finds the maximum regret value associated with each possible action, identifies the minimum among these maxima, and chooses the action associated with this minimum of maximum regret values.

   The above criteria can be criticized for a number of reasons, including undue sensitivity of the results to irrelevant factors, undue reliance on extreme values, and, of course, a total lack of consideration for the probabilities of events.

4. A form of decision making under uncertainty that employs probabilities, but only those available prior to the gathering of new experimental or sample evidence about the likelihood of alternative future events, is called *prior analysis.* Once again, a number of alternative criteria can be employed.

   a. According to the *maximum-likelihood criterion,* a decision maker simply ignores all the events that might occur except the most likely one and selects the action that produces the optimal result (maximum benefit or minimum cost) associated with this most likely event.

   b. A decision maker using the *expected-monetary-value criterion* determines an expected monetary value for each possible action and selects the action with the optimal expected monetary value (the optimal *EMV* is the largest one if the objective is to maximize some benefit; the optimal *EMV* is the smallest one if the objective is to minimize some cost). This criterion is often employed with the help of a decision tree; expected monetary values are then computed (from ultimate-payoff values and event-branch probabilities) for each fork of the tree by a process termed *backward induction.*

c. A decision maker using the *expected-opportunity-loss criterion* determines an expected opportunity loss for each possible action and selects the action with the smallest of these values. This criterion necessarily always yields the same result as the expected-monetary-value criterion.

Observation shows that people making decisions under uncertainty often fail to follow the *EMV* criterion, a fact that can be explained by studying people's attitudes toward risk. For a risk-averse person, the marginal utility of money declines with larger amounts of money; for a risk-neutral person, it is constant; for a risk-seeking person, it rises. Except for the risk-neutral person, the optimization of expected monetary value does not produce maximum welfare. For the risk averter and the risk seeker, therefore, a different decision criterion is superior—namely that of *maximizing expected utility.*

5. A form of decision making under uncertainty that starts out with prior probabilities, proceeds to gather additional or sample evidence about event probabilities, and then uses this new evidence to transform the set of prior probabilities, by means of Bayes's theorem, into a revised set of posterior probabilities (with the help of which a final decision is reached) is called *posterior analysis.*

6. The maximum amount a decision maker can be expected to pay for obtaining perfect information about future events—and, thus, for eliminating uncertainty completely—is called the *expected value of perfect information,* EVPI. It always equals the expected opportunity loss of the optimal decision under uncertainty. The maximum amount a decision maker can be expected to pay for obtaining admittedly imperfect information about future events—and, thus, for reducing (rather than eliminating) uncertainty—is called the *expected value of sample information,* EVSI. The ratio of the *EVSI* to the (prior) *EVPI,* multiplied by 100, measures the efficiency of sample information, or *ESI.*

7. A form of decision making under uncertainty that decides the worth of obtaining additional (experimental or sample) information before proceeding to prior and posterior analysis is called *preposterior analysis.* This analysis can be performed by using either one of two alternative approaches that always yield the same result: (a) the decision-tree approach *(extensive-form analysis)* and (b) the tabular approach *(normal-form analysis).*

8. A logical extension of preposterior analysis is *sequential sampling,* in which the sample size is not fixed in advance but in which a decision is made, after each bit of information is received, either to continue sampling or to put a stop to it.

9. With the development of decision theory (and its widespread use of subjective probabilities), a great controversy among statisticians has emerged. Classicists wish to rely exclusively on objective probabilities and view statistics as a means to the making of inferences; Bayesians freely incorporate personal judgment by means of subjective probabilities and are apt to view statistics as a collection of techniques for the making of decisions in the face of uncertainty.

Key Terms

action branches
action point
actions
backward induction
Bayes's postulate
Bayes's strategy
criterion of insufficient reason
decision branches
decision fork
decision making under uncertainty
decision node
decision point
decision theory

decision tree
dominant action
efficiency of sample information *(ESI)*
equal-likelihood criterion
event branches
event point
events
expected monetary value *(EMV)*
expected-monetary-value criterion
expected opportunity loss *(EOL)*
expected-opportunity-loss criterion
expected regret value
expected utility *(EU)*

expected-utility criterion
expected value of perfect information *(EVPI)*
expected value of sample information *(EVSI)*
extensive-form analysis
fair gamble
inadmissible action
Laplace criterion
marginal utility of money
maximax
maximin
maximum-likelihood criterion
minimax
minimax regret
minimin
more-than-fair gamble
normal-form analysis
opportunity loss *(OL)*
optimal strategy
payoffs
payoff table
posterior analysis

posterior expected value of perfect information
   (posterior *EVPI*)
preposterior analysis
prior analysis
prior expected value of perfect information
   (prior *EVPI*)
regret
regret table
risk aversion
risk neutrality
risk seeking
sequential sampling
state-of-nature branches
state-of-nature fork
state-of-nature node
state-of-nature point
states of nature
total utility of money
unfair gamble
utility-of-money function

Questions and Problems

The computer program noted at the end of the chapter can be used to work some of the subsequent problems.

1. Depending on whether future demand is low, moderate, or high, a real-estate developer foresees profits (in millions of dollars) of 4, 5, or 6 from a small project; of 1, 6, or 10 from a medium-sized project; and of −5, 0, or 30 from a large project. Determine the best action under the criterion of
a. maximin.
b. maximax.

2. Depending on whether future demand is low, moderate, or high, a producer of aircraft ELTs (emergency locator transmitters) foresees profits (in thousands of dollars) of 15, 20, or 25 from a small production run; of 6, 20, or 35 from a medium-sized run; and of −10, 0, or 69 from a large run. Determine the best action under the criterion of
a. maximin.
b. maximax.

3. Table 17.A shows profits (in million dollars) that are associated with 5 actions and 6 events. Determine the best action under the criterion of
a. maximin.
b. maximax.

4. Table 17.B shows profits (in million dollars) that are associated with 4 actions and 5 events. Determine the best action under the criterion of
a. maximin.
b. maximax.

5. Table 17.C shows benefits that are associated with 5 actions and 4 events. Determine the best action under the criterion of
a. maximin.
b. maximax.

6. The Zoning Board of Appeals has granted permission to a builder to erect between 1 and 5 homes on a certain lot, but whatever the chosen number is, the homes must all be built now. The builder's cost is $100,000 per home; up to the number demanded, homes built can be sold at 25 percent above cost; any surplus homes can be sold to a real-estate firm at 60 percent of cost. How many homes should the profit-seeking builder erect, given that 1 home is the minimum and the criterion is
a. maximin.
b. maximax.

Table 17.A

Events Actions	E_1	E_2	E_3	E_4	E_5	E_6
A_1	0	-15	-35	-50	23	67
A_2	-15	-35	-50	23	67	102
A_3	-35	-50	23	67	102	139
A_4	-49	23	67	102	139	150
A_5	23	67	102	139	145	-100

Table 17.B

Events Actions	E_1	E_2	E_3	E_4	E_5
A_1	93	75	60	-10	0
A_2	0	17	65	107	2
A_3	5	6	7	79	-5
A_4	-53	89	0	-12	112

Table 17.C

Events Actions	E_1	E_2	E_3	E_4
A_1	22	39	56	-17
A_2	33	34	36	36
A_3	12	-50	-1	79
A_4	-60	12	19	70
A_5	0	0	12	62

7. Rework problem 6, but let the builder's cost be $60,000 per home, while regular sales are at 50 percent above cost and surplus sales at 80 percent of cost.

8. Rework problem 6, but let the builder's cost be $120,000 per home, while regular sales are at 15 percent above cost and surplus sales at 50 percent of cost.

9. An advertising manager believes that a given volume of sales can be achieved at different costs, depending on whether the ads are disseminated by means of direct mail, newspapers, or television.

However, it is also believed that the state of the economy (recession, inflation, stagflation) will influence the cost figures such that the cost (in thousands of dollars) of a direct-mail campaign will be 75, 20, or 40; of a newspaper campaign will be 50, 60, or 30; and of a TV campaign will be 55, 40, or 45. Determine the best action under the criterion of
a. minimax.
b. minimin.

10. A given volume of output can be produced by one of three different methods, but the associated cost will vary with the state of the economy (recession, inflation, stagflation) because prices of different productive inputs will be affected differently. The projected costs of method 1 are (in thousands of dollars) 15, 19, 29; the corresponding figures for method 2 are 29, 19, 14 and for method 3 are 25, 17, 25. Determine the best action under the criterion of
a. minimax.
b. minimin.

11. Consider Table 17.A and assume the entries are costs. Determine the best action under the criterion of
a. minimax.
b. minimin.

12. Consider Table 17.B and assume the entries are costs. Determine the best action under the criterion of
a. minimax.
b. minimin.

13. Consider Table 17.C and assume the entries are costs. Determine the best action under the criterion of
a. minimax.
b. minimin.

14. Reconsider problem 1, but this time determine the best action under the criterion of minimax regret.

15. Reconsider problem 2, but this time determine the best action under the criterion of minimax regret.

16. Reconsider problem 3, but this time determine the best action under the criterion of minimax regret.

17. Reconsider problem 4, but this time determine the best action under the criterion of minimax regret.

18. Reconsider problem 5, but this time determine the best action under the criterion of minimax regret.

19. Reconsider problem 6, but this time determine the best action under the criterion of minimax regret.

20. Reconsider problem 7, but this time determine the best action under the criterion of minimax regret.

21. Reconsider problem 9, but this time determine the best action under the criterion of minimax regret.

22. Reconsider problem 10, but this time determine the best action under the criterion of minimax regret.

23. Assume the entries in Table 17.A are costs. Determine the best action under the criterion of minimax regret.

24. Assume the entries in Table 17.B are costs. Determine the best action under the criterion of minimax regret.

25. Assume the entries in Table 17.C are costs. Determine the best action under the criterion of minimax regret.

26. Reconsider problem 3, but determine the best action under the maximum-likelihood criterion, assuming that $p(E_1) = .1$, $p(E_2) = .2$, $p(E_3) = .3$, $p(E_4) = .1$, $p(E_5) = .05$, and $p(E_6) = .25$.

27. Reconsider problem 4, but determine the best action under the maximum-likelihood criterion, assuming that $p(E_1) = .4$, $p(E_2) = .2$, $p(E_3) = .05$, $p(E_4) = .05$, and $p(E_5) = .3$.

28. Assume the entries in Table 17.C are costs. Determine the best action under the maximum-likelihood criterion, assuming that $p(E_1) = .4$, $p(E_2) = .5$, $p(E_3) = .02$, and $p(E_4) = .08$.

29. A business executive has determined that keeping the business in its present location is a disaster and that it must be moved to one of two new locations, East or West. Moving East will produce future net earnings with a present value of either $50 million or $10 million; moving the business West will produce future net earnings with a present value of either $100 million or zero—in each case, depending on whether the attempt to break into the new market will be a success or a failure.[3] Assuming that the chances for success or failure at the new locations are judged to be equal, determine the best action under the criterion of expected monetary value.

30. Reconsider problem 2. Determine the best action under the expected-monetary-value criterion, assuming $p(E_1) = .1$, $p(E_2) = .7$, and $p(E_3) = .2$.

31. Reconsider problem 3. Determine the best action under the expected-monetary-value criterion, assuming $p(E_1) = .1$, $p(E_2) = .2$, $p(E_3) = .3$, $p(E_4) = .1$, $p(E_5) = .05$, and $p(E_6) = .25$.

---

[3]In order to make comparable dollars of revenue and cost that pertain to different periods of time, it is customary to calculate payoffs in terms of "present value." For a detailed discussion of this concept, see Heinz Kohler, *Intermediate Microeconomics: Theory and Applications,* 2nd ed. (Glenview, Ill.: Scott, Foresman and Company, 1986), chapter 15.

32. Reconsider problem 4. Determine the best action under the expected-monetary-value criterion, assuming $p(E_1) = .4$, $p(E_2) = .2$, $p(E_3) = .05$, $p(E_4) = .05$, and $p(E_5) = .3$.

33. Reconsider problem 5. Determine the best action under the expected-monetary-value criterion, assuming $p(E_1) = .4$, $p(E_2) = .5$, $p(E_3) = .02$, and $p(E_4) = .08$.

34. Analytical Example 17.2 ends with a question. What is the answer?

35. An independent producer of TV films wants to market a new soap opera. The rights can be sold to a distributor for $125 million now, or the program can be offered to a TV network for review with these possible results (in the judgment of the film producer): a 60-percent chance of rejection (which ruins all further chances of a sale to anyone and spells a $30 million loss) or a 40-percent chance of getting a contract (which means a $300 million profit). Use a decision tree to determine the optimal action, given the desire to maximize expected monetary value.

36. An investor is thinking of buying a bankrupt private airport and its associated resort for $5 million (which would involve an annual cost of $500,000). An additional cost of $200,000 per year is expected to run the new venture. At a contemplated landing fee of $100, the probabilities are .1, .5, and .4, respectively, that 6,000 planes, 7,000 planes, or 8,000 planes will land per year. Use a decision tree to determine the optimal action, given the desire to maximize expected monetary value and given the option not to make the investment at all.

37. An investor is thinking of publishing a new magazine, *The Aviation Consumer*. The monthly fixed cost will be $400,000; the variable cost will be 25 cents per magazine. At a contemplated price of $1.50 per magazine, the probabilities are .1, .2, .3, and .4, respectively, that 500,000 or 400,000 or 300,000 or 250,000 monthly magazines will be sold. Use a decision tree to determine the optimal

action, given the desire to maximize expected monetary value and given the option not to make the investment at all.

38. Reconsider problem 29. Determine the best action under the criterion of expected opportunity loss.

39. Rework problem 30, but apply the expected-opportunity-loss criterion.

40. Rework problem 31, but apply the expected-opportunity-loss criterion.

41. Rework problem 32, but apply the expected-opportunity-loss criterion.

42. Rework problem 33, but apply the expected-opportunity-loss criterion.

43. Reconsider problem 29. Determine the best action under the criterion of expected utility, assuming the executive has a utility function of $U = \sqrt{\$}$.

44. Rework problem 30, but apply the expected-utility criterion, assuming $U = \sqrt{\$}$.

45. Rework problem 31, but apply the expected-utility criterion, assuming $U = \sqrt{\$}$.

46. Rework problem 32, but apply the expected-utility criterion, assuming $U = \sqrt{\$}$.

47. Reconsider problem 35. Suppose that a consulting firm is willing to offer advice on the network's likely reaction for a $1 million fee. The consulting firm's "track record" is given in Table 17.D. Use a decision tree to determine the optimal strategy if the film producer definitely decides to buy the advice and to then take the action (sell the rights or offer the film to the network) that maximizes expected monetary value.

48. Study problems 35 and 47. Determine the (prior) expected value of perfect information.

49. Study problems 35 and 47. Determine the expected value of the sample information.

50. Study problems 35 and 47. Determine the efficiency of the sample information.

51. Reread the chapter discussion refering to Figure 17.6, then redo the decision-tree analysis under the assumption that the survey costs $1 million rather than $100,000.

Table 17.D

Advice Given	Subsequent Events	
	E_1 = Rejection	E_2 = Contract Offer
R_1 = network will reject	$p(R_1\|E_1) = .8$	$p(R_1\|E_2) = .3$
R_2 = network will offer contract	$p(R_2\|E_1) = .2$	$p(R_2\|E_2) = .7$

Table 17.E

	Subsequent Events	
Report Issued	E_1 = oil is found	E_2 = oil is not found
R_1 = oil present R_2 = no oil present	$p(R_1|E_1) = .6$ $p(R_2|E_1) = .4$	$p(R_1|E_2) = .2$ $p(R_2|E_2) = .8$

52. Consider the answer to problem 51. If no survey is made, what is the (prior) expected value of perfect information, assuming $p(R_1) = p(E_1)$ and also that $p(R_2) = p(E_2)$?

53. Consider the answer to problem 51. What is the expected value of the sample information?

54. Conduct an extensive-form (preposterior) analysis for an oil company, given the following information. It can drill for oil on a given site now (and the prior probabilities of E_1 = finding oil or E_2 = not finding oil are considered equal) or sell its rights to the site for $10 million. If it drills and finds oil, a net payoff of $100 million is expected; if no oil is found, that payoff equals − $15 million. It can instead contract for a seismic test. If the test predicts oil, it can drill (and get a net payoff of $90 million if oil is in fact found and of − $25 million if none is found) or sell the rights (for $20 mil-

lion). If the test predicts no oil, it can drill (and get a net payoff of $90 million if oil is in fact found and of − $25 million if none is found) or sell the rights for $1 million. The "track record" of the seismic-test company is given in Table 17.E.

55. Consider the answer to problem 54. If no test is made, what is the (prior) expected value of perfect information, assuming $p(R_1) = p(E_1)$ and also that $p(R_2) = p(E_2)$?

56. Consider the answer to problem 54. What is the expected value of the sample information?

57. Table 17.19 gives expected payoffs for six different strategies, but the computation is shown only for strategies S_1 and S_6 (in Tables 17.17 and 17.18). Make a similar computation for S_2.

58. Rework problem 57, but for S_3.

59. Rework problem 57, but for S_4.

60. Rework problem 57, but for S_5.

Selected Readings

Aitchison, John. *Choice Against Chance: An Introduction to Statistical Decision Theory.* Reading, Mass.: Addison-Wesley, 1970.

Bayes, Thomas. "An Essay Towards Solving a Problem in the Doctrine of Chances." *Philosophical Transactions of the Royal Society* 53(1763):370–418. The posthumous publication containing the Bayes's theorem. Now most easily available in: Thomas Bayes. *Facsimiles of Two Papers by Bayes.* New York: Hafner, 1963.

Bell, D. E., R. L. Keeney, H. Raiffa. *Conflicting Objectives in Decisions.* New York: Wiley, 1977.

Berkson, Joseph. "My Encounter With Neo-Bayesianism." *International Statistical Review* 45(1977):1–8. Presents arguments against the Bayesian approach.

Box, George E. P. and George C. Tiao. *Bayesian Inference in Statistical Analysis.* Reading, Mass.: Addison-Wesley, 1973.

Brown, Rex V., Andrew S. Kahr, C. Peterson. *Decision Analysis for the Manager.* New York: Holt, Rinehart and Winston, 1974.

Efron, B. "Why Isn't Everyone a Bayesian?" *The American Statistician,* February 1986, pp. 1–11. A discussion, with numerous comments by critics, of the Fisher versus Neyman-Pearson-Wald controversy.

Fisher, Ronald A. *Statistical Methods and Scientific Inference,* 2nd ed., rev. New York: Hafner, 1959. In this, his last book, Fisher strongly rejects the Bayesian approach to statistics as well as the view of statistics as a decision-making science traceable to Wald.

Harvard Business Review. *Statistical Decision Series,* Parts I-IV. Boston: 1951–70.

Kempthorne, Oscar, and Leroy Folks. *Probability, Statistics, and Data Analysis.* Ames: Iowa State University Press, 1971.

Kohler, Heinz. *Intermediate Microeconomics: Theory and Applications,* 2nd. ed. (Glenview, Ill.: Scott, Foresman and Company, 1986). Includes presentations of basic utility theory (Chapter 3), the economics of uncertainty (Chapter 16), and (in the accompanying Student Workbook) game theory.

Lindley, Dennis V. *Bayesian Statistics: A Review.* Philadelphia: Society for Industrial and Applied Mathematics, 1972.

Luce, Robert D., and Howard Raiffa. *Games and Decisions.* New York: Wiley, 1958.

Menges, G. "Inference and Decision." *Selecta Statistica Canadiana,* vol. I. New York: Wiley, 1973. On the Fisher-Wald controversy concerning the nature of statistical science.

Morris, W. T. *Management Science: A Bayesian Introduction.* Englewood Cliffs, N.J.: Prentice-Hall, 1968.

Neumann, John von, and Oskar Morgenstern. *The Theory of Games and Economic Behavior,* rev. ed. Princeton: Princeton University Press, 1953. The classic work on *game theory* that studies decision making under uncertainty that is complicated by a conscious conflict of wills so that the payoff to an action depends not only on the decision maker's choice and "nature," but also on the conscious choices made by other people.

Raiffa, Howard. *Decision Analysis: Introductory Lectures on Choices Under Uncertainty.* Reading, Mass.: Addison-Wesley, 1968.

Savage, Leonard J. "The Theory of Statistical Decision." *Journal of the American Statistical Association* (March 1951), pp. 55–67. A classic article, including a stimulating exposition of Wald's work.

Schlaifer, Robert. *Probability and Statistics for Business Decisions: An Introduction to Managerial Economics Under Uncertainty.* New York: McGraw-Hill, 1959. On Bayesian inference, with emphasis on decision theory.

Schlaifer, Robert. *Analysis of Decisions Under Uncertainty.* New York: McGraw-Hill, 1969.

Stigler, Stephen M. "Who Discovered Bayes's Theorem?" *The American Statistician* (November 1983), pp. 290–96. A fascinating article that suggests that Bayes, perhaps, was not the first to discover the theorem bearing his name.

Tsokos, Chris P., and I. N. Shimi, eds. *The Theory and Applications of Reliability With Emphasis on Bayesian and Nonparametric Methods.* New York: Academic Press, 1977.

Wald, Abraham. *Sequential Analysis.* New York: Wiley, 1947, and *Statistical Decision Functions.* New York: Wiley, 1950. This book discusses some crucial landmarks in the development of statistics as a decision making science.

Winkler, Robert L. *Introduction to Bayesian Inference and Decision.* New York: Holt, Rinehart and Winston, 1972.

Zellner, Arnold. *An Introduction to Bayesian Inference in Econometrics.* New York: Wiley, 1971.

Computer Programs

The DAYSAL-2 personal-computer diskettes contain one program of interest to this chapter:

16. Expected Monetary Values helps you calculate expected monetary values for decision-tree analysis.

APPENDIX TABLES

Appendix Table A
Random Numbers

71274	84346	75444	85690	35384	87841	97411	78698	46796	33552
64017	01373	14665	31891	80997	14321	47741	59980	87739	38174
43747	17686	11045	15549	52779	65135	00275	95434	36337	24041
59688	48689	41591	47042	83615	93034	25077	64835	67798	30547
95016	73467	11447	59500	94921	15166	69217	26267	11316	22651
65207	30591	65947	58339	00952	32111	45459	14986	57395	34492
34510	78657	08883	49489	85619	52912	01662	49854	78354	30631
56299	60624	91572	31734	18159	18927	31314	59682	41320	88602
02113	12579	86172	03819	69968	02616	72687	42699	04792	16510
00884	87979	45184	61572	20086	14498	29640	94263	90964	29278
89367	53577	97412	19603	57234	63055	49059	35761	72007	22751
99781	56740	42659	46617	21828	99831	45987	63450	66919	78252
92024	12100	76013	12587	86340	74880	79979	35906	38122	64917
82861	09215	87342	72789	76132	24468	93065	78968	03321	48081
11286	13011	67982	74101	44961	25468	14247	95934	50711	24492
68674	24686	14460	61242	92310	86810	87702	69811	53996	99517
24882	20749	94139	28785	74402	18561	79069	56838	30020	99707
21740	51134	39298	92203	66230	30636	58169	78982	50057	39908
46901	34825	28673	10404	97777	30782	04680	15319	84125	40937
92686	81702	74149	76326	01101	96278	90855	55145	89705	51199
98743	59366	94797	96803	69876	87533	19675	59246	65348	06606
11463	25619	38107	91053	58416	02720	86563	27443	99598	04074
76975	18636	54975	67422	57101	68857	35389	35641	34505	71552
67359	50379	81053	97357	00717	59504	34480	77127	23243	48682
34084	48031	27227	48912	10797	88917	93126	52945	79457	66528
94553	44441	83166	40056	73935	52103	13972	56781	31900	95037
20545	43211	34500	92233	53497	39401	78535	82360	57410	32060
05646	01152	13235	39168	69214	83852	00144	08105	94247	37189
54022	58295	96122	87620	74774	06884	81689	68392	25776	08748
32291	30700	32561	83579	94582	77930	06826	96855	97751	42664
66212	75061	65891	96896	15107	12985	97616	64241	08592	81036
98059	44951	23078	92793	80756	52799	20340	62969	81775	31065
23157	52179	24394	39833	89427	58771	26992	00649	25475	50888
28911	81929	91368	49372	43335	44465	43257	66893	34761	60423
85959	90369	02100	28727	83001	84166	20473	35305	38088	54795
34459	31400	58760	17157	73816	55527	54133	24605	56153	35354
04073	57781	36894	93000	57834	29343	98195	58425	97275	71392
22126	91330	95667	75737	36869	55209	41663	04943	59401	17039
10288	61685	25302	84097	13088	86840	04020	43046	01043	43157
75431	32853	72907	99432	65482	23011	70466	87386	67471	77629
90800	17425	28042	53770	98924	31863	84115	82488	23239	82185
19083	89475	05207	41284	83405	55825	31117	59821	96455	63796
10405	67911	77238	46262	42766	07215	02391	47316	78724	41170
34711	77325	99768	63455	44335	91028	27740	86163	81474	08159
73334	61941	16883	05012	63191	35763	60157	09617	25501	44989
79452	68381	71937	23274	60273	47091	82876	24641	03825	50894
13864	28746	32434	88325	99996	96130	39471	74020	56077	22133
73082	50271	83240	80065	09328	02940	41686	32758	89467	73553
43060	88221	35010	79829	71520	80453	95049	66352	77495	83256
15172	42061	33264	63832	48528	23258	13520	83222	45659	39074

Source: The Rand Corporation, *A Million Random Digits with 100,000 Normal Deviates* (Glencoe, Illinois: The Free Press, 1955), p. 377.

Appendix Table B
The Greek Alphabet

A	α	alpha	H	η	eta	N	ν	nu	T	τ	tau
B	β	beta	Θ	θ	theta	Ξ	ξ	xi	Y	υ	upsilon
Γ	γ	gamma	I	ι	iota	O	o	omicron	Φ	φ	phi
Δ	δ	delta	K	κ	kappa	Π	π	pi	X	χ	chi
E	ε	epsilon	Λ	λ	lambda	P	ρ	rho	Ψ	ψ	psi
Z	ζ	zeta	M	μ	mu	Σ	σ	sigma	Ω	ω	omega

Appendix Table C
Binomial Probabilities for Individual Values of x

Entries in the table give the probability of x successes in n trials of a binomial experiment, where π is the probability of a success in one trial. For example, with two trials and $\pi = .40$, the probability of two successes is .1600.

							π					
n	x	.05	.10	.15	.20	.25	.30	.35	.40	.45	.50
1	0	.9500	.9000	.8500	.8000	.7500	.7000	.6500	.6000	.5500	.5000
	1	.0500	.1000	.1500	.2000	.2500	.3000	.3500	.4000	.4500	.5000
2	0	.9025	.8100	.7225	.6400	.5625	.4900	.4225	.3600	.3025	.2500
	1	.0950	.1800	.2550	.3200	.3750	.4200	.4550	.4800	.4950	.5000
	2	.0025	.0100	.0225	.0400	.0625	.0900	.1225	.1600	.2025	.2500
3	0	.8574	.7290	.6141	.5120	.4219	.3430	.2746	.2160	.1664	.1250
	1	.1354	.2430	.3251	.3840	.4219	.4410	.4436	.4320	.4084	.3750
	2	.0071	.0270	.0574	.0960	.1406	.1890	.2389	.2880	.3341	.3750
	3	.0001	.0010	.0034	.0080	.0156	.0270	.0429	.0640	.0911	.1250
4	0	.8145	.6561	.5220	.4096	.3164	.2401	.1785	.1296	.0915	.0625
	1	.1715	.2916	.3685	.4096	.4219	.4116	.3845	.3456	.2995	.2500
	2	.0135	.0486	.0975	.1536	.2109	.2646	.3105	.3456	.3675	.3750
	3	.0005	.0036	.0115	.0256	.0469	.0756	.1115	.1536	.2005	.2500
	4	.0000	.0001	.0005	.0016	.0039	.0081	.0150	.0256	.0410	.0625
5	0	.7738	.5905	.4437	.3277	.2373	.1681	.1160	.0778	.0503	.0312
	1	.2036	.3280	.3915	.4096	.3955	.3602	.3124	.2592	.2059	.1562
	2	.0214	.0729	.1382	.2048	.2637	.3087	.3364	.3456	.3369	.3125
	3	.0011	.0081	.0244	.0512	.0879	.1323	.1811	.2304	.2757	.3125
	4	.0000	.0004	.0022	.0064	.0146	.0284	.0488	.0768	.1128	.1562
	5	.0000	.0000	.0001	.0003	.0010	.0024	.0053	.0102	.0185	.0312
6	0	.7351	.5314	.3771	.2621	.1780	.1176	.0754	.0467	.0277	.0156
	1	.2321	.3543	.3993	.3932	.3560	.3025	.2437	.1866	.1359	.0938
	2	.0305	.0984	.1762	.2458	.2966	.3241	.3280	.3110	.2780	.2344
	3	.0021	.0146	.0415	.0819	.1318	.1852	.2355	.2765	.3032	.3125
	4	.0001	.0012	.0055	.0154	.0330	.0595	.0951	.1382	.1861	.2344
	5	.0000	.0001	.0004	.0015	.0044	.0102	.0205	.0369	.0609	.0938
	6	.0000	.0000	.0000	.0001	.0002	.0007	.0018	.0041	.0083	.0156
7	0	.6983	.4783	.3206	.2097	.1335	.0824	.0490	.0280	.0152	.0078
	1	.2573	.3720	.3960	.3670	.3115	.2471	.1848	.1306	.0872	.0547
	2	.0406	.1240	.2097	.2753	.3115	.3177	.2985	.2613	.2140	.1641
	3	.0036	.0230	.0617	.1147	.1730	.2269	.2679	.2903	.2918	.2734
	4	.0002	.0026	.0109	.0287	.0577	.0972	.1442	.1935	.2388	.2734
	5	.0000	.0002	.0012	.0043	.0115	.0250	.0466	.0774	.1172	.1641
	6	.0000	.0000	.0001	.0004	.0013	.0036	.0084	.0172	.0320	.0547
	7	.0000	.0000	.0000	.0000	.0001	.0002	.0006	.0016	.0037	.0078

Source: The National Bureau of Standards, *Tables of the Binomial Probability Distribution*, Applied Mathematics Series, no. 6 (Washington, D.C.: U.S. Government Printing Office, 1949). The original contains probabilities for values of n up to 49, and for values of π in increments of .01. All entries were calculated by the formula in Box 6.B.

Appendix Table C (continued)

n	x	.05	.10	.15	.20	.25	π .30	.35	.40	.45	.50
8	0	.6634	.4305	.2725	.1678	.1001	.0576	.0319	.0168	.0084	.0039
	1	.2793	.3826	.3847	.3355	.2670	.1977	.1373	.0896	.0548	.0312
	2	.0515	.1488	.2376	.2936	.3115	.2965	.2587	.2090	.1569	.1094
	3	.0054	.0331	.0839	.1468	.2076	.2541	.2786	.2787	.2568	.2188
	4	.0004	.0046	.0185	.0459	.0865	.1361	.1875	.2322	.2627	.2734
	5	.0000	.0004	.0026	.0092	.0231	.0467	.0808	.1239	.1719	.2188
	6	.0000	.0000	.0002	.0011	.0038	.0100	.0217	.0413	.0703	.1094
	7	.0000	.0000	.0000	.0001	.0004	.0012	.0033	.0079	.0164	.0312
	8	.0000	.0000	.0000	.0000	.0000	.0001	.0002	.0007	.0017	.0039
9	0	.6302	.3874	.2316	.1342	.0751	.0404	.0207	.0101	.0046	.0020
	1	.2985	.3874	.3679	.3020	.2253	.1556	.1004	.0605	.0339	.0176
	2	.0629	.1722	.2597	.3020	.3003	.2668	.2162	.1612	.1110	.0703
	3	.0077	.0446	.1069	.1762	.2336	.2668	.2716	.2508	.2119	.1641
	4	.0006	.0074	.0283	.0661	.1168	.1715	.2194	.2508	.2600	.2461
	5	.0000	.0008	.0050	.0165	.0389	.0735	.1181	.1672	.2128	.2461
	6	.0000	.0001	.0006	.0028	.0087	.0210	.0424	.0743	.1160	.1641
	7	.0000	.0000	.0000	.0003	.0012	.0039	.0098	.0212	.0407	.0703
	8	.0000	.0000	.0000	.0000	.0001	.0004	.0013	.0035	.0083	.0176
	9	.0000	.0000	.0000	.0000	.0000	.0000	.0001	.0003	.0008	.0020
10	0	.5987	.3487	.1969	.1074	.0563	.0282	.0135	.0060	.0025	.0010
	1	.3151	.3874	.3474	.2684	.1877	.1211	.0725	.0403	.0207	.0098
	2	.0746	.1937	.2759	.3020	.2816	.2335	.1757	.1209	.0763	.0439
	3	.0105	.0574	.1298	.2013	.2503	.2668	.2522	.2150	.1665	.1172
	4	.0010	.0112	.0401	.0881	.1460	.2001	.2377	.2508	.2384	.2051
	5	.0001	.0015	.0085	.0264	.0584	.1029	.1536	.2007	.2340	.2461
	6	.0000	.0001	.0012	.0055	.0162	.0368	.0689	.1115	.1596	.2051
	7	.0000	.0000	.0001	.0008	.0031	.0090	.0212	.0425	.0746	.1172
	8	.0000	.0000	.0000	.0001	.0004	.0014	.0043	.0106	.0229	.0439
	9	.0000	.0000	.0000	.0000	.0000	.0001	.0005	.0016	.0042	.0098
	10	.0000	.0000	.0000	.0000	.0000	.0000	.0000	.0001	.0003	.0010
11	0	.5688	.3138	.1673	.0859	.0422	.0198	.0088	.0036	.0014	.0005
	1	.3293	.3835	.3248	.2362	.1549	.0932	.0518	.0266	.0125	.0054
	2	.0867	.2131	.2866	.2953	.2581	.1998	.1395	.0887	.0513	.0269
	3	.0137	.0710	.1517	.2215	.2581	.2568	.2254	.1774	.1259	.0806
	4	.0014	.0158	.0536	.1107	.1721	.2201	.2428	.2365	.2060	.1611
	5	.0001	.0025	.0132	.0388	.0803	.1321	.1830	.2207	.2360	.2256
	6	.0000	.0003	.0023	.0097	.0268	.0566	.0985	.1471	.1931	.2256
	7	.0000	.0000	.0003	.0017	.0064	.0173	.0379	.0701	.1128	.1611
	8	.0000	.0000	.0000	.0002	.0011	.0037	.0102	.0234	.0462	.0806
	9	.0000	.0000	.0000	.0000	.0001	.0005	.0018	.0052	.0126	.0269
	10	.0000	.0000	.0000	.0000	.0000	.0000	.0002	.0007	.0021	.0054
	11	.0000	.0000	.0000	.0000	.0000	.0000	.0000	.0000	.0002	.0005

Appendix Table C *(continued)*

n	x	.05	.10	.15	.20	.25	π .30	.35	.40	.45	.50
12	0	.5404	.2824	.1422	.0687	.0317	.0138	.0057	.0022	.0008	.0002
	1	.3413	.3766	.3012	.2062	.1267	.0712	.0368	.0174	.0075	.0029
	2	.0988	.2301	.2924	.2835	.2323	.1678	.1088	.0639	.0339	.0161
	3	.0173	.0853	.1720	.2362	.2581	.2397	.1954	.1419	.0923	.0537
	4	.0021	.0213	.0683	.1329	.1936	.2311	.2367	.2128	.1700	.1208
	5	.0002	.0038	.0193	.0532	.1032	.1585	.2039	.2270	.2225	.1934
	6	.0000	.0005	.0040	.0155	.0401	.0792	.1281	.1766	.2124	.2256
	7	.0000	.0000	.0006	.0033	.0115	.0291	.0591	.1009	.1489	.1934
	8	.0000	.0000	.0001	.0005	.0024	.0078	.0199	.0420	.0762	.1208
	9	.0000	.0000	.0000	.0001	.0004	.0015	.0048	.0125	.0277	.0537
	10	.0000	.0000	.0000	.0000	.0000	.0002	.0008	.0025	.0068	.0161
	11	.0000	.0000	.0000	.0000	.0000	.0000	.0001	.0003	.0010	.0029
	12	.0000	.0000	.0000	.0000	.0000	.0000	.0000	.0000	.0001	.0002
13	0	.5133	.2542	.1209	.0550	.0238	.0097	.0037	.0013	.0004	.0001
	1	.3512	.3672	.2774	.1787	.1029	.0540	.0259	.0113	.0045	.0016
	2	.1109	.2448	.2937	.2680	.2059	.1388	.0836	.0453	.0220	.0095
	3	.0214	.0997	.1900	.2457	.2517	.2181	.1651	.1107	.0660	.0349
	4	.0028	.0277	.0838	.1535	.2097	.2337	.2222	.1845	.1350	.0873
	5	.0003	.0055	.0266	.0691	.1258	.1803	.2154	.2214	.1989	.1571
	6	.0000	.0008	.0063	.0230	.0559	.1030	.1546	.1968	.2169	.2095
	7	.0000	.0001	.0011	.0058	.0186	.0442	.0833	.1312	.1775	.2095
	8	.0000	.0000	.0001	.0011	.0047	.0142	.0336	.0656	.1089	.1571
	9	.0000	.0000	.0000	.0001	.0009	.0034	.0101	.0243	.0495	.0873
	10	.0000	.0000	.0000	.0000	.0001	.0006	.0022	.0065	.0162	.0349
	11	.0000	.0000	.0000	.0000	.0000	.0001	.0003	.0012	.0036	.0095
	12	.0000	.0000	.0000	.0000	.0000	.0000	.0000	.0001	.0005	.0016
	13	.0000	.0000	.0000	.0000	.0000	.0000	.0000	.0000	.0000	.0001
14	0	.4877	.2288	.1028	.0440	.0178	.0068	.0024	.0008	.0002	.0001
	1	.3593	.3559	.2539	.1539	.0832	.0407	.0181	.0073	.0027	.0009
	2	.1229	.2570	.2912	.2501	.1802	.1134	.0634	.0317	.0141	.0056
	3	.0259	.1142	.2056	.2501	.2402	.1943	.1366	.0845	.0462	.0222
	4	.0037	.0349	.0998	.1720	.2202	.2290	.2022	.1549	.1040	.0611
	5	.0004	.0078	.0352	.0860	.1468	.1963	.2178	.2066	.1701	.1222
	6	.0000	.0013	.0093	.0322	.0734	.1262	.1759	.2066	.2088	.1833
	7	.0000	.0002	.0019	.0092	.0280	.0618	.1082	.1574	.1952	.2095
	8	.0000	.0000	.0003	.0020	.0082	.0232	.0510	.0918	.1398	.1833
	9	.0000	.0000	.0000	.0003	.0018	.0066	.0183	.0408	.0762	.1222
	10	.0000	.0000	.0000	.0000	.0003	.0014	.0049	.0136	.0312	.0611
	11	.0000	.0000	.0000	.0000	.0000	.0002	.0010	.0033	.0093	.0222
	12	.0000	.0000	.0000	.0000	.0000	.0000	.0001	.0005	.0019	.0056
	13	.0000	.0000	.0000	.0000	.0000	.0000	.0000	.0001	.0002	.0009
	14	.0000	.0000	.0000	.0000	.0000	.0000	.0000	.0000	.0000	.0001

Appendix Table C (continued)

n	x	.05	.10	.15	.20	.25	.30	.35	.40	.45	.50
						π					
15	0	.4633	.2059	.0874	.0352	.0134	.0047	.0016	.0005	.0001	.0000
	1	.3658	.3432	.2312	.1319	.0668	.0305	.0126	.0047	.0016	.0005
	2	.1348	.2669	.2856	.2309	.1559	.0916	.0476	.0219	.0090	.0032
	3	.0307	.1285	.2184	.2501	.2252	.1700	.1110	.0634	.0318	.0139
	4	.0049	.0428	.1156	.1876	.2252	.2186	.1792	.1268	.0780	.0417
	5	.0006	.0105	.0449	.1032	.1651	.2061	.2123	.1859	.1404	.0916
	6	.0000	.0019	.0132	.0430	.0917	.1472	.1906	.2066	.1914	.1527
	7	.0000	.0003	.0030	.0138	.0393	.0811	.1319	.1771	.2013	.1964
	8	.0000	.0000	.0005	.0035	.0131	.0348	.0710	.1181	.1647	.1964
	9	.0000	.0000	.0001	.0007	.0034	.0116	.0298	.0612	.1048	.1527
	10	.0000	.0000	.0000	.0001	.0007	.0030	.0096	.0245	.0515	.0916
	11	.0000	.0000	.0000	.0000	.0001	.0006	.0024	.0074	.0191	.0417
	12	.0000	.0000	.0000	.0000	.0000	.0001	.0004	.0016	.0052	.0139
	13	.0000	.0000	.0000	.0000	.0000	.0000	.0001	.0003	.0010	.0032
	14	.0000	.0000	.0000	.0000	.0000	.0000	.0000	.0000	.0001	.0005
	15	.0000	.0000	.0000	.0000	.0000	.0000	.0000	.0000	.0000	.0000
16	0	.4401	.1853	.0743	.0281	.0100	.0033	.0010	.0003	.0001	.0000
	1	.3706	.3294	.2097	.1126	.0535	.0228	.0087	.0030	.0009	.0002
	2	.1463	.2745	.2775	.2111	.1336	.0732	.0353	.0150	.0056	.0018
	3	.0359	.1423	.2285	.2463	.2079	.1465	.0888	.0468	.0215	.0085
	4	.0061	.0514	.1311	.2001	.2252	.2040	.1553	.1014	.0572	.0278
	5	.0008	.0137	.0555	.1201	.1802	.2099	.2008	.1623	.1123	.0667
	6	.0001	.0028	.0180	.0550	.1101	.1649	.1982	.1983	.1684	.1222
	7	.0000	.0004	.0045	.0197	.0524	.1010	.1524	.1889	.1969	.1746
	8	.0000	.0001	.0009	.0055	.0197	.0487	.0923	.1417	.1812	.1964
	9	.0000	.0000	.0001	.0012	.0058	.0185	.0442	.0840	.1318	.1746
	10	.0000	.0000	.0000	.0002	.0014	.0056	.0167	.0392	.0755	.1222
	11	.0000	.0000	.0000	.0000	.0002	.0013	.0049	.0142	.0337	.0667
	12	.0000	.0000	.0000	.0000	.0000	.0002	.0011	.0040	.0115	.0278
	13	.0000	.0000	.0000	.0000	.0000	.0000	.0002	.0008	.0029	.0085
	14	.0000	.0000	.0000	.0000	.0000	.0000	.0000	.0001	.0005	.0018
	15	.0000	.0000	.0000	.0000	.0000	.0000	.0000	.0000	.0001	.0002
	16	.0000	.0000	.0000	.0000	.0000	.0000	.0000	.0000	.0000	.0000
17	0	.4181	.1668	.0631	.0225	.0075	.0023	.0007	.0002	.0000	.0000
	1	.3741	.3150	.1893	.0957	.0426	.0169	.0060	.0019	.0005	.0001
	2	.1575	.2800	.2673	.1914	.1136	.0581	.0260	.0102	.0035	.0010
	3	.0415	.1556	.2359	.2393	.1893	.1245	.0701	.0341	.0144	.0052
	4	.0076	.0605	.1457	.2093	.2209	.1868	.1320	.0796	.0411	.0182
	5	.0010	.0175	.0668	.1361	.1914	.2081	.1849	.1379	.0875	.0472
	6	.0001	.0039	.0236	.0680	.1276	.1784	.1991	.1839	.1432	.0944
	7	.0000	.0007	.0065	.0267	.0668	.1201	.1685	.1927	.1841	.1484
	8	.0000	.0001	.0014	.0084	.0279	.0644	.1134	.1606	.1883	.1855
	9	.0000	.0000	.0003	.0021	.0093	.0276	.0611	.1070	.1540	.1855
	10	.0000	.0000	.0000	.0004	.0025	.0095	.0263	.0571	.1008	.1484
	11	.0000	.0000	.0000	.0001	.0005	.0026	.0090	.0242	.0525	.0944
	12	.0000	.0000	.0000	.0000	.0001	.0006	.0024	.0081	.0215	.0472
	13	.0000	.0000	.0000	.0000	.0000	.0001	.0005	.0021	.0068	.0182
	14	.0000	.0000	.0000	.0000	.0000	.0000	.0001	.0004	.0016	.0052
	15	.0000	.0000	.0000	.0000	.0000	.0000	.0000	.0001	.0003	.0010
	16	.0000	.0000	.0000	.0000	.0000	.0000	.0000	.0000	.0000	.0001
	17	.0000	.0000	.0000	.0000	.0000	.0000	.0000	.0000	.0000	.0000

Appendix Table C (continued)

n	x	.05	.10	.15	.20	.25	π .30	.35	.40	.45	.50
18	0	.3972	.1501	.0536	.0180	.0056	.0016	.0004	.0001	.0000	.0000
	1	.3763	.3002	.1704	.0811	.0338	.0126	.0042	.0012	.0003	.0001
	2	.1683	.2835	.2556	.1723	.0958	.0458	.0190	.0069	.0022	.0006
	3	.0473	.1680	.2406	.2297	.1704	.1046	.0547	.0246	.0095	.0031
	4	.0093	.0700	.1592	.2153	.2130	.1681	.1104	.0614	.0291	.0117
	5	.0014	.0218	.0787	.1507	.1988	.2017	.1664	.1146	.0666	.0327
	6	.0002	.0052	.0301	.0816	.1436	.1873	.1941	.1655	.1181	.0708
	7	.0000	.0010	.0091	.0350	.0820	.1376	.1792	.1892	.1657	.1214
	8	.0000	.0002	.0022	.0120	.0376	.0811	.1327	.1734	.1864	.1669
	9	.0000	.0000	.0004	.0033	.0139	.0386	.0794	.1284	.1694	.1855
	10	.0000	.0000	.0001	.0008	.0042	.0149	.0385	.0771	.1248	.1669
	11	.0000	.0000	.0000	.0001	.0010	.0046	.0151	.0374	.0742	.1214
	12	.0000	.0000	.0000	.0000	.0002	.0012	.0047	.0145	.0354	.0708
	13	.0000	.0000	.0000	.0000	.0000	.0002	.0012	.0045	.0134	.0327
	14	.0000	.0000	.0000	.0000	.0000	.0000	.0002	.0011	.0039	.0117
	15	.0000	.0000	.0000	.0000	.0000	.0000	.0000	.0002	.0009	.0031
	16	.0000	.0000	.0000	.0000	.0000	.0000	.0000	.0000	.0001	.0006
	17	.0000	.0000	.0000	.0000	.0000	.0000	.0000	.0000	.0000	.0001
	18	.0000	.0000	.0000	.0000	.0000	.0000	.0000	.0000	.0000	.0000
19	0	.3774	.1351	.0456	.0144	.0042	.0011	.0003	.0001	.0000	.0000
	1	.3774	.2852	.1529	.0685	.0268	.0093	.0029	.0008	.0002	.0000
	2	.1787	.2852	.2428	.1540	.0803	.0358	.0138	.0046	.0013	.0003
	3	.0533	.1796	.2428	.2182	.1517	.0869	.0422	.0175	.0062	.0018
	4	.0112	.0798	.1714	.2182	.2023	.1491	.0909	.0467	.0203	.0074
	5	.0018	.0266	.0907	.1636	.2023	.1916	.1468	.0933	.0497	.0222
	6	.0002	.0069	.0374	.0955	.1574	.1916	.1844	.1451	.0949	.0518
	7	.0000	.0014	.0122	.0443	.0974	.1525	.1844	.1797	.1443	.0961
	8	.0000	.0002	.0032	.0166	.0487	.0981	.1489	.1797	.1771	.1442
	9	.0000	.0000	.0007	.0051	.0198	.0514	.0980	.1464	.1771	.1762
	10	.0000	.0000	.0001	.0013	.0066	.0220	.0528	.0976	.1449	.1762
	11	.0000	.0000	.0000	.0003	.0018	.0077	.0233	.0532	.0970	.1442
	12	.0000	.0000	.0000	.0000	.0004	.0022	.0083	.0237	.0529	.0961
	13	.0000	.0000	.0000	.0000	.0001	.0005	.0024	.0085	.0233	.0518
	14	.0000	.0000	.0000	.0000	.0000	.0001	.0006	.0024	.0082	.0222
	15	.0000	.0000	.0000	.0000	.0000	.0000	.0001	.0005	.0022	.0074
	16	.0000	.0000	.0000	.0000	.0000	.0000	.0000	.0001	.0005	.0018
	17	.0000	.0000	.0000	.0000	.0000	.0000	.0000	.0000	.0001	.0003
	18	.0000	.0000	.0000	.0000	.0000	.0000	.0000	.0000	.0000	.0000
	19	.0000	.0000	.0000	.0000	.0000	.0000	.0000	.0000	.0000	.0000
20	0	.3585	.1216	.0388	.0115	.0032	.0008	.0002	.0000	.0000	.0000
	1	.3774	.2702	.1368	.0576	.0211	.0068	.0020	.0005	.0001	.0000
	2	.1887	.2852	.2293	.1369	.0669	.0278	.0100	.0031	.0008	.0002
	3	.0596	.1901	.2428	.2054	.1339	.0716	.0323	.0123	.0040	.0011
	4	.0133	.0898	.1821	.2182	.1897	.1304	.0738	.0350	.0139	.0046
	5	.0022	.0319	.1028	.1746	.2023	.1789	.1272	.0746	.0365	.0148
	6	.0003	.0089	.0454	.1091	.1686	.1916	.1712	.1244	.0746	.0370
	7	.0000	.0020	.0160	.0545	.1124	.1643	.1844	.1659	.1221	.0739
	8	.0000	.0004	.0046	.0222	.0609	.1144	.1614	.1797	.1623	.1201
	9	.0000	.0001	.0011	.0074	.0271	.0654	.1158	.1597	.1771	.1602
	10	.0000	.0000	.0002	.0020	.0099	.0308	.0686	.1171	.1593	.1762
	11	.0000	.0000	.0000	.0005	.0030	.0120	.0336	.0710	.1185	.1602
	12	.0000	.0000	.0000	.0001	.0008	.0039	.0136	.0355	.0727	.1201
	13	.0000	.0000	.0000	.0000	.0002	.0010	.0045	.0146	.0366	.0739
	14	.0000	.0000	.0000	.0000	.0000	.0002	.0012	.0049	.0150	.0370
	15	.0000	.0000	.0000	.0000	.0000	.0000	.0003	.0013	.0049	.0148
	16	.0000	.0000	.0000	.0000	.0000	.0000	.0000	.0003	.0013	.0046
	17	.0000	.0000	.0000	.0000	.0000	.0000	.0000	.0000	.0002	.0011
	18	.0000	.0000	.0000	.0000	.0000	.0000	.0000	.0000	.0000	.0002
	19	.0000	.0000	.0000	.0000	.0000	.0000	.0000	.0000	.0000	.0000
	20	.0000	.0000	.0000	.0000	.0000	.0000	.0000	.0000	.0000	.0000

Appendix Table D
Binomial Probabilities for Cumulative Values of x

Entries in this table give the probability of *x* or fewer successes in *n* trials of a binomial experiment, where π is the probability of a success in one trial. For example, with $n = 5$ trials and $\pi = .20$, the probability of two or fewer successes is .9421. (Probability values in blank spaces equal 1.0000.)

n	x	.05	.10	.20	.30	.40	.50
1	0	0.9500	0.9000	0.8000	0.7000	0.6000	0.5000
	1	1.0000	1.0000	1.0000	1.0000	1.0000	1.0000
2	0	0.9025	0.8100	0.6400	0.4900	0.3600	0.2500
	1	0.9975	0.9900	0.9600	0.9100	0.8400	0.7500
	2	1.0000	1.0000	1.0000	1.0000	1.0000	1.0000
3	0	0.8574	0.7290	0.5120	0.3430	0.2160	0.1250
	1	0.9927	0.9720	0.8960	0.7840	0.6480	0.5000
	2	0.9999	0.9990	0.9920	0.9730	0.9360	0.8750
	3	1.0000	1.0000	1.0000	1.0000	1.0000	1.0000
4	0	0.8145	0.6561	0.4096	0.2401	0.1296	0.0625
	1	0.9860	0.9477	0.8192	0.6517	0.4752	0.3125
	2	0.9995	0.9963	0.9728	0.9163	0.8208	0.6875
	3	1.0000	0.9999	0.9984	0.9919	0.9744	0.9375
	4		1.0000	1.0000	1.0000	1.0000	1.0000
5	0	0.7738	0.5905	0.3277	0.1681	0.0778	0.0313
	1	0.9774	0.9185	0.7373	0.5282	0.3370	0.1875
	2	0.9988	0.9914	0.9421	0.8369	0.6826	0.5000
	3	1.0000	0.9995	0.9933	0.9692	0.9130	0.8125
	4		1.0000	0.9997	0.9976	0.9898	0.9688
	5			1.0000	1.0000	1.0000	1.0000
6	0	0.7351	0.5314	0.2621	0.1176	0.0467	0.0156
	1	0.9672	0.8857	0.6554	0.4202	0.2333	0.1094
	2	0.9978	0.9841	0.9011	0.7443	0.5443	0.3438
	3	0.9999	0.9987	0.9830	0.9295	0.8208	0.6563
	4	1.0000	0.9999	0.9984	0.9891	0.9590	0.8906
	5		1.0000	0.9999	0.9993	0.9959	0.9844
	6			1.0000	1.0000	1.0000	1.0000
7	0	0.6983	0.4783	0.2097	0.0824	0.0280	0.0078
	1	0.9556	0.8503	0.5767	0.3294	0.1586	0.0625
	2	0.9962	0.9743	0.8520	0.6471	0.4199	0.2266
	3	0.9998	0.9973	0.9667	0.8740	0.7102	0.5000
	4	1.0000	0.9998	0.9953	0.9712	0.9037	0.7734
	5		1.0000	0.9996	0.9962	0.9812	0.9375
	6			1.0000	0.9998	0.9984	0.9922
	7				1.0000	1.0000	1.0000
8	0	0.6634	0.4305	0.1678	0.0576	0.0168	0.0039
	1	0.9428	0.8131	0.5033	0.2553	0.1064	0.0352
	2	0.9942	0.9619	0.7969	0.5518	0.3154	0.1445
	3	0.9996	0.9950	0.9437	0.8059	0.5941	0.3633
	4	1.0000	0.9996	0.9896	0.9420	0.8263	0.6367
	5		1.0000	0.9988	0.9887	0.9502	0.8555
	6			0.9999	0.9987	0.9915	0.9648
	7			1.0000	0.9999	0.9993	0.9961
	8				1.0000	1.0000	1.0000

Source: For values of *n* up to 20, the National Bureau of Standards, *Tables of the Binomial Probability Distribution*, Applied Mathematics Series, No. 6 (Washington D.C.: U.S. Government Printing Office, 1949); for *n* = 50 and *n* = 100, Lawrence L. Lapin, *Statistics for Modern Business Decisions*, 3rd ed. (New York: Harcourt, Brace, Jovanovich, 1982), pp. 830–32.

Appendix Table D *(continued)*

n	x	.05	.10	.20	π .30	.40	.50
9	0	0.6302	0.3874	0.1342	0.0404	0.0101	0.0020
	1	0.9288	0.7748	0.4362	0.1960	0.0705	0.0195
	2	0.9916	0.9470	0.7382	0.4628	0.2318	0.0898
	3	0.9994	0.9917	0.9144	0.7297	0.4826	0.2539
	4	1.0000	0.9991	0.9804	0.9012	0.7334	0.5000
	5		0.9999	0.9969	0.9747	0.9006	0.7461
	6		1.0000	0.9997	0.9957	0.9750	0.9102
	7			1.0000	0.9996	0.9962	0.9805
	8				1.0000	0.9997	0.9980
	9					1.0000	1.0000
10	0	0.5987	0.3487	0.1074	0.0282	0.0060	0.0010
	1	0.9139	0.7361	0.3758	0.1493	0.0464	0.0107
	2	0.9885	0.9298	0.6778	0.3828	0.1673	0.0547
	3	0.9990	0.9872	0.8791	0.6496	0.3823	0.1719
	4	0.9999	0.9984	0.9672	0.8497	0.6331	0.3770
	5	1.0000	0.9999	0.9936	0.9526	0.8338	0.6230
	6		1.0000	0.9991	0.9894	0.9452	0.8281
	7			0.9999	0.9999	0.9877	0.9453
	8			1.0000	1.0000	0.9983	0.9893
	9					0.9999	0.9990
	10					1.0000	1.0000
11	0	0.5688	0.3138	0.0859	0.0198	0.0036	0.0005
	1	0.8981	0.6974	0.3221	0.1130	0.0302	0.0059
	2	0.9848	0.9104	0.6174	0.3127	0.1189	0.0327
	3	0.9984	0.9815	0.8369	0.5696	0.2963	0.1133
	4	0.9999	0.9972	0.9496	0.7897	0.5328	0.2744
	5	1.0000	0.9997	0.9883	0.9218	0.7535	0.5000
	6		1.0000	0.9980	0.9784	0.9006	0.7256
	7			0.9998	0.9957	0.9707	0.8867
	8			1.0000	0.9994	0.9941	0.9673
	9				1.0000	0.9993	0.9941
	10					1.0000	0.9995
	11						1.0000
12	0	0.5404	0.2824	0.0687	0.0138	0.0022	0.0002
	1	0.8816	0.6590	0.2749	0.0850	0.0196	0.0032
	2	0.9804	0.8891	0.5583	0.2528	0.0834	0.0193
	3	0.9978	0.9744	0.7946	0.4925	0.2253	0.0730
	4	0.9998	0.9957	0.9274	0.7237	0.4382	0.1938
	5	1.0000	0.9995	0.9806	0.8821	0.6652	0.3872
	6		0.9999	0.9961	0.9614	0.8418	0.6128
	7		1.0000	0.9994	0.9905	0.9427	0.8062
	8			0.9999	0.9983	0.9847	0.9270
	9			1.0000	0.9998	0.9972	0.9807
	10				1.0000	0.9997	0.9968
	11					1.0000	0.9998
	12						1.0000
13	0	0.5133	0.2542	0.0550	0.0097	0.0013	0.0001
	1	0.8646	0.6213	0.2336	0.0637	0.0126	0.0017
	2	0.9755	0.8661	0.5017	0.2025	0.0579	0.0112
	3	0.9969	0.9658	0.7473	0.4206	0.1686	0.0461
	4	0.9997	0.9935	0.9009	0.6543	0.3530	0.1334
	5	1.0000	0.9991	0.9700	0.8346	0.5744	0.2905
	6		0.9999	0.9930	0.9376	0.7712	0.5000
	7		1.0000	0.9988	0.9818	0.9023	0.7095
	8			0.9998	0.9960	0.9679	0.8666
	9			1.0000	0.9993	0.9922	0.9539
	10				0.9999	0.9987	0.9888
	11				1.0000	0.9999	0.9983
	12					1.0000	0.9999
	13						1.0000

Appendix Table D (continued)

n	x	.05	.10	.20	.30	.40	.50
14	0	0.4877	0.2288	0.0440	0.0068	0.0008	0.0001
	1	0.8470	0.5846	0.1979	0.0475	0.0081	0.0009
	2	0.9699	0.8416	0.4481	0.1608	0.0398	0.0065
	3	0.9958	0.9559	0.6982	0.3552	0.1243	0.0287
	4	0.9996	0.9908	0.8702	0.5842	0.2793	0.0898
	5	1.0000	0.9985	0.9561	0.7805	0.4859	0.2120
	6		0.9998	0.9884	0.9067	0.6925	0.3953
	7		1.0000	0.9976	0.9685	0.8499	0.6047
	8			0.9996	0.9917	0.9417	0.7880
	9			1.0000	0.9983	0.9825	0.9102
	10				0.9998	0.9961	0.9713
	11				1.0000	0.9994	0.9935
	12					0.9999	0.9991
	13					1.0000	0.9999
	14						1.0000
15	0	0.4633	0.2059	0.0352	0.0047	0.0005	0.0000
	1	0.8290	0.5490	0.1671	0.0353	0.0052	0.0005
	2	0.9638	0.8159	0.3980	0.1268	0.0271	0.0037
	3	0.9945	0.9444	0.6482	0.2969	0.0905	0.0176
	4	0.9994	0.9873	0.8358	0.5155	0.2173	0.0592
	5	0.9999	0.9978	0.9389	0.7216	0.4032	0.1509
	6	1.0000	0.9997	0.9819	0.8689	0.6098	0.3036
	7		1.0000	0.9958	0.9500	0.7869	0.5000
	8			0.9992	0.9848	0.9050	0.6964
	9			0.9999	0.9963	0.9662	0.8491
	10			1.0000	0.9993	0.9907	0.9408
	11				0.9999	0.9981	0.9824
	12				1.0000	0.9997	0.9963
	13					1.0000	0.9995
	14						1.0000
16	0	0.4401	0.1853	0.0281	0.0033	0.0003	0.0000
	1	0.8108	0.5147	0.1407	0.0261	0.0033	0.0003
	2	0.9571	0.7892	0.3518	0.0994	0.0183	0.0021
	3	0.9930	0.9316	0.5981	0.2459	0.0651	0.0106
	4	0.9991	0.9830	0.7982	0.4499	0.1666	0.0384
	5	0.9999	0.9967	0.9183	0.6598	0.3288	0.1051
	6	1.0000	0.9995	0.9733	0.8247	0.5272	0.2272
	7		0.9999	0.9930	0.9256	0.7161	0.4018
	8		1.0000	0.9985	0.9743	0.8577	0.5982
	9			0.9998	0.9929	0.9417	0.7728
	10			1.0000	0.9984	0.9809	0.8949
	11				0.9997	0.9951	0.9616
	12				1.0000	0.9991	0.9894
	13					0.9999	0.9979
	14					1.0000	0.9997
	15						1.0000
17	0	0.4181	0.1668	0.0225	0.0023	0.0002	0.0000
	1	0.7922	0.4818	0.1182	0.0193	0.0021	0.0001
	2	0.9497	0.7618	0.3096	0.0774	0.0123	0.0012
	3	0.9912	0.9174	0.5489	0.2019	0.0464	0.0064
	4	0.9988	0.9779	0.7582	0.3887	0.1260	0.0245
	5	0.9999	0.9953	0.8943	0.5968	0.2639	0.0717
	6	1.0000	0.9992	0.9623	0.7752	0.4478	0.1662
	7		0.9999	0.9891	0.8954	0.6405	0.3145
	8		1.0000	0.9974	0.9597	0.8011	0.5000
	9			0.9995	0.9873	0.9081	0.6855
	10			0.9999	0.9968	0.9652	0.8338
	11			1.0000	0.9993	0.9894	0.9283
	12				0.9999	0.9975	0.9755
	13				1.0000	0.9995	0.9936
	14					0.9999	0.9988
	15					1.0000	0.9999
	16						1.0000

The column headers are grouped under π.

Appendix Table D *(continued)*

n	x	.05	.10	.20	π .30	.40	.50
18	0	0.3972	0.1501	0.0180	0.0016	0.0001	0.0000
	1	0.7735	0.4503	0.0991	0.0142	0.0013	0.0001
	2	0.9419	0.7338	0.2713	0.0600	0.0082	0.0007
	3	0.9891	0.9018	0.5010	0.1646	0.0328	0.0038
	4	0.9985	0.9718	0.7164	0.3327	0.0942	0.0154
	5	0.9998	0.9936	0.8671	0.5344	0.2088	0.0481
	6	1.0000	0.9988	0.9487	0.7217	0.3743	0.1189
	7		0.9998	0.9837	0.8593	0.5634	0.2403
	8		1.0000	0.9957	0.9404	0.7368	0.4073
	9			0.9991	0.9790	0.8653	0.5927
	10			0.9998	0.9939	0.9424	0.7597
	11			1.0000	0.9986	0.9797	0.8811
	12				0.9997	0.9942	0.9519
	13				1.0000	0.9987	0.9846
	14					0.9998	0.9962
	15					1.0000	0.9993
	16						0.9999
	17						1.0000
19	0	0.3774	0.1351	0.0144	0.0011	0.0001	0.0000
	1	0.7547	0.4203	0.0829	0.0104	0.0008	0.0000
	2	0.9335	0.7054	0.2369	0.0462	0.0055	0.0004
	3	0.9868	0.8850	0.4551	0.1332	0.0230	0.0022
	4	0.9980	0.9648	0.6733	0.2822	0.0696	0.0096
	5	0.9998	0.9914	0.8369	0.4739	0.1629	0.0318
	6	1.0000	0.9983	0.9324	0.6655	0.3081	0.0835
	7		0.9997	0.9767	0.8180	0.4878	0.1796
	8		1.0000	0.9933	0.9161	0.6675	0.3238
	9			0.9984	0.9674	0.8139	0.5000
	10			0.9997	0.9895	0.9115	0.6762
	11			0.9999	0.9972	0.9648	0.8204
	12			1.0000	0.9994	0.9884	0.9165
	13				0.9999	0.9969	0.9682
	14				1.0000	0.9994	0.9904
	15					0.9999	0.9978
	16					1.0000	0.9996
	17						1.0000
20	0	0.3585	0.1216	0.0115	0.0008	0.0000	0.0000
	1	0.7358	0.3917	0.0692	0.0076	0.0005	0.0000
	2	0.9245	0.6769	0.2061	0.0355	0.0036	0.0002
	3	0.9841	0.8670	0.4114	0.1071	0.0160	0.0013
	4	0.9974	0.9568	0.6296	0.2375	0.0510	0.0059
	5	0.9997	0.9887	0.8042	0.4164	0.1256	0.0207
	6	1.0000	0.9976	0.9133	0.6080	0.2500	0.0577
	7		0.9996	0.9679	0.7723	0.4159	0.1316
	8		0.9999	0.9900	0.8867	0.5956	0.2517
	9		1.0000	0.9974	0.9520	0.7553	0.4119
	10			0.9994	0.9829	0.8725	0.5881
	11			0.9999	0.9949	0.9435	0.7483
	12			1.0000	0.9987	0.9790	0.8684
	13				0.9997	0.9935	0.9423
	14				1.0000	0.9984	0.9793
	15					0.9997	0.9941
	16					1.0000	0.9987
	17						0.9998
	18						1.0000

Appendix Table D **(continued)**

n	x	.05	.10	.20	π .30	.40	.50
50	0	0.0769	0.0052	0.0000	0.0000	0.0000	0.0000
	1	0.2794	0.0338	0.0002	0.0000	0.0000	0.0000
	2	0.5405	0.1117	0.0013	0.0000	0.0000	0.0000
	3	0.7604	0.2503	0.0057	0.0000	0.0000	0.0000
	4	0.8964	0.4312	0.0185	0.0002	0.0000	0.0000
	5	0.9622	0.6161	0.0480	0.0007	0.0000	0.0000
	6	0.9882	0.7702	0.1034	0.0025	0.0000	0.0000
	7	0.9968	0.8779	0.1904	0.0073	0.0001	0.0000
	8	0.9992	0.9421	0.3073	0.0183	0.0002	0.0000
	9	0.9998	0.9755	0.4437	0.0402	0.0008	0.0000
	10	1.0000	0.9906	0.5836	0.0789	0.0022	0.0000
	11		0.9968	0.7107	0.1390	0.0057	0.0000
	12		0.9990	0.8139	0.2229	0.0133	0.0002
	13		0.9997	0.8894	0.3279	0.0280	0.0005
	14		0.9999	0.9393	0.4468	0.0540	0.0013
	15		1.0000	0.9692	0.5692	0.0955	0.0033
	16			0.9856	0.6839	0.1561	0.0077
	17			0.9937	0.7822	0.2369	0.0164
	18			0.9975	0.8594	0.3356	0.0325
	19			0.9991	0.9152	0.4465	0.0595
	20			0.9997	0.9522	0.5610	0.1013
	21			0.9999	0.9749	0.6701	0.1611
	22			1.0000	0.9877	0.7660	0.2399
	23				0.9944	0.8438	0.3359
	24				0.9976	0.9022	0.4439
	25				0.9991	0.9427	0.5561
	26				0.9997	0.9686	0.6641
	27				0.9999	0.9840	0.7601
	28				1.0000	0.9924	0.8389
	29					0.9966	0.8987
	30					0.9986	0.9405
	31					0.9995	0.9675
	32					0.9998	0.9836
	33					0.9999	0.9923
	34					1.0000	0.9967
	35						0.9987
	36						0.9995
	37						0.9998
	38						1.0000

Appendix Table D *(continued)*

n	x	.05	.10	.20	.30	.40	.50	n	x	.20	.30	.40	.50
100	0	0.0059	0.0000	0.0000	0.0000	0.0000	0.0000	100	36	0.9999	0.9201	0.2386	0.0033
	1	0.0371	0.0003	0.0000	0.0000	0.0000	0.0000		37	1.0000	0.9470	0.3068	0.0060
	2	0.1183	0.0019	0.0000	0.0000	0.0000	0.0000		38		0.9660	0.3822	0.0105
	3	0.2578	0.0078	0.0000	0.0000	0.0000	0.0000		39		0.9790	0.4621	0.0176
	4	0.4360	0.0237	0.0000	0.0000	0.0000	0.0000		40		0.9875	0.5433	0.0284
	5	0.6160	0.0576	0.0000	0.0000	0.0000	0.0000						
	6	0.7660	0.1172	0.0001	0.0000	0.0000	0.0000		41		0.9928	0.6225	0.0443
	7	0.8720	0.2061	0.0003	0.0000	0.0000	0.0000		42		0.9960	0.6967	0.0666
	8	0.9369	0.3209	0.0009	0.0000	0.0000	0.0000		43		0.9979	0.7635	0.0967
	9	0.9718	0.4513	0.0023	0.0000	0.0000	0.0000		44		0.9989	0.8211	0.1356
	10	0.9885	0.5832	0.0057	0.0000	0.0000	0.0000		45		0.9995	0.8689	0.1841
	11	0.9957	0.7030	0.0126	0.0000	0.0000	0.0000		46		0.9997	0.9070	0.2421
	12	0.9985	0.8018	0.0253	0.0000	0.0000	0.0000		47		0.9999	0.9362	0.3086
	13	0.9995	0.8761	0.0469	0.0001	0.0000	0.0000		48		0.9999	0.9577	0.3822
	14	0.9999	0.9274	0.0804	0.0002	0.0000	0.0000		49		1.0000	0.9729	0.4602
	15	1.0000	0.9601	0.1285	0.0004	0.0000	0.0000		50			0.9832	0.5398
	16		0.9794	0.1923	0.0010	0.0000	0.0000		51			0.9900	0.6178
	17		0.9900	0.2712	0.0022	0.0000	0.0000		52			0.9942	0.6914
	18		0.9954	0.3621	0.0045	0.0000	0.0000		53			0.9968	0.7579
	19		0.9980	0.4602	0.0089	0.0000	0.0000		54			0.9983	0.8159
	20		0.9992	0.5595	0.0165	0.0000	0.0000		55			0.9991	0.8644
	21		0.9997	0.6540	0.0288	0.0000	0.0000		56			0.9996	0.9033
	22		0.9999	0.7389	0.0479	0.0001	0.0000		57			0.9998	0.9334
	23		1.0000	0.8109	0.0755	0.0003	0.0000		58			0.9999	0.9557
	24			0.8686	0.1136	0.0006	0.0000		59			1.0000	0.9716
	25			0.9125	0.1631	0.0012	0.0000		60				0.9824
	26			0.9442	0.2244	0.0024	0.0000		61				0.9895
	27			0.9658	0.2964	0.0046	0.0000		62				0.9940
	28			0.9800	0.3768	0.0084	0.0000		63				0.9967
	29			0.9888	0.4623	0.0148	0.0000		64				0.9982
	30			0.9939	0.5491	0.0248	0.0000		65				0.9991
	31			0.9969	0.6331	0.0398	0.0001		66				0.9996
	32			0.9984	0.7107	0.0615	0.0002		67				0.9998
	33			0.9993	0.7793	0.0913	0.0004		68				0.9999
	34			0.9997	0.8371	0.1303	0.0009		69				1.0000
	35			0.9999	0.8839	0.1795	0.0018						

Appendix Table E
Exponential Functions

μ	e^{μ}	$e^{-\mu}$	μ	e^{μ}	$e^{-\mu}$
0.00	1.0000	1.000000	3.00	20.086	.049787
0.10	1.1052	.904837	3.10	22.198	.045049
0.20	1.2214	.818731	3.20	24.533	.040762
0.30	1.3499	.740818	3.30	27.113	.036883
0.40	1.4918	.670320	3.40	29.964	.033373
0.50	1.6487	.606531	3.50	33.115	.030197
0.60	1.8221	.548812	3.60	36.598	.027324
0.70	2.0138	.496585	3.70	40.447	.024724
0.80	2.2255	.449329	3.80	44.701	.022371
0.90	2.4596	.406570	3.90	49.402	.020242
1.00	2.7183	.367879	4.00	54.598	.018316
1.10	3.0042	.332871	4.10	60.340	.016573
1.20	3.3201	.301194	4.20	66.686	.014996
1.30	3.6693	.272532	4.30	73.700	.013569
1.40	4.0552	.246597	4.40	81.451	.012277
1.50	4.4817	.223130	4.50	90.017	.011109
1.60	4.9530	.201897	4.60	99.484	.010052
1.70	5.4739	.182684	4.70	109.95	.009095
1.80	6.0496	.165299	4.80	121.51	.008230
1.90	6.6859	.149569	4.90	134.29	.007447
2.00	7.3891	.135335	5.00	148.41	.006738
2.10	8.1662	.122456	5.10	164.02	.006097
2.20	9.0250	.110803	5.20	181.27	.005517
2.30	9.9742	.100259	5.30	200.34	.004992
2.40	11.023	.090718	5.40	221.41	.004517
2.50	12.182	.082085	5.50	244.69	.004087
2.60	13.464	.074274	5.60	270.43	.003698
2.70	14.880	.067206	5.70	298.87	.003346
2.80	16.445	.060810	5.80	330.30	.003028
2.90	18.174	.055023	5.90	365.04	.002739
3.00	20.086	.049787	6.00	403.43	.002479

Additional values for this table can be found by interpolation or by noting the following:

$$e = 2.71828$$
$$e^{\mu} = 2.71828^{\mu}$$
$$e^{-\mu} = \frac{1}{e^{\mu}} = \frac{1}{2.71828^{\mu}}$$

With the help of logarithms, one can easily solve these expressions for any given μ. For example, if $\mu = 2.33$ (a value not found in the table), we can write

$$e^{2.33} = 2.71828^{2.33}$$
$$\log e^{2.33} = 2.33(\log 2.71828) = 2.33(.434294) = 1.011905$$
$$e^{2.33} = \text{antilog } 1.011905 = 10.277976$$
$$e^{-2.33} = \frac{1}{10.277976} = .097295$$

Source: George F. Becker and C.E. Van Orstrand, *Smithsonian Mathematical Tables: Hyperbolic Functions* (Washington, D.C.: Smithsonian Institution, 1909), pp 226–58.

Appendix Table F
Poisson Probabilities for Individual Values of x

Entries in this table give the probability of x occurrences for a Poisson process with a mean of μ. For example, when $\mu = 1.5$, the probability of $x = 4$ occurrences is .0471.

x	μ 0.1	0.2	0.3	0.4	0.5	0.6	0.7	0.8	0.9	1.0
0	.9048	.8187	.7408	.6703	.6065	.5488	.4966	.4493	.4066	.3679
1	.0905	.1637	.2222	.2681	.3033	.3293	.3476	.3595	.3659	.3679
2	.0045	.0164	.0333	.0536	.0758	.0988	.1217	.1438	.1647	.1839
3	.0002	.0011	.0033	.0072	.0126	.0198	.0284	.0383	.0494	.0613
4	.0000	.0001	.0002	.0007	.0016	.0030	.0050	.0077	.0111	.0153
5	.0000	.0000	.0000	.0001	.0002	.0004	.0007	.0012	.0020	.0031
6	.0000	.0000	.0000	.0000	.0000	.0000	.0001	.0002	.0003	.0005
7	.0000	.0000	.0000	.0000	.0000	.0000	.0000	.0000	.0000	.0001

x	μ 1.1	1.2	1.3	1.4	1.5	1.6	1.7	1.8	1.9	2.0
0	.3329	.3012	.2725	.2466	.2231	.2019	.1827	.1653	.1496	.1353
1	.3662	.3614	.3543	.3452	.3347	.3230	.3106	.2975	.2842	.2707
2	.2014	.2169	.2303	.2417	.2510	.2584	.2640	.2678	.2700	.2707
3	.0738	.0867	.0998	.1128	.1255	.1378	.1496	.1607	.1710	.1804
4	.0203	.0260	.0324	.0395	.0471	.0551	.0636	.0723	.0812	.0902
5	.0045	.0062	.0084	.0111	.0141	.0176	.0216	.0260	.0309	.0361
6	.0008	.0012	.0018	.0026	.0035	.0047	.0061	.0078	.0098	.0120
7	.0001	.0002	.0003	.0005	.0008	.0011	.0015	.0020	.0027	.0034
8	.0000	.0000	.0001	.0001	.0001	.0002	.0003	.0005	.0006	.0009
9	.0000	.0000	.0000	.0000	.0000	.0000	.0001	.0001	.0001	.0002

x	μ 2.1	2.2	2.3	2.4	2.5	2.6	2.7	2.8	2.9	3.0
0	.1225	.1108	.1003	.0907	.0821	.0743	.0672	.0608	.0550	.0498
1	.2572	.2438	.2306	.2177	.2052	.1931	.1815	.1703	.1596	.1494
2	.2700	.2681	.2652	.2613	.2565	.2510	.2450	.2384	.2314	.2240
3	.1890	.1966	.2033	.2090	.2138	.2176	.2205	.2225	.2237	.2240
4	.0992	.1082	.1169	.1254	.1336	.1414	.1488	.1557	.1622	.1680
5	.0417	.0476	.0538	.0602	.0668	.0735	.0804	.0872	.0940	.1008
6	.0146	.0174	.0206	.0241	.0278	.0319	.0362	.0407	.0455	.0504
7	.0044	.0055	.0068	.0083	.0099	.0118	.0139	.0163	.0188	.0216
8	.0011	.0015	.0019	.0025	.0031	.0038	.0047	.0057	.0068	.0081
9	.0003	.0004	.0005	.0007	.0009	.0011	.0014	.0018	.0022	.0027
10	.0001	.0001	.0001	.0002	.0002	.0003	.0004	.0005	.0006	.0008
11	.0000	.0000	.0000	.0000	.0000	.0001	.0001	.0001	.0002	.0002
12	.0000	.0000	.0000	.0000	.0000	.0000	.0000	.0000	.0000	.0001

Source: Richard S. Burington and Donald C. May, *Handbook of Probability and Statistics with Tables* (Sandusky, Ohio: Handbook Publishers, 1953), pp. 259–62. All entries were calculated by the formula in Box 6.F.

Appendix Table F (continued)

x	3.1	3.2	3.3	3.4	3.5	μ 3.6	3.7	3.8	3.9	4.0
0	.0450	.0408	.0369	.0334	.0302	.0273	.0247	.0224	.0202	.0183
1	.1397	.1304	.1217	.1135	.1057	.0984	.0915	.0850	.0789	.0733
2	.2165	.2087	.2008	.1929	.1850	.1771	.1692	.1615	.1539	.1465
3	.2237	.2226	.2209	.2186	.2158	.2125	.2087	.2046	.2001	.1954
4	.1734	.1781	.1823	.1858	.1888	.1912	.1931	.1944	.1951	.1954
5	.1075	.1140	.1203	.1264	.1322	.1377	.1429	.1477	.1522	.1563
6	.0555	.0608	.0662	.0716	.0771	.0826	.0881	.0936	.0989	.1042
7	.0246	.0278	.0312	.0348	.0385	.0425	.0466	.0508	.0551	.0595
8	.0095	.0111	.0129	.0148	.0169	.0191	.0215	.0241	.0269	.0298
9	.0033	.0040	.0047	.0056	.0066	.0076	.0089	.0102	.0116	.0132
10	.0010	.0013	.0016	.0019	.0023	.0028	.0033	.0039	.0045	.0053
11	.0003	.0004	.0005	.0006	.0007	.0009	.0011	.0013	.0016	.0019
12	.0001	.0001	.0001	.0002	.0002	.0003	.0003	.0004	.0005	.0006
13	.0000	.0000	.0000	.0000	.0001	.0001	.0001	.0001	.0002	.0002
14	.0000	.0000	.0000	.0000	.0000	.0000	.0000	.0000	.0000	.0001

x	4.1	4.2	4.3	4.4	4.5	μ 4.6	4.7	4.8	4.9	5.0
0	.0166	.0150	.0136	.0123	.0111	.0101	.0091	.0082	.0074	.0067
1	.0679	.0630	.0583	.0540	.0500	.0462	.0427	.0395	.0365	.0337
2	.1393	.1323	.1254	.1188	.1125	.1063	.1005	.0948	.0894	.0842
3	.1904	.1852	.1798	.1743	.1687	.1631	.1574	.1517	.1460	.1404
4	.1951	.1944	.1933	.1917	.1898	.1875	.1849	.1820	.1789	.1755
5	.1600	.1633	.1662	.1687	.1708	.1725	.1738	.1747	.1753	.1755
6	.1093	.1143	.1191	.1237	.1281	.1323	.1362	.1398	.1432	.1462
7	.0640	.0686	.0732	.0778	.0824	.0869	.0914	.0959	.1002	.1044
8	.0328	.0360	.0393	.0428	.0463	.0500	.0537	.0575	.0614	.0653
9	.0150	.0168	.0188	.0209	.0232	.0255	.0280	.0307	.0334	.0363
10	.0061	.0071	.0081	.0092	.0104	.0118	.0132	.0147	.0164	.0181
11	.0023	.0027	.0032	.0037	.0043	.0049	.0056	.0064	.0073	.0082
12	.0008	.0009	.0011	.0014	.0016	.0019	.0022	.0026	.0030	.0034
13	.0002	.0003	.0004	.0005	.0006	.0007	.0008	.0009	.0011	.0013
14	.0001	.0001	.0001	.0001	.0002	.0002	.0003	.0003	.0004	.0005
15	.0000	.0000	.0000	.0000	.0001	.0001	.0001	.0001	.0001	.0002

x	5.1	5.2	5.3	5.4	5.5	μ 5.6	5.7	5.8	5.9	6.0
0	.0061	.0055	.0050	.0045	.0041	.0037	.0033	.0030	.0027	.0025
1	.0311	.0287	.0265	.0244	.0225	.0207	.0191	.0176	.0162	.0149
2	.0793	.0746	.0701	.0659	.0618	.0580	.0544	.0509	.0477	.0446
3	.1348	.1293	.1239	.1185	.1133	.1082	.1033	.0985	.0938	.0892
4	.1719	.1681	.1641	.1600	.1558	.1515	.1472	.1428	.1383	.1339
5	.1753	.1748	.1740	.1728	.1714	.1697	.1678	.1656	.1632	.1606
6	.1490	.1515	.1537	.1555	.1571	.1584	.1594	.1601	.1605	.1606
7	.1086	.1125	.1163	.1200	.1234	.1267	.1298	.1326	.1353	.1377
8	.0692	.0731	.0771	.0810	.0849	.0887	.0925	.0962	.0998	.1033
9	.0392	.0423	.0454	.0486	.0519	.0552	.0586	.0620	.0654	.0688
10	.0200	.0220	.0241	.0262	.0285	.0309	.0334	.0359	.0386	.0413
11	.0093	.0104	.0116	.0129	.0143	.0157	.0173	.0190	.0207	.0225
12	.0039	.0045	.0051	.0058	.0065	.0073	.0082	.0092	.0102	.0113
13	.0015	.0018	.0021	.0024	.0028	.0032	.0036	.0041	.0046	.0052
14	.0006	.0007	.0008	.0009	.0011	.0013	.0015	.0017	.0019	.0022
15	.0002	.0002	.0003	.0003	.0004	.0005	.0006	.0007	.0008	.0009
16	.0001	.0001	.0001	.0001	.0001	.0002	.0002	.0002	.0003	.0003
17	.0000	.0000	.0000	.0000	.0000	.0001	.0001	.0001	.0001	.0001

Appendix Table F (continued)

x	6.1	6.2	6.3	6.4	6.5	μ 6.6	6.7	6.8	6.9	7.0
0	.0022	.0020	.0018	.0017	.0015	.0014	.0012	.0011	.0010	.0009
1	.0137	.0126	.0116	.0106	.0098	.0090	.0082	.0076	.0070	.0064
2	.0417	.0390	.0364	.0340	.0318	.0296	.0276	.0258	.0240	.0223
3	.0848	.0806	.0765	.0726	.0688	.0652	.0617	.0584	.0552	.0521
4	.1294	.1249	.1205	.1162	.1118	.1076	.1034	.0992	.0952	.0912
5	.1579	.1549	.1519	.1487	.1454	.1420	.1385	.1349	.1314	.1277
6	.1605	.1601	.1595	.1586	.1575	.1562	.1546	.1529	.1511	.1490
7	.1399	.1418	.1435	.1450	.1462	.1472	.1480	.1486	.1489	.1490
8	.1066	.1099	.1130	.1160	.1188	.1215	.1240	.1263	.1284	.1304
9	.0723	.0757	.0791	.0825	.0858	.0891	.0923	.0954	.0985	.1014
10	.0441	.0469	.0498	.0528	.0558	.0588	.0618	.0649	.0679	.0710
11	.0245	.0265	.0285	.0307	.0330	.0353	.0377	.0401	.0426	.0452
12	.0124	.0137	.0150	.0164	.0179	.0194	.0210	.0227	.0245	.0264
13	.0058	.0065	.0073	.0081	.0089	.0098	.0108	.0119	.0130	.0142
14	.0025	.0029	.0033	.0037	.0041	.0046	.0052	.0058	.0064	.0071
15	.0010	.0012	.0014	.0016	.0018	.0020	.0023	.0026	.0029	.0033
16	.0004	.0005	.0005	.0006	.0007	.0008	.0010	.0011	.0013	.0014
17	.0001	.0002	.0002	.0002	.0003	.0003	.0004	.0004	.0005	.0006
18	.0000	.0001	.0001	.0001	.0001	.0001	.0001	.0002	.0002	.0002
19	.0000	.0000	.0000	.0000	.0000	.0000	.0000	.0001	.0001	.0001

x	7.1	7.2	7.3	7.4	7.5	μ 7.6	7.7	7.8	7.9	8.0
0	.0008	.0007	.0007	.0006	.0006	.0005	.0005	.0004	.0004	.0003
1	.0059	.0054	.0049	.0045	.0041	.0038	.0035	.0032	.0029	.0027
2	.0208	.0194	.0180	.0167	.0156	.0145	.0134	.0125	.0116	.0107
3	.0492	.0464	.0438	.0413	.0389	.0366	.0345	.0324	.0305	.0286
4	.0874	.0836	.0799	.0764	.0729	.0696	.0663	.0632	.0602	.0573
5	.1241	.1204	.1167	.1130	.1094	.1057	.1021	.0986	.0951	.0916
6	.1468	.1445	.1420	.1394	.1367	.1339	.1311	.1282	.1252	.1221
7	.1489	.1486	.1481	.1474	.1465	.1454	.1442	.1428	.1413	.1396
8	.1321	.1337	.1351	.1363	.1373	.1382	.1388	.1392	.1395	.1396
9	.1042	.1070	.1096	.1121	.1144	.1167	.1187	.1207	.1224	.1241
10	.0740	.0770	.0800	.0829	.0858	.0887	.0914	.0941	.0967	.0993
11	.0478	.0504	.0531	.0558	.0585	.0613	.0640	.0667	.0695	.0722
12	.0283	.0303	.0323	.0344	.0366	.0388	.0411	.0434	.0457	.0481
13	.0154	.0168	.0181	.0196	.0211	.0227	.0243	.0260	.0278	.0296
14	.0078	.0086	.0095	.0104	.0113	.0123	.0134	.0145	.0157	.0169
15	.0037	.0041	.0046	.0051	.0057	.0062	.0069	.0075	.0083	.0090
16	.0016	.0019	.0021	.0024	.0026	.0030	.0033	.0037	.0041	.0045
17	.0007	.0008	.0009	.0010	.0012	.0013	.0015	.0017	.0019	.0021
18	.0003	.0003	.0004	.0004	.0005	.0006	.0006	.0007	.0008	.0009
19	.0001	.0001	.0001	.0002	.0002	.0002	.0003	.0003	.0003	.0004
20	.0000	.0000	.0001	.0001	.0001	.0001	.0001	.0001	.0001	.0002
21	.0000	.0000	.0000	.0000	.0000	.0000	.0000	.0000	.0001	.0001

Appendix Table F (continued)

x	8.1	8.2	8.3	8.4	8.5	8.6	8.7	8.8	8.9	9.0
0	.0003	.0003	.0002	.0002	.0002	.0002	.0002	.0002	.0001	.0001
1	.0025	.0023	.0021	.0019	.0017	.0016	.0014	.0013	.0012	.0011
2	.0100	.0092	.0086	.0079	.0074	.0068	.0063	.0058	.0054	.0050
3	.0269	.0252	.0237	.0222	.0208	.0195	.0183	.0171	.0160	.0150
4	.0544	.0517	.0491	.0466	.0443	.0420	.0398	.0377	.0357	.0337
5	.0882	.0849	.0816	.0784	.0752	.0722	.0692	.0663	.0635	.0607
6	.1191	.1160	.1128	.1097	.1066	.1034	.1003	.0972	.0941	.0911
7	.1378	.1358	.1338	.1317	.1294	.1271	.1247	.1222	.1197	.1171
8	.1395	.1392	.1388	.1382	.1375	.1366	.1356	.1344	.1332	.1318
9	.1256	.1269	.1280	.1290	.1299	.1306	.1311	.1315	.1317	.1318
10	.1017	.1040	.1063	.1084	.1104	.1123	.1140	.1157	.1172	.1186
11	.0749	.0776	.0802	.0828	.0853	.0878	.0902	.0925	.0948	.0970
12	.0505	.0530	.0555	.0579	.0604	.0629	.0654	.0679	.0703	.0728
13	.0315	.0334	.0354	.0374	.0395	.0416	.0438	.0459	.0481	.0504
14	.0182	.0196	.0210	.0225	.0240	.0256	.0272	.0289	.0306	.0324
15	.0098	.0107	.0116	.0126	.0136	.0147	.0158	.0169	.0182	.0194
16	.0050	.0055	.0060	.0066	.0072	.0079	.0086	.0093	.0101	.0109
17	.0024	.0026	.0029	.0033	.0036	.0040	.0044	.0048	.0053	.0058
18	.0011	.0012	.0014	.0015	.0017	.0019	.0021	.0024	.0026	.0029
19	.0005	.0005	.0006	.0007	.0008	.0009	.0010	.0011	.0012	.0014
20	.0002	.0002	.0002	.0003	.0003	.0004	.0004	.0005	.0005	.0006
21	.0001	.0001	.0001	.0001	.0001	.0002	.0002	.0002	.0002	.0003
22	.0000	.0000	.0000	.0000	.0001	.0001	.0001	.0001	.0001	.0001

x	9.1	9.2	9.3	9.4	9.5	9.6	9.7	9.8	9.9	10
0	.0001	.0001	.0001	.0001	.0001	.0001	.0001	.0001	.0001	.0000
1	.0010	.0009	.0009	.0008	.0007	.0007	.0006	.0005	.0005	.0005
2	.0046	.0043	.0040	.0037	.0034	.0031	.0029	.0027	.0025	.0023
3	.0140	.0131	.0123	.0115	.0107	.0100	.0093	.0087	.0081	.0076
4	.0319	.0302	.0285	.0269	.0254	.0240	.0226	.0213	.0201	.0189
5	.0581	.0555	.0530	.0506	.0483	.0460	.0439	.0418	.0398	.0378
6	.0881	.0851	.0822	.0793	.0764	.0736	.0709	.0682	.0656	.0631
7	.1145	.1118	.1091	.1064	.1037	.1010	.0982	.0955	.0928	.0901
8	.1302	.1286	.1269	.1251	.1232	.1212	.1191	.1170	.1148	.1126
9	.1317	.1315	.1311	.1306	.1300	.1293	.1284	.1274	.1263	.1251
10	.1198	.1210	.1219	.1228	.1235	.1241	.1245	.1249	.1250	.1251
11	.0991	.1012	.1031	.1049	.1067	.1083	.1098	.1112	.1125	.1137
12	.0752	.0776	.0799	.0822	.0844	.0866	.0888	.0908	.0928	.0948
13	.0526	.0549	.0572	.0594	.0617	.0640	.0662	.0685	.0707	.0729
14	.0342	.0361	.0380	.0399	.0419	.0439	.0459	.0479	.0500	.0521
15	.0208	.0221	.0235	.0250	.0265	.0281	.0297	.0313	.0330	.0347
16	.0118	.0127	.0137	.0147	.0157	.0168	.0180	.0192	.0204	.0217
17	.0063	.0069	.0075	.0081	.0088	.0095	.0103	.0111	.0119	.0128
18	.0032	.0035	.0039	.0042	.0046	.0051	.0055	.0060	.0065	.0071
19	.0015	.0017	.0019	.0021	.0023	.0026	.0028	.0031	.0034	.0037
20	.0007	.0008	.0009	.0010	.0011	.0012	.0014	.0015	.0017	.0019
21	.0003	.0003	.0004	.0004	.0005	.0006	.0006	.0007	.0008	.0009
22	.0001	.0001	.0002	.0002	.0002	.0002	.0003	.0003	.0004	.0004
23	.0000	.0001	.0001	.0001	.0001	.0001	.0001	.0001	.0002	.0002
24	.0000	.0000	.0000	.0000	.0000	.0000	.0000	.0001	.0001	.0001

Appendix Table F *(continued)*

x	11	12	13	14	15	μ 16	17	18	19	20
0	.0000	.0000	.0000	.0000	.0000	.0000	.0000	.0000	.0000	.0000
1	.0002	.0001	.0000	.0000	.0000	.0000	.0000	.0000	.0000	.0000
2	.0010	.0004	.0002	.0001	.0000	.0000	.0000	.0000	.0000	.0000
3	.0037	.0018	.0008	.0004	.0002	.0001	.0000	.0000	.0000	.0000
4	.0102	.0053	.0027	.0013	.0006	.0003	.0001	.0001	.0000	.0000
5	.0224	.0127	.0070	.0037	.0019	.0010	.0005	.0002	.0001	.0001
6	.0411	.0255	.0152	.0087	.0048	.0026	.0014	.0007	.0004	.0002
7	.0646	.0437	.0281	.0174	.0104	.0060	.0034	.0018	.0010	.0005
8	.0888	.0655	.0457	.0304	.0194	.0120	.0072	.0042	.0024	.0013
9	.1085	.0874	.0661	.0473	.0324	.0213	.0135	.0083	.0050	.0029
10	.1194	.1048	.0859	.0663	.0486	.0341	.0230	.0150	.0095	.0058
11	.1194	.1144	.1015	.0844	.0663	.0496	.0355	.0245	.0164	.0106
12	.1094	.1144	.1099	.0984	.0829	.0661	.0504	.0368	.0259	.0176
13	.0926	.1056	.1099	.1060	.0956	.0814	.0658	.0509	.0378	.0271
14	.0728	.0905	.1021	.1060	.1024	.0930	.0800	.0655	.0514	.0387
15	.0534	.0724	.0885	.0989	.1024	.0992	.0906	.0786	.0650	.0516
16	.0367	.0543	.0719	.0866	.0960	.0992	.0963	.0884	.0772	.0646
17	.0237	.0383	.0550	.0713	.0847	.0934	.0963	.0936	.0863	.0760
18	.0145	.0256	.0397	.0554	.0706	.0830	.0909	.0936	.0911	.0844
19	.0084	.0161	.0272	.0409	.0557	.0699	.0814	.0887	.0911	.0888
20	.0046	.0097	.0177	.0286	.0418	.0559	.0692	.0798	.0866	.0888
21	.0024	.0055	.0109	.0191	.0299	.0426	.0560	.0684	.0783	.0846
22	.0012	.0030	.0065	.0121	.0204	.0310	.0433	.0560	.0676	.0769
23	.0006	.0016	.0037	.0074	.0133	.0216	.0320	.0438	.0559	.0669
24	.0003	.0008	.0020	.0043	.0083	.0144	.0226	.0328	.0442	.0557
25	.0001	.0004	.0010	.0024	.0050	.0092	.0154	.0237	.0336	.0446
26	.0000	.0002	.0005	.0013	.0029	.0057	.0101	.0164	.0246	.0343
27	.0000	.0001	.0002	.0007	.0016	.0034	.0063	.0109	.0173	.0254
28	.0000	.0000	.0001	.0003	.0009	.0019	.0038	.0070	.0117	.0181
29	.0000	.0000	.0001	.0002	.0004	.0011	.0023	.0044	.0077	.0125
30	.0000	.0000	.0000	.0001	.0002	.0006	.0013	.0026	.0049	.0083
31	.0000	.0000	.0000	.0000	.0001	.0003	.0007	.0015	.0030	.0054
32	.0000	.0000	.0000	.0000	.0001	.0001	.0004	.0009	.0018	.0034
33	.0000	.0000	.0000	.0000	.0000	.0001	.0002	.0005	.0010	.0020
34	.0000	.0000	.0000	.0000	.0000	.0000	.0001	.0002	.0006	.0012
35	.0000	.0000	.0000	.0000	.0000	.0000	.0000	.0001	.0003	.0007
36	.0000	.0000	.0000	.0000	.0000	.0000	.0000	.0001	.0002	.0004
37	.0000	.0000	.0000	.0000	.0000	.0000	.0000	.0000	.0001	.0002
38	.0000	.0000	.0000	.0000	.0000	.0000	.0000	.0000	.0000	.0001
39	.0000	.0000	.0000	.0000	.0000	.0000	.0000	.0000	.0000	.0001

Appendix Table G
Poisson Probabilities for Cumulative Values of x

Entries in this table give the probability of *x* or fewer occurrences for a Poisson process with a mean of μ. For example, when μ = 2.9, the probability of 4 or fewer occurrences is .8318.

x	0.1	0.2	0.3	0.4	0.5	μ 0.6	0.7	0.8	0.9	1.0
0	0.9048	0.8187	0.7408	0.6703	0.6065	0.5488	0.4966	0.4493	0.4066	0.3679
1	0.9953	0.9825	0.9631	0.9384	0.9098	0.8781	0.8442	0.8088	0.7725	0.7358
2	0.9998	0.9989	0.9964	0.9921	0.9856	0.9769	0.9659	0.9526	0.9371	0.9197
3	1.0000	0.9999	0.9997	0.9992	0.9982	0.9966	0.9942	0.9909	0.9865	0.9810
4	1.0000	1.0000	1.0000	0.9999	0.9998	0.9996	0.9992	0.9986	0.9977	0.9963
5	1.0000	1.0000	1.0000	1.0000	1.0000	1.0000	0.9999	0.9998	0.9997	0.9994
6	1.0000	1.0000	1.0000	1.0000	1.0000	1.0000	1.0000	1.0000	1.0000	0.9999
7	1.0000	1.0000	1.0000	1.0000	1.0000	1.0000	1.0000	1.0000	1.0000	1.0000

x	1.1	1.2	1.3	1.4	1.5	μ 1.6	1.7	1.8	1.9	2.0
0	0.3329	0.3012	0.2725	0.2466	0.2231	0.2019	0.1827	0.1653	0.1496	0.1353
1	0.6990	0.6626	0.6268	0.5918	0.5578	0.5249	0.4932	0.4628	0.4338	0.4060
2	0.9004	0.8795	0.8571	0.8335	0.8088	0.7834	0.7572	0.7306	0.7037	0.6767
3	0.9743	0.9662	0.9569	0.9463	0.9344	0.9212	0.9068	0.8913	0.8747	0.8571
4	0.9946	0.9923	0.9893	0.9857	0.9814	0.9763	0.9704	0.9636	0.9559	0.9473
5	0.9990	0.9985	0.9978	0.9968	0.9955	0.9940	0.9920	0.9896	0.9868	0.9834
6	0.9999	0.9997	0.9996	0.9994	0.9991	0.9987	0.9981	0.9974	0.9966	0.9955
7	1.0000	1.0000	0.9999	0.9999	0.9998	0.9997	0.9996	0.9994	0.9992	0.9989
8	1.0000	1.0000	1.0000	1.0000	1.0000	1.0000	0.9999	0.9999	0.9998	0.9998
9	1.0000	1.0000	1.0000	1.0000	1.0000	1.0000	1.0000	1.0000	1.0000	1.0000

x	2.1	2.2	2.3	2.4	2.5	μ 2.6	2.7	2.8	2.9	3.0
0	0.1225	0.1108	0.1003	0.0907	0.0821	0.0743	0.0672	0.0608	0.0550	0.0498
1	0.3796	0.3546	0.3309	0.3084	0.2873	0.2674	0.2487	0.2311	0.2146	0.1991
2	0.6496	0.6227	0.5960	0.5697	0.5438	0.5184	0.4936	0.4695	0.4460	0.4232
3	0.8386	0.8194	0.7993	0.7787	0.7576	0.7360	0.7141	0.6919	0.6696	0.6472
4	0.9379	0.9275	0.9162	0.9041	0.8912	0.8774	0.8629	0.8477	0.8318	0.8153
5	0.9796	0.9751	0.9700	0.9643	0.9580	0.9510	0.9433	0.9349	0.9258	0.9161
6	0.9941	0.9925	0.9906	0.9884	0.9858	0.9828	0.9794	0.9756	0.9713	0.9665
7	0.9985	0.9980	0.9974	0.9967	0.9958	0.9947	0.9934	0.9919	0.9901	0.9881
8	0.9997	0.9995	0.9994	0.9991	0.9989	0.9985	0.9981	0.9976	0.9969	0.9962
9	0.9999	0.9999	0.9999	0.9998	0.9997	0.9996	0.9995	0.9993	0.9991	0.9989
10	1.0000	1.0000	1.0000	1.0000	0.9999	0.9999	0.9999	0.9998	0.9998	0.9997
11	1.0000	1.0000	1.0000	1.0000	1.0000	1.0000	1.0000	1.0000	0.9999	0.9999
12	1.0000	1.0000	1.0000	1.0000	1.0000	1.0000	1.0000	1.0000	1.0000	1.0000

x	3.1	3.2	3.3	3.4	3.5	μ 3.6	3.7	3.8	3.9	4.0
0	0.0450	0.0408	0.0369	0.0334	0.0302	0.0273	0.0247	0.0224	0.0202	0.0183
1	0.1847	0.1712	0.1586	0.1468	0.1359	0.1257	0.1162	0.1074	0.0992	0.0916
2	0.4012	0.3799	0.3594	0.3397	0.3208	0.3027	0.2854	0.2689	0.2531	0.2381
3	0.6248	0.6025	0.5803	0.5584	0.5366	0.5152	0.4942	0.4735	0.4533	0.4335
4	0.7982	0.7806	0.7626	0.7442	0.7254	0.7064	0.6872	0.6678	0.6484	0.6288
5	0.9057	0.8946	0.8829	0.8705	0.8576	0.8441	0.8301	0.8156	0.8006	0.7851
6	0.9612	0.9554	0.9490	0.9421	0.9347	0.9267	0.9182	0.9091	0.8995	0.8893
7	0.9858	0.9832	0.9802	0.9769	0.9733	0.9692	0.9648	0.9599	0.9546	0.9489
8	0.9953	0.9943	0.9931	0.9917	0.9901	0.9883	0.9863	0.9840	0.9815	0.9786
9	0.9986	0.9982	0.9978	0.9973	0.9967	0.9960	0.9952	0.9942	0.9931	0.9919
10	0.9996	0.9995	0.9994	0.9992	0.9990	0.9987	0.9984	0.9981	0.9977	0.9972
11	0.9999	0.9999	0.9998	0.9998	0.9997	0.9996	0.9995	0.9994	0.9993	0.9991
12	1.0000	1.0000	1.0000	0.9999	0.9999	0.9999	0.9999	0.9998	0.9998	0.9997
13	1.0000	1.0000	1.0000	1.0000	1.0000	1.0000	1.0000	1.0000	0.9999	0.9999
14	1.0000	1.0000	1.0000	1.0000	1.0000	1.0000	1.0000	1.0000	1.0000	1.0000

Source: Lawrence L. Lapin. *Statistics for Modern Business Decisions*, 3rd ed. (New York: Harcourt, Brace, Jovanovich, 1982), pp. 843–47.

Appendix Table G *(continued)*

x	4.1	4.2	4.3	4.4	4.5	μ 4.6	4.7	4.8	4.9	5.0
0	0.0166	0.0150	0.0136	0.0123	0.0111	0.0101	0.0091	0.0082	0.0074	0.0067
1	0.0845	0.0780	0.0719	0.0663	0.0611	0.0563	0.0518	0.0477	0.0439	0.0404
2	0.2238	0.2102	0.1974	0.1851	0.1736	0.1626	0.1523	0.1425	0.1333	0.1247
3	0.4142	0.3954	0.3772	0.3595	0.3423	0.3257	0.3097	0.2942	0.2793	0.2650
4	0.6093	0.5898	0.5704	0.5512	0.5321	0.5132	0.4946	0.4763	0.4582	0.4405
5	0.7693	0.7531	0.7367	0.7199	0.7029	0.6858	0.6684	0.6510	0.6335	0.6160
6	0.8786	0.8675	0.8558	0.8436	0.8311	0.8180	0.8046	0.7908	0.7767	0.7622
7	0.9427	0.9361	0.9290	0.9214	0.9134	0.9049	0.8960	0.8867	0.8769	0.8666
8	0.9755	0.9721	0.9683	0.9642	0.9597	0.9549	0.9497	0.9442	0.9382	0.9319
9	0.9905	0.9889	0.9871	0.9851	0.9829	0.9805	0.9778	0.9749	0.9717	0.9682
10	0.9966	0.9959	0.9952	0.9943	0.9933	0.9922	0.9910	0.9896	0.9880	0.9863
11	0.9989	0.9986	0.9983	0.9980	0.9976	0.9971	0.9966	0.9960	0.9953	0.9945
12	0.9997	0.9996	0.9995	0.9993	0.9992	0.9990	0.9988	0.9986	0.9983	0.9980
13	0.9999	0.9999	0.9998	0.9998	0.9997	0.9997	0.9996	0.9995	0.9994	0.9993
14	1.0000	1.0000	1.0000	0.9999	0.9999	0.9999	0.9999	0.9999	0.9998	0.9998
15	1.0000	1.0000	1.0000	1.0000	1.0000	1.0000	1.0000	1.0000	0.9999	0.9999
16	1.0000	1.0000	1.0000	1.0000	1.0000	1.0000	1.0000	1.0000	1.0000	1.0000

x	5.1	5.2	5.3	5.4	5.5	μ 5.6	5.7	5.8	5.9	6.0
0	0.0061	0.0055	0.0050	0.0045	0.0041	0.0037	0.0033	0.0030	0.0027	0.0025
1	0.0372	0.0342	0.0314	0.0289	0.0266	0.0244	0.0224	0.0206	0.0189	0.0174
2	0.1165	0.1088	0.1016	0.0948	0.0884	0.0824	0.0768	0.0715	0.0666	0.0620
3	0.2513	0.2381	0.2254	0.2133	0.2017	0.1906	0.1801	0.1700	0.1604	0.1512
4	0.4231	0.4061	0.3895	0.3733	0.3575	0.3422	0.3272	0.3127	0.2987	0.2851
5	0.5984	0.5809	0.5635	0.5461	0.5289	0.5119	0.4950	0.4783	0.4619	0.4457
6	0.7474	0.7324	0.7171	0.7017	0.6860	0.6703	0.6544	0.6384	0.6224	0.6063
7	0.8560	0.8449	0.8335	0.8217	0.8095	0.7970	0.7842	0.7710	0.7576	0.7440
8	0.9252	0.9181	0.9106	0.9026	0.8944	0.8857	0.8766	0.8672	0.8574	0.8472
9	0.9644	0.9603	0.9559	0.9512	0.9462	0.9409	0.9352	0.9292	0.9228	0.9161
10	0.9844	0.9823	0.9800	0.9775	0.9747	0.9718	0.9686	0.9651	0.9614	0.9574
11	0.9937	0.9927	0.9916	0.9904	0.9890	0.9875	0.9859	0.9840	0.9821	0.9799
12	0.9976	0.9972	0.9967	0.9962	0.9955	0.9949	0.9941	0.9932	0.9922	0.9912
13	0.9992	0.9990	0.9988	0.9986	0.9983	0.9980	0.9977	0.9973	0.9969	0.9964
14	0.9997	0.9997	0.9996	0.9995	0.9994	0.9993	0.9991	0.9990	0.9988	0.9986
15	0.9999	0.9999	0.9999	0.9998	0.9998	0.9998	0.9997	0.9996	0.9996	0.9995
16	1.0000	1.0000	1.0000	0.9999	0.9999	0.9999	0.9999	0.9999	0.9999	0.9998
17	1.0000	1.0000	1.0000	1.0000	1.0000	1.0000	1.0000	1.0000	1.0000	0.9999
18	1.0000	1.0000	1.0000	1.0000	1.0000	1.0000	1.0000	1.0000	1.0000	1.0000

Appendix Table G *(continued)*

x	6.1	6.2	6.3	6.4	6.5	6.6	6.7	6.8	6.9	7.0
0	0.0022	0.0020	0.0018	0.0017	0.0015	0.0014	0.0012	0.0011	0.0010	0.0009
1	0.0159	0.0146	0.0134	0.0123	0.0113	0.0103	0.0095	0.0087	0.0080	0.0073
2	0.0577	0.0536	0.0498	0.0463	0.0430	0.0400	0.0371	0.0344	0.0320	0.0296
3	0.1425	0.1342	0.1264	0.1189	0.1119	0.1052	0.0988	0.0928	0.0871	0.0818
4	0.2719	0.2592	0.2469	0.2351	0.2237	0.2127	0.2022	0.1920	0.1823	0.1730
5	0.4298	0.4141	0.3988	0.3837	0.3690	0.3547	0.3407	0.3270	0.3137	0.3007
6	0.5902	0.5742	0.5582	0.5423	0.5265	0.5108	0.4953	0.4799	0.4647	0.4497
7	0.7301	0.7160	0.7018	0.6873	0.6728	0.6581	0.6433	0.6285	0.6136	0.5987
8	0.8367	0.8259	0.8148	0.8033	0.7916	0.7796	0.7673	0.7548	0.7420	0.7291
9	0.9090	0.9016	0.8939	0.8858	0.8774	0.8686	0.8596	0.8502	0.8405	0.8305
10	0.9531	0.9486	0.9437	0.9386	0.9332	0.9274	0.9214	0.9151	0.9084	0.9015
11	0.9776	0.9750	0.9723	0.9693	0.9661	0.9627	0.9591	0.9552	0.9510	0.9466
12	0.9900	0.9887	0.9873	0.9857	0.9840	0.9821	0.9801	0.9779	0.9755	0.9730
13	0.9958	0.9952	0.9945	0.9937	0.9929	0.9920	0.9909	0.9898	0.9885	0.9872
14	0.9984	0.9981	0.9978	0.9974	0.9970	0.9966	0.9961	0.9956	0.9950	0.9943
15	0.9994	0.9993	0.9992	0.9990	0.9988	0.9986	0.9984	0.9982	0.9979	0.9976
16	0.9998	0.9997	0.9997	0.9996	0.9996	0.9995	0.9994	0.9993	0.9992	0.9990
17	0.9999	0.9999	0.9999	0.9999	0.9998	0.9998	0.9998	0.9997	0.9997	0.9996
18	1.0000	1.0000	1.0000	1.0000	0.9999	0.9999	0.9999	0.9999	0.9999	0.9999
19	1.0000	1.0000	1.0000	1.0000	1.0000	1.0000	1.0000	1.0000	1.0000	0.9999
20	1.0000	1.0000	1.0000	1.0000	1.0000	1.0000	1.0000	1.0000	1.0000	1.0000

x	7.1	7.2	7.3	7.4	7.5	7.6	7.7	7.8	7.9	8.0
0	0.0008	0.0007	0.0007	0.0006	0.0006	0.0005	0.0005	0.0004	0.0004	0.0003
1	0.0067	0.0061	0.0056	0.0051	0.0047	0.0043	0.0039	0.0036	0.0033	0.0030
2	0.0275	0.0255	0.0236	0.0219	0.0203	0.0188	0.0174	0.0161	0.0149	0.0138
3	0.0767	0.0719	0.0674	0.0632	0.0591	0.0554	0.0518	0.0485	0.0453	0.0424
4	0.1641	0.1555	0.1473	0.1395	0.1321	0.1249	0.1181	0.1117	0.1055	0.0996
5	0.2881	0.2759	0.2640	0.2526	0.2414	0.2307	0.2203	0.2103	0.2006	0.1912
6	0.4349	0.4204	0.4060	0.3920	0.3782	0.3646	0.3514	0.3384	0.3257	0.3134
7	0.5838	0.5689	0.5541	0.5393	0.5246	0.5100	0.4956	0.4812	0.4670	0.4530
8	0.7160	0.7027	0.6892	0.6757	0.6620	0.6482	0.6343	0.6204	0.6065	0.5926
9	0.8202	0.8096	0.7988	0.7877	0.7764	0.7649	0.7531	0.7411	0.7290	0.7166
10	0.8942	0.8867	0.8788	0.8707	0.8622	0.8535	0.8445	0.8352	0.8257	0.8159
11	0.9420	0.9371	0.9319	0.9265	0.9208	0.9148	0.9085	0.9020	0.8952	0.8881
12	0.9703	0.9673	0.9642	0.9609	0.9573	0.9536	0.9496	0.9453	0.9409	0.9362
13	0.9857	0.9841	0.9824	0.9805	0.9784	0.9762	0.9739	0.9714	0.9687	0.9658
14	0.9935	0.9927	0.9918	0.9908	0.9897	0.9886	0.9873	0.9859	0.9844	0.9827
15	0.9972	0.9968	0.9964	0.9959	0.9954	0.9948	0.9941	0.9934	0.9926	0.9918
16	0.9989	0.9987	0.9985	0.9983	0.9980	0.9978	0.9974	0.9971	0.9967	0.9963
17	0.9996	0.9995	0.9994	0.9993	0.9992	0.9991	0.9989	0.9988	0.9986	0.9984
18	0.9998	0.9998	0.9998	0.9997	0.9997	0.9996	0.9996	0.9995	0.9994	0.9993
19	0.9999	0.9999	0.9999	0.9999	0.9999	0.9999	0.9998	0.9998	0.9998	0.9997
20	1.0000	1.0000	1.0000	1.0000	1.0000	0.9999	0.9999	0.9999	0.9999	0.9999
21	1.0000	1.0000	1.0000	1.0000	1.0000	1.0000	1.0000	1.0000	1.0000	1.0000

Appendix Table G (continued)

					μ					
x	8.1	8.2	8.3	8.4	8.5	8.6	8.7	8.8	8.9	9.0
0	0.0003	0.0003	0.0002	0.0002	0.0002	0.0002	0.0002	0.0002	0.0001	0.0001
1	0.0028	0.0025	0.0023	0.0021	0.0019	0.0018	0.0016	0.0015	0.0014	0.0012
2	0.0127	0.0118	0.0109	0.0100	0.0093	0.0086	0.0079	0.0073	0.0068	0.0062
3	0.0396	0.0370	0.0346	0.0323	0.0301	0.0281	0.0262	0.0244	0.0228	0.0212
4	0.0941	0.0887	0.0837	0.0789	0.0744	0.0701	0.0660	0.0621	0.0584	0.0550
5	0.1822	0.1736	0.1653	0.1573	0.1496	0.1422	0.1352	0.1284	0.1219	0.1157
6	0.3013	0.2896	0.2781	0.2670	0.2562	0.2457	0.2355	0.2256	0.2160	0.2068
7	0.4391	0.4254	0.4119	0.3987	0.3856	0.3728	0.3602	0.3478	0.3357	0.3239
8	0.5786	0.5647	0.5508	0.5369	0.5231	0.5094	0.4958	0.4823	0.4689	0.4557
9	0.7041	0.6915	0.6788	0.6659	0.6530	0.6400	0.6269	0.6137	0.6006	0.5874
10	0.8058	0.7955	0.7850	0.7743	0.7634	0.7522	0.7409	0.7294	0.7178	0.7060
11	0.8807	0.8731	0.8652	0.8571	0.8487	0.8400	0.8311	0.8220	0.8126	0.8030
12	0.9313	0.9261	0.9207	0.9150	0.9091	0.9029	0.8965	0.8898	0.8829	0.8758
13	0.9628	0.9595	0.9561	0.9524	0.9486	0.9445	0.9403	0.9358	0.9311	0.9262
14	0.9810	0.9791	0.9771	0.9749	0.9726	0.9701	0.9675	0.9647	0.9617	0.9585
15	0.9908	0.9898	0.9887	0.9875	0.9862	0.9847	0.9832	0.9816	0.9798	0.9780
16	0.9958	0.9953	0.9947	0.9941	0.9934	0.9926	0.9918	0.9909	0.9899	0.9889
17	0.9982	0.9979	0.9976	0.9973	0.9970	0.9966	0.9962	0.9957	0.9952	0.9947
18	0.9992	0.9991	0.9990	0.9989	0.9987	0.9985	0.9983	0.9981	0.9978	0.9976
19	0.9997	0.9996	0.9996	0.9995	0.9995	0.9994	0.9993	0.9992	0.9991	0.9989
20	0.9999	0.9999	0.9998	0.9998	0.9998	0.9997	0.9997	0.9997	0.9996	0.9996
21	1.0000	0.9999	0.9999	0.9999	0.9999	0.9999	0.9999	0.9999	0.9998	0.9998
22	1.0000	1.0000	1.0000	1.0000	1.0000	1.0000	1.0000	0.9999	0.9999	0.9999
23	1.0000	1.0000	1.0000	1.0000	1.0000	1.0000	1.0000	1.0000	1.0000	1.0000

					μ					
x	9.1	9.2	9.3	9.4	9.5	9.6	9.7	9.8	9.9	10.0
0	0.0001	0.0001	0.0001	0.0001	0.0001	0.0001	0.0001	0.0001	0.0001	0.0000
1	0.0011	0.0010	0.0009	0.0009	0.0008	0.0007	0.0007	0.0006	0.0005	0.0005
2	0.0058	0.0053	0.0049	0.0045	0.0042	0.0038	0.0035	0.0033	0.0030	0.0028
3	0.0198	0.0184	0.0172	0.0160	0.0149	0.0138	0.0129	0.0120	0.0111	0.0103
4	0.0517	0.0486	0.0456	0.0429	0.0403	0.0378	0.0355	0.0333	0.0312	0.0293
5	0.1098	0.1041	0.0987	0.0935	0.0885	0.0838	0.0793	0.0750	0.0710	0.0671
6	0.1978	0.1892	0.1808	0.1727	0.1650	0.1575	0.1502	0.1433	0.1366	0.1301
7	0.3123	0.3010	0.2900	0.2792	0.2687	0.2584	0.2485	0.2388	0.2294	0.2202
8	0.4426	0.4296	0.4168	0.4042	0.3918	0.3796	0.3676	0.3558	0.3442	0.3328
9	0.5742	0.5611	0.5479	0.5349	0.5218	0.5089	0.4960	0.4832	0.4705	0.4579
10	0.6941	0.6820	0.6699	0.6576	0.6453	0.6330	0.6205	0.6080	0.5955	0.5830
11	0.7932	0.7832	0.7730	0.7626	0.7520	0.7412	0.7303	0.7193	0.7081	0.6968
12	0.8684	0.8607	0.8529	0.8448	0.8364	0.8279	0.8191	0.8101	0.8009	0.7916
13	0.9210	0.9156	0.9100	0.9042	0.8981	0.8919	0.8853	0.8786	0.8716	0.8645
14	0.9552	0.9517	0.9480	0.9441	0.9400	0.9357	0.9312	0.9265	0.9216	0.9165
15	0.9760	0.9738	0.9715	0.9691	0.9665	0.9638	0.9609	0.9579	0.9546	0.9513
16	0.9878	0.9865	0.9852	0.9838	0.9823	0.9806	0.9789	0.9770	0.9751	0.9730
17	0.9941	0.9934	0.9927	0.9919	0.9911	0.9902	0.9892	0.9881	0.9869	0.9857
18	0.9973	0.9969	0.9966	0.9962	0.9957	0.9952	0.9947	0.9941	0.9935	0.9928
19	0.9988	0.9986	0.9985	0.9983	0.9980	0.9978	0.9975	0.9972	0.9969	0.9965
20	0.9995	0.9994	0.9993	0.9992	0.9991	0.9990	0.9989	0.9987	0.9986	0.9984
21	0.9998	0.9998	0.9997	0.9997	0.9996	0.9996	0.9995	0.9995	0.9994	0.9993
22	0.9999	0.9999	0.9999	0.9999	0.9998	0.9998	0.9998	0.9998	0.9997	0.9997
23	1.0000	1.0000	1.0000	0.9999	0.9999	0.9999	0.9999	0.9999	0.9999	0.9999
24	1.0000	1.0000	1.0000	1.0000	1.0000	1.0000	1.0000	1.0000	0.9999	0.9999
25	1.0000	1.0000	1.0000	1.0000	1.0000	1.0000	1.0000	1.0000	1.0000	1.0000

Appendix Table G (continued)

x	11.0	12.0	13.0	14.0	15.0	μ 16.0	17.0	18.0	19.0	20.0
0	0.0000	0.0000	0.0000	0.0000	0.0000	0.0000	0.0000	0.0000	0.0000	0.0000
1	0.0002	0.0001	0.0000	0.0000	0.0000	0.0000	0.0000	0.0000	0.0000	0.0000
2	0.0012	0.0005	0.0002	0.0001	0.0000	0.0000	0.0000	0.0000	0.0000	0.0000
3	0.0049	0.0023	0.0011	0.0005	0.0002	0.0001	0.0000	0.0000	0.0000	0.0000
4	0.0151	0.0076	0.0037	0.0018	0.0009	0.0004	0.0002	0.0001	0.0000	0.0000
5	0.0375	0.0203	0.0107	0.0055	0.0028	0.0014	0.0007	0.0003	0.0002	0.0001
6	0.0786	0.0458	0.0259	0.0142	0.0076	0.0040	0.0021	0.0010	0.0005	0.0003
7	0.1432	0.0895	0.0540	0.0316	0.0180	0.0100	0.0054	0.0029	0.0015	0.0008
8	0.2320	0.1550	0.0998	0.0621	0.0374	0.0220	0.0126	0.0071	0.0039	0.0021
9	0.3405	0.2424	0.1658	0.1094	0.0699	0.0433	0.0261	0.0154	0.0089	0.0050
10	0.4599	0.3472	0.2517	0.1757	0.1185	0.0774	0.0491	0.0304	0.0183	0.0108
11	0.5793	0.4616	0.3532	0.2600	0.1847	0.1270	0.0847	0.0549	0.0347	0.0214
12	0.6887	0.5760	0.4631	0.3585	0.2676	0.1931	0.1350	0.0917	0.0606	0.0390
13	0.7813	0.6815	0.5730	0.4644	0.3632	0.2745	0.2009	0.1426	0.0984	0.0661
14	0.8540	0.7720	0.6751	0.5704	0.4656	0.3675	0.2808	0.2081	0.1497	0.1049
15	0.9074	0.8444	0.7636	0.6694	0.5681	0.4667	0.3714	0.2866	0.2148	0.1565
16	0.9441	0.8987	0.8355	0.7559	0.6641	0.5660	0.4677	0.3750	0.2920	0.2211
17	0.9678	0.9370	0.8905	0.8272	0.7489	0.6593	0.5640	0.4686	0.3784	0.2970
18	0.9823	0.9626	0.9302	0.8826	0.8195	0.7423	0.6549	0.5622	0.4695	0.3814
19	0.9907	0.9787	0.9573	0.9235	0.8752	0.8122	0.7363	0.6509	0.5606	0.4703
20	0.9953	0.9884	0.9750	0.9521	0.9170	0.8682	0.8055	0.7307	0.6472	0.5591
21	0.9977	0.9939	0.9859	0.9711	0.9469	0.9108	0.8615	0.7991	0.7255	0.6437
22	0.9989	0.9969	0.9924	0.9833	0.9672	0.9418	0.9047	0.8551	0.7931	0.7206
23	0.9995	0.9985	0.9960	0.9907	0.9805	0.9633	0.9367	0.8989	0.8490	0.7875
24	0.9998	0.9993	0.9980	0.9950	0.9888	0.9777	0.9593	0.9317	0.8933	0.8432
25	0.9999	0.9997	0.9990	0.9974	0.9938	0.9869	0.9747	0.9554	0.9269	0.8878
26	1.0000	0.9999	0.9995	0.9987	0.9967	0.9925	0.9848	0.9718	0.9514	0.9221
27	1.0000	0.9999	0.9998	0.9994	0.9983	0.9959	0.9912	0.9827	0.9687	0.9475
28	1.0000	1.0000	0.9999	0.9997	0.9991	0.9978	0.9950	0.9897	0.9805	0.9657
29	1.0000	1.0000	1.0000	0.9999	0.9996	0.9989	0.9973	0.9940	0.9881	0.9782
30	1.0000	1.0000	1.0000	0.9999	0.9998	0.9994	0.9985	0.9967	0.9930	0.9865
31	1.0000	1.0000	1.0000	1.0000	0.9999	0.9997	0.9992	0.9982	0.9960	0.9919
32	1.0000	1.0000	1.0000	1.0000	0.9999	0.9999	0.9996	0.9990	0.9978	0.9953
33	1.0000	1.0000	1.0000	1.0000	1.0000	0.9999	0.9998	0.9995	0.9988	0.9973
34	1.0000	1.0000	1.0000	1.0000	1.0000	1.0000	0.9999	0.9997	0.9994	0.9985
35	1.0000	1.0000	1.0000	1.0000	1.0000	1.0000	0.9999	0.9999	0.9997	0.9992
36	1.0000	1.0000	1.0000	1.0000	1.0000	1.0000	1.0000	0.9999	0.9998	0.9996
37	1.0000	1.0000	1.0000	1.0000	1.0000	1.0000	1.0000	1.0000	0.9999	0.9998
38	1.0000	1.0000	1.0000	1.0000	1.0000	1.0000	1.0000	1.0000	1.0000	0.9999
39	1.0000	1.0000	1.0000	1.0000	1.0000	1.0000	1.0000	1.0000	1.0000	0.9999
40	1.0000	1.0000	1.0000	1.0000	1.0000	1.0000	1.0000	1.0000	1.0000	1.0000

Appendix Table H
Standard Normal Curve Areas

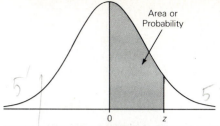

Area or Probability

Entries in this table give the area under the curve between the mean and z standard deviations above the mean. For example, for z = 2.25, the area under the curve between the mean and z is .4878

z	.00	.01	.02	.03	.04	.05	.06	.07	.08	.09
.0	.0000	.0040	.0080	.0120	.0160	.0199	.0239	.0279	.0319	.0359
.1	.0398	.0438	.0478	.0517	.0557	.0596	.0636	.0675	.0714	.0753
.2	.0793	.0832	.0871	.0910	.0948	.0987	.1026	.1064	.1103	.1141
.3	.1179	.1217	.1255	.1293	.1331	.1368	.1406	.1443	.1480	.1517
.4	.1554	.1591	.1628	.1664	.1700	.1736	.1772	.1808	.1844	.1879
.5	.1915	.1950	.1985	.2019	.2054	.2088	.2123	.2157	.2190	.2224
.6	.2257	.2291	.2324	.2357	.2389	.2422	.2454	.2486	.2518	.2549
.7	.2580	.2612	.2642	.2673	.2704	.2734	.2764	.2794	.2823	.2852
.8	.2881	.2910	.2939	.2967	.2995	.3023	.3051	.3078	.3106	.3133
.9	.3159	.3186	.3212	.3238	.3264	.3289	.3315	.3340	.3365	.3389
1.0	.3413	.3438	.3461	.3485	.3508	.3531	.3554	.3577	.3599	.3621
1.1	.3643	.3665	.3686	.3708	.3729	.3749	.3770	.3790	.3810	.3830
1.2	.3849	.3869	.3888	.3907	.3925	.3944	.3962	.3980	.3997	.4015
1.3	.4032	.4049	.4066	.4082	.4099	.4115	.4131	.4147	.4162	.4177
1.4	.4192	.4207	.4222	.4236	.4251	.4265	.4279	.4292	.4306	.4319
1.5	.4332	.4345	.4357	.4370	.4382	.4394	.4406	.4418	.4429	.4441
1.6	.4452	.4463	.4474	.4484	.4495	.4505	.4515	.4525	.4535	.4545
1.7	.4554	.4564	.4573	.4582	.4591	.4599	.4608	.4616	.4625	.4633
1.8	.4641	.4649	.4656	.4664	.4671	.4678	.4686	.4693	.4699	.4706
1.9	.4713	.4719	.4726	.4732	.4738	.4744	.4750	.4756	.4761	.4767
2.0	.4772	.4778	.4783	.4788	.4793	.4798	.4803	.4808	.4812	.4817
2.1	.4821	.4826	.4830	.4834	.4838	.4842	.4846	.4850	.4854	.4857
2.2	.4861	.4864	.4868	.4871	.4875	.4878	.4881	.4884	.4887	.4890
2.3	.4893	.4896	.4898	.4901	.4904	.4906	.4909	.4911	.4913	.4916
2.4	.4918	.4920	.4922	.4925	.4927	.4929	.4931	.4932	.4934	.4936
2.5	.4938	.4940	.4941	.4943	.4945	.4946	.4948	.4949	.4951	.4952
2.6	.4953	.4955	.4956	.4957	.4959	.4960	.4961	.4962	.4963	.4964
2.7	.4965	.4966	.4967	.4968	.4969	.4970	.4971	.4972	.4973	.4974
2.8	.4974	.4975	.4976	.4977	.4977	.4978	.4979	.4979	.4980	.4981
2.9	.4981	.4982	.4982	.4983	.4984	.4984	.4985	.4985	.4986	.4986
3.0	.4986	.4987	.4987	.4988	.4988	.4989	.4989	.4989	.4990	.4990
3.1	.4990	.4991	.4991	.4991	.4992	.4992	.4992	.4992	.4993	.4993
3.2	.4993	.4993	.4994	.4994	.4994	.4994	.4994	.4995	.4995	.4995
3.3	.4995	.4995	.4995	.4996	.4996	.4996	.4996	.4996	.4996	.4997
3.4	.4997	.4997	.4997	.4997	.4997	.4997	.4997	.4997	.4997	.4998
3.6	.4998	.4998	.4999	.4999	.4999	.4999	.4999	.4999	.4999	.4999
3.9	.5000									

Source: The National Bureau of Standards, *Tables of Normal Probability Functions,* Applied Mathematics Series, no. 23 (Washington, D.C.: U.S. Government Printing Office, 1953). The original contains probabilities for values of z from 0 to 8.285, mostly in increments of .0001, and for areas from $\mu - z$ to $\mu + z$.

Appendix Table I
Exponential Probabilities for Cumulative Values of x

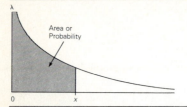

Entries in this table give the area under the curve between zero and x. For example, for $\lambda = 4.2$ and $x = .9$, the area under the curve is .9772, because $\lambda \cdot x = 3.78$.

λx	.00	.01	.02	.03	.04	.05	.06	.07	.08	.09
0.0	0.0000	0.0100	0.0198	0.0296	0.0392	0.0488	0.0582	0.0676	0.0769	0.0861
0.1	0.0952	0.1042	0.1131	0.1219	0.1306	0.1393	0.1479	0.1563	0.1647	0.1730
0.2	0.1813	0.1894	0.1975	0.2055	0.2134	0.2212	0.2289	0.2366	0.2442	0.2517
0.3	0.2592	0.2666	0.2739	0.2811	0.2882	0.2953	0.3023	0.3093	0.3161	0.3229
0.4	0.3297	0.3363	0.3430	0.3495	0.3560	0.3624	0.3687	0.3750	0.3812	0.3874
0.5	0.3935	0.3995	0.4055	0.4114	0.4173	0.4231	0.4288	0.4345	0.4401	0.4457
0.6	0.4512	0.4566	0.4621	0.4674	0.4727	0.4780	0.4831	0.4883	0.4934	0.4984
0.7	0.5034	0.5084	0.5132	0.5181	0.5229	0.5276	0.5323	0.5370	0.5416	0.5462
0.8	0.5507	0.5551	0.5596	0.5640	0.5683	0.5726	0.5768	0.5810	0.5852	0.5893
0.9	0.5934	0.5975	0.6015	0.6054	0.6094	0.6133	0.6171	0.6209	0.6247	0.6284
1.0	0.6321	0.6358	0.6394	0.6430	0.6465	0.6501	0.6535	0.6570	0.6604	0.6638
1.1	0.6671	0.6704	0.6737	0.6770	0.6802	0.6834	0.6865	0.6896	0.6927	0.6958
1.2	0.6988	0.7018	0.7048	0.7077	0.7106	0.7135	0.7163	0.7192	0.7220	0.7247
1.3	0.7275	0.7302	0.7329	0.7355	0.7382	0.7408	0.7433	0.7459	0.7484	0.7509
1.4	0.7534	0.7559	0.7583	0.7607	0.7631	0.7654	0.7678	0.7701	0.7724	0.7746
1.5	0.7769	0.7791	0.7813	0.7835	0.7856	0.7878	0.7899	0.7920	0.7940	0.7961
1.6	0.7981	0.8001	0.8021	0.8041	0.8060	0.8080	0.8099	0.8118	0.8136	0.8155
1.7	0.8173	0.8191	0.8209	0.8227	0.8245	0.8262	0.8280	0.8297	0.8314	0.8330
1.8	0.8347	0.8363	0.8380	0.8396	0.8412	0.8428	0.8443	0.8459	0.8474	0.8489
1.9	0.8504	0.8519	0.8534	0.8549	0.8563	0.8577	0.8591	0.8605	0.8619	0.8633
2.0	0.8647	0.8660	0.8673	0.8687	0.8700	0.8713	0.8725	0.8738	0.8751	0.8763
2.1	0.8775	0.8788	0.8800	0.8812	0.8823	0.8835	0.8847	0.8858	0.8870	0.8881
2.2	0.8892	0.8903	0.8914	0.8925	0.8935	0.8946	0.8956	0.8967	0.8977	0.8987
2.3	0.8997	0.9007	0.9017	0.9027	0.9037	0.9046	0.9056	0.9065	0.9074	0.9084
2.4	0.9093	0.9102	0.9111	0.9120	0.9128	0.9137	0.9146	0.9154	0.9163	0.9171
2.5	0.9179	0.9187	0.9195	0.9203	0.9211	0.9219	0.9227	0.9235	0.9242	0.9250
2.6	0.9257	0.9265	0.9272	0.9279	0.9286	0.9293	0.9301	0.9307	0.9314	0.9321
2.7	0.9328	0.9335	0.9341	0.9348	0.9354	0.9361	0.9367	0.9373	0.9380	0.9386
2.8	0.9392	0.9398	0.9404	0.9410	0.9416	0.9422	0.9427	0.9433	0.9439	0.9444
2.9	0.9450	0.9455	0.9461	0.9466	0.9471	0.9477	0.9482	0.9487	0.9492	0.9497
3.0	0.9502	0.9507	0.9512	0.9517	0.9522	0.9526	0.9531	0.9536	0.9540	0.9545
3.1	0.9550	0.9554	0.9558	0.9563	0.9567	0.9571	0.9576	0.9580	0.9584	0.9588
3.2	0.9592	0.9596	0.9600	0.9604	0.9608	0.9612	0.9616	0.9620	0.9624	0.9627
3.3	0.9631	0.9635	0.9638	0.9642	0.9646	0.9649	0.9653	0.9656	0.9660	0.9663
3.4	0.9666	0.9670	0.9673	0.9676	0.9679	0.9683	0.9686	0.9689	0.9692	0.9695
3.5	0.9698	0.9701	0.9704	0.9707	0.9710	0.9713	0.9716	0.9718	0.9721	0.9724
3.6	0.9727	0.9729	0.9732	0.9735	0.9737	0.9740	0.9743	0.9745	0.9748	0.9750
3.7	0.9753	0.9755	0.9758	0.9760	0.9762	0.9765	0.9767	0.9769	0.9772	0.9774
3.8	0.9776	0.9779	0.9781	0.9783	0.9785	0.9787	0.9789	0.9791	0.9793	0.9796
3.9	0.9798	0.9800	0.9802	0.9804	0.9806	0.9807	0.9809	0.9811	0.9813	0.9815
4.0	0.9817	0.9834	0.9850	0.9864	0.9877	0.9889	0.9899	0.9909	0.9918	0.9926
5.0	0.9933	0.9939	0.9945	0.9950	0.9955	0.9959	0.9963	0.9967	0.9970	0.9973
6.0	0.9975	0.9978	0.9980	0.9982	0.9983	0.9985	0.9986	0.9988	0.9989	0.9990
7.0	0.9991	0.9992	0.9993	0.9993	0.9994	0.9994	0.9995	0.9995	0.9996	0.9996
8.0	0.9997	0.9997	0.9997	0.9998	0.9998	0.9998	0.9998	0.9998	0.9998	0.9999
9.0	0.9999	0.9999	0.9999	0.9999	0.9999	0.9999	0.9999	0.9999	0.9999	0.9999

All the entries in this table have been calculated with the formula given in Box 7.F of this text.

Appendix Table J
Critical Normal Deviate Values

(A) For Statistical Estimation		(B) For Hypothesis Testing			
		Two-Tailed Tests		One-Tailed Tests	
Confidence Level, C	Normal Deviate, z	Significance Level, α	Normal Deviate, $z_{\alpha/2}$	Significance Level, α	Normal Deviate, z_{α}
.80	1.2817	.10	1.645	.10	1.2817
.90	1.645	.05	1.96	.05	1.645
.95	1.96	.025	2.24	.025	1.96
.98	2.3267	.01	2.575	.01	2.3267
.99	2.575	.005	2.81	.005	2.575
.998	3.08	.001	3.27	.001	3.08
.999	3.27				

(A) For Statistical Estimation: Area = C or confidence level; $\mu - z\sigma_{\overline{X}}$, μ, $\mu + z\sigma_{\overline{X}}$

Two-Tailed Tests: Area = $\alpha/2$; $-z_{\alpha/2}$, 0, $+z_{\alpha/2}$
E.g.
$-1.96 \leq z \leq +1.96$

One-Tailed Tests, Lower-tailed: Area = α; $-z_{\alpha}$, 0
E.g.
$z \geq -1.645$

One-Tailed Tests, Upper-tailed: Area = α; 0, $+z_{\alpha}$
E.g.
$z \leq +1.645$

Source: The National Bureau of Standards, *Tables of Normal Probability Functions,* Applied Mathematics Series, No. 23 (Washington, D.C.: U.S. Government Printing Office, 1953).

Appendix Table K
Student _t_ Distributions

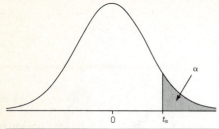

The following table provides the values of t_α that correspond to a given upper-tail area α and a specified number of degrees of freedom. For example, for an upper-tail area of .05 and 9 degrees of freedom, the critical value of $t_\alpha = 1.833$.

Degrees of Freedom	Critical tail areas (= α for one-tailed tests, = $\alpha/2$ for two-tailed tests)									
	.4	.25	.1	.05	.025	.01	.005	.0025	.001	.0005
1	0.325	1.000	3.078	6.314	12.706	31.821	63.657	127.32	318.31	636.62
2	.289	.816	1.886	2.920	4.303	6.965	9.925	14.089	22.327	31.598
3	.277	.765	1.638	2.353	3.182	4.541	5.841	7.453	10.214	12.924
4	.271	.741	1.533	2.132	2.776	3.747	4.604	5.598	7.173	8.610
5	0.267	0.727	1.476	2.015	2.571	3.365	4.032	4.773	5.893	6.869
6	.265	.718	1.440	1.943	2.447	3.143	3.707	4.317	5.208	5.959
7	.263	.711	1.415	1.895	2.365	2.998	3.499	4.029	4.785	5.408
8	.262	.706	1.397	1.860	2.306	2.896	3.355	3.833	4.501	5.041
9	.261	.703	1.383	1.833	2.262	2.821	3.250	3.690	4.297	4.781
10	0.260	0.700	1.372	1.812	2.228	2.764	3.169	3.581	4.144	4.587
11	.260	.697	1.363	1.796	2.201	2.718	3.106	3.497	4.025	4.437
12	.259	.695	1.356	1.782	2.179	2.681	3.055	3.428	3.930	4.318
13	.259	.694	1.350	1.771	2.160	2.650	3.012	3.372	3.852	4.221
14	.258	.692	1.345	1.761	2.145	2.624	2.977	3.326	3.787	4.140
15	0.258	0.691	1.341	1.753	2.131	2.602	2.947	3.286	3.733	4.073
16	.258	.690	1.337	1.746	2.120	2.583	2.921	3.252	3.686	4.015
17	.257	.689	1.333	1.740	2.110	2.567	2.898	3.222	3.646	3.965
18	.257	.688	1.330	1.734	2.101	2.552	2.878	3.197	3.610	3.922
19	.257	.688	1.328	1.729	2.093	2.539	2.861	3.174	3.579	3.883
20	0.257	0.687	1.325	1.725	2.086	2.528	2.845	3.153	3.552	3.850
21	.257	.686	1.323	1.721	2.080	2.518	2.831	3.135	3.527	3.819
22	.256	.686	1.321	1.717	2.074	2.508	2.819	3.119	3.505	3.792
23	.256	.685	1.319	1.714	2.069	2.500	2.807	3.104	3.485	3.767
24	.256	.685	1.318	1.711	2.064	2.492	2.797	3.091	3.467	3.745
25	0.256	0.684	1.316	1.708	2.060	2.485	2.787	3.078	3.450	3.725
26	.256	.684	1.315	1.706	2.056	2.479	2.779	3.067	3.435	3.707
27	.256	.684	1.314	1.703	2.052	2.473	2.771	3.057	3.421	3.690
28	.256	.683	1.313	1.701	2.048	2.467	2.763	3.047	3.408	3.674
29	.256	.683	1.311	1.699	2.045	2.462	2.756	3.038	3.396	3.659
30	0.256	0.683	1.310	1.697	2.042	2.457	2.750	3.030	3.385	3.646
40	.255	.681	1.303	1.684	2.021	2.423	2.704	2.971	3.307	3.551
60	.254	.679	1.296	1.671	2.000	2.390	2.660	2.915	3.232	3.460
120	.254	.677	1.289	1.658	1.980	2.358	2.617	2.860	3.160	3.373
∞	.253	.674	1.282	1.645	1.960	2.326	2.576	2.807	3.090	3.291

Confidence Level C = 1 = α										
one-tailed test	.60	.75	.90	.95	.975	.99	.995	.9975	.999	.9995
two-tailed test	.20	.50	.80	.90	.95	.98	.99	.995	.998	.999

Note: A significance level α is associated with, for example, an upper-tail area α in an upper-tailed test, but with an upper- _as well as_ lower-tail area $\alpha/2$ in a two-tailed test. The confidence level always equals C = 1 − α. Therefore, a significance level of, say, $\alpha = .05$, goes with an upper-tail area of .05 in an upper-tailed test (C = .95), but with an upper-tail area of .025 in a two-tailed test (C = .95).

Source: E. S. Pearson and H. O. Hartley, _Biometrika Tables for Statisticians_, vol. I. (Cambridge: Cambridge University Press, 1966), p. 146.

Appendix Table L
Chi-Square Distributions

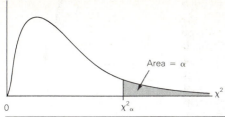

Area = α

χ^2

0   χ^2_α

This table provides values of χ^2_α that correspond to a given upper-tail area, α, and a specified number of degrees of freedom. For example, for an upper-tail area of .05 and 4 degrees of freedom, the critical value of $\chi^2_{.05}$ equals 9.488.

Degrees of Freedom	Critical Values for Upper-Tail Area, α													
	.99	.98	.95	.90	.80	.70	.50	.30	.20	.10	.05	.02	.01	.001
1	.0³157	.0³628	.00393	.0158	.0642	.148	.455	1.074	1.642	2.706	3.841	5.412	6.635	10.827
2	.0201	.0404	.103	.211	.446	.713	1.386	2.408	3.219	4.605	5.991	7.824	9.210	13.815
3	.115	.185	.352	.584	1.005	1.424	2.366	3.665	4.642	6.251	7.815	9.837	11.345	16.266
4	.297	.429	.711	1.064	1.649	2.195	3.357	4.878	5.989	7.779	9.488	11.668	13.277	18.467
5	.554	.752	1.145	1.610	2.343	3.000	4.351	6.064	7.289	9.236	11.070	13.388	15.086	20.515
6	.872	1.134	1.635	2.204	3.070	3.828	5.348	7.231	8.558	10.645	12.592	15.033	16.812	22.457
7	1.239	1.564	2.167	2.833	3.822	4.671	6.346	8.383	9.803	12.017	14.067	16.622	18.475	24.322
8	1.646	2.032	2.733	3.490	4.594	5.527	7.344	9.524	11.030	13.362	15.507	18.168	20.090	26.125
9	2.088	2.532	3.325	4.168	5.380	6.393	8.343	10.656	12.242	14.684	16.919	19.679	21.666	27.877
10	2.558	3.059	3.940	4.865	6.179	7.267	9.342	11.781	13.442	15.987	18.307	21.161	23.209	29.588
11	3.053	3.609	4.575	5.578	6.989	8.148	10.341	12.899	14.631	17.275	19.675	22.618	24.725	31.264
12	3.571	4.178	5.226	6.304	7.807	9.034	11.340	14.011	15.812	18.549	21.026	24.054	26.217	32.909
13	4.107	4.765	5.892	7.042	8.634	9.926	12.340	16.985	16.985	19.812	22.362	25.472	27.688	34.528
14	4.660	5.368	6.571	7.790	9.467	10.821	13.339	16.222	18.151	21.064	23.685	26.873	29.141	36.123
15	5.229	5.985	7.261	8.547	10.307	11.721	14.339	17.322	19.311	22.307	24.996	28.259	30.578	37.697
16	5.812	6.614	7.962	9.312	11.152	12.624	15.338	18.418	20.465	23.542	26.296	29.633	32.000	39.252
17	6.408	7.255	8.672	10.085	12.002	13.531	16.338	19.511	21.615	24.769	27.587	30.995	33.409	40.790
18	7.015	7.906	9.390	10.865	12.857	14.440	17.338	20.601	22.760	25.989	28.869	32.346	34.805	42.312
19	7.633	8.567	10.117	11.651	13.716	15.352	18.338	21.689	23.900	27.204	30.144	33.687	36.191	43.820
20	8.260	9.237	10.851	12.443	14.578	16.266	19.337	22.775	25.038	28.412	31.410	35.020	37.566	45.315
21	8.897	9.915	11.591	13.240	15.445	17.182	20.337	23.858	26.171	29.615	32.671	36.343	38.932	46.797
22	9.542	10.600	12.338	14.041	16.314	18.101	21.337	24.939	27.301	30.813	33.924	37.659	40.289	48.268
23	10.196	11.293	13.091	14.848	17.187	19.021	22.337	26.018	28.429	32.007	35.172	38.968	41.638	49.728
24	10.856	11.992	13.848	15.659	18.062	19.943	23.337	27.096	29.553	33.196	36.415	40.270	42.980	51.179
25	11.524	12.697	14.611	16.473	18.940	20.867	24.337	28.172	30.675	34.382	37.652	41.566	44.314	52.620
26	12.198	13.409	15.379	17.292	19.820	21.792	25.336	29.246	31.795	35.563	38.885	42.856	45.642	54.052
27	12.879	14.125	16.151	18.114	20.703	22.719	26.336	30.319	32.912	36.741	40.113	44.140	46.963	55.476
28	13.565	14.847	16.928	18.939	21.588	23.647	27.336	31.391	34.027	37.916	41.337	45.419	48.278	56.893
29	14.256	15.574	17.708	19.768	22.475	24.577	28.336	32.461	35.139	39.087	42.557	46.693	49.588	58.302
30	14.953	16.306	18.493	20.599	23.364	25.508	29.336	33.530	36.250	40.256	43.773	47.962	50.892	59.703

Source: Ronald A. Fisher and Frank Yates, *Statistical Tables for Biological, Agricultural and Medical Research*, 6th ed. (New York, Hafner, 1963), p. 47.

Note: When the number of degrees of freedom exceeds 30, the χ^2 distribution can be approximated by the normal distribution, but the original Fisher/Yates table provides values up to 70 degrees of freedom.

Appendix Table M

Critical Values of D for the Kolmogorov-Smirnov Maximum Deviation Test for Goodness of Fit

This table provides the critical values D_α that correspond to an upper-tail probability, α, of the test statistic, D. For example, for an upper-tail area of .05 and a sample of $n = 10$, the critical value of $D_{.05} = .36866$. Thus, the probability is .05 that an observed value of $D \geq .36866$.

	Critical Values for Upper-Tail Area, α				
n	.10	.05	.025	.01	.005
1	.90000	.95000	.97500	.99000	.99500
2	.68377	.77639	.84189	.90000	.92929
3	.56481	.63604	.70760	.78456	.82900
4	.49265	.56522	.62394	.68887	.73424
5	.44698	.50945	.56328	.62718	.66853
6	.41037	.46799	.51926	.57741	.61661
7	.38148	.43607	.48342	.53844	.57581
8	.35831	.40962	.45427	.50654	.54179
9	.33910	.38746	.43001	.47960	.51332
10	.32260	.36866	.40925	.45662	.48893
11	.30829	.35242	.39122	.43670	.46770
12	.29577	.33815	.37543	.41918	.44905
13	.28470	.32549	.36143	.40362	.43247
14	.27481	.31417	.34890	.38970	.41762
15	.26588	.30397	.33760	.37713	.40420
16	.25778	.29472	.32733	.36571	.39201
17	.25039	.28627	.31796	.35528	.38086
18	.24360	.27851	.30936	.34569	.37062
19	.23735	.27136	.30143	.33685	.36117
20	.23156	.26473	.29408	.32866	.35241
21	.22617	.25858	.28724	.32104	.34427
22	.22115	.25283	.28087	.31394	.33666
23	.21645	.24746	.27490	.30728	.32954
24	.21205	.24242	.26931	.30104	.32286
25	.20790	.23768	.26404	.29516	.31657
26	.20399	.23320	.25907	.28962	.31064
27	.20030	.22898	.25438	.28438	.30502
28	.19680	.22497	.24993	.27942	.29971
29	.19348	.22117	.24571	.27471	.29466
30	.19032	.21756	.24170	.27023	.28987
31	.18732	.21412	.23788	.26596	.28530
32	.18445	.21085	.23424	.26189	.28094
33	.18171	.20771	.23076	.25801	.27677
34	.17909	.20472	.22743	.25429	.27279
35	.17659	.20185	.22425	.25073	.26897
36	.17418	.19910	.22119	.24732	.26532
37	.17188	.19646	.21826	.24404	.26180
38	.16966	.19392	.21544	.24089	.25843
39	.16753	.19148	.21273	.23786	.25518
40	.16547	.18913	.21012	.23494	.25205
41	.16349	.18687	.20760	.23213	.24904
42	.16158	.18468	.20517	.22941	.24613
43	.15974	.18257	.20283	.22679	.24332
44	.15796	.18053	.20056	.22426	.24060
45	.15623	.17856	.19837	.22181	.23798
46	.15457	.17665	.19625	.21944	.23544
47	.15295	.17481	.19420	.21715	.23298
48	.15139	.17302	.19221	.21493	.23059
49	.14987	.17128	.19028	.21277	.22828
50	.14840	.16959	.18841	.21068	.22604

Source: Leslie H. Miller, "Table of Percentage Points of Kolmogorov Statistics," *Journal of the American Statistical Association* 51 (1956):113–15.

Appendix Table M (continued)

	Critical Values for Upper-Tail Area, α				
n	.10	.05	.025	.01	.005
51	.14697	.16796	.18659	.20864	.22386
52	.14558	.16637	.18482	.20667	.22174
53	.14423	.16483	.18311	.20475	.21968
54	.14292	.16332	.18144	.20289	.21768
55	.14164	.16186	.17981	.20107	.21574
56	.14040	.16044	.17823	.19930	.21384
57	.13919	.15906	.17669	.19758	.21199
58	.13801	.15771	.17519	.19590	.21019
59	.13686	.15639	.17373	.19427	.20844
60	.13573	.15511	.17231	.19267	.20673
61	.13464	.15385	.17091	.19112	.20506
62	.13357	.15263	.16956	.18960	.20343
63	.13253	.15144	.16823	.18812	.20184
64	.13151	.15027	.16693	.18667	.20029
65	.13052	.14913	.16567	.18525	.19877
66	.12954	.14802	.16443	.18387	.19729
67	.12859	.14693	.16322	.18252	.19584
68	.12766	.14587	.16204	.18119	.19442
69	.12675	.14483	.16088	.17990	.19303
70	.12586	.14381	.15975	.17863	.19167
71	.12499	.14281	.15864	.17739	.19034
72	.12413	.14183	.15755	.17618	.18903
73	.12329	.14087	.15649	.17498	.18776
74	.12247	.13993	.15544	.17382	.18650
75	.12167	.13901	.15442	.17268	.18528
76	.12088	.13811	.15342	.17155	.18408
77	.12011	.13723	.15244	.17045	.18290
78	.11935	.13636	.15147	.16938	.18174
79	.11860	.13551	.15052	.16832	.18060
80	.11787	.13467	.14960	.16728	.17949
81	.11716	.13385	.14868	.16626	.17840
82	.11645	.13305	.14779	.16526	.17732
83	.11576	.13226	.14691	.16428	.17627
84	.11508	.13148	.14605	.16331	.17523
85	.11442	.13072	.14520	.16236	.17421
86	.11376	.12997	.14437	.16143	.17321
87	.11311	.12923	.14355	.16051	.17223
88	.11248	.12850	.14274	.15961	.17126
89	.11186	.12779	.14195	.15873	.17031
90	.11125	.12709	.14117	.15786	.16938
91	.11064	.12640	.14040	.15700	.16846
92	.11005	.12572	.13965	.15616	.16755
93	.10947	.12506	.13891	.15533	.16666
94	.10889	.12440	.13818	.15451	.16579
95	.10833	.12375	.13746	.15371	.16493
96	.10777	.12312	.13675	.15291	.16408
97	.10722	.12249	.13606	.15214	.16324
98	.10668	.12187	.13537	.15137	.16242
99	.10615	.12126	.13469	.15061	.16161
100	.10563	.12067	.13403	.14987	.16081

Note: For larger samples, critical values can be calculated as follows:

$\alpha = .10$:	$\alpha = .05$:	$\alpha = .025$:	$\alpha = .01$:	$\alpha = .005$:
$\dfrac{1.22}{\sqrt{n}}$	$\dfrac{1.36}{\sqrt{n}}$	$\dfrac{1.48}{\sqrt{n}}$	$\dfrac{1.63}{\sqrt{n}}$	$\dfrac{1.73}{\sqrt{n}}$

Appendix Table N F Distribution

Entries in the table give F_α values, where α is the area or probability in the upper tail of the F distribution. For example, with 12 numerator degrees of freedom, 15 denominator degrees of freedom, and a .05 area in the upper tail, $F_{.05} = 2.48$.

$F_{.05}$ Values

Denominator Degrees of Freedom	Numerator Degrees of Freedom																		
	1	2	3	4	5	6	7	8	9	10	12	15	20	24	30	40	60	120	∞
1	161.4	199.5	215.7	224.6	230.2	234.0	236.8	238.9	240.5	241.9	243.9	245.9	248.0	249.1	250.1	251.1	252.2	253.3	254.3
2	18.51	19.00	19.16	19.25	19.30	19.33	19.35	19.37	19.38	19.40	19.41	19.43	19.45	19.45	19.46	19.47	19.48	19.49	19.50
3	10.13	9.55	9.28	9.12	9.01	8.94	8.89	8.85	8.81	8.79	8.74	8.70	8.66	8.64	8.62	8.59	8.57	8.55	8.53
4	7.71	6.94	6.59	6.39	6.26	6.16	6.09	6.04	6.00	5.96	5.91	5.86	5.80	5.77	5.75	5.72	5.69	5.66	5.63
5	6.61	5.79	5.41	5.19	5.05	4.95	4.88	4.82	4.77	4.74	4.68	4.62	4.56	4.53	4.50	4.46	4.43	4.40	4.36
6	5.99	5.14	4.76	4.53	4.39	4.28	4.21	4.15	4.10	4.06	4.00	3.94	3.87	3.84	3.81	3.77	3.74	3.70	3.67
7	5.59	4.74	4.35	4.12	3.97	3.87	3.79	3.73	3.68	3.64	3.57	3.51	3.44	3.41	3.38	3.34	3.30	3.27	3.23
8	5.32	4.46	4.07	3.84	3.69	3.58	3.50	3.44	3.39	3.35	3.28	3.22	3.15	3.12	3.08	3.04	3.01	2.97	2.93
9	5.12	4.26	3.86	3.63	3.48	3.37	3.29	3.23	3.18	3.14	3.07	3.01	2.94	2.90	2.86	2.83	2.79	2.75	2.71
10	4.96	4.10	3.71	3.48	3.33	3.22	3.14	3.07	3.02	2.98	2.91	2.85	2.77	2.74	2.70	2.66	2.62	2.58	2.54
11	4.84	3.98	3.59	3.36	3.20	3.09	3.01	2.95	2.90	2.85	2.79	2.72	2.65	2.61	2.57	2.53	2.49	2.45	2.40
12	4.75	3.89	3.49	3.26	3.11	3.00	2.91	2.85	2.80	2.75	2.69	2.62	2.54	2.51	2.47	2.43	2.38	2.34	2.30
13	4.67	3.81	3.41	3.18	3.03	2.92	2.83	2.77	2.71	2.67	2.60	2.53	2.46	2.42	2.38	2.34	2.30	2.25	2.21
14	4.60	3.74	3.34	3.11	2.96	2.85	2.76	2.70	2.65	2.60	2.53	2.46	2.39	2.35	2.31	2.27	2.22	2.18	2.13
15	4.54	3.68	3.29	3.06	2.90	2.79	2.71	2.64	2.59	2.54	2.48	2.40	2.33	2.29	2.25	2.20	2.16	2.11	2.07
16	4.49	3.63	3.24	3.01	2.85	2.74	2.66	2.59	2.54	2.49	2.42	2.35	2.28	2.24	2.19	2.15	2.11	2.06	2.01
17	4.45	3.59	3.20	2.96	2.81	2.70	2.61	2.55	2.49	2.45	2.38	2.31	2.23	2.19	2.15	2.10	2.06	2.01	1.96
18	4.41	3.55	3.16	2.93	2.77	2.66	2.58	2.51	2.46	2.41	2.34	2.27	2.19	2.15	2.11	2.06	2.02	1.97	1.92
19	4.38	3.52	3.13	2.90	2.74	2.63	2.54	2.48	2.42	2.38	2.31	2.23	2.16	2.11	2.07	2.03	1.98	1.93	1.88
20	4.35	3.49	3.10	2.87	2.71	2.60	2.51	2.45	2.39	2.35	2.28	2.20	2.12	2.08	2.04	1.99	1.95	1.90	1.84
21	4.32	3.47	3.07	2.84	2.68	2.57	2.49	2.42	2.37	2.32	2.25	2.18	2.10	2.05	2.01	1.96	1.92	1.87	1.81
22	4.30	3.44	3.05	2.82	2.66	2.55	2.46	2.40	2.34	2.30	2.23	2.15	2.07	2.03	1.98	1.94	1.89	1.84	1.78
23	4.28	3.42	3.03	2.80	2.64	2.53	2.44	2.37	2.32	2.27	2.20	2.13	2.05	2.01	1.96	1.91	1.86	1.81	1.76
24	4.26	3.40	3.01	2.78	2.62	2.51	2.42	2.36	2.30	2.25	2.18	2.11	2.03	1.98	1.94	1.89	1.84	1.79	1.73
25	4.24	3.39	2.99	2.76	2.60	2.49	2.40	2.34	2.28	2.24	2.16	2.09	2.01	1.96	1.92	1.87	1.82	1.77	1.71
26	4.23	3.37	2.98	2.74	2.59	2.47	2.39	2.32	2.27	2.22	2.15	2.07	1.99	1.95	1.90	1.85	1.80	1.75	1.69
27	4.21	3.35	2.96	2.73	2.57	2.46	2.37	2.31	2.25	2.20	2.13	2.06	1.97	1.93	1.88	1.84	1.79	1.73	1.67
28	4.20	3.34	2.95	2.71	2.56	2.45	2.36	2.29	2.24	2.19	2.12	2.04	1.96	1.91	1.87	1.82	1.77	1.71	1.65
29	4.18	3.33	2.93	2.70	2.55	2.43	2.35	2.28	2.22	2.18	2.10	2.03	1.94	1.90	1.85	1.81	1.75	1.70	1.64
30	4.17	3.32	2.92	2.69	2.53	2.42	2.33	2.27	2.21	2.16	2.09	2.01	1.93	1.89	1.84	1.79	1.74	1.68	1.62
40	4.08	3.23	2.84	2.61	2.45	2.34	2.25	2.18	2.12	2.08	2.00	1.92	1.84	1.79	1.74	1.69	1.64	1.58	1.51
60	4.00	3.15	2.76	2.53	2.37	2.25	2.17	2.10	2.04	1.99	1.92	1.84	1.75	1.70	1.65	1.59	1.53	1.47	1.39
120	3.92	3.07	2.68	2.45	2.29	2.17	2.09	2.02	1.96	1.91	1.83	1.75	1.66	1.61	1.55	1.50	1.43	1.35	1.25
∞	3.84	3.00	2.60	2.37	2.21	2.10	2.01	1.94	1.88	1.83	1.75	1.67	1.57	1.52	1.46	1.39	1.32	1.22	1.00

Area or Probability = α

F_α

Appendix Table N (continued)

F_{.025} Values

Denominator Degrees of Freedom	Numerator Degrees of Freedom																		
	1	2	3	4	5	6	7	8	9	10	12	15	20	24	30	40	60	120	∞
1	647.8	799.5	864.2	899.6	921.8	937.1	948.2	956.7	963.3	968.6	976.7	984.9	993.1	997.2	1001	1006	1010	1014	1018
2	38.51	39.00	39.17	39.25	39.30	39.33	39.36	39.37	39.39	39.40	39.41	39.43	39.45	39.46	39.46	39.47	39.48	39.49	39.50
3	17.44	16.04	15.44	15.10	14.88	14.73	14.62	14.54	14.47	14.42	14.34	14.25	14.17	14.12	14.08	14.04	13.99	13.95	13.90
4	12.22	10.65	9.98	9.60	9.36	9.20	9.07	8.98	8.90	8.84	8.75	8.66	8.56	8.51	8.46	8.41	8.36	8.31	8.26
5	10.01	8.43	7.76	7.39	7.15	6.98	6.85	6.76	6.68	6.62	6.52	6.43	6.33	6.28	6.23	6.18	6.12	6.07	6.02
6	8.81	7.26	6.60	6.23	5.99	5.82	5.70	5.60	5.52	5.46	5.37	5.27	5.17	5.12	5.07	5.01	4.96	4.90	4.85
7	8.07	6.54	5.89	5.52	5.29	5.12	4.99	4.90	4.82	4.76	4.67	4.57	4.47	4.42	4.36	4.31	4.25	4.20	4.14
8	7.57	6.06	5.42	5.05	4.82	4.65	4.53	4.43	4.36	4.30	4.20	4.10	4.00	3.95	3.89	3.84	3.78	3.73	3.67
9	7.21	5.71	5.08	4.72	4.48	4.32	4.20	4.10	4.03	3.96	3.87	3.77	3.67	3.61	3.56	3.51	3.45	3.39	3.33
10	6.94	5.46	4.83	4.47	4.24	4.07	3.95	3.85	3.78	3.72	3.62	3.52	3.42	3.37	3.31	3.26	3.20	3.14	3.08
11	6.72	5.26	4.63	4.28	4.04	3.88	3.76	3.66	3.59	3.53	3.43	3.33	3.23	3.17	3.12	3.06	3.00	2.94	2.88
12	6.55	5.10	4.47	4.12	3.89	3.73	3.61	3.51	3.44	3.37	3.28	3.18	3.07	3.02	2.96	2.91	2.85	2.79	2.72
13	6.41	4.97	4.35	4.00	3.77	3.60	3.48	3.39	3.31	3.25	3.15	3.05	2.95	2.89	2.84	2.78	2.72	2.66	2.60
14	6.30	4.86	4.24	3.89	3.66	3.50	3.38	3.29	3.21	3.15	3.05	2.95	2.84	2.79	2.73	2.67	2.61	2.55	2.49
15	6.20	4.77	4.15	3.80	3.58	3.41	3.29	3.20	3.12	3.06	2.96	2.86	2.76	2.70	2.64	2.59	2.52	2.46	2.40
16	6.12	4.69	4.08	3.73	3.50	3.34	3.22	3.12	3.05	2.99	2.89	2.79	2.68	2.63	2.57	2.51	2.45	2.38	2.32
17	6.04	4.62	4.01	3.66	3.44	3.28	3.16	3.06	2.98	2.92	2.82	2.72	2.62	2.56	2.50	2.44	2.38	2.32	2.25
18	5.98	4.56	3.95	3.61	3.38	3.22	3.10	3.01	2.93	2.87	2.77	2.67	2.56	2.50	2.44	2.38	2.32	2.26	2.19
19	5.92	4.51	3.90	3.56	3.33	3.17	3.05	2.96	2.88	2.82	2.72	2.62	2.51	2.45	2.39	2.33	2.27	2.20	2.13
20	5.87	4.46	3.86	3.51	3.29	3.13	3.01	2.91	2.84	2.77	2.68	2.57	2.46	2.41	2.35	2.29	2.22	2.16	2.09
21	5.83	4.42	3.82	3.48	3.25	3.09	2.97	2.87	2.80	2.73	2.64	2.53	2.42	2.37	2.31	2.25	2.18	2.11	2.04
22	5.79	4.38	3.78	3.44	3.22	3.05	2.93	2.84	2.76	2.70	2.60	2.50	2.39	2.33	2.27	2.21	2.14	2.08	2.00
23	5.75	4.35	3.75	3.41	3.18	3.02	2.90	2.81	2.73	2.67	2.57	2.47	2.36	2.30	2.24	2.18	2.11	2.04	1.97
24	5.72	4.32	3.72	3.38	3.15	2.99	2.87	2.78	2.70	2.64	2.54	2.44	2.33	2.27	2.21	2.15	2.08	2.01	1.94
25	5.69	4.29	3.69	3.35	3.13	2.97	2.85	2.75	2.68	2.61	2.51	2.41	2.30	2.24	2.18	2.12	2.05	1.98	1.91
26	5.66	4.27	3.67	3.33	3.10	2.94	2.82	2.73	2.65	2.59	2.49	2.39	2.28	2.22	2.16	2.09	2.03	1.95	1.88
27	5.63	4.24	3.65	3.31	3.08	2.92	2.80	2.71	2.63	2.57	2.47	2.36	2.25	2.19	2.13	2.07	2.00	1.93	1.85
28	5.61	4.22	3.63	3.29	3.06	2.90	2.78	2.69	2.61	2.55	2.45	2.34	2.23	2.17	2.11	2.05	1.98	1.91	1.83
29	5.59	4.20	3.61	3.27	3.04	2.88	2.76	2.67	2.59	2.53	2.43	2.32	2.21	2.15	2.09	2.03	1.96	1.89	1.81
30	5.57	4.18	3.59	3.25	3.03	2.87	2.75	2.65	2.57	2.51	2.41	2.31	2.20	2.14	2.07	2.01	1.94	1.87	1.79
40	5.42	4.05	3.46	3.13	2.90	2.74	2.62	2.53	2.45	2.39	2.29	2.18	2.07	2.01	1.94	1.88	1.80	1.72	1.64
60	5.29	3.93	3.34	3.01	2.79	2.63	2.51	2.41	2.33	2.27	2.17	2.06	1.94	1.88	1.82	1.74	1.67	1.58	1.48
120	5.15	3.80	3.23	2.89	2.67	2.52	2.39	2.30	2.22	2.16	2.05	1.94	1.82	1.76	1.69	1.61	1.53	1.43	1.31
∞	5.02	3.69	3.12	2.79	2.57	2.41	2.29	2.19	2.11	2.05	1.94	1.83	1.71	1.64	1.57	1.48	1.39	1.27	1.00

Appendix Table N *(continued)*

F_{01} Values

Denominator Degrees of Freedom	1	2	3	4	5	6	7	8	9	10	12	15	20	24	30	40	60	120	∞
1	4052	4999.5	5403	5625	5764	5859	5928	5981	6022	6056	6106	6157	6209	6235	6261	6287	6313	6339	6366
2	98.50	99.00	99.17	99.25	99.30	99.33	99.36	99.37	99.39	99.40	99.42	99.43	99.45	99.46	99.47	99.47	99.48	99.49	99.50
3	34.12	30.82	29.46	28.71	28.24	27.91	27.67	27.49	27.35	27.23	27.05	26.87	26.69	26.60	26.50	26.41	26.32	26.22	26.13
4	21.20	18.00	16.69	15.98	15.52	15.21	14.98	14.80	14.66	14.55	14.37	14.20	14.02	13.93	13.84	13.75	13.65	13.56	13.46
5	16.26	13.27	12.06	11.39	10.97	10.67	10.46	10.29	10.16	10.05	9.89	9.72	9.55	9.47	9.38	9.29	9.20	9.11	9.02
6	13.75	10.92	9.78	9.15	8.75	8.47	8.26	8.10	7.98	7.87	7.72	7.56	7.40	7.31	7.23	7.14	7.06	6.97	6.88
7	12.25	9.55	8.45	7.85	7.46	7.19	6.99	6.84	6.72	6.62	6.47	6.31	6.16	6.07	5.99	5.91	5.82	5.74	5.65
8	11.26	8.65	7.59	7.01	6.63	6.37	6.18	6.03	5.91	5.81	5.67	5.52	5.36	5.28	5.20	5.12	5.03	4.95	4.86
9	10.56	8.02	6.99	6.42	6.06	5.80	5.61	5.47	5.35	5.26	5.11	4.96	4.81	4.73	4.65	4.57	4.48	4.40	4.31
10	10.04	7.56	6.55	5.99	5.64	5.39	5.20	5.06	4.94	4.85	4.71	4.56	4.41	4.33	4.25	4.17	4.08	4.00	3.91
11	9.65	7.21	6.22	5.67	5.32	5.07	4.89	4.74	4.63	4.54	4.40	4.25	4.10	4.02	3.94	3.86	3.78	3.69	3.60
12	9.33	6.93	5.95	5.41	5.06	4.82	4.64	4.50	4.39	4.30	4.16	4.01	3.86	3.78	3.70	3.62	3.54	3.45	3.36
13	9.07	6.70	5.74	5.21	4.86	4.62	4.44	4.30	4.19	4.10	3.96	3.82	3.66	3.59	3.51	3.43	3.34	3.25	3.17
14	8.86	6.51	5.56	5.04	4.69	4.46	4.28	4.14	4.03	3.94	3.80	3.66	3.51	3.43	3.35	3.27	3.18	3.09	3.00
15	8.68	6.36	5.42	4.89	4.56	4.32	4.14	4.00	3.89	3.80	3.67	3.52	3.37	3.29	3.21	3.13	3.05	2.96	2.87
16	8.53	6.23	5.29	4.77	4.44	4.20	4.03	3.89	3.78	3.69	3.55	3.41	3.26	3.18	3.10	3.02	2.93	2.84	2.75
17	8.40	6.11	5.18	4.67	4.34	4.10	3.93	3.79	3.68	3.59	3.46	3.31	3.16	3.08	3.00	2.92	2.83	2.75	2.65
18	8.29	6.01	5.09	4.58	4.25	4.01	3.84	3.71	3.60	3.51	3.37	3.23	3.08	3.00	2.92	2.84	2.75	2.66	2.57
19	8.18	5.93	5.01	4.50	4.17	3.94	3.77	3.63	3.52	3.43	3.30	3.15	3.00	2.92	2.84	2.76	2.67	2.58	2.49
20	8.10	5.85	4.94	4.43	4.10	3.87	3.70	3.56	3.46	3.37	3.23	3.09	2.94	2.86	2.78	2.69	2.61	2.52	2.42
21	8.02	5.78	4.87	4.37	4.04	3.81	3.64	3.51	3.40	3.31	3.17	3.03	2.88	2.80	2.72	2.64	2.55	2.46	2.36
22	7.95	5.72	4.82	4.31	3.99	3.76	3.59	3.45	3.35	3.26	3.12	2.98	2.83	2.75	2.67	2.58	2.50	2.40	2.31
23	7.88	5.66	4.76	4.26	3.94	3.71	3.54	3.41	3.30	3.21	3.07	2.93	2.78	2.70	2.62	2.54	2.45	2.35	2.26
24	7.82	5.61	4.72	4.22	3.90	3.67	3.50	3.36	3.26	3.17	3.03	2.89	2.74	2.66	2.58	2.49	2.40	2.31	2.21
25	7.77	5.57	4.68	4.18	3.85	3.63	3.46	3.32	3.22	3.13	2.99	2.85	2.70	2.62	2.54	2.45	2.36	2.27	2.17
26	7.72	5.53	4.64	4.14	3.82	3.59	3.42	3.29	3.18	3.09	2.96	2.81	2.66	2.58	2.50	2.42	2.33	2.23	2.13
27	7.68	5.49	4.60	4.11	3.78	3.56	3.39	3.26	3.15	3.06	2.93	2.78	2.63	2.55	2.47	2.38	2.29	2.20	2.10
28	7.64	5.45	4.57	4.07	3.75	3.53	3.36	3.23	3.12	3.03	2.90	2.75	2.60	2.52	2.44	2.35	2.26	2.17	2.06
29	7.60	5.42	4.54	4.04	3.73	3.50	3.33	3.20	3.09	3.00	2.87	2.73	2.57	2.49	2.41	2.33	2.23	2.14	2.03
30	7.56	5.39	4.51	4.02	3.70	3.47	3.30	3.17	3.07	2.98	2.84	2.70	2.55	2.47	2.39	2.30	2.21	2.11	2.01
40	7.31	5.18	4.31	3.83	3.51	3.29	3.12	2.99	2.89	2.80	2.66	2.52	2.37	2.29	2.20	2.11	2.02	1.92	1.80
60	7.08	4.98	4.13	3.65	3.34	3.12	2.95	2.82	2.72	2.63	2.50	2.35	2.20	2.12	2.03	1.94	1.84	1.73	1.60
120	6.85	4.79	3.95	3.48	3.17	2.96	2.79	2.66	2.56	2.47	2.34	2.19	2.03	1.95	1.86	1.76	1.66	1.53	1.38
∞	6.63	4.61	3.78	3.32	3.02	2.80	2.64	2.51	2.41	2.32	2.18	2.04	1.88	1.79	1.70	1.59	1.47	1.32	1.00

Numerator Degrees of Freedom

Source: E. S. Pearson and H. O. Hartley, *Biometrika Tables for Statisticians,* vol. 1 (Cambridge: Cambridge University Press, 1966), pp. 171–73.

Appendix Table O
Values of q_α in Tukey's *HSD* Test

Error d.f.	α	2	3	4	5	6	7	8	9	10	11	12
		\multicolumn{11}{c}{k = Number of means or number of steps between ordered means}										
5	.05	3.64	4.60	5.22	5.67	6.03	6.33	6.58	6.80	6.99	7.17	7.32
	.01	5.70	6.98	7.80	8.42	8.91	9.32	9.67	9.97	10.24	10.48	10.70
6	.05	3.46	4.34	4.90	5.30	5.63	5.90	6.12	6.32	6.49	6.65	6.79
	.01	5.24	6.33	7.03	7.56	7.97	8.32	8.61	8.87	9.10	9.30	9.48
7	.05	3.34	4.16	4.68	5.06	5.36	5.61	5.82	6.00	6.16	6.30	6.43
	.01	4.95	5.92	6.54	7.01	7.37	7.68	7.94	8.17	8.37	8.55	8.71
8	.05	3.26	4.04	4.53	4.89	5.17	5.40	5.60	5.77	5.92	6.05	6.18
	.01	4.75	5.64	6.20	6.62	6.96	7.24	7.47	7.68	7.86	8.03	8.18
9	.05	3.20	3.95	4.41	4.76	5.02	5.24	5.43	5.59	5.74	5.87	5.98
	.01	4.60	5.43	5.96	6.35	6.66	6.91	7.13	7.33	7.49	7.65	7.78
10	.05	3.15	3.88	4.33	4.65	4.91	5.12	5.30	5.46	5.60	5.72	5.83
	.01	4.48	5.27	5.77	6.14	6.43	6.67	6.87	7.05	7.21	7.36	7.49
11	.05	3.11	3.82	4.26	4.57	4.82	5.03	5.20	5.35	5.49	5.61	5.71
	.01	4.39	5.15	5.62	5.97	6.25	6.48	6.67	6.84	6.99	7.13	7.25
12	.05	3.08	3.77	4.20	4.51	4.75	4.95	5.12	5.27	5.39	5.51	5.61
	.01	4.32	5.05	5.50	5.84	6.10	6.32	6.51	6.67	6.81	6.94	7.06
13	.05	3.06	3.73	4.15	4.45	4.69	4.88	5.05	5.19	5.32	5.43	5.53
	.01	4.26	4.96	5.40	5.73	5.98	6.19	6.37	6.53	6.67	6.79	6.90
14	.05	3.03	3.70	4.11	4.41	4.64	4.83	4.99	5.13	5.25	5.36	5.46
	.01	4.21	4.89	5.32	5.63	5.88	6.08	6.26	6.41	6.54	6.66	6.77
15	.05	3.01	3.67	4.08	4.37	4.59	4.78	4.94	5.08	5.20	5.31	5.40
	.01	4.17	4.84	5.25	5.56	5.80	5.99	6.16	6.31	6.44	6.55	6.66
16	.05	3.00	3.65	4.05	4.33	4.56	4.74	4.90	5.03	5.15	5.26	5.35
	.01	4.13	4.79	5.19	5.49	5.72	5.92	6.08	6.22	6.35	6.46	6.56
17	.05	2.98	3.63	4.02	4.30	4.52	4.70	4.86	4.99	5.11	5.21	5.31
	.01	4.10	4.74	5.14	5.43	5.66	5.85	6.01	6.15	6.27	6.38	6.48
18	.05	2.97	3.61	4.00	4.28	4.49	4.67	4.82	4.96	5.07	5.17	5.27
	.01	4.07	4.70	5.09	5.38	5.60	5.79	5.94	6.08	6.20	6.31	6.41
19	.05	2.96	3.59	3.98	4.25	4.47	4.65	4.79	4.92	5.04	5.14	5.23
	.01	4.05	4.67	5.05	5.33	5.55	5.73	5.89	6.02	6.14	6.25	6.34
20	.05	2.95	3.58	3.96	4.23	4.45	4.62	4.77	4.90	5.01	5.11	5.20
	.01	4.02	4.64	5.02	5.29	5.51	5.69	5.84	5.97	6.09	6.19	6.28
24	.05	2.92	3.53	3.90	4.17	4.37	4.54	4.68	4.81	4.92	5.01	5.10
	.01	3.96	4.55	4.91	5.17	5.37	5.54	5.69	5.81	5.92	6.02	6.11
30	.05	2.89	3.49	3.85	4.10	4.30	4.46	4.60	4.72	4.82	4.92	5.00
	.01	3.89	4.45	4.80	5.05	5.24	5.40	5.54	5.65	5.76	5.85	5.93
40	.05	2.86	3.44	3.79	4.04	4.23	4.39	4.52	4.63	4.73	4.82	4.90
	.01	3.82	4.37	4.70	4.93	5.11	5.26	5.39	5.50	5.60	5.69	5.76
60	.05	2.83	3.40	3.74	3.98	4.16	4.31	4.44	4.55	4.65	4.73	4.81
	.01	3.76	4.28	4.59	4.82	4.99	5.13	5.25	5.36	5.45	5.53	5.60
120	.05	2.80	3.36	3.68	3.92	4.10	4.24	4.36	4.47	4.56	4.64	4.71
	.01	3.70	4.20	4.50	4.71	4.87	5.01	5.12	5.21	5.30	5.37	5.44
∞	.05	2.77	3.31	3.63	3.86	4.03	4.17	4.29	4.39	4.47	4.55	4.62
	.01	3.64	4.12	4.40	4.60	4.76	4.88	4.99	5.08	5.16	5.23	5.29

Source: E. S. Pearson and H. O. Hartley, *Biometrika Tables for Statisticians*, vol. 1, (Cambridge: Cambridge University Press, 1966), pp. 192–93.

Appendix Table P
Values of d_L and d_U for the Durbin-Watson Test

						$\alpha = .05$					
	$m = 1$		$m = 2$		$m = 3$		$m = 4$		$m = 5$	
n	d_L	d_U	d_L	d_U	d_L	d_U	d_L	d_U	d_L	d_U
15	1.08	1.36	0.95	1.54	0.82	1.75	0.69	1.97	0.56	2.21
16	1.10	1.37	0.98	1.54	0.86	1.73	0.74	1.93	0.62	2.15
17	1.13	1.38	1.02	1.54	0.90	1.71	0.78	1.90	0.67	2.10
18	1.16	1.39	1.05	1.53	0.93	1.69	0.82	1.87	0.71	2.06
19	1.18	1.40	1.08	1.53	0.97	1.68	0.86	1.85	0.75	2.02
20	1.20	1.41	1.10	1.54	1.00	1.68	0.90	1.83	0.79	1.99
21	1.22	1.42	1.13	1.54	1.03	1.67	0.93	1.81	0.83	1.96
22	1.24	1.43	1.15	1.54	1.05	1.66	0.96	1.80	0.86	1.94
23	1.26	1.44	1.17	1.54	1.08	1.66	0.99	1.79	0.90	1.92
24	1.27	1.45	1.19	1.55	1.10	1.66	1.01	1.78	0.93	1.90
25	1.29	1.45	1.21	1.55	1.12	1.66	1.04	1.77	0.95	1.89
26	1.30	1.46	1.22	1.55	1.14	1.65	1.06	1.76	0.98	1.88
27	1.32	1.47	1.24	1.56	1.16	1.65	1.08	1.76	1.01	1.86
28	1.33	1.48	1.26	1.56	1.18	1.65	1.10	1.75	1.03	1.85
29	1.34	1.48	1.27	1.56	1.20	1.65	1.12	1.74	1.05	1.84
30	1.35	1.49	1.28	1.57	1.21	1.65	1.14	1.74	1.07	1.83
31	1.36	1.50	1.30	1.57	1.23	1.65	1.16	1.74	1.09	1.83
32	1.37	1.50	1.31	1.57	1.24	1.65	1.18	1.73	1.11	1.82
33	1.38	1.51	1.32	1.58	1.26	1.65	1.19	1.73	1.13	1.81
34	1.39	1.51	1.33	1.58	1.27	1.65	1.21	1.73	1.15	1.81
35	1.40	1.52	1.34	1.58	1.28	1.65	1.22	1.73	1.16	1.80
36	1.41	1.52	1.35	1.59	1.29	1.65	1.24	1.73	1.18	1.80
37	1.42	1.53	1.36	1.59	1.31	1.66	1.25	1.72	1.19	1.80
38	1.43	1.54	1.37	1.59	1.32	1.66	1.26	1.72	1.21	1.79
39	1.43	1.54	1.38	1.60	1.33	1.66	1.27	1.72	1.22	1.79
40	1.44	1.54	1.39	1.60	1.34	1.66	1.29	1.72	1.23	1.79
45	1.48	1.57	1.43	1.62	1.38	1.67	1.34	1.72	1.29	1.78
50	1.50	1.59	1.46	1.63	1.42	1.67	1.38	1.72	1.34	1.77
55	1.53	1.60	1.49	1.64	1.45	1.68	1.41	1.72	1.38	1.77
60	1.55	1.62	1.51	1.65	1.48	1.69	1.44	1.73	1.41	1.77
65	1.57	1.63	1.54	1.66	1.50	1.70	1.47	1.73	1.44	1.77
70	1.58	1.64	1.55	1.67	1.52	1.70	1.49	1.74	1.46	1.77
75	1.60	1.65	1.57	1.68	1.54	1.71	1.51	1.74	1.49	1.77
80	1.61	1.66	1.59	1.69	1.56	1.72	1.53	1.74	1.51	1.77
85	1.62	1.67	1.60	1.70	1.57	1.72	1.55	1.75	1.52	1.77
90	1.63	1.68	1.61	1.70	1.59	1.73	1.57	1.75	1.54	1.78
95	1.64	1.69	1.62	1.71	1.60	1.73	1.58	1.75	1.56	1.78
100	1.65	1.69	1.63	1.72	1.61	1.74	1.59	1.76	1.57	1.78

Source: J. Durbin and G. S. Watson, "Testing for Serial Correlation in Least Squares Regression II," *Biometrika* 38(June 1951):173–75.

Note: The α levels are for one-tailed tests; for two-tailed tests, consult $\alpha/2$.

Appendix Table P (continued)

		$\alpha = .025$
	$m = 1$		$m = 2$		$m = 3$		$m = 4$		$m = 5$	
n	d_L	d_U	d_L	d_U	d_L	d_U	d_L	d_U	d_L	d_U
15	0.95	1.23	0.83	1.40	0.71	1.61	0.59	1.84	0.48	2.09
16	0.98	1.24	0.86	1.40	0.75	1.59	0.64	1.80	0.53	2.03
17	1.01	1.25	0.90	1.40	0.79	1.58	0.68	1.77	0.57	1.98
18	1.03	1.26	0.93	1.40	0.82	1.56	0.72	1.74	0.62	1.93
19	1.06	1.28	0.96	1.41	0.86	1.55	0.76	1.72	0.66	1.90
20	1.08	1.28	0.99	1.41	0.89	1.55	0.79	1.70	0.70	1.87
21	1.10	1.30	1.01	1.41	0.92	1.54	0.83	1.69	0.73	1.84
22	1.12	1.31	1.04	1.42	0.95	1.54	0.86	1.68	0.77	1.82
23	1.14	1.32	1.06	1.42	0.97	1.54	0.89	1.67	0.80	1.80
24	1.16	1.33	1.08	1.43	1.00	1.54	0.91	1.66	0.83	1.79
25	1.18	1.34	1.10	1.43	1.02	1.54	0.94	1.65	0.86	1.77
26	1.19	1.35	1.12	1.44	1.04	1.54	0.96	1.65	0.88	1.76
27	1.21	1.36	1.13	1.44	1.06	1.54	0.99	1.64	0.91	1.75
28	1.22	1.37	1.15	1.45	1.08	1.54	1.01	1.64	0.93	1.74
29	1.24	1.38	1.17	1.45	1.10	1.54	1.03	1.63	0.96	1.73
30	1.25	1.38	1.18	1.46	1.12	1.54	1.05	1.63	0.98	1.73
31	1.26	1.39	1.20	1.47	1.13	1.55	1.07	1.63	1.00	1.72
32	1.27	1.40	1.21	1.47	1.15	1.55	1.08	1.63	1.02	1.71
33	1.28	1.41	1.22	1.48	1.16	1.55	1.10	1.63	1.04	1.71
34	1.29	1.41	1.24	1.48	1.17	1.55	1.12	1.63	1.06	1.70
35	1.30	1.42	1.25	1.48	1.19	1.55	1.13	1.63	1.07	1.70
36	1.31	1.43	1.26	1.49	1.20	1.56	1.15	1.63	1.09	1.70
37	1.32	1.43	1.27	1.49	1.21	1.56	1.16	1.62	1.10	1.70
38	1.33	1.44	1.28	1.50	1.23	1.56	1.17	1.62	1.12	1.70
39	1.34	1.44	1.29	1.50	1.24	1.56	1.19	1.63	1.13	1.69
40	1.35	1.45	1.30	1.51	1.25	1.57	1.20	1.63	1.15	1.69
45	1.39	1.48	1.34	1.53	1.30	1.58	1.25	1.63	1.21	1.69
50	1.42	1.50	1.38	1.54	1.34	1.59	1.30	1.64	1.26	1.69
55	1.45	1.52	1.41	1.56	1.37	1.60	1.33	1.64	1.30	1.69
60	1.47	1.54	1.44	1.57	1.40	1.61	1.37	1.65	1.33	1.69
65	1.49	1.55	1.46	1.59	1.43	1.62	1.40	1.66	1.36	1.69
70	1.51	1.57	1.48	1.60	1.45	1.63	1.42	1.66	1.39	1.70
75	1.53	1.58	1.50	1.61	1.47	1.64	1.45	1.67	1.42	1.70
80	1.54	1.59	1.52	1.62	1.49	1.65	1.47	1.67	1.44	1.70
85	1.56	1.60	1.53	1.63	1.51	1.65	1.49	1.68	1.46	1.71
90	1.57	1.61	1.55	1.64	1.53	1.66	1.50	1.69	1.48	1.71
95	1.58	1.62	1.56	1.65	1.54	1.67	1.52	1.69	1.50	1.71
100	1.59	1.63	1.57	1.65	1.55	1.67	1.53	1.70	1.51	1.72

Appendix Table P *(continued)*

					$\alpha = .01$					
	$m = 1$		$m = 2$		$m = 3$		$m = 4$		$m = 5$	
n	d_L	d_U	d_L	d_U	d_L	d_U	d_L	d_U	d_L	d_U
15	0.81	1.07	0.70	1.25	0.59	1.46	0.49	1.70	0.39	1.96
16	0.84	1.09	0.74	1.25	0.63	1.44	0.53	1.66	0.44	1.90
17	0.87	1.10	0.77	1.25	0.67	1.43	0.57	1.63	0.48	1.85
18	0.90	1.12	0.80	1.26	0.71	1.42	0.61	1.60	0.52	1.80
19	0.93	1.13	0.83	1.26	0.74	1.41	0.65	1.58	0.56	1.77
20	0.95	1.15	0.86	1.27	0.77	1.41	0.68	1.57	0.60	1.74
21	0.97	1.16	0.89	1.27	0.80	1.41	0.72	1.55	0.63	1.71
22	1.00	1.17	0.91	1.28	0.83	1.40	0.75	1.54	0.66	1.69
23	1.02	1.19	0.94	1.29	0.86	1.40	0.77	1.53	0.70	1.67
24	1.04	1.20	0.96	1.30	0.88	1.41	0.80	1.53	0.72	1.66
25	1.05	1.21	0.98	1.30	0.90	1.41	0.83	1.52	0.75	1.65
26	1.07	1.22	1.00	1.31	0.93	1.41	0.85	1.52	0.78	1.64
27	1.09	1.23	1.02	1.32	0.95	1.41	0.88	1.51	0.81	1.63
28	1.10	1.24	1.04	1.32	0.97	1.41	0.90	1.51	0.83	1.62
29	1.12	1.25	1.05	1.33	0.99	1.42	0.92	1.51	0.85	1.61
30	1.13	1.26	1.07	1.34	1.01	1.42	0.94	1.51	0.88	1.61
31	1.15	1.27	1.08	1.34	1.02	1.42	0.96	1.51	0.90	1.60
32	1.16	1.28	1.10	1.35	1.04	1.43	0.98	1.51	0.92	1.60
33	1.17	1.29	1.11	1.36	1.05	1.43	1.00	1.51	0.94	1.59
34	1.18	1.30	1.13	1.36	1.07	1.43	1.01	1.51	0.95	1.59
35	1.19	1.31	1.14	1.37	1.08	1.44	1.03	1.51	0.97	1.59
36	1.21	1.32	1.15	1.38	1.10	1.44	1.04	1.51	0.99	1.59
37	1.22	1.32	1.16	1.38	1.11	1.45	1.06	1.51	1.00	1.59
38	1.23	1.33	1.18	1.39	1.12	1.45	1.07	1.52	1.02	1.58
39	1.24	1.34	1.19	1.39	1.14	1.45	1.09	1.52	1.03	1.58
40	1.25	1.34	1.20	1.40	1.15	1.46	1.10	1.52	1.05	1.58
45	1.29	1.38	1.24	1.42	1.20	1.48	1.16	1.53	1.11	1.58
50	1.32	1.40	1.28	1.45	1.24	1.49	1.20	1.54	1.16	1.59
55	1.36	1.43	1.32	1.47	1.28	1.51	1.25	1.55	1.21	1.59
60	1.38	1.45	1.35	1.48	1.32	1.52	1.28	1.56	1.25	1.60
65	1.41	1.47	1.38	1.50	1.35	1.53	1.31	1.57	1.28	1.61
70	1.43	1.49	1.40	1.52	1.37	1.55	1.34	1.58	1.31	1.61
75	1.45	1.50	1.42	1.53	1.39	1.56	1.37	1.59	1.34	1.62
80	1.47	1.52	1.44	1.54	1.42	1.57	1.39	1.60	1.36	1.62
85	1.48	1.53	1.46	1.55	1.43	1.58	1.41	1.60	1.39	1.63
90	1.50	1.54	1.47	1.56	1.45	1.59	1.43	1.61	1.41	1.64
95	1.51	1.55	1.49	1.57	1.47	1.60	1.45	1.62	1.42	1.64
100	1.52	1.56	1.50	1.58	1.48	1.60	1.46	1.63	1.44	1.65

GLOSSARY

α risk in hypothesis testing, the probability of making a type I error **(9)**

absolute class frequency the absolute number of observations that fall into a given class **(3)**

absolute frequency distribution a tabular summary of a data set showing the absolute class frequencies in each of several collectively exhaustive and mutually exclusive classes **(3)**

acceptance number see **critical value**

acceptance region in a hypothesis test, that range of possible values of the test statistic which signals the acceptance of the null hypothesis **(9)**

action branches branches emanating from an action point in a decision-tree diagram and representing the possible actions of the decision maker **(17)**

action point in a decision-tree diagram, a point of choice (at which the decision maker is in control), symbolized by a square **(17)**

actions in decison theory, the mutually exclusive decision alternatives open to a decision maker **(17)**

addition law a law of probability theory that is used to compute the probability for the occurrence of a *union* of two or more events; according to the general law, for any two events $p(A \text{ } or \text{ } B) = p(A) + p(B) - p(A \text{ } and \text{ } B)$; according to the special law for two mutually exclusive events, $p(A \text{ } or \text{ } B) = p(A) + p(B)$ **(5)**

additive time-series model a time-series model that expresses the actual value of a time series as the sum of its components—for example, $Y = T + C + S + I$ **(15)**

adjusted sample coefficient of multiple determination a measure, denoted by $\overline{R}^2$ (and equal to R^2 adjusted for degrees of freedom), that is an unbiased estimator of the corresponding population coefficient **(14)**

alternative hypothesis this hypothesis is the second of two opposing hypotheses in a hypothesis test, symbolized by H_A, and often (but not necessarily) representing that proposition about an unknown population parameter that is tentatively assumed to be false **(9)**

analysis of covariance (ANOCOVA) a combination of (a) regression analysis that incorporates quantitative and dummy variables and (b) an analysis of variance **(14)**

analysis of residuals a careful study of the differences, $e = Y - \hat{Y}$, between actual observations, Y, concerning the dependent variable and the associated estimates, $\hat{Y}$, made by the regression equation **(14)**

analysis of variance (ANOVA) a statistical technique specially designed to test whether the means of more than two quantitative populations are equal **(12)**

analytical statistics a branch of the discipline that is concerned with developing and utilizing techniques for properly analyzing (or drawing inferences from) numerical information **(1)**

ANOVA table in the analysis of variance, a table that shows, for each source of variation, the sum of squares, the degrees of freedom and their ratio, the mean square (which is the desired estimate of the population variance), and more **(12)**

anticipatory surveys surveys of various economic actors (such as households, businesses, and governments) to determine their economic plans (for example, with regard to spending) **(15)**

arithmetic mean (μ or $\overline{X}$) a measure of central tendency obtained from ungrouped data by adding the values observed and dividing their sum by the number of observations **(4)**

array see **ordered array**

autocorrelation see **serial correlation**

β risk in hypothesis testing, the probability of making a type II error **(9)**

backward-elimination method a method of developing a multiple regression equation that starts with *all* potential independent *X* variables in a model designed to explain the behavior of the dependent *Y* variable and then eliminates one *X* at a time—always the one the coefficient of which has that *t* value, among all insignificant ones, that is closest to zero **(14)**

backward induction a process by which a decision problem is solved with the help of a decision-tree diagram and that involves the computation, with the help of ultimate payoff values and event-branch probability values, of expected payoff values at each fork of the tree **(17)**

bar chart a graphical portrayal of qualitative or discrete quantitative variables by a series of noncontiguous horizontal bars the lengths of which are proportional to the values that are to be depicted **(3)**

barometric indicators qualitative indicators of business and economic conditions, such as the results of anticipatory surveys and a variety of leading economic indicators **(15)**

base period in index construction, the arbitrarily chosen period to which the magnitudes of the prices (or quantities) of other periods are being compared **(16)**

base shifting converting an index-number series to a new base by successively dividing each value of the original series by the original index number for the year that is to become the new base and then multiplying each result by 100 **(16)**

basic outcome any one of a random experiment's possible outcomes the occurrence of which rules out the occurrence of all the alternative outcomes **(5)**

Bayesian estimation a type of estimation procedure that views the parameter being estimated not as a constant, but, just like the estimator, as a random variable **(8)**

Bayes's postulate see **equal-likelihood criterion**

Bayes's strategy the optimal strategy selected in decision-making problems that are solved by optimizing the expected payoff **(17)**

Bayes's theorem a formula for revising an initial subjective (prior) probability value on the basis of results obtained by an empirical investigation and for, thus, obtaining a new (posterior) probability value **(5)**

Bernoulli process a sequence of *n* identical trials of a random experiment such that each trial (a) produces one of two possible complementary outcomes that are conventionally called *success* and *failure* and (b) is independent of any other trial so that the probability of success or failure is constant from trial to trial **(6)**

bias see **systematic error**

bimodal frequency distribution a frequency distribution in which two different values occur with the highest frequency, or almost that **(4)**

binary variables see **dummy variables**

binomial coefficient (c_x^n) a part of the binomial formula that gives the number of permutations of *x* successes and $n - x$ failures that can be achieved in *n* trials that satisfy the conditions of a Bernoulli process **(6)**

binomial formula a formula for calculating the probability of *x* successes in *n* trials of a random experiment that satisfies the conditions of a Bernoulli process **(6)**

binomial probability distribution a probability distribution that shows the probabilities associated with possible values of a discrete random variable that are generated by a Bernoulli process **(6)**

binomial probability tables tables that list binomial probabilities (probabilities of *x* successes in a Bernoulli process) for various combinations of possible values of *n* (number of trials) and π (probability of success in any one trial) **(6)**

binomial random variable the number of successes achieved in a Bernoulli process **(6)**

bivariate data set a data set containing information on two variables **(2)**

blocking a procedure that eliminates the effects of extraneous factors during an experiment by forming blocks of experimental units within each of which all units are as alike as possible with respect to those factors **(2)**

blocks mean square *(BMS)* in the analysis of variance, the population variance estimate, s_A^2, based on the observed variation among the blocks samples **(12)**

blocks sum of squares *(BSS)* in the analysis of variance, the sum of the squared deviations between each block sample mean and the grand mean, multiplied by the number of observations made for each block multiplied by the number of observations per cell (which may well equal 1) **(12)**

blocks variation in the analysis of variance, variation among block sample means that is attributable to inherent differences among blocks of experimental units **(12)**

census a complete survey, in which observations are made about one or more characteristics of interest for every elementary unit that exists **(2)**

central limit theorem the theorem that states: if $\overline{X}$ is the mean of (or P is the proportion in) a random sample taken from a population and if the population values are *not* normally distributed, the sampling distribution of $\overline{X}$ (or P) nevertheless approaches a normal distribution as sample size increases—provided sample size, n, remains small relative to population size, N; this approximation is near-perfect for samples of $n \geq 30$ but $n < .05N$ (or when $n \cdot \pi$ as well as $n(1 - \pi) \geq 5$) **(8)**

chain indexes series of index numbers each of which uses the immediately preceding period as a base rather than using a common base **(16)**

chance error see **random error**

chance variable see **random variable**

Chebyshev's theorem the theorem that states: regardless of the shape of a population's frequency distribution, the proportion of observations falling within k standard deviations of the mean is at least $1 - (1/k^2)$, given that k is 1 or more **(4)**

chi-square distribution the sampling distribution of the χ^2 statistic **(10)**

chi-square statistic the sum of all the ratios that can be constructed by taking the difference between each observed and expected frequency (in, for example, a contingency table), squaring the difference, and dividing this squared deviation by the expected frequency **(10)**

circularity test one of several index-number quality tests; a price-index formula, for instance, is said to pass the test if the price index of year 1 with year 0 base, when multiplied by the price index of year 2 with year 1 base and divided by 100, equals the independently calculated price index of year 2 with year 0 base **(16)**

classes groupings of data **(3)**

class frequency see **absolute, relative,** and **cumulative class frequency**

classical time-series model a model that attempts to explain the pattern observed in an actual time series by the presence of four components: trend, cyclical, seasonal, and irregular components **(15)**

class mark the midpoint of a class width **(3)**

class width the difference between the numerical lower and upper limit of a class of quantitative data **(3)**

clustered random sample a subset of a frame, or of an associated population, chosen by taking separate censuses in a randomly chosen subset of geo-graphically distinct clusters into which the frame or population is naturally subdivided **(2)**

coefficient of kurtosis (K or k) a measure of a frequency curve's shape that classifies such curves as platykurtic, mesokurtic, or leptokurtic **(4)**

coefficient of skewness see **Pearson's coefficient of skewness**

coefficient of variation (V or v) a measure of relative dispersion; equal to the ratio of the standard deviation to the arithmetic mean **(4)**

coincident economic indicators time series that precisely coincide with the business cycle; they turn down when the GNP reaches its peak and turn up when it reaches its trough **(15)**

collectively exhaustive events different random events that jointly contain all the basic outcomes in the sample space **(5)**

column chart a graphical portrayal of qualitative or discrete quantitative variables by a series of non-contiguous vertical columns the lengths of which are proportional to the values that are to be depicted **(3)**

combinations different selections of items such that possible alternative sequences among the components of any one such selection are considered immaterial **(5)**

compatible events different random events that have at least some basic outcomes in common **(5)**

complementary events two random events such that precisely all those basic outcomes that are not contained in one are contained in the other **(5)**

completely randomized design see **randomized group design**

composite event any combination of two or more basic outcomes **(5)**

composite price index a number that compares a set of prices for a variety of items at one time or place with such a set for the same items at another time or place **(16)**

conditional mean of Y in regression analysis, the mean, $\mu_{Y \cdot X}$, of a conditional probability distribution of Y **(13)**

conditional probability a measure that a particular event occurs, given that another event has already occurred or is certain to occur **(5)**

conditional probability distribution of Y in regression analysis, the normal curve for a given population of Y values that is associated with a specified value of X **(13)**

conditional standard deviation of Y in regression analysis, the standard deviation, $\sigma_{Y \cdot X}$, of a conditional probability distribution of Y **(13)**

confidence band in regression analysis, a band of values lying above and below an estimated regression line and indicating the limits of the prediction interval for each $\mu_{Y \cdot X}$ **(13)**

confidence intervals intervals within which population parameters presumably lie **(8)**

confidence level the percentage of interval estimates (obtained from repeated samples, each of size *n,* taken from a given population) that can be expected to contain the actual value of the parameter being estimated **(8)**

confidence level of the hypothesis test the probability (equal to $1 - \alpha$) of avoiding a type I error when the null hypothesis is true **(9)**

confidence limits the two limits that define the interval within which a population parameter that is being estimated presumably lies **(8)**

consistent estimator a sample statistic such that the probability of its being close to the parameter being estimated gets ever larger (and, therefore, approaches unity) as the sample size increases **(8)**

consumer price index (CPI) an index measuring the average change in U.S. consumer prices over time **(16)**

contingency table a table that classifies data according to two or more categories associated with each of two qualitative variables that may or may not be statistically independent **(10)**

continuous quantitative variable a quantitative variable observations about which can assume values at all points on a scale of values, with no breaks between possible values **(2)**

control charts graphical devices that highlight the average performance of data series and the dispersion around this average so that average and dispersion of the past become standards for controlling performance in the present **(4)**

control group a set of experimental units that is not exposed to anything new **(2)**

convenience sample a subset of a frame, or of an associated population, resulting when expediency is the primary consideration in selecting elementary units for observation and simply the most easily accessible units are chosen **(2)**

correlation analysis a statistical method the central focus of which is the establishment of an *index* that provides, in a single number, an instant picture of the strength of association between two variables **(13)**

correlation matrix a table showing simple correlation coefficients between each possible pair of variables included in a regression model **(14)**

criterion of insufficient reason see **equal-likelihood criterion**

critical value in hypothesis testing, the value of a test statistic that divides all possible values into acceptance and rejection regions **(9)**

cross-section data numerical values pertaining to units of different populations that have been observed simultaneously at the same point in time (in the case of a stock variable) or during the same period of time (in the case of a flow variable) **(15)**

cumulative class frequency the sum of the (absolute or relative) class frequencies for all classes up to and including the class in question, beginning at either end of the frequency distribution **(3)**

cumulative frequency distribution a tabular summary of a data set showing for each of several collectively exhaustive and mutually exclusive classes the absolute number or proportion of observations that are less than or equal to the upper limits of the classes in question (LE type) or that are more than or equal to their lower limits (ME type) **(3)**

cumulative probability distribution a probability distribution that shows the probabilities of a random variable being less than or equal to any given possible value **(6)**

cutoff point see **critical value**

cyclical component in time-series analysis, wide up-and-down swings of the variable of interest around the trend, with the swings lasting from one to several years each and typically differing in length and amplitude from one cycle to the next **(15)**

data set a collection of observations about one or more characteristics of interest, for one or more elementary units **(2)**

data transformation a systematic conversion of all the numerical values of some variable, as by squaring them, taking their reciprocals, or finding their logarithms **(13)**

datum a single observation about a specified characteristic of interest possessed by an elementary unit **(2)**

deciles values in a data set that divide the total of observations into ten parts, each of which contains .10 (or 10 percent) of the observed values **(4)**

decision branches see **action branches**

decision fork see **action point**

decision making under uncertainty any situation in which the ultimate outcome of a decision maker's choice depends on chance **(17)**

decision node see **action point**

decision point see **action point**

decision rule a hypothesis-testing rule that specifies in advance, for all possible values of a test statistic that might be observed in a sample, when the null hypothesis should be accepted and when it should be rejected in favor of the alternative hypothesis **(9)**

decision theory a variety of methods that can be employed in the systematic analysis and solution of decision-making problems that arise because uncertainty exists about future events over which the decision maker has no control, but which are bound to influence the ultimate outcome of a decision **(17)**

decision tree a summary of a decision-making situation in the context of uncertainty, showing graphically and in chronological order from left to right every potential action, event, and payoff **(17)**

deductive reasoning the drawing of inferences about an unknown part from a known whole **(1)**

deflation a process that removes the effect of price changes from a current-dollar time series and, thus, restates the series in terms of constant dollars **(16)**

degrees of freedom *(d.f.)* the number of values that can be freely chosen when calculating a statistic, equal to sample size minus 1 in the case of a single sample **(8)**

dependent events two random events such that the probability of one event is affected by the occurrence of the other event; hence, $p(A) \neq p(A|B)$ **(5)**

dependent variable in regression analysis, the variable (denoted by Y) the value of which is assumed unknown and that is being explained or predicted on the basis of its relationship with one or more other variables **(13)**

descriptive statistics a branch of the discipline that is concerned with developing and utilizing techniques for the careful collection and effective presentation of numerical information **(1)**

deseasonalized time series the original values of a time series divided by seasonal indexes and, thus, reflecting only trend, cyclical, and irregular components **(15)**

deterministic relationship a relationship between any two variables, X and Y, such that the value of Y is uniquely determined whenever the value of X is specified **(13)**

dichotomous qualitative variable a qualitative variable about which observations can be made in only two categories **(2)**

diffusion index an index indicating the percentage of selected leading economic indicators that are rising in a given month **(15)**

direct relationship a relationship between any two variables, X and Y, such that the values of Y increase with larger values of X **(13)**

discrete quantitative variable a quantitative variable such that observations about it can assume values only at specific points on a scale of values, with gaps between **(2)**

disjoint events see **mutually exclusive events**

distribution-free tests nonparametric tests **(11)**

dominant action in decision theory, an action that is unambiguously superior to an alternative action because it produces payoffs that are as good as or better than those of the alternative action no matter which event occurs **(17)**

double-blind experiments experiments in which response bias is controlled by letting neither the subjects nor the judges know who is receiving which type of treatment **(2)**

dummy variables variables that can take on two values only—namely, 0 or 1—and that are used to indicate the absence or presence of a particular qualitative characteristic **(14)**

Durbin-Watson test a test for the detection of serial correlation **(14)**

econometric model typically, a system of regression equations estimated from past time-series data and used to show the effect of various independent variables on various dependent variables **(15)**

Edgeworth price index a weighted aggregative price index that uses the sum of base-year and current-year quantities as weights **(16)**

efficiency of sample information *(ESI)* the ratio of the expected value of sample information to the (prior) expected value of perfect information, multiplied by 100 **(17)**

efficient estimator that sample statistic, among all the available unbiased estimators, which has the smallest variance for a given sample size **(8)**

elementary event see **simple event**

elementary unit a person or object possessing a characteristic in which a statistician is interested **(2)**

equal-likelihood criterion in decision theory, the assignment of equal prior probabilities to all possible events when absolutely nothing is known about the likelihood of occurrence of these events **(17)**

error see **experimental error, random error,** and **unexplained variation**

error mean square *(EMS)* in the analysis of variance, the population variance estimate, s_w^2, based on the observed variation within samples; the denominator of the F statistic **(12)**

error of acceptance see **type II error**

error of rejection see **type I error**

error sum of squares *(ESS)* in the analysis of variance, the sum of all samples' sums of squared deviations of individual observations from their sample mean; the total sum of squares minus explained variation **(12,13)**

escalator clauses clauses in legal contracts according to which nominal monetary payments are raised periodically and automatically in accordance with increases in the consumer price index so as to preserve the real purchasing power of these payments **(16)**

estimated partial-regression coefficient the coefficient of an independent variable in an estimated multiple-regression equation; it gives the net change in Y for a unit change in that independent variable, while holding other independent variables constant **(14)**

estimated regression coefficients the values of a and b found in the equation of the estimated regression line, $\hat{Y}_X = a + bX$ **(13)**

estimated regression line the regression line calculated from sample data by the method of least squares **(13)**

estimation the process of inferring the values of unknown population parameters from known sample statistics **(8)**

estimator the type of sample statistic that serves the purpose of making inferences about a given type of parameter **(8)**

event branches branches emanating from an event point in a decision-tree diagram and representing the possible events confronting the decision maker **(17)**

event point in a decision-tree diagram, a point of chance (at which the decision maker exercises no control but at which "nature" is in charge), symbolized by a circle **(17)**

events in decision theory, the mutually exclusive future occurrences that will affect the outcome of any present action taken but that are not under the control of a decision maker **(17)**

exact hypothesis a hypothesis that specifies a single value for an unknown parameter **(9)**

expected monetary value *(EMV)* the sum of the weighted payoffs associated with an action, the weights being the probabilities of the alternative events that produce the various possible payoffs **(17)**

expected-monetary-value criterion one of several probabilistic criteria for making decisions under uncertainty according to which a decision maker determines an expected monetary value for each possible action and selects the action with the optimal expected monetary value (the largest *EMV* is optimal if the objective is to maximize some benefit; the smallest *EMV* is optimal if the objective is to minimize some cost) **(17)**

expected opportunity loss *(EOL)* the sum of the weighted opportunity-loss values associated with an action, the weights being the probabilities of the alternative events that produce the various possible opportunity losses **(17)**

expected-opportunity-loss criterion one of several probabilistic criteria for making decisions under uncertainty according to which a decision maker determines an expected opportunity loss for each possible action and selects the action with the smallest of these values **(17)**

expected regret value see **expected opportunity loss**

expected utility *(EU)* the sum of the weighted utilities associated with an action, the weights being the probabilities of the alternative events that produce the various possible utility payoffs **(17)**

expected-utility criterion one of several probabilistic criteria for making decisions under uncertainty according to which a decision maker determines the expected utility for each possible action and selects the action that maximizes it **(17)**

expected value *(E_R)* the weighted arithmetic mean of a random variable; that value of a random variable which one can expect to find on the average by numerous repetitions of the random experiment that generates the variable's actual value **(6)**

expected value of perfect information *(EVPI)* the maximum amount a decision maker can be expected to pay for obtaining perfect information about future events and, thus, for eliminating uncertainty completely (given the objective of maximizing some benefit, it equals the difference between the expected benefit when the optimal decision is made with perfect information and the expected benefit when that decision is made under uncertainty; given the objective of minimizing some cost, it equals the difference between the expected cost when the optimal decision is made under uncertainty and the expected cost when that decision is made with perfect information) **(17)**

expected value of sample information *(EVSI)* the maximum amount a decision maker can be expected to pay for obtaining admittedly imperfect information about future events and, thus, for reducing, rather than eliminating, uncertainty (given the objective of maximizing some benefit, it equals the difference between the expected benefit when an optimal decision is made with the sample information and before paying for it and the expected benefit when that decision is made without it; given the objective of minimizing some cost, it equals the difference between the expected cost when the optimal decision is made without the sample information and the expected cost when that decision is made with it, but before paying for it) **(17)**

experiment the collection of data from elementary units, while exercising control over some or all factors that may make these units different from one another and that may, therefore, affect the characteristic of interest being observed **(2)**

experimental design a plan for assigning treatments to experimental units under controlled conditions and, thus, for generating valid data **(2)**

experimental error the difference between the value of a variable obtained by performing a single experiment and the value obtained by averaging the results of a large number of identical experiments **(2)**

experimental group a set of experimental units that is exposed to something new **(2)**

experimental units elementary units; the recipients of treatments in controlled experiments **(2)**

explained deviation of Y in regression analysis, the difference between the regression estimate of an individual Y, or $\hat{Y}_X$, and the mean, $\overline{Y}$, of all the observations of the dependent variable **(13)**

explained variable see **dependent variable**

explained variance in the analysis of variance, the numerator of the F statistic **(12)**

explained variation in the analysis of variance, the variation among sample means; it is attributed to inherent differences among the sampled populations and equals the treatments sum of squares (1-way ANOVA) or the treatments plus blocks sums of squares (2-way ANOVA without interaction) or the treatments plus blocks plus interactions sums of squares (2-way ANOVA with interaction) **(12)**

explained variation of Y in regression analysis, the sum of the squares of all the explained deviations

of Y, or $\Sigma(\hat{Y}_X - \overline{Y})^2$, also called the **regression sum of squares** **(13)**

explanatory variable see **independent variable**

exponential probability distribution a probability density function for a continuous random variable that measures intervals of time or space between consecutive events in a Poisson process **(7)**

exponential random variable the uncertain time or space between any two consecutive events in a Poisson process **(7)**

exponential smoothing a forecasting procedure that produces self-correcting forecasts by means of a built-in adjustment mechanism that corrects for earlier forecasting errors **(15)**

extensive-form analysis a form of preposterior analysis that uses backward induction on a decision tree to identify the optimal strategy **(17)**

factorial the product of a series of positive whole numbers that descends from a given number, n, down to 1 **(5)**

factorial product see **factorial**

factor-reversal test one of several index-number quality tests; a quantity-weighted price-index formula, for instance, is said to pass the test if division of the price index by 100, and subsequent multiplication by a corresponding quantity index, produces a value equal to an independently derived value index **(16)**

fair gamble a gamble in which the expected monetary value of what is given up is precisely equal to the expected monetary value of what is received **(17)**

F distributions the different members of the continuous probability distribution family of the F statistic that are distinguished by different numerator and denominator degrees of freedom associated with the two variance estimates making up this statistic **(12)**

fiducial interval in the theory of estimation, a range of values that can be trusted, with a specified probability, to contain the value of some parameter **(9)**

first-order serial correlation serial correlation such that e_t is correlated with e_{t-1} **(14)**

first quartile the .25 fractile or 25th percentile in a data set below which a quarter of all observations lie **(4)**

Fisher's ideal index a weighted aggregative (price or quantity) index equal to the geometric mean of the Laspeyres and Paasche indexes **(16)**

forecasting making statements about unknown, uncertain, and, typically, future events **(15)**

forward-selection method a method of developing a multiple regression equation that starts with no independent X variables in a model designed to explain the behavior of the dependent Y variable and then adds one X at a time—always the X that provides the greatest decrease in the hitherto unexplained variation in Y **(14)**

fractile a value in a data set below which a specified proportion of all values is found **(4)**

frame a complete listing of all elementary units relevant to a statistical investigation **(2)**

frequency curve a graphical portrayal, by a smooth curve, of an absolute or relative frequency distribution of a continuous quantitative variable **(3)**

frequency density the ratio of class frequency to class width **(3)**

frequency distribution see **absolute, relative,** and **cumulative frequency distribution**

frequency histogram a graphical portrayal, as a series of contiguous rectangles, of an absolute or relative frequency distribution of a continuous quantitative variable in such a way that the areas of the rectangles correspond to the frequencies being depicted **(3)**

frequency polygon a graphical portrayal (as a many-sided figure) of an absolute or relative frequency distribution of a continuous quantitative variable **(3)**

F statistic in the analysis of variance, the ratio of the explained to the unexplained variance **(12)**

goodness-of-fit test a statistial test to determine the likelihood that sample data have been generated from a population that conforms to a specified type of probability distribution **(10)**

grand mean ($\bar{\bar{X}}$) the arithmetic mean of two or more means **(12)**

heteroscedasticity the characteristic of different statistical populations having unequal variances **(14)**

histogram see **frequency histogram**

homoscedasticity the characteristic of different statistical populations having equal variances **(12)**

hypergeometric formula a formula for calculating the probability of x successes when a random sample of n is drawn without replacement from a population of N within which S units exist with the characteristic denoting success **(6)**

hypergeometric probability distribution a probability distribution that shows the probabilities associated with possible values of a discrete random variable in situations in which these values are generated by sampling without replacement and in which the probability of success, therefore, changes from one trial to the next **(6)**

hypergeometric random variable the number of successes achieved when a random variable of n is drawn without replacement from a population of N within which S units exist with the characteristic denoting success **(6)**

hypothesis a proposition tentatively advanced as possibly true **(9)**

hypothesis testing a systematic approach to assessing beliefs about reality; confronting a belief with evidence and deciding whether that belief can be maintained as reasonable or must be discarded as untenable **(9)**

implicit price deflator a price index found by dividing any value of a current-dollar time series by the corresponding value of a constant-dollar time series, and multiplying by 100 **(16)**

inadmissible action in decision theory, an action that is unambiguously inferior to an alternative action because it produces payoffs that are at best as good as and often worse than those of the alternative action no matter which event occurs **(17)**

incompatible events see **mutually exclusive events**

independent events two random events such that the probability of one event is not affected by the occurrence of the other event; hence, $p(A) = p(A|B)$ **(5)**

independent variable in regression analysis, the variable (denoted by X) the value of which is assumed known and that is being used to explain or predict the value of the other variable **(13)**

index-number problem the problem of being able to construct a different index for a given phenomenon depending on which one of several equally logical sets of weights is being employed **(16)**

index numbers numbers that measure the magnitude of a variable at one time or place relative to the magnitude of the same variable at another time or place **(16)**

index of industrial production an index measuring changes over time in the physical volume of industrial goods produced in the United States **(16)**

indicator variables see **dummy variables**

inductive reasoning the drawing of inferences about an unknown whole from a known part **(1)**

inexact hypothesis a hypothesis that specifies a range of values for an unknown parameter **(9)**

inferential statistics see **analytical statistics**

interactions mean square (IMS) in the analysis of variance, the population variance estimate, s_A^2, based on the observed interactions between treatments and blocks **(12)**

interactions sum of squares (ISS) in the analysis of variance, the number of observations per cell times the sum of squared interactions **(12)**

interfractile ranges distance measures of dispersion; differences between two values, called fractiles, in a data set **(4)**

interquartile range a distance measure of dispersion; the difference between the third and first quartile in a data set, containing the middle 50 percent of all values observed **(4)**

intersection of two events all the basic outcomes simultaneously contained in both one random event *and* another **(5)**

interval data numbers that by their size rank observations in order of importance and between which *intervals* or distances are comparable, while their ratios are meaningless **(2)**

interval estimate the estimate of a population parameter, expressed as a range of values within which the unknown parameter presumably lies **(8)**

inverse relationship a relationship between any two variables, X and Y, such that the values of Y decrease with larger values of X **(13)**

irregular component in time-series analysis, random movements of the variable of interest around the trend/cyclical/seasonal components **(15)**

joint probability a measure of the likelihood of the simultaneous occurrence of two or more events **(5)**

judgment sample a subset of a frame, or of an associated population, resulting when "expert" judgment plays a major role in selecting elementary units for observation in such a way that the chosen ones are "representative" of the whole **(2)**

Kolmogorov-Smirnov one-sample test an alternative to the χ^2 test for goodness of fit **(11)**

Kruskal-Wallis test an extension of the Wilcoxon rank-sum test from two to more than two statistical populations **(11)**

kurtosis a frequency curve's degree of peakedness **(4)**

lagging economic indicators time series that lag behind business-cycle turns; they turn down only after the GNP has begun its cyclical decline and turn up only after the GNP has begun to rise **(15)**

Laplace criterion see **equal-likelihood criterion**

Laspeyres price index a weighted aggregative price index that uses base-year quantity weights **(16)**

law of large numbers the law that states that: the probability of a significant deviation of an empirically determined probability value from a theoretically determined one is smaller the larger is the number of repetitions of the random experiment in question **(5)**

leading economic indicators time series that anticipate business-cycle turns; they turn down just before the GNP reaches its business-cycle peak and turn up just before the GNP reaches its trough **(15)**

lower-tailed test a one-tailed hypothesis test in which the entire rejection region is found in the lower tail **(9)**

Mann-Whitney test (U test) a nonparametric test equivalent to the Wilcoxon rank-sum test **(11)**

marginal probability see **unconditional probability**

marginal utility of money the change in the total utility of money that is associated with a unit change in the quantity of money **(17)**

matched-pairs design a randomized block design with blocks of two experimental units **(2)**

matched-pairs sample a sample taken after each experimental unit in population A has been matched with a "twin" from population B so that any sample observation about a unit in population A automatically yields an associated observation about a unit in population B **(8)**

maximax one of several nonprobabilistic criteria for decision making under uncertainty according to which a decision maker who seeks to maximize some benefit finds the maximum benefit associated with each possible action, identifies the maximum among these maxima, and chooses the action associated with this maximum of maxima **(17)**

maximin one of several nonprobabilistic criteria for decision making under uncertainty according to which a decision maker who seeks to maximize some benefit finds the minimum benefit associated with each possible action, identifies the maximum among these minima, and chooses the action associated with this maximum of minima **(17)**

maximum-likelihood criterion one of several probabilistic criteria for making decisions under uncertainty according to which a decision maker simply ignores all the events that might occur except the most likely one and selects the action that produces the optimal result (maximum benefit or minimum cost) associated with this most likely event **(17)**

mean see **arithmetic mean**

mean absolute deviation (MAD) an average deviation measure of dispersion, obtained by averaging the absolute values of deviations of individual observations from the mean **(4)**

mean of squared deviations see **variance**

mean squared error the sum of an estimator's squared bias plus its variance **(8)**

measurement the assignment of numbers to characteristics that are being observed **(2)**

measures of central tendency (or location) values around which observations tend to cluster and that describe the location of the "center" of a data set **(4)**

measures of dispersion (or variability) numbers that indicate the spread or scatter of observations; they show the extent to which individual values in a data set differ from one another and, hence, differ from their central location **(4)**

measures of shape numbers that indicate either the degree of asymmetry or that of peakedness in a frequency distribution **(4)**

median (M or m) a measure of central tendency; that value in an ordered array of data above and below which an equal number of observations can be found **(4)**

median class the class among grouped data that contains the median **(4)**

method of least squares in regression analysis, the method of fitting a line to data displayed in a scatter diagram in such a way that the sum of the squares of the vertical deviations between the line and the individual data plots is minimized **(13)**

minimax one of several nonprobabilistic criteria for making decisions under uncertainty according to which a decision maker who seeks to minimize some cost finds the maximum cost associated with each possible action, identifies the minimum among these maxima, and chooses the action associated with this minimum of maxima **(17)**

minimax regret one of several nonprobabilistic criteria for making decisions under uncertainty according to which a decision maker finds the maximum regret value associated with each possible action, identifies the minimum among these max-

ima, and chooses the action associated with this minimum of maximum regret values **(17)**

minimin one of several nonprobabilistic criteria for making decisions under uncertainty according to which a decision maker who seeks to minimize some cost finds the minimum cost associated with each possible action, identifies the minimum among these minima, and chooses the action associated with this minimum of minima **(17)**

mixed time-series model a time-series model that combines features of the multiplicative and additive models—for example, $Y = T \cdot C \cdot I + S$ **(15)**

modal class the class among grouped data that contains the mode **(4)**

mode (Mo or mo) a measure of central tendency; the most frequently occurring value in a set of data **(4)**

Monte Carlo method a procedure that simulates a real-life process in order to generate a body of data that is then viewed as a "historical record" to be subjected to statistical analysis **(7)**

more-than-fair gamble a gamble in which the expected monetary value of what is given up is less than the expected monetary value of what is received **(17)**

moving-averages series a series of numbers obtained by successively averaging overlapping groups of two or more consecutive values in a time series and replacing the central value in each group by the group's average **(15)**

multicollinearity a high correlation between or among the independent variables in a regression model **(14)**

multimodal frequency distribution a frequency distribution in which two or more values occur with the highest frequency, or almost that **(4)**

multinomial qualitative variable a qualitative variable observations about which can be made in more than two categories **(2)**

multiple correlation analysis correlation analysis that measures the strength of association among more than two variables **(14)**

multiple regression analysis regression analysis in which several (independent) variables are used to estimate the value of an unknown (dependent) variable **(13,14)**

multiplication law a law of probability theory that is used to compute the probability for an *intersection* of two or more events; according to the general law for any two events, $p(A \text{ and } B) = p(A) \cdot p(B|A)$ and also $p(A \text{ and } B) = p(B) \cdot p(A|B)$; according to the special law for independent events, $p(A \text{ and } B) = p(A) \cdot p(B)$ **(5)**

multiplicative time-series model a time-series model, such as the classical one, that expresses the actual value of a time series as the product of its components—for example, $Y = T \cdot C \cdot S \cdot I$ **(15)**

multivariate data set a data set containing information on more than two variables **(2)**

mutually exclusive events different random events that have no basic outcomes in common **(5)**

negative serial correlation serial correlation such that a small value (Y below its mean and $e < 0$) follows a large one or a large one follows a small one **(14)**

nominal data numbers that merely *name* or label differences in kind **(2)**

nonparametric statistics test statistics used in nonparametric tests **(11)**

nonparametric tests hypothesis tests that make no assumptions whatsoever about the nature of underlying population distributions or their parameters **(11)**

nonresponse bias a systematic tendency for selected elementary units with particular characteristics not to contribute data in a survey, while other such units, with other characteristics, do **(2)**

nonsampling error see **systematic error**

nonsense correlation see **spurious correlation**

normal-form analysis a tabular form of preposterior analysis that systematically computes an expected payoff value for every conceivable strategy and selects the strategy with the best of these values as the optimal one **(17)**

normal probability distribution a probability density function that is (a) single-peaked above the random variable's mean, median, and mode, (b) perfectly symmetric about this central value, and (c) characterized by tails extending indefinitely in both directions from the center, approaching but never touching the horizontal axis **(7)**

not statistically significant result in hypothesis testing, any sample result that leads to the continued acceptance of the null hypothesis because it has a high probability of occurring when that hypothesis is true **(9)**

null hypothesis the first of two opposing hypotheses in a hypothesis test, symbolized by H_0, and often (but not necessarily) representing that proposition about an unknown population parameter that is tentatively assumed to be true **(9)**

number-of-runs test a nonparametric procedure designed to discover whether the elements in a sequence appear in random order **(11)**

objective probability a numerical measure of chance that estimates the likelihood of a specific occurrence (event A) in a repeatable random experiment (according to the classical approach, it is equal to the number of equally likely basic outcomes favorable to the occurrence of event A divided by the number of equally likely basic outcomes possible; according to the empirical approach, it is equal to the number of times event A did occur in the past during a large number of experiments divided by the maximum number of times event A could have occurred during these experiments) **(5)**

observational study see **survey**

observed significance level see *p* **value**

OC curve see **operating-characteristic curve**

ogive a graphical portrayal of a cumulative frequency distribution **(3)**

one-factor ANOVA see **one-way ANOVA**

one-sided hypothesis an alternative hypothesis that holds for deviations from the null hypothesis in one direction only **(9)**

one-tailed test a hypothesis test in which the alternative hypothesis is one-sided and in which the null hypothesis, therefore, is rejected only for very low (or very high) values of the test statistic located entirely in one tail of that statistic's sampling distribution **(9)**

one-way ANOVA one of several versions of the analysis of variance (it is performed with data derived from experiments based on the completely randomized design that controls extraneous factors by creating one treatment group for each treatment and assigning each experimental unit to one of these groups by a random process; thus, only one factor (the treatment) is considered as possibly affecting the variable of interest) **(12)**

open-ended classes classes that have only one stated end point, the upper or lower limit **(3)**

operating-characteristic curve in hypothesis testing, a graph showing, for all possible values of a population parameter that contradict the null hypothesis, the probability, β, of erroneously accepting it (type II error), given sample size and a specified maximum α risk **(9)**

opportunity loss (OL) when a decision maker maximizes benefit, the difference between (a) the optimal payoff for a given event and (b) the actual payoff achieved as a result of taking a specified action and the subsequent occurrence of that event; when a decision maker minimizes cost, the difference between (a) the actual cost incurred as a result of taking a specified action and the subsequent occurrence of an event and (b) the minimum cost achievable for that event **(17)**

optimal strategy in decision theory, a complete plan specifying the actions to be taken at each conceivable action point, if the expected (monetary or utility) payoff is to be the best one available **(17)**

ordered array a listing of numerical data in order of ascending or descending magnitude **(3)**

ordinal data numbers that by their size *order* or rank observations on the basis of importance, while intervals between the numbers, or ratios of such numbers, are meaningless **(2)**

outcome space see **sample space**

outliers wild values or maverick data that are not believable because they differ greatly from the majority of observed values **(2)**

Paasche price index a weighted aggregative price index that uses current-year quantity weights **(16)**

parameter a summary measure based on population data **(4)**

parametric statistic a test statistic used in a parametric test **(1)**

parametric test a hypothesis test that is based on certain specific assumptions about the probability distribution of population values or the sizes of population parameters **(11)**

payoffs in decision theory, the positive or negative net benefits that are associated with each possible action/event combination and that are, thus, the joint outcome of choice and chance **(17)**

payoff table a summary of a decision-making situation in the context of uncertainty, consisting of a tabular listing of the payoffs associated with all possible combinations of actions and events **(17)**

Pearson's coefficient of skewness (*Sk* or *sk*) a measure of a frequency curve's shape that relates the difference between the mode and the mean to the standard deviation **(4)**

percentile a value in a data set below which a specified percentage of all values is found **(4)**

permutations distinguishable ordered arrangements of items all of which have been drawn from a given group of items **(5)**

pictogram a set of symbols used to depict data **(3)**

pie chart a portrayal of divisions of some aggregate by a segmented circle in such a way that the sector areas are proportional to the sizes of the divisions in question **(3)**

point estimate the estimate of a population parameter, expressed as a single numerical value **(8)**

Poisson formula a formula for calculating the probability, within a specified time or space, of x

occurrences of a specified event that satisfy the conditions of a Poisson process **(6)**

Poisson probability tables tables that list Poisson probabilities (probabilities of x occurrences in a Poisson process) for various values of μ (the mean number of occurrences) **(6)**

Poisson process the occurrence of a series of events of a given type in a random (and, hence, unpredictable) pattern over time or space such that: (a) the number of occurrences within a specified time or space can equal any integer between zero and infinity, (b) the number of occurrences within one unit of time or space is independent of that within any other such unit, and (c) the probability of an occurrence is the same in all such units **(6)**

Poisson process rate the mean number of occurrences of an event per unit of time or space during the Poisson process **(6)**

Poisson random variable the number of occurrences of a specified event within a specified time or space **(6)**

polygon see **frequency polygon**

pooled variance the weighted average of two sample variances (the weights being the degrees of freedom associated with each variance), used to estimate the variance (known to be identical) in each of two populations from which the samples were taken **(9)**

population the set of all possible observations about a specified characteristic of interest **(2)**

population coefficient of correlation an index of the degree of association between two variables, denoted by ρ and equal to the square root of the population coefficient of determination **(13)**

population coefficient of determination the most important measure of how well a true regression line fits the population data on which it is based; this coefficient is denoted by ρ^2 **(13)**

population regression line see **true regression line**

population standard error of the estimate in regression analysis, the standard deviation, $\sigma_{Y \cdot X}$, of a conditional probability distribution of Y **(13)**

positive serial correlation serial correlation such that one large value (Y above its mean and $e > 0$) follows another one or one small value follows another **(14)**

posterior analysis a form of decision making under uncertainty that starts out with prior probabilities, proceeds to gather additional experimental or sample evidence about event probabilities, and then uses this new evidence to transform the set of prior probabilities, by means of Bayes's theorem, into a revised set of posterior probabilities with the help of which a final decision is reached   **(17)**

posterior expected value of perfect information (posterior *EVPI*)   the *EVPI* that is computed when the expected value of the optimal action under uncertainty is based on posterior probabilities **(17)**

posterior probability   a prior probability modified on the basis of new information   **(5)**

power curve   in hypothesis testing, a graph showing, for all possible values of a population parameter that contradict the null hypothesis, the probability, $1 - \beta$, of correctly rejecting it, given sample size and a specified maximum α risk   **(9)**

power of the hypothesis test   the probability (equal to $1 - \beta$) of avoiding a type II error when the null hypothesis is false   **(9)**

predicted variable   see **dependent variable**

prediction interval   in regression analysis, a confidence interval within which a population parameter that is being estimated presumably lies   **(13)**

predictor variable   see **independent variable**

preposterior analysis   a form of decision making under uncertainty that decides the worth of obtaining additional (experimental or sample) information before proceeding to prior or posterior analysis **(17)**

price indexes   numbers that measure the level of either a single price or a set of prices at one time or place relative to another time or place   **(16)**

price relative   the ratio of the price of one period or place to the price of another (base) period or place   **(16)**

price-relatives index   see **simple price index**

prior analysis   decision making under uncertainty that employs probabilities but only those available prior to the gathering of new experimental or sample evidence about the likelihood of alternative future events   **(17)**

prior expected value of perfect information (prior *EVPI*)   the *EVPI* that is computed when the expected value of the optimal action under uncertainty is based on prior probabilities   **(17)**

prior probability   a person's initial subjective estimate of the likelihood of an event, prior to any empirical investigation   **(5)**

probability density function   a smooth frequency curve that describes the probability distribution of a continuous random variable   **(7)**

probability distribution   a systematic listing of each possible value of a random variable, together with the associated likelihood of its occurrence **(6)**

probability sample   see **random sample**

probability space   see **sample space**

producer price index   an index measuring average price changes over time in U.S. primary (nonretail) markets   **(16)**

proportion (π or P)   a number that describes the frequency of observations in a particular category as a fraction of all observations made   **(4)**

***p* value**   in hypothesis testing, the probability that a value of the test statistic as extreme as or more extreme than the one actually observed could have occurred by chance, assuming that a hypothesized parameter value is true **(9)**; in regression analysis, the probability that a *t* value as extreme as or more extreme than the one actually observed could have occurred by chance, assuming the true regression coefficient, β, equals zero   **(14)**

qualitative variable   a variable that is normally not expressed numerically because it differs in kind rather than degree among elementary units   **(2)**

quantitative variable   a variable that is normally expressed numerically because it differs in degree rather than kind among elementary units   **(2)**

quantity indexes   numbers that measure the magnitude of either a single quantity or of a set of quantities at one time or place relative to another time or place   **(16)**

quartiles   values in a data set that divide the total of observation into four quarters, each of which contains .25 (or 25 percent) of the observed values **(4)**

random error   in surveys, the difference between the value of a variable obtained by taking a single random sample and the value obtained by taking a census (or by averaging the results of all possible random samples of like size); in comparative experiments, see **experimental error**   **(2)**

random event   any subset of the sample space   **(5)**

random experiment   any activity that will result in one and only one of several well-defined outcomes but that does not allow us to tell in advance which of these will prevail in any particular instance   **(5)**

randomization   a procedure that lets extraneous factors operate during an experiment but that assures, by virtue of the random assignment of experimental units to experimental and control groups, that each treatment has an equal chance to be enhanced or handicapped by those factors   **(2)**

randomized block design   an experimental plan that divides the available experimental units into blocks of fairly homogeneous units (each block containing as many units as there are treatments or some multiple of that number) and then matches each treatment with one or more units within each block by a random process **(2)**

randomized group design an experimental plan that creates one treatment group for each treatment and assigns each experimental unit to one of these groups by a random process **(2)**

random-numbers table a listing of numbers generated by a random process in such a way that each possible digit is equally likely to precede or follow any other one **(2)**

random sample a subset of a frame, or of an associated population, chosen by a random process that gives each unit of the frame or associated population a known positive (but not necessarily equal) chance to be selected **(2)**

random variable any quantitative variable the numerical value of which is determined by a random experiment and, thus, by chance **(6)**

range a distance measure of dispersion; the difference between the largest and smallest observation in a set of ungrouped data or between the upper limit of the largest class and the lower limit of the smallest class for grouped data **(4)**

rank-correlation test see **Spearman rank-correlation test**

rank-sum test see **Wilcoxon rank-sum test**

ratio data numbers that by their size rank observations in order of importance and between which intervals as well as *ratios* are meaningful **(2)**

ratio-to-moving-average method a method of isolating the seasonal/irregular component of a time series and, ultimately, of computing seasonal indexes **(15)**

rectangular probability distribution see **uniform probability distribution**

reference period see **base period**

regressand see **dependent variable**

regression analysis a statistical method the central focus of which is the establishment of an *equation* that allows the unknown value of one variable to be estimated from the known value of one or more other variables **(13)**

regression effect the tendency of those members of any population who, with respect to a given characteristic, are in an extreme position (below or above the population mean) at one time to be in a less extreme position at a later time (either personally or by means of their offspring) **(13)**

regression fallacy the incorrect attribution of the regression effect to the operation of some important unseen factor **(13)**

regression hyperplane the four- and higher-dimensional equivalent of a regression line **(14)**

regression line in regression analysis, a line that summarizes the stochastic relationship between an independent variable, X, and a dependent variable, Y, while also minimizing the errors made when the equation of that line is employed to estimate Y from X **(13)**

regression plane the three-dimensional equivalent of a regression line that minimizes the sum of the squared vertical deviations between the sample points suspended in Y vs. X_1 vs. X_2 space and their associated multiple regression equation estimates—all of which lie on this plane **(14)**

regression sum of squares (RSS) see **explained variation of Y**

regressor see **independent variable**

regret see **opportunity loss**

regret table in decision theory, a table showing the opportunity-loss values associated with each possible action/event combination **(17)**

rejection region in a hypothesis test, that range of possible values of the test statistic which signals the rejection of the null hypothesis **(9)**

relative class frequency the proportion of all observations that fall into a given class **(3)**

relative frequency distribution a tabular summary of a data set showing the relative class frequencies in each of several collectively exhaustive and mutually exclusive classes **(3)**

residual plots a graphical display of residuals, defined as differences between values obtained by observation and corresponding values anticipated on the basis of a theoretical model or previous experience **(3)**

residual variation see **unexplained variation**

response bias a tendency for answers to survey questions to be wrong in some systematic way **(2)**

risk aversion an attitude according to which a person considers the utility of a certain prospect of money to be higher than the expected utility of an uncertain prospect of equal expected monetary value **(17)**

risk neutrality an attitude according to which a person considers the utility of a certain prospect of money to be equal to the expected utility of an uncertain prospect of equal expected monetary value **(17)**

risk seeking an attitude according to which a person considers the utility of a certain prospect of money to be lower than the expected utility of an uncertain prospect of equal expected monetary value **(17)**

robustness the degree of sensitivity of a statistical test to errors in basic assumptions **(11)**

run the occurrence of an unbroken succession of like observations in a sequence containing potentially different types of observations **(11)**

sample a subset of a population or of the frame from which it is derived **(2)**

sample coefficient of alienation an index of the degree of linear association between two variables, denoted by k and equal to the square root of the sample coefficient of nondetermination **(13)**

sample coefficient of correlation an index of the degree of linear association between two variables, denoted by r and equal to the square root of the sample coefficient of determination **(13)**

sample coefficient of determination the most important measure of how well an estimated regression line fits the sample data on which it is based, denoted by r^2 and equal to the proportion of the total variation in the values of Y that can be explained by the association of Y with X as measured by the estimated regression line **(13)**

sample coefficient of multiple correlation an index of the degree of linear association between more than two variables, denoted by R and equal to the square root of the sample coefficient of multiple determination **(14)**

sample coefficient of multiple determination the most important measure of how well an estimated regression plane (or hyperplane) fits the sample data on which it is based, denoted by R^2 and equal to the proportion of the total variation in the values of the dependent variable Y that is explained by the multiple regression of Y on X_1, X_2, and possibly additional independent variables (X_3, X_4, and so on) **(14)**

sample coefficient of nondetermination a measure of how well an estimated regression line fits the sample data on which it is based, denoted by k^2 and equal to the proportion of the total variation in the values of Y that *cannot* be explained by the association of Y with X as measured by the estimated regression line **(13)**

sample coefficient of partial correlation the square root of the sample coefficient of partial determination **(14)**

sample coefficient of partial determination in multiple regression analysis, an index of association between the dependent variable and just *one* of the independent variables, given a prior accounting of the effects of one or more other independent variables **(14)**

sample covariance the tendency of paired sample observations to vary concurrently; measured by the simple correlation coefficient **(14)**

sample point see **simple event**

sample regression line see **estimated regression line**

sample space a listing of all the basic outcomes of a random experiment **(5)**

sample standard error of the estimate in regression analysis, the standard deviation, such as $s_{Y \cdot X}$ or $s_{Y \cdot X_1 X_2}$, of the values of Y observed in a sample around the least-squares regression line (or plane) estimated from this sample **(13,14)**

sample survey a partial survey in which observations are made about one or more characteristics of interest for only a subset of all existing elementary units **(2)**

sampling distribution a probability distribution that shows the likelihood of occurrence associated with all the possible values of a sample statistic, which values would be obtained when drawing all possible simple random samples of a given size from a population **(8)**

sampling error see **random error**

scatter diagram a graphical display used in regression analysis, consisting of a scatter of dots, each dot of which represents one value of the independent variable (measured along the horizontal axis) and an associated value of the dependent variable (measured along the vertical axis) **(13)**

seasonal component in time-series analysis, narrow up-and-down swings of the variable of interest around the trend/cyclical components, with the swings predictably repeating each other within periods of one year or less **(15)**

seasonal indexes indices showing the seasonal component of a time series **(15)**

seasonally adjusted time series see **deseasonalized time series**

second-order serial correlation serial correlation such that e_t is correlated with e_{t-2} **(14)**

second quartile the .50 fractile or 50th percentile in a data set below which half of all observations lie; the median **(4)**

selection bias a systematic tendency to favor the inclusion in a survey of selected elementary units with particular characteristics, while excluding other such units with other characteristics **(2)**

sequential sampling a form of sampling in which the sample size is not fixed in advance but in which a decision is made, after each bit of information is received, either to continue sampling or to put a stop to it **(17)**

serial correlation in regression analysis, a situation in which the value of Y (and the associated error term, e) that is observed at one time, t, is correlated with that observed at an earlier time, such as $t - 1$ or $t - 2$ **(14)**

signed-rank test see **Wilcoxon signed-rank test**

significance level of the hypothesis test among all the sample results that are possible when the null hypothesis is true, the (arbitrary) maximum proportion, α, that is considered sufficiently unusual to reject the null hypothesis **(9)**

sign test a nonparametric test based on a matched-pairs sample and designed to determine whether two statistical populations are identical to or different from one another **(11)**

simple correlation analysis correlation analysis that measures the strength of association between two variables only **(13)**

simple event any single one of a random experiment's basic outcomes **(5)**

simple price index a number that compares the price of a single item at one time or place with the price of the same item at another time or place **(16)**

simple random sample a subset of a frame, or of an associated population, chosen in such a fashion that every possible subset of like size has an equal chance of being selected **(2)**

simple regression analysis regression analysis in which a single variable is used to estimate the value of an unknown variable **(13)**

single-parameter exponential smoothing one of several exponential-smoothing techniques **(15)**

size of the hypothesis test see **significance level of the hypothesis test**

skewness a frequency distribution's degree of distortion from horizontal symmetry **(4)**

Slutsky-Yule effect the appearance of artificially created and, therefore, misleading waves in a moving-averages series even when there are no waves in the original time series **(15)**

smoothing constant a parameter employed in the exponential-smoothing formula **(15)**

smoothing technique a statistical procedure, such as the construction of a moving-averages series, that dampens (or averages out) fluctuations in a time series **(15)**

Spearman rank-correlation coefficient the test statistic used in the Spearman rank-correlation test, symbolized by ρ **(11)**

Spearman rank-correlation test a nonparametric test that measures the degree of association between two variables for which only rank-order data are available **(11)**

specification error the exclusion of an explanatory variable that theory suggests should be included in a regression or the inclusion of an explanatory variable that theory suggests to be irrelevant **(14)**

splicing uniting two index-number time series (that cover different spans of time) into a single and longer series **(16)**

spurious correlation high positive or negative correlation between two variables that have no logical connection with one another **(13)**

standard deviation (σ or s) an average deviation measure of dispersion, equal to the positive square root of the variance **(4)**

standard error the standard deviation of any statistic **(8)**

standard error of the proportion the standard deviation calculated from a sampling distribution of the proportion **(8)**

standard error of the sample mean the standard deviation, $\sigma_{\bar{X}}$, of the sampling distribution of the sample mean, $\bar{X}$ **(8)**

standard error of the sample proportion the standard deviation, σ_P, of the sampling distribution of the sample proportion, P **(8)**

standard normal curve a normal probability density function with a mean of zero and a standard deviation of one **(7)**

standard normal deviate the deviation from the mean, μ_R, of any given value, x, of a normally distributed random variable, measured in units of standard deviations, σ_R; symbolized by z **(7)**

standard scores the number of standard deviations that particular observations lie below or above the mean of all observations in a data set **(4)**

state-of-nature branches see **event branches**

state-of-nature fork see **event point**

state-of-nature node see **event point**

state-of-nature point see **event point**

states of nature see **events**

statistic a summary measure based on sample data **(4)**

statistically significant result in hypothesis testing, any sample result that leads to the rejection of the null hypothesis because it has a low probability of occurring when that hypothesis is true **(9)**

statistical map a portrayal of data for areal units by such devices as differentiated cross-hatching or shading of these units on a geographic map **(3)**

statistics a branch of mathematics that is concerned with facilitating wise decision making in the face of uncertainty and that, therefore, develops and utilizes techniques for the careful collection, effective presentation, and proper analysis of numerical information **(1)**

stem-and-leaf diagram   an unusual type of diagram that combines the features of an ordered array and a frequency histogram   **(3)**

step-down method   see **backward-elimination method**

step-up method   see **forward-selection method**

stepwise multiple regression   a procedure that develops a multiple regression equation in carefully delineated steps, either by means of the forward-selection method or the backward-elimination method   **(14)**

stochastic relationship   a relationship between any two variables, X and Y, such that many possible values of Y can be associated with any one value of X **(13)**

stratified random sample   a subset of a frame, or of an associated population, chosen by dividing the latter into mutually exclusive and clearly distinguishable strata (which are internally homogeneous but differ greatly from one another with respect to some characteristic of interest) and then taking separate (simple or systematic) random samples from every stratum, often in such a way that the sizes of the separate samples vary with the importance of the different strata   **(2)**

Student's *t* distribution   a sampling distribution for a random variable, t, derived from a normally distributed population and defined as $(\overline{X} - \mu)/(s/\sqrt{n})$ **(8)**

subjective probability   a numerical measure of chance that expresses a purely personal degree of belief in the likelihood of a specific occurrence in a unique random experiment   **(5)**

survey   the collection of data from elementary units without exercising any particular control over factors that may make these units different from one another and that may, therefore, affect the characteristic of interest being observed   **(2)**

systematic error   in surveys, the difference between the value of a variable obtained by taking a census (or by averaging the results of all random samples of a given size) and the true value; in comparative experiments, the difference between the differential value obtained by averaging the results of many replications of an experiment and the true difference   **(2)**

systematic random sample   a subset of a frame, or of an associated population, chosen by randomly selecting one of the first k elements and then including every kth element thereafter   **(2)**

test statistic   the type of statistic to be computed from a simple random sample taken from the population of interest in a hypothesis test and to be

used for establishing the truth or falsity of the null hypothesis   **(9)**

theory of probability   the calculus of the likelihood of specific occurrences   **(5)**

third quartile   the .75 fractile of 75th percentile in a data set below which three quarters of all observations lie   **(4)**

three-parameter exponential smoothing   one of several exponential-smoothing techniques that includes seasonal as well as trend smoothing   **(15)**

time-reversal test   one of several index-number quality tests; a price-index formula, for instance, is said to pass the test if, in a comparison of prices at two dates, the same result is obtained regardless of which of the two dates is chosen as a base   **(16)**

time-series analysis   a statistical procedure that employs time-series data, usually for the purpose of explaining past or forecasting future events   **(15)**

time-series data   numerical values pertaining to units of a given population that have been observed repeatedly at different points in time (in the case of a stock variable) or during different periods of time (in the case of a flow variable)   **(14,15)**

time-series line graph   a graphical portrayal, by a continuous line, of data that are linked with time   **(3)**

tolerable error level   the maximum amount a point estimate should, in the opinion of the statistician, extend above or below the parameter being estimated   **(8)**

total deviation of *Y*   in regression analysis, the difference between any particular observation of Y and the mean, $\overline{Y}$, of all such observations   **(13)**

total sum of squares *(Total SS)*   in the analysis of variance, the sum of the squared deviations of each individual sample observation from the mean of all observations   **(12,13)**

total utility of money   the overall welfare a person derives from the possession of a given quantity of money   **(17)**

total variation of *Y*   in regression analysis, the sum of the squares of all the total deviations of Y, or $\Sigma(Y - \overline{Y})^2$, also called the **total sum of squares**   **(13)**

***t* ratio**   see ***t* value**

treatments   stimuli applied to experimental units in controlled experiments   **(2)**

treatments mean square *(TMS)*   in the analysis of variance, the population variance estimate, s_A^2, based on the observed variation among the treatments samples   **(12)**

treatments sum of squares *(TSS)*   in the analysis of variance, the sum of the squared deviations between each treatment sample mean and the grand mean, multiplied by the number of observations made for each treatment, multiplied by the number of observations per cell (which may well equal 1) **(12)**

treatments variation in the analysis of variance, variation among treatment sample means that is attributable to inherent differences among treatment populations **(12)**

tree diagram a device for defining the sample space of multiple-step random experiments **(5) see also decision tree (17)**

trend component in time-series analysis, a relatively smooth, progressively upward or downward movement of the variable of interest over an extended period of time **(15)**

trend smoothing constant a parameter employed in the multiple-parameter exponential smoothing formula **(15)**

true regression coefficients the values of α and β found in the equation of the true regression line, $E(Y) = \mu_{Y \cdot X} = \alpha + \beta X$ **(13)**

true regression line the regression line calculated from census data by the method of least squares **(13)**

Tukey's HSD test a test, employed as a follow-up to ANOVA, that seeks out "honestly significant differences" among paired sample means **(12)**

t value the ratio of an estimated partial regression coefficient to its standard error **(14)**

two-factor ANOVA see **two-way ANOVA**

two-parameter exponential smoothing one of several exponential-smoothing techniques that includes trend smoothing **(15)**

two-sided hypothesis an alternative hypothesis that holds for deviations from the null hypothesis in either direction **(9)**

two-tailed test a hypothesis test in which the alternative hypothesis is two-sided and in which the null hypothesis, therefore, is rejected for values of the test statistic located in either tail of that statistic's sampling distribution **(9)**

two-way ANOVA one of several versions of the analysis of variance [performed with data derived from experiments based on the randomized block design that controls extraneous factors by dividing the available experimental units into distinctly different, but internally homogeneous, blocks and then randomly matching each treatment with one or more units within any given block; thus, two factors (treatments and blocks) are considered as possibly affecting the variable of interest] **(12)**

type I error in hypothesis testing, the erroneous rejection of a null hypothesis that is in fact true **(9)**

type II error in hypothesis testing, the erroneous acceptance of a null hypothesis that is in fact false **(9)**

typical-year aggregative price index a weighted aggregative price index that uses arbitrary, "typical-year" quantity weights **(16)**

unbiased estimator a sample statistic that, on the average, across many samples, takes on a value equal to the population parameter that is being estimated with its help **(8)**

unconditional probability a measure of the likelihood that a particular event occurs, regardless of whether another event occurs **(5)**

unexplained deviation of Y in regression analysis, the difference between any particular observation of Y and its regression estimate, $\hat{Y}_X$ **(13)**

unexplained variance in the analysis of variance, the error mean square; the denominator of the F statistic **(12)**

unexplained variation in the analysis of variance, the variation of the sample data *within* each of the samples about the respective sample means; it is attributed to chance and equals the total sum of squares minus explained variation **(12)**

unexplained variation of Y in regression analysis, the sum of the squares of all the unexplained deviations of Y, or $\Sigma(Y - \hat{Y}_X)^2$, also called the **error sum of squares (13)**

unfair gamble a gamble in which the expected monetary value of what is given up exceeds the expected monetary value of what is received **(17)**

uniform probability distribution a probability density function for a random variable that is equally likely to take on any of the values within a given range **(7)**

uniform random variable a random variable that has an equal chance of assuming any value within a specified range along a continuous scale **(7)**

union of two events all the basic outcomes contained in either one random event *or* another, or, conceivably, in both **(5)**

univariate data set a data set containing information on one variable only **(2)**

universe see **population**

unweighted aggregative price index a composite price index constructed by summing all the prices of the current year (for which the index is desired) and relating that sum to a similar sum of all the base-year prices **(16)**

upper-tailed test a one-tailed hypothesis test in which the entire rejection region is found in the upper tail **(9)**

U **test** see **Mann-Whitney test**

utility-of-money function the relationship between alternative amounts of money a person might possess and the different utility totals associated with these amounts **(17)**

variables characteristics possessed by elementary units **(2)**

variance (σ^2 or s^2) an average deviation measure of dispersion, obtained by averaging the squares of deviations of individual observations from the mean **(4)**

Venn diagrams graphical devices that depict sample spaces and random events symbolically **(5)**

weighted aggregative price indexes composite price indexes that compute a sum of quantity-weighted prices for the current year and then relate that sum to a similar one for the base year **(16)**

weighted mean (μ_w or $\overline{X}_w$) a measure of central tendency; a form of arithmetic mean that gives different observations unequal weights in accordance with their unequal relative importance **(4)**

Wilcoxon rank-sum test (W test) a nonparametric test based on two independent simple random samples and designed to determine whether two statistical populations of continuous values are identical to or different from one another **(11)**

Wilcoxon signed-rank test a nonparametric test based on a matched-pairs sample and designed to determine whether two statistical populations are identical to or different from one another **(11)**

work sampling a method of collecting sample data that assumes that the proportion of time spent by a worker in the pursuit of a given activity during a period is identical to the relative frequency with which momentary observations of this worker engaged in this activity have been made **(8)**

W test see **Wilcoxon rank-sum test**

ANSWERS TO ODD-NUMBERED QUESTIONS AND PROBLEMS

Chapter 2

1. a. Seven (as given in the headings of columns 2–8).

  b. Qualitative: the headings of columns 2, 3, 4, and 7.
  Quantitative: the headings of columns 5, 6, and 8.

  c. Qualitative: all multinomial
  Quantitative: all continuous

3. qualitative and dichotomous: *c, e*
  qualitative and multinomial: *f*
  quantitative and discrete: *a, b*
  quantitative and continuous: *b, d*

  Note: Item *b* is listed twice because dollar figures that measure the value of coins and paper bills in our pockets might be considered *discrete* because numbers are necessarily separated from each other by distances of one penny. One could find numbers such as $17.23 and $17.24, but never a $17.238. Dollar figures in other contexts, however, might be considered *continuous.* A tax rate might be expressed as 66.75 mills, which equals 6.675 cents on the dollar; an exchange rate might be stated as $8.9776 per unit of foreign currency; and the average cost of production might equal $17.5396 per unit of output.

5. Qualitative: *c, e, f*
  Quantitative: *a, b, d*

7. multinomial: *c, e, f*
  continuous: *a, b, d*

9. Nominal: *a, d, f;* ordinal: *c;* ratio: *b, e*

11. Nominal: *b, d*

13. Census: *a* (if small firm); sample: *a* (if large firm), *b* (the population is infinite), *c* for all of the reasons for sampling given in the text).

15. Answers can vary, but this is how the A. C. Nielsen Co. does it: Electronic devices, called *audimeters,* are attached to all TV sets owned by a sample of 1,700 homes and connected to a Nielsen computer via special telephone lines. The devices record when the TV sets are on or off, as well as the channels tuned. A Nielson rating of 25 means that 25 percent of sample households were tuned to the rated show for at least 6 minutes.

17. Convenience sample: *a;* judgment sample: *b;* probability sample: *c.*

19. Problem 16: none; problem 17c: clustered random sample; problem 18a: clustered random sample; problem 18b: systematic random sample.

21. See footnote 1 of this chapter. For $n = 5$ (and $N = 100$), the answer is 75,287,520. For $n = 50$, it is 100,891,344,545,564,200,000,000,000,000, a remarkable number calculated by computer. (The footnote formula here reappears in Chapter 5 as Box 5F.)

23. See footnote 1 of this chapter. For $n = 3$ (and $N = 9$), the answer is 84. For $n = 1$, it is 9.

25. 13 Burroughs, 41 Getty Oil, 39 General Motors, 27 Digital Equipment, 02 American Cyanamid, 31 Exxon, 04 American Home Products, 20 Colgate-Palmolive, 43 Goodyear, 26 Deere.

27. 06 Evergreen Fund, 02 Amcap Fund, 15 Nicholas Fund, 04 American Capital Venture Fund, 05 Claremont Capital Corp., 08 Fidelity Destiny Fund.

29. 13 Mass Cap Development Fund, 20 Sigma Venture Shares, 22 Twentieth Century Select, 16 Over-the-Counter Securities Fund, 04 American Capital Venture Fund, 03 American Capital Pace Fund.

31. 25 Dart and Kraft, 41 Getty Oil, 30 Eastman Kodak, 93 Union Carbide, 57 Kmart, 31 Exxon.

33. Answers will vary, but students should have followed a procedure like the one discussed in the text in connection with the selection of 10 multinationals.

35. The value of k is (100/4) = 25; the sample is 15 Chase Manhattan, 40 General Telephone and Electric, 65 J. P. Morgan, 90 Texas Instruments.

37. By taking every 50th bill, starting with one chosen at random from serial numbers 00–49. If that random number were 17, the sample would consist of bills numbered 17, 67, 117, and so on.

39. A sample of one.

41. The numbers are 3 3 5 5 2 3 8 1 7 4. Hence the sample is 03 American Express, 13 Burroughs, 25 Dart and Kraft, 35 Fluor, 42 Gillette, 53 International Harvester, 68 NCR, 71 PepsiCo., 87 Sun Co., 94 Union Oil California.

43. *Cluster 17:* 85 Standard Oil California, 86 Standard Oil Indiana, 87 Sun Co., 88 Tenneco, 89 Texaco. *Cluster 05:* 25 Dart and Kraft, 26 Deere, 27 Digital Equipment, 28 Dow Chemical, 29 DuPont de Nemours.

45. Selection bias: *d;* response bias: *a, e;* nonresponse bias: *b, c.*

47. Stratified random sample; random error.

49. Answers will vary. See the text section "How Bias Can Enter Surveys."

51. Answers will vary. Review the text section noted in answer 49. Telephone numbers should be selected by some type of random process, presumably from a university directory. (Some national phone surveys generate telephone numbers from a random numbers table, then dial them.)

53. Answers will vary, but consider the types of samples discussed in the text.

55. Answers will vary. Many such polls involve convenience sampling; some by design, others by default (as a random sample is ruined by huge nonresponse bias).

57. One possible answer is the randomized block design embodied in Table 12.4.

59. Once again the randomized block design can help: one can separately record income data for the 3 education levels and also for the 4 occupations, making for 12 observations. This procedure is discussed in Chapter 12 under "Two-way ANOVA."

Chapter 3

1. The *stem-and-leaf diagram* is given in Table A.1. The *ordered array* is given in Table A.2.

3. The *ordered array* is given in Table A.3.

5. For $n = 80$: $k = 1 + 3.3 \log 80 = 1 + 3.3(1.9031) = 7.28$. Seven classes. For $n = 400$: $k = 1 + 3.3 \log 400 = 1 + 3.3(2.6021) = 9.59$. Ten classes.

7. Arrangement 1 fails to make the classes *collectively exhaustive;* it does not cover the whole range of possible data; some of them, therefore, could not be classified (for example, 15.5 or 15.59). Arrangement 2 fails to make the classes *mutually exclu-*

Table A.1

Stems						Leafs											
15.3	3	9															
15.4	4																
15.5	1	3	7														
15.6	2	9															
15.7	1	2	3	4	5	5	7	7	9								
15.8	1	2	3	4	4	5	9										
15.9	0	0	0	1	1	1	2	2	2	3	3	4	5	6	8	8	8	9
16.0	0	0	0	1	1	1	4	5	7	8 .	9						
16.1	0	1	2	3	7	9											
16.2	0	1	4	5	7	9											
16.3	1	7	8														
16.4	0	2	3	7	9												
16.5	3																
16.6	0	3	5	8													
16.7																	
16.8	1	8															

Table A.2

15.33	15.71	15.79	15.90	15.92	15.98	16.04	16.13	16.29	16.49
15.39	15.72	15.81	15.90	15.93	15.99	16.05	16.17	16.31	16.53
15.44	15.73	15.82	15.90	15.93	16.00	16.07	16.19	16.37	16.60
15.51	15.74	15.83	15.91	15.94	16.00	16.08	16.20	16.38	16.63
15.53	15.75	15.84	15.91	15.95	16.00	16.09	16.21	16.40	16.65
15.57	15.75	15.84	15.91	15.96	16.01	16.10	16.24	16.42	16.68
15.62	15.77	15.85	15.92	15.98	16.01	16.11	16.25	16.43	16.81
15.69	15.77	15.89	15.92	15.98	16.01	16.12	16.27	16.47	16.88

Table A.3

15,971	19,900	21,486	23,380	25,829
17,356	20,153	21,536	24,110	25,861
18,329	20,209	21,705	24,378	25,920
18,577	20,452	22,064	24,435	26,300
18,954	20,607	22,141	24,456	26,709
19,014	20,691	22,520	24,520	27,384
19,180	20,934	22,600	24,577	28,401
19,563	21,057	22,737	24,628	28,621
19,690	21,208	22,924	25,125	29,000
19,700	21,307	23,089	25,610	39,751
19,800				

sive; classes overlap; some data, therefore, have two places to go (for example, 15.5).

9. a. See Table A.4.

Table A.4

Class (net weight in ounces) (1)	Absolute Class Frequency (number of cans in class) (2)
15.3 to under 15.4	2
15.4 to under 15.5	1
15.5 to under 15.6	3
15.6 to under 15.7	2
15.7 to under 15.8	9
15.8 to under 15.9	7
15.9 to under 16.0	18
16.0 to under 16.1	11
16.1 to under 16.2	6
16.2 to under 16.3	6
16.3 to under 16.4	3
16.4 to under 16.5	5
16.5 to under 16.6	1
16.6 to under 16.7	4
16.7 to under 16.8	0
16.8 to under 16.9	2
Total	80

b. See Table A.5.
11. See Table A.6.

Table A.5

Class (net weight in ounces) (1)	Absolute Class Frequency (number of cans in class) (2)
15.3 to under 16.1	53
16.1 to under 16.9	27
Total	80

Table A.6

Class (net weight in ounces) (1)	Absolute Class Frequency (number of cans in class) (2)	Relative Class Frequency (proportion of all cans in class) (3)
15.3 to under 15.5	3	.0375
15.5 to under 15.7	5	.0625
15.7 to under 15.9	16	.2000
15.9 to under 16.1	29	.3625
16.1 to under 16.3	12	.1500
16.3 to under 16.5	8	.1000
16.5 to under 16.7	5	.0625
16.7 to under 16.9	2	.0250
Totals	80	1.0000

13. Sturgess's rule suggests seven classes ($1 + 3.3 \log 51 = 6.64$). Dividing the data range of \$39,751 − \$15,971 = \$23,780 by 7, suggests class widths of

$3,500 (rounded as specified). The result is Table A.7.

Table A.7

Class (salary) (1)	Absolute Class Frequency (number of states in class) (2)	Relative Class Frequency (proportion of states in class) (3)
15,500 to under 19,000	5	.0980
19,000 to under 22,500	21	.4118
22,500 to under 26,000	18	.3529
26,000 to under 29,500	6	.1176
29,500 to under 33,000	0	0
33,000 to under 36,500	0	0
36,500 to under 40,000	1	.0196
Total	51	1.0000

15. See Table A.8.

Table A.8

Class (net weight in ounces) (1)	Cumulative Absolute Class Frequency (number of cans in class or lower ones) (2)	Cumulative Relative Class Frequency (proportion of cans in class or lower ones) (3)
15.3 to under 15.5	3	.0375
15.5 to under 15.7	8	.1000
15.7 to under 15.9	24	.3000
15.9 to under 16.1	53	.6625
16.1 to under 16.3	65	.8125
16.3 to under 16.5	73	.9125
16.5 to under 16.7	78	.9750
16.7 to under 16.9	80	1.0000

17. See Table A.9.

Table A.9

Classes (salary) (1)	Cumulative Absolute Class Frequency (number of states in class or higher ones) (2)	Cumulative Relative Class Frequency (proportion of states in class or higher ones) (3)
15,500 to under 19,000	51	1.0000
19,000 to under 22,500	46	.9020
22,500 to under 26,000	25	.4902
26,000 to under 29,500	7	.1373
29,500 to under 33,000	1	.0196
33,000 to under 36,500	1	.0196
36,500 to under 40,000	1	.0196

19. The missing values are 469.5, 843.1, 260.5, 140.5, totaling 1,713.6.

21. See Figures A.1 and A.2, respectively, depicting the cases of too many and too few classes.

Figure A.1

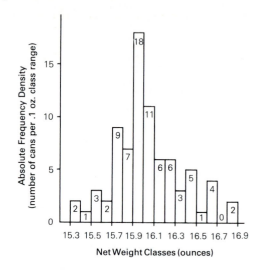

Figure A.2

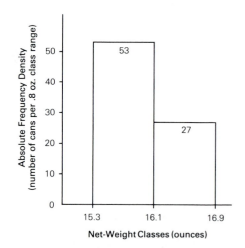

23. The answers can vary, but some likely ones are as follows: skewed to the right—a, c, e; skewed to the left—d, f, h; rectangular—b, g, i.

25. Quite possibly the two machines are not producing parts of the same average thickness, as the double peak in the histogram suggests.

27. See Figures A.3 through A.5. Compare the answer to the answer you got for problem 21.

Figure A.3

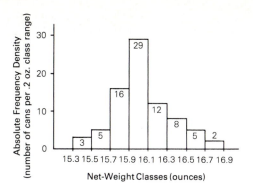

Figure A.4

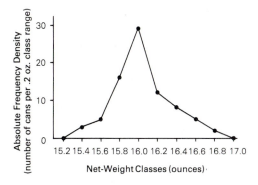

Figure A.5

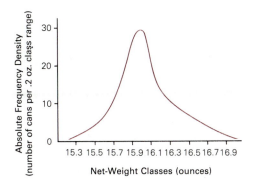

29. See Figure A.6.

31. Table 3G, "U.S. Net Import Reliance, 1981," is graphed in Figure A.7. Note: It is customary to order the bars by size in ascending or descending order. Any category named other is, however, always shown last, if it exists. (In this example such a category does not exist.)

33. See Figure A.8.

Figure A.6

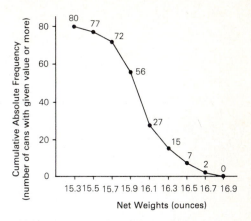

Figure A.7

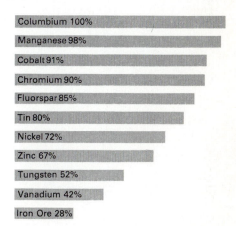

Figure A.8   State and Local Government Taxes—Percent Distribution by Type: 1965 to 1982

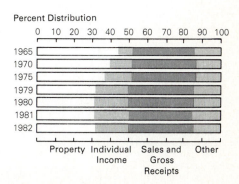

Source: U. S. Bureau of the Census.

Figure A.9   Retail Store Sales by Kind of Business: 1982 and 1983

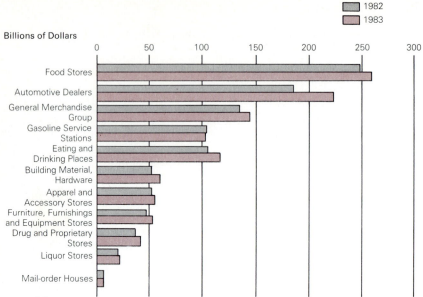

Source: U. S. Bureau of the Census.

35. See Figure A.9.
37. See Figure A.10.

Figure A.10

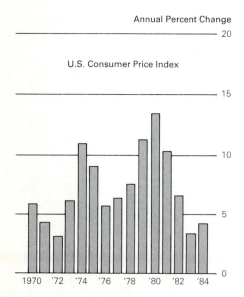

Source: U. S. Bureau of the Census.

39. Table 3.O, "U.S. Business Inventories," is graphed in Figure A.11.

Figure A.11

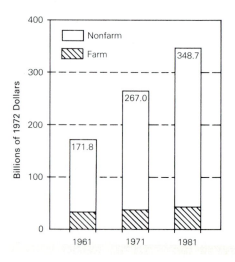

41. Table 3P, "Personnel Problems at City Motors, 1986," is graphed in Figure A.12.

Figure A.12

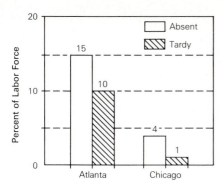

43. Table 3R, "Car-Rental Market Shares, March 1982," is graphed in Figure A.13. Note: The 360 degrees of the circle are split in accordance with the percentages given in the table: 39.7 percent of $360° = 143°$ going to Hertz, 23.4 percent of $360° = 84°$ to Avis, and so on.

Figure A.13

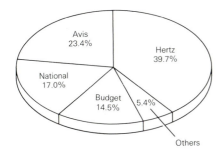

45. See Figure A.14.

Figure A.14

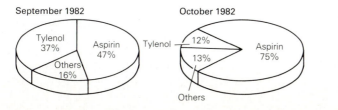

47. See Figure A.15. The angles are 97°, 76°, 68°, 40°, 40°, 40° (rounded).

Figure A.15

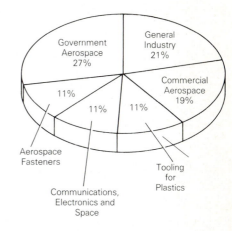

49. See Figure A.16. The angles are 169°, 144°, 47°.

Figure A.16

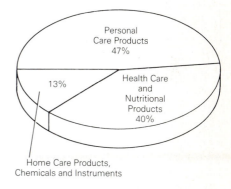

51. See Figure A.17.

Figure A.17

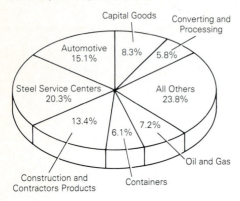

Steel Shipments by Market—1981

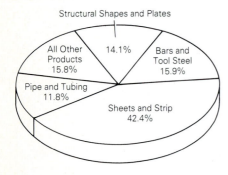

Steel Shipments by Product Group—1981

53. See Figure A.18.

Figure A.18

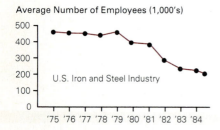

Average Number of Employees (1,000's)

55. See Figure A.19.
57. Table 3T, "Fewer Farmers Feeding More People," is graphed in Figure A.20.

Figure A.19

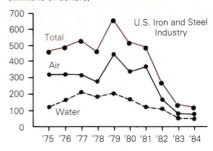

Capital Expenditures for Environmental Quality (millions of dollars)

Figure A.20

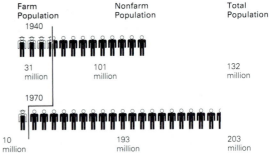

59. a. See Table A.10.

Table A.10

Age Groups	Relative Frequency A	B
10–19	.005	.004
20–29	.323	.070
30–39	.501	.109
40–49	.126	.160
50–59	.036	.264
60–69	.008	.383
70–79	.001	.010
	1.000	1.000

b. Baby food: magazine A, because a larger percentage of subscribers is found in the relevant age groups (between 20 and 39, 82.4% vs. 17.9% for magazine B). Denture cleaners: magazine B, because a larger percentage of subscribers is found in the relevant age groups (between 50 and 79, 65.7% vs. 4.5% for magazine A).

Chapter 4

1. a. List 1: the sample mean is $\overline{X}$ = $912.33 million; the sample median is m = $631 million; there is no mode (mo). List 2: the sample mean is $\overline{X}$ = 15.53%; the sample median is m = 15.2%; there is no mode (mo). List 3: the sample mean is $\overline{X}$ = $10,208.93 million; the sample median is m = $9,039; there is no mode (mo).

b. No, except that the population mean, median, and mode, respectively, would be symbolized by μ, **M**, and **Mo**.

3. The sample mean is $\overline{X}$ = 16.03 ounces; the sample median is m = 15.98 ounces; there are 6 modes: 15.9, 15.91, 15.92, 15.98, 16, and 16.01 ounces.

5. a. The population mean is μ = $22,832.34; the population median is **M** = $22,141; there is no mode.

b. Presumably, the U.S. average given in the table is a weighted average that weights each state average by that state's school-teacher population.

7. The population mean is μ = $22,808.83; the population median is **M** = $22,416.67; the population mode is **Mo** = $21,947.37.

9. The sample mean is $\overline{X}$ = $65.08; the sample median is m = $60.60; the sample mode is mo = $61.19.

11. We can make use of the relative frequency distribution found in the Chapter 3, problem 13 answer (Table A.7) and draw the graph, labelled A.21, that follows.

13. Answers can vary, but here is a possible one: $278,000; 17,000; 15,000; 12,000; 11,000; $9,000; 2,000; 2,000; 2,000; 2,000. Note how the *mean* equals $35,000; the *median* is $10,000; the *mode* is $2,000. Presumably, the real-estate agent trying to sell houses will use the first of these "averages"; the mayor looking for a federal grant will probably use the last one.

15. Mean: *iii, v, viii*; median: *ii* (the distributions are skewed), *iv, vii*; mode: *i, vi, vii*.

17.

$$8.62 \quad (21.5) = 185.33$$
$$57.51 \quad (7.9) = 454.33$$
$$\underline{42.91 \quad (7.4) = 317.53}$$
$$\Sigma wX = 957.19$$

$$\frac{\Sigma wX}{\Sigma w} = \frac{957.19}{109.04} = 8.8 \text{ percent}$$

This weighted average compares with an unweighted average of 12.3 percent.

19. The weighted mean is μ_w = 3.54 percent. The unweighted mean is μ = 3.51 percent.

21. a. Range: 16.88 − 15.33 = 1.55 ounces

b. 1st quartile: 15.835 ounces; 3rd quartile: 16.205 ounces.

c. 2nd decile: 15.78 ounces; 7th decile: 16.125 ounces.

d. See Table A.11 and Figure A.22. Note: Can you see why less-than and more-than ogives, when plotted in the same graph, would intersect above the median?

Figure A.21

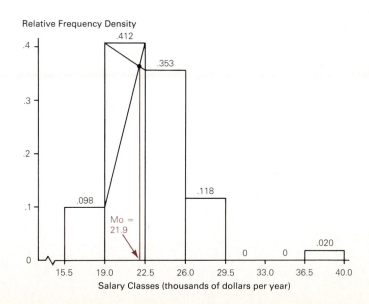

Table A.11

Class (net weight in ounces)	Less-than-or-Equal Cumulative Relative Class Frequency (proportion of cans in class or lower ones)
15.3 to under 15.5	.04
15.5 to under 15.7	.10
15.7 to under 15.9	.30
15.9 to under 16.1	.66
16.1 to under 16.3	.81
16.3 to under 16.5	.91
16.5 to under 16.7	.97
16.7 to under 16.9	1.00

Figure A.22

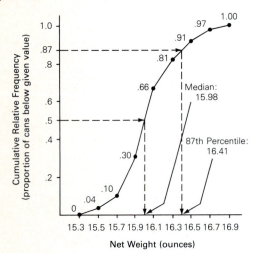

23. a. Range
List 1: $2,250 million − $347 million = $1,903 million
List 2:   22.1% − 11.8% = 10.3%
List 3: $15,060 − $8,210 = $6,850
b. Interquartile Range
List 1: $1,612 million − $503 million = $1,109 million
List 2:   17.2% − 13.1% = 4.1%
List 3: $11,130 − $8,357 = $2,773
25. a. Range
Table 4.F data: 17.5 percent − 4.7 percent = 12.8 percent
Table 4.G data: 3.95 percent − 2.32 percent = 1.63 percent
Table 4.H data: 12.3 percent − 2.4 percent = 9.9 percent

b. Interquartile Range
Table 4.F data: 17 percent − 9.3 percent = 7.7 percent
Table 4.G data: 3.79 percent − 3.19 percent = 0.60 percent
Table 4.H data: 10.35 percent − 3.95 percent = 6.4 percent
27. a. MAD = (31.6/15) = 2.1 percent. See Table A.12.

Table A.12

$|X - \overline{X}|$
6.6
3.4
2.3
1.7
1.5
.5
.2
.3
.7
1.1
1.5
2.4
2.8
2.9
3.7
$\Sigma|X - \overline{X}| = 31.6$

b. See Table A.13.

Table A.13

$(X - \overline{X})^2$	X^2
43.56	488.41
11.56	357.21
5.29	316.84
2.89	295.84
2.25	289.00
.25	256.00
.04	234.09
.09	231.04
.49	219.04
1.21	207.36
2.25	196.00
5.76	171.61
7.84	161.29
8.41	158.76
13.69	139.24
$\Sigma(X - \overline{X})^2 = 105.58$	$\Sigma X^2 = 3,721.73$

Using the *long method:*

$$s^2 = \frac{105.58}{14} = 7.54 \text{ percent squared.}$$

Using the *short method:*

$$s^2 = \frac{3,721.73 - 15(15.52667)^2}{14}$$

$$= \frac{3,721.73 - 3,616.16}{14} = \frac{105.57}{14}$$

$$= 7.54 \text{ percent squared.}$$

c. $s = \sqrt{7.54 \text{ percent squared}} = 2.746 \text{ percent.}$

29. Table 3.A data: MAD $= .244$ ounces; $s^2 = .1016$ squared ounces; $s = .3188$ ounces. Table 3.B data: MAD $= 13,200$ miles; $s^2 = 239,015,400$ squared miles; $s = 15,460$ miles. Table 3.C data: MAD $= \$2,850.85$; $\sigma^2 = 14,696,250$ squared dollars; $\sigma = \$3,833.57$. (Data treated as population data.)

31. See Table A.14.

33. MAD $= \$559.677$; $s^2 = 1,398,516$ squared dollars; $s = \$1,182.589$. If these were population data, one would estimate $\sigma^2 = 1,398,286$ squared dollars and $\sigma = \$1,182.491$.

35. MAD $= 8.845$ months; $\sigma^2 = 442.948$ squared months; $\sigma = 21.046$ months. If these were sample data, one would estimate $s^2 = 443.135$ squared months and $s = 21.051$ months.

37. The lower control limit is $\mu - 2\sigma = 15.394004$ ounces; the upper control limit is $\mu + 2\sigma = 16.661256$ ounces. The sample contains one read-

ing below the lower limit and two above the upper limit. Thus, there is reason for concern.

39. A normal frequency distribution was assumed because it would place 95.4 percent of all observations within 2 standard deviations of the mean. If the assumption did not hold, a more conservative estimate might be made using Chebyshev's theorem: At least 75 percent of the lightbulbs would have lifetimes within the range indicated by the manager.

41. Relative dispersion can be measured by the coefficient of variation, $v = (s/\overline{X})$. The three sample means were calculated in problem 1a as \$912.33 million, 15.53 percent, and \$10,208.93 million. The sample standard deviation for List 2 was calculated in problem 27c as 2.746 percent; the remaining two have to be calculated as shown in Table A.15. Thus, the three coefficients of variation are the following:

$$v_1 = \frac{\$598.1733 \text{ million}}{\$912.33 \text{ million}} = .656$$

$$v_2 = \frac{2.746\%}{15.53\%} = .177$$

$$v_3 = \frac{\$2,580.3949}{\$10,208.93} = .253$$

Dispersion is least by the criterion of R&D spending as a percent of sales; it is highest by the criterion of total dollar spending on R&D.

Table A.14

f	X	fX	X^2	fX^2
3	15.4	46.2	237.16	711.48
5	15.6	78.0	243.36	1,216.80
16	15.8	252.8	249.64	3,994.24
29	16.0	464.0	256.00	7,424.00
12	16.2	194.4	262.44	3,149.28
8	16.4	131.2	268.96	2,151.68
5	16.6	83.0	275.56	1,377.80
2	16.8	33.6	282.24	564.48
$n = 80$		$\Sigma fX = 1,283.2$		$\Sigma fX^2 = 20,589.76$

$$\overline{X} = \frac{1,283.2}{80} = 16.04 \text{ ounces.}$$

$$s^2 = \frac{20,589.76 - 80(16.04)^2}{79} = \frac{20,589.76 - 20,582.53}{79} = \frac{7.23}{79} = .09 \text{ ounces squared.}$$

$$s = \sqrt{.09 \text{ ounces squared}} = .3 \text{ ounces.}$$

Table A.15

List 1	List 3
X^2	X^2
5,062,500	226,803,600
2,951,524	220,552,201
2,842,596	217,857,600
2,598,544	123,876,900
712,336	113,998,329
662,596	94,517,284
541,696	88,228,449
398,161	81,703,521
396,900	72,795,024
378,225	71,605,444
276,676	70,862,724
253,009	69,839,449
163,216	68,840,209
136,161	67,667,076
120,409	67,404,100

$$\Sigma X^2 = 17,494,549$$
$$s^2 = \frac{17,494,549 - 15(912.33)^2}{14}$$
$$= \frac{5,009,359}{14}$$
$$= \$357,811.35 \text{ million squared}$$
$$s = \$598.1733 \text{ million}$$

$$\Sigma X^2 = 1,656,551,910$$
$$s^2 = \frac{1,656,551,910 - 15(10,208.93)^2}{14}$$
$$= \frac{93,218,134}{14}$$
$$= \$16,658,438.1 \text{ squared}$$
$$s = \$2,580.3949$$

43. Greater variability in reading scores. The coefficients of variation were

for math: $V_M = (\sigma/\mu) = \dfrac{11.53634}{55.36364} = .2083739.$

for reading: $V_R = (\sigma/\mu) = \dfrac{12.72023}{50.30606} = .2528568.$

45. With the help of answers to Problems 1a and 27c, we can determine that the *skewness* is positive:

$$sk = \frac{3(\overline{X} - m)}{s} = \frac{3(15.53 - 15.2)}{2.746} = .361$$

With the help of Table A.12, we can develop Table A.16 and determine that *kurtosis* is mesokurtic:

$$k = \frac{\dfrac{\Sigma(X - \overline{X})^4}{n}}{s^4} = \frac{\dfrac{2,432.12}{15}}{(2.746)^4} = 2.852.$$

Table A.16

$|X - \overline{X}|$	$(X - \overline{X})^4$
6.6	1,897.47
3.4	133.63
2.3	27.98
1.7	8.35
1.5	5.06
.5	.06
.2	.00
.3	.01
.7	.24
1.1	1.46
1.5	5.06
2.4	33.18
2.8	61.47
2.9	70.73
3.7	187.42
	$\begin{matrix} 2,432.12 \\ \Sigma(X - \overline{X})^4 \end{matrix}$

47. Column 5 data: $sk = 1.002587$; $k = 5.847568$
Column 8 data: $sk = .5761308$ or $.7803349$ (depending on use of average mode or median in formula); $k = 5.351316$.

49. Table 3.C data: $Sk = .5410109$; $K = 8.33559$. Table 3.V (Town A): $Sk = .1108403$ or $.3325209$ (depending on whether formula using mode or median is selected); $K = 4.425422$.

Table 3.V (Town B): $Sk = -.5687124$ or $-.6108391$ (depending on whether formula using mode or median is selected); $K = 1.710968$.

Table 4.I (math): $Sk = .7856598$ or $.3026008$ (depending on whether formula using mode or median is selected); $K = 2.26701$.

Table 4.I (reading): $Sk = .661794$; $K = 2.370322$.

51. a. The previous answer to Problem 31 determined $\overline{X} = 16.04$ ounces and $s = .3$ ounces. An answer to Problem 6 determined $m = 16.01$ ounces and mo $= 15.99$ ounces. Thus *skewness* is positive:

$$sk = \frac{\overline{X} - mo}{s} = \frac{16.04 - 15.99}{.3} = .167 \text{ or}$$

$$sk = \frac{3(\overline{X} - m)}{s} = \frac{3(16.04 - 16.01)}{.3} = .3.$$

b. With the help of Table A.17, we can also determine that *kurtosis* is leptokurtic:

$$k = \frac{\frac{\Sigma f(X - \overline{X})^4}{\Sigma f}}{s^4} = \frac{\frac{2.0452}{80}}{(.3)^4} = 3.156$$

Table A.17

f	$X - \overline{X}$	$(X - \overline{X})^4$	$f(X - \overline{X})^4$
3	$-.64$	.1678	.5034
5	$-.44$	.0375	.1875
16	$-.24$	.0033	.0528
29	$-.04$	.0000	.0000
12	.16	.0007	.0084
8	.36	.0168	.1344
5	.56	.0983	.4915
2	.76	.3336	.6672
$\Sigma f = 80$			$\Sigma f(X - \overline{X})^4$ 2.0452

53. Previous answers to Problems 9 and 34 can help. They established $\overline{X} = \$65.08$, $m = \$60.60$, mo $= \$61.19$, $s = \$50.114$. Thus

$$sk = \frac{\overline{X} - mo}{s} = \frac{\$65.08 - 61.19}{\$50.114} = .0776 \text{ or}$$

$$sk = \frac{3(\overline{X} - m)}{s} = \frac{3(\$65.08 - 60.60)}{\$50.114} = .2682.$$

55. The proportions are .010, .037, .015, .041, and .052 for shipments A through E, respectively. Thus, shipments D and E would be rejected.

57. Aggressive growth: $4/23 = .17$
Small company growth: $6/23 = .26$

59. The range of $\overline{X} \pm 2.5s$ extends from $16.03 - (2.5 \times .3188) = 15.233$ ounces to $16.03 + (2.5 \times .3188) = 16.827$ ounces. It covers 79 of the 80 data, or 98.75 percent of them. Chebyshev's theorem gives a minimum proportion of

$$1 - \frac{1}{(2.5)^2} = 1 - \frac{1}{6.25} = 1 - .16 = .84;$$

that is, of 84 percent. The result is consistent with Chebyshev's theorem.

Chapter 5

1. a. There are 38 univariate basic outcomes.
b. See Figure A.23. There are 8 multivariate basic outcomes.

Figure A.23

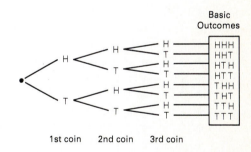

3. See the four parts of Figure A.24 on the next page.

5. Mutually exclusive: *i, ii*; compatible: *iii, iv, v*; collectively exhaustive: *ii, iv*; not collectively exhaustive: *i, iii, v*.

7. (i) Not finding defective items in the group. (ii) Inflation not above 10 percent per year. (iii) Real GNP growth below 3 percent per year. (iv) Not drawing an ace from a deck of cards. (v) Finding at least one defective item in the group.

9. Unions: A or C; B or $\overline{C}$
Intersections: A and B; $\overline{B}$ and C
Complements: $\overline{A}$ (a nonbrick house)

11. Objective, classical: *iii, viii, x*; objective, empirical: *vi*; subjective: *i, ii, iv, v, vii, ix*.

13. Four outcomes are possible, as shown in panel (a) of Figure 5.2 in the text, if "1st toss" and "2nd

Figure A.24

Case i   Text Figure B

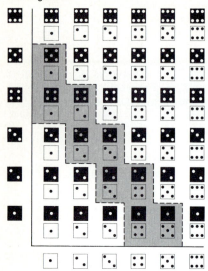

Case ii   Text Figure C

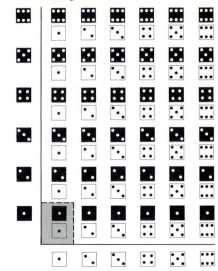

Case iii   Text Figure D

	Clubs	Diamonds	Hearts	Spades
King	♣ K	♦ K	♥ K	♠ K
Queen	♣ Q	♦ Q	♥ Q	♠ Q
Jack	♣ J	♦ J	♥ J	♠ J
10	♣ 10	♦ 10	♥ 10	♠ 10
9	♣ 9	♦ 9	♥ 9	♠ 9
8	♣ 8	♦ 8	♥ 8	♠ 8
7	♣ 7	♦ 7	♥ 7	♠ 7
6	♣ 6	♦ 6	♥ 6	♠ 6
5	♣ 5	♦ 5	♥ 5	♠ 5
4	♣ 4	♦ 4	♥ 4	♠ 4
3	♣ 3	♦ 3	♥ 3	♠ 3
Deuce	♣ 2	♦ 2	♥ 2	♠ 2
Ace	♣ A	♦ A	♥ A	♠ A

Case iv   Text Figure E

	Clubs	Diamonds	Hearts	Spades
King	♣ K	♦ K	♥ K	♠ K
Queen	♣ Q	♦ Q	♥ Q	♠ Q
Jack	♣ J	♦ J	♥ J	♠ J
10	♣ 10	♦ 10	♥ 10	♠ 10
9	♣ 9	♦ 9	♥ 9	♠ 9
8	♣ 8	♦ 8	♥ 8	♠ 8
7	♣ 7	♦ 7	♥ 7	♠ 7
6	♣ 6	♦ 6	♥ 6	♠ 6
5	♣ 5	♦ 5	♥ 5	♠ 5
4	♣ 4	♦ 4	♥ 4	♠ 4
3	♣ 3	♦ 3	♥ 3	♠ 3
Deuce	♣ 2	♦ 2	♥ 2	♠ 2
Ace	♣ A	♦ A	♥ A	♠ A

toss," respectively, are replaced by "1st coin" and "2nd coin." Each person, therefore, has an *equal* chance to win. In the long run, the person proposing the bet will get $2 for every $1 paid out.

15. a. p(Avis customer) = .234, assuming that the distribution of customers equals that of their expenditures on which Figure A.13 is presumably based.

  b. p(net weight 15.5 to under 15.7 oz) = .06. (The answer is *not* .10 because the table shows a *cumulative* distribution.)

17. $p(<\$100) = (900/6,066) = .148$; $p(\$100$ to under $\$500) = (4,264/6,066) = .703$; $p(>\$500) = (902/6,066) = .149$.

19. $P_1^2 = 2$
  $P_4^6 = 360$
  $P_2^{15} = 210$
  $P_7^{25} = 2,422,728,000$
  P_{32}^{25} is an impossibility, because Formula 5D requires $x \leq n$.

21. $P_{5,5,5}^{15} = 756,756$; $P_{2,2,3,3}^{10} = 25,200$.

23. See Table A.18.

Table A.18

					Number of Combinations (x at a time out of n distinct items; no repetitions allowed.)						
x n	0	1	2	3	4	5	6	7	8	9	10
0	1										
1	1	1									
2	1	2	1								
3	1	3	3	1							
4	1	4	6	4	1						
5	1	5	10	10	5	1					
6	1	6	15	20	15	6	1				
7	1	7	21	35	35	21	7	1			
8	1	8	28	56	70	56	28	8	1		
9	1	9	36	84	126	126	84	36	9	1	
10	1	10	45	120	210	252	210	120	45	10	1
11	1	11	55	165	330	462	462	330	165	55	11
12	1	12	66	220	495	792	924	792	495	220	66
13	1	13	78	286	715	1,287	1,716	1,716	1,287	715	286
14	1	14	91	364	1,001	2,002	3,003	3,432	3,003	2,002	1,001
15	1	15	105	455	1,365	3,003	5,005	6,435	6,435	5,005	3,003
16	1	16	120	560	1,820	4,368	8,008	11,440	12,870	11,440	8,008
17	1	17	136	680	2,380	6,188	12,376	19,448	24,310	24,310	19,448
18	1	18	153	816	3,060	8,568	18,564	31,824	43,758	48,620	43,758
19	1	19	171	969	3,876	11,628	27,132	50,388	75,582	92,378	92,378
20	1	20	190	1,140	4,845	15,504	38,760	77,520	125,970	167,960	184,756
21	1	21	210	1,330	5,985	20,349	54,264	116,280	203,490	293,930	352,716
22	1	22	231	1,540	7,315	26,334	74,613	170,544	319,770	497,420	646,646
23	1	23	253	1,771	8,855	33,649	100,947	245,157	490,314	817,190	1,144,066
24	1	24	276	2,024	10,626	42,504	134,596	346,104	735,471	1,307,504	1,961,256
25	1	25	300	2,300	12,650	53,130	177,100	480,700	1,081,575	2,042,975	3,268,760
26	1	26	325	2,600	14,950	65,780	230,230	657,800	1,562,275	3,124,550	5,311,735
27	1	27	351	2,925	17,550	80,730	296,010	888,030	2,220,075	4,686,825	8,436,285
28	1	28	378	3,276	20,475	98,280	376,740	1,184,040	3,108,105	6,906,900	13,123,110
29	1	29	406	3,654	23,751	118,755	475,020	1,560,780	4,292,145	10,015,005	20,030,010
30	1	30	435	4,060	27,405	142,506	593,775	2,035,800	5,852,925	14,307,150	30,045,015
31	1	31	465	4,495	31,465	169,911	736,281	2,629,575	7,888,725	20,160,075	44,352,165
32	1	32	496	4,960	35,960	201,376	906,192	3,365,856	10,518,300	28,048,800	64,512,240
33	1	33	528	5,456	40,920	237,336	1,107,568	4,272,048	13,884,156	38,567,100	92,561,040
34	1	34	561	5,984	46,376	278,256	1,344,904	5,379,616	18,156,204	52,451,256	131,128,140
35	1	35	595	6,545	52,360	324,632	1,623,160	6,724,520	23,535,820	70,607,460	183,579,396
36	1	36	630	7,140	58,905	376,992	1,947,792	8,347,680	30,260,340	94,143,280	254,186,856
37	1	37	666	7,770	66,045	435,897	2,324,784	10,295,472	38,608,020	124,403,620	348,330,136
38	1	38	703	8,436	73,815	501,942	2,760,681	12,620,256	48,903,492	163,011,640	472,733,756
39	1	39	741	9,139	82,251	575,757	3,262,623	15,380,937	61,523,748	211,915,132	635,745,396
40	1	40	780	9,880	91,390	658,008	3,838,380	18,643,560	76,904,685	273,438,880	847,660,528
41	1	41	820	10,660	101,270	749,398	4,496,388	22,481,940	95,548,245	350,343,565	1,121,099,408
42	1	42	861	11,480	111,930	850,668	5,245,786	26,978,328	118,030,185	445,891,810	1,471,442,973
43	1	43	903	12,341	123,410	962,598	6,096,454	32,224,114	145,008,513	563,921,995	1,917,334,783
44	1	44	946	13,244	135,751	1,086,008	7,059,052	38,320,568	177,232,627	708,930,508	2,481,256,778
45	1	45	990	14,190	148,995	1,221,759	8,145,060	45,379,620	215,553,195	886,163,135	3,190,187,286
46	1	46	1,035	15,180	163,185	1,370,754	9,366,819	53,524,680	260,932,815	1,101,716,330	4,076,350,421
47	1	47	1,081	16,215	178,365	1,533,939	10,737,573	62,891,499	314,457,495	1,362,649,145	5,178,066,751
48	1	48	1,128	17,296	194,580	1,712,304	12,271,512	73,629,072	377,348,994	1,677,106,640	6,540,715,896
49	1	49	1,176	18,424	211,876	1,906,884	13,983,816	85,900,584	450,978,066	2,054,455,634	8,217,822,536
50	1	50	1,225	19,600	230,300	2,118,760	15,890,700	99,884,400	536,878,650	2,505,433,700	10,272,278,170

25.

$$\text{Entrees: } C_2^5 = \frac{5!}{2!(5-2)!} = 10.$$

$$\text{Vegetables: } C_2^4 = \frac{4!}{2!(4-2)!} = 6.$$

$$\text{Desserts: } C_2^7 = \frac{7!}{2!(7-2)!} = 21.$$

Dinners: $C_2^3 \cdot C_2^5 \cdot C_2^4 \cdot C_2^6 \cdot C_2^7$
$$= 3 \cdot 10 \cdot 6 \cdot 15 \cdot 21 = 56,700.$$

27. $C_3^{10} \cdot C_2^{10}$

$$= \left(\frac{10!}{3!(10-3)!} \right)\left(\frac{10!}{2!(10-2)!} \right)$$

$$= \left(\frac{10 \cdot 9 \cdot 8}{3 \cdot 2 \cdot 1} \right)\left(\frac{10 \cdot 9}{2 \cdot 1} \right) = 120(45)$$

$$= 5,400.$$

29. *Favorable outcomes* $= C_{11}^{50}$ because in every hand with the desired cards, there are 11 other cards that can be selected in C_{11}^{50} ways from the remaining 50 cards. *Possible outcomes* $= C_{13}^{52}$ because there are 52 cards in a deck and 13 cards are to be selected. Thus, the probability is

$$\frac{C_{11}^{50}}{C_{13}^{52}} = \frac{\dfrac{50!}{11!39!}}{\dfrac{52!}{13!39!}} = \frac{50!}{11!39!} \times \frac{13!39!}{52!}$$

$$= \frac{13 \cdot 12}{52 \cdot 51} = \frac{156}{2,652} \cong .059$$

31. a. The events are mutually exclusive; thus, the special law applies:

$$p(5 \text{ or } 6) = \frac{4}{36} + \frac{5}{36} = \frac{9}{36} = \frac{1}{4}$$

This result was noted in problems 3i and 12d as well.

b. The events are not mutually exclusive; thus, the general law must be used. $p(\text{black } or \text{ ace}) = p(\text{black}) + p(\text{ace}) - p(\text{black } and \text{ ace}) =$

$$\frac{26}{52} + \frac{4}{52} - \frac{2}{52} = \frac{28}{52} = \frac{7}{13}.$$

This result was noted in problems 3iv and 12d as well.

c. The events are not mutually exclusive; thus,

$$p(\text{inflation } or \text{ recession}) = .8 + .2 - .1 = .9.$$

d. The events are not mutually exclusive; thus, the general law must be used, but the information is insufficient for determining an answer.

33. $p(A \text{ or } B \text{ or } C) = p(A) + p(B) + p(C) - p(A \text{ and } B) - p(A \text{ and } C) - p(B \text{ and } C) + p(A \text{ and } B \text{ and } C) = .2 + .1 + .05 - .08 - .02 - .07 + .05 = .23.$

35.

$$p(G|H) = \frac{240}{300} = .8$$

Alternative:

$$p(G|H) = \frac{p(H \text{ and } G)}{p(H)} = \frac{.60}{.75} = .8$$

$$p(B|H) = \frac{60}{300} = .2$$

Alternative:

$$p(B|H) = \frac{p(H \text{ and } B)}{p(H)} = \frac{.15}{.75} = .2$$

$\left. \right\} 1.0$

$$p(G|L) = \frac{10}{100} = .1$$

Alternative:

$$p(G|L) = \frac{p(L \text{ and } G)}{p(L)} = \frac{.025}{.25} = .1$$

$$p(B|L) = \frac{90}{100} = .9$$

Alternative:

$$p(B|L) = \frac{p(L \text{ and } B)}{p(L)} = \frac{.225}{.25} = .9$$

$\left. \right\} 1.0$

$$p(H|G) = \frac{240}{250} = .96$$

Alternative:

$$p(H|G) = \frac{p(H \text{ and } G)}{p(G)} = \frac{.60}{.625} = .96$$

$$p(L|G) = \frac{10}{250} = .04$$

Alternative:

$$p(L|G) = \frac{p(L \text{ and } G)}{p(G)} = \frac{.025}{.625} = .04$$

$\left. \right\} 1.0$

$$p(H|B) = \frac{60}{150} = .4$$

Alternative:

$$p(H|B) = \frac{p(H \text{ and } B)}{p(B)} = \frac{.15}{.375} = .4$$

$$p(L|B) = \frac{90}{150} = .6$$

Alternative:

$$p(L|B) = \frac{p(L \text{ and } B)}{p(B)} = \frac{.225}{.375} = .6$$

$\left. \right\} 1.0$

37. (i) $p(U|S) = (30/630) \cong .048$; conditional probability. (ii) $p(A|F) = (90/100) = .9$; conditional probability. (iii) $p(N \ and \ H) = (100/900) \cong .11$; joint probability. (iv) $p(N|H) = (100/400) = .25$; conditional probability. (v) $p(F|S) = (200/500) = .4$; conditional probability. (vi) $p(U \ and \ E) = (1,000/5,100) \cong .2$; joint probability. (vii) $p(A) = (3,400/5,100) \cong .67$; marginal probability.

39. Table 5.B: $p(H \ and \ B) = p(H) \cdot p(B|H) = .75(.2) = .15$
Proof: $(60/400) = .15$
Table 5.C: $p(S \ and \ U) = p(S) \cdot p(U|S) = .9(\approx.048) \cong .043$
Proof: $(30/700) \cong .043$
Table 5.D: $p(M \ and \ B) = p(M) \cdot p(B|M) = (\approx.67)(.9) = .6$
Proof: $(180/300) = .6$
Table 5.E: $p(S \ and \ F) = p(S) \cdot p(F|S) = (\approx.56)(.4) \cong .22$
Proof: $(200/900) \cong .22$
Table 5.F: $p(U \ and \ E) = p(U) \cdot p(E|U) = (\approx.33)(\approx.59) = .196$
Proof: $(1,000/5,100) = .196$

41. From the information given, we can set up Table A.19.

Table A.19 Number of Union Members Who . . .

	Voted Yes, Y	Voted No, N	Total
were from Northeast, NE	5,000	15,000	20,000
were from Southeast, SE	4,000	3,000	7,000
were from Northwest, NW	4,000	6,000	10,000
were from Southwest, SW	13,000	8,000	21,000
Total	26,000	32,000	58,000

Accordingly, the probabilities are the following:
a. $p(NE \ and \ N) = (15,000/58,000) \cong .26$; joint probability
b. $p(SW|N) = (8,000/32,000) = .25$; conditional probability
c. $p(Y|NW \ or \ SW) = (17,000/31,000) \cong .55$; conditional probability
d. $p(Y) = (26,000/58,000) \cong .45$; marginal probability

43. $p(A \ and \ C) = p(A) \cdot p(C|A) = .\overline{55}(.6) = .\overline{33}$
$p(A \ and \ D) = p(A) \cdot p(D|A) = .\overline{55}(.4) = .\overline{22}$
$p(B \ and \ C) = p(B) \cdot p(C|B) = .\overline{44}(.25) = .\overline{11}$
$p(B \ and \ D) = p(B) \cdot p(D|B) = .\overline{44}(.75) = .\overline{33}$

45. $p(S \ or \ A) = p(S) + p(A) - p(S \ and \ A) = .9 + (\approx.91) - (\approx.86) = .95$
$p(M \ or \ A) = p(M) + p(A) - p(M \ and \ A) = (\approx.67) + (\approx.37) - (\approx.07) = (\approx.97)$
$p(N \ or \ H) = p(N) + p(H) - p(N \ and \ H) = (\approx.44) + (\approx.44) - (\approx.11) = (\approx.77)$
$p(U \ or \ L) = p(U) + p(L) - p(U \ and \ L) = (\approx.33) + (\approx.41) - (\approx.14) = (\approx.61)$

47. Table 5.B: H and B are dependent. Proof: $p(H) = .75 \neq p(H|B) = .40$.
Table 5.C: S and U are dependent. Proof: $p(S) = .9 \neq p(S|U) = .46$.
Table 5.D.: M and B are dependent. Proof: $p(M) = .67 \neq p(M|B) = .95$.
Table 5.E: S and F are dependent. Proof: $p(S) = .56 \neq p(S|F) = .40$.
Table 5.F: U and E are independent. Proof: $p(U) = .33 = p(U|E) = .33$.

49. The events are independent; on each mission, there is a 98-percent chance of survival. The special multiplication law applies:

$$p(\text{survival of 50 missions}) = .98^{50} = .36,$$

and the probability of being shot down is .64.

51. The events are independent; at each barrier, the missile has an 80 percent chance of going through. The special multiplication law applies:

$$p(\text{passing all five barriers}) = .80^5 = .33,$$

and the probability of destroying the missile is .67.

53. From the information given, we can set up Table A.20.

Table A.20 Number of People Who . . .

	Get the Flu, F	Avoid the Flu, A	Total
were vaccinated, V	80,000	320,000	400,000
were unvaccinated, U	240,000	360,000	600,000
Total	320,000	680,000	1,000,000

Thus, the answers are the following:
a. $p(F) = .32$
b. 320,000
c. $p(A) = .68$, but $p(A|V) = .8$. The events are dependent.

55. See Figure A.25 on the next page.

Figure A.25

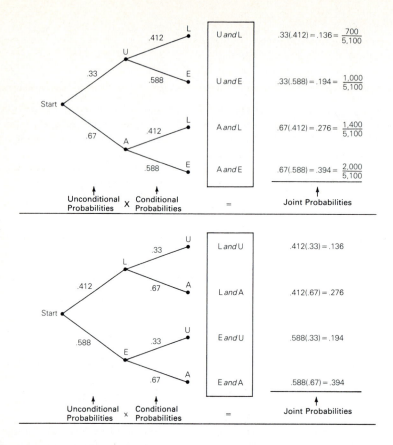

57. When E is the event in question, $\overline{E}$ its complement, and R the empirical result, we have

$$p(E|R) = \frac{p(E) \cdot p(R|E)}{p(E) \cdot p(R|E) + p(\overline{E}) \cdot p(R|\overline{E})}.$$

Thus,

$$p(E|R) = \frac{.3(.75)}{.3(.75) + (.7)(.1)} = \frac{.225}{.225 + .07} = .76$$

if praised and

$$p(E|R) = \frac{.3(.25)}{.3(.25) + (.7)(.9)} = \frac{.075}{.075 + .63} = .11$$

if criticized.

59. The solution is analogous to problem 57:

$$p(E|R) = \frac{.5(.9)}{.5(.9) + .5(.4)} = \frac{.45}{.45 + .2} = .69.$$

Chapter 6

1. See Table A.21. For determining the probability distribution, you may wish to take another look at panel (c) of Figure 5.1.

Table A.21

Probability Distribution	
Point Value, x	Probability $p(R = x)$
5	36/108
10	48/108
20	16/108
50	4/108
100	4/108
	$\Sigma p = 108/108$ $= 1.00$

Table A.22

Probability Distribution Case a		Probability Distribution Case f	
x	$p(x) = (x/10)$	x	$p(x) = (10 - x)/40$
1	1/10	0	10/40
2	2/10	1	9/40
3	3/10	2	8/40
4	4/10	3	7/40
		4	6/40
	$\Sigma p = 10/10$ $= 1.00$		$\Sigma p = 40/40$ $= 1.00$
$\mu_R = 3; \sigma^2 = 1; \sigma_R = 1$		$\mu_R = 1.75; \sigma_R^2 = 1.9375; \sigma = 1.3919$	

Table A.23

Probability Distribution		Summary Measures			
Number of Points, x	Probability, $p(R = x)$	$p \cdot x$	$x - \mu_R$	$(x - \mu_R)^2$	$p(x - \mu_R)^2$
1	1/6	1/6	-2.5	6.25	1.042
2	1/6	2/6	-1.5	2.25	.375
3	1/6	3/6	$-.5$	.25	.042
4	1/6	4/6	.5	.25	.042
5	1/6	5/6	1.5	2.25	.375
6	1/6	6/6	2.5	6.25	1.042
	$\Sigma p = (6/6)$ $= 1.00$	$\mu_R = (21/6)$ $= 3.50$			$\sigma_R^2 = 2.918$ $\sigma_R = 1.708$

3. The following could *not* be probability distributions: *b* (implies negative probability); *c* or *e* (imply sums of probabilities of less than unity); *d* (implies a sum of probabilities of more than unity). This leaves *a* and *f*. See Table A.22.

5. The Box 6.A formulas apply. See Table A.23.

7. The summary measures are given in Table A.24. $\mu_R = 3$ lawn mowers will be ordered.

Table A.24

$p \cdot x$	$x - \mu_R$	$(x - \mu_R)^2$	$p(x - \mu_R)^2$
0	-3	9	.45
.10	-2	4	.40
.40	-1	1	.20
.90	0	0	0
.80	1	1	.20
.50	2	4	.40
.30	3	9	.45
$\mu_R = 3.00$			$\sigma_R^2 = 2.1$ $\sigma_R = 1.45$

9. The implied probability distribution and the expected value of the random variable (cars coming in) is given in Table A.25. Hence the expected revenue is $310.

Table A.25

Probability Distribution	
x	$p(x)$
20	(79/313) = .2524
25	(121/313) = .3866
32	(19/313) = .0607
40	(25/313) = .0799
47	(39/313) = .1246
51	(30/313) = .0958
	$\Sigma p = 1.0000$

$$\mu_R = \Sigma p(x) \cdot x = 30.59$$

11. Using the class midpoints, one can develop Table A.26 on the next page.

Table A.26

\multicolumn{2}{c}{Probability Distribution}	
x	$p(x)$
5,500	.05
8,500	.07
11,500	.11
14,500	.54
17,500	.11
20,500	.07
23,500	.05
	$\Sigma p = 1.00$

$\mu_R = \Sigma p(x) \cdot x = 14{,}500$

$\sigma_R = 3{,}888.44$

13. a. The answer, from Appendix Table C, is given in Table A.27.

Table A.27

x	p
0	.3585
1	.3774
2	.1887
3	.0596
4	.0133
5	.0022
6	.0003
7	.0000

b. Since 5 percent of a sample of 20 cars equals 1 car, the answer equals the probability of finding more than 1 defective car, or $1 - .3585 - .3774 = .2641$.

15. a. According to Pascal's triangle, $p(5 \text{ boys, } 5 \text{ girls}) = (252/1{,}024) = .2461$. According to Ap-

pendix Table C, for $n = 10$, $x = 5$, and $\pi = .5$, $p = .2461$. What seems probable, may not be!

b. According to Appendix Table C, for $n = 15$, $x = 15$, and $\pi = .5$, $p = .0000$. According to the Box 6.B formula, $p = .00003052$.

17. See Figure A.26. Q = quits, S = stays. Thus the probability distribution is given in Table A.28. This is confirmed in Appendix Table C for $n = 3$, $\pi = .1$.

Table A.28

x	$p(x)$
0	.729
1	.243
2	.027
3	.001
	$\Sigma p = 1.000$

19. See Table A.29 for answers (from Appendix Table C, $n = 15$, $\pi = .25$).

a. For medium-sized cars, read the table directly.

b. For compacts, the table must be read in reverse: $x = 0$ for medium-sized cars corresponds to $x = 15$ for compact cars; $x = 1$ for medium-sized cars corresponds to $x = 14$ for compacts, and so on.

Table A.29

x	p	x	p
0	.0134	7	.0393
1	.0668	8	.0131
2	.1559	9	.0034
3	.2252	10	.0007
4	.2252	11	.0001
5	.1651	12	.0000
6	.0917		

Figure A.26

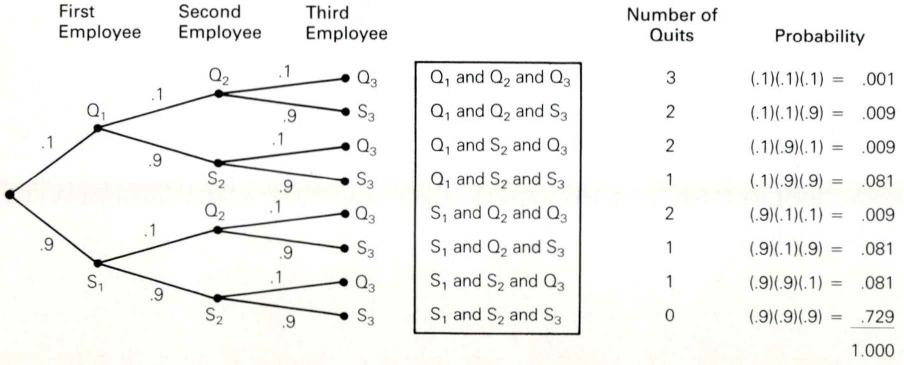

First Employee	Second Employee	Third Employee		Number of Quits	Probability	
			Q_1 and Q_2 and Q_3	3	$(.1)(.1)(.1) =$	.001
			Q_1 and Q_2 and S_3	2	$(.1)(.1)(.9) =$	.009
			Q_1 and S_2 and Q_3	2	$(.1)(.9)(.1) =$	.009
			Q_1 and S_2 and S_3	1	$(.1)(.9)(.9) =$	.081
			S_1 and Q_2 and Q_3	2	$(.9)(.1)(.1) =$	.009
			S_1 and Q_2 and S_3	1	$(.9)(.1)(.9) =$	.081
			S_1 and S_2 and Q_3	1	$(.9)(.9)(.1) =$	.081
			S_1 and S_2 and S_3	0	$(.9)(.9)(.9) =$	.729
						1.000

21. The answers, from Appendix Table D for $n = 100$ and $\pi = .40$, are (i) .0000, (ii) .0248 − .0148 = .01, (iii), .4621, (iv) 1 − .8689 = .1311, (v) 1 − .9729 = .0271, (vi) = .5433 − .0148 = .5285.

23. Answers (from Appendix Table C), for $n = 5$ and $\pi = .20$, are: (i) .0003, (ii) .3277, (iii) 1 − .3277 = .6723, (iv) .0512 + .0064 + .0003 = .0579, (v) .2048, (vi) .2048 + .0512 + .0064 = .2624.

25. Referring to Appendix Table D (for $n = 50$ and $\pi = .30$), which gives cumulative probabilities for tails coming up, (i) 44 or more heads is equivalent to 6 or fewer tails, $p = .0025$; (ii) exactly 40 heads is equivalent to exactly 10 tails, $p = .0789 − .0402 = .0387$; (iii) more than 37 heads is equivalent to 12 or fewer tails, $p = .2229$; (iv) 31 or fewer heads is equivalent to 19 or more tails, $p = 1 − .8594 = .1406$; (v) fewer than 28 heads is equivalent to more than 22 tails, $p = 1 − .9877 = .0123$; (vi) between 30 and 40 heads, inclusive, is equivalent to between 20 and 10 tails, inclusive,

$$p = .9522 − .0402 = .9120.$$

27. According to the Box 6.C formulas,

$$\mu_R = n \cdot \pi = 5(.20) = 1;$$
$$\sigma_R^2 = n \cdot \pi(1 − \pi)$$
$$= 5(.20)(.80) = .80;$$
$$\sigma_R = \sqrt{.80} = .89.$$

The skewness is positive because $\pi < .5$. Positive skewness is also evident by looking at the entire probability distribution in Appendix Table C, in the section where $n = 5$ and $\pi = .20$.

29. According to Appendix Table D, for $n = 100$ and $\pi = .30$, a. $p(R \leq 20) = .0165$; b. $p(R = 31) = p(R \leq 31) − p(R \leq 30) = .6331 − .5491 = .084$; c. $p(R \geq 47) = 1 − p(R \leq 46) = 1 − .9997 = .0003$; d. $\mu_R = 30$, $\sigma_R^2 = 21$, $\sigma_R = 4.58$, skewness positive.

31. According to Appendix Tables C and D, for $n = 7$ and $\pi = .05$, these answers apply: (i) $p(\text{all executed}) = p(\text{none late}) = .6983$; (ii) $p(\text{none executed}) = p(7 \text{ late}) = .0000$; (iii) $p(5 \text{ or more executed}) = p(2 \text{ or fewer late}) = .9962$; (iv) $p(6 \text{ or fewer executed}) = p(1 \text{ or more late}) = 1 − p(0 \text{ late}) = 1 − .6983 = .3017$.

33. As Table A.30 shows, the hypergeometric formula can be employed. Caution: In this answer, S stands for *success* (here *defective* units found), not for satisfactory units found.

35. The binomial probability distribution can be used as an approximation in this case. Appendix Table C, for $n = 3$ and $\pi = .25$, gives probability values of .4219, .4219, .1406, and .0156, respectively, for x values of 0, 1, 2, and 3.

37. See Table A.31 on the next page. The problem is analogous to Problem 36, part ii.

$$\mu_P = \pi = .4$$
$$\sigma_P^2 = \frac{\pi(1 − \pi)}{n}\left(\frac{N − n}{N − 1}\right) = \frac{.4(.6)}{3}\left(\frac{10 − 3}{10 − 1}\right)$$
$$= .08\left(\frac{7}{9}\right) = .06\overline{22}$$
$$\sigma_P = .25$$

Table A.30

Number of Successes (Defective Units Found), x	$c_x^S = \dfrac{S!}{x!(S − x)!}$	$c_{n-x}^{N-S} = \dfrac{(N − S)!}{(n − x)![(N − S) − (n − x)]!}$	$c_n^N = \dfrac{N!}{n!(N − n)!}$	$p(R = x \mid n = 3,\ N = 16,\ S = 4)$
0	$\dfrac{4!}{0!\ 4!} = 1$	$\dfrac{12!}{3!\ 9!} = 220$	$\dfrac{16!}{3!\ 13!} = 560$	$\dfrac{220}{560} = .3929$
1	$\dfrac{4!}{1!\ 3!} = 4$	$\dfrac{12!}{2!\ 10!} = 66$	$\dfrac{16!}{3!\ 13!} = 560$	$\dfrac{264}{560} = .4714$
2	$\dfrac{4!}{2!\ 2!} = 6$	$\dfrac{12!}{1!\ 11!} = 12$	$\dfrac{16!}{3!\ 13!} = 560$	$\dfrac{72}{560} = .1286$
3	$\dfrac{4!}{3!\ 1!} = 4$	$\dfrac{12!}{0!\ 12!} = 1$	$\dfrac{16!}{3!\ 13!} = 560$	$\dfrac{4}{560} = .0071$
				1.0000

Table A.31

Proportion of Successes, $P = (x/n)$	Probability, $p\left(P = \dfrac{x}{n}\right)$
0/3 = 0	.1667
1/3 = .33	.5000
2/3 = .67	.3000
3/3 = 1.00	.0333
	Σp = 1.0000

39. The hypergeometric formula suggests a probability (for $N = 20$, $S = 10$, $n = 10$, $x = 10$) of .0000054125.

41. Given a hypergeometric distribution (with $N = 10$, $S = 5$, $n = 4$), $\mu_R = 2$, $\sigma_R^2 = .6666$, $\sigma_R = .8165$.

43. Given a hypergeometric distribution (with $N = 20$, $S = 4$, $n = 4$), $\mu_R = .8$, $\sigma_R^2 = .5389$, $\sigma_R = .7341$.

45. Given a hypergeometric distribution (with $N = 100$, $S = 35$, $n = 10$), $\mu_R = 3.5$; $\sigma_R^2 = 2.0681$; $\sigma_R = 1.4381$.

47. With the help of Appendix Table F, the graphs in Figure A.27 can be drawn.

49. Assuming a Poisson process is occurring, the formula in Box 6.F applies. See Table A.32.

51. Assuming a Poisson process is occurring, Appendix Table F (for $\mu = 1$ and $x = 0$ or $x = 1$) suggests $p = .3679$ in either case. (The binomial formula suggests probabilities of .3660 and .3697, respectively.)

53. Assuming a Poisson process, $\mu = 1.9$, and the expected number per year is $12(1.9) = 22.8$. Appendix Tables F and G show $p(R = 0) = .1496$ and $p(R < 3) = .7037$.

55. a. False. According to Box 6.C, $\mu_R = n \cdot \pi$, and $\sigma_R^2 = n \cdot \pi(1 - \pi)$. Hence, $(\sigma_R^2/\mu_R) = 1 - \pi$. In this case, $1 - \pi = (.8/4) = .2$. Thus, $\pi = .8$.

b. True. According to Box 6.C, $\mu_R = n \cdot \pi$. Hence, $(\mu_R/\pi) = n$. In this case, $n = (4/.8) = 5$.

c. True. See Appendix Table C for $n = 5$ and read the $\pi = .20$ column backwards. Because .20 is the probability of *failure,* 0 failures implies 5 successes, and so on.

d. True. See Boxes 6.C and 6.E; μ_R is the same for the binomial and hypergeometric probability distributions.

Figure A.27

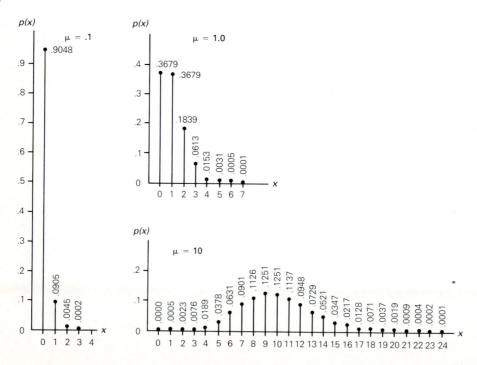

Table A.32

Number of Occurrences of Specified Event (customer arrival), x	$e^{-\mu}$ (from Appendix Table E)	μ^x	$x!$	$p(R = x\|\mu)$
Saturday: $\lambda = 50$ customers per hour, $t = .1$ hour; hence, $\mu = 5$				
0		$5^0 = 1$	$0! = 1$	.0067
1		$5^1 = 5$	$1! = 1$	.0337
2		$5^2 = 25$	$2! = 2$	.0842
3		$5^3 = 125$	$3! = 6$	.1404
4		$5^4 = 625$	$4! = 24$	.1755
5	.006738	$5^5 = 3,125$	$5! = 120$	.1755
6		$5^6 = 15,625$	$6! = 720$	.1462
7		$5^7 = 78,125$	$7! = 5,040$	.1044
8		$5^8 = 390,625$	$8! = 40,320$	.0653
9		$5^9 = 1,953,125$	$9! = 362,880$	.0363
10		$5^{10} = 9,765,625$	$10! = 3,628,800$	.0181
11		$5^{11} = 48,828,125$	$11! = 39,916,800$	.0082
Monday: $\lambda = 2$ customers per hour, $t = .1$ hour; hence, $\mu = .2$				
0		$.2^0 = 1$	$0! = 1$	.8187
1		$.2^1 = .2$	$1! = 1$	.1637
2		$.2^2 = .04$	$2! = 2$	.0164
3	.818731	$.2^3 = .008$	$3! = 6$	.0011
4		$.2^4 = .0016$	$4! = 24$	.0001
5		$.2^5 = .00032$	$5! = 120$	.0000
6		$.2^6 = .000064$	$6! = 720$	.0000

57. a. False. See Appendix Table G for $\lambda = 10$ and $t = .5$ (hence, $\mu = 5$), under $x \le 10$: it shows $p = .9863$.

b. True. See Appendix Table G for $\lambda = .5$ and $t = 40$ (hence, $\mu = 20$), under $x \le 9$: it shows $p = .005$. Thus, the desired answer equals $1 - .005 = .995$.

c. False. See Appendix Table F for $\lambda = 2$ and $t = .5$ (hence, $\mu = 1$), under $x = 1$ and $x = 2$. The sum of the probabilities equals $.3679 + .1839 = .5518$.

d. True. See Appendix Table G for $\lambda = 2$ and $t = .25$ (hence, $\mu = .50$), under $x \le 2$: it shows $p = .9856$. Thus, the desired answer equals $1 - .9856 = .0144$.

59. a. False. The probabilities are, respectively, .9179 and .9231. Appendix Table G for $\lambda = 5$ and $t = .5$—hence $\mu = 2.5$—gives $p = .0821$ for $x = 0$; hence, $p = 1 - .0821 = .9179$ for $x \ge 1$. Appendix Table D for $n = 50$ and $\pi = .05$ gives $p = .0769$ for $x = 0$; hence, $p = 1 - .0769 = .9231$ for $x \ge 1$.

b. True. See Appendix Table C, section $n = 6$, $\pi = .5$, for $x = 4$.

c. True. See Appendix Table D, section $n = 20$, $\pi = .2$, for $x \le 3$.

d. False. See Appendix Table C, section $n = 10$, $\pi = .1$, for $x = 1$. The answer is .3874 (picking 1 unsatisfactory item).

Chapter 7

1. Given the assumption of a normally distributed random variable, Appendix Table H can be used. *The 10th percentile* is that time in which 10 percent or fewer of all checks are cashed. Thus, 10 percent of the normal curve area must lie to the left of this value, and 90 percent must lie to the right of it. This means also that 40 percent of the area must lie between the 10th percentile and the (higher) mean (which is the 50th percentile). In Appendix Table H, an entry of .4000 lies between .3997 and .4015, corresponding to z values of 1.28 and 1.29, respectively. Interpolating, we find the z value corresponding to .4000 as 1.2817. Given that $z = (x - \mu)/\sigma$, it follows that $x = (\sigma \cdot z) + \mu$. In this case, with $\mu = 30$, $\sigma = 10$, and z negative (it lies *below* the mean), we have $x = 10(-1.2817) + 30 = 17.183$. Thus, 10 percent or fewer of all checks are cashed in 17.2 seconds. *The 75th percentile* is that time in which 75 percent or fewer of all checks are cashed. Thus, 75 percent of the normal-curve area must lie to the left of this value, and 25 percent of the area lies between the mean and the 75th percentile. In Appendix Table H, an entry of .2500 lies between .2486 and .2518, corresponding to z values of .67 and .68, respectively. Interpolating, we find that the z value for .2500 is .6744. Hence, $x = 10(.6744) + 30 = 36.74$. Thus, 75 percent or fewer of all checks are cashed in 36.7 seconds.

3. A raw score of 140 corresponds to a z value of $[(140 - 100)/16] = 2.5$. Appendix Table H lists a corresponding probability of .4938 for values between the mean and 2.5z. Thus, $p(z \le 2.5) = .5000 + .4938 = .9938$; and the answer equals $1 - .9938 = .0062$.

5. The 75th percentile has to be found, as in problem 1. The z value is $+.6744$; hence, $x = 100(.6744) + 500 = 567.44$. This is also the lowest acceptable score.
The 95th percentile has to be found, as in problem 2. The z value is $+1.645$; hence, $x = 12(1.645) + 60 = 79.74$. The cutoff time should be 80 minutes.

7. The 99th percentile has to be found. It corresponds to a *positive* z value that places 49 percent of the normal curve area between itself and the (lower) mean. As in answer 6, the z value is 2.3267. Hence, $x = 2.45(2.3267) + 25.9 = 31.6$. The rating must be 31.6 miles per gallon or better.

9. **a.** z values are 0 to 1.5; $p = .4332$.
   b. z values are -2 to 0; $p = .4772$.
   c. z values are -1 to $+2$; $p = .3413 + .4772 = .8185$.
   d. z values are $+3$ to ∞; $p = .5000 - .4986 = .0014$.
   e. z values are $-\infty$ to -3.9; $p = .5000 - .4999 = .0001$.
   f. z values are $+1$ to $+2$; $p = .4772 - .3413 = .1359$.
   g. z values are -3 to -2; $p = .4986 - .4772 = .0214$.
   h. z values are $-\infty$ to $+3.7$; $p = .5000 + .4999 = .9999$.
   i. z values are -1.7 to ∞; $p = .4554 + .5000 = .9554$.

11. The relevant z value is $[(140,000 - 122,000)/10,000] = 1.8$; $p = .5000 - .4641 = .0359$. Only in about 4 percent of all weeks will the workload exceed 140,000 checks.

13. **a.** The relevant z value is $[(680,000 - 700,000)/50,000] = -.4$; the relevant range for $p(\text{tolls} \ge 680,000)$ is $-.4 \le z \le \infty$; thus $p = .1554 + .5000 = .6554$.
   b. The multiplication law for independent events (Box 5.J) can be used; $p(\text{toll} \ge 680,000$ in year 1 *and* toll $\ge 680,000$ in year 2) $= .6554(.6554) = .4295$.
   c. The 75th percentile has to be found. According to the procedure discussed in Answer 1, $x = 50,000(.6744) + 700,000 = 733,720$. In the best 25 percent of years, revenue will exceed $733,720.

15. **a.** The relevant z value is $z = [(400 - 375)/40] = .625$; thus, $p(>400 \text{ students}) = p(.625 \le z \le \infty) = .5000 - .2341 = .2659$. (The value of .2341 is found in Appendix Table H by interpolating between .2324 and .2357.)
   b. The relevant z values are $z = [(350 - 375)/40] = -.625$ and, as before, $+.625$. Thus, $p(350 \le \text{students} \le 400) = p(-.625 \le z \le .625) = .2341 + .2341 = .4682$.
   c. From (a) and (b) we know that $p(350 \ge \text{students}) = p(-\infty \le z \le -.625) = .5000 - .2341 = .2659$. It is no accident that the three answers add to unity; the three events are collectively exhaustive.

17. The probability for acceptable parts is $p(2.87 \le \text{length} \le 3.10) = p\left(\dfrac{2.87 - 3}{.15} \le z \le \dfrac{3.10 - 3}{.15}\right) = p(-.87 \le z \le .67) = .3078 + .2486 = .5564$. Thus the probability sought is $1 - .5564 = .4436$.

19. $p(\text{breakage at 73 or less}) = p\left(z \le \dfrac{73 - 80}{7}\right) = p(z \le -1) = .5000 - .3413 = .1587$; $p(\text{breakage at 100 or more}) = p\left(z \ge \dfrac{100 - 80}{7}\right) = p(z \ge 2.86) = .5000 - .4979 = .0021$. Thus, $50:.1587 = x:.0021$, and $x = .66$. About a single pane could take such a force.

21. $p(\text{income} \ge 127,000) = p\left(z \ge \dfrac{127,000 - 89,000}{19,000}\right) = p(z \ge 2) = .5000 - .4772 = .0228$. Thus, $3:.0228 = x:1$, and $x = 131.6$. There are 132 psychiatrists.

23. Assume the normal distribution applies. Some 39.25 percent of workers produce between 17.3 and 18.1 tons; the z value corresponding to 18.1 tons is found in Appendix Table H as 1.24. But $\sigma = [(x - \mu)/z] = [(18.1 - 17.3)/1.24] = .645$ tons. Thus, the z value for 16 tons is $z = [(x - \mu)/\sigma] = [(16 - 17.3)/.645] = -2.02$. The associated probability is $p(\geq 16 \text{ tons}) = p(-2.02 \leq z \leq \infty) = .4783 + .5000 = .9783$. The answer is 97.83 percent.

25. a. $p(\leq 1 \text{ lb}) = p\left(-\infty \leq z \leq \dfrac{1 - 1.5}{.3}\right) =$
$p(-\infty \leq z \leq -1.67) = .5000 - .4525 = .0474.$

b. $p(\leq 2 \text{ lb}) = p\left(-\infty \leq z \leq \dfrac{2 - 1.5}{.3}\right) =$
$p(-\infty \leq z \leq 1.67) = .5000 + .4525 = .9525.$

c. $p(\geq 1.4 \text{ lb}) = p\left(\dfrac{1.4 - 1.5}{.3} \leq z \leq \infty\right) =$
$p(-.33 \leq z \leq \infty) = .1293 + .5000 = .6293.$

d. $p(1.6 \leq \text{weight gain} \leq 1.7) = p\left(\dfrac{1.6 - 1.5}{.3} \leq\right.$
$\left. z \leq \dfrac{1.7 - 1.5}{.3}\right) = p(.33 \leq z \leq .66) = .2454 -$
$.1293 = .1161.$

27. $p(\geq 63) = p\left(\dfrac{63 - 52}{10} \leq z \leq \infty\right) = p(1.1 \leq z \leq$
$\infty) = .5000 - .3643 = .1357.$ Also $p(\geq 93) =$
$p\left(\dfrac{93 - 52}{10} \leq z \leq \infty\right) = p(4.1 \leq z \leq \infty) =$
$.5000 - .5000 = 0.$ At least 13.57 percent of employees can be expected to take 63 or more hours, like A. The case of B is, indeed, highly unusual.

29. a. $p(\text{time} \leq 11) = p\left(-\infty \leq z \leq \dfrac{11 - 14}{1.3}\right) =$
$p(-\infty \leq z \leq -2.31) = .5000 - .4896 = .0104.$ Not impossible, but only 1 percent of electricians can do it.

b. $p(\text{time} \geq 17) = p\left(\dfrac{17 - 14}{1.3} \leq z \leq \infty\right) =$
$p(2.31 \leq z \leq \infty) = .5000 - .4896 = .0104.$ Only 1 percent of all electricians need this much time or more.

31. The *95th percentile* places 5 percent of possible outflow values beyond the sought value of x. It also places 45 percent of all possible values between itself and the (lower) mean; thus Appendix Table H suggests a z value of 1.645. Since $x = (\sigma \cdot z) + \mu, x = [(400)(1.645)] + 1,500 = 2,158.$ The answer: \$2,158.

33. $p(\text{insider loans} \geq 1.9) = p\left(\dfrac{1.9 - .437}{.129} \leq z \leq\right.$
$\left. \infty\right) = p(11.34 \leq z \leq \infty) = .5000 - .5000 = 0.$
Yes, there is practically no chance of another bank like this.

35. $p(W \geq 500) = p\left(\dfrac{500 - 400}{30} \leq z \leq \infty\right) =$
$p(3.33 \leq z \leq \infty) = .5000 - .4996 = .0004.$
$p(W \leq 350) = p\left(-\infty \leq z \leq \dfrac{350 - 400}{30}\right) =$
$p(-\infty \leq z \leq -1.67) = .5000 - .4525 = .0475.$
$p(410 \leq W \leq 430) = p\left(\dfrac{410 - 400}{30} \leq z \leq\right.$
$\left. \dfrac{430 - 400}{30}\right) = p(.33 \leq z \leq 1) = .3413 -$
$.1293 = .212.$

37. The *10th percentile* has to be found; it places 40 percent of all P/E ratios between itself and the (higher) mean. Appendix Table H suggests a z value of -1.2817. Since $x = (\sigma \cdot z) + \mu, x = [(2.1)(-1.2817)] + 9.6 = 6.9.$

39. a. The approximation of the binomial by the normal distribution can be used. In this example, $n = 20$, $\pi = .25$, and $1 - \pi = .75$. The rule of thumb [that $n \cdot \pi$ as well as $n(1 - \pi)$ equal or exceed 5] is met; the approximation, therefore, should be close. The necessary calculations are organized in Table A.33. As noted in the text, z scores are calculated as

$$z = \frac{x - (n \cdot \pi)}{\sqrt{n \cdot \pi(1 - \pi)}}$$

which in this case equals

$$z = \frac{x - [20(.25)]}{\sqrt{20(.25).75}} = \frac{x - 5}{1.9365}.$$

The corresponding probabilities come from Appendix Table H. The normal approximation to binomial probabilities is calculated from the probabilities associated with the two z scores as follows: in rows 0–4 of Table A.33 by deducting the probability of the upper z score from that of the lower z score, because the relevant area under the normal curve lies to the left of the mean; in row 5, by adding the two probabilities, because the relevant area under the normal curve straddles the mean; in the remaining rows, by deducting the probability of the lower z score from that of the upper z score, because the relevant area under the normal curve now lies to the right of the mean. The probabilities for x values above 12 are still positive, but tiny; note that the probabilities given for $x \leq 12$ add to only .9976. An interesting comparison can also be made by consulting Appendix Table C and finding the precise binomial probabilities directly in the section for $n = 20$ and $\pi = .25$. The values so found are given in the last column of Table A.33. As a comparison of the last two columns of Table A.33 shows, the normal approximation of the binomial probablities is extremely close.

b. Using the normal approximation, the answer is .0079 + .0249 + .0634 + .1221 = .2183. The precise value, found by adding the corresponding binomial probabilities, is .2251.

41. Use the normal approximation to the binomial distribution, noting that $n \cdot \pi = 500(.9) = 450$ as well as $n(1 - \pi) = 500(.1) = 50$ exceed 5. Here $\mu = n\pi = 450$; $\sigma = \sqrt{n \cdot \pi(1 - \pi)} = \sqrt{500(.1)(.9)} = 6.7082.$

a. $p(\text{attend} > 400) = p\left(\dfrac{400.5 - 450}{6.7082} \leq z \leq \infty\right)$
$= p(-7.38 \leq z \leq \infty) = .5000 + .5000 = 1.$

b. $p(\text{attend} < 450) = p\left(-\infty \leq z \leq \dfrac{449.5 - 450}{6.7082}\right) = p(-\infty \leq z \leq -.07) =$
.5000 − .0279 = .4721. (Remember this is an approximation.)

c. $p(413 \leq \text{attendees} \leq 463) =$
$p\left(\dfrac{412.5 - 450}{6.7082} \leq z \leq \dfrac{463.5 - 450}{6.7082}\right) =$
$p(-5.59 \leq z \leq 2.01) = .5000 + .4778 =$
.9778.

43. Use the normal approximation to the binomial distribution, noting that $n \cdot \pi$ as well as $n \cdot \pi(1 - \pi)$ exceed 5. Here $\mu = n \cdot \pi = 100(.1) = 10$, and $\sigma = \sqrt{n \cdot \pi(1 - \pi)} = \sqrt{100(.1)(.9)} = 3.$

$p(\text{errors} < 5) = p\left(-\infty \leq z \leq \dfrac{4.5 - 10}{3}\right) =$

$p(-\infty \leq z \leq -1.83) = .5000 - .4664 = .0336.$

45. See Table A.34. In this example, $n = 5$, $N = 50$, $S = 40$; hence, $\pi = (S/N) = .8$. This time, z

Table A.33

Discrete x Value	Corresponding Interval x Scores	Corresponding Interval z Scores	Probabilities Lower z Score	Probabilities Upper z Score	Normal Approximation	Precise Binomial Probabilities
0	−.5 to .5	−2.84 to −2.32	.4977	.4898	.0079	.0032
1	.5 to 1.5	−2.32 to −1.81	.4898	.4649	.0249	.0211
2	1.5 to 2.5	−1.81 to −1.29	.4649	.4015	.0634	.0669
3	2.5 to 3.5	−1.29 to −.77	.4015	.2794	.1221	.1339
4	3.5 to 4.5	−.77 to −.26	.2794	.1026	.1768	.1897
5	4.5 to 5.5	−.26 to .26	.1026	.1026	.2052	.2023
6	5.5 to 6.5	.26 to .77	.1026	.2794	.1768	.1686
7	6.5 to 7.5	.77 to 1.29	.2794	.4015	.1221	.1124
8	7.5 to 8.5	1.29 to 1.81	.4015	.4649	.0634	.0609
9	8.5 to 9.5	1.81 to 2.32	.4649	.4898	.0249	.0271
10	9.5 to 10.5	2.32 to 2.84	.4898	.4977	.0079	.0099
11	10.5 to 11.5	2.84 to 3.36	.4977	.4996	.0019	.0030
12	11.5 to 12.5	3.36 to 3.87	.4996	.4999	.0003	.0008
13	12.5 to 13.5	3.87 to 4.39	.4999	.4999		.0002
14	13.5 to 14.5	4.39 to 4.91	.4999	.4999		
15	14.5 to 15.5	4.91 to 5.42	.4999	.4999		
16	15.5 to 16.5	5.42 to 5.94	.4999	.4999	.0000	
17	16.5 to 17.5	5.94 to 6.45	.4999	.4999		.0000
18	17.5 to 18.5	6.45 to 6.97	.4999	.4999		
19	18.5 to 19.5	6.97 to 7.49	.4999	.4999		
20	19.5 to 20.5	7.49 to 8.00	.4999	.4999		
					.9976	1.0000

Table A.34

Discrete *x* Value	Corresponding Interval		Probabilities		Normal Approximation	Precise Hypergeometric Probabilities
	x Scores	*z* Scores	Lower *z* Score	Upper *z* Score		
0	−.5 to .5	−5.25 to −4.08	.4999	.4999	.0000	.0001
1	.5 to 1.5	−4.08 to −2.92	.4999	.4982	.0017	.0040
2	1.5 to 2.5	−2.92 to −1.75	.4982	.4599	.0383	.0442
3	2.5 to 3.5	−1.75 to −.58	.4599	.2190	.2409	.2098
4	3.5 to 4.5	−.58 to .58	.2190	.2190	.4380	.4313
5	4.5 to 5.5	.58 to 1.75	.2190	.4599	.2409	.3106
					.9598	1.0000

scores must be calculated, in accordance with the formulas in Box 6.E, as

$$z = \frac{x - (n \cdot \pi)}{\sqrt{n \cdot \pi(1 - \pi)\left(\dfrac{N - n}{N - 1}\right)}}$$

$$= \frac{x - [5(.8)]}{\sqrt{5(.8)(.2)\left(\dfrac{50 - 5}{50 - 1}\right)}} = \frac{x - 4}{.8571}.$$

The precise probabilities from Table 6.12 have been reproduced in the last column of Table A.34 for comparison. Note once again that the normal probabilities do not add to 1 because the normal distribution assigns positive, though tiny, probabilities even to intervals between very large *x* values. Note also that the match of the last two columns is not perfect, but then the rule of thumb for using the normal approximation (let $n \cdot \pi$ and $n(1 - \pi) \geq 5$) has been violated; in this case, $n \cdot \pi = 4$ and $n(1 - \pi) = 1$. (This is merely an exercise.)

47. In this example, $\mu_R = 2$.

a. Using the formula in Box 7.F, along with Appendix Table E,

$$p(R < 1) = 1 - e^{-(x/\mu)} = 1 - e^{-(1/2)} = 1 - e^{-.5}$$
$$= 1 - .6065 = .3935.$$

b. $p(R < 4) = 1 - e^{-2} = 1 - .1353 = .8647.$

c. Using the formula in Box 7.G,

$$p(2 < R < 6) = e^{-(2/2)} - e^{-(6/2)} = e^{-1} - e^{-3}$$
$$= .3679 - .0498 = .3181.$$

d. Using the formula in Box 7.E,
$$p(R > 5) = e^{-(5/2)} = e^{-2.5} = .0821.$$

49. Given $1/\lambda$ of 500 hours, $\lambda = (1/500) = .002$ failures per hour.

a. $p(R \leq 10) = .0198.$ (Appendix Table I for $\lambda \cdot x = .002(10) = .02.$)

b. If the generators work independently, $p = .0198^2 = .000392.$

c. The chance of failure during a 5-hour blackout is given in Appendix Table I for $\lambda \cdot x = .002(5) = .01$ as .01. Thus, the chance of no failure or of success is $\pi = 1 - .01 = .99$. Using the binomial formula,

$$p(R = 10 | n = 10, \pi = .99)$$
$$= \frac{10!}{10! \, 0!} (.99)^{10}(.01)^0 = .9044.$$

d. The chance of failure during a 5-hour blackout now is $.01^2 = .0001$; hence, $\pi = 1 - .0001 = .9999$. The binomial formula now gives us

$$p(R = 10 | n = 10, \pi = .9999)$$
$$= \frac{10!}{10! \, 0!} (.9999)^{10}(.0001)^0 = .999.$$

51. Given $\mu = 2, \lambda = .5$

a. From Box 7E and Appendix Table E: $p(R > 2) = e^{-.5(2)} = e^{-1} = .3679.$

b. $p(R > 3) = e^{-.5(3)} = e^{-1.5} = .2231$

c. From Box 7F and Appendix Table I: $p(R < .5) = 1 - e^{-.5(.5)} = 1 - e^{-.25} = .2212.$

d. From Box 7G and Appendix Table E: $p(2 < R < 3) = e^{-.5(2)} - e^{-.5(3)} = e^{-1} - e^{-1.5} = .3679 - .2231 = .1448.$ The answer can also be found in Appendix Table I for $\lambda \cdot x = 1.5$ and 1, respectively, as $.7769 - .6321 = .1448.$

53. Given $\mu = 66, \lambda = .0\overline{15}$. All questions can be answered with the help of Appendix Table I.

a. $p(R \leq 30) = 1 - e^{-.0\overline{15}(30)} = 1 - e^{-.45} = .3624$

b. $p(R \leq 66) = 1 - e^{-.0\overline{15}(66)} = 1 - e^{-1} = .6321$

c. $p(R \leq 90) = 1 - e^{-.0\overline{15}(90)} = 1 - e^{-1.36} = .7433$

d. $p(120 < R < 180) = e^{-.0\overline{15}(120)} - e^{-.0\overline{15}(180)} = e^{-1.82} - e^{-2.73} = (1 - .8380) - (1 - .9348) = .1620 - .0652 = .0968.$

55. You don't. According to Appendix Table I, the probability value of .0488 is associated with $\lambda \cdot x = .05$. Given $x = 10$, $\lambda = .005$; therefore, $\mu = (1/.005) = 200$. The firm's product would have to have a mean lifetime of 200 years, which is highly doubtful. Either claim (a) or (b) is untrue.

57. The uniform probability distribution applies. In this case, $a = 0$ and $b = 1$ (see Box 7H and 7I)

a. $p(R < .25) = \dfrac{.25 - 0}{1 - 0} = .25.$

b. $p(R < .79) = \dfrac{.79 - 0}{1 - 0} = .79.$

c. $p(R > .79) = 1 - p(R < .79) = 1 - .79 = .21.$

d. $p(.25 < R < .79) = p(R < .79) - p(R < .25) = .79 - .25 = .54.$

59. The uniform probability distribution applies. In this case, $a = 0$ and $b = 30$ (see Boxes 7H–7J).

a. $\mu = \dfrac{a + b}{2} = 15$ minutes.

b. $p(R > 25) = 1 - p(R < 25) = 1 - \dfrac{25 - 0}{30 - 0} = 1 - .8\overline{333} = .1\overline{666}.$

c. $p(R < 5) = \dfrac{5 - 0}{30 - 0} = .1\overline{666}.$

d. $p(7 < R < 10) = p(R < 10) - p(R < 7) = \dfrac{10 - 0}{30 - 0} - \dfrac{7 - 0}{30 - 0} = .\overline{3333} - .\overline{2333} = .1.$

Chapter 8

1. From Box 8.A, we can determine $\sigma_{\overline{X}} = (\sigma/\sqrt{n}) = (50/5) = 10$. Given $\sigma_{\overline{X}} = 10$ and a normal sampling distribution (because the population is normal), we can employ Appendix Table H to find z values and our answers: (a) Between $z = [(140 - 150)/10] = -1$ and $z = [(160 - 150)/10] = +1$, the probability is $.3413 + .3413 = .6826$. (b) Above $z = [(140 - 150)/10] = -1$, the probability is $.3413 + .5000 = .8413$. (c) Below $z = [(130 - 150)/10] = -2$, the probability is $.5000 - .4772 = .0228$. (d) Between $z = [(160 - 150)/10] = +1$ and $z = [(170 - 150)/10] = +2$, the probability is $.4772 - .3413 = .1359$. (e) Above $z = [(155 - 150)/10] = +.5$, the probability is $.5000 - .1915 = .3085$.

3. The formulas of Box 8.A apply.

a. $\mu_{\overline{X}} = 30$; $\sigma_{\overline{X}} = (3/\sqrt{10}) = .9487.$

b. $\mu_{\overline{X}} = 50$; $\sigma_{\overline{X}} = (2/\sqrt{15}) = .5164.$

c. $\mu_{\overline{X}} = 100$; $\sigma_{\overline{X}} = (10/\sqrt{30}) = 1.8257.$

d. $\mu_{\overline{X}} = 400$; $\sigma_{\overline{X}} = (8/\sqrt{100}) = .8.$

5. $\mu_{\overline{X}} = 1,200$ and $\sigma_{\overline{X}} = (250/\sqrt{36}) = 41.\overline{66}.$

$$p(1,150 \leq \overline{X} \leq 1,250) = p\left(\dfrac{1,150 - 1,200}{41.\overline{66}} \leq z \leq \dfrac{1,250 - 1,200}{41.\overline{66}}\right) = p(-1.2 \leq z \leq 1.2) = .3849 + .3849 = .7698.$$

7. See Table A.35, which proceeds from the fact that the possible number of equally likely samples is

$$C_2^5 = \dfrac{5!}{2!(5 - 2)!} = 10.$$

Table A.35

Sample Number	Sample Ages	Possible Sample Means, $\overline{X}$	Probability $p(R = \overline{X})$	The Sampling Distribution of the Sample Mean	
				$\overline{X}$	$p(R = \overline{X})$
1	34, 29	31.5	.1	23.5	.1
2	34, 22	28.0	.1	25.5	.1
3	34, 30	32.0	.1	26.0	.1
4	34, 25	29.5	.1	27.0	.1
5	29, 22	25.5	.1	27.5	.1
6	29, 30	29.5	.1	28.0	.1
7	29, 25	27.0	.1	29.5	.2
8	22, 30	26.0	.1	31.5	.1
9	22, 25	23.5	.1	32.0	.1
10	30, 25	27.5	.1		
		$\Sigma\overline{X} = 280$			1.0

Using the Box 6.A formulas, $\mu_{\bar{X}} = 280/10 = 28$; $\sigma_{\bar{X}}^2 = \Sigma p(\bar{X} - \mu_{\bar{X}})^2 = 6.45$; $\sigma_{\bar{X}} = 2.54$.

9. The solution here requires the conversion of the probability distribution for possible *numbers* of females (given in Figure 6.4) into the one for possible *proportions* of females given in Table A.36. The *summary measures* for the sampling distribution in this case must employ the finite population correction factor of $(N - n)/(N - 1)$, as shown in Box 8.B. Thus, $\mu_P = .6$ (because there were $S = 6$ females among the $N = 10$ names in the personnel manager's file),

$$\sigma_P^2 = \frac{.6(.4)}{4} \cdot \left(\frac{10 - 4}{10 - 9}\right) = .06(6) = .36,$$

and $\sigma_P = \sqrt{.36} = .6$.

Table A.36

Possible Sample Proportions, $P = \dfrac{x}{n}$	Probability, $p(R = P)$
4/4	(360/5040) = .071
3/4	(1920/5040) = .381
2/4	(2160/5040) = .429
1/4	(576/5040) = .114
0/4	(24/5040) = .005
	1.000

11. **a.** $\pi = .61$.
   b. $\pi = .6$.
   c. $\mu = \$243$.
13. This problem corresponds to Practice Problem 6 in the text: $n \geq 30$, $n < .05N$, σ is known. Hence,

$$\sigma_{\bar{X}} = \frac{\sigma}{\sqrt{n}} = \frac{4}{\sqrt{50}} = .566.$$

The required z value equals 1.96. Thus, the 95-percent confidence interval goes from 7.99 to 10.21 minutes, because $\mu = \bar{X} \pm z\sigma_{\bar{X}} = 9.1 \pm 1.96(.566) = 9.1 \pm 1.11$.

15. This problem corresponds to Practice Problem 7 in the text: $n \geq 30$, $n \geq .05N$, the population is normal, σ is known. Hence,

$$\sigma_{\bar{X}} = \frac{\sigma}{\sqrt{n}}\sqrt{\frac{N - n}{N - 1}} = \frac{99}{\sqrt{36}}\sqrt{\frac{75 - 36}{74}} = 11.98.$$

The required z value equals 3.08. Thus, the 99.8-percent confidence interval goes from \$699.10 to \$772.90, because $\mu = \bar{X} \pm z\sigma_{\bar{X}} = 736 \pm 3.08(11.98) = 736 \pm 36.90$.

17. This problem corresponds to Practice Problem 8 in the text: $n \geq 30$, $n < .05N$, σ unknown. Hence,

$$\sigma_{\bar{X}} \cong \frac{s}{\sqrt{n}} = \frac{3,600}{\sqrt{100}} = 360.$$

The required z value equals 1.96. Thus the 95-percent confidence interval goes from \$14,044.40 to \$15,455.60, because $\mu = \bar{X} \pm z\sigma_{\bar{X}} = 14,750 \pm 1.96(360) = 14,750 \pm 705.60$.

19. In this problem, $n \geq 30$, $n \geq .05N$, σ unknown. Hence,

$$\sigma_{\bar{X}} \cong \frac{s}{\sqrt{n}}\sqrt{\frac{N - n}{N - 1}} = \frac{.231}{\sqrt{50}}\sqrt{\frac{200 - 50}{200 - 1}} = .0284.$$

The required z value equals 1.28. Thus the 80-percent confidence interval goes from \$1.303 to \$1.375, because $\mu = \bar{X} \pm z\sigma_{\bar{X}} = 1.339 \pm 1.28(.0284) = 1.339 \pm .036$.

21. This problem corresponds to Practice Problem 8 in the text: $n \geq 30$, $n < .05N$, σ unknown. Hence,

$$\sigma_{\bar{X}} \cong \frac{s}{\sqrt{n}} = \frac{5.2}{\sqrt{81}} = .5777.$$

The required z value equals 3.08. Thus the 99.8-percent confidence interval goes from 12.12 to 15.68 seats, because $\mu = \bar{X} \pm z\sigma_{\bar{X}} = 13.9 \pm 3.08(.5777) = 13.9 \pm 1.7796$.

23. This problem corresponds to Practice Problem 8 in the text: $n \geq 30$, $n < .05N$, σ unknown. Hence,

$$\sigma_{\bar{X}} \cong \frac{s}{\sqrt{n}} = \frac{2.01}{\sqrt{100}} = .201.$$

The required z value (from Appendix Table H) is 2.24. Thus the 97.5-percent confidence interval goes from 5.30 lb. to 6.20 lb., because $\mu = \bar{X} \pm z\sigma_{\bar{X}} = 5.75 \pm 2.24(.201) = 5.75 \pm .45$.

25. $$\pi = P + (z\sigma_P) \cong P \pm \left(1.96\sqrt{\frac{P(1 - P)}{n}}\right)$$

$$1960: \quad .51 \pm \left(1.96\sqrt{\frac{.51(.49)}{1,500}}\right) = .51 \pm .025;$$

range is .485 to .535.

1964: $.64 \pm \left(1.96\sqrt{\dfrac{.64(.36)}{1,500}}\right) = .64 \pm .024;$

range is .616 to .664.

1968: $.50 \pm \left(1.96\sqrt{\dfrac{.5(.5)}{1,500}}\right) = .50 \pm .025;$

range is .475 to .525.

1972: $.38 \pm \left(1.96\sqrt{\dfrac{.38(.62)}{1,500}}\right) = .38 \pm .025;$

range is .355 to .405.

27. $\pi = .39 \pm \left(1.96\sqrt{\dfrac{.39(.61)}{100}}\right) = .39 \pm .10;$ range

is .29 to .49.

29. $\pi = P \pm (z\sigma_P)$

$\cong P \pm \left(z\sqrt{\dfrac{P(1 - P)}{n}}\sqrt{\dfrac{N - n}{N - 1}}\right)$

$= .7 \pm \left(2.57\sqrt{\dfrac{.7(.3)}{30}}\sqrt{\dfrac{100 - 30}{99}}\right)$

$= .7 \pm .18;$ range is .52 to .88.

31. $\pi = .38 \pm \left(1.22\sqrt{\dfrac{.38(.62)}{50}}\right) = .38 \pm .084;$ range

is .30 to .46.

33. $\pi = .26 \pm \left(2.51\sqrt{\dfrac{.26(.74)}{50}}\sqrt{\dfrac{150 - 50}{150 - 1}}\right) = .26 \pm$

.1276; the range is .13 to .39.

35. $\pi = .12 \pm \left(2.33\sqrt{\dfrac{.12(.88)}{100}}\right) = .12 \pm .0757;$ the

range is .04 to .20. The auditor believes that be-tween 4 and 20 percent of all bank statements mailed to customers are in error; the procedure employed can be relied upon to give the correct answer 98 times out of 100.

37. This problem corresponds to Practice Problem 10 in the text: $n < 30$, $n \geq .05N$, the population is normal, σ is known. Hence,

$$\sigma_{\bar{X}} = \dfrac{\sigma}{\sqrt{n}}\sqrt{\dfrac{N - n}{N - 1}} = \dfrac{3.7}{\sqrt{10}}\sqrt{\dfrac{35 - 10}{34}} = 1.00.$$

The required z value equals 2.054 (from Appendix Table H on page T–25, looking for an inside value

of $.96/2 = .4800$). Thus, the 96-percent confidence interval goes from 17.946 to 22.054 gallons, be-cause $\mu = \bar{X} \pm z\sigma_{\bar{X}} = 20 \pm 2.054(1)$.

39. Part (i): This problem corresponds to Practice Problem 12 in the text: $n < 30$, $n < .05N$, the pop-ulation is normal, σ is unknown. The required t value (for 19 degrees of freedom and an upper-tail area, $\alpha = .025$) is 2.093. The 95-percent confi-dence interval goes from 6.83 to 9.17 days, because

$$\mu = \bar{X} \pm t\dfrac{s}{\sqrt{n}} = 8 \pm \left[2.093\left(\dfrac{2.5}{\sqrt{20}}\right)\right]$$
$$= 8 \pm 1.17.$$

Part (ii): This problem corresponds to Practice Problem 11 in the text: $n < 30$, $n < .05N$, the pop-ulation is *not* normal, σ is known. As in the text, $k = 4.472$. The 95-percent confidence interval now goes from 5 to 11 days, because

$$\mu = \bar{X} \pm \left(k\dfrac{\sigma}{\sqrt{n}}\right)$$
$$= 8 \pm \left[4.472\left(\dfrac{3}{\sqrt{20}}\right)\right] = 8 \pm 3.$$

41. The same procedure as in part (i) of problem 39 applies. The required t value (for 4 degrees of freedom and an upper-tail area, $\alpha = .005$) is 4.604. The 99-percent confidence interval goes from 47.3 to 56.7 knots, because

$$\mu = \bar{X} \pm \left(t\dfrac{s}{\sqrt{n}}\right) = 52 \pm \left(4.604\dfrac{2.3}{\sqrt{5}}\right)$$
$$= 52 \pm 4.7356.$$

43. The procedure used in problem 35 can be adapted; the required t value (for 24 degrees of freedom and an upper-tail area, $\alpha = .005$) is 2.797. Thus

$$\pi \cong P \pm \left(t\sqrt{\dfrac{P(1 - P)}{n}}\right) = .68$$
$$\pm \left(2.797\sqrt{\dfrac{.68(.32)}{25}}\right) = .68$$
$$\pm .26;\text{ the range is .42 to .94.}$$

45. This case involves two large, independent samples; we can apply the Box 8.E formula to the following data: $n_A = 290$; $n_B = 333$; $\overline{X}_A = 59$, $\overline{X}_B = 71$; $s_A = 20$; $s_B = 30$; $z = 2.57$. Hence,

$$\mu_A - \mu_B \cong -12 \pm \left(2.57\sqrt{\frac{400}{290} + \frac{900}{333}} \right)$$
$$= -12 \pm [2.57(2.02)]$$
$$= -12 \pm 5.19.$$

The difference between the mean intensities lies between -17.19 and -6.81 miles per hour. Storms are that much less intense in region A.

47. The same procedure as in problem 45 can be applied; but to these data: $n_A = 42$; $n_B = 32$; $\overline{X}_A = 59.05$; $\overline{X}_B = 70.08$; $s_A = 7.92$; $s_B = 25.39$; $z = 1.64$. Hence,

$$\mu_A - \mu_B \cong -11.03$$
$$\pm \left(1.64\sqrt{\frac{(7.92)^2}{42} + \frac{(25.39)^2}{32}} \right)$$
$$= -11.03 \pm 7.63.$$

The range goes from $-\$18.66$ to $-\$3.40$, suggesting an average NYSE price this much smaller.

49. This case involves two small, independent samples, with $n_A = n_B = 15$. The t value (for $\alpha = .025$ and 28 degrees of freedom) equals 2.048. Thus,

$$\mu_A - \mu_B \cong 24 \pm \left(2.048\sqrt{\frac{441}{15} + \frac{2704}{15}} \right)$$
$$= 24 \pm (2.048\sqrt{209.667})$$
$$= 24 \pm 29.65.$$

The population difference ranges from -5.65 to $+53.65$ houses; thus, it is not clear that one camera is better than the other.

51. The case corresponds to Practice Problem 17 in the text, but $t = 2.65$ (upper-tail area $\alpha = .01$; 13 degrees of freedom). Hence,

$$\mu_A - \mu_B \cong \overline{D} \pm \left(t\frac{s_D}{\sqrt{n}} \right)$$
$$= 133 \pm \left(2.65\frac{41}{\sqrt{14}} \right) = 133 \pm 29.04.$$

The daily sales of restaurant A are higher, somewhere between $103.96 and $162.04.

53. The case involves large and independent samples; hence the formula of Box 8.G applies.

$$\pi_A - \pi_B \cong (.68 - .22)$$
$$\pm \left(2.33\sqrt{\frac{(.68)(.32)}{50} + \frac{(.22)(.78)}{50}} \right)$$
$$= .46 \pm .206.$$

The difference in proportions ranges from .254 to .666.

55. The same procedure as in problem 53 can be applied.

$$\pi_D - \pi_N \cong (.116 - .092)$$
$$\pm \left(2.33\sqrt{\frac{(.116)(.884)}{4,900} + \frac{(.092)(.908)}{3,900}} \right)$$
$$= .024 \pm .015.$$

The interval goes from .009 to .039.

57. a. Applying the formula in Box 8.H,

$$n = \left(\frac{1.96(.5)}{.1} \right)^2 = 96.04.$$

A sample of 97 computer users is needed.

b. Applying the formula in Box 8.I,

$$n = \frac{(2.33)^2(.25)(.75)}{(.05)^2} = 407.17.$$

and

$$n = \frac{(2.33)^2(.5)(.5)}{(.05)^2} = 542.89.$$

A sample of 408 interviewees is needed in the first instance; a sample of 543 interviewees is needed in the second instance.

c. Applying the formula in Box 8.H,

$$n = \left(\frac{2.57(5)}{2} \right)^2 = 41.28.$$

and

$$n = \left(\frac{1.96(5)}{2} \right)^2 = 24.01.$$

The sample size must equal 42 applicants in the first case and 25 in the second.

59. a. Applying the formula in Box 8.H,

$$n = \left(\frac{1.96(2)}{1}\right)^2 = 15.36.$$

A sample of 16 air specimens is needed.

b. Applying the formula in Box 8.H,

$$n = \left(\frac{3.27(3)}{.1}\right)^2 = 9,623.6$$

A sample of 9,624 is needed.

c. Applying the formula in Box 8.I,

$$n = \frac{(1.96)^2(.5)(.5)}{(.01)^2} = 9,604.$$

A sample of 9,604 is needed.

d. Applying the formula in Box 8.I,

$$n = \frac{(2.33)^2(.12)(.88)}{(.0757)^2} = 100.0.$$

A sample of 100 is needed. (Compare this answer with the answer to problem 35.)

Chapter 9

1. For $\alpha = .001$, this being a two-tailed test, the critical values in Practice Problem 1 would have been $\pm z_{\alpha/2} = \pm 3.27$ (see Appendix Table J). Hence, the decision rule would have been "Accept H_0 if $-3.27 \leq z \leq +3.27$." Given the computed value of the test statistic ($z = -1$), the null hypothesis would have been *accepted*. For $\alpha = .25$, we would have found $\pm z_{\alpha/2} = \pm 1.15$ (see Appendix Table H), with the same result.

3. *Step 1:* H_0: $\mu = 33$. H_A: $\mu \neq 33$
Step 2:

$$z = \frac{\overline{X} - \mu_0}{\sigma_{\overline{X}}}$$

Step 3: The critical value is: $\pm z_{\alpha/2} = \pm 2.24$. The decision rule is "Accept H_0 if $-2.24 \leq z \leq +2.24$."
Step 4:

$$z = \frac{34 - 33}{2/\sqrt{100}} = 5$$

H_0 should be *rejected:* the sleeves are longer.

5. *Step 1:* H_0: $\mu = 12,357$. H_A: $\mu \neq 12,357$
Step 2:

$$z = \frac{\overline{X} - \mu_0}{\sigma_{\overline{X}}}$$

Step 3: The critical value is: $\pm z_{\alpha/2} = \pm 3.27$. The decision rule is "Accept H_0 if $-3.27 \leq z \leq +3.27$."
Step 4:

$$z = \frac{13,950 - 12,357}{3,972/\sqrt{36}} = 2.41$$

H_0 should be *accepted.*

7. *Step 1:* H_0: $\mu = 500$. H_A: $\mu \neq 500$
Step 2:

$$z = \frac{\overline{X} - \mu_0}{\sigma_{\overline{X}}}$$

Step 3: The critical value can be found in Appendix Table H (by noting that $\alpha/2 = .01$, that $.5 - .01 = .49$, and that $.49$ corresponds to $z = 2.327$) as: $\pm z_{\alpha/2} = \pm 2.327$. The decision rule is "Accept H_0 if $-2.327 \leq z \leq +2.327$."
Step 4:

$$z = \frac{592 - 500}{101/\sqrt{32}} = 5.15$$

H_0 should be *rejected;* the cost is higher.

9. *Step 1:* H_0: $\mu \geq 60$. H_A: $\mu < 60$
Step 2:

$$z = \frac{\overline{X} - \mu_0}{\sigma_{\overline{X}}}$$

Step 3: The critical value is: $-z_{\alpha} = -1.64$. The decision rule is "Accept H_0 if $z \geq -1.64$."
Step 4:

$$z = \frac{101 - 60}{21/\sqrt{50}} = 13.81$$

H_0 should be *accepted.*

11. *Step 1:* H_0: $\mu \leq 50$. H_A: $\mu > 50$
Step 2:

$$z = \frac{\overline{X} - \mu_0}{\sigma_{\overline{X}}}$$

Step 3: The critical value is: $z_\alpha = 3.08$. The decision rule is "Accept H_0 if $z \leq 3.08$."

Step 4:

$$z = \frac{71 - 50}{30/\sqrt{100}} = 7$$

H_0 should be *rejected;* the mean was higher.

13. In each case, the interval limits are $\mu = \overline{X} \pm z_\alpha \sigma_{\overline{X}}$.

Problem 7: 98-percent confidence inverval, $z = 2.33$.

$$\mu = 592 \pm \left[2.33\left(\frac{101}{\sqrt{32}}\right)\right] = 592 \pm 41.60.$$

The range is $550.40 to $633.60.

Problem 8: 99-percent confidence interval, $z = 2.57$.

$$\mu = 399 \pm \left[2.57\left(\frac{25}{\sqrt{50}}\sqrt{\frac{450 - 50}{450 - 1}}\right)\right] = 399 \pm 8.576.$$

The range is $390.42 to $407.58.

Problem 9: 95-percent confidence interval, $z = 1.96$.

$$\mu = 101 \pm \left[1.96\left(\frac{21}{\sqrt{50}}\right)\right] = 101 \pm 5.82.$$

The range is 95.18 minutes to 106.82 minutes.

Problem 10: 97.5-percent confidence interval, $z = 2.24$.

$$\mu = 4.2 \pm \left[2.24\left(\frac{1}{\sqrt{64}}\right)\right] = 4.2 \pm .28.$$

The range is 3.92 ppm to 4.48 ppm.

15. *Step 1:* H_0: $\pi = .25$. H_A: $\pi \neq .25$

Step 2:

$$z = \frac{P - \pi_0}{\sigma_P}$$

Step 3: The critical value is: $\pm z_{\alpha/2} = \pm 2.57$. The decision rule is "Accept H_0 if $-2.57 \leq z \leq +2.57$."

Step 4:

$$\sigma_P \cong \sqrt{\frac{P(1 - P)}{n}} = \sqrt{\frac{.19(.81)}{200}} = .0277$$

$$z = \frac{.19 - .25}{.0277} = -2.16$$

H_0 should be *accepted.*

17. a. *Step 1:* H_0: $\pi \geq .50$. H_A: $\pi < .50$

Step 2:

$$z = \frac{P - \pi_0}{\sigma_P}$$

Step 3: The critical value is: $-z_\alpha = -1.64$. The decision rule is "Accept H_0 if $z \geq -1.64$."

Step 4:

$$\sigma_P \cong \sqrt{\frac{P(1 - P)}{n}} = \sqrt{\frac{.4(.6)}{25}} = .098$$

$$z = \frac{.4 - .5}{.098} = -1.02$$

H_0 should be *accepted.*

b. Now $\sigma_P = .049$ and $z = -2.04$.

H_0 should be *rejected* in this case.

19. *Step 1:* H_0: $\pi \leq .12$. H_A: $\pi > .12$.

Step 2:

$$z = \frac{P - \pi_0}{\sigma_P}$$

Step 3: The critical value is: $z_\alpha = 1.64$. The decision rule is "Accept H_0 if $z \leq 1.64$."

Step 4:

$$\sigma_P \cong \sqrt{\frac{P(1 - P)}{n}} = \sqrt{\frac{.108(.892)}{250}} = .01963$$

$$z = \frac{.108 - .12}{.01963} = -.61$$

H_0 should be *accepted.*

21. *Step 1:* H_0: $\pi = .53$. H_A: $\pi \neq .53$

Step 2:

$$z = \frac{P - \pi_0}{\sigma_P}$$

Step 3: The critical value is: $\pm z_{\alpha/2} = \pm 1.96$. The decision rule is "Accept H_0 if $-1.96 \leq z \leq +1.96$."

Step 4:

$$\sigma_P \cong \sqrt{\frac{.51(.49)}{1,500}} = .01291$$

$$z = \frac{.51 - .53}{.01291} = -1.55$$

H_0 should be *accepted*. Note how, in the Chapter 8, problem 25 answer, a proportion of .53 also lay within the 95-percent confidence interval.

23. *Step 1:* H_0: $\mu_A - \mu_B = 0$; H_A: $\mu_A - \mu_B \neq 0$.
Step 2:

$$z = \frac{\overline{X}_A - \overline{X}_B}{\sigma_d}.$$

Step 3: The critical value is: $\pm z_{\alpha/2} = \pm 2.24$. The decision rule is "Accept H_0 if $-2.24 \leq z \leq +2.24$."
Step 4:

$$\sigma_d \cong \sqrt{\frac{s_A^2}{n_A} + \frac{s_B^2}{n_B}}$$

$$= \sqrt{\frac{(200)^2}{100} + \frac{(100)^2}{80}} = 22.9129.$$

$$z = \frac{500 - 410}{22.9129} = 3.93.$$

H_0 should be *rejected*. New York wages are higher.

25. *Step 1:* H_0: $\mu_A - \mu_B = 0$. H_A: $\mu_A - \mu_B \neq 0$
Step 2:

$$z = \frac{\overline{X}_A - \overline{X}_B}{\sigma_d}$$

Step 3: The critical value is: $\pm z_{\alpha/2} = \pm 2.81$. The decision rule is "Accept H_0 if $-2.81 \leq z \leq +2.81$."
Step 4:

$$\sigma_d \cong \sqrt{\frac{s_A^2}{n_A} + \frac{s_B^2}{n_B}} = \sqrt{\frac{64}{36} + \frac{9}{49}} = 1.4005$$

$$z = \frac{93 - 87}{1.4005} = 4.28$$

H_0 should be *rejected*. Center A scores are higher.

27. *Step 1:* H_0: $\mu_A - \mu_B \geq 0$. H_A: $\mu_A - \mu_B < 0$.
Step 2:

$$z = \frac{\overline{X}_A - \overline{X}_B}{\sigma_d}$$

Step 3: The critical value is: $-z_\alpha = -1.64$. The decision rule is "Accept H_0 if $z \geq -1.64$."
Step 4:

$$\sigma_d \cong \sqrt{\frac{s_A^2}{n_A} + \frac{s_B^2}{n_B}} = \sqrt{\frac{(18.2)^2}{100} + \frac{(3.2)^2}{100}} = 1.8479$$

$$z = \frac{48.25 - 39.95}{1.8479} = 4.49$$

H_0 should be *accepted*.

29. *Step 1:* H_0: $\pi_A - \pi_B = 0$. H_A: $\pi_A - \pi_B \neq 0$
Step 2:

$$z = \frac{P_A - P_B}{\sigma_d}$$

Step 3: The critical value is: $\pm z_{\alpha/2} = \pm 1.64$. The decision rule is "Accept H_0 if $-1.64 \leq z \leq +1.64$."
Step 4:

$$P_P = \frac{25(.72) + 171(.269)}{196} = .3265$$

$$\sigma_d \cong \sqrt{.3265(.6735)\left(\frac{1}{25} + \frac{1}{171}\right)} = .1004$$

$$z = \frac{.72 - .269}{.1004} = 4.49$$

H_0 should be *rejected*. Having a committee member's district involved made a big difference.

31. *Step 1:* H_0: $\pi_A - \pi_B \geq 0$. H_A: $\pi_A - \pi_B < 0$
Step 2:

$$z = \frac{P_A - P_B}{\sigma_d}$$

Step 3: The critical value is: $-z_\alpha = -2.33$. The decision rule is "Accept H_0 if $z \geq -2.33$."
Step 4:

$$P_P = \frac{200(.18) + 200(.19)}{400} = .185$$

$$\sigma_d \cong \sqrt{.185(.815)\left(\frac{1}{200} + \frac{1}{200}\right)} = .03883$$

$$z = \frac{.18 - .19}{.03883} = -.26.$$

H_0 should be *accepted*.

33. *Step 1:* H_0: $\pi_A \leq \pi_B$. H_A: $\pi_A > \pi_B$.
Step 2:

$$z = \frac{P_A - P_B}{\sigma_d}.$$

Step 3: The critical value is: $z_\alpha = 2.57$. The decision rule is "Accept H_0 if $z \leq 2.57$."
Step 4:

$$P_P = \frac{100(.9) + 150(.7)}{250} = .78$$

$$\sigma_d \cong \sqrt{.78(.22)\left(\frac{1}{100} + \frac{1}{150}\right)} = .0535$$

$$z = \frac{.9 - .7}{.0535} = 3.74.$$

H_0 should be *rejected:* The quality of the foreign components is better.

35. *Step 1:* H_0: $\mu = 5$. H_A: $\mu \neq 5$. Assume normal distribution of lifetimes.
Step 2:

$$t = \frac{\overline{X} - \mu_0}{s/\sqrt{n}}$$

Step 3: The critical value is: $\pm t_{.0005,15} = \pm 4.073$ (Appendix Table K, 15 degrees of freedom). The decision rule is "Accept H_0 if $-4.073 \leq t \leq +4.073$."
Step 4:

$$t = \frac{4.1 - 5}{1.6/\sqrt{16}} = -2.25$$

H_0 should be *accepted:* The seemingly low sample result can be attributed to sampling error; rats with artificial blood do live normal life spans.

37. *Step 1:* H_0: $\mu = 190$. H_A: $\mu \neq 190$. Assume normal distribution of temperatures.
Step 2:

$$t = \frac{\overline{X} - \mu_0}{s/\sqrt{n}}$$

Step 3: The critical value is: $\pm t_{.025,24} = \pm 2.064$ (Appendix Table K, 24 degrees of freedom). The decision rule is "Accept H_0 if $-2.064 \leq t \leq +2.064$."

Step 4:

$$t = \frac{193 - 190}{3/\sqrt{25}} = 5$$

H_0 should be *rejected.* The engine *is* running hot.

39. *Step 1:* H_0: $\mu \leq 105$. H_A: $\mu > 105$. Assume normal distribution of crankcase weights.
Step 2:

$$t = \frac{\overline{X} - \mu_0}{s/\sqrt{n}}$$

Step 3: The critical value is $t_\alpha = t_{.01,19} = 2.539$ (Appendix Table K, 19 degrees of freedom). The decision rule is "Accept H_0 if $t \leq 2.539$."
Step 4:

$$t = \frac{107 - 105}{.1/\sqrt{20}} = 89.44$$

H_0 should be *rejected.* The crankcases are unsuitable.

41. *Step 1:* H_0: $\mu \geq 60$. H_A: $\mu < 60$. Assume normal distribution of top speeds.
Step 2:

$$t = \frac{\overline{X} - \mu_0}{s/\sqrt{n}}$$

Step 3: The critical value is: $-t_\alpha = -t_{.01,4} = -3.747$ (Appendix Table K, 4 degrees of freedom). The decision rule is "Accept H_0 if $t \geq -3.747$."
Step 4:

$$t = \frac{52 - 60}{2.3/\sqrt{5}} = -7.78.$$

H_0 should be *rejected.* The speed is lower. (Note how the postulated 60 knots or more lies outside the Chapter 8, problem 41 confidence interval as well.)

43. *Step 1:* H_0: $\mu_A - \mu_B = 0$. H_A: $\mu_A - \mu_B \neq 0$
Step 2:

$$t = \frac{\overline{X}_A - \overline{X}_B}{\sqrt{s_P^2\left(\frac{1}{n_A} + \frac{1}{n_B}\right)}}$$

Step 3: The critical value is: $\pm t_{\alpha/2} = \pm t_{.05,40} = \pm 1.684$ (Appendix Table K with $n_A + n_B - 2$ degrees of freedom). The decision rule is "Accept H_0 if $-1.684 \le t \le +1.684$."

Step 4:

$$s_P^2 = \frac{(n_A - 1)s_A^2 + (n_B - 1)s_B^2}{n_A + n_B - 2}$$

$$= \frac{15(.2)^2 + 25(2.3)^2}{40} = 3.3213$$

$$t = \frac{23 - 24}{\sqrt{3.3213\left(\frac{1}{16} + \frac{1}{26}\right)}} = -1.73$$

H_0 should be *rejected*. Yarn B is stronger.

45. Step 1: H_0: $\mu_A - \mu_B = 0$. H_A: $\mu_A - \mu_B \ne 0$
Step 2:

$$t = \frac{\overline{D}}{s_D/\sqrt{n}}$$

Step 3: The critical value is: $\pm t_{\alpha/2} = \pm t_{.025,15} = \pm 2.131$ (see Appendix Table K with $n - 1 = 15$ degrees of freedom). The decision rule is "Accept H_0 if $-2.131 \le t \le +2.131$."

Step 4:

$$t = \frac{100}{59/\sqrt{16}} = 6.78$$

H_0 should be *rejected*. Salaried workers make larger sales.

47. Step 1: H_0: $\mu_A - \mu_B = 0$. H_A: $\mu_A - \mu_B \ne 0$
Step 2:

$$t = \frac{\overline{D}}{s_D/\sqrt{n}}$$

Step 3: The critical value is: $\pm t_{\alpha/2} = \pm t_{.025,3} = \pm 3.182$ (Appendix Table K, with $n - 1 = 3$ degrees of freedom). The decision rule is "Accept H_0 if $-3.182 \le t \le +3.182$."

Step 4:

$$t = \frac{-57}{12/\sqrt{4}} = -9.5$$

H_0 should be *rejected*. The fertilizer *reduced* the yield.

49. Step 1: H_0: $\mu_A - \mu_B \ge 0$. H_A: $\mu_A - \mu_B < 0$.
Step 2:

$$t = \frac{\overline{X}_A - \overline{X}_B}{\sqrt{s_P^2\left(\frac{1}{n_A} + \frac{1}{n_B}\right)}}$$

Step 3: The critical value is: $-t_\alpha = -t_{.10,23} = -1.319$ (see Appendix Table K with $n_A + n_B - 2 = 23$ degrees of freedom). The decision rule is "Accept H_0 if $t \ge -1.319$."

Step 4:

$$s_P^2 = \frac{8(25)^2 + 15(22)^2}{23} = 393.91$$

$$t = \frac{56 - 29}{\sqrt{393.91\left(\frac{1}{9} + \frac{1}{16}\right)}} = 3.26$$

H_0 should be *accepted*. Coffee before bed keeps occasional coffee drinkers up longer.

51. a. Step 1: H_0: $\mu \ge 25$. H_A: $\mu < 25$.
Step 2: $\overline{X} = \mu_0 + (z\sigma_{\overline{x}})$
Step 3: The critical value is: $-z_\alpha = -1.96$. Thus the decision rule is "Accept H_0 if $\overline{X} \ge \mu_0 - (1.96\,\sigma_{\overline{x}}) = \mu_0 - [1.96(\sigma/\sqrt{n})]$ or if $\overline{X} \ge 25 - [1.96(5/\sqrt{36})] = 23.37$."

b. The sketch of the situation in Figure A.28, which is similar to text Figure 9.3 but now involves a lower-tailed test, helps us to visualize the situation. When $\mu = 22$, given $\alpha = .025$ and $\sigma_{\overline{x}} = (5/\sqrt{36}) = .8333$, the critical value of $\overline{X} = 23.37$ is shown to be higher than the mean of the true sampling distribution; the corresponding z value is $+1.64$. To the right of this value we find .0505 of the area under the normal curve. So $\beta = .0505$. Similarly, for $\mu = 24$, we find $\beta = .7764$; but for $\mu = 26$ the null hypothesis is *true*, so $\beta = 0$ (one can't make a type II error when H_0 is true).

c. For $\alpha = .10$, the critical z equals $z_\alpha = -1.28$; thus the critical $\overline{X} = 25 - 1.28\,(5/\sqrt{36}) = 23.93$. Then, for $\mu = 22$, $z_\beta = [(23.93 - 22)/.8333] = 2.32$ and $\beta = .5000 - .4898 = .0102$. Similarly, for $\mu = 24$, $z_\beta = [(23.93 - 24)/.8333] = -.084$ and $\beta = .0335 + .5000 = .5335$. Note how this confirms the trade-off between α and β: the higher is α, the lower is β.

d. In this case, the rejection region for one sampling distribution (centered on $\mu = 25$) must have a lower-tail area of .025 and, hence, a critical z value of $z_\alpha = -1.96$, while the accep-

Figure A.28

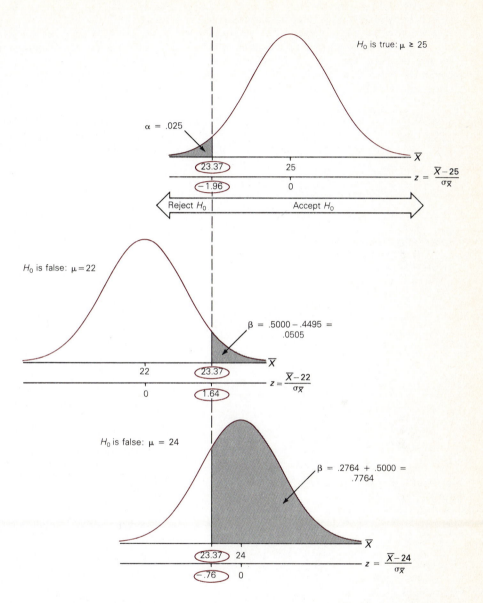

tance region for the other sampling distribution (centered on $\mu = 24$) must have an upper-tail area of $\beta = .10$ and, hence, a critical value of $z_\beta = +1.28$. Hence,

$$n = \left[\frac{\sigma(z_\alpha - z_\beta)}{(\mu_1 - \mu_0)}\right]^2 = \left[\frac{5(-1.96 - 1.28)}{24 - 25}\right]^2 = 262.44.$$

A sample of 263 is required.

53. a. $\alpha = .05, .3954, .8808$, respectively. (In each case, find the Appendix Table H entry corresponding to $z_{\alpha/2}$, subtract it from .5000, and double the result.)

b. $\alpha = .025, .1056, .4404$, respectively. (In each case, find the Appendix Table H entry corresponding to z_α and subtract it from .5000.)

55. The answer is that you would want α to be *small* to minimize the chance for an erroneous rejection of H_0.

57. a. See Figure A.29. In this example, $\alpha = .05$ and is split between the two tails. Similarly, the p value is split. Thus, if the observed test statistic is $z = -.92$, the crosshatched plus dotted area on the left represents only half the p value; combined with a similar area on the right, the p value equals .3576.

Figure A.29

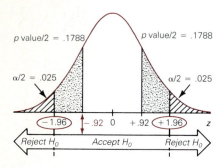

p value/2 = .1788 p value/2 = .1788

$\alpha/2 = .025$ $\alpha/2 = .025$

−1.96 −.92 0 +.92 +1.96 z

Reject H_0 Accept H_0 Reject H_0

b. The p values are given in Table A.37.

Table A.37

Practice Problems	p value	End-of-Chapter Problem	p value	End-of-Chapter Problem	p value
1	.3174	3	.0000	22	.0000
4	.0124	4	.0000	23	.0000
6	.0000	5	.0160	25	.0000
8	.0000	6	.0000	26	.0000
10	.1388	7	.0000	28	.3078
AE 9.2	.6170	14	.0750	29	.0000
	.2542	15	.0308	52a	.3472
	.0012	21	.1212	52b	.0000
	.0004				
	.0910				

Note: Depending on the size of the observed test statistic, the p value can exceed α (as in Figure A.29), and H_0 is accepted, or it can fall short of α (being smaller than the crosshatched area in Figure A.29), and H_0 is rejected.

59. The p values are given in Table A.38.

Table A.38

Practice Problems	p value	End-of-Chapter Problems	p value
13(9 d.f.)	$.001 \geq p \geq .0005$	38 (15 d.f.)	$.0005 \geq p$
15(25 d.f.)	$.4 \geq p \geq .25$	39 (19 d.f.)	$.0005 \geq p$
AE 9.3(16 d.f.)	$.005 \geq p \geq .0025$	40 (10 d.f.)	$.1 \geq p \geq .05$
		41 (4 d.f.)	$.001 \geq p \geq .0005$
		48 (9 d.f.)	$.99 \geq p \geq .975$
		49 (23 d.f.)	$.999 \geq p \geq .9975$

Chapter 10

1. *Step 1:*

H_0: The two variables are independent.
H_A: The two variables are dependent.

Step 2:

$$\chi^2 = \sum \frac{(f_0 - f_e)^2}{f_e}$$

Step 3: For $\alpha = .02$ and 4 degrees of freedom, the critical value is $\chi^2_{.02,4} = 11.668$. The decision rule is "Accept H_0 if $\chi^2 \leq 11.668$."

Step 4: The χ^2 value is computed in Table A.39. H_0 should be *accepted.* Severity and location are independent variables in the case of automobile accidents.

Table A.39

Row, Column	f_0	f_e	$f_0 - f_e$	$(f_0 - f_e)^2$	$\dfrac{(f_0 - f_e)^2}{f_e}$
1,1	10	15	−5	25	1.667
1,2	20	20	0	0	0
1,3	20	15	5	25	1.667
2,1	10	7.5	2.5	6.25	.833
2,2	10	10	0	0	0
2,3	5	7.5	−2.5	6.25	.833
3,1	10	7.5	2.5	6.25	.833
3,2	10	10	0	0	0
3,3	5	7.5	−2.5	6.25	.833
Total	100	100	0		$\chi^2 = 6.666$

3. *Steps 1* and *2* are formulated as in problem 1. *Step 3:* For $\alpha = .01$ and 10 degrees of freedom, the critical value is $\chi^2_{.01,10} = 23.209$. The decision rule is: "Accept H_0 if $\chi^2 \leq 23.209$." *Step 4:* According to the χ^2 value computed in Table A.40, H_0 should be *rejected.* Source of supply and type of defect are dependent variables.

5. a. $d.f. = 9$, critical $\chi^2_{.001} = 27.877$. *Reject H_0.*
b. $d.f. = 1$, critical $\chi^2_{.01} = 6.635$. *Accept H_0.*
c. $d.f. = 27$, critical $\chi^2_{.02} = 44.140$. *Reject H_0.*
d. $d.f. = 22$, critical $\chi^2_{.05} = 33.924$. *Reject H_0.*

Table A.40

Row, Column	f_o	f_e	$f_e - f_e$	$(f_o - f_e)^2$	$\dfrac{(f_o - f_e)^2}{f_e}$
1,A	5	8.33	−3.33	11.09	1.33
1,B	0	11.67	−11.67	136.19	11.67
1,C	3	13.33	−10.33	106.71	8.01
1,D	12	25.00	−13.00	169.00	6.76
1,E	50	11.67	38.33	1,469.19	125.89
1,F	30	30.00	0	0	0
2,A	18	16.67	1.33	1.77	.11
2,B	32	23.33	8.67	75.17	3.22
2,C	42	26.67	15.33	235.01	8.81
2,D	68	50.00	18.00	324.00	6.48
2,E	0	23.33	−23.33	544.29	23.33
2,F	40	60.00	−20.00	400.00	6.67
3,A	27	25.00	2.00	4.00	.16
3,B	38	35.00	3.00	9.00	.26
3,C	35	40.00	−5.00	25.00	.63
3,D	70	75.00	−5.00	25.00	.33
3,E	20	35.00	−15.00	225.00	6.43
3,F	110	90.00	20.00	400.00	4.44
Total	600	600	0		$\chi^2 = 214.53$

7. *Steps 1* and *2* are formulated as in problem 1. *Step 3:* For $\alpha = .02$ and 4 degrees of freedom, the critical value is $\chi^2_{.02,4} = 11.668$. The decision rule is "Accept H_0 if $\chi^2 \leq 11.668$." *Step 4:* Actual $\chi^2 = 95.66$ *Reject H_0.* Stock price and dividend yield are dependent variables.

9. *Steps 1* and *2* are formulated as in problem 1. *Step 3:* For $\alpha = .05$ and 4 degrees of freedom, the critical value is $\chi^2_{.05,4} = 9.488$. The decision rule is "Accept H_0 if $\chi^2 \leq 9.488$." *Step 4:* Actual $\chi^2 = 96.03$. *Reject H_0.* Awareness of ad and magazine type are dependent variables.

11. *Steps 1* and *2* are formulated as in problem 1. *Step 3:* For $\alpha = .10$ and 2 degrees of freedom, the critical $\chi^2_{.10,2} = 4.605$. The decision rule is "Accept H_0 if $\chi^2 \leq 4.605$." *Step 4:* Actual $\chi^2 = 138.898$. *Reject H_0.* The type of chicken consumed and watching "Falcon Crest" are dependent variables.

13. The test must be based on the data in Table A.41:

Table A.41 Drivers in Accidents (Massachusetts sample, 1986)

Accident Type	Seat Belt Worn		
	Yes	No	Total
Serious	49	114	163
Non-Serious	222	165	387
Total	271	279	550

Steps 1 and 2 are formulated as in problem 1. *Step 3:* For $\alpha = .01$ and 1 degree of freedom, the critical $\chi^2_{.01,1} = 6.635$. The decision rule is "Accept H_0 if $\chi^2 \leq 6.635$." *Step 4:* Actual $\chi^2 = 34.206$. *Reject H_0.* Severity of accidents and wearing seat belts are dependent variables.

15. *Steps 1* and *2* are formulated as in problem 1. *Step 3:* For $\alpha = .01$ and 6 degrees of freedom, the critical $\chi^2_{.01,6} = 16.812$. The decision rule is "Accept H_0 if $\chi^2 \leq 16.812$." *Step 4:* Actual $\chi^2 = 136.959$. *Reject H_0.* Type of policy sold and geographic region are dependent variables.

17. See Table A.42.

Table A.42

Problem	E_{χ^2}	$\sigma^2_{\chi^2}$	σ_{χ^2}
12	3	6	2.449
13	1	2	1.414
14	3	6	2.449
15	6	12	3.464

19. *Step 1:* H_0: $\pi_A = \pi_B = \pi_C = \pi_D = .35$

Step 2: $\chi^2 = \Sigma \dfrac{(f_o - f_e)^2}{f_e}$

Step 3: Given $\alpha = .10$ and 3 degrees of freedom, the critical value is $\chi^2_{.10,3} = 6.251$ (Appendix Table L), making the decision rule: "Accept H_0 if $\chi^2 \leq 6.251$." *Step 4:* According to the χ^2 value computed in Table A.43, H_0 should be *rejected.* The proportions differ.

Table A.43

Row, Column	f_o	f_e	$f_o - f_e$	$(f_o - f_e)^2$	$\dfrac{(f_o - f_e)^2}{f_e}$
1,A	10	17.5	−7.5	56.25	3.21
1,B	0	17.5	−17.5	306.25	17.50
1,C	20	17.5	2.5	6.25	.36
1,D	40	17.5	22.5	506.25	28.93
2,A	40	32.5	7.5	56.25	1.73
2,B	50	32.5	17.5	306.25	9.42
2,C	30	32.5	−2.5	6.25	.19
2,D	10	32.5	−22.5	506.25	15.58
Total	200	200	0		$\chi^2 = 76.92$

21. *Step 1:* H_0: $\pi_P = \pi_N = \pi_D = .5$. *Step 2:* χ^2 is formulated as in problem 19. *Step 3:* Given $\alpha = .05$ and 2 degrees of freedom (a contingency table would have 2 rows—heads, tails—and 3 columns—penny, nickel, dime), the critical value is $\chi^2_{.05,2} = 5.991$. The decision rule is "Accept H_0 if $\chi^2 \le 5.991$." *Step 4:* Actual $\chi^2 = 4$. *Accept H_0.* The coins are fair.

23. *Step 1:* H_0: $\pi_A = \pi_B = .066$. *Step 2:* χ^2 is formulated as in problem 19. *Step 3:* Given $\alpha = .01$ and 1 degree of freedom (2 rows—defective, acceptable—and 2 columns—A and B), the critical value is $\chi^2_{.01,1} = 6.635$. The decision rule is "Accept H_0 if $\chi^2 \le 6.635$." *Step 4:* Actual $\chi^2 = .389$. *Accept H_0.* The proportions are the same.

25. *Step 1:* H_0: $\pi_A = \pi_B = \pi_C = \pi_D = .551$. *Step 2:* χ^2 is formulated as in problem 19. *Step 3:* Given $\alpha = .001$ and 3 degrees of freedom (2 rows—last week, not last week—and 4 columns—companies A–D), the critical value is $\chi^2_{.001,3} = 16.266$. The decision rule is "Accept H_0 if $\chi^2 \le 16.266$." *Step 4:* Actual $\chi^2 = 22.276$. *Reject H_0.* The proportion differs among the companies.

27. H_0: $\pi_A = .2$, $\pi_B = .5$, $\pi_C = .1$, $\pi_D = .2$. Given $\alpha = .10$ and 3 degrees of freedom, the critical $\chi^2_{.10,3} = 6.251$. The decision rule is "Accept H_0 if $\chi^2 \le 6.251$." The *computation of χ^2* is given in Table A.44. H_0 should be *rejected*. (At $\alpha = .01$, the critical $\chi^2_{.01,3}$ would have been 11.345 and H_0 would

have been *accepted*. The educational level of jurors would then have been deemed to reflect the educational level of the county population.)

29. Dividing the year into quarters beginning with the birthday, there were 215 deaths in the first quarter, 145 in the second, 95 in the third, 45 in the fourth (preceding the *next* birthday). Without any linkage between time of death and birthday, one would expect an equal number of deaths in each quarter, here $(500/4) = 125$. Thus:
Step 1: H_0: $\pi_I = \pi_{II} = \pi_{III} = \pi_{IV} = .25$.
Step 2:

$$\chi^2 = \sum \frac{(f_o - f_e)^2}{f_e}$$

Step 3: Given $\alpha = .01$ and 3 degrees of freedom, the critical $\chi^2_{.01,3} = 11.345$. The decision rule is "Accept H_0 if $\chi^2 \le 11.345$." *Step 4:* Actual $\chi^2 = 126.4$. *Reject H_0.* There *is* a strange linkage between time of death and birthday. (See also Close-Up 9.2 on page 382.)

31. Given 29 degrees of freedom and $\alpha = .04$, Appendix Table L yields $\chi^2_U = \chi^2_{.02,29} = 46.693$ and $\chi^2_L = \chi^2_{.98,29} = 15.574$. According to Box 10.F, the interval is

$$\frac{127.69(29)}{46.693} \le \sigma^2 \le \frac{127.69(29)}{15.574}$$

or

$$79.31 \le \sigma^2 \le 237.77.$$

33. Given 9 degrees of freedom and $\alpha = .02$, Appendix Table L yields $\chi^2_U = \chi^2_{.01,9} = 21.666$ and $\chi^2_L = \chi^2_{.99,9} = 2.088$. According to Box 10.F, the interval is

$$\frac{26(9)}{21.666} \le \sigma^2 \le \frac{26(9)}{2.088}$$

Table A.44

f_o	f_e	$f_o - f_e$	$(f_o - f_e)^2$	$\dfrac{(f_o - f_e)^2}{f_e}$
12	20	−8	64	3.20
64	50	14	196	3.92
12	10	2	4	.40
12	20	−8	64	3.20
100	100	0		$\chi^2 = 10.72$

or
$$10.800 \le \sigma^2 \le 112.069.$$

The altimeters do not meet specifications.

35. a. Lower-tailed test. Critical $\chi^2_{.99,25} = 11.524$.

$$\text{Computed } \chi^2 = \frac{10(25)}{64} = 3.91.$$

H_0 should be *rejected*.

b. Lower-tailed test. Critical $\chi^2_{.98,14} = 5.368$.

$$\text{Computed } \chi^2 = \frac{.1(14)}{.64} = 2.19.$$

H_0 should be *rejected*.

c. Upper-tailed test. Critical $\chi^2_{.001,4} = 18.467$.

$$\text{Computed } \chi^2 = \frac{110(4)}{100} = 4.4.$$

H_0 should be *accepted*.

37. *Step 1:* H_0: $\sigma^2 = 100$. H_A: $\sigma^2 \ne 100$

Step 2: $\chi^2 = \dfrac{s^2(n-1)}{\sigma_0^2}$

Step 3: For $\alpha = .04$ and 13 degrees of freedom, critical χ^2 values are $\chi^2_{.98,13} = 4.765$ and $\chi^2_{.02,13} = 25.472$. The decision rule is "Accept H_0 if $4.765 \le \chi^2 \le 25.472$."

Step 4: To compute χ^2, we need the Table A.45 computations.

Table A.45

X	X^2
−17	289
22	484
5	25
0	0
8	64
−3	9
13	169
20	400
25	625
−41	1,681
0	0
−4	16
8	64
13	121
$\Sigma X = 49$	$\Sigma X^2 = 3{,}947$

$$\overline{X} = 3.5$$

Thus,

$$s^2 = \frac{\Sigma X^2 - n\overline{X}^2}{n-1}$$
$$= \frac{3{,}947 - 14(3.5)^2}{13} = 290.42$$

and

$$\chi^2 = \frac{s^2(n-1)}{\sigma_0^2} = \frac{290.42(13)}{100} = 37.75$$

H_0 should be *rejected*. The variance is now larger.

39. *Step 1:* H_0: $\sigma^2 \ge 80$. H_A: $\sigma^2 < 80$.

Step 2: $\chi^2 = \dfrac{s^2(n-1)}{\sigma_0^2}$

Step 3: Given $\alpha = .05$ and 24 degrees of freedom, Appendix Table L suggests a critical value of $\chi^2_{.95,24} = 13.848$. The decision rule is "Accept H_0 if $\chi^2 \ge 13.848$."

Step 4: $\chi^2 = \dfrac{64(24)}{80} = 19.2$

H_0 should be *accepted*. The claim is justified.

41. *Step 1:* $H_0 = \sigma^2 \ge 200$. H_A: $\sigma^2 < 200$.

Step 2: $\chi^2 = \dfrac{s^2(n-1)}{\sigma_0^2}$

Step 3: Given $\alpha = .10$ and 19 degrees of freedom, Appendix Table L suggests a critical value of $\chi^2_{.90,19} = 11.651$. The decision rule is "Accept H_0 if $\chi^2 \ge 11.651$."

Step 4: $\chi^2 = \dfrac{198(19)}{200} = 18.81$

H_0 should be *accepted*. The belief is justified.

43. *Step 1:* H_0: $\sigma^2 \le .01$. H_A: $\sigma^2 > .01$

Step 2: $\chi^2 = \dfrac{s^2(n-1)}{\sigma_0^2}$

Step 3: Given $\alpha = .02$ and 29 degrees of freedom, Appendix Table L suggests a critical value of $\chi^2_{.02,29} = 46.693$. The decision rule is "Accept H_0 if $\chi^2 \le 46.693$."

Step 4: $\chi^2 = \dfrac{.0576(29)}{.01} = 167.04$

H_0 should be *rejected*. The variance is larger.

45. See Table A.46 on the next page. Degrees of freedom $= k - 1 - 0 = 4$. For $\alpha = .02$, the critical value is $\chi^2_{.02,4} = 11.668$. H_0 should be *accepted*. In the population of all flights, the number of engines requiring oil after a 7-hour flight is a binomially distributed random variable with a probability of success in any one trial of $\pi = .5$.

Table A.46

Number of Engines Requiring Oil After 7-Hour Flight	f_o	Binomial Probability, p, for $n = 4$ and $\pi = .5$	$f_e = 200p$	$f_o - f_e$	$(f_o - f_e)^2$	$\dfrac{(f_o - f_e)^2}{f_e}$
0	15	.0625	12.5	2.5	6.25	.50
1	44	.2500	50.0	−6.0	36.00	.72
2	75	.3750	75.0	0	0	0
3	56	.2500	50.0	6.0	36.00	.72
4	10	.0625	12.5	−2.5	6.25	.50
Total	200	1.0000	200	0		$\chi^2 = 2.44$

Table A.47

Number of Acceptances	f_o	Binomial Probability, p, for $n = 3$ and $\pi = .25$	$f_e = 100p$	$f_o - f_e$	$(f_o - f_e)^2$	$\dfrac{(f_o - f_e)^2}{f_e}$
0	24	.4219	42.19	−18.19	330.88	7.84
1	36	.4219	42.19	−6.19	38.32	.91
2	26	.1406	14.06	} 24.38	594.38	38.05
3	14	.0156	1.56			
Total	100	1.0000	100	0		$\chi^2 = 46.80$

Table A.48

Number of Yearly Auto Accidents Per Driver	f_o	Poisson Probability, p, for $\mu = .1$	$f_e = 112{,}956p$	$f_o - f_e$	$\dfrac{(f_o - f_e)^2}{f_e}$
0	103,628	.9048	102,202.58	1,425.42	19.880
1	7,389	.0905	10,222.52	−2,833.52	785.407
2	1,910	.0045	508.30	1,401.70	3,865.361
3	29	.0002	22.59	6.41	1.819
Total	112,956	1.0000	112,955.99*	.01*	4,672.467 χ^2

*rounding error of .01

47. See Table A.47. Degrees of freedom $= k - 1 - 0 = 2$. For $\alpha = .05$, the critical value is $\chi^2_{.05,2} = 5.991$. H_0 should be *rejected*. The number of acceptances is either not binomially distributed or, if it is, the value of $\pi \neq .25$.

49. See Table A.48. There were 11,296 accidents by 112,956 drivers. Accordingly, the mean number of occurrences is estimated from the given data as $\mu = .1$. For $4 - 1 - 1 = 2$ degrees of freedom and $\alpha = .10$, the critical $\chi^2_{.10,2} = 4.605$. Accordingly, H_0 should be *rejected*.

51. See Table A.49. There were a minimum of 179 cancellations in 96 days. Accordingly, one can estimate $\mu = 1.9$. For $5 - 1 - 1 = 3$ degrees of freedom and $\alpha = .02$, the critical $\chi^2_{.02,3} = 9.837$. Accordingly, H_0 should be *accepted*.

53. One would first have to calculate $\overline{X}$ and s and use these statistics as estimates of the missing parameters. This calculation could be accomplished with the help of (Chapter 10's) Table L data (repeated in the first and third columns of Table A.50 and a more or less arbitrary estimate of the midpoints of

Table A.49

Number of Cancellations per Day	f_o	Poisson Probability, p, for $\mu = 1.9$	$f_e = 96p$	$f_o - f_e$	$\dfrac{(f_o - f_e)^2}{f_e}$
0	22	.1496	14.36	7.64	4.065
1	30	.2842	27.28	2.72	.271
2	22	.2700	25.92	−3.92	.593
3	10	.1710	16.42	−6.42	2.510
4	5	.0812	7.80		
5	2	.0309	2.97		
6	0	.0098	.94		
7	1	.0027	.26	−.04	.000
8	1	.0006	.06		
9	0	.0001	.01		
10 and more	3	.0000	0		
Total	96	1.0000	96.02*	−.02*	$\chi^2 = 7.439$

*rounding error of .02

Table A.50

Number of Futures Contracts Traded per Day (millions)	Class Midpoint, X	Observed Frequency, f_o	$f_o X$	X^2	$f_o X^2$
under 10	5	5	25	25	125
10 − under 20	15	9	135	225	2,025
20 − under 30	25	15	375	625	9,375
30 − under 40	35	23	805	1,225	28,175
40 − under 50	45	20	900	2,025	40,500
50 − under 60	55	8	440	3,025	24,200
60 − under 70	65	6	390	4,225	25,350
70 − under 80	75	3	225	5,625	16,875
80 and above	85	1	85	7,225	7,225
		90	3,380		153,850
		$\Sigma f_o = n$	$\Sigma f_o X$		$\Sigma f_o X^2$

the two open-ended classes (given in the second column of Table A.50).

$$\overline{X} = \frac{\Sigma f_o X}{n} = \frac{3,380}{90} = 37.56.$$

$$s = \sqrt{\frac{\Sigma f_o X^2 - n\overline{X}^2}{n - 1}}$$

$$= \sqrt{\frac{153,850 - 90(37.56)^2}{89}} = 17.379.$$

These two values could be inserted in the null hypothesis, making Chapter 10's Table M look like Table A.51 on the next page. After combining classes as needed, we have $k = 7$ classes. Since two parameters were estimated, we have $7 - 1 - 2 = 4$ degrees of freedom. With $\alpha = .02$, this $d.f.$ gives us a critical value of $\chi^2_{.02,4} = 11.668$. Thus, the decision rule must be "Accept H_0 if $\chi^2 \leq 11.668$." The actual value of χ^2 is calculated in Table A.52 on the next page. Accordingly, the null hypothesis, as amended, should be *accepted*.

Table A.51

Number of Futures Contracts Traded per Day (millions): Upper Class Limit, x	Normal Deviate, $z = \dfrac{x - \mu}{\sigma}$ $= \dfrac{x - 37.56}{17.379}$	Area Under Standard Normal Curve Left of x	Area of Class Interval, p	Expected Frequency (if H_0 is true), $f_e = 90p$
10	-1.59	.0559	.0559	5.031
20	-1.01	.1562	.1003	9.027
30	$-.44$	.3300	.1738	15.642
40	.14	.5557	.2257	20.313
50	.72	.7642	.2085	18.765
60	1.29	.9015	.1373	12.357
70	1.87	.9693	.0678	6.102 ⎫
80	2.44	.9927	.0234	2.106 ⎬ 8.865
∞	∞	1.0000	.0073	.657 ⎭
			1.0000	90

Table A.52

Number of Futures Contracts Traded per Day (millions)	f_o	f_e	$f_o - f_e$	$\dfrac{(f_o - f_e)^2}{f_e}$
under 10	5	5.031	$-.031$	.0002
10 — under 20	9	9.027	$-.027$	.0001
20 — under 30	15	15.642	$-.642$	.0263
30 — under 40	23	20.313	2.687	.3554
40 — under 50	20	18.765	1.235	.0813
50 — under 60	8	12.357	-4.357	1.5363
60 and above	10	8.865	1.135	.1453
Total	90	90	0	$\chi^2 = 2.1449$

55. H_0: The preferences are uniformly distributed among the 5 brands. There are $5 - 1 - 0 = 4$ degrees of freedom; the critical value is $\chi^2_{10,4} = 7.779$. According to the value of χ^2 calculated in Table A.53, H_0 should be *accepted*.

57. H_0: The 6 faces of the die are uniformly distributed in the population of all possible rolls. There are $6 - 1 - 0 = 5$ degrees of freedom; given $\alpha = .05$, the critical value is $\chi^2_{05,5} = 11.070$. According to the value of χ^2 calculated in Table A.54, H_0 should be *rejected*.

Table A.53

Most Preferred Brand	f_o	f_e	$f_o - f_e$	$\dfrac{(f_o - f_e)^2}{f_e}$
A	91	100	-9	.81
B	109	100	9	.81
C	85	100	-15	2.25
D	100	100	0	0
E	115	100	15	2.25
Total	500	500	0	$\chi^2 = 6.12$

Table A.54

Face Value of Die	f_o	f_e	$f_o - f_e$	$\dfrac{(f_o - f_e)^2}{f_e}$
1	5	20	-15	11.25
2	36	20	16	12.80
3	8	20	-12	7.20
4	9	20	-11	6.05
5	7	20	-13	8.45
6	55	20	35	61.25
Total	120	120	0	$\chi^2 = 107$

Table A.55

Opinion Expressed About the President's Performance	f_o (after 3 years)	f_e (1.5 times f_o after 1 year)	$f_o - f_e$	$\dfrac{(f_o - f_e)^2}{f_e}$
Does superb job	68	75	-7	.653
Does good job	33	37.5	-4.5	.540
Does fair job	22	22.5	$-.5$	.011
Does lousy job	27	15	12.0	9.600
Total	150	150	0	$\chi^2 = 10.804$

59. H_0: Public opinion 3 years after inauguration is the same as 1 year after inauguration. There being $4 - 1 - 0 = 3$ degrees of freedom, the critical value is $\chi^2_{.02, 3} = 9.837$. According to the χ^2 value calculated in Table A.55, H_0 should be *rejected*.

Chapter 11

1. *The Wilcoxon rank-sum test* is appropriate. *Step 1:* H_0: Starting salaries for A graduates are the same as for B graduates. *Step 2:* Use the normal deviate for the Wilcoxon rank-sum test. *Step 3:* Given $\alpha = .10$ and a two-tailed test, Appendix Table J suggests critical values of $z_{\alpha/2} = \pm 1.64$. Hence, the decision rule is: "Accept H_0 if $-1.64 \leq z \leq +1.64$." *Step 4:* See Table A.56.

Table A.56

Ordering of Original Data	Sample of Origin	Ranks	Sample A Ranks
$1,200	A	1	1
1,250	B	2	
1,390	B	3	
1,410	B	4	
1,450	A	5	5
1,490	B	6	
1,500	A	7	7
1,600	A	8	8
1,710	B	9	
1,750	B	10	
1,800	A	11	11
1,850	A	12	12
1,900	A	13	13
2,000	A	14	14
2,050	B	15	
2,100	B	16	
2,200	B	17.5	
2,200	A	17.5	17.5
2,300	A	19	19
2,350	B	20	
			$W = 107.5$

$$\mu_W = \frac{10(10 + 10 + 1)}{2}$$
$$= 105;$$

$$\sigma_W = \sqrt{\frac{10 \cdot 10(10 + 10 + 1)}{12}}$$
$$= 13.2288;$$

$$z = \frac{107.5 - 105}{13.2288}$$
$$= .19.$$

Thus, H_0 should be *accepted*. Starting salaries are the same for graduates of the two institutions.

3. The *Wilcoxon rank-sum test* is appropriate. *Step 1:* H_0: Output per worker under plan A is the same as or less than output under plan B. *Step 2:* Use the normal deviate for the Wilcoxon rank-sum test. *Step 3:* Given $\alpha = .01$ and a one-tailed test, Appendix Table J suggests a critical value of $z_\alpha = +2.33$. Hence, the decision rule is "Accept H_0 if $z \leq 2.33$." *Step 4:* See Table A.57 on the next page.

$$\mu_W = \frac{10(10 \cdot \ \ ) + 1)}{} = 105;$$

$$\sigma_W = \sqrt{\frac{10 \cdot 10(10 + 10 + 1)}{12}}$$
$$= 13.2288;$$

$$z = \frac{85.5 - 105}{13.2288} = -1.47.$$

H_0 should be *accepted*. Output per worker under the piece-work plan is better than or equal to output under the hourly plan.

Table A.57

Ordering of Original Data	Sample of Origin	Ranks	Sample A Ranks
199	B	1	
241	A	2	2
252	B	3	
274	A	4	4
278	A	5	5
492	A	6.5	6.5
492	B	6.5	
510	A	8	8
517	B	9	
547	A	10	10
552	A	11	11
602	A	12	12
631	A	13	13
651	A	14	14
707	B	15	
751	B	16	
908	B	17	
917	B	18	
937	B	19	
981	B	20	

$$W = 85.5$$

5. The *Wilcoxon rank-sum test* is appropriate. *Step 1:* Model A cars last as long as or longer than Model B cars. *Step 2:* Use the normal deviate for the Wilcoxon rank-sum test. *Step 3:* Given $\alpha = .005$ and a lower-tailed test, Appendix Table J suggests "Accept H_0 if $z \geq -2.57$." *Step 4:* $W = 159.5$, $\mu_w = 175.5$, $\sigma_w = 19.5$, $z = -.82$. H_0 should be *accepted*.

7. The *Wilcoxon rank-sum test* is appropriate. *Step 1:* H_0: Assessed values are the same in the two neighborhoods. *Step 2:* Use the normal deviate for the Wilcoxon rank-sum test. *Step 3:* Given $\alpha = .025$ and a two-tailed test, Appendix Table J suggests "Accept H_0 if $-2.24 \leq z \leq +2.24$." *Step 4:* $W = 153.5$, $\mu_w = 169$, $\sigma_w = 18.38478$, $z = -.84$. H_0 should be *accepted*.

9. For the *U test,* Step 1 is identical to Problem 3. *Step 2:* Use the normal deviate for the Mann-Whitney

test. *Step 3:* Given $\alpha = .01$, an upper-tailed test (hence the cautionary note in the text applies), and Appendix Table J, "Accept H_0 if $-z \leq +2.33$." *Step 4:*

$$U = \left[10 \cdot 10 + \frac{10(10 + 1)}{2} \right] - 85.5 = 69.5$$

$$\mu_U = \frac{10 \cdot 10}{2} = 50$$

$$\sigma_U = \sigma_w = 13.22876$$

$$z = \frac{69.5 - 50}{13.22876} = 1.47$$

Since $-1.47 \leq +2.33$, H_0, again, should be *accepted*.

11. *Step 1:* H_0: Test scores are at least as high for high-school dropouts as for high-school graduates. *Step 2:* Use the normal deviate for the Mann-Whitney test. *Step 3:* Given $\alpha = .05$ and a lower-tailed test (hence the cautionary note in the text applies), Appendix Table J suggests "Accept H_0 if $-z \geq -1.64$," *Step 4:* $U = 92$, $\mu_U = 77$, $\sigma_U = 18.26654$, $z = .82$. Since $-.82 \geq -1.64$, H_0 should be *accepted*.

13. *Step 1:* H_0: Planes purchase equal amounts of fuel at the two airports. *Step 2:* Use the normal deviate for the Mann-Whitney test. *Step 3:* Given $\alpha = .01$ and a two-tailed test, Appendix Table J suggests "Accept H_0 if $-2.57 \leq z \leq +2.57$." *Step 4:* $U = 88.5$, $\mu_U = 70$, $\sigma_U = 17.07825$, $z = 1.08$. H_0 should be *accepted*.

15. *Step 1:* H_0: Career earnings of baseball players are at most equal to those of golfers. *Step 2:* Use the normal deviate for the Mann-Whitney test. *Step 3:* Given $\alpha = .05$ and an upper-tailed test (hence the cautionary note in the text applies), Appendix Table J suggests "Accept H_0 if $-z \leq +1.64$." *Step 4:* $U = 33$, $\mu_U = 71.5$, $\sigma_U = 17.26026$, $z = -2.23$. Since $+2.23 > +1.64$, H_0 should be *rejected*. Baseball players earn more.

17. See Table A.58.

Table A.58

Problem	W	μ_w	σ_w	z
10	156	156	16.12452	0
11	128	143	18.26654	-.82
12	108.5	105	13.22876	.26
13	106.5	125	17.07825	-1.08
14	155	232.5	24.10913	-3.21
15	176	137.5	17.26026	2.23

19. The *sign test* is appropriate, lower-tailed. H_0: Scores under A equal or exceed those under B. The decision rule is "Accept H_0 if $z \geq -1.64$." Given $n = 30$, $S = 20$, $\mu_S = 15$, $\sigma_S = 2.7386$,

$$z = \frac{20 - 15}{2.7386} = 1.83.$$

H_0 should be *accepted*.

21. The *sign test* is appropriate, two-tailed. H_0: Sales are the same under scheme A as under scheme B. The decision rule is "Accept H_0 if $-2.24 \leq z \leq +2.24$." Given $n = 11$, $S = 7$, $\mu_S = 5.5$, $\sigma_S = 1.6583$,

$$z = \frac{7 - 5.5}{1.6583} = .90.$$

H_0 should be *accepted*.

23. The *sign test* is appropriate, upper-tailed. H_0: Realtor A is slower to sell houses than realtor B. The decision rule is "Accept H_0 if $z \leq +1.64$." Given $n = 98$, $S = 82$, $\mu_S = 49$, $\sigma_S = 4.9497$,

$$z = \frac{82 - 49}{4.9497} = 6.67.$$

H_0 should be *rejected*. Realtor A is faster, not slower.

25. The *sign test* is appropriate, two-tailed. H_0: The population median is 800 service hours. The deci-

sion rule is "Accept H_0 if $-1.96 \leq z \leq +1.96$." Given $n = 50$, $S = 12$, $\mu_S = 25$, $\sigma_S = 3.5355$,

$$z = \frac{12 - 25}{3.5355} = -3.68.$$

H_0 should be *rejected*. The median is lower.

27. The *Wilcoxon signed-rank test* is appropriate (and preferable to the *sign test*). *Step 1:* H_0: The shelf life with the traditional preservative is shorter than or at best equal to that with the new preservative. *Step 2:* Use the normal deviate for the Wilcoxon signed-rank test. *Step 3:* Given $\alpha = .01$, and a one-tailed test, Appendix Table J suggests a critical value of $z_\alpha = +2.33$. Hence, the decision rule is: "Accept H_0 if $z \leq +2.33$." *Step 4:* See Table A.59. $n = 10$, hence

$$\sigma_T = \sqrt{\frac{10(10 + 1)(20 + 1)}{6}} = 19.62$$

and

$$z = \frac{6}{19.62} = .31.$$

H_0 should be *accepted*. The new preservative produces a shelf life for meat equal to or longer than the traditional one.

29. The *Wilcoxon signed-rank test* is appropriate (and preferable to the *sign test*). *Step 1:* H_0: Salaries in

Table A.59

| Supplier | Shelf Life | | Difference, (A) − (B) | Absolute Value of Nonzero Difference | Rank | Signed Rank |
	Old (A)	New (B)				
A	4.2	4.8	−.6	.6	8.5	−8.5
B	6.1	6.3	−.2	.2	2	−2
C	5.3	5.0	+.3	.3	4.5	+4.5
D	3.9	4.5	−.6	.6	8.5	−8.5
E	4.5	4.2	+.3	.3	4.5	+4.5
F	6.0	5.5	+.5	.5	7	+7
G	5.8	5.9	−.1	.1	1	−1
H	4.9	4.9	0	—	—	—
I	5.3	5.0	+.3	.3	4.5	+4.5
J	5.5	5.5	0	—	—	—
K	5.2	5.5	−.3	.3	4.5	−4.5
L	4.8	4.0	+.8	.8	10	+10

$$T = +6$$

academic and nonacademic employment are the same. *Step 2:* Use the normal deviate for the Wilcoxon signed-rank test. *Step 3:* Given $\alpha = .025$ and a two-tailed test, Appendix Table J suggests:"Accept H_0 if $-2.24 \leq z \leq +2.24$." *Step 4:* $T = -108$, $\sigma_T = 35.21$, $z = -3.07$. H_0 should be *rejected*. Academic salaries are lower.

31. The *number-of-runs test* is appropriate. H_0: The sequence is random. For $\alpha = .05$, the decision rule is "Accept H_0 if $-1.96 \leq z \leq +1.96$."

$$n_A = 35, \quad n_B = 25, \quad R_A = 5$$

$$\mu_{R_A} = \frac{35(25 + 1)}{35 + 25} = 15.17$$

$$\sigma_{R_A} = \sqrt{\frac{35(25 + 1)(35 - 1)}{(35 + 25)^2}\left(\frac{25}{35 + 25 - 1}\right)}$$
$$= 1.9083$$

$$z = \frac{5 - 15.17}{1.9083} = -5.33$$

H_0 should be *rejected*.

33. The *number-of-runs test* is appropriate. H_0: The sequence is random. For $\alpha = .025$, the decision rule is "Accept H_0 if $-2.24 \leq z \leq +2.24$."

$$n_M = 41, \, n_F = 19, R_M = 10$$

$$\mu_{R_M} = \frac{41(19 + 1)}{41 + 19} = 13.67.$$

$$\sigma_{R_M} = \sqrt{\frac{41(19 + 1)(41 - 1)}{(41 + 19)^2}\left(\frac{19}{41 + 19 - 1}\right)}$$
$$= 1.7129$$

$$z = \frac{10 - 13.67}{1.7129} = -2.14.$$

H_0 should be *accepted*. A test at the 5-percent significance level would have *rejected* H_0 because the critical z values would then have been ± 1.96.

35. The *number-of-runs test* is appropriate. H_0: The sequence is random. For $\alpha = .05$ and a two-tailed test, Appendix Table J suggests "Accept H_0 if $-1.96 \leq z \leq +1.96$."

$$n_W = 16, \, n_L = 14, R_W = 5$$
$$\mu_{R_W} = 8, \sigma_{R_W} = 1.3896, z = -2.16.$$

H_0 should be *rejected*.

37. The *number-of-runs test* is appropriate. H_0: The sequence is random. For $\alpha = .10$ and a two-tailed test, Appendix Table J suggests "Accept H_0 if $-1.64 \leq z \leq +1.64$."

$$n_{(1)} = 11, n_{(9)} = 16, R_{(1)} = 5$$
$$\mu_{R_{(1)}} = 6.9259, \sigma_{R_{(1)}} = 1.2564, z = -1.53.$$

H_0 should be *accepted*.

39. The *Kruskal-Wallis test* is appropriate. *Step 1:* H_0: The test-score populations are identical; H_A: The test-score populations differ. *Step 2:* Select the K statistic. *Step 3:* For $\alpha = .05$ and $x = 4$ samples (3 degrees of freedom), Appendix Table L shows the critical value of $\chi^2_{.05,3} = 7.815$. Hence, the decision rule is "Accept H_0 if $K \leq 7.815$." *Step 4:* See Table A.60.

$$K = \frac{12}{27(27 + 1)}\left[\frac{(121.5)^2}{7} + \frac{(42)^2}{6}\right.$$
$$\left. + \frac{(64.5)^2}{7} + \frac{(150)^2}{7}\right] - 3(27 + 1)$$
$$= 14.595.$$

H_0 should be *rejected*. The training is not equally effective at the four locations.

41. The *Kruskal-Wallis test* is appropriate. *Step 1:* H_0: The price-increase populations are identical. *Step 2:* Select the K statistic. *Step 3:* For $\alpha = .02$ and $x = 3$ samples (2 degrees of freedom), Appendix

Table A.60

A		B		C		D	
Observation	Rank	Observation	Rank	Observation	Rank	Observation	Rank
96	27	65	3	60	1	95	26
82	11.5	74	8	73	7	93	25
88	18	72	6	85	13.5	90	21
70	5	66	4	61	2	88	18
90	21	79	9.5	79	9.5	91	23.5
91	23.5	82	11.5	85	13.5	87	15.5
87	15.5			88	18	90	21
	$W_A = 121.5$		$W_B = 42$		$W_C = 64.5$		$W_D = 150$

Table L suggests "Accept H_0 if $K \leq 7.824$." *Step 4:* $W_A = 64$, $W_B = 41$, $W_C = 15$

$$K = \frac{12}{15(15 + 1)}\left[\frac{(64)^2}{5} + \frac{(41)^2}{5} + \frac{(15)^2}{5}\right]$$
$$- 3(15 + 1) = 12.02$$

H_0 should be *rejected*. The rate of appreciation does differ with house quality.

43. The *Kruskal-Wallis test* is appropriate. *Step 1: H_0:* The salary populations are identical. *Step 2:* Select the K statistic. *Step 3:* For $\alpha = .10$ and $x = 3$ samples (2 degrees of freedom), Appendix Table L suggests: "Accept H_0 if $K \leq 4.605$." *Step 4:* $W_A = 50$, $W_B = 58$, $W_G = 63$

$$K = \frac{12}{18(18 + 1)}\left[\frac{(50)^2}{7} + \frac{(58)^2}{5} + \frac{(63)^2}{6}\right]$$
$$- 3(18 + 1) = 2.349.$$

H_0 should be *accepted*.

45. The *Kruskal-Wallis test* is appropriate. *Step 1: H_0:* The evaluation-score populations are identical. *Step 2:* Select the K statistic. *Step 3:* For $\alpha = .05$ and $x = 3$ samples (2 degrees of freedom), Appendix Table L suggests: "Accept H_0 if $K \leq 5.991$." *Step 4:* $W_A = 66.5$, $W_B = 65.5$, $W_C = 39$.

$$K = \frac{12}{18(18 + 1)}\left[\frac{(66.5)^2}{6} + \frac{(65.5)^2}{6} + \frac{(39)^2}{6}\right]$$
$$- 3(18 + 1) = 2.845.$$

H_0 should be *accepted*.

47. The *Kolmogorov-Smirnov one-sample test* is appropriate. *Step 1: H_0:* The number of promising wells per site in the population of all possible sites is binomially distributed with a probability of success in any one trial of $\pi = .2$. *H_A:* The number of promising wells per site in the population of all possible sites is not correctly described by H_0. *Step 2: D* = max. $|F_o - F_e|$. *Step 3:* Given $\alpha = .01$ and $n = 500$, we compute a critical $D_{.01} = (1.63/\sqrt{500}) = .07290$. The decision rule is "Accept H_0 if $D \leq .07290$." *Step 4:* See Table A.61.

Given $D = .012$, H_0 should be *accepted*. [Note: This result is the same as for problem 7 in Chapter 10 of the *Student Workbook,* wherein the same problem is subjected to a χ^2 test.]

49. *Step 1: H_0:* The population of yearly automobile accidents per driver is Poisson distributed with $\mu = .1$. *Step 2: D* = max. $|F_o - F_e|$. *Step 3:* Given $\alpha = .10$ and $n = 112,956$, we compute a critical $D_{.10} = (1.22/\sqrt{n}) = .00363$. (Appendix Table M). The decision rule is "Accept H_0 if $D \leq .00363$." *Step 4:* See Table A.62. Given $D = .0126$, H_0 should be *rejected* (just as in Chapter 10's χ^2 test).

Table A.61

Number of Promising Wells per Site	f_o	$\dfrac{f_o}{500}$	F_o	Binomial Probability, p, for $n = 3$ and $\pi = .2$ (from Appendix Table C)	F_e	$|F_o - F_e|$
0	250	.500	.500	.5120	.5120	.012 = D
1	198	.396	.896	.3840	.8960	0
2	47	.094	.990	.0960	.9920	.002
3	5	.010	1.000	.0080	1.0000	0
Total	500	1.000		1.0000		

Table A.62

Number of Yearly Auto Accidents per Driver	f_o	$\dfrac{f_o}{112,956}$	F_o	Poisson Probability for $\mu = .1$	F_e	$|F_o - F_e|$
0	103,628	.9174	.9174	.9048	.9048	.0126 = D
1	7,389	.0654	.9828	.0905	.9953	.0125
2	1,910	.0169	.9993	.0045	.9998	.0005
3	29	.0003	1.0000	.0002	1.0000	0
Total	112,956	1.0000		1.0000		

51. *Step 1:* H_0: The population of annual incomes of lawyers is normally distributed with $\mu = \$45,000$ and $\sigma = \$10,000$. *Step 2:* $D = $ max. $|F_o - F_e|$. *Step 3:* Given $\alpha = .05$ and $n = 100$, the critical $D_{.05} = .12067$ (Appendix Table M). The decision rule is "Accept H_0 if $D \le .12067$." *Step 4:* See Table A.63. Given $D = .1385$, H_0 should be *rejected* (just as in Chapter 10's χ^2 test).

53. *Step 1:* H_0: The numbers of service requests are uniformly distributed over the six business days. *Step 2:* $D = $ max. $|F_o - F_e|$. *Step 3:* Given $\alpha = .05$ and $n = 11,016$, the critical $D_{.05} = (1.36/\sqrt{11,016}) = .01296$ (Appendix Table M). The decision rule is "Accept H_0 if $D \le .01296$." *Step 4:* See Table A.64. Given $D = .1033$, H_0 should be *rejected* (just as in Chapter 10's χ^2 test).

55. *Step 1:* H_0: There is no correlation between the two sets of rankings. *Step 2:* Select the normal deviate for Spearman's rank-correlation coefficient (Box 11.T). *Step 3:* Given $\alpha = .10$ and a two-tailed test, Appendix Table J suggests "Accept H_0 if $-1.64 \le z \le +1.64$." *Step 4:* See Table A.65a; it defines ranking at time of hiring as x_i and ranking after 2 years as y_i. $n = 10$; $\sigma_\rho = .33$.

$$\rho = 1 - \frac{6(0)}{10(10^2 - 1)} = +1;$$

$$z = \frac{+1}{.33} = +3.$$

H_0 should be *rejected*. (A perfect *positive* rank correlation exists.)

Table A.63

Annual Incomes of Lawyers (thousands of dollars)	f_o	$\dfrac{f_o}{100}$	F_o	Normal Probability for $\mu = 45,000$ and $\sigma = 10,000$	F_e	$|F_o - F_e|$
under 20	2	.02	.02	.0062	.0062	.0138
20–under 30	5	.05	.07	.0606	.0668	.0032
30–under 40	10	.10	.17	.2417	.3085	.1385 = D
40–under 50	60	.60	.77	.3830	.6915	.0785
50–under 60	10	.10	.87	.2417	.9332	.0632
60–under 70	5	.05	.92	.0606	.9938	.0738
70–under 80	3	.03	.95	.0060	.9998	.0498
80–under 90	0	0	.95	.0001	.9999	.0499
90 and over	5	.05	1.00	.0001	1.0000	0
Total	100	1.00				

Table A.64

Day of the Week When Service Was Requested	f_o	$\dfrac{f_o}{11,016}$	F_o	Uniform Probability	F_e	$|F_o - F_e|$
Monday	1,251	.1136	.1136	.1667	.1667	.0531
Tuesday	1,830	.1661	.2797	.1667	.3334	.0537
Wednesday	1,675	.1521	.4318	.1667	.5000	.0682
Thursday	1,450	.1316	.5634	.1667	.6667	.1033 = D
Friday	1,905	.1729	.7363	.1667	.8334	.0971
Saturday	2,905	.2637	1.0000	.1667	1.0000	0
	11,016	1.0000		1.0000		

Table A.65a

Salesperson	A	B	C	D	E	F	G	H	I	J	
$d_i = x_i - y_i$	0	0	0	0	0	0	0	0	0	0	$\Sigma d_i = 0$
d_i^2	0	0	0	0	0	0	0	0	0	0	$\Sigma d_i^2 = 0$

57. *Steps 1* to *3* are formulated as in Problem 55. *Step 4:* See Table A.65b; it defines monthly rankings as x_i and priority-number rankings as y_i.

Table A.65b

Month	Priority Number	$d_i = x_i - y_i$	d_i^2
1	8	-7	49
2	9	-7	49
3	12	-9	81
4	10	-6	36
5	11	-6	36
6	7	-1	1
7	5	$+2$	4
8	4	$+4$	16
9	3	$+6$	36
10	6	$+4$	16
11	2	$+9$	81
12	1	$+11$	121
			$\Sigma d_i^2 = 526$

$n = 12;$

$\sigma_p = .3015;$

$\rho = 1 - \dfrac{6(526)}{12(12^2 - 1)} = -.8392;$

$z = \dfrac{-.8392}{.3015} = -2.78.$

H_0 should be *rejected*. (A strong negative correlation exists: the higher the month, the lower the draft-priority number tends to be.)

59. Wilcoxon rank-sum test (or Mann-Whitney test): *a;* sign test: *b, d, e;* Wilcoxon signed-rank test: *e;* number-of-runs test: *c.*

Chapter 12

1. The sample means are: $\overline{X}_1 = 3.2, \overline{X}_2 = 3, \overline{X}_3 = 2.4, \overline{X}_4 = 2.8, \overline{X}_5 = 3.6;$ the grand mean is $\overline{\overline{X}} = 3.$ $TSS = r\Sigma(\overline{X}_j - \overline{\overline{X}})^2 = 5[(3.2 - 3)^2 + (3 - 3)^2 + (2.4 - 3)^2 + (2.8 - 3)^2 + (3.6 - 3)^2] = 4.$ ESS can be calculated from Table A.66 as $ESS = 4.8 + 10 + 5.2 + 24.8 + 5.2 = 50.$ The ANOVA table is given in Table A.67. The critical $F_{.025}$ for $d.f. = (4,20)$ equals 3.51. H_0 should be *accepted*. The mean delivery times are the same.

3. H_0: The mean housing prices are the same regardless of which of 3 air pollution levels prevails: $\mu_L = \mu_M = \mu_H; H_A$: At least one of the population means is different from the others. The sample means are $\overline{X}_L = 81.2, \overline{X}_M = 77, \overline{X}_H = 55.4;$ the grand mean is $\overline{\overline{X}} = 71.2.$ $TSS = r\Sigma(\overline{X}_j - \overline{\overline{X}})^2 = 5[(81.2 - 71.2)^2 + (77 - 71.2)^2 + (55.4 - 71.2)^2] = 1,916.4.$ ESS can be calculated from Table A.68 on the next page as: $ESS = 3,570.8 + 1,682 + 729.2 = 5,982.$ The ANOVA table is given in

Table A.66

Observation	1 $(X_{i1} - \overline{X}_1)^2$	2 $(X_{i2} - \overline{X}_2)^2$	3 $(X_{i3} - \overline{X}_3)^2$	4 $(X_{i4} - \overline{X}_4)^2$	5 $(X_{i5} - \overline{X}_5)^2$
1	$(3 - 3.2)^2 = .04$	$(4 - 3)^2 = 1$	$(2 - 2.4)^2 = .16$	$(7 - 2.8)^2 = 17.64$	$(2 - 3.6)^2 = 2.56$
2	$(3 - 3.2)^2 = .04$	$(1 - 3)^2 = 4$	$(4 - 2.4)^2 = 2.56$	$(1 - 2.8)^2 = 3.24$	$(3 - 3.6)^2 = .36$
3	$(2 - 3.2)^2 = 1.44$	$(3 - 3)^2 = 0$	$(3 - 2.4)^2 = .36$	$(1 - 2.8)^2 = 3.24$	$(4 - 3.6)^2 = .16$
4	$(5 - 3.2)^2 = 3.24$	$(5 - 3)^2 = 4$	$(1 - 2.4)^2 = 1.96$	$(2 - 2.8)^2 = .64$	$(5 - 3.6)^2 = 1.96$
5	$(3 - 3.2)^2 = .04$	$(2 - 3)^2 = 1$	$(2 - 2.4)^2 = .16$	$(3 - 2.8)^2 = .04$	$(4 - 3.6)^2 = .16$
	4.8	10	5.2	24.8	5.2

Table A.67

Source of Variation	Sum of Squares (1)	Degrees of Freedom (2)	Mean Square (3) = (1) ÷ (2)	Test Statistic (4)
Treatments	$TSS = 4$	$c - 1 = 4$	$TMS = \dfrac{4}{4} = 1$	$F = \dfrac{TMS}{EMS} = \dfrac{1}{2.5} = .4$
Error	$ESS = 50$	$(r - 1)c = 20$	$EMS = \dfrac{50}{20} = 2.5$	
Total	$Total\ SS = 54$	$rc - 1 = 24$		

Table A.68

Observation	Low $(X_{iL} - \overline{X}_L)^2$	Moderate $(X_{iM} - \overline{X}_M)^2$	High $(X_{iH} - \overline{X}_H)^2$
1	$(120 - 81.2)^2 = 1{,}505.44$	$(61 - 77)^2 = 256$	$(40 - 55.4)^2 = 237.16$
2	$(68 - 81.2)^2 = 174.24$	$(59 - 77)^2 = 324$	$(55 - 55.4)^2 = .16$
3	$(40 - 81.2)^2 = 1{,}697.44$	$(110 - 77)^2 = 1{,}089$	$(73 - 55.4)^2 = 309.76$
4	$(95 - 81.2)^2 = 190.44$	$(75 - 77)^2 = 4$	$(45 - 55.4)^2 = 108.16$
5	$(83 - 81.2)^2 = 3.24$	$(80 - 77)^2 = 9$	$(64 - 55.4)^2 = 73.96$
	$3{,}570.8$	$1{,}682$	729.2

Table A.69

Source of Variation	Sum of Squares (1)	Degrees of Freedom (2)	Mean Square (3) = (1) ÷ (2)	Test Statistic (4)
Treatments	$TSS = 1{,}916.4$	$c - 1 = 2$	$TMS = 958.2$	$F = \dfrac{TMS}{EMS} = \dfrac{958.2}{498.5} = 1.92$
Error	$ESS = 5{,}982.0$	$(r - 1)c = 12$	$EMS = 498.5$	
Total	$Total\ SS = 7{,}898.4$	$(rc) - 1 = 14$		

Table A.70

Source of Variation	Sum of Squares	Degrees of Freedom	Mean Square	Test Statistic
Treatments	3,078.11	2	1,539.06	$F = 38.58$
Error	598.33	15	39.89	
Total	3,676.44	17		

Table A.71

Source of Variation	Sum of Squares	Degrees of Freedom	Mean Square	Test Statistic
Treatments	.67	2	.33	$F = .0018$
Error	3,322.29	18	184.57	
Total	3,322.95	20		

Table A.69. The critical $F_{.025}$ for $d.f. = (2,12) = 5.10$ (Appendix Table N). H_0 should be *accepted.*

5. H_0: $\mu_A = \mu_B = \mu_C$. The ANOVA table is Table A.70. Given $\alpha = .025$ and $d.f. = (2,15)$, Appendix Table N suggests "Accept H_0 if $F \leq 4.77$." H_0 should be *rejected.* Worker productivity differs. The results are valid if the three output-per-worker populations are normal and have equal variances.

7. H_0: $\mu_A = \mu_B = \mu_C$. The ANOVA table is Table A.71. Given $\alpha = .05$ and $d.f. = (2,18)$, Appendix Table N suggests "Accept H_0 if $F \leq 3.55$." H_0 should be *accepted.* The results are valid if the three reading-score populations are normal and have equal variances.

9. H_0: $\mu_A = \mu_B = \mu_C$. The ANOVA table is Table A.72. Given $\alpha = .01$ and $d.f. = (2,15)$, Appendix Table N suggests "Accept H_0 if $F \leq 6.36$." H_0 should be *accepted.* The results are valid if the three consumption populations are normal and have equal variances.

Table A.72

Source of Variation	Sum of Squares	Degrees of Freedom	Mean Square	Test Statistic
Treatments	107.11	2	53.56	$F = .36$
Error	2,208.67	15	147.24	
Total	2,315.78	17		

Table A.73

Source of Variation	Sum of Squares	Degrees of Freedom	Mean Square	Test Statistic
Treatments	1,065.88	3	355.29	$F = 10.56$
Error	1,211.10	36	33.64	
Total	2,276.98	39		

Table A.74

Source of Variation	Sum of Squares	Degrees of Freedom	Mean Square	Test Statistic
Treatments	1.70	3	.568	$F = 8.80$
Error	1.29	20	.065	
Total	3.00	23		

11. $H_0: \mu_A = \mu_B = \mu_C = \mu_D$. The ANOVA table is Table A.73.
Given $\alpha = .025$ and $d.f. = (3,36)$, Appendix Table N suggests "Accept H_0 if $F \leq 3.51$ (interpolated)." H_0 should be *rejected*. The incidence of crime differs among the districts. The results are valid if the four police-call populations are normal and have equal variances.

13. $H_0: \mu_A = \mu_B = \mu_C = \mu_D$. The ANOVA table is Table A.74.
Given $\alpha = .05$ and $d.f. = (3,20)$, Appendix Table N suggests "Accept H_0 if $F \leq 3.10$." H_0 should be *rejected*. The average debt-to-equity ratios differ among the industries. The results are valid if the four debt-to-equity ratio populations are normal and have equal variances.

15. The sample means are: $\overline{X}_1 = 13$, $\overline{X}_2 = 11.33$, $\overline{X}_3 = 9.67$, $\overline{X}_4 = 15.33$, $\overline{X}_5 = 10.67$, $\overline{X}_A = 12.4$, $\overline{X}_B = 12.6$, $\overline{X}_C = 11$; the grand mean is $\overline{\overline{X}} = 12$. The hypothesis for fertilizer treatments is H_0: The mean yields are the same for all types of fertilizer. The hypothesis for soil blocks is H_0: The mean yields are the same for all 3 soil qualities. $TSS = r\Sigma(\overline{X}_j - \overline{\overline{X}})^2 = 3[(13 - 12)^2 + (11.33 - 12)^2 + (9.67 - 12)^2 + (15.33 - 12)^2 + (10.67 -$

$12)^2] = 59.33.$ $BSS = c\Sigma(\overline{X}_i - \overline{\overline{X}})^2 = 5[(12.4 - 12)^2 + (12.6 - 12)^2 + (11 - 12)^2] = 7.6.$ *Total* $SS = (15 - 12)^2 + (13 - 12)^2 + \ldots + (12 - 12)^2 = 136.$ $ESS = Total\ SS - TSS - BSS = 136 - 59.33 - 7.6 = 69.07.$ The ANOVA table is given in Table A.75 on the next page. The critical $F_{.025}$ for $d.f. = (4,8) = 5.05$; for $d.f. = (2,8) = 6.06$. Both null hypotheses should be *accepted*.

17. For paint treatments, H_0: The mean quality score is the same for all types of paint. For surface blocks, H_0: The mean quality score is the same for all types of surfaces. The ANOVA table is Table A.76 on the next page. According to Appendix Table N, given $\alpha = .025$, the critical F for $d.f. = (4,16)$ is 3.73; both null hypotheses should be *accepted*.

19. For temperature treatments, $H_0: \mu_1 = \mu_2 = \mu_3 = \mu_4$. For pressure blocks, $H_0: \mu_A = \mu_B = \mu_C$. The ANOVA table is Table A.77, the third table on the following page. According to Appendix Table N, given $\alpha = .05$, the critical F for $d.f. = (3,6)$ is 4.76; for $d.f. = (2,6)$, it is 5.14. Both null hypotheses should be *accepted*.

21. For area treatments, $H_0: \mu_N = \mu_E = \mu_S = \mu_W$. For station-size blocks, $H_0: \mu_A = \mu_B = \mu_C$. The ANOVA table is Table A.78 on the next page.

Table A.75

Source of Variation	Sum of Squares (1)	Degrees of Freedom (2)	Mean Square (3) = (1) ÷ (2)	Test Statistic (4)
Treatments	$TSS = 59.33$	$c - 1 = 4$	$TMS = \dfrac{59.33}{4} = 14.83$	$F_T = \dfrac{TMS}{EMS} = \dfrac{14.83}{8.63} = 1.72$
Blocks	$BSS = 7.60$	$r - 1 = 2$	$BMS = \dfrac{7.6}{2} = 3.80$	$F_B = \dfrac{BMS}{EMS} = \dfrac{3.80}{8.63} = .44$
Error	$ESS = 69.07$	$(r - 1)(c - 1) = 8$	$EMS = \dfrac{69.07}{8} = 8.63$	
Total	$Total\ SS = 136$	$rc - 1 = 14$		

Table A.76

Source of Variation	Sum of Squares	Degrees of Freedom	Mean Square	Test Statistic
Treatments	315.2	4	78.80	$F_T = .72$
Blocks	1,333.6	4	333.40	$F_B = 3.04$
Error	1,753.2	16	109.58	
Total	3,402.0	24		

Table A.77

Source of Variation	Sum of Squares	Degrees of Freedom	Mean Square	Test Statistic
Treatments	60,000	3	20,000	$F_T = \overline{.66}$
Blocks	20,000	2	10,000	$F_B = \overline{.33}$
Error	180,000	6	30,000	
Total	260,000	11		

Table A.78

Source of Variation	Sum of Squares	Degrees of Freedom	Mean Square	Test Statistic
Treatments	130,466.7	3	43,488.9	$F_T = 2.82$
Blocks	389,616.7	2	194,808.4	$F_B = 12.65$
Error	92,383.3	6	15,397.2	
Total	612,466.7	11		

Table A.79

Source of Variation	Sum of Squares	Degrees of Freedom	Mean Square	Test Statistic
Treatments	120.66	2	60.33	$F_T = 3.72$
Blocks	172.29	3	57.43	$F_B = 3.54$
Error	97.25	6	16.21	
Total	390.20	11		

Table A.80

Source of Variation	Sum of Squares	Degrees of Freedom	Mean Square	Test Statistic
Treatments	1,716.93	2	858.46	$F_T = 47.41$
Blocks	827.81	2	413.91	$F_B = 22.86$
Interactions	6,006.96	4	1,501.74	$F_I = 82.94$
Error	814.83	45	18.11	
Total	9,366.54	53		

According to Appendix Table N, given $\alpha = .01$, the critical F for $d.f. = (3,6)$ is 9.78; for $d.f. = (2,6)$, it is 10.92. H_0 should be *accepted* for treatments; *rejected* for blocks.

23. For method treatments, H_0: $\mu_1 = \mu_2 = \mu_3$. For laboratory blocks, H_0: $\mu_A = \mu_B = \mu_C = \mu_D$. The ANOVA table is Table A.79.

According to Appendix Table N, given $\alpha = .025$, the critical F for $d.f. = (2,6)$ is 7.26; for $d.f. = (3,6)$, it is 6.60. Both null hypotheses should be *accepted*.

25. The null hypotheses are the following:
for treatments, H_0: $\mu_1 = \mu_2 = \mu_3$
for blocks, H_0: $\mu_A = \mu_B = \mu_C$
for interactions, H_0: There is no interaction between treatments, j, and blocks, i.
The ANOVA table is Table A.80.
According to Appendix Table N, given $\alpha = .05$, the critical values of F are
for $d.f. = (2,45) = 3.21$ (treatments).
for $d.f. = (2,45) = 3.21$ (blocks).
for $d.f. = (4,45) = 2.59$ (interactions).
Thus, H_0 should be *rejected* in all three cases.

27. The null hypotheses are the following:
for treatments, H_0: $\mu_C = \mu_F = \mu_H = \mu_P = \mu_W$
for blocks, H_0: $\mu_C = \mu_G = \mu_L$

for interactions, H_0: There is no interaction between treatments, j, and blocks, i.
The ANOVA table is Table A.81 on the next page.
According to Appendix Table N, given $\alpha = .01$, the critical values of F are
for $d.f. = (4,45) = 3.78$ (treatments).
for $d.f. = (2,45) = 5.13$ (blocks).
for $d.f. = (8,45) = 2.95$ (interactions).
Thus, H_0 should be *rejected* for treatments and blocks, but *accepted* for interactions.

29. The null hypotheses are the following:
for treatments, H_0: $\mu_U = \mu_S = \mu_H$.
for blocks, H_0: $\mu_A = \mu_B = \mu_C$.
for interactions, H_0: There is no interaction between treatments, j, and blocks, i.
The ANOVA table is Table A.82.
According to Appendix Table N, given $\alpha = .025$, the critical values of F are
for $d.f. = (2,9) = 5.71$ (treatments).
for $d.f. = (2,9) = 5.71$ (blocks).
for $d.f. = (4,9) = 4.72$ (interactions).
Thus, H_0 should be *accepted* for treatments and interactions, *rejected* for blocks.

31. The null hypotheses are the following:
for treatments, H_0: $\mu_R = \mu_S = \mu_H$
for blocks, H_0: $\mu_M = \mu_F$

Table A.81

Source of Variation	Sum of Squares	Degrees of Freedom	Mean Square	Test Statistic
Treatments	156.71	4	39.18	$F_T = 11.17$
Blocks	131.45	2	65.73	$F_B = 18.74$
Interactions	66.98	8	8.37	$F_I = 2.39$
Error	157.85	45	3.51	
Total	512.99	59		

Table A.82

Source of Variation	Sum of Squares	Degrees of Freedom	Mean Square	Test Statistic
Treatments	140.33	2	70.17	$F_T = 2.14$
Blocks	2,791.00	2	1,395.50	$F_B = 42.65$
Interactions	70.67	4	17.67	$F_I = .54$
Error	294.50	9	32.72	
Total	3,296.50	17		

for interactions, H_0: There is no interaction between treatments, *j*, and blocks, *i*.
The ANOVA table is Table A.83.
According to Appendix Table N, given $\alpha = .05$, the critical values of F are
for *d.f.* $= (2,12) = 3.89$ (treatments).
for *d.f.* $= (1,12) = 4.75$ (blocks).
for *d.f.* $(2,12)$ 3.89 (interactions.)
Thus, H_0 should be *accepted* in all three cases.

33. The null hypotheses are the following:
for treatments, H_0: $\mu_L = \mu_M = \mu_H$
for blocks, H_0: $\mu_A = \mu_B = \mu_C$
for interactions, H_0: There is no interaction between treatments, *j*, and blocks, *i*.
The ANOVA table is Table A.84.
According to Appendix Table N, given $\alpha = .01$, the critical values of F are
for *d.f.* $= (2,18) = 6.01$ (treatments).
for *d.f.* $= (2,18) = 6.01$ (blocks).
for *d.f.* $= (4,18) = 4.58$ (interactions).
Thus, H_0 should be *rejected* in all three cases.

35. The null hypotheses are the following:
for treatments, H_0: $\mu_G = \mu_M = \mu_N$

for blocks, H_0: $\mu_J = \mu_F = \mu_M = \mu_A$
for interactions, H_0: There is no interaction between treatments, *j*, and blocks, *i*.
The ANOVA table is Table A.85.
According to Appendix Table N, given $\alpha = .025$, the critical values of F are
for *d.f.* $= (2,24) = 4.32$ (treatments).
for *d.f.* $= (3,24) = 3.72$ (blocks).
for *d.f.* $= (6,24) = 2.99$ (interactions).
Thus, H_0 should be *rejected* for treatments and interactions, *accepted* for blocks.

37. The null hypotheses are the following:
for treatments, H_0: $\mu_A = \mu_C = \mu_W$
for blocks, H_0: $\mu_R = \mu_U$
for interactions, H_0: There is no interaction between treatments, *j*, and blocks, *i*.
The ANOVA table is Table A.86.
According to Appendix Table N, given $\alpha = .05$, the critical values of F are
for *d.f.* $= (2,12) = 3.89$ (treatments).
for *d.f.* $= (1,12) = 4.75$ (blocks).
for *d.f.* $= (2,12) = 3.89$ (interactions).
Thus, H_0 should be *rejected* in all three cases.

Table A.83

Source of Variation	Sum of Squares	Degrees of Freedom	Mean Square	Test Statistic
Treatments	560.78	2	280.39	$F_T = 2.17$
Blocks	264.50	1	264.50	$F_B = 2.05$
Interactions	247.00	2	123.50	$F_I = .95$
Error	1,552.00	12	129.33	
Total	2,624.28	17		

Table A.84

Source of Variation	Sum of Squares	Degrees of Freedom	Mean Square	Test Statistic
Treatments	2,260.22	2	1,130.11	$F_T = 87.43$
Blocks	5,984.22	2	2,992.11	$F_B = 231.48$
Interactions	7,771.56	4	1,942.89	$F_I = 150.31$
Error	232.66	18	12.93	
Total	16,248.67	26		

Table A.85

Source of Variation	Sum of Squares	Degrees of Freedom	Mean Square	Test Statistic
Treatments	68,893.71	2	34,446.85	$F_T = 104.83$
Blocks	1,735.23	3	578.41	$F_B = 1.76$
Interactions	13,325.61	6	2,220.94	$F_I = 6.76$
Error	7,885.99	24	328.58	
Total	91,840.53	35		

Table A.86

Source of Variation	Sum of Squares	Degrees of Freedom	Mean Square	Test Statistic
Treatments	2,170,767.0	2	1,085,384	$F_T = 70.12$
Blocks	1,727,941.0	1	1,727,941	$F_B = 111.63$
Interactions	243,622.3	2	121,811.2	$F_I = 7.87$
Error	185,745.7	12	15,478.8	
Total	4,328,076	17		

39. The Box 12.B formula applies, the appropriate t value being $t_{.025,18}$. Thus,

$$\mu_A = 77.43 \pm \left(2.101\sqrt{\frac{184.57}{7}}\right) = 77.43 \pm 10.79.$$

$\mu_B = 77.57 \pm 10.79.$
$\mu_C = 77.86 \pm 10.79.$

Hence,

$$66.64 \leq \mu_A \leq 88.22$$
$$66.78 \leq \mu_B \leq 88.36$$
$$67.07 \leq \mu_C \leq 88.65$$

41. The Box 12.B formula applies, the appropriate t value being $t_{.005,15}$. Thus,

$$\mu_A = 18 \pm \left(2.947\sqrt{\frac{147.24}{6}}\right) = 18 \pm 14.60.$$

$\mu_B = 23.33 \pm 14.60.$
$\mu_C = 18.33 \pm 14.60.$

Hence,

$$3.40 \leq \mu_A \leq 32.60$$
$$8.73 \leq \mu_B \leq 37.93$$
$$3.73 \leq \mu_C \leq 32.93$$

43. The Box 12.B formula applies, the appropriate t value being $t_{.025,20}$. Thus,

$$\mu_A = .150 \pm \left(2.086\sqrt{\frac{.065}{6}}\right) = .150 \pm .217.$$

$\mu_B = .300 \pm .217.$
$\mu_C = .117 \pm .217.$
$\mu_D = .783 \pm .217.$

Hence,

$$-.067 \leq \mu_A \leq .367$$
$$.083 \leq \mu_B \leq .517$$
$$-.100 \leq \mu_C \leq .334$$
$$.566 \leq \mu_D \leq 1.000$$

45. The Box 12.B formula applies, the appropriate t value being $t_{.025,6}$. Thus,

$$\mu_1 = 700 \pm \left(2.447\sqrt{\frac{30,000}{3}}\right) = 700 \pm 244.7.$$

$\mu_2 = 800 \pm 244.7.$
$\mu_3 = 900 \pm 244.7.$
$\mu_4 = 800 \pm 244.7.$

$$\mu_A = 750 \pm \left(2.447\sqrt{\frac{30,000}{4}}\right) = 750 \pm 211.92$$

$\mu_B = 850 \pm 211.92.$
$\mu_C = 800 \pm 211.92.$

Hence,

$$455.3 \leq \mu_1 \leq 944.7.$$
$$555.3 \leq \mu_2 \leq 1,044.7.$$
$$655.3 \leq \mu_3 \leq 1,144.7.$$
$$555.3 \leq \mu_4 \leq 1,044.7.$$

$$538.08 \leq \mu_A \leq 961.92$$
$$638.08 \leq \mu_B \leq 1,061.92$$
$$588.08 \leq \mu_C \leq 1,011.92$$

47. The Box 12.B formula applies, the appropriate t value being $t_{.005,6}$. Thus,

$$\mu_N = 266.67 \pm \left(3.707\sqrt{\frac{15,397.22}{3}}\right)$$
$$= 266.67 \pm 265.57.$$
$\mu_E = 333.33 \pm 265.57.$
$\mu_S = 143.33 \pm 265.57.$
$\mu_W = 430 \pm 265.57.$

$$\mu_A = 122.5 \pm \left(3.707\sqrt{\frac{15,397.22}{4}}\right) = 122.5 \pm 229.99$$

$\mu_B = 215 \pm 229.99.$
$\mu_C = 542.5 \pm 229.99.$

Hence,

$$1.10 \leq \mu_N \leq 532.24$$
$$67.76 \leq \mu_E \leq 598.90$$
$$-122.24 \leq \mu_S \leq 408.90$$
$$164.43 \leq \mu_W \leq 695.57$$

$$-107.49 \leq \mu_A \leq 352.49$$
$$-14.99 \leq \mu_B \leq 444.99$$
$$312.51 \leq \mu_C \leq 772.49$$

49. A correct answer to Problem 2 reveals sample means of $\overline{X}_1 = 70.57$, $\overline{X}_2 = 72.34$, $\overline{X}_3 = 58.04$, as well as an error mean square and error degrees of freedom of $EMS = 31.52$ and $d.f. = 21$, respectively. There were $n = 8$ observations per sample. Appendix Table K provides the appropriate value of $t_{.025,21} = 2.08$. Accordingly, from Box 12.C, *for Boston vs. Chicago means,* $\mu_1 - \mu_2 \cong (70.57 - 72.34) \pm 2.08\sqrt{[2(31.52)]/8} = -1.77 \pm 5.84$; $-7.61 \leq (\mu_1 - \mu_2) \leq 4.07$. The means are the same. *For Boston vs. Los Angeles means,* $\mu_1 - \mu_3 \cong (70.57 - 58.04) \pm 5.84 = 12.53 \pm 5.84$; $6.69 \leq (\mu_1 - \mu_3) \leq 18.37$. The Boston mean is higher. *For Chicago vs. Los Angeles means,* $\mu_2 - \mu_3 \cong (72.34 - 58.04) \pm 5.84 = 14.30 \pm 5.84$; $8.46 \leq (\mu_2 - \mu_3) \leq 20.14$. The Chicago mean is higher.

51. The Box 12.C formula applies, the appropriate t value being $t_{.025,45}$. Thus,

$$\mu_1 - \mu_2 = (33.22 - 25) \pm \left(2.016 \sqrt{\frac{2(18.11)}{18}} \right)$$
$$= 8.22 \pm 2.86.$$

$\mu_1 - \mu_3 = (33.22 - 19.5) \pm 2.86 = 13.72 \pm 2.86.$
$\mu_2 - \mu_3 = (25 - 19.5) \pm 2.86 = 5.5 \pm 2.86.$

$$\mu_A - \mu_B = (24.78 - 31.17) \pm \left(2.016 \sqrt{\frac{2(18.11)}{18}} \right)$$
$$= -6.39 \pm 2.86$$

$\mu_A - \mu_C = (24.78 - 21.78) \pm 2.86 = 3 \pm 2.86$
$\mu_B - \mu_C = (31.17 - 21.78) \pm 2.86 = 9.39 \pm 2.86$

Hence,

$5.36 \le (\mu_1 - \mu_2) \le 11.08.$ μ_1 is larger than μ_2.
$10.86 \le (\mu_1 - \mu_3) \le 16.58.$ μ_1 is larger than μ_3.
$2.64 \le (\mu_2 - \mu_3) \le 8.36.$ μ_2 is larger than μ_3.

$-9.25 \le (\mu_A - \mu_B) \le -3.53.$ μ_A is smaller than μ_B.
$.14 \le (\mu_A - \mu_C) \le 5.86.$ μ_A is larger than μ_C.
$6.53 \le (\mu_B - \mu_C) \le 12.25.$ μ_B is larger than μ_C.

53. The Box 12.C formula applies, the appropriate t value being $t_{.005,45}$. Thus,

$$\mu_C - \mu_G = (4.36 - 6.23) \pm \left(2.693 \sqrt{\frac{2(3.51)}{20}} \right)$$
$$= -1.87 \pm 1.60.$$

$\mu_C - \mu_L = (4.36 - 2.61) \pm 1.60 = 1.75 \pm 1.60.$
$\mu_G - \mu_L = (6.23 - 2.61) \pm 1.60 = 3.62 \pm 1.60.$

Hence,

$-3.47 \le (\mu_C - \mu_G) \le -.27.$ μ_C is smaller than μ_G.
$.15 \le (\mu_C - \mu_L) \le 3.35.$ μ_C is larger than μ_L.
$2.02 \le (\mu_G - \mu_L) \le 5.22.$ μ_G is larger than μ_L.

55. The Box 12.C formula applies, the appropriate t value being $t_{.025,12}$. Thus

$$\mu_R - \mu_S = (88.33 - 74.67) \pm \left(2.179 \sqrt{\frac{2(129.33)}{6}} \right)$$
$$= 13.66 \pm 14.31.$$

$\mu_R - \mu_H = (88.33 - 81.17) \pm 14.31$
$\quad = 7.16 \pm 14.31.$
$\mu_S - \mu_H = (74.67 - 81.17) \pm 14.31$
$\quad = -6.50 \pm 14.31.$

$$\mu_M - \mu_F = (77.56 - 85.22) \pm \left(2.179 \sqrt{\frac{2(129.33)}{9}} \right)$$
$$= -7.66 \pm 11.68.$$

Hence,

$-.65 \le (\mu_R - \mu_S) \le 27.97.$
The means are the same.

$-7.15 \le (\mu_R - \mu_H) \le 21.47.$
The means are the same.

$-20.81 \le (\mu_S - \mu_H) \le 7.81.$
The means are the same.

$-19.34 \le (\mu_M - \mu_F) \le 4.02.$
The means are the same.

57. The absolute differences in the sample means can be calculated from text Table 12.6 and are given in Table A.87 on the next page, which can be read like a mileage chart. Table 12.6 also shows $n = 2$, while Table 12.7 shows $EMS = 7.5$. For $k = 12$, $\alpha = .01$, and error $d.f. = 12$, Appendix Table O gives $q_\alpha = 7.06$. Thus,

$$HSD = 7.06 \sqrt{\frac{7.5}{2}} = 13.67.$$

Hence, all the encircled differences in Table A.87 (all of which exceed 13.67) can be viewed as significant.

59. For $k = 3$, $\alpha = .01$, and error $d.f. = 18$, Appendix Table O suggests $q_\alpha = 4.70$. Thus,

$$HSD = 4.70 \sqrt{\frac{12.93}{9}} = 5.63.$$

The absolute differences in the sample means are

$|\overline{X}_L - \overline{X}_M| = |36.67 - 58.56| = 21.89.$
$|\overline{X}_L - \overline{X}_H| = |36.67 - 43.44| = 6.77.$
$|\overline{X}_M - \overline{X}_H| = |58.56 - 43.44| = 15.12.$

$|\overline{X}_A - \overline{X}_B| = |60.44 - 25.67| = 34.77.$
$|\overline{X}_A - \overline{X}_C| = |60.44 - 52.56| = 7.88.$
$|\overline{X}_B - \overline{X}_C| = |25.67 - 52.56| = 26.89.$

Conclusion: In each case, the population means differ.

Table A.87

	$\overline{X}_{A1}$	$\overline{X}_{A2}$	$\overline{X}_{A3}$	$\overline{X}_{B1}$	$\overline{X}_{B2}$	$\overline{X}_{B3}$	$\overline{X}_{C1}$	$\overline{X}_{C2}$	$\overline{X}_{C3}$	$\overline{X}_{D1}$	$\overline{X}_{D2}$	$\overline{X}_{D3}$
$\overline{X}_{A1}$	—	5	5	8	(14)	10	4	13	8	10	0	7
$\overline{X}_{A2}$		—	0	13	9	5	9	8	3	5	5	2
$\overline{X}_{A3}$			—	13	9	5	9	8	3	5	5	2
$\overline{X}_{B1}$				—	(22)	(18)	4	(21)	(16)	(18)	8	(15)
$\overline{X}_{B2}$					—	4	(18)	1	6	4	(14)	7
$\overline{X}_{B3}$						—	(14)	3	2	0	10	3
$\overline{X}_{C1}$							—	(17)	12	(14)	4	11
$\overline{X}_{C2}$								—	5	3	13	6
$\overline{X}_{C3}$									—	2	8	1
$\overline{X}_{D1}$										—	10	3
$\overline{X}_{D2}$											—	7
$\overline{X}_{D3}$												—

Chapter 13

1. a. See Figure A.30 for the scatter diagram.

b. $\hat{Y}_X = 52.89 - 1.12\,X$.

c. The relationship does make sense. According to the estimated regression line, the incumbent party can expect to capture 52.89 percent of the vote if there is no change in the election-year unemployment rate (point B) but to gain (or lose) about one percentage point in the vote for every one percentage point difference between the election-year unemployment rate and that of the previous year. If unemployment is down 3 percentage points, the predicted vote is 56.3 percent (point A); if unemployment is up 3 percentage points, the predicted vote is 49.5 percent (point C).

3. $\hat{Y}_X = 452.6558 + 96.69197\,X$.

5. $\hat{Y}_X = 57.08909 + 17.37734\,X$.

7. $\hat{Y}_X = 17.46059 - .07219\,X$.

9. $\hat{Y}_X = 10.47237 - .33201\,X$.

11. $\hat{Y}_X = .35432 + .02364\,X$.

13. $\hat{Y}_X = 228.8908 + 19.70607\,X$.

15. $\hat{Y}_X = 1.23145 + .35778\,X$.

17. $\hat{Y}_X = 36.94287 + 4.21079\,X$.

19. $\hat{Y}_X = -72.75798 + .68403\,X$.

21. The *point estimate* is $\hat{Y}_5 = 52.89 - 1.12(5) = 47.29$. For the *98-percent prediction intervals,* the Box 13.B and 13.D formulas apply (with $t_{.01,22} = 2.508$ and $s_{Y\cdot X} = 6.021974$):

$$\mu_{Y\cdot 5} = 47.29$$

$$\pm \left[2.508(6.021974)\sqrt{\frac{1}{24} + \frac{(5 - .13)^2}{159.86 - 24(.0169)}} \right]$$

$$= 47.29 \pm 6.59$$

Hence, $40.7 \le \mu_{Y\cdot 5} \le 53.88$.

$$I_{Y\cdot 5} = 47.29 \pm 16.48$$

Hence, $30.81 \le I_{Y\cdot 5} \le 63.77$.

23. The *point estimate* is $\hat{Y}_5 = 452.6558 + 96.69197(5) = 936.12$. For the *90-percent prediction intervals,* the Box 13.B and 13.D formulas apply (with $t_{.05,8} = 1.86$ and $s_{Y\cdot X} = 263.3122$):

$$\mu_{Y\cdot 5} = 936.12$$

$$\pm \left[1.86(263.3122)\sqrt{\frac{1}{10} + \frac{(5 - 5.7)^2}{433 - 10(5.7)^2}} \right]$$

$$= 936.12 \pm 158.35$$

Hence, $777.77 \le \mu_{Y\cdot 5} \le 1{,}094.47$.

$$I_{Y\cdot 5} = 936.12 \pm 514.72$$

Hence, $421.40 \le I_{Y\cdot 5} \le 1{,}450.84$.

Figure A.30

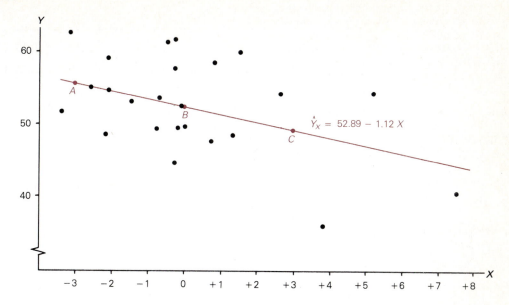

25. The *point estimate* is $\hat{Y}_{11} = 57.08909 + 17.37734(11) = 248.24$. For the *98-percent prediction intervals,* the Box 13.B and 13.D formulas apply (with $t_{.01,12} = 2.681$ and $s_{Y\cdot X} = 56.83893$):

$$\mu_{Y\cdot 11} = 248.24$$
$$\pm \left[2.681(56.83893)\sqrt{\frac{1}{14} + \frac{(11 - 7.71)^2}{1,016 - 14(7.71)^2}} \right]$$
$$= 248.24 \pm 55.01.$$

Hence, $193.23 \le \mu_{Y\cdot 11} \le 303.25$.

$$I_{Y\cdot 11} = 248.24 \pm 162.01$$

Hence, $86.23 \le I_{Y\cdot 11} \le 410.25$.

27. The *point estimate* is $\hat{Y}_{140} = 17.46059 - .07219(140) = 7.35$. For the *99.5-prediction intervals,* the Box 13.B and 13.D formulas apply (with $t_{.0025,14} = 3.326$ and $s_{Y\cdot X} = 1.41038$):

$$\mu_{Y\cdot 140} = 7.35$$
$$\pm \left[3.326(1.41038)\sqrt{\frac{1}{16} + \frac{(140 - 74.69)^2}{114,377 - 16(74.69)^2}} \right]$$
$$= 7.35 \pm 2.26$$

Hence, $5.09 \le \mu_{Y\cdot 140} \le 9.61$.

$$I_{Y\cdot 140} = 7.35 \pm 5.21$$

Hence, $2.14 \le I_{Y\cdot 140} \le 12.56$.

29. The *point estimte* is $\hat{Y}_{14} = 10.47237 - .33201(14) = 5.82$. For the *99.9-percent prediction intervals,* the Box 13.B and 13.D formulas apply (with $t_{.0005,7} = 5.408$ and $s_{Y\cdot X} = .8641118$):

$$\mu_{Y\cdot 14} = 5.82$$
$$\pm \left[5.408(.8641118)\sqrt{\frac{1}{9} + \frac{(14 - 9.69)^2}{896.54 - 9(9.69)^2}} \right]$$
$$= 5.82 \pm 3.21$$

Hence, $2.61 \le \mu_{Y\cdot 14} \le 9.03$.

$$I_{Y\cdot 14} = 5.82 \pm 5.67$$

Hence, $.15 \le I_{Y\cdot 14} \le 11.49$.

No, the relationship doesn't make sense. One would expect, all else being equal, that higher interest rates lower investment demand, hence aggregate demand, hence output and employment. Thus higher interest rates should be associated with higher, not lower unemployment. The value of slope b should be positive. Presumably, other things were not equal as this reasoning assumes.

31. The *point estimate* is $\hat{Y}_{25} = .35432 + .02364(25) = .945$. For the *95-percent prediction intervals,* the Box 13.B and 13.D formulas apply (with $t_{.025,8} = 2.306$ and $s_{Y\cdot X} = .07485795$):

$$\mu_{Y\cdot 25} = .945$$
$$\pm \left[2.306(.07485795)\sqrt{\frac{1}{10} + \frac{(25 - 22.49)^2}{6,923.354 - 10(22.49)^2}} \right]$$
$$= .945 \pm .056$$

Hence, $.889 \leq \mu_{Y\cdot 25} \leq 1.001$.

$$I_{Y\cdot 25} = .945 \pm .181$$

Hence, $.764 \leq 1.126$.

33. The Box 13.F and 13.G formulas apply. Given $t_{.025,22} = 2.074$,

$$\beta = -1.118 \pm \left(2.074 \frac{6.021974}{\sqrt{159.86 - 24(.13)^2}} \right)$$
$$= -1.118 \pm .989$$

Hence, $-2.107 \leq \beta \leq -.129$.

$$\alpha = 52.891 \pm \left(.989 \sqrt{\frac{159.86}{24}} \right)$$
$$= 52.891 \pm 2.552$$

Hence, $50.339 \leq \alpha \leq 55.443$.

35. See Table A.88.

Table A.88

Problem	95-percent Confidence Interval For	
	β	α
3	$38.29 \leq \beta \leq 155.09$	$68.36 \leq \alpha \leq 836.95$
5	$8.22 \leq \beta \leq 26.54$	$-20.94 \leq \alpha \leq 135.11$
7	$-.09 \leq \beta \leq -.05$	$15.85 \leq \alpha \leq 19.07$

37. See Table A.89.

Table A.89

Problem	95-percent Confidence Interval For	
	β	α
9	$-.62 \leq \beta \leq -.05$	$7.63 \leq \alpha \leq 13.31$
11	$.02 \leq \beta \leq .03$	$.25 \leq \alpha \leq .46$
13	$8.64 \leq \beta \leq 30.77$	$11.30 \leq \alpha \leq 446.48$

39. H_0: $\beta = 0$ and H_A: $\beta \neq 0$. Given $n - 2 = 22$ degrees of freedom, Appendix Table K implies a critical value of $\pm t_{.025,22} = \pm 2.074$. Hence the decision rule is "Accept H_0 if $-2.074 \leq t \leq +2.074$." The actual value of t is

$$t = \frac{b}{\frac{s_{Y\cdot X}}{\sqrt{\Sigma X^2 - n\bar{X}^2}}} = \frac{-1.11836}{\frac{6.021974}{\sqrt{159.86 - 24(.13)^2}}}$$
$$= -2.345.$$

H_0 should be *rejected*.

41. See Table A.90.

43. See Table A.91.

Table A.90

Problem	b	Actual t	Critical t	Result
3	96.692	3.818	± 2.306	Reject H_0.
5	17.377	4.134	± 2.179	Reject H_0.
7	$-.072$	-8.114	± 2.145	Reject H_0.
9	$-.332$	-2.762	± 2.365	Reject H_0.
11	.024	13.644	± 2.306	Reject H_0.

Table A.91

Problem	b	Actual t	Critical t	Result
13	19.706	4.106	± 2.306	Reject H_0.
15	.358	13.122	± 2.365	Reject H_0.
17	4.211	7.596	± 2.306	Reject H_0.
19	.684	58.030	± 2.074	Reject H_0.

45. See Table A.92.

Table A.92

Problem	$r^2 = \dfrac{RSS}{Total\ SS}$	$r = \sqrt{r^2}$
1	.20	−.45
3	.65	.80
5	.59	.77
7	.82	−.91
9	.52	−.72

Problem 1: Only 20 percent of the variation in Y (the presidential voting percentage captured by the incumbent party) can be explained by the association of Y with X (the change in the election-year unemployment rate for which the incumbent party might be blamed).

47. See Table A.93.

Table A.93

Problem	$r^2 = \dfrac{RSS}{Total\ SS}$	$r = \sqrt{r^2}$
11	.96	.98
13	.68	.82
15	.96	.98
17	.88	.94
19	.994	.997

49. a. For $F_{.05}$ and $d.f. = (1,22)$, Appendix Table N suggests "Accept H_0 if $F \leq 4.30$." The ANOVA table is Table A.94.

Table A.94

Source of Variation	Sum of Squares	Degrees of Freedom	Mean Square	Test Statistic
Regression	199.41	1	199.41	$F = 5.50$
Error	797.81	22	36.26	
Total	997.22	23		

Reject H_0 for problem 1, as in problem 39.

b. For $F_{.05}$ and $d.f. = (1,8)$, Appendix Table N suggests "Accept H_0 if $F \leq 5.32$." The ANOVA table is Table A.95.

Table A.95

Source of Variation	Sum of Squares	Degrees of Freedom	Mean Square	Test Statistic
Regression	1,010,663.0	1	1,010,663.00	$F = 14.58$
Error	554,666.6	8	69,333.33	
Total	1,565,329.6	9		

Reject H_0 for problem 3, as in problem 41.

51. a. For $F_{.05}$ and $d.f. = (1,12)$, Appendix Table N suggests "Accept H_0 if $F \leq 4.75$." The ANOVA table is Table A.96.

Table A.96

Source of Variation	Sum of Squares	Degrees of Freedom	Mean Square	Test Statistic
Regression	55,217.72	1	55,217.72	$F = 17.09$
Error	38,767.97	12	3,230.66	
Total	93,985.69	13		

Reject H_0 for problem 5, as in problem 41.

b. For $F_{.05}$ and $d.f. = (1,14)$, Appendix Table N suggests "Accept H_0 if $F \leq 4.60$." The ANOVA table is Table A.97.

Table A.97

Source of Variation	Sum of Squares	Degrees of Freedom	Mean Square	Test Statistic
Regression	130.95	1	130.95	$F = 65.83$
Error	27.85	14	1.99	
Total	158.79	15		

Reject H_0 for problem 7, as in problem 41.

53. a. For $F_{.05}$ and $d.f. = (1,7)$, Appendix Table N suggests "Accept H_0 if $F \leq 5.59$." The ANOVA table is Table A.98.

Table A.98

Source of Variation	Sum of Squares	Degrees of Freedom	Mean Square	Test Statistic
Regression	5.70	1	5.70	$F = 7.63$
Error	5.23	7	.75	
Total	10.92	8		

Reject H_0 for problem 9, as in problem 41.

b. For $F_{.05}$ and $d.f. = (1,8)$, Appendix Table N suggests "Accept H_0 if $F \leq 5.32$." The ANOVA table is Table A.99.

Table A.99

Source of Variation	Sum of Squares	Degrees of Freedom	Mean Square	Test Statistic
Regression	1.04	1	1.04	$F = 186.16$
Error	.05	8	.0056	
Total	1.09	9		

Reject H_0 for problem 11, as in problem 41.
c. For $F_{.05}$ and $d.f. = (1,8)$, Appendix Table N suggests "Accept H_0 if $F \leq 5.32$." The ANOVA table is Table A.100.

Table A.100

Source of Variation	Sum of Squares	Degrees of Freedom	Mean Square	Test Statistic
Regression	243,094.1	1	243,094.1	$F = 16.86$
Error	115,336.3	8	14,417.04	
Total	358,430.4	9		

Reject H_0 for problem 13, as in problem 43.
55. a. For $F_{.05}$ and $d.f. = (1,7)$, Appendix Table N suggests "Accept H_0 if $F \leq 5.59$." The ANOVA table is Table A.101.

Table A.101

Source of Variation	Sum of Squares	Degrees of Freedom	Mean Square	Test Statistic
Regression	1,721.38	1	1,721.38	$F = 172.17$
Error	69.98	7	10.00	
Total	1,791.36	8		

Reject H_0 for problem 15, as in problem 43.
b. For $F_{.05}$ and $d.f. = (1,8)$, Appendix Table N suggests "Accept H_0 if $F \leq 5.32$." The ANOVA table is Table A.102.

Table A.102

Source of Variation	Sum of Squares	Degrees of Freedom	Mean Square	Test Statistic
Regression	14,922.21	1	14,922.21	$F = 57.71$
Error	2,068.71	8	258.59	
Total	16,990.92	9		

Reject H_0 for problem 17, as in problem 43.

c. For $F_{.05}$ and $d.f. = (1,22)$, Appendix Table N suggests "Accept H_0 if $F \leq 4.30$." The ANOVA table is Table A.103.

Table A.103

Source of Variation	Sum of Squares	Degrees of Freedom	Mean Square	Test Statistic
Regression	712,449.3	1	712,449.3	$F = 3,367.5$
Error	4,654.5	22	211.6	
Total	717,103.7	23		

Reject H_0 for problem 19, as in problem 43.
57. $\hat{Y}_X = -26.79262 + 15.14805 X$.
$s_{Y \cdot X} = 21.38662$
$r^2 = .8072$
For $F_{.05}$ and $d.f. = (1,16)$, Appendix Table N suggests "Accept H_0 if $F \leq 4.49$." The ANOVA table is Table A.104.

Table A.104

Source of Variation	Sum of Squares	Degrees of Freedom	Mean Square	Test Statistic
Regression	30,648.28	1	30,648.28	$F = 67.01$
Error	7,318.20	16	457.39	
Total	37,966.48	17		

H_0 should be *rejected*.
59. See Table A.105. Differences are due to rounding.

Table A.105

XY	X^2	$\hat{Y}_X$	$Y - \hat{Y}_X$	Y^2
220.5	2,025	4.67	.23	24.01
330.4	3,481	6.01	−.41	31.36
697.0	6,724	8.20	.30	72.25
616.0	6,400	8.01	−.31	59.29
504.1	5,041	7.15	−.05	50.41
366.0	3,600	6.10	0	37.21
319.0	3,025	5.63	.17	33.64
489.9	4,761	6.96	.14	50.41
600.4	6,241	7.92	−.32	57.76
921.5	9,025	9.44	.26	94.09
5,064.8 $= \Sigma XY$	50,323 $= \Sigma X^2$		$\Sigma(Y - \hat{Y}_X) = 0$, except for rounding	510.43 $= \Sigma Y^2$

$$\Sigma X = 695, \overline{X} = 69.5, \Sigma Y = 70.1, \overline{Y} = 7.01.$$

$$b = \frac{5{,}064.8 - 10(69.5)(7.01)}{50{,}323 - 10(4{,}830.25)} = \frac{192.85}{2{,}020.5} = .0954$$

$$a = 7.01 - (.0954)(69.5) = .3797$$

From Box 13.J,

$$r^2 = \frac{.3797(70.1) + .0954(5,064.8) - 10(7.01)^2}{510.43 - 10(7.01)^2}$$

$$= .967.$$

Chapter 14

1. $\hat{Y} = 27.3595 + .0337 X_1 - 1.2383 X_2$
    (4.4477) (3.5919) (-3.7682)
$R^2 = .8027$ $\bar{R}^2 = .7869$
The ANOVA table is Table A.106.

Table A.106

Source of Variation	Sum of Squares	Degrees of Freedom	Mean Square	Test Statistic
Regression	370.01	2	185.00	$F = 50.85$
Error	90.95	25	3.64	
Total	460.96	27		

Critical $F_{.05}$ for $d.f. = (2,25) = 3.39$. *Reject* H_0: $\beta_1 = \beta_2 = 0$.

3. $\hat{Y} = 4.0231 + .0848 X_1 - 3.8750 D_1$
    (1.2556) (3.9120) (-2.1685)
$R^2 = .8012$ $\bar{R}^2 = .7651$.
The ANOVA table is Table A.107.

Table A.107

Source of Variation	Sum of Squares	Degrees of Freedom	Mean Square	Test Statistic
Regression	332.37	2	166.19	$F = 22.17$
Error	82.46	11	7.50	
Total	414.83	13		

Critical $F_{.05}$ for $d.f. = (2,11) = 3.98$. *Reject* H_0: $\beta_1 = \beta_2 = 0$.

5. $\hat{Y} = 313.5316 + 1.5120 X_1 - 4.9085 X_2$
    (2.6985) (1.9284) (-2.0893)
$R^2 = .7014$ $\bar{R}^2 = .6161$
The ANOVA table is Table A.108.

Table A.108

Source of Variation	Sum of Squares	Degrees of Freedom	Mean Square	Test Statistic
Regression	67,121.21	2	33,560.60	$F = 8.22$
Error	28,568.80	7	4,081.26	
Total	95,690.01	9		

Critical $F_{.05}$ for $d.f. = (2,7) = 4.74$. *Reject* H_0: $\beta_1 = \beta_2 = 0$.

7. $\hat{Y} = -29.7684 + 6.6804 X_1 + 9.9322 X_2$
    ($-.3222$)   (.9656)   (21.5428)
$R^2 = .9752$ $\bar{R}^2 = .9710$
The ANOVA table is Table A.109.

Table A.109

Source of Variation	Sum of Squares	Degrees of Freedom	Mean Square	Test Statistic
Regression	14,071,500	2	7,035,750	$F = 235.60$
Error	358,358.8	12	29,863.23	
Total	14,429,858.8	14		

Critical $F_{.05}$ for $d.f. = (2,12) = 3.89$. *Reject* H_0: $\beta_1 = \beta_2 = 0$.

9. $\hat{Y} = -111.4061 + .9286 X_1 + .6844 X_2$
    (-2.2822)   (.8427)   (1.8663)
$R^2 = .8824$ $\bar{R}^2 = .8354$.
The ANOVA table is Table A.110.

Table A.110

Source of Variation	Sum of Squares	Degrees of Freedom	Mean Square	Test Statistic
Regression	1,275.99	2	638.00	$F = 18.76$
Error	170.01	5	34.00	
Total	1,446.00	7		

Critical $F_{.05}$ for $d.f. = (2,5) = 5.79$. *Reject* H_0: $\beta_1 = \beta_2 = 0$.

11. $\hat{Y} = 96.2144 + .0891 X_1 + 43.2533 D_1$
    (1.8197) (.2389)   (1.6825)
$R^2 = .3695$ $\bar{R}^2 = .2293$.
The ANOVA table is Table A.111.

Table A.111

Source of Variation	Sum of Squares	Degrees of Freedom	Mean Square	Test Statistic
Regression	4,399.75	2	2,199.88	$F = 2.64$
Error	7,508.92	9	834.32	
Total	11,908.67	11		

Critical $F_{.05}$ for $d.f. = (2,9) = 4.26$. *Accept* H_0: $\beta_1 = \beta_2 = 0$.

13. $\hat{Y} = 142.4233 - 1.8289X_1 - 127.3630D_1$
$\quad\quad (3.3380)\ (-.1294)\quad\quad (-2.4030)$
$R^2 = .4963\quad\quad \overline{R}^2 = .3523$
The ANOVA table is Table A.112.

Table A.112

Source of Variation	Sum of Squares	Degrees of Freedom	Mean Square	Test Statistic
Regression	40,544.16	2	20,272.08	$F = 3.45$
Error	41,156.00	7	5,879.43	
Total	81,700.16	9		

Critical $F_{.05}$ for $d.f. = (2,7) = 4.74$. *Accept* H_0:
$\beta_1 = \beta_2 = 0$.

15. $\hat{Y} = 11.4382 - .0328\,X_1 + 1.0701\,D_1$
$\quad\quad (2.5444)(-.5995)\quad\quad (.7003)$
$R^2 = .1092\quad \overline{R}^2 = -.3362$
The ANOVA table is Table A.113.

Table A.113

Source of Variation	Sum of Squares	Degrees of Freedom	Mean Square	Test Statistic
Regression	.51	2	.26	$F = .25$
Error	4.20	4	1.05	
Total	4.71	6		

Critical $F_{.05}$ for $d.f. = (2,4) = 6.94$. *Accept* H_0:
$\beta_1 = \beta_2 = 0$.

17. $\hat{Y} = 500 - 1\,X_1 - .00027\,D_1$
$\quad\quad (\infty)\quad\ (\infty)\,(-2.66185)$
$R^2 = 1\quad \overline{R}^2 = 1$
The ANOVA table is Table A.114.

Table A.114

Source of Variation	Sum of Squares	Degrees of Freedom	Mean Square	Test Statistic
Regression	170,930.8	2	85,465.4	$F = 5.7$ trillion
Error	.0	9	.0	
Total	170,930.8	11		

Critical $F_{.05}$ for $d.f. = (2,9) = 4.26$. *Reject* H_0:
$\beta_1 = \beta_2 = 0$.

19. $\hat{Y} = 230.6815 - 4.1441X_1 + 144.7237D_1$
$\quad\quad (1.4056)\quad (-.3970)\quad\quad (.8267)$
$R^2 = .1171\quad \overline{R}^2 = -.1352$
The ANOVA table is Table A.115.

Table A.115

Source of Variation	Sum of Squares	Degrees of Freedom	Mean Square	Test Statistic
Regression	67,376.22	2	33,688.11	$F = .46$
Error	508,152.20	7	72,593.18	
Total	575,528.42	9		

Critical $F_{.025}$ for $d.f. = (2,7) = 6.54$. *Accept* H_0:
$\beta_1 = \beta_2 = 0$.

21. $\hat{Y} = 23.1063 - .0138X_1 + .0143X_2$
$\quad\quad (2.4007)\ (-.7788)\quad (.0611)$

$\quad\quad + .0114X_3 - 1.6115\,X_4 + 20.7406\ D_1$
$\quad\quad\ (.6883)\ (-1.6771)\quad\quad (2.5624)$

$R^2 = .9121\quad \overline{R}^2 = .8022$
The ANOVA table is Table A.116.

Table A.116

Source of Variation	Sum of Squares	Degrees of Freedom	Mean Square	Test Statistic
Regression	2,307.62	5	461.52	$F = 8.30$
Error	222.38	4	55.59	
Total	2,530	9		

Critical $F_{.025}$ for $d.f. = (5,4) = 9.36$. *Accept* H_0:
$\beta_1 = \beta_2 = \beta_3 = \beta_4 = \beta_5 = 0$.

23. $\hat{Y} = 101.7044 - 10.1697X_1 - .6381X_2$
$\quad\quad (4.5223)\quad (-2.5110)\ (-.0966)$
$R^2 = .4800\quad \overline{R}^2 = .3315$
The ANOVA table is Table A.117.

Table A.117

Source of Variation	Sum of Squares	Degrees of Freedom	Mean Square	Test Statistic
Regression	3,285.95	2	1,642.98	$F = 3.23$
Error	3,559.10	7	508.44	
Total	6,845.05	9		

Critical $F_{.01}$ for $d.f. = (2,7) = 9.55$. *Accept* H_0:
$\beta_1 = \beta_2 = 0$.

25. See Table A.118.
27. See Table A.119.
29. a. Box 14.B applies. The DAYSAL-2 computer program shows a point estimate of 60.76 and an interval of $41.81 \leq \mu_{Y\cdot12} \leq 79.71$.

Table A.118

Problem	Sample Standard Error of the Estimate of Y
1	1.9074
2	281.3748
3	2.7380
4	46.9302
5	63.8847
6	1.2469
7	172.8098
8	.8179

Table A.119

Problem	Sample Standard Error of the Estimate of Y
17	.000122
18	16.2215
19	269.4312
20	20.6976
21	7.4562
22	.1293
23	22.5487
24	28.9192

b. Box 14.C applies. The DAYSAL-2 computer program shows a point estimate of 60.76 and an interval of $4.18 \leq I_{Y \cdot 12} \leq 117.35$.

31. Using the same procedure as in problem 29, the point estimate is 42.83, and the intervals are $23.39 \leq \mu_{Y \cdot 12345} \leq 62.26$ and $14.44 \leq I_{Y \cdot 12345} \leq 71.22$.

33. Using the same procedure as in problem 29, the point estimate is 180.95, and the intervals are $-79.32 \leq \mu_{Y \cdot 12} \leq 441.22$ and $-507.36 \leq I_{Y \cdot 12} \leq 869.26$.

35. Box 14.D applies. The answer to problem 17 implies sample standard errors of $s_{b_1} = 0$ and $s_{b_2} = .0001$ (coefficient divided by its t value); a critical $t_{\alpha/2}$ for 9 degrees of freedom is 2.262 (Appendix Table K). Hence
$\beta_1 = 1 \pm [2.262(0)]$ and $1 \leq \beta_1 \leq 1$.
$\beta_2 = -.00027 \pm [2.262(.0001)]$ and $-.0005 \leq \beta_2 \leq -.00004$.

37. Using the same procedure as in problem 35,
$\beta_1 = -.0328 \pm [2.776(.0547)]$ and $-.1846 \leq \beta_1 \leq +.1190$.
$\beta_2 = 1.0701 \pm [2.776(1.5281)]$ and $-3.1719 \leq \beta_2 \leq +5.3121$.

39. Given the Box 14.I formula,

$$r^2_{Y1 \cdot 2} = \frac{.845 - .072}{1 - .072} = .833.$$

$$r^2_{Y2 \cdot 1} = \frac{.845 - .715}{1 - .715} = .456.$$

41. The best alternative is the one with the highest regression sum of squares and, thus, the lowest error sum of squares. Table A.120 shows the crucial data for the two regressions.

43. $\hat{Y} = -7.0969 + 1.9334 X_1 + .6494 X_2$
$\quad\quad (-1.8591) \quad (8.8643) \quad (3.6991)$

$R^2_{Y \cdot 12} = .8450.$

$\hat{Y} = 1.5398 + 1.8440 X_1 + .0156 X_3$
$\quad\quad (.2950) \quad (6.3188) \quad (.1696)$

$R^2_{Y \cdot 13} = .7157.$

45. See Table A.121.

Table A.120

Regressing Y against X_1 and	Regression Sum of Squares	Error Sum of Squares	Total Sum of Squares
X_2	1,538.226	282.2003	1,820.426
X_3	1,302.936	517.4903	1,820.426

Table A.121

Source of Variation	Sum of Squares	Degrees of Freedom	Mean Square	Test Statistic
Regression	$RSS = 237.797$	$m = 1$	$RMS = 237.797$	$F = \dfrac{RMS}{EMS} = 22.6$
Error	$ESS = 52.5$	$n - m - 1 = 5$	$EMS = 10.5$	
Total	290.297	$n - 1 = 7$		

Given the critical $F_{.025}$ for *d.f.* = (1,5) = 10.01, the result is highly significant. (In fact, the *p* value is .005). The value of r^2 is (237.797/290.297) = .8192 (which agrees with the chapter, except for rounding).

47. For α = .01, n = 25, and m = 4, we find d_L = .83 and d_U = 1.52. Because $4 - 1.52 \le 2.95 \le 4 - .83$, we are uncertain.

49. See Table A.122. Note the strong correlation (in excess of .80) between the independent variables in problems 9 and 10, which suggests multicollinearity. Problem 9 also involves a high R^2 = .88, yet insignificant *t* values. The other signs of multicollinearity, such as extreme values and meaningless signs of coefficients, are generally absent.

Table A.122

Problem	Correlation Between X_1 and X_2	Problem	Correlation Between X_1 and X_2
1	−.73	7	.08
2	−.27	8	.73
3	−.57	9	.90
4	−.61	10	.88
5	−.51	11	.75
6	−.66	12	.70

51. a. *Caution:* The parenthetical values are standard errors, thus, *t* values must be calculated by dividing the coefficients by these standard errors. In the order given, these ratios equal 1.59, −1.07, .39, .06, −.28, .67. They are hardly significant. Indeed, the computer shows corresponding *p* values of .073, .156, .353, .476, .394, .260. (The probabilities of getting the results by pure chance are 7.3 percent, 15.6 percent, and so on.)

b. We can calculate the *F* statistic and compare it with a critical value of, say $F_{.05}^{5,9}$ = 3.48 (Appendix Table N). (There are m = 5 degrees of freedom associated with the numerator, or *RSS*, and $n - m - 1 = 15 - 5 - 1 = 9$ degrees of freedom associated with the denominator, or *ESS*.) Thus,

$$RSS = 9,773 - 8,337 = 1,436$$
$$RMS = (1,436/5) = 287.2$$
$$EMS = (8,337/9) = 926.\overline{33}$$
$$F = (287.2/926.\overline{33}) = .31$$

This result does not show statistical significance. Indeed, the associated *p* value is .897.

53. a. $\hat{C} = -11.46 + .9589\ DPI$; *t* values in parentheses.
$$(-.95)\ (11.7010)$$

b. It is the marginal propensity to consume. For each \$1 change in disposable personal income, personal consumption expenditures change by about 96 cents.

c. See Figure A.31.

Figure A.31

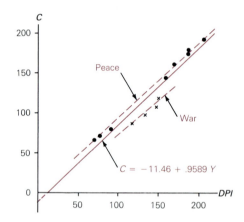

d. The equation for *war years only* is

$$\hat{C}_w = -12.62 + .8565\ DPI;\ t\ \text{values in parentheses.}$$
$$(-.53)\ (4.8781)$$

The equation for *peace years only* is

$$\hat{C}_p = -1.39 + .9413\ DPI;\ t\ \text{values in parentheses.}$$
$$(-.37)\ (38.7450)$$

The aggregate consumption function's intercept as well as slope declined during the war. (See again Figure A.31).

e. $s_{C\cdot Y}$ = 12.23 (all years), 4.473 (war years only), 3.561 (peace years only).

f. The ANOVA tables are given in Parts A–C of Table A.123.
All three regressions are highly significant; r^2 = (*RSS*/Total *SS*) = .93 (all years), .92 (war years), and .996 (peace years).

55. a. The *ratios* of *b/t* are 8.563, .491, .498, 1.053, .517, .181, .0001117, .01309, .01399, .03963. (These

Table A.123

Source of Variation	Sum of Squares	Degrees of Freedom	Mean Square	Test Statistic
A. All Years				
Regression	20,489.693	1	20,489.693	$F = 136.91$
Error	1,496.547	10	149.6547	
Total	21,986.24	11		

Given a critical $F_{.05}$ for $d.f. = (1,10) = 4.96$, *reject* H_0: $\beta_1 = 0$.

		B. War Years		
Regression	476.17637	1	476.17637	$F = 23.80$
Error	40.02113	2	20.010565	
Total	516.1975	3		

Given a critical $F_{.05}$ for $d.f. = (1,2) = 18.51$, *reject* H_0: $\beta_1 = 0$.

		C. Peace Years		
Regression	19,039.721	1	19,039.721	$F = 1,501.17$
Error	76.09928	6	12.683213	
Total	19,115.82	7		

Given a critical $F_{.05}$ for $d.f. = (1,6) = 5.99$, *reject* H_0: $\beta_1 = 0$

are computer printouts; your answer may differ slightly because of rounding.)

b. 52.2 percent, as shown by R^2.

57. a. PDR_t would *rise* by $.1(70) = 7$ to 12 per 100,000.

b. P_{LIB}/P_{IND} would *rise* by $.0071(.5) = .0355$ to 10.0355.

59. a. $SPEND/TAX$ will *rise* by $.20(25) = 5$.

b. Yes, they would. Each of the alternative measures of decentralization (R, E, and L) has a negative sign: more decentralization, smaller public sector; less decentralization, larger public sector.

Chapter 15

1. The missing values are: $I_t = .94$, $Y_t = \$26,647.50$, $S_t = .40$, $C_t = .89$. Because, in this model, $Y_t = T_t \cdot C_t \cdot S_t \cdot I_t$, the missing I_t (row 1) is found as $Y_t/(T_t \cdot C_t \cdot S_t)$. The Y_t equation applies directly in row 2, while the missing S_t (row 3) is found as $Y_t/(T_t \cdot C_t \cdot I_t)$ and the missing C_t (row 4) as $Y_t/(T_t \cdot S_t \cdot I_t)$.

3. a. The 10-year moving-averages series (going from 1944–1981 in successive *columns*) is

172.47	278.89	402.535
193	283.34	428.7
213.2	287.88	462.21
232.085	292.62	501.13
247.25	297.895	544.995
256.755	304.175	594.915
260.805	312.125	651.22
261.61	320.595	716.425
262.465	329.145	797.585
264.615	339.56	897.7851
267.76	352.09	1017.16
271.22	366.67	1154.125
274.8	383.02	

b. Between 1980 and 1981, the debt rose at a rate of 136.965 per year; hence a 1989 forecast is $1,154.125 + [8(136.965)] = 2.249.845$ (billions of dollars).

5. a. The 6-year moving-averages series (going from 1962–1981 in successive *columns*) is

1480.333	1618.083	1669.833
1468.333	1778	1641.667
1456.417	1823.333	1632.417
1457.167	1779.917	1586.833
1450.25	1759.083	1523.083
1431.333	1758.75	1475.25
1471.583	1723.25	

b. Between 1980 and 1981, housing starts *fell* at a rate of 47.833; hence a 1989 forecast is $1{,}475.25 - [8(47.833)] = 1{,}092.586$ (thousand units).

7. a. The 14-year moving-averages series (going from 1954–1976 in successive *columns*) is

67.6536	92.6357	127.525
69.9214	96.6429	132.7286
72.3929	100.575	138.275
75.08923	104.6678	143.9893
77.9964	109.1071	149.8
81.2	113.6929	155.6571
84.7607	118.2036	161.5214
88.6250	122.7143	

b. Between 1975 and 1976, the service-industry contribution to GNP rose at a rate of 5.8643; hence a 1989 forecast is $161.5214 + [13(5.8643)] = 237.7573$ (billion 1972 dollars).

9. a. The 4-year moving-averages series (going from 1949–1981 in successive *columns*) is

8.825	12.55	18.225
9.3375	12.7	20.75
9.7375	12.9	25.25
10.25	13.1	30.925
10.725	13.375	37.1375
11.4125	13.7375	42.825
12.1875	14.1625	48.800
12.7375	14.725	58.925
12.9125	15.475	76.1625
12.8	16.25	95.9125
12.6	17.0625	110.75

b. Between 1980 and 1981, the mining industry contribution to GNP rose at a rate of 14.8375 per year; hence a 1989 forecast is $110.75 + [8(14.8375)] = 229.45$ (billion 1972 dollars).

11. a. The 4-year moving averages series (going from 1962–1984 in successive *columns*) is

24.8	57.7	184.325
26.25	66.6875	207.8125
27.6875	77.35	236.925
29.8625	91.1625	270.2125
33.1125	108.8375	301.0375
37.5625	128.475	325.2125
43	148.0125	345.2625
49.7125	165.925	

b. Between 1983 and 1984, the transfer payments rose at a rate of 20.05; hence a 1989 forecast is $345.2625 + [5(20.05)] = 445.5125$ (billion dollars).

13. The 1987 forecast is 2,015.02 billion dollars. The use of $\alpha = .8$ is preferable: The error sum of squares is 2,774,340 when $\alpha = .2$ is used; it is only 427,407.3 when $\alpha = .8$ is used.

15. The 1985 forecast is 1,740.84 thousand units. The use of $\alpha = .9$ is preferable: The error sum of squares is 2,932,269 when $\alpha = .1$ is used; it is only 2,620,507 when $\alpha = .9$ is used.

17. The 1984 forecast is 204.28 billions of 1972 dollars. The use of $\alpha = .7$ is preferable: The error sum of squares is 7,467.688 when $\alpha = .3$ is used; it is only 1,668.609 when $\alpha = .7$ is used.

19. The 1984 forecast is 113.58 billions of 1972 dollars. The use of $\alpha = .5$ is preferable: The error sum of squares is 7,887.674 when $\alpha = .4$ is used; it is only 6,105.464 when $\alpha = .5$ is used.

21. The 1987 forecast is 373.38 billion dollars. The use of $\alpha = .8$ is preferable: The error sum of squares is 174,257.9 when $\alpha = .15$ is used; it is only 12,445.12 when $\alpha = .8$ is used.

23. The 1985 forecast is 1,948.88 thousand units. Note the last row of Table A.124. (If you have done it, see problem 14 for comparison).

25. The 1984 forecast is 147.76 billions of 1972 dollars. Note the last row of Table A.125 on page A-72. (If you have done it, see problem 18 for comparison.)

27. a. The ANOVA table is Table A.126 on page A-72. Given a critical $F_{.05}$ for *d.f.* = $(1,18) = 4.41$, reject H_0: $\beta_1 = 0$.
b. $r^2 = .69$.

29. A scatter diagram suggesting a linear regression, is shown in Figure A.32 on page A-73 as $\hat{C}_X = 677.27 + 25.2X$ (with $X = 0$ at mid-1970). A 1988 (or $X = 18$) forecast comes to 1,130.87 (billion 1972 dollars).

Table A.124

Year, t	Actual Value, Y_t	Forecast Value, F_t	Error, $Y_t - F_t$
1959	1554.00		
1960	1296.00		
1961	1365.00	1296.00	69.00
1962	1493.00	1047.66	445.34
1963	1635.00	854.77	780.23
1964	1561.00	726.57	834.43
1965	1510.00	637.18	872.82
1966	1196.00	586.53	609.47
1967	1322.00	533.93	788.07
1968	1545.00	530.71	1014.29
1969	1500.00	590.69	909.31
1970	1469.00	676.54	792.46
1971	2085.00	782.40	1302.60
1972	2379.00	991.38	1387.62
1973	2058.00	1264.37	793.63
1974	1353.00	1509.70	−156.70
1975	1171.00	1653.74	−482.74
1976	1548.00	1745.86	−197.86
1977	2002.00	1858.55	143.45
1978	2036.00	2011.11	24.89
1979	1760.00	2152.81	−392.81
1980	1313.00	2237.03	−924.03
1981	1100.00	2231.17	−1131.17
1982	1072.00	2159.34	−1087.34
1983	1713.00	2048.41	−335.41
1984	1751.00	1999.25	−248.25
1985		1948.88	

31. The regression (with t values in parentheses) is

$$\hat{Y}_X = 113.7242 + 25.6180\,X \quad \text{and } r^2 = .64.$$
$$(-1.4677)\quad(9.0193)$$

The 1989 ($X = 50$) forecast is 1,167.1758.

33. a. $s_{Y \cdot X_1} = 272.6152$

b. $-265.74 \le \alpha \le 38.29$
$20.05 \le \beta \le 31.19$
Note: This is a *large-sample* test, use $z_{\alpha/2} = 1.96$.

c. The critical $z_{\alpha/2} = \pm 1.96$. The coefficients, standard errors, and t values are
for a: $-113.72,\ 77.48,\ -1.47.\ H_0$ is accepted.
for b: $25.62,\ 2.84,\ 9.02.\ H_0$ is rejected.

35. a. The ANOVA table is Table A.127 on page A-73.
Given a critical $F_{.05}$ for $d.f. = (1,24) = 4.26$, *accept* H_0: $\beta = 0$.

b. $1,241.21 \le \alpha \le 1,769.30$
$-13.04 \le \beta \le 23.19$

37. a. $29.85 \le \alpha \le 42.10$
$3.98 \le \beta \le 4.56$
Note: This is a *large-sample* test; use $z_{\alpha/2} = \pm 1.96$.

b. The ANOVA table is Table A.128 on page A-73.
Given a critical $F_{.05}$ for $d.f. = (1,35) = 4.13$, *reject* H_0: $\beta = 0$.

39. a. The ANOVA table is Table A.129 on page A-74.
Given a critical $F_{.05}$ for $d.f. = (1,35) = 4.13$, *reject* H_0: $\beta = 0$.

b. The critical $z_{\alpha/2} = \pm 1.96$. The coefficients, standard errors, and t values are
for a: $-12.59,\ 7.26,\ -1.73.\ H_0$ is accepted.
for b: $2.35,\ .35,\ 6.78.\ H_0$ is rejected.

41. The regression (with t values in parentheses) is

$$\hat{Y}_X = -49.0503 + 14.4811X \text{ and } r^2 = .89.$$
$$(-3.2359)\,(14.4777)$$

The 1989 ($X = 29$) forecast is 370.90.

43. The regression (with t values in parentheses) is

$$\hat{Y}_X = 97.7312 + 1.5754X \text{ and } r^2 = .56.$$
$$(25.4104)\,(5.0250)$$

The 1990 ($X = 25$) forecast is 137.12.

Table A.125

Year, t	Actual Value, Y_t	Forecast Value, F_t	Error, $Y_t - F_t$
1947	6.80		
1948	9.40		
1949	8.10	9.40	−1.30
1950	9.30	11.06	−1.76
1951	10.20	11.98	−1.78
1952	10.10	12.32	−2.22
1953	10.60	11.77	−1.17
1954	10.90	11.27	−0.37
1955	12.40	10.97	1.43
1956	13.40	11.85	1.55
1957	13.50	13.27	0.23
1958	12.40	14.24	−1.84
1959	12.30	13.79	−1.49
1960	12.60	13.00	−0.40
1961	12.70	12.52	0.18
1962	12.80	12.33	0.47
1963	13.10	12.41	0.69
1964	13.40	12.79	0.61
1965	13.50	13.34	0.16
1966	14.20	13.76	0.44
1967	14.60	14.43	0.17
1968	15.30	15.05	0.25
1969	16.10	15.78	0.32
1970	17.60	16.64	0.96
1971	17.40	18.06	−0.66
1972	19.00	18.63	0.37
1973	21.70	19.72	1.98
1974	32.20	22.09	10.11
1975	38.80	30.95	7.85
1976	43.00	41.42	1.58
1977	47.40	49.88	−2.48
1978	52.00	55.93	−3.93
1979	66.80	60.14	6.66
1980	96.00	70.72	25.28
1981	132.30	96.83	35.47
1982	125.10	138.37	−13.27
1983	112.40	156.17	−43.77
1984		147.76	

Table A.126

Source of Variation	Sum of Squares	Degrees of Freedom	Mean Square	Test Statistic
Regression	1,469,408	1	1,469,408	$F = 40.13$
Error	659,140.5	18	36,618.92	
Total	2,128,548.5	19		

Figure A.32

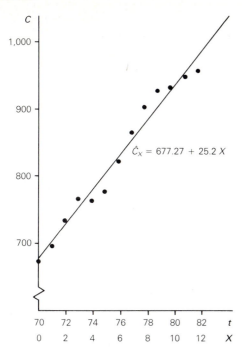

$\hat{C}_X = 677.27 + 25.2\,X$

45. a. The regression (with t values in parentheses) is

$$\hat{Y}_X = 91.90 \quad -4\overline{.66}X \text{ and } r^2 = .69.$$
$$(15.47)\ (-4.19)$$

The 1988 ($X = 13$) forecast is 31.23.

b. The ANOVA table is Table A.130.

Given a critical $F_{.05}$ for *d.f.* $= (1,8) = 5.32$, reject H_0: $\beta = 0$.

47. We require X values measured in months from a zero base, each of them centered in the middle of a month. If $X = 0$ were placed in the middle of the first month of 1968, or at mid-January, this would shift the base backward in time by 5.5 months (the original base was mid-1968, at the end of the 6th month). The original equation shows sales rising annually by $b = 47$ units; hence, they are rising by $(b/12) = (47/12)$ units per month, which is the slope of the new regression line for monthly data. Hence, the new intercept equals 200 (or mid-1968 trend sales) minus 5.5 (47/12), or 178.46 units. The transformed regression line (measuring sales monthly at annual rates) is $\hat{Y}_X = 178.46 + 3.92X$ (where X is measured in months and $X = 0$ is located at mid-January, 1968). Note how this equation accurately predicts the

Table A.127

Source of Variation	Sum of Squares	Degrees of Freedom	Mean Square	Test Statistic
Regression	37,711.38	1	37,711.38	$F = .33$
Error	2,703,212	24	112,633.8	
Total	2,740,923.38	25		

Table A.128

Source of Variation	Sum of Squares	Degrees of Freedom	Mean Square	Test Statistic
Regression	76,858.81	1	76,858.81	$F = 818.56$
Error	3,286.35	35	93.90	
Total	80,145.16	36		

Table A.129

Source of Variation	Sum of Squares	Degrees of Freedom	Mean Square	Test Statistic
Regression	23,295.78	1	23,295.78	$F = 45.91$
Error	17,758.88	35	507.40	
Total	41,054.66	36		

Table A.130

Source of Variation	Sum of Squares	Degrees of Freedom	Mean Square	Test Statistic
Regression	1,796.67	1	1,796.67	$F = 17.58$
Error	817.67	8	102.21	
Total	2,614.34	9		

Table A.131

Year	X	$Y_t = T_t \cdot C_t \cdot I_t$	T_t	$\dfrac{Y_t}{T_t} = C_t \cdot I_t$
1976	0	200,000	200,000	1.000
1977	1	300,000	250,000	1.200
1978	2	400,000	300,000	1.333
1979	3	400,000	350,000	1.143
1980	4	390,000	400,000	.975
1981	5	370,000	450,000	.822
1982	6	350,000	500,000	.700
1983	7	330,000	550,000	.600
1984	8	300,000	600,000	.500
1985	9	250,000	650,000	.385

200-unit annual trend sales of 1968 that constitute the intercept of the annual-data regression line in Figure 15.10: $\hat{Y}_{5.5} = 178.46 + 3.92(5.5) = 200$.

49. $\hat{Y}_X = 97.1404 + .3939X$ (with $X = 0$ at mid-February 1965).

51. See Table A.131. The annual data do not contain a seasonal component; thus, $Y_t = T_t \cdot C_t \cdot I_t$. The trend can be calculated by means of the equation. Thus, the ratio Y_t/T_t equals $C_t \cdot I_t$; most of its variation is probably cyclical.

53. See Table A.132. Note: $\Sigma(Y_X - \hat{Y}_X)$ should equal zero but equals -6.56 due to rounding. Thus,

$$d = \frac{2,283.3924}{5,156.7718} = .44.$$

Assuming we test H_0: "No serial correlation" against H_A: "Positive serial correlation exists," and use $\alpha = .05$, we consult Appendix Table P and find, for $n = 27$ and $m = 1$, $d_L = 1.32$ and $d_U = 1.47$. Since $d < d_L$, we reject H_0 and conclude that (positive) serial correlation does exist.

55. See Table A.133; $d = 1.68$.
Given $\alpha/2 = .025$, $m = 1$, and $n = 10$, we find no entry in Appendix Table P. Using the closest entries for $n = 15$, $d_L = .95$ and $d_U = 1.23$, one would accept H_0: No serial correlation.

57. See Table A.134. The grouping of $S \cdot I$ data is given in Table A.135.

59. Following the procedure explained in the answer to Problem 57, one can derive Table A.136.

Table A.132

Year, X	Actual Observation, Y_X	Regression Estimate, $\hat{Y}_X = 33.07(1.081)^X$	Error, $e = Y_X - \hat{Y}_X$	$e_X - e_{X-1}$	$(e_X - e_{X-1})^2$	e_X^2
0	34	33.07	.93	—	—	.8649
1	38	35.75	2.25	1.32	1.7424	5.0625
2	40	38.64	1.36	−.89	.7921	1.8496
3	40	41.77	−1.77	−3.13	9.7969	3.1329
4	44	45.16	−1.16	.61	.3721	1.3456
5	45	48.82	−3.82	−2.66	7.0756	14.5924
6	45	52.77	−7.77	−3.95	15.6025	60.3729
7	47	57.04	−10.04	−2.27	5.1529	100.8016
8	53	61.67	−8.67	1.37	1.8769	75.1689
9	61	66.66	−5.66	3.01	9.0601	32.0356
10	70	72.06	−2.06	3.60	12.9600	4.2436
11	79	77.90	1.10	3.16	9.9856	1.2100
12	97	84.21	12.79	11.69	136.6561	163.5841
13	119	91.03	27.97	15.18	230.4324	782.3209
14	126	98.40	27.60	−.37	.1369	761.7600
15	123	106.37	16.63	−10.97	120.3409	276.5569
16	124	114.99	9.01	−7.62	58.0644	81.1801
17	137	124.30	12.70	3.69	13.6161	161.2900
18	145	134.37	10.63	−2.07	4.2849	112.9969
19	148	145.25	2.75	−7.88	62.0944	7.5625
20	147	157.02	−10.02	−12.77	163.0729	100.4004
21	160	169.74	−9.74	.28	.0784	94.8676
22	172	183.48	−11.48	−1.74	3.0276	131.7904
23	196	198.35	−2.35	9.13	83.3569	5.5225
24	212	214.41	−2.41	−.06	.0036	5.8081
25	222	231.78	−9.78	−7.37	54.3169	95.6484
26	205	250.55	−45.55	−35.77	1,279.4929	2.074.8025
					2,283.3924	5,156.7718

Table A.133

e_t	$e_t - e_{t-1}$	$(e_t - e_{t-1})^2$	e_t^2
−12.00	—	—	144.00
−0.33	11.67	136.19	.11
−1.27	−.94	.88	1.61
9.80	11.07	122.54	96.04
13.77	3.97	15.76	189.61
0.13	−13.64	186.05	.02
9.70	9.57	91.58	94.09
−15.93	−25.63	656.90	253.76
−5.87	10.06	101.20	34.46
2.00	7.87	61.94	4.00
		1,373.05	817.70

Table A.134

| Time, t | Actual Carloadings, $Y_t = T \cdot C \cdot S \cdot I$ | 12-Months Moving Average | | Ratio of Actual to Moving Average (adj.) $S \cdot I$ |
		Unadjusted	Adjusted, $T \cdot C$	
1979 Jan.	392	—	—	—
Feb.	407	—	—	—
Mar.	584	—	—	—
Apr.	478	—	—	—
May	494	—	—	—
June	611	—	—	—
		476		
July	446		478	.93
		479		
Aug.	476		480	.99
		481		
Sept.	476		476	1.00
		471		
Oct.	488		470	1.04
		468		
Nov.	459		466	.98
		463		
Dec.	399		456	.88
		449		
1980 Jan.	426		447	.95
		445		
Feb.	439		443	.99
		441		
Mar.	457		440	1.04
		438		
Apr.	442		437	1.01
		435		
May	435		434	1.00
		433		
June	450		434	1.04
		434		
July	387		434	.89
		433		
Aug.	434		433	1.00
		433		
Sept.	436		433	1.01
		433		
Oct.	454		431	1.05
		428		
Nov.	439		426	1.03
		423		
Dec.	410		422	.97
		421		
1981 Jan.	410		422	.97
		423		
Feb.	437		424	1.03
		424		
Mar.	456		424	1.08
		424		
Apr.	382		423	.90
		422		
May	385		421	.91
		419		
June	424		417	1.02
		415		

Table A.134 (continued)

Time, t	Actual Carloadings, $Y_t = T \cdot C \cdot S \cdot I$	12-Months Moving Average		Ratio of Actual to Moving Average (adj.) $S \cdot I$
		Unadjusted	Adjusted, $T \cdot C$	
July	411	—	—	—
Aug.	439	—	—	—
Sept.	435	—	—	—
Oct.	441	—	—	—
Nov.	403	—	—	—
Dec.	360	—	—	—

Table A.135

Month	Ratio of Actual to Moving Average			S_t	
	1979	1980	1981	Unadjusted[a]	Adjusted[b]
Jan.	—	.95	.97	.97	.98
Feb.	—	.99	1.03	1.01	1.02
Mar.	—	1.04	1.08	1.06	1.07
Apr.	—	1.01	.90	.955	.97
May	—	1.00	.91	.955	.97
June	—	1.04	1.02	1.03	1.04
July	.93	.89	—	.91	.92
Aug.	.99	1.00	—	.995	1.01
Sept.	1.00	1.01	—	1.005	1.02
Oct.	1.04	1.05	—	1.045	1.06
Nov.	.98	1.03	—	1.005	1.02
Dec.	.88	.97	—	.925	.94
			Average:	.98875	1.00

[a]Median of row data.
[b]Unadjusted S_t times 1/.98875, equal to the seasonal indices sought.

Table A.136

Month	Ratio of Actual to Moving Average			S_t	
	1985	1986	1987	Unadjusted[a]	Adjusted[b]
Jan.	—	.999	.997	.998	1.00
Feb.	—	1.000	.998	.999	1.00
Mar.	—	.999	.992	.9955	1.00
Apr.	—	.982	.974	.978	.98
May	—	.964	.974	.969	.97
June	—	.986	.981	.9835	.99
July	.999	.996	—	.9975	1.00
Aug.	1.003	1.005	—	1.004	1.01
Sept.	1.013	1.012	—	1.0125	1.02
Oct.	.998	.997	—	.9975	1.00
Nov.	.999	.998	—	.9985	1.00
Dec.	1.000	.996	—	.998	1.00
			Average:	.99425	1.00

[a]Median of row data.
[b] Unadjusted S_t times 1/.99425; equal to the seasonal indices sought.

Chapter 16

1. a. Formula 16.A applies:

$$\text{gasoline: } \frac{\$1.40}{\$1.30}(100) = 107.7$$

$$\text{oil: } \frac{\$1.00}{\$1.20}(100) = 83.3$$

$$\text{repairs: } \frac{\$20}{\$25}(100) = 80$$

$$\text{tires: } \frac{\$30}{\$30}(100) = 100$$

b. Formula 16.B applies: $I_{84,85} = 371/4 = 92.75$. 1984 automobile operating costs were 92.75 percent of 1985 costs.

c. See Table A.137. The formula in Box 16.C applies: $I_{84,85} = 189{,}492.8/1{,}824 = 103.9$. 1984 automobile operating costs were 103.9 percent of 1985 costs. The discrepancy is explained by the heavy weighting of gasoline, the only item with a *higher* 1984 price.

Table A.137

Item	$\dfrac{P_{84}}{P_{85}} \cdot 100$	V_{84}	$\left(\dfrac{P_{84}}{P_{85}} \cdot 100\right)V_{84}$
Gasoline	107.7	1,484	159,826.8
Oil	83.3	20	1,666.0
Repairs	80	200	16,000.0
Tires	100	120	12,000.0
		1,824	189,492.8

3. a. auto: 125.8; credit card: 103.7; mortgage: 85.8; signature: 98.4.

b. 103.4

c. 94.7.

5. Formula 16.D applies:

$$I_{84,85} = \frac{\Sigma P_{85}}{\Sigma P_{84}} \cdot 100 = \frac{57.50}{52.40} \cdot 100 = 109.7.$$

7. $I_{87,86} = (58.9/58.3) \cdot 100 = 101.0$

9. See Table A.138. Formula 16.E applies. From the value and price data given, the needed 1984 quantities can be computed. Hence,

$$I_{85,84}^{P(\text{Laspeyres})} = \frac{1{,}772}{1{,}824} \cdot 100 = 97.1.$$

11. $I_{86,85}^{P(\text{Laspeyres})} = \left(\dfrac{262{,}921.80}{274{,}571.10}\right) \cdot 100 = 95.8.$

13. $I_{87,84}^{P(\text{Laspeyres})} = \left(\dfrac{230{,}715.75}{349{,}347.00}\right) \cdot 100 = 66.0.$

15. $I_{87,86}^{P(\text{Paasche})} = \left(\dfrac{764{,}450.55}{878{,}045.20}\right) \cdot 100 = 87.1.$

17. $I_{85,84}^{P(\text{Paasche})} = \left(\dfrac{274{,}571.1}{347{,}550}\right) \cdot 100 = 79.0.$

19. See Table A.139. Formula 16.G applies. From the quantity data computed in problems 14 and 9, one can find the average quantities here. Hence,

$$I_{85,84}^{P(T-Y)} = \frac{2{,}071}{2{,}109.5} \cdot 100 = 98.2.$$

Compare the answer with that of problem 9 (97.1) and problem 14 (99.0), which used 1984 and 1985 quantities, respectively.

21. Formula 16.H applies, but it is equivalent to using current-year and base-year average quantities as weights. Hence, the answer is, in this case, the same as for problem 19.

23. $I_{86,85}^{P(\text{Edgeworth})} = 91.4.$

25. $I_{87,84}^{P(\text{Edgeworth})} = 63.6.$

27. $I_{87,86}^{P(\text{Fisher})} = 88.6.$

29. $I_{85,84}^{P(\text{Fisher})} = 77.4.$

31. Using 1979 price weights:

$$I_{80,79}^{Q(\text{Laspeyres})} = \frac{775}{870} \cdot 100 = 89$$

$$I_{81,79}^{Q(\text{Laspeyres})} = \frac{870}{870} \cdot 100 = 100$$

$$I_{82,79}^{Q(\text{Laspeyres})} = \frac{970}{870} \cdot 100 = 111$$

33. $I_{86,85}^{Q(\text{Laspeyres})} = \left(\dfrac{470{,}185.75}{274{,}571.1}\right) \cdot 100 = 171.2.$

35. Using current price weights:

$$I_{80,79}^{Q(\text{Paasche})} = \frac{789}{900} \cdot 100 = 88$$

$$I_{81,79}^{Q(\text{Paasche})} = \frac{849}{888} \cdot 100 = 96$$

$$I_{82,79}^{Q(\text{Paasche})} = \frac{765}{732} \cdot 100 = 105$$

Indexes other than Laspeyres' involve more complex calculations.

37. $I_{86,85}^{Q(\text{Paasche})} = \left(\dfrac{417{,}954.7}{262{,}921.8}\right) \cdot 100 = 159.0.$

Table A.138

Item	P_{84}	P_{85}	$Q_{84} = \dfrac{V_{84}}{P_{84}}$	$P_{85} \cdot Q_{84}$	$P_{84} \cdot Q_{84}$
Gasoline	1.40	1.30	1,060	1,378	1,484
Oil	1.00	1.20	20	24	20
Repairs	20.00	25.00	10	250	200
Tires	30.00	30.00	4	120	120
				1,772	1,824

Table A.139

Item	P_{84}	P_{85}	$Q_a = \dfrac{Q_{84} + Q_{85}}{2}$	$P_{85} \cdot Q_a$	$P_{84} \cdot Q_a$
Gasoline	1.40	1.30	1,180	1,534	1,652
Oil	1.00	1.20	22.5	27	22.5
Repairs	20.00	25.00	15	375	300
Tires	30.00	30.00	4.5	135	135
				2,071	2,109.5

39. $I^{Q(\text{Fisher})}_{80,79} = \sqrt{89 \cdot 88} = 88$

$I^{Q(\text{Fisher})}_{81,79} = \sqrt{100 \cdot 96} = 98$

$I^{Q(\text{Fisher})}_{82,79} = \sqrt{111 \cdot 105} = 108$

41. $I^{Q(\text{Fisher})}_{86,85} = \sqrt{171.2(159.0)} = 165.0.$

See problems 33 and 37.

43. Consider the product of the 1985 indexes, based on 1984: The Laspeyres price index · Paasche quantity index =

$$\left(\frac{\Sigma P_{85} \cdot Q_{84}}{\Sigma P_{84} \cdot Q_{84}} \cdot 100 \right) \cdot \left(\frac{\Sigma Q_{85} \cdot P_{85}}{\Sigma Q_{84} \cdot P_{85}} \cdot 100 \right).$$

Canceling terms and dividing by 100 does yield the value index:

$$\frac{\Sigma P_{85} \cdot Q_{85}}{\Sigma P_{84} \cdot Q_{84}} \cdot 100.$$

Now consider that the Paasche price index · Laspeyres quantity index =

$$\left(\frac{\Sigma P_{85} \cdot Q_{85}}{\Sigma P_{84} \cdot Q_{85}} \cdot 100 \right) \cdot \left(\frac{\Sigma Q_{85} \cdot P_{84}}{\Sigma Q_{84} \cdot P_{84}} \cdot 100 \right).$$

Canceling terms and dividing by 100 does yield the same value index:

$$\frac{\Sigma P_{85} \cdot Q_{85}}{\Sigma P_{84} \cdot Q_{84}} \cdot 100.$$

45. See Table A.140. The formula in Box 16.C applies:

$$I_{84,85} = \frac{239,512}{2,370} = 101.1.$$

1984 automobile operating costs were 101.1 percent of 1985 costs, according to this calculation. This contradicts the equally logical computation in (1c) with 1984 weights.

Table A.140

Item	$\dfrac{P_{84}}{P_{85}} \cdot 100$	V_{85}	$\left(\dfrac{P_{84}}{P_{85}} \cdot 100 \right) V_{85}$
Gas	107.7	1,690	182,013
Oil	83.3	30	2,499
Repairs	80	500	40,000
Tires	100	150	15,000
		2,370	239,512

47. Any price set in which food prices were four times as large as machinery prices would do the trick.

49. Time-reversal test: yes

$$I_{85,84} = \frac{57.50}{52.40} \cdot 100 = 109.7$$

$$I_{84,85} = \frac{52.40}{57.50} \cdot 100 = 91.1$$

$$\frac{109.7}{100} \cdot \frac{91.1}{100} = 1.00$$

Factor-reversal test: not applicable
Circularity test: yes

$$I_{85,84} = \frac{57.50}{52.40} \cdot 100 = 109.73$$

$$I_{86,85} = \frac{67.00}{57.50} \cdot 100 = 116.52$$

$$I_{86,84} = \frac{67.00}{52.40} \cdot 100 = 127.86$$

$$\frac{I_{85,84} \cdot I_{86,85}}{100} = \frac{109.73(116.52)}{100} = 127.86$$

51. No. While $I_{87,86}^{P(\text{Paasche})} = 87.1$, $I_{86,87}^{P(\text{Paasche})} = \left(\frac{417,954.7}{376,414.85}\right) \cdot 100 = 111.0$. But $.871(1.11) = .967 \neq 1$.

53. No. While $I_{87,86}^{P(\text{Laspeyres})} = 90.1$ and $I_{87,86}^{Q(\text{Laspeyres})} = 210.1$, $I_{87,86}^{V} = \left(\frac{764,450.55}{417,954.7}\right) \cdot 100 = 182.9$, but $[90.1(210.1)]$: $100 = 189.3$.

55. Yes. While $I_{87,86}^{P(\text{Fisher})} = 88.6$ and $I_{87,86}^{Q(\text{Fisher})} = 206.6$, $I_{87,86}^{V} = \left(\frac{764,450.55}{417,954.7}\right) \cdot 100 = 182.9$, just as $[88.6(206.6)]$: $100 = 183.0$, the same answer, except for rounding.

57. No. Note how $\frac{I_{85,84}^{P(\text{Paasche})} \times I_{86,85}^{P(\text{Paasche})} \times I_{87,86}^{P(\text{Paasche})}}{100 \times 100} = 61.2$ and *not* 62.9 (which is $I_{87,84}^{P(\text{Paasche})}$).

59. a. 92, 95, 98, 100, 105, 111, 97, 95, 103
b. 95, 98, 101, 103, 108, 114, 100, 98, 106
c. 125, 138, 174, 150, 100, 75

Chapter 17

1. See Table A.141. The maximin action is A_1; the maximax action is A_3.
3. The maximin action is A_4; the maximax action is A_4 also.
5. The maximin action is A_2; the maximax action is A_3.
7. See Table A.142. (The cell entries show revenue minus cost equaling profit, in thousands of dol-

Table A.141

Actions	Events			Row Minimum	Row Maximum
	E_1 = low demand	E_2 = moderate demand	E_3 = high demand		
A_1 = small project	4	5	6	④ maximin	6
A_2 = medium project	1	6	10	1	10
A_3 = large project	−5	0	30	−5	㉚ maximax

Table A.142

Actions	Events					
	E_0 Demand = 0	E_1 Demand = 1	E_2 Demand = 2	E_3 Demand = 3	E_4 Demand = 4	E_5 Demand = 5
A_1 = build 1	48 − 60 = −12	90 − 60 = 30	90 − 60 = 30	90 − 60 = 30	90 − 60 = 30	90 − 60 = 30
A_2 = build 2	96 − 120 = −24	138 − 120 = 18	180 − 120 = 60	180 − 120 = 60	180 − 120 = 60	180 − 120 = 60
A_3 = build 3	144 − 180 = −36	186 − 180 = 6	228 − 180 = 48	270 − 180 = 90	270 − 180 = 90	270 − 180 = 90
A_4 = build 4	192 − 240 = −48	234 − 240 = −6	276 − 240 = 36	318 − 240 = 78	360 − 240 = 120	360 − 240 = 120
A_5 = build 5	240 − 300 = −60	282 − 300 = −18	324 − 300 = 24	366 − 300 = 66	408 − 300 = 108	450 − 300 = 150

lars.) The maximin action is A_1; the maximax action is A_5.

9. See Table A.143. The minimax action is A_3; the minimin action is A_1.
11. The minimax action is A_1; the minimin action is A_5.
13. The minimax action is A_2; the minimin action is A_4.
15. See Table A.144. The minimax-regret action is A_3.
17. The minimax-regret action is A_2 (which brings a maximum regret of 110).
19. The minimax-regret action is A_2 (which brings a maximum regret of 75.)

21. See Table A.145, the third table on this page. The minimax-regret action is A_3. Caution: Unlike the payoffs in problems 14–20, the payoffs involve costs, not benefits.
23. The minimax-regret action is A_1 (which brings a maximum regret of 167).
25. The minimax-regret action is A_5 (which brings a maximum regret of 79).
27. The maximum-likelihood action is A_1.
29. See Table A.146. The expected-monetary-value action is A_2.

Table A.143

Actions	Events			Row Maximum	Row Minimum
	E_1 = Recession	E_2 = Inflation	E_3 = Stagflation		
A_1 = direct-mail campaign	75	20	40	75	⑳ Minimin
A_2 = newspaper campaign	50	60	30	60	30
A_3 = TV campaign	55	40	45	㊵ Minimax	40

Table A.144

Actions	Events			Row Maximum
	E_1 = low demand	E_2 = moderate demand	E_3 = high demand	
A_1 = small run	$15 - 15 = 0$	$20 - 20 = 0$	$69 - 25 = 44$	44
A_2 = medium run	$15 - 6 = 9$	$20 - 20 = 0$	$69 - 35 = 34$	34
A_3 = large run	$15 - (-10) = 25$	$20 - 0 = 20$	$69 - 69 = 0$	㉕

Table A.145

Actions	Events			Row Maximum
	E_1 = Recession	E_2 = Inflation	E_3 = Stagflation	
A_1 = Direct-mail campaign	$75 - 50 = 25$	$20 - 20 = 0$	$40 - 30 = 10$	25
A_2 = newspaper campaign	$50 - 50 = 0$	$60 - 20 = 40$	$30 - 30 = 0$	40
A_3 = TV campaign	$55 - 50 = 5$	$40 - 20 = 20$	$45 - 30 = 15$	⑳ Minimax Regret

Table A.146

Actions	Events		EMV
	E_1 = Success at New Location; $p(E_1)$ = .5	E_2 = Failure at New Location; $p(E_2)$ = .5	
A_1 = move East	50	10	30
A_2 = move West	100	0	㊿ Optimum

31. *EMV* for $A_1 = -.6$, for $A_2 = 7.65$, for $A_3 = 39.95$, for $A_4 = 74.45$, and for $A_5 = 42.45$. Thus, A_4 is best.

33. *EMV* for $A_1 = 28.06$, for $A_2 = 33.8$, for $A_3 = -13.9$, for $A_4 = -12.02$, and for $A_5 = 5.2$. Thus, A_2 is best.

35. The optimal action is to sell the rights for $125 million, as Figure A.33 shows.

37. The optimal action is to invest *or* not to invest. It's a toss-up; see Figure A.34.

39. *EOL* for $A_1 = 8.8$, for $A_2 = 7.7$, and for $A_3 = 16.5$. Thus, A_2 again is best.

41. *EOL* for $A_1 = 42.5$, for $A_2 = 84.6$, for $A_3 = 91.2$, and for $A_4 = 67.6$. Thus, A_1 again is best.

43. See Table A.147. The expected-utility action is A_1.

45. *EU* for $A_1 = -.97$, for $A_2 = -.28$, for $A_3 = 3.70$, for $A_4 = 7.38$, and for $A_5 = 4.43$. Thus, A_4 is best. (The square root has been taken of the absolute values of negative dollar figures and the resultant utility values were considered negative.)

Figure A.34

Figure A.33

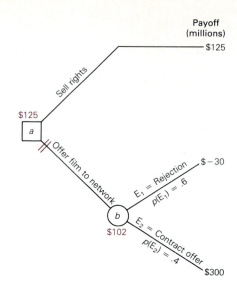

Figure A.34

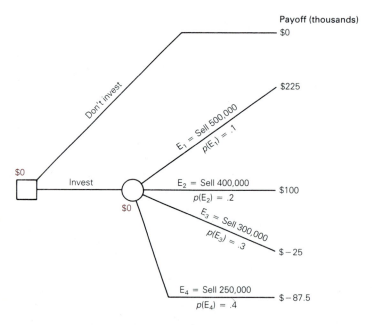

Table A.147

Actions	Events		EU
	E_1 = Success at New Location; $p(E_1) = .5$	E_2 = Failure at New Location; $p(E_2) = .5$	
A_1 = move East	$\sqrt{50} = 7.07$	$\sqrt{10} = 3.16$	5.12 Optimum
A_2 = move West	$\sqrt{100} = 10.00$	$\sqrt{0} = 0$	5.00

47. Given prior probabilities of $p(E_1) = .6$ and $p(E_2) = .4$, we can compute the following. *If the advice is R_1:*

$$p(E_1|R_1) = \frac{.6(.8)}{.6(.8) + .4(.3)}$$
$$= .8$$
$$p(E_2|R_1) = \frac{.4(.3)}{.4(.3) + .6(.8)}$$
$$= .2$$
$$p(R_1) = p(E_1 \text{ and } R_1)$$
$$\quad + p(E_2 \text{ and } R_1)$$
$$= .6(.8) + .4(.3)$$
$$= .6$$

If the advice is R_2:

$$p(E_1|R_2) = \frac{.6(.2)}{.6(.2) + .4(.7)}$$
$$= .3$$
$$p(E_2|R_2) = \frac{.4(.7)}{.4(.7) + .6(.2)}$$
$$= .7$$
$$p(R_2) = p(E_1 \text{ and } R_2)$$
$$\quad + p(E_2 \text{ and } R_2)$$
$$= .6(.2) + .4(.7)$$
$$= .4$$

On the basis of the preceding calculations, we can construct Figure A.35.

Figure A.35

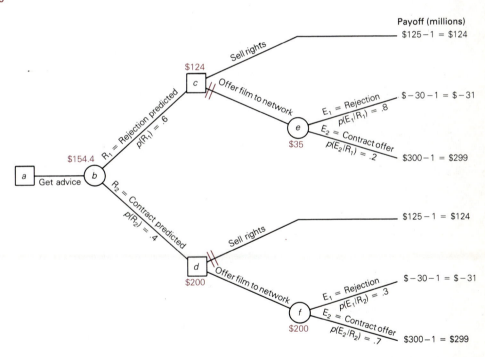

The optimal strategy is: if R_1 is received, sell the rights (and take the $125 million minus the $1 million fee); if R_2 is received, offer the film to the network (and earn an expected $200 million).

49. The expected gross benefit with the information is $154.4 million + $1.0 million = $155.4 million;

hence, the $EVSI = $155.4 million $-$ $125 million $=$ $30.4 million. The $1 million fee was cheap advice.

51. See Figure A.36 on the next page.

53. $EVSI = $7.06 million (at b) + $1 million (fee) $-$ $7.10 million (at e) $=$ $960,000. A $1 million fee

Figure A.36

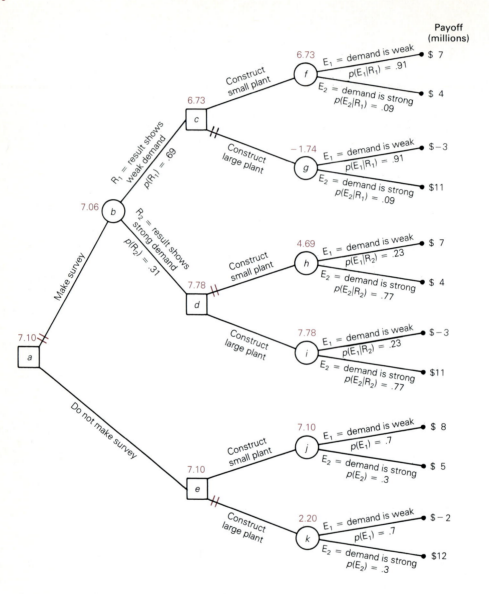

Payoff
(millions)

is not worth it. Hence the pruning scissors in Figure A.36 at the back of the text are applied above point of choice *a,* not below it (as in text Figure 17.6, "Preposterior Decision-Tree Analysis," which can be found on page 757).

55. *EVPI* = $55 million − $42.5 million = $12.5 million, where $55 million is computed as .5($100 million) + .5($10 million).
57. See Table A.148 (in millions of dollars).
59. See Table A.149 (in millions of dollars).

Table A.148

First-Stage Event Probability	Second-Stage Event Probability	Joint Probability	Payoff	Joint Probability × Payoff
$p(R_1) = .69$	$p(E_1|R_1) = .91$	$.69(.91) = .63$	$7.9	$.63(7.9) = \$4.98$
$p(R_1) = .69$	$p(E_2|R_1) = .09$	$.69(.09) = .06$	4.9	$.06(4.9) = .29$
$p(R_2) = .31$	$p(E_1|R_2) = .23$	$.31(.23) = .07$	-2.1	$.07(-2.1) = -.15$
$p(R_2) = .31$	$p(E_2|R_2) = .77$	$.31(.77) = .24$	11.9	$.24(11.9) = 2.86$
		1.00		Expected payoff (S_2): 7.98

Table A.149

First-Stage Event Probability	Second-Stage Event Probability	Joint Probability	Payoff	Joint Probability × Payoff
$p(R_1) = .69$	$p(E_1|R_1) = .91$	$.69(.91) = .63$	$-\$2.1$	$.63(-2.1) = -\$1.32$
$p(R_1) = .69$	$p(E_2|R_1) = .09$	$.69(.09) = .06$	11.9	$.06(11.9) = .71$
$p(R_2) = .31$	$p(E_1|R_2) = .23$	$.31(.23) = .07$	-2.1	$.07(-2.1) = -.15$
$p(R_2) = .31$	$p(E_2|R_2) = .77$	$.31(.77) = .24$	11.9	$.24(11.9) = 2.86$
		1.00		Expected payoff (S_4): $2.10

ACKNOWLEDGMENTS

Cover design by Anthony Ma.

Table 2.2 on p. 19 and Table 3.1 on p. 50: From "The 100 largest U.S. multinationals," *Forbes,* July 5, 1982, pp. 126–28. Copyright © 1982 by Forbes, Inc. Reprinted by permission.

Table 3.C on p. 32 from "Teacher Salaries—1984–85" from *The New York Times,* August 31, 1985. Copyright © 1985 by The New York Times Company. Reprinted by permission.

Table on p. 35 from "A Study of Vaccination Against the Common Cold" by Sir Austin Bradford Hill in *Principles of Medical Statistics.* Copyright © 1966 by The Lancet Ltd. Reprinted by permission of The Lancet Ltd.

Tables on p. 36 from "The Fluorescein Experiment." Reprinted with permission from *The Science Teacher,* a publication of the National Science Teachers Association, Vol. 35, No. 8, November 1968. Copyright © 1968 by the National Science Teachers Association.

Figures on p. 37 from "The Salk Vaccine Study." Reprinted with permission from *The Science Teacher,* a publication of the National Science Teachers Association, Vol. 35, No. 8, November 1968. Copyright © 1968 by the National Science Teachers Association.

Photo on p. 41: *Biometrika.*

Table 2.A on p. 44 from "1985 Mutual Funds 'Honor Roll.'" Excerpted by permission of *Forbes* magazine, September 16, 1985. Copyright © Forbes Inc., 1985.

Table on p. 123: From *National Income and Its Composition,* 1919–1938 by Simon Kuznets, pp. 512–13. Copyright 1941 by National Bureau of Economic Research, Inc. Reprinted by permission.

Table 4.A on p. 127 reprinted from the July 5, 1982 issue of *Business Week* by special permission, Copyright © 1982 by McGraw-Hill, Inc.

Table 4.A on p. 127 from "Three Measures of the Top 15 in R&D spending." Reprinted from the July 5, 1982 issue of *Business Week* by special permission, Copyright © 1982 by McGraw-Hill, Inc.

Table 4.F on p. 129: From "The 500 Largest Industrial Corporations Outside the U.S." in *Fortune,* August 23, 1982, p. 183. Copyright © 1982 by Time, Inc. Reprinted by permission.

Table 4.G on p. 129 from "Public Utility Finance and Economic Waste" by Glenn P. Jenkins, *Canadian Journal of Economics,* August 1985. Reprinted by permission.

Table 4.I on p. 131 from "New York Public Schools-Mathematics and Reading Scores" from *The New York Times,* June 20, 1982. Copyright © 1982 by The New York Times Company. Reprinted by permission.

Photo on p. 179: The Granger Collection.

Photo on p. 230: The Bettmann Archive.

Photo on p. 231: Offentliche Bibliothek der Universitat, Basel, Switzerland.

Table on p. 329 from *The Design and Analysis of Industrial Experiments,* edited by Owen L. Davies, 1960. Reprinted by permission of the publishers, The Longman Group Limited.

Photo on p. 330: The Granger Collection.

Table on p. 380: From "Antitrust Pork Barrel" by Roger L. Faith, Donald R. Leavens, and Robert D. Tollison, *Journal of Law and Economics,* October 1982, pp. 349–42. Copyright © 1982 by the University of Chicago. All rights reserved. Reprinted by permission of the University of Chicago Press.

Photo on p. 383: Courtesy of the Statistical Laboratory, Iowa State University, Ames, Iowa 50011.

Photo on p. 384: Courtesy, University of California, Berkeley, Department of Statistics.

Photo on p. 385: Keystone Press Agency, Ltd.

Table A on p. 427: From "Leukemia in Man Following Exposure to Ionizing Radiation" by A. Bertrand Brill, Masanobu Tomonaga, and Robert M. Heyssel in *Annals of Internal Medicine* 56, 4 (April 1962). Copyright © 1962 by The American College of Physicians. Reprinted by permission.

Table B on p. 429 from "On a Distribution Law for Word Frequencies" by H. S. Sichel, *Journal of the American Statistical Association,* September 1975. Copyright © 1975 by Journal of the American Statistical Association. Reprinted by permission.

Photo on p. 430: *Biometrika.*

First photo on p. 477 by W. G. Smith, Jr., Boyce Thompson Institute.

Second photo on p. 477: Keystone Press Agency, Ltd.

Table on p. 515: From "The Fairness of Earnings Differentials" by Joanne Martin in *The Journal of Human Resources* 17 (Winter 1982): 21. Reprinted by permission of The University of Wisconsin Press.

SUBJECT INDEX

NAME INDEX

GLOSSARY OF SYMBOLS

a the vertical intercept of the sample regression line (estimator of α)

$A \cup B$ the union of events A, B (A *or* B (or both) occur)

$A \cap B$ the intersection of events A, B (A *and* B occur simultaneously)

b the slope of the sample regression line (estimator of β)

b_1 and b_2 estimated partial regression coefficients

BMS blocks mean square (variance explained by blocks)

BSS blocks sum of squares (variation explained by blocks)

C_x^n combinations for x at a time out of n distinct items

c_x^n binomial coefficient; number of permutations of x successes and $n - x$ failures in n Bernoulli trials

C_t the cyclical component in time-series analysis, measured against time, t

D the Kolmogorov-Smirnov one-sample test statistic; dummy variable; the difference between 2 sample observations in a matched-pairs sample

$\overline{D}$ the average of differences between 2 sample observations in a matched-pairs sample

d the difference between 2 sample means (or proportions) from independent samples; the Durbin-Watson test statistic

d_1 the difference between the frequency density of one class and the preceding class

d_2 the difference between the frequency density of one class and the following class

$d.f.$ degrees of freedom

D-W the Durbin-Watson test statistic

E the statistic that is being used as an estimator

e the constant 2.71828; the tolerable error level; a sampling error

EMS error mean square (unexplained variance)

EMV expected monetary value

EOL expected opportunity loss

ESI efficiency of sample information *(EVSI/EVPI)*

ESS error sum of squares (unexplained variation)

EU expected utility

$EVPI$ expected value of perfect information

$EVSI$ expected value of sample information

E_R expected value of random variable, R

F the sum of frequencies up to, but not including those of the median class; the test statistic that is the ratio of explained variance to unexplained variance (e.g., TMS/EMS or RMS/EMS)

f class frequency

$f(x)$ any function of x

$f(z)$ the standard normal probability density function

f_e expected frequency

f_o observed frequency

F_t the current period's forecast value (of a time series)

F_{t+1} next period's forecast value (of a time series)

HSD Tukey's honestly significant difference

H_0 the null hypothesis

H_A the alternative hypothesis

IMS interactions mean square (variance explained by interactions)

ISS interactions sum of squares (variation explained by interactions)

I_t the irregular component in time-series analysis, measured against time, t

$I_{t,0}$ intertemporal (price or quantity) index for period t with period 0 base

$I_{X,b}$ interspatial (price or quantity) index for place X with place b base

$I_{Y \cdot X}$ expected individual value of Y, given X

$I_{Y \cdot 12}$ expected individual value of Y, given X_1 and X_2

K population coefficient of kurtosis; the Kruskal-Wallis test statistic

k sample coefficient of kurtosis; sample coefficient of alienation; desirable number of classes in Sturgess's rule; number of standard deviations in Chebyshev's theorem; number of observations per cell in two-way ANOVA; number of means in Tukey's HSD test

k^2 sample coefficient of nondetermination *(ESS/ Total SS)*

L lower limit of median (or modal) class

m sample median; number of independent variables in a regression

MAD mean absolute deviation

Mo population mode

mo sample mode

N the total number of observations in a population

n the total number of observations in a sample; number of trials in a random experiment

$n!$ n factorial

OL opportunity loss

p a probability value that measures the likelihood that a test-statistic value could occur by chance, given that the null hypothesis is true

$p(A)$ probability of event A

$p(A|B)$ probability of event A, given B

$p(R = x)$ probability of random variable (R) being equal to a given possible value, x

P sample proportion

P_P pooled estimator of the population proportion

P_t current-year price

P_x^n permutations for x at a time out of n distinct items

$P_{x_1, x_2, \ldots x_k}^n$ permutations for n at a time out of n nondistinct items, there being k kinds of items and x_1 units of kind *1*, x_2 units of kind *2*, and so forth.

P_0 year 0 price

9_α key number in Tukey's HSD test

Q_t current-year quantity

Q_0 year 0 quantity

RSS regression sum of squares (variation explained by regression)

RMS regression mean square (variance explained by regression)

R name of some random variable; sample coefficient of multiple correlation; the number-of-runs test statistic

r sample coefficient of correlation

R^2 sample coefficient of multiple determination *(RSS/Total SS* in multiple regression)

r^2 sample coefficient of determination *(RSS/Total SS* in simple regression)

$\overline{R}^2$ adjusted sample coefficient of multiple determination